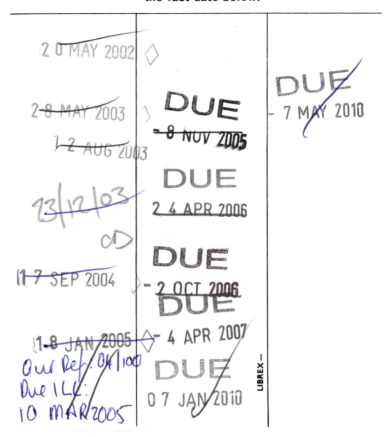

THE ELECTRICAL ENGINEERING
AND SIGNAL PROCESSING SERIES
Edited by Alexander Poularikas and Richard C. Dorf

The Advanced Signal Processing Handbook: Theory and Implementation for Radar, Sonar, and Medical Imaging Real-Time Systems
Stergios Stergiopoulos

The Transform and Data Compression Handbook
K.R. Rao and P.C. Yip

Forthcoming Titles

Handbook of Antennas in Wireless Communications
Lal Chand Godara

Propagation Data Handbook for Wireless Communications
Robert Crane

The Digital Color Imaging Handbook
Guarav Sharma

Handbook of Neural Network Signal Processing
Yu Hen Hu and Jeng-Neng Hwang

Handbook of Multisensor Data Fusion
David Hall

Applications in Time Frequency Signal Processing
Antonia Papandreou-Suppappola

Noise Reduction in Speech Applications
Gillian Davis

Signal Processing in Noise
Vyacheslav Tuzlukov

Electromagnetic Radiation and the Human Body: Effects, Diagnosis and Therapeutic Technologies
Nikolaos Uzunoglu and Konstantina S. Nikita

ADVANCED SIGNAL PROCESSING HANDBOOK

Theory and Implementation for Radar, Sonar, and Medical Imaging Real-Time Systems

Edited by

STERGIOS STERGIOPOULOS

CRC Press
Boca Raton London New York Washington, D.C.

Library of Congress Cataloging-in-Publication Data

Advanced signal processing handbook : theory and implementation for radar, sonar, and medical imaging real-time systems / edited by Stergios Stergiopoulos.
p. cm. — (Electrical engineering and signal processing series)
Includes bibliographical references and index.
ISBN 0-8493-3691-0 (alk. paper)
1. Signal processing—Digital techniques. 2. Diagnostic imaging—Digital techniques. 3. Image processing—Digital techniques. I. Stergiopoulos, Stergios. II. Series.

TK5102.9 .A383 2000
621.382′2—dc21 00-045432
CIP

Direct all inquiries to CRC Press LLC, 2000 N.W. Corporate Blvd., Boca Raton, Florida 33431.

International Standard Book Number 0-8493-3691-0
Library of Congress Card Number 00-045432
Printed in the United States of America 1 2 3 4 5 6 7 8 9 0
Printed on acid-free paper

Preface

Recent advances in digital signal processing algorithms and computer technology have combined to provide the ability to produce real-time systems that have capabilities far exceeding those of a few years ago. The writing of this handbook was prompted by a desire to bring together some of the recent theoretical developments on advanced signal processing, and to provide a glimpse of how modern technology can be applied to the development of current and next-generation active and passive real-time systems.

The handbook is intended to serve as an introduction to the principles and applications of advanced signal processing. It will focus on the development of a generic processing structure that exploits the great degree of processing concept similarities existing among the radar, sonar, and medical imaging systems. A high-level view of the above real-time systems consists of a high-speed *Signal Processor* to provide mainstream signal processing for detection and initial parameter estimation, a *Data Manager* which supports the data and information processing functionality of the system, and a *Display Sub-System* through which the system operator can interact with the data structures in the data manager to make the most effective use of the resources at his command.

The *Signal Processor* normally incorporates a few fundamental operations. For example, the sonar and radar signal processors include beamforming, "matched" filtering, data normalization, and image processing. The first two processes are used to improve both the signal-to-noise ratio (SNR) and parameter estimation capability through spatial and temporal processing techniques. Data normalization is required to map the resulting data into the dynamic range of the display devices in a manner which provides a CFAR (constant false alarm rate) capability across the analysis cells.

The processing algorithms for spatial and temporal spectral analysis in real-time systems are based on conventional FFT and vector dot product operations because they are computationally cheaper and more robust than the modern non-linear high resolution adaptive methods. However, these non-linear algorithms trade robustness for improved array gain performance. Thus, the challenge is to develop a concept which allows an appropriate mixture of these algorithms to be implemented in practical real-time systems.

The non-linear processing schemes are adaptive and synthetic aperture beamformers that have been shown experimentally to provide improvements in array gain for signals embedded in partially correlated noise fields. Using system image outputs, target tracking, and localization results as performance criteria, the impact and merits of these techniques are contrasted with those obtained using the conventional processing schemes. The reported real data results show that the advanced processing schemes provide improvements in array gain for signals embedded in anisotropic noise fields. However, the same set of results demonstrates that these processing schemes are not adequate enough to be considered as a replacement for conventional processing. This restriction adds an additional element in our generic signal processing structure, in that the conventional and the advanced signal processing schemes should run in parallel in a real-time system in order to achieve optimum use of the advanced signal processing schemes of this study.

The handbook also includes a generic concept for implementing successfully adaptive schemes with near-instantaneous convergence in 2-dimensional (2-D) and 3-dimensional (3-D) arrays of sensors, such as planar, circular, cylindrical, and spherical arrays. It will be shown that the basic step is to minimize the number of degrees of freedom associated with the adaptation process. This step will minimize the adaptive scheme's convergence period and achieve near-instantaneous convergence for integrated active and passive sonar applications. The reported results are part of a major research project, which includes the definition of a generic signal processing structure that allows the implementation of adaptive and synthetic aperture signal processing schemes in real-time radar, sonar, and medical tomography (CT, MRI, ultrasound) systems that have 2-D and 3-D arrays of sensors.

The material in the handbook will bridge a number of related fields: detection and estimation theory; filter theory (Finite Impulse Response Filters); 1-D, 2-D, and 3-D sensor array processing that includes conventional, adaptive, synthetic aperture beamforming and imaging; spatial and temporal spectral analysis; and data normalization. Emphasis will be placed on topics that have been found to be particularly useful in practice. These are several interrelated topics of interest such as the influence of medium on array gain system performance, detection and estimation theory, filter theory, space-time processing, conventional, adaptive processing, and model-based signal processing concepts. Moveover, the system concept similarities between sonar and ultrasound problems are identified in order to exploit the use of advanced sonar and model-based signal processing concepts in ultrasound systems.

Furthermore, issues of information post-processing functionality supported by the Data Manager and the Display units of real-time systems of interest are addressed in the relevant chapters that discuss normalizers, target tracking, target motion analysis, image post-processing, and volume visualization methods.

The presentation of the subject matter has been influenced by the authors' practical experiences, and it is hoped that the volume will be useful to scientists and system engineers as a textbook for a graduate course on sonar, radar, and medical imaging digital signal processing. In particular, a number of chapters summarize the state-of-the-art application of advanced processing concepts in sonar, radar, and medical imaging X-ray CT scanners, magnetic resonance imaging, and 2-D and 3-D ultrasound systems. The focus of these chapters is to point out their applicability, benefits, and potential in the sonar, radar, and medical environments. Although an all-encompassing general approach to a subject is mathematically elegant, practical insight and understanding may be sacrificed. To avoid this problem and to keep the handbook to a reasonable size, only a modest introduction is provided. In consequence, the reader is expected to be familiar with the basics of linear and sampled systems and the principles of probability theory. Furthermore, since modern real-time systems entail sampled signals that are digitized at the sensor level, our signals are assumed to be discrete in time and the subsystems that perform the processing are assumed to be digital.

It has been a pleasure for me to edit this book and to have the relevant technical exchanges with so many experts on advanced signal processing. I take this opportunity to thank all authors for their responses to my invitation to contribute. I am also greatful to CRC Press LLC and in particular to Bob Stern, Helena Redshaw, Naomi Lynch, and the staff in the production department for their truly professional cooperation. Finally, the support by the European Commission is acknowledged for awarding Professor Uzunoglu and myself the Fourier Euroworkshop Grant (HPCF-1999-00034) to organize two workshops that enabled the contributing authors to refine and coherently integrate the material of their chapters as a handbook on advanced signal processing for sonar, radar, and medical imaging system applications.

Stergios Stergiopoulos

Editor

Stergios Stergiopoulos received a B.Sc. degree from the University of Athens in 1976 and the M.S. and Ph.D. degrees in geophysics in 1977 and 1982, respectively, from York University, Toronto, Canada. Presently he is an Adjunct Professor at the Department of Electrical and Computer Engineering of the University of Western Ontario and a Senior Defence Scientist at Defence and Civil Institute of Environmental Medicine (DCIEM) of the Canadian DND. Prior to this assignment and from 1988 and 1991, he was with the SACLANT Centre in La Spezia, Italy, where he performed both theoretical and experimental research in sonar signal processing. At SACLANTCEN, he developed jointly with Dr. Sullivan from NUWC an acoustic synthetic aperture technique that has been patented by the U.S. Navy and the Hellenic Navy. From 1984 to 1988 he developed an underwater fixed array surveillance system for the Hellenic Navy in Greece and there he was appointed senior advisor to the Greek Minister of Defence. From 1982 to 1984 he worked as a research associate at York University and in collaboration with the U.S. Army Ballistic Research Lab (BRL), Aberdeen, MD, on projects related to the stability of liquid-filled spin stabilized projectiles. In 1984 he was awarded a U.S. NRC Research Fellowship for BRL. He was Associate Editor for the *IEEE Journal of Oceanic Engineering* and has prepared two special issues on Acoustic Synthetic Aperture and Sonar System Technology. His present interests are associated with the implementation of non-conventional processing schemes in multi-dimensional arrays of sensors for sonar and medical tomography (CT, MRI, ultrasound) systems. His research activities are supported by Canadian-DND Grants, by Research and Strategic Grants (NSERC-CANADA) ($300K), and by a NATO Collaborative Research Grant. Recently he has been awarded with European Commission-ESPRIT/IST Grants as technical manager of two projects entitled "New Roentgen" and "MITTUG." Dr. Stergiopoulos is a Fellow of the Acoustical Society of America and a senior member of the IEEE. He has been a consultant to a number of companies, including Atlas Elektronik in Germany, Hellenic Arms Industry, and Hellenic Aerospace Industry.

Contributors

Dimos Baltas
Department of Medical Physics and Engineering
Strahlenklinik, Städtische Kliniken Offenbach
Offenbach, Germany

Institute of Communication and Computer Systems
National Technical University of Athens
Athens, Greece

Klaus Becker
FGAN Research Institute for Communication, Information Processing, and Ergonomics (FKIE)
Wachtberg, Germany

James V. Candy
Lawrence Livermore National Laboratory
University of California
Livermore, California, U.S.A.

G. Clifford Carter
Naval Undersea Warfare Center
Newport, Rhode Island, U.S.A.

N. Ross Chapman
School of Earth and Ocean Sciences
University of Victoria
Victoria, British Columbia, Canada

Ian Cunningham
The John P. Robarts Research Institute
University of Western Ontario
London, Ontario, Canada

Konstantinos K. Delibasis
Institute of Communication and Computer Systems
National Technical University of Athens
Athens, Greece

Amar Dhanantwari
Defence and Civil Institute of Environmental Medicine
Toronto, Ontario, Canada

Reza M. Dizaji
School of Earth and Ocean Sciences
University of Victoria
Victoria, British Columbia, Canada

Donal B. Downey
The John P. Robarts Research Institute
University of Western Ontario
London, Ontario, Canada

Geoffrey Edelson
Advanced Systems and Technology
Sanders, A Lockheed Martin Company
Nashua, New Hampshire, U.S.A.

Aaron Fenster
The John P. Robarts Research Institute
University of Western Ontario
London, Ontario, Canada

Dimitris Hatzinakos
Department of Electrical and Computer Engineering
University of Toronto
Toronto, Ontario, Canada

Simon Haykin
Communications Research Laboratory
McMaster University
Hamilton, Ontario, Canada

Grigorios Karangelis
Department of Cognitive Computing and Medical Imaging
Fraunhofer Institute for Computer Graphics
Darmstadt, Germany

R. Lynn Kirlin
School of Earth and Ocean Sciences
University of Victoria
Victoria, British Columbia, Canada

Wolfgang Koch
FGAN Research Institute for Communciation, Information Processing, and Ergonomics (FKIE)
Wachtberg, Germany

Christos Kolotas
Department of Medical Physics and Engineering
Strahlenklinik, Städtische Kliniken Offenbach
Offenbach, Germany

Harry E. Martz, Jr.
Lawrence Livermore National Laboratory
University of California
Livermore, California, U.S.A.

George K. Matsopoulos
Institute of Communication
and Computer Systems
National Technical University
of Athens
Athens, Greece

Charles A. McKenzie
Cardiovascular Division
Beth Israel Deaconess Medical Center
and Harvard Medical School
Boston, Massachusetts, U.S.A.

Bernard E. McTaggart
Naval Undersea Warfare Center
(retired)
Newport, Rhode Island, U.S.A.

Sanjay K. Mehta
Naval Undersea Warfare Center
Newport, Rhode Island, U.S.A.

Natasa Milickovic
Department of Medical Physics
and Engineering
Strahlenklinik, Städtische
Kliniken Offenbach
Offenbach, Germany

Gerald R. Moran
Lawson Research Institute and
Department of Medical
Biophysics
University of Western Ontario
London, Ontario, Canada

Nikolaos A. Mouravliansky
Institute of Communication
and Computer Systems
National Technical University
of Athens
Athens, Greece

Arnulf Oppelt
Siemens Medical Engineering Group
Erlangen, Germany

Kostantinos N. Plataniotis
Department of Electrical
and Computer Engineering
University of Toronto
Toronto, Ontario, Canada

Andreas Pommert
Institute of Mathematics and
Computer Science in Medicine
University Hospital Eppendorf
Hamburg, Germany

Frank S. Prato
Lawson Research Institute
and Department
of Medical Biophysics
University of Western Ontario
London, Ontario, Canada

John M. Reid
Department of Biomedical
Engineering
Drexel University
Philadelphia, Pennsylvania, U.S.A.

Department of Radiology
Thomas Jefferson University
Philadelphia, Pennsylvania, U.S.A.

Department of Bioengineering
University of Washington
Seattle, Washington, U.S.A.

Georgios Sakas
Department of Cognitive Computing
and Medical Imaging
Fraunhofer Institute
for Computer Graphics
Darmstadt, Germany

Daniel J. Schneberk
Lawrence Livermore
National Laboratory
University of California
Livermore, California, U.S.A.

Stergios Stergiopoulos
Defence and Civil Institute
of Environmental Medicine
Toronto, Ontario, Canada

Department of Electrical
and Computer Engineering
University of Western Ontario
London, Ontario, Canada

Edmund J. Sullivan
Naval Undersea Warfare Center
Newport, Rhode Island, U.S.A.

Rebecca E. Thornhill
Lawson Research Institute and
Department of Medical
Biophysics
University of Western Ontario
London, Ontario, Canada

Nikolaos Uzunoglu
Department of Electrical
and Computer Engineering
National Technical University
of Athens
Athens, Greece

Nikolaos Zamboglou
Department of Medical Physics
and Engineering
Strahlenklinik, Städtische
Kliniken Offenbach
Offenbach, Germany

Institute of Communication
and Computer Systems
National Technical University
of Athens
Athens, Greece

Dedication

To my lifelong companion Vicky, my son Steve, and my daughter Erene

Contents

SECTION II Sonar and Radar System Applications

SECTION III Medical Imaging System Applications

1

Signal Processing Concept Similarities among Sonar, Radar, and Medical Imaging Systems

Stergios Stergiopoulos
Defence and Civil Institute of Environmental Medicine

University of Western Ontario

1.1 Introduction

Several review articles on sonar,[1,3–5] radar,[2,3] and medical imaging[3,6–14] system technologies have provided a detailed description of the mainstream signal processing functions along with their associated implementation considerations. The attempt of this handbook is to extend the scope of these articles by introducing an implementation effort of non-mainstream processing schemes in real-time systems. To a large degree, work in the area of *sonar and radar system technology* has traditionally been funded either directly or indirectly by governments and military agencies in an attempt to improve the capability of anti-submarine warfare (ASW) sonar and radar systems. A secondary aim of this handbook is to promote, where possible, wider dissemination of this military-inspired research.

1.2 Overview of a Real-Time System

In order to provide a context for the material contained in this handbook, it would seem appropriate to briefly review the basic requirements of a high-performance real-time system. Figure 1.1 shows one possible high-level view of a generic system.[15] It consists of an array of sensors and/or sources; a high-speed *signal*

0-8493-3691-0/01/$0.00+$.50

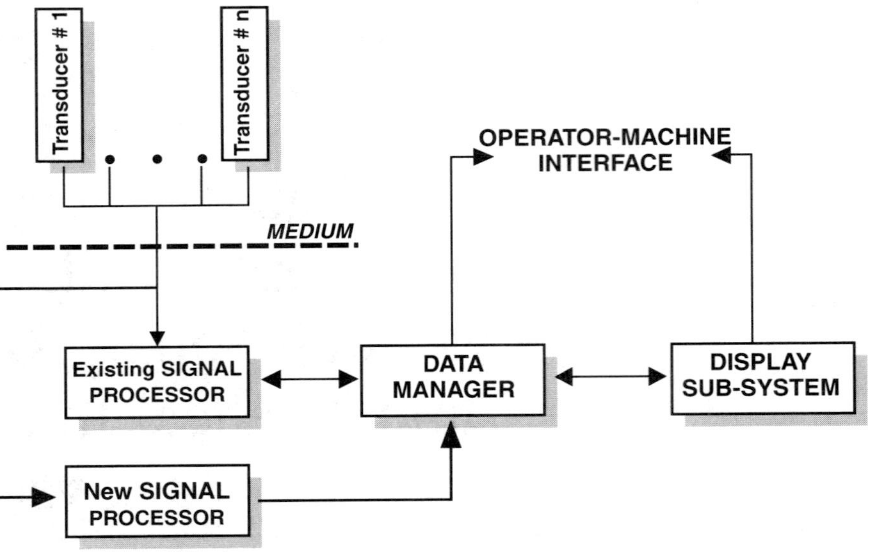

FIGURE 1.1 Overview of a generic real-time system. It consists of an array of transducers, a *signal processor* to provide mainstream signal processing for detection and initial parameter estimation; a *data manager*, which supports the data, information processing functionality, and data fusion; and a *display sub-system* through which the system operator can interact with the manager to make the most effective use of the information available at his command.

processor to provide mainstream signal processing for detection and initial parameter estimation; a *data manager*, which supports the data and information processing functionality of the system; and a *display sub-system* through which the system operator can interact with the data structures in the data manager to make the most effective use of the resources at his command.

In this handbook, we will be limiting our attention to the *signal processor*, the *data manager*, and *display sub-system*, which consist of the algorithms and the processing architectures required for their implementation. *Arrays of sources and sensors* include devices of varying degrees of complexity that illuminate the medium of interest and sense the existence of signals of interest. These devices are arrays of transducers having cylindrical, spherical, planar, or linear geometric configurations, depending on the application of interest. Quantitative estimates of the various benefits that result from the deployment of arrays of transducers are obtained by the *array gain* term, which will be discussed in Chapters 6, 10, and 11. Sensor array design concepts, however, are beyond the scope of this handbook and readers interested in transducers can refer to other publications on the topic.[16–19]

The *signal processor* is probably the single, most important component of a real-time system of interest for this handbook. In order to satisfy the basic requirements, the processor normally incorporates the following fundamental operations:

- Multi-dimensional beamforming
- Matched filtering
- Temporal and spatial spectral analysis
- Tomography image reconstruction processing
- Multi-dimensional image processing

The first three processes are used to improve both the signal-to-noise ratio (SNR) and parameter estimation capability through spatial and the temporal processing techniques. The next two operations are image reconstruction and processing schemes associated mainly with image processing applications. As indicated in Figure 1.1, the replacement of the *existing signal processor* with a *new signal processor*, which would include advanced processing schemes, could lead to improved performance functionality of a real-time system of interest, while the associated development cost could be significantly lower than using other hardware (H/W) alternatives. In a sense, this statement highlights the future trends of state-of-the-art investigations on advanced real-time signal processing functionalities that are the subject of the handbook.

Furthemore, post-processing of the information provided by the previous operations includes mainly the following:

- Signal tracking and target motion analysis
- Image post-processing and data fusion
- Data normalization
- OR-ing

These operations form the functionality of the *data manager* of sonar and radar systems. However, identification of the processing concept similarities between sonar, radar, and medical imaging systems may be valuable in identifying the implementation of these operations in other medical imaging system applications. In particular, the operation of data normalization in sonar and radar systems is required to map the resulting data into the dynamic range of the display devices in a manner which provides a constant false alarm rate (CFAR) capability across the analysis cells. The same operation, however, is required in the display functionality of medical ultrasound imaging systems as well.

In what follows, each sub-system, shown in Figure 1.1, is examined briefly by associating the evolution of its functionality and characteristics with the corresponding signal processing technological developments.

1.3 Signal Processor

The implementation of signal processing concepts in real-time systems is heavily dependent on the computing architecture characteristics, and, therefore, it is limited by the progress made in this field. While the mathematical foundations of the signal processing algorithms have been known for many years, it was the introduction of the microprocessor and high-speed multiplier-accumulator devices in the early 1970s which heralded the turning point in the development of digital systems. The first systems were primarily fixed-point machines with limited dynamic range and, hence, were constrained to use conventional beamforming and filtering techniques.[1,4,15] As floating-point central processing units (CPUs) and supporting memory devices were introduced in the mid to late 1970s, multi-processor digital systems and modern signal processing algorithms could be considered for implementation in real-time systems. This major breakthrough expanded in the 1980s into massively parallel architectures supporting multi-sensor requirements.

The limitations associated with these massively parallel architectures became evident by the fact that they allow only fast-Fourier-transform (FFT), vector-based processing schemes because of efficient implementation and of their very cost-effective throughput characteristics. Thus, non-conventional schemes (i.e., adaptive, synthetic aperture, and high-resolution processing) could not be implemented in these types of real-time systems of interest, even though their theoretical and experimental developments suggest that they have advantages over existing conventional processing approaches.[2,3,15,20–25] It is widely believed that these advantages can address the requirements associated with the difficult operational problems that next generation real-time sonar, radar, and medical imaging systems will have to solve.

New scalable computing architectures, however, which support both scalar and vector operations satisfying high input/output bandwidth requirements of large multi-sensor systems, are becoming available.[15] Recent frequent announcements include successful developments of super-scalar and massively parallel signal processing computers that have throughput capabilities of hundred of billions of floating-point operations per second (GFLOPS).[31] This resulted in a resurgence of interest in algorithm development of new covariance-based, high-resolution, adaptive[15,20–22,25] and synthetic aperture beamforming algorithms,[15,23] and time-frequency analysis techniques.[24]

Chapters 2, 3, 6, and 11 discuss in some detail the recent developments in adaptive, high-resolution, and synthetic aperture array signal processing and their advantages for real-time system applications. In particular, Chapter 2 reviews the basic issues involved in the study of adaptive systems for signal processing. The virtues of this approach to statistical signal processing may be summarized as follows:

- The use of an adaptive filtering algorithm, which enables the system to adjust its free parameters (in a supervised or unsupervised manner) in accordance with the underlying statistics of the environment in which the system operates, hence, avoiding the need for determining the statistical characteristics of the environment
- Tracking capability, which permits the system to follow statistical variations (i.e., non-stationarity) of the environment
- The availability of many different adaptive filtering algorithms, both linear and non-linear, which can be used to deal with a wide variety of signal processing applications in radar, sonar, and biomedical imaging
- Digital implementation of the adaptive filtering algorithms, which can be carried out in hardware or software form

In many cases, however, special attention is required for non-linear, non-Gaussian signal processing applications. Chapter 3 addresses this topic by introducing a Gaussian mixture approach as a model in such problems where data can be viewed as arising from two or more populations mixed in varying proportions. Using the Gaussian mixture formulation, problems are treated from a global viewpoint that readily yields and unifies previous, seemingly unrelated results. Chapter 3 introduces novel signal processing techniques applied in applications problems, such as target tracking in polar coordinates and interference rejection in impulsive channels. In other cases these advanced algorithms, introduced in Chapters 2 and 3, trade robustness for improved performance.[15,25,26] Furthermore, the improvements achieved are generally not uniform across all signal and noise environments of operational scenarios. The challenge is to develop a concept which allows an appropriate mixture of these algorithms to be implemented in practical real-time systems. The advent of new adaptive processing techniques is only the first step in the utilization of *a priori* information as well as more detailed information for the mediums of the propagating signals of interest. Of particular interest is the rapidly growing field of matched field processing (MFP).[26] The use of linear models will also be challenged by techniques that utilize higher order statistics,[24] neural networks,[27] fuzzy systems,[28] chaos, and other non-linear approaches. Although these concerns have been discussed[27] in a special issue of the *IEEE Journal of Oceanic Engineering* devoted to sonar system technology, it should be noted that a detailed examination of MFP can be found also in the July 1993 issue of this journal which has been devoted to detection and estimation of MFP.[29]

The discussion in Chapter 4 focuses on the class of problems for which there is some information about the signal propagation model. From the basic formalism of blind system identification process, signal processing methods are derived that can be used to determine the unknown parameters of the medium transfer function and to demonstrate its performance for estimating the source location and the environmental parameters of a shallow water waveguide. Moreover, the system concept similarities between sonar and ultrasound systems are analyzed in order to exploit the use of model-based sonar signal processing concepts in ultrasound problems.

The discussion on model-based signal processing is extended in Chapter 5 to determine the most appropriate signal processing approaches for measurements that are contaminated with noise and underlying uncertainties. In general, if the SNR of the measurements is high, then simple non-physical techniques such as Fourier transform-based temporal and spatial processing schemes can be used to extract the desired information. However, if the SNR is extremely low and/or the propagation medium is uncertain, then more of the underlying propagation physics must be incorporated somehow into the processor to extract the information. These are issues that are discussed in Chapter 5, which introduces a generic development of model-based processing schemes and then concentrates specifically on those designed for sonar system applications.

Thus, Chapters 2, 3, 4, 5, 6, and 11 address a major issue: the implementation of advanced processing schemes in real-time systems of interest. The starting point will be to identify the signal processing concept similarities among radar, sonar, and medical imaging systems by defining a generic signal processing structure integrating the processing functionalities of the real-time systems of interest. The definition of a generic signal processing structure for a variety of systems will address the above continuing interest that is supported by the fact that synthetic aperture and adaptive processing techniques provide new gain.[2,15,20,21,23] This kind of improvement in array gain is equivalent to improvements in system performance.

In general, improvements in system performance or array gain improvements are required when the noise environment of an operational system is non-isotropic, such as the noise environment of (1) atmospheric noise or clutter (radar applications), (2) cluttered coastal waters and areas with high shipping density in which sonar systems operate (sonar applications), and (3) the complexity of the human body (medical imaging applications). An alternative approach to improve the array gain of a real-time system requires the deployment of very large aperture arrays, which leads to technical and operational implications. Thus, the implementation of non-conventional signal processing schemes in operational systems will minimize very costly H/W requirements associated with array gain improvements.

Figure 1.2 shows the configuration of a generic signal processing scheme integrating the functionality of radar, sonar, ultrasound, medical tomography CT/X-ray, and magnetic resonance imaging (MRI) systems. There are five major and distinct processing blocks in the generic structure. Moreover, reconfiguration of the different processing blocks of Figure 1.2 allows the application of the proposed concepts to a variety of active or passive digital signal processing (DSP) systems.

The **first point** of the generic processing flow configuration is that its implementation is in the frequency domain. The **second point** is that with proper selection of filtering weights and careful data partitioning, the frequency domain outputs of conventional or advanced processing schemes can be made equivalent to the FFT of the broadband outputs. This equivalence corresponds to implementing finite impulse response (FIR) filters via circular convolution with the FFT, and it allows spatial-temporal processing of narrowband and broadband types of signals,[2,15,30] as defined in Chapter 6. Thus, each processing block in the generic DSP structure provides continuous time series; this is the central point of the implementation concept that allows the integration of quite diverse processing schemes, such as those shown in Figure 1.2.

More specifically, the details of the generic processing flow of Figure 1.2 are discussed very briefly in the following sections.

1.3.1 Signal Conditioning of Array Sensor Time Series

The block titled *Signal Conditioning for Array Sensor Time Series* in Figure 1.2 includes the partitioning of the time series from the receiving sensor array, their initial spectral FFT, the selection of the signal's frequency band of interest via bandpass FIR filters, and downsampling. The output of this block provides continuous time series at a reduced sampling rate for improved temporal spectral resolution. In many system applications including moving arrays of sensors, array shape estimation or the sensor coordinates would be required to be integrated with the signal processing functionality of the system, as shown in this block.

Typical system requirements of this kind are towed array sonars,[15] which are discussed in Chapters 6, 10, and 11; CT/X-ray tomography systems,[6–8] which are analyzed in Chapters 15 and 16; and ultrasound imaging systems deploying long line or planar arrays,[8–10] which are discussed in Chapters 6, 7, 13, and 14.

The processing details of this block will be illustrated in schematic diagrams in Chapter 6. The FIR band selection processing of this block is typical in all the real-time systems of interest. As a result, its output can be provided as input to the blocks named *Sonar, Radar & Ultrasound Systems* or *Tomography Imaging Systems.*

1.3.2 Tomography Imaging CT/X-Ray and MRI Systems

The block at the right-hand side of Figure 1.2, which is titled *Tomography Imaging Systems*, includes image reconstruction algorithms for medical imaging CT/X-ray and MRI systems. The processing details of these

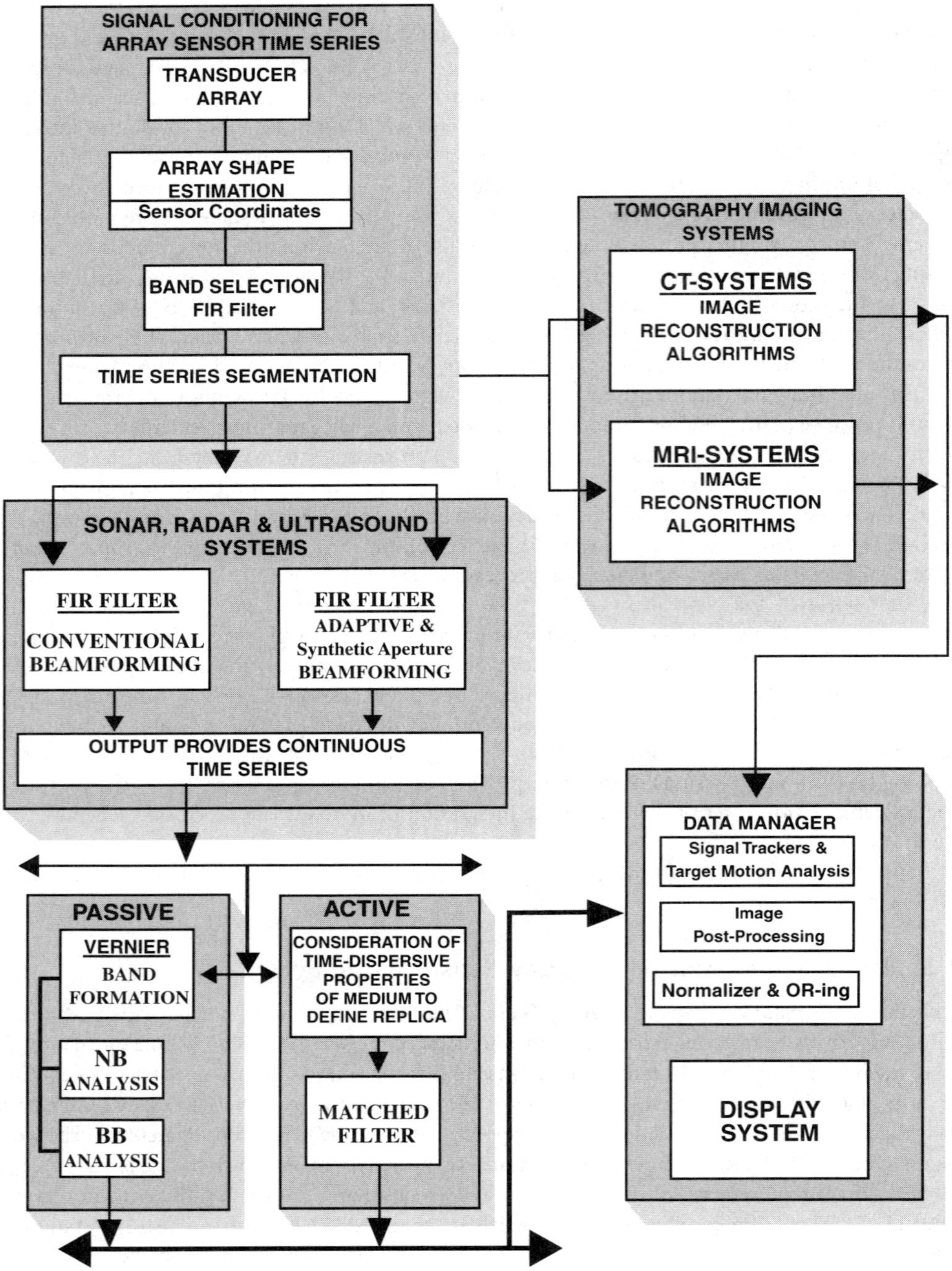

FIGURE 1.2 A generic signal processing structure integrating the signal processing functionalities of sonar, radar, ultrasound, CT/X-ray, and MRI medical imaging systems.

algorithms will be discussed in Chapters 15 through 17. In general, image reconstruction algorithms[6,7,11–13] are distinct processing schemes, and their implementation is practically efficient in CT and MRI applications. However, tomography imaging and the associated image reconstruction algorithms can be applied in other system applications such as diffraction tomography using ultrasound sources[8] and acoustic tomography of the ground using various acoustic frequency regimes. Diffraction tomography is not practical for medical imaging applications because of the very poor image resolution and the very high absorption rate of the acoustic energy by the bone structure of the human body. In geophysical applications, however, seismic waves can be used in tomographic imaging procedures to detect and classify very large buried objects. On the other hand, in working with higher acoustic frequencies, a better image resolution would allow detection and classification of small, shallow buried objects such as anti-personnel land mines,[41] which is a major humanitarian issue that has attracted the interest of U.N. and the highly industrialized countries in North America and Europe. The rule of thumb in acoustic tomography imaging applications is that higher frequency regimes in radiated acoustic energy would provide better image resolution at the expense of higher absorption rates for the radiated energy penetrating the medium of interest. All these issues and the relevant industrial applications of computed tomography imaging are discussed in Chapter 15.

1.3.3 Sonar, Radar, and Ultrasound Systems

The underlying signal processing functionality in sonar, radar, and modern ultrasound imaging systems deploying linear, planar, cylindrical, or spherical arrays is beamforming. Thus, the block in Figure 1.2 titled *Sonar, Radar & Ultrasound Systems* includes such sub-blocks as *FIR Filter/Conventional Beamforming* and *FIR Filter/Adaptive & Synthetic Aperture Beamforming* for multi-dimensional arrays with linear, planar, circular, cylindrical, and spherical geometric configurations. The output of this block provides continuous, directional beam time series by using the FIR implementation scheme of the spatial filtering via circular convolution. The segmentation and overlap of the time series at the input of the beamformers take care of the wraparound errors that arise in fast-convolution signal processing operations. The overlap size is equal to the effective FIR filter's length.[15,30] Chapter 6 will discuss in detail the conventional, adaptive, and sythetic aperture beamformers that can be implemented in this block of the generic processing structure in Figure 1.2. Moreover, Chapters 6 and 11 provide some real data output results from sonar systems deploying linear or cylindrical arrays.

1.3.4 Active and Passive Systems

The blocks named *Passive* and *Active* in the generic structure of Figure 1.2 are the last major processes that are included in most of the DSP systems. Inputs to these blocks are continuous beam time series, which are the outputs of the conventional and advanced beamformers of the previous block. However, continuous sensor time series from the first block titled *Signal Conditioning for Array Sensor Time Series* can be provided as the input of the *Active* and *Passive* blocks for temporal spectral analysis. The block titled *Active* includes a *Matched Filter* sub-block for the processing of active signals. The option here is to include the medium's propagation characteristics in the replica of the active signal considered in the matched filter in order to improve detection and gain.[15,26] The sub-blocks *Vernier/Band Formation, NB* (Narrowband) *Analysis*, and *BB* (Broadband) *Analysis* include the final processing steps of a temporal spectral analysis for the beam time series. The inclusion of the *Vernier* sub-block is to allow the option for improved frequency resolution. Chapter 11 discusses the signal processing functionality and system-oriented applications associated with active and passive sonars. Furthermore, Chapter 13 extends the discussion to address the signal processing issues relevant with ultrasound medical imaging systems.

In summary, the strength of the generic processing structure in Figure 1.2 is that it identifies and exploits the processing concept similarities among radar, sonar, and medical imaging systems. Moreover, it enables the implementation of non-linear signal processing methods, adaptive and synthetic aperture, as well as the equivalent conventional approaches. This kind of parallel functionality for conventional and advanced processing schemes allows for a very cost-effective evaluation of any type of improvement during the concept demonstration phase.

As stated above, the derivation of the effective filter length of an FIR adaptive and synthetic aperture filtering operation is very essential for any type of application that will allow simultaneous NB and BB signal processing. This is a non-trivial problem because of the dynamic characteristics of the adaptive algorithms, and it has not as yet been addressed.

In the past, attempts to implement matrix-based signal processing methods such as adaptive processing were based on the development of systolic array H/W because systolic arrays allow large amounts of parallel computation to be performed efficiently since communications occur locally. Unfortunately, systolic arrays have been much less successful in practice than in theory. Systolic arrays big enough for real problems cannot fit on one board, much less on one chip, and interconnects have problems. A two-dimensional (2-D) systolic array implementation will be even more difficult. Recent announcements, however, include successful developments of super-scalar and massively parallel signal processing computers that have throughput capabilities of hundred of billions of GFLOPS.[40] It is anticipated that these recent computing architecture developments would address the computationally intensive scalar and matrix-based operations of advanced signal processing schemes for next-generation real-time systems.

Finally, the block *Data Manager* in Figure 1.2 includes the display system, normalizers, target motion analysis, image post-processing, and OR-ing operations to map the output results into the dynamic range of the display devices. This will be discussed in the next section.

1.4 Data Manager and Display Sub-System

Processed data at the output of the mainstream signal processing system must be stored in a temporary database before they are presented to the system operator for analysis. Until very recently, owing to the physical size and cost associated with constructing large databases, the *data manager* played a relatively small role in the overall capability of the aforementioned systems. However, with the dramatic drop in the cost of solid-state memories and the introduction of powerful microprocessors in the 1980s, the role of the *data manager* has now been expanded to incorporate post-processing of the signal processor's output data. Thus, post-processing operations, in addition to the traditional display data management functions, may include

- For sonar and radar systems
  - Normalization and OR-ing
  - Signal tracking
  - Localization
  - Data fusion
  - Classification functionality
- For medical imaging systems
  - Image post-processing
  - Normalizing operations
  - Registration and image fusion

It is apparent from the above discussion that for a next-generation DSP system, emphasis should be placed on the degree of interaction between the operator and the system through an operator-machine interface (OMI), as shown schematically in Figure 1.1. Through this interface, the operator may selectively proceed with localization, tracking, diagnosis, and classification tasks.

A high-level view of the generic requirements and the associated technologies of the data manager of a next-generation DSP system reflecting the above concerns could be as shown in Figure 1.3. The central point of Figure 1.3 is the operator that controls two kinds of displays (the processed information and tactical displays) through a continuous interrogation procedure. In response to the operator's request, the units in the *data manager* and *display sub-system* have a continuous interaction including data flow and requests for processing that include localization, tracking, classification for sonar-radar systems (Chapters 8 and 9), and diagnostic images for medical imaging systems (Chapter 7). Even though the processing steps of radar and airborne systems associated with localization, tracking, and classification have conceptual similarities with those of a sonar system, the processing techniques that have been successfully applied in airborne systems have not been successful with sonar systems. This

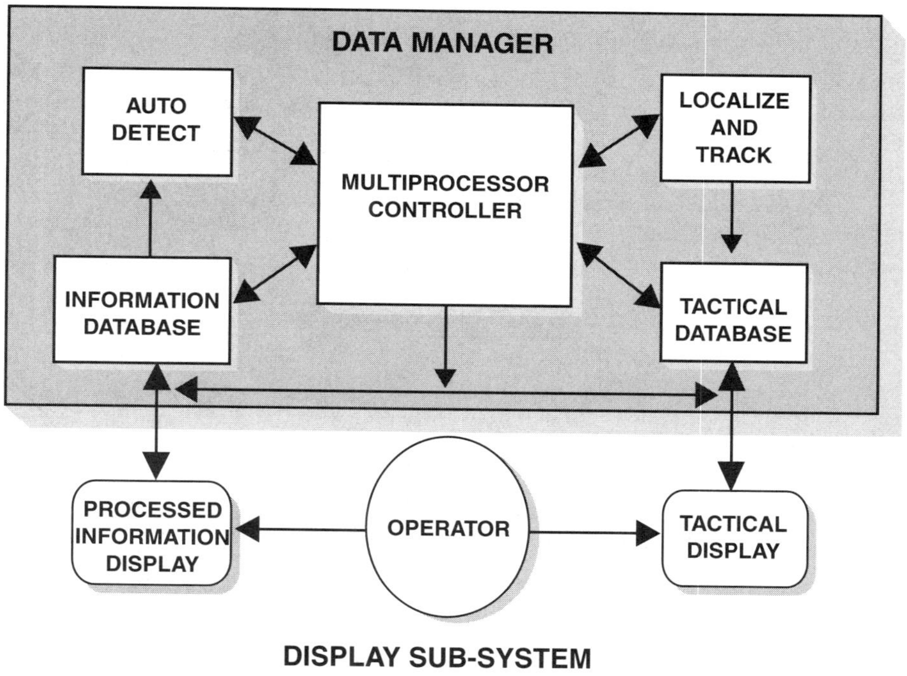

FIGURE 1.3 Schematic diagram for the generic requirements of a data manager for a next-generation, real-time DSP system.

is a typical situation that indicates how hostile, in terms of signal propagation characteristics, the underwater environment is with respect to the atmospheric environment. However, technologies associated with *data fusion, neural networks, knowledge-based systems,* and *automated parameter estimation* will provide solutions to the very difficult operational sonar problem regarding localization, tracking, and classification. These issues are discussed in detail in Chapters 8 and 9. In particular, Chapter 8 focuses on target tracking and sensor data processing for active sensors. Although active sensors certainly have an advantage over passive sensors, nevertheless, passive sensors may be prerequisite to some tracking solution concepts, namely, passive sonar systems. Thus, Chapter 9 deals with a class of tracking problems for passive sensors only.

1.4.1 Post-Processing for Sonar and Radar Systems

To provide a better understanding of these differences, let us examine the levels of information required by the data management of sonar and radar systems. Normally, for sonar and radar systems, the processing and integration of information from sensor level to a command and control level include a few distinct processing steps. Figure 1.4 shows a simplified overview of the integration of four different levels of information for a sonar or radar system. These levels consist mainly of

- Navigation and non-sensor array data
- Environmental information and estimation of propagation characteristics in order to assess the medium's influence on sonar or radar system performance
- Signal processing of received sensor signals that provide parameter estimation in terms of bearing, range, and temporal spectral estimates for detected signals
- Signal following (tracking) and localization that monitors the time evolution of a detected signal's estimated parameters

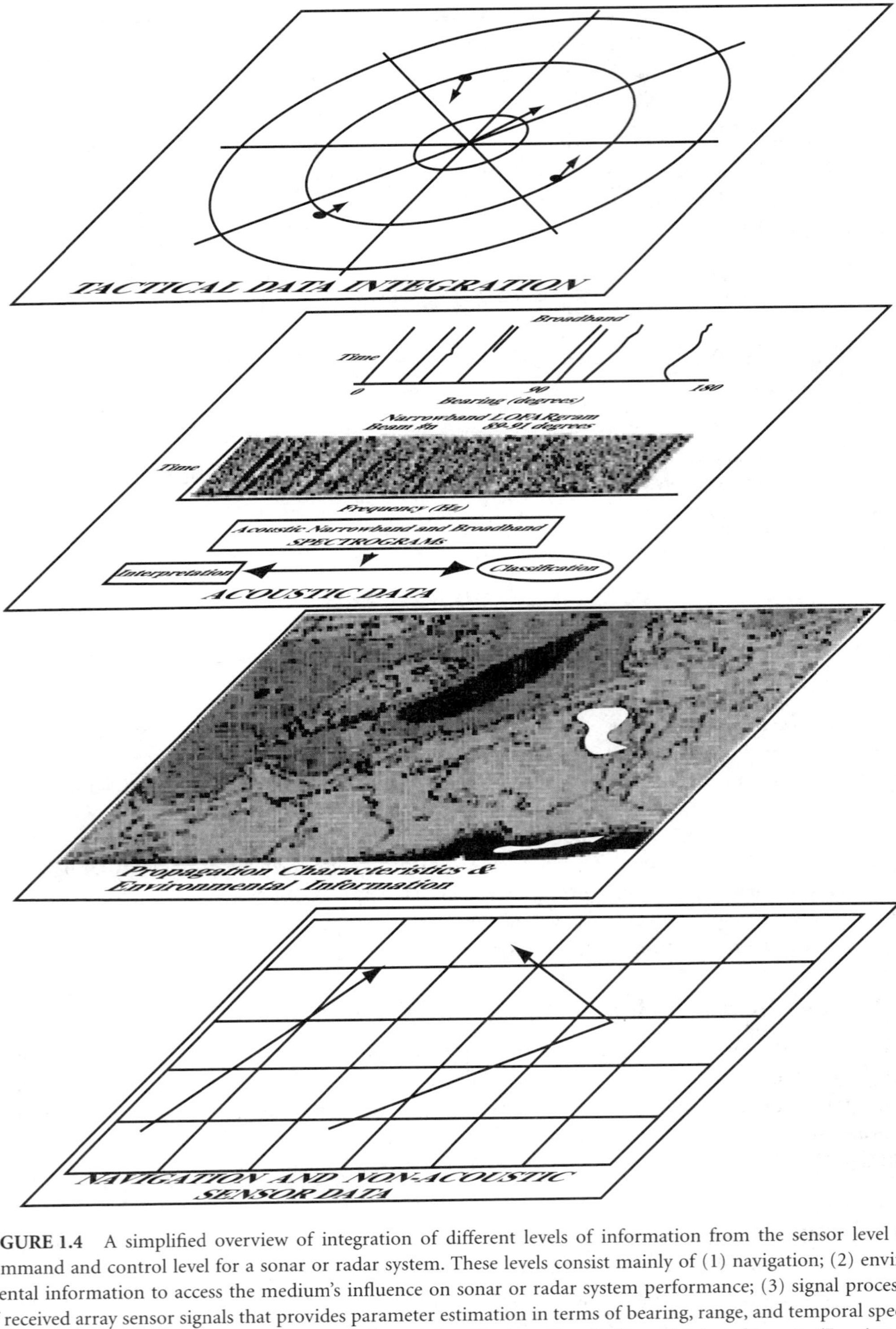

FIGURE 1.4 A simplified overview of integration of different levels of information from the sensor level to a command and control level for a sonar or radar system. These levels consist mainly of (1) navigation; (2) environmental information to access the medium's influence on sonar or radar system performance; (3) signal processing of received array sensor signals that provides parameter estimation in terms of bearing, range, and temporal spectral estimates for detected signals; and (4) signal following (tracking) and localization of detected targets. (Reprinted by permission of IEEE ©1998.)

This last tracking and localization capability[32,33] allows the sonar or radar operator to rapidly assess the data from a multi-sensor system and carry out the processing required to develop an array sensor-based tactical picture for integration into the platform level command and control system, as shown later by Figure 1.9.

In order to allow the databases to be searched effectively, a high-performance OMI is required. These interfaces are beginning to draw heavily on modern workstation technology through the use of windows, on-screen menus, etc. Large, flat panel displays driven by graphic engines which are equally adept at pixel manipulation as they are with 3-D object manipulation will be critical components in future systems. It should be evident by now that the term data manager describes a level of functionality which is well beyond simple data management. The data manager facility applies technologies ranging from relational databases, neural networks,[26] and fuzzy systems[27] to expert systems.[15,26] The problems it addresses can be variously characterized as signal, data, or information processing.

1.4.2 Post-Processing for Medical Imaging Systems

Let us examine the different levels of information to be integrated by the data manager of a medical imaging system. Figure 1.5 provides a simplified overview of the levels of information to be integrated by a current medical imaging system. These levels include

- The system structure in terms of array-sensor configuration and computing architecture
- Sensor time series signal processing structure
- Image processing structure
- Post-processing for reconstructed image to assist medical diagnosis

In general, current medical imaging systems include very limited post-processing functionality to enhance the images that may result from mainstream image reconstruction processing. It is anticipated, however, that next-generation medical imaging systems will enhance their capabilities in post-processing functionality by including image post-processing algorithms that are discussed in Chapters 7 and 14.

More specifically, although modern medical imaging modalities such as CT, MRA, MRI, nuclear medicine, 3-D ultrasound, and laser con-focal microscopy provide "slices of the body," significant differences exist between the image content of each modality. Post-processing, in this case, is essential with special emphasis on data structures, segmentation, and surface- and volume-based rendering for visualizing volumetric data. To address these issues, the first part of Chapter 7 focuses less on explaining algorithms and rendering techniques, but rather points out their applicability, benefits, and potential in the medical environment. Moreover, in the second part of Chapter 7, applications are illustrated from the areas of craniofacial surgery, traumatology, neurosurgery, radiotherapy, and medical education. Furthermore, some new applications of volumetric methods are presented: 3-D ultrasound, laser con-focal data sets, and 3D-reconstruction of cardiological data sets, i.e., vessels as well as ventricles. These new volumetric methods are currently under development, but due to their enormous application potential they are expected to be clinically accepted within the next few years.

As an example, Figures 1.6 and 1.7 present the results of image enhancement by means of post-processing on images that have been acquired by current CT/X-ray and ultrasound systems. The left-hand-side image of Figure 1.6 shows a typical X-ray image of a human skull provided by a current type of CT/X-ray imaging system. The right-hand-side image of Figure 1.6 is the result of post-processing the original X-ray image. It is apparent from these results that the right-hand-side image includes imaging details that can be valuable to medical staff in minimizing diagnostic errors and interpreting image results. Moreover, this kind of post-processing image functionality may assist in cognitive operations associated with medical diagnostic applications.

Ultrasound medical imaging systems are characterized by poor image resolution capabilities. The three images in Figure 1.7 (top left and right images, bottom left-hand-side image) provide pictures of the skull of a fetus as provided by a conventional ultrasound imaging system. The bottom right-hand-side image of Figure 1.7 presents the resulting 3-D post-processed image by applying the processing algorithms discussed in Chapter 7. The 3-D features and characteristics of the skull of the fetus are very pronounced in this case,

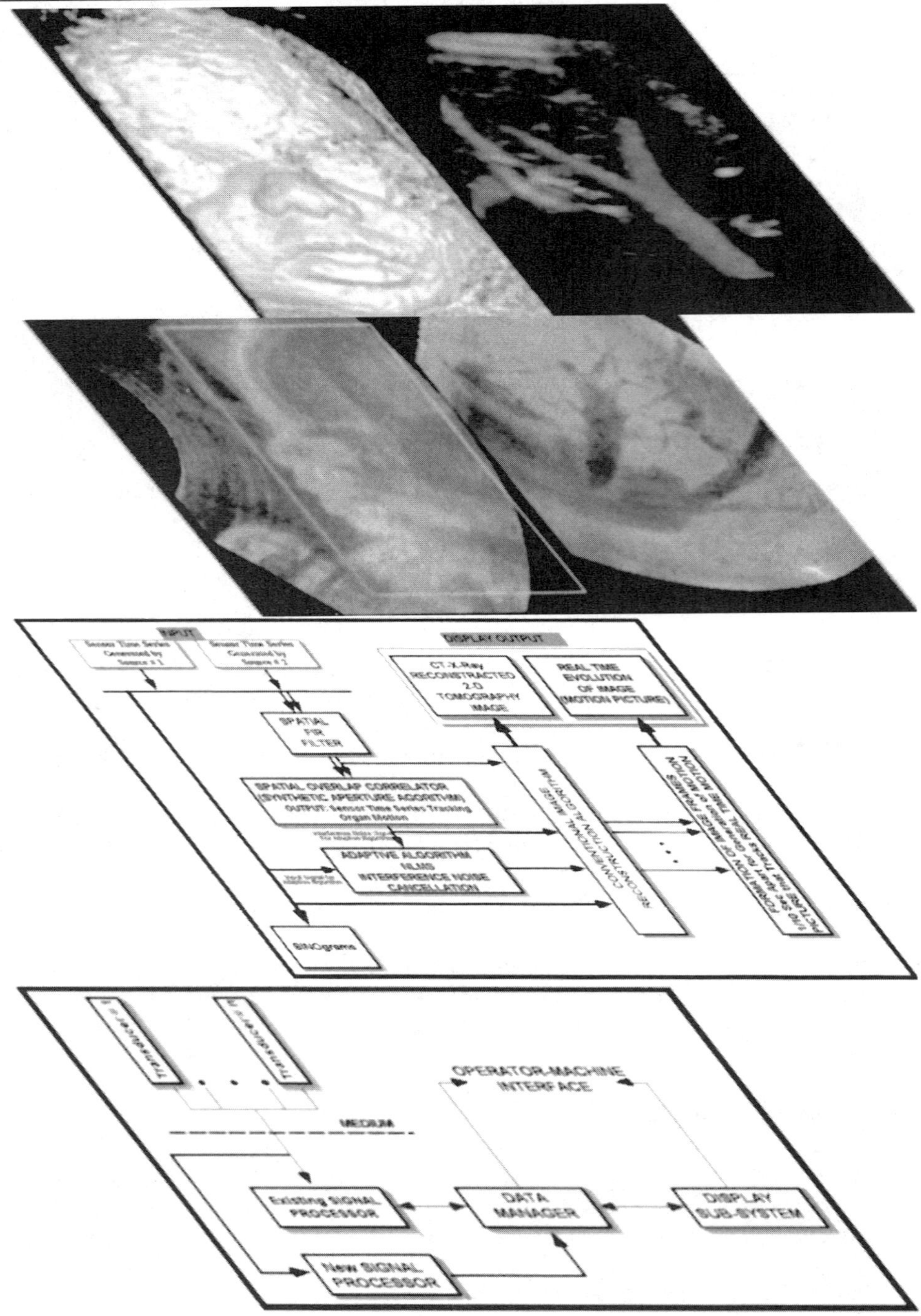

FIGURE 1.5 A simplified overview of the integration of different levels of information from the sensor level to a command and control level for a medical imaging system. These levels consist mainly of (1) sensor array configuration, (2) computing architecture, (3) signal processing structure, and (4) reconstructed image to assist medical diagnosis.

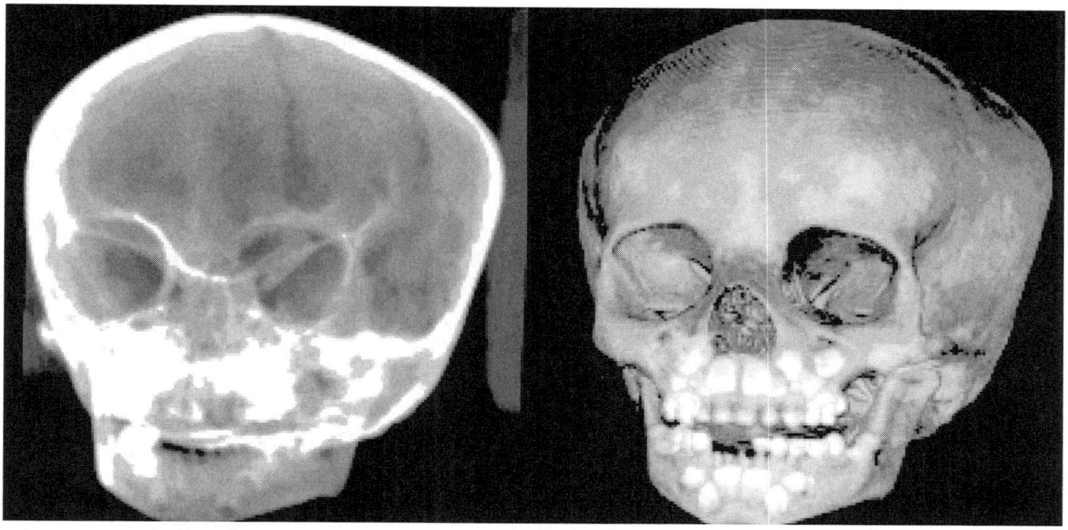

FIGURE 1.6 The left-hand-side is an X-ray image of a human skull. The right-hand-side image is the result of image enhancement by means of post-processing the original X-ray image. (Courtesy of Prof. G. Sakas, Fraunhofer IDG, Durmstadt, Germany.)

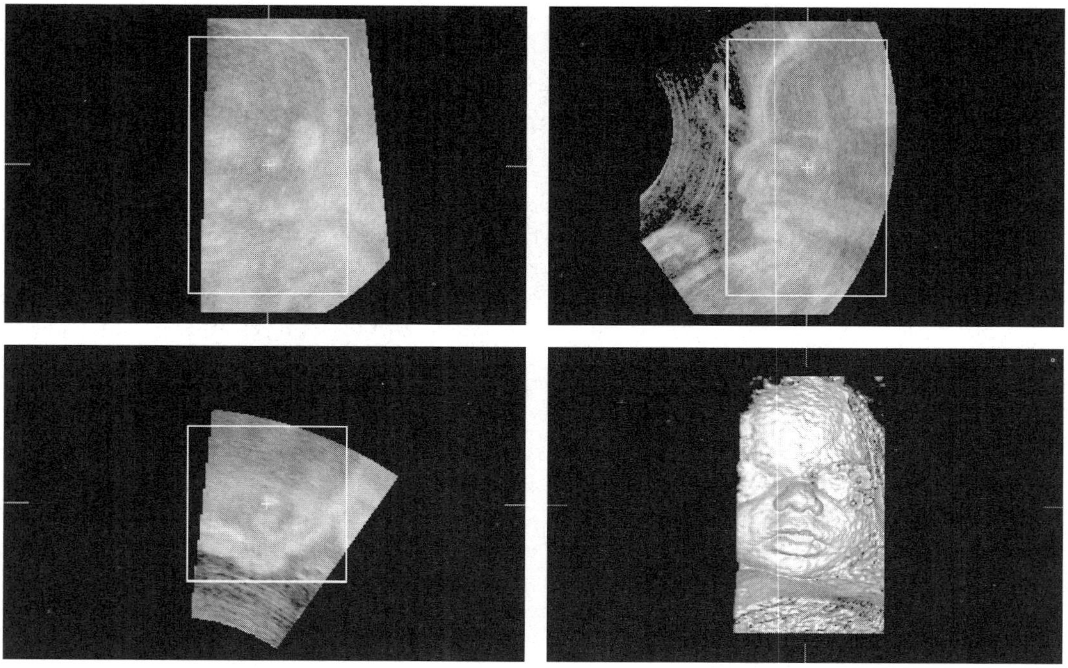

FIGURE 1.7 The two top images and the bottom left-hand-side image provide details of a fetus' skull using convetional medical ultrasound systems. The bottom right-hand-side 3-D image is the result of image enhancement by means of post-processing the original three ultrasound images. (Courtesy of Prof. G. Sakas, Fraunhofer IDG, Durmstadt, Germany.)

although the clarity is not as good as in the case of the CT/X-ray image in Figure 1.6. Nevertheless, the image resolution characteristics and 3-D features that have been reconstructed in both cases, shown in Figures 1.6 and 1.7, provide an example of the potential improvements in the image resolution and cognitive functionality that can be integrated in the next-generation medical imaging systems.

Needless to say, the image post-processing functionality of medical imaging systems is directly applicable in sonar and radar applications to reconstruct 2-D and 3-D image details of detected targets. This kind of image reconstruction post-processing capability may improve the difficult classification tasks of sonar and radar systems.

At this point, it is also important to re-emphasize the significant differences existing between the image content and system functionality of the various medical imaging systems mainly in terms of sensor-array configuration and signal processing structures. Undoubtedly, a generic approach exploiting the conceptually similar processing functionalities among the various configurations of medical imaging systems will simplify OMI issues that would result in better interpretation of information of diagnostic importance. Moreover, the integration of data fusion functionality in the data manager of medical imaging systems will provide better diagnostic interpretation of the information inherent at the output of the medical imaging systems by minimizing human errors in terms of interpretation.

Although these issues may appear as exercises of academic interest, it becomes apparent from the above discussion that system advances made in the field of sonar and radar systems may be applicable in medical imaging applications as well.

1.4.3 Signal and Target Tracking and Target Motion Analysis

In sonar, radar, and imaging system applications, single sensors or sensor networks are used to collect information on time-varying signal parameters of interest. The individual output data produced by the sensor systems result from complex estimation procedures carried out by the *signal processor* introduced in Section 1.3 (**sensor signal processing**). Provided the quantities of interest are related to moving point-source objects or small extended objects (radar targets, for instance), relatively simple statistical models can often be derived from basic physical laws, which describe their temporal behavior and thus define the underlying dynamical system. The formulation of adequate dynamics models, however, may be a difficult task in certain applications. For an efficient exploitation of the sensor resources as well as to obtain information not directly provided by the individual sensor reports, appropriate data association and estimation algorithms are required (**sensor data processing**). These techniques result in tracks, i.e., estimates of state trajectories, which statistically represent the quantities or objects considered along with their temporal history. Tracks are initiated, confirmed, maintained, stored, evaluated, fused with other tracks, and displayed by the *tracking system* or *data manager*. The tracking system, however, should be carefully distinguished from the underlying sensor systems, though there may exist close interrelations, such as in the case of multiple target tracking with an agile-beam radar, increasing the complexity of sensor management.

In contrast to the target tracking via active sensors, discussed in Chapter 8, Chapter 9 deals with a class of tracking problems that use passive sensors only. In solving tracking problems, active sensors certainly have an advantage over passive sensors. Nevertheless, passive sensors may be a prerequisite to some tracking solution concepts. This is the case, e.g., whenever active sensors are not feasible from a technical or tactical point of view, as in the case of passive sonar systems deployed by submarines and surveillance naval vessels. An important problem in passive target tracking is the target motion analysis (TMA) problem. The term TMA is normally used for the process of estimating the state of a radiating target from noisy measurements collected by a single passive observer. Typical applications can be found in passive sonar, infrared (IR), or radar tracking systems.

For signal followers, the parameter estimation process for tracking the bearing and frequency of detected signals consists of peak picking in a region of bearing and frequency space sketched by fixed gate sizes at the outputs of the conventional and non-conventional beamformers depicted in Figure 1.2. Figure 1.8 provides a schematic interpretation of the signal followers functionality in tracking the time-varying frequency and bearing estimates of detected signals in sonar and radar applications. Details about this

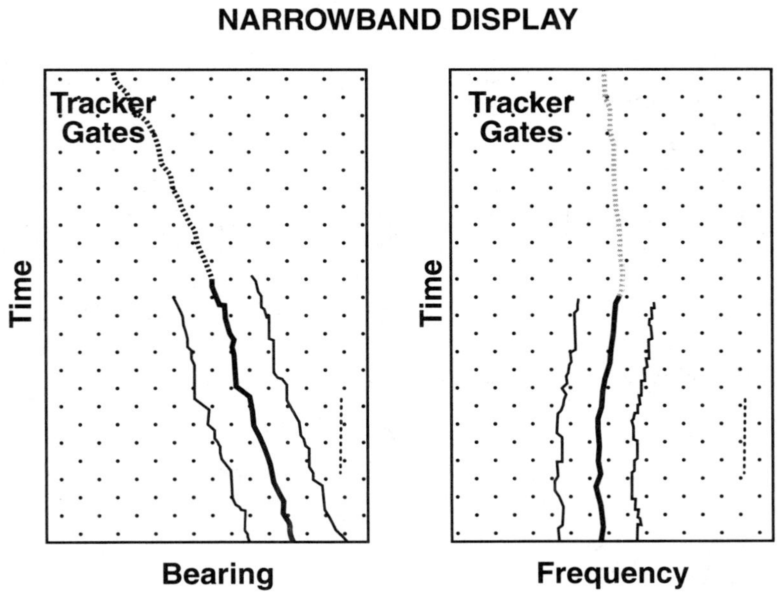

FIGURE 1.8 Signal following functionality in tracking the time-varying frequency and bearing of a detected signal (target) by a sonar or radar system. (Courtesy of William Cambell, Defence Research Establishment Atlantic, Dartmouth, NS, Canada.)

estimation process can be found in Reference 34 and in Chapters 8 and 9 of this handbook. Briefly, in Figure 1.8, the choice of the gate sizes was based on the observed bearing and frequency fluctuations of a detected signal of interest during the experiments. Parabolic interpolation was used to provide refined bearing estimates.[35] For this investigation, the bearings-only tracking process described in Reference 34 was used as an NB tracker, providing unsmoothed time evolution of the bearing estimates to the localization process.[32,36]

Tracking of the time-varying bearing estimates of Figure 1.8 forms the basic processing step to localize a distant target associated with the bearing estimates. This process is called localization or TMA, which is discussed in Chapter 9. The output results of a TMA process form the tactical display of a sonar or radar system, as shown in Figures 1.4 and 1.8. In addition, the temporal-spatial spectral analysis output results and the associated display (Figures 1.4 and 1.8) form the basis for classification and the target identification process for sonar and radar systems. In particular, data fusion of the TMA output results with those of temporal-spatial spectral analysis output results outline an integration process to define the tactical picture for sonar and radar operations, as shown in Figure 1.9. For more details, the reader is referred to Chapters 8 and 9, which provide detailed discussions of target tracking and TMA operations for sonar and radar systems.[32–36]

It is apparent from the material presented in this section that for next-generation sonar and radar systems, emphasis should be placed on the degree of interaction between the operator and the system, through an OMI as shown schematically in Figures 1.1 and 1.3. Through this interface, the operator may selectively proceed with localization, tracking, and classification tasks, as depicted in Figure 1.7.

In standard computed tomography (CT), image reconstruction is performed using projection data that are acquired in a time sequential manner.[6,7] Organ motion (cardiac motion, blood flow, lung motion due to respiration, patient's restlessness, etc.) during data acquisition produces artifacts, which appear as a blurring effect in the reconstructed image and may lead to inaccurate diagnosis.[14] The intuitive solution to this problem is to speed up the data acquisition process so that the motion effects become negligible. However, faster CT scanners tend to be significantly more costly, and, with current X-ray tube technology, the scan times that are required are simply not realizable. Therefore, signal processing algorithms to account for organ motion artifacts are needed. Several mathematical techniques have been proposed as a solution

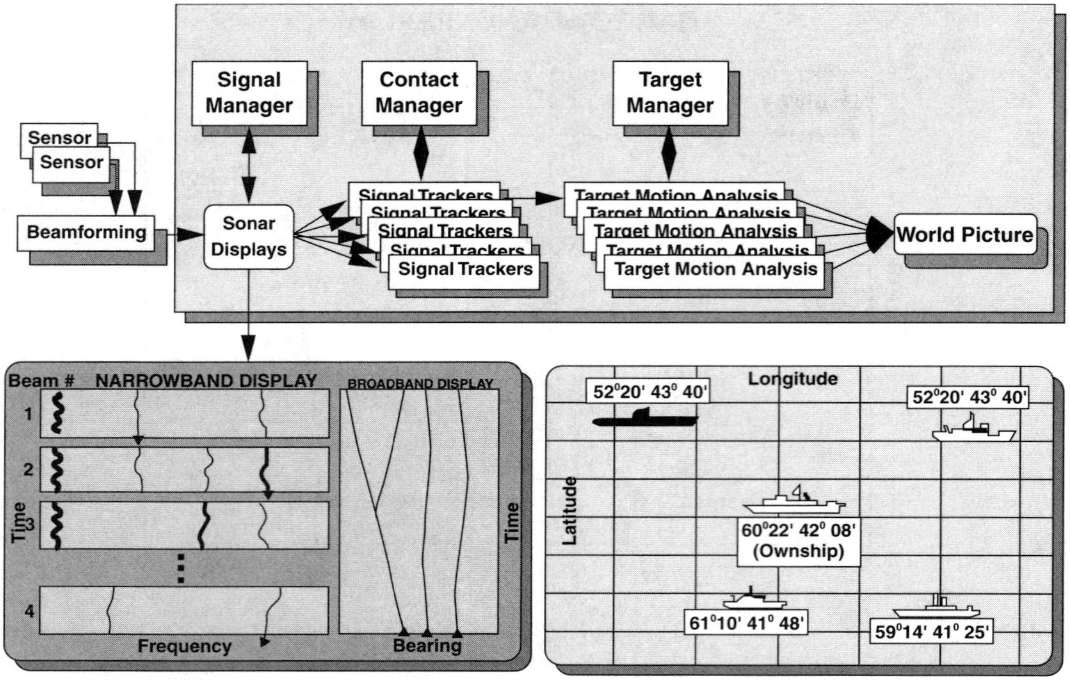

FIGURE 1.9 Formation of a tactical picture for sonar and radar systems. The basic operation is to integrate by means of data fusion the signal tracking and localization functionality with the temporal-spatial spectral analysis output results of the generic signal processing structure of Figure 1.2. (Courtesy of Dr. William Roger, Defence Research Establishment Atlantic, Dartmouth, NS, Canada.)

to this problem. These techniques usually assume a simplistic linear model for the motion, such as translational, rotational, or linear expansion.[14] Some techniques model the motion as a periodic sequence and take projections at a particular point in the motion cycle to achieve the effect of scanning a stationary object. This is known as a retrospective electrocardiogram (ECG)-gating algorithm, and projection data are acquired during 12 to 15 continuous 1-s source rotations while cardiac activity is recorded with an ECG. Thus, the integration of ECG devices with X-ray CT medical tomography imaging systems becomes a necessity in cardiac imaging applications using X-ray CT and MRI systems. However, the information provided by the ECG devices to select in-phase segments of CT projection data can be available by signal trackers that can be applied on the sensor time series of the CT receiving array. This kind of application of signal trackers on CT sensor time series will identify the in-phase motion cycles of the heart under a similar configuration as the ECG-gating procedure. Moreover, the application of the signal trackers in cardiac CT imaging systems will eliminate the use of the ECG systems, thus making the medical imaging operations much simpler. These issues will be discussed in some detail in Chapter 16.

It is anticipated, however, that radar, sonar, and medical imaging systems will exhibit fundamental differences in their requirements for information post-processing functionality. Furthermore, bridging conceptually similar processing requirements may not always be an optimum approach in addressing practical DSP implementation issues; rather it should be viewed as a source of inspiration for the researchers in their search for creative solutions.

In summary, past experience in DSP system development that "improving the signal processor of a sonar or radar or medical imaging system was synonymous with the development of new signal processing algorithms and faster hardware" has changed. While advances will continue to be made in these areas, future developments in data (contact) management represent one of the most exciting avenues of research in the development of high-performance systems.

In sonar, radar, and medical imaging systems, an issue of practical importance is the operational requirement by the operator to be able to rapidly assess numerous images and detected signals in terms of localization, tracking, classification, and diagnostic interpretation in order to pass the necessary information up through the chain of command to enable tactical or medical diagnostic decisions to be made in a timely manner. Thus, an assigned task for a *data manager* would be to provide the operator with quick and easy access to both the output of the *signal processor*, which is called processed data display, and the tactical display, which will show medical images and localization and tracking information through graphical interaction between the processed data and tactical displays.

1.4.4 Engineering Databases

The design and integration of engineering databases in the functionality of a *data manager* assist the identification and classification process, as shown schematically in Figure 1.3. To illustrate the concept of an engineering database, we will consider the land mine identification process, which is a highly essential functionality in humanitarian demining systems to minimize the false alarm rate. Although a lot of information on land mines exists, often organized in electronic databases, there is nothing like a CAD engineering database. Indeed, most databases serve either documentation purposes or are land mine signatures related to a particular sensor technology. This wealth of information must be collected and organized in such a way so that it can be used online, through the necessary interfaces to the sensorial information, by each one of the future identification systems. Thus, an engineering database is intended to be the common core software applied to all future land mine detection systems.[41] It could be built around a specially engineered database storing all available information on land mines. The underlying idea is, using techniques of cognitive and perceptual sciences, to extract the particular features that characterize a particular mine or a class of mines and, successively, to define the sensorial information needed to detect these features in typical environments. Such a land mine identification system would not only trigger an alarm for every suspect object, but would also reconstruct a comprehensive model of the target. Successively, it would compare the model to an existing land mine engineering database deciding or assisting the operator to make a decision as to the nature of the detected object.

A general approach of the engineering database concept and its applicability in the aforementioned DSP systems would assume that an effective engineering database will be a function of the available information on the subjects of interest, such as underwater targets, radar targets, and medical diagnostic images. Moreover, the functionality of an engineering database would be highly linked with the multi-sensor data fusion process, which is the subject of discussion in the next section.

1.4.5 Multi-Sensor Data Fusion

Data fusion refers to the acquisition, processing, and synergistic combination of information from various knowledge sources and sensors to provide a better understanding of the situation under consideration.[39] Classification is an information processing task in which specific entities are mapped to general categories. For example, in the detection of land mines, the fusion of acoustic,[41] electromagnetic (EM), and IR sensor data is in consideration to provide a better land mine field picture and minimize the false alarm rates. The discussion of this section has been largely influenced by the work of Kundur and Hatzinakos[39] on "Blind Image Deconvolution" (for more details the reader is referred to Reference 39).

The process of multi-sensor data fusion addresses the issue of system integration of different type of sensors and the problems inherent in attempting to fuse and integrate the resulting data streams into a coherent picture of operational importance. The term *integration* is used here to describe operations wherein a sensor input may be used independently with respect to other sensor data in structuring an overall solution. ***Fusion*** is used to describe the result of joint analysis of two or more originally distinct data streams.

More specifically, while multi-sensors are more likely to correctly identify positive targets and eliminate false returns, using them effectively will require ***fusing*** the incoming data streams, each of which may have a different character. This task will require solutions to the following engineering problems:

- Correct combination of the multiple data streams in the same context
- Processing multiple signals to eliminate false positives and further refine positive returns

For example, in humanitarian demining, a positive return from a simple metal detector might be combined with a ground penetrating radar (GPR) evaluation, resulting in the classification of the target as a spent shell casing and allowing the operator to safely pass by in confidence.

Given a design that can satisfy the above goals, it will then be possible to design and implement computer-assisted or automatic recognition in order to positively identify the nature, position, and orientation of a target. Automatic recognition, however, will be pursued by the engineering database, as shown in Figure 1.3.

In data fusion, another issue of equal importance is the ability to deal with conflicting data, producing interim results that the algorithm can revise as more data become available. In general, the data interpretation process, as part of the functionality of data fusion, consists briefly of the following stages:[39]

- Low-level data manipulation
- Extraction of features from the data either using signal processing techniques or physical sensor models
- Classification of data using techniques such as Bayesian hypothesis testing, fuzzy logic, and neural networks
- Heuristic expert system rules to guide the previous levels, make high-level control decisions, provide operator guidance, and provide early warnings and diagnostics

Current research and development (R&D) projects in this area include the processing of localization and identification of data from various sources or type of sensors. The systems combine features of modern multi-hypothesis tracking methods and correlation. This approach, to process all available data regarding targets of interest, allows the user to extract the maximum amount of information concerning target location from the complex "sea" of available data. Then a correlation algorithm is used to process large volumes of data containing localization and to attribute information using multiple hypothesis methods.

In image classification and fusion strategies, many inaccuracies often result from attempting to fuse data that exhibit motion-induced blurring or defocusing effects and background noise.[37,38] Compensation for such distortions is inherently sensor dependent and non-trivial, as the distortion is often time varying and unknown. In such cases, blind image processing, which relies on partial information about the original data and the distorting process, is suitable.[39]

In general, multi-sensor data fusion is an evolving subject, which is considered to be highly essential in resolving the sonar, radar detection/classification, and diagnostic problems in medical imaging systems. Since a single sensor system with an acceptable very low false alarm rate is rarely available, current developments in sonar, radar, and medical imaging systems include multi-sensor configurations to minimize the false alarm rates. Then the multi-sensor data fusion process becomes highly essential. Although data fusion and databases have not been implemented yet in medical imaging systems, their potential use in this area will undoubtedly be a rapidly evolving R&D subject in the near future. Then system experience in the areas of sonar and radar systems would be a valuable asset in that regard. For medical imaging applications, the data and image fusion processes will be discussed in detail in Chapter 19.

Finally, Chapter 20 concludes the material of this handbook by providing clinical data and discussion on the role of medical imaging in radiotherapy treatment planning.

References

1. W.C. Knight, R.G. Pridham, and S.M. Kay, Digital signal processing for sonar, *Proc. IEEE,* 69(11), 1451–1506, 1981.
2. B. Windrow et al., Adaptive antenna systems, *Proc. IEEE,* 55(12), 2143–2159, 1967.
3. B. Windrow and S.D. Stearns, *Adaptive Signal Processing*, Prentice-Hall, Englewood Cliffs, NJ, 1985.
4. A. A. Winder, Sonar system technology, *IEEE Trans. Sonic Ultrasonics*, SU-22(5), 291–332, 1975.
5. A.B. Baggeroer, Sonar signal processing, in *Applications of Digital Signal Processing*, A.V. Oppenheim, Ed., Prentice-Hall, Englewood Cliffs, NJ, 1978.
6. H.J. Scudder, Introduction to computer aided tomography, *Proc. IEEE*, 66(6), 628–637, 1978.
7. A.C. Kak and M. Slaney, *Principles of Computerized Tomography Imaging*, IEEE Press, New York, 1992.
8. D. Nahamoo and A.C. Kak, Ultrasonic Diffraction Imaging, TR-EE 82–80, Department of Electrical Engineering, Purdue University, West Lafayette, IN, August 1982.
9. S.W. Flax and M. O'Donnell, Phase-aberration correction using signals from point reflectors and diffuse scatterers: basic principles, *IEEE Trans. Ultrasonics, Ferroelectrics Frequency Control*, 35(6), 758–767, 1988.
10. G.C. Ng, S.S. Worrell, P.D. Freiburger, and G.E. Trahey, A comparative evaluation of several algorithms for phase aberration correction, *IEEE Trans. Ultrasonics, Ferroelectrics Frequency Control*, 41(5), 631–643, 1994.
11. A.K. Jain, *Fundamentals of Digital Image Processing*, Prentice-Hall, Englewood Cliffs, NJ, 1990.
12. Q.S. Xiang and R.M. Henkelman, K-space description for the imaging of dynamic objects, *Magn. Reson. Med.*, 29, 422–428, 1993.
13. M.L. Lauzon, D.W. Holdsworth, R. Frayne, and B.K. Rutt, Effects of physiologic waveform variability in triggered MR imaging: theoretical analysis, *J. Magn. Reson. Imaging*, 4(6), 853–867, 1994.
14. C.J. Ritchie, C.R. Crawford, J.D. Godwin, K.F. King, and Y. Kim, Correction of computed tomography motion artifacts using pixel-specific back-projection, *IEEE Trans. Medical Imaging*, 15(3), 333–342, 1996.
15. S. Stergiopoulos, Implementation of adaptive and synthetic-aperture processing schemes in integrated active-passive sonar systems, *Proc. IEEE*, 86(2), 358–396, 1998.
16. D. Stansfield, *Underwater Electroacoustic Transducers*, Bath University Press and Institute of Acoustics, 1990.
17. J.M. Powers, Long range hydrophones, in *Applications of Ferroelectric Polymers,* T.T. Wang, J.M. Herbert, and A.M. Glass, Eds., Chapman & Hall, New York, 1988.
18. P.B. Boemer, W.A. Edelstein, C.E. Hayes, S.P. Souza, and O.M. Mueller, The NMR phased array, *Magn. Reson. Med.*, 16, 192–225, 1990.
19. P.S. Melki, F.A. Jolesz, and R.V. Mulkern, Partial RF echo planar imaging with the FAISE method. I. Experimental and theoretical assessment of artifact, *Magn. Reson. Med.*, 26, 328–341, 1992.
20. N.L. Owsley, *Sonar Array Processing*, S. Haykin, Ed., Signal Processing Series, A.V. Oppenheim, Series Ed., p. 123, Prentice-Hall, Englewood Cliffs, NJ, 1985.
21. B. Van Veen and K. Buckley, Beamforming: a versatile approach to spatial filtering, *IEEE ASSP Mag.*, 4–24, 1988.
22. A.H. Sayed and T. Kailath, A state-space approach to adaptive RLS filtering, *IEEE SP Mag.*, July, 18–60, 1994.
23. E.J. Sullivan, W.M. Carey, and S. Stergiopoulos, Editorial special issue on acoustic synthetic aperture processing, *IEEE J. Oceanic Eng.*, 17(1), 1–7, 1992.
24. C.L. Nikias and J.M. Mendel, Signal processing with higher-order spectra, *IEEE SP Mag.*, July, 10–37, 1993.
25. S. Stergiopoulos and A.T. Ashley, Guest Editorial for a special issue on sonar system technology, *IEEE J. Oceanic Eng.*, 18(4), 361–365, 1993.

26. A.B. Baggeroer, W.A. Kuperman, and P.N. Mikhalevsky, An overview of matched field methods in ocean acoustics, *IEEE J. Oceanic Eng.*, 18(4), 401–424, 1993.
27. "Editorial" special issue on neural networks for oceanic engineering systems, *IEEE J. Oceanic Eng.*, 17, 1–3, October 1992.
28. A. Kummert, Fuzzy technology implemented in sonar systems, *IEEE J. Oceanic Eng.*, 18(4), 483–490, 1993.
29. R.D. Doolitle, A. Tolstoy, and E.J. Sullivan, Editorial special issue on detection and estimation in matched field processing, *IEEE J. Oceanic Eng.*, 18, 153–155, 1993.
30. A. Antoniou, *Digital Filters: Analysis, Design, and Applications, 2nd Ed.*, McGraw-Hill, New York, 1993.
31. Mercury Computer Systems, Inc., *Mercury News Jan-97*, Mercury Computer Systems, Inc., Chelmsford, MA, 1997.
32. Y. Bar-Shalom and T.E. Fortman, *Tracking and Data Association*, Academic Press, Boston, MA, 1988.
33. S.S. Blackman, *Multiple-Target Tracking with Radar Applications*, Artech House Inc., Norwood, MA, 1986.
34. W. Cambell, S. Stergiopoulos, and J. Riley, Effects of bearing estimation improvements of non-conventional beamformers on bearing-only tracking, Proc. Oceans '95 MTS/IEEE, San Diego, CA, 1995.
35. W.A. Roger and R.S. Walker, Accurate estimation of source bearing from line arrays, Proc. Thirteen Biennial Symposium on Communications, Kingston, Ontario, Canada, 1986.
36. D. Peters, Long Range Towed Array Target Analysis — Principles and Practice, DREA Memorandum 95/217, Defence Research Establishment Atlantic, Dartmouth, NS, Canada, 1995.
37. A.H.S. Solberg, A.K. Jain, and T. Taxt, A Markov random field model for classification of multi-source satellite imagery, *IEEE Trans. Geosci. Remote Sensing*, 32, 768–778, 1994.
38. L.J. Chipman et al., Wavelets and image fusion, *Proc. SPIE*, 2569, 208–219, 1995.
39. D. Kundur and D. Hatzinakos, Blind image deconvolution, *Signal Processing Magazine*, 13, 43–64, May 1996.
40. Mercury Computer Systems, Inc., *Mercury News Jan-98*, Mercury Computer Systems, Inc., Chelmsford, MA, 1998.
41. S. Stergiopoulos, R. Alterson, D. Havelock, and J. Grodski, Acoustic Tomography Methods for 3D Imaging of Shallow Buried Objects, 139th Meeting of the Acoustical Society of America, Atlanta, GA, May 2000.

I

General Topics on Signal Processing

2

Adaptive Systems for Signal Process*

Simon Haykin
McMaster University

2.1 The Filtering Problem

The term "filter" is often used to describe a device in the form of a piece of physical hardware or software that is applied to a set of noisy data in order to extract information about a prescribed quantity of interest. The noise may arise from a variety of sources. For example, the data may have been derived by means of noisy sensors or may represent a useful signal component that has been corrupted by transmission through a communication channel. In any event, we may use a filter to perform three basic information-processing tasks.

1. *Filtering* means the extraction of information about a quantity of interest at time t by using data measured up to and including time t.
2. *Smoothing* differs from filtering in that information about the quantity of interest need not be available at time t, and data measured later than time t can be used in obtaining this information. This means that in the case of smoothing there is a *delay* in producing the result of interest. Since

* The material presented in this chapter is based on the author's two textbooks: (1) *Adaptive Filter Theory* (1996) and (2) *Neural Networks: A Comprehensive Foundation* (1999), Prentice-Hall, Englewood Cliffs, NJ.

0-8493-3691-0/01/$0.00+$.50

in the smoothing process we are able to use data obtained not only up to time t, but also data obtained after time t, we would expect smoothing to be more accurate in some sense than filtering.

3. *Prediction* is the forecasting side of information processing. The aim here is to derive information about what the quantity of interest will be like at some time $t + \tau$ in the future, for some $\tau > 0$, by using data measured up to and including time t.

We may classify filters into linear and nonlinear. A filter is said to be *linear* if the filtered, smoothed, or predicted quantity at the output of the device is a *linear function of the observations applied to the filter input.* Otherwise, the filter is *nonlinear.*

In the statistical approach to the solution of the *linear filtering problem* as classified above, we assume the availability of certain statistical parameters (i.e., *mean and correlation functions*) of the useful signal and unwanted additive noise, and the requirement is to design a linear filter with the noisy data as input so as to minimize the effects of noise at the filter output according to some statistical criterion. A useful approach to this filter-optimization problem is to minimize the mean-square value of the *error* signal that is defined as the difference between some desired response and the actual filter output. For stationary inputs, the resulting solution is *commonly* known as the *Wiener filter,* which is said to be *optimum in the mean-square sense.* A plot of the mean-square value of the error signal vs. the adjustable parameters of a linear filter is referred to as the *error-performance surface.* The minimum point of this surface represents the Wiener solution.

The Wiener filter is inadequate for dealing with situations in which *nonstationarity* of the signal and/or noise is intrinsic to the problem. In such situations, the optimum filter has to assume a *time-varying form.* A highly successful solution to this more difficult problem is found in the *Kalman filter,* a powerful device with a wide variety of engineering applications.

Linear filter theory, encompassing both Wiener and Kalman filters, has been developed fully in the literature for *continuous-time* as well as *discrete-time* signals. However, for technical reasons influenced by the wide availability of digital computers and the ever-increasing use of digital signal-processing devices, we find in practice that the discrete-time representation is often the preferred method. Accordingly, in this chapter, we only consider the discrete-time version of Wiener and Kalman filters. In this method of representation, the input and output signals, as well as the characteristics of the filters themselves, are all defined at discrete instants of time. In any case, a continuous-time signal may always be represented by a *sequence of samples* that are derived by observing the signal at uniformly spaced instants of time. No loss of information is incurred during this conversion process provided, of course, we satisfy the well-known *sampling theorem,* according to which the sampling rate has to be greater than twice the highest frequency component of the continuous-time signal (assumed to be of a low-pass kind). We may thus represent a continuous-time signal $u(t)$ by the sequence $u(n)$, $n = 0, \pm1, \pm2, \ldots$, where for convenience we have normalized the sampling period to unity, a practice that we follow throughout this chapter.

2.2 Adaptive Filters

The design of a Wiener filter requires *a priori* information about the statistics of the data to be processed. The filter is optimum only when the statistical characteristics of the input data match the *a priori* information on which the design of the filter is based. When this information is not known completely, however, it may not be possible to design the Wiener filter or else the design may no longer be optimum. A straightforward approach that we may use in such situations is the "estimate and plug" procedure. This is a two-stage process whereby the filter first "estimates" the statistical parameters of the relevant signals and then "plugs" the results so obtained into a *nonrecursive* formula for computing the filter parameters. For a *real-time operation,* this procedure has the disadvantage of requiring excessively elaborate and costly hardware. A more efficient method is to use an adaptive filter. By such a device we mean one that is *self-designing* in that the adaptive filter relies on a *recursive algorithm* for its operation, which makes it possible for the filter to perform satisfactorily in an environment where complete knowledge of the relevant signal characteristics is not available. The algorithm starts from some predetermined set of *initial conditions,* representing whatever we know about the environment. Yet, in a stationary environment, we find that after successive iterations of the algorithm it *converges* to the optimum Wiener solution in some statistical sense. In a nonstationary environment, the algorithm offers a *tracking* capability, in that it can track time variations in the statistics of the input data, provided that the variations are sufficiently slow.

As a direct consequence of the application of a recursive algorithm whereby the parameters of an adaptive filter are updated from one iteration to the next, the parameters become *data dependent.* This, therefore, means that an adaptive filter is in reality a *nonlinear device, in the sense that it does not obey the principle of superposition.* Notwithstanding this property, adaptive filters are commonly classified as linear or nonlinear. An adaptive filter is said to be *linear* if the estimate of quantity of interest is computed adaptively (at the output of the filter) as a *linear combination of the available set of observations applied to the filter input.* Otherwise, the adaptive filter is said to be *nonlinear.*

A wide variety of recursive algorithms have been developed in the literature of the operation of linear adaptive filters. In the final analysis, the choice of one algorithm over another is determined by one or more of the following factors:

- *Rate of convergence* — This is defined as the number of iterations required for the algorithm, in response to stationary inputs, to converge "close enough" to the optimum Wiener solution in the mean-square sense. A fast rate of convergence allows the algorithm to adapt rapidly to a stationary environment of unknown statistics.
- *Misadjustment* — For an algorithm of interest, this parameter provides a quantitative measure of the amount by which the final value of the mean-squared error, averaged over an ensemble of adaptive filters, deviates from the minimum mean-squared error that is produced by the Wiener filter.
- *Tracking* — When an adaptive filtering algorithm operates in a nonstationary environment, the algorithm is required to *track* statistical variations in the environment. The tracking performance of the algorithm, however, is influenced by two contradictory features: (1) the rate of convergence and (b) the steady-state fluctuation due to algorithm noise.
- *Robustness* — For an adaptive filter to be *robust,* small disturbances (i.e., disturbances with small energy) can only result in small estimation errors. The disturbances may arise from a variety of factors internal or external to the filter.
- *Computational requirements* — Here, the issues of concern include (1) the number of operations (i.e., multiplications, divisions, and additions/subtractions) required to make one complete iteration of the algorithm, (2) the size of memory locations required to store the data and the program, and (3) the investment required to program the algorithm on a computer.
- *Structure* — This refers to the structure of information flow in the algorithm, determining the manner in which it is implemented in hardware form. For example, an algorithm whose structure exhibits high modularity, parallelism, or concurrency is well suited for implementation using very large-scale integration (VLSI).*
- *Numerical properties* — When an algorithm is implemented numerically, inaccuracies are produced due to *quantization errors.* The quantization errors are due to analog-to-digital conversion of the input data and digital representation of internal calculations. Ordinarily, it is the latter source of quantization errors that poses a serious design problem. In particular, there are two basic issues

* VLSI technology favors the implementation of algorithms that possess high modularity, parallelism, or concurrency. We say that a structure is *modular* when it consists of similar stages connected in cascade. By *parallelism,* we mean a large number of operations being performed side by side. By *concurrency,* we mean a large number of *similar* computations being performed at the same time. For a discussion of VLSI implementation of adaptive filters, see Shabhag and Parhi (1994). This book emphasizes the use of *pipelining,* an architectural technique used for increasing the throughput of an adaptive filtering algorithm.

of concern: numerical stability and numerical accuracy. *Numerical stability* is an inherent characteristic of an adaptive filtering algorithm. *Numerical accuracy,* on the other hand, is determined by the number of *bits* (i.e., **bi**nary digi**ts** used in the numerical representation of data samples and filter coefficients). An adaptive filtering algorithm is said to be numerically robust when it is insensitive to variations in the word length used in its digital implementation.

These factors, in their own ways, also enter into the design of nonlinear adaptive filters, except for the fact that we now no longer have a well-defined frame of reference in the form of a Wiener filter. Rather, we speak of a nonlinear filtering algorithm that may converge to a local minimum or, hopefully, a global minimum on the error-performance surface.

In the sections that follow, we shall first discuss various aspects of linear adaptive filters. Discussion of nonlinear adaptive filters is deferred to Section 2.6.

2.3 Linear Filter Structures

The operation of a linear adaptive filtering algorithm involves two basic processes: (1) a *filtering* process designed to produce an output in response to a sequence of input data, and (2) an *adaptive* process, the purpose of which is to provide mechanism for the *adaptive control* of an *adjustable* set of parameters used in the filtering process. These two processes work interactively with each other. Naturally, the choice of a structure for the filtering process has a profound effect on the operation of the algorithm as a whole.

There are three types of filter structures that distinguish themselves in the context of an adaptive filter with *finite memory* or, equivalently, *finite-duration impulse response.* The three filter structures are transversal filter, lattice predictor, and systolic array.

2.3.1 Transversal Filter

The transversal filter,* also referred to as a *tapped-delay line filter,* consists of three basic elements, as depicted in Figure 2.1: (1) a *unit-delay element,* (2) a *multiplier,* and (3) an *adder.* The number of delay elements used in the filter determines the finite duration of its impulse response. The number of delay elements, shown as $M - 1$ in Figure 2.1, is commonly referred to as the filter order. In Figure 2.1, the delay elements are each identified by the unit-delay operator z^{-1}. In particular, when z^{-1} operates on the

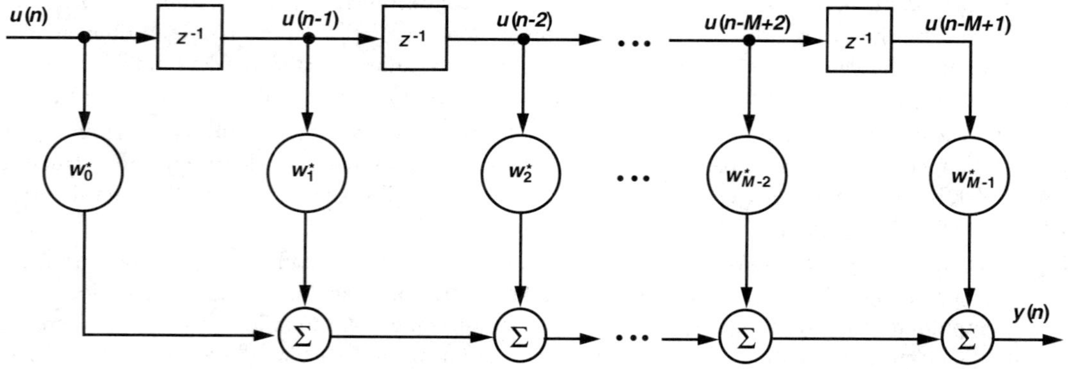

FIGURE 2.1 Transversal filter.

* The transversal filter was first described by Kallmann as a continuous-time device whose output is formed as a linear combination of voltages taken from uniformly spaced taps in a nondispersive delay line (Kallmann, 1940). In recent years, the transversal filter has been implemented using digital circuitry, charged-coupled devices, or surface-acoustic wave devices. Owing to its versatility and ease of implementation, the transversal filter has emerged as an essential signal-processing structure in a wide variety of applications.

input $u(n)$, the resulting output is $u(n-1)$. The role of each multiplier in the filter is to multiply the *tap input,* to which it is connected by a filter coefficient referred to as a *tap weight.* Thus, a multiplier connected to the kth tap input $u(n-k)$ produces the scalar version of the *inner product,* $w_k^* u(n-k)$, where w_k is the respective tap weight and $k = 0, 1, \ldots, M-1$. The asterisk denotes *complex conjugation,* which assumes that the tap inputs and, therefore, the tap weights are all *complex valued.* The combined role of the adders in the filter is to sum the individual multiplier outputs and produce an overall filter output. For the transversal filter described in Figure 2.1, the filter output is given by

$$y(n) = \sum_{k=0}^{m-1} w_k^* u(n-k) \tag{2.1}$$

Equation 2.1 is called a finite *convolution sum* in the sense that it convolves the finite-duration impulse response of the filter, w_n^*, with the filter input $u(n)$ to produce the filter output $y(n)$.

2.3.2 Lattice Predictor

A *lattice predictor** is *modular* in structure in that it consists of a number of individual stages, each of which has the appearance of a lattice, hence, the name "lattice" as a structural descriptor. Figure 2.2 depicts a lattice predictor consisting of $M-1$ stages; the number $M-1$ is referred to as the *predictor order.* The mth stage of the lattice predictor in Figure 2.2 is described by the pair of input-output relations (assuming the use of complex-valued, wide-sense stationary input data):

$$f_m(n) = f_{m-1}(n) + \kappa_m^* b_{m-1}(n-1) \tag{2.2}$$

$$b_m(n) = b_{m-1}(n-1) + \kappa_m f_{m-1}(n) \tag{2.3}$$

where $m = 1, 2, \ldots, M-1$, and $M-1$ is the final predictor order. The variable $f_m(n)$ is the mth *forward prediction error,* and $b_m(n)$ is the mth *backward prediction error.* The coefficient κ_m is called the mth *reflection coefficient.* The forward prediction error $f_m(n)$ is defined as the difference between the input $u(n)$ and its *one-step predicted* value; the latter is based on the set of m past inputs $u(n-1), \ldots, u(n-m)$. Correspondingly, the backward prediction error $b_m(n)$ is defined as the difference between the input $u(n-m)$ and its "backward" prediction based on the set of m "future" inputs $u(n), \ldots, u(n-m+1)$. Considering the conditions at the input of stage 1 in Figure 2.2, we have

$$f_0(n) = b_o(n) = u(n) \tag{2.4}$$

where $u(n)$ is the lattice predictor input at time n. Thus, starting with the *initial conditions* of Equation 2.4 and given the set of reflection coefficients $\kappa_1, \kappa_2, \ldots, \kappa_{M-1}$, we may determine the final pair of outputs $f_{M-1}(n)$ and $b_{M-1}(n)$ by moving through the lattice predictor, stage by stage.

For a *correlated* input sequence $u(n), u(n-1), \ldots, u(n-M+1)$ drawn from a stationary process, the backward prediction errors $b_0, b_1(n), \ldots, b_{M-1}(n)$ form a sequence of *uncorrelated* random variables. Moreover, there is a one-to-one correspondence between these two sequences of random variables in the sense that if we are given one of them, we may uniquely determine the other and vice versa. Accordingly, a linear combination of the backward prediction errors $b_0, b_1(n), \ldots, b_{M-1}(n)$ may be used to provide an *estimate* of some desired response $d(n)$, as depicted in the lower half of Figure 2.2. The arithmetic difference between $d(n)$ and the estimate so produced represents the estimation error $e(n)$. The process described herein is referred to as a *joint-process estimation.* Naturally, we may use the original input sequence $u(n), u(n-1), \ldots, u(n-M+1)$ to produce an estimate of the desired response $d(n)$ directly. The indirect method depicted in Figure 2.2, however, has the advantage of simplifying the computation

* The development of the lattice predictor is credited to Itakura and Saito (1972).

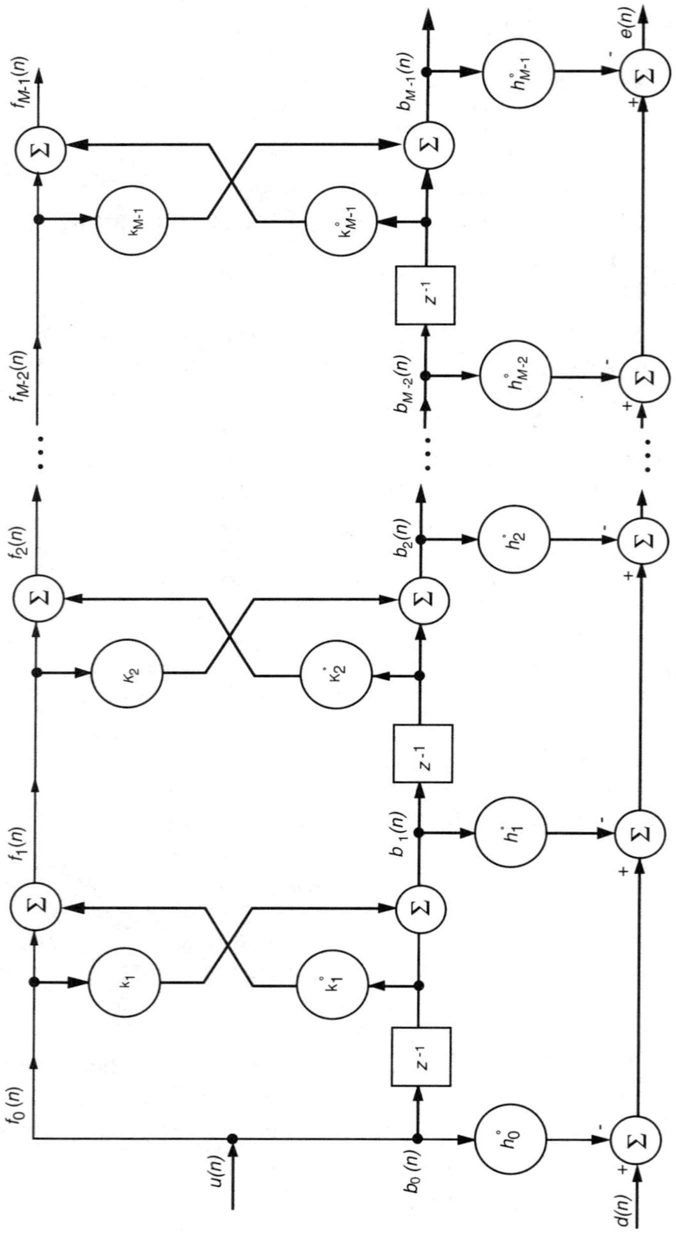

FIGURE 2.2 Multistage lattice filter.

of the tap weights $h_0, h_1(n), \ldots, h_{M-1}$ by exploiting the uncorrelated nature of the corresponding backward prediction errors used in the estimation.

2.3.3 Systolic Array

A *systolic array** represents a *parallel computing* network ideally suited for *mapping* a number of important linear algebra computations, such as *matrix multiplication, triangularization,* and *back substitution.* Two basic types of processing elements may be distinguished in a systolic array: *boundary cells* and *internal cells.* Their functions are depicted in Figures 2.3a and 2.3b, respectively. In each case, the parameter r represents a value *stored* within the cell. The function of the boundary cell is to produce an output equal to the input u divided by the number r stored in the cell. The function of the internal cell is twofold: (1) to multiply the input z (coming in from the top) by the number r stored in the cell, subtract the product rz from the second input (coming in from the left), and thereby produce the difference $u - rz$ as an output from the right-hand side of the cell; and (2) to transmit the first z downward without alteration.

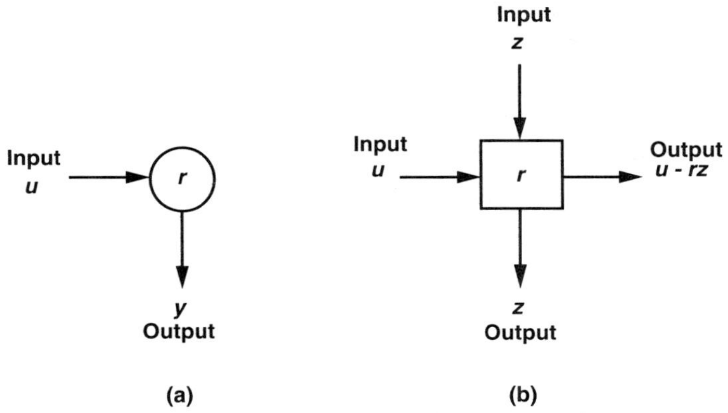

FIGURE 2.3 Two basic cells of a systolic array: (a) boundary cell and (b) internal cell.

Consider, for example, the 3 × 3 triangular array shown in Figure 2.4. This systolic array involves a combination of boundary and internal cells. In this case, the triangular array computes an output vector **y** related to the input vector **u** as follows:

$$y = \mathbf{R}^{-T}\mathbf{u} \tag{2.5}$$

where the $\mathbf{R}^{-T}$ is the *inverse* of the transposed matrix $\mathbf{R}^T$. The elements of $\mathbf{R}^T$ are the respective cell contents of the triangular array. The zeros added to the inputs of the array in Figure 2.4 are intended to provide the delays necessary for pipelining the computation described in Equation 2.5.

A systolic array architecture, as described herein, offers the desirable features of *modularity, local interconnections,* and highly *pipelined* and *synchronized* parallel processing; the synchronization is achieved by means of a global *clock.*

We note that the transversal filter of Figure 2.1, the joint-process estimator of Figure 2.2 based on a lattice predictor, and the triangular systolic array of Figure 2.4 have a common property: all three of

* The systolic array was pioneered by Kung and Leiserson (1978). In particular, the use of systolic arrays has made it possible to achieve a high throughput, which is required for many advanced signal-processing algorithms to operate in *real time.*

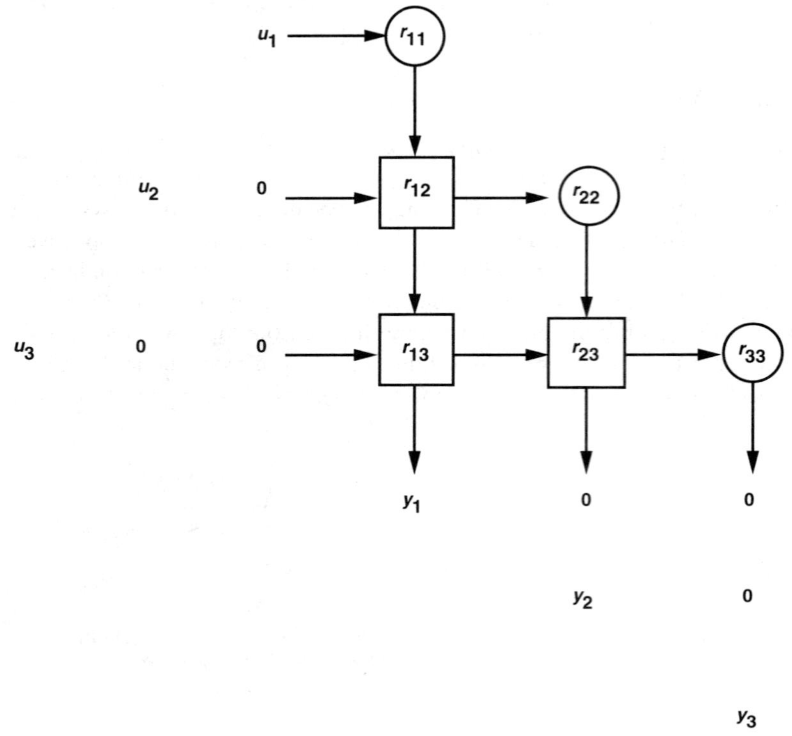

FIGURE 2.4 Triangular systolic array.

them are characterized by an impulse response of finite duration. In other words, they are examples of a *finite-duration impulse response* (*FIR*) *filter,* whose structures contain *feedforward* paths only. On the other hand, the filter structure shown in Figure 2.5 is an example of an *infinite-duration impulse response* (*IIR*) *filter.* The feature that distinguishes an IIR filter from an FIR filter is the inclusion of *feedback* paths. Indeed, it is the presence of feedback that makes the duration of the impulse response of an IIR filter infinitely long. Furthermore, the presence of feedback introduces a new problem, namely, that of *stability.* In particular, it is possible for an IIR filter to become unstable (i.e., break into oscillation), unless special precaution is taken in the choice of feedback coefficients. By contrast, an FIR filter in inherently *stable.* This explains the reason for the popular use of FIR filters, in one form or another, as the structural basis for the design of linear adaptive filters.

2.4 Approaches to the Development of Linear Adaptive Filtering Algorithms

There is no unique solution to the linear adaptive filtering problem. Rather, we have a "kit of tools" represented by a variety of recursive algorithms, each of which offers desirable features of its own. (For complete detailed treatment of linear adaptive filters, see the book by Haykin [1996].) The challenge facing the user of adaptive filtering is (1) to understand the capabilities and limitations of various adaptive filtering algorithms and (2) to use this understanding in the selection of the appropriate algorithm for the application at hand.

Basically, we may identify two distinct approaches for deriving recursive algorithms for the operation of linear adaptive filters, as discussed next.

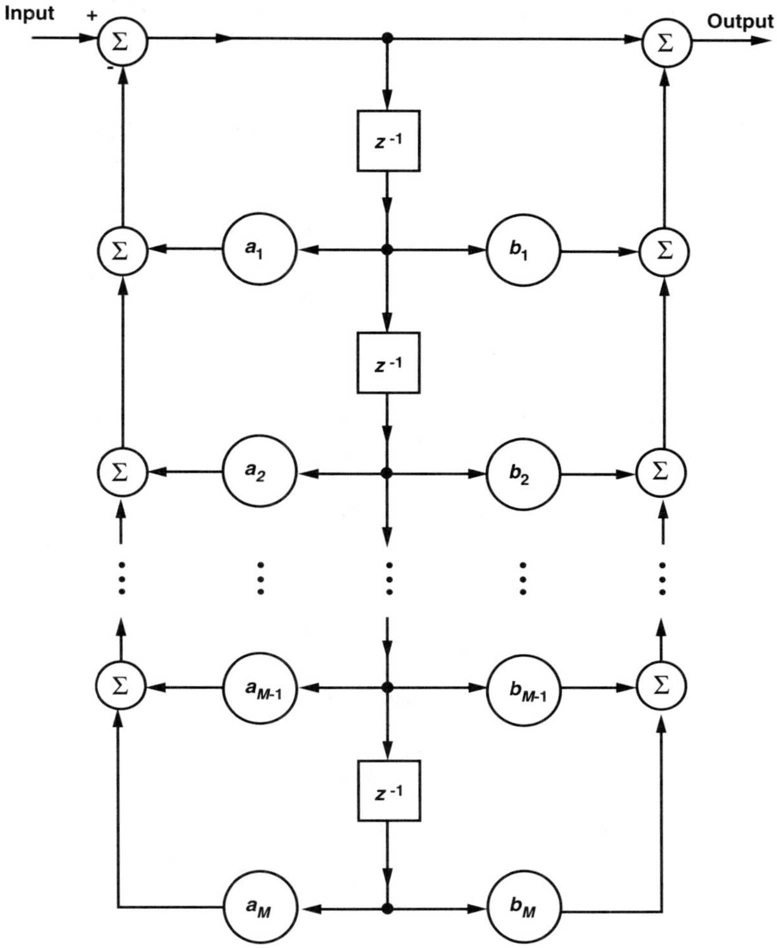

FIGURE 2.5 IIR filter.

2.4.1 Stochastic Gradient Approach

Here, we may use a tapped-delay line or transversal filter as the structural basis for implementing the linear adaptive filter. For the case of stationary inputs, the *cost function*,* also referred to as the *index of performance*, is defined as the *mean-squared error* (i.e., the mean-square value of the difference between the desired response and the transversal filter output). This cost function is precisely a second-order function of the tap weights in the transversal filter. The dependence of the mean-squared error on the unknown tap weights may be viewed to be in the form of a *multidimensional paraboloid* (i.e., punch bowl) with a uniquely defined bottom or *minimum point*. As mentioned previously, we refer to this paraboloid as the *error-performance surface;* the tap weights corresponding to the minimum point of the surface define the optimum Wiener solution.

To develop a recursive algorithm for updating the tap weights of the adaptive transversal filter, we proceed in two stages. We first modify the system of *Wiener-Hopf equations* (i.e., the matrix equation defining the optimum Wiener solution) through the use of the *method of steepest descent,* a well-known technique in

* In the general definition of a function, we speak of a transformation from a vector space into the space of real (or complex) scalars (Luenberger, 1969; Dorny, 1975). A cost function provides a quantitative measure for assessing the quality of performance and, hence, the restriction of it to a real scalar.

optimization theory. This modification requires the use of a *gradient vector,* the value of which depends on two parameters: the *correlation matrix* of the tap inputs in the transversal filter and the *cross-correlation vector* between the desired response and the same tap inputs. Next, we use instantaneous values for these correlations so as to derive an *estimate* for the gradient vector, making it assume a *stochastic* character in general. The resulting algorithm is widely known as the *least-mean-square* (*LMS*) *algorithm,* the essence of which may be described in words as follows for the case of a transversal filter operating on real-valued data:

$$\begin{pmatrix}\text{updated value}\\ \text{of tap-weight}\\ \text{vector}\end{pmatrix} = \begin{pmatrix}\text{old value}\\ \text{of tap-weight}\\ \text{vector}\end{pmatrix} + \begin{pmatrix}\text{learning-}\\ \text{rate}\\ \text{parameter}\end{pmatrix}\begin{pmatrix}\text{tap-}\\ \text{input}\\ \text{vector}\end{pmatrix}\begin{pmatrix}\text{error}\\ \text{signal}\end{pmatrix}$$

where the error signal is defined as the difference between some desired response and the actual response of the transversal filter produced by the tap-input vector.

The LMS algorithm, summarized in Table 2.1, is simple and yet capable of achieving satisfactory performance under the right conditions. Its major limitations are a relatively slow rate of convergence and a sensitivity to variations in the condition number of the correlation matrix of the tap inputs; the *condition number* of a Hermitian matrix is defined as the ratio of its largest eigenvalue to its smallest eigenvalue. Nevertheless, the LMS algorithm is highly popular and widely used in a variety of applications.

TABLE 2.1 Summary of the LMS Algorithm

Notations:		
	$\mathbf{u}(n)$	= tap-input vector at time n
		= $[u(n), u(n-1), \ldots, u(n-M+1)]^T$
	M	= number of tap inputs
	$d(n)$	= desired response at time n
	$\hat{\mathbf{w}}(n)$	= $[w_0(n), w_1(n), \ldots, w_{M-1}(n)]^T$
		= tap-weight vector at time n
	$y(n)$	= actual response of the tapped-delay line filter
		= $\hat{\mathbf{w}}^H(n)\mathbf{u}(n)$, where superscript H denotes Hermitian transposition
	$e(n)$	= error signal
		= $d(n) - y(n)$
Parameters:		
	M	= number of taps
	m	= step-size parameter
		$0 < \mu < \dfrac{2}{\text{tap-input power}}$
		tap-input power = $\sum_{k=0}^{M-1} E[\lvert u(n-k)\rvert^2]$
Initialization: If prior knowledge on the tap-weight vector $\hat{\mathbf{w}}(n)$ is available, use it to select an appropriate value for $\hat{\mathbf{w}}(0)$. Otherwise, set $\hat{\mathbf{w}}(0) = \mathbf{0}$.		
Date:		
Given:		$\mathbf{u}(n) = M \times 1$ tap-input vector at time n
		$d(n)$ = desired response at time n
To be computed:		$\hat{\mathbf{w}}(n+1)$ = estimate of tap-weight vector at time $n+1$
Computation:		For $n = 0, 1, 2, \ldots$, compute
		$e(n) = d(n) - \hat{\mathbf{w}}^H(n)\mathbf{u}(n)$
		$\hat{\mathbf{w}}(n+1) = \hat{\mathbf{w}}(n) = \mu\mathbf{u}(n)e^*(n)$

In a nonstationary environment, the orientation of the error-performance surface varies continuously with time. In this case, the LMS algorithm has the added task of continually *tracking* the bottom of the error-performance surface. Indeed, tracking will occur provided that the input data vary slowly compared to the *learning rate* of the LMS algorithm.

The stochastic gradient approach may also be pursued in the context of a lattice structure. The resulting adaptive filtering algorithm is called the *gradient adaptive lattice (GAL) algorithm.* In their own individual ways, the LMS and GAL algorithms are just two members of the *stochastic gradient family* of linear adaptive filters, although it must be said that the LMS algorithm is by far the most popular member of this family.

2.4.2 Least-Squares Estimation

The second approach to the development of linear adaptive filtering algorithms is based on the *method of least squares.* According to this method, we minimize a cost function or index of performance that is defined as the *sum of weighted error squares,* where the *error* or *residual* is itself defined as the difference between some desired response and the actual filter output. The method of least squares may be formulated with *block estimation* or *recursive estimation* in mind. In block estimation, the input data stream is arranged in the form of blocks of equal length (duration), and the filtering of input data proceeds on a block-by-block basis. In recursive estimation, on the other hand, the estimates of interest (e.g., tap weights of a transversal filter) are *updated* on a sample-by-sample basis. Ordinarily, a recursive estimator requires less storage than a block estimator, which is the reason for its much wider use in practice.

Recursive least-squares (RLS) estimation may be viewed as a special case of Kalman filtering. A distinguishing feature of the Kalman filter is the notion of *state,* which provides a measure of all the inputs applied to the filter up to a specific instant of time. Thus, at the heart of the Kalman filtering algorithm we have a recursion that may be described in words as follows:

$$\begin{pmatrix}\text{updated value}\\ \text{of the}\\ \text{state}\end{pmatrix} = \begin{pmatrix}\text{old value}\\ \text{of the}\\ \text{state}\end{pmatrix} + \begin{pmatrix}\text{Kalman}\\ \text{gain}\end{pmatrix}\begin{pmatrix}\text{innovation}\\ \text{vector}\end{pmatrix}$$

where the *innovation vector* represents new information put into the filtering process at the time of the computation. For the present, it suffices to say that there is indeed a one-to-one correspondence between the Kalman variables and RLS variables. This correspondence means that we can tap the vast literature on Kalman filters for the design of linear adaptive filters based on RLS estimation.

We may classify the *RLS family* of linear adaptive filtering algorithms into three distinct categories, depending on the approach taken:

1. *Standard RLS algorithm* assumes the use of a transversal filter as the structural basis of the linear adaptive filter. Table 2.2 summarizes the standard RLS algorithm. Derivation of this algorithm relies on a basic result in linear algebra known as the *matrix inversion lemma.* Most importantly, it enjoys the same virtues and suffers from the same limitations as the standard Kalman filtering algorithm. The limitations include lack of numerical robustness and excessive computational complexity. Indeed, it is these two limitations that have prompted the development of the other two categories of RLS algorithms, described next.
2. *Square-root RLS algorithms* are based on *QR decomposition* of the incoming data matrix. Two well-known techniques for performing this decomposition are the *Householder transformation* and the *Givens rotation,* both of which are data adaptive transformations. At this point in the discussion, we need to merely say that RLS algorithms based on the Householder transformation or Givens rotation are numerically stable and robust. The resulting linear adaptive filters are referred to as *square-root adaptive filters,* because in a matrix sense they represent the square-root forms of the standard RLS algorithm.

TABLE 2.2 Summary of the RLS Algorithm

Notations:

$\mathbf{u}(n)$	= tap-input vector at time n
	$= [u(n), u(n-1), \ldots, u(n-M+1)]^T$
M	= number of tap inputs
$d(n)$	= desired response at time n
$\hat{\mathbf{w}}(n)$	$= [\hat{w}_0(n), \hat{w}_1(n), \ldots, \hat{w}_{M-1}(n)]^T$
	= tap-weight vector at time n
$\xi(n)$	= innovation (i.e., *a priori* error signal) at time n
1	= exponential weighting factor
$\mathbf{k}(n)$	= gain vector at time n
$\mathbf{P}(n)$	= weight-error correlation matrix

Initialize the algorithm by setting

$$\mathbf{P}(0) = \delta\angle^{1}\mathbf{I}, \ \delta = \text{small positive constant}$$

$$\mathbf{w}(0) = 0$$

For each instant of time, n - 1, 2, …, compute

$$\mathbf{k}(n) = \frac{\lambda^{-1}\mathbf{P}(n-1)\mathbf{u}(n)}{1+\lambda^{-1}\mathbf{u}^H(n)\mathbf{P}(n-1)\mathbf{u}(n)}$$

$$\xi(n) = d(n) - \hat{\mathbf{w}}^H(n-1)\mathbf{u}(n)$$

$$\hat{\mathbf{w}}(n) = \hat{\mathbf{w}}(n-1) + \mathbf{k}(n)\xi^*(n)$$

$$\lambda^{-1}\mathbf{P}(n-1) - \lambda^{-1}\mathbf{k}(n)\mathbf{u}(n)\mathbf{P}(n-1)$$

3. *Fast RLS algorithms,* which include the standard RLS algorithm and square-root RLS algorithms, have a computational complexity that increases as the square of M, where M is the number of adjustable weights (i.e., the number of degrees of freedom) in the algorithm. Such algorithms are often referred to as $O(M^2)$ algorithms, where $O(\cdot)$ denotes "order of." By contrast, the LMS algorithm is an $O(M)$ algorithm, in that its computational complexity increases linearly with M. When M is large, the computational complexity of $O(M^2)$ algorithms may become objectionable from a hardware implementation point of view. There is therefore a strong motivation to modify the formulation of the RLS algorithm in such a way that the computational complexity assumes an $O(M)$ form. This objective is indeed achievable, in the case of temporal processing, first by virtue of the inherent *redundancy* in the *Toeplitz structure* of the input data matrix and second by exploiting this redundancy through the use of *linear least-squares prediction in both the forward and backward directions.* The resulting algorithms are known collectively as fast RLS algorithms; they combine the desirable characteristics of recursive linear least-squares estimation with an $O(M)$ computational complexity. Two types of fast RLS algorithms may be identified, depending on the filtering structure employed:
   - *Order-recursive adaptive filters,* which are based on a lattice-like structure for making linear forward and backward predictions
   - *Fast transversal filters,* in which the linear forward and backward predictions are performed using separate transversal filters

Certain (but not all) realizations of order-recursive adaptive filters are known to be numerically stable, whereas fast transversal filters suffer from a numerical stability problem and, therefore, require some form of stabilization for them to be of practical use.

An introductory discussion of linear adaptive filters would be incomplete without saying something about their tracking behavior. In this context, we note that stochastic gradient algorithms such as the LMS algorithm are *model independent;* generally speaking, we would expect them to exhibit good tracking behavior, which indeed they do. In contrast, RLS algorithms are *model dependent;* this, in turn, means that their tracking behavior may be inferior to that of a member of the stochastic gradient family, unless care is taken to minimize the mismatch between the mathematical model on which they are based and the underlying physical process responsible for generating the input data.

2.4.3 How to Choose an Adaptive Filter

Given the wide variety of adaptive filters available to a system designer, how can a choice be made for an application of interest? Clearly, whatever the choice, it has to be *cost effective.* With this goal in mind, we may identify three important issues that require attention: *computational cost, performance,* and *robustness.* The use of computer simulation provides a good first step in undertaking a detailed investigation of these issues. We may begin by using the LMS algorithm as an adaptive filtering tool for the study. The LMS algorithm is relatively simple to implement. Yet it is powerful enough to evaluate the practical benefits that may result from the application of adaptivity to the problem at hand. Moreover, it provides a practical frame of reference for assessing any further improvement that may be attained through the use of more sophisticated adaptive filtering algorithms. Finally, the study must include tests with real-life data, for which there is no substitute.

Practical applications of adaptive filtering are very diverse, with each application having peculiarities of its own. The solution for one application may not be suitable for another. Nevertheless, to be successful we have to develop a physical understanding of the environment in which the filter has to operate and thereby relate to the realities of the application of interest.

2.5 Real and Complex Forms of Adaptive Filters

In the development of adaptive filtering algorithms, regardless of their origin, it is customary to assume that the input data are in baseband form. The term "baseband" is used to designate the band of frequencies representing the original (message) signal as generated by the source of information.

In such applications as communications, radar, and sonar, the information-bearing signal component of the receiver input typically consists of a message signal *modulated* onto a carrier wave. The bandwidth of the message signal is usually small compared to the carrier frequency, which means that the modulated signal is a *narrowband signal.* To obtain the baseband representation of a narrowband signal, the signal is translated down in frequency in such a way that the effect of the carrier wave is completely removed, yet the information content of the message signal is fully preserved. In general, the baseband signal so obtained is *complex.* In other words, a sample $u(n)$ of the signal may be written as

$$u(n) = u_{\mathrm{I}}(n) + ju_{\mathrm{Q}}(n) \tag{2.6}$$

where $u_{\mathrm{I}}(n)$ is the *in-phase* (real) *component,* and $u_{\mathrm{Q}}(n)$ is the *quadrature* (imaginary) *component.* Equivalently, we may express $u(n)$ as

$$u(n) = |u(n)|e^{j\phi(n)} \tag{2.7}$$

where $|u(n)|$ is the *magnitude,* and $\phi(n)$ is the *phase angle.*

The LMS and RLS algorithms summarized in Tables 2.1 and 2.2 assume the use of complex signals. The adaptive filtering algorithm so described is said to be in *complex form.* The important virtue of complex adaptive filters is that they preserve the mathematical formulation and elegant structure of complex signals encountered in the aforementioned areas of application.

If the signals to be processed are *real,* we naturally use the *real form* of the adaptive filtering algorithm of interest. Given the complex form of an adaptive filtering algorithm, it is straightforward to deduce the corresponding real form of the algorithm. Specifically, we do two things:

1. The operation of *complex conjugation,* wherever in the algorithm, is simply removed.
2. The operation of *Hermitian transposition* (i.e., conjugate transposition) of a matrix, wherever in the algorithm, is replaced by ordinary transposition.

Simply put, complex adaptive filters include real adaptive filters as special cases.

2.6 Nonlinear Adaptive Systems: Neural Networks

The theory of linear optimum filters is based on the mean-square error criterion. The Wiener filter that results from the minimization of such a criterion, and which represents the goal of linear adaptive filtering for a stationary environment, can only relate to second-order statistics of the input data and no higher. This constraint limits the ability of a linear adaptive filter to extract information from input data that are non-Gaussian. Despite its theoretical importance, the existence of Gaussian noise is open to question (Johnson and Rao, 1990). Moreover, non-Gaussian processes are quite common in many signal processing applications encountered in practice. The use of a Wiener filter or a linear adaptive filter to extract signals of interest in the presence of such non-Gaussian processes will therefore yield suboptimal solutions. We may overcome this limitation by incorporating some form of *nonlinearity* in the structure of the adaptive filter to take care of higher order statistics. Although, by so doing, we no longer have the Wiener filter as a frame of reference and so complicate the mathematical analysis, we would expect to benefit in two significant ways: improving learning efficiency and a broadening of application areas.

In this section, we describe an important class of the nonlinear adaptive system commonly known as artificial neural networks or just simply *neural networks.* This terminology is derived from analogy with biological neural networks that make up the human brain.

A *neural network* is a massively parallel distributed processor that has a natural propensity for storing experiential knowledge and making it available for use. It resembles the brain in two respects:

1. Knowledge is acquired by the network through a learning process.
2. Interconnection strengths known as synaptic weights are used to store the knowledge.

Basically, learning is a process by which the free parameters (i.e., synaptic weights and bias levels) of a neural network are adapted through a continuing process of stimulation by the environment in which the network is embedded. The type of learning is determined by the manner in which the parameter changes take place. Specifically, learning machines may be classified as follows:

- Learning with a teacher, also referred to as supervised learning
- Learning without a teacher

This second class of learning machines may also be subdivided into

- Reinforcement learning
- Unsupervised learning or self-organized learning

In the subsequent sections of this chapter, we will describe the important aspects of these learning machines and highlight the algorithms involved in their designs. For a detailed treatment of the subject, see Haykin (1999); this book has an up-to-date bibliography that occupies 41 pages of references.

In the context of adaptive signal-processing applications, neural networks offer the following advantages:

- *Nonlinearity,* which makes it possible to account for the nonlinear behavior of physical phenomena responsible for generating the input data
- The ability to *approximate any prescribed input-output mapping* of a continuous nature
- *Weak statistical assumptions* about the environment, in which the network is embedded
- *Learning* capability, which is accomplished by undertaking a training session with input-output examples that are representative of the environment
- *Generalization,* which refers to the ability of the neural network to provide a satisfactory performance in response to *test data* never seen by the network before
- *Fault tolerance,* which means that the network continues to provide an acceptable performance despite the failure of some neurons in the network
- *VLSI implementability,* which exploits the massive parallelism built into the design of a neural network

This is indeed an impressive list of attributes, which accounts for the widespread interest in the use of neural networks to solve signal-processing tasks that are too difficult for conventional (linear) adaptive filters.

2.6.1 Supervised Learning

This form of learning assumes the availability of a labeled (i.e., ground truthed) set of training data made up of N input-output examples:

$$T = \{(\mathbf{x}_i, d_i)\}_{i=1}^{N} \tag{2.8}$$

where $\mathbf{x}_i$ = input vector of ith example
d_i = desired (target) response of ith example, assumed to be scalar for convenience of presentation
N = sample size

Given the training sample T, the requirement is to compute the free parameters of the neural network so that the actual output y_i of the neural network due to $\mathbf{x}_i$ is close enough to d_i for all i in a statistical sense. For example, we may use the mean-squared error

$$E(n) = \frac{1}{N}\sum_{i=1}^{N}(d_i - y_i)^2 \tag{2.9}$$

as the index of performance to be minimized.

2.6.1.1 Multilayer Perceptrons and Back-Propagation Learning

The back-propagation algorithm has emerged as the workhorse for the design of a special class of layered feedforward networks known as *multilayer perceptrons.* As shown in the block diagram of Figure 2.6, a multilayer perceptron consists of the following:

- *Input layer* of nodes, which provide the means for connecting the neural network to the source(s) of signals driving the network
- One or more *hidden layers* of processing units, which act as "feature detectors"
- *Output layer* of processing units, which provide one final stage of computation and thereby produce the response of the network to the signals applied to the input layer

The processing units are commonly referred to as artificial neurons or just *neurons.* Typically, a neuron consists of a linear combiner with a set of adjustable synaptic weights, followed by a nonlinear activation function, as depicted in Figure 2.7.

Two commonly used forms of the activation function $\varphi(\cdot)$ are shown in Figure 2.8. The first one, shown in Figure 2.8a, is called the *hyperbolic function,* which is defined by

$$\begin{aligned}\varphi(v) &= \tanh(v) \\ &= \frac{1-\exp(-2v)}{1+\exp(-2v)}\end{aligned} \tag{2.10}$$

The second one, shown in Figure 2.8b, is called the *logistic function,* which is defined by

$$\varphi(v) - \frac{1}{1+\exp(-v)} \tag{2.11}$$

From these definitions, we readily see that the logistic function is of a unipolar form that is nonsymmetric, whereas the hyperbolic function is bipolar that is antisymmetric.

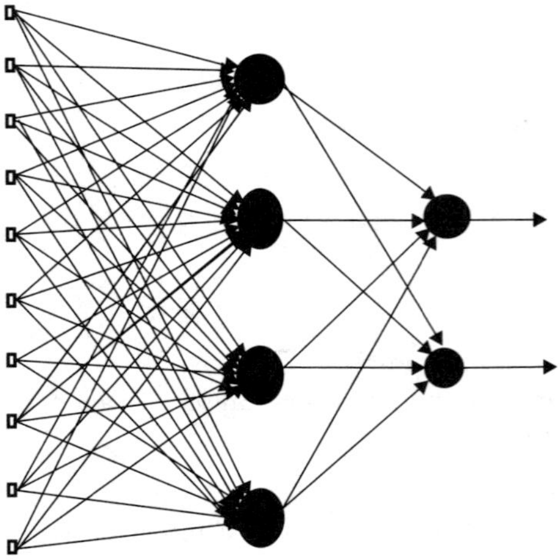

FIGURE 2.6 Fully connected feedforward of acyclic network with one hidden layer and one output layer.

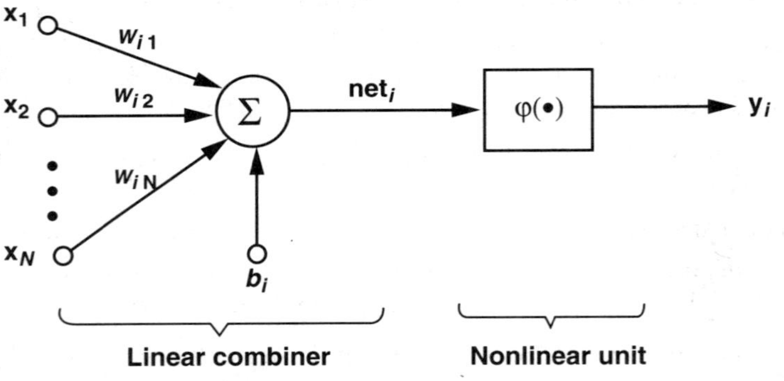

FIGURE 2.7 Simplified model of a neuron.

The training of an MLP is usually accomplished by using the *back-propagation (BP) algorithm* that involves two phases (Werbos, 1974; Rumelhart et al., 1986):

- *Forward phase:* During this phase, the free parameters of the network are fixed, and the input signal is propagated through the network of Figure 2.6 layer by layer. The forward phase finishes with the computation of an error signal defined as the difference between a desired response and the actual output produced by the network in response to the signals applied to the input layer.
- *Backward phase:* During this second phase, the error signal e_i is propagated through the network of Figure 2.6 in the backward direction, hence the name of the algorithm. It is during this phase that adjustments are applied to the free parameters of the network so as to minimize the error e_i in a statistical sense.

BP learning may be implemented in one of two basic ways, as summarized here:

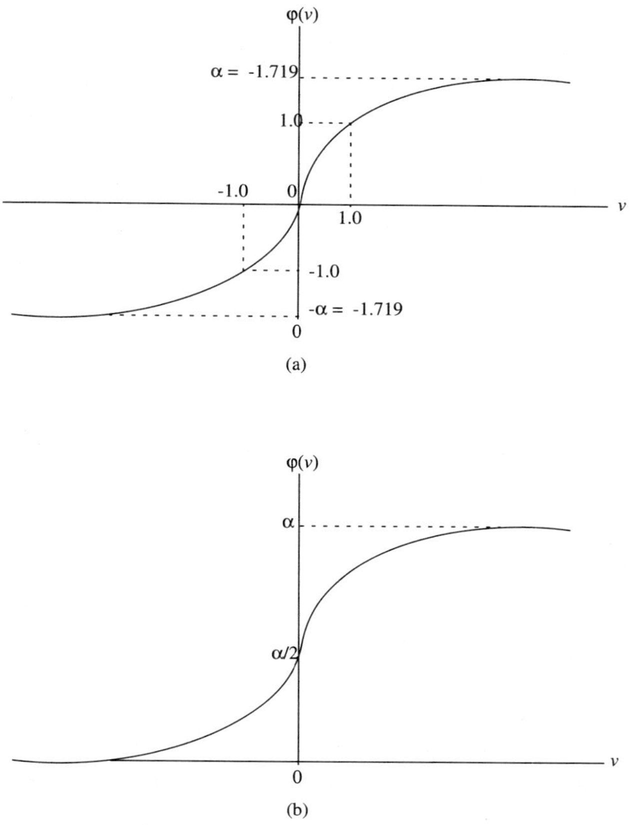

FIGURE 2.8 (a) Antisymmetric activation function; (b) nonsymmetric activation function.

1. *Sequential mode* (also referred to as the pattern mode, on-line mode, or stochastic mode): In this mode of BP learning, adjustments are made to the free parameters of the network on an example-by-example basis. The sequential mode is best suited for pattern classification.
2. *Batch mode:* In this second mode of BP learning, adjustments are made to the free parameters of the network on an epoch-by-epoch basis, where each epoch consists of the entire set of training examples. The batch mode is best suited for nonlinear regression.

The BP learning algorithm is simple to implement and computationally efficient in that its complexity is linear in the synaptic weights of the network. However, a major limitation of the algorithm is that it can be excruciatingly slow, particularly when we have to deal with a difficult learning task that requires the use of a large network.

Traditionally, the derivation of the BP algorithm is done for real-valued data. This derivation may be extended to complex-valued data by permitting the free parameters of the multilayer perceptron to assume complex values. However, in the latter case, care has to be exercised in how the activation function is handled for complex-valued inputs. For a detailed derivation of the complex BP algorithm, see Haykin (1996).

In any event, we may try to make BP learning perform better by invoking the following list of neuristics:

- Use neurons with antisymmetric activation functions (e.g., hyperbolic tangent function) in preference to nonsymmetric activation functions (e.g., logistic function). Figure 2.8 shows examples of these two forms of activation functions.
- Shuffle the training examples after the presentation of each epoch; an epoch involves the presentation of the entire set of training examples to the network.

- Follow an easy-to-learn example with a difficult one.
- Preprocess the input data so as to remove the mean and decorrelate the data.
- Arrange for the neurons in the different layers to learn at essentially the same rate. This may be attained by assigning a learning-rate parameter to neurons in the last layers that is smaller than those at the front end.
- Incorporate prior information into the network design whenever it is available.

One other heuristic that deserves to be mentioned relates to the size of the training set, N, for a pattern classification task. Given a multilayer perceptron with a total number of synaptic weights including bias levels, denoted by W, a rule of thumb for selecting N is

$$N = O\left(\frac{W}{\varepsilon}\right) \tag{2.12}$$

where O denotes "the order of," and ε denotes the fraction of classification errors permitted on test data. For example, with an error of 10%, the number of training examples needed should be about ten times the number of synaptic weights in the network.

Supposing that we have chosen a multilayer perceptron to be trained with the BP algorithm, how do we determine when it is "best" to stop the training session? How do we select the size of individual hidden layers of the MLP? The answers to these important questions may be obtained through the use of a statistical technique known as *cross-validation,* which proceeds as follows:

- The set of training examples is split into two parts:
  1. Estimation subset used for training of the model
  2. Validation subset used for evaluating the model performance
- The network is finally tuned by using the entire set of training examples and then tested on test data not seen before.

2.6.1.2 Radial-Basis Function (RBF) Networks

Another popular layered feedforward network is the radial-basis function (RBF) network, whose structure is shown in Figure 2.9. RBF networks use memory-based learning for their design. Specifically, learning is viewed as a curve-fitting problem in high-dimensional space (Broomhead and Lowe, 1988; Poggio and Girosi, 1990).

1. Learning is equivalent to finding a surface in a multidimensional space that provides a best fit to the training data.
2. Generalization (i.e., response of the network to input data not seen before) is equivalent to the use of this multidimensional surface to interpolate the test data.

A commonly used formulation of the RBFs, which constitute the hidden layer, is based on the *Gaussian function.* To be specific, let $\mathbf{u}$ denote the signal vector applied to the input layer and $\mathbf{u}_i$ denote the center of the Gaussian function assigned to hidden unit i. We may then define the corresponding RBF as

$$\varphi(\|\mathbf{u}-\mathbf{u}_i\|) = \exp\left(-\frac{\|\mathbf{u}-\mathbf{u}_i\|^2}{2\sigma^2}\right), \qquad i = 1, 2, \ldots, K \tag{2.13}$$

where the symbol $\|\mathbf{u}-\mathbf{u}_i\|$ denotes the Euclidean distance between the vectors $\mathbf{u}$ and $\mathbf{u}_i$ and σ^2 is the width common to all K RBFs. (Each RBF may also be permitted to have a different width, but such a generalization results in increased complexity.) On this basis, we may define the input-output mapping realized by the RBF network (assuming a single output) to be

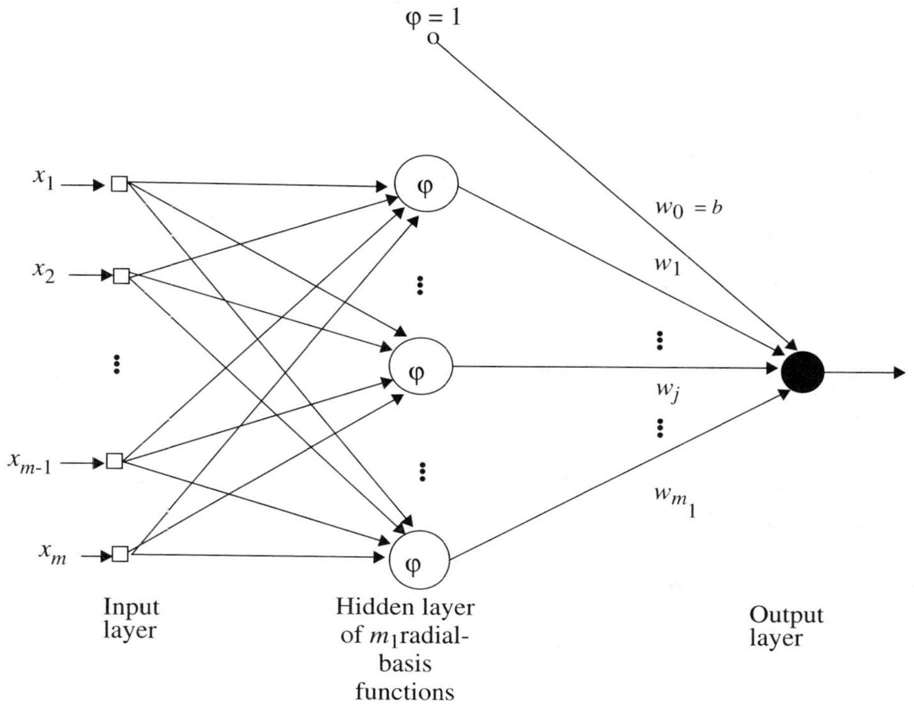

FIGURE 2.9 RBF network.

$$
\begin{aligned}
y &= \sum_{i=1}^{K} w_i \varphi(\|\mathbf{u} - \mathbf{u}_i\|) \\
&= \sum_{i=1}^{K} w_i \exp\left(-\frac{\|\mathbf{u} - \mathbf{u}_i\|^2}{2\sigma^2}\right)
\end{aligned}
\tag{2.14}
$$

where the set of weights $\{w_i\}_{i=1}^{K}$ constitutes the output layer. Equation 2.14 represents a *linear mixture* of Gaussian functions.

RBF networks differ from the multilayer perceptrons in some fundamental respects:

- RBF networks are local approximators, whereas multilayer perceptrons are global approximators.
- RBF networks have a single hidden layer, whereas multilayer perceptrons can have any number of hidden layers.
- The output layer of a RBF network is always linear, whereas in a multilayer perceptron it can be linear or nonlinear.
- The activation function of the hidden layer in an RBF network computes the Euclidean distance between the input signal vector and a parameter vector of the network, whereas the activation function of a multilayer perceptron computes the inner product between the input signal vector and the pertinent synaptic weight vector.

The use of a linear output layer in an RBF network may be justified in light of *Cover's theorem* on the separability of patterns. According to this theorem, provided that the transformation from the input space to the feature (hidden) space is nonlinear and the dimensionality of the feature space is high compared to that of the input (data) space, then there is a high likelihood that a nonseparable pattern classification task in the input space is transformed into a linearly separable one in the feature space.

Design methods for RBF networks include the following:

1. Random selection of fixed centers (Broomhead and Lowe, 1988)
2. Self-organized selection of centers (Moody and Darken, 1989)
3. Supervised selection of centers (Poggio and Girosi, 1990)
4. Regularized interpolation exploiting the connection between an RBF network and the Watson-Nadaraya regression kernel (Yee, 1998)

2.6.1.3 Support Vector Machines

Support vector machine (SVM) theory provides the most principled approach to the design of neural networks, eliminating the need for domain knowledge (Vapnik, 1998). SVM theory applies to pattern classification, regression, or density estimation using an RBF network (depicted in Figure 2.9) or an MLP with a single hidden layer (depicted in Figure 2.6).

Unlike BP learning, different cost functions are used for pattern classification and regression. Most importantly, the use of SVM learning eliminates the need for how to select the size of the hidden layer in an MLP or RBF network. In the latter case, it also eliminates the need for how to specify the centers of the RBF units in the hidden layer.

Simply stated, support vectors are those data points (for the linearly separable case) that are the most difficult to classify and are optimally separated from each other.

In an SVM, the selection of basis functions is required to satisfy *Mercer's theorem;* that is, each basis function is in the form of a positive, definite, inner-product kernel (assuming real-valued data):

$$K(\mathbf{x}_i, \mathbf{x}_j) = \underset{\sim}{\varphi}^T(\mathbf{x}_i)\underset{\sim}{\varphi}(\mathbf{x}_j) \tag{2.15}$$

where $\mathbf{x}_i$ and $\mathbf{x}_j$ are input vectors for examples i and j, and $\underset{\sim}{\varphi}(\mathbf{x}_i)$ is the vector of hidden-unit outputs for inputs $\mathbf{x}_i$. The hidden (feature) space is chosen to be of high dimensionality so as to transform a nonlinear, separable, pattern classification problem into a linearly separable one. Most importantly, however, in a pattern classification task, for example, the support vectors are selected by the SVM learning algorithm so as to maximize the margin of separation between classes.

The curse-of-dimensionality problem, which can plague the design of multilayer perceptrons and RBF networks, is avoided in SVMs through the use of quadratic programming. This technique, based directly on the input data, is used to solve for the linear weights of the output layer (Vapnik, 1998).

2.6.2 Unsupervised Learning

Turning next to unsupervised learning, adjustment of synaptic weights may be carried through the use of neurobiological principles such as Hebbian learning and competitive learning or information-theoretic principles. In this section we will describe specific applications of these three approaches.

2.6.2.1 Principal Components Analysis

According to *Hebb's postulate of learning,* the change in synaptic weight Δw_{ji} of a neural network is defined by (for real-valued data)

$$\Delta w_{ji} = \eta x_i y_i \tag{2.16}$$

where η = learning-rate parameter
x_i = input (presynaptic) signal
y_j = output (postsynaptic) signal

Principal component analysis (PCA) networks use a modified form of this self-organized learning rule. To begin with, consider a linear neuron designed to operate as a maximum eigenfilter; such a neuron is referred to as *Oja's neuron* (Oja, 1982). It is characterized as follows:

$$\Delta w_{ji} = \eta y_j(x_i - y_j w_{ji}) \tag{2.17}$$

where the term $-\eta y_j^2 w_{ji}$ is added to stabilize the learning process. As the number of iterations approaches infinity, we find the following:

1. The synaptic weight vector of neuron j approaches the eigenvector associated with the largest eigenvalue λ_{max} of the correlation matrix of the input vector (assumed to be of zero mean).
2. The variance of the output of neuron j approaches the largest eigenvalue λ_{max}.

The generalized Hebbian algorithm (GHA), due to Sanger (1989), is a straightforward generalization of Oja's neuron for the extraction of any desired number of principal components.

2.6.2.2 Self-Organizing Maps

In a self-organizing map (SOM), due to Kohonen (1997), the neurons are placed at the nodes of a lattice, and they become selectively tuned to various input patterns (vectors) in the course of a competitive learning process. The process is characterized by the formation of a topographic map in which the spatial locations (i.e., coordinates) of the neurons in the lattice correspond to intrinsic features of the input patterns. Figure 2.10 illustrates the basic idea of an SOM, assuming the use of a two-dimensional lattice of neurons as the network structure.

In reality, the SOM belongs to the class of vector coding algorithms (Luttrell, 1989); that is, a fixed number of code words are placed into a higher dimensional input space, thereby facilitating data compression.

An integral feature of the SOM algorithm is the neighborhood function centered around a neuron that wins the competitive process. The neighborhood function starts by enclosing the entire lattice initially and is then allowed to shrink gradually until it encompasses the winning neuron.

The algorithm exhibits two distinct phases in its operation:

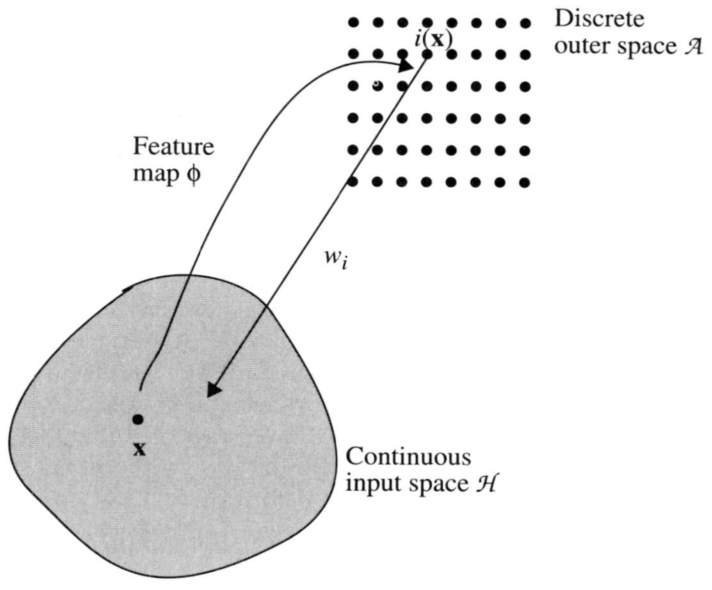

FIGURE 2.10 Illustration of the relationship between feature map ϕ and weight vector $\mathbf{w}_i$ of winning neuron i.

1. *Ordering phase,* during which the topological ordering of the weight vectors takes place
2. *Convergence phase,* during which the computational map is fine tuned

The SOM algorithm exhibits the following properties:

1. Approximation of the continuous input space by the weight vectors of the discrete lattice
2. Topological ordering exemplified by the fact that the spatial location of a neuron in the lattice corresponds to a particular feature of the input pattern
3. The feature map computed by the algorithm reflects variations in the statistics of the input distribution
4. SOM may be viewed as a nonlinear form of principal components analysis

2.6.3 Information-Theoretic Models

Mutual information, defined in accordance with Shannon's information theory, provides the basis of a powerful approach for self-organized learning. The theory is embodied in the maximum mutual information (Infomax) principle, due to Linsker (1988), which may be stated as follows:

> The transformation of a random vector **X** observed in the input layer of a neural network to a random vector **Y** produced in the output layer should be chosen so that the activities of the neurons in the output layer jointly maximize information about the activities in the input layer. The objective function to be maximized is the mutual information $I(\mathbf{Y};\mathbf{X})$ between **X** and **Y**.

The Infomax principle finds applications in the following areas:

- Design of self-organized models and feature maps (Linsker, 1989)
- Discovery of properties of a noisy sensory input exhibiting coherence across both space and time (first variant of Infomax due to Becker and Hinton, 1992)
- Dual-image processing designed to maximize the spatial differentiation between the corresponding regions of two separate images (views) of an environment of interest as in radar polarimetry (second variant of Infomax due to Ukrainec and Haykin, 1996)
- Independent components analysis (ICA) for blind source separation (due to Barlow, 1989); see also Comon (1994); ICA may be viewed as the third variant of Infomax (Haykin, 1999)

2.6.4 Temporal Processing Using Feedforward Networks

Time is an essential dimension of learning. We may incorporate time into the design of a neural network implicitly or explicitly. A straightforward method of implicit representation of time is to add a short-term memory structure at the input end of a static neural network (e.g., multilayer perceptron), as illustrated in Figure 2.11. This configuration is called a *focused time-lagged feedforward network* (TLFN). Focused TLFNs are limited to stationary dynamical processes.

To deal with nonstationary dynamical processes, we may use distributed TLFNs where the effect of time is distributed at the synaptic level throughout the network. One way in which this may be accomplished is to use FIR filters to implement the synaptic connections of an MLP. The training of a distributed TLFN is naturally a more difficult proposition than the training of a focused TLFN. Whereas we may use the ordinary BP algorithm to train a focused TLFN, we have to extend the BP algorithm to cope with the replacement of a synaptic weight in the ordinary MLP by a synaptic weight vector. This extension is referred to as the temporal BP algorithm due to Wan (1994).

2.6.5 Dynamically Driven Recurrent Networks

Another practical way of accounting for time in a neural network is to employ feedback at the local or global level. Neural networks so configured are referred to as recurrent networks.

We may identify two classes of recurrent networks:

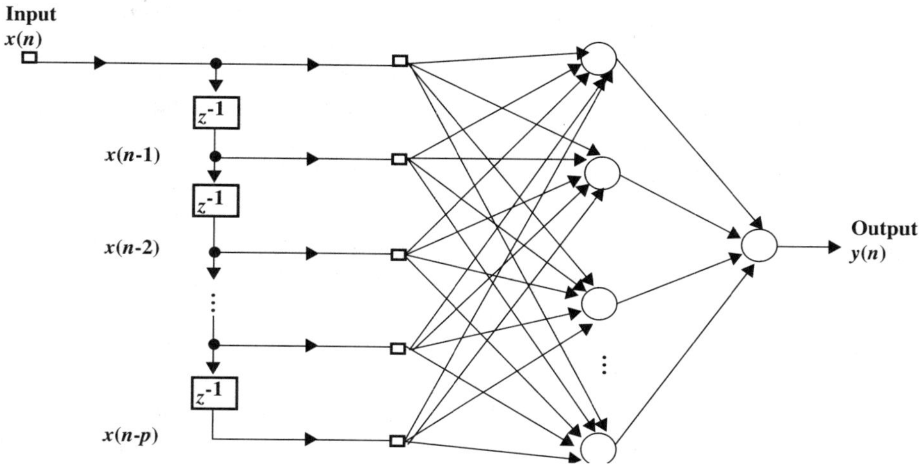

FIGURE 2.11 TLFN — the bias levels have been omitted for the convenience of presentation.

1. Autonomous recurrent networks exemplified by the Hopfield network (Hopfield, 1982) and brain-state-in-a-box (BSB) model. These networks are well suited for building associative memories, each with its own domain of applications. Figure 2.12 shows an example of a Hopfield network involving the use of four neurons.
2. Dynamically driven recurrent networks are well suited for input-output mapping functions that are temporal in character.

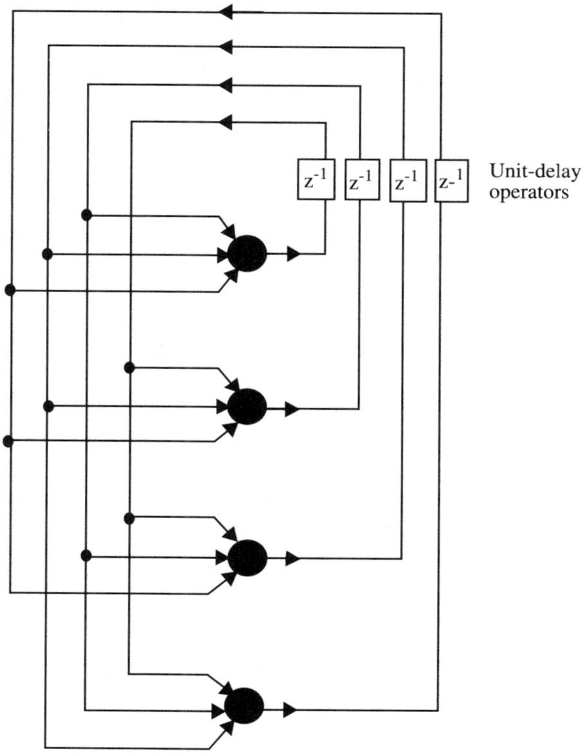

FIGURE 2.12 Recurrent network with no self-feedback loops and no hidden neurons.

A powerful approach for the design of dynamically driven recurrent networks with the goal of solving an input-output mapping task is to build on the state-space approach of modern control theory (Sontag, 1990). Such an approach is well suited for the two network configurations shown in Figures 2.13 and 2.14, which are respectively referred to as a *nonlinear autoregressive with exogeneous inputs (NARX) model* and a *recurrent multilayer perceptron (RMLP)*.

To design a dynamically driven recurrent network for input-output mapping, we may use any one of the following approaches:

- *Back-propagation through time* (BPTT) involves unfolding the temporal operation of the recurrent network into a layered feedforward network (Werbos, 1990). This unfolding facilitates the application of the ordinary BP algorithm.
- *Real-time recurrent learning* adjustments are made (using a gradient-descent method) to the synaptic weights of a fully connected recurrent network in real time (Williams and Zipser, 1989).
- An *extended Kalman filter* (EKF) builds on the classic Kalman filter theory to compute the synaptic weights of the recurrent network. Two versions of the algorithm are available (Feldkamp and Puskorius, 1998):
  - Decoupled EKF
  - Global EKF

The decoupled EKF algorithm is computationally less demanding but somewhat less accurate than the global EKF algorithm.

A serious problem that can arise in the design of a dynamically driven recurrent network is the *vanishing gradients problem.* This problem pertains to the training of a recurrent network to produce a desired response at the current time that depends on input data in the distant past (Bengio et al., 1994). It makes the learning of long-term dependencies in gradient-based training algorithms difficult if not impossible in certain cases. To overcome the problem, we may use the following methods:

1. EKF (encompassing second-order information) for training
2. Elaborate optimization methods such as pseudo-Newton and simulated annealing (Bengio et al., 1994)
3. Use of long time delays in the network architecture (Giles et al., 1997)
4. Hierarchically structuring of the network in multiple levels associated with different time scales (El Hihi and Bengio, 1996)
5. Use of gating units to circumvent some of the nonlinearities (Hochreiter and Schmidhuber, 1997)

2.7 Applications

The ability of an adaptive filter to operate satisfactorily in an unknown environment and track time variations of input statistics make the adaptive filter a powerful device for signal-processing and control applications. Indeed, adaptive filters have been successfully applied in such diverse fields as communications, radar, sonar, seismology, and biomedical engineering. Although these applications are indeed quite different in nature, nevertheless, they have one basic common feature: an input vector and a desired response are used to compute an estimation error, which is in turn used to control the values of a set of adjustable filter coefficients. The adjustable coefficients may take the form of tap weights, reflection coefficients, rotation parameters, or synaptic weights, depending on the filter structure employed. However, the essential difference between the various applications of adaptive filtering arises in the manner in which the desired response is extracted. In this context, we may distinguish four basic classes of adaptive filtering applications, as depicted in Figure 2.15. For convenience of presentation, the following notations are used in Figure 2.15:

u = input applied to the adaptive filter
y = output of the adaptive filter
d = desired response
$e = d - y$ = estimation error

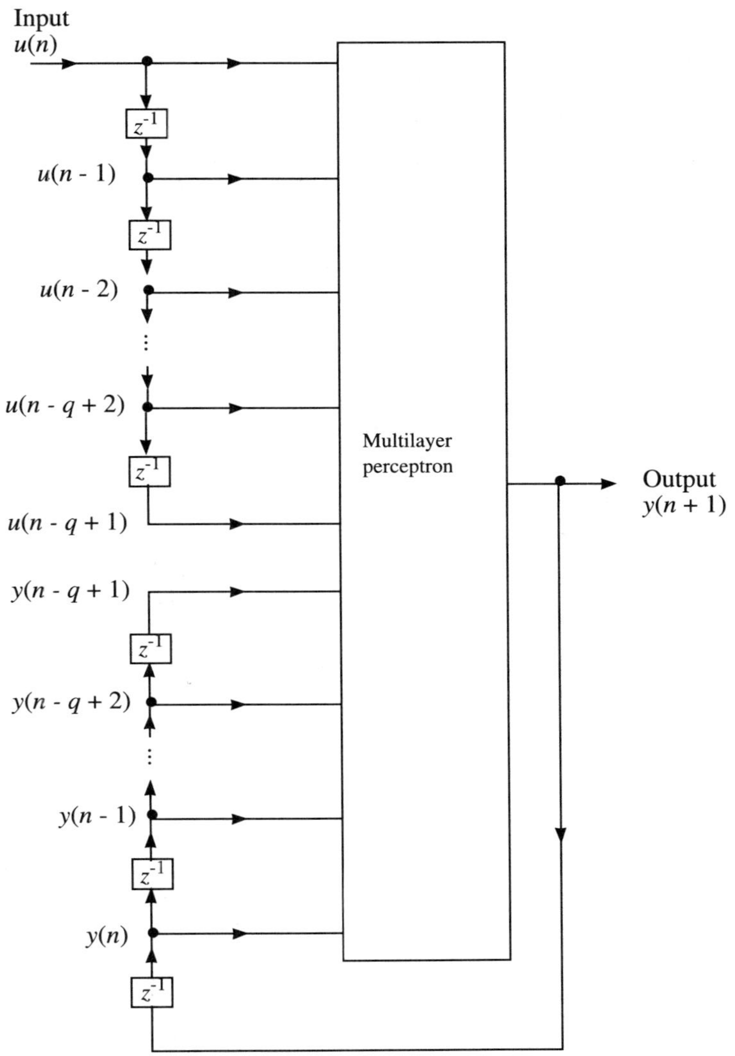

FIGURE 2.13 NARX model.

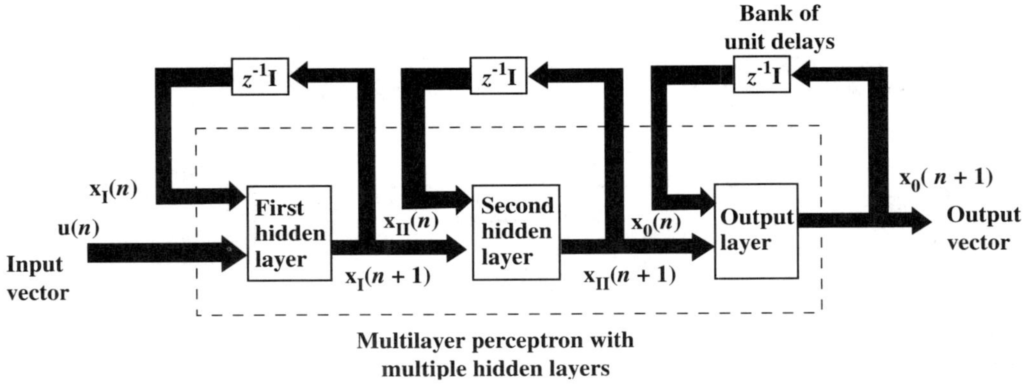

FIGURE 2.14 Recurrent multilayer perceptron.

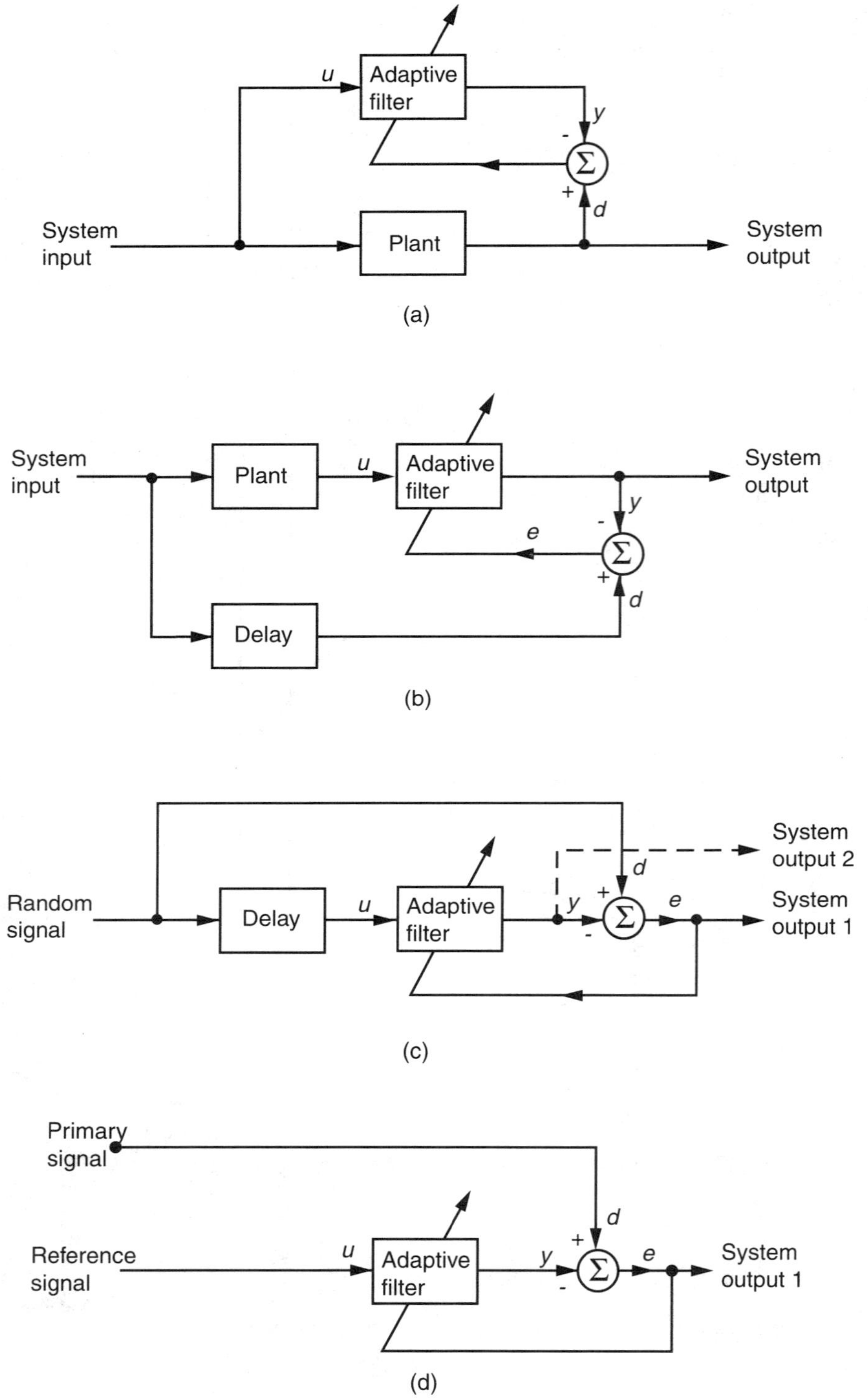

FIGURE 2.15 Four basic classes of adaptive filtering applications: (a) class I, identification; (b) class II, inverse modeling; (c) class III, prediction; (d) class IV, interference canceling.

The functions of the four basic classes of adaptive filtering applications depicted herein are as follows:

Class I. *Identification* (Figure 2.15a): The notion of a *mathematical model* is fundamental to sciences and engineering. In the class of applications dealing with identification, an adaptive filter is used to provide a linear model that represents the best fit (in some sense) to an *unknown plant*. The plant and the adaptive filter are driven by the same input. The plant output supplies the desired response for the adaptive filter. If the plant is dynamic in nature, the model will be time varying.

Class II. *Inverse modeling* (Figure 15.2b): In this second class of applications, the function of the adaptive filter is to provide an *inverse model* that represents the best fit (in some sense) to an *unknown noisy plant*. Ideally, in the case of a linear system, the inverse model has a transfer function equal to the *reciprocal* (*inverse*) of the plant's transfer function, such that the combination of the two constitutes an ideal transmission medium. A delayed version of the plant (system) input constitutes the desired response for the adaptive filter. In some applications, the plant input is used without delay as the desired response.

Class III. *Prediction* (Figure 2.15c): Here, the function of the adaptive filter is to provide the best *prediction* (in some sense) of the present value of a random signal. The present value of the signal thus serves the purpose of a desired response for the adaptive filter. Past values of the signal supply the input applied to the adaptive filter. Depending on the application of interest, the adaptive filter output or the estimation (prediction) error may serve as the system output. In the first case, the system operates as a predictor; in the latter case, it operates as a *prediction-error filter*.

Class IV. *Interference canceling* (Figure 2.15d): In this final class of applications, the adaptive filter is used to cancel *unknown interference* contained (alongside an information-bearing signal component) in a *primary signal*, with the cancellation being optimized in some sense. The primary signal serves as the desired response for the adaptive filter. A *reference* (*auxiliary*) *signal* is employed as the input to the adaptive filter. The reference signal is derived from a sensor or set of sensors located in relation to the sensor(s) supplying the primary signal in such a way that the information-bearing signal component is weak or essentially undesirable.

In Table 2.3, we have listed some applications that are illustrative of the four basic classes of adaptive filtering applications. These applications, totaling twelve, are drawn from the fields of control systems, seismology, electrocardiography, communications, and radar. A selected number of these applications are described individually in the remainder of this section.

2.7.1 System Identification

System identification is the experimental approach to the modeling of a process or a plant (Goodwin and Rayne, 1977; Ljung and Söderström, 1983; Åström and Eykhoff, 1971). It involves the following steps:

TABLE 2.3 Applications of Adaptive Filters

Class of Adaptive Filtering	Application
I. Identification	System identification
	Layered earth modeling
II. Inverse modeling	Predictive deconvolution
	Adaptive equalization
	Blind equalization
III. Prediction	Linear predictive coding
	Adaptive differential pulse-code modulation
	Autoregressive spectrum analysis
	Signal detection
IV. Interference canceling	Adaptive noise canceling
	Echo cancelation
	Adaptive beamforming

experimental planning, the selection of a model structure, parameter estimation, and model validation. The procedure of system identification, as pursued in practice, is iterative in nature in that we may have to go back and forth between these steps until a satisfactory model is built. Here, we discuss briefly the idea of adaptive filtering algorithms for estimating the parameters of an unknown plant modeled as a transversal filter.

Suppose we have an unknown dynamic plant that is linear and time varying. The plant is characterized by a *real-valued* set of discrete-time measurements that describe the variation of the plant output in response to a known stationary input. The requirement is to develop an *on-line transferal filter model* for this plant, as illustrated in Figure 2.16. The model consists of a finite number of unit-delay elements and a corresponding set of adjustable parameters (tap weights).

Let the available input signal at time n be denoted by the set of samples: $u(n)$, $u(n-1)$, ..., $u(n-M+1)$, where M is the number of adjustable parameters in the model. This input signal is applied simultaneously to the plant and the model. Let their respective outputs be denoted by $d(n)$ and $y(n)$. The plant output $d(n)$ serves the purpose of a desired response for the adaptive filtering algorithm employed to adjust the model parameters. The model output is given by

$$y(n) = \sum_{k=0}^{M-1} \hat{w}_k(n)u(n-k) \tag{2.18}$$

where $\hat{w}_0(n)$, $\hat{w}_1(n)$, ... and $\hat{w}_{M-1}(n)$ are the estimated model parameters. The model output $y(n)$ is compared with the plant output $d(n)$. The difference between them, $d(n) - y(n)$, defines the modeling (estimation) error. Let this error be denoted by $e(n)$.

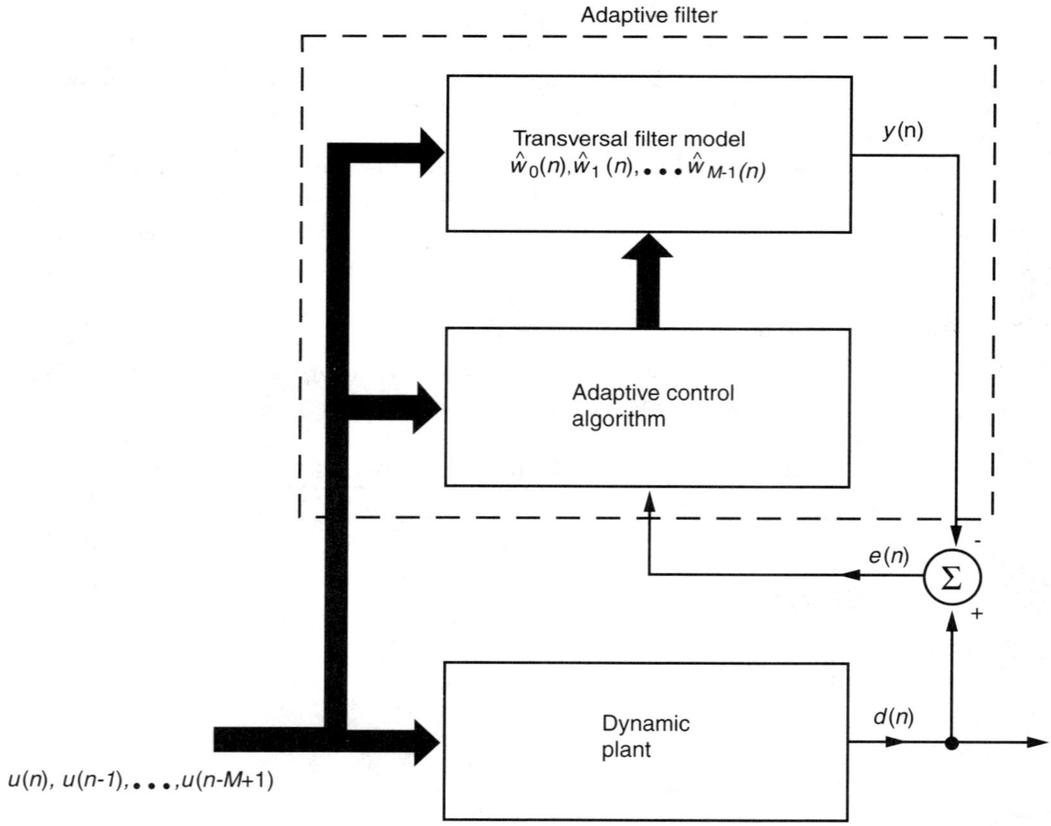

FIGURE 2.16 System identification.

Typically, at time n, the modeling error $e(n)$ is non-zero, implying that the model deviates from the plant. In an attempt to account for this deviation, the error $e(n)$ is applied to an *adaptive control algorithm*. The samples of the input signal, $u(n), u(n-1), \ldots, u(n-M+1)$, are also applied to the algorithm. The combination of the transversal filter and the adaptive control algorithm constitutes the adaptive filtering algorithm. The algorithm is designed to control the adjustments made in the values of the model parameters. As a result, the model parameters assume a new net of values for use on the next iteration. Thus, at time $n+1$, a new model output is computed, and with it a new value for the modeling error. The operation described is then repeated. This process is continued for a sufficiently large number of iterations (starting from time $n = 0$), until the deviation of the model from the plant, measured by the magnitude of the modeling error $e(n)$, becomes sufficiently small in some statistical sense.

When the plant is time varying, the plant output is *nonstationary* and so is the desired response presented to the adaptive filtering algorithm. In such a situation, the adaptive filtering algorithm has the task of not only keeping the modeling error small, but also continually tracking the time variations in the dynamics of the plant.

2.7.2 Spectrum Estimation

The *power spectrum* provides a quantitative measure of the second-order statistics of a discrete-time stochastic process as a function of frequency. In *parametric spectrum analysis,* we evaluate the power spectrum of the process by assuming a *model* for the process. In particular, the process is modeled as the output of a linear filter that is excited by a *white-noise process,* as in Figure 2.17. By definition, a white-noise process has a constant power spectrum. A model that is of practical utility is the *autoregressive (AR) model,* in which the transfer function of the filter is assumed to consist of poles only. Let this transfer function be denoted by

$$H(e^{j\omega}) = \frac{1}{1 + a_1 e^{j\omega} + \ldots + a_M e^{-jM\omega}} = \frac{1}{1 + \sum_{k=1}^{M} a_k e^{-jk\omega}} \tag{2.19}$$

where the a_k are called the *AR parameters,* and M is the model order. Let σ_v^2 denote the constant power spectrum of the white-noise process $v(n)$ applied to the filter input. Accordingly, the power spectrum of the filter output $u(n)$ equals

$$S_{\text{AR}}(\omega) = \sigma_v^2 |H(e^{j\omega})|^2 \tag{2.20}$$

We refer to $S_{\text{AR}}(\omega)$ as the *AR power spectrum.* Equation 2.19 assumes that the AR process $u(n)$ is real, in which case the AR parameters themselves assume real values.

When the AR model is time varying, the model parameters become time dependent, as shown by $a_1(n), a_2(n), \ldots, a_M(n)$. In this case, we express the power spectrum of the time-varying AR process as

$$S_{\text{AR}}(\omega, n) = \frac{\sigma_v^2}{\left|1 + \sum_{k=1}^{M} a_k e^{-jk\omega}\right|} \tag{2.21}$$

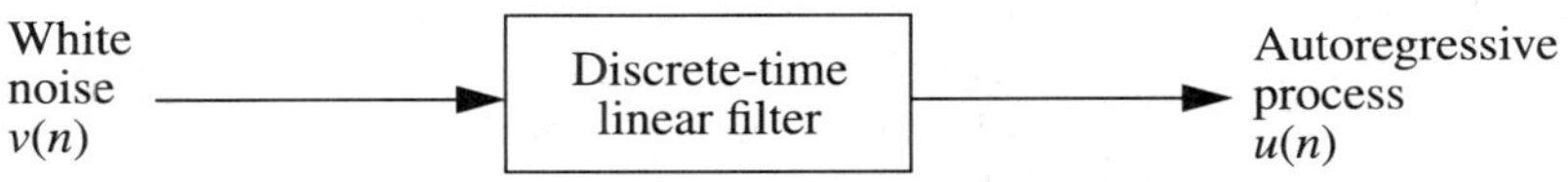

FIGURE 2.17 Black box representation of a stochastic model.

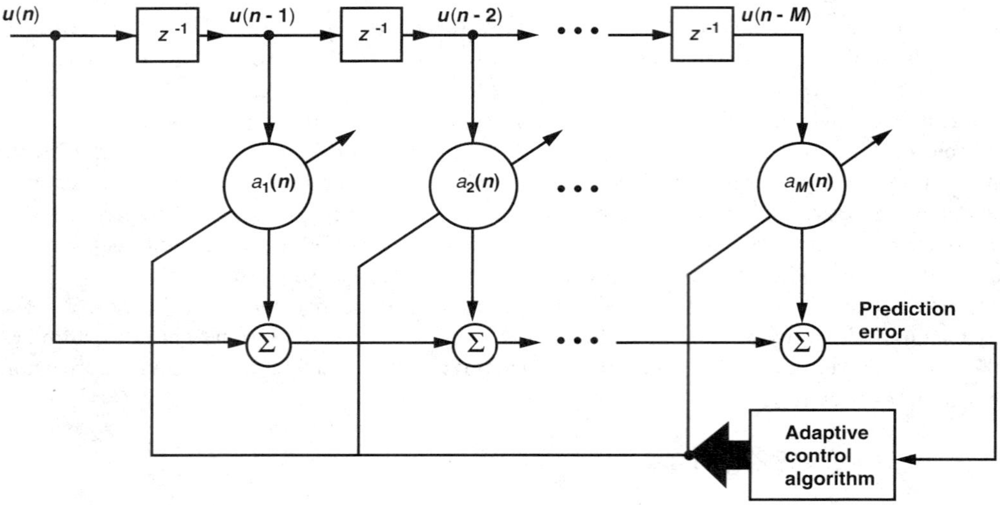

FIGURE 2.18 Adaptive prediction-error filter for real-valued data.

We may determine the AR parameters of the time-varying model by applying $u(n)$ to an *adaptive prediction-error filter,* as indicated in Figure 2.18. The filter consists of a transversal filter with adjustable tap weights. In the adaptive scheme of Figure 2.18, the prediction error produced at the output of the filter is used to control the adjustments applied to the tap weights of the filter. The *adaptive AR model* provides a practical means for measuring the *instantaneous frequency* of a frequency-modulated process. In particular, we may do this by measuring the frequency at which the AR power spectrum $S_{AR}(\omega,n)$ attains its peak value for varying time n.

2.7.3 Signal Detection

The *detection problem,* that is, the problem of detecting an information-bearing signal in noise, may be viewed as one of *hypothesis testing* with deep roots in *statistical decision theory* (Van Trees, 1968). In the statistical formulation of hypothesis testing, there are two criteria of most interest: the *Bayes criterion* and the *Neyman-Pearson criterion.* In the Bayes test, we minimize the *average cost* or *risk* of the experiment of interest, which incorporates two sets of parameters: (1) *a priori probabilities* that represent the observer's information about the source of information before the experiment is conducted, and (2) a set of *costs* assigned to the various possible courses of action. As such, the Bayes criterion is directly applicable to digital communications. In the Neyman-Pearson test, on the other hand, we maximize the *probability of detection* subject to the constraint that the *probability of false alarm* does not exceed some preassigned value. Accordingly, the Neyman-Pearson criterion is directly applicable to radar or sonar. An idea of fundamental importance that emerges in hypothesis testing is that for a Bayes criterion or Neyman-Pearson criterion, the optimum test consists of two distinct operations: (1) processing the observed data to compute a test statistic called the *likelihood ratio* and (2) computing the likelihood ratio with a *threshold* to make a *decision* in favor of one of the two hypotheses. The choice of one criterion or the other merely affects the value assigned to the threshold. Let H_1 denote the hypothesis that the observed data consist of noise alone, and let H_2 denote the hypothesis that the data consist of signal plus noise. The likelihood ratio is defined as the ratio of two maximum likelihood functions, with the numerator assuming that hypothesis H_2 is true and the denominator assuming that hypothesis H_1 is true. If the likelihood ratio exceeds the threshold, the decision is made in favor of hypothesis H_2; otherwise, the decision is made in favor of hypothesis H_1.

In simple binary hypothesis testing, it is assumed that the signal is known and the noise is both white and Gaussian. In this case, the likelihood ratio test yields a *matched filter* (matched in the sense that its impulse response equals the time-reversed version of the known signal). When the additive noise is a *colored Gaussian noise* of known mean and correlation matrix, the likelihood ratio test yields a filter that consists of two sections: a *whitening filter* that transforms the colored noise component at the input into a white Gaussian noise process and a *matched filter* that is matched to the new version of the known signal as modified by the whitening filter.

However, in some important operational environments such as *communications, radar,* and *active sonar,* there may be inadequate information on the signal and noise statistics to design a fixed optimum detector. For example, in a sonar environment it may be difficult to develop a precise model for the received sonar signal, one that would account for the following factors completely:

- Loss in the signal strength of a *target echo* from an object of interest (e.g., enemy vessel), due to oceanic propagation effects and reflection loss at the target
- Statistical variations in the additive *reverberation* component, produced by reflections of the transmitted signal from scatterers such as the ocean surface, ocean floor, biologies, and in homogeneities within the ocean volume
- Potential sources of *noise* such as biological, shipping, oil drilling, and seismic and oceanographic phenomena

In situations of this kind, the use of adaptivity offers a powerful approach to solve difficult signal detection problems. The particular application we have chosen for our present discussion is the detection of a small radar target in sea clutter (i.e., radar backscatter from the ocean surface). The radar target is a small piece of ice called a *growler;* the portion of which is visible above the sea surface is about the size of a grand piano. Recognizing that 90% of the volume of ice lies inside the water, a growler can indeed pose a threat to navigation in ice-infested waters as on the east coast of Canada.

The detection problem described herein is further compounded by the nonstationary nature of both sea clutter and the target echo from the growler. The strategy we have chosen to solve this difficult signal detection problem reformulates it into a pattern classification problem for which neural networks are well suited.

Figure 2.19 shows a block diagram of the detection strategy described in Haykin and Thomson, 1998, and Haykin and Bhattacharya, 1997. It consists of three functional units:

- *Time-frequency analyzer,* which converts the time-varying waveform of the input signal into a picture with two coordinates, namely, time and frequency
- *Feature extractor,* the purpose of which is to compress the two-dimensional data produced by the time-frequency analyzer by extracting a set of features that retain the essential frequency content of the original signal
- *Pattern classifier,* which is trained to categorize the set of features applied to its input into two classes:
  1. No target present (i.e., the input signal consists of clutter only)
  2. Target present (i.e., the input signal consists of target echo plus clutter)

A signal-processing tool that is well suited for the application described herein is the *Wigner-Ville distribution* (WVD) (Cohen, 1995). The time-frequency map produced by this method is highly dependent on the nature of the input signal. If the input signal consists of clutter only, the resulting WVD picture is determined entirely by the time-frequency characteristics of sea clutter. Figure 2.20a shows a typical WVD picture due to sea clutter acting alone. On the other hand, if the input signal consists of a target echo plus sea clutter, the resulting WVD picture consists of three components: one

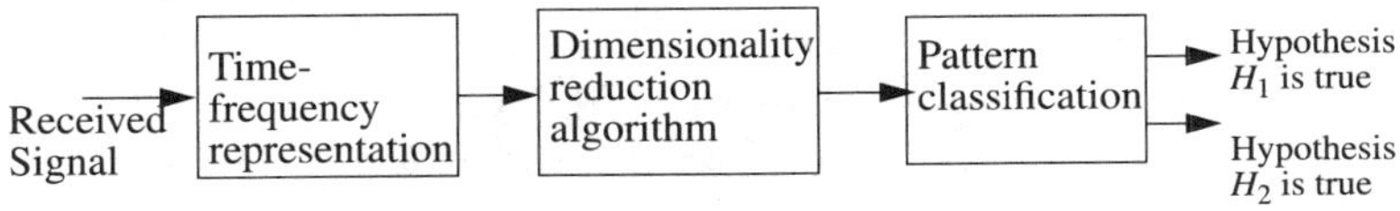

FIGURE 2.19 Block diagram of the detection strategy used in a nonstationary environment.

due to the target echo, one due to clutter in the background, and one (commonly referred to as a "cross-product term") due to the interaction between these two components. Ordinarily, the cross-product terms are viewed as highly undesirable, as they tend to complicate the spectral interpretation of WVD pictures; indeed, much effort has been expended in the literature to reduce the effects of cross-product terms. However, in the application of WVD to target detection described herein, the cross-product terms perform a useful service by enhancing the detection power of the method. In particular, cross-product terms are there to be seen only when a target is present; they disappear when the input signal is target free. Figure 2.20b shows a typical WVD picture pertaining to the combined presence of sea clutter and radar echo from a small growler. The zebra-like pattern (consisting of an alternating set of dark and light stripes) is due to the cross-product terms. The point to note here is that the target echo in the original input signal is hardly visible; yet is shows up ever so clearly in the WVD picture of Figure 2.20b.

For the feature extraction, we may use PCA, which was briefly described in Section 2.6.2.1. Finally, for pattern classification, we may use a multilayer perceptron trained with the BP algorithm. Design details of these two functional units and those of the WVD are presented elsewhere (Haykin and Thomson, 1998; Haykin and Bhattacharya, 1997). For the present discussion, it suffices to compare the receiver operating characteristics of this new radar detection strategy against those of an ordinary constant false alarm rate (CFAR) receiver. Figure 2.21 presents the results of this comparison using real-life data, which were collected at a site on the east coast of Canada by means of an instrument-quality radar known as the IPIX radar (designed and built at McMaster University, Hamilton, Ontario). From Figure 2.21, we see that for probability of false alarm $P_{FA} = 10^{-3}$, we have the following values for probability of detection:

$$P_D = \begin{cases} 0.91 & \text{for adaptive receiver based on the detection strategy of Figure 2.19} \\ 0.71 & \text{for Doppler CFAR receiver} \end{cases}$$

2.7.4 Target Tracking

The objective of *tracking* is to estimate the *state* of a target of interest by processing measurements obtained from the target through the use of sensors and other means. The measurements are *noise-corrupted observables,* which are related to the current state of the target. Typically, the state consists of kinematic components such as the position, velocity, and acceleration of a moving target.

To state the tracking problem in mathematical terms, let the vector $\mathbf{x}(n)$ denote the state of a target, the vector $\mathbf{u}(n)$ denote the (known) input or control signal, and the vector $\mathbf{y}(n)$ denote the corresponding measurements obtained from the target. We may then express the state-space equations of the system in its most generic setting as follows:

$$\mathbf{x}(n+1) = \mathbf{f}(n, x(n), \mathbf{u}(n), \mathbf{v}_1(n)) \tag{2.22}$$

$$\mathbf{y}(n) = \mathbf{h}(n, \mathbf{x}(n), \mathbf{v}_2(n)) \tag{2.23}$$

where $\mathbf{f}(\cdot)$ and $\mathbf{h}(\cdot)$ are vector-valued functions, and $\mathbf{v}_1(n)$ and $\mathbf{v}_2(n)$ are noise vectors. The time argument n indicates that both $\mathbf{f}(\cdot)$ and $\mathbf{g}(\cdot)$ are time varying. Equations 2.22 and 2.23 are referred to as the process and measurement equations, respectively. The issue of interest is to estimate the state vector $\mathbf{x}(n)$, given the measurement vector $\mathbf{y}(n)$.

When the process and measurement equations are both linear, and the process noise vector $\mathbf{v}_1(n)$ and measurement noise vector $\mathbf{v}_2(n)$ are both modeled as zero-mean, white, Gaussian processes that are statistical independent, the Kalman filter provides the optimum estimate of the state $\mathbf{x}(n)$, given $\mathbf{y}(n)$ (Bar-Shalom and Fortmann, 1988). Optimality here refers to minimization of the mean-square error between the actual motion of the target and the *track* (i.e., state trajectory) estimated from the measurements associated with the target.

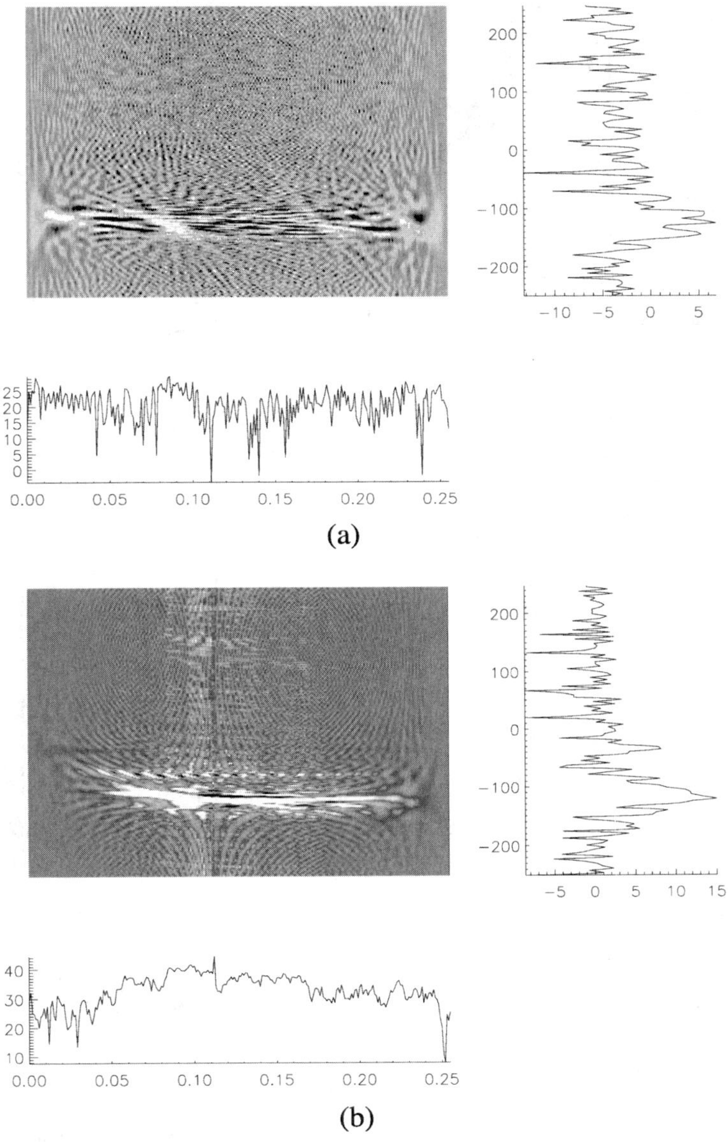

FIGURE 2.20 (a) WVD for sea clutter; (b) WVD for a barely visible growler. For the images, the horizontal axes are time in seconds and the vertical axes are frequency in Hertz. Horizontal axes of power spectra are in decibels.

Unfortunately, in many of the radar and sonar target-tracking problems encountered in practice, the process and measurement equations are nonlinear, and the noise processes corrupting the state and measured data are non-Gaussian. The traditional approach for dealing with nonlinear dynamics is to use the EKF, the derivation of which assumes knowledge of the nonlinear functions $\mathbf{f}(\cdot)$ and $\mathbf{h}(\cdot)$ and maintains the Gaussian assumptions about the process and noise vectors. The EKF closely resembles a Kalman filter except for the fact that each step of the standard Kalman filtering algorithm is replaced by its linearized equivalent (Bar-Shalom and Fortmann, 1988). However, the EKF approach to target tracking suffers from the following drawbacks when it is applied to an environment with nonlinear dynamics:

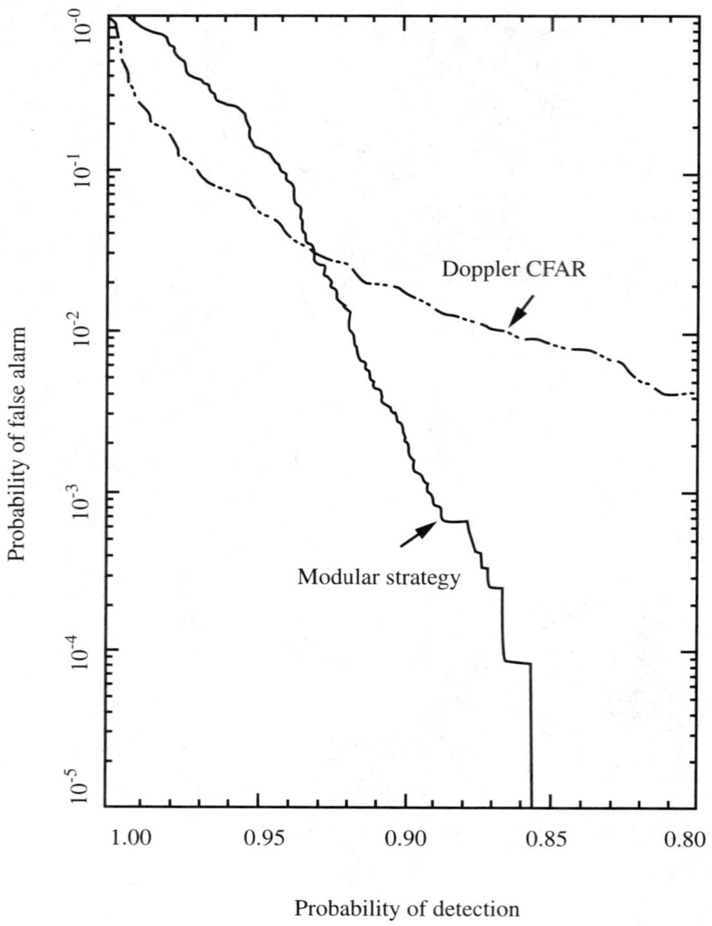

FIGURE 2.21 Composite receiver operating characteristics.

1. Linearization of the vector-valued functions $\mathbf{f}(\cdot)$ and $\mathbf{h}(\cdot)$ can produce system instability if the time steps are not sufficiently short in duration.
2. Linearization of the underlying dynamics requires the determination of two Jacobians (i.e., matrices of partial derivatives):
   - Jacobian of the vector-valued function $\mathbf{f}(\cdot)$, evaluated at the latest filtered estimate of the state at time n
   - Jacobian of the vector-valued function $\mathbf{h}(\cdot)$, evaluated at the one-step predicted estimate of the state at time $n + 1$

   The determination of these two Jacobians may lead to computational difficulties.
3. The use of a short-time step to avoid system instability, combined with the determination of these two Jacobians, may impose a high computational overload on the system.

To overcome these shortcomings of the EKF, we may use the *unscented Kalman filter* (UKF) (Julier and Uhlmann, 1997; Wan et al., 1999), which is a generalization of the standard linear Kalman filter to systems whose process and measurement models are nonlinear. The UKF is preferable to the EKF for solving nonlinear filtering problems for two reasons:

1. The UKF is accurate to the third order for Gaussian-distributed process and measurement errors. For non-Gaussian distributions, the UKF is accurate to at least the second order. Accordingly, the UKF provides better performance than the traditional EKF.

2. Unlike the EKF, the UKF does not require the computation of Jacobians pertaining to process and measurement equations. It is therefore simpler than the EKF in computational terms.

These are compelling reasons to reconsider the design of tracking systems for radar and sonar systems using the UKF.

2.7.5 Adaptive Noise Canceling

As the name implies, adaptive noise canceling relies on the use of *noise canceling* by subtracting noise from a received signal, an operation controlled in an *adaptive* manner for the purpose of improved signal-to-noise ratio. Ordinarily, it is inadvisable to subtract noise from a received signal, because such an operation could produce disastrous results by causing an increase in the average power of the output noise. However, when proper provisions are made, and filtering and subtraction are controlled by an adaptive process, it is possible to achieve a superior system performance compared to direct filtering of the received signal (Widrow et al., 1975b; Widrow and Stearns, 1985).

Basically, an adaptive noise canceler is a *dual-input, closed-loop adaptive feedback system* as illustrated in Figure 2.22. The two inputs of the system are derived from a pair of sensors: a *primary sensor* and a *reference (auxiliary) sensor.* Specifically, we have the following:

1. The primary sensor receives an information-bearing signal $s(n)$ corrupted by additive noise $v_0(n)$, as shown by

$$d(n) = s(n) + v_0(n) \tag{2.24}$$

The signal $s(n)$ and the noise $v_0(n)$ are uncorrelated with each other; that is,

$$E[s(n)v_1(n-k)] = 0 \text{ for all } k \tag{2.25}$$

where $s(n)$ and $v_0(n)$ are assumed to be real valued.

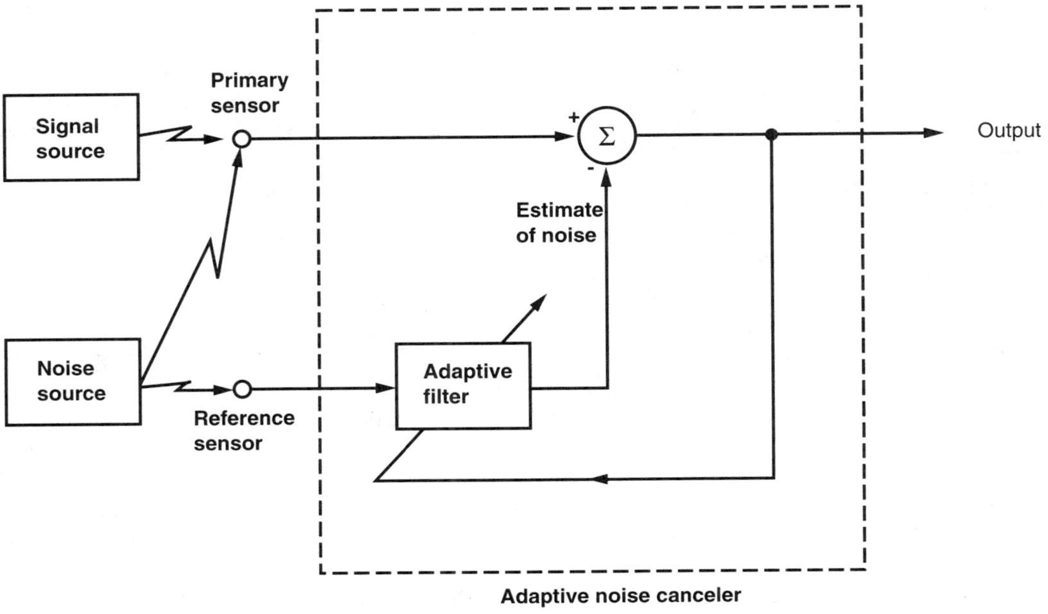

FIGURE 2.22 Adaptive noise cancelation.

2. The reference sensor receives a noise $v_1(n)$ that is uncorrelated with the signal $s(n)$, but correlated with the noise $v_0(n)$ in the primary sensor output in an unknown way; that is,

$$E[s(n)v_1(n-k)] = 0 \text{ for all } k \tag{2.26}$$

and

$$E[v_0(n)v_1(n-k)] = p(k) \tag{2.27}$$

where, as before, the signals are real valued, and $p(k)$ is an unknown cross-correlation for lag k.

The reference signal $v_1(n)$ is processed by an adaptive filter to produce the output signal

$$y(n) = \sum_{k=0}^{M-1} \hat{w}_k(n)v_1(n-k) \tag{2.28}$$

where $\hat{w}_k(n)$ is the adjustable (real) tap weights of the adaptive filter. The filter output $y(n)$ is subtracted from the primary signal $d(n)$, serving as the "desired" response for the adaptive filter. The error signal is defined by

$$e(n) = d(n) - y(n) \tag{2.29}$$

Thus, substituting Equation 2.22 into Equation 2.28, we get

$$e(n) = s(n) + v_0(n) - y(n) \tag{2.30}$$

The error signal is, in turn, used to adjust the tap weights of the adaptive filter, and the control loop around the operations of filtering and subtraction is thereby closed. Note that the information-bearing signal $s(n)$ is indeed part of the error signal $e(n)$, as indicated in Equation 2.30.

The error signal $e(n)$ constitutes the overall system output. From Equation 2.30, we see that the noise component in the system output is $v_0(n) - y(n)$. Now the adaptive filter attempts to minimize the mean-square value (i.e., average power) of the error signal $e(n)$. The information-bearing signal $s(n)$ is essentially unaffected by the adaptive noise canceler. Hence, minimizing the mean-square value of the error signal $e(n)$ is equivalent to minimizing the mean-square value of the output noise $v_0(n) - y(n)$. With the signal $s(n)$ remaining essentially constant, it follows that the *minimization of the mean-square value of the error signal is indeed the same as the maximization of the output signal-to-noise ratio of the system.*

The signal-processing operation described herein has two limiting cases that are noteworthy:

1. The adaptive filtering operation is *perfect* in the sense that

$$y(n) = v_0(n)$$

In this case, the system output is *noise free*, and the noise cancelation is perfect. Correspondingly, the output signal-to-noise ratio is infinitely large.

2. The reference signal $v_1(n)$ is completely uncorrelated with both the signal and noise components of the primary signal $d(n)$; that is,

$$E[d(n)v_1(n-k)] = 0 \text{ for all } k$$

In this case, the adaptive filter "switches itself off," resulting in a zero value for the output $y(n)$. Hence, the adaptive noise canceler has *no* effect on the primary signal $d(n)$, and the output signal-to-noise ratio remains unaltered.

The effective use of adaptive noise canceling therefore requires that we place the reference sensor in the noise field of the primary sensor with two specific objectives in mind. First, the information-bearing signal component of the primary sensor output is *undetectable* in the reference sensor output. Second, the reference sensor output is *highly correlated* with the noise component of the primary sensor output. Moreover, the adaptation of the adjustable filter coefficients must be near optimum.

In the remainder of this section, we described three useful applications of the adaptive noise canceling operation:

1. *Canceling 60-Hz interference in electrocardiography:* In electrocardiography (ECG) commonly used to monitor heart patients, an *electrical discharge* radiates energy through human *tissue* and the resulting output is received by an *electrode.* The electrode is usually positioned in such a way that the received energy is maximized. Typically, however, the electrical discharge involves very low potentials. Correspondingly, the received energy is very small. Hence, extra care has to be exercised in minimizing signal degradation due to external *interference.* By far, the strongest form of interference is that of a 60-Hz periodic waveform picked up by the receiving electrode (acting like an antenna) from nearby electrical equipment (Huhta and Webster, 1973). Needless to say, this interference has undesirable effects in the interpretation of electrocardiograms. Widrow et al. (1975b) have demonstrated the use of adaptive noise canceling (based on the LMS algorithm) as a method for reducing this form of interference. Specifically, the primary signal is taken from the ECG preamplifier, and the reference signal is taken from a wall outlet with proper attenuation. Figure 2.23 shows a block diagram of the adaptive noise canceler used by Widrow et al. (1975b). The adaptive filter has two adjustable weights, $\hat{w}_0(n)$ and $\hat{w}_1(n)$. One weight, $\hat{w}_0(n)$, is fed directly from the reference point. The second weight, $\hat{w}_1(n)$, is fed from a 90° phase-shifted version of the reference input. The sum of the two weighted versions of the reference signal is then subtracted from the ECG output to produce an error signal. This error signal together with the weighted inputs are applied to the LMS algorithm, which, in turn, controls the adjustments applied to the two weights. In this application, the adaptive noise canceler acts as a variable "notch filter." The frequency of the sinusoidal interference in the ECG output is presumably the same as that of the sinusoidal reference signal. However, the amplitude and phase of the sinusoidal interference in the ECG output are unknown. The two weights, $\hat{w}_0(n)$ and $\hat{w}_1(n)$, provide the *two degrees of freedom* required to control the amplitude and phase of the sinusoidal reference signal so as to cancel the 60-Hz interference contained in the ECG output.
2. *Reduction of acoustic noise in speech:* At a noisy site (e.g., the cockpit of a military aircraft), voice communication is affected by the presence of *acoustic noise.* This effect is particularly serious when linear predictive coding (LPC) is used for the digital representation of voice signals at low bit rates; LPC was discussed earlier. To be specific, high-frequency acoustic noise severely affects the estimated LPC spectrum in both the low- and high-frequency regions. Consequently, the intelligibility of digitized speech using LPC often falls below the minimum acceptable level. Kang and Fransen (1987) describe the use of an adaptive noise canceler, based on the LMS algorithm, for reducing acoustic noise in speech. The noise-corrupted speech is used as the primary signal. To provide the reference signal (noise only), a reference microphone is placed in a location where there is sufficient isolation from the source of speech (i.e., the known location of the speaker's mouth). In the experiments described by Kang and Fransen, a reduction of 10 to 15 dB in the acoustic noise floor is achieved without degrading voice quality. Such a level of noise reduction is significant in improving voice quality, which may be unacceptable otherwise.
3. *Adaptive speech enhancement:* Consider the situation depicted in Figure 2.24. The requirement is to listen to the voice of the desired speaker in the presence of background noise, which may be satisfied through the use of the adaptive noise canceling. Specifically, *reference microphones* are added at locations far enough away from the desired speaker such that their outputs

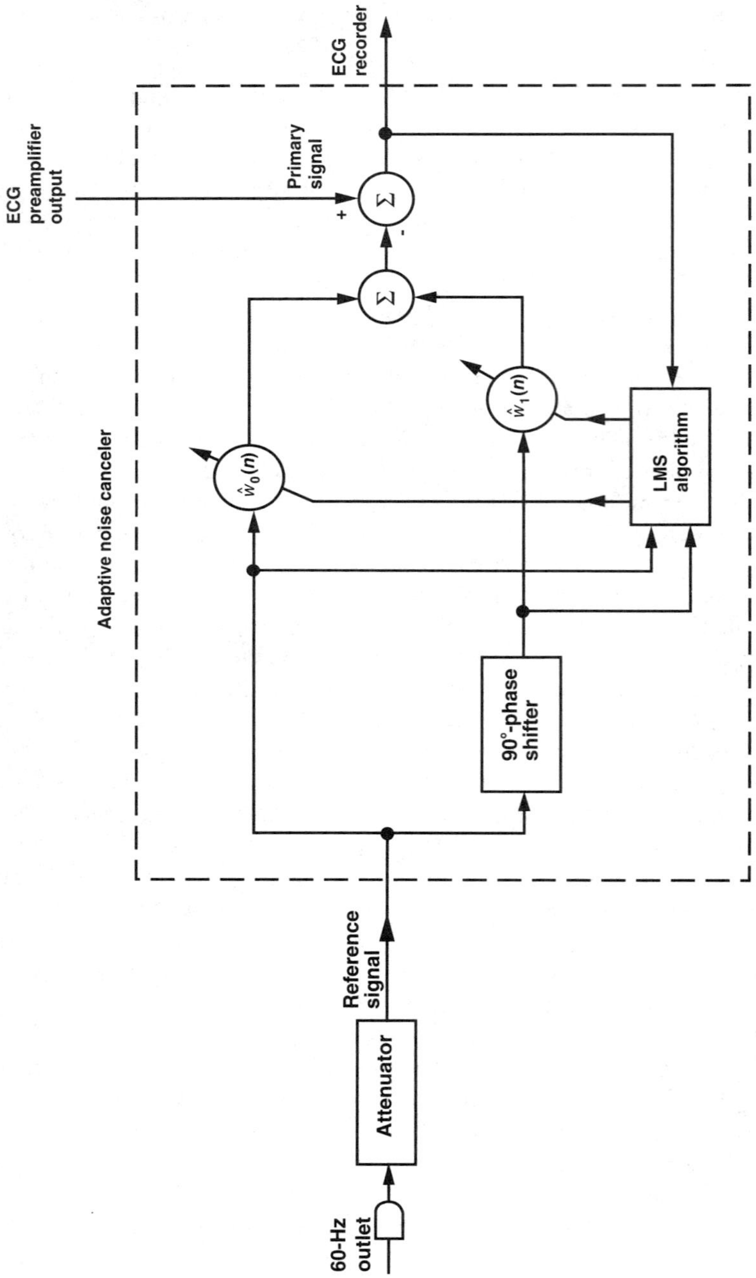

FIGURE 2.23 Adaptive noise canceler for suppressing 60-Hz interference in ECG. (After Widrow et al., 1975b.)

contain *only* noise. As indicated in Figure 2.24, a weighted sum of the auxiliary microphone outputs is subtracted from the output of the desired speech-containing microphone, and an adaptive filtering algorithm (e.g., the LMS algorithm) is used to adjust the weights so as to minimize the average output power. A useful application of the idea described herein is in the adaptive noise cancelation for heating aids* (Chazan et al., 1988). The so-called "cocktail party effect" severely limits the usefulness of hearing aids. The cocktail party phenomenon refers to the ability of a person with normal hearing to focus on a conversation taking place at a distant location in a crowded room. This ability is lacking in a person who wears hearing aids because of extreme sensitivity to the presence of *background noise.* This sensitivity is attributed to two factors: (1) the loss of directional cues and (2) the limited channel capacity of the ear caused by the reduction in both dynamic range and frequency response. Chazan et al. (1988) describe an adaptive noise-canceling technique aimed at overcoming this problem. The technique involves the use of an *array of microphones* that exploit the difference in spatial characteristics between the desired signal and the noise in a crowded room. The approach taken by Chazan et al. is based on the fact that each microphone output may be viewed as the sum of the signals produced by the individual speakers engaged in conversations in the room. Each signal contribution in a particular microphone output is essentially the result of a speaker's speech signal having passed through the *room filter.* In other words, each speaker (including the desired speaker) produces a signal at the microphone output that is the sum of the direct transmission of his/her speech signal and its reflections from the walls of the room. The requirement is to reconstruct the desired speaker signal, including its room reverberations, while canceling out the source of noise. In general, the transformation undergone by the speech signal from the desired speaker is not known. Also, the characteristics of the background noise are variable. We thus have a signal-processing problem for which adaptive noise canceling offers a feasible solution.

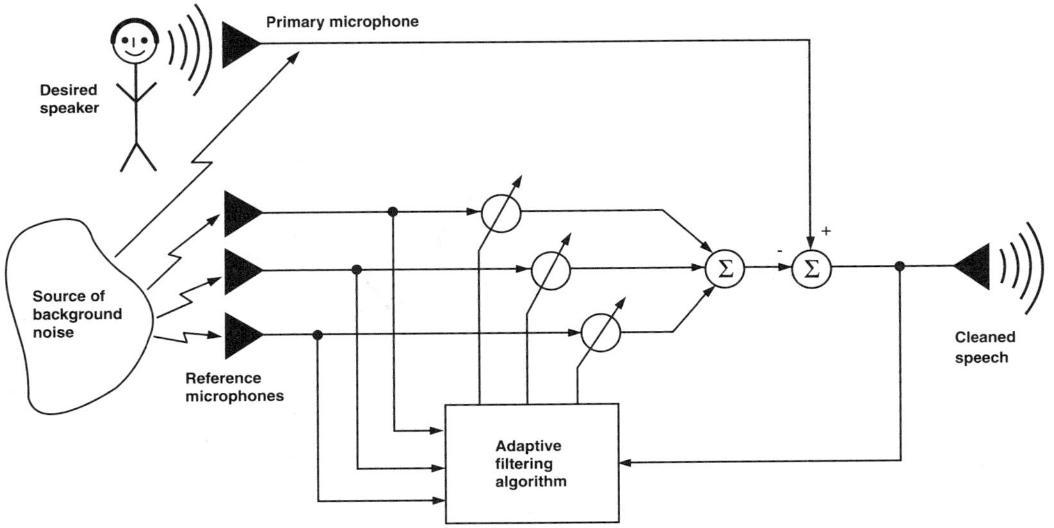

FIGURE 2.24 Block diagram of an adaptive noise canceler for speech.

* This idea is similar to that of adaptive spatial filtering in the context of antennas, which is considered later in this section.

2.7.6 Adaptive Beamforming

For our last application, we describe a *spatial* form of adaptive signal processing that finds practical use in radar, sonar, communications, geophysical exploration, astrophysical exploration, and biomedical signal processing.

In the particular type of spatial filtering of interest to us in this book, a number of independent *sensors* are placed at different points in space to "listen" to the received signal. In effect, the sensors provide a means of *sampling* the received signal in *space.* The set of sensor outputs collected at a particular instant of time constitutes a *snapshot.* Thus, a snapshot of data in spatial filtering (for the case when the sensors lie uniformly on a straight line) plays a role analogous to that of a set of consecutive tap inputs that exist in a transversal filter at a particular instant of time.*

In radar, the sensors consist of antenna elements (e.g., dipoles, horns, slotted waveguides) that respond to incident electromagnetic waves. In sonar, the sensors consist of hydrophones designed to respond to acoustic waves. In any event, spatial filtering, known as *beamforming,* is used in these systems to distinguish between the spatial properties of signal and noise. The device used to do the beamforming is called a *beamformer.* The term "beamformer" is derived from the fact that the early forms of antennas (spatial filters) were designed to form *pencil beams,* so as to receive a signal radiating from a specific direction and attenuate signals radiating from other directions of no interest (Van Veen and Buckley, 1988). Note that the beamforming applies to the radiation (transmission) or reception of energy.

In a primitive type of spatial filtering, known as the *delay-and-sum beamformer,* the various sensor outputs are delayed (by appropriate amounts to align spatial components coming from the direction of a target) and then summed, as in Figure 2.25. Thus, for a single target, the average power at the output of the delay-and-sum beamformer is maximized when it is steered toward the target. A major limitation of the delay-and-sum beamformer, however, is that it has no provisions for dealing with sources of *interference.*

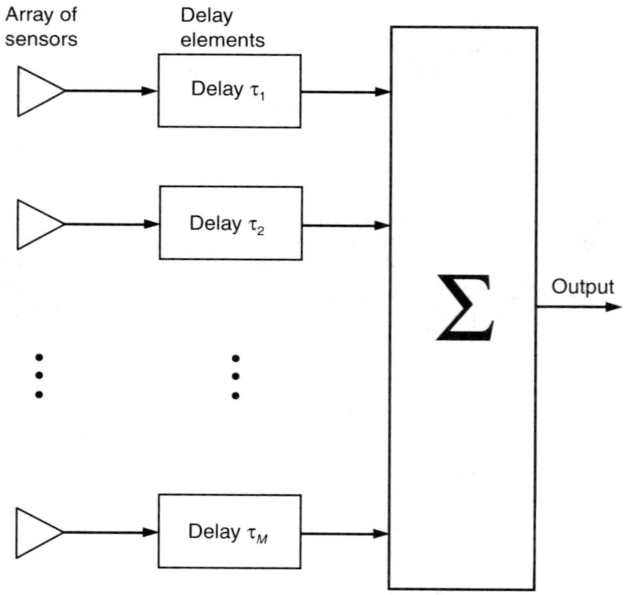

FIGURE 2.25 Delay-and-sum beamformer.

* For a discussion of the analogies between time- and space-domain forms of signal processing, see Bracewell (1978) and Van Veen and Buckley (1988).

In order to enable a beamformer to respond to an unknown interference environment, it has to be made *adaptive* in such a way that it places *nulls* in the direction(s) of the source(s) of interference automatically and in real time. By so doing, the output signal-to-noise ratio of the system is increased, and the *directional response* of the system is thereby improved. In the next section, we consider two examples of *adaptive beamformers* that are well suited for use with narrowband signals in radar and sonar systems.

2.7.6.1 Adaptive Beamformer with Minimum-Variance Distortionless Response

Consider an adaptive beamformer that uses a linear array of M identical sensors, as in Figure 2.26. The individual sensor outputs, assumed to be in *baseband* form, are weighted and then summed. The beamformer has to satisfy two requirements: (1) a *steering* capability whereby the target signal is always protected, and (2) the effects of sources of interference whereby the effects are minimized. One method of providing for these two requirements is to minimize the variance (i.e., average power) of the beamformer output, subject to the *constraint* that during the process of adaptation the weights satisfy the condition

$$\mathbf{w}^H(n)\mathbf{s}(\phi) = 1 \qquad \text{for all } n \text{ and } \phi = \phi_t \tag{2.31}$$

where $\mathbf{w}(n)$ is the $M \times 1$ weight vector, and $\mathbf{s}(\phi)$ is an $M \times 1$ steering vector. The superscript H denotes Hermitian transposition (i.e., transposition combined with complex conjugation). In this application, the baseband data are complex valued, hence the need for complex conjugation. The value of the *electrical*

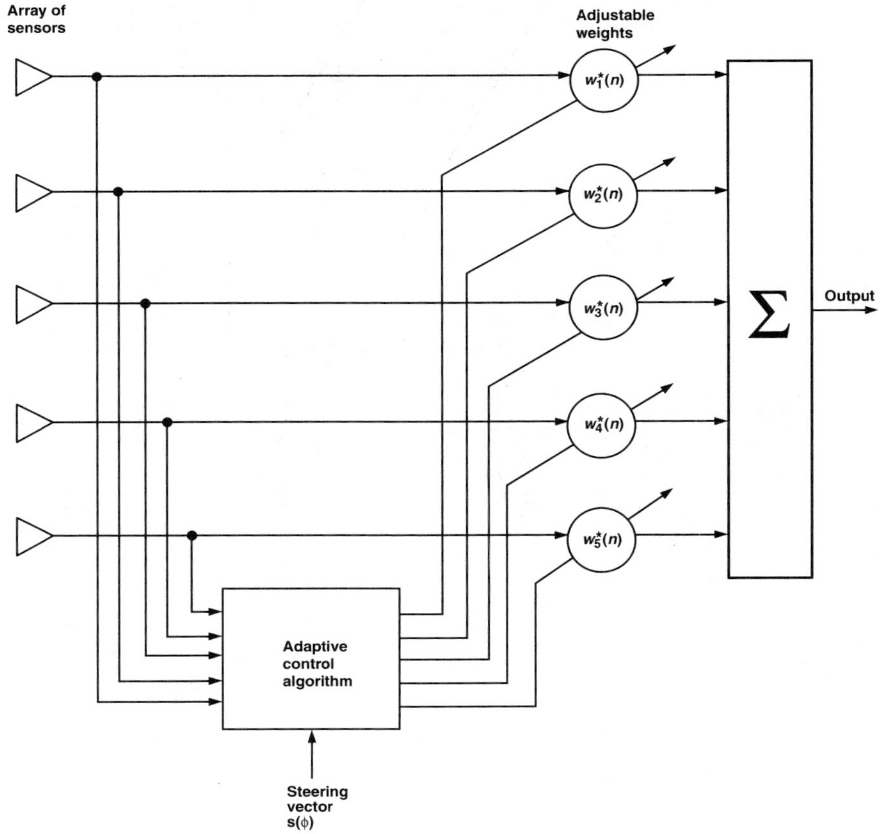

FIGURE 2.26 Adaptive beamformer for an array of five sensors. The sensor outputs (in baseband form) are complex valued; hence the weights are complex valued.

angle $\phi = \phi_t$ is determined by the direction of the target. The angle ϕ is itself measured with sensor 1 (at the top end of the array) treated as the point of reference.

The dependence of vector $\mathbf{s}(\phi)$ on the angle ϕ is defined by

$$\mathbf{s}(\phi) = [1, e^{-j\phi}, \ldots, e^{-j(M-1)\phi}]^T$$

The angle ϕ is itself related to incidence angle θ of a plane wave, measured with respect to the normal to the linear array, as follows:*

$$\phi = \frac{2\pi d}{\lambda}\sin\theta \tag{2.32}$$

where d is the spacing between adjacent sensors of the array and λ is the wavelength (see Figure 2.27). The incidence angle θ lies inside the range $-\pi/2$ to $\pi/2$. The permissible values that the angle ϕ may

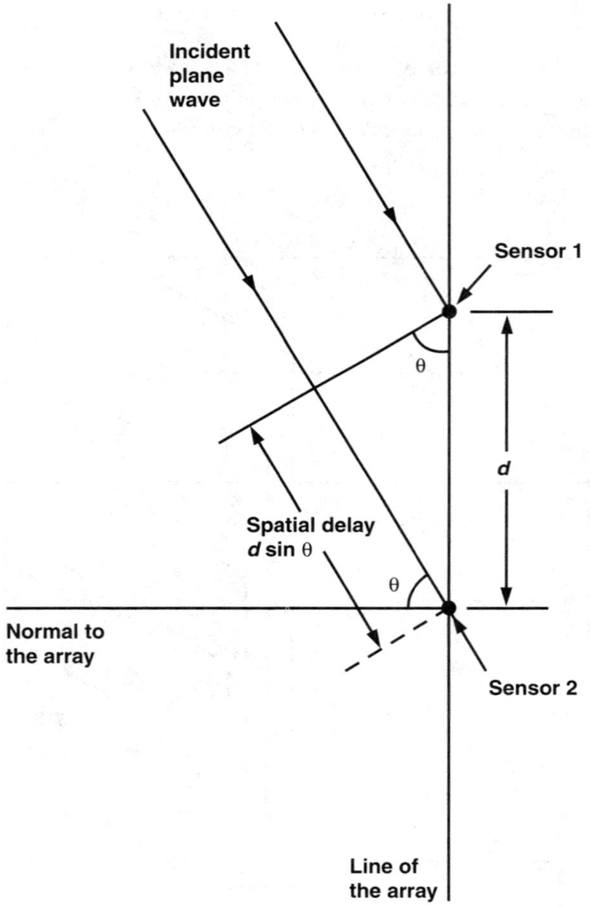

FIGURE 2.27 Spatial delay incurred when a plane wave impinges on a linear array.

* When a plane wave impinges on a linear array as in Figure 2.27, there is a spatial delay of $d \sin\theta$ between the signals received at any pair of adjacent sensors. With a wavelength of λ, this spatial delay is translated into an electrical angular difference defined by $\phi = 2\pi(d \sin\theta/\lambda)$.

assume lie inside the range $-\pi$ to π. This means that we must choose the spacing $d < \lambda/2$, so that there is a one-to-one correspondence between the values of θ and ϕ without ambiguity. The condition $d < \lambda/2$ may be viewed as the spatial analog of the sampling theorem.

The imposition of the *signal-protection constraint* in Equation 2.31 ensures that, for a prescribed look direction, the response of the array is maintained as constant (i.e., equal to 1), no matter what values are assigned to the weights. An algorithm that minimizes the variance of the beamformer output, subject to this constraint, is therefore referred to as the *minimum-variance distortionless response (MVDR) beamforming algorithm* (Capon, 1969; Owsley, 1985). The imposition of the constraint described in Equation 2.31 reduces the number of "degrees of freedom" available to the MVDR algorithm to $M - 2$, where M is the number of sensors in the array. This means that the number of independent nulls produced by the MVDR algorithm (i.e., the number of independent interferences that can be canceled) is $M - 2$.

The MVDR beamforming is a special case of *linearly constrained minimum variance (LCMV) beamforming.* In the latter case, we minimize the variance of the beamformer output, subject to the constraint

$$\mathbf{w}^H(n)\mathbf{s}(\phi) = g \qquad \text{for all } n \text{ and } \phi = \phi_t \tag{2.33}$$

where g is a complex constant. The LCMV beamformer linearly constraints the weights, such that any signal coming from electrical angle ϕ_t is passed to the output with response (gain) g. Comparing the constraint of Equation 2.31 with that of Equation 2.33, we see that the MVDR beamformer is indeed a special case of the LCMV beamformer for $g = 1$.

2.7.6.2 Adaptation in Beam Space

The MVDR beamformer performs adaptation directly in the *data space.* The adaptation process for interference cancelation may also be performed in *beam space.* To do so, the input data (received by the array of sensors) are transformed into the beam space by means of an *orthogonal multiple beamforming network,* as illustrated in the block diagram of Figure 2.28. The resulting output is processed by a *multiple sidelobe canceler* so as to cancel interference(s) from unknown directions.

The beamforming network is designed to generate a set of *orthogonal* beams. The multiple outputs of the beamforming network are referred to as *beam ports.* Assume that the sensor outputs are equally weighted and have a *uniform* phase. Under this condition, the response of the array produced by an incident plane wave arriving at the array along direction θ, measured with respect to the normal to the array, is given by

$$A(\phi, \alpha) = \sum_{n=-N}^{N} e^{jn\varphi}e^{-jn\alpha} \tag{2.34}$$

where $M = (2N + 1)$ is the total number of sensors in the array, with the sensor at the midpoint of the array treated as the point of reference. The electrical angle ϕ is related to θ by Equation 2.32, and α is a constant called the *uniform phase factor.* The quantity $A(\phi, \alpha)$ is called the array pattern. For $\phi = \lambda/2$, we find from Equation 2.31 that

$$\phi = \pi\sin\theta$$

Summing the geometric series in Equation 2.34, we may express the array pattern as

$$A(\phi, \sigma) = \frac{\sin\left[\frac{1}{2}(2N+1)(\phi - \alpha)\right]}{\sin\left[\frac{1}{2}(\phi - \alpha)\right]} \tag{2.35}$$

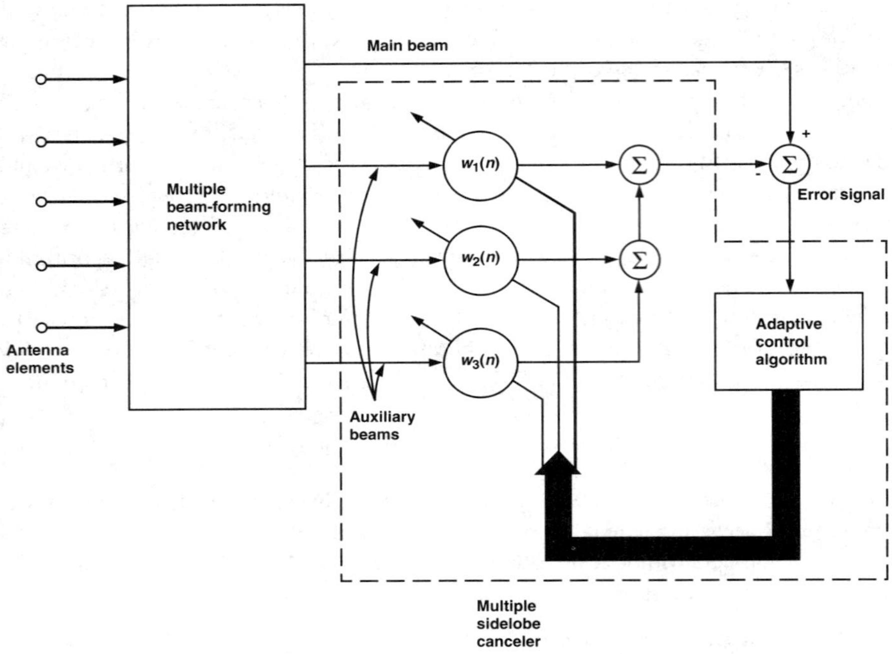

FIGURE 2.28 Block diagram of adaptive combiner with fixed beams; owing to the symmetric nature of the multiple beamforming network, final values of the weights are real valued.

By assigning different values to α, the main beam of the antenna is thus scanned across the range $-\pi < \phi \le \pi$. To generate an orthogonal set of beams equal to $2N$ in number, we assign the following discrete values to the uniform phase factor

$$\alpha = \frac{\pi}{2N+1}k, \qquad k = \pm 1, \pm 3, \ldots, \pm 2N-1 \tag{2.36}$$

Figure 2.29 illustrates the variations of the magnitude of the array pattern $A(\phi,\alpha)$ with ϕ for the case of $2N + 1 = 5$ elements and $\alpha = \pm 3\pi/5, \pm 3\pi/5$. Note that owing to the symmetric nature of the beamformer, the final values of the weights are real valued.

The orthogonal beams generated by the beamforming network represent $2N$ independent *look directions,* one per beam. Depending on the target direction of interest, a particular beam in the set is identified as the *main beam* and the remainder are viewed as *auxiliary beams.* We note from Figure 2.29 that each of the auxiliary beams has a *null in the look direction of the main beam.* The auxiliary beams are adaptively weighted by the multiple sidelobe canceler so as to form a cancelation beam that is subtracted from the main beam. The resulting estimation error is fed back to the multiple sidelobe canceler so as to control the corrections applied to its adjustable weights.

Since all the auxiliary beams have nulls in the look direction of the main beam, and the main beam is excluded from the multiple sidelobe canceler, the overall output of the adaptive beamformer is constrained to have a constant response in the look direction of the main beam (i.e., along the direction of the target). Moreover, with $(2N - 1)$ degrees of freedom (i.e., the number of available auxiliary beams), the system is capable of placing up to $(2N - 1)$ nulls along the (unknown) directions of independent interferences.

Note that with any array of $(2N + 1)$ sensors, we may produce a beamforming network with $(2N + 1)$ orthogonal beam ports by assigning the uniform phase factor the following set of values:

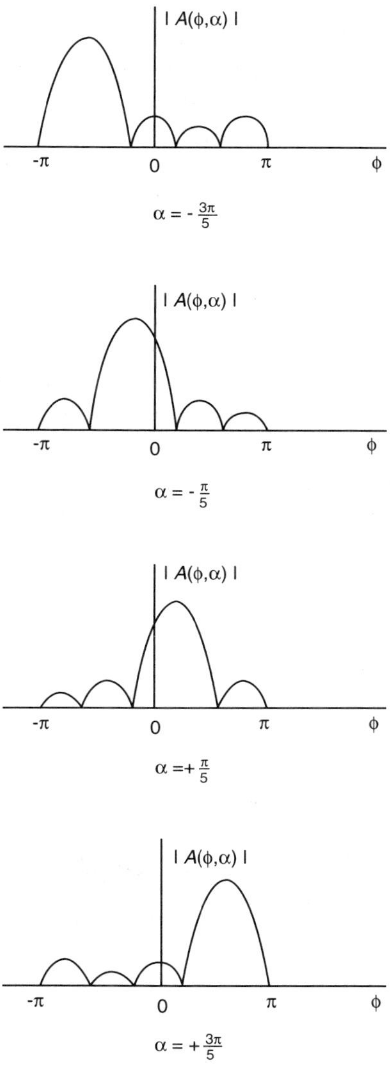

FIGURE 2.29 Variations of the magnitude of the array pattern, $A(\phi,\alpha)$ with ϕ and α.

$$\alpha = \frac{k\pi}{2N-1}, \qquad k = 0, \pm 2, \ldots, \pm 2N \tag{2.37}$$

In this case, a small fraction of the main lobe of the beam port at either end lies in the nonvisible region. Nevertheless, with one of the beam ports providing the main beam and the remaining $2N$ ports providing the auxiliary beams, the adaptive beamformer is now capable of producing up to $2N$ independent nulls.

2.8 Concluding Remarks

Adaptive signal processing, be it in time, space, or space time, is essential to the design of modern radar, sonar, and biomedical imaging systems. We say so for the following reasons:

- The underlying statistics of the signals of interest may be unknown, which makes it difficult, if not impossible, to design optimum filters by classical methods. An adaptive signal processor overcomes this difficulty by learning the underlying statistics of the environment in an on-line fashion, off-line fashion, or combination thereof.
- Signals generated by radar, sonar, and biomedical systems are inherently *nonstationary.* Adaptive signal processing provides an elegant approach to deal with nonstationary phenomena by adjusting the free parameters of a filter in accordance with prescribed algorithms.

In this chapter, we have presented a guided tour of adaptive systems for signal processing by focusing attention on the following issues:

- Linear adaptive systems, exemplified by the LMS and RLS algorithms. The LMS algorithm is simple to design but slow to converge. The RLS algorithm, in contrast, is complex but fast to converge. When the adaptive filtering is of a temporal kind, the complexity of the RLS algorithm can be reduced significantly by exploiting the time-shifting property of the input signals. For details, see Haykin (1996).
- Nonlinear adaptive systems, exemplified by neural networks. This class of adaptive systems takes many different forms. The network can be of the feedforward type, which is exemplified by multilayer perceptrons and RBF networks. For the design of multilayer perceptrons, we can use the BP algorithm, which is a generalization of the LMS algorithm. A more principled approach for the design of multilayer perceptrons and RBF networks is to use SVM theory pioneered by Vapnik and co-workers (Vapnik, 1998). Another important type of neural network involves the abundant use of feedback, which is exemplified by Hopfield networks and dynamically driven recurrent networks. The state-space approach of the modern control theory provides a powerful basis for the design of dynamically driven recurrent networks, so as to synthesize input-output mappings of interest.

The choice of one of these adaptive systems over another can only be determined by the application of interest.

To motive the study of these two broadly defined classes of adaptive systems, we described five different applications:

- System identification
- Spectrum estimation
- Noise cancellation
- Signal detection
- Beamforming

The need for each and every one of these applications arises in radar, sonar, and medical imaging systems in a variety of different ways, as illustrated in the subsequent chapters of the book.

References

1. Åström, K.J. and P. Eykhoff (1971). System identification — a survey, *Automatica,* 7, 123–162.
2. Barlow, H.B. (1989). Unsupervised learning, *Neural Computat.,* 1, 295–311.
3. Bar-Shalom, Y. and T.E. Fortmann (1988). *Tracking and Data Association,* Academic Press, New York.
4. Becker, S. and G.E. Hinton (1992). A self-organizing neural network that discovers surfaces in random-dot stereograms, *Nature* (*London*), 355, 161–163.
5. Bengio, Y., P. Simard, and P. Frasconi (1994). Learning long-term dependencies with gradient descent is difficult, *IEEE Trans. Neural Networks,* 5, 157–166.
6. Bracewell, R.N. (1978). *The Fourier Transform and Its Applications,* 2nd ed., McGraw-Hill, New York.

7. Broomhead, D.S. and D. Lowe (1988). Multivariable functional interpolation and adaptive networks, *Complex Syst.,* 2, 321–355.
8. Capon, J. (1969). High-resolution frequency-wavenumber spectrum analysis, *Proc. IEEE,* 57, 1408–1418.
9. Chazan, D., Y. Medan, and U. Shvadron (1988). Noise cancellation for hearing aids, *IEEE Trans. Acoust. Speech Signal Process.,* ASSP-36, 1697–1705.
10. Cohen, L. (1995). *Time-Frequency Analysis,* Prentice-Hall, Englewood Cliffs, NJ.
11. Comon, P. (1994). Independent component analysis: A new concept?, *Signal Process.,* 36, 287–314.
12. El Hihi, S. and Y. Bengio (1996). Hierarchical recurrent neural networks for long-term dependencies, *Adv. Neural Inf. Process. Syst.,* 8, 493–499.
13. Feldkamp, L.A. and G.V. Puskorius (1988). A signal processing framework based on dynamic neural networks with application to problems in adaptation, filtering and classification, Special issue, *Proc. IEEE Intelligent Signal Process.,* 86, November.
14. Goodwin, G.C. and R.L. Rayne (1977). *Dynamic System Identification: Experiment Design and Data Analysis,* Academic Press, New York.
15. Haykin, S. (1996). *Adaptive Filter Theory,* 3rd ed., Prentice-Hall, Englewood Cliffs, NJ.
16. Haykin, S. (1999). *Neural Networks: A Comprehensive Foundation,* 2nd ed., Prentice-Hall, Englewood Cliffs, NJ.
17. Haykin, S. and T. Bhattacharya (1997). Modular learning strategy for signal detection in a nonstationary environment, *IEEE Trans. Signal Process.,* 45, 1619–1637.
18. Haykin, S. and D. Thomson (1998). Signal detection in a nonstationary environment reformulated as an adaptive pattern classification problem, *Proc. IEEE,* 86(11), 2325–2344.
19. Hochreiter, S. and J. Schmidhuber (1997). LSTM can solve hard long time lag problems, *Adv. Neural Inf. Process. Syst.,* 9, 473–479.
20. Itakura, F. and S. Saito (1972). On the optimum quantization of feature parameters in the PARCOR speech synthesizer, in *IEEE 1972 Conf. Speech Commun. Process.,* New York, pp. 434–437.
21. Johnson, D.H. and P.S. Rao (1990). On the existence of Gaussian noise, in The 1990 Digital Signal Processing Workshop, New Paltz, NY, sponsored by *IEEE Signal Processing Society,* pp. 8.13.1–8.14.2.
22. Julier, J.J. and J.K. Uhlmann (1997). A New Extension of the Kalman Filter to Nonlinear Systems, in *Proc. Eleventh International Symposium on Aerospace/Defence Sensing, Simulation, and Controls,* Orlando, FL.
23. Kallmann, H.J. (1940). Transversal filters, *Proc. IRE,* 28, 302–310.
24. Kang, G.S. and L.J. Fransen (1987). Experimentation with an adaptive noise-cancellation filter, *IEEE Trans. Circuits Syst.,* CAS-34, 753–758.
25. Kohonen, T. (1997). *Self-Organizing Maps,* 2nd ed., Springer-Verlag, Berlin.
26. Kung, H.T. and C.E. Leiserson (1978). Systolic arrays (for VLSI), Sparse Matrix Proc. 1978, Soc. Ind. Appl. Math., pp. 256–282.
27. Ljung, L. and T. Söderström (1983). *Theory and Practice of Recursive Identification,* MIT Press, Cambridge, MA.
28. Linsker, R. (1988). Towards an organizing principle for a layered perceptual network, in *Neural Information Processing Systems,* D.Z. Anderson, Ed., American Institute of Physics, New York, pp. 485–494.
29. Linsker, R. (1989). How to generate ordered maps by maximizing the mutual information between input and output signals, *Neural Computat.,* 1, 402–411.
30. Luenberger, D.G. (1969). *Optimization by Vector Space Methods,* Wiley, New York; C.N. Dorny (1975). *A Vector Space Approach to Models and Optimization,* Wiley-Interscience, New York.
31. Luttrell, S.P. (1989). Self-organization: a derivation from first principle of a class of learning algorithms, *IEEE Conf. Neural Networks,* pp. 495–498.
32. Moody and Darken (1989). Fast learning in networks of locally-tuned processing units, *Neural Computat.,* 1, 281–294.

33. Oja, E. (1982). A simplified neuron model as a principal component analysis, *J. Math. Biol.,* 15, 267–273.
34. Owsley, N.L. (1985). Sonary array processing, in *Array Signal Processing,* S. Haykin, Ed., Prentice-Hall, Englewood Cliffs, NJ, pp. 115–193.
35. Poggio, T. and F. Girosi (1990). Networks for approximation and learning, *Proc. IEEE,* 78, 1481–1497.
36. Rumelhart, D.E., G.E. Hinton, and R.J. Williams (1986). *Learning Internal Representations by Error Propagation,* Vol. 1, Chap. 8, D.E. Rumelhart and J.L. McCleland, Eds., MIT Press, Cambridge, MA.
37. Sanger, T.D. (1989). An optimality principle for unsupervised learning, *Adv. Neural Inf. Process. Syst.,* 1, 11–19.
38. Shabhag, N.R. and K.K. Parhi (1994). *Pipelined Adaptive Digital Filters,* Kluwer, Boston, MA.
39. Sontag, E.D. (1990). *Mathematical Control Theory: Deterministic Finite Dimensional Systems,* Springer-Verlag, New York.
40. Ukrainec, A.M. and S. Haykin (1996). A modular neural network for enhancement of cross-polar radar targets, *Neural Networks,* 9, 143–168.
41. Van Trees, H.L. (1968). *Detection, Estimation and Modulation Theory, Part I,* Wiley, New York.
42. Van Veen, B.D. and K.M. Buckley (1988). Beamforming: a versatile approach to spatial filtering, *IEEE ASSP Mag.,* 5, 4–24.
43. Vapnik, V.N. (1998). *Statistical Learning Theory,* Wiley, New York.
44. Wan, E.A. (1994). Time series prediction by using a connectionist network with internal delay lines, in *Time Series Prediction: Forecasting the Future and Understanding the Past,* A.S. Weigend and N.A. Gershenfield, Eds., Addison-Wesley, Reading, MA, pp. 195–217.
45. Wan, E.A., R. vander Merwe, and A.T. Nelson (1999). Dual estimation and the unsented transformation, *Neural Inf. Process. Syst.* (*NIPS*), Denver, CO.
46. Werbos, P.J. (1974). Beyond Regression: New Tools for Prediction and Analysis in the Behavioral Sciences, Ph.D. thesis, Harvard University, Cambridge, MA.
47. Werbos, P.J. (1990). Backpropagation through time: what it does and how to do it, *Proc. IEEE,* 78, 1550–1560.
48. Widrow, B. et al. (1975b). Adaptive noise cancelling: principles and applications, *Proc. IEEE,* 63, 1692–1716.
49. Widrow, B. and S.D. Stearns (1985). *Adaptive Signal Processing,* Prentice-Hall, Englewood Cliffs, NJ.
50. Williams, R.J. and D. Zipser (1989). A learning algorithm for continually running fully recurrent neural networks, *Neural Computat.,* 1, 270–280.
51. Yee, P.V. (1998). Regularized Radial Basis Function Networks: Theory and Applications to Probability Estimation, Classification, and Time Series Prediction, Ph.D. thesis, McMaster University, Hamilton, Ontario.

3

Gaussian Mixtures and Their Applications to Signal Processing

Kostantinos N. Plataniotis
University of Toronto

Dimitris Hatzinakos
University of Toronto

Nomenclature

Φ	Family of distributions
$F(x\|\hat{\theta})$	Conditional distribution
$\hat{\theta}$	Unknown parameter
$\hat{\theta}$	Estimated value
$N(x; \mu_i, \Sigma_i)$	d-Dimensional Gaussian
μ	Mean value
Σ	Covariance
w_i	Mixing coefficient
ML	Maximum likelihood
EM	Expectation maximization
$\log\Lambda$	Log likelihood
Z	Missing data (EM algorithm)
LRT	Likelihood ratio test
N_g	Number of mixture components
RBF	Radial-basis functions
PNN	Probabilistic neural network
$x(k)$	State vector

0-8493-3691-0/01/$0.00+$.50

$Z^k = z(1), z(2), \ldots, z(k), \ldots$	Observation record
$\hat{x}(k\|k) = E(x(k)\|Z^k)$	Mean-squared-error filtered estimate
EKF	Extended Kalman filter
$J_h(x(k))$	Jacobian matrix
AGSF	Adaptive Gaussian sum filter
CMKF	Converted measurement Kalman filter
ε-mixture	ε-Contaminated Gaussian mixture model
$i(k)$	Inter-symbol inference
$n(k)$	Thermal noise
NMSE	Normalized mean square error
DS/SS	Direct-sequence spread-spectrum
IIR	Infinite-duration impulse response

Abstract

There are a number of engineering applications in which a function should be estimated from data. Mixtures of distributions, especially Gaussian mixtures, have been used extensively as models in such problems where data can be viewed as arising from two or more populations mixed in varying proportions.[1–3] The objective of this chapter is to highlight the use of mixture models as a way to provide efficient and accurate solutions to problems of important engineering significance. Using the Gaussian mixture formulation, problems are treated from a global viewpoint that readily yields and unifies previous, seemingly unrelated results. This chapter reviews the existing methodologies, examines current trends, provides connections with other methodologies and practices, and discusses application areas.

3.1 Introduction

Central to unsupervised learning in adaptive signal processing, stochastic estimation, and pattern recognition is the determination of the underlying probability density function of the quantity of interest based on available measurement data.[4] If no *a priori* knowledge of the functional form of the requested density is available, non-parametric techniques should be used. Therefore, over the years a number of techniques ranging from data histograms and kernel estimators to neural network and fuzzy system-based approximators have been proposed.[4,7,8] On the other hand, if some impartial *a priori* knowledge regarding the data characteristics is available, the requested probability function is assumed to be of a known functional form, but with a set of unknown parameters. The parameterized function provides a partial description where the full knowledge of the underlying phenomenon is achieved through the specific values of the parameters.

Let x be a d-dimensional vector with a probability distribution $F(x)$ and a probability density $f(x)$. In most engineering problems, a density such as the multi-dimensional Gaussian is assumed. More often, families of parametric distributions are used.[14] In this case, the family is considered to be a linear combination of given distributions. This family is often called parametric since its members can be characterized by a finite number of parameters.[9–12,15,18,23–26]

The family of distributions considered in this chapter can be defined as

$$\Phi = [F(x|\theta);\ \theta \in \Theta] \tag{3.1}$$

Suppose that a sequence of random identically distributed observation's $x_1, x_2, \ldots, x_n$ are drawn from $F(x|\hat{\theta})$ with $\hat{\theta}$ unknown to the observer. An estimate of the unknown parameter $\hat{\theta}$ which can be obtained as a function of the observations can be used to completely characterize the mixture.[10,16,18]

Let us assume that associated with each one of the random samples $x_1, x_2, \ldots$ is a probability distribution with the possibility of some of the samples being from $F(x|\theta^1)$, some from $F(x|\theta^2)$, etc., where θ^1, θ^2 are different realizations of the unknown parameter θ. In other words, any sample x could be from any of the member distributions in the parametric family Φ.[5] Defining a mixing distribution $G(\theta)$, which describes the probability that point θ characterizes the mixture, the sample x can be considered as having a distribution

$$H(x) = \int F(x|\theta)dG(\theta) \tag{3.2}$$

which is called a mixture. In most engineering applications, a finite number of points $\theta^1, \theta^2, \ldots, \theta^g$ are assumed. Then the mixing distribution is expressed as

$$G(\theta) = \sum_{i=1}^{Ng} P(\theta^i)\delta(\theta - \theta^i) \tag{3.3}$$

Substituting Equation 3.3 into the mixture expression of Equation 3.2, the finite mixture

$$H(x) = \sum_{i=1}^{Ng} F(x|\theta^i)P(\theta^i) \tag{3.4}$$

can be obtained. The parameter points used to discretize the mixture can be known *a priori* with the only unknown elements in the Equation 3.4 being the mixing parameters $P(\theta^i)$. In such a scenario, the distributions used in the mixture (basis functions) are determined *a priori*. Thus, only the mixture coefficients are fit to the observations, usually through the minimization of an error criterion. Alternatively, the basis functions themselves (through their parameters) are adapted to the data in addition to the mixing coefficients. In such a case, the optimization of the mixture parameters becomes a difficult non-linear problem, and the type of the basis function selected as well as the type of the optimization strategy used becomes very important. Because of their simplicity, Gaussian densities are most often used as basis functions.[5]

The discussion in this chapter is intended to provide a perspective on the Gaussian mixture approach to developing solutions and methodologies for signal processing problems. We will discuss in detail a number of engineering areas of application of finite Gaussian mixtures. In engineering applications, the finite mixture representation can be used to (1) directly represent the underlying physical phenomenon, e.g., tracking in a multi-target environment, medical diagnosis, etc., and (2) indirectly model underlying phenomena that do not necessarily have a direct physical interpretation, e.g., outlier modeling in communication channels. The problem of tracking a target using polar coordinate measurements is used here to demonstrate the applicability of the Gaussian mixture model to model an actual physical phenomenon. The process of tracking a target involves the reception and processing of received signals. The Gaussian mixture model is used to approximate the densities involved in the derivation of the optimal Bayesian estimator needed to provide reliable and cost-effective estimates of the state of the system. In addition, we also discuss in detail the problem of narrowband interference suppression as an example of indirect application of the Gaussian mixture model. Spread-spectrum communication systems often use estimation techniques to reject narrowband interference. The basic assumption is that the direct sequence spread-spectrum signal along with the background noise can be viewed as non-Gaussian measurement noise. The Gaussian mixture framework is then used to model the non-Gaussian measurement channels. Similar treatment of signals can easily be extended to any application subject to non-linear effects of non-Gaussian measurements, e.g., biomedical systems. For example, Gaussian mixtures have been used to model random noise, magnetic field inhomogeneities, and biological variations of the tissue in magnetic resonance imaging (MRI) as well as computerized tomography (CT).[27–30]

After a brief review of the mathematical aspects of Gaussian mixtures, three methodologies for estimating mixture parameters are discussed. Particular emphasis is placed on the expectation/maximization (EM) algorithm and its applicability to the problem of adaptive mixture parameter determination. Computational issues are also analyzed with emphasis on the computer generation of mixture variables. Then the framework is applied to two problems, and numerical results are presented. The results included in this chapter are meant to be illustrative rather than exhaustive. Finally, to demonstrate the versatility and the powerful nature of the framework, connections with other research areas are drawn, with particular emphasis on the connection between Gaussian mixtures and the radial-basis functions (RBF) networks.

3.2 Mathematical Aspects of Gaussian Mixtures

3.2.1 The Approximation Theorem

In an adaptive signal processing, unsupervised learning environment, the usefulness of the Gaussian mixture model depends on two factors: (1) whether or not the approximation is sufficiently powerful to represent a broad class of density functions, most notably those that are encountered in engineering applications, and (2) if such an approximation can be obtained in a reasonable manner through a parameter estimation scheme which allows the user to compute the optimal values of the mixture parameters from a finite set of data samples.[6,13,51–53]

Regarding the first factor, a Gaussian mixture can be constructed to approximate any given density. This can be proven by utilizing the Wiener's theorem of approximation or by considering delta functions of a positive type. This methodology, first presented in References 8 and 51, is reviewed in this chapter. The resulting class of density functions is rich enough to approximate all density functions of engineering interests.[8,51]

We start reviewing the methodology by briefly discussing the characteristics and properties of delta functions. Delta families of positive type are families of functions which converge to a delta (impulse) function as a parameter characterizing the family converging to a limit value. Specifically, let δ_λ be a family of functions on the interval $(-\infty, \infty)$ which are integrable over every interval. This is called a delta family of positive type if the following conditions are satisfied:

1. $\int_{-a}^{a} \delta_\lambda(x)dx \to \lambda$ as $\lambda \to \lambda_0$ for some a.
2. For every constant $\gamma > 0$, δ_λ tends to zero uniformly for $\gamma \le |x| \le \infty$ as $\lambda \to \lambda_0$.
3. $\delta_\lambda(x) \ge 0$ for all x and λ.

If such a function is required to satisfy the condition that

$$\int_{-\infty}^{\infty} \delta_\lambda(x)dx = 1 \tag{3.5}$$

then it defines a probability density function for all λ. It can seen by inspection that the Gaussian density tends to the delta function as the variance tends to zero and, therefore, can be used as a basis function for approximation purposes.[51,73]

Using the delta families, the following result can be used for the approximation of an arbitrary density function p.

The sequence $p_\lambda(x)$, which is formed by the convolution of δ_λ and p,

$$p_\lambda(x) = \int_{-\infty}^{\infty} \delta_\lambda(x-u)p(u)du \tag{3.6}$$

converges uniformly to $p(x)$ on every interior subinterval of $(-\infty, \infty)$.

When the density p has a finite number of discontinuities, Equation 3.6 holds true except at the points of discontinuity. Since the Gaussian density can be used as a delta family of positive type, the approximation p_λ can be written as follows:

$$p_\lambda(x) = \int_{-\infty}^{\infty} N_\lambda(x-u)p(u)du \tag{3.7}$$

which forms the basis for the Gaussian sum approximation. The term $\delta_\lambda(x - u)p(u)$ is integrable on $(-\infty, \infty)$, and it is at least piecewise continuous. Thus, $p_\lambda(x)$ itself can be approximated on any finite interval by a *Riemann* sum. In particular, if a bounded interval (a, b) is considered, the function is given as

$$p_{\lambda,n}(x) = \frac{1}{k}\sum_{i=1}^{n} N_\lambda(x - x_i)[\xi_i - \xi_{i-1}] \tag{3.8}$$

where the interval (a, b) is divided into n subintervals by selecting points such that

$$a = \xi_0 < \xi_1 < \xi_2 < \ldots < \xi_n = b \tag{3.9}$$

In each such subinterval, a point x_i is chosen such as

$$p(x_i)[\xi_i - \xi_{i-1}] = \int_{-\xi_{i-1}}^{\xi_i} p(x)dx \tag{3.10}$$

which is possible by the mean value theorem. The normalization constant k ensures that the density $p_{\lambda,n}$ is a density function.

Consequently, an approximation of p_λ over some bounded interval (a, b) can be written as

$$p_{\lambda,n}(x) = \sum_{i=1}^{n} w_i N_{\sigma i}(x - x_i) \tag{3.11}$$

where $\sum_{i=1}^{n} w_i = 1$ and $w_i \geq 0$ for all i.

The relation between Equations 3.10 and 3.11 is obvious by inspection. However, in Equation 3.11, the variance σ_i can vary from one term to another. This has been done to obtain greater flexibility for an approximation using Gaussian mixtures with a finite number of terms. As the number of terms in the mixture increases, it is necessary to require that σ_i become equal and vanish.

Under this framework, an unknown d-dimensional distribution (density function) can be expressed as a linear combination of Gaussian terms. The form of the approximation is as follows:

$$p(x) = \sum_{i=1}^{Ng} \omega_i N(x; \mu_i, \Sigma_i) \tag{3.12}$$

where $N(.)$ represents a d-dimensional Gaussian density defined as

$$N(x; \mu, \Sigma) = \frac{1}{(2\pi)^{0.5}|\Sigma|^{0.5}} \exp(-0.5(x - \mu)^\tau \Sigma^{-1}(x - \mu)) \tag{3.13}$$

where μ, Σ are the mean and covariance of the Gaussian basis functions and w_i in Equation 3.12 is the ith mixing coefficient (weight) with the assumption that $\omega_i \geq 0$, $\forall_i = 1, 2, \ldots, Ng$, and $\sum_{i=1}^{Ng} \omega_i = 1$.

3.2.2 The Identifiability Problem

The problem most often encountered in the context of finite mixtures is that of identifiability meaning the uniqueness of representation in the mixture.[20–22,31] If the Gaussian mixture of Equation 3.12 is identifiable, then

$$\sum_{i=1}^{M} w_i F(x|\theta^i) = \sum_{j=1}^{M'} w_j' F(x|\theta^j) \tag{3.14}$$

implies that

1. $M = M'$
2. For each i, $1 \leq i \leq M$, there exists uniquely j, $1 \leq j \leq M'$ such that $w_i = w_j'$ and $F(x|\theta^i) = F(x|\theta^j)$

There exists extensive literature on the problem of mixture identifiability. A necessary and sufficient condition that the class Φ of all finite mixtures be identifiable is that Φ be a linearly independent set over the field of real numbers.[5,20,31] Using the above conditions, the identifiability of several common distribution functions has been investigated. Among the class of all finite mixtures, that of Gamma distributions, the one-dimensional Cauchy distribution, the one-dimensional Gaussian family, and the multi-dimensional Gaussian family are identifiable.

The following theorem discusses the identifiability problem.[21,31]

Theorem

A necessary and sufficient condition that the class of all finite mixtures of the family $\aleph$ be identifiable is that F be a linearly independent set over the field of real numbers.

Proof

Necessity: Let $\sum_{i=1}^{M} \alpha_i F_i = 0 \ \forall x$, where α_i real numbers are a linear relation in $\aleph$. Assume that the α_i's are subscripted so that $\alpha_i < 0$ if $i < N$. We then have

$$\sum_{i=1}^{N} \alpha_i F_i + \sum_{i=N+1}^{M} \alpha_i F_i = 0 \rightarrow \sum_{i=1}^{N} |\alpha_i| F_i = \sum_{i=N+1}^{M} |\alpha_i| F_i$$

Since the F_i are distribution functions **d.f** or **c.d.f**, $F_i(\infty) = 1$. Thus,

$$\sum_{i=1}^{N} |\alpha_i| = \sum_{i=N+1}^{M} |\alpha_i| = b > 0$$

Therefore, if we define $w_i = |\alpha_i|/b$, we have

$$\sum_{i=1}^{N} w_i^1 F_i = \sum_{i=N+1}^{M} w_i F_i$$

Since by definition $w_i > 0$ and $\sum_{i=1}^{N} w_i^1 = \sum_{i=N+1}^{M} w_i = 1$, the coefficients satisfy the requirements for mixing parameters.

The relation $\sum_{i=1}^{N} w_i^1 F_i = \sum_{i=N+1}^{M} w_i F_i$ asserts that there exist two distinct representations of a finite mixture so that *leph* cannot be identifiable.

Since the proof of necessity requires that $\aleph$ is identifiable, we are led to a contradiction which follows from assuming that the members of the family are linearly dependent. Consequently, the conclusion follows that the members of the family form a linearly independent set over the field of real numbers.

Sufficiency: If a given mixture is a linear independent set, then it can be considered as a basis which spans the family $\aleph$. If there were two distinct representations of the same mixture, this would contradict the unique representation property of a basis. This does not mean that there exists only one representation of the mixture, but rather that given a basis which spans the family consisting of $(F_i)_{i=1}^{M}$, the relation $\sum_{i=1}^{N} w_i^1 F_i = \sum_{i=N+1}^{M} w_i F_i$ implies always that $w_i^1 = w_i$. The unique representation property of a basis allows the conclusion that if F is a linearly independent set, then it is sufficient for identifiability.

The problem of identifiability is of significant practical importance in all practical applications of mixtures. Without resolving the problem of the unique characterization of the mixture model, a reliable estimation procedure to determine its parameters cannot be defined. There are many classes of mixture models in which we are unable to define a unique representation. A simple example of such a non-identifiable mixture is the uniform distribution which can be expressed as a mixture of two other uniform distributions, e.g., $U(x; 0.5, 0.5) = 0.5U(x; 0.25, 0.25) + 0.5U(x; 0.75, 0.25)$. However, by utilizing the theorems summarized above, it has been proven that the class of all finite mixtures of Gaussian (normal) distributions is identifiable.[1,5]

3.3 Methodologies for Mixture Parameter Estimation

The problem of determining the parameters of the mixture to best approximate a given density function can be solved in more than one way. There exists considerable literature on mixture parameter estimation with a variety of different approaches ranging from the moments method,[46] to the moment generation function,[3] graphical methods,[47] Bayesian methods,[9] and the different variations of the maximum likelihood method.[1,4,10,32] In this chapter, we will concentrate on the maximum likelihood approach.

There are two different methodologies in estimating the parameters of the Gaussian mixture by using the maximum likelihood principle. The first approach is the iterative one, in which the parameter values are refined by processing the data iteratively. Alternatively, one can use a recursive approach, refining the mixture parameter values with each new available data value. A recursive procedure requires that the latest value of a parameter within the mixture model depends only on the previous value of the estimate and the current data sample. Generally speaking, an iterative procedure will produce better results than a recursive one. On the other hand, the recursive parameter estimator is usually much faster than the iterative one. In the case of Gaussian mixture approximation, we are interested in estimating from the data the mixing coefficients (weights) and, if needed, the first two moments of the Gaussian basis functions.

The method of choice for the estimation of the Gaussian mixture parameters is currently the EM algorithm.[32,33] This is an iterative procedure which starts with an initial estimate of the mixture's parameters. Based on that initial guess, the method constructs a sequence of estimates by first evaluating the expectation of the log likelihood of the current estimate and then proceeds by determining the new parameter value which maximizes this expectation. Although the EM methodology is most often used, we continue our analysis by reviewing first the classical maximum likelihood approach to the problem of mixture parameter estimation. In this approach, estimates of the mixture parameters are obtained by maximizing the marginal likelihood function of (n) independent observations drawn from the mixture. A detailed description of the method follows in the next section.

3.3.1 The Maximum Likelihood Approach

Let us assume that a set of unlabeled data samples $(x_1, x_2, \ldots, x_n)$ are drawn from a Gaussian mixture density

$$p(x) = \sum_{i=1}^{Ng} p(\omega_i)p(x|\omega_i) = \sum_{i=1}^{Ng} \omega_i N(x, \theta_i) \tag{3.15}$$

with $\sum_{i=1}^{Ng} \omega_i = 1, \omega_i \geq 0$ for all i, and θ the unknown parameter vector which summarizes the uncertainty on the mean value and the variance (covariance) of the Gaussian basis function. By applying the Bayes rule, the following relation holds:

$$p(\omega_i|x) = \frac{p(\omega_i)p(x|\omega_i)}{p(x)} = \frac{\omega_i N(x, \theta_i)}{\sum_{j=1}^{N} \omega_j N(x, \theta_j)} \tag{3.16}$$

We are seeking parameters θ and ω which minimize the log likelihood of the available samples:

$$\log\Lambda = \sum_{k=1}^{n} \log p(x_k) \tag{3.17}$$

Using Lagrange multipliers, Equation 3.17 can be rewritten as follows:

$$\log\hat{\Lambda} = \sum_{k=1}^{n} \log p(x_k) - \lambda\left(\sum_{i=1}^{n} \omega_i - 1\right) \tag{3.18}$$

Taking the partial derivative with respect to ω_i, and setting it equal to 0, we have the following expression:

$$\hat{\omega}_i = \frac{1}{\lambda}\sum_{k=1}^{n} p(\omega_i|x_k) \tag{3.19}$$

To obtain estimates of the generic basis parameter θ, the partial derivative with respect to θ is set equal to 0:

$$\sum_{k=1}^{n} p(\omega_i|x_k)\frac{\partial}{\partial\theta_i}N(x_k, \theta_i) = 0 \tag{3.20}$$

For the case of a multi-dimensional Gaussian density, the parameter vector θ is comprised of the mean value and the covariance matrix. Taking together the partial derivatives of the logarithm with respect to their elements, we have the following relations:

$$\hat{\mu}_i = \frac{\sum_{k=1}^{n} p(\omega_i|x_k)x_k}{\sum_{k=1}^{n} p(\omega_i|x_k)} \tag{3.21}$$

$$\hat{\Sigma}_i = \frac{\sum_{k=1}^{n} p(\omega_i|x_k)(x_k - \hat{\mu}_i)^{\tau}(x_k - \hat{\mu}_i)}{\sum_{k=1}^{n} p(\omega_i|x_k)} \tag{3.22}$$

The systems of Equations 3.16, 3.21, and 3.22 can be solved using iterative methods. However, when such an approach is used, singular solutions may occur since a component density centered on a single design sample may have a likelihood that approaches infinity as the variance (covariance) of the component approaches zero. The simplest way to avoid this problem is to utilize a new set of design data samples for each iteration of the solution, making it impossible for a single data sample to dominate the whole component density.

Although simple in concept, this method does not work well in practice. Therefore, alternative solutions have been developed to alleviate the problem. Among them is the stochastic gradient descent solution reviewed in the next section.

3.3.2 The Stochastic Gradient Descent Approach

Let us start with the generic, parametric update formula devised through the utilization of the maximum likelihood solution. For both the man and the variance (covariance), the update equation has the following form:

$$\theta_n = \frac{\sum_{k=1}^{n} p(\omega|x_k)\theta(x_k)}{\sum_{k=1}^{n} p(\omega|x_k)} \tag{3.23}$$

After some simple algebraic manipulation, a recursive expression for the θ_{n+1} as a function of θ_n can be obtained as

$$\theta_{n+1} = \theta_n + \gamma_{n+1}(\theta_{n+1} - \theta_n) \tag{3.24}$$

with

$$\gamma_{n+1} = \frac{p(\omega|x_{n+1})}{\sum_{k=1}^{n+1} p(\omega|x_{n+1})} \tag{3.25}$$

Equation 3.25 can also be formulated in a recursive format. However, the denominator for the calculation of the correction term is not bounded for growing data sets (n), and thus, such an estimation procedure would require infinite memory. Therefore, if we assume only a finite sample set with samples drawn unbiased from the unknown distribution and with the fixed set size (n) large, then the correction factor can be approximately calculated as follows:

$$\gamma_{n+1} \approx \frac{p(\omega|x_{n+1})}{(n+1)p(\omega)} \tag{3.26}$$

By utilizing Equations 3.24 and 3.26, explicit time update equations for the parameters of the Gaussian mixture can be written. Although it may be impossible to obtain convergence from only one iteration if the design set is too small, acceptable estimates can be obtained if the data samples are drawn with replacement until a stable solution is obtained.

3.3.3 The EM Approach

As before, we assume that a set of unlabeled data samples ($x_1, x_2, \ldots, x_n$) are drawn from a Gaussian mixture density

$$p(x) = \sum_{i=1}^{Ng} p(\omega_i)p(x|\omega_i) = \sum_{i=1}^{Ng} \omega_i N(x, \theta_i) \tag{3.27}$$

with $\sum_{i=1}^{Ng} \omega_i = 1$, $\omega_i \geq 0$ for all i, and θ_i is the unknown parameter vector consisting of the elements of the mean value μ_i and the distinct elements of the covariance (variance) Σ_i of the Gaussian basis function $N(x; \theta_i)$. The EM algorithm utilizes the concept of missing data, which in our case is the knowledge of which Gaussian function of each data sample is coming from. Let us assume that the variable Z_j provides the density membership for the jth sample available. In other words, if $Z_{ij} = 1$, then x_j has a density $N(x, \theta_i)$. The values of Z_{ij} are unknown and are treated by EM as missing information to be estimated along with the parameters θ and ω of the mixture model. The likelihood of the model parameters θ, w, given the joint distribution of the data set and the missing values Z, can be defined as

$$\log L(\theta, w|((x_1, x_2, \ldots, x_n), Z)) = \sum_{i=1}^{n}\sum_{j=1}^{Ng} Z_{ij}\log(p(x_i|\theta_j)\omega_j) \tag{3.28}$$

The EM algorithm iteratively maximizes the expected log likelihood over the conditional distribution of the missing data Z given (1) the observed data $x_1, x_2, \ldots, x_n$ and (2) the current estimates of the mixture model parameters θ and ω. This is achieved by repeatedly applying the E-step and the M-step of the algorithm. The E-step of EM finds the expected value of the log likelihood over the values of the missing data Z given the observed data and the current parameters $\theta = \theta^0$ and $\omega = \omega^0$.

It can be shown that the following equation holds true:

$$Z_{ij}^0 = \frac{p(x_i|\theta_j^0)\omega_j^0}{\sum_{t=1}^{g} pp(x_i|theta_t^0)\omega_t^0} \tag{3.29}$$

with $i = 1, 2, \ldots, n$ and $t = 1, 2, \ldots, N$. The M-step of the EM algorithm maximizes the log likelihood over θ and ω in order to find the next estimates for them, the so-called θ^1 and ω^1.

The maximization over ω leads to a solution

$$\omega_{ji}^1 = \sum_{i=1}^{n} \frac{Z_{ij}^n}{n} \tag{3.30}$$

We can then maximize over the parameters θ by maximizing the terms of the log likelihood separately over each θ_j with $j = 1, 2, \ldots, g$. Therefore, evaluation of this step means calculations of the

$$\theta_j^1 = \max_{\theta_j} \sum_{i=1}^{n} Z_{ij}^0 \log(p(x_i|\theta_j)) \tag{3.31}$$

For the case of Gaussian mixtures, the solution to the M-step of the algorithm exists in closed form. Thus, at the $(k + 1)^{\text{th}}$ iteration, the current estimates for the mixture coefficients, the elemental means, and the covariance matrices are given as

$$\omega_j(k+1) = \sum_{i=1}^{n} \frac{\hat{\tau}_j(k+1)}{n} \tag{3.32}$$

$$\hat{\tau}(k+1) = \frac{\omega_j(k)}{N}(x_j;\, \theta_i(k)) \sum_{i=1}^{g} \omega_j(kN(x_j;\, \theta_i(k))) \tag{3.33}$$

$$\mu_i(k+1) = \frac{\sum_{j=1}^{n} \hat{\tau}_j(k+1)x_j}{n\omega_i(k+1)} \tag{3.34}$$

$$\Sigma_i(k+1) = \frac{\sum_{j=1}^{n} \hat{\tau}_j(k+1)(x_j - \mu_i(k+1))(x_j - \mu_i(k+1))^{\tau}}{n\omega_i(k+1)} \tag{3.35}$$

The EM algorithm increases the likelihood function of the data at each iteration and, under suitable regularity conditions, converges to a stationary parameter vector.[32] The convergence properties of the EM algorithm have been discussed extensively in the literature. The EM algorithm produces a monotonic increasing sequence of likelihoods, thus if the algorithm converges it will reach a stationary point in the likelihood function, which can be different from the global maximum. However, like any other optimization algorithm, the EM algorithm depends on the provided initial values to determine the solution. Given a specific test of initial conditions, it may converge to the optimal solution, while for another set of initial parameters it may find only a suboptimal one. The final set of values, as well as the number of iterations needed for the convergence of the EM algorithm, is thus greatly affected from the initial parameter values. Therefore, the initial placement of the Gaussian components are of paramount importance for the convergence of the EM algorithm.

In the problem of function approximation or distribution modeling in which the EM algorithm is used to guide the function approximation, a good starting point for the elemental Gaussian terms may be near the means of the actual underlying component Gaussian terms. To this end, many different techniques have been devised over the years. Among the clustering techniques, the different variants of the K-means algorithm are the most popular.[4] In this approach, the components of the underlying distributions which generate the data are considered as data clusters, and pattern recognition techniques are used to identify them. When the K-means algorithm is used to identify initial values for the EM algorithm, the number of Gaussian functions in the mixture (clusters) has to be specified in advance. Having the number of clusters predefined, an iterative procedure is invoked to move the cluster centers in order to minimize the mean square error between cluster centers and available data points. The procedure can be described as follows:

1. Randomly select N_g data points as the initial starting locations of the elemental Gaussian terms (clusters).
2. Assign a novel data point x_j to cluster center μ_i if $|x_j - \mu_i| \leq |x_j - \mu_l|$ for all $l = 1, \ldots, N_g, l \neq j$.
3. Calculate the new mean value of the data points associated with the center μ_i.
4. Repeat Steps 1 and 2 until the centers are stationary.

Although this algorithm is simple and works well in many practical applications, it has several drawbacks. The procedure itself depends on the initial conditions, and it can converge to different solutions depending on which initial data points were selected as initial cluster centers. Thus, if it is used as the initial starting point for the EM algorithm, then the varying final configuration of the cluster centers produced by the K-means algorithm may lead to variations in the final Gaussian mixture generated by the EM algorithm.[38]

Alternatively, scale-space techniques can be utilized to determine the Gaussian term parameters from the available data samples. Such techniques initially motivated by the use of Gaussian filters for edge detection can provide constructing descriptions of signals and functions by decomposing the data histogram into sums of Gaussian distributions.[39,40] The scale-space description of a given data set indicates the zero-crossing points of the second derivatives of the data at varying resolutions.[41] When scale-space techniques are used to determine the parameters of a Gaussian mixture, we are particularly interested in the location of zero crossings in the second derivative and the sign of the third derivative at the zero crossing. By determining the second derivatives of the data waveform and locating the zero-crossing points, the number of Gaussian terms present in the approximating Gaussian mixture can be identified. The sign of the waveform's second derivative can be used to determine where the function is convex or concave.[40]

In general, to determine an (N_g) component's normal mixture, $(3N_g - 1)$ parameters must be estimated. The direct calculation of these parameters as a function of the location of the zero-crossing points form a system of $(3N_g - 1)$ simultaneous non-linear equations. To overcome the computational complexity of a direct estimation, a two-stage procedure was proposed.[40] In this approach, a rough estimate of the parameter values are obtained based on the zero-crossing locations. With this initial set as a starting point, the EM algorithm is utilized to provide the final set of Gaussian mixture parameters. The procedure can be summarized as follows:

- At any scale, sign changes will alternate left to right. Odd (even)-numbered zero crossings will thus correspond to lower (upper) turning points.
- Given the locations of upper and lower turning points, the point halfway between the turning point pair is used to provide the initial estimate of the mean μ_i.
- Half the distance between turning point pairs is used as an estimate of the standard deviation (covariance).
- Given these initial estimates of the parameters which determine the mixture, the EM algorithm is used to calculate the optimal set of parameters.

In summary, clustering techniques such as the K-means algorithm or scale-space filters can be used to provide initial values for the EM algorithm. Changes in the initial conditions will result in varying final Gaussian mixtures, and although there is no guarantee that the final mixture chosen is optimal, those which are based on initial sets selected from these algorithms are usually better.

Finally, to improve the properties of the EM algorithm, a stochastic version of the algorithm, the so-called stochastic EM (SEM) algorithm, has been proposed in the literature.[42] Stochastic perturbation and sampling methodologies are used in the context of SEM to reduce the dependence on the initial values and to speed up convergence. If the initial parameters are sufficiently close to the actual values, the convergence is exponential for Gaussian mixtures. Although the dependence on the initial values is largely reduced in the SEM algorithm, SEM seems inappropriate for small sample records.[43]

3.3.4 The EM Algorithm for Adaptive Mixtures

The problem of determining the number (N_g) of components in the Gaussian mixture when the mixing coefficients, means, and variances (covariances) of the elemental Gaussian terms are also unknown parameters to be determined from the data is a difficult but important one. Most of the studies undertaken in the past concern the problem of testing the hypothesis of ($N_g = g_1$) vs. the alternative ($N_g = g_2$) with the two numbers $1 \leq g_1 \leq g_2$. If the classical likelihood approach is utilized to determine the rest of the parameters in the mixture, the maximum likelihood ratio test (LRT) can be used to determine the actual number of the components in the mixture. The LRT test rejects the hypothesis H^n_{g1} and decides for H^n_{g2} whether the likelihood ratio

$$\lambda = \frac{\Lambda_{g1}}{\Lambda_{g2}} \leq 1$$

is too small or, equivalently, the log likelihood statistic is too large.[43]

Recently, adaptive versions of the EM algorithm have also appeared in the literature in an attempt to circumvent the problem of determining the number of components in the mixture. The so-called adaptive mixture is essentially a recursively calculated Gaussian mixture with the ability to create new terms or drop existing terms as dictated by the data. In the case of multivariate Gaussian basis functions examined here, a recursive formulation of the EM algorithm can be used to evaluate the number of basis functions as well as their parameters at every time instant. The parameter update equations are summarized as

$$\hat{\tau}_i(k+1) = \frac{\omega_j(k)N(x_{k+1};\, \theta_i(k))}{\sum_{i=1}^{N} \omega_j(k)N(x_{k+1};\, \theta_i(k))} \tag{3.36}$$

$$\omega_j(k+1) = \omega_j(k) + \frac{1}{n}(\hat{\tau}(k+1) - \omega_j(k)) \tag{3.37}$$

$$\mu_i(k+1) = \mu_i(k) + \frac{\hat{\tau}(k+1)}{n\omega_j(k)}(x_{k+1} - \mu_i(k)) \tag{3.38}$$

$$\Sigma_i(k+1) = \Sigma_i(k) + \frac{\hat{\tau}(k+1)}{n\omega_j(k)}((x_{k+1} - \mu_i(k+1))(x_{k+1} - \mu_i(k+1))^{\tau}) - \Sigma_i(k) \tag{3.39}$$

with the time index k defined over the interval $k = 1, 2, \ldots, n$. Given a new data point at a certain time instant k, the algorithm either updated the parameters of the existing basis on the mixture by utilizing the equations above or added a new term to the mixture. The addition of a new term should be based on the utilization of an appropriate measure as to the likelihood that the current measurement has been drawn from the existing model. One such measure proposed is the Mahalanobis distance between the observation and each of the existing bases in the Gaussian mixture. For the Gaussian basis mixtures considered here, the square Mahalanobis distance between a data point x_j and a Gaussian basis function with mean value μ_i and covariance Σ_i is given as $d^2_M = (x_j - \mu_i)^{\tau}\, \Sigma_i^{-1}\, (x_j - mu_i)$. Thus, if the distance between a new point and each basis function in the current Gaussian mixture exceeds a predefined threshold, then a new term is created with its mean value given by the location of the point and a covariance which is based on the covariances of the surrounding terms and their mixing coefficients.[34] After the insertion of the new term, the mixing (weighting) coefficients of the Gaussian basis functions are renormalized appropriately.

3.4 Computer Generation of Mixture Variables

It is of paramount importance in many practical applications to generate random variables which can be described in terms of mixtures. The availability of such techniques will not only help the practitioner to understand the applications of mixtures to a variety of engineering problems, but it can also provide insights useful for modifying or extending mixture methodologies.

Let us assume that the mixture model $f(.) = \sum_{j=1}^{Ng} \omega_j f_j(.)$ is available. It can be seen that the mixture is defined in terms of three distinguishable steps:

1. The number of elements present in the mixtures Ng (typically a finite number is selected)
2. The mixture weights ω_j, j = 1, 2, …, Ng which regulate the contribution of each element in the final outcome
3. The elements (elemental density functions) $f_j(.)$, $j = 1, 2, \ldots, Ng$ of the mixture

To generate a random variable X from a given mixture, the following steps should be performed:

1. Generate an element identifier $J = P(J = j) = \omega_j$. In most applications, the number N_g of mixture elements is chosen to be 2, in which case the identifier can be generated simply as a result of a comparison of a uniform (0, 1) variable with ω_j. In the case of $g > 2$, the identifier may be generated by one of several discrete variable generating techniques.
2. Generate realizations X_j, $f_j(.)$ for $j = 1, 2, \ldots, g$.
3. Using Steps 1 and 2, calculate X, $f(.)$.

By the application of this method, the resulting random variable X has the desired distribution $f(.)$ since construction follows the distribution:

$$\sum_{j=1}^{Ng} f_j(.)P(J=j) = \sum_{j=1}^{Ng} f_j(.)\omega_j = f(.) \tag{3.40}$$

The above-described methodology can be utilized to generate random variables from a given mixture model and is used in the simulation studies reported in this chapter.

In this section, an application example is used to demonstrate the applicability of the above generation method. The problem selected is that of "glint noise generation." In radar target applications, the observation noise is highly non-Gaussian. It is well documented in the literature that the so-called "glint noise" possesses the characteristics of a long-tailed distribution.[63–65] Conventional minimum mean square estimators can be seriously degraded if non-Gaussian noise is present. Therefore, it is of paramount importance to have accurate modeling of the non-Gaussian noise phenomenon prior to the development of any efficient tracking algorithm. Many different models have been used for the non-Gaussian glint noise present in target tracking applications. Among them is a mixture approach, originally proposed by Hewer et al.,[64] which argues that the radar glint noise can be modeled as a mixture of background Gaussian noise with outliers. Their results were based on the analysis of the *QQ-plots* of glint noise records.[65] Examination of such records reveals that the glint *QQ-plot* is fairly linear around the origin, an indication that the distribution is Gaussian-like around its mean. However, in the tail region, the plot deviates from linearity and indicates a non-Gaussian, long-tailed character. The data in the tail region are essentially associated with the glint spikes and are considered to be outliers. These outliers have a considerable influence on conventional target tracking filters, such as the Kalman filter which is quite non-robust. The effect of the glint spikes is even greater on the sample variance (covariance) used in the derivation of the filter's gain. It is not difficult to see that variances (covariances) which are quadratic functions of the data are more sensitive to outliers than the sample means. Therefore, the glint spikes can be modeled as a Gaussian noise with large variance (covariance), resulting in an overall glint noise model which can be considered as a Gaussian mixture with the two components used to model the background (thermal) Gaussian noise and the glint spikes, respectively. The weighting coefficients in the mixture (percentage of contamination) can be used to model the non-Gaussian nature of the glint spikes. Therefore, the glint noise model can be generated as the mixture of two Gaussian distributions, each with zero mean and with fixed variance (covariance).

In most studies, the variances (covariances) are proportional to each other. Assuming that the Gaussian terms are denoted as $N_1(0, \sigma_1)$ and $N_2(0, \sigma_2)$, the mixture distribution has the following form:

$$f(k, \sigma_1, \sigma_2) = (1 - k)N_1(0, \sigma_1) + kN_2(0, \sigma_2) \tag{3.41}$$

with $0 < k < 1$. A random variable X of this distribution can be generated by first selecting uniformly a sample U from the interval $[0, 1]$. If $U > k$, then X is generated by an independent sample from $N_1(0, \sigma_1)$. Otherwise, the requested variable X is a sample from $N_2(0, \sigma_2)$. In a first experiment, it is assumed that the regulatory coefficient is the unknown parameter in the mixture. The weighting coefficient assumes the values of $k = 0.1$, $k = 0.2$, and $k = 0.3$, respectively. The variances of the two components are given by $\sigma_1 = 1.0$ and $\sigma_2 = 100.0$. The resulting noise profiles can be seen in Figure 3.1. In a second experiment, we assume that the weighting coefficient in the mixture is known and the only parameter is the variance of the second component in the mixture. We assume that the variance of the first component is fixed, $\sigma_1 = 1.0$. The variance σ_2 of the second component assumes the values of 10.0, 100.0, and 1000.0. By varying

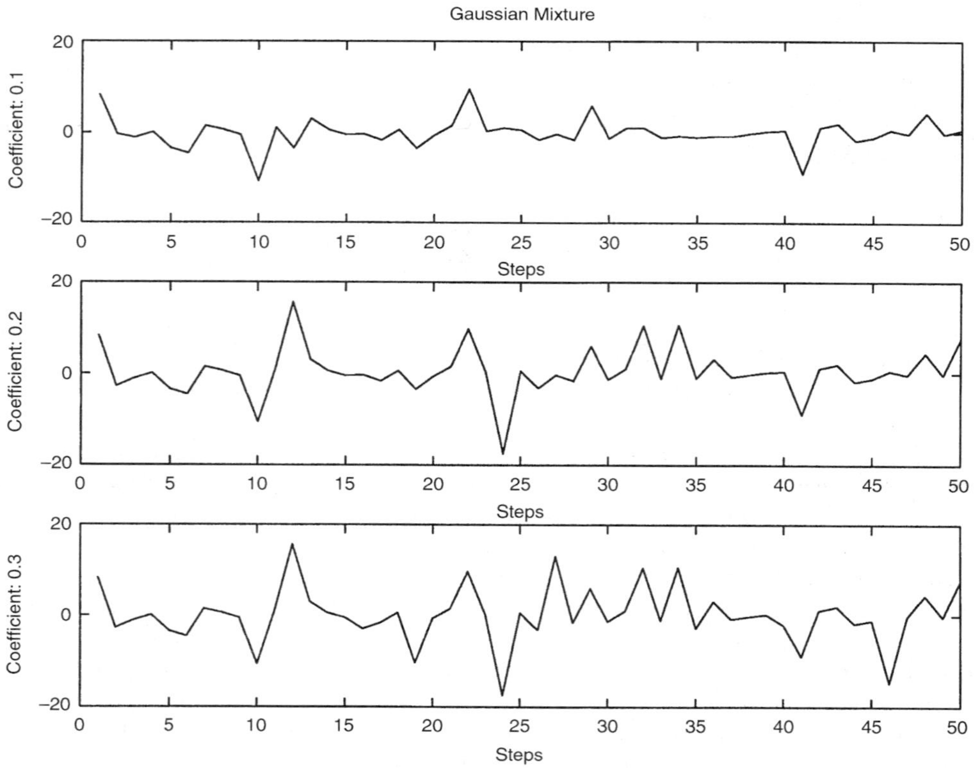

FIGURE 3.1 Gaussian mixture generation: the effect of the weighting coefficient.

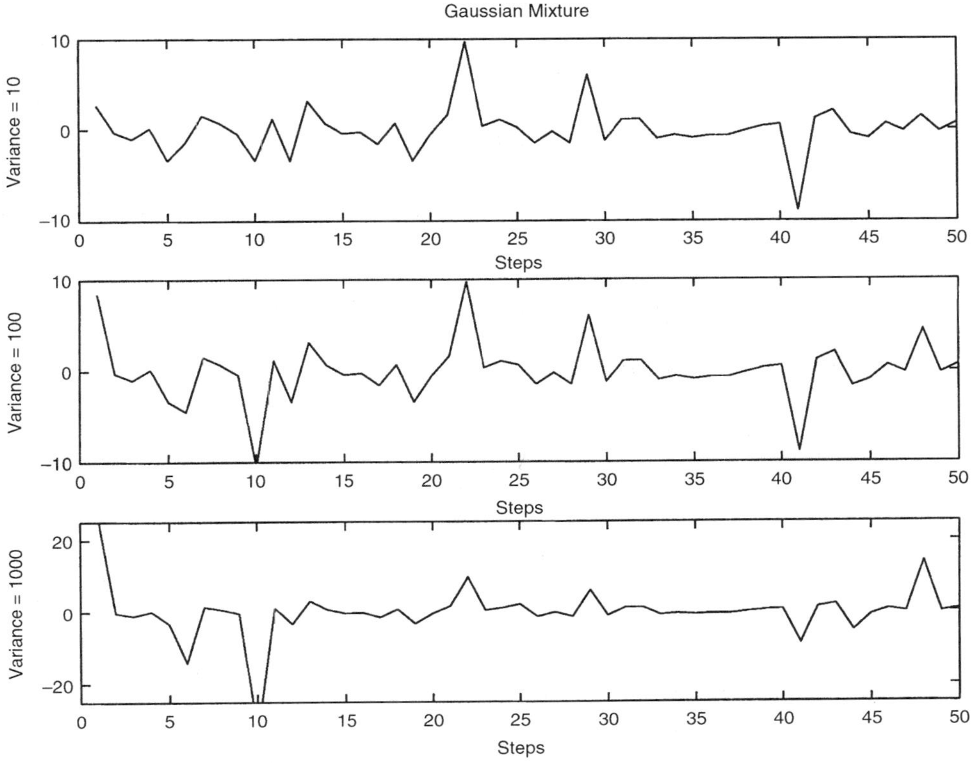

FIGURE 3.2 Gaussian mixture generation: the effect of the variance.

the parameters of the second term in the mixture, a different noise profile can be obtained. It is evident from Figure 3.2 that by increasing the contribution or the variance of the second component in the mixture, the resulting profile deviates more from Gaussian, becoming increasingly non-Gaussian.

3.5 Mixture Applications

In this section, we will describe, in detail, three areas of application of Gaussian mixture models where the model is used either to represent the underlying physical phenomenon or to assist in the development of an efficient and cost-effective algorithmic solution. The application areas considered are those of target tracking in polar coordinates and stochastic estimation for non-linear systems, non-Gaussian (impulsive) noise modeling and inter-symbol interference rejection, and neural networks for function approximation. These applications were selected mainly due to their importance to the signal processing community. It should be emphasized at this point that Gaussian mixtures have been applied to a number of different areas. Such areas include electrophoresis, medical diagnosis and prognosis, econometric applications such as switching econometric models, astronomy, geophysical applications, and applications in agriculture. The interested reader should refer to the extensive summary of references to Gaussian mixture applications provided in Reference 1 for mixture applications.

Apart from that, Gaussian mixture models are essential tools in other literature, such as neural networks where RBF networks and probabilistic neural networks (PNN) are based on Gaussian mixture models, fuzzy systems where fuzzy basis functions are often constructed to imitate the Gaussian mixture model, and image processing/computer vision where Gaussian mixtures can be used to model image intensities and to assist in the estimation of the optical flow.[23,25,66–75]

In the next few paragraphs, we consider three case studies illustrating the effectiveness of the Gaussian mixture approach in solving difficult signal processing problems. The problem of target tracking in polar coordinates is considered in the next section.

3.5.1 Applications to Non-Linear Filtering

Estimation (filtering) theory has received considerable attention in the past four decades, primarily due to its practical significance in solving engineering and scientific problems. As a result of the combined research efforts of many scientists in the field, numerous estimation algorithms have been developed. These can be classified into two major categories, namely, linear and non-linear filtering algorithms corresponding to linear (or linearized) physical dynamic models with Gaussian noise statistics and to non-linear or non-Gaussian physical models.[48,49] The most challenging problem arising in stochastic estimation and control is the development of an efficient estimation (filtering) algorithm which can provide estimates of the state of a dynamical system when non-linear dynamic models coupled with non-Gaussian statistics are assumed. We seek, therefore, the optimal, in the minimum mean square sense, estimator of the state vector x(k) of a dynamic system which can be described by the following set of equations:

$$x(k+1) = f(x(k), v(k), k) \tag{3.42}$$

$$z(k+1) = h(x(k), w(k), k) \tag{3.43}$$

where $f(.)$ is the non-linear function which describes the state evolution over time, and $v(k)$ is the state process noise which can be of a non-Gaussian nature. In most cases, the state noise is modeled as additive white Gaussian noise with covariance $Q(k)$. The only information available about this system is a sequence of measurements $z(1), z(2), \ldots, z(k), \ldots$ obtained at discrete time intervals. The measurement Equation 3.43 describes the observation model which transforms the plant state vector into the measurement space. Most often the observation matrix $h(.)$, it is assumed to be non-linear with additive measurement noise $w(k)$. The additive measurement noise is considered to be white Gaussian with noise covariance $R(k)$ and uncorrelated to the state noise process. The initial state vector $x(0)$, which is generally unknown, is modeled as a random variable which is Gaussian distributed with mean value $\hat{x}(0)$ and covariance $P(0)$. It is considered uncorrelated to the noise processes $\forall k > 0$.

Given the set of measurements $Z^k = [z(1), z(2), \ldots, z(k-1), z(k)]$, we desire the mean-squared-error optimal filtered estimate $\hat{x}(k|k)$ of $x(k)$:

$$\hat{x}(k|k) = E(x(k)|Z^k) \tag{3.44}$$

of the system state.

For the case of linear dynamics and additive Gaussian noise, the problem was first solved by Kalman through his well-known filter.[49] The so-called Kalman filter is the optimal recursive estimator for this case. However, if the dynamics of the system are non-linear and/or the noise processes in Equations 3.42 and 3.43 are non-Gaussian, the degradation in the performance of the Kalman filter will be rather dramatic.[50]

The requested state estimate in Equation 3.44 can be obtained recursively through the application of the Bayes theorem as follows:

$$\hat{x}(k|k) = E(x(k)|Z^k) = \int_{-\infty}^{\infty} x(k)f(x(k)|Z^k)dz \tag{3.45}$$

$$f(x(k), z(k)|Z^{k-1}) = f(x(k)|z(k), Z^{k-1})f(z(k)|Z^{k-1}) = f(z(k)|x(k), Z^{k=1})f(x(k)|Z^{k-1}) \tag{3.46}$$

$$f(x(k)|z(k), Z^{k-1}) = \frac{f(z(k)|x(k), Z^{k-1})f(x(k)|Z^{k-1})}{f(z(k)|Z^{k-1})} = \frac{f(z(k)|x(k), Z^{k-1})f(x(k)|Z^{k-1})}{\int f(z(k)|x(k), Z^{k-1})f(x(k)|Z^{k-1})dx(k)} \quad (3.47)$$

Based on the assumptions of the model, the density function $f(x(k)|z(k))$ can be considered as Gaussian with mean value $h(x(k))$ and covariance $R(k)$:

$$f(x(k)|z(k)) = \frac{1}{(2\pi)^m}|R(k)|^{-0.5}\exp(-0.5\|z(k) - h(x(k))\|^2_{R^{-1}(k)}) \quad (3.48)$$

In a similar manner, the density $f(x(k)|x(x-1))$ can be considered Gaussian with mean value $f(x(k-1))$ and covariance $Q(k-1)$. Given the fact that the initial conditions are assumed Gaussian and thus,

$$f(x(0)|z(0)) = \frac{f(x(0))f(z(0)|x(0))}{f(z(0))} \quad (3.49)$$

a set of equations which can be used to recursively evaluate the state estimate is now available.[48–55]

The above estimation problem is solvable only when the density $f(x(k)|z(k))$ can be evaluated for all k. However, this is possible only for a linear state-space model and if the *a priori* noise and state distributions are Gaussian in nature. In this case, the relations describing the conditional mean and covariance are the well-known Kalman filter equations.[54] To overcome the difficulties associated with the determination of the integrals in Equations 3.46 and 3.47, suboptimal estimation procedures have been developed over the years.[51–55] The most commonly used involves the assumption that the *a priori* distributions are Gaussian and that the non-linear system can be linearized relative to the latest available state estimate resulting in a Kalman-like filter, the so-called "extended" Kalman filter (EKF). Although EKF performs well in many practical applications, there are numerous situations in which unsatisfactory results have been reported. Thus, a number of different methodologies have appeared in the literature. Among them is the Gaussian sum filter which utilizes the approximation theorem reported in Section 3.2.1 to approximate Equations 3.46 and 3.47. This estimation procedure utilizes a Gaussian mixture to approximate the posterior density $f(x(k)|z(k), Z^{k-1})$ in conjunction with the linearization procedure used in EKF. This so-called Gaussian sum approach assumes that at a certain time instant k the one step-ahead predicted density $f(x(k)|Z^{k-1})$ can be written in the form of a Gaussian mixture.[51,52,54]

Then, given the next available measurement and the non-linear model, the filtering density $f(x(k)|z(k), Z^{k-1})$ is calculated as

$$f(x(k)|z(k), Z^{k-1}) = c(k)\sum_{i=1}^{Ng}\omega_i N((x(k) - a_i), B_i)f((z(k) - h(x(k)))) \quad (3.50)$$

Parallelizing the EKF operation, the Gaussian sum filter linearizes $h(x(k))$ relative to a_i so that $f((z(k) - h(x(k))))$ can be approximated by a Gaussian-like function in the region around each a_i. Once the *a posteriori* density $f(x(k)|z(k), Z^{k-1})$ is in the form of a Gaussian mixture, the prediction step of the non-linear estimator can be performed in the same manner by linearizing $f(x(k+1)|x(k))$ about each term in the Gaussian mixture defined to approximate $f(x(k)|z(k), Z^{k-1})$.

In this review, a non-linear filter based on Gaussian mixture models is utilized to provide an efficient, computationally attractive solution to the radar target tracking problem. In tracking applications, the target motion is usually best modeled in a simple fashion using Cartesian coordinates. However, the target position measurements are provided in polar coordinates (range and azimuth) with respect to the sensor location. Due to the geometry of the problem and the non-linear relationship between the two coordinate systems, tracking in Cartesian coordinates using polar measurements can be seen as a non-linear estimation problem, which is described in terms of the following non-linear state-space model:

$$x(k+1) = F(k+1, k)x(k) + G(k+1, k)v(k) \tag{3.51}$$

where $x(k)$ is the vector of Cartesian coordinates target states, $F(.)$ is the state transition matrix, $G(.)$ is the noise gain matrix, and $v(k)$ is the system noise process which is modeled as a zero-mean white Gaussian random process with covariance matrix $Q(k)$.

The polar coordinate measurement of the target position is related to the Cartesian coordinate target state as follows:

$$z(k) = h(x(k)) + w(k) \tag{3.52}$$

where $z(k)$ is the vector of polar coordinates measurements, $h(\cdot)$ is the Cartesian-to-polar coordinate transformation, and $w(k)$ is the observation noise process which is assumed to be a zero-mean white Gaussian noise process with covariance matrix $R(k)$. Thus, target tracking becomes the problem of estimating the target states $x(k)$ from the noisy polar measurements $z(k)$, $k = 1, 2, \ldots$.

A Gaussian mixture model can be used to approximate the densities involved in the derivation of the optimal Bayesian estimator of Equations 3.45 to 3.47 when it is applied to the tracking problem.

To evaluate the state prediction density $p(x(k)|Z^{k-1})$ efficiently, we will assume the conditional density $p(x(k-1)|Z^{k-1})$ to be Gaussian with mean $\hat{x}(k-1|k-1)$ and covariance matrix $P(k-1|k-1)$. Based on this assumption, the state prediction density is a Gaussian density with

$$\hat{x}(k|k-1) = F\hat{x}(k-1|k-1) \tag{3.53}$$

$$P(k|k-1) = FP(k-1|k-1)F^T + GQ(k)G^T \tag{3.54}$$

Given the state-space model of the problem, the function $p(z(k)|x(k))$ can be defined by the measurement equation and the known statistics of the measurement noise $w(k)$:

$$\begin{aligned} p(z(k)|x(k)) &= \int p(z(k)|x(k), w(k))p(w(k)|x(k))dw(k) \\ &= \int \delta(x(k) - h(x(k)) - w(k))p_w(w(k))dw(k) \\ &= p_w(x(k) - h(x(k))) \end{aligned} \tag{3.55}$$

Thus, the function $p(z(k)|x(k))$ can be obtained by applying the transformation $w(k) = z(k) - h(\underset{\sim}{x}(0))$ to the density function $p_w(w(k))$. Utilizing this observation, we select some initial parameters $\tilde{\alpha}_{k,i}$, $\tilde{m}_{k,i}$, and $\tilde{B}_{k,i}$ from the known statistics of the noise $w(k)$, transform these parameters from the $w(k)$ space to the $f(k)$ space based on the transformation $w(k) = z(k) - h(x(k))$ $z(k) - h(x(k))$, and, finally, collect them as a Gaussian mixture approximation for the function $p(z(k)|x(k))$ (see Figure 3.3).

The Gaussian mixture procedure used to approximate the non-linear prediction density $p(x(k)|Z^{k-1})$ is summarized as follows:

1. For initialization, select the parameters $\tilde{\alpha}_{k,i}$, $\tilde{m}_{k,i}$, and $\tilde{B}_{k,i}$ for a prescribed value of N such that the following sum-of-squared error is minimized:

$$\sum_{j=1}^{K} \left| p_w(w_{k,jj}) - \sum_{i=1}^{N} \tilde{\alpha}_{k,i} \mathcal{N}(w_{k,j} - \tilde{m}_{k,i}, \tilde{B}_{k,i}) \right|^2 < \varepsilon \tag{3.56}$$

where $w_{k,j}$:$j = 1, \ldots, K$ is the set of uniformly spaced points distributed through the region containing non-negligible probability, and ε is the prescribed accuracy.

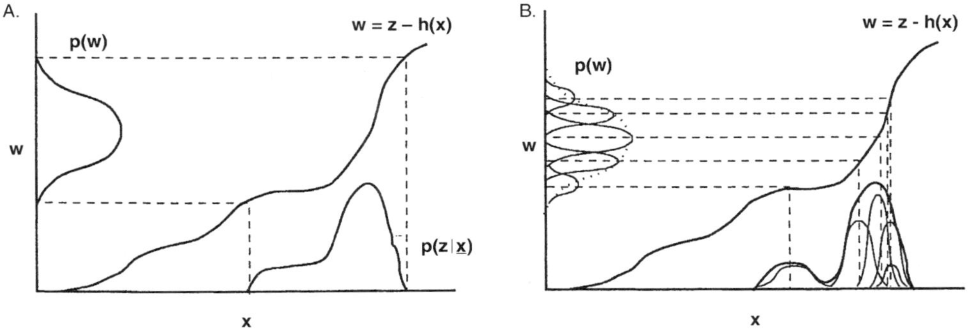

FIGURE 3.3 Gaussian mixture approximation of $p(z(k)|x(k))$.

2. For each new measurement $z(k)$, update the new parameters $\alpha_{k,i}$, $m_{k,i}$, and $B_{k,i}$ such that

$$p(z(k)|x(k)) \approx \sum_{i=1}^{N} \alpha_{k,i} N(m_{k,i} - D(x(k)), B_{k,i}) \tag{3.57}$$

where

$$m_{k,i} = h^{-1}(\overline{m}_{k,i}) \tag{3.58}$$

$$\overline{m}_{k,i} = z(k) - \overline{m}_{k,i} \tag{3.59}$$

$$B_{k,i} = [J_h(m_{k,i})^T \tilde{B}_{k,i}^{-1} J_h(m_{k,i})]^{-1} \tag{3.60}$$

$$\beta_{k,i} = |J_h(m_{k,i})| \tag{3.61}$$

$$\alpha_{k,i} = \beta_{k,i}\tilde{\alpha}_{k,i} \tag{3.62}$$

Here, we assume the function is invertible; however, if the inverse does not exist, then we must choose $m_{k,i}$ to be the most likely solution given $m_{k,i} = h(\tilde{m}_{k,i})$. Moreover, $J_h(x(k))$, $He_h(m_{k,i})$ are the Jacobian and the Hessian of the function $h(x(k))$, respectively, evaluated as

$$\begin{aligned} J_{F_i}(m_{k,i}) &= \left.\frac{\partial F_i(x_n)}{\partial x_n}\right|_{x_n = m_{n,i}} \\ &= \frac{1}{2} J_h(m_{n,i})^T \tilde{B}_{n,i}^{-1}(m_{n,i} - h(m_{n,i})) \end{aligned} \tag{3.63}$$

$$\begin{aligned} He_{F_i}(m_{n,i}) &= \left.\frac{\partial^2 F_i(x_n)}{\partial x_n \partial x_n^T}\right|_{x_n = m_{n,i}} \\ &= -[He_h(m_{n,i})^T \tilde{B}_{n,i}^{-1}(m_{n,i} - h(m_{n,i})) + J_h(m_{n,i})^T \tilde{B}_{n,i}^{-1} J_h(m_{n,i})] \end{aligned} \tag{3.64}$$

Given the form of the approximation, the algorithmic description of the non-linear adaptive Gaussian sum filter (AGSF) for one processing cycle is as follows (see Figure 3.4):

1. Assume that at time k the mean $\hat{x}(k-1|k-1)$ and the associated covariance matrix $P(k-1|k-1)$ of the conditional density $p(x(k-1)|Z^{k-1})$ are available.

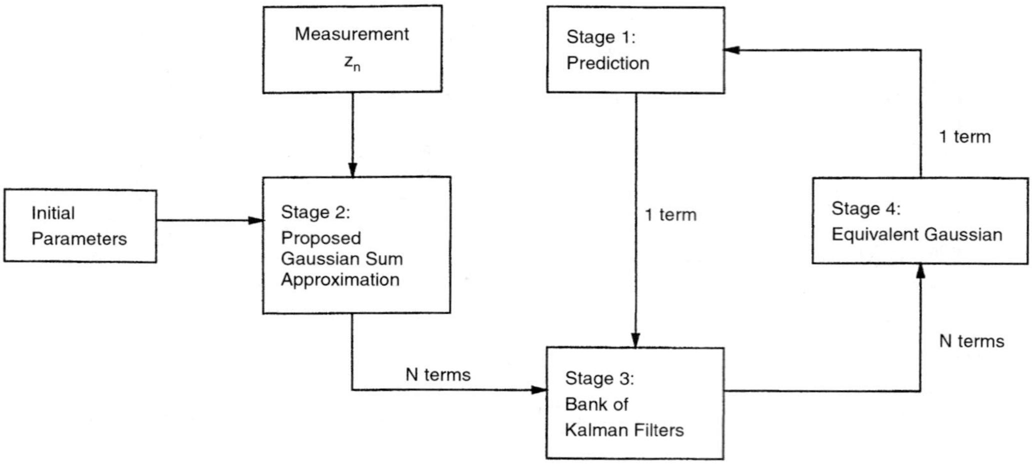

FIGURE 3.4 The adaptive Gaussian sum filter (AGSF).

The predictive mean $\hat{x}(k|k-1)$ and the corresponding covariance matrix $P(k|k-1)$ of the predictive density $p(x(k)|Z^{k-1})$ are determined through Equations 3.53 and 3.54 using the state equation of the model.

2. The density $p(x(k)|Z^k)$ is approximated systematically by a weighed sum of Gaussian terms.
3. The Gaussian terms in the mixture are passed to a bank of N Kalman filters which evaluate the parameters for the Gaussian mixture approximation for the density $p(x(k)|Z^k)$.
4. The Gaussian mixture approximation for the density $p(x(k)|Z^k)$ is collapsed into one equivalent Gaussian term with mean $\hat{x}(k|k)$ and covariance $P(k|k)$.

A two-dimensional, long-range target tracking application is simulated to demonstrate the performance of the AGSF on target state estimation. The target trajectory is modeled by the second-order kinematic model of Equation 3.51 with a process noise of standard variation 0.01 m/s^2 in each coordinate. The measurements are modeled according to Equation 3.52. The standard deviations for range errors are assumed to be 50 m, and two standard deviations of bearing error are used: $\sigma_\theta = 2.5°$ and 5.73°. The parameters of the model are defined as follows:

$$\mathbf{x}_{n+1} = \begin{bmatrix} 1 & 1 & 0 & 0 \\ 0 & 1 & 0 & 0 \\ 0 & 0 & 1 & 1 \\ 0 & 0 & 0 & 1 \end{bmatrix} \mathbf{x}_n + \begin{bmatrix} 1/2 & 0 \\ 1 & 0 \\ 0 & 1/2 \\ 0 & 1 \end{bmatrix} \mathbf{w}_n \tag{3.65}$$

$$\mathbf{z}_n = \begin{bmatrix} \sqrt{x_n^2 + y_n^2} \\ \tan^{-1} y_n / x_n \end{bmatrix} + \mathbf{v}_n$$

$$\mathbf{Q} = \begin{bmatrix} 0.0001 & 0 \\ 0 & 0.0001 \end{bmatrix} \tag{3.66}$$

$$\mathbf{R} = (1)\begin{bmatrix} 2500 & 0 \\ 0 & 0.037 \end{bmatrix}, (2)\begin{bmatrix} 2500 & 0 \\ 0 & 0.01 \end{bmatrix}$$

The AGSF is compared with the EKF and the converted measurement Kalman filter (CMKF) in this experiment. All these filters are initialized with the same initial filtered estimate $\hat{\mathbf{x}}_{0|0}$ and the same initial error covariance $\mathbf{P}_{0|0}$ based on the first two measurements. The initial number of Gaussian terms in the preprocessing stage is 30. After preprocessing, the number of the Gaussian terms used in the implementation of the AGSF is 9. The results presented here are based on 1500 measurements averaged over 1000 independent Monte Carlo realizations of the experiment with the sampling interval of 1 s and with two different measurement noise levels. In order to generate the measurement record, the initial state x_0 is assumed Gaussian with an average range of 50 km and an average velocity of 20 m/s. For each Monte Carlo realization of the experiment, the initial value is chosen randomly from the assumed Gaussian distribution.

The position errors and the velocity errors for the three filters are shown in Figures 3.5 and 3.6, respectively, for $\sigma_\theta = 2.5°$. The error is defined as the root mean square of the difference between the actual value and the estimated value. The Gaussian sum approach converges faster and yields estimates of smaller error than the EKF and the CMKF. For $\sigma_\theta = 2.5°$, the CMKF converges faster than the EKF initially, but it ceases to converge after the first 400 measurements. The EKF, on the other hand, is very steady and consistent. As σ_θ increases to 5.72° (0.1 rad), the EKF starts to diverge due to the fact that the EKF is extremely sensitive to the initial filter conditions. When the cross-error gets too large, the wrong set of initial conditions can lead to divergence. The CMKF, however, seems to be more robust to inconsistent initial conditions. The AGSF, due to its parallel nature and the fact that the Bayes rule operates as a correcting/adjusting mechanism, is also in position to compensate for inconsistent initial conditions.

3.5.2 Non-Gaussian Noise Modeling

The Gaussian mixture density approximation has been extensively used to accomplish practical models for non-Gaussian noise sources in a variety of applications. The appearance of the noise and its effect are related to its characteristics. Noise signals can be either periodic in nature or random. Usually, noise

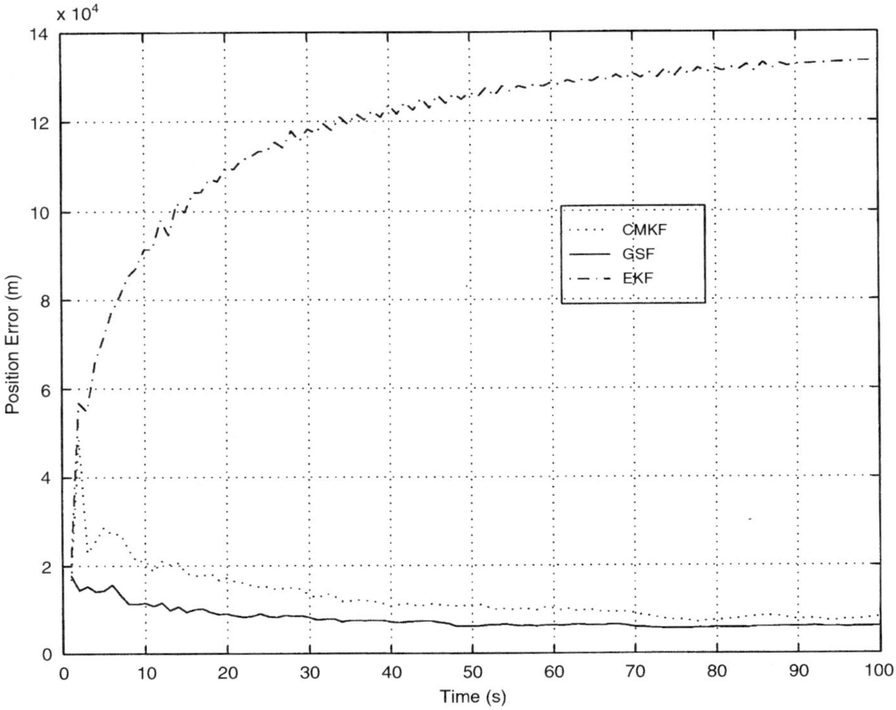

FIGURE 3.5 Target tracking: comparison of position errors.

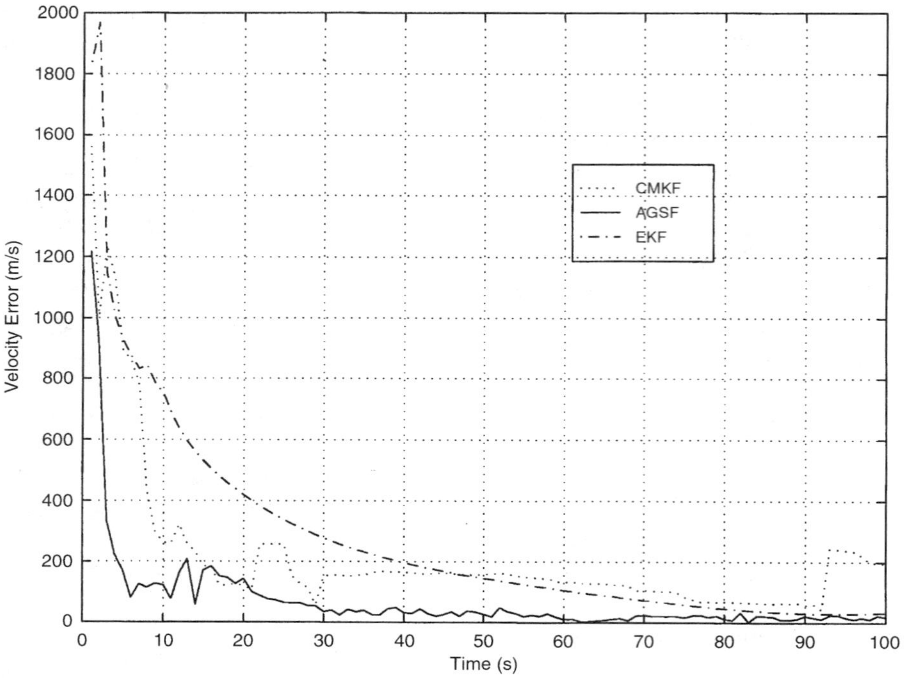

FIGURE 3.6 Target tracking: comparison of velocity errors.

signals introduced during signal transmission are random in nature, resulting in abrupt local changes in the transmitting sequence. These noise signals cannot be adequately described in terms of the commonly used Gaussian noise model. Rather, they can be characterized in terms of impulsive sequences (interferences) which occur in the form of short time duration, high energy spikes attaining large amplitudes with probability higher than the probability predicted by a Gaussian density model.

There are various sources that can generate such non-Gaussian noise signals. Among others, some are man-made phenomena, such as car ignition systems, industrial machines in the vicinity of the signal receiver, switching transients in power lines, and various unprotected electric switches. In addition, natural causes such as lightning in the atmosphere and ice cracking in the Antarctic region also generate non-Gaussian, long-tailed types of noise.

Several models have been used to date to model non-Gaussian noise environments. Some of these models have been developed directly from the underlying physical phenomenon. On the other hand, empirically devised noise models have been used over the years to approximate many non-Gaussian noise distributions. Based on the density approximation theorem presented above, any non-Gaussian noise distribution can be expressed as, or approximated sufficiently well by, a finite sum of known Gaussian probability density functions (pdfs). The Gaussian sum model has been used in the development of approximate empirical distributions which relate to many physical non-Gaussian phenomena.

The most commonly used empirical model is the ε-mixture or ε-contaminated Gaussian mixture model in which the noise pdf has the form of

$$f(x) = (1 - \varepsilon)f_b(x) + \varepsilon f_o(x) \qquad (3.67)$$

where $\varepsilon \in [0, 1]$ is the mixture weighting coefficient. The mixing parameter ε regulates the contribution of the non-Gaussian component, and usually it varies between 0.01 to 0.25.

The $f_b(x)$ pdf is usually taken to be a Gaussian pdf representing background noise. Among the choices for the contaminating pdf are various "heavy-tailed" distributions such as the Laplacian or the double exponential. However, most often f_o is taken to be Gaussian with variance σ_o^2 taken to be many times the variance of f_o, σ_b^2. The ratio $k = \sigma_o^2/\sigma_b^2$ has generally been taken to be between 1 and 10,000. Although the parameters of the mixture model are not directly related to the underlying physical phenomenon, the model is widely used in a variety of applications, primarily due to its analytical simplicity. The flexibility of the model allows for the approximation of many different, naturally occurring noise distribution shapes. This approach has been used to model non-Gaussian measurement channels in narrowband interference suppression, a problem of considerable engineering interest.[60]

Spread-spectrum communication systems often use estimation techniques to reject narrowband interference. Recently, the interference rejection problem has been formulated as a non-linear estimation problem using a state-space representation.[58] Following the state-space approach, the narrowband interference is modeled as the state trajectory, and the combination of the direct-sequence spread-spectrum signal with the background noise is treated as non-Gaussian measurement noise.

The basic idea is to spread the bandwidths of transmitting signals so that they are much greater than the information rate. The problem of interest is the suppression of a narrowband interferer in a direct-sequence spread-spectrum (DS/SS) system operating as an Nth-order autoregressive process of the form:

$$i_k = \sum_{n=1}^{N} \Phi_n i_{k-n} + e_k \tag{3.68}$$

where e_k is a zero-mean white Gaussian noise process, and $\Phi_i, \Phi_2, \ldots, \Phi_{N-1}, \Phi_N$ are the autoregressive parameters known to the receiver.

The discrete time model arises when the received continuous time signal is passed through an integrate-and-dump filter operating at the chip rate.[59]

The DS/SS modulation waveform is written as

$$m(t) = \sum_{k=0}^{N_c-1} c_k q(t - k\tau_c) \tag{3.69}$$

where N_c is the pseudo-noise chip sequence used to spread the transmitted signal, and $q(.)$ is a rectangular pulse of duration τ_c. The transmitter signal can be then expressed as

$$s(t) = \sum_k b_k m(t - kT_b) \tag{3.70}$$

where $b(k)$ is the binary information sequence, and $T_b = N_c\tau_c$ is the bit duration. Based on that, the received signal is defined as

$$z(t) = as(t-\tau) + n(t) + i(t) \tag{3.71}$$

where a is an attenuation factor, τ is a delay offset, $n(t)$ is wideband Gaussian noise, and $i(t)$ is narrowband interference. Assuming that $n(t)$ is band limited and hence white after sampling, with $\tau = 0$ and $a = 1$ for simplicity, if the received signal is chip matched and sampled at the chip rate of the pseudo-noise sequence, the discrete time sequence resulting from the continuous model above can be rewritten as follows:

$$z(k) = s(k) + n(k) + i(k) \tag{3.72}$$

The system noise contains an interference component $i(k)$ and a thermal noise component $n(k)$. We assume binary signaling and a processing gain of K chips/bit so that during each bit interval, a pseudo-random code sequence of length K is transmitted. The code sequences can be denoted as

$$S^K = [s_1(1), s_1(2), \ldots, s_1(K)] \tag{3.73}$$

with $s_1 \in (+1, -1)$.

Based on this, a state-spacer representation for the received signal and the interference can be constructed as follows:

$$\begin{aligned} x(k) &= \Phi x(k-1) + v(k) \\ z(k) &= Hx(k) + w(k) \end{aligned} \tag{3.74}$$

with $x(k) = [i_k, i_{k-1}, \ldots, i_{k-N+1}]^{\tau}$, $v(k) = [e_k, 0, \ldots, 0]^{\tau}$, $H = [1, 0, \ldots, 0]$, and

$$\Phi = \begin{vmatrix} \Phi_1 & \Phi_2 & \ldots & \Phi_N \\ 1. & 0. & \ldots & 0. \\ \ldots & \ldots & \ldots & \ldots \\ 0. & 0. & \ldots & 1. \end{vmatrix}$$

The additive observation noise $w(k)$ in the state-space model is defined as

$$v(k) = n(k) + s(k)$$

Since the first component of the system state $x(k)$ is the interference $i(k)$, an estimate of the state contains an estimate of $i(k)$ which can be subtracted from the received signal in order to increase the system's performance. The additive observation (measurement) noise $v(k)$ is the sum of two independent variables: one is Gaussian distributed and the other takes on values –1 or –1 with equal probability. Therefore, its density is the weighted sum of two Gaussian densities (Gaussian sum):[59,60]

$$f(w(k)) = (1-\varepsilon)N(\mu, \sigma_n^2) + \varepsilon N(-\mu, \lambda\sigma_n^2) \tag{3.75}$$

with $\varepsilon = 0.5$ and $\mu = 1$.

In summary, the narrowband interference is modeled as the state trajectory, and the combination of the DS/SS signal and additive Gaussian noise is treated as non-Gaussian measurement noise. Non-linear statistical estimators can be used then to estimate the narrowband interference and to subtract it from the received signal. Due to the nature of the non-Gaussian measurement noise, a non-linear filter should be used to provide the estimates. The non-linear filter takes advantage of the Gaussian mixture representation of the measurement noise to provide online estimates of the inter-symbol interference. By collapsing the Gaussian mixture at every step through the utilization of the Bayes theorem, a Kalman-like recursive filter with constant complexity can be devised.

For the state-space model of Equation 3.72, if the measurement noise is expressed in terms of the Gaussian mixture of Equation 3.75, an estimate $\hat{x}(k|k)$ of the system state $x(k)$ at time instant k can be computed recursively by an AGSF as follows:

$$\hat{x}(k|k) = \hat{x}(k|k-1) + K(k)(z(k) - \hat{z}(k|k-1)) \tag{3.76}$$

$$P(k|k) = (I - K(k)H(k))P(k|k-1) \tag{3.77}$$

$$\hat{x}(k|k-1) = \Phi(k, k-1)\hat{x}(k-1|k-1) \tag{3.78}$$

$$P(k|k-1) = \Phi(k, k-1)P(k-1|k-1)\Phi(k, k-1)^{\tau} + Q(k-1) \tag{3.79}$$

with initial conditions $\hat{x}(0|0) = \hat{x}(0)$ and $P(0|0) = P(0)$.

$$K(k) = P(k|k-1)H^{\tau}(k|k-1)P_z^{-1}(k|k-1) \tag{3.80}$$

$$\hat{z}(k|k-1) = \sum_{i=1}^{Ng} \omega_i(k)\hat{z}_i(k|k-1) \tag{3.81}$$

$$\hat{z}_i(k|k-1) = H(k)\hat{x}(k|k-1) + \mu_i \tag{3.82}$$

$$P_{zi}(k|k-1) = H(k)P(k|k-1)H^{\tau}(k) + R_i \tag{3.83}$$

In Equation 3.75, $Ng = 2$, with $\mu_i = \mu$ and $R_1 = \sigma_n^2$, $R_2 = \lambda\sigma_n^2$.

The corresponding innovation covariance and the posterior weights used in the Bayesian decision module are defined as

$$P_z(k|k-1) = \sum_{i=1}^{Ng} (P_{zi}(k|k-1) + (\hat{z}(k|k-1) - \hat{z}_i(k|k-1))(\hat{z}(k|k-1) - \hat{z}_i(k|k-1))^{\tau})\omega_i(k) \tag{3.84}$$

$$\omega_i(k) = \frac{((2\pi)^{-m}|P_{zi}|^{-1}\exp(-0.5(\|z(k) - \hat{z}_i(k|k-1)\|^2_{P_{zi}^{-1}(k|k-1)})))a_i}{c(k)} \tag{3.85}$$

where $|.|$ denotes the determinant of the matrix, and $||.||$ denotes inner product. The parameter a_i is the initial weighting coefficient used in Gaussian mixture which describes the additive measurement noise. In Equation 3.75, $a_1 = (1 - \varepsilon)$ and $a_2 = \varepsilon$.

Finally, the normalization factor $c(k)$ is calculated recursively as follows:

$$c(k) = \sum_{i=1}^{Ng} ((2\pi)^{-m}|P_{zi}|^{-1}\exp(-0.5(\|z(k) - \hat{z}_i(k|k-1)\|^2_{P_{zi}^{-1}(k|k-1)})))a_i \tag{3.86}$$

Simulation results are included here to demonstrate the effectiveness of such an approach. In this study, the interferer is found by channeling white noise through a second-order infinite-duration impulse response (IIR) with two poles at 0.99:

$$i_k = 1.98i_{k-1} - 0.9801i_{k-2} + e_k \tag{3.87}$$

where e_k is zero-mean white Gaussian noise with variance 0.01. The regulatory coefficient ε used in the Gaussian mixture of Equation 3.75 is set to be $\varepsilon = 0.2$, and the ratio λ is taken to be $\lambda = 10$ or $\lambda = 10{,}000$ with σ_n 1.0. The non-Gaussian measurement noise profile, for a single run, is depicted in Figure 3.7 ($\lambda = 10$) and Figure 3.10 ($\lambda = 10{,}000$).

The normalized mean square error (NMSE) is utilized for filter comparison purposes in all experiments. The data were averaged through Monte Carlo techniques. Given the form of the state vector, the first component of $x(k)$ is used in the evaluation analysis. The NMSE is therefore defined as

$$NMSE = \frac{1}{MCRs}\left(\sum_{k=1}^{MCRs} \frac{(x_{1r}^k - \hat{x}_{1j}^k)^2}{x_{1r}^{k^2}}\right)$$

Where MCRs is the number of Monte Carlo runs, x_{1r} is the actual value, and $\hat{x}_{1j}$ is the outcome of the j filter under consideration.

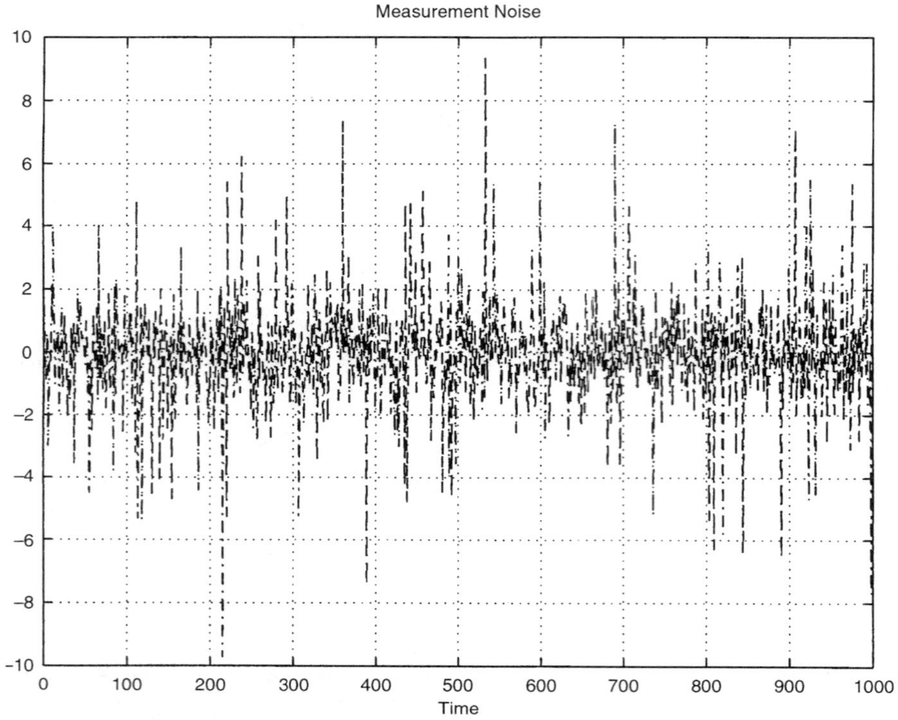

FIGURE 3.7 Inter-symbol interference-I: measurement noise profile.

In this experiment, 100 independent runs (MCRs), each 1000 samples in length, were considered. Due to its high complexity and the unavailability of suitable non-linear transformation for the "score function," the Masreliez filter was not included in these simulation studies.

Two different plot types are reported in this chapter. First, state estimation plots for single MCRs are included to facilitate the performance of the different estimation schemes (Figures 3.8 and 3.11). In addition, the NMSE plots for all the simulation studies are also reported (Figures 3.9 and 3.12). From the plots included in this chapter, we can clearly see the improvement accomplished by the utilization of the new filter vs. the Kalman filter and the Masreliez filter. The effects have appeared more pronounced at more dense non-Gaussian (impulsive) environments. This trend was also verified during the error analysis utilizing the Monte Carlo error plots (Figures 3.9 and 3.12).

3.5.3 Radial-Basis Functions (RBF) Networks

Although Gaussian mixtures have been used for many years in adaptive signal processing, stochastic estimation, statistical pattern recognition, Bayesian analysis, and decision theory, only recently have they been considered by the neural networks community as a valuable tool for the development of a rich class of neural nets, the so-called RBF networks.[72] RBF networks can be used to provide an effective and computationally efficient solution to the interpolation problem. In other words, given a sequence of (n) available data points $X = (x_1, x_2, \ldots, x_n)$ (which can be vectors) and the corresponding (n) measurement values $Y(y_1, y_2, \ldots, y_n)$, the objective is to define a function F satisfying the interpolation condition $F(x_i) = y_i$, $i = 1, 2, \ldots, n$. The RBF neural approach consists of choosing F from a linear space of dimension (n) which depends on the data points x_i.[73] The basis of this linear space is chosen to be the set of radial functions. Radial functions are a special class of functions in which their response decreases or increases monotonically with distance from a central point. The central point, the distance scale, as well as the shape of the radial function are parameters of the RBF neural model. Although many radical functions

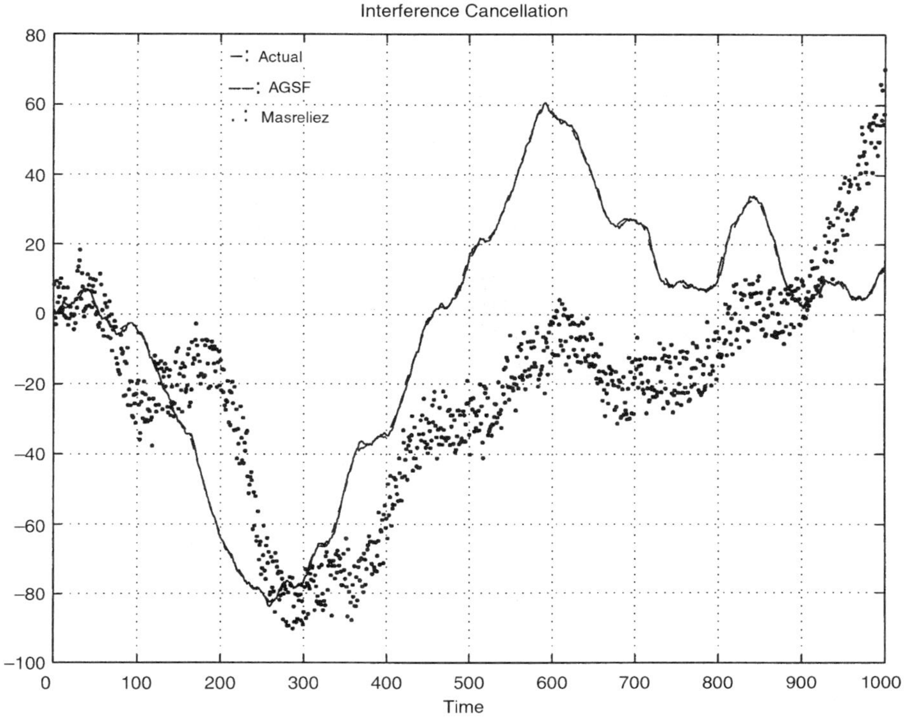

FIGURE 3.8 Inter-symbol interference-I: performance comparison.

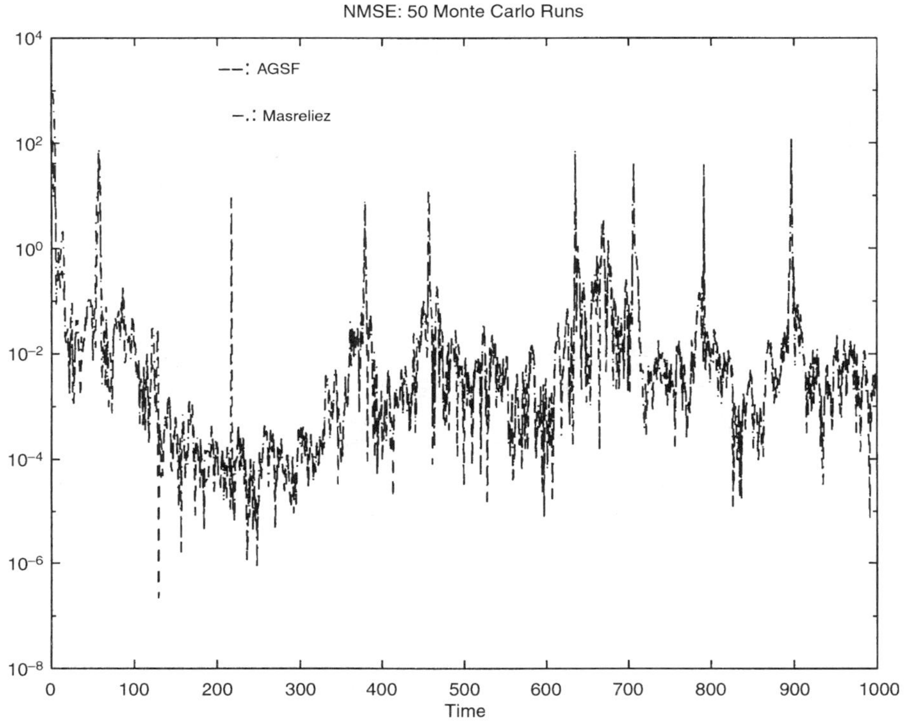

FIGURE 3.9 Inter-symbol interference-I: Monte Carlo evaluation.

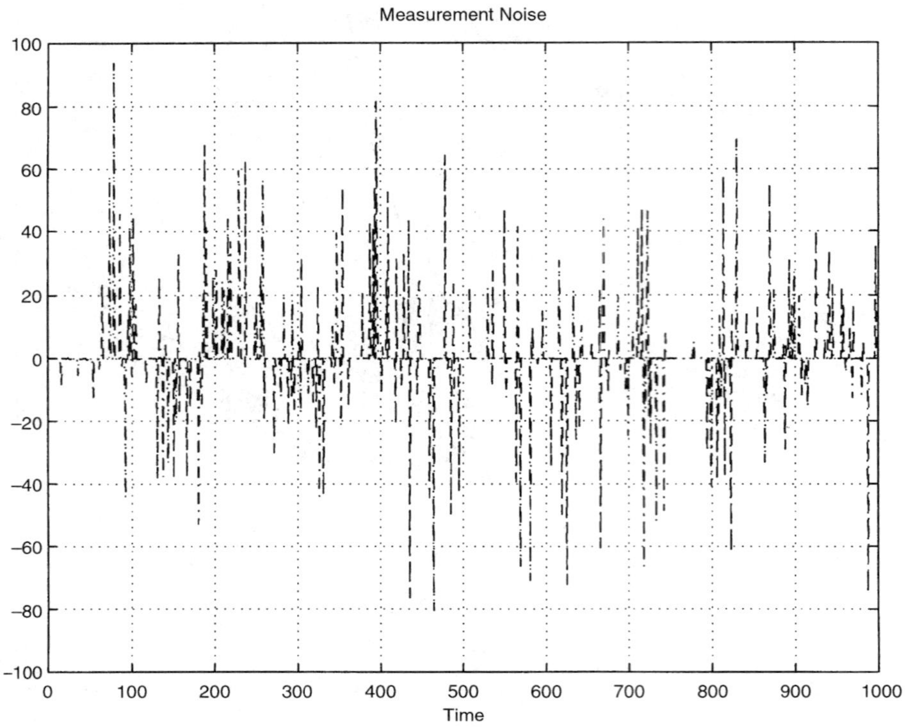

FIGURE 3.10 Inter-symbol interference-II: measurement noise profile.

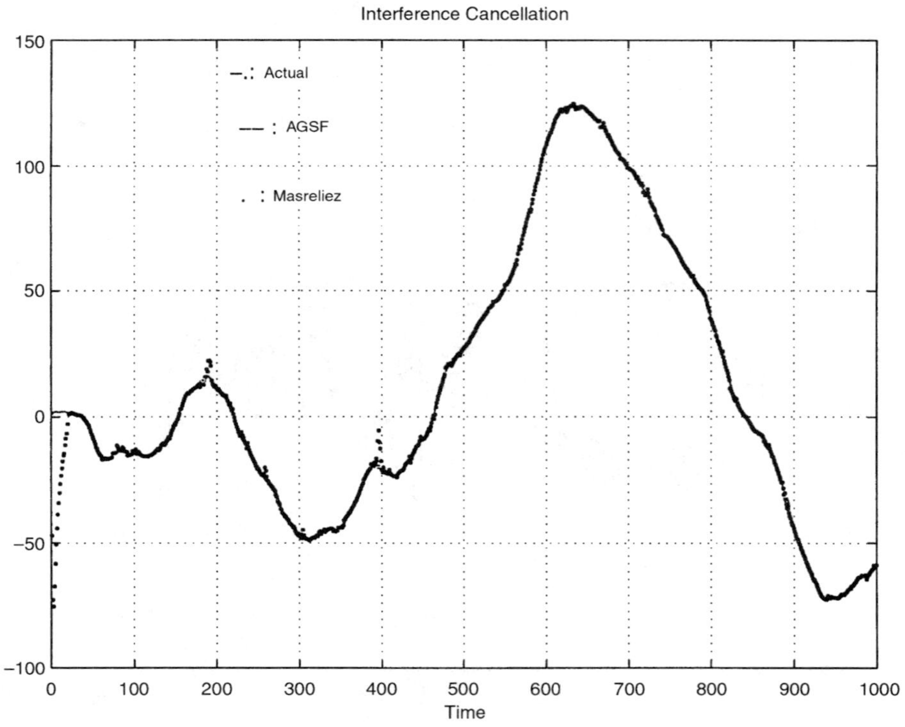

FIGURE 3.11 Inter-symbol interference-II: performance comparison.

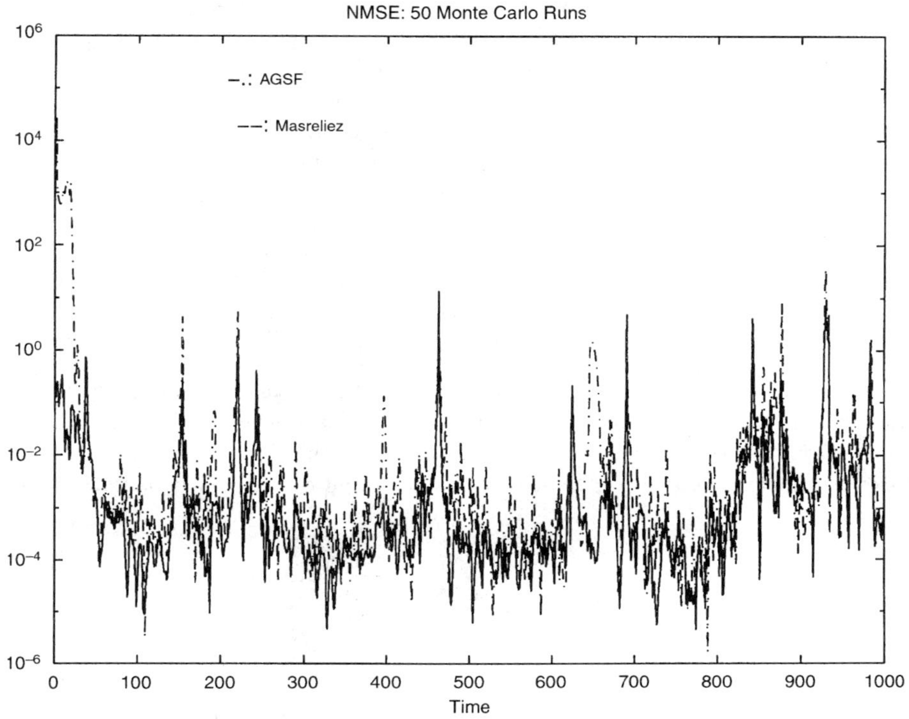

FIGURE 3.12 Inter-symbol interference-II: Monte Carlo evaluation.

have been defined and used in the literature, the typical one is the Gaussian which, in the case of a scalar input, is defined as

$$f(x;\, c,\, r) \;=\; \exp\left(-\frac{(x-c)^2}{r^2}\right) \tag{3.88}$$

with parameters the center c and the radius r. A single layer network consisting of such Gaussian basis functions is usually called RBF net in the neural network literature. Optimization techniques can be used to adjust the parameters of the basis functions in order to achieve better results. Assuming that the number of basis (Gaussian) functions is fixed, the interpolation problem is formulated as follows:

$$F(x) \;=\; \sum_{i=1}^{Ng} \omega_i f(x;\, c_i,\, r) \tag{3.89}$$

Although the number Ng of elemental Gaussian terms in the mixture expression can be defined *a priori*, it can also be considered as a parameter. In such a case, the smallest possible number of Gaussian bases is targeted.

In this setting, the problem is the equivalent of solving for a set of $(3Ng)$ non-linear equations using (n) data points. Thus, the problem is to determine the Gaussian centers and radius along with the mixture parameters from the sample data set.

One of the most convenient ways to implement this is to start with an initial set of parameters and then iteratively modify them until a local minimum is reached in the error function between the available data set and the approximating Gaussian mixture.

However, defining a smooth curve from available data is an ill-posed problem in the sense that the information in the data may not be sufficient to uniquely reconstruct the function mapping in regions where data samples are not available. Moreover, if the available data set is subject to measurement errors or stochastic variations, additional steps such as introduction of penalty terms in the error function are needed in order to guarantee good results. In a general d-dimensional space, the Gaussian radial basis can be $f(x) = \exp(-0.5\|x - \mu_i\|_{\Sigma_i^{-1}})$, where μ_i and Σ_i represent the mean vector and the covariance matrix of the ith RBF.

The quadratic term in the Gaussian basis function form can be written as an expanded form

$$\|x - \mu_i\|_{\Sigma_i^{-1}} = \sum_{k=1}^{d}\sum_{j=1}^{d} \lambda_{ikj}(x_j - \mu_{ij})(x_k - \mu_{ik}) \tag{3.90}$$

with μ_{ij} as the jth element of the mean vector μ_i, and λ_{kj} as the (j, k) element of the shape matrix Σ_i^{-1}. The elements of the shape function can be evaluated in terms of the marginal standard deviations σ_{ij}, σ_{ik} and the correlation coefficient. Assuming that the shape matrix is a positive diagonal, a much simpler expression can be obtained. In such a case, the output of the ith Gaussian basis function can be defined as

$$o_i = \exp\left(-0.5\sum_{k=1}^{d}\frac{(x_k - \mu_{ik})^2}{\sigma_{ik}}\right) \tag{3.91}$$

with $1 \le i \le Ng$.

The output of the ith Gaussian basis function forms a hyper-ellipsoid in the d-dimensional space with the mean and the variance being the parameters which determine the geometric shape and the position of that hyper-ellipsoid. Therefore, the radial-basis network consists of an array of Gaussian functions determined by some parameter vectors.[68]

$$F(x) = \sum_{i=1}^{Ng}\omega_i \exp\left(-0.5\sum_{k=1}^{d}\frac{(x_k - \mu_{ik})^2}{\sigma_{ik}}\right) \tag{3.92}$$

RBF networks have been used extensively to approximate non-linear functions.[78] In most cases, single, hidden layer structures with Gaussian units are used due to their simplicity and fast training. To demonstrate the function approximation capabilities of the RBF network, a simple scalar example is considered. The RBF network consists of five Gaussian units equally weighted. Figure 3.13 depicts the initial placement of the five Gaussian terms, as well as the overall function to be approximated. It can be seen from the plot that the basis functions are equally distributed on the interval 50 to 200. Figure 3.14 depicts the final location of the Gaussian basis functions. The unequal weights and the shifted placement of the basis functions provide an efficient and cost effective approximation to the original function.

The deterministic function approximation approach is probably not the best way to characterize an RBF network when the relationship between the input and output parameters is a statistical rather a deterministic one. It was suggested in Reference 79 that in this case it is better to consider the input and output pair x, $F(x)$ as realizations of random vectors which are statistically dependent. In such a case, if a complete statistical description of the data is available, the output value can be estimated given only the input values. However, since a complete statistical description is seldom available in most cases, the optimal statistical estimator cannot be realized. One way to overcome the problem is to assume a certain parametric model and use the data to construct a model which fits the data reasonably well.[80] A number of different neural networks based on parametric modeling of data have been proposed in the literature. Among them are the so-called probabilistic neural networks (PNN)[25,73]

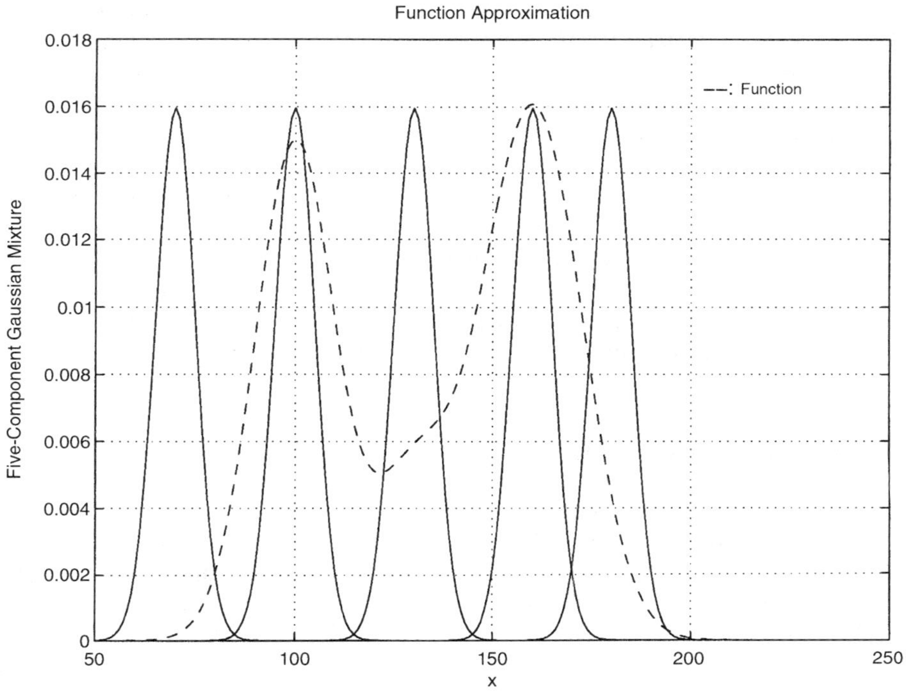

FIGURE 3.13 Function approximation via RBF nets: initial placement of the Gaussian terms.

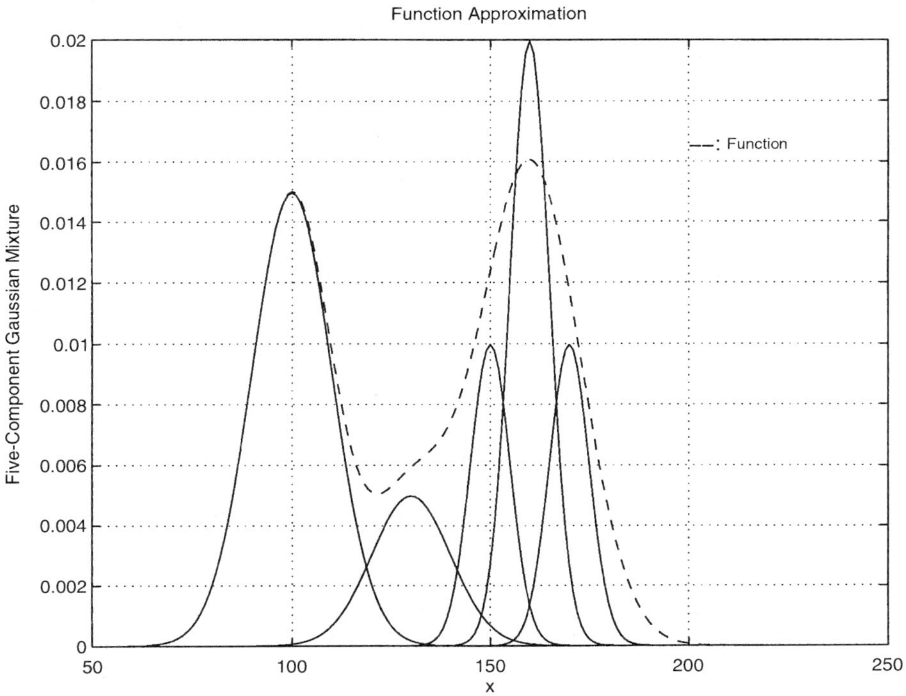

FIGURE 3.14 Function approximation via RBF nets: final placement of the Gaussian terms.

and the Gaussian mixture (GM) model of References 80 and 81. The GM model is a parametric probabilistic model based on the GM model discussed throughout this chapter. In the context of GM, it is assumed that the available input/output pairs result from a mixture of N_g populations of Gaussian random vectors, each one with a probability of occurrence of ω_i, $i = 1, \ldots, N_g$. Given that assumption, a GM basis function network (GMBFN)[80] can be used to provide an estimate of the output variable given a set of input values and the set of N_g Gaussian bases. The GMBFN parallelizes the GM models used in the development of non-linear statistical estimators. Parameter estimation techniques, such as the EM algorithm discussed in this chapter, can be used to estimate the parameters of the GMBFN model during training. The GMBFN network can be viewed as the link between the GM models used in statistical signal processing and the RBF networks used for function approximation. This type of network has been shown to have good approximation capabilities in non-linear mappings and has been proven to provide efficient solutions in application problems such as channel equalization and image restoration.

3.6 Concluding Remarks

In this chapter we reviewed some of the issues related to the GM approach and its applications to signal processing. Due to the nature of the GM model, special attention was given to non-linear, non-Gaussian signal processing applications. Novel signal processing techniques were developed to provide effective, simple, and computationally attractive solutions in important application problems, such as target tracking in polar coordinates and interference rejection in impulsive channels. Emphasis was also given on theoretical results, such as the approximation theorem and the EM algorithm for mixture parameter estimation. Although these issues are not related to any particular practical application, they can provide the practitioner with the necessary tools needed to support a successful application of GMs.

The authors' intention was to illustrate the applicability of the GM methodology in signal processing applications and to highlight the similarities between GM models used in statistical signal processing and neural network methodologies such as RBF used in function approximation and optimization. Since mixture model analysis yields a large number of theorems, methods, applications, and test procedures, there is much pertinent theoretical work as well as research on GM applications which has been omitted for reasons of space and time.

Apart from the practical problems discussed here, there is a large class of problems that appear to be amenable to solution by GMs. Among them are emerging areas of significant importance, such as data mining, estimation of video flow, and modeling of (computer) communication channels. It is the authors' belief that GM models provide effective tools for these emerging signal processing applications, and, thus, surveys on GM analysis and applications can contribute to further advances in these emerging research areas.

References

1. D.M. Tittirington, A.F.M. Smith, and U.F. Makov, *Statistical Analysis of Finite Distributions*, Wiley, New York, 1985.
2. G.J. McLachlam and K.E. Basford, *Mixture Models: Inference and Applications to Clustering*, Marcel Dekker, New York, 1988.
3. B.S. Everitt and D.J. Hand, *Finite Mixture Distributions*, Chapman & Hall, London, 1981.
4. R.O. Duda and P.E. Hart, *Pattern Recognition and Scene Analysis*, Wiley, New York, 1973.
5. E.A. Patrick, *Fundamentals of Pattern Recognition*, Prentice-Hall, Englewood Cliffs, NJ, 1972.
6. R.S. Bucy and P.D. Joseph, *Filtering for Stochastic Process with Application to Guidance*, Interscience, New York, 1968.
7. J. Koreyaar, *Mathematical Methods, Vol. 1*, pp. 330–333, Academic Press, New York, 1968.

8. W. Feller, *An Introduction to Probability and Its Applications, Vol. II*, p. 249, John Wiley & Sons, New York, 1966.
9. M. Aitkin and D.B. Rubin, Estimation and hypothesis testing in finite mixture models, *J. R. Stat. Soc. B*, 47, 67–75, 1985.
10. K.E. Basford and G.J. MacLachlan, Likelihood estimation with normal mixture models, *Appl. Stat.*, 34, 282–289, 1985.
11. J. Behboodian, On a mixture of normal distributions, *Biometrika*, 57, 215–217, 1970.
12. J.G. Fryer and C.A. Robertson, A comparison of some methods for estimating mixed normal distributions, *Biometrika*, 59, 639–648, 1972.
13. A.S. Zeevi and R. Meir, Density estimation through convex combination of densities: Approximation and estimation bounds, *Neural Networks*, 10(1), 99–109, 1997.
14. S.S. Gupta and W.T. Huang, On mixtures of distributions: A survey and some new results on ranking and selection, *Sankhya Ser. B*, 43, 245–290, 1981.
15. V. Husselblad, Estimation of finite mixtures of distributions from exponential family, *J. Am. Stat. Assoc.*, 64, 1459–1471, 1969.
16. R.J. Hathaway, Another interpretation of the EM algorithm for mixture distributions, *Stat. Probability Lett.*, 4, 53–56.
17. R.A. Maronna, Robust M-estimators of multivariate location and scatter, *Ann. Stat.*, 4, 51–67, 1976.
18. R.A. Render and H.F. Walker, Mixture densities, maximum likelihood and the EM algorithm, *SIAM Rev.*, 26(2), 195–293, 1984.
19. T. Hastie and R. Tibshirani, Discriminant analysis by Gaussian mixtures, *J. R. Stat. Soc.* B, 58(1), 155–176, January 1996.
20. S.J. Yakowitz and J.D. Sprangins, On the identifiability of finite mixtures, *Ann. Math. Stat.*, 39, 209–214, 1968.
21. S.J. Yakowitz, Unsupervised learning and the identification of finite mixtures, *IEEE Trans. Inf. Theory*, IT-16, 258–263, 1970.
22. S.J. Yakowitz, A consistent estimator for the identification of finite mixtures, *Ann. Math. Stat.*, 40, 1728–1735, 1968.
23. H.G.C. Traven, A neural network approach to statistical pattern classification by semiparametric estimation of probability density function, *IEEE Trans. Neural Networks*, 2(3), 366–377, 1991.
24. H. Amindavar and J.A. Ritchey, Pade approximations of probability functions, *IEEE Trans. Aerosp. Electron. Syst.*, AES-30, 416–424, 1994.
25. D.F. Specht, Probabilistic neural networks, *Neural Networks*, 3, 109–118, 1990.
26. T.Y. Young and G. Copaluppi, Stochastic estimation of a mixture of normal density functions using an information criterion, *IEEE Trans. Inf. Theory*, IT-16, 258–263, 1970.
27. J.C. Rajapakse, J.N. Gieldd, and J.L. Rapaport, Statistical approach to segmentation of single-channel cerebral MR images, *IEEE Trans. Med. Imaging*, 16(2), 176–186, 1997.
28. P. Schroeter, J.M. Vesin, T. Langenberger, and R. Meuli, Robust parameter estimation of intensity distributions for brain magnetic resonance images, *IEEE Trans. Med. Imaging*, 27(2), 172–186, 1998.
29. J.C. Rajapakse and F. Kruggel, Segmentation of MR images with intensity inhomogeneities, *Image Vision Comput.*, 16(3), 165–180, 1998.
30. S.G. Sanjay and T.J. Hebert, Bayesian pixel classification using spatially variant finite mixtures and the generalized EM algorithm, *IEEE Trans. Image Process.*, 7(7), 1024–1028, 1998.
31. H. Teicher, Identifiability of finite mixtures, *Ann. Stat.*, 34, 1265–1269, 1963.
32. A. P. Dempster, N.M. Laird, and D.B. Rubin, Maximum likelihood from incomplete data via the EM algorithm, *J.R. Stat. Soc. B*, 39, 1–38, 1977.
33. T.K. Moon, The expectation-maximization algorithm, *IEEE Signal Process. Mag.*, 16(2), 47–60, 1997.
34. C.E. Priebe, Adaptive mixtures, *J. Am. Stat. Assoc.*, 89, 796–806, 1994.

35. J. Diebolt and C. Robert, Estimation of finite mixture distributions through Bayesian sampling, *J. R. Stat. Soc. B*, 56, 363–375, 1994.
36. M. Escobar and M. West, Bayesian density estimation and inference using mixtures, *J. Am. Stat. Assoc.*, 90, 577–588, 1995.
37. D.M. Titterington, Some recent research in the analysis of mixture distribution, *Statistics*, 21, 619–640, 1990.
38. P. McKenzie and M. Alder, Initializing the EM algorithm for use in Gaussian mixture modelling, in *Pattern Recognition in Practice IV*, E.S. Gelsema and L.N. Kanal, Eds., Elsevier Science, New York, pp. 91–105, 1994.
39. A. Witkin, Scale space filtering, *Proc. Int. J. Conf. Artificial Intelligence*, IJCAI-83, 1019–1022, 1983.
40. M.J. Carlotto, Histogram analysis using a scale space approach, *IEEE Trans. Pattern Recognition Mach. Intelligence*, PAMI-9(1), 121–129, 1987.
41. A. Goshtasby and W.D. O'Neill, Curve fitting by a sum of Gaussians', *Graphical Models Image Process.*, 56(4), 281–288, 1994.
42. G. Celeux and J. Diebolt, The SEM algorithm: A probabilistic teacher algorithm derived from the EM algorithm for the mixture problem, *Computat. Stat. Q.*, 2, 35–52, 1986.
43. H.H. Bock, Probability models and hypotheses testing in partitioning cluster analysis, in *Clustering and Classification*, P. Arabie, L.J. Hubert, and G. DeSoete, Eds., pp. 377–453, World Scientific Publishers, Singapore, 1996.
44. J.L. Solka, W.L. Poston, E.J. Wegman, and B.C. Wallet, A new iterative adaptive mixture type estimator, *Proc. 28th Symp. Interface*, in press.
45. C.E. Priebe and D.M. Marchette, Adaptive mixtures: Recursive nonparametric pattern recognition, *Pattern Recognition*, 24, 1197–1209, 1991.
46. D.B. Cooper and P.W. Cooper, Nonsupervised adaptive signal detection and pattern recognition, *Inf. Control*, 7, 416–444, 1964.
47. B.S. Everitt, *Graphical Techniques for Multivariate Data*, Heinemann, London, 1978.
48. R.S. Bucy, Liner and non-linear filtering, *Proc. IEEE*, 58, 854–864, 1970.
49. D.G. Lainiotis, Partitioning: A unifying framework for adaptive systems I: Estimation, *Proc. IEEE*, 64, 1126–1143, 1976.
50. R.S. Bucy and K.D. Senne, Digital synthesis of non-linear filters, *Automatica*, 7, 287–298, 1971.
51. H.W. Sorenson and D.L. Alspach, Recursive Bayesian estimation using Gaussian sums, *Automatica*, 7, 465–479, 1971.
52. H.W. Sorenson and A.R. Stubberud, Nonlinear filtering by approximation of a-posteriori density, *Int. J. Control*, 18, 33–51, 1968.
53. T. Numera and A.R. Stubberud, Gaussian sum approximation for non-linear fixed point prediction, *Int. J. Control*, 38, 1047–1053, 1983.
54. D.L. Alspach, Gaussian sum approximations in nonlinear filtering and control, *Inf. Sci.*, 7, 271–290, 1974.
55. T.S. Rao and M. Yar, Linear and non-linear filters for linear, but non-Gaussian processes, *Int. J. Control*, 39, 235–246, 1983.
56. D. Lerro and Y. Bar-Shalom, Tracking with debiased consistent converted measurements versus EKF, *IEEE Trans Aerosp. Electron. Syst.*, 29(3), 1015–1022, 1993.
57. W.-I. Tam, K.N. Plataniotis, and D. Hatzinakos, An adaptive Gaussian sum algorithm for target tracking, *Signal Process.*, 77(1), 85–104, August 1999.
58. R. Vijayan and H.V. Poor, Nonlinear techniques for interference suppression in spread-spectrum systems, *IEEE Trans. Commun.*, COM-38, 1060–1065, 1990.
59. K.S. Vastola, Threshold detection in narrowband non-Gaussian noise, *IEEE Trans. Commun.*, COM-32, 134–139, 1984.
60. L.M. Garth and H.V. Poor, Narrowband interference suppression in impulsive environment, *IEEE Trans. Aerosp. Electron. Syst.*, AES-28, 15–33, 1992.

61. C.J. Masreliez, Approximate non-Gaussian filtering with linear state and observation relations, *IEEE Trans. Autom. Control*, AC-20, 107–110, 1975.
62. W.R. Wu and A. Kundu, Recursive filtering with non-Gaussian noises, *IEEE Trans. Signal Process.*, 44(4), 1454–1468, 1996.
63. W.R. Wu and P.P. Cheng, A nonlinear IMM algorithm for maneuvering target tracking, *IEEE Trans. Aerosp. Electron. Syst.*, AES-30, 875–885, 1994.
64. G.A. Hewer, R.D. Martin, and J. Zeh, Robust preprocessing for Kalman filtering of glint noise, *IEEE Trans. Aerosp. Electron. Syst.*, AES-23, 120–128, 1987.
65. Z.M. Durovic and B.D. Kovacevic, QQ-plot approach to robust Kalman filtering, *Int. J. of Control*, 61(4), 837–857, 1994.
66. A.R. Webb, Functional approximation by feed-forward networks: A least squares approach to generalization, *IEEE Trans. Neural Networks*, 5, 363–371, 1994.
67. J. Mooddy and C.J. Darken, Fast learning in networks of locally-tuned processing units, *Neural Computat.*, 1, 281–294, 1989.
68. L. Jin, M.M. Gupta, and P.N. Nikiforuk, Neural networks and fuzzy basis functions for functional approximation, in *Fuzzy Logic and Intelligent Systems*, H. Li and M.M. Gupta, Eds., Kluwer Academic Publishers, Dordrecht, 1996.
69. T. Poggio and F. Girosi, Networks for approximation and learning, *Proc. IEEE*, 78, 1481–1497, 1990.
70. D.A. Cohn, Z. Ghrahramani, and M.I. Jordan, Active learning with statistical models, *J. Artificial Intelligence Res.*, 4, 129–145, 1996.
71. M.I. Jordan and C.M. Bishop, *Neural Networks*, A.I. Memo No. 1562, Massachusetts Institute of Technology, Boston, 1996.
72. B. Mulgrew, Applying radial basis functions, *IEEE Signal Process. Mag.*, 13(2), 50–65, 1996.
73. H.M. Kim and J.M. Mendel, Fuzzy basis functions: Comparisons with other basis functions, *IEEE Trans. Fuzzy Syst.*, 3, 158–168, 1995.
74. A. Jepson and M. Black, Mixture Models for Image Representation, Technical Report ARK96-PUB-54, Department of Computer Science, University of Toronto, Canada, March 1996.
75. P. Kontkane, P. Myllymaki, and H. Tirri, Predictive data mining with finite mixtures, in *Proceedings of the 2nd International Conference on Knowledge Discovery and Data Mining*, IEEE Computer Society Press, Portland, OR, pp. 176–182, 1996.
76. I. Caballero, C.J. Pantaleon-Prieto, and A. Artes-Rodriguez, Sparce deconvolution using adaptive mixed-Gaussian models, *Signal Process.*, 54, 161–172, 1996.
77. Y. Zhao, X. Zhuang, and S.J. Ting, Gaussian mixture density modeling of non-Gaussian source for autoregressive process, *IEEE Trans. Signal Process.*, 43(4), 894–903, 1995.
78. J. Park and I.W. Sandberg, Universal approximation using radial-basis function networks, *Neural Computat.*, 3, 246–257, 1991.
79. I. Cha and S.A. Kassam, Gaussian-mixture basis function networks for nonlinear signal processing, *Proc. 1995 Workshop Nonlinear Signal Process.*, 1, 44–47, 1995.
80. I. Cha and S.A. Kassam, RBNF restoration of nonlinear degraded images, *IEEE Trans Image Process.*, 5(6), 964–975, 1996.
81. R.A. Render, R.J. Hathaway, and J.C. Bezdeck, Estimating the parameters of mixture models with modal estimators, *Commun. Stat. Part A: Theory and Methods*, 16, 2639–2660, 1987.

4

Matched Field Processing — A Blind System Identification Technique

N. Ross Chapman
University of Victoria

Reza M. Dizaji
University of Victoria

R. Lynn Kirlin
University of Victoria

4.1 Introduction

In underwater acoustics, there has been an intensive research effort over the past 20 years to develop model based signal processing methods[1–3] and system-theoretical approaches[4] for use in advanced sonar design. One of the techniques, known as matched field processing (MFP), has gained widespread use. MFP was described in the underwater acoustics literature initially as a generalized beamforming method for source localization with an array of sensors.[1–3] More recently, MFP has been applied as an inversion method to estimate either the source location or the environmental parameters of the ocean waveguide from measurements of the acoustic field.[5–12] The technique has been remarkably successful, and there is now extensive literature on various applications. Readers can refer to the review paper by Baggeroer et al.[6] and the monograph by Tolstoy[9] for information on various matched field (MF) processors that are in use. Applications for inversion of source location and waveguide model parameters are addressed in recent special issues of the *IEEE Journal of Oceanic Engineering*[13] and *Journal of Computational Acoustics*.[14]

MFP can be considered as a sub-category of a more general approach known as blind system identification (BSI). Blind system identification is a fundamental signal processing technique for estimating both unknown source and unknown system parameters when only the system output data are known.[15]

0-8493-3691-0/01/$0.00+$.50

The technique has widespread application in a number of different areas, such as speech recognition, cancellation of reverberation, image restoration, data communication, and seismic and underwater acoustic signal processing. In many instances, especially in sonar and seismic applications, the transfer function, or signal propagation model, is nearly known, and the desired result is not the transfer function itself, but the unknown parameters of the signal propagation model. For example, in underwater acoustics numerical methods based on ray theory,[16] wave-number integration,[17] parabolic equation,[18,19] and normal modes[20,21] are available for calculating the acoustic field in an arbitrary waveguide to very high accuracy; the task is instead to find the unknown parameters of the waveguide by modeling the acoustic field.

In this chapter, we focus on the class of problems for which there is some information about the signal propagation model. From the basic formalism of BSI, we derive methods that can be used to determine the unknown parameters of the transfer function. We show that the widely used Bartlett family of MF processors can be obtained from this formalism. We then introduce a cross-relation (CR) MFP technique and demonstrate its performance for estimating the source location and the environmental parameters of a shallow water waveguide. The source is assumed to be either broadband or narrowband random noise. However, estimation formulas are derived for deterministic, non-stationary (NS), and wide-sense stationary (WSS) random sources. For the NS case, two formulations are proposed, one of which is based on an evolutive spectrum concept that obtains the advantages of time-frequency analysis. For each formulation, two estimation methods are proposed, based on a self-CR and a cross-CR processor (defined according to the specific output channel signal that is used to construct the estimator). All the preceding formulations are derived as second-order MF processors. We extend the CR concept to higher order and discuss the application to real data.

4.2 Blind System Identification

4.2.1 Basic Concept and Formulation

The model for a multi-channel, single input multiple output (SIMO) system is shown in Figure 4.1.

When the system is linear and time invariant, the system output for the ith channel, $x_i(t)$, is given by

$$x_i(t) = y_i(t) + w_i(t) \tag{4.1}$$

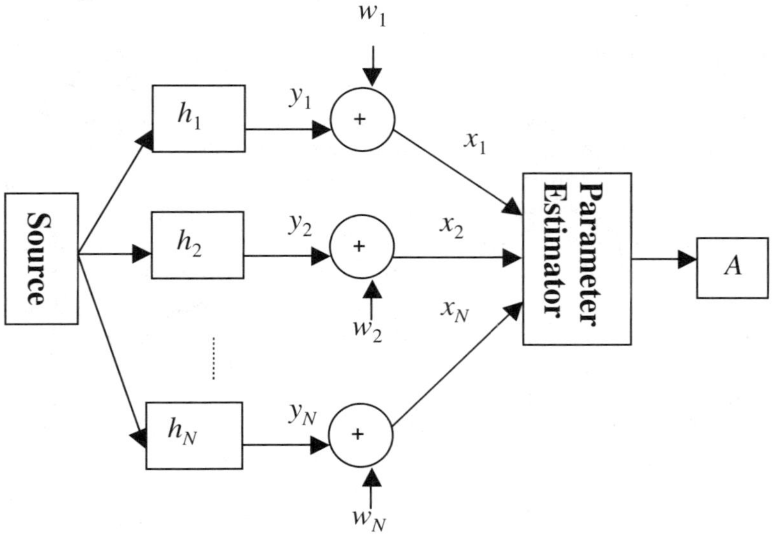

FIGURE 4.1 The model for a SIMO system.

where $y_i(t) = s(t) * h_i(t)$. Here, $s(t)$ is the unknown input source, $h_i(t)$ is the transfer function for the ith channel, and $w_i(t)$ is additive noise. For underwater acoustics, the transfer function $h_i(t)$ corresponds to the paths traveled by acoustic waves from the source to the ith sensor in an array.

The analysis is convenient in the frequency domain using a matrix form in which all channel quantities are stacked into a single vector. Thus, the measured signal vector representing the system output is given by

$$X = [X_{f1} X_{f2} \ldots X_{fM}] \tag{4.2}$$

with $X_{f_i} = [x_1(f_i)\ x_2(f_i) \ldots\ x_N(f_i)]^H$ and $(.)^H$ representing the Hermitian operator. Based on the system block diagram in Figure 4.1,

$$X = Y + W \tag{4.3}$$

where Y is the received signal vector given by

$$Y = HD_u \tag{4.4}$$

and W is additive noise that is assumed to be spatially and spectrally independent, zero-mean, Gaussian noise having a covariance matrix for each spectral component

$$C_{W_{f_i}}^{N \times N} = \sigma^2 I^{N \times N}, i = 1, 2, \ldots, N \tag{4.5}$$

D_u is the signal input vector given by

$$D_u = diag(u_{f1} u_{f2} \ldots u_{fM}) \tag{4.6}$$

The source generates either a deterministic or a random signal that can be narrowband or broadband with M frequency components f_p, $p = 1, \ldots, M$.

H is a generalized Sylvester matrix[22]

$$H = [H_{f1:A} H_{f2:A} \ \ldots\ H_{fM:A}] \tag{4.7}$$

where $H_{f_{1:A}} = [H_1(f_i;A) H_2(f_i;A)\ \ldots\ H_N(f_i;A)]^H$ whose components $H_j(f;A), j = 1, \ldots, N$ are the Fourier transforms of the transfer functions h_j, $j = 1, 2, \ldots, N$. The vector A corresponds to the set of unknown channel parameters, for instance, the set of source location or waveguide model parameters. The columns of matrix H are linearly independent, since each of them corresponds to a distinct frequency.

4.2.2 Identifiability

A system is considered to be completely identifiable if all unknown system parameters can be determined uniquely from output signals. However, from Equations 4.3 and 4.4, it is clear that a measured output X can only imply an input source D_u or a system transfer function H to within an unknown scalar constant. Identifiability conditions have been studied in detail by Hua and Wax[23] and further by Abed-Meraim et al.[24]

The necessary and sufficient conditions for identifiability ensure the following intuitive requirements for SIMO systems:

1. All channels in the system must be sufficiently different from each other. For example, measured outputs cannot be the same for any two channels.
2. The number of channels and size of the finite impulse response (FIR) transfer functions are known *a priori*.

3. There must be a sufficient number of output data samples; the number of data samples cannot be less than the number of unknown system parameters.
4. The input source must be sufficiently complex. For example, it cannot be zero, a constant, or a single sinusoidal.

The last three assumptions are based on the condition that an FIR structure is used. In practice, especially in underwater acoustic approaches, we often know the sinusoidal model of signal propagation, and the task is to estimate unknown parameters of the model. For the specific problem of estimating unknown parameters of the transfer function, the question of identifiability is somewhat difficult. In this case, the necessary conditions are satisfied, but the sufficient conditions are generally not satisfied.

There are two powerful techniques in widespread use for identifying the system transfer function in Equation 4.7.[15] The first is the classic maximum likelihood method that is applicable to any estimation problem for which the probability density function of the available data is known. The second method, the channel subspace method, is easily adapted to the problem of estimating unknown parameters of the transfer function.

4.2.3 Bartlett Matched Field Processor Family

The Bartlett MF processor family is the most widely used for source localization and environmental parameter estimation. Some well-known MF processors like minimum variance, multiple constraint, and matched mode processors are members of this family.[6,9]

From Equation 4.3, the covariance matrix of the measured data (for deterministic signals) is expressed as

$$\begin{aligned} C_X &= C_Y + C_W = HD_u D_u^H H^H + C_W = HD_{|u|^2} H^H + C_W \\ D_{|u|^2} &= diag(|u_{f1}|^2, |u_{f2}|^2, \ldots, |u_{fM}|^2) \end{aligned} \tag{4.8}$$

With a random source, Equation 4.8 becomes

$$\begin{aligned} C_X &= C_Y + C_W = HE(D_u D_u^H) H^H + C_W = HD_{S_u} H^H + C_W \\ D_{S_u} &= diag(S_u(f_1), S_u(f_2), \ldots, S_u(f_M)) \end{aligned} \tag{4.9}$$

where $S_u(f_i)$, $i = 1, 2, \ldots, M$ is the power spectral density of the random source at the ith frequency. Equations 4.8 and 4.9 represent a relationship between the signal and noise subspaces of C_X and the column vectors of H based on the following theorem.

Theorem

There is a linear relationship between the eigenvectors of the received signal covariance matrix C_Y (which spans the signal subspace of C_X) and the non-zero columns of the transfer function matrix H if, and only if, $D_{|u|^2}$ is full rank.

Proof[25]

Let the singular value decomposition (SVD) of C_Y be given by

$$C_Y = U_Y \Lambda_Y U_Y^H = HD_{S_u} H^H \tag{4.10}$$

where U_Y and Λ_Y are matrices of eigenvectors and eignvalues of the covariance matrix of C_Y, respectively. Now, define K as

$$K = \Lambda_Y^{1/2} \Omega^H D_{S_u}^{-1/2} \tag{4.11}$$

where Ω can be any $M \times M$ unitary matrix and $D_{|u|^2}$ is full rank. Substituting Equation 4.11 in Equation 4.10 we obtain the following equation:

$$\begin{aligned} C_Y &= U_Y \Lambda_Y U_Y^H = U_Y \Lambda_Y^{1/2} (\Lambda_Y^{1/2})^H U_Y^H = U_Y K D_{S_u}^{1/2} \Omega \Omega^H (D_{S_n}^{1/2})^H K^H U_Y^H \\ &= U_Y K D_{S_u} K^H U_Y^H = H D_{S_u} H^H \end{aligned} \tag{4.12}$$

Equation 4.12 implies the following linear relationship between the eigenvectors of C_Y (which span the signal subspace of C_X) and the columns of the transfer function matrix H:

$$U_Y K = H \tag{4.13}$$

The above theorem implies that

1. $D_{|u|^2}$ must be full rank.
2. Since U_Y is not a null matrix, then $H_i(f)$, $i = 1, 2, \ldots, N$ should not share common zeros at all frequencies.

The uniqueness conditions for random sources are the same as the conditions for deterministic sources, except that we have an expectation operator rather than a deterministic measure. For random sources, the theorem implies that D_{S_u} should be full rank, so the power spectral density (PSD) of the source must be non-zero over the frequency band.

Using the theorem, we will describe two formulations based on the multiple signal classification (MUSIC) concept[26] to estimate the unknown transfer functions. These formulations are categorized as channel subspace (CS) methods in the BSI literature. The theorem provides both necessary and sufficient conditions for estimation of transfer functions. However, for source localization and environmental parameter estimation from field data, the theorem only provides the necessary conditions for the existence of solutions for the parameters.

In the first formulation we consider the fact that the space spanned by the columns of the matrix H is orthogonal to the null subspace of C_X. Therefore, we have

$$\hat{H}_{f_i;A}^{MUSIC} = \arg \min_{\|H_{f_i;A}\| = 1} \left\| H_{f_i;A}^H E_n \right\|^2, \; i = 1, 2, \ldots, M \tag{4.14}$$

where E_n is a matrix with columns consisting of eigenvectors of the noise subspace of C_X. The reason for using C_X instead of C_Y is that in reality there is no way to measure C_Y. C_X is full rank with ordered eigenvalues as below:

$$\begin{gathered} \lambda_{s_1} \geq \lambda_{s_2} \geq \ldots, \lambda_{s_T} \geq \lambda_{n_1} \geq \lambda_{n_2} = \ldots = \lambda_{n_{(N-T)}} = \lambda \\ T\text{: \# of sources} \end{gathered} \tag{4.15}$$

Replacing C_Y with C_X gives an acceptable estimation of the transfer function if the ratio λ_{s_i}/λ is large enough for all $1 \leq i \leq T$.

In Equation 4.14 it is assumed that the minimizing procedure results in the global minimum point. In the MFP literature, the inverse of Equation 4.14 is known as the eigenvector processor.[27] Tolstoy[9] has commented that the eigenvector processor is not appropriate for estimating source or spectral intensities since it is seeking the zeros of Equation 4.14, whereas we have shown theoretically that the processor gives us a unique answer for transfer functions if the conditions of the theorem are satisfied. The MUSIC algorithm has been widely used in source localization for horizontal arrays, and there is a large body of work published on this technique, most of which has assumed plane wave fields and mutually uncorrelated sources. The plane wave assumption is to assure that the sufficiency conditions are satisfied for the uniqueness of the source localization. The approximate orthogonal processor (AOP)[2] is another processor using the same concept as MUSIC. Fizell[2] has commented that the disadvantages of the AOP include its high false alarm rate and its instability in the presence of noise. The instability comes from the inability to truly separate the signal subspace from the noise subspace when the signal-to-noise ratio (SNR) is low. In this case, we can use some model order estimators such as the Akaike[28] or the minimum description length (MDL)[29] to overcome this problem.

In the second formulation the projection of the column vectors of H onto the signal subspace of C_X is maximized. Thus, we have

$$\hat{H}_{f_i;A}^{MUSIC} = \underset{\|H_{f_i;A}\| = 1}{\arg\max} \left\| H_{f_i;A}^H E_y \right\|^2, \; i = 1, 2, \ldots, M \tag{4.16}$$

where E_y is a matrix whose columns are the eigenvectors of the null subspace of C_X.

The formulation in Equation 4.16 is applicable to any vector located in the signal subspace of C_X, and, thus, includes the received signal vector, i.e., Y_{f_i} $i = 1, 2, \ldots, M$, so that we may write

$$\hat{H}_{f_i;A}^{Bartlett} = \underset{\|H_{f_i;A}\| = 1}{\arg\max} \left| H_{f_i;A}^H Y_{f_i} \right|^2, \; i = 1, \ldots, M \tag{4.17}$$

where Y_{f_i} is the received signal vector at the frequency f_i.

Equation 4.17 is known as the narrowband Bartlett MF processor, from which the source location or environmental parameters of the transfer functions are estimated. The formulation can be extended to multiple frequencies in either the coherent form,

$$\hat{H}_{A}^{Bartlett} = \underset{\|H_{f_i;A}\| = 1}{\arg\max} \left| \sum_{f_i} H_{f_i;A}^H Y_{f_i} \right|^2, \; i = 1, \ldots, M \tag{4.18}$$

or the incoherent form,

$$\hat{H}_{A}^{Bartlett} = \underset{\|H_{f_i;A}\| = 1}{\arg\max} \sum_{f_i} \left| H_{f_i;A}^H Y_{f_i} \right|^2, \; i = 1, \ldots, M \tag{4.19}$$

For random sources, Equations 4.18 and 4.19 become

$$\hat{H}_{A}^{Bartlett} = \underset{\|H_{f_i;A}\| = 1}{\arg\max} \sum_{f_i} \sum_{f_j} (H_{f_i;A}^H E(Y_{f_i} Y_{f_j}) H_{f_j;A}) = \underset{\|H_{f_i}\| = 1}{\arg\max} \sum_{f_i} \sum_{f_j} (H_{f_i;A}^H C_{Y_{f_i,f_j}} H_{f_j;A}), \; i, j = 1, 2, \ldots, M \tag{4.20}$$

for the coherent form of the Bartlett processor and

$$\hat{H}_{A}^{Bartlett} = \underset{\|H_{f_i;A}\| = 1}{\arg\max} \sum_{f_i} (H_{f_i;A}^H E(Y_{f_i} Y_{f_i}^H) H_{f_i;A}) = \underset{\|H_{f_i}\| = 1}{\arg\max} \sum_{f_i} (H_{f_i;A}^H C_{Y_{f_i}} H_{f_i;A}), \; i = 1, 2, \ldots, M \tag{4.21}$$

for the incoherent form, where

$$C_{Y_{f_i}} = [S_{y_p, y_q}(f_i)]_{p, q = 1, \ldots, N} \text{ and } C_{Y_{f_i, f_j}} = [S_{y_p, y_q}(f_i, f_j)]_{p, q = 1, \ldots, N} \tag{4.22}$$

$S_{y_p, y_q}(f_i)$ and $S_{y_p, y_q}(f_i, f_j)$ are the PSD and cross PSD of $y_p(f_i)$ and $y_q(f_i)$, respectively. In practice, $C_{Y_{f_i}}$ and $C_{Y_{f_i, f_j}}$ $i, j = 1, 2, \ldots, M$ are obtained using the periodogram technique.[30]

As mentioned before, we have in practice no access to the received signals, so use of the measured signals is inevitable. This suggests a sub-optimum formulation. In this case, assuming that the SNR is high enough, we find the maximum likelihood measure of the estimated transfer functions to be

$$\hat{H}_{f_i;A}^{Bartlett} \approx \underset{\|H_{f_i;A}\|=1}{\arg\max}(H_{f_i;A}^H C_{X_{f_i}} H_{f_i;A}),\ i = 1, \ldots, M \tag{4.23}$$

where

$$C_{X_{f_i}} = E(X_{f_i} X_{f_i}^H)$$

and X_{f_i} is the measured signal vector at the frequency f_i.

The non-coherent form of the processor for a broadband signal is

$$\hat{H}_{f_i;A}^{Bartlett} \approx \underset{\|H_{f_i;A}\|=1,\, i=1,\ldots,M}{\arg\max} \sum_{i=1}^{M} (H_{f_i;A}^H C_{X_{f_i}} H_{f_i;A}),\ i = 1, \ldots, M \tag{4.24}$$

Although the processor in Equation 4.24 is a significant advance over conventional plane wave processors, it is not perfect because of its sidelobes and lack of resolution at the source location. However, this kind of processor (linear processor) has the least sensitivity to the mismatch between the waveguide model and the real environment.

In an effort to improve the Bartlett processor performance, the minimum variance (MV) processor has been developed. It has been designed to be optimum in the sense that the output noise power is minimized subject to the constraint that the signal be undistorted by the filter. The processor is defined as

$$\hat{H}_{f_i;A}^{MV} = \underset{\|H_{f_i;A}\|=1}{\arg\max} \frac{1}{(H_{f_i;A}^H C_{X_{f_i}}^{-1} H_{f_i;A})},\ i = 1, \ldots M \tag{4.25}$$

There are two possible non-coherent formulations for wideband MV processors. The first one sums the denominator terms for different frequencies, giving the reciprocal of sums

$$\hat{H}_A^{MV} = \underset{\|H_{f_i;A}\|=1,\, i=1,2,\ldots,M}{\arg\max} \frac{1}{\sum_{i=1}^{M} (H_{f_i;A}^H C_{X_{f_i}}^{-1} H_{f_i;A})} \tag{4.26}$$

The other formulation adds a term for each frequency, giving the sum of reciprocals

$$\hat{H}_A^{MV} = \underset{\|H_{f_i;A}\|=1,\, i=1,2,\ldots,M}{\arg\max} \sum_{i=1}^{M} \frac{1}{(H_{f_i;A}^H C_{X_{f_i}}^{-1} H_{f_i;A})} \tag{4.27}$$

The reciprocal of the sum formulation (Equation 4.26) is closer to the non-coherent concept than the sum of reciprocals (Equation 4.27). The reason is that in the reciprocal sum formulation the maximum point is obtained only if all denominator values corresponding to different frequencies have small values, while in the sum of reciprocal formulation the maximum point can be obtained if only one denominator term (corresponding to one frequency) is small. We would expect the reciprocal of the sum formulation to be more stable and accurate.

The covariance matrix of the received signal should be full rank, given that all conditions mentioned for the Bartlett processor are satisfied. Sometimes it may be necessary to diagonally load the matrix, i.e., add some small quantity to the diagonal. The performance of the MV processor degrades rapidly in the presence of errors in the model estimates of the field as well as under mismatch conditions. This sensitivity requires that quantitative knowledge of the environmental parameters must be extraordinarily accurate. In addition, the propagation model used must also be highly accurate, a difficult requirement if range and depth change rapidly in space. Finally, the MV performance mimics that of the Bartlett processor in low signal-to-noise conditions if noise is temporally and spatially white Gaussian.

In order to overcome the high sensitivity of the MV processor, Schmidt et al.[31] have introduced a multiple constraint (MC) processor. The principle behind the approach is to design a neighborhood response rather than a single point response, e.g., near the precise source range and depth, for which the signal is passed without distortion. The derivation of the MC processor is very similar to that of the MV processor, except that the constraint condition which optimized the response only at a single point is extended by a system of constraints imposed at L neighborhood points. For the vector d with L entries corresponding to constraints at L points, the processor is given by

$$\hat{H}_{f_i;A}^{MC} = \underset{\|H_{f_i;A}\|=1}{\arg\max} \frac{d^H (H_{f_i;A}^H C_{X_{f_i}}^{-1} H_{f_i;A})^{-1} d}{d^H (H_{f_i;A}^H H_{f_i;A})^{-1} d}, \; i = 1, \ldots, M \tag{4.28}$$

For wideband MC processors, the non-coherent formulation is given by

$$\hat{H}_A^{MC} = \underset{\|H_{f_i;A}\|=1,\, i=1,2,\ldots,M}{\arg\max} \sum_{i=1}^{M} \frac{d^H (H_{f_i;A}^H C_{X_{f_i}}^{-1} H_{f_i;A})^{-1} d}{d^H (H_{f_i;A}^H H_{f_i;A})^{-1} d} \tag{4.29}$$

Schmidt el al. suggest that the number of constraints be $L = 2N_{dim} + 1$, where N_{dim} is the number of dimensions or parameters in the problem.[31]

The matched mode processor (MMP) proposed by Shang[32] and Yang[33] operates in modal space (recall the normal mode solution of wave propagation) in contrast to the Bartlett processor that operates in field space, using the signals recorded by hydrophones. The key advantage of MMP is that prior to processing the data can be filtered to eliminate modes which degrade the localization, e.g., poorly modeled or noise-dominated modes. The technique requires that the number of hydrophones N be greater than or equal to the number of effective modes L at the array range. The non-coherent, wideband MMP is given by

$$\hat{H}_A^{MP} = \underset{\|H_{f_i;A}\|=1,\, i=1,2,\ldots,M}{\arg\max} \sum_{i=1}^{M} \left| \sum_{l=1}^{L} \hat{a}_l^* a_l \right|^2 \tag{4.30}$$

where $\hat{a}_l$ is the lth modal excitation inferred from the data; that is,

$$X_k(f_i) = \sum_{l=1}^{L} a_l(f_i) \Psi_l(z_k, f_i), \; i = 1, 2, \ldots, M; \; k = 1, 2, \ldots, N$$

and $\hat{a}_l$ is the model prediction for the lth modal excitation; that is,

$$H_k(f_i) = \sum_{l=1}^{L} \hat{a}_l(f_i) \Psi_l(z_k, f_i), \; i = 1, 2, \ldots, M; \; k = 1, 2, \ldots, N$$

where Ψ_l, $l = 1, \ldots, L$ are excited modes, and M is the number of frequency components. The performance of the processor is highly dependent upon the accuracy with which the mode excitation a_l can be inferred from the data, particularly for finite aperture vertical arrays that discretely sample the field. MMP and Bartlett processors are equivalent if the vertical array fully samples the effective modes composing the field at the array range.

4.3 Cross-Relation Matched Field Processor

This section introduces the CR based MF processor technique for estimating the source location and environmental parameters in shallow water. The source is assumed to be either broadband or narrowband random noise. However, the processor can be applicable for broadband and narrowband deterministic sources. The estimation formulas are derived for deterministic and random sources including NS, and WSS random sources. For the NS case, two formulations are proposed: one is based on a time-varying correlation function, and another is based on an evolutive spectrum concept to obtain the advantages of time-frequency analysis. For each estimation formula, two estimation methods are proposed: one is the self-CR, and the other is a cross-CR named according to which channel output signal is used to construct the estimator. All the above formulations derive a second-order MF processor.

We extend the second-order CR concept to introduce higher order MF processors. The higher order characteristic of these processors provides the ability of canceling the effect of Gaussian random sources (either white or non-white) since the third and some higher order moments of Gaussian random signals are zero.

4.3.1 Cross-Relation Concept

Let us consider the geometry of the measurement system in shallow water using a vertical linear array with N sensors. A schematic diagram to demonstrate the CR concept for each pair of sensors is shown in Figure 4.2. The cross-relation in Equation 4.31 between the transfer function and the measured signal for any pair of sensors follows from the linearity of the transfer functions.

$$\begin{cases} y_p(n) = h_p(n;\alpha)*S(n) \\ y_q(n) = h_q(n;\alpha)*S(n) \end{cases} \Rightarrow h_p(n;\alpha)*y_q(n;\alpha)*y_p(n) \qquad (4.31)$$

$$p, q = 1, 2, \ldots, N;\ p \neq q,\ n = 1, 2, \ldots, L$$

Equation 4.31 shows that the outputs of each channel pair are related by their channel responses. It gives a relationship that allows, under certain identifiability conditions, identification of multi-channel systems based on only channel outputs. Consequently, the method can be classified as a blind identification technique.

4.3.2 Deterministic Sources

For the deterministic source, Equation 4.31 can be rewritten in the following matrix form to solve for all channel responses simultaneously.

$$Yh = 0 \qquad (4.32)$$

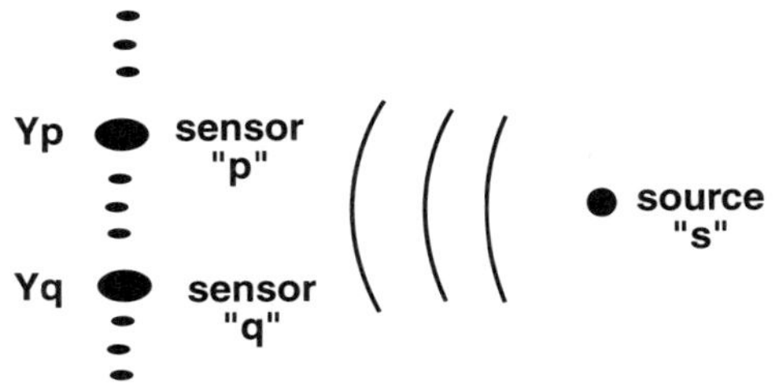

FIGURE 4.2 A scheme to demonstrate the CR concept for each pair of sensors.

where

$$h = [h_1^T, h_2^T, \ldots, h_N^T]^T, h_i = [h_i(L-1), \ldots h_i(0)]^T, i = 1, 2, \ldots, N$$

$$Y = \Big[\underbrace{Y_{1,2}, \ldots, Y_{p,q;p\neq q}^T, \ldots, Y_{N,N-1}^T}_{\frac{N(N-1)}{2} blocks}\Big]^T$$

$$Y_{p,q,p\neq q}^{\frac{M}{L}\times NL} = \Big[\underbrace{0}_{\frac{M}{L}\times(p-1)L} \quad \underbrace{Y_p}_{\frac{M}{L}\times L} \quad \underbrace{0}_{\frac{M}{L}\times(q-p-1)L} \quad \underbrace{-Y_q}_{\frac{M}{L}\times L} \quad 0\Big]$$

$$Y_p = \begin{bmatrix} y_p(0) & \ldots & y_p(L-1) \\ \ldots & \ldots & \ldots \\ y_p(M-L-1) & \ldots & y_p(M-1) \end{bmatrix}$$

The above multi-channel system can be identified uniquely (up to a scalar) if, and only if, the null space dimension of the data matrix Y is one. This is a general necessary and sufficient condition for channel identification. Xu et al.[22] have given theorems and explicit expressions that provide insight into the characteristics of the channels and input signal. The proof is given in the time domain; however, it is easier to prove it in the frequency domain. We begin by writing Equation 4.31 in the frequency domain:

$$\begin{cases} Y_p(f_i) = H_p(f_i;A)S_{f_i} \\ Y_q(f_i) = H_q(f_i;A)S_{f_i} \end{cases} \Rightarrow H_p(f_i;A)Y_q(f_i) = H_q(f_i;A)Y_p(f_i)$$
$$p, q = 1, 2, \ldots, N \quad p \neq q;\ i = 1, \ldots, L \tag{4.33}$$

Rewriting Equation 4.33 for all channels, we have

$$Y_F H_F = 0^{L\times L}$$
$$Y_F^{2Lr\times L} = \Big[\underbrace{Y_{1,2}^T, \ldots, Y_{p,q,p\neq q}^T, \ldots, Y_{N,N-1}^T}_{r = \frac{N(N-1)}{2}}\Big]^T, H_F^{2Lr\times L} = \Big[\underbrace{H_{1,2}, \ldots, H_{p,q,p\neq q}, \ldots, H_{N,N-1}}_{r = \frac{N(N-1)}{2}}\Big]^T \tag{4.34}$$
$$p = 1, 2, \ldots, N \quad q = p+1, \ldots, N$$

where

$$Y_{p,q}^{L\times 2L} = \begin{bmatrix} Y_p(f_1) & -Y_q(f_1) & 0 & 0 & \ldots & 0 & 0 \\ 0 & 0 & Y_p(f_2) & -Y_q(f_2) & \ldots & 0 & 0 \\ 0 & 0 & 0 & 0 & \ldots & 0 & 0 \\ 0 & 0 & 0 & 0 & \ldots & 0 & 0 \\ 0 & 0 & 0 & 0 & \ldots & Y_p(f_L) & -Y_q(f_L) \end{bmatrix}$$

$$H_{p,q}^{L\times 2L} = \begin{bmatrix} H_q(f_1, A) & H_p(f_1, A) & 0 & 0 & \ldots & 0 & 0 \\ 0 & 0 & H_q(f_2, A) & H_p(f_2, A) & \ldots & 0 & 0 \\ 0 & 0 & 0 & 0 & \ldots & 0 & 0 \\ 0 & 0 & 0 & 0 & \ldots & 0 & 0 \\ 0 & 0 & 0 & 0 & \ldots & H_q(f_L, A) & H_p(f_L, A) \end{bmatrix}$$

We first state the following lemma.

Lemma

The matrix H_F is full column rank if, and only if, for all the frequency bands f_i, $i = 1, 2, \ldots, L$ the transfer functions $H_p(f_i;A)$, $p = 1, \ldots, N$ are not zero.

Proof

The proof of lemma is straightforward since $H_p(f_i;A)$, $p = 1, \ldots, N$ are columns of H_F. Now, we state the following theorem that gives the identifiability condition.

Theorem

The multi-channel system (in the frequency domain) is uniquely identified up to a scalar if, and only if, the null space dimension of Y_F is L while H_F is assumed to be full column rank (i.e., the lemma is satisfied).

Proof

The proof of the theorem is straightforward and can be found in Reference 25.

In order to have the null space dimension of Y_F equal to L, assuming the condition in the lemma is satisfied, S_{f_i}, $i = 1, 2, \ldots, L$ should be non-zero (see Equation 4.31). This implies that the source signal should be sufficiently complex, another form of condition 2 given in Reference 22.

Equation 4.34 describes the noise-free case. For the case where channels are corrupted by noise, the following MF processor is proposed based on the least square criterion:

$$\hat{H}_F^{CR} = \underset{\|H_p(f;A)\| = 1,\, p = 1, \ldots, N}{\arg\min} E(\|X_F H_F\|^2) \tag{4.35}$$

X_F is the matrix of the measured signal with dimensions $2Lr \times L$ where $r = (N(N-1))/2$.

$$X_F = \left[\underbrace{X_{1,2}^T, \ldots, X_{p,q,p \neq q}^T, \ldots, X_{N,N-1}^T}_{r = \frac{N(N-1)}{2}} \right]^T, p = 1, 2, \ldots, N \quad q = p+1, \ldots, N$$

$$X_{p,q} = \begin{bmatrix} X_p(f_1) & -X_q(f_1) & 0 & 0 & \cdots & 0 & 0 \\ 0 & 0 & X_p(f_2) & -X_q(f_2) & \cdots & 0 & 0 \\ 0 & 0 & 0 & 0 & \cdots & 0 & 0 \\ 0 & 0 & 0 & 0 & \cdots & 0 & 0 \\ 0 & 0 & 0 & 0 & \cdots & X_p(f_L) & -X_q(f_L) \end{bmatrix}_{L \times 2L}$$

In MF inversion, ensuring identifiability conditions for the transfer functions h does not guarantee a unique solution. The sufficiency conditions are highly dependent on the particular model for the ocean waveguide and the specific parameters. Since the ocean environment is very complicated and there is generally no exact analytical relationship between the model parameters and the acoustic field, ocean parameters cannot be fully interpreted by the model, and it is not practical to analytically determine the sufficiency. In order to satisfy the sufficiency, we should carefully set the bounds of parameter variations. The bounds are usually obtained from complementary information provided by other sources such as ground truth data.

4.3.3 Non-Stationary Random Sources

Let us multiply both sides of Equation 4.31 by the conjugate of $y_p(n_1)$ or $y_q(n_1)$ to produce the self-CR equation or by $y_k(n_1)$, $k = 1, \ldots, N$, $k \neq p, q$ to produce the cross-CR equation, and then apply the expectation operator to this product. Here, n_1 is the time index, independent of n. A scheme to demonstrate the cross-CR technique is shown in Figure 4.3. The results are given in Equation 4.36 for the self-CR estimator and Equation 4.37 for the cross-CR estimator.

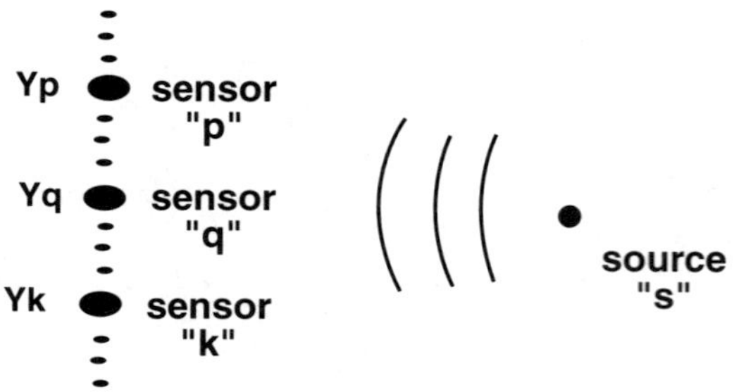

FIGURE 4.3 A scheme to demonstrate the cross-CR technique.

$$\begin{aligned} h_p(n;\alpha)*E(y_q(n)y_p^*(n_1)) &= h_q(n;\alpha)*E(y_p(n)y_p^*(n_1)) \\ h_p(n;\alpha)*R_{y_q,y_p}(n,n_1) &= h_q(n;\alpha)*R_{y_p}(n,n_1) \\ p,q &= 1,2,\ldots,N; p\neq q \end{aligned} \tag{4.36}$$

$$\begin{aligned} h_p(n;A)*E(y_q(n)y_k^*(n_1)) &= h_q(n;\alpha)*E(y_p(n)y_k^*(n_1)) \\ h_p(n;\alpha)*R_{y_q,y_p}(n,n_1) &= h_q(n;\alpha)*R_{y_p,y_k}(n,n_1) \\ p,q &= 1,2,\ldots,N;\ k = 1,\ldots,N,\ k\neq p,q \end{aligned} \tag{4.37}$$

For an NS source, Equations 4.36 and 4.37 can be rewritten in the following matrix form to solve for all channel responses simultaneously.

$$R_{y,self\text{-}CR}(n_1)h = 0,\ R_{y,cross\text{-}CR}(n_1)h = 0 \tag{4.38}$$

where

$$h = [h_1^T, h_2^T, \ldots, h_N^T]^T,\ h_i = [h_i(L-1), \ldots, h_i(0)]^T,\ i = 1,2,\ldots,N$$

$$R_{y,\,self\text{-}CR}(n_1) = \left[\underbrace{R_{1,2}^T(n_1), \ldots, R_{p,q;p\neq q}^T(n_1), \ldots, R_{N,N-1}^T(n_1)}_{\frac{N(N-1)}{2}blocks}\right]$$

$$R_{y,\,cross\text{-}CR}(n_1) = \left[\underbrace{R_{1,2,3}^T(n_1), \ldots, R_{p,q,k;k\neq p,q}^T(n_1), \ldots, R_{N,N-1,N-2}^T(n_1)}_{\frac{N(N-1)(N-2)}{3}blocks}\right]$$

$$R_{p,q;p\neq q}^{\frac{M}{L}\times NL}(n_1) = \left[\underbrace{0}_{\frac{M}{L}\times(p-1)L} \quad \underbrace{R_{y_q,y_p}(n_1)}_{\frac{M}{L}\times L} \quad \underbrace{0}_{\frac{M}{L}\times(q-p-1)L} \quad \underbrace{-R_{y_p}(n_1)}_{\frac{M}{L}\times L} \quad 0\right]$$

$$R_{p,q,k;k\neq p,q}^{\frac{M}{L}\times NL}(n_1) = \left[\underbrace{0}_{\frac{M}{L}\times(p-1)L} \quad \underbrace{R_{y_q,y_k}(n_1)}_{\frac{M}{L}\times L} \quad \underbrace{0}_{\frac{M}{L}\times(q-p-1)L} \quad \underbrace{-R_{y_p,y_k}(n_1)}_{\frac{M}{L}\times L} \quad 0\right]$$

$$R_{y_q, y_p}(n_1) = \begin{bmatrix} R_{y_q, y_p}(0, n_1) & \dots & R_{y_q, y_p}(L-1, n_1) \\ \dots & \dots & \dots \\ R_{Y_q, y_p}(M-L-1, n_1) & \dots & R_{y_q, y_p}(M-1, n_1) \end{bmatrix}$$

$$R_{y_p}(n_1) = \begin{bmatrix} R_{y_p}(0, n_1) & \dots & R_{y_p}(L-1, n_1) \\ \dots & \dots & \dots \\ R_{Y_p}(M-L-1, n_1) & \dots & R_{y_p}(M-1, n_1) \end{bmatrix}$$

The identifiability condition requires that the null space dimension of matrices $R_{y,\, self\text{-}CR}$ and $R_{y,\, cross\text{-}CR}$ should be one, while the vector h should be non-zero, implying that h_i, $i = 1, 2, \dots, N$ should not be zeros. To give more explicit expressions and provide more insight into the characteristics of the channels and the source signal (similar to what we had for deterministic sources, but here we have a statistical sense), the following conditions are given:

1. The polynomials $\{H_i(z)\}$, $i = 1, \dots, N$ are coprime, i.e., they do not share any common roots.
2. The linear complexity of the expected value of the NS source signal is greater than $2L + 1$ (for sufficient condition) and not less than $L + 1$ (for necessary condition).

As shown above, the *cross-CR* estimator only has cross-correlation components, so in the presence of spatially white noise, this estimator gives better performance than the self-CR estimator which contains noise autocorrelation.

For the case where channels are corrupted by noise, the following least square estimators are proposed:

$$P_{R,\alpha}^{self\text{-}CR} = \sum_{n_1} \|R_{x,\, self\text{-}CR}(n_1)h\|^{-1},\; P_{R,\alpha}^{cross\text{-}CR} = \sum_{n_1} \|R_{x,\, cross\text{-}CR}(n_1)h\|^{-1} \tag{4.39}$$

Ideally, at solutions, the inverse terms in the series go to zero, and for $P_{R,\alpha}^{cross\text{-}CR}$, they become infinite; however, due to the noise autocorrelation, $P_{R,\alpha}^{self\text{-}CR}$ remains finite at solutions.

4.3.4 Wide-Sense Stationary Random Sources

For WSS sources, Equation 4.36 for self-CR and Equation 4.37 for cross-CR become

$$\begin{aligned} h_p(n;\alpha) * R_{y_q, y_p}(n - n_1) &= h_q(n;\alpha) * R_{y_p}(n - n_1) \\ H_p(mF;\alpha) S_{y_q, y_p}(mF) &= H_q(mF;\alpha) S_{y_p}(mF) \\ p, q &= 1, 2, \dots, N; p \neq q \end{aligned} \tag{4.40}$$

$$\begin{aligned} h_p(n;\alpha) * R_{y_q, y_k}(n - n_1) &= h_q(n;\alpha) * R_{y_p, y_k}(n - n_1) \\ H_p(mF;\alpha) S_{y_q, y_k}(mF) &= H_q(mF;\alpha) S_{y_p, y_k}(mF) \\ p, q &= 1, 2, \dots, N;\; k = 1, \dots, N,\; k \neq p, q \end{aligned} \tag{4.41}$$

where $S_{y_p}(f)$ is the PSD of y_p, and $S_{y_p, y_q}(f)$ is the cross PSD of y_p and y_q. Equations 4.40 and 4.41 can be written in the following matrix form to solve for all channel responses simultaneously:

$$S_{y,\, self\text{-}CR} H = 0,\; S_{y,\, cross\text{-}CR} H = 0 \tag{4.42}$$

where

$$H = [H_1^T, H_2^T, \ldots, H_N^T]^T, H_i = [H_i(0), H_i(F), \ldots, H_i((L-1)F)]^T, i = 1, 2, \ldots, N$$

$$S_{y,\,self\text{-}CR} = \left[\underbrace{S_{1,2}^T, \ldots, S_{p,q;p \neq q}^T, \ldots, S_{N,N-1}^T}_{\frac{N(N-1)}{2} blocks} \right]$$

$$S_{y,\,cross\text{-}CR} = \left[\underbrace{S_{1,2,3}^T, \ldots, S_{p,q,k;k \neq p,q}^T, \ldots, S_{N,N-1,N-2}^T}_{\frac{N(N-1)(N-2)}{3} blocks} \right]$$

$$S_{p,q;p \neq q}^{1 \times NL} = \left[\underbrace{0}_{1 \times (p-1)L} \quad \underbrace{S_{y_q,y_p}}_{1 \times L} \quad \underbrace{0}_{1 \times (q-p-1)L} \quad \underbrace{-S_{y_p}}_{1 \times L} \quad 0 \right]$$

$$S_{p,q,k;k \neq p,q}^{1 \times NL} = \left[\underbrace{0}_{1 \times (p-1)L} \quad \underbrace{S_{y_q,y_k}}_{1 \times L} \quad \underbrace{0}_{1 \times (q-p-1)L} \quad \underbrace{-S_{y_p,y_k}}_{1 \times L} \quad 0 \right]$$

$$S_{y_q y_p} = [S_{y_q,y_p}(0) \ldots S_{y_q,y_p}((L-1)F)],\ S_{y_p} = [S_{y_p}(0) \ldots S_{y_p}((L-1)F)]$$

The identifiability condition is that the null space dimension of matrices $S_{y,\,self\text{-}CR}$ and $S_{y,\,cross\text{-}CR}$ should be one and that H_i, $i = 1, 2, \ldots, N$ should not be zero.

Again, the *cross-CR* estimator only has signal cross-spectrum density components, so in the presence of spatially white noises this estimator performs better than the self-CR estimator, which has both PSD and cross PSD.

The unknown parameter is estimated by maximizing the following equations when the channels are corrupted by noise:

$$P_{S,\alpha}^{self\text{-}CR} = \|S_{x,\,self\text{-}CR} H\|^{-1},\ P_{S,\alpha}^{cross\text{-}CR} = \|S_{x,\,self\text{-}CR} H\|^{-1} \tag{4.43}$$

Equation 4.43 can be rewritten in the following form to give more explicit expressions:

$$P_{\alpha}^{self\text{-}CR,\,Linear} = \frac{1}{\sum_{f_k} \sum_{i=1}^{N} \sum_{j=1, j \neq i}^{N} \left| S_{x_i}(f_k) H_{j;\alpha}(f_k) - S_{x_i,x_j}(f_k) H_{i;\alpha}(f_k) \right|^2} \tag{4.44}$$

$$P_{\alpha}^{cross\text{-}CR,\,Linear} = \frac{1}{\sum_{f_i} \sum_{i=1}^{N} \sum_{j=1, j \neq i} \sum_{k=1, k \neq i,j} \left| S_{x_k,x_i}(f_k) H_{j;\alpha}(f_k) - S_{x_k,x_j}(f_k) H_{i;\alpha}(f_k) \right|^2} \tag{4.45}$$

4.4 Time-Frequency Matched Field Processor

4.4.1 Background Theory

Time-frequency representations (TFR) of signals map a one-dimensional signal of time, $x(t)$, into a two-dimensional function of time and frequency, $T_x(t, f)$. The values of the TFR surface in the time-frequency plane give an indication as to which spectral components are present at each point in time.

TFRs have been applied to analyze, modify, and synthesize NS or time-varying signals. Three-dimensional plots of TFR surfaces have been used as pictorial representations enabling a signal analyst to determine which spectral components of a signal or system vary with time.[34–36] TFRs are divided into two major groups: linear and quadratic (bilinear). The short-time Fourier transform (STFT), Gabor transform, and the time-frequency version of the wavelet transform (WT) are members of linear TFR group. All linear TFRs satisfy the superposition or linearity principle which states that if $x(t)$ is a linear combination of some signal components, then the TFR of $x(t)$ is the same linear combination of the TFRs of each of the signal components. The Wigner distribution and ambiguity function[34] are the most important members of the energetic and correlative interpretations of the quadratic group, respectively. The Wigner distribution (Equation 4.46) and ambiguity function (Equation 4.47) are given, respectively, as

$$W_{x,y}(t,f) = \int_{\tau} x\left(t+\frac{\tau}{2}\right)y^*\left(t-\frac{\tau}{2}\right)e^{-j2\pi f\tau}d\tau = \int_{\nu} X\left(f+\frac{\nu}{2}\right)Y^*\left(f-\frac{\nu}{2}\right)e^{j2\pi t\nu}d\nu \tag{4.46}$$

$$A_{x,y}(\tau,\nu) = \int_{t} x\left(t+\frac{\tau}{2}\right)y^*\left(t-\frac{\tau}{2}\right)e^{-j2\pi\nu t}dt = \int_{\nu} X\left(f+\frac{\nu}{2}\right)X^*\left(f-\frac{\nu}{2}\right)e^{j2\pi t\nu}d\nu \tag{4.47}$$

The Wigner distribution and ambiguity function are dual in the sense that they are a Fourier transform pair:

$$A_{x,y}(\tau,\nu) = \int_t\int_f W_{x,y}(t,f)e^{-j2\pi(\nu t-\tau f)}dtdf \tag{4.48}$$

For random signals the expectation of Equation 4.48 is considered. In this case, the Wigner distribution is called an *evolutive spectrum*. For the energetic TFR, we seek to combine information from the instantaneous power $p_x(t) = |x(t)|^2$ and the spectral energy density $P_x(f) = |X(f)|^2$; while in the correlative TFR representation we seek to combine information from the temporal correlation $r_x(\tau) = \int_t x(t+\tau)x^*(t)dt$ and the spectral correlation $R_x(\nu) = \int_f X(f+\nu)X^*(f)df$. Two prominent examples of the energetic form are the spectrogram and the scalogram, defined as the squared magnitudes of the STFT and WT, respectively. Two fundamental classes of energetic TFRs are the classical Cohen class and the affine class.[34] The Cohen class includes all time-frequency, shift-invariant, quadratic TFRs in which the shift of signal in time and/or frequency results in the shift in TFR by the same time delay and /or modulation frequency. Every member of the Cohen class, Tx, is interpreted as a two-dimensional filtered Wigner distribution (an evolutive power spectrum for a random signal),[34] i.e.,

$$T_x \in \textit{Cohen class} \Leftrightarrow T_x(t,f) = \int_{t'}\int_{f'} \Psi_T(t-t', f-f')W_x(t',f')dt'df' \tag{4.49}$$

where $W_x(t, f)$ is the Wigner distribution of $x(t)$. Each member of Cohen's class is associated with a unique, signal-independent, kernel function $\Psi_T(t, f)$.

The affine class includes all energetic, quadratic TFRs which preserve time scaling and time shift. Any TFR which is an element of the affine class can be derived from the Wigner transform by means of an affine transformation,[34] i.e.,

$$T_x \in \textit{Affine class} \Leftrightarrow T_x(t,f) = \int_{t'}\int_{f'} \chi_T\left(f(t-t'), \frac{f'}{f}\right)W_x(t',f')dt'df' \tag{4.50}$$

where $\chi_T(\alpha, \beta)$ is a two-dimensional kernel function. The scalogram is the most famous member of this group.

4.4.2 Formulation

Let us now derive a time-frequency based MF processor. We multiply both sides of the Fourier transform of Equation 4.31 by $y_p^*(n_1T)e^{j2\pi n_1 mTF}$ or $y_q^*(n_1T)e^{j2\pi n_1 mTF}$ for the self-CR case and by $y_k^*(n_1T)e^{j2\pi n_1 mTF}$, $k = 1, \ldots, N$, $k \neq p, q$ for the cross-CR case, and then apply the expectation operator to produce

$$\underbrace{H_p(mF;\alpha)E(Y_q(mF)y_p^*(n_1T)e^{j2\pi n_1 mTF})}_{RD_{y_q, y_p}(mF, n_1T)} = \underbrace{H_q(mF;\alpha)E(Y_p(mF)y_p^*(n_1T)e^{j2\pi n_1 mTF})}_{RD_{y_p}(mF, n_1T)}$$
$$p, q = 1, 2, \ldots, N; p \neq q \tag{4.51}$$

$$\underbrace{H_p(mF;\alpha)E(Y_q(mF)y_k^*(n_1T)e^{j2\pi n_1 mTF})}_{RD_{y_k, y_q}(mF, n_1T)} = \underbrace{H_q(mF;\alpha)E(Y_p(mF)y_k^*(n_1T)e^{j2\pi n_1 mTF})}_{RD_{y_k, y_p}(mF, n_1T)}$$
$$p, q, k = 1, 2, \ldots, N; k \neq p, q \tag{4.52}$$

where RD_{y_p} is the *self-Rihaczek* distribution of y_p, and RD_{y_k, y_q} is the *cross-Rihaczek* distribution of y_p and y_q.[34,37] This distribution is a bilinear time-frequency distribution and a member of the Cohen class.[34] The *self-Rihaczek* and *cross-Rihaczek* distributions have the following relationship with the ambiguity function:[34]

$$RD_x(t, f) = \int_\tau\int_\nu [e^{j\pi\tau\nu}A_x(\tau, \nu)]e^{j2\pi(t\nu - f\tau)}d\tau d\nu$$
$$RD_{x,y}(t, f) = \int_\tau\int_\nu [e^{j\pi\tau\nu}A_{x,y}(\tau, \nu)]e^{j2\pi(t\nu - f\tau)}d\tau d\nu \tag{4.53}$$

where $A_x(\tau, \nu)$ is the ambiguity function of $x(t)$, and $A_{x,y}(\tau, \nu)$ is the cross-ambiguity function of $x(t)$ and $y(t)$.

The relationship between the ambiguity function and the evolutive spectrum is given in Equation 4.48. The *Rihaczek* distribution exhibits the following properties of bilinear time-frequency representation:

1. Time shift: $RD_{\tilde{x}}(t, f) = RD_x(t - t_0, f)$ for $\tilde{x}(t) = x(t - t_0)$.
3. Frequency shift: $RD_{\tilde{x}}(t, f) = RD_x(t, f - f_0)$ for $\tilde{x}(t) = x(t)e^{j2\pi f_0 t}$.
4. Time marginal: $\int_f RD_x(t, f)df = |x(t)|^2$.
5. Frequency marginal: $\int_t RD_x(t, f)dt = |X(f)|^2$.
6. Time moments: $\int_t\int_f t^n RD_x(t, f)dtdf = \int_t t^n|x(t)|^2 dt$.
7. Frequency moments: $\int_t\int_f f^n RD_x(t, f)dtdf = \int_f f^n|X(f)|^2 df$.
8. Time-frequency scaling: $RD_{\tilde{x}}(t, f) = RD_x\left(at\frac{f}{a}\right)$ for $\tilde{x}(t) = \sqrt{|a|}x(at)$.
9. Finite time support: $RD_x(t, f) = 0$ for t outside $[t_1, t_2]$ if $x(t) = 0$ outside $[t_1, t_2]$.
10. Finite frequency support: $RD_x(t, f) = 0$ for f outside $[f_1, f_2]$ if $X(f) = 0$ outside $[f_1, f_2]$.
11. Moyal's formula (unitarity): $(RD_{x_1, y_1}(t, f), RD_{x_2, y_2}(t, f)) = (x_1, x_2)(y_1, y_2)^*$.
12. Convolution: $RD_{\tilde{x}}(t, f) = \int_t RD_h(t - t', f)RD_x(t', f)dt'$ for $\tilde{x}(t) = \int_t h(t - t')x(t')dt'$.
13. Multiplication: $RD_{\tilde{x}}(t, f) = \int_f RD_h(t, f - f')RD_x(t, f')df'$ for $\tilde{x}(t) = h(t)x(t)$.

Equations 4.51 and 4.52 can be rewritten in the following matrix forms to solve for all channel responses simultaneously:

$$RD_{y,\,self\text{-}CR}(n_1)H = 0,\ RD_{y,\,cross\text{-}CR}(n_1)H = 0 \tag{4.54}$$

where

$$H = [H_1^T, H_2^T, \ldots, H_N^T]^T, H_i = [H_i(0), H_i(F), \ldots, H_i((M-1)F)]^T, i = 1, 2, \ldots, N$$

$$RD_{y,\,self\text{-}CR}(n_1) = \left[\underbrace{RD_{1,2}^T(n_1), \ldots, RD_{p,\,q;p\neq q}^T(n_1), \ldots, RD_{N,\,N-1}^T(n_1)}_{\frac{N(N-1)}{2} blocks}\right]$$

$$RD_{y,\,cross\text{-}CR}(n_1) = \left[\underbrace{RD_{1,2,3}^T(n_1), \ldots, RD_{p,\,q,\,k;k\neq p,\,q}^T(n_1), \ldots, RD_{N,\,N-1,\,N-2}^T(n_1)}_{\frac{N(N-1)(N-2)}{3} blocks}\right]$$

$$RD_{p,\,q;p\neq q}^{1\times NM}(n_1) = \left[\underbrace{0}_{1\times(p-1)M} \quad \underbrace{RD_{y_q,\,y_p}(n_1)}_{1\times M} \quad \underbrace{0}_{1\times(q-p-1)M} \quad \underbrace{-RD_{y_p}(n_1)}_{1\times M} \quad 0\right]$$

$$RD_{p,\,q,\,k;k\neq p,\,q}^{1\times NM}(n_1) = \left[\underbrace{0}_{1\times(p-1)M} \quad \underbrace{RD_{y_q,\,y_k}(n_1)}_{1\times M} \quad \underbrace{0}_{1\times(q-p-1)M} \quad \underbrace{-RD_{y_p,\,y_k}(n_1)}_{1\times M} \quad 0\right]$$

$$RD_{y_q,\,y_p}(n_1) = [RD_{y_q,\,y_p}(0, n_1T)\ldots RD_{y_q,\,y_p}((M-1)F, n_1T)]$$

$$RD_{y_p}(n_1) = [RD_{y_p}(0, n_1T)\ldots RD_{y_p}((M-1)F, n_1T)]$$

The identifiability condition requires that the null space dimension of matrices $RD_{y,\,self\text{-}CR}$ and $RD_{y,\,cross\text{-}CR}$ should be unity and that the vector H be non-zero, i.e., H_i, $i = 1, 2, \ldots, N$ should not be zero for all frequencies. To give more explicit expressions and provide more insights into the characteristics of the channels and the source signal (for input signals with time-frequency signature), the following conditions are given:

1. For the frequency band f_l, $l = 1, 2, \ldots, M$, the transfer functions H_i, $i = 1, 2, \ldots, N$ should not be zero.
2. In order to have the null space dimension of $RD_{y,\,self\text{-}CR}$ or $RD_{y,\,cross\text{-}CR}$ equal to unity, assuming the condition in Item 1 is satisfied, the *Rihaczek* distribution of the source $RD_s(n_1, f_l)$, $i = 1, 2, \ldots, L$ should be non-zero for all time indices n_1. This implies that the source signal should be sufficiently complex in the time-frequency sense.

As shown above, the *cross-CR* estimator only has signal cross-correlation components, so in the presence of spatially white noises, this estimator performs better than the self-CR estimator which retains noise autocorrelations.

For channels corrupted by noise, the following least square estimators are proposed:

$$P_{RD,\,\alpha}^{self\text{-}CR} = \sum_{n_1} \|RD_{x,\,self\text{-}CR}(n_1)H\|^{-1}, \; P_{RD,\,\alpha}^{cross\text{-}CR} = \sum_{n_1} \|RD_{x,\,cross\text{-}CR}(n_1)H\|^{-1} \tag{4.55}$$

At solution, each term in $P_{RD,\,\alpha}^{cross\text{-}CR}$ approaches infinity, while each term in $P_{RD,\,\alpha}^{self\text{-}CR}$ is large but bounded.

4.5 Higher Order Matched Field Processors

4.5.1 Background Theory

Higher order statistics have shown wide applicability in many diverse fields such as sonar, radar, and seismic signal processing; data analysis; and system identification.[38,39,40] Specific higher order statistics known as cumulants and their associated Fourier transforms known as polyspectra reveal not only amplitude information, but also phase information. This is an important distinction from the well-known second order statistics such as autocorrelation which are phase blind.

Cumulants, on the other hand, are blind to any kind of Gaussian process; that is, they automatically null the effects of colored Gaussian measurement noise, whereas correlation-based methods do not. By considering the process distribution we can choose the appropriate cumulant to reach our goals. For example, if a random process is symmetrically distributed as are Laplace, uniform, Gaussian, and Bernoulli-Gaussian distributions, then its third order cumulant equals zero. Thus, in order to obtain non-zero information about such a process, we should use at least a fourth order cumulant. For non-symmetric distributions such as exponential, Rayleigh, and k-distributions, the third order cumulant is not zero. In underwater acoustics, ship noise has a complex distribution with non-zero statistics higher than second order, so an MF processor based on higher order statistics will yield more information.

Higher order statistics are applicable when we are dealing with non-Gaussian processes. Many real-world applications are truly non-Gaussian. The greatest drawbacks to the use of higher order statistics are that they require longer data records and much more computation than do correlation based methods. Longer data lengths are needed in order to reduce the variance associated with estimating the higher order statistics from real data using sample averaging techniques.

Let $v = [v_1, v_2, \ldots, v_k]^T$ and $x = [x_1, x_2, \ldots, x_k]T$, where x denotes a collection of random variables. The kth order cumulant of these random variables is defined as the coefficient of $(v_1, v_2, \ldots, v_k)$ in the Taylor series expansion of the cumulant-generating function,[38] i.e.,

$$K(v) = \ln E(e^{jv'x}) \tag{4.56}$$

The kth order cumulant is defined in terms of its joint moments of orders up to k and vise versa. The moment-to-cumulant formula is

$$C_x(I) = \sum_{\cup_{p=1}^{q} I_p = I} (-1)^{q-1}(q-1)! \prod_{p=1}^{q} m_x(I_p) \tag{4.57}$$

where $\cup_{p=1}^{q} I_p = I$ denotes summation over all partitions of set I.[38] Set I contains the indices of the components of vector x. The partition of the set I is the unordered collection of non-intersecting non-empty sets I_p such that $\cup_{p=1}^{q} I_p = I$ where q is the number of partitions sets I_p. $m_x(I_p)$ stands for the moment of the partition x corresponding to set I_p, i.e., $m_x(I_p) = E(x_1 x_2 \ldots x_p)$. As examples, for zero-mean real random variables, the second, third, and fourth order cumulants are given by

$$cum(x_1, x_2) = E(x_1 x_2),\ cum(x_1, x_2, x_3) = E(x_1 x_2 x_3) \tag{4.58}$$

$$cum(x_1, x_2, x_3, x_4) = E(x_1 x_2 x_3 x_4) - E(x_1 x_2)E(x_3 x_4) - E(x_1 x_3)E(x_2 x_4) - E(x_1 x_4)E(x_2 x_3) \tag{4.59}$$

The cumulant-to-moment formula is

$$m_x(I) = \sum_{\cup_{p=1}^{q} I_p = I} C_x(I_p) \tag{4.60}$$

The most important properties of cumulants are listed.

1. If λ_i, $i = 1, 2, \ldots, k$ are constants, and x_i, $i = 1, 2, \ldots, k$ are random variables, then

$$cum(\lambda_1 x_1, \ldots, \lambda_k x_k) = \left(\prod_{i=1}^{k} \lambda_i\right) cum(x_1, \ldots, x_k) \tag{4.61}$$

2. Cumulants are symmetric in their arguments.

3. Cumulants are additive in their arguments, i.e.,

$$cum(x_0 + y_0, z_1, \ldots, z_k) = cum(x_0, z_1, \ldots, z_k) + cum(y_0, z_1, \ldots, z_k) \tag{4.62}$$

4. If α is a constant, then $cum(\alpha + z_1, z_2, \ldots, z_k) = cum(z_1, \ldots, z_k)$.
5. If the random variables $\{x_i\}$, $i = 1, \ldots, k$ are independent of the random variables $\{y_i\}$, $i = 1, \ldots, k$, then $cum(x_1 + y_1, \ldots, x_k + y_k) = cum(x_1, \ldots, x_k) + cum(y_1, \ldots, y_k)$. If $\{y_i\}$, $i = 1, \ldots, k$ is taken from Gaussian (colored or white) and $k \geq 3$, then $cum(y_1, \ldots, y_k)$. This makes the higher order statistics more robust to additive measurement noise than correlation, even if the noise is colored.
6. If a subset of the k random variables $\{x_i\}$, $i = 1, \ldots, k$ is independent of the rest, then $cum(x_1, \ldots, x_k) = 0$.
7. Cumulants of an independent, identically distributed random sequences are delta functions.

In many practical applications, we are given data and want to calculate cumulants from the data. Cumulants involve expectations, and, as in the case of correlation, they cannot be computed in an exact manner from real data; they must be approximated in much the same way that correlation is approximated. Cumulants are approximated by replacing expectations by sample averages.

4.5.2 Formulation

Now, we derive an MF processor based on higher order statistics by multiplying both sides of Equation 4.31 by a subset of $Y_k(F)$, $k = 1, \ldots, N$, $k \neq p, q$, and then applying the expectation operator to this product to produce a T order CR equation ($T \leq N$):

$$\begin{aligned} H_p(F;\alpha)E(Y_q(F)Y_m(F)\ldots Y_{m+T-2}(F)) &= H_q(F;\alpha)E(Y_p(F)Y_m(F)\ldots Y_{m+T-2}(F)) \\ H_p(F;\alpha)m_Y(I_q) &= H_q(F;\alpha)m_Y(I_p) \\ p, q, m = 1, 2, \ldots, N; p \neq q, \; MFP\ order{:}\ T, I_q &= \{q, m, m+1, \ldots, m+T-2\} \\ I_p &= \{p, m, m+1, \ldots, m+T-2\} \end{aligned} \tag{4.63}$$

If we replace moments by cumulants in Equation 4.63 we obtain

$$H_p(F;\alpha)\left(\sum_{\cup_{r=1}^{s} I_r = I_p} C_{Y(F)}(I_r)\right) = H_q(F;\alpha)\left(\sum_{\cup_{r=1}^{s} I_r = I_q} C_{Y(F)}(I_r)\right) \tag{4.64}$$

Equations 4.63 and 4.64 can be written in the following matrix form to solve for all channel responses simultaneously:

$$CUM_Y H = 0 \tag{4.65}$$

where

$$H = [H_1^T, H_2^T, \ldots, H_N^T]^T, H_i = [H_i(0), H_i(F), \ldots, H_i((L-1)F)]^T, i = 1, 2, \ldots, N$$

$$CUM_Y = \left[\underbrace{CUM_{1,2}^T, \ldots, CUM_{p,q;k_1,\ldots,k_{T-2}}^T, \ldots}_{\binom{N}{T} blocks}\right]$$

$$cum_{p,q,k_1,\ldots,k_{T-2}}^{1\times NL} = \left[\underbrace{0}_{1\times(p-1)L} \quad \underbrace{cum_{Y_p, Y_{k_1}, \ldots, Y_{k_{T-2}}}}_{1\times L} \quad \underbrace{0}_{1\times(q-p-1)L} \quad \underbrace{cum_{Y_p, Y_{k_1}, \ldots, Y_{k_{T-2}}}}_{1\times L} \quad 0\right]$$

$$cum_{Y_p, Y_{k_1}, \ldots, Y_{k_{T-2}}} = \left[\sum_{\cup_{r=1}^{s} I_r = I_p} C_{Y_p(0), Y_{k_1}(0), \ldots, Y_{k_{T-2}}(0)}(I_r) \ldots \sum_{\cup_{r=1}^{s} I_r = I_p} C_{Y_p((L-1)F), Y_{k_1}((L-1)F), \ldots, Y_{k_{T-2}}((L-1)F)}(I_r)\right]$$

The identifiability condition is that the null space dimension of matrix CUM_Y should be 1 and H_i, $i = 1, 2, \ldots, N$ should not be zero. To give more explicit expressions and provide more insights into the characteristics of the channels and the source signal, the following conditions are given:

1. For all frequencies band, f_i, $i = 1, 2, \ldots, M$, transfer functions H_i, $i = 1, 2, \ldots, N$, should not be zero.
2. In order to have the null space dimension of CUM_i equal to one, and assuming the condition mentioned above is satisfied, the source T order moment should be non-zero for all frequencies.

For the case where channels are corrupted by noise, the least square estimator, referred to as the high order CR based MF processor, is

$$P_{Y,\alpha}^{H\_CR} = \|CUM_Y H\|^{-1} \tag{4.66}$$

Equation 4.66 can be rewritten in the following form to give a more explicit expression of the processor:

$$\begin{aligned} P_{Y;\alpha}^{H\_CR} &= \frac{1}{\sum_p\sum_q\cdots\sum_m \left| H_p(F;\alpha)E\begin{pmatrix} (Y_q(F)Y_m(F)\ldots Y_{m+L-2}(F)) \\ -H_q(F;\alpha)E(Y_p(F)Y_m(F)\ldots Y_{m+L-2}(F)) \end{pmatrix}\right|^2} \\ &= \frac{1}{\sum_p\sum_q\cdots\sum_m |H_p(F;\alpha)m_Y(I_q) - H_q(F;\alpha)m_Y(I_p)|^2} \\ &= \frac{1}{\sum_p\sum_q\cdots\sum_m \left| H_p(F;\alpha)\sum_{\bigcup_{r=1}^{s} I_r = I_p} C_{Y(F)}(I_r) - H_q(F;\alpha)\sum_{\bigcup_{r=1}^{s} I_r = I_q} C_{Y(F)}(I_r)\right|^2} \end{aligned} \tag{4.67}$$

where p, q, m, and r are not equal, chosen from set $\{1, 2, \ldots, N\}$. The MF processor's order is T.

As an example, let us consider the third order case as follows:

$$\begin{aligned} P_{Y;\alpha}^{H\_CR} &= \frac{1}{\sum_p\sum_q\sum_m\sum_r |H_p(F;\alpha)E(Y_q(F)Y_m(F)Y_r(F)) - H_q(F;\alpha)E(Y_q(F)Y_m(F)Y_r(F))|^2} \\ &= \frac{1}{\sum_p\sum_q\sum_m\sum_r |H_p(F;\alpha)cum(Y_q(F)Y_m(F)Y_r(F)) - H_q(F;\alpha)E(Y_q(F)Y_m(F)Y_r(F))|^2} \end{aligned} \tag{4.68}$$

In the case of a true match, we write the output signals in terms of transfer function and source signal (as given in Equation 4.31),

$$P_{Y;\alpha}^{H\_CR} = \frac{1}{\begin{pmatrix} |cum(S^3(F))|^2 \sum_p\sum_q\sum_m\sum_r |H_p(F;\alpha)H_q(F)H_m(F)H_r(F) \\ -H_q(F;\alpha)H_p(F)H_m(F)H_r(F)|^2 \end{pmatrix}} \tag{4.69}$$

For zero-mean, Gaussian random sources (white or non-white), we have $cum(S^3(F)) = 2\sigma^2\bar{S} = 0$. A Gaussian source cannot be localized with a third order moment because its third cumulant is zero and does not satisfy the second part of the identifiability condition. However, this same feature gives the third order MF processor the ability of discriminating between Gaussian and non-Gaussian signals such as those radiated by noise from ships.

The effect of zero-mean, spatially white, non-Gaussian interference is canceled since each output signal does not appear more than once in the cumulants.

Let us assume that a deviation in the true source location or environmental parameter has occurred. The CR term (Equation 4.63) takes the form

$$CR_{pq} = E(S^T)H_m(F)\ldots H_{m+T-2}(F)\left(\underbrace{H_p(F;\alpha)H_q(F) - H_q(F;\alpha)H_p(F)}_{\mu_{pq}}\right) \tag{4.70}$$

For parameters with low sensitivity to the pressure field, there is no considerable change in the amplitude of the transfer function. In this case we mainly focus on the transfer functions phase. Moreover, let us assume that the array length is small enough in comparison to the water depth, so with good approximation we can assume that the amplitudes of the transfer functions appearing in the formulation are the same. Equation 4.70 can be simplified to

$$|CR_{pq}| \approx |E(S^T(F))||H(F)|^{T-1}|\mu_{pq}| \tag{4.71}$$

By substituting Equation 4.71 in the MF processor formulation (Equation 4.67) we have

$$P_{Y,\alpha}^{H\_CR} = \frac{1}{M(T)|H(F)|^{2(T-1)}|E(S^T(F))|^2\sum_p\sum_q|\mu_{pq}|^2} \tag{4.72}$$

where $M(L)$ is a constant multiplier that shows the number of CR terms in the MF processor formulation. For an array with N sensors we have

$$M(T) = \binom{N-2}{T-1}$$

To obtain a simpler equation, let us assume that the deviation due to the mismatch is independent of the sensors p and q:

$$P_{Y,\alpha}^{H\_CR} \approx \frac{1}{M(T)|H(F)|^{2(T-1)}|E(S^T(F))|^2\binom{N}{2}|\mu|^2} \tag{4.73}$$

Now, define the MF processor sensitivity function S as

$$S = [S_1, S_2, \ldots, S_q] \tag{4.74}$$

where

$$S_i = \left|\frac{\partial P_{Y,\alpha}^{H\_CR}}{\partial \alpha_i}\right|; i = 1, \ldots, q \tag{4.75}$$

The sensitivity function from Equation 4.73 becomes

$$S_i^T \approx \frac{2}{M(T)|H(F)|^{2(T-1)}|E(S^T(F))|^2\binom{N}{2}|\mu|^3}\frac{\partial \mu}{\partial \alpha_i} \tag{4.76}$$

To see how the MF processor sensitivity changes with order increasing from T to $T+1$, we obtain

$$S_i^{T+1} = \frac{M(T)|E(S^T(F))|^2}{M(T+1)|H(F)|^2|E(S^{T+1}(F))|^2}S_i^T = \frac{T|E(S^T(F))|^2}{|H(F)|^2|E(S^{T+1}(F))|^2(N-T-1)}S_i^T \tag{4.77}$$

The transfer function norms represent the transmission loss from the source to the vertical array sensors that are relatively small because of the high ocean attenuation. This fact potentially causes the sensitivity function to have a large value for higher order MF processors; however, in order to calculate the sensitivity function we need to know the relative value of moments, i.e., $(M(T))/(M(T+1))$

4.6 Simulation and Experimental Examples

This section presents an evaluation of CR MF processors for source localization and compares their performance with that for other MF processors. The evaluation is based on examples from underwater acoustics, using both simulation and experimental data. Environmental parameters of the waveguide are assumed known in the simulation, and for the real data we rely on values obtained from seismic ground truth data in the region of experiment. The simulation results are given in Section 4.6.1 for a random broadband source. The performance of the two different kinds of CR MF processors (self and cross) are compared with that for the Bartlett and MV processors.

In Section 4.6.2, source localization results using radiated ship noise are shown from an experiment in shallow water off the West Coast of Vancouver Island in the Northeast Pacific Ocean. The ship noise data were obtained using a 16-element vertical line array with a 15-m hydrophone spacing that was deployed in a 400-m ocean waveguide.

We refer readers to Reference 41 for simulation and experimental results for time-frequency and higher order MF processor.

4.6.1 Simulation Results

The performance of the Bartlett family processors and the CR processors is compared using an example from underwater acoustics: localization of a source radiating a broadband random signal. The example simulates a typical shallow water environment in which a noise source with an SNR value of –20 dB is operating at an unknown range and depth. The true source depth and range are 106 m and 3.6 km in a 400-m ocean waveguide. Ambiguity surfaces are calculated over a grid that spans the range from 50 m to 10 km with a resolution of 110 m, and the depth from 1 to 400 m with a resolution of 20 m. The localization results are compared for the Bartlett, MV, self-CR, and cross-CR in Figures 4.4 to 4.7. The replica or modeled fields used in experiments are calculated using Westwood et al.'s normal mode model, ORCA.[21]

Figures 4.4 to 4.7 demonstrate that all estimators have detected the true source location. The cross-CR gives superior performance with respect to resolution and sidelobe levels since it has only cross-spectrum components in its structure and the measurement white noise is canceled. The Bartlett and MV estimators show nearly the same performance as each other. This fact was discussed and proved in Reference 9, where the received random signals are contaminated by white measurement noise. In this example the MV-MF processor performance is degraded sharply, compared to that for the noise-free condition, and approaches the Bartlett performance. Self-CR gives a performance close to that of the Bartlett because of the effect of white noise that contaminates the self-spectrum elements.

Cross-sections of the ambiguity surface in range where the depth is 106 m and in depth where the range is 3.6 km are shown in Figures 4.8 and 4.9, respectively. Lower sidelobe levels and sharper mainlobes around the source location are obtained for the cross-CR MF processor in comparison to that for other processors.

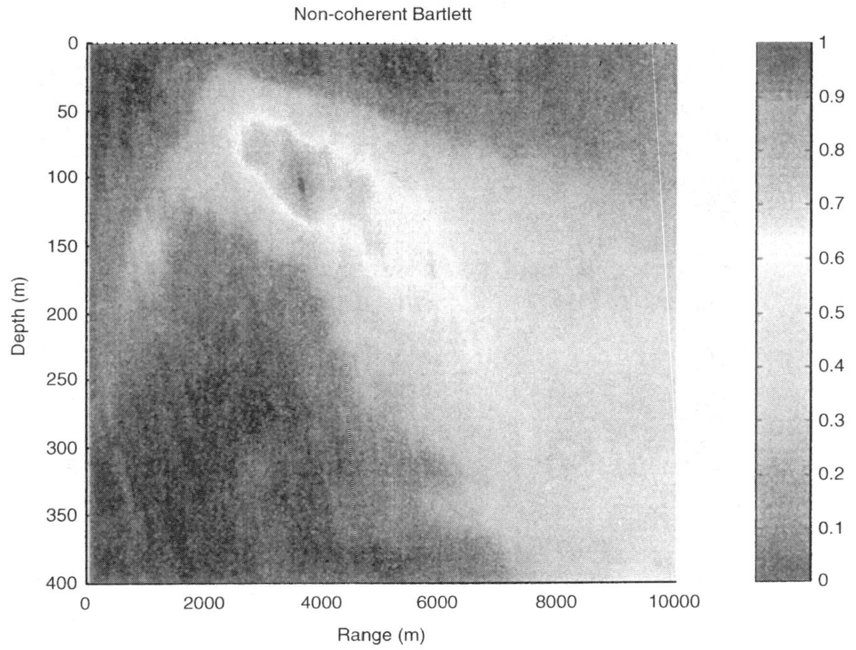

FIGURE 4.4 Ambiguity surface for Bartlett processor (SNR = –20 dB).

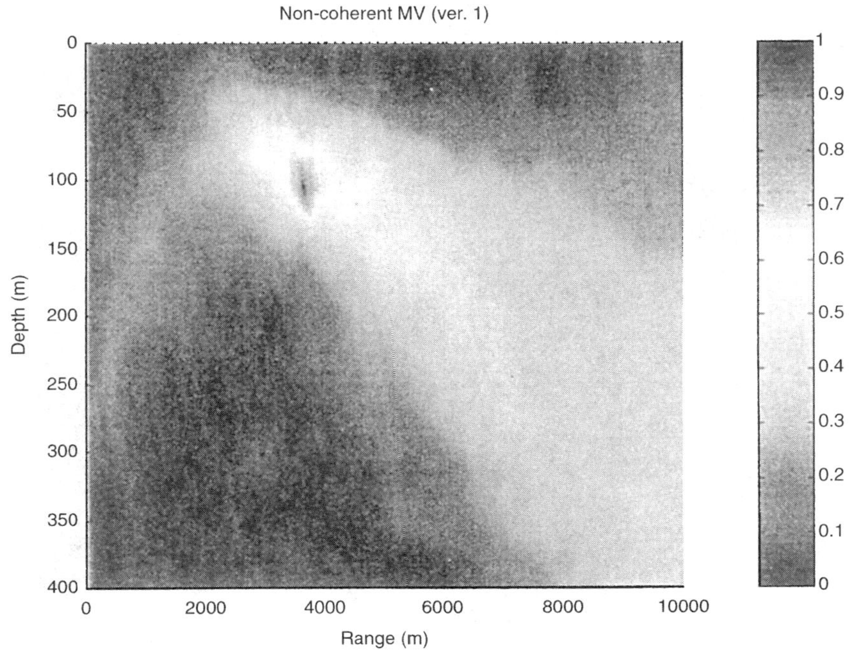

FIGURE 4.5 Ambiguity surface for MV processor, version one (SNR = –20 dB).

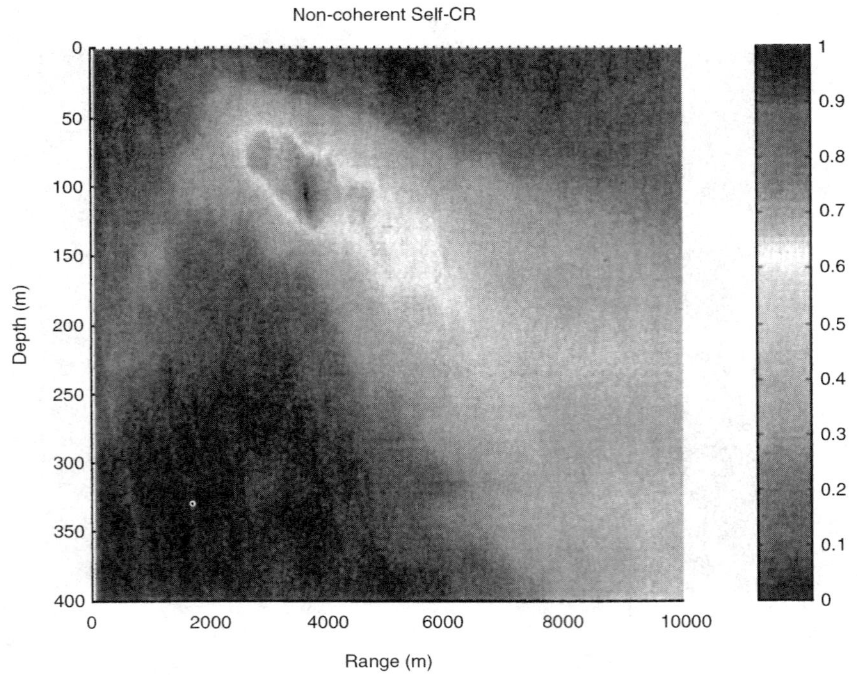

FIGURE 4.6 Ambiguity surface for self-CR processor (SNR = –20 dB).

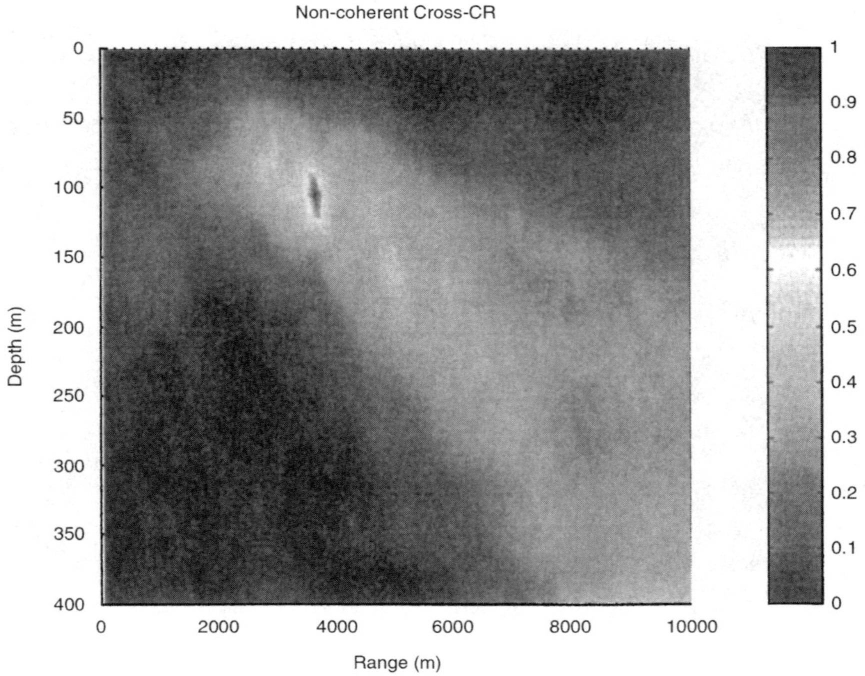

FIGURE 4.7 Ambiguity surface for cross-CR processor (SNR = –20 dB).

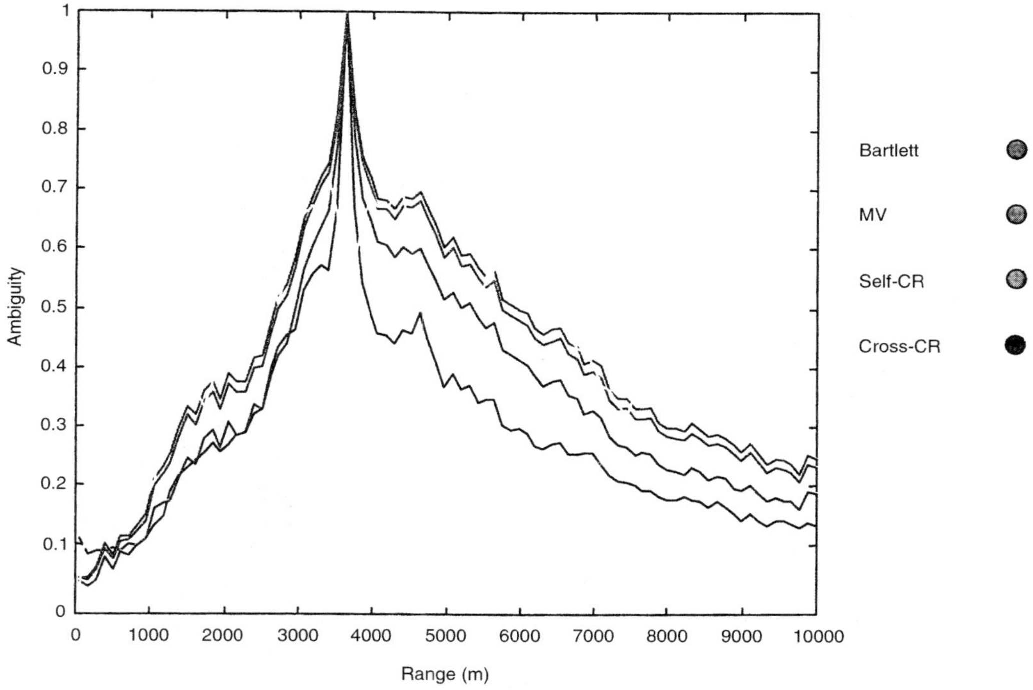

FIGURE 4.8 Performance of the different MF processors in range for a depth of 106 m.

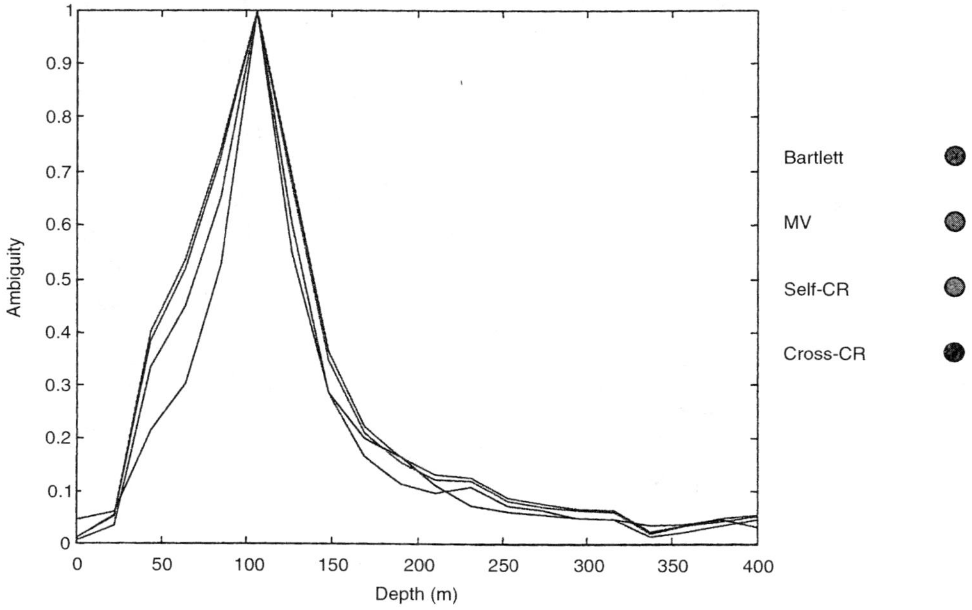

FIGURE 4.9 Performance of the different MF processors in depth for a range of 3.6 km.

4.6.2 Experimental Results

In this section, experimental data representing radiated ship noise in the band 73 to 133 Hz are used to demonstrate source localization performance. The true ship position was at a range of 3.3 km from the vertical array. The data were obtained in a portion of the experimental track where the properties of the waveguide were not varying with range. The ambiguity surface of the Bartlett and the two CR-MF processors are given in Figures 4.10 to 4.12. The grid search spans ranges from 50 m to 10 km, with resolution around 110 m, and depths from 1 to 400 m, with a resolution of 10 m. Along with the recorded ship noise, there are two tonal signals at 45 and 70 Hz generated by a sound source towed behind the ship. We show an expanded portion of the ambiguity surface around the source position in Figures 4.13 to 4.15.

The cross-CR processor has clearly localized both the ship and the towed source. The appearance of the continuous wave (CW) source in the ambiguity surface is due to incomplete cancelation of the harmonics of the towed source. The Bartlett processor has localized the source, but with poor resolution. The main peak includes both the towed source and ship noise. The CR based MF processors give considerable improvement in sidelobe level reduction and sharpness of the mainlobe width in comparison with the Bartlett MF processor. The self-CR processor shows both CW and ship locations, but with a weak value at the ship location. The cross-CR processor shows both ship and CW source locations with strong ambiguity values. The ship is localized at a range of 3.35 km, very close to the true value measured by GPS. The ship depth is estimated around 10 m, while we expect it to be close to the ocean surface. The error can be due to the poor depth resolution of the ambiguity surface and possible mismatches in the model of the ocean waveguide.

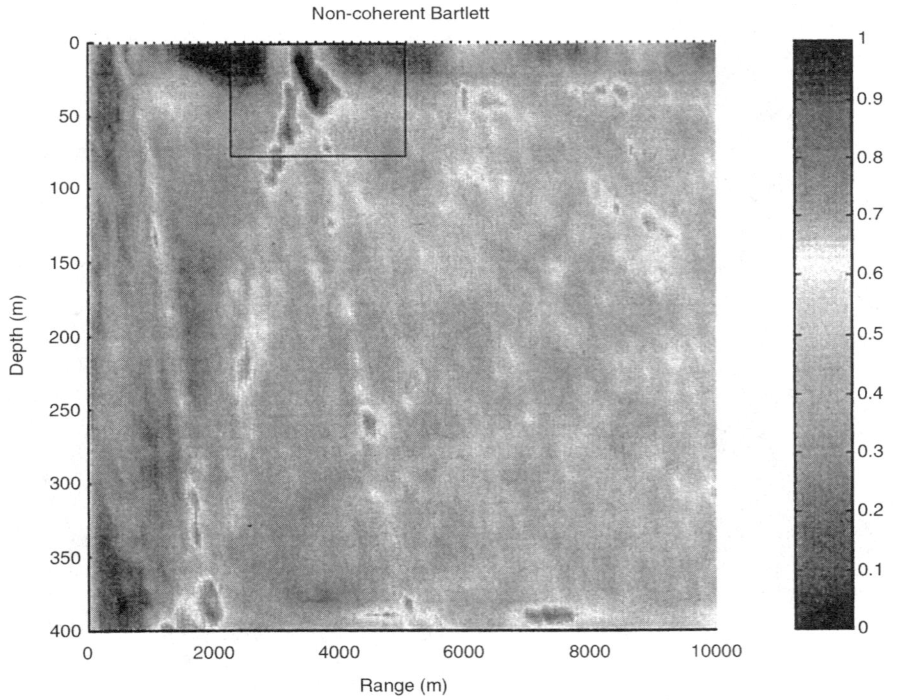

FIGURE 4.10 Ambiguity surface for Bartlett processor (73–133 Hz).

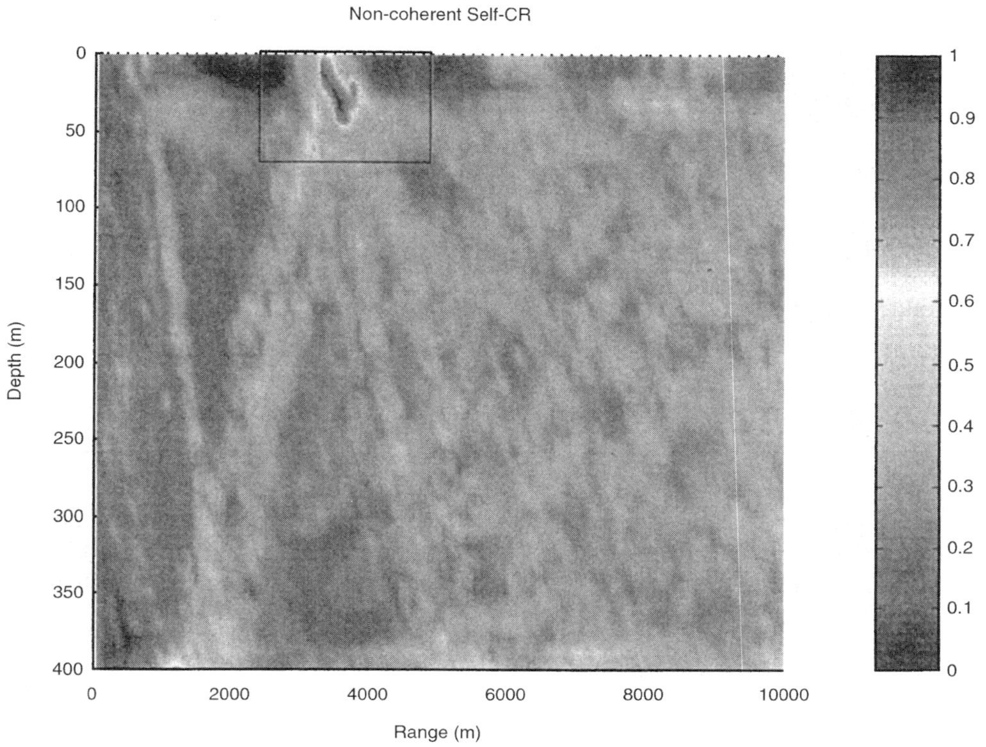

FIGURE 4.11 Ambiguity surface for self-CR processor (73–133 Hz).

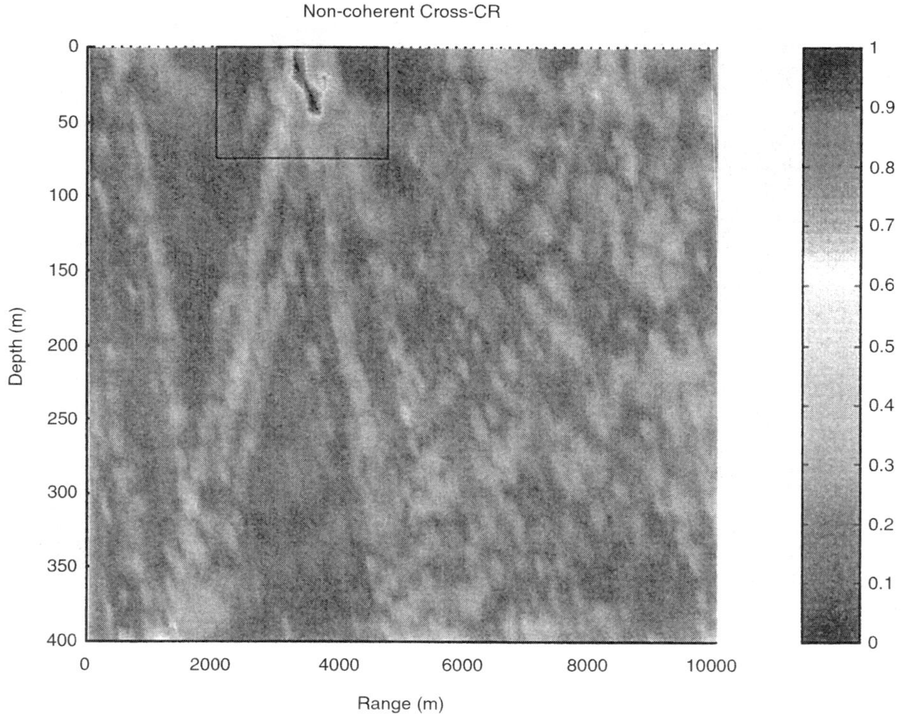

FIGURE 4.12 Ambiguity surface for cross-CR processor (73–133 Hz).

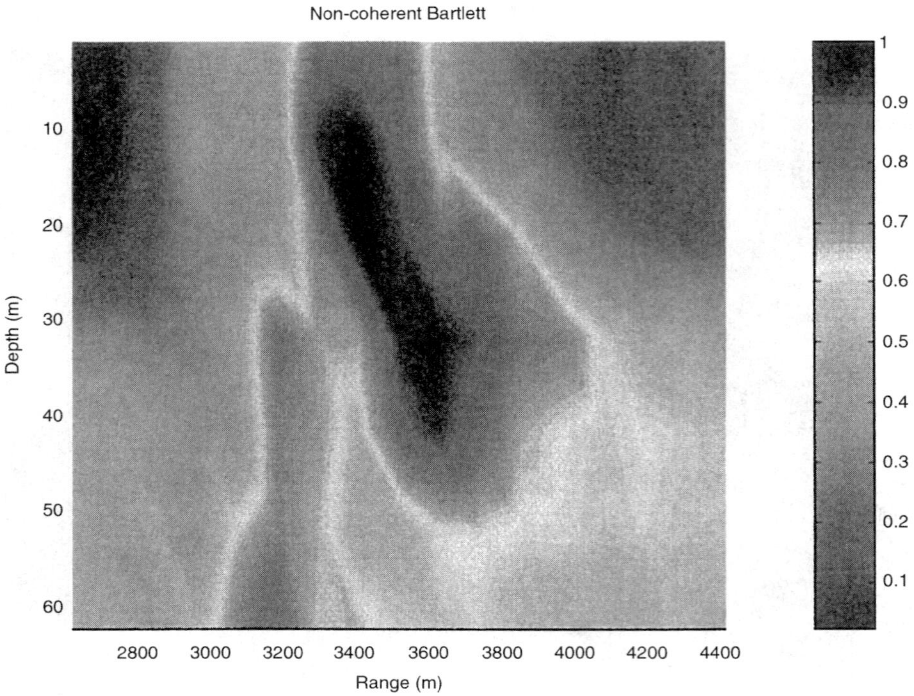

FIGURE 4.13 Ambiguity surface for Bartlett processor (73–133 Hz).

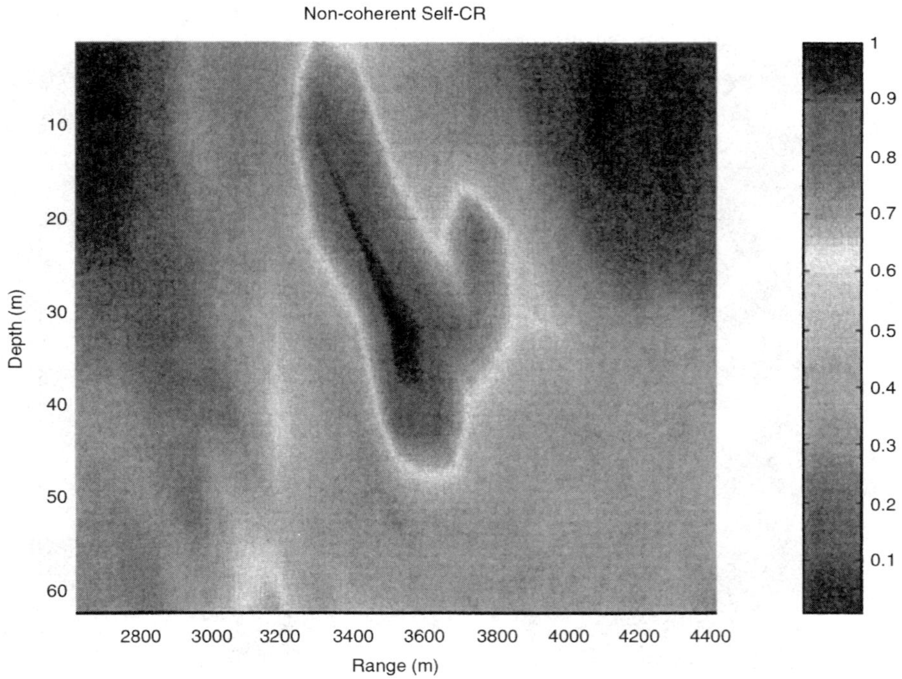

FIGURE 4.14 Ambiguity surface for self-CR processor (73–133 Hz).

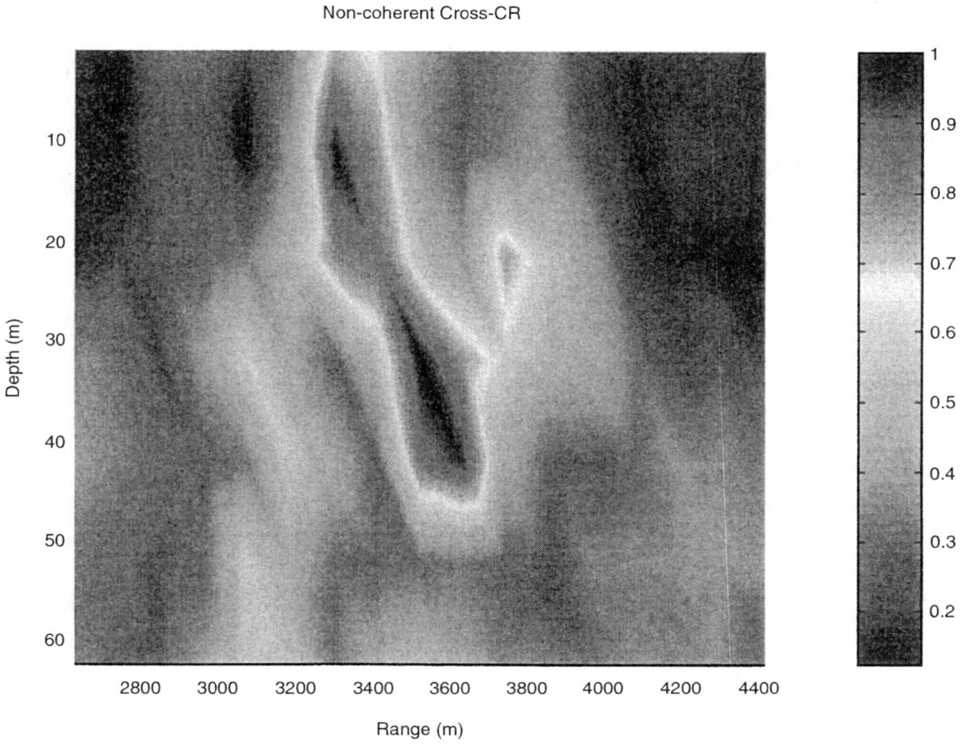

FIGURE 4.15 Ambiguity surface for cross-CR processor (73–133 Hz).

References

1. Bucker, H. P., Use of calculated sound fields and matched field detection to locate sound sources in shallow water, *J. Acoust. Soc. Am.*, 59(2), 368–373, 1976.
2. Fizell, R. G., Application of high-resolution processing to range and depth estimation using ambiguity function methods, *J. Acoust. Soc. Am.*, 82(2), 606–613, 1987.
3. Baggeroer, A.B., Kuperman, W.A., and Schmidt, H., Matched field processing: Source localization in correlated noise as an optimum parameter estimation problems, *J. Acoust. Soc. Am.*, 83, 571–587, 1988.
4. Candy, J. V. and Sullivan, E. J., Sound velocity profile estimation: A system theoretic approach, *IEEE J. Oceanic Eng.*, 18(3), 240–252, 1993.
5. Richardson, A. M. and Nolte, W., A posteriori probability source localization in an uncertain sound speed deep ocean environment, *J. Acoust. Soc. Am.*, 89(5), 2280–2284, 1991.
6. Baggeroer, A.B., Kuperman, W.A., and Mikhalevsky, N., An overview of matched field methods in ocean acoustic, *IEEE J. Oceanic Eng.*, 18(4), 401–424, 1993.
7. Middleton, D. and Sullivan, E. J., Estimation and detection issues in matched field processing, *IEEE J. Oceanic Eng.*, 18(3), 156–167, 1993.
8. Porter, M. B., Acoustic models and sonar systems, *IEEE J. Oceanic Eng.*, 18(4), 425–437, 1993.
9. Tolstoy, A., *Matched Field Processing for Underwater Acoustics*, World Scientific, Singapore, 1993.
10. Collins, M. D., Kuperman, W. A., and Schmidt, H., Nonlinear inversion for ocean bottom properties, *J. Acoust. Soc. Am.*, 92, 2770–2783, 1992.

11. Lindsay, C. E. and Chapman, N. R., Matched field inversion for geoacoustic model parameters using adaptive simulated annealing, *IEEE J. Oceanic Eng.*, 18(3), 224–231, 1993.
12. Gerstoft, P., Inversion of seismo-acoustic data using genetic algorithms and a posteriori probability distributions, *J. Acoust. Soc. Am.*, 95, 770–782, 1994.
13. Wilson, J. H. and Rajan, S. D., Eds., Special issue on inversions and the variability of sound propagation in shallow water, *IEEE J. Oceanic Eng.*, 21(4), 321–504, 1996.
14. Chapman, R. and Tolstoy, A., Eds., Special issue on benchmarking geoacoustic inversion methods, *J. Comp. Acoust.*, 6(1, 2), 1–289, 1998.
15. Abed-Meraim, K., Qiu, W., and Hua, Y., Blind system identification, *Proc. IEEE*, 85, 1310–1322, 1997.
16. Westwood, E. K. and Vidmar, P. J., Eigenary finding and time series simulation in a layered bottom, *J. Acoust. Soc. Am.*, 81, 912–924, 1987.
17. Schmidt, H. and Jensen, F. B., A full wave solution for propagation in multilayered viscoelastic media with application to Gaussian beam reflection at fluid-solid interfaces, *J. Acoust. Soc. Am.*, 77, 813–825, 1985.
18. Thomson, D. J. and Chapman, N. R., A wide-angle split-step algorithm for the parabolic equation, *J. Acoust. Soc. Am.*, 74, 1848–1854, 1983.
19. Collins, M. D., Applications and time-domain solution of higher-order parabolic equations in underwater acoustics, *J. Acoust. Soc. Am.*, 86, 1097–1102, 1989.
20. Porter, M. B. and Reiss, E. L., A numerical method for bottom interacting ocean acoustic normal modes, *J. Acoust. Soc. Am.*, 77, 1760–1767, 1985.
21. Westwood, E. K., Tindle, C. T., and Chapman, N. R., A normal mode model for acousto-elastic ocean environments, *J. Acoust. Soc. Am.*, 100, 3631–3645, 1996.
22. Xu, G., Liu, H., Tong, L., and Kailath, T., A least squares approach to blind channel identification, *IEEE Trans. SP*, 43, 2982–2993, 1995.
23. Hua, Y. and Wax, M., Strict identifiability of multiple FIR channels driven by an unknown arbitrary sequence, *IEEE Trans. SP*, 44, 756–759, 1996.
24. Abed-Meraim, K., Cardoso, J. F., Gorokhov, A. Y., and Loubaton, P., On subspace methods for blind identification of single-input multiple-output FIR systems, *IEEE Trans. SP*, 45, 42–55, 1997.
25. Kirlin, R. L., Kaufhold, B., and Dizaji, R., Blind system identification using normalized Fourier coefficient gradient vectors obtained from time-frequency entropy-based blind clustering of data wavelets, *Digital Signal Process. J.*, 9, 18–35, 1999.
26. Justice, J. H. et al. (Haykin, S., Ed.), *Array Signal Processing*, Prentice-Hall, Englewood Cliffs, NJ, 1985.
27. Bienvenu, G. and Kopp, L., Optimality of high resolution array processing using the eigensystem approach, *IEEE Trans. ASSP*, 31, 1235–1247, 1983.
28. Akaike, M., A new look at the statistical model identification, *IEEE Trans. Automat. Control*, 19, 716–737, 1974.
29. Wax, M., Model based processing in sensor arrays, in *Advances in Spectrum Analysis and Array Processing*, Vol. 3, S. Haykin, Ed., 1–47, Prentice-Hall, Englewood, Cliffs, NJ.
30. Papoulis, A., *Probability, Random Variables, and Stochastic Processes*, McGraw-Hill, New York, 1990.
31. Schmidt, H., Baggeroer, W., Kuperman, A., and Scheer, E. K., Environmentally tolerant beamforming for high resolution matched field processing: Deterministic mismatch, *J. Acoust. Soc. Am.*, 88, 1851–1862, 1990.
32. Shang, E. C., Source depth estimation in waveguides, *J. Acoust. Soc. Am.*, 77, 1413–1418, 1985.
33. Yang, T. C., A method of range and depth estimation by modal decomposition, *J. Acoust. Soc. Am.*, 82(5), 1736–1745, 1987.
34. Hlawatsch, F. and Boudreaux-Bartles, G. F., Linear and quadratic time-frequency signal representations, *IEEE SP Mag.*, 21–67, April 1992.

35. Papandreou-Suppappola, A., Hlawatsch, F., and Boudreaux-Bartels, F., Quadratic time-frequency representations with scale covariance and generalized time-shift covariance: A unified framework for the affine, hyperbolic, and power classes, *Digital Signal Process. J.*, 8, 3–48, 1998.
36. Boashash, B., Ed., *Time-Frequency Signal Analysis, Methods and Applications*, John Wiley & Sons, New York, 1992.
37. Rihaczeck, W., Signal energy distribution in time and frequency, *IEEE Trans. Inf. Theory*, 14(3), 369–374, 1968.
38. Mendel, J. M., Tutorial on higher-order statistics (spectra) in signal processing and system theory: Theoretical results and some applications, *Proc. IEEE*, 79, 278–305, 1991.
39. Nikias, C. L. and Raghuveer, M., Bispectrum estimation: a digital signal processing framework, *Proc. IEEE*, 75, 869–891, 1987.
40. Pan, R. and Nikias, C. L., The complex cepstrum of higher-order moments, *IEEE Trans. ASSP*, 36, 186–205, 1988.
41. Dizaji, R., Matched Field Processing, a Blind System Identification Technique, Ph.D. dissertation, University of Victoria, Victoria, B.C., Canada, 2000.

5

Model-Based Ocean Acoustic Signal Processing

James V. Candy
University of California

Edmund J. Sullivan
Naval Undersea Warfare Center

Abstract

Signal processing can simply be defined as a technique or set of techniques to extract the useful information from noisy measurement data while rejecting the extraneous. These techniques can range from simple, non-physical representations of the measurement data such as the Fourier or wavelet transforms to parametric black-box models used for data prediction to lumped mathematical physical representations usually characterized by ordinary differential equations to full physical partial differential equation models capturing the critical details of wave propagation in a complex medium. The determination of which approach is the most appropriate is usually based on how severely contaminated the measurements are with noise and underlying uncertainties. If the signal-to-noise (SNR) of the measurements is high, then simple non-physical techniques can be used to extract the desired information. However, if the SNR is extremely low and/or the propagation medium is uncertain, then more of the underlying propagation physics must be incorporated somehow into the processor to extract the information. Model-based signal processing is an approach that incorporates propagation, measurement, and noise/uncertainty models into the processor to extract the required signal information while rejecting the extraneous data even in

0-8493-3691-0/01/$0.00+$.50

highly uncertain environments — like the ocean. This chapter outlines the motivation and development of model-based processors (MBP) for ocean acoustic applications. We discuss the generic development of MBP schemes and then concentrate specifically on those designed for application in the hostile ocean environment. Once the MBP is characterized, we then discuss a set of ocean acoustic applications demonstrating this approach.

5.1 Introduction

The detection and localization of an acoustic source has long been the motivation of early sonar systems. With the advent of quieter and quieter submarines due to new manufacturing technologies and the recent proliferation of small non-nuclear powered vessels, the need for more sophisticated processing techniques has been apparent for quite some time. It has often been contemplated that the incorporation of ocean acoustic propagation models into signal processing schemes can offer more useful information necessary to improve overall processor performance and to achieve the desired enhancement/detection/localization even under the most hostile of conditions. Model-based techniques offer high expectations of performance, since a processor based on the predicted physical phenomenology that inherently has generated the measured signal must produce a better (minimum error variance) estimate then one that does not.[1,2] The uncertainty of the ocean medium also motivates the use of stochastic models to capture the random and often non-stationary nature of the phenomena ranging from ambient noise and scattering to distant shipping noise. Therefore, processors that do not take these effects into account are susceptible to large estimation errors. This uncertainty was discussed by Tolstoy[3] in the work of Carey and Moseley[4] when investigating space-time processing and in the overview by Sullivan and Middleton[5] and Baggeroer et al.[6] Therefore, if the model embedded in this process is inaccurate or for that matter incorrect, then the model-based processor (MBP) can actually perform worse. Hence, it is necessary, as part of the MBP design procedure, to estimate/update the model parameters either through separate experiments or jointly (adaptively) while performing the required processing.[7] Note that the introduction of a recursive, on-line MBP can offer a dramatic detection improvement in a tactical passive or active sonar-type system, especially when a rapid environmental assessment is required.[8]

Incorporating a propagation model into a signal processing scheme was most probably initiated by the work of Hinich,[9] who applied it to the problem of source depth estimation. However, as early as 1966, Clay[10] suggested matching the modal functions of an acoustic waveguide to estimate source depth. The concept of matched-field processing (MFP), which compares the measured pressure field to that predicted by a propagation model to estimate source range and depth, was introduced by Bucker[11] in 1976. In MFP, the localization problem is solved by exhaustively computing model predictions of the field at the array for various assumed source positions. The final position estimate is the one achieving maximum correlation with the measured field at the array. Many papers have been written exploiting and improving on the MFP and are best summarized in the text of Tolstoy,[3] the special issues of Doolittle[12] and Stergiopoulos and Ashley,[13] as well as the recent text by Diachok et al.[14] Other approaches to solve the localization problem have also evolved, with the most noteworthy being the simulated annealing approach of Kuperman et al.,[15] the maximum a posteriori estimator of Richardson and Nolfe,[16] and the empirical eigenfunctions of Krolik.[17] All of these works contain most of the references therein to the effort performed in MFP over the past 25 years. However, matched field is mainly aimed at the localization problem; indeed most estimators implemented by MFP are focused on seeking an estimation of localization parameters.

In ocean acoustics, there are many problems of interest other than localization that are governed by propagation models of varying degrees of sophistication. Here, we are interested primarily in a shallow water environment characterized by a normal-mode model, and, therefore, our development will eventually lead to adaptively adjusting parameters of the propagation model to "fit" the ever-changing ocean environment encompassing our sensor array. In fact, one way to think about this processor is that it passively listens to the ocean environment and "learns" or adapts to its changes. It is clear that the resulting processor will be much more sensitive to changes than one that is not, thereby providing current information and processing. Once recent paper utilizes such a processor as the heart of its model-based localization scheme.[18]

With this background in mind, we investigate the development of an "MBP," that is, a processor that incorporates a mathematical representation of the ocean acoustic propagation and can be used to perform various signal processing functions ranging from simple filtering or signal enhancement to adaptively adjusting model parameters to localization to tracking to sound speed estimation or inversion. In all of these applications, the heart of the processing lies in the development of the MBP and its variants. Clearly, each of the MFP methods described above can be classified as model based, for instance, the MFP incorporates a fixed (parametrically) propagation model. However, in this chapter, we will investigate the state-space forward propagation scheme of Candy and Sullivan[7] and apply it to various ocean acoustic signal processing problems. We choose to differentiate between the terms model-based processing and matched-field processing, primarily to emphasize the fact that this work is based on the existing state-space framework that enables access to all of the statistical properties inherited through this formalism, such as the predicted conditional means and covariances.[1,2] This approach also enables us access to the residual or innovation sequence associated with MBPs (Kalman filter estimator/identifiers), permitting us to monitor the performance of the embedded models in representing the phenomenology (ocean acoustics, noise, etc.) as well as the on-line potential of refining these models adaptively using the innovations.[7,8] The state-space formalism can be considered as a general framework that *already* contains the signal processing algorithms, and it is the task of the modeler to master the art of embedding his models of interest. Thus, in this sense, the modeler is not practicing signal processing per se, but actually dealing with the problem of representing his models within the state-space framework. Furthermore, this framework is not limited to localization, but because of its flexibility, tomographic reconstructions can be performed to directly attack the *mismatch problem* that plagues MFP.[3,12,19–21] This can be accomplished by constructing an "adaptive" MBP that allows continuous updating of the model parameters and is easily implemented by augmenting them into the current state vector. That is, unlike the conventional view of the inverse problem, where the functional relationship between the measurements and the parameters of interest must be invertible, here we simply treat these parameters as quantities to be estimated by augmenting them into the state vector, creating an adaptive joint estimation problem. In MFP, most of the techniques employed to "correct" this mismatch problem usually achieve their result by a desensitization of the algorithm (multiple constraint minimum variance distortionless response [MVDR][17]) or by multiple parameter estimation (e.g., simulated annealing[15]). Simulated annealing is a sub-optimal multiple parameter estimator capable of estimating bottom parameters, but it has difficulty in dealing with functions such as the modal functions, since it would require them to be parameterized in some arbitrary manner. An adaptive MBP does not sacrifice any potential information available in the model, but actually can refine it, as will be seen by *adaptively*, i.e., recursively, updating parameters. This, of course, enlarges the dimension of the state space. In this way, the original states and the augmented states are updated by the recursive processor in a self-consistent manner. The fact that the relationship between the original states and the parameters of interest may be complicated and/or non-linear is not an issue here, since only the "forward" problem is explicitly used in each recursion via the measurement relations. Thus, the usual complications of the inverse problem are avoided at the expense of creating a higher dimensional state space. All that is necessary is that the parameters of interest be observable or identifiable in the system theoretic sense.[8,22,23]

Much of the formalism for this model-based signal processing has been worked out.[7,8,22–25] Model-based processing is concerned with the incorporation of environmental (propagation, seabed, sound speed, etc.), measurement (sensor arrays), and noise (ambient, shipping, surface, etc.) models long with measured data into a sophisticated processing algorithm capable of detecting, filtering (enhancing), and localizing an acoustic source (target) in the complex ocean environment. This technique offers a well-founded statistical approach for comparing propagation/noise models to measured data and is *not* constrained to a stationary environment which is essential in the hostile ocean. Not only does the processor offer a means of estimating various quantities of high interest (modes, pressure field, sound speed, etc.), but it also provides a methodology to statistically evaluate its performance on-line, which is especially useful for model validation experiments.[22,24] Although model-based techniques have been around for quite a while,[1,2] they have just recently found their way into ocean acoustics. Some of the major advantages of MBPs are that they (1) are recursive; (2) are statistical, incorporating both noise and parameter uncertainties; (3) are not constrained to only stationary statistics; (4) are capable of being extended to incorporate both linear/non-linear space-time varying models; (5) are capable of on-line processing of the measured data at each recursion; (6) are capable of filtering the pressure field as well as simultaneously estimating the modal functions and/or sound speed (inversion/tomography); (7) are capable of monitoring their own performance by testing the residual between the measurement and its prediction; and (8) are easily extended to perform adaptively. However, a drawback (in some cases) is the increased computational load. This feature will not be much of a problem with the constant improvement in the speed of modern computer systems and the reduced computations required in a low frequency, shallow water environment.

Let us examine the inherent structure of the MBP. Model-based processing is a direct approach that uses *in situ* measurements. More specifically, the acoustic measurements are combined with a set of preliminary sound speed and other model parameters usually obtained from *a priori* information or a sophisticated simualtor[26–29] that solves the underlying boundary value problem to extract the initial parameters/states in order to construct the forward propagator and initialize the algorithm. The algorithm then uses the incoming data to adaptively update the parameter set jointly with the acoustic signal processing task (detection, enhancement, and localization). In principle, any propagation model can be included in this method;[29] however, in this chapter, our designs are all based on the normal-mode model of propagation. We define the MBP as a Kalman filter whose estimated states are the modal functions $\hat{\phi}(z_\ell)$ and states representing the models of the ocean estimated acoustic parameters $\hat{\theta}(z_\ell)$ that have been augmented into the processor. The basic processor is shown in Figure 5.1. The inputs to the MBP can be either raw data $[\{p(z_\ell)\}, \{c(z_\ell)\}]$ or a combination of raw data and outputs $\theta(z_\ell)$ of a "modal solver" (such as SNAP,[26] the SACLANT normal-mode propagation model). There are basically three advantages to this type of processor. First, it is recursive and, therefore, can continuously update the estimates of the sonar and environmental parameters. Second, it can include the system and measurement noise in a self-consistent manner. By noise, it is meant not only acoustic noise, but also errors in the input parameters of the model. Third, one of the outputs of the MBP is the so-called "innovation sequence," $\varepsilon(z_\ell)$, which provides an on-line test of the "goodness of fit" of the model to the data. This innovation sequence plays a major role in the recursive nature of this processor by providing information that can be used to adaptively correct the processor and the propagation model itself, as well as the input to a sequential detector.[30] Along with the ability of this processing scheme to self-consistently estimate parameters of interest along with the signal processing task, stand-alone estimators can also be used to provide refined inputs to the model. For example, by using data from a towed array, the horizontal wave numbers can be directly estimated as a spatial spectral analysis problem.[31] Further, these estimates can be refined by use of new towed array processing schemes.[32–36]

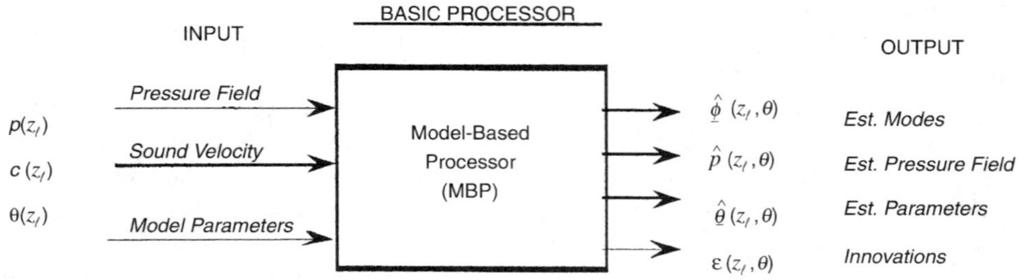

FIGURE 5.1 Model-based ocean acoustic processor: the basic processor.

In ocean acoustics, we are also concerned with an environmental model of the ocean and how it effects the propagation of sound through this noisy, complex environment. The problem of estimating the environmental parameters characterizing the ocean medium is called the *ocean tomography* or, equivalently, the *environmental inversion* problem and has long been a concern because of its detrimental effect on various detection/localization schemes.[3,19–21,37] Much of the work accomplished on this problem has lacked quantitative measures of the mismatch of the model with its environment. In a related work,[24] it was shown how to quantify "modeling errors" both with and without a known ocean environmental model available. In the first case, standard errors were calculated for modal/pressure-field estimates, as well as an overall measure of fit based on the innovations or residuals between the measured and predicted pressure field. In the second case, only the residuals were used. These results quantify the mismatch between the embedded models and the actual measurements both on simulated as well as experimental data.

It has already been shown that the state-space representation can be utilized for signal enhancement to spatially propagate both modal and range functions as discussed in Candy and Sullivan.[7] Specifically, using the normal-mode model of the acoustic field and a known source location, the modal functions and the pressure field can be estimated from noisy array measurements. In the stochastic case, a Gauss-Markov model evolves, allowing the inclusion of stochastic phenomena such as noise and modeling errors in a consistent manner. The Gauss-Markov representation includes the second-order statistics of the measurement noise and the modal uncertainty. In our case, the measurement noise can be "lumped" to represent the near-field acoustic noise field, flow noise on the hydrophones, and electronic noise, whereas the modal/range uncertainty can also be lumped to represent sound speed profile (SSP) errors, noise from distant shipping, errors in the boundary conditions, sea state effects, and ocean inhomogeneities. It should also be noted that adaptive forms of the MBP are also available to provide a realizable solution to the so-called *mismatch problem*, where the model and it underlying parameters do not faithfully represent the measured pressure-field data.[3,37] References 8 and 22 address the mismatch problem and its corresponding solution using MBPs.

Clearly, it is not possible to discuss all of the details of MBP designs for various ocean acoustic applications. In this chapter, we first develop the "concept" of model-based signal processing in Section 5.2 and compare it to conventional beamforming. We show, simply, how this approach can be extended even further to solve an important towed array problem. We discuss the basic approach to "minimum-variance" design, which is the basis of MBP design and performance analysis. With the MBP background complete, we briefly discuss the shallow water ocean environment, normal-mode modeling, and the evolution of state-space forward propagators in Section 5.3. The development of MBP schemes to solve various applications are surveyed in Section 5.4. Conclusions and summaries are in Section 5.5. Keep in mind that the primary purpose of this chapter is to introduce the concept of model-based signal processing, develop a basic understanding of processor design and performance, and demonstrate its applicability in the hostile ocean environment by presenting various solutions developed exclusively for this application.

5.2 Model-Based Processing

5.2.1 Motivation

In this section, we discuss the basics of the model-based approach to signal processing. Formally, the model-based approach is simply "incorporating mathematical models of both physical phenomenology and the measurement process (including noise) into the processor to extract the desired information." This approach provides a mechanism to incorporate knowledge of the underlying physics or dynamics in the form of mathematical propagation models along with measurement system models and accompanying uncertainties such as instrumentation noise or ambient noise as well as model uncertainties directly into the resulting processor. In this way, the MBP enables the interpretation of results directly in terms of the physics. The MBP is really a modeler's tool, enabling the incorporation of any *a priori*

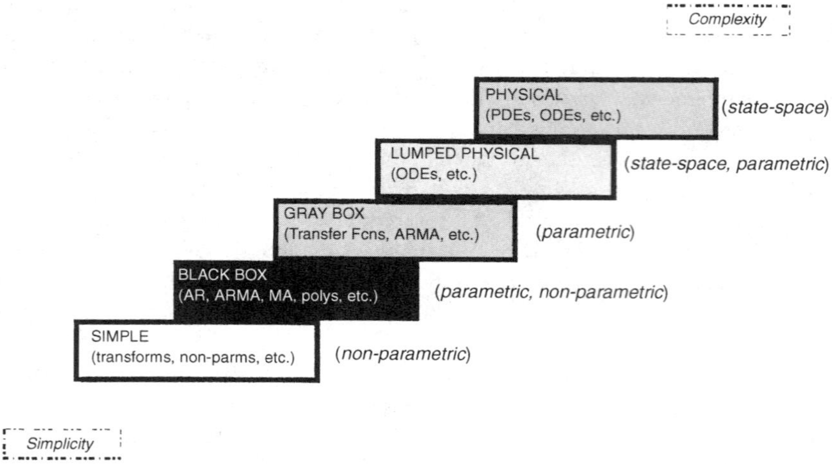

FIGURE 5.2 Fidelity of the embedded model determines the complexity of the resulting MBP required to achieve the desired SNR: simple implied model (Fourier, wavelet, etc.), black-box model (data prediction model), gray-box model (implied physical model), lumped physical model (differential equations), full physical model (partial differential equations).

information about the problem to extract the desired information. The fidelity of the model incorporated into the processor determines the complexity of the MBP. These models can range from simple, implied non-physical representations of the measurement data, such as the Fourier or wavelet transforms, to parametric black-box models used for data prediction to lumped mathematical physical representations usually characterized by ordinary differential equations to full physical partial differential equation models capturing the critical details of wave propagation in a complex medium. The dominating factor of which model is the most appropriate is usually determined by how severely contaminated the measurements are with noise and underlying uncertainties. If the signal-to-noise ratio (SNR) of the measurements is high, then simple non-physical techniques can be used to extract the desired information. This approach of selecting the appropriate model is depicted in Figure 5.2, where we note that as we progress up the "modeling" steps to increase the SNR, the complexity of the model increases to achieve the desired results.

For our problem in ocean acoustics, the model-based approach is shown in Figure 5.3. The underlying physics are represented by an acoustic propagation model depicting how the sound propagates from a source or target to the sensor measurement array of hydrophones. Noise in the form of background or ambient noise, shipping noise, uncertainty in the model parameters, etc. is shown in the figure as input to both the propagation and measurement system models. Besides the model parameters and initial conditions, the raw measurement data is input to the model with the output being the filtered or enhanced signal.

Before we develop the MBP for ocean acoustic applications, let us motivate the approach with a simple example taken from ocean acoustics. Suppose we have a plane wave signal characterized by

$$s_k(t) = ae^{i\beta_k \sin\theta_o - i\omega_o t} \tag{5.1}$$

where $s_k(t)$ is the space-time signal measured by the kth sensor, and a is the plane wave amplitude factor with β, θ_o, ω_o as the respective wavenumber, bearing, and temporal frequency parameters. Let us further assume that the signal is measured by a horizontal array. A simple, but important example in ocean acoustics is that of a 50 Hz plane wave source (target) at a bearing of 45° impinging on a 2-element array at a 10 dB SNR. We would we like to solve two basic ocean acoustic processing problems: (1) signal

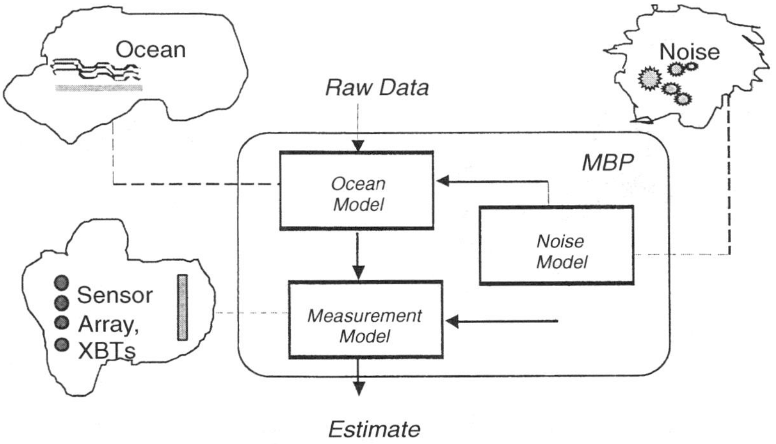

FIGURE 5.3 Model-based approach: structure of the MBP showing the incorporation of propagation (ocean), measurement (sensor array), and noise (ambient) models.

enhancement, and (2) extraction of the source bearing, θ_o, and temporal frequency, ω_o, parameters. The basic problem geometry and synthesized measurements are shown in Figure 5.4.

The signal enhancement problem can be solved classically by constructing a 50 Hz bandpass filter with a narrow 1 Hz bandwidth and filtering each channel, while the model-based approach would be to define the various models as described in Figure 5.3 and incorporate them into the processor structure. For the *plane wave enhancement problem*, we have the following models:

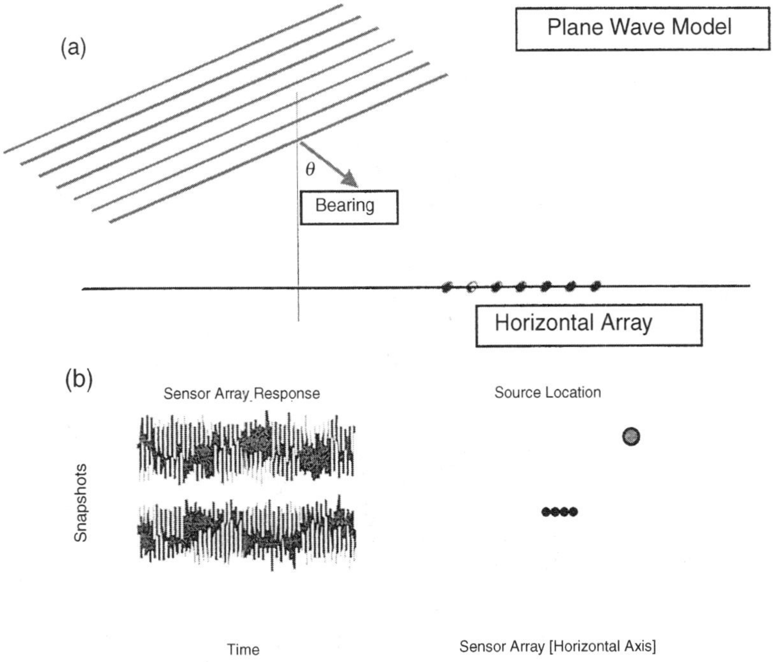

FIGURE 5.4 Plane wave propagation: (a) problem geometry; (b) synthesized 50 Hz, 45°, plane wave impinging on a 2-element sensor array at 10 dB SNR.

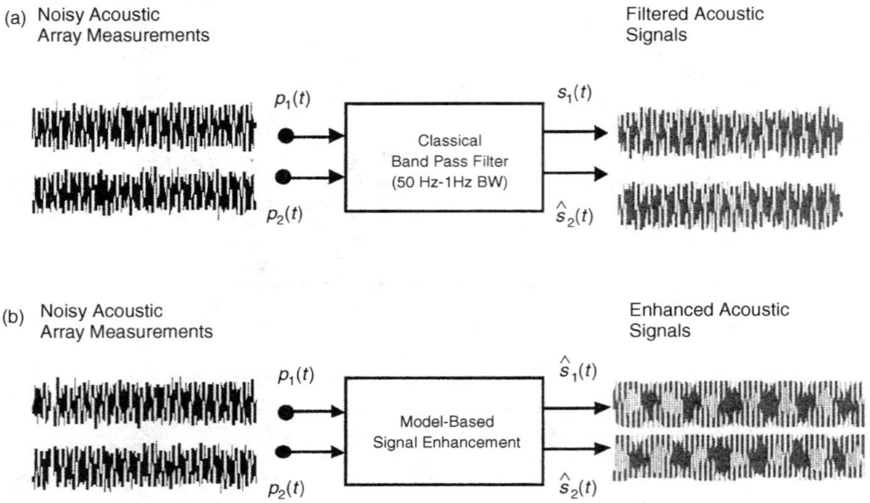

FIGURE 5.5 Plane wave signal enhancement problem: (a) classical bandpass filter (50 Hz to 1 Hz BW) approach, (b) MBP using 50 Hz, 45°, plane wave model impinging on a 2-element sensor array.

Signal model:

$$s_k(t) = ae^{i\beta_k \sin\theta_o - i\omega_o t} \tag{5.2}$$

Measurement/noise model:

$$p_k(t) = s_k(t) + n_k(t) \tag{5.3}$$

The results of the classical and MBP outputs are shown in Figure 5.5. In Figure 5.5a, the classical bandpass filter design is by no means optimal, as noted by the random amplitude fluctuations created by the additive measurement noise process discussed above. The output of the optimal MBP is, in fact, optimal for this problem, since it is incorporating the correct propagation (plane wave), measurement (hydrophone array), and noise (white, Gaussian) into its internal structure. We observed the optimal outputs in Figure 5.5b.

Summarizing this simple example, we see that the classical processor design is based on *a priori* knowledge of the desired signal frequency (50 Hz), but cannot incorporate knowledge of the propagation physics or noise into its design. On the other hand, the MBP *uses* the *a priori* information about the plane wave propagation signal and sensor array measurements along with any *a priori* knowledge of the noise process in the form of mathematical models embedded in its processing scheme to achieve optimal performance. We will discuss the concept of optimal processor performance in Section 5.2.7. Note that conceptually there are other processors that could be used to represent the classical design (e.g., autoregressive filters) solution to this filtering problem, here we are just attempting to tutorially motivate the application of the MBP. The next variant to this problem is even more compelling.

Consider now the same plane wave, the same noisy hydrophone measurements, with the more realistic objective of estimating the bearing and temporal frequency of the target. In essence, this is a problem of estimating a set of parameters, $\{\theta_o, \omega_o\}$, from noisy array measurements, $\{p_k(t)\}$. The classical approach to this problem is to first take either sensor channel and perform spectral analysis on the filtered time series to estimate the temporal frequency, ω_o. The bearing can be estimated independently by performing classical beamforming[38] on the array data. A beamformer can be considered a spatial spectral estimator which is scanned over bearing angle, indicating the true source location at maximum power. The results of applying this approach to our problem are shown in Figure 5.6a, showing the outputs of both spectral estimators peaking at the correct frequency and angle parameters.

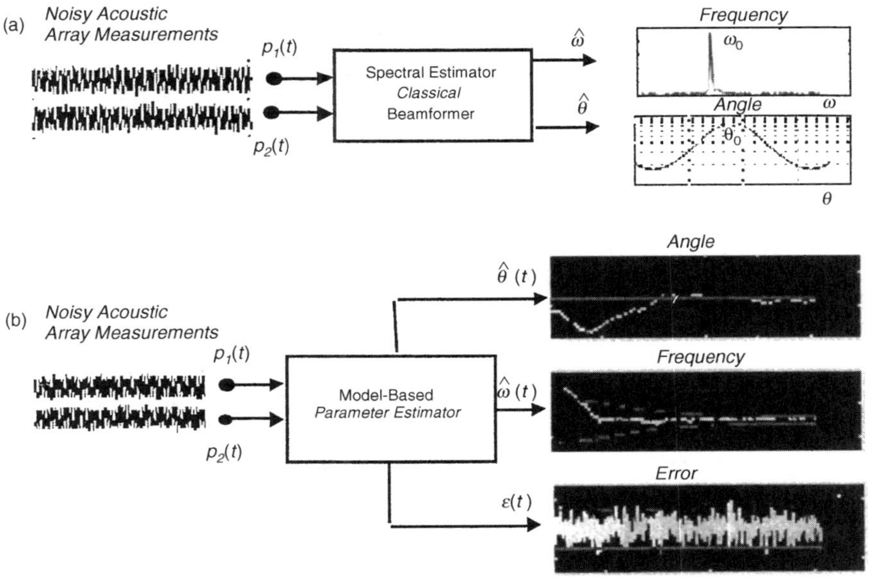

FIGURE 5.6 Plane wave impinging on a 2-element sensor array — Frequency and Bearing Estimation Problem: (a) classical spectral (temporal and spatial) estimation approach; (b) model-based approach using a parametric adaptive (non-linear) processor to estimate bearing angle, temporal frequency, and the corresponding residual or innovations sequence.

The MBP is implemented as before by incorporating the plane wave propagation, hydrophone array, and noise models; however, the temporal frequency and bearing angle parameters are now unknown and must be estimated along with simultaneous enhancement of the signals. The solution to this problem is performed by "augmenting" the unknown parameters into the MBP structure and solving the so-called joint estimation problem.[1,2] This is the *parameter adaptive* form of the MBP used in most ocean acoustic applications. Here, the problem becomes non-linear due to the augmentation and is more computationally intensive; however, the results are appealing, as shown in Figure 5.6b. Here, we see the bearing angle and temporal frequency estimates as a function of time eventually converging to the true values (ω_o = 50 Hz, θ_o = 45°). The MBP also produces the "residual sequence" (shown in Figure 5.6b), which is used in determining its overall performance. We will discuss this in Section 5.2.2.

Next, we summarize the classical and model-based solutions to the temporal frequency and bearing angle estimation problem. The classical approach simply performs spectral analysis temporally and spatially (beamforming) to extract the parameters from noisy data, while the model-based approach embeds the unknown parameters into its propagation, measurement, and noise models through augmentation, enabling a solution to the joint estimation problem. The MBP also enables a monitoring of its performance by analyzing the statistics of its residual or innovations sequence. It is this sequence that indicates the optimality of the MBP outputs. This completes the simple example, next, we provide a brief overview of model-based signal processing which will eventually lead us to solutions of various ocean acoustic problems.

5.2.2 Overview

Model-based signal processing algorithms are based on a well-defined procedure. First, the relevant phenomenology, which can be described by a deterministic model, is placed in the form of a Gauss-Markov model which, among other things, allows modeling and measurement uncertainties to be represented by stochastic components. This allows the states, or in the shallow water ocean case, the modal functions, and the associated horizontal wave numbers of the normal-mode model to be estimated by a recursive MBP. Note that since the model is Gauss-Markov, the resulting optimal MBP is the well-known Kalman filter.[1,2] Usually, in ocean acoustics, we are not only interested in the estimates of the states, but also in estimates of model parameters. These are found by augmenting them into the Gauss-Markov model, that is, the parameter vector is augmented into the state vector, thereby enlarging the overall dimension of the system. This process, the inclusion of the model parameters as states, is commonly referred to as *parameter adaptive estimation* or *identification*. Finally, since the states depend upon the parameters in a non-linear manner, it is necessary to replace the standard linear Kalman filter algorithm with the so-called *extended Kalman filter* (EKF).[1,2] It is the intent of Sections 5.2.3 to 5.2.7 to provide a (somewhat abbreviated) description of this procedure as a preparation for the applications in Section 5.4 in which we will apply the model-based approach to various problems of the parameter adaptive processing in the hostile ocean environment.

5.2.3 Gauss-Markov Model

The first step is to place the "dynamic" model into state-space form. The main advantage is that it decomposes higher order differential equations to a set of first-order equations. As will be seen, this enables the dynamics to be represented in the form of a first-order Markov process. Generically, we will be considering a "spatial" processor. The model, which is driven by its initial conditions, is placed in state-space form as

$$\frac{d}{dz}\mathbf{x}(z) = \mathbf{A}(z)\mathbf{x}(x) \tag{5.4}$$

where z is the spatial variable. This model leads to the following Gauss-Markov representation of the model when additive white, Gaussian noise or uncertainty is assumed:

$$\frac{d}{dz}\mathbf{x}(z) = \mathbf{A}(z)\mathbf{x}(z) + \mathbf{w}_x(z) \tag{5.5}$$

where $\mathbf{w}_x(z)$ is additive, zero-mean, white, Gaussian noise with corresponding covariance matrix, $\mathbf{R}_{w_x w_x}$. The system or process matrix is defined by $\mathbf{A}(z)$. Also, the corresponding measurement model can be written as

$$y(z) = \mathbf{C}(z)\mathbf{x}(z) + v(z)\,, \tag{5.6}$$

where y is the measurement, and C is the measurement matrix. The random noise terms in the model $\mathbf{w}_x(z)$ and $v(z)$ are assumed to be Gaussian and zero mean with respective covariance matrices, $\mathbf{R}_{w_x w_x}$ and $\mathbf{R}_{vv}$. In ocean acoustics, the measurement noise, $v(z)$, can be used to represent the lumped effects of near-field acoustic noise field, flow noise on the hydrophone, and electronic noise. The process noise, $\mathbf{w}_x(z)$, can be used to represent the lumped uncertainty of sound speed errors, distant shipping noises, errors in the boundary conditions, sea state effects, and ocean inhomogeneities that propagate through the ocean acoustic system dynamics (normal-mode model). Having our Gauss-Markov model in hand, we are now in a position to implement a recursive estimator in the form of a Kalman filter. Kalman filtering theory is a study in itself, and the interested reader is referred to References 1 and 2 for more detailed information. For our purposes, however, it will be sufficient to point out that the Kalman filter seeks a minimum error variance estimate of the states and measurement.

5.2.4 Model-Based Processor (Kalman Filter)

Since we will be concerned with discrete sensor arrays that spatially sample the pressure field, we choose to discretize the differential state equations using a finite (first) difference approach (central differences can also be used for improved numerical stability), that is,

$$\frac{d}{dz}\mathbf{x}(z) \approx \frac{\mathbf{x}(z_{\ell+1}) - \mathbf{x}(z_\ell)}{\Delta z} \tag{5.7}$$

where $\Delta z = z_{\ell+1} - z_\ell$ and z_ℓ is the location of the ℓth sensor array element. Substituting Equation 5.7 into Equation 5.5 leads to the discrete form of our Gauss-Markov model given by

$$\mathbf{x}(z_{\ell+1}) = [I + \Delta z \mathbf{A}(z_\ell)]\,\mathbf{x}(z_\ell) + \Delta z \mathbf{w}_x(z_\ell) \tag{5.8}$$

Once the Gauss-Markov model is established, we are in a position to define our recursive model-based (Kalman) processor for the state vector $\mathbf{x}$ as

$$\hat{\mathbf{x}}(z_{\ell+1 z_l}) = \mathbf{A}_x(z_\ell)\hat{\mathbf{x}}(z_\ell|z_\ell) \tag{5.9}$$

where $\mathbf{A}_x(z_\ell) = [I + \Delta z\mathbf{A}(z_\ell)]$ and the "hat" signifies estimation. Note that the notation has been generalized to reflect the recursive nature of the process. $\hat{\mathbf{x}}(z_{\ell+1}|z_\ell)$ is the predicted estimate of the state vector $\mathbf{x}$ at depth $z_{\ell+1}$ based on the data up to depth z_ℓ, and $\hat{\mathbf{x}}(z_\ell|z_\ell)$ is the corrected or "filtered" value of the estimate of $\mathbf{x}$ at depth z_ℓ. We now have everything in place to outline the actual recursive MBP algorithm shown in Table 5.1.

Prediction
Given the initial or trial value of the state vector, $\hat{\mathbf{x}}(z_\ell|z_\ell)$, Equation 5.9 is used to provide the predicted estimate $\hat{\mathbf{x}}(z_{\ell+1}|z_\ell)$. This constitutes a prediction of the state estimate at depth $z_{\ell+1}$ based on the data up to depth z_ℓ.

Innovation
The residual or innovation, $\varepsilon(z_{\ell+1})$, is then computed as the difference between the new measurement, $p(z_{\ell+1})$, that is, the measurement taken at $z_{\ell+1}$, and the predicted measurement $\hat{p}(z_{\ell+1})$, obtained by substituting $\hat{\mathbf{x}}(z_{\ell+1}|z_\ell)$ into the measurement equation. Thus,

$$\varepsilon(z_{\ell+1}) = p(z_{\ell+1}) - \hat{p}(z_{\ell+1}) = p(z_{\ell+1}) - \mathbf{C}^T(z_s)\hat{\mathbf{x}}(z_{\ell+1}|z_\ell) \tag{5.10}$$

TABLE 5.1 Spatial Kalman Filter Algorithm (Predictor/Corrector Form)

Prediction	
$\hat{x}(z_\ell\|z_{\ell-1}) = A(z_{\ell-1})\hat{x}(z_{\ell-1}\|z_{\ell-1})$	(state prediction)
$P(z_\ell\|z_{\ell-1}) = A(z_{\ell-1})\tilde{P}(z_{\ell-1}\|z_{\ell-1})A'(z_{\ell-1}) + W(z_{\ell-1})R_{ww}(z_{\ell-1})W'(z_{\ell-1})$	(covariance prediction)
Innovation	
$\varepsilon(z_\ell) = y(z_\ell) - \hat{y}(z_\ell\|z_{\ell-1}) = y(z_\ell) - C(z_\ell)\hat{x}(z_\ell\|z_{\ell-1})$	(innovation)
$R_{\varepsilon\varepsilon}(z_\ell) = C(z_\ell)\tilde{P}(z_\ell\|z_{\ell-1})C'(z_\ell) + R_{vv}(z_\ell)$	(innovation covariance)
Gain	
$K(z_\ell) = \tilde{P}(z_\ell\|z_{\ell-1})C'(z_\ell)R_{\varepsilon\varepsilon}^{-1}(z_\ell)$	(Kalman gain or weight)
Correction	
$\hat{x}(z_\ell\|z_\ell) = \hat{x}(z_\ell\|z_{\ell-1}) + K(z_\ell)e(z_\ell)$	(state correction)
$\tilde{P}(z_\ell\|z_\ell) = [I - K(z_\ell)C(z_\ell)]\tilde{P}(z_\ell\|z_{\ell-1})$	(covariance correction)
Initial Conditions	
$\hat{x}(0\|0) \quad \tilde{P}(0\|0)$	

Gain

Next, the Kalman gain or weight is computed. It is given by

$$\mathbf{K}(z_{\ell+1}) = \tilde{\mathbf{P}}(z_{\ell+1}|z_\ell)\mathbf{C}(z_s)\mathbf{R}_{\varepsilon\varepsilon}^{-1}(z_{\ell+1}) \tag{5.11}$$

where $\tilde{\mathbf{P}}$ is the error covariance of the state estimate, and $\mathbf{R}_{\varepsilon\varepsilon}$ is the innovation covariance.

Correction

The Kalman gain is then used in the correction equation as follows:

$$\hat{\mathbf{x}}(z_{\ell+1}|z_{\ell+1}) = \hat{\mathbf{x}}(z_{\ell+1}|z_\ell) + \mathbf{K}(z_{\ell+1})\varepsilon(z_{\ell+1}) \tag{5.12}$$

This corrected estimate or filtered value of the state vector is then inserted into the right-hand side of the prediction equation, thereby initiating the next recursion.

The algorithm described above is based on the *linear* Kalman filter. In our parameter adaptive problem, the system is non-linear. This leads to the EKF, which we will discuss subsequently. The EKF algorithm formally resembles the same steps as the linear algorithm outlined above and is given in detail in Tables 5.1 and 5.2, respectively.

5.2.5 Augmented Gauss-Markov Model

In order to streamline the notation somewhat, we shall denote the unknown parameters to be estimated as θ. Referring to Equation 5.4, we see that our *augmented* state vector for the result in the *augmented* Gauss-Markov model is given by

$$\frac{d}{dz}\underline{\mathbf{x}}(z) = \underline{\mathbf{A}}(z, \theta)\mathbf{x}(z) + \underline{\mathbf{w}}(z) \tag{5.13}$$

where $\underline{\mathbf{w}}(z)$ is an additive Gaussian noise term. The dependence of $\mathbf{A}$ on θ denotes the inclusion of the unknown model parameters in the system matrix. Modeling the unknown parameters as constants in $\underline{\mathbf{A}}(z, \theta)$ simply means that there are no "dynamics" associated with the parameter vector θ or the parameter is modeled as a random walk.[1,2] Thus, the Gauss-Markov model for θ can be written as

$$\frac{d}{dz}\theta(z) = \mathbf{O} + \mathbf{w}_\theta(z) \tag{5.14}$$

where the depth dependence of θ is purely stochastic, modeled by $\mathbf{w}_\theta(z)$. The new measurement model corresponding to the augmentation is now non-linear, since the measurement can no longer be written as a linear function of the state vector $\underline{\mathbf{x}}(z)$. This means that the linear Kalman filter algorithm (see References 1 and 2) outlined previously is no longer valid, which leads us to the non-linear EKF.

5.2.6 Extended Kalman Filter

The non-linear measurement equation is now written as

$$p(z) = \mathbf{c}[\underline{\mathbf{x}}(z)] + v(z) \tag{5.15}$$

where the $\mathbf{c}[\bullet]$ is a non-liner vector function of x.

The EKF is based on approximating the non-linearities by a first-order Taylor's series expansion. This means that we will require the first derivatives of the elements of the augmented state vector and the measurements with respect to these state vector elements. These derivatives form the so-called *Jacobian* matrices.

TABLE 5.2 Discrete Extended Kalman Filter Algorithm (Predictor/Corrector Form)

Predictor

$$\hat{x}(z_{\ell+1}|z_\ell) = a[\hat{x}(z_{\ell+1}|z_\ell)] + b[u(z_\ell)] \quad \text{(state prediction)}$$

$$\tilde{P}(z_{\ell+1}|z_\ell) = A[\hat{x}(z_{\ell+1}|z_\ell)]\tilde{P}(z_{\ell+1}|z_\ell)A^T[\hat{x}(z_{\ell+1}|z_\ell)] + R_{ww}(z_\ell) \quad \text{(covariance prediction)}$$

Innovation

$$\varepsilon(z_{\ell+1}) = y(z_{\ell+1}) - \hat{y}(z_{\ell+1}|z_\ell) = y(z_{\ell+1}) - c[\hat{x}(z_{\ell+1}|z_\ell)] \quad \text{(innovation)}$$

$$R_{\varepsilon\varepsilon}(z_{\ell+1}) = C[\hat{x}(z_{\ell+1}|z_\ell)]\tilde{P}(z_{\ell+1}|z_\ell)C^T[\hat{x}(z_{\ell+1}|z_\ell)] + R_{vv}(z_{\ell+1}) \quad \text{(innovation covariance)}$$

Gain

$$K(z_{\ell+1}) = \tilde{P}(z_{\ell+1}|z_\ell)C^T[\hat{x}(z_{\ell+1}|z_\ell)]R_{\varepsilon\varepsilon}^{-1}(z_{\ell+1}) \quad \text{(Kalman gain or weight)}$$

Correction

$$\hat{x}(z_{\ell+1}|z_{\ell+1}) = \hat{x}(z_{\ell+1}|z_\ell) + K(z_{\ell+1})\varepsilon(z_{\ell+1}) \quad \text{(state correction)}$$

$$P(z_{\ell+1}|z_{\ell+1}) = [I - K(z_{\ell+1})C[\hat{x}(z_{\ell+1}|z_{\ell+1})]\tilde{P}(z_{\ell+1}|z_\ell)] \quad \text{(covariance correction)}$$

Initial Conditions

$$x(0|0) \qquad P(0|0) \qquad A[\hat{x}(z_{\ell+1}|z_\ell)] \equiv \frac{\partial}{\partial x}a[x]\bigg|_{x=\hat{x}(z_{\ell+1}|z_\ell)} \qquad C[\hat{x}(z_{\ell+1}|z_\ell)] \equiv \frac{\partial}{\partial x}c[x]\bigg|_{x=\hat{x}(z_{\ell+1}|z_\ell)}$$

As previously mentioned, the actual algorithm for the EKF formally resembles the algorithm for the linear Kalman filter of Table 5.1 and is shown in Table 5.2. Since the derivation of the EKF algorithm is somewhat complicated and would detract from the main point of this Chapter, the interested reader is directed to References 1 and 2.

5.2.7 Model-Based Processor Design Methodology

Here we discuss design methodology: the design of an MBP requires the following basic steps: (1) model development, (2) simulation/minimum-variance design, (3) application ("tuning") to data sets, and (4) performance analysis.

In the modeling first step, we develop the mathematical models of the propagation dynamics, measurement system (sensor array), and noise sources (assumed Gaussian). Once the state-space models have been specified, we next search for a set of parameters/data to initialize the processor. The parameters/data can come from a historical database of the region, from work performed by others in the same or equivalent environment, from previous detailed "truth" model computer simulations, or merely from educated guesses based on any *a priori* information and experience of the modeler/designer.

After gathering all of the parameters/data information and employing it in our state-space models, we perform a Gauss-Markov simulation, the second step, where the use of *additive noise sources* in this formulation enable us to "lump" the uncertainties evolving from initial conditions, $\mathbf{x}(0) \sim N(\hat{\mathbf{x}}(0), \hat{\mathbf{P}}(0))$; propagation dynamics, $\mathbf{w} \sim N(0, \mathbf{R}_{ww})$; and measurement noise, $\mathbf{v} \sim N(0, \mathbf{R}_{vv})$. Note that we are not attempting to "precisely" characterize the uncertainties as such specific entities as modal noise, flow noise, shipping noise, etc., because building a model for each would involve a separate Gauss-Markov representation which would then eventually be augmented into the processor.[1,2] Here, we simply admit to our "ignorance" and lump the uncertainty into additive Gaussian noise processes controlled by their inherent covariances. Once we have specified the parameters/data, we are now ready to perform Gauss-Markov simulations. This procedure is sometimes referred to as "sanity" testing. We typically use the following definitions of SNR in deciding what signal/noise levels to perform the simulations at:

$$\mathbf{SNR}_{in} \equiv \frac{\mathbf{P}_{m,\,m}}{diag[\mathbf{R}_{ww}]} \tag{5.16}$$

and

$$\mathbf{SNR}_{out} \equiv \frac{\mathbf{C}(z_\ell)\mathbf{P}_{m,\,m}\mathbf{C}^T(z_\ell)}{diag[\mathbf{R}_{vv}]},\ \ell = 1, \ldots, L \tag{5.17}$$

where $\mathbf{P}_{m,\,m} = diag[\mathbf{P}]$ is the state covariance ($Cov(\mathbf{x}_m)$), $\mathbf{C}(z_\ell)$ is the measurement model of Equation 5.6, and $\mathbf{R}_{ww}$ and $\mathbf{R}_{vv}$ are the noise covariances discussed in the previous section.

Note also that when we augment the unknown parameters into the state vector to construct our parameter adaptive processor, we assume that they are also random (walks) with our precomputed initial values specified (initial conditions or means) and their corresponding covariances are used to bound their uncertainty (2σ confidence bounds). In fact, if we know more about how they evolve dynamically or we would like to constrain their values, we can place "hard" limits directly into the processor. For instance, suppose we know that the states are constrained; therefore, we can limit the corresponding parameter excursions practically to (θ_{min}, θ_{max}) to avoid these erroneous states values by using an augmented parameter model such as

$$\theta(z_{\ell+1}) = \theta(z_\ell),\ \theta_{min} < \theta(z_\ell) < \theta_{max} \tag{5.18}$$

thereby constraining any excursions to remain within this interval. Of course, the random walk model certainly can provide "soft" constraints in the simulation, since the parameter is modeled as Gauss-Markov, implying that 95% of the samples must lie within confidence limits controlled by ($\pm 1.96\sqrt{\mathbf{P}_{m,\,m}}$). This constitutes a *soft* statistical constraint of the parameter variations. However, this approach does not guarantee that the processor will remain within this bound; therefore, hard constraints may offer a better alternative.

Once the Gauss-Markov simulation is complete, the processor is designed to achieve a minimum-variance estimate in the linear model case or a best mean-squared-error estimate in the non-linear model case — this is called *minimum-variance design*. Since the models used in the processor are "exactly" those used to perform the simulation and synthesize the design data, we will have as output the minimum-variance or best mean-squared-error estimates. For the non-linear case, the estimates are deemed approximate minimum variance because the EKF processor uses linearization techniques which approximate the non-linear functions in order to obtain the required estimates. We expect the processor to perform well using the results of the various Gauss-Markov simulations to bound its overall expected performance on real data. We essentially use simulations at various SNRs to obtain a "feel" for learning how to tune (adjusting noise covariances etc.) the processor using our particular model sets and parameters.

Having completed the simulations and studied the sensitivity of the processor to various parameter values, we are ready to attack the actual data set. With real data, the processor can only perform as well as the dynamical models being used represent the underlying phenomenology generating the data. Poor models used in the processor can actually cause the performance to degrade substantially, and enhancement may not be possible at all in contrast to the simulation step where we have assured that the model "faithfully" represents the data. In practice, it is never possible to accurately model everything. Thus, the goal of minimum-variance design is to achieve as close to the optimal design as possible by investigating the consistency of the processor relative to the measured data. This is accomplished by utilizing the theoretical properties of the processor,[1,2] that is, the processor is deemed *optimal* if, and only if, the residual/innovations sequence is zero mean and statistically white (uncorrelated). This approach to performance analysis is much like the results in time series/regression analysis which implies that when the model "explains" or fits the data, nothing remains and the residuals are uncorrelated.[39,40] Therefore, when applying the processor to real data, it is necessary to adjust or "tune" the model parameters (usually the elements of the process noise covariance matrix, $\mathbf{R}_{ww}$; see References 1 and 2 for details) until the innovations are zero mean/white. If it is not possible to achieve this property, then the models are deemed inadequate and must be improved by incorporating more of the phenomenology. The important point here is that the model-based schemes enable the modeler to assess how well the model is performing on real data and decide where it may be improved. For instance, for the experimental data, we can statistically test the innovations and show that they are white, but when we visually observed the sample correlation function estimate used in the test, it is clear that there still remains some correlation in the innovations. This leads us, as modelers, to believe that we have not captured all of the phenomenology that has generated the data, and therefore, we must improve the model or explain why the model is inadequate.

Care must be taken when using these statistical tests as noted in References 2 and 39 to 44, because if the models are non-linear or non-stationary, then the usual whiteness/zero-mean tests, that is, testing that 95% of the sample (normalized) innovation correlations lie within the bounds given by

$$\left[\hat{c}_{\varepsilon\varepsilon}(k) \pm \frac{1.96}{\sqrt{N}}\right], \hat{c}_{\varepsilon\varepsilon}(k) = \frac{R_{\varepsilon\varepsilon}(k)}{R_{\varepsilon\varepsilon}(0)} \tag{5.19}$$

and testing for zero mean as

$$\left[\hat{m}_{\varepsilon}(k) < 1.96\sqrt{\frac{\hat{R}_{\varepsilon\varepsilon}(k)}{N}}\right] \tag{5.20}$$

rely on quasi-stationary assumptions and sample statistics to estimate the required correlations. However, it can be argued heuristically that when the estimator is tuned, the non-stationarities are being tracked by the MBP even in the non-linear case, and therefore, the innovations should be covariance stationary.

When data are non-stationary, then a more reliable statistic to use is the *weighted sum-squared residual* (WSSR), which is a measure of the overall global estimation performance for the MBP processor, determining the "whiteness" of the innovations sequence.[2,39] It essentially aggregates all of the information available in the innovation vector and tests whiteness by requiring that the decision function lies below the specified threshold to be deemed statistically white. If the WSSR statistic does lie beneath the calculated threshold, then theoretically, the estimator is tuned and said to converge. That is, for sensor array measurements, we test that the corresponding innovations sequence is zero mean/white by performing a statistical hypothesis test against the threshold. The WSSR statistic essentially aggregates all of the information available in the innovation vector over some finite window of N samples, that is, the WSSR defined by $\rho(k)$ is

$$\rho(k) \equiv \sum_{k=\ell-N+1}^{\ell} \varepsilon^{\dagger}(t_k)\mathbf{R}_{\varepsilon\varepsilon}^{-1}(k)\varepsilon(t_k) \qquad \ell \geq N \tag{5.21}$$

$$\rho(k) \underset{< H_0}{\overset{> H_1}{}} \quad \tau \tag{5.22}$$

where H_0 is the hypothesis that there is no model "mismatch" (white innovations), while H_1 is the hypothesis that there is mismatch specified by non-zero-mean, non-white innovations. Under the zero-mean assumption, the WSSR statistic is equivalent to testing that the vector innovation sequence is white. Under H_0, $\rho(k)$ is distributed as $\chi^2(NL)$. It is possible to show[2,39] for a large $NL > 30$ (L the number of hydrophones) and a level of significance of $\alpha = 0.05$, that

$$\tau = NL + 1.96\sqrt{2NL} \tag{5.23}$$

and the WSSR statistic must lie below the calculated threshold, τ.

Here, the window is designed to slide through the innovations data and estimate its whiteness. Even in the worst case, where these estimators may not prove to be completely consistent, the processor (when tuned) predicts the non-stationary innovations covariance, $\mathbf{R}_{\varepsilon\varepsilon}(z_\ell)$, enabling a simple (varying with ℓ) confidence interval to be constructed and used for testing. This confidence interval is

$$[\varepsilon(z_\ell) \pm 1.96\sqrt{\mathbf{R}_{\varepsilon\varepsilon}(z_\ell)}],\ \ell = 1, \ldots, L \tag{5.24}$$

Thus, overall performance of the processor can be assessed by analyzing the statistical properties of the innovations. There are other tests that can be used with real data to check the consistency of the processor, and we refer the reader to Chapter 5, Reference 2, for more details.

5.3 State-Space Ocean Acoustic Forward Propagators

The ocean is an extremely hostile environment compared to other media. It can be characterized by random, non-stationary, non-linear, space-time varying parameters that must be estimated to "track" its dynamics. As shown in Figure 5.7, not only does the basic ocean propagation medium depend directly on its changing temperature variations effecting the sound speed directly, but also on other forms of noise and clutter that create uncertainty and contaminate any array measurements. The need for a processor is readily apparent. However, the hostile operational environment places unusual demands on it. For instance, the processor should be capable of (1) "learning" about its own operational environment including clutter, (2) "detecting" a target with minimal target information, (3) "enhancing" the target signal while removing both clutter and noise, (4) "localizing" the target position, and (5) "tracking" the target as it moves. An MBP is capable of satisfying these requirements, but before we motivate the processor, we must characterize the shallow ocean environment. In this section, we investigate the development of state-space signal processing models from the corresponding ocean acoustic normal-mode solutions to the wave equation. The state-space models will eventually be employed as "forward" propagators in model-based signal processing schemes.[2] Note that this approach does not offer a new solution to the resulting boundary value problem, but, in fact, requires that a solution be available *a priori* in order to propagate the normal modes recursively in an initial value scheme.

For our propagation model, we assume a horizontally stratified ocean of depth h with a *known* source position (x, y, z). We assume that the acoustic energy from a point source propagating over a long range, $r\ (r >> h)$, toward a receiver can be modeled as a trapped wave characterized by a waveguide phenomenon. For a layered waveguide model with sources on the z (or vertical)axis, the pressure field p is symmetric about z (with known source bearing) and, therefore, is governed by the cylindrical wave equations which is given by

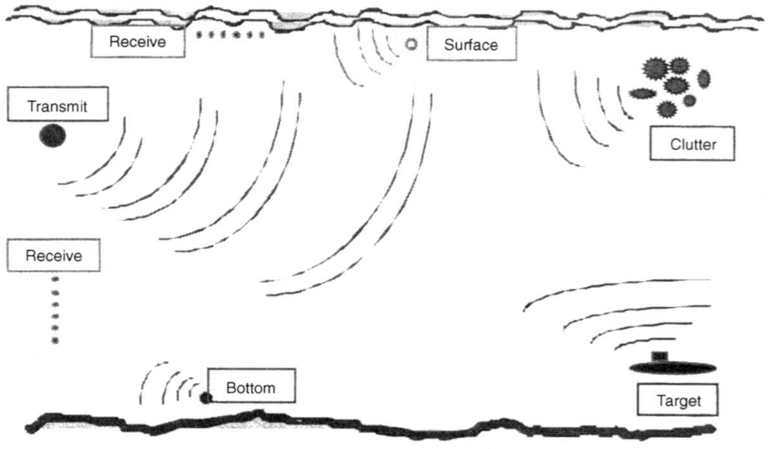

FIGURE 5.7 Complex ocean acoustic environment: transmit/receive arrays, surface scattering, bottom interaction, clutter, target, and ambient noise.

$$\frac{\partial^2}{\partial r^2}p(r, z, t) + \frac{1}{r}\frac{\partial}{\partial r}p(r, z, t) + \frac{\partial^2}{\partial z^2}p(r, z, t) = \frac{1}{c^2}\frac{\partial^2}{\partial t^2}p(r, z, t). \tag{5.25}$$

The solution to this equation is accomplished by using the separation of variables technique, that is,

$$p(r, z, t) = \mu(r)\phi(z)T(t) \tag{5.26}$$

Substituting Equation 5.26 into Equation 5.25, assuming a harmonic source

$$T(t) = e^{j\omega t} \tag{5.27}$$

and defining separation constants κ_z, κ_r, we obtain the following set of ordinary differential equations:

$$\begin{aligned} \frac{d^2}{dr^2}\mu(r) + \frac{1}{r}\frac{d}{dr}\mu(r) &= -\kappa_r^2\mu(r) \\ \frac{d^2}{dz^2}\phi(z) &= -\kappa_z^2\phi(z) \\ \kappa^2 &= \frac{\omega^2}{c^2(z)} \\ \kappa^2 &= \kappa_r^2 + \kappa_z^2 \end{aligned} \tag{5.28}$$

where solutions to each of these relations describe the propagation of the acoustic pressure in cylindrical coordinates assuming the harmonic source of Equation 5.27 and the speed of sound a function of depth, $c = c(z)$.

An approximation to the solution of the range-dependent relation of Equation 5.28 is given by the Hankel function approximation as

$$\mu(r) = \frac{1}{\sqrt{2\pi\kappa_r r}}e^{-j\left(\kappa_r r - \frac{\pi}{4}\right)} \tag{5.29}$$

and shows that the waves spread radially from the source and are attenuated by $1/\sqrt{r}$.

The depth relation of Equation 5.28 is an eigenvalue equation in z with

$$\frac{d^2}{dz^2}\phi_m(z) + \kappa_z(m)\phi_m(z) = 0, m = 1, \ldots, M \tag{5.30}$$

whose eigensolutions $\{\phi_m(z)\}$ are the *modal functions*, and κ_z is the wavenumber in the z-direction. These solutions depend on the sound speed profile, $c(z)$, and the boundary conditions at the surface and bottom.

Using the orthogonality property of the modal functions with a known point source located at z_s, we obtain the total wavenumber as

$$\kappa^2 = \frac{\omega^2}{c^2(z)} = \kappa_r^2(m) + \kappa_z^2(m), m = 1, \ldots, M \tag{5.31}$$

where κ_r, κ_z are the respective wave numbers in the r and z directions with c the depth-dependent sound velocity profile and ω the harmonic source frequency.

For our purpose, we are concerned with the estimation of the pressure field, therefore, we remove the time dependence, normalize units, and obtain the *acoustic pressure propagation model*,

$$p(r,z) = q\sum_{m=1}^{M}\phi_m(z_s)\phi_m(z)\frac{e^{-\alpha_r(m)r}}{\sqrt{\kappa_r(m)r}}e^{j\kappa_r(m)r} \tag{5.32}$$

where p is the acoustic pressure, q is the source intensity, α is the attenuation coefficient, ϕ_m is the mth modal function at z and z_s, $\kappa_r(m)$ is the horizontal wavenumber associated with the mth mode, and r is the horizontal range.

To develop the state-space forward propagation model, we use the cylindrical wave equation and make the appropriate assumptions leading to the normal-mode solutions of Equation 5.28. Recall that the depth equation is an eigenvalue equation whose eigensolutions $\phi_m(z)$ are the *modal functions* and κ_z is the wavenumber in the z-direction depending on the sound velocity profile, $c(z)$, and the boundary conditions at the surface and bottom. As before, the mth mode satisfies the depth relation of Equation 5.30.

Since the depth relation is a linear, space-varying coefficient (for each layer) differential equation, it can easily be placed in state-space form where the state vector is defined as $\underline{x}_m := [\phi_m(z)\ (d/dz)\phi_m(z)]'$. Examining mode propagation in more detail, we see that each mode is characterized by a set of second-order, ordinary differential equations which can be written as

$$\frac{d}{dz}\underline{x}_m(z) = A_m(z)\underline{x}_m(z) = \begin{bmatrix} 0 & 1 \\ -\kappa_z^2(m) & 0 \end{bmatrix}\underline{x}_m(z) \tag{5.33}$$

The solution to this equation is governed by the state-transition matrix, $\Phi(z, z_0)$, where the state equation is solved by

$$\underline{x}_m(z) = \Phi_m(z, z_0)\underline{x}_m(z_0) \qquad m = 1, \ldots, M \tag{5.34}$$

and the transition matrix satisfies

$$\frac{d}{dz}\Phi_m(z, z_0) = A_m(z)\Phi_m(z, z_0) \qquad m = 1, \ldots, M \tag{5.35}$$

with $\Phi_m(z_0, z_0) = I$.

If we include all of the M modes in the model, then we obtain

$$\frac{d}{dz}\underline{x}(z) = \begin{bmatrix} A_1(z) & \ldots & O \\ \vdots & 0 & \vdots \\ O & \ldots & A_M(z) \end{bmatrix}\underline{x}(z) \tag{5.36}$$

or simply

$$\frac{d}{dz}\underline{x}(z) = A(z)\underline{x}(z) \tag{5.37}$$

Next we develop the state-space form for the range propagator. The range equation is given in Equation 5.28, but since we have expanded the depth relation in terms of normal modes, then Equation 5.31 shows that the horizontal wavenumbers $\{\kappa_r(m)\}$, $m = 1, \ldots, M$ are also a function of mode number m. Thus, we can rewrite Equation 5.28 as

$$\frac{d^2}{dr^2}\mu_m(r) + \frac{1}{r}\frac{d}{dr}\mu_m(r) = -\kappa_r^2(m)\mu_m(r) \qquad m = 1, \ldots, M \tag{5.38}$$

and now define the range state vector as $\underline{x}_m := [\mu_m(r) \; (d/dr)\mu_m(r)]'$. Again, the state equation corresponding to the mth mode is

$$\frac{d^2}{dr}\underline{x}_m(r) = A_m(r)\underline{x}_m(r) = \begin{bmatrix} 0 & 1 \\ -\kappa_r^2(m) & -\frac{1}{r} \end{bmatrix}\underline{x}_m(r) \tag{5.39}$$

The range state-space solutions can be calculated using numerical integration techniques as before, that is,

$$\underline{x}_m(r) = \Phi_m(r, r_0)\underline{x}_m(r_0) \qquad m = 1, \ldots, M \tag{5.40}$$

and similarly

$$\frac{d}{dr}\Phi_m(r, r_0) = A_m(r)\Phi_m(r, r_0) \qquad m = 1, \ldots, M \tag{5.41}$$

with $\Phi_m(r_0, r_0) = \mathrm{I}$. Including all of the M wavenumbers, we obtain the same form of the block decoupled system matrix of Equations 5.36 or simply

$$\frac{d}{dr}\underline{x}(r) = A(r)\underline{x}(r) \tag{5.42}$$

So, we see that the modal and range propagators consist of two "decoupled" state spaces, each of which is excited by the separable point source driving function

$$u(r, z) = \delta(r - r_s, z - z_s) = \delta(r - r_s)\delta(z - z_s) \tag{5.43}$$

which can be modeled in terms of the initial state vectors. The coupling of the range and depth is through the vertical wavenumbers, $\kappa_z(m)$, and the corresponding dispersion relation of Equation 5.31. These wavenumbers are obtained as the solution to the boundary value problem in depth and are in fact that resulting eigenvalues. Thus, the boundary values, source intensity, etc. are "built" into these wavenumbers, and for this reason we are able to forward propagate or reproduce the modal solutions from the state-space (initial value) solutions. The final parameters that must be resolved are those which depend on the point source $[q, \{\phi_m(z_s)\}]$ and the corresponding energy constraints on the modal functions. We will include these parameters in the measurement or output equations associated with the state-space propagator.

Next we develop the pressure measurement model. If we assume that the pressure field is measured, remove the time dependence, etc. and use the two-dimensional model

$$p(r, z) = \mu(r)\phi(z) \tag{5.44}$$

then we can derive the corresponding pressure measurement model. Assuming the normal-mode solution, we must sum over all of the modes as in the acoustic propagation model of Equation 5.32. That is,

$$p(r, z) = \sum_{i=1}^{M} w_i\mu_i(r)\phi_i(z) \tag{5.45}$$

where w_i is a weighting function given by

$$w_i = \frac{q\phi_i(z_s)}{\int_0^L \phi_i^2(z)dz} \tag{5.46}$$

In terms of our state-space propagation models, recall that

$$\underline{x}_m(z) = \begin{bmatrix} x_{m1}(z) \\ x_{m2}(z) \end{bmatrix} = \begin{bmatrix} \phi_m(z) \\ \frac{d}{dz}\phi_m(z) \end{bmatrix} \text{ and } \underline{x}_m(r) = \begin{bmatrix} x_{m1}(r) \\ x_{m2}(r) \end{bmatrix} = \begin{bmatrix} \mu_m(r) \\ \frac{d}{dr}\mu_m(r) \end{bmatrix} \tag{5.47}$$

and that

$$p(r, z_\ell) = \sum_{i=1}^{M} w_i x_{i1}(r) x_{i1}(z_\ell) \tag{5.48}$$

which is the value of the pressure field at the ℓth sensor and clearly a non-linear function of the states. If we employ a complete vertical array of hydrophones as our measurements, the depth must be discretized over the L_z elements, so that $z_1 \ldots, z_{L_z}$ and

$$\underline{p}(r, z) = \begin{bmatrix} x_{11}(z_1) & x_{21}(z_1) & \ldots & x_{M1}(z_1) \\ \vdots & \vdots & \vdots & \vdots \\ x_{11}(z_{L_z}) & z_{21}(z_{L_z}) & \ldots & x_{M1}(z_{L_z}) \end{bmatrix} \begin{bmatrix} w_1 x_{11}(r) \\ \vdots \\ w_M x_{M1}(r) \end{bmatrix} \tag{5.49}$$

or simply

$$\underline{p} = \chi(z)\underline{x}_w(r) \tag{5.50}$$

where $\underline{p} \in C^{L_z \times 1}$ is the resulting pressure field. It is possible to extend the measurement model to a horizontal array by merely discretizing over r, that is, $r_1, \ldots, r_{L_r}$, and interchanging the roles of $x_{ij}(z) \rightarrow x_{ij}(r_\ell)$ in Equation 5.49. It is important to understand that the recursive nature of the state-space formulation causes the simulation and eventual processing to occur in a sequential manner in depth (range) for this vertical (horizontal) array, yielding values of all modal functions at a particular depth z_ℓ corresponding to the ℓth sensor. To see this, rewrite Equation 5.45 in vector form as

$$p(r, z_\ell) = [w_1 x_{11}(r)\ 0\ w_2 x_{21}(r)\ 0 \ldots w_M x_{M1}(r)\ 0] \begin{bmatrix} x_{11}(z_\ell) \\ x_{12}(z_\ell) \\ --- \\ \vdots \\ --- \\ x_{M1}(z_\ell) \\ x_{M2}(z_\ell) \end{bmatrix}, \ell = 1, \ldots, L_z \tag{5.51}$$

which can be viewed as sequentially sampling the modal (range) functions at the given array sensor depth (range) location. That is, all modal functions are sampled at the ℓth sensor located at depth z_ℓ, next all the modal functions are sampled at $z_{\ell+1}$, and so on until the final sensor at z_{L_z}. For the design of large arrays, this sequential approach can offer considerable computational savings.

Next let us consider the representation of a two-dimensional, equally spaced array with sensors located at $\{(r_i, z_j)\}$, $i = 1, \ldots, L_r$; $j = 1, \ldots, L_z$. We write the equations by "stacking" the vertical equation of Equation 5.48 into $L_r - L_z$ vectors, that is,

$$\begin{bmatrix} \underline{p}(r_1) \\ \vdots \\ \underline{p}(r_{L_r}) \end{bmatrix} = \begin{bmatrix} \chi_1(z) & \ldots & 0 \\ \vdots & & \vdots \\ 0 & \ldots & \chi_{L_r}(z) \end{bmatrix} \begin{bmatrix} \underline{x}_w(r_1) \\ \vdots \\ \underline{x}_w(r_{L_r}) \end{bmatrix} \tag{5.52}$$

or

$$\underline{\mathbf{p}} = \chi(z)\mathbf{X}_w(r) \tag{5.53}$$

for $\mathbf{p} \in \mathbf{C}^{L_rL_z \times 1}$, $\chi \in \mathbf{C}^{L_rL_z \times ML_r}$, and $\mathbf{X}_w \in \mathbf{C}^{ML_r \times 1}$. Note also that for the case of equally spaced vertical line arrays with the identical number of sensors, we have $\chi_1(z) = \ldots \chi_{L_z}(z) = \chi(z)$. Also, recall that the pressure field will be processed sequentially due to the recursive (in depth or range) nature of the state-space approach. Note also that a two-dimensional thinned array can also be characterized by defining $L_z = \sum_{j=1}^{L_r} L_{z_j}$ is the number of vertical sensors in the jth line.

Clearly, the array pressure measurement is a non-linear vector function of $x_{ij}(r)$ and $x_{ij}(z_\ell)$, that is,

$$\underline{p}(r, z) = C(\underline{x}(r), \underline{x}(z)) \tag{5.54}$$

Note that we will restrict our development to the vertical line array of Equation 5.49.

In summary, we have just shown that the separable solutions to the modal and range equations can be characterized *individually* in "linear" state-space form with a non-linear pressure measurement system. We summarize the state-space formulation of the "normal-mode" model for a *vertical line array* of L_z elements as

$$\textit{Modal Model: } \frac{d}{dz}\underline{x}(z) = A(z)\underline{x}(z) \tag{5.55}$$

$$\textit{Range Model: } \frac{d}{dr}\underline{x}(r) = A(r)\underline{x}(r) \tag{5.56}$$

$$\textit{Pressure Measurement Model: } \underline{p}(r, z) = C(\underline{x}(r), \underline{x}(z)) \tag{5.57}$$

where $\underline{x}(z)$, $\underline{x}(r)$, $B \in R^{2M \times 1}$, $A \in R^{2M \times 2M}$, $\underline{p}$, $C \in R^{L_z \times 1}$, and

$$A(z) = \begin{bmatrix} A_1(z) & \ldots & O \\ \vdots & & \vdots \\ O & \ldots & A_M(z) \end{bmatrix}, A(r) = \begin{bmatrix} A_1(r) & \ldots & O \\ \vdots & & \vdots \\ O & \ldots & A_M(r) \end{bmatrix} \tag{5.58}$$

with

$$A_m(z) = \begin{bmatrix} 0 & 1 \\ -\kappa_z^2(m) & 0 \end{bmatrix}, A_m(r) = \begin{bmatrix} 0 & 1 \\ -\kappa_r^2(m) & -\frac{1}{r} \end{bmatrix} \tag{5.59}$$

and also

$$C(\underline{x}(z), \underline{x}(r)) = \left[\sum_{i=1}^{M} w_i x_{i1}(r) x_{i1}(z_1) \ldots \sum_{i=1}^{M} w_i x_{i1}(r) x_{i1}(z_{L_z}) \right]' \tag{5.60}$$

$$u(r, z) = \delta(r - r_s)\delta(z - z_s),\ w_i = \frac{q\phi_i(z_s)}{\int_0^L \phi_i^2(z)dz} \tag{5.61}$$

This constitutes a complete deterministic representation of the normal-mode model in state-space form. However, since propagation in the ocean is affected by inhomogeneities in the water, slow time variations in the sound speed, and motion of the surface, the model must be modified to include these effects. This can be done in a natural way by placing the model into a Gauss-Markov representation which includes the second-order statistics of the measurement as well as the modal/range noise. Note that since the pressure measurement model is non-linear, this model is only approximately Gauss-Markov because the associated non-linearities are linearized about their mean. The measurement noise can represent the near-field acoustic noise field, flow noise on the hydrophone, and electronic noise. The modal/range noise can represent sound speed errors, distant shipping noise, errors in the boundary conditions, sea state effects, and ocean inhomogeneities. Besides the ability to lump the various noise terms into the model, the Gauss-Markov representation provides a framework in which the various statistics associated with the model, such as the means and their associated covariances, can be computed. Upon simplification of the notation used above, the general Gauss-Markov propagation model for our problem is shown in Table 5.3.

Here we note some interesting features of this representation. First, there is no constraint of stationary statistics because the vector functions and associated matrices are functions of the index variable ℓ which is (r, z) in our problem. Note that the covariance evolution equations must be calculated through integration techniques. Also, this model is non-linear, as observed from the measurement model of Equation 5.48, resulting in the *approximate* Gauss-Markov representation due to linearization about the mean in the measurement covariance equations of Table 5.3. These statistics are quite useful in simulation because confidence intervals can be constructed about the mean for validation purposes.

Note also that in this framework well-known notions are easily captured. For instance, the solution to the state equation is given in terms of the state transition matrix, $\Phi(\ell, \tau)$, which is related to the Green's function of the propagation medium. In fact, an interesting way in which the process noise, w, enters the Gauss-Markov model is similar to the shipping noise. This can be observed through the state solutions of Equation 5.39 or 5.42,

$$\underline{x}(\ell) = \Phi(\ell, \ell_0)\underline{x}(\ell_0) + \int_{\ell_0}^{\ell} \Phi(\ell, \alpha)W(\alpha)\underline{w}(\alpha)d\alpha \tag{5.62}$$

where we see that the process noise propagates through the same medium as the source. Note also that the covariance contribution of the noise is part of the state variance/covariance terms in Table 5.3.

For our normal-mode representation, we can specify the model in terms of the previously derived relationships; thus, we define the *approximate Gauss-Markov ocean propagation model* as

$$\begin{bmatrix} \frac{d}{dz}\underline{x}(z) \\ -- \\ \frac{d}{dr}\underline{x}(r) \end{bmatrix} = \begin{bmatrix} A(z) & | & O \\ & -- & \\ O & | & A(r) \end{bmatrix} \begin{bmatrix} \underline{x}(z) \\ -- \\ \underline{x}(r) \end{bmatrix} + \begin{bmatrix} W(z) & | & O \\ & -- & \\ O & | & W(r) \end{bmatrix} \begin{bmatrix} \underline{w}(z) \\ -- \\ \underline{w}(r) \end{bmatrix} \tag{5.63}$$

$$\underline{y}(r, z) = C(\underline{x}(r), \underline{x}(z)) + \underline{v}(r, z) \tag{5.64}$$

where the model parameters $\{A(z), A(r), W(z), W(r), C(\underline{x}(r), \underline{x}(z))\}$ are defined above. We also assume that $\{\underline{w}(z), \underline{w}(r)\}$ are zero mean, Gaussian with respective covariances, and $R_{w_z w_z}$ and $R_{w_r w_r}$ with $\underline{v}(r, z)$

TABLE 5.3 Approximate Continuous-Discrete Gauss-Markov Representation

State propagation

$$\frac{d}{d\ell}x(\ell) = A(\ell)x(\ell) + W(\ell)w(\ell) \tag{5.65}$$

State mean propagation

$$\frac{d}{d\ell}m_x(\ell) = A(\ell)m_x(\ell) \tag{5.66}$$

State variance/covariance propagation

$$\frac{d}{d\ell}P(\ell) = A(\ell)P(\ell) + P(\ell)A'(\ell) + W(\ell)R_{ww}(\ell)W'(\ell)$$

$$\frac{d}{d\ell}P(\ell,\tau) = \begin{cases} \Phi(\ell,\tau)P(\ell) & t \geq \tau \\ P(\ell)\Phi'(\ell,\tau) & t \leq \tau \end{cases}$$

Measurement propagation

$$y(\ell) = C(x(\ell)) + v(\ell) \tag{5.67}$$

Measurement mean propagation

$$m_y(\ell) = C(m_x(\ell)) \tag{5.68}$$

Measurement variance/covariance propagation

$$R_{yy}(\ell) = \frac{\partial}{\partial x}C[m_x(\ell)]P(\ell)\frac{\partial}{\partial x}C[m_x(\ell)]' + R_{vv}(\ell)$$

$$R_{yy}(\ell,\tau) = \frac{\partial}{\partial x}C[m_x(\ell)]P(\ell,\tau)\frac{\partial}{\partial x}C[m_x(\tau)]' + R_{vv}(\ell,\tau)$$

State transition propagation

$$\frac{d}{d\ell}\Phi(\ell,\tau) = A(\ell)\Phi(\ell,\tau) \tag{5.69}$$

$$\frac{\partial}{\partial x}C[m_y(\ell)] := \frac{\partial}{\partial x}C[x(\ell)]\Big|_{x(\ell)=m_x(\ell)} \tag{5.70}$$

are also zero mean, Gaussian with covariance R_{vv}. Also, the initial state vectors $\underline{x}(z_0) \sim N(m_z, P_z)$ and $\underline{x}(r_0) \sim N(m_r, P_r)$ are Gaussian.

This completes the development of the state-space forward propagators for model-based signal processing. But before we close, we emphasize that we are not actually solving the differential equation in the usual sense. The MBP is a recursive estimation scheme which requires initial values for the state vector. The processor then propagates sequentially down the vertical array using the forward propagator, which evolves from the wave equation. This is somewhat similar to an initial value problem, which attempts to directly solve the differential equation, but does not incorporate the data. In our case, we used SNAP[26,27,29] to provide the initial values as the solution to the two point boundary value problem. Given these initial values, the forward propagator then sequentially marches down the array, and at each step (hydrophone) the predicted measurement is compared to the actual measurement (innovation) to generate the model-based correction. The effectiveness of this approach is dependent upon these initial values. That is, the initial solution (in this case provided by SNAP) must be reasonably close to the truth, otherwise the estimates will converge slowly or not at all. However, since the data are used as an integral part of the processing, the initial solution need not be exact, since any errors in the initial values are dealt with by the correction stage of the processor at each step in the recursion. We discuss this further in subsequent sections.

5.4 Ocean Acoustic Model-Based Processing Applications

In this section, we discuss various applications of the optimal MBP to ocean acoustic data. In each application, we have applied the MBP design methodology of Section 5.2 by following the basic steps outlined: (1) model development, (2) simulation/minimum-variance design, (3) application ("tuning") to data sets, and (4) performance analysis. Here, we will concentrate on Step 3 of the application to ocean acoustic data with the understanding that Steps 1 and 2 have been completed. For more details and the step-by-step processor designs, see the appropriate papers referenced. We will briefly discuss the following applications to a portion of the ocean known as the Hudson Canyon, which has been used extensively for ocean acoustic sea testing. After describing the Hudson Canyon data, we proceed to develop the following MBP applications.

5.4.1 Ocean Acoustic Data: Hudson Canyon Experimental Data

Here, we discuss a set of experimental data that was granted over a representative part of the ocean which captures many of the features captured in various experimental scenarios that could be envisioned. The Hudson Canyon experiment was performed in 1988 in the Atlantic Ocean. It was led by Dr. W. Carey of the Naval Undersea Warfare Center (NUWC), with the primary goal of investigating acoustic propagation (transmission and attenuation) using continuous wave data. Hudson Canyon is located off the coast of New Jersey in the region of the Atlantic Margin Coring project (AMCOR) borehole 6010. The seismic and coring data are combined with sediment properties measured at that site.[45,46] Excellent agreement was achieved between the model and data as recently reported by Rogers,[47] indicating a well-known, well-documented shallow water experiment with bottom interaction and yielding ideal data sets for investigating the applicability of an MBP to measured ocean acoustic data. All of the required measurements were performed carefully to obtain environmental information such as sonar depth soundings, CTD, and sound speed profile measurements along the acoustic track.

The Hudson Canyon is topologically characterized by two clearly distinct bottoms: a flat bottom and a sloping bottom. A vertical array of 24 hydrophones was anchored at the bottom, and an acoustic source was driven both away and toward the array on the predefined environmental tracks. The Hudson Canyon experiment was performed at low frequencies (50 to 600 Hz) in shallow water at a depth of 73 m during a period of calm sea state. A calibrated acoustic source was towed at roughly 36 m depth radially to distances of 0.5 to 26 km. The ship speed was between 2 and 4 km. The fixed vertical hydrophone array consisted of 24 phones spaced 2.5 m apart extending from the seafloor up to a depth of about 14 m below the surface. CTD and SSP measurements were made at regular intervals, and the data were collected under carefully controlled conditions in the ocean environment. We note that, experimentally, the spectrum of the time series data collected at 50 Hz is dominated by five modes occurring at wave numbers between 0.14 to 0.21 $m-^{1}$, with relative amplitudes increasing with increased wave number. As seen from the work of Rogers,[47] the SNAP[26] simulation was performed and the results agree quite closely, indicating a well-understood ocean environment.

In order to construct the state-space propagator, we require the set of experimental parameters discussed above which were obtained from the measurements and processing (wave number spectra). The horizontal wave number spectra were estimated using synthetic aperture processing.[32–36] Eight temporal frequencies were employed: four on the inbound (75, 275, 575, and 600 Hz) and four on the outbound (50, 175, 375, and 425 Hz). In this application, we will confine our investigation to the 50 Hz case, which is well documented, and to horizontal ranges from 0.5 to 4 km. The raw measured data were processed (sampled, corrected, filtered, etc.) by Mr. J Doutt of the Woods Hole Oceanographic Institute (WHOI) and supplied for this investigation.

Due to the shallow ocean (73 m) and low temporal frequency (50 Hz), only five modes were supported by the water column. The horizontal wave numbers and relative amplitudes used in the propagator were those obtained from the measured spectrum. We used the same array geometry as the experiment, that is, a vertical array with sensors uniformly spaced at 2.5 m, spanning the water column from the bottom

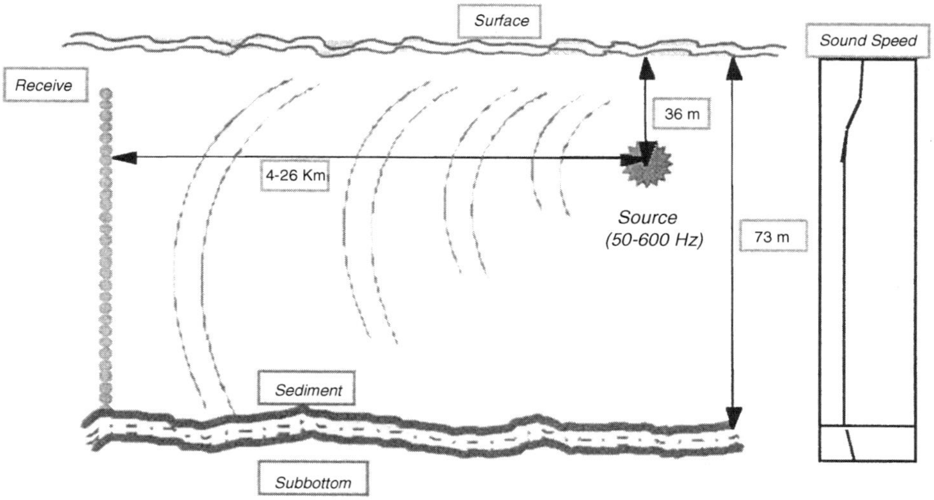

FIGURE 5.8 Hudson canyon experiment with vertical array and sound source.

up to 14.5 m below the surface. Note that when we perform the simulation and model-based processing, all the modal functions will "start" at 14.5 m because of the initial array depth. We will use a piecewise linear approximation of the SSP obtained from the measured SSP of Figure 5.8 and confine our investigation to the 50 Hz case.

The linear space-varying "depth only" Gauss-Markov model of Section 5.3 is employed along with its corresponding statistics. The model takes on the following form for a 24-element vertical sensor array with 2.5 m vertical spacing located at (r_0, z_0):

$$\frac{d}{dz}\begin{bmatrix} \phi_1(z) \\ \frac{d}{dz}\phi_1(z) \\ --- \\ \vdots \\ --- \\ \phi_5(z) \\ \frac{d}{dz}\phi_5(z) \end{bmatrix} = \left[\begin{array}{cc|cc} 0 & 1 & O & \\ -k_z^2(1) & 0 & & \\ \hline & & \ddots & \\ O & & 0 & 1 \\ & & -k_z^2(5) & 0 \end{array}\right]\begin{bmatrix} \phi_1(z) \\ \frac{d}{dz}\phi_1(z) \\ --- \\ \vdots \\ --- \\ \phi_5(z) \\ \frac{d}{dz}\phi_5(z) \end{bmatrix} + \underline{w}(z) \tag{5.71}$$

$$p(r_s, z_\ell) = [\beta_1(r_s, z_\ell)\ 0\ \beta_2(r_s, z_\ell)\ 0\ \ldots\ \beta_5(r_s, z_\ell)\ 0]\begin{bmatrix} \phi_1(z) \\ \frac{d}{dz}\phi_1(z) \\ --- \\ \vdots \\ --- \\ \phi_5(z) \\ \frac{d}{dz}\phi_5(z) \end{bmatrix} + v(z_\ell) \qquad \ell = 1, \ldots, L \tag{5.72}$$

where, in this case, we use $\beta_i(r_s, z) = (q\phi_i(z_s))/\int_0^L \phi_i^2(z)dz\ H_o(k_r(i)r_s)$, and H_o is the Hankel function solution to the range propagator.[29] We model the SSP as piecewise linear with

$$c(z) = a(z)z + b(z) \qquad a, b \text{ known} \tag{5.73}$$

where $a(z)$ is the slope and $b(z)$ is the corresponding intercept, these are given by

$$a(z) = \frac{c(z_\ell) - c(z_{\ell-1})}{z_\ell - z_{\ell-1}} \qquad z_{\ell-1} < z < z_\ell \tag{5.74}$$

and

$$b(z) = c(z_\ell) \tag{5.75}$$

where these parameters are estimated directly from the measured profile. This SSP is embedded in the $\kappa_z(m)$ parameter of the system matrix $A_m(z)$, and the $\beta_i(r_0, z)$ are the relative modal amplitudes estimated from the horizontal wave number spectrum. This completes the development of the state-space propagator developed for the Hudson Canyon experiment. Next we discuss the design of the MBP for the noisy measured experimental data.

5.4.2 Ocean Acoustic Application: Adaptive Model-Based Signal Enhancement

Here we briefly discuss the development of a model-based signal enhancer or filter designed to extract the desired signals, which in this application is the estimated modal functions and pressure field. A model-based approach is developed to solve an adaptive ocean acoustic signal processing problem, Here, we investigate the design of a model-based identifier for a normal-mode model developed for the Hudson Canyon shallow water ocean experiment and apply it to a set of experimental data, demonstrating the feasibility of this approach. In this application, we show how the parametric adaptive processor can be structured to estimate the horizontal wave numbers directly from measured pressure field and sound speed. Improvements can be achieved by developing processors that incorporate knowledge of the surrounding ocean environment and noise into their processing schemes.[9–13] However, as mentioned previously, it is well known that if the incorporated model is inaccurate either parametrically or from the basic principles, then the processor can actually perform worse in the sense that the predicted error variance is greater than that of the raw measurements.[1,2] In fact, one way to choose the "best" model or processor is based on comparing predicted error variances — the processor achieving the smallest wins. In practice, the usual procedure to check for model adequacy is to analyze the statistical properties of the resulting residual or innovations sequence, that is, the difference between the measured and predicted measurements. Here again, the principle of minimum (residual) variance is applied to decide on the best processor or, equivalently, the best embedded model.[7] Other sophisticated statistical tests have been developed for certain classes of models with high success to make this decision.[2,7,18,24] In any case, the major problem with *model-based* signal processing schemes is assuring that the model incorporated in the algorithm is adequate for the proposed application, that is, it can faithfully represent the ongoing phenomenology. Therefore, it is necessary, as part of the MBP design procedure, to estimate/update the model parameters either through separate experiments or jointly (adaptively) while performing the required processing.[18,23] The introduction of a recursive, on-line MBP can offer a dramatic detection improvement in a tactical passive or active sonar-type system, especially when a rapid environmental assessment is required.[8,21] In this section, we discuss the development of a processor capable of adapting to the ever-changing ocean environment and providing the required signal enhancement for eventual detection and localization.

Here, we investigate the development of an adaptive MBP which we define as the model-based identifier (MBID). The MBID incorporates an initial mathematical representation of the ocean acoustic propagation model into its framework and adapts, on-line, its parameters as the ocean changes environmentally. Here, we are interested primarily in a shallow water environment characterized by a normal-mode model, and therefore, our development will concentrate on adaptively adjusting parameters of the normal-mode

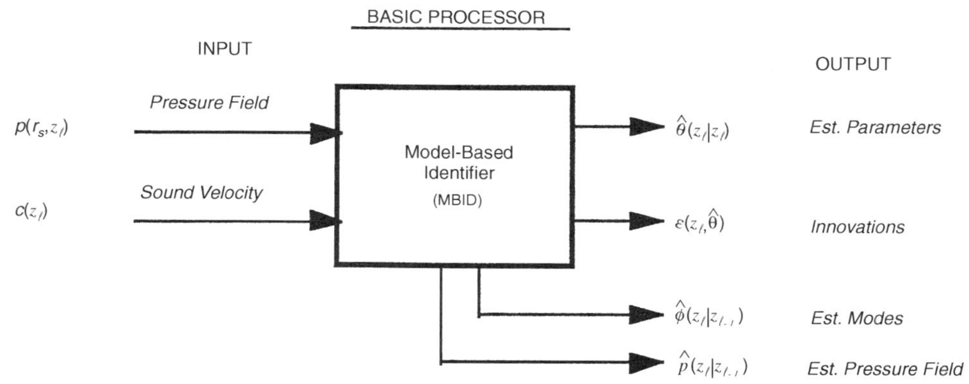

FIGURE 5.9 Adaptive model-based signal enhancement: the basic processor.

propagation model to "fit" the ocean surrounding our sensor array. In fact, one way to think about this processor is that is passively listens to the ocean environment and "learns" or adapts to its changes. It is clear that the resulting processor will be much more sensitive to changes than one that does not adapt, thereby providing current information and processing. In the following, we define the MBID as a Kalman filter whose estimated states are the modal functions $\hat{\phi}(z_\ell)$ and states representing the estimated ocean acoustic parameters $\hat{\theta}(z_\ell)$ that have been augmented into the processor. The basic processor is shown in Figure 5.9. The inputs to the MBID are raw data $[\{p(z_\ell), \{c(z_\ell)\}]$, and its outputs are $\theta(z_\ell)$, the set of parameters of interest. There are advantages to this type of processing. First, it is recursive and, therefore, can adaptively update the estimates of the sonar and environmental parameters. Second, it can include the system and measurement noise or uncertainty in a self-consistent manner. By uncertainty, it is meant errors in the input parameters of the model. Third, one of the outputs of the MBID is the innovation sequence, $e(z_\ell)$, which provides an on-line test of the goodness of fit of the model to the data.[7,24] This innovation sequence plays a major role in the iterative nature of this processor by providing information that can be used to adaptively correct the processor and the propagation model itself.[23]

The application of this adaptive approach to other related problems of interest is apparent. For *signal enhancement, the* adaptive MBP or MBID can provide enhanced signal estimates of modal functions (modal filtering), pressure-field estimates (measurement filtering), and parameters of interest (parameter estimation) such as wave numbers, range-depth functions, sound speed, etc.[18,23] For *model monitoring* and *source detection* purposes, the MBID provides estimates of the residuals or innovations sequence, which can be statistically tested for adequacy[7] or used to calculate a decision function.[7,30] For *localization*, the MBID provides estimates of the enhanced range-depth and modal (modal filter) functions used for model-based localization as discussed in Reference 18. It can also be used to provide enhanced modes/pressure field for MMP or the MFP. In fact, for rapid assessment of the ocean environment — a definite requirement in a tactical situation — it is possible to perform *model-based inversion* on-line, using the MBID scheme to adaptively estimate the changing parameters characterizing the sound speed profile.[8,22] Thus, the MBID provides a technique capable of "listening and learning."

5.4.2.1 Model-Based Signal Enhancement: Parametrically Adaptive Model

Next, we briefly develop the MBID for use with the normal-mode, ocean acoustic propagation model. System identification is typically concerned with the estimation of a model and its associated parameters from noisy measurement data. Usually, the model structure is predefined (as in our case) and then a parameter estimator is developed to fit parameters according to some error criterion. After completion or during this estimation, the quality of the estimates must be evaluated to decide if the processor performance is satisfactory or, equivalently, if the model adequately represents the data. There are various types (criteria) of identifiers employing many different model (usually linear) structures.[40–44] Since our efforts are primarily aimed at ocean acoustics in which the models and parameters are usually non-linear, we will concentrate

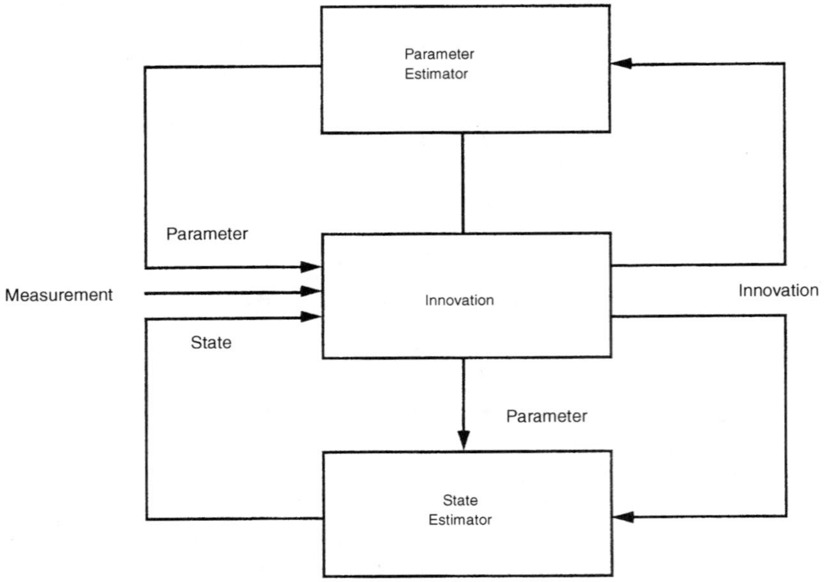

FIGURE 5.10 MBID: the processor structure.

on developing a parameter estimator capable of on-line (shipboard) operations and non-linear dynamics. From our previous discussions, it is clear that the EKF identifier will satisfy these constraints nicely. The general non-linear identifier or, equivalently, parameter estimator structure can be derived directly from the EKF algorithm[1,2] in discrete form, which we showed in Table 5.2 previously. We note that this algorithm is *not* implemented in this fashion; it is implemented in the numerically stable UD-factorized form as in SSPACK_PC,[48] a toolbox in MATLAB.[49] Here, we are just interested in the overall internal structure of the algorithm and the decomposition that evolves. The simplified structure of the EKF parameter estimator is shown in Figure 5.10. The basic structure of the MBID consists of two distinct, yet coupled, processors: a parameter estimator and a state estimator (filter). The parameter estimator provides estimates that are corrected by the corresponding innovations during each recursion. These estimates are then provided to the state estimator (EKF) to update the model parameters used in the estimator. After both state and parameters estimates are calculated, a new measurement is processed and the procedure continues.

For propagation in a shallow water environment, we choose the normal-mode model which can easily be placed in state-space form (see Section 5.3 for details). We choose the "depth only" structure and assume a vertical array which yields a linear space-varying formulation, then we develop the identifier — a non-linear processor. Next, we investigate the performance of the processor on the Hudson Canyon data set discussed above.

Recall from Section 5.3 that the pressure-field measurement model is given by

$$p(r_s, z) = \mathbf{C}^T(r_s, z)\phi(z) + \mathbf{v}(z) \tag{5.76}$$

where

$$\mathbf{C}^T(r_s, z) = [\beta_1(r_s, z_s)\ 0|\beta_2(r_s, z_s)\ 0|\ldots|\beta_M(r_s, z_s)\ 0] \tag{5.77}$$

with

$$\beta_m(r_s, z_s) = \frac{q\phi_m(z_s)}{\int_o^h \phi_m^2(z)dz} H_0(k_r(m)r_s)$$

The random noise vector $\mathbf{v}$ is assumed Gaussian, zero mean with respective covariance matrix, R_{vv}. Our array spatially samples the pressure field, therefore, we choose to discretize the differential state equations using a finite (first) difference approach. Since a vertical line sensor array was used to measure the pressure field, the measurement model for the mth mode becomes

$$p_m(r_s, z_\ell) = \beta_m(r_s, z_s)\phi_{m1}(z_\ell) + v_m(z_\ell) \tag{5.78}$$

It is this model that we employ in our MBID. Next, suppose we assume that the horizontal wave numbers, $\{\kappa_r(m)\}$, are unknown and we would like to estimate them directly from the pressure-field measurements. Note that the horizontal wave numbers are *not* a function of depth, they are constant or invariant over depth. Once estimated, the horizontal wave numbers along with the known sound speed can be used to determine the vertical wave numbers directly from the dispersion relation of Equation 5.31 (see Section 5.3).

The basic form of the coupled modal equations follow from Equation 5.8 with $\kappa_t \to \theta$ and $m = 1, \ldots, M$:

$$\begin{aligned}
\phi_{m1}(z_\ell) &= \phi_{m1}(z_{\ell-i}) + \Delta z_\ell \phi_{m2}(z_{\ell-1}) \\
\phi_{m2}(z_\ell) &= -\Delta z_\ell\left(\frac{\omega^2}{c^2(z_{\ell-1})} - \theta_m^2(z_{\ell-1})\right)\phi_{m1}(z_{\ell-1}) + \phi_{m2}(z_{\ell-1}) \\
\theta_m(z_\ell) &= \theta_m(z_{\ell-1})
\end{aligned} \tag{5.79}$$

and corresponding measurement model

$$p_m(r_s, z_\ell) = \beta_m(r_s, z_s)\phi_{m1}(z_\ell) \tag{5.80}$$

The information required to construct the adaptive processor is derived from the above Gauss-Markov process and measurement functions using the augmented approach to design the parametrically adaptive processor (see Section 5.2). The details can also be found in Reference 23. The overall MBID relations used to enhance both modal and pressure-field measurements are given by the *prediction equations for* the mth mode and wave number:

$$\begin{aligned}
\hat{\phi}_{m1}(z_\ell|z_{\ell-1}) &= \hat{\phi}_{m1}(z_\ell|z_{\ell-1}) + \Delta z_\ell \hat{\phi}_{m2}(z_{\ell-1}|z_{\ell-1}) \\
\hat{\phi}_{m2}(z_\ell|z_{\ell-1}) &= -\Delta z_\ell\left(\frac{\omega^2}{c^2(z_\ell)} - \hat{\theta}_m^2(z_{\ell-1}|z_{\ell-1})\right)\hat{\phi}_{m1}(z_{\ell-1}|z_{\ell-1}) + \hat{\phi}_{m2}(z_{\ell-1}|z_{\ell-1}) \\
\hat{\theta}_m(z_\ell|z_{\ell-1}) &= \hat{\theta}_m(z_{\ell-1}|z_{\ell-1})
\end{aligned} \tag{5.81}$$

The corresponding innovations (parameterized by θ) are given by

$$e(z_\ell, \theta) = p(r_s, z_\ell) - \sum_{m=1}^{M} c_m[\phi, \theta]\hat{\phi}_{m1}(z_\ell|z_{\ell-1}, \theta) \tag{5.82}$$

with the correction equations

$$\begin{aligned}
\hat{\phi}(z_\ell|z_\ell) &= \hat{\phi}(z_\ell|z_{\ell-1}) + K_\phi(z_\ell)e(z_\ell, \theta) \\
\hat{\theta}(z_\ell|z_\ell) &= \hat{\theta}_m(z_\ell|z_{\ell-1}) + K_\theta(z_\ell)e(z_\ell, \theta)
\end{aligned} \tag{5.83}$$

Note the signal enhancements produced by the adaptive processor are the sets of modal and pressure-field estimates, $[\{\hat{\phi}_m(z_\ell\}), \{\hat{p}_m(z_\ell)\}]$, $m = 1, \ldots, M$.

Next we discuss the application of the MBID to noisy experimental measurements from the Hudson Canyon data set discussed above. We investigate the results of the MBID performance on the experimental hydrophone measurements from the Hudson Canyon. Here, we have the 24-element vertical array and initialize the MBID with the average set of horizontal wave numbers: {0.208, 0.199, 0.183, 0.175, 0.142} m^{-1} for the five modes supporting the water column from a 36 m deep, 50 Hz source at 0.5 km range (see References 8 and 33 for more details). The performance of the processor is best analyzed by the results in Figures 5.11 and 5.12, where we see that the residual or innovations sequence which lies within the $\pm 2\sqrt{R_{\varepsilon\varepsilon}}$ bounds and the associated *zero-mean/whiteness tests* are also shown. Recall that it is necessary for the innovations sequence to be zero mean and white for the processor to be deemed as tracking for the modes and associated parameters (Figure 5.11). Thus, the processor is successfully tracking and the model is *valid* for this data set. Note that the whiteness test is limited to stationary processes, since it employs a sample covariance estimator. However, it can be argued heuristically that when the estimator is tuned, the non-stationarities are being tracked by the MBP, and, therefore, the innovations should be covariance stationary. The associated WSSR statistic (also shown in Figure 5.12) essentially aggregates

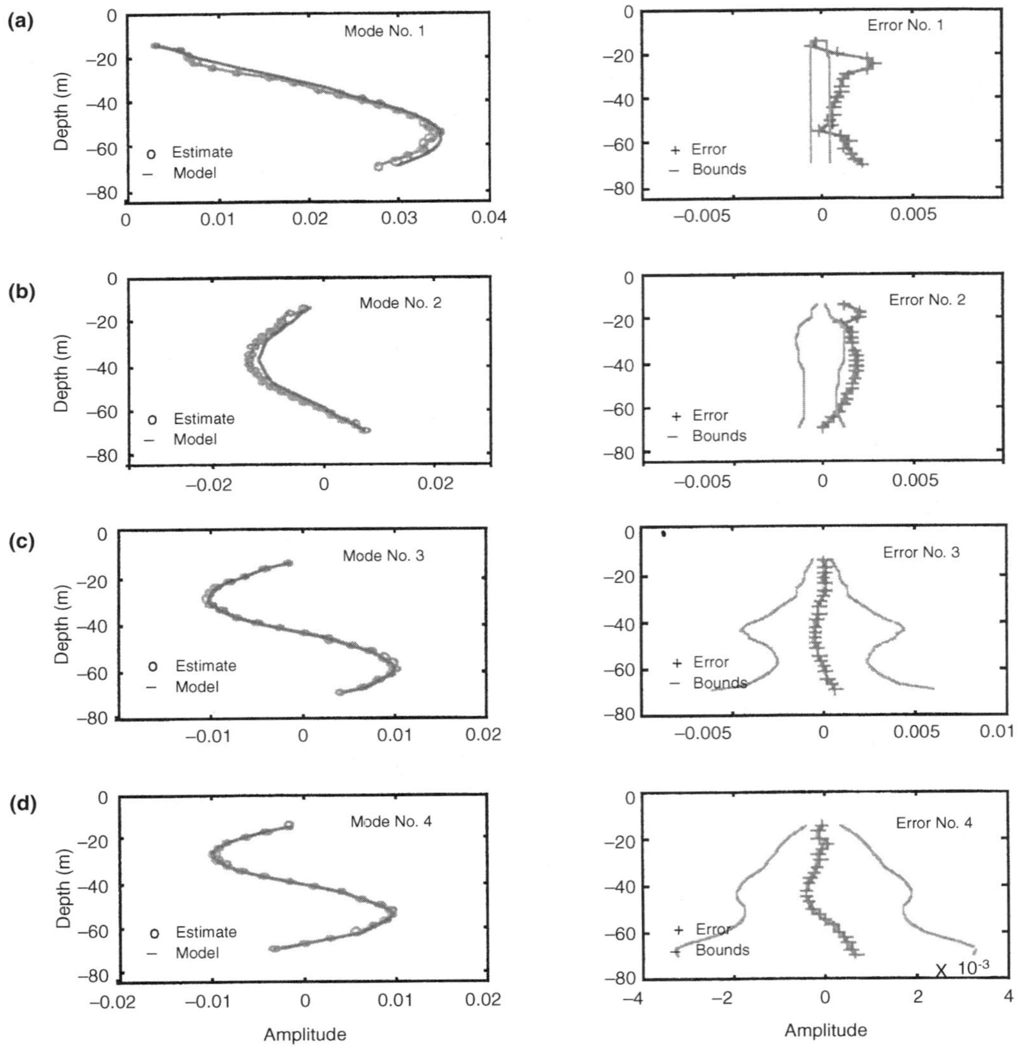

FIGURE 5.11 MBID of the Hudson Canyon experiment (0.5 km): (a) mode 1 and error (91% out), (b) mode 2 and error (83% out), (c) mode 3 and error (0% out), and (d) mode 4 and error (0% out).

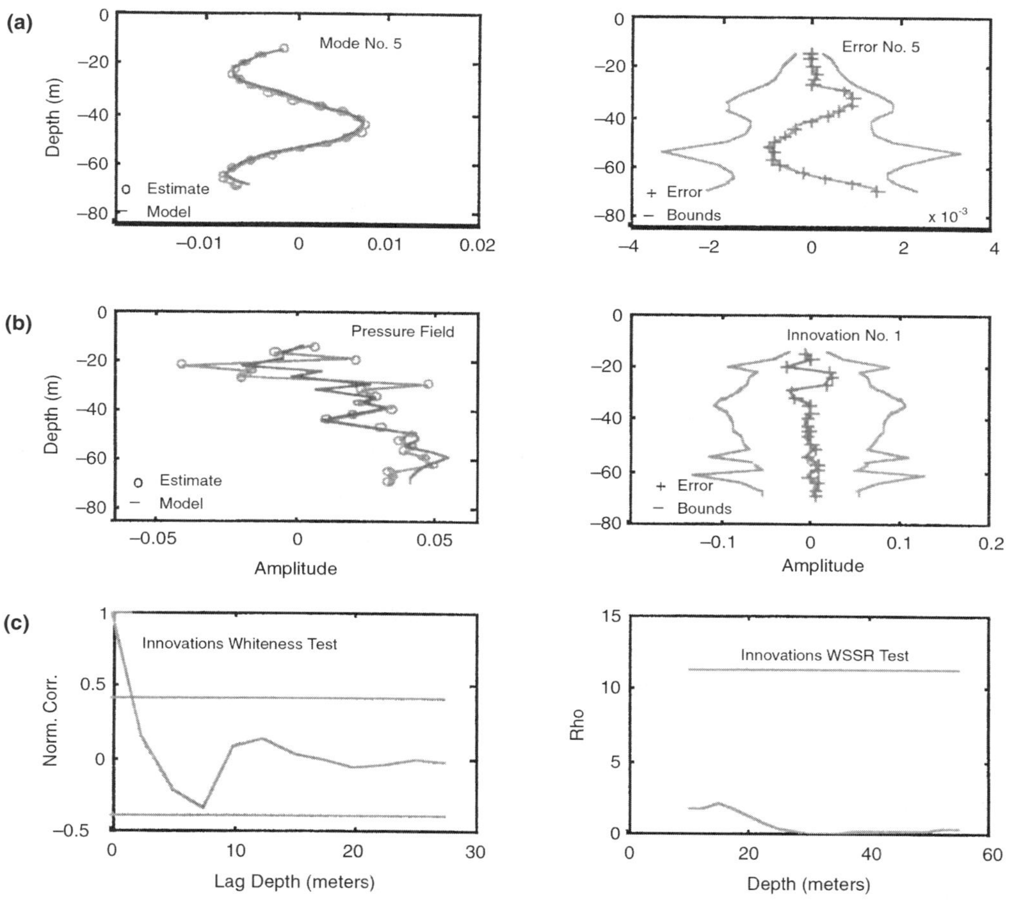

FIGURE 5.12 MBID of the Hudson Canyon experiment (0.5 km): (a) mode 5 and error (0% out), (b) pressure field and innovation (2–0% out), and (c) whiteness test and WSSR (0% out).

all of the information available in the innovation vector (see Reference 2 for details) and tests whiteness by requiring the decision function $\rho(z)$ which lies below the specified threshold to be white. Note that the WSSR statistic is not confined to a stationary process. The resulting estimates are quite reasonable, as shown in Figure 5.12. Note that although there is a little difficulty tracking the first couple of modes, the results actually appear better than those reported previously for this data set (see References 24). The results for the higher order modes follow those predicted by the model as observed in Figure 5.12 and corresponding estimation errors. From Figure 5.12 we see that the reconstructed pressure field and innovations are also quite reasonable, indicating a "tuned" processor with its zero mean ($1.9 \times 10^{-3} < 6.7 \times 10^{-3}$) and white (~0% out and WSSR $< \tau$) innovation sequence.

The final parameter estimates are shown in Figure 5.13 with the predicted error statistics for these data, which are also included in Tables 5.4 and 5.5 for comparison to the simulated. We note that the parameter estimates continue to adapt to the changing ocean environment based on the pressure-field measurements. We initially start the wave numbers at their averages and then allow them to adapt to the measured sensor data. The first wave number estimate appears to converge (approximately) to the average with a slight bias, but the others adapt to other values due to changes in the data. We see that the MBID appears to perform better than the MBP with the augmented parameter estimator simply because the horizontal wave numbers are "adaptively" estimated on-line, providing a superior fit to the raw data. Thus, we see that the use of the MBID in conjunction with vertical array measurements enables us to enhance the modal and pressure-field measurements even in the ever-changing ocean

TABLE 5.4 MBID: Wave Number Estimation

Hudson Canyon Experiment			
Wave Numbers	Model	Simulation	Experiment
κ_1	0.2079	0.2105 ± 0.0035	0.2076 ± 0.0043
κ_2	0.1991	0.1993 ± 0.0052	0.1978 ± 0.0036
κ_3	0.1827	0.1846 ± 0.0359	0.1817 ± 0.0251
κ_4	0.1746	0.1770 ± 0.0149	0.1746 ± 0.0098
κ_5	0.1423	0.1466 ± 0.0385	0.1479 ± 0.0288

TABLE 5.5 MBID: Wave Number Estimation

Hudson Canyon Experiment: Modal Modeling Error		
Mode No.	Fixed MBP	Adaptive Wave No.
1	1.8×10^{-3}	1.2×10^{-3}
2	1.4×10^{-3}	1.9×10^{-3}
3	1.9×10^{-4}	3.0×10^{-4}
4	5.8×10^{-4}	3.2×10^{-4}
5	5.4×10^{-4}	6.7×10^{-4}

environment. In this particular application, we see how the MBID is employed to adaptively estimate the wave numbers (horizontal) from noisy pressure-field and sound speed measurements evolving from a vertical array or hydrophones. This completes the section on applying the identifier to a critical ocean acoustic estimation problem.

We have developed an on-line, parametrically adaptive, model-based solution to the ocean acoustic signal processing problem based on coupling the normal-mode propagation model to a vertical sensor array. The algorithm employed was the non-linear EKF identifier/parameter estimator in predictor/corrector form which evolved as the solution to the minimum-variance estimation problem when the models were placed in state-space form. It was shown that the MBID follows quite naturally from the MBP. In fact, a horizontal wave number identifier was constructed to investigate the underlying structure of the processor and apply it to both simulated and Hudson Canyon experimental data, yielding enhancement results better than those reported previously[24] in the sense that the estimated modal functions track those predicted by propagation models more closely (smaller variances etc.).

5.4.3 Ocean Acoustic Application: Adaptive Environmental Inversion

In this section, a model-based approach to invert or estimate the SSP from noisy pressure-field measurements is discussed. The resulting MBP is based on the state-space representation of the normal-mode propagation model. Using data obtained from the Hudson Canyon experiment, the adaptive processor is designed, and the results are compared to the data. It is shown that the MBP is capable of predicting the sound speed quite well.

In ocean acoustics, we are usually concerned with an environmental model of the ocean and how it effects the propagation of sound through this noisy, hostile environment. The problem of estimating the environmental parameters characterizing the ocean medium is called the *ocean tomography* or, equivalently, the *environmental inversion* problem and has long been a concern because of its detrimental effect on various detection/localization schemes.[3,19–21] Much of the work accomplished on this problem has lacked quantitative measures of the mismatch of the model with its environment. In a related work,[24] it was shown how to quantify "modeling errors" both with and without a known ocean environmental model available. In the first case, it was shown how to calculate standard errors for modal/pressure-field estimates, as well as an overall measure of fit based on the innovations or residuals between the measured and predicted pressure field. In the second case, only the residuals were used. These results quantify the

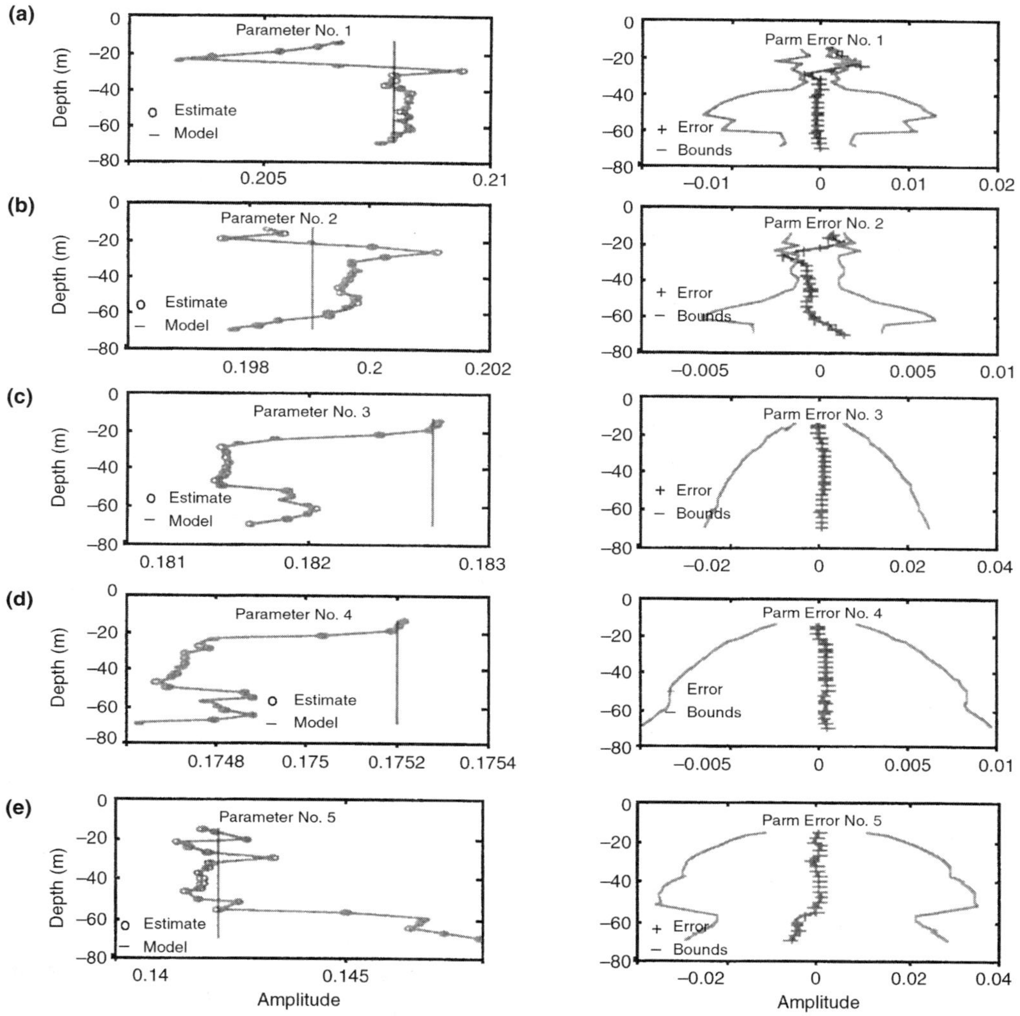

FIGURE 5.13 MBID of the Hudson Canyon experiment (0.5 km): (a) parameter 1 and error (8.7% out), (b) parameter 2 and error (4.4% out), (c) parameter 3 and error (0% out), (d) parameter 4 and error (0% out), and (e) parameter 5 and error (0% out).

mismatch between the embedded models and the actual measurements both on simulated as well as experimental data.

Here, we concentrate on the design of an adaptive MBP to solve the environmental inversion or oceanographic tomography problem while jointly estimating the underlying signals — which we term *model-based inversion*. That is, the MBP is designed to estimate the SSP as well as enhance the corresponding modal/pressure-field signals with its accompanying performance statistics quantified using the corresponding residuals. More specifically, we are concerned with estimating the SSP from noisy hydrophone measurements in a real, hostile ocean acoustic experimental environment. Theoretical work on the design of the MBP for this problem has been accomplished, indicating that a solution exists.[8] Here we apply these techniques for inversion to measured data from the Hudson Canyon experiment. Note that there is presently a growing literature on oceanographic tomography, since it can deal with the estimation of many different parameters relative to ocean acoustics. For more information on these issues, see References 3 and 12 to 14 and references therein.

5.4.3.1 Adaptive Environmental Inversion: Augmented Gauss-Markov Model

The normal-mode solutions can easily be placed in state-space form, as discussed above in presenting the Hudson Canyon experiment. Recall that the measurement noise can represent the near-field acoustic noise field, flow noise on the hydrophone, and electronic noise. The modal or process noise can represent SSP errors, distant shipping noise, errors in the boundary conditions, sea state effects, and ocean inhomogeneities. By assuming that the horizontal range of the source r_s is known *a priori*, we can use the Hankel function $H_0(\kappa_r r_s)$, which is the source range solution; therefore, we reduce the state-space model to that of depth only of Equation 5.71 as before.

The vertical wave numbers are functions of the SSP through the dispersion relationship of Equation 5.31 and can be further analyzed through knowledge of the SSP. Since our processor will be sequential, it is recursing over depth. We would like it to improve the estimation of the SSP "in-between" sensors. With a state-space processor, we can employ two spatial increments in z simultaneously: one for the measurement system $\Delta z_\ell := z_\ell - z_{\ell-1}$ and one for the modal state space $\Delta z_j = (\Delta z_\ell)/(N_\Delta)$ where N_Δ is an integer. Therefore, in order to propagate the states (modes) at Δz_j and measurements at Δz_ℓ, we must have values of the SSP at each Δz_j (sub-interval) as well. Suppose we expand $c(z)$ in a Taylor series about a nominal depth, say z_0, then we obtain

$$c(z) \approx \theta_0(z_0) + \theta_1(z_0)(z-z_0) + \ldots + \theta_N(z_0)\frac{(z-z_0)^N}{N!} \tag{5.84}$$

where

$$\theta_i(z_0) := \left.\frac{\partial^i c(z)}{\partial z^i}\right|_{z = z_0} \qquad i = 0, \ldots, N$$

In this formulation, therefore, we have a "model" of the SSP of the form $\hat{c}_N(z_\ell) = \underline{\Delta}_N^T(z_\ell)\theta(z_\ell)$, where

$$\underline{\Delta}_N^T(z_\ell) = \left[1 \ \Delta z_{\ell-1} \ldots \frac{\Delta z_{\ell-1}^N}{N!}\right]$$

and the set of $\{\theta_i(z_\ell)\}$ are only known at a sparse number of depths, $\ell = j, j+1, \ldots, j+N_\theta$. More simply, we have a set of measurements of the SSP measured *a priori* at specific depths — not necessarily corresponding to all sensor locations $\{z_\ell\}$ — therefore, we use these values as initial values to the MBP, enabling it to sequentially update the set of parameters $\{\theta_i(z_\ell)\}$ over the layer $z_{\ell-1} \le z_j < z_\ell$ until a new value of $\theta_i(z_j)$ becomes available, then we re-initialize the parameter estimator with this value and continue our SSP estimation until we have recursed through each sensor location. In this way we can utilize our measured SSP information in the form of a parameter update and improve the estimates using the processor. Thus, we can characterize this SSP representation in an approximate Gauss-Markov model, which is non-linear, when we constrain the SSP parameters to the set $\{\theta(z_\ell)\}$, $z_\ell = z_j, \ldots, z_j + N_\theta$, that is,

$$\underline{\theta}_N(z_{\ell+1}) = \begin{cases} \underline{\theta}_N(z_\ell) + \Delta z_\ell \underline{w}_\theta(z_\ell) & z_\ell < z_j < z_{\ell+1} \\ \underline{\theta}_N(z_\ell)\delta(z_\ell - z_j) & z_\ell = z_j \end{cases} \tag{5.85}$$

where we have $w_\theta \sim N(0, R_{w_\theta w_\theta})$, and $\underline{\theta}_N(z_0) \sim N(\underline{\theta}_N(0), P_\theta(0))$.

It is this model that we use in our adaptive MBP to estimate the sound speed and solve the environmental inversion problem. We can now "augment" this SSP representation into our normal-mode/pressure-field propagation model to obtain an overall system model. The augmented *Gauss-Markov (approximate) model* for M-modal functions in the interval $z_\ell < z_j < z_{\ell+1}$ is given by

$$\begin{bmatrix} \underline{x}(z_{\ell+1}) \\ --- \\ \underline{\theta}_N(z_{\ell+1}) \end{bmatrix} = \begin{bmatrix} A(z_\ell, \theta) & | & 0 \\ - & - & - \\ 0 & | & I_{N+1} \end{bmatrix} \begin{bmatrix} \underline{x}(z_\ell) \\ --- \\ \underline{\theta}_N(z_\ell) \end{bmatrix} + \begin{bmatrix} \underline{w}(z_\ell) \\ --- \\ \underline{w}_\theta(z_\ell) \end{bmatrix}$$

with the corresponding measurement model

$$p(r_s, z_\ell) = [\underline{C}^T(r_s, z_\ell)|\underline{0}] \begin{bmatrix} \underline{x}(z_\ell) \\ ---- \\ \underline{\theta}(z_\ell) \end{bmatrix} + v_p(z_\ell) \tag{5.86}$$

This completes the development of the state-space forward propagator for the experiment. Next we discuss the design of the MBP for the Hudson Canyon data.

5.4.3.2 Adaptive Environmental Inversion: Sound Speed Estimation

Next we develop a solution to the environmental inversion problem by designing an MBP to estimate the sound speed, on-line, from noisy pressure-field measurements. The processor is based on the augmented model above. We briefly discuss the approach and then the algorithm and apply it to the Hudson Canyon experimental data for a 500 m range at a 50 Hz temporal frequency.

The environmental inversion problem can be defined in terms of our previous models as the following:

GIVEN a set of noisy acoustic (pressure-field) measurements $\{p(r_0, z_\ell)\}$ and a set of sound speed parameters $\{\underline{\theta}(z_\ell)\}$, **FIND** the best (minimum-variance) estimate of the SSP, $\hat{c}(z_\ell)$.

For this problem, we have a sparse set of SSP measurements available at $N_\theta = 9$ depths with a complete set of pressure-field measurements. The solution to the inversion problem can be obtained using the parametrically adaptive form of the EKF algorithm[1,2] discussed in the previous application employed as a joint state/parameter estimator. Here, we choose the discrete EKF available in SSPACK_PC.[48]

The experimental measurements consist of sound speed in the form of discrete data pairs $\{c(z_j), z_j\}$ which can be utilized in the estimator for correction as it processes the acoustic data. We use the first two terms ($N = 2$) of the Taylor series expansion of the SSP (piecewise linear) for our model, where both θ_0 and θ_1 are space-varying, Gaussian random functions with specified means and variances, and, therefore, through linearity, so is $c(z_\ell)$. Thus, our Gauss-Markov model for this problem is given by Equation 5.86. We will use a spatial sampling interval of $\Delta z_j = (\Delta z_\ell)/10$ in the state propagation equations as discussed previously. The corresponding EKF estimator evolves from the algorithm (see Reference 2) with all of the appropriate functions and Jacobians.

We observe the performance of the model-based SSP processor. Here, we use only the acoustic measurements and the nine sound speed data values $\{c(z_j)\}$, $j = 1, \ldots, 9$ to set hard constraints on the parameter estimator and force it to meet these values only when the appropriate depth is achieved. The results of the runs are shown in Figure 5.14. Here, we see the estimated SSP parameters and reconstructed SSP. The estimator appears to track the SSP parameters as well as the profile. The standard rms modeling errors (see Reference 24 for details) for the SSP parameters and profile are, respectively, 1.6×10^{-4}, 2.7×10^{-5}, and 1.0×10^{-2}. Since we are using a joint estimator, the enhanced estimates of both modal functions and the pressure field are in excellent agreement with all of the modes predicted by the validated SNAP propagation model. The standard rms modeling errors for each mode, respectively, are 1.0×10^{-2}, 1.3×10^{-3}, 2.3×10^{-4}, 2.2×10^{-4}, and 3.5×10^{-4}. The innovations or residuals are shown in Figure 5.15, where we see that they are zero mean and reasonably white (8.3% out of bounds). The rms standard error for the residuals is given by 2.9×10^{-3}. So we see that the processor is clearly capable of jointly estimating the SSP and enhancing the modal/pressure field.

In this section, we have developed an on-line, adaptive, model-based solution to the environmental inversion problem, that is, an SSP estimation scheme based on coupling the normal-mode propagation

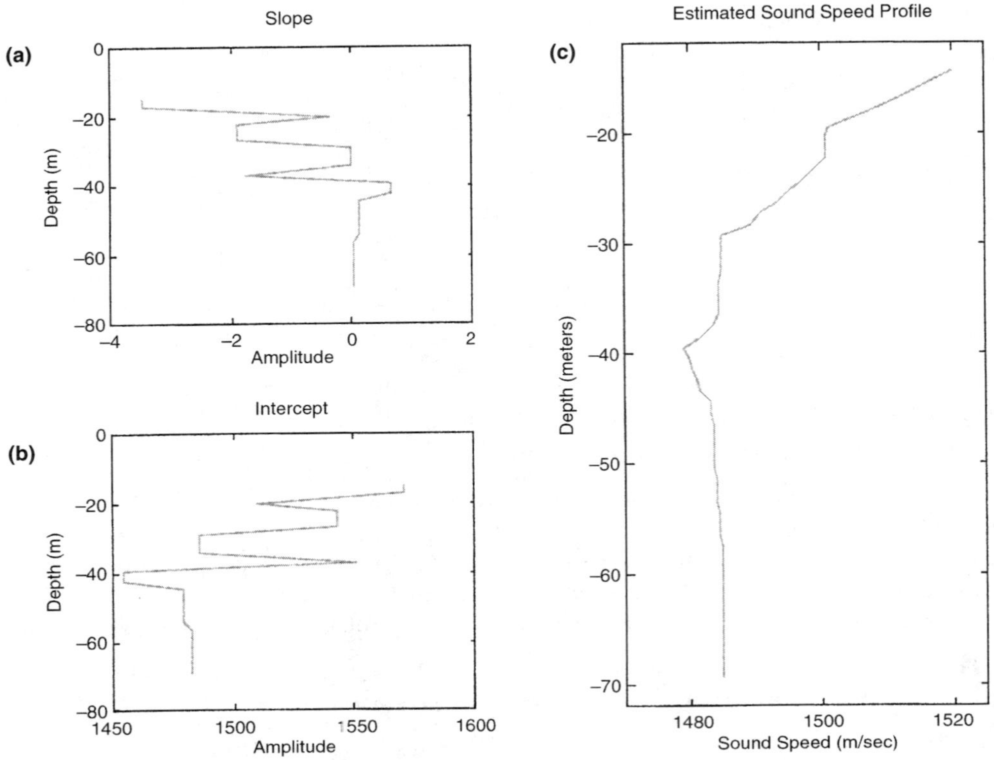

FIGURE 5.14 Model-based inversion: (a) slope estimation, (b) intercept estimation, and (c) sound speed estimation.

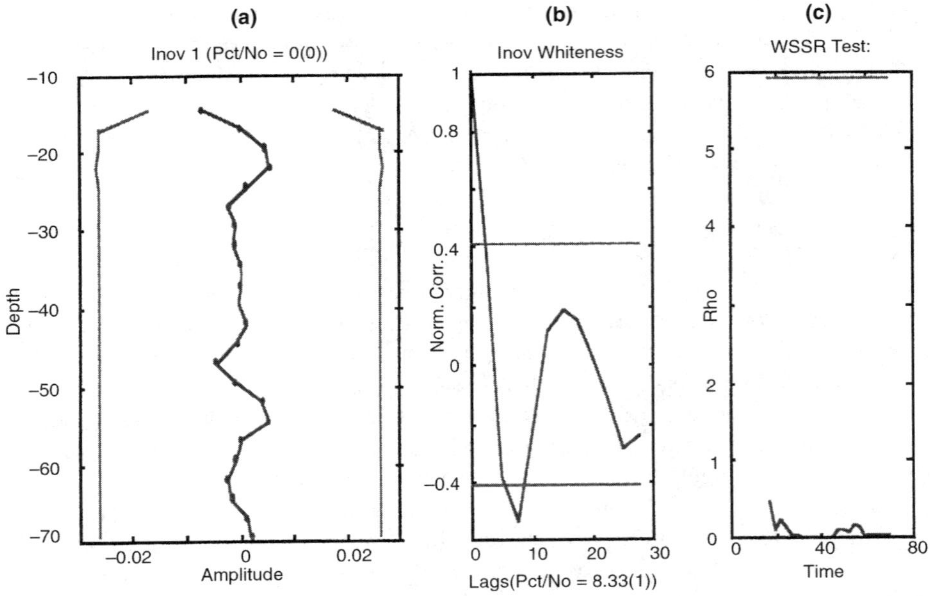

FIGURE 5.15 Model-based residuals: (a) innovation/residual, (b) whiteness test (8.3% out), and (c) WSSR test (passed).

model to a functional model of the SSP evolving from Taylor series expansion about the most current sound speed measurement available. The algorithm employed was the parametrically adaptive EKF, which evolved as the solution to the minimum-variance estimation problem when the augmented models were placed in state-space form.

5.4.4 Ocean Acoustic Application: Model-Based Localization

In this section, a parametrically adaptive, model-based approach is developed to solve the passive localization problem in ocean acoustics using the state-space formulation. It is shown that the inherent structure of the resulting processor consists of a parameter estimator coupled to a non-linear optimization scheme. We design the parameter estimator or more appropriately the adaptive MBID for a propagation model developed from the Hudson Canyon shallow water ocean experiment.

Let us examine the inherent structure of the adaptive model-based localizer shown in Figure 5.16. Here, we see that it consists of two distinct parts: a parameter estimator implemented using an MBID as discussed above and a non-linear optimizer to estimate the source position. We see that the primary purpose of the parameter estimator is to provide estimates of the inherent localization functions that then must be solved (implicitly) for the desired position. In this application, we will show that the parameter estimator (or identifier) will be model-based, incorporating the ocean acoustic propagation model. Thus, we see that it is, in fact, the adaptive MBP or, in this case, the MBID that provides the heart of the model-based localization scheme.

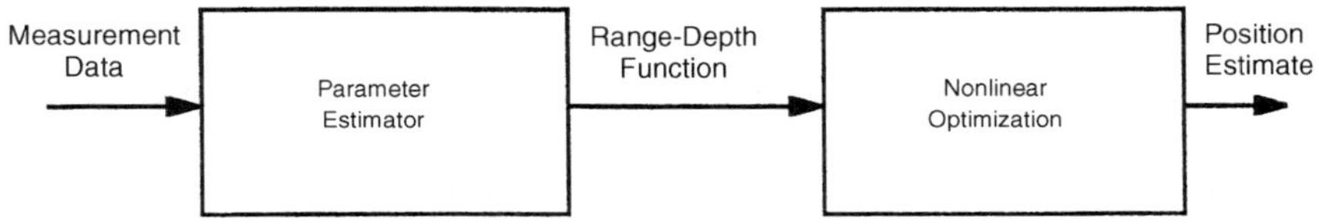

FIGURE 5.16 Model-based localization: the basic processor.

We develop the model-based localizer (MBL) for our ocean acoustic problem and show how it is realized by utilizing an MBID coupled to a non-linear optimizer. It will also be shown that the MBID provides an enhanced estimate of the required range-depth function, which is essentially the scaled modal coefficients that are supplied to the optimal position estimator. Next, we briefly outline the complete processor and show the structure of the embedded MBID. The MBL is applied to the Hudson Canyon experimental data, demonstrating the impact of using the MBID for signal enhancement prior to localization.

5.4.4.1 Model-Based Localization: Non-Linear Optimizer

First, we develop an MBL for use with the normal-mode model and choose the "depth only" structure with a vertical array as before in Equation 5.71. Recall that the acoustic pressure propagation model is

$$p(r_s, z) = q\sum_{m=1}^{M} H_0(\kappa_r(m)r_s)\phi_m(z_s)\phi_m(z) \tag{5.87}$$

where p is the acoustic pressure, q is the source amplitude, ϕ_m is the mth modal function at z and z_s, $\kappa_r(m)$ is the horizontal wavenumber associated with the mth mode, r_s is the source range, and $H_0(\kappa_r r_s)$ is the zeroth-order Hankel function which provides the range solution. The localization problem solution evolves from the measurement equation of the Gauss-Markov model where we can write the sampled pressure field in terms of range-depth-dependent terms as

$$p(r_s, z_\ell) = \sum_{m=1}^{M} \beta_m(r_s, z_s)\phi_{m1}(z_\ell) + \mathbf{v}(z_\ell) \tag{5.88}$$

For the two-dimensional localization problem, we can decompose the pressure measurement further as

$$p(r_s, z_\ell) = \sum_{m=1}^{M} \gamma_m(r, z)\theta_m(r_s, z_s)\phi_{m1}(z_\ell) + \mathbf{v}(z_\ell) \tag{5.89}$$

where γ_m represents the known function and $\theta_m(r_s, z_s)$ the *unknown* function of position. Equating these functions with β_m from Equation 5.88 we have

$$\gamma_m(r, z) = \frac{q}{\int_0^h \phi_m^2(z)dz} \tag{5.90}$$

and

$$\theta_m(r_s, z_s) = H_0(k_r(m)r_s)\phi_m(z_s) \tag{5.91}$$

an implicit, separable function of $\mathbf{r}_s$ and $\mathbf{z}_s$ which we will call the source range-depth function. With these definitions in mind, it is now possible to define (simply) the *MBL problem* as the following:

GIVEN a set of noisy pressure-field and sound speed measurements, $[\{p(r_s, z_\ell)\}, \{c(z_\ell)\}]$, and the normal-mode model, **FIND** the best (minimum-error variance) estimate of the source position (r_s, z_s), that is, find $\hat{r}_s$ and $\hat{z}_s$.

In order to solve this problem, we must first estimate the "unknown" range-depth function $\theta_m(r_s, z_s)$ from the noisy pressure-field measurement model and then use numerical optimization techniques to perform the localization (r_s, z_s). We discuss the MBP used to perform the required parameter estimation in Section 5.4.4.2; here we concentrate on the localization problem and the related range-depth functions.

In the design of a localizer, we choose a non-linear least squares approach.[50] Thus, the optimization problem is to find the source position (r_s, z_s) that minimizes

$$J(r_s, z_s) := \frac{1}{M}\sum_{m=1}^{M} (\theta_m(r_s, z_s) - H_0(\kappa_r(m)r_s)\phi_m(z_s))^2 \tag{5.92}$$

Since we know from our previous analysis[18] that a unique optimum does exist, we choose to use a brute force, *direct search* method for our localizer primarily because it requires the minimal amount of *a priori* information and should slowly converge to the global optimum. For an on-line application, more rapidly convergent algorithms requiring *a priori* information (gradient and Hessian) should be investigated,[50] but here we use an off-line search to investigate the feasibility of the MBL.

The "direct search" localization algorithm follows the *polytope* method of Nelder and Meade.[51] At each stage of iteration, $N + 1$ points, say, $\alpha_1, \ldots, \alpha_{N+1}$, are retained together with the function of these values, that is,

$$\alpha_n := (r_n, z_n), \qquad J(\alpha_n), \qquad for \qquad n = 1, 2, \ldots, N+1 \tag{5.93}$$

where the functions are ordered such that

$$J(\alpha_{N+1}) \geq J(\alpha_N) \geq \ldots \geq J(\alpha_1) \tag{5.94}$$

and constitute the vertices of the polytope in N space. At each iteration, a new polytope is generated, producing a new point to replace the "worst" point α_{N+1} — the point with the largest function value. If we define $c(\alpha)$ as the centroid of the best N vertices of $\alpha_1, \ldots, \alpha_N$ given by

$$c(\alpha) = \frac{1}{N}\sum_{n=1}^{N} \alpha_n \tag{5.95}$$

then at the beginning of the nth iteration a search or trial point is constructed by a single reflection step using

$$\alpha_r = c(\alpha) + (c(\alpha) - \alpha_{N+1})\Delta_r \tag{5.96}$$

where Δ_r is the reflection coefficient ($\Delta_r > 0$). The function is evaluated at α_r, giving $J(\alpha_r)$ and yielding three possibilities:

1. $J(\alpha_1) \le J(\alpha_r) \le J(\alpha_N)$ and, therefore, $\alpha_r \to \alpha_{N+1}$.
2. $J(\alpha_r) < J(\alpha_1)$ and $\alpha_r \to \alpha_1$ a new best point, since we are minimizing J. The direction Δ_r is assumed correct, and we then *expand* the polytope by defining

$$\alpha_e = c(\alpha) + (\alpha_r - c(\alpha))\Delta_e \tag{5.97}$$

   where Δ_e is the expansion coefficient ($\Delta_e > 1$). If $J(\alpha_e) \le J(\alpha_r)$, $\alpha_e \to \alpha_{N+1}$, otherwise $\alpha_r \to \alpha_{N+1}$.
3. In $J(\alpha_r) > J(\alpha_N)$, the polytope is too large and we must "contract" it using

$$\alpha_c = \begin{cases} \alpha_1 + (\alpha_{N+1} - \alpha_1)\Delta_c & \textit{for} \quad J(\alpha_r) \ge J(\alpha_{N+1}) \\ \alpha_1 + (\alpha_r - \alpha_1)\Delta_c & \textit{for} \quad J(\alpha_r) < J(\alpha_{N+1}) \end{cases} \tag{5.98}$$

   where Δ_c is the contraction coefficient. If $J(\alpha_c) < \min\{J(\alpha_r), J(\alpha_{N+1})\}$, then $\alpha_c \to \alpha_{N+1}$.

Using the MATLAB Optimization Toolbox,[52] we apply the polytope algorithm to the shallow water experimental data discussed in Section 5.4.4.2. Before we discuss the details of the MBID, let us see how the model-based approach is used to implement the localizer. From the cost function $J(r_s, z_s)$ of Equation 5.92, we see that we must have an estimate of the range-depth function, $\theta_m(r_s, z_s)$, and this is provided by our MBID. However, we must also have estimates of the associated Hankel function, $H_0(\kappa_r r_n)$, and the corresponding modal functions evaluated at the current iterate depth, z_n as $\phi_{m1}(z_n)$. The MBID provides us with estimates of these modal functions $\{\hat{\phi}_{m1}(z_\ell)\}$, $m = 1, \ldots, M$, $\ell = 1, \ldots, L$ at each sensor location (in depth). Since the optimizer requires a finer mesh (in depth) than the modal function estimates at each sensor to perform its search, we use the state-space forward propagator to generate the estimates at a finer depth sampling interval

$$\Delta z_n := \frac{\Delta z_\ell}{p} \qquad p \in \mathbf{I} \tag{5.99}$$

Thus, for a given value of "search" depth z_n, we find the closest available depths from the estimator (array geometry) to bracket the target depth, $z_{\ell-1} < z_n < z_\ell$, and use the lower bound $z_{\ell-1}$ to select the initial condition vector for our propagator. We then forward propagate the modal function at the finer Δz_n to obtain the desired estimate at $\hat{\phi}_{m1}(z_n)$.

Note that the propagator evolves simply by discretizing the differential equation using first differences

$$\frac{d}{dz}\phi(z) \approx \frac{\phi(z_n) - \phi(z_{n-1})}{\Delta z_n} \tag{5.100}$$

which leads to the corresponding *state-space propagator* given by[7]

$$\hat{\phi}(z_n) = [\mathbf{I} - \Delta z_n A(z_n)]\hat{\phi}(z_{n-1}) \qquad \textit{for} \qquad \hat{\phi}(z_{n-1}) = \phi(z_{\ell-1}) \tag{5.101}$$

In this way, the state-space forward propagator is used to provide functional estimates to the nonlinear optimizer for localization, so we see that the MBID (Section 5.4.4.2) is designed to not only provide estimates of the range-depth function, but also to provide enhanced estimates of the modal functions at each required depth interation, that is,

$$[\{\hat{\theta}_m(r_s, z_s)\}, \{\hat{\phi}_{m1}(z_\ell)\}] \rightarrow [\{\hat{\phi}_n(z_n)\}, (\hat{r}_s, \hat{z}_s)] \tag{5.102}$$

From an estimation viewpoint, it is important to realize the ramifications of the output of the processor and its relationship to the position estimates. The respective range-depth and modal estimates $\hat{\theta}$ and $\hat{\phi}$ provided by the MBID are minimum-variance estimates (approximately). In the case of Gaussian noise, they are, if fact, the maximum likelihood (maximum *a posteriori*) estimates and, therefore, the corresponding maximum likelihood *invariance theorem* guarantees that the solutions for the (r_s, z_s) are also the maximum likelihood estimates of position.[5] This completes the description of the localizer. Next we discuss how the range-depth and modal functions are estimated from noisy pressure-field measurements by developing the MBID.

5.4.4.2 Model-Based Localization: Parametrically Adaptive Processor

Next we develop the adaptive parameter estimator or, more appropriately the MBID which provides the basis of our eventual localizer design (see Figure 5.17). However, before we can provide a solution to the localization problem, we must develop the identifier to extract the desired range-depth function of the previous section as well as provide the necessary enhancement required for localization. From our previous work, it is clear that the EKF identifier will satisfy these constraints nicely.[1,2,18,23] It is also clear from the localization discussion in Section 5.4.4.1 that we must estimate the vector source range-depth function $\theta(r_s, z_s)$ directly from the measured data as well as the required modal functions. The basic approach we take, therefore, is to realize that at a given source depth the implicit range-depth function is fixed; therefore, we can assume that $\theta(r_s, z_s)$ is a constant ($\dot{\theta} = 0$) or a random walk with a discrete Gauss-Markov representation given by

$$\theta(r_s, z_\ell) = \theta(r_s, z_{\ell-1}) + w_\theta(z_{\ell-1}) \tag{5.103}$$

Therefore, the underlying model for our ocean acoustic problem becomes the normal-mode propagation model (in discrete form) with an augmented parameter space as discussed in Section 5.2 (ignoring the noise sources):

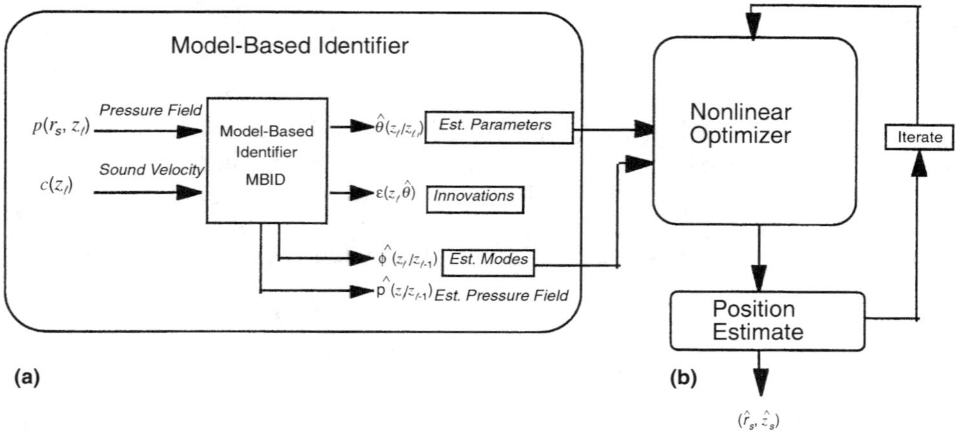

FIGURE 5.17 MBL processor structure: (a) MBID and (b) optimizer.

$$
\begin{aligned}
\phi_{m1}(z_\ell) &= \phi_{m1}(z_{\ell-1}) + \Delta z_\ell \phi_{m2}(z_{\ell-1}) \\
\phi_{m2}(z_\ell) &= -\Delta z_\ell \kappa_z^2(m)\phi_{m1}(z_{\ell-1}) + \phi_{m2}(z_{\ell-1}),\ m = 1, \ldots, M \\
\theta_1(r_s, z_\ell) &= \theta_1(r_s, z_{\ell-1}) \\
\vdots \qquad & \qquad \vdots \\
\theta_M(r_s, z_\ell) &= \theta_M(r_s, z_{\ell-1})
\end{aligned}
\tag{5.104}
$$

and the corresponding measurement model is given by

$$
p(r_s, z_\ell) = \sum_{m=1}^{M} \gamma_m(r, z_\ell)\theta_m(r_s, z_\ell)\phi_{m1}(z_\ell) + v(z_\ell) \tag{5.105}
$$

We choose this general representation where the set $\{\theta_m(r_s, z_\ell)\}$, $m = 1, \ldots, M$ is the *unknown* implicit function of range and depth, while the parameters $\{\gamma_m(r, z_\ell)\}$ represent the *known a priori* information that is included in the processor. The basic *prediction* estimates for the mth mode from the MBID are

$$
\begin{aligned}
\hat{\phi}_{m1}(z_\ell|z_{\ell-1}) &= \hat{\phi}_{m1}(z_{\ell-1}|z_{\ell-1}) + \Delta z_\ell \hat{\phi}_{m2}(z_\ell|z_{\ell-1}) \\
\hat{\phi}_{m2}(z_\ell|z_{\ell-1}) &= -\Delta z_\ell \kappa_z^2(m)\hat{\phi}_{m1}(z_{\ell-1}|z_{\ell-1}) + \hat{\phi}_{m2}(z_{\ell-1}|z_{\ell-1}),\ m = 1, \ldots, M \\
\hat{\theta}_1(r_s, z_\ell|z_{\ell-1}) &= \hat{\theta}_1(r_s, z_{\ell-1}|z_{\ell-1}) \\
\vdots \qquad & \qquad \vdots \\
\hat{\theta}_M(r_s, z_\ell|z_{\ell-1}) &= \hat{\theta}_M(r_s, z_{\ell-1}|z_{\ell-1})
\end{aligned}
\tag{5.106}
$$

and the corresponding innovations (parameterized by θ) are given by

$$
\varepsilon(z_\ell, \theta) = p(r_s, z_\ell) - \sum_{m=1}^{M} \gamma_m(r, z_s)\hat{\theta}_m(r_s, z_\ell|z_{\ell-1})\hat{\phi}_{m1}(z_\ell|z_{\ell-1}, \theta) \tag{5.107}
$$

with the vector *correction equations*

$$
\begin{aligned}
\hat{\phi}(z_\ell|z_\ell) &= \hat{\phi}(z_\ell|z_{\ell-1}) + K_\phi(z_\ell)\varepsilon(z_\ell, \theta) \\
\hat{\theta}(r_s, z_\ell|z_\ell) &= \hat{\theta}(r_s, z_\ell|z_{\ell-1}) + K_\theta(z_\ell)\varepsilon(z_\ell, \theta)
\end{aligned}
\tag{5.108}
$$

So we see (simply) how the unknown range-depth or scaled modal coefficient parameters are augmented into the MBID algorithm to enhance the required signals and extract the desired parameters. We summarize the detailed structure of the MBL incorporating the MBID in Figure 5.17.

5.4.4.3 Model-Based Localization: Application to Hudson Canyon Data

Again, we use the Hudson Canyon experimental data to analyze the localizer performance. Recall that a 23-element vertical array was deployed from the bottom with 2.5 m separation to measure the pressure field and through spectral analysis the following average horizontal wave numbers: {0.28, 0.199, 0.183, 0.175, 0.142} m^{-1} for the five modes supporting the water column from a 36 m deep, 50 Hz source at 0.5 km range (see References 43 to 45 for more details) resulted. Using SSPACK_PC,[48] a toolbox available in MATLAB, we investigate the design using the experimental hydrophone measurements from the Hudson Canyon. Here, we initialize the MBID with the average set of horizontal wave numbers as before. The resulting estimates are quite reasonable, as shown in Figures 5.18 and 5.19. The results are better than those reported previously for this data set (see Reference 24), primarily because we have allowed the processor to dynamically adapt (parameter estimator) to the changing parameters. The results for the higher order modes follow those predicted by the model as observed in Figure 5.19 and by the

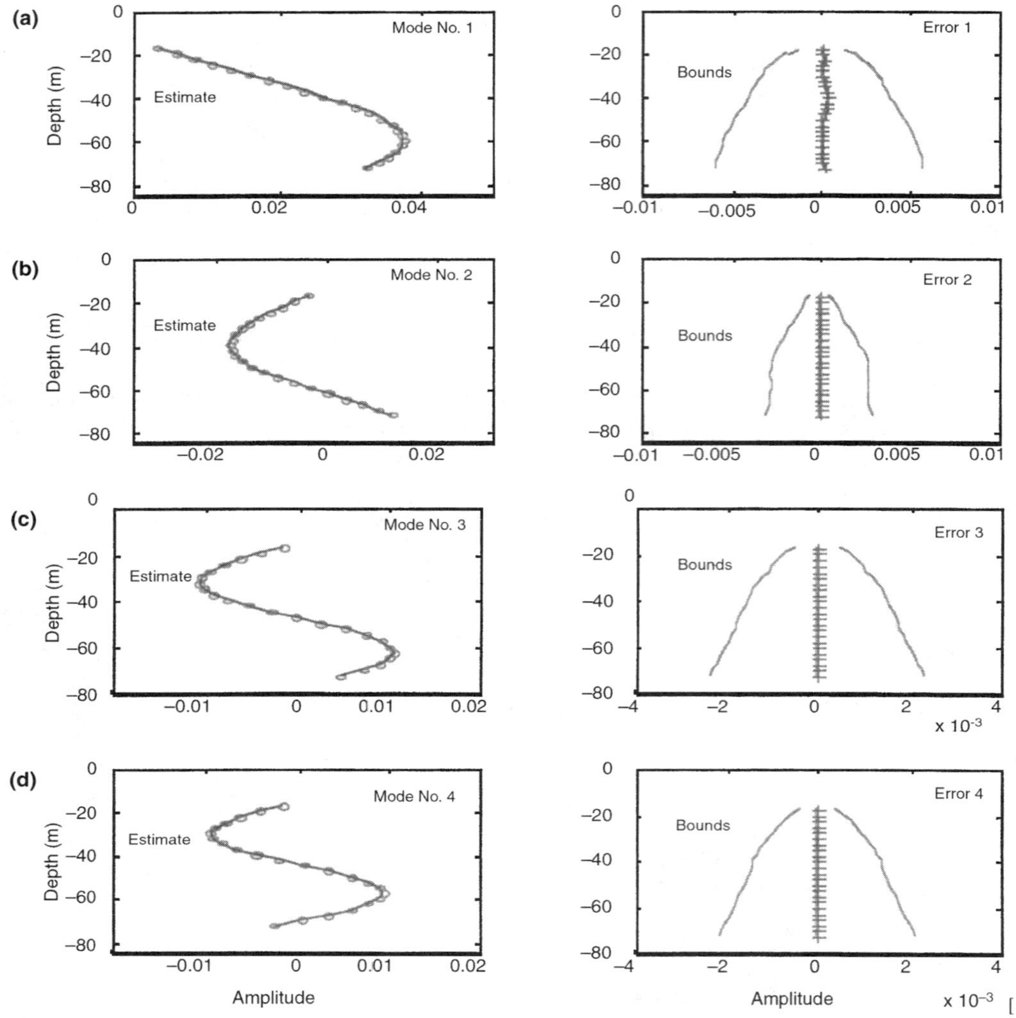

FIGURE 5.18 MBID of the Hudson Canyon experiment (0.5 km): (a) mode 1 and error (0% out), (b) mode 2 and error (0% out), (c) mode 3 and error (0% out), and (d) mode 4 and error (0% out).

corresponding estimation errors. The reconstructed pressure field and innovations are also quite reasonable as shown in Figure 5.19 and indicate a "tuned" processor with its zero mean ($1.0 \times 10^{-3} < 2.6 \times 10^{-3}$) and white innovations (~8.3% out and WSSR < τ). The final parameter estimates with predicted error statistics for this data are also included in Table 5.6 for comparison to the simulation. We see again that the MBID appears to perform better than the fixed MBP simply because the range-depth parameters or scaled modal coefficients are "adaptively" estimated on-line, providing a superior fit to the raw data as long as we have reasonable estimates to initialize the processor. This completes the discussion on the design of the MBID. Next we take these intermediate results and apply the non-linear optimizer Section 5.4.4.2 to obtain a solution to the localization problem as depicted in Figure 5.17.

Next we consider the application of the MBL to the experimental Hudson Canyon data sets. Here, we use the MBID to provide estimates of $[\{\theta_m(r_s, z_s)\}, \{\hat{\phi}_{m1}(z_\ell)\}]$. And then use the polytope search algorithm along with the state-space propagator of Equation 5.101 to provide the localization (Equations 5.93 to 9.96) discussed previously. We applied the optimizer to the resulting range-depth parameters estimated by the MBID. The results of the localization are shown in Figure 5.20. Here, we see the range-depth parameter estimates from the MBID, the true values from the simulator, and the estimates developed by

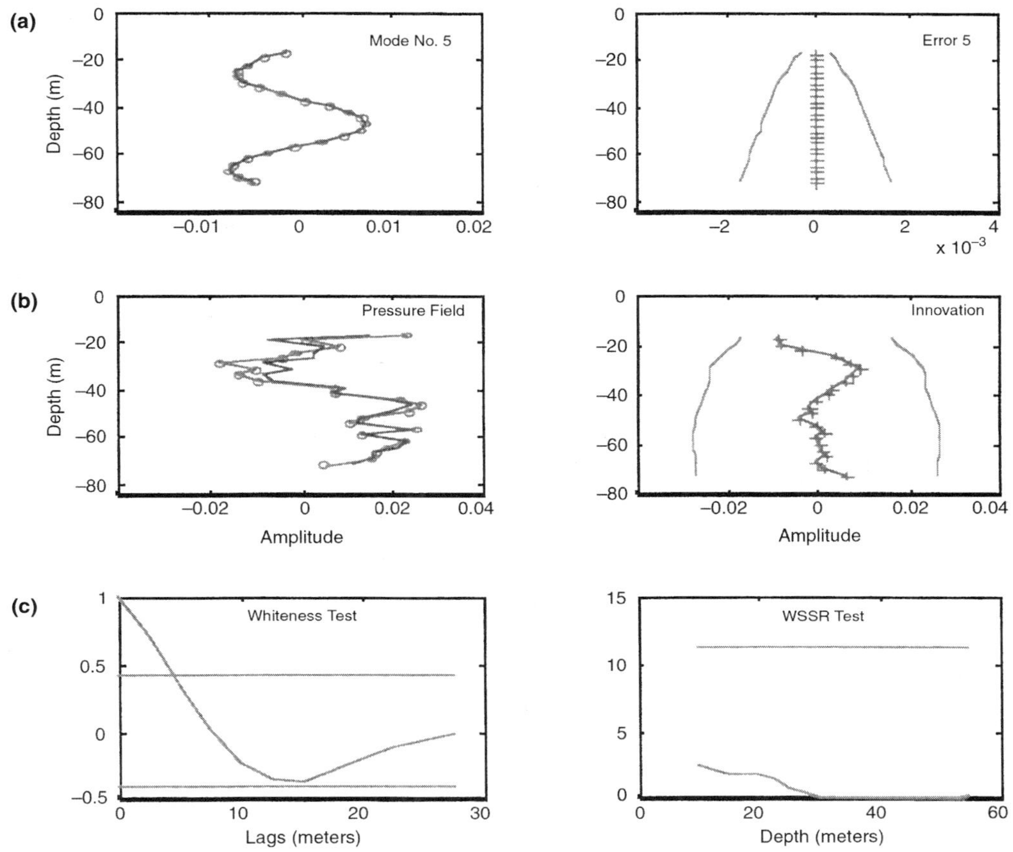

FIGURE 5.19 MBID of the Hudson Canyon experiment (0.5 km): (a) mode 5 and error (0% out), (b) pressure field and innovation (0–0% out), and (c) whiteness test and WSSR (8.3% out).

TABLE 5.6 MBID: Range-Depth Parameter Estimation

	Hudson Canyon Experiment		
Wave Numbers	Model Prediction	Simulation Est./Err.	Experiment Est./Err.
θ_1	1.000	1.014 ± 0.235	1.015 ± 0.362
θ_2	0.673	0.701 ± 0.238	0.680 ± 0.364
θ_3	0.163	0.127 ± 0.430	0.163 ± 0.393
θ_4	0.166	0.138 ± 0.476	0.166 ± 0.393
θ_5	0.115	0.141 ± 0.463	0.116 ± 0.364

the optimizer given, respectively, by the +, x, o, characters on the plots. The corresponding mean-squared errors are also shown, indicating the convergence of the optimizer after about 30 iterations as well as the actual range-depth search (position iterates) with the *true* (500 m, 36 m) and *estimated* (500.2 m, 35.7 m) position estimates shown. The algorithm converges quite readily. It appears that the MBL is able to perform quite well over this data set.

To understand why we can achieve this quality of localization performance, we observe that the effect of the MBID is to enhance these noisy measurements, enabling the optimizer to converge to the correct position. The parametrically adaptive MBID enhancement capability is clear from Figures 5.18 and 5.19. The effect of the MBID leads very closely to the true depth function estimate, since the modal function estimates are quite good. Thus, we can think of the MBID as providing the necessary enhancements in

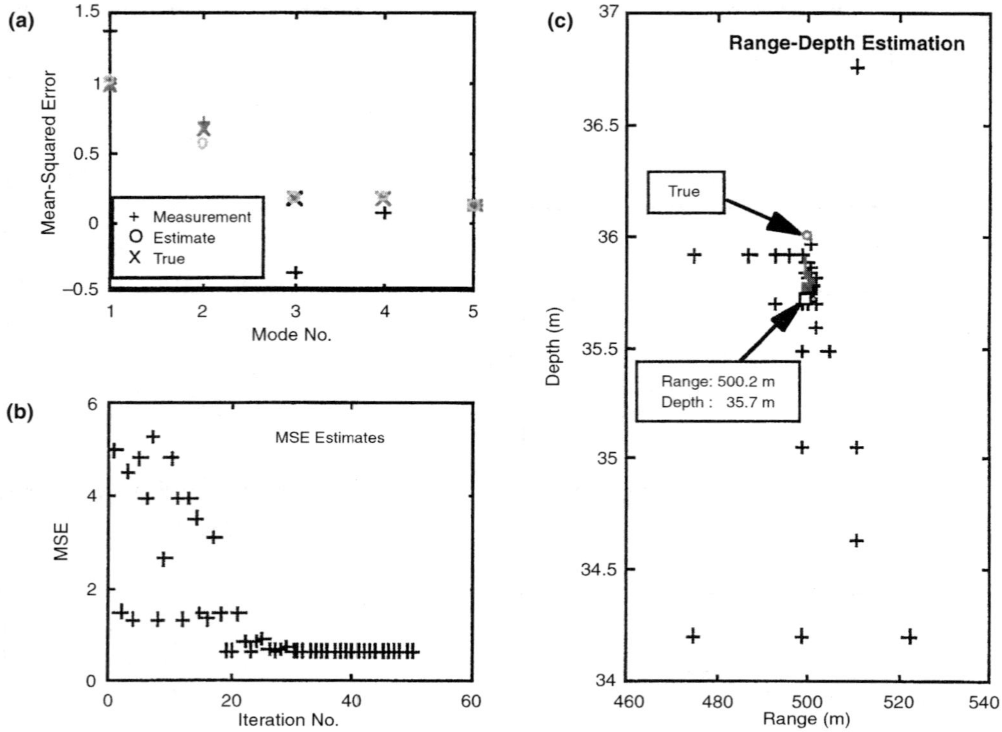

FIGURE 5.20 Hudson Canyon experiment localization: (a) estimated range-depth parameters, (b) mean-squared error, and (c) localization: true and estimated (500.2 m, 35.7 m).

SNR as well as decreasing the dimensionality of the search space to that of the modal space by using the estimated range-depth functions from the MBID as the "raw" measurement data input (along with the modal estimates) to the polytope optimizer.

In summary, we have developed an on-line, parametrically adaptive, model-based solution to the localization problem, that is, a source position location estimation scheme based on coupling the normal-mode propagation model to a functional model of position. The algorithm employed was the non-linear EKF identifier/parameter estimator coupled to a direct search optimizer using the polytope approach. We showed that the MBL evolves quite naturally from the MBID. Thus, the results of applying the MBL scheme to the raw experimental data from the Hudson Canyon experiment were quite good.

5.4.5 Ocean Acoustic Application: Model-Based Towed Array Processor

In this section, we discuss the final application of model-based processing techniques to the development of a processor capable of estimating the bearing of a fixed source from data acquired from a towed array. The signal and measurement systems are placed into state-space form, thereby allowing the unknown parameters of the model, such as multiple source bearings, to be estimated by an adaptive MBID. It is shown that the method outperforms the conventional beamforming approach by providing a continuously time coherent process that avoids the need for spatial and temporal discrete Fourier transforms. A major advantage of the method is that there is no inherent limitation to the degree of sophistication of the models used, and therefore, it can deal with other than plane wave models, such as cylindrically or spherically spreading propagation models as well as more sophisticated representations such as the normal mode and the parabolic equation propagation models.

Here, we consider a simple plane wave propagation model (developed in Section 5.2) and apply it to the problem of multiple source bearing estimation. We will see that this multichannel, adaptive approach casts the bearing estimation problem into a rather general form that *eliminates* the need for an explicit beamformer, while evolving into a passive synthetic aperture (PASA) structure in a natural way.[31–36] In model-based array processing,[31] it will become apparent that PASA forms a natural framework in which to view this approach, since it is a time-evolving spatial process involving a moving array. The issue of the motion is an important one, since, as shown by Edelson,[35] the Cramer-Rao lower bound (CRLB) on the bearing estimate for a moving line array is less than that for the same physical array when not moving. In particular, the ratio of the CRLB for the moving array to that for the stationary array, which we denote by R, is given by

$$R = CRLB_{moving}/CRLB_{fixed} = \frac{1}{1 + (3/2)(D/L) + (D/L)^2} \tag{5.109}$$

Here, D is the "dynamic aperture," that is, the speed of motion of the array times the total time, and L is the length of the physical aperture. As can be seen from this relation, there is potential for a highly significant improvement in performance since R dramatically decreases as D increases.

5.4.5.1 Model-Based Towed Array Processor: Adaptive Processing

Next we develop the model-based solution to the space-time array processing problem by developing a general form of the MBP design with various sets of unknown parameters. We define the *acoustic array space-time processing problem* as the following:

> **GIVEN** a set of noisy pressure-field measurements and a horizontal array of L sensors, **FIND** the best (minimum error variance) estimate of source bearings, temporal frequencies, amplitudes, and array speed.

We use the following non-linear pressure-field measurement model for M monochromatic plane wave sources. We will characterize each of the sources by a corresponding set of temporal frequencies, bearings, and amplitudes, $[\{\omega_m\}, \{\theta_m\}, \{\alpha_m\}]$. That is,

$$p(x, t_k) = \sum_{m=1}^{M} a_m e^{i\omega_m t_k - \beta(t_k)\sin\theta_m} + n(t_k) \tag{5.110}$$

where

$$\beta(t_k) = k_o(x_o + vt_k) \tag{5.111}$$

and $k_o = (2\pi)/\lambda$, x is the current spatial position along the x-axis in meters, v is the array speed (m/sec), and n is additive random noise. The inclusion of the motion in the generalized wave number, β, is critical to the improvement of the processing, since the synthetic aperture effect is actually created through the motion.

If we further assume that the single sensor Equation 5.110 is expanded to include an array of L sensors, then $x \rightarrow x_\ell$, $\ell = 1, \ldots, L$, and we obtain

$$p(x_\ell, t_k) = \sum_{m=1}^{M} a_m e^{i\omega_m t_k - \beta(t_k)\sin\theta_m} + n_\ell(t_k) \tag{5.112}$$

This expression can be written in a more concise form as

$$\mathbf{p}(t_k) = \mathbf{c}(t_k, \Theta) + \mathbf{n}(t_k) \tag{5.113}$$

where the vector Θ represents the parameters of the plane wave sources and

$$\mathbf{p}(t_k) = [p(x_1, t_k)p(x_2, t_k), \ldots, p(x_L, t_k)]^T \tag{5.114}$$

Since we model these parameters as constants, then the *augmented* Gauss-Markov state-space model evolves as

$$\begin{bmatrix} \theta(t_k) \\ --- \\ \omega(t_k) \\ --- \\ \mathbf{a}(t_k) \end{bmatrix} = \begin{bmatrix} \theta(t_{k-1}) \\ --- \\ \omega(t_{k-1}) \\ --- \\ \mathbf{a}(t_{k-1}) \end{bmatrix} + \mathbf{w}(t_{k-1}) \tag{5.115}$$

where $\theta := [\theta_1 \dots \theta_M]^T$, $\omega := [\omega_1 \dots \omega_M]^T$, $\mathbf{a} := [a_1 \dots a_M]^T$, and $\mathbf{w}$ is a zero-mean, Gaussian random vector with covariance R_{ww}.

Defining the composite parameter vector Θ as

$$\Theta := \begin{bmatrix} \theta \\ --- \\ \omega \\ --- \\ \mathbf{a} \end{bmatrix} \tag{5.116}$$

for $\Theta \in R^{3M}$, the following *augmented* state prediction equation evolves for our adaptive MBP:

$$\hat{\Theta}(t_k|t_{k-1}) = \hat{\Theta}(t_{k-1}|t_{k-1}) + \Delta t_k \mathbf{w}(t_{k-1}) \tag{5.117}$$

with the associated measurement equation.

Since the state-space model is linear with no explicit dynamics, the prediction relations are greatly simplified while the correction equations become non-linear due to the plane wave measurement model. This leads to the EKF solution, wherein the non-linearities are approximated with a first-order Taylor series expansion. Here, we require the measurement Jacobian,

$$C(t_k, \Theta) := \frac{\partial \mathbf{c}(t_k, \Theta)}{\partial \Theta} \tag{5.118}$$

an $L \times 3M$ matrix. We simplify the notation by defining the following time-varying coefficient,

$$\alpha_m(t_k) := a_m e^{i\omega_m t_k} \tag{5.119}$$

then the mth plane wave source measured by the ℓth pressure-field sensor is simply given by

$$p_\ell(t_k) = \sum_{m=1}^{M} \alpha_m(t_k) e^{-i\beta_\ell(t_k)\sin\theta_m} \tag{5.120}$$

Given Equations 5.116, 5.118, and 5.119, we are now in a position to compute a recursive (predictor/corrector form) EKF estimate of the state vector Θ. The steps of the algorithm based on Table 5.2 are as follows:

1. Given an initial or trial value of the state estimate, $\hat{\Theta}(t_{k-1}|t_{k-1})$, the parameter prediction equation is used to predict the value of $\hat{\Theta}(t_k|t_{k-1})$. This constitutes a prediction of the state vector for $t = t_k$ based on the data up to $t = t_{k-1}$ as shown in Equation 5.117.

2. The *innovation*, $\varepsilon(t_k)$, is then computed as the difference between the new measurement taken at $t = t_k$ and the predicted measurement obtained by substituting $\hat{\Theta}(t_k|t_{k-1})$ into the measurement equation, that is,

$$\varepsilon(t_k) = \mathbf{p}(t_k) - \hat{\mathbf{p}}(t_k|t_{k-1}) = \mathbf{p}(t_k) - \mathbf{c}(t_k, \hat{\Theta}) \tag{5.121}$$

3. Next, the Kalman gain or weight, $\mathbf{K}(t_k)$, is computed (see Table 5.2).
4. The Kalman gain is then used in the correction stage, producing $\hat{\Theta}(t_k|t_k)$, the corrected estimate from

$$\hat{\Theta}(t_k|t_k) = \hat{\Theta}(t_k|t_{k-1}) + \mathbf{K}(t_k)\varepsilon(t_k) = \mathbf{p}(t_k) - \mathbf{c}(t_k, \hat{\Theta}) \tag{5.122}$$

5. This corrected estimate is then substitute into the right-hand side of the prediction equation, thereby initiating the next iteration.

This completes the discussion of the adaptive model-based array processor. Next we present some examples based on synthesized data.

5.4.6 Model-Based Towed Array Processor: Application to Synthetic Data

Here, we will evaluate the performance of the adaptive MBP to synthesized data assuming that there are two plane wave sources. Then, we will assume that the two sources are both operating at the same frequency. Although, in principle, the speed of the array's motion, v, is observable, sensitivity calculations have shown that the algorithm is sufficiently insensitive to small variations in v to the extent that measured ship speed will suffice as an input value. Finally, we will reduce the number of amplitude parameters, $\{a_m\}$, $m = 1, 2, \ldots, M$, from two to one by rewriting Equation 5.119 as

$$p(x_\ell, t_k) = a_1 e^{i\omega_1 t_k}[e^{-i\beta_\ell(t_k)\sin\theta_1} + \delta e^{-i\beta_\ell(t_k)\sin\theta_2}] + n_\ell(t_k) \tag{5.123}$$

Here, $\delta = a_2/a_1$ and $\omega_1 = \omega_2 = \omega_0$. The parameter a_1 appearing outside the square brackets can be considered to be a data scaling parameter. Consequently, we have four parameters to deal with so that our measurement equation becomes

$$p(x_\ell, t_k) = e^{i\omega_1 t_k + \beta_\ell(t_k)\sin\theta_1} + \delta e^{i\omega_o t_k + \beta_\ell(t_k)\sin\theta_2} + n_\ell(t_k) \tag{5.124}$$

and Equation 5.115 simplifies to

$$\Theta = \begin{bmatrix} \theta_1 \\ \theta_2 \\ --- \\ \omega_0 \\ --- \\ \delta \end{bmatrix} \tag{5.125}$$

We now have all the necessary equations to implement the predictor/corrector form of the Kalman filter algorithm. The calculations are carried out in MATLAB using the SSPACK_PC Toolbox.[48]

In all of the following examples, we assume that the two sources are radiating narrowband energy at a frequency of 50 Hz. The true values of the two bearings, θ_1 and θ_2, are 45° and –10°, respectively. The true amplitude ratio δ is 2. The corresponding initial values for θ_1, θ_2, $f_0 = \omega_0/2\pi$, and δ are 43°, –8°, 50.1 Hz, and 2.5, respectively.

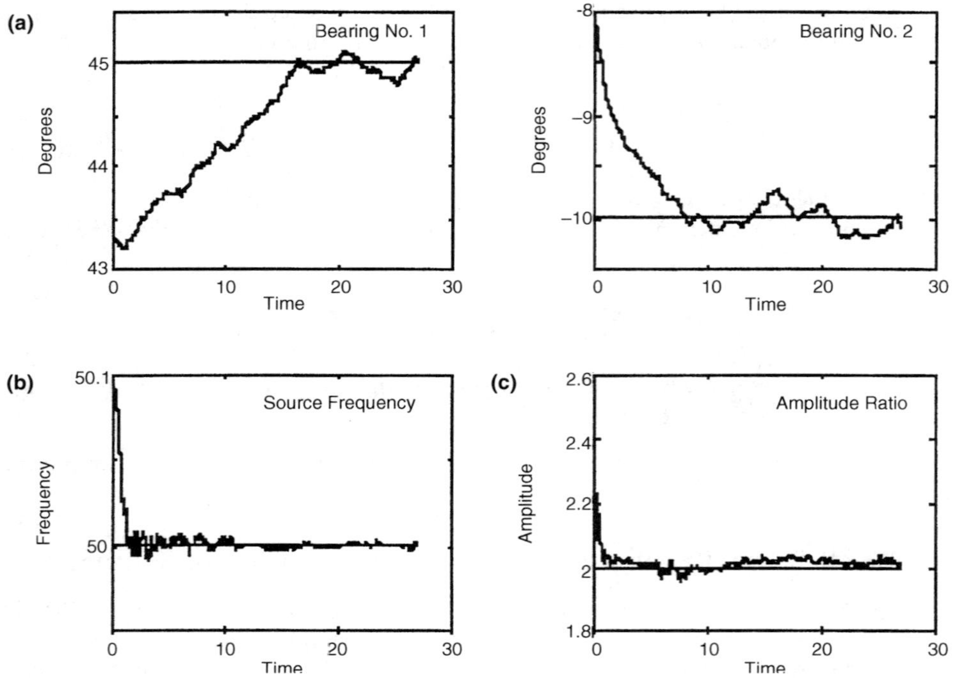

FIGURE 5.21 Case 1: four-element array at 0 dB SNR: (a) two source bearing estimates, (b) temporal frequency estimate, and (c) source amplitude ratio.

Case 1. The number of hydrophones is four, the array's speed of motion is 5 m/sec, and the SNR on the unit amplitude hydrophone is 0 dB. Since the duration of the signal is 27 sec, the array traces out an aperture of 6λ, a factor of four increase over the 1.5λ physical aperture. The parameters being estimated are θ_1, θ_2, $f_0 = \omega_0/2\pi$, and δ. The results are shown in Figure 5.21.

Case 2. This is the same as Case 1 except that the number of hydrophones has been increased to eight. As can be seen in Figure 5.22, as would be expected, the quality of the estimates is significantly improved over the four-hydrophone case.

Case 3. This example is the same as Case 2 except that the speed v is set to zero. From the values of the corresponding predicted variances in Table 5.7, it is clear that the performance has degraded with respect to the $v = 5$ m/sec case as shown in Figure 5.23.

The predicted variances of the estimates for our method are given in Table 5.7, where it is seen that, for the eight-hydrophone moving array $v = 5$ m/sec, the variance on the estimate of θ_1, that is, the bearing associated with the signal with SNR = 0 dB, is 2.5×10^{-5} deg^2, whereas for the $v = 0$ case the predicted

TABLE 5.7 Predicted Variances of the Three Test Cases

Towed Array Synthesis Experiment		
Case Number	Bearing Variance (θ_1) (deg^2):	Bearing Variance (θ_2) (deg^2)
1	7.5×10^{-5}	24.0×10^{-6}
2	2.5×10^{-5}	6.0×10^{-6}
3	14.0×10^{-5}	33.0×10^{-6}

Note: In all cases, the initial values of the parameters, θ_1, θ_2, f_0, and δ are 42°, –8°, and 50.1 Hz, respectively.

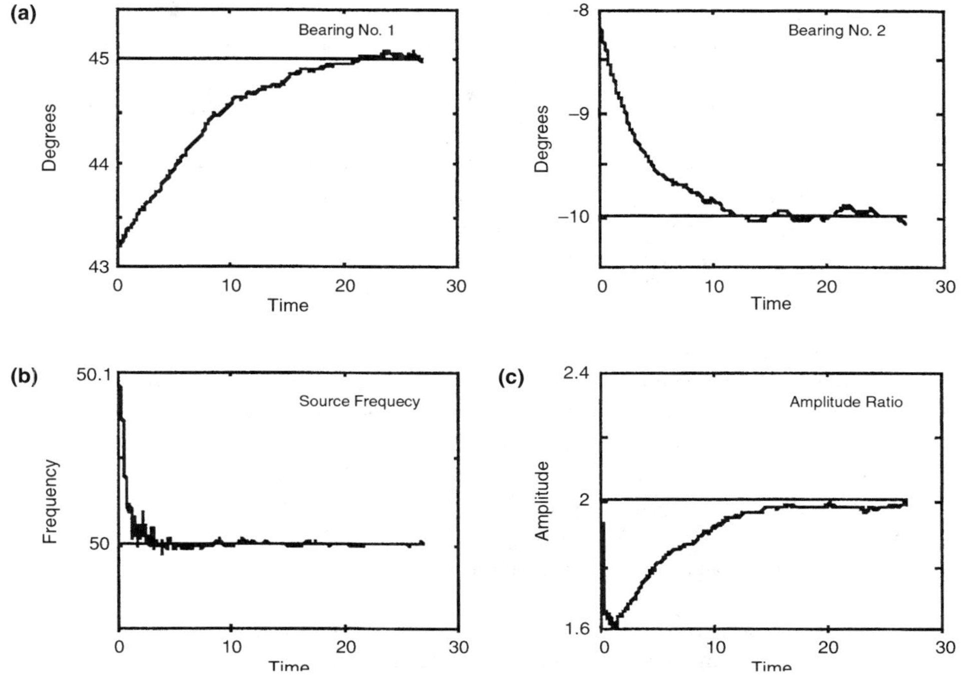

FIGURE 5.22 Case 2: eight-element array at 0 dB SNR: (a) two source bearing estimates, (b) temporal frequency estimate, and (c) source amplitude ratio.

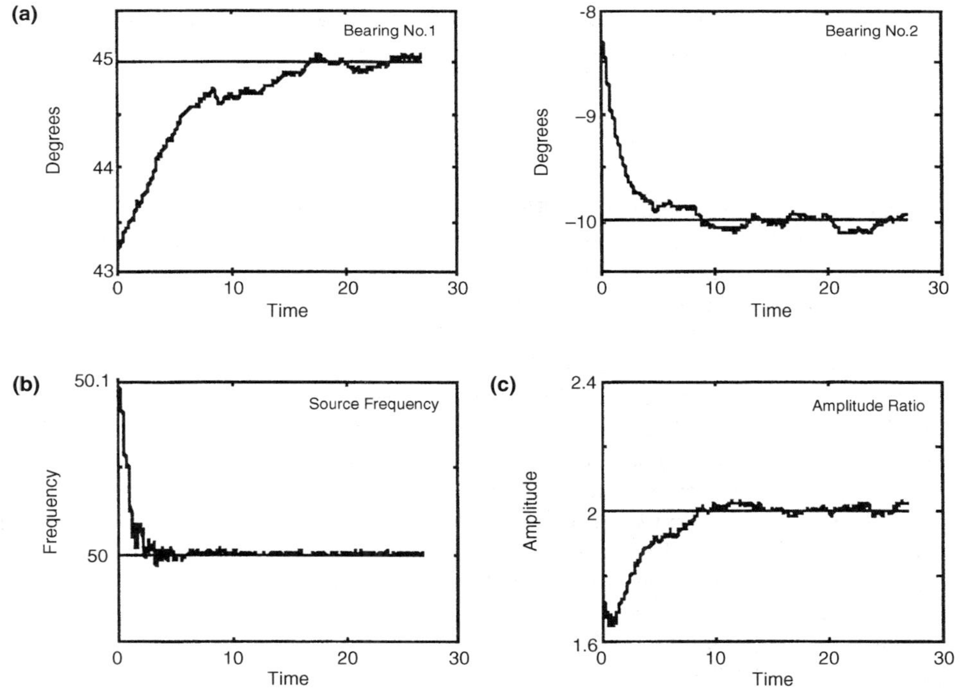

FIGURE 5.23 Case 3: eight-element array at 0 dB SNR and no motion: (a) two source bearing estimates, (b) temporal frequency estimate, and (c) source amplitude ratio.

variance increases to 14.0×10^{-5} deg^2. This is an improvement of approximately a factor of five in the moving array case over that where the array is stationary. Standard acoustic array bearing estimators (beamformers) do not take advantage of the motion of the array and, therefore, cannot comprise efficient estimators of the source bearing(s). A popular form of bearing estimator is the so-called $k - \omega$ beamformer wherein the beamformer takes the form of a discrete spatial Fourier transform. However, this introduces a limit on the bearing accuracy, since the finite bin size of the spatial transform limits the bearing resolution, unless the size of the transform can somehow be increased in an adaptive manner. A further issue is that most conventional beamformers are based on a time domain fast Fourier transform (FFT), which results in the incoherent concatenation of a sequence of coherent processes, whereas our model-based algorithm provides a continuous coherent process in the time domain.

A model-based approach to space-time acoustic array processing has been presented. By explicitly including the motion of the array in the signal model, improved bearing estimation and resolution performance is obtained. The technique is shown to be equivalent to a passive synthetic aperture processor which allows the motion of the array to effectively increase its useful aperture, thereby providing all of the associated improvement in performance. The advantage of this approach is that there is essentially no limit to the degree of sophistication allowed for the particular models chosen for the processor. In this work, we have chosen the signal model to be a sum of plane waves. However, the method can easily be generalized to more general models such as signals with spherical or cylindrical wavefronts which, for example, would permit a wavefront curvature ranging scheme to be implemented that could exploit the large apertures available.

A unique aspect of this processor is that it performs bearing estimation without the necessity of introducing an explicit beamformer structure. The advantage of this is that all of the limitations of standard beamformers, such as finite beam bin sizes, limited time domain coherence due to time domain FFT processing, and the limitation imposed by a predetermined number of beams, are avoided. A further advantage of our MBP is that the innovations sequence provided by the Kalman estimator carries information regarding the performance of the model. Although we have not exploited this in this work, there remains the potential of using this information to monitor and update the model on-line. Thus, deviations from plane wave signals and distortions due to array motion could, in principle, be compensated for in a self-consistent manner. This completes the application.

5.5 Summary

In this chapter, we have described the model-based approach to ocean acoustic signal processing. This approach offers a mechanism to incorporate any *a priori* knowledge or historical information into the processor to extract and enhance the desired information. This *a priori* information is usually in the form of mathematical models which describe the propagation, measurement, and noise processes. The application of the resulting MBP in various forms has led to some very enlightening applications.

The MBP and its variants were discussed briefly in Section 5.2 where a simple plane wave source example was synthesized to illuminate the concepts. It was discussed how various forms of the underlying Gauss-Markov models evolve based on the particular form of the problem under investigation. Linear problems were solved with the standard, linear Kalman filter algorithm for the stationary data case, while time- or space-varying structures led to the non-stationary case which could also be handled by the linear algorithm. When an adaptive form for the ocean acoustic problem had to be developed due the changing nature of the hostile ocean environment, then the non-linear processor evolved in the form of the extended Kalman algorithm and its parametrically adaptive form, performing jointly both state and parameter estimation.

In Section 5.3, the normal-mode propagation model was developed and placed in state-space form for both range and depth. These synthesizers were termed "forward propagators" due to their similarity to the marching method used in solving boundary value problems. The extension of these representations to include uncertainty and noise led finally to the Gauss-Markov model which is the underlying basis of the Kalman solutions to the optimal estimation problem. Here, these results provided the underlying basis for model-based signal processing in the ocean.

In Section 5.4, several ocean acoustic applications were discussed based on a set of ocean acoustic data obtained from the Hudson Canyon experiments performed in 1988. This data set represents a well-known, well-defined region which has extensively been modeled and used in many algorithm applications. We posed, solved, and demonstrated the results for model-based adaptive signal enhancement, model-based inversion, model-based localization, and model-based space-time processing.

In summary, we re-emphasize several points regarding the advantages of MBP. First, it enhances the SNR of the signals of interest. This, in itself, is of major importance and does not seem to be generally recognized. Second, it basically solves the so-called mismatch problem that plagues MFP without desensitizing the performance, but, indeed, improving it. Third, since it is recursive, it can easily deal with adaptive and non-stationary problems, as demonstrated in all of oceanic applications. Last, the MBP is capable of monitoring its own performance, thus providing information on the fidelity of the models employed.

References

1. A. Jazwinski, *Stochastic Processes and Filtering Theory.* New York: Academic Press, 1970.
2. J. V. Candy, *Signal Processing: The Model-Based Approach.* New York: McGraw-Hill, 1986.
3. A. Tolstoy, *Matched Field Processing for Ocean Acoustics.* New Jersey: World Scientific Publishing Co., 1993.
4. W. M. Carey and W. B. Moseley, Space-time processing, environmental-acoustic effects, *IEEE J. Oceanic Eng.*, 16(3), 285–301, 1991.
5. E. J. Sullivan and D. Middleton, Estimation and detection issues in matched-field processing, *IEEE Trans. Oceanic Eng.*, 18(3), 156–167, 1993.
6. A. B. Baggeroer, W. A., Kuperman, and H. Schmidt, Matched-field processing: source localization in correlated noise as an optimum parameter estimation problem, *J. Acoust. Soc. Am.*, 83(2), 571–587, 1988.
7. J. V. Candy and E. J. Sullivan, Ocean acoustic signals processing: a model-based approach, *J. Acoust. Soc. Am.*, 92(12), 3185–3201, 1992.
8. J. V. Candy and E. J. Sullivan, Sound velocity profile estimation: a system theoretic approach, *IEEE Trans. Oceanic Eng.*, 18(3), 240–252, 1993.
9. M. J. Hinich, Maximum likelihood signal processing for a vertical array, *J. Acoust. Soc. Am.*, 54, 499–503, 1973.
10. C. S. Clay, Use of arrays for acoustic transmission in a noisy ocean, *Res. Geophys.*, 4(4), 475–507, 1966.
11. H. P. Bucker, Use of calculated sound fields and matched-field detection to locate sound in shallow water, *J. Acoust. Soc. Am.*, 59, 329–337, 1976.
12. R. D. Doolittle, A. Tolstoy, and E. J. Sullivan, Eds., Special issue on detection and estimation in matched-field processing, *IEEE J. Oceanic Eng.*, 18(3), 153–357, 1993.
13. S. Stergiopoulos and A. T. Ashley, Eds., Special issue on sonar system technology, *IEEE J. Oceanic Eng.*, 18(4), 1993.
14. O. Diachok, A. Caiti, P. Gerstoft, and H. Schmidt, Eds., *Full Field Inversion Methods in Ocean and Seismo-Acoustics.* Boston: Kluwer, 1995.
15. W. A. Kuperman, M. D. Collins, J. S. Perkins, and N. R. Davis, Optimum time domain beamforming with simulated annealing including application of a-priori information, *J. Acoust. Soc. Am.*, 88, 1802–1810, 1990.
16. A. M. Richardson and L. W. Nolte, A posteriori probability source localization in an uncertain sound speed, deep ocean environment, *J. Acoust. Soc. Am.*, 89(6), 2280–2284, 1991.
17. J. L. Krolik, Matched-field minimum variance beamforming in a random ocean channel, *J. Acoust. Soc. Am.*, 92(3), 1408–1419, 1992.

18. J. V. Candy and E. J. Sullivan, Passive localization in ocean acoustics: a model-based approach, *J. Acoust. Soc. Am.*, 98(3), 1455–1471, 1995.
19. C. Feuillade, D. Del Balzo, and M. Rowe, Environmental mismatch in shallow-water matched-field processing: geoacoustic parameter variability, *J. Acoust. Soc. Am.*, 85(6), 2354–2364, 1989.
20. R. Hamson and R. Heitmeyer, Environmental and system effects on source localization in shallow water by the matched-field processing of a vertical array, *J. Acoust. Soc. Am.*, 86, 1950–1959, 1989.
21. E. J. Sullivan and K. Rameau, Passive ranging as an inverse problem: a sensitivity study, SACLANT-CEN Report SR-118, SACLANT Undersea Research Centre, La Spezia, Italy, 1987.
22. J. V. Candy and E. J. Sullivan, Model-based environmental inversion: a shallow water ocean application, *J. Acoust. Soc. Am.*, 98(3), 1446–1454, 1995.
23. J. V. Candy and E. J. Sullivan, Model-based identification: an adaptive approach to ocean acoustic processing, *IEEE Trans. Oceanic Eng.*, 21(3), 273–289, 1996.
24. J. V. Candy and E. J. Sullivan, Model-based processor design for a shallow water ocean acoustic experiment, *J. Acoust. Soc. Am.*, 95(4), 2038–2051, 1994.
25. J. V. Candy and E. J. Sullivan, Model-based processing of a large aperture array, *IEEE Trans. Oceanic Eng.*, 19(4), 519–528, 1994.
26. F. B. Jensen and M. C. Ferla, SNAP: the SACLANTCEN normal-model propagation model, SACLANTCEN Report SM-121, SACLANT Undersea Research Centre, La Spezia, Italy, 1979.
27. M. B. Porter, The KRACKEN normal mode program, SACLANTCEN Report SM-245, SACLANT Undersea Research Centre, La Spezia, Italy, 1991.
28. H. Schmidt, SAFARI: Seismo-acoustic fast field algorithm for range independent environments, SACLANTCEN Report SM-245, SACLANT Undersea Research Centre, La Spezia, Italy, 1987.
29. F. B. Jensen, W. A. Kuperman, M. B. Porter, and H. Schmidt, *Computational Ocean Acoustics*. New York: Am. Inst. Physics Press, 1994.
30. J. V. Candy and E. J. Sullivan, Monitoring the ocean environment: a model-based detection approach, 5th European Conf. Underwater Acoustics, Lyon, France, July 2000.
31. J. V. Candy and E. J. Sullivan, Model-based passive ranging, *J. Acoust. Soc. Am.*, 85(6), 2472–2480, 1989.
32. E. J. Sullivan, W. Carey, and S. Stergiopoulos, Eds., Editorial, Special issue on acoustic synthetic aperture processing, *IEEE Trans. Oceanic Eng.*, 17, 1–7, 1993.
33. S. Stergiopoulos and E. J. Sullivan, Extended towed array processing by an overlap correlator, *J. Acoust. Soc. Am.*, 86, 158–171, 1989.
34. N. Yen and W. Carey, Application of synthetic aperture processing to towed array data, *J. Acoust. Soc. Am.*, 86, 754–765, 1989.
35. G. S. Edelson, On the Estimation of Source Location Using a Passive Towed Array, Ph.D. dissertation, University of Rhode Island, Kingston, 1993.
36. E. J. Sullivan and J. V. Candy, Space-time array processing: the model-based approach, *J. Acoust. Soc. Am.*, 102(5), 2809–2820, 1997.
37. E. J. Sullivan, Passive localization using propagation models, SACLANTCEN Report SR-117, SACLANT Undersea Research Centre, La Spezia, Italy, 1987.
38. D. Johnson and R. Mersereau, *Array Signal Processing*. Englewood Cliffs, NJ: Prentice-Hall, 1993.
39. L. J. Ljung, Asymptotic behavior of the extended Kalman filter as a parameter estimator for linear systems, *IEEE Trans. Autom. Control*, AC-24, 36–50, 1979.
40. L. J. Ljung, *System Identification: Theory for the User*. Englewood Cliffs, NJ: Prentice-Hall, 1987.
41. T. Soderstrom and P. Stoica, *System Identification*. Englewood Cliffs, NJ: Prentice-Hall, 1989.
42. J. P. Norton, *An Introduction to Identification*. New York: Academic Press, 1986.
43. L. J. Ljung and T. Soderstrom, *Theory and Practice of Recursive Identification*. Boston: MIT Press, 1983.
44. G. C. Goodwin and K. S. Sin, *Adaptive Filtering, Prediction and Control*. Englewood Cliffs, NJ: Prentice-Hall, 1984.

45. W. M. Carey, J. Doutt, R. Evans, and L. Dillman, Shallow water transmission measurements taken on the New Jersey continental shelf, *IEEE J. Oceanic Eng.*, 20(4), 321–336, 1995.
46. W. Carey, J. Doutt, and L. Maiocco, Shallow water transmission measurements taken on the New Jersey continental shelf, *J. Acoust. Soc. Am.*, 89, 1981(A), 1991.
47. A. R. Rogers, T. Yamamoto, and W. Carey, Experimental investigation of sediment effect on acoustic wave propagation in shallow water, *J. Acoust. Soc. Am.*, 93, 1747–1761, 1993.
48. J. V. Candy and P. M. Candy, SSPACK_PC: A model-based signal processing package on personal computers, *DSP Appl.*, 2(3), 33–42, 1993.
49. Math Works, *MATLAB*, Boston: The MathWorks, 1990.
50. P. E. Gill, W. Murray, and M. H. Wright, *Practical Optimization*. New York: Academic Press, 1981.
51. J. A. Nelder and R. Meade, A simplex method for function minimization, *Comput. J.*, 7, 308–313, 1965.
52. A. Grace, *Optimization Toolbox for Use with MATLAB*, Boston: The MathWorks, 1992.

6

Advanced Beamformers

Stergios Stergiopoulos
Defence and Civil Institute of Environmental Medicine

University of Western Ontario

Abbreviations and Symbols

$(\ )^*$	Complex conjugate transpose operator
$A_s(f_i)$	Power spectral density of signal $s(t_i)$
BW	Signal bandwidth
$B(f, \theta_s)$	Beamforming plane wave response in frequency domain for a line array steered at azimuth angle θ_s and expressed by $B(f, \theta_s) = \bar{D}^*(f, \theta_s)\bar{X}(f)$
$b(t_i, \theta_s)$	Beam time series of a conventional or adaptive plane wave beamformer of a line array steered at azimuth angle θ_s and expressed by $b(t_i, \theta_s, \phi_s) = \mathrm{IFFT}\{B(f_i, \theta_s, \phi_s)\}$
$b(f_i, \theta_s, \phi_s)$	Beam time series for conventional or adaptive plane wave beamformers of a multi-dimensional array steered at azimuth angle θ_s and elevation angle ϕ_s
$B(f_i, \theta_s, \phi_s)$	Plane wave response in frequency domain for a line array steered at azimuth angle θ_s and elevation angle ϕ_s, expressed by $B(f_i, \theta_s, \phi_s) = \bar{D}^*(f, \theta_s, \phi_s)W(\theta_s)\bar{X}(f)$

0-8493-3691-0/01/$0.00+$.50

CFAR	Constant false alarm rate
C	Signal blocking matrix in GSC adaptive algorithm
CW	Continuous wave; narrowband pulse signal
c	Speed of sound in the underwater sea environment
δ	Sensor spacing for a line array receiver
$\delta_{nm} = (n - m)\delta$	Sensor spacing between the nth and m^{th} sensors of a line array
δ_z	Denotes distance between each ring along z-axis of a cylindrical array receiver
$\bar{D}(f_i, \theta)$	Steering vector for a line array having its nth term for the plane wave arrival with angle θ being expressed by $d_n(f_i, \theta) = \exp\left[j2\pi\frac{(i-1)f_s}{M}\tau_n(\theta)\right]$
$\bar{D}(f, \theta_s, \phi_s))$	Steering vector for a circular array with the nth term being expressed by $d_n(f, \theta_s, \phi_s) = \exp(j2\pi fR\sin\phi_s\cos(\theta_s - \theta_n)/c)$
$\bar{\varepsilon}$	Noise vector component with n th element $\varepsilon_n(t_i)$ for sensor outputs (i.e., $\bar{x} = \bar{s} + \bar{\varepsilon}$)
E{...}	Expectation operator
ETAM	Extended towed array measurements
θ	Azimuth angle of plane wave arrival with respect to a line or multi-dimensional array
$\theta_n = 2\pi m/M$	Angular location of the mth sensor of a M sensor circular array, with $m = 0, 1, \ldots, M-1$
f	Frequency in hertz (Hz)
f_s	Sampling frequency
FM	Frequency modulated active pulse
ϕ	Elevation angle of plane wave arrival with respect to a multi-dimensional array
GSC	Generalized sidelobe canceller
G	Total number of sub-apertures for sub-aperture adaptive beamforming, $g = 1, 2, \ldots, G$
HFM	Hyperbolic frequency modulated pulse
i	Index of time samples of sensor time series, $\{x_n(t_i), i = 1, 2, \ldots, M_s\}$
I	Unit matrix
$\bar{k}$	Wavenumber parameter
k	Iteration number of adaptation process
λ	Wavelength of acoustic signal with frequency f, where $c = f\lambda$
L	Size of line array expressed by $L = (N-1)\delta$
LCMV	Linearly constrained minimum variance beamformers
MVDR	Minimum variance distortionless response
M	M is the number of sensors in a circular array
M_s	M_s is the number of temporal samples of a sensor time series
μ	Convergence controlling parameter or "step size" for the NLMS algorithm
$\aleph = NM$	Number of sensors of a multi-dimensional array that can be decomposed into circular and line array beamformers, where N is the number of circular rings and M is the number of sensors in each ring
N	Number of sensors in a line array receiver, where $\{x_n(t_i), n = 1, 2, \ldots, N\}$, or number of rings in a cylindrical array
NLMS	Normalized least mean square

n	Index for space samples of line array sensor time series $\{x_n(t_i), n = 1,2, \ldots, N\}$
$P(f, \theta_s)$	Beam power pattern in the frequency domain for a line array steered at azimuth angle θ_s and expressed by $P(f, \theta_s) = B(f, \theta_s) \times B^*(f, \theta_s)$
π	3.14159
r	Index for the rth ring of a cylindrical or spherical array of sensors
R	Radius of a receiving circular array
$R(f_i)$	Spatial correlation matrix with elements $R_{nm}(f, d_{nm})$ for received sensor time series
$\rho_{nm}(f, \delta_{nm})$	Cross-correlation coefficients given from $\rho_{nm}(f, d_{nm}) = R_{nm}(f, d_{nm})/\bar{X}^2(f)$
$\bar{S}$	Signal vector whose n^{th} element is expressed by $s_n(t_i) = s_n[t_i + \tau_n(\theta)]$
S	Spatial correlation matrix for the plane wave signal $s_n(t_i)$
$S(f_i, \theta)$	Spatial correlation matrix for the plane wave signal in the frequency domain; it has as its nth row and m^{th} column defined by, $S_{nm}(f_i, \theta) = A_s(f_i) d_n(f_i, \theta) d^*_m(f_i, \theta)$
STCM	Steered covariance matrix
STMV	Steered minimum variance
SVD	Singular value decomposition method
$\sigma_n^2(f_i)$	Power spectral density of noise, $\varepsilon_n(t_i)$
$\bar{X}^*$	Row vector of received N sensor time series $\{x_n(t_i), n = 1, 2, \ldots, N\}$
$X_n(f)$	Fourier transform of $x_n(t_i)$
$X_n(f_i, \theta_s)$	Pre-steered sensor time series in frequency domain
$x_n(t_i, \tau_n(\theta_s))$	Pre-steered sensor time series in the time domain
$X^2(f)$	Mean acoustic intensity of sensor time sequences at frequency bin f
$\tau_n(\theta, \phi)$	Time delay between $(n-1)^{st}$ and n^{th} sensor of a multi-dimensional array for incoming plane waves with direction of propagation of azimuth angle θ and elevation angle ϕ
TL	Propagation loss for the range separating the source (reflected signals) and the array
$\tau_n(\theta)$	Time delay between the first and the nth sensor of the line array for an incoming plane wave with direction of propagation θ
$W(\theta_s)$	Diagonal matrix with the off diagonal terms being zero and the diagonal terms being the weights of a spatial window to reduce the sidelobe structure of a circular array beamformer
$w_{r,m}$	The $(r, m)^{th}$ term of a 3-D spatial window of a multi-dimensional plane wave beamformer
ω	Frequency in radians/second
$\bar{Z}(f_i, \theta_s)$	Result of the signal blocking matrix $\mathbf{C}$ of the GSC adaptive line array beamformer being applied to pre-steered sensor time series $\bar{X}(f_i, \theta_s)$
$\bar{Z}(f_i, \theta)$	Line array adaptive beamforming weights or solution to the constrained minimization problem that allows signals from the look direction θ to pass with a specified gain

6.1 Introduction

The aim of this chapter is to bring together some of the recent theoretical developments on beamformers and to provide suggestions of how modern technology can be applied to the development of current and next-generation ultrasound systems and integrated active and passive sonars. It will focus on the development of an advanced beamforming structure that allows the implementation of adaptive and synthetic aperture signal processing techniques in ultrasound systems and integrated active-passive sonars deploying multi-dimensional arrays of sensors.

The concept of implementing successfully adaptive schemes in 2-dimensional (2-D) and 3-dimensional (3-D) arrays of sensors, such as planar, circular, cylindrical, and spherical arrays, is similar to that of line arrays. In particular, the basic step is to minimize the number of degrees of freedom associated with the adaptation process. The material of this chapter is focused on the definition of a generic beamforming structure that decomposes the beamforming process of 2-D and 3-D sensor arrays into subsets of coherent processes. The approach is to fractionate the computationally intensive multi-dimensional beamformer into two simple modules: linear and circular array beamformers. As a result of the decomposition process, application of spatial shading to reduce the sidelobe structures can now be easily incorporated in 2-D and 3-D beamformers of real-time ultrasound, sonar, and radar systems that include arrays with hundreds of sensors. Then the next step is to define a generic sub-aperture scheme for 2-D and 3-D sensor arrays. The multi-dimensional generic sub-aperture structure leads to minimization of the associated convergence period and makes the implementation of adaptive schemes with near-instantaneous convergence practically feasible.

The reported real data results show that the adaptive processing schemes provide improvements in array gain for signals embedded in a partially correlated noise field. For ultrasound medical imaging systems, practically realizable angular resolution improvements have been quantitatively assessed to be equivalent with those provided by the conventional beamformer of a three-time longer physical aperture and for broadband frequency modulation (FM) and CW type of active pulses. The same set of results also demonstrate that the combined implementation of a synthetic aperture and the sub-aperture adaptive scheme suppresses significantly the sidelobe structure of CW pulses for medical imaging applications. In summary, the reported development of the generic, multi-dimensional beamforming structure has the capability to include several algorithms (adaptive, synthetic aperture, conventional beamfomers, matched filters, and spectral analyzers) working in synergism.

Section 6.2 presents very briefly a few issues of *space-time signal processing* related to detection and sources' parameters estimation procedures. Section 6.3 introduces advanced beamforming processing schemes and the practical issues associated with their implementation in systems deploying multi-dimensional receiving arrays. Our intent here is not to be exhaustive, but only to be illustrative of how the receiving array, the underwater or human body medium, and the subsequent signal processing influence the performance of systems of interest. Issues of practical importance related to system-oriented applications are also addressed, and generic approaches are suggested that could be considered for the development of next-generation array signal processing concepts. Then, these generic approaches are applied to the central problem that the ultrasound and sonar systems deal with — detection and estimation.

Section 6.6 introduces the development of a realizable generic processing scheme that allows the implementation and testing of *adaptive processing techniques* in a wide spectrum of real-time systems. The computing architecture requirements for future ultrasound and sonar systems are addressed in the same section. It identifies the matrix operations associated with high-resolution and adaptive signal processing and discusses their numerical stability and implementation requirements. The mapping onto signal processors of matrix operations includes specific topics such as QR decomposition, Cholesky factorization, and singular value decomposition for solving least-squares and eigensystem problems.[1,29] Schematic diagrams also illustrate the mapping of the signal processing flow for the advanced beamformers in real-time computing architectures. Finally, a concept demonstration of the above developments is presented in Section 6.7, which provides real and synthetic data outputs from an advanced beamforming structure incorporating adaptive and synthetic aperture beamformers.

6.2 Background

In general, the mainstream conventional signal processing of current sonar and ultrasound systems consists of a selection of temporal and spatial processing algorithms.[2–6] These algorithms are designed to increase the signal-to-noise ratio for improved signal delectability while simultaneously providing parameter estimates such as frequency, time delay, Doppler, and bearing for incorporation into localization, classification, and signal tracking algorithms. Their implementation in real-time systems has been directed at providing high-quality, artifact-free conventional beamformers currently used in operational ultrasound and sonar systems. However, aberration effects associated with ultrasound system operations and the drastic changes in the target acoustic signatures associated with sonars suggest that fundamentally new concepts need to be introduced into the signal processing structure of next-generation ultrasound and sonar systems.

To provide a context for the material contained in this chapter, it would seem appropriate to review briefly the basic requirements of high-performance sonar systems deploying multi-dimensional arrays of sensors. Figure 6.1 shows one possible high-level view of a generic warfare sonar system. The upper part of Figure 6.1 presents typical sonar mine-hunting operations carried out by naval platforms (i.e., surface vessels). The lower left-hand side of Figure 6.1 provides a schematic representation of the coordinate system for a hull-mounted cylindrical array of an active sonar.[7,8] The lower right-hand side of Figure 6.1 provides a schematic representation of the coordinate system for a variable depth active sonar deploying a spherical array of sensors for mine warfare operations.[9] In particular, it is assumed that the sensors form a cylindrical or spherical array that allows for beam steering across 0 to 360° in azimuth and a 180° angular searching sector in elevation along the vertical axis of the coordinate system.

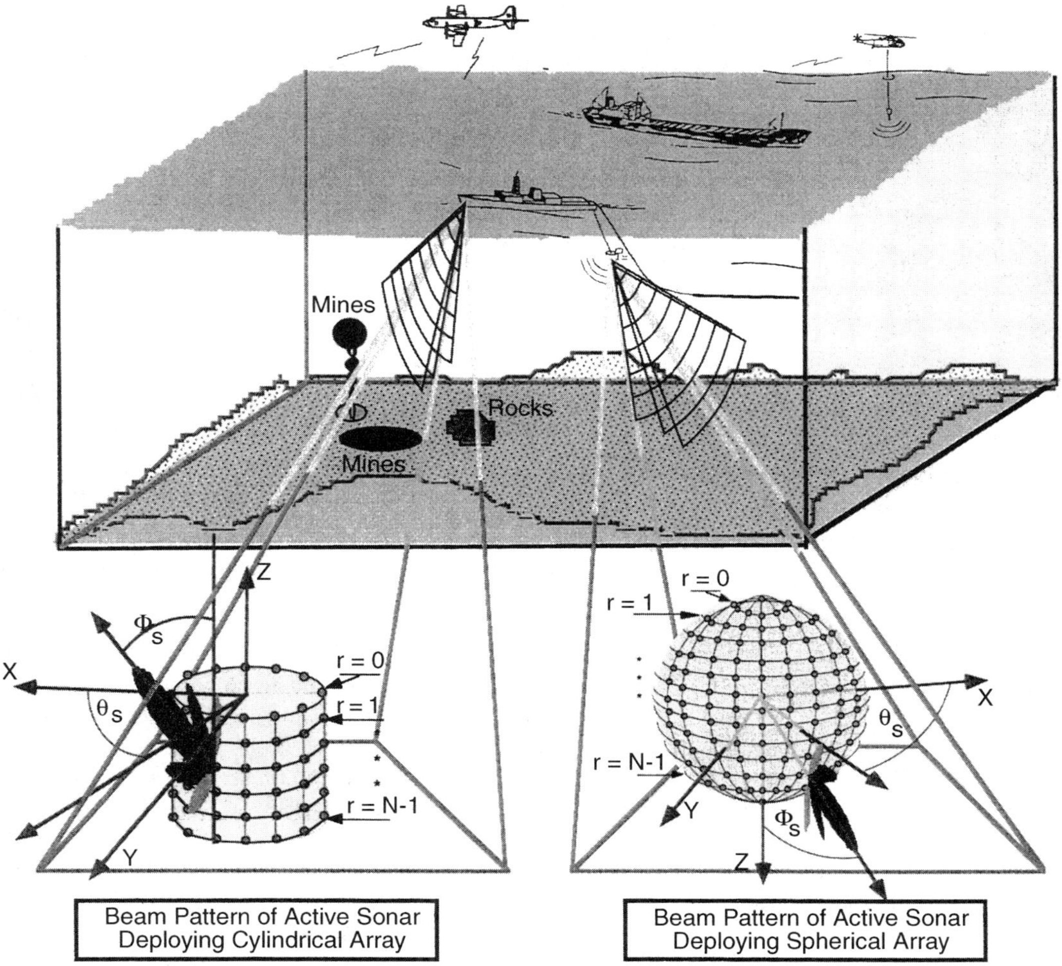

FIGURE 6.1 Mine warefare sonar operations (top). Schematic representation of the coordinate system for a hull mounted cylindrical array of an active sonar (bottom left). Schematic representation of the coordinate system for a variable depth spherical array of an active sonar (bottom right).

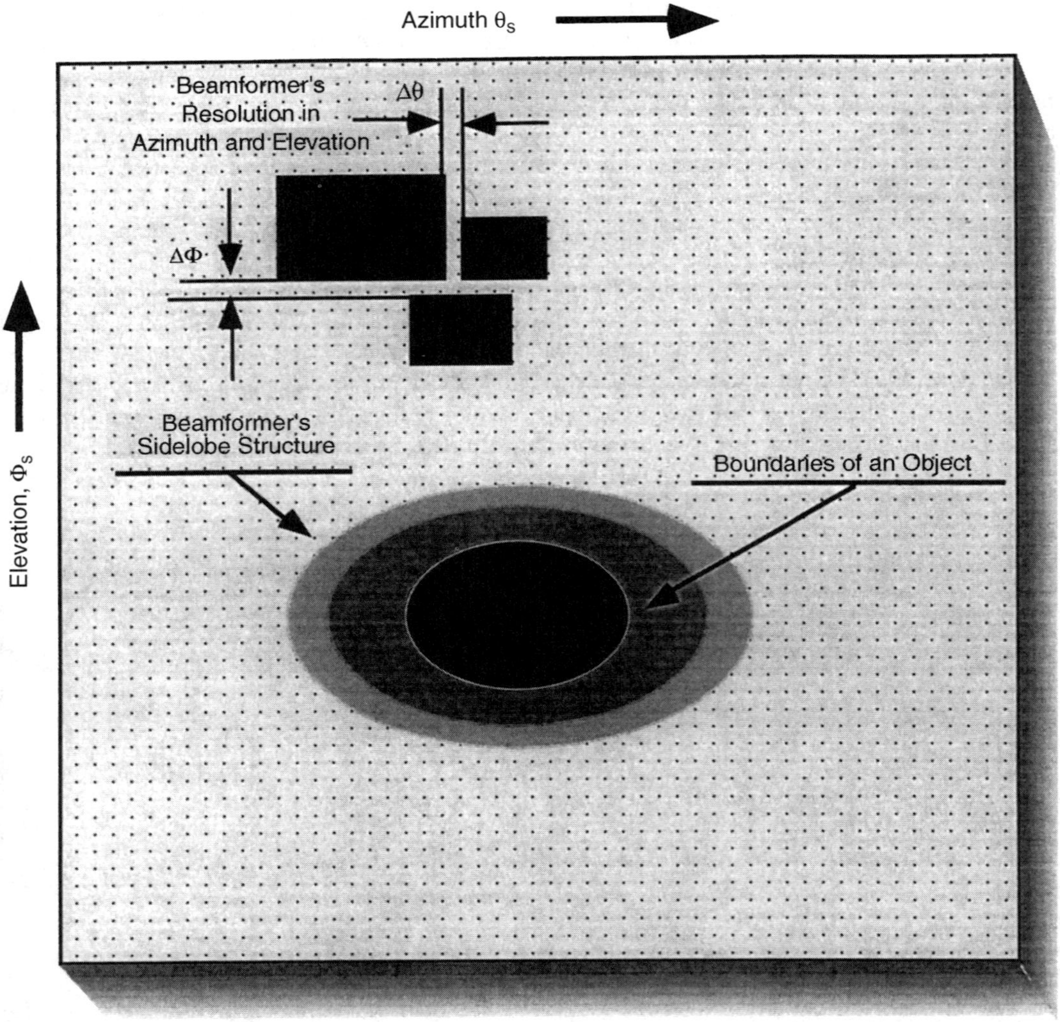

FIGURE 6.2 Angular resolution performance in terms of azimuth and elevation beam steering and the effects of a beamformer's sidelobe structure for mine warfare sonar operations.

Thus, for effective sonar operations, the beam width and the sidelobe structure of the beam steering patterns (shown in the lower part of Figure 6.1 for a given azimuth θ_s and elevation Φ_s beam steering) should be very small to allow for high image and spatial resolution of detected mines that are in close proximity with other objects. More specifically, the beam steering pattern characteristics of a mine-hunting sonar define its performance in terms of image and spatial resolution characteristics, as shown schematically in Figure 6.2. For a given angular resolution in azimuth and elevation, shown schematically in the upper left-hand corner of Figure 6.2, a mine-hunting sonar would not be able to distinguish detected objects and mines that are closer than the angular resolution performance limits. Moreover, the beam steering sidelobe structure would affect the image resolution performance of the system, as depicted in the lower part of Figure 6.2. Thus, for a high-performance sonar, it is desirable that the system should provide the highest possible angular resolution in azimuth and elevation as well as the lowest possible levels of sidelobe structures, properties that are defined by the aperture size of the receiving array. The above arguments are equally valid for ultrasound system operations since the beamforming process for ultrasound imaging assumes plane wave arrivals.

Because the increased angular resolution means longer sensor arrays with consequent technical and operational implications, many attempts have been made to increase the effective array length by synthesizing additional sensors (i.e., synthetic aperture processing)[1,6,11–16] or using adaptive beam processing techniques.[1–5,17–24]

In previous studies, the impact and merits of these techniques have been assessed for towed array[1,4,5,10–20] and cylindrical array hull-mounted[2,4,7,25,26] sonars and contrasted with those obtained using the conventional beamformer. The present material extends previous investigations and further assesses the performance characteristics of ultrasound and sonar systems that are assumed to include adaptive processing schemes integrated with a plane wave conventional beamforming structure.

6.3 Theoretical Remarks

Sonar operations can be carried out by a wide variety of naval platforms, as shown in Figure 6.1. This includes surface vessels, submarines, and airborne systems such as airplanes and helicopters. Shown also in Figure 6.1 is a schematic representation of active and passive sonar operations in an underwater sea environment. Active sonar and ultrasound operations involve the transmission of well-defined acoustic signals, called replicas, which illuminate targets in an underwater sea or human body medium, respectively. The reflected acoustic energy from a target or body organ provides the array receiver with a basis for detection and estimation. Passive sonar operations base their detection and estimation on acoustic sounds, which emanate from submarines and ships. Thus, in passive systems, only the receiving sensor array is under the control of the sonar operators. In this case, major limitations in detection and classification result from imprecise knowledge of the characteristics of the target radiated acoustic sounds.

The passive sonar concept can be made clearer by comparing sonar systems with radars, which are always active. Another major difference between the two systems arises from the fact that sonar system performance is more affected than that of radar systems by the underwater medium propagation characteristics. All these issues have been discussed in several review articles[1–6] that form a good basis for interested readers to become familiar with "mainstream" sonar signal processing developments. Therefore, discussions of issues of conventional sonar signal processing, detection, estimation, and influence of medium on sonar system performance are beyond the scope of this chapter. Only a very brief overview of these issues will be highlighted in this section in order to define the basic terminology required for the presentation of the main theme of this chapter. Let us start with a basic system model that reflects the interrelationships between the target, the underwater sea environment or the human body (medium), and the receiving sensor array of a sonar or an ultrasound system.

A schematic diagram of this basic system is shown in Figure 6.3, where array signal processing is shown to be two dimensional[1,5,10,12,18] in the sense that it involves both temporal and spatial spectral analysis. The temporal processing provides spectral characteristics that are used for target classification, and the spatial processing provides estimates of the directional characteristics, (i.e., bearing and possibly range) of a detected signal. Thus, *space-time processing* is the fundamental processing concept in sonar and ultrasound systems, and it will be the subject of discussion in the next section.

6.3.1 Space-Time Processing

For geometrical simplicity and without any loss of generality, we consider here a combination of N equally spaced acoustic transducers in a linear array, which may form a towed or hull-mounted array system that can be used to estimate the directional properties of echoes and acoustic signals. As shown in Figure 6.3, a direct analogy between sampling in space and sampling in time is a natural extension of the sampling theory in space-time signal representation, and this type of space-time sampling is the basis in array design that provides a description of an array system response. When the sensors are arbitrarily distributed, each element will have an added degree of freedom, which is its position along the axis of the array. This is analogous to non-uniform temporal sampling of a signal. In this chapter we extend our discussion to multi-dimensional array systems.

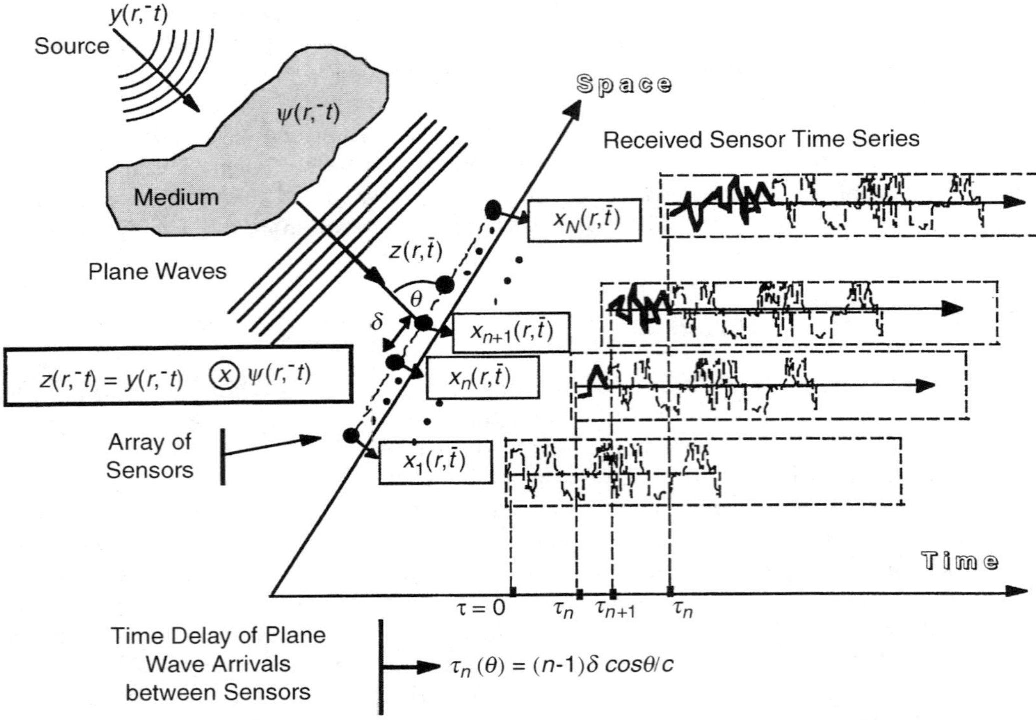

FIGURE 6.3 A model of space-time signal processing. It shows that ultrasound and sonar signal processing is two dimensional in the sense that it involves both temporal and spatial spectral analysis. The temporal processing provides characteristics for target classification, and the spatial processing provides estimates of the directional characteristics (bearing, range depth) of detected echoes (active case) or signals of interest (passive case). (Reprinted by permission of IEEE ©1998.)

Sources of sound that are of interest in sonar and ultrasound system applications are harmonic narrowband, and broadband, and these sources satisfy the wave equation.[2,10] Furthermore, their solutions have the property that their associated temporal-spatial characteristics are separable.[10] Therefore, measurements of the pressure field $z(\bar{r}, t)$, which is excited by acoustic source signals, provide the spatial-temporal output response, designated by $x(\bar{r}, t)$, of the measurement system. The vector $\bar{r}$ refers to the source-sensor relative position, and t is the time. The output response $x(\bar{r}, t)$ is the convolution of $z(\bar{r}, t)$ with the line array system response $h(\bar{r}, t)$:[10,30]

$$x(\bar{r}, t) = z(\bar{r}, t) \otimes h(\bar{r}, t) \tag{6.1}$$

where ⊗ refers to convolution. Since $z(\bar{r}, t)$ is defined at the input of the receiver, it is the convolution of the source's characteristics $y(\bar{r}, t)$ with the underwater medium's response $\Psi(\bar{r}, t)$,

$$z(\bar{r}, t) = y(\bar{r}, t) \otimes \Psi(\bar{r}, t). \tag{6.2}$$

Fourier transformation of Equation 6.1 provides

$$X(\omega, \bar{k}) = \{Y(\omega, \bar{k}) \cdot \Psi(\omega, \bar{k})\}H(\omega, \bar{k}), \tag{6.3}$$

where $\omega, \bar{k}$ are the frequency and wavenumber parameters of the temporal and spatial spectrums of the transform functions in Equations 6.1 and 6.2. Signal processing, in terms of beamforming operations, of the receiver's output, $x(\bar{r}, t)$, provides estimates of the source bearing and possibly of the source range. This is a well-understood concept of the forward problem, which is concerned with determining the parameters of the received signal $x(\bar{r}, t)$, given that we have information about the other two functions $z(\bar{r}, t)$ and $h(\bar{r}, t)$.[5] The inverse problem is concerned with determining the parameters of the impulse response of the medium $\Psi(\bar{r}, t)$ by extracting information from the received signal $x(\bar{r}, t)$, assuming that the function $x(\bar{r}, t)$ is known.[5] The ultrasound and sonar problems, however, are quite complex and include both forward and inverse problem operations. In particular, detection, estimation, and tracking-localization processes of sonar and ultrasound systems are typical examples of the forward problem, while target classification for passive-active sonars and diagnostic ultrasound imaging are typical examples of the inverse problem. In general, the inverse problem is a computationally costly operation, and typical examples in acoustic signal processing are seismic deconvolution and acoustic tomography.

6.3.2 Definition of Basic Parameters

This section outlines the context in which the sonar or the ultrasound problem can be viewed in terms of simple models of acoustic signals and noise fields. The signal processing concepts that are discussed in this chapter have been included in sonar and radar investigations with sensor arrays having circular, planar, cylindrical, and spherical geometric configurations.[7,25,26,28] Thus, we consider a multi-dimensional array of equally spaced sensors with spacing δ. The output of the nth sensor is a time series denoted by $x_n(t_i)$, where $(i = 1, \ldots, M_s)$ are the time samples for each sensor time series. $*$ denotes complex conjugate transposition so that $\bar{X}^*$ is the row vector of the received $\aleph$ sensor time series $\{x_n(t_i), n = 1, 2, \ldots, \aleph\}$.

Then $x_n(t_i) = s_n(t_i) + \varepsilon_n(t_i)$, where $s_n(t_i)$ and $\varepsilon_n(t_i)$ are the signal and noise components in the received sensor time series. $\bar{S}$ and $\bar{\varepsilon}$ denote the column vectors of the signal and noise components of the vector $\bar{X}$ of the sensor outputs (i.e., $\bar{x} = \bar{s} + \bar{\varepsilon}$).

$$X_n(f) = \sum_{i=1}^{M_s} x_n(t_i)\exp(-j2\pi f t_i) \tag{6.4}$$

is the Fourier transform of $x_n(t_i)$ at the signal with frequency f, $c = f\lambda$ is the speed of sound in the underwater or human-body medium, and λ is the wavelength of the frequency f. $S = E\{\bar{S}\,\bar{S}^*\}$ is the spatial correlation matrix of the signal vector $\bar{S}$, whose n^{th} element is expressed by

$$s_n(t_i) = s_n[t_i + \tau_n(\theta, \phi)], \tag{6.5}$$

$E\{\ldots\}$ denotes expectation, and $\tau_n(\theta, \phi)$ is the time delay between the $(n-1)^{st}$ and the nth sensor of the array for an incoming plane wave with direction of propagation of azimuth angle θ and an elevation angle ϕ, as depicted in Figure 6.3. In frequency domain, the spatial correlation matrix **S** for the plane wave signal $s_n(t_i)$ is defined by

$$S(f_i, \theta, \phi) = A_s(f_i)\bar{D}(f_i, \theta, \phi)\bar{D}^*(f_i, \theta, \phi) \tag{6.6}$$

where $A_s(f_i)$ is the power spectral density of $s(t_i)$ for the ith frequency bin, and $\bar{D}(f, \theta, \phi)$ is the steering vector with the nth term being denoted by $d_n(f, \theta, \phi)$. Then matrix $\mathbf{S}(f_i, \theta, \phi)$ has its nth row and m^{th} column defined by $S_{nm}(f_i, \theta, \phi) = A_s(f_i)d_n(f_i, \theta, \phi)d^*_m(f_i, \theta, \phi)$. Moreover, $R(f_i)$ is the spatial correlation matrix of received sensor time series with elements $R_{mn}(f, d_{nm})$. $R_\varepsilon(f_i) = \sigma_n^2(f_i)R_\varepsilon(f_i)$ is the spatial correlation matrix of the noise for the ith frequency bin with $\sigma_n^2(f_i)$ being the power spectral density of the noise, $\varepsilon_n(t_i)$. In what is considered as an estimation procedure in this chapter, the associated problem of detection is defined in the classical sense as a hypothesis test that provides a detection probability and a probability of false alarm.[31–33] This choise of definition is based on the standard CFAR (constant false alarm rate) processor, which is based on the Neyman-Pearson (N-P) criterion.[31] The CFAR processor provides an estimate of the ambient noise or clutter level so that the threshold can be varied dynamically to stabilize the false alarm rate. Ambient noise estimates for the CFAR processor are provided mainly by noise normalization techniques[34] that account for the slowly varying changes in the background noise or clutter. The above estimates of ambient noise are based upon the average value of the received signal, the desired probability of detection, and the probability of false alarms.

At this point, a brief discussion on the fundamentals of detection and estimation process is required to address implementation issues of signal processing schemes in sonar and ultrasound systems.

6.3.3 Detection and Estimation

In passive systems, in general, we do not have the *a priori* probabilities associated with the hypothesis H_1 that the signal is assumed present and the null hypothesis H_0 that the received time series consists only of noise. As a result, costs cannot be assigned to the possible outcomes of the experiment. In this case, the N-P criterion[31] is applied because it requires only a knowledge of the signal's and noise's probability density functions (pdf).

Let $x_{n=1}(t_i)$, $(i = 1, \ldots, M)$ denote the received vector signal by a single sensor. Then for hypothesis H_1, which assumes that the signal is present, we have

$$H_1{:}x_{n=1}(t_i) = s_{n=1}(t_i) + \varepsilon_{n=1}(t_i), \tag{6.7}$$

where $s_{n=1}(t_i)$ and $\varepsilon_{n=1}(t_i)$ are the signal and noise vector components in the received signal, and $p_1(x)$ is the pdf of the received signal $x_{n=1}(t_i)$ given that H_1 is true. Similarly, for hypothesis H_0,

$$H_0{:}\ x_{n=1}(t_i) = \varepsilon_{n=1}(t_i) \tag{6.8}$$

and $p_0(x)$ is the pdf of the received signal given that H_0 is true. The N-P criterion requires maximization of probability of detection for a given probability of false alarm. So, there exists a non-negative number η such that if hypothesis H_1 is chosen, then

$$\lambda(x) = \frac{p_1(x)}{p_o(x)} \geq \eta, \tag{6.9}$$

which is the likelihood ratio. By using the analytic expressions for $p_0(x)$ (the pdf for H_0) and $p_1(x)$ (the pdf for H_1) in Equation 6.9 and by taking the $ln\ [\lambda(x)]$, we have,[31]

$$\lambda_\tau = \ln[\lambda(x)] = \bar{s}^* R'_\varepsilon \bar{x} \tag{6.10}$$

where λ_τ is the log likelihood ratio and R_ε' is the covariance matrix of the noise vector, as defined in the Section 6.3.2. For the case of white noise with $R_\varepsilon' = \sigma_n^2 I$ and I being the unit matrix, the test statistic in Equation 6.10 is simplified into a simple correlation receiver (or replica correlator)

$$\lambda_\tau = \bar{s}^* \otimes \bar{x}. \tag{6.11}$$

For the case of anisotropic noise, however, an optimum detector should include the correlation properties of the noise in the correlation receiver as defined in Equation 6.10.

For plane wave arrivals that are observed by an N sensor array receiver, the test statistics are[31]

$$\lambda_\tau = \sum_{i=1}^{\frac{M_s}{2}-1} \overline{X^*}(f_i) \cdot R'_\varepsilon(f_i) \cdot S(f_i, \phi, \theta) \cdot [S(f_i, \phi, \theta) + R'_\varepsilon(f_i)]^{-1} \cdot \bar{X}(f_i), \tag{6.12}$$

where the above statistics are for the frequency domain with parameters defined in Equations 6.5 and 6.6 in the Section 6.3.2. Then, for the case of an array of sensors receiving plane wave signals, the log likelihood ratio λ_τ in Equation 6.12 is expressed by the following equation, which is the result of simple matrix manipulations based on the frequency domain Equation 6.5 and 6.6 and their parameter definitions presented in Section 6.3.2. Thus,

$$\lambda_\tau = \sum_{i=1}^{\frac{M}{2}-1} \left|\varphi(f_i)\overline{D}^*(f_i, \phi, \theta)R'_\varepsilon(f_i)^{-1}\overline{X}(f_i)\right|^2, \tag{6.13}$$

where[31]

$$\varphi^2(f_i) = \frac{A_s(f_i)/\sigma_n^2(f_i)}{1 + A_s(f_i)\overline{D}^*(f_i, \phi, \theta)R'^{-1}_\varepsilon(f_i)\overline{D}(f_i)/\sigma_n^2(f_i)}. \tag{6.14}$$

Equation 6.13 can be written also as follows:

$$\lambda_\tau = \sum_{i=1}^{\frac{M}{2}-1} \left[\sum_{n=1}^{N} \zeta_n^*(f_i, \phi, \theta)X_n(f_i)\right]^2. \tag{6.15}$$

This last expression in Equation 6.15 of the log likelihood ratio indicates that an optimum detector in this case requires the filtering of each one of the N sensor received time series $X_n(f_i)$ with a set of filters being the elements of the vector,

$$\overline{\zeta}(f_i, \phi, \theta) = \varphi(f_i)\overline{D}^*(f_i, \phi, \theta)R'_\varepsilon(f_i)^{-1}. \tag{6.16}$$

Then, the summation of the filtered sensor outputs in the frequency domain according to Equation 6.16 provides the test statistics for optimum detection. For the simple case of white noise $R'_\varepsilon = \sigma_n^2 I$ and for a line array receiver, the filtering operation in Equation 6.16 indicates plane wave conventional beamforming in the frequency domain,

$$\lambda_\tau = \sum_{i=1}^{\frac{M}{2}-1} \left[\Psi \sum_{n=1}^{N} d_n^*(f_i, \theta)X_n(f_i)\right]^2, \tag{6.17}$$

where $\Psi = \zeta/(1 + N\zeta)$ is a scalar, which is a function of the signal-to-noise ratio $\zeta = A_s^2/\sigma_n^2$.

For the case of narrowband signals embedded in spatially and/or temporally correlated noise or interferences, it has been shown[13] that the deployment of very long arrays or application of acoustic synthetic aperture will provide sufficient array gain and will achieve optimum detection and estimation for the parameters of interest.

For the general case of broadband and narrowband signals embedded in a spatially anisotropic and temporally correlated noise field, Equation 6.17 indicates that the filtering operation for optimum detection and estimation requires adaptation of the sonar and ultrasound signal processing according to the ambient noise's and human body's noise characteristics, respectively. The family of algorithms for optimum beamforming that uses the characteristics of the noise are called *adaptive beamformers*[3,17–20,22,23] and *a detailed definition of an adaptation process requires knowledge of the correlated noise's covariance matrix* $R'_\varepsilon(f_i)$. However, if the required knowledge of the noise's characteristics is inaccurate, the performance of the optimum beamformer will degrade dramatically.[18,23] As an example, the case of cancellation of the desired signal is often typical and significant in adaptive beamforming applications.[18,24] This suggests that the implementation of useful adaptive beamformers in real-time operational systems is not a trivial task. The existence of numerous articles on adaptive beamforming suggests the dimensions of the difficulties associated with this kind of implementation. In order to minimize the generic nature of the problems associated with adaptive beamforming, the concept of partially adaptive beamformer design was introduced. This concept reduces the degrees of freedom, which results in lowering the computational requirements and often improving the adaptive response time.[17,18] However, the penalty associated with the reduction of the degrees of freedom in partially adaptive beamformers is that they cannot converge to the same optimum solution as the fully adaptive beamformer.

Although a review of the various adaptive beamformers would seem relevant at this point, we believe that this is not necessary since there are excellent review articles[3,17,18,21] that summarize the points that have been considered for the material of this chapter. There are two main families of adaptive beamformers: the generalized sidelobe cancellers (GSC)[44,45] and the linearly constrained minimum variance (LCMV) beamformers.[18] A special case of the LCMV is Capon's maximum likelihood method,[22] which is called minimum variance distortionless response (MVDR).[17,18,22,23,38,39] This algorithm has proven to be one of the more robust of the adaptive array beamformers, and it has been used by numerous researchers as a basis to derive other variants of MVDR.[18] In this chapter, we will address implementation issues for various partially adaptive variants of the MVDR and a GSC adaptive beamformer,[1] which are discussed in Section 6.5.2.

In summary, the classical estimation problem assumes that the *a priori* probability of the signal's presense $p(H_1)$ is unity.[31–33] However, if the signal's parameters are not known *a priori* and $p(H_1)$ is known to be less than unity, then a series of detection decisions over an exhaustive set of source parameters constitutes a detection procedure, where the results incidentally provide an estimation of the source's parameters. As an example, we consider the case of a matched filter, which is used in a sequential manner by applying a series of matched filter detection statistics to estimate the range and speed of the target, which are not known *a priori*. This kind of estimation procedure is not optimal since it does not constitute an appropriate form of Bayesian minimum variance or minimum mean square error procedure.

Thus, the problem of detection[31–33] is much simpler than the problem of estimating one or more parameters of a detected signal. Classical decision theory[31–33] treats signal detection and signal estimation as separate and distinct operations. A detection decision as to the presence or absence of the signal is regarded as taking place independently of any signal parameter or waveform estimation that may be indicated as the result of a detection decision. However, interest in joint or simultaneous detection and estimation of signals arises frequently. Middleton and Esposito[46] have formulated the problem of simultaneous optimum detection and estimation of signals in noise by viewing *estimation* as a generalized detection process. Practical considerations, however, require different cost functions for each process.[46] As a result, it is more effective to retain the usual distinction between detection and estimation.

Estimation, in passive sonar and ultrasound systems, includes both the temporal and spatial structure of an observed signal field. For active systems, correlation processing and Doppler (for moving target indications) are major concerns that define the critical distinction between these two approaches (i.e., *passive, active*) to sonar and ultrasound processing. In this chapter, we restrict our discussion only to topics related to spatial signal processing for estimating signal parameters. However, spatial signal processing has a direct representation that is analogous to the frequency domain representation of temporal signals. Therefore, the spatial signal processing concepts discussed here have direct applications to temporal spectral analysis.

6.3.4 Cramer-Rao Lower Bound (CRLB) Analysis

Typically, the performance of an estimator is represented as the variance in the estimated parameters. Theoretical bounds associated with this performance analysis are specified by the Cramer-Rao bound,[31–33] and this has led to major research efforts by the sonar signal processing community to define the idea of an optimum processor for discrete sensor arrays.[12,16,56–60] If the *a priori* probability of detection is close to unity, then the minimum variance achievable by any unbiased estimator is provided by the *Cramer-Rao lower bound* (CRLB).[31,32,46]

More specifically, let us consider that the received signal by the nth sensor of a receiving array is expressed by

$$x_n(t_i) = s_n(t_i) + \varepsilon_n(t_i) \tag{6.18}$$

where $s_n(t_i, \bar{\Theta}) = s_n[t_i + \tau_n(\theta, \phi)]$ defines the received signal model, with $\tau_n(\theta, \phi)$ being the time delay between the $(n-1)^{st}$ and the nth sensor of the array for an incoming plane wave with the direction of propagation of azimuth angle θ and an elevation angle ϕ, as depicted in Figure 6.3. The vector $\bar{\Theta}$ includes all the unknown parameters considered in Equation 6.18. Let $\sigma^2_{\theta_i}$ denote the variance of an unbiased estimate of an unknown parameter θ_i in the vector $\bar{\Theta}$. The Cramer-Rao[31–33] bound states that the best unbiased estimate $\tilde{\Theta}$ of the parameter vector $\bar{\Theta}$ has the covariance matrix

$$\text{cov}\tilde{\Theta} \geq J(\bar{\Theta})^{-1}, \tag{6.19}$$

where J is the Fisher information matrix whose elements are

$$J_{ij} = -E\left(\frac{\vartheta^2 \ln P\langle \bar{X}|\bar{\Theta}\rangle}{\vartheta\theta_i \vartheta\theta_j}\right). \tag{6.20}$$

In Equation 6.20, $P\langle \bar{X}|\bar{\Theta}\rangle$ is the pdf governing the observations

$$\bar{X} = [x_1(t_i), x_2(t_i), x_3(t_i), \ldots, x_N(t_i)]^*$$

for each of the N and M_s independent spatial and temporal samples, respectively, that are described by the model in Equation 6.18. The variance of the unbiased estimates $\tilde{\Theta}$ has a lower bound (called the CRLB), which is given by the diagonal elements of Equation 6.19. This CRLB is used as the standard of performance and provides a good measure for the performance of signal processing algorithms which gives unbiased estimates $\tilde{\Theta}$ for the parameter vector $\bar{\Theta}$. In this case, if there exists a signal processor to achieve the CRLB, it will be the maximum likelihood estimation (MLE) technique. The above requirement associated with the *a priori* probability of detection is very essential because if it is less than one, then the estimation is biased and the theoretical CRLBs do not apply. This general framework of optimality is very essential in order to account for Middleton's[32] warning that a system optimized for the one function (detection or estimation) may not be necessarily optimized for the other.

For a given model describing the received signal by a sonar or ultrasound system, the CRLB analysis can be used as a tool to define the information inherent in a sonar system. This is an important step related to the development of the signal processing concept for a sonar system as well as in defining the optimum sensor configuration arrangement under which we can achieve, in terms of system performance, the optimum estimation of signal parameters of our interest. This approach has been applied successfully to various studies related to the present development.[12,15,56–60]

As an example, let us consider the simplest problem of one source with the bearing θ_1 being the unknown parameter. Following Equation 6.20, the results of the variance $\sigma^2_{\theta_i}$ in the bearing estimates are

$$\sigma^2_{\theta_i} = \frac{3}{2\Psi N}\left(\frac{B_w}{\pi \sin\theta_1}\right)^2, \tag{6.21}$$

where $\Psi = M_s A_1^2/\sigma_N^2$, and the parameter $B_w = \lambda/(N-1)\delta$ gives the beam width of the physical aperture that defines the angular resolution associated with the estimates of θ_1. The signal-to-noise ratio (SNR) at the sensor level is SNR $= 10 \times \log_{10}(\Psi)$ or

$$\text{SNR} = 20 \times \log_{10}(A_1/\sigma_1) + 10 \times \log_{10}(M_s). \tag{6.22}$$

It is obvious from Equations 6.21 and 6.22 that the variance of the bearing $\sigma^2_{\theta_i}$ can get smaller when the observation period, $T = M_s/f_s$, becomes long and the receiving array size, $L = (N - 1)\lambda$, gets very long.

The next question needed to be addressed is about the unbiased estimator that can exploit this available information and provide results asymptotically reaching the CRLBs. For each estimator, it is well known that there is a range of SNR in which the variance of the estimates rises very rapidly as SNR decreases. This effect, which is called the *threshold effect of the estimator*, determines the range of the SNR of received signals for which the parameter estimates can be accepted. In passive sonar systems, the SNR of signals of interest are often quite low and probably below the threshold value of an estimator. In this case, high-frequency resolution in both time and spatial domains for the parameter estimation of narrowband signals is required. In other words, the threshold effect of an estimator determines the frequency resolution for processing and the size of the array receivers required in order to detect and estimate signals of interest that have very low SNR.[12-14,53,61,62] The CRLB analysis has been used in many studies to evaluate and compare the performance of the various non-conventional processing schemes[17,18,55] that have been considered for implementation in the generic beamforming structure to be discussed in Section 6.5.1. In general, array signal processing includes a large number of algorithms for a variety of systems that are quite diverse in concept. There is a basic point that is common in all of them, however, and this is the beamforming process, which we will examine in Section 6.4.

6.4 Optimum Estimators for Array Signal Processing

Sonar signal processing includes mainly estimation (after detection) of the source's bearing, which is the main concern in sonar array systems because in most of the sonar applications the acoustic signal's wavefronts tend to be planar, which assumes distant sources. Passive ranging by measurement of wavefront curvature is not appropriate for the far-field problem. The range estimate of a distant source, in this case, must be determined by various target motion analysis methods discussed in Reference 1 and Chapter 9, which address the localization tracking performance of non-conventional beamformers with real data.

More specifically, a one-dimensional (1-D) device such as a line sensor array satisfies the basic requirements of a spatial filter. It provides direction discrimination, at least in a limited sense, and an SNR improvement relative to an omni-directional sensor. Because of the simplified mathematics and reduced number of the involved sensors, relative to multi-dimensional arrays, most of the researchers have focused on the investigation of the line sensor arrays in system applications.[1-6] Furthermore, implementation issues of synthetic aperture and adaptive techniques in real time systems have been extensively investigated for line arrays as well.[1,5,6,12,17,19,20] However, the configuration of the array depends on the purpose for which it is to be designed. For example, if a wide range of horizontal angles is to be observed, a circular configuration may be used, giving rise to beam characteristics that are independent of the direction of steering. Vertical direction may be added by moving into a cylindrical configuration.[8] In a more general case, where both vertical and horizontal steering are to be required and where a large range of angles is to be covered, a spherically symmetric array would be desirable.[9] In modern ultrasound imaging systems, planar arrays are required to reconstruct real-time 3-D images. However, the huge computational load required for multi-dimensional conventional and adaptive beamformers makes the applications of these 2-D and 3-D arrays in real-time systems non-feasible.

Furthermore, for modern sonar and radar systems, it has become a necessity these days that all possible active and passive modes of operation should be exploited under an integrated processing structure that reduces redundancy and provides cost-effective, real-time system solutions.[6] Similarly, the implementation of computationally intensive data adaptive techniques in real-time systems is also an issue of equal practical importance. However, when these systems include multi-dimensional (2-D, 3-D) arrays with hundreds of sensors, then the associated beamforming process requires very large memory and very intensive throughput characteristics, things that make its implementation in real-time systems a very expensive and difficult task.

To counter this implementation problem, this chapter introduces a generic approach of implementing conventional beamforming processing schemes with integrated passive and active modes of operations in systems that may include planar, cylindrical, or spherical arrays.[25-28] This approach decomposes the 2-D and 3-D beamforming process into sets of linear and/or circular array beamformers. Because of the decomposition process, the fully multi-dimensional beamformer can now be divided into subsets of coherent processes that can be implemented in small size CPUs that can be integrated under the parallel configuration of existing computing architectures. Furthermore, application of spatial shading for multi-dimensional beamformers to control sidelobe structures can now be easily incorporated. This is because the problem of spatial shading for linear arrays has been investigated thoroughly,[36] and the associated results can be integrated into a circular and a multi-dimensional beamformer, which can be decomposed now into coherent subsets of linear and/or circular beamformers of the proposed generic processing structure.

As a result of the decomposition process provided by the generic processing structure, the implementation effort for adaptive schemes is reduced to implementing adaptive processes in linear and circular arrays. Thus, a multi-dimensional adaptive beamformer can now be divided into two coherent modular steps which lead to efficient system-oriented implementations. In summary, the proposed approach demonstrates that the incorporation of adaptive schemes with near-instantaneous convergence in multi-dimensional arrays is feasible.[7,25–28]

At this point, it is important to note that the proposed decomposition process of 2-D and 3-D conventional beamformers into sets of linear and/or circular array beamformers is an old concept that has been exploited over the years by sonar system designers. Thus, references on this subject may exist in U.S. Navy-labs' and industrial institutes' technical reports that are not always readily available, and the authors of this chapter are not aware of any kind of reports in this area. Previous efforts had attempted to address practical implementation issues and had focused on cylindrical arrays. As an example, a cylindrical array beamformer is decomposed into time-delay line array beamformers, providing beams along elevation angles of the cylindrical array. These are called staves. Then, the beam time series associated with a particular elevation steering of interest are provided at the input of a circular array beamformer.

In this chapter, the attempt is to provide a higher degree of development than the one discussed above for cylindrical arrays. The task is to develop a generic processing structure that integrates the decomposition process of multi-dimensional planar, cylindrical, and spherical array beamformers into line and/or circular array beamformers. Furthermore, the proposed generic processing structure integrates passive and active modes of operation into a single signal processing scheme.

6.4.1 Generic, Multi-Dimensional Conventional Beamforming Structure

6.4.1.1 Linear Array Conventional Beamformer

Consider an N sensor linear array receiver with uniform sensor spacing δ, shown in Figure 6.4, receiving plane wave arrivals with direction of propagation θ. Then, as a follow-up of the parameter definition in Section 6.3.2,

$$\tau_n(\theta) = (n-1)\delta\cos\theta/c \tag{6.23}$$

is the time delay between the 1st and the nth sensor of the line array for an incoming plane wave with direction θ, as illustrated in Figure 6.4.

$$d_n(f_i, \theta) = \exp\left[j2\pi\frac{(i-1)f_s}{M}\tau_n(\theta)\right] \tag{6.24}$$

is the nth term of the steering vector $\overline{D}(f, \theta)$. Moreover, because of Equations 6.16 and 6.17 the plane wave response of the N-sensor line array steered at a direction θ_s can be expressed by

$$B(f, \theta_s) = \overline{D}^*(f, \theta_s)\overline{X}(f). \tag{6.25}$$

Previous studies[1] have shown that for a single source this conventional beamformer without shading is an optimum processing scheme for bearing estimation. The sidelobe structure can be suppressed at

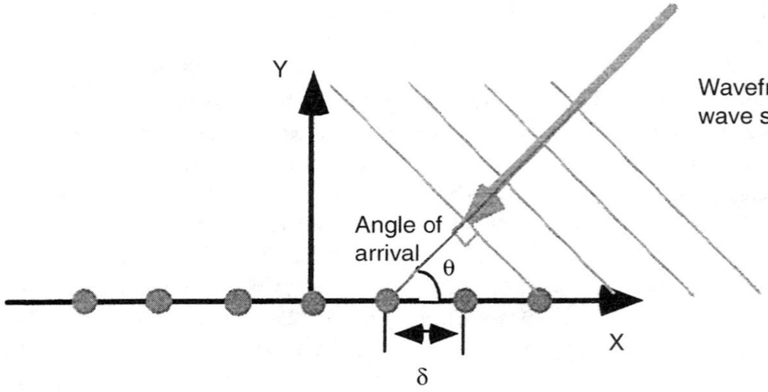

FIGURE 6.4 Geometric configuration and coordinate system for a line array of sensors.

the expense of a beam width increase by applying different weights (i.e., spatial shading window).[36] The angular response of a line array is ambiguous with respect to the angle θ_s, responding equally to targets at angles θ_s and $-\theta_s$ where θ_s varies over $[0, \pi]$.

Equation 6.25 is basically a mathematical interpretation of Figure 6.3 and shows that a line array is basically a spatial filter because by steering a beam in a particular direction we spatially filter the signal coming from that direction, as illustrated in Figure 6.3. On the other hand, Equation 6.25 is fundamentally a discrete Fourier transform relationship between the sensor weightings and the beam pattern of the line array, and, as such, it is computationally a very efficient operation. However, Equation 6.25 can be generalized for non-linear 2-D and 3-D arrays, and this is discussed in Section 6.4.2.

As an example, let us consider a distant monochromatic source. Then the plane wave signal arrival from the direction θ received by an N hydrophone line array is expressed by Equation 6.24. The beam power pattern $P(f, \theta_s)$ is given by $P(f, \theta_s) = B(f, \theta_s) \times B^*(f, \theta_s)$ that takes the form

$$P(f, \theta_s) = \sum_{n=1}^{N} \sum_{m=1}^{N} X_n(f) X_m^*(f) \exp\left[\frac{j2\pi f \delta_{nm} \cos\theta_s}{c}\right], \tag{6.26}$$

where δ_{nm} is the spacing $\delta(n - m)$ between the nth and m^{th} sensors. As a result of Equation 6.26, the expression for the power beam pattern $P(f, \theta_s)$ is reduced to

$$P(f, \theta_s) = \left\{ \frac{\sin\left[N\frac{\pi\delta}{\lambda}(\sin\theta_s - \sin\theta)\right]}{\sin\left[\frac{\pi\delta}{\lambda}(\sin\theta_s - \sin\theta)\right]} \right\}^2. \tag{6.27}$$

Let us consider for simplicity the source bearing θ to be at array broadside, $\delta = \lambda/2$, and $L = (N-1)\delta$ is the array size. Then Equation 6.27 is modified as[4,10]

$$P(f, \theta_s) = \frac{N^2 \sin^2\left[\frac{\pi L \sin\theta_s}{\lambda}\right]}{\left(\frac{\pi L \sin\theta_s}{\lambda}\right)^2}, \tag{6.28}$$

which is the far-field radiation or directivity pattern of the line array as opposed to near-field regions. The results in Equations 6.27 and 6.28 are for a perfectly coherent incident acoustic signal, and an increase in array size L results in additional power output and a reduction in beam width, which are similar arguments with those associated with the CRLB analysis expressed by Equation 6.21. The sidelobe structure of the directivity pattern of a line array, which is expressed by Equation 6.27, can be suppressed at the expense of a beam width increase by applying different weights. The selection of these weights will act as spatial filter coefficients with optimum performance.[5,17,18] There are two different approaches to select these weights: **pattern optimization** and **gain optimization**. For pattern optimization, the desired array response pattern $P(f, \theta_s)$ is selected first. A desired pattern is usually one with a narrow main lobe and low sidelobes. The weighting or shading coefficients in this case are real numbers from well-known window functions that modify the array response pattern. Harris' review[36] on the use of windows in discrete Fourier transforms and temporal spectral analysis is directly applicable in this case to spatial spectral analysis for towed line array applications.

Using the approximation $sin\theta \cong \theta$ for small θ at array broadside, the first null in Equation 6.25 occurs at $\pi L \sin\theta/\lambda = \pi$ or $\Delta\theta \, x L/\lambda \cong 1$. The major conclusion drawn here for line array applications is that[4,10]

$$\Delta\theta \approx \frac{\lambda}{L} \text{ and } \Delta f \times T = 1 \tag{6.29}$$

where $T = M_s/F_s$ is the sensor time series length. Both of the relations in Equation 6.29 express the well-known temporal and spatial resolution limitations in line array applications that form the driving force and motivation for adaptive and synthetic aperture signal processing that we will discuss later.

An additional constraint for sonar and ultrasound applications requires that the frequency resolution Δf of the sensor time series for spatial spectral analysis that is based on fast Fourier transform (FFT) beamforming processing must be such that

$$\Delta f \times \frac{L}{c} \ll 1 \tag{6.30}$$

in order to satisfy *frequency quantization* effects associated with discrete frequency domain beamforming following the FFT of sensor data.[17,42] This is because, in conventional beamforming, **f**inite-duration **i**mpulse **r**esponse (FIR) filters are used to provide realizations in designing digital phase shifters for beam steering. Since fast-convolution signal processing operations are part of the processing flow of a sonar signal processor, the effective beamforming filter length needs to be considered as the overlap size between successive snapshots. In this way, the overlap process will account for the wraparound errors that arise in the fast-convolution processing.[1,40–42] It has been shown[42] that an approximate estimate of the effective beamforming filter length is provided by Equations 6.28 and 6.30.

Because of the linearity of the conventional beamforming process, an exact equivalence of the frequency domain narrowband beamformer with that of the time domain beamformer for broadband signals can be derived.[42,43] Based on the model of Figure 6.3, the time domain beamformer is simply a time delaying[43] and summing process across the sensors of the line array, which is expressed by

$$b(\theta_s, t_i) = \sum_{n=1}^{N} x_n(t_i - \tau_s) . \tag{6.31}$$

Since

$$b(\theta_s, t_i) = \text{IFFT}\{B(f, \theta_s)\}, \tag{6.32}$$

by using FFTs and fast-convolution procedures, continuous beam time sequences can be obtained at the output of the frequency domain beamformer.[42] This is a very useful operation when the implementation of beamforming processors in sonar systems is considered.

The beamforming operation in Equation 6.31 is not restricted only for plane wave signals. More specifically, consider an acoustic source at the near field of a line array with r_s as the source range and θ as its bearing. Then the time delay for steering at θ is

$$\tau_s = (r_s^2 + d_{nm}^2 - 2r_s d_{nm}\cos\theta)^{1/2}/c. \quad (6.33)$$

As a result of Equation 6.33, the steering vector $d_n(f, \theta_s) = \exp[j2\pi f\tau_s]$ will include two parameters of interest, the bearing θ and range r_s of the source. In this case, the beamformer is called *focused beamformer*, which is used mainly in ultrasound system applications There are, however, practical considerations restricting the application of the focused beamformer in passive sonar line array systems, and these have to do with the fact that effective range focusing by a beamformer requires extremely long arrays.

6.4.1.2 Circular Array Conventional Beamformer

Consider M sensors distributed uniformly on a ring of radius R receiving plane wave arrivals at an azimuth angle θ and an elevation angle ϕ as shown in Figure 6.5. The plane wave response of this circular array for azimuth steering θ_s and an elevation steering ϕ_s can be written as follows:

$$B(f, \theta_s, \phi_s) = \bar{D}^*(f, \theta_s, \phi_s)W(\theta_s)\bar{X}(f), \quad (6.34)$$

where $\bar{D}(f, \theta_s, \phi_s)$ is the steering vector with the nth term being expressed by $d_n(f, \theta_s, \phi_s) = \exp(j2\pi/R\sin\phi_s\cos(\theta_s - \theta_n)/c)$, and $\theta_m = 2\pi m/M$ is the angular location of the mth sensor with $m = 0, 1, \ldots, M-1$. $W(\theta_s)$ is a diagonal matrix with the off diagonal terms being zero and the diagonal terms being the weights of a spatial window to reduce the sidelobe structure.[36] This spatial window, in general, is not uniform and depends on the sensor location (θ_n) and the beam steering direction (θ_s). The beam power pattern $P(f, \theta_s, \phi_s)$ is given by $P(f, \theta_s, \phi_s) = B(f, \theta_s, \phi_s) \times B^*(f, \theta_s, \phi_s)$. The azimuth angular response of the circular array covers the range$[0, 2\pi]$, and, therefore, there is no ambiguity with respect to the azimuth angle θ.

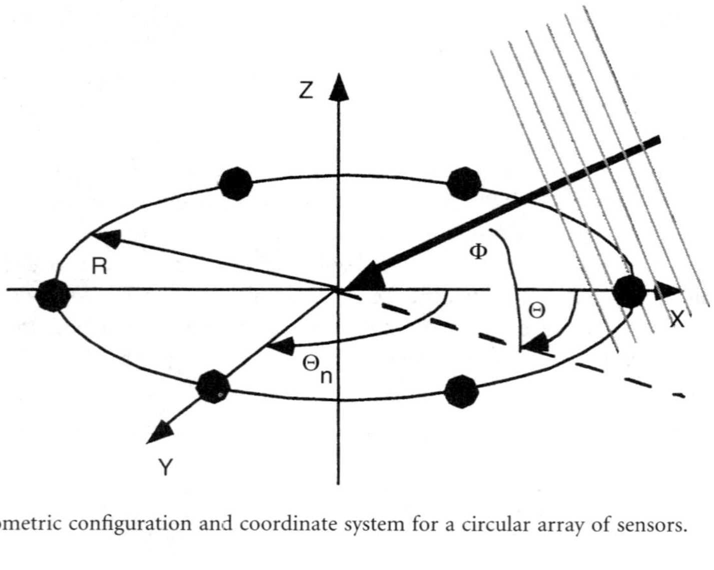

FIGURE 6.5 Geometric configuration and coordinate system for a circular array of sensors.

6.4.2 Multi-Dimensional (3-D) Array Conventional Beamformer

Presented in this section is a generic approach to decompose the planar, cylindrical, and spherical array beamformers into coherent subsets of linear and/or circular array beamformers. In this chapter, we will restrict the discussion to 3-D arrays with cylindrical and planar geometric configuration. The details of the decomposition process for spherical arrays are similar and can be found in References 7, and 25 to 28.

6.4.2.1 Decomposition Process for 2-D and 3-D Sensor Array Beamformers

6.4.2.1.1 Cylindrical Array Beamformer

Consider the cylindrical array shown in Figure 6.6 with $\aleph$ sensors and $\aleph = NM$, where N is the number of circular rings and M is the number of sensors on each ring. The angular response of this cylindrical array to a steered direction at (θ_s, ϕ_s)can be expressed as

$$B(f, \theta_s, \phi_s) = \sum_{r=0}^{N-1}\sum_{m=0}^{M-1} w_{r,m}X_{r,m}(f)d^*_{r,m}(f, \theta_s, \phi_s), \tag{6.35}$$

where $w_{r,m}$ is the $(r, m)^{th}$ term of a 3-D spatial window, $X_{r,m}(f)$ is the $(r, m)^{th}$ term of the matrix $\underline{X}(f)$, or $X_{r,m}(f)$ is the Fourier transform of the signal received by the mth sensor on the rth ring, and $d_{r,m}(f, \theta_s, \phi_s) = \exp\{j2\pi f[(r\delta_z \cos\phi_s + R\sin\phi_s \cos(\theta_s - \theta_m)/c]\}$ is the $(r, m)^{th}$ steering term of $\bar{D}(f, \theta_s, \phi_s)$. R is the radius of the ring, δ_z is the distance between each ring along the z-axis, r is the index for the rth ring, and $\theta_m = 2\pi m/M$, $m = 0, 1, \ldots, M-1$. Assuming $w_{r,m} = w_r\, x\, w_m$, Equation 6.35 can be re-arranged as follows:

$$B(f, \theta_s, \phi_s) = \sum_{r=0}^{N-1} w_r d^*_r(f, \theta_s, \phi_s)\left[\sum_{m=0}^{M-1} X_{r,m}(f) w_m d^*_m(f, \theta_s, \phi_s)\right], \tag{6.36}$$

where $d_r(f, \theta_s, \phi_s) = \exp\{j2\pi f[(r\delta_z \cos\phi_s/c)]\}$ is the rth term of the steering vector for line array beamforming, w_r is the rth term of a spatial window for line array spatial shading, $d_m(f, \theta_s, \phi_s) = \exp\{j2\pi f(R\sin\phi_s \cos(\theta_s - \theta_m)/c)\}$ is the mth term of the steering vector for a circular beamformer discussed in Section 6.4.1, and w_m is the mth term of a spatial window for circular array shading.

Thus, Equation 6.36 suggests the decomposition of the cylindrical array beamformer into two steps, which is a well-known process in array theory. The first step is to perform circular array beamforming for each of the N rings with M sensors on each ring. The second step is to perform line array beamforming along the z-axis on the N beam time series outputs of the first step. This kind of implementation, which is based on the decomposition of the cylindrical beamformer into line and circular array beamformers, is shown in Figure 6.6. The coordinate system is identical to that shown in Figure 6.5. The decomposition process of Equation 6.36 also makes the design and incorporation of 3-D spatial windows much simpler. Non-uniform shading windows can be applied to each circular beamformer to improve the angular response with respect to the azimuth angle θ. A uniform shading window can then be applied to the line array beamformer to improve the angular response with respect to the elevation angle ϕ. Moreover, the decomposition process, shown in Figure 6.6, leads to an efficient implementation in computing architectures based on the following two factors:

- The number of sensors for each of these circular and line array beamformers is much less than the total number of sensors, $\aleph$, of the cylindrical array. This kind of decomposition process for the 3-D beamformer eliminates the need for very large memory and CPUs with very high throughput requirements on one board for real-time system applications.
- All the circular and line array beamformers can be executed in parallel, which allows their implementation in much simpler parallel architectures with simpler CPUs, which is a practical requirement for real-time system applications.

Thus, under the restriction $w_{r,m} = w_r \times w_m$ for 3-D spatial shading, the decomposition process provides equivalent beam time series with those that would have been provided by a 3-D cylindrical beamformer, as shown by Equations 6.35 and 6.36.

6.4.2.1.2 Planar Array Beamformer

Consider the discrete planar array shown in Figure 6.7 with $\aleph$ sensors where $\aleph = NM$ and M, N are the number of sensors along x-axis and y-axis, respectively. The angular response of this planar array to a steered direction (θ_s, ϕ_s) can be expressed as

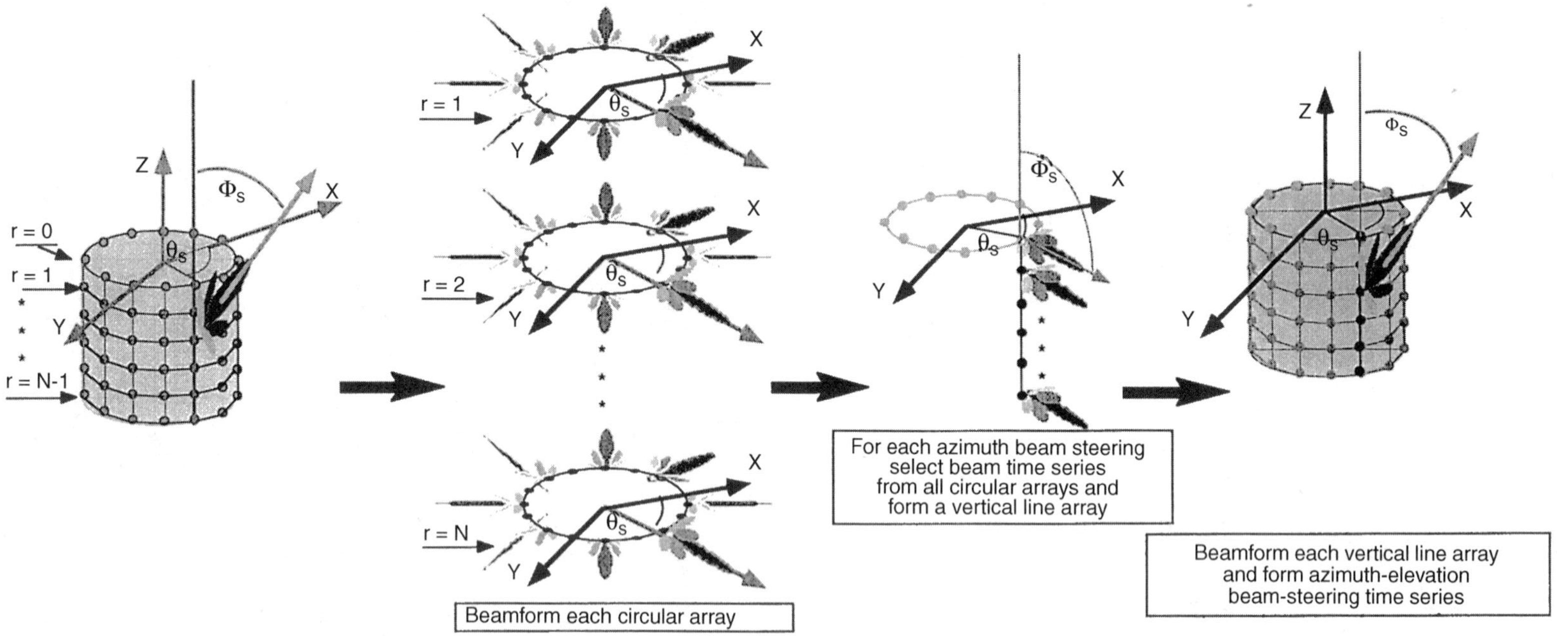

FIGURE 6.6 Coordinate system and geometric representation of the concept of decomposing a cylindrical array beamformer. The $\aleph = NM$ sensor cylindrical array beamformer consists of N circular arrays with M being the number of sensors in each circular array. Then, the beamforming structure for cylindrical arrays is reduced into coherent subsets of circular (for 0° to 360° azimuth bearing estimates) and line array (for 0° to 180° angular elevation bearing estimates) beamformers.

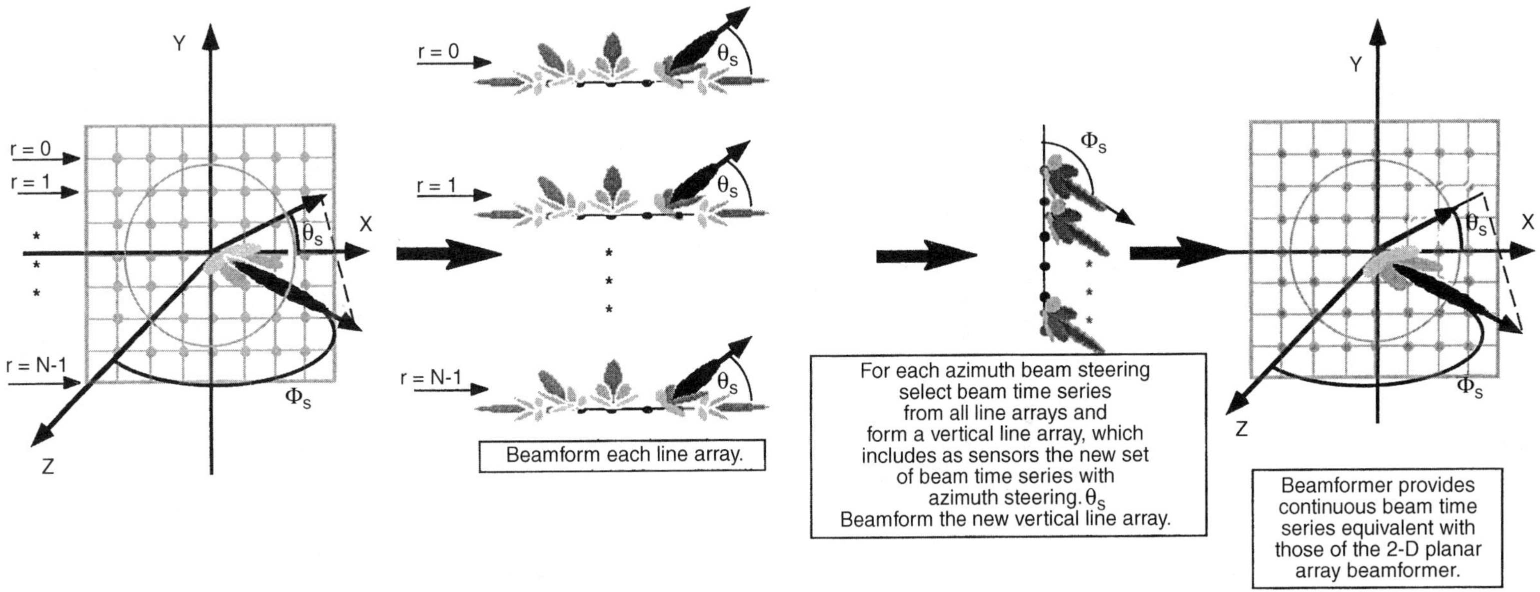

FIGURE 6.7 Coordinate system and geometric representation of the concept of decomposing a planar array beamformer. The $\aleph = NM$ sensor planar array beamformer consists of N line arrays with M being the number of sensors in each line array. Then, the beamforming structure for planar arrays is reduced into coherent subsets of line (for 0° to 180° azimuth bearing estimates) and line array (for 0° to 180° elevation bearing estimates) beamformers.

$$B(f, \theta_s, \phi_s) = \sum_{r=0}^{N-1} \sum_{m=0}^{M-1} w_{r,m} X_{r,m}(f) d^*_{r,m}(f, \theta_s, \phi_s), \quad (6.37)$$

where $w_{r,m}$ is the $(r, m)^{th}$ term of matrix $W(\theta, \phi)$ including the weights of a 2-D spatial window, and $X_{r,m}(f)$ is the $(r, m)^{th}$ term of the matrix $\underline{X}(f)$ including the Fourier transform of the received signal by the $(m, r)^{th}$ sensor along the x-axis and y-axis, respectively. $\underline{D}(f, \theta_s, \phi_s)$ is the steering matrix having its $(r, m)^{th}$ term defined by

$$d_{r,m}(f, \theta_s, \phi_s) = \exp(j2\pi f(m\delta_x \sin\theta_s + r\delta_y \cos\theta_s \cos\phi_s)/c).$$

Assuming that the matrix of spatial shading (weighting) $W(\theta, \phi)$ is separable [i.e., $\underline{W}(\theta, \phi) = \underline{W}_1(\theta)\underline{W}_2(\phi)$], Equation 6.37 can be simplified as follows:

$$B(f, \theta_s, \phi_s) = \sum_{r=0}^{N-1} w_{1,r} d^*_r(f, \theta_s, \phi_s) \left[\sum_{m=0}^{M-1} w_{2,m} X_{r,m}(f) d^*_m(f, \theta_s, \phi_s) \right] \quad (6.38a)$$

where $d_r(f, \theta_s, \phi_s) = \exp(j2\pi fr\delta_y \cos\theta_s \cos\phi_s/c)$ is the rth term of the steering vector $\bar{D}_y(f, \theta_s, \phi_s)$ and $d_m(f, \theta_s, \phi_s) = \exp(j2\pi fm\delta_x \sin\theta_s/c)$ is the mth term of the steering vector $\bar{D}_x(f, \theta_s, \phi_s)$. The summation term enclosed by parentheses in Equation 6.38a is equivalent to the response of a line array beamformer along the x-axis. Then all the steered beams from this summation term form a vector denoted by $\bar{B}_y(f, \theta_s)$. This vector defines a line array with directional sensors, which are the beams defined by the second summation process of Equation 6.38a. Therefore, Equation 6.38a can be expressed as

$$B(f, \theta_s, \phi_s) = \bar{D}^*_y(f, \theta_s, \phi_s) \underline{W}_1(\theta) \bar{B}_y(f, \theta_s). \quad (6.38b)$$

Equation 6.38b suggests that the 2-D planar array beamformer can be decomposed into two line array beamforming steps. The first step includes a line array beamforming along the x-axis and will be repeated N times to get the vector $\bar{B}_y(f, \theta_s)$ that includes the beam times series $b_r(f, \theta_s)$, where the index $r = 0, 1, \ldots, N-1$ is along the y-axis. The second step includes line array beamforming along the y-axis and will be done only once by treating the vector $\bar{B}_y(f, \theta_s)$ as the input signal for the line array beamformer to get the output $B(f, \theta_s, \phi_s)$. The separable spatial windows can now be applied separately on each line array beamformer to suppress sidelobe structures. Figure 6.7 shows the involved steps of decomposing the 2-D planar array beamformer into two steps of line array beamformers. The coordinate system is identical to that shown in Figure 6.4. The decomposition of the planar array beamformer into these two line-array beamforming steps leads to an efficient implementation based on the following two factors. First, the number of the involved sensors for each of these line array beamformers is much less than the total number of sensors, $\aleph$, of the planar array. This kind of decomposion process for the 2-D beamformer eliminates the need for very large memory and CPUs with very high throughput requirements on one board for real-time system applications. Second, all these line array beamformers can be executed in parallel, which allows their implementation in much simpler parallel architectures with simpler CPUs, which is a practical requirement for real-time system applications. Besides the advantage of the efficient implementation, the proposed decomposition approach makes the application of the spatial window much simpler to be incorporated.

6.4.3 Influence of the Medium's Propagation Characteristics on the Performance of a Receiving Array

In ocean acoustics and medical ultrasound imaging, the wave propagation problem is highly complex due to the spatial properties of the non-homogeneous underwater and human body mediums. For stationary source and receiving arrays, the space-time properties of the acoustic pressure fields include a limiting resolution imposed by these mediums. This limitation is due either to the angular spread of the incident energy about a single arrival as a result of the scattering phenomena or to the multipaths and their variation over the aperture of the receiving array.

More specifically, an acoustic signal that propagates through anisotropic mediums will interact with the transmitting medium microstructure and the rough boundaries, resulting in a net field that is characterized by irregular spatial and temporal variations. As a consequence of these interactions, a point source detected by a high-angular resolution receiver is perceived as a source of finite extent. It has been suggested[47] that due to the above spatial variations the sound field consists not of parallel, but of superimposed wavefronts of different directions of propagation. As a result, coherence measurements of this field by a receiving array give an estimate for the spatial coherence function. In the model for the spatial uncertainty of the above study,[47] the width of the coherence function is defined as the coherence length of the medium, and its reciprocal value is a measure of the angular uncertainty caused by the scattered field of the underwater environment.

By the *coherence* of acoustic signals in the sea or the human body, we mean the degree to which the acoustic pressures are the same at two points in the medium of interest located a given distance and direction apart. Pressure sensors placed at these two points will have phase coherent outputs if the received acoustic signals are perfectly coherent; if the two sensor outputs, as a function of space or time, are totally dissimilar, the signals are said to be incoherent. Thus, the loss of spatial coherence results in an upper limit on the useful aperture of a receiving array of sensors.[10] Consequently, knowledge of the angular uncertainty of the signal caused by the medium is considered essential in order to determine quantitatively the influence of the medium on the array gain, which is also influenced significantly by a partially directive anisotropic noise background. Therefore, for a given non-isotropic medium, it is desirable to estimate the optimum array size and achievable array gain for sonar and ultrasound array applications.

For geometrical simplicity and without any loss of generality, we consider the case of a receiving line array. Quantitative estimates of the spatial coherence for a receiving line array are provided by the cross-spectral density matrix in the frequency domain between any set of two sensor time series of the line array. An estimate of the cross-spectral density matrix $R(f)$ with its nm^{th} term is defined by

$$R_{nm}(f, \delta_{nm}) = E[X_n(f)X_m^*(f)]\,. \tag{6.39}$$

The above space-frequency correlation function in Equation 6.39 can be related to the angular power directivity pattern of the source, $\Psi_s(f, \theta)$, via a Fourier transformation by using a generalization of Bello's concept[48] of time-frequency correlation function $[t \Leftrightarrow 2\pi f]$ into space $[\delta_{nm} \Leftrightarrow 2\pi f \sin\theta / c]$, which gives

$$R_{nm}(f, \delta_{nm}) = \int_{-\pi/2}^{\pi/2} \Psi_s(f, \theta) \exp\left[\frac{-j2\pi f \delta_{nm}\theta}{c}\right] d\theta\,, \tag{6.40}$$

or

$$\Psi_s(f, \theta) = \int_{-N\delta/2}^{N\delta/2} R_{nm}(f, \delta_{nm}) \exp\left[\frac{j2\pi f \delta_{nm}\theta}{c}\right] d(\delta_{nm})\,. \tag{6.41}$$

The above transformation can be converted into the following summation:

$$R_{nm}(f_o, \delta_{nm}) = \Delta\theta \sum_{g=-G/2}^{G/2} \Psi_s(f_o, \theta_g) \exp\left[\frac{-j2\pi f_o \delta_{nm} \sin(g\Delta\theta)}{c}\right] \cos(g\Delta\theta)\,, \tag{6.42}$$

where $\Delta\theta$ is the angle increment for sampling the angular power directivity pattern $\theta_g = g\Delta\theta$, g is the index for the samples, and G is the total number of samples.

For line array applications, the power directivity pattern (calculated for a homogeneous free space) due to a distant source, which is treated as a point source, should be a delta function. Estimates, however, of the source's directivity from a line array operating in an anisotropic ocean are distorted by the underwater medium. In other words, the directivity pattern of the received signal is the convolution of the original pattern and the angular directivity of the medium (i.e., the angular scattering function of the underwater environment). As a result of this the angular pattern of the received signal, by a receiving line array system, is the scattering function of the medium.

In this chapter, the concept of spatial coherence is used to determine the statistical response of a line array to the acoustic field. This response is the result of the multipath and scattering phenomena discussed above, and there are models[10,47] to relate the spatial coherence with the physical parameters of an anisotropic medium for measurement interpretation. In these models, the interaction of the acoustic signal with the transmitting medium is considered to result in superimposed wavefronts of different directions of propagation. Then Equations 6.24 and 6.25, which define a received sensor signal from a distant source, are expressed by

$$x_n(t_i) = \sum_{l=1}^{J} A_l \exp\left[-j2\pi f_l\left(t_i - \frac{\delta(n-1)}{c}\theta_l\right)\right] + \varepsilon_{n,i}(0, \sigma_e), \tag{6.43}$$

where $l = 1, 2, \ldots, J$, and J is the number of superimposed waves. As a result, a generalized form of the cross-correlation function between two sensors, which has been discussed by Carey and Moseley,[10] is

$$R_{nm}(f, \delta_{nm}) = \bar{X}^2(f)\exp\left[-\left(\frac{\delta_{nm}}{L_c}\right)^k\right], k = 1, \quad \text{or} \quad 1.5 \quad \text{or} \quad 2, \tag{6.44}$$

where L_c is the correlation length and $\bar{X}^2(f)$ is the mean acoustic intensity of a received sensor time sequence at the frequency bin f. A more explicit expression for the Gaussian form of Equation 6.44 is given in[47]

$$R_{nm}(f, \delta_{nm}) \approx \bar{X}^2(f)\exp\left[-\left(\frac{2\pi f \delta_{nm}\sigma_\theta}{c}\right)^2 / 2\right], \tag{6.45}$$

and the cross-correlation coefficients are given from

$$\rho_{nm}(f, \delta_{nm}) = R_{nm}(f, \delta_{nm}) / \bar{X}^2(f). \tag{6.46}$$

At the distance $L_c = c/(2\pi f \sigma_\theta)$, called "*the coherence length*," the correlation function in Equation 6.46 will be 0.6. This critical length is determined from experimental coherence measurements plotted as a function of δ_{nm}. Then a connection between the medium's angular uncertainty and the measured coherence length is derived as

$$\sigma_\theta = 1/L_c, \text{ and } L_c = 2\pi\delta_{1m} f/c. \tag{6.47}$$

Here, δ_{1m} is the critical distance between the first and the mth sensors at which the coherence measurements get smaller than 0.6. Using the parameter definition in Equation 6.47, the effective aperture size and array gain of a deployed towed line array can be determined[10,47] for a specific underwater ocean environment.

Since the correlation function for a Gaussian acoustic field is given by Equation 6.45, the angular scattering function $\Phi(f, \theta)$ of the medium can be derived. Using Equation 6.41 and following a rather simple analytical integral evaluation, we have

$$\Phi(f, \theta) = \frac{1}{\sigma_\theta\sqrt{2\pi}}\exp\left[-\frac{\theta^2}{2\sigma_\theta^2}\right], \tag{6.48}$$

where $\sigma_\theta = c/(2\pi f\delta_{nm})$. This is an expression for the angular scattering function of a Gaussian underwater ocean acoustic field.[10,47]

It is apparent from the above discussion that the estimates of the cross-correlation coefficients $\rho_{nm}(f_i,\delta_{nm})$ are necessary to define experimentally the coherence length of an underwater or human body medium. For details on experimental studies on coherence estimation for underwater sonar applications, the reader may review References 10 and 30.

6.4.4 Array Gain

The performance of a line array to an acoustic signal embodied in a noise field is characterized by the "*array gain*" parameter, **AG**. The mathematical relation of this parameter is defined by

$$\mathrm{AG} = 10\log\frac{\sum_{n=1}^{N}\sum_{m=1}^{N}\tilde{\rho}_{nm}(f,\delta_{nm})}{\sum_{n=1}^{N}\sum_{m=1}^{N}\tilde{\rho}_{\varepsilon,nm}(f,\delta_{nm})}, \tag{6.49}$$

where $\rho_{nm}(f_i,\delta_{nm})$ and $\rho_{\varepsilon,nm}(f,\delta_{nm})$ denote the normalized cross-correlation coefficients of the signal and noise field, respectively. Estimates of the correlation coefficients are given by Equation 6.46.

If the noise field is isotropic, that is, not partially directive, then the denominator in Equation 6.49 is equal to N, i.e.,

$$\sum_{n=1}^{N}\sum_{m=1}^{N}\tilde{\rho}_{\varepsilon,nm}(f,\delta_{nm}) = N,$$

because the non-diagonal terms of the cross-correlation matrix for the noise field are negligible. Then Equation 6.49 simplifies to

$$\mathrm{AG} = 10\log\frac{\sum_{n=1}^{N}\sum_{m=1}^{N}\tilde{\rho}_{nm}(f,\delta_{nm})}{N}. \tag{6.50}$$

For perfect spatial coherence across the line array, the normalized cross-correlation coefficients are $\rho_{nm}(f,\delta_{nm}) \cong 1$, and the expected values of the array gain estimates are $\mathrm{AG} = 10\times\log N$. For the general case of isotropic noise and for frequencies smaller than the towed array's design frequency, the term **AG** is reduced to the quantity called the directivity index,

$$DI = 10\times\log[(N-1)\delta/(\lambda/2)]. \tag{6.51}$$

When $\delta << \lambda$ and the conventional beamforming processing is employed, Equation 6.29 indicates that the deployment of very long line arrays is required in order to achieve sufficient AG and angular resolution for precise bearing estimates. Practical deployment considerations, however, usually limit the overall dimensions of a hull-mounted line or towed array. In addition, the medium's spatial coherence[10,30] sets an upper limit on the effective towed array length. In general, the medium's spatial coherence length is of the order of $O(10^2)\lambda$.[10,30] In addition, for sonar systems, very long towed arrays suffer degradation in the AG due to array shape deformation and increased levels of self noise.[49–53]

Alternatives to large aperture sonar arrays are signal processing schemes discussed in Reference 1. Theoretical and experimental investigations have shown that bearing resolution and detectability of weak signals in the presence of strong interferences can be improved by applying non-conventional beamformers such as adaptive beamforming,[1–5,17–24] or acoustic synthetic aperture processing[1,11–16] to the sensor time series of deployed short sonar and ultrasound arrays, which are discussed in the next section.

6.5 Advanced Beamformers

6.5.1 Synthetic Aperture Processing

Various synthetic aperture techniques have been investigated to increase signal gain and improve angular resolution for line array systems. While these techniques have been successfully applied to aircraft and satellite-active radar systems, they have not been successful with sonar and ultrasound systems. In this section, we will review synthetic aperture techniques that have been tested successfully with real data.[11–16] They are summarized in terms of their experimental implementation and the basic approach involved.

Let us start with a few theoretical remarks. The plane wave response of a line array to a distant monochromatic signal, received by the nth element of the array, is expressed by Equations 6.23, 6.24, and 6.25. In these expressions, the frequency f includes the Doppler shift due to a combined movement of the receiving array and the source (or object reflecting the incoming acoustic wavefront) radiating signal. Let υ denote the relative speed; it is assumed here that the component of the source's velocity along its bearing is negligible. If f_o is the frequency of the stationary field, then the frequency of the received signal is expressed by

$$f = f_o(1 \pm \upsilon \sin\theta / c) \tag{6.52}$$

and an approximate expression for the received sensor time series in Equations 6.18 and 6.43 is given by

$$x_n(t_i) = A\exp\left[j2\pi f_o\left(t_i - \frac{\upsilon t_i + (n-1)\delta}{c}\sin\theta\right)\right] + \varepsilon_{n,i}. \tag{6.53}$$

τ seconds later, the relative movement between the receiving array and the radiated source is $\upsilon\tau$. By proper choice of the parameters υ and τ, we have $\upsilon\tau = q\delta$, where q represents the number of sensor positions that the array has moved, and the received signal $x_n(t_i + \tau)$ is expressed by

$$x_n(t_i + \tau) = \exp(j2\pi f_o\tau)A\exp\left[j2\pi f_o\left(t_i - \frac{\upsilon t_i + (q+n-1)\delta}{c}\sin\theta\right)\right] + \varepsilon^{\tau}_{n,i}. \tag{6.54}$$

As a result, we have the Fourier transform of $x_n(t_i + \tau)$, as

$$\tilde{X}_n(f)_\tau = \exp(j2\pi f_o\tau)\tilde{X}_n(f), \tag{6.55}$$

where $\tilde{X}_n(f)_\tau$ and $\tilde{X}_n(f)$ are the DFTs of $x_n(t_i + \tau)$ and $x_n(t_i)$, respectively. If the phase term $\exp(-j2\pi f_o\tau)$ is used to correct the line array measurements shown in Equation 6.55, then the spatial information included in the successive measurements at $t = t_i$ and $t = t_i + \tau$ is equivalent to that derived from a line array of $(q + N)$ sensors. When idealized conditions are assumed, the phase correction factor for Equation 6.52 in order to form a synthetic aperture is $\exp(-j2\pi f_o\tau)$. However, this phase correction estimate requires *a priori* knowledge of the source receiver relative speed υ and accurate estimates for the frequency f of the received signal. An additional restriction is that the synthetic aperture processing techniques have to compensate for the disturbed paths of the receiving array during the integration period that the synthetic aperture is formed. Moreover, the temporal coherence of the source signal should be greater or at least equal to the integration time of the synthetic aperture.

At this point, it is important to review a few fundamental physical arguments associated with passive synthetic aperture processing. In the past,[13] there was a conventional wisdom regarding synthetic aperture techniques which held that practical limitations prevent them from being applicable to real-world systems. The issues were threshold.

1. Since passive synthetic aperture can be viewed as a scheme that converts temporal gain to spatial gain, most signals of interest do not have sufficient temporal coherence to allow a long spatially coherent aperture to be synthesized.
2. Since past algorithms required *a priori* knowledge of the source frequency in order to compute the phase correction factor, as shown by Equations 6.52 to 6.55, the method was essentially useless in any bearing estimation problem since Doppler would introduce an unknown bias on the frequency observed at the receiver.
3. Since synthetic aperture processing essentially converts temporal gain to spatial gain, there was no "new" gain to be achieved and, therefore, no point to the method.

Recent work[12–16] has shown that there can be realistic conditions under which all of these objections are either not relevant or do not constitute serious impediments to practical applications of synthetic aperture processing in operational systems.[1] Theoretical discussions have shown[13] that the above three arguments are valid for cases that include the formation of synthetic aperture in mediums with isotropic noise characteristics. However, when the noise characteristics of the received signal are non-isotropic and the receiving array includes more than one sensor, then there is spatial gain available from passive synthetic aperture processing; this has been discussed analytically in Reference 13. Recently, there have been only two passive synthetic aperture techniques[11–16,54] and an MLE estimator[12] published in the open literature that deal successfully with the above restrictions. In this section, they are summarized in terms of their experimental implementation for sonar and ultrasound applications. For more details about these techniques, the reader may review References 11 to 16, 54, and 55.

6.5.1.1 FFT Based Synthetic Aperture Processing (FFTSA Method)

Shown in the upper part of Figure 6.8, under *Physical Aperture*, are the basic processing steps of Equations 6.23 to 6.26 for conventional beamforming applications including line arrays. This processing includes the generation of the aperture function of the line array via FFT transformation (i.e., Equation 6.25), with the beamforming done in the frequency domain. The output (i.e., Equation 6.26) provides the directionality power pattern of the acoustic signal/noise field received by the N sensors of the line array. As an example, the theoretical response of the power pattern for a 64-sensor line array is given in Figure 6.9. In the lower part of Figure 6.8, under *Synthetic Aperture*, the concept of an FFT-based synthetic aperture technique, called FFTSA,[55] is presented. The experimental realization of this method includes

1. The time series acquisition, using the N sensor line array, of a number M of snapshots of the acoustic field under surveillance taken every τ seconds
2. The generation of the aperture function for each of the M snapshots
3. The beamforming in the frequency domain of each generated aperture function

This beamforming processor provides M beam patterns with N beams each. For each beam of the beamforming output, there are M time-dependent samples with a τ second sampling interval.

The FFT transformation in the time domain of the M time-dependent samples of each beam provides the synthetic aperture output, which is expressed analytically by Equation 6.56. For more details, please refer to Reference 55.

$$P(f, \theta_s)_M = \left\{ \frac{\sin\left[N\frac{\pi\delta}{\lambda}(\sin\theta_s)\right]}{\sin\left[\frac{\pi\delta}{\lambda}(\sin\theta_s)\right]} \cdot \frac{\sin\left[M\frac{N\pi\delta}{2\,\lambda}(\sin\theta_s)\right]}{\sin\left[\frac{N\pi\delta}{2\,\lambda}(\sin\theta_s)\right]} \right\}^2 . \tag{6.56}$$

Equation 6.56 assumes that $\upsilon\tau = (N\delta)/2$, which indicates that there is a 50% spatial overlap between two successive sets of M measurements and that the source bearing of θ is approximately at the boresight.

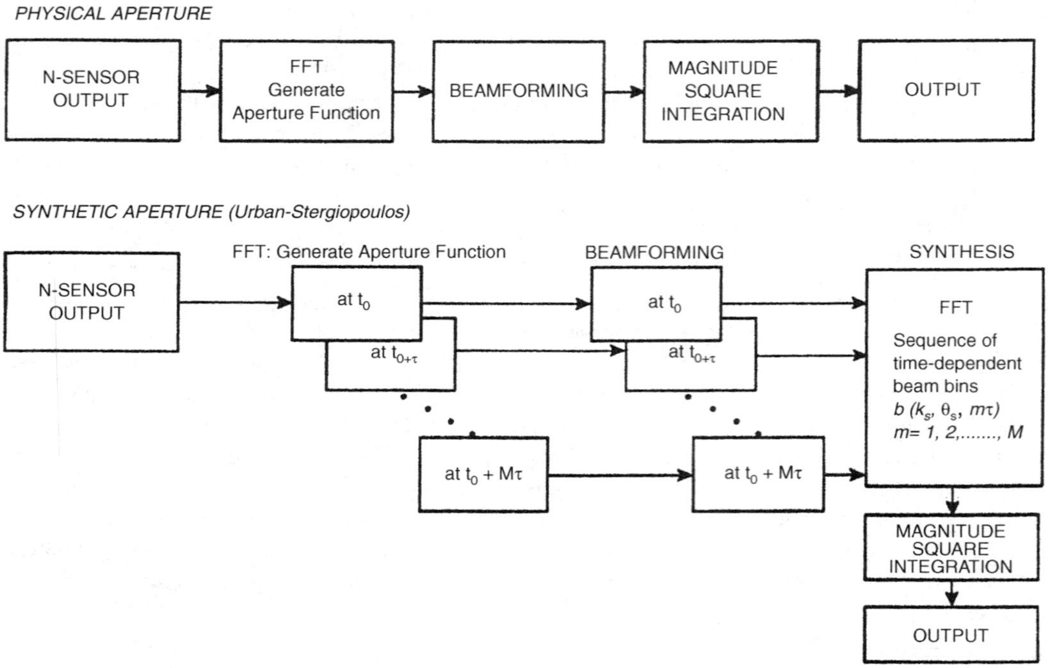

FIGURE 6.8 Shown under the title *Physical Aperture* is conventional beamforming processing in the frequency domain for a physical line array. Presented under the title *Synthetic Aperture* is the signal processing concept of the FFTSA method. (Reprinted by permission of IEEE ©1992.)

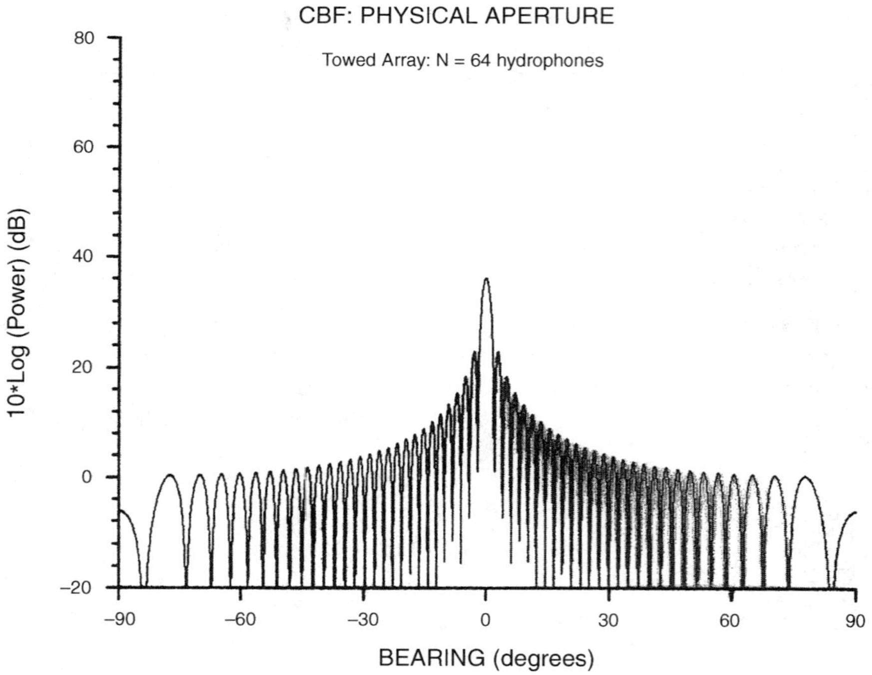

FIGURE 6.9 The azimuth power pattern from the beamforming output of the 64-sensor line array considered for the synthetic aperture processing in Figure 6.10. (Reprinted by permission of IEEE ©1992.)

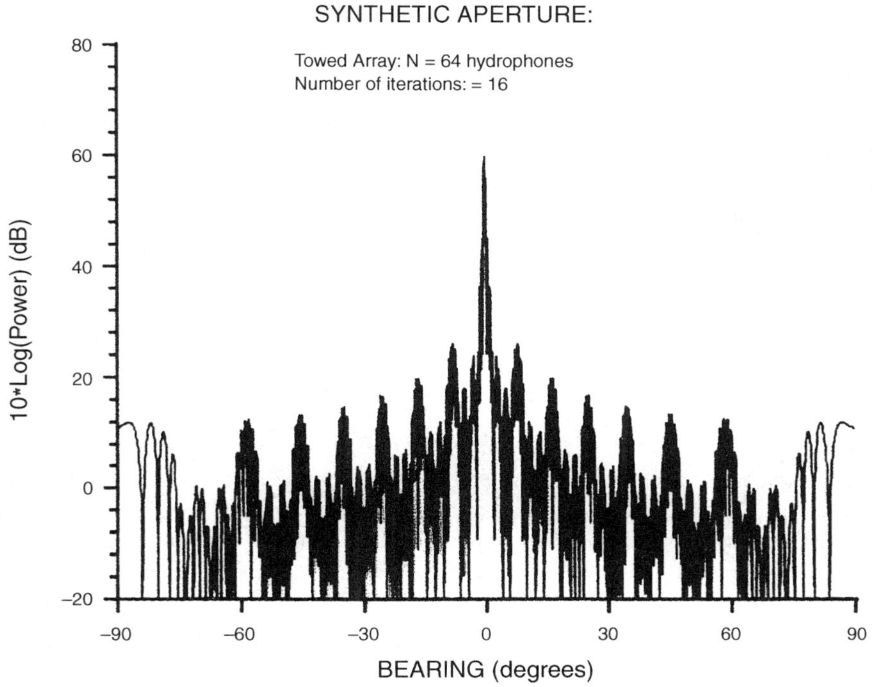

FIGURE 6.10 The azimuth power pattern from the beamforming output of the FFTSA method. (Reprinted by permission of IEEE ©1992.)

The azimuthal power pattern of Equation 6.56 for the beamforming output of the FFTSA method is shown in Figure 6.10.

6.5.1.2 Yen and Carey's Synthetic Aperture Method

The concept of the experimental implementation of Yen and Carey's synthetic aperture method[54] is shown in Figure 6.11 and is also expressed by the following relation,

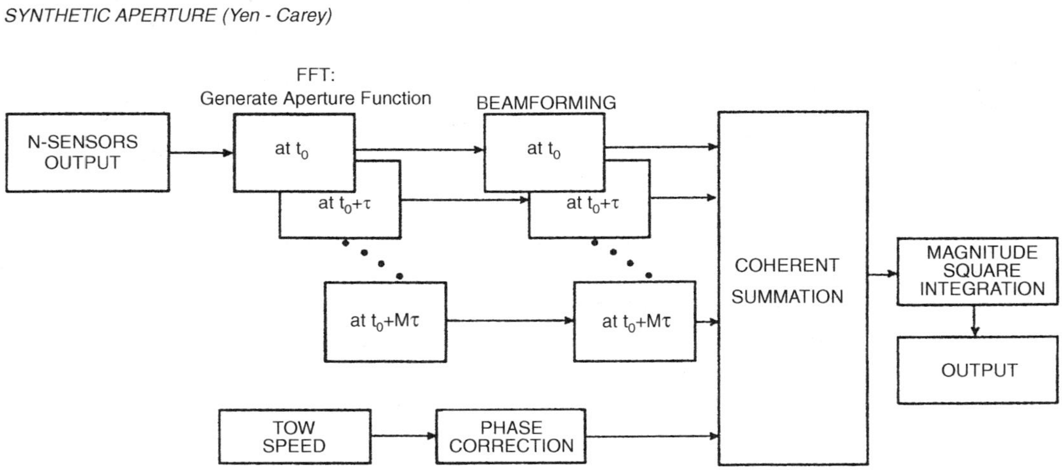

FIGURE 6.11 The concept of the experimental implementation of Yen and Carey's synthetic aperture method is shown under the same arrangement as the FFTSA method for comparison. (Reprinted by permission of IEEE ©1992.)

$$B(f_o, \theta_s)_M = \sum_{m=1}^{M} b(f_o, \theta_s)_{m\tau} \exp(-j\phi_m), \tag{6.57}$$

which assumes that estimates of the phase corrector ϕ_m require knowledge of the relative source receiver speed υ or the velocity filter concept, introduced by Yen and Carey.[54] The basic difference of this method[54] with the FFTSA[55] technique is the need to estimate a phase correction factor ϕ_m in order to synthesize the M time-dependent beam patterns. Estimates of ϕ_m are given by

$$\phi_m = 2\pi f_o(1 \pm \upsilon \sin\theta_s / c) m\tau, \tag{6.58}$$

and the application of a velocity filter concept for estimating the relative source receiver speed, υ. This method has been successfully applied to experimental sonar data, including CW signals, and the related application results have been reported.[10,54]

6.5.1.3 Nuttall's MLE Method for Synthetic Aperture Processing

It is also important to mention here the development by Nuttall[12] of an MLE estimator for synthetic aperture processing. This MLE estimator requires the acquisition of very long sensor time series over an interval T which corresponds to the desired length υT of the synthetic aperture. This estimator includes searching for the values of ϕ and ω that maximize the term

$$MLE(\omega, \phi) = \left| \Delta t \sum_{n=0}^{N-1} \exp(-jn\phi) \left[\sum_{m=1}^{M} x_n(m\Delta t) \exp(-jm\Delta t\omega) \right] \right|, \tag{6.59}$$

where

$$\omega = 2\pi f_o\left(1 - \frac{\upsilon \sin\theta}{c}\right), \qquad \phi = \frac{\delta}{c} 2\pi f_o \sin\theta. \tag{6.60}$$

Equations 6.59 and 6.60 indicate that the N complex vectors $X_n(\omega) = \sum_{m=1}^{M} x_n(t_m) \exp(-j\omega\Delta t)$, which give the spectra for the very long sensor time series $x_n(t)$ at ω, are phased together by searching ϕ over $(-p, p)$, until the largest vector length occurs in Equation 6.59. Estimates of (θ, f) are determined from Equation 6.60 using the values of ϕ and ω that maximize Equation 6.59. The MLE estimator has been applied on real sonar data sets, and the related application results have been reported.[12]

A physical interpretation of the above synthetic aperture methods is that the realistic conditions for effective acoustic synthetic aperture processing can be viewed as schemes that convert temporal gain to spatial gain. Thus, a synthetic aperture method requires that successive snapshots of the received acoustic signal have good cross-correlation properties in order to synthesize an extended aperture and that the speed fluctuations are successfully compensated by means of processing. It has also been suggested[11–16] that the prospects for successfully extending the physical aperture of a line array require algorithms which are not based on the synthetic aperture concept used in active radars. The reported results in References 10 to 16, 54, and 55 have shown that the problem of creating an acoustic synthetic aperture is centered on the estimation of a phase correction factor, which is used to compensate for the phase differences between sequential line array measurements in order to coherently synthesize the spatial information into a synthetic aperture. When the estimates of this phase correction factor are correct, then the information inherent in the synthetic aperture is the same as that of an array with an equivalent physical aperture.[11–16]

6.5.1.4 Spatial Overlap Correlator for Synthetic Aperture Processing (ETAM Method)

Recent theoretical and experimental studies have addressed the above concerns and indicated that the space and time coherence of the acoustic signal in the sea[10–16] appears to be sufficient to extend the physical aperture of a moving line array. In the above studies, the fundamental question related to the angular resolution capabilities of a moving line array and the amount of information inherent in a received signal have been addressed. These investigations include the use of the CRLB analysis and show that for long observation intervals of the order of 100 s, the additional information provided by a moving line array over a stationary array is expressed as a large increase in angular resolution, which is due to the Doppler caused by the movement of the array (see Figure 3 in Reference 12). A summary of these research efforts has been reported in a special issue of the *IEEE Journal of Oceanic Engineering*.[13] The synthetic aperture processing scheme that has been used in broadband sonar applications[1] is based on the extended towed array measurements (ETAM) algorithm, which was invented by Stergiopoulos and Sullivan.[11] The basic concept of this algorithm is a phase correction factor that is used to combine coherently successive measurements of the towed array to extend the effective towed array length.

Shown in Figure 6.12 is the experimental implementation of the ETAM algorithm in terms of the line array speed and sensor positions as a function of time and space. Between two successive positions of the N sensor line array with sensor spacing δ, there are $(N\text{-}q)$ pairs of space samples of the acoustic field that have the same spatial information, their difference being a phase factor[11–12,55] related to the time delay these measurements were taking. By cross-correlating the $(N\text{-}q)$ pairs of the sensor time series that overlap, the desired phase correction factor is derived, which compensates for the time delay between these measurements and the phase fluctuations caused by irregularities of the tow path of the physical array or relative speed between source and receiver; this is called the *overlap correlator.* Following the above, the key parameter in the ETAM algorithm is the time increment $\tau = q\delta/\upsilon$ between two successive sets of measurements, where υ is the tow speed and q represents the number of sensor positions that the towed array has moved during the τ seconds or the number of sensors to which the physical aperture of the array is extended at each successive set of measurements. The optimum overlap size $(N\text{-}q)$, which is related to the variance of the phase correction estimates, has been shown[13] to be $N/2$. The total number of sets of measurements required to achieve a desired extended aperture size is then defined by $J = (2/N)(T\upsilon/\delta)$, where T is the period taken by the towed array to travel a distance equivalent to the desired length of the synthetic aperture.

Then, for the frequency bin f_i and between two successive j^{th} and $(j+1)^{\text{th}}$ snapshots, the phase correction factor estimate is given by,

$$\Psi_j(f_i) = \arg\left\{\frac{\sum_{n=1}^{N/2} X_{j,\left(\frac{n}{2}+n\right)}(f_i) \times X^*_{(j+1),n}(f_i) \times \rho_{j,n}(f_i)}{\sum_{n=1}^{N/2} \rho_{j,n}(f_i)}\right\} \tag{6.61}$$

where, for a frequency band with central frequency f_i and observation bandwidth Δf or $f_i - \Delta f/2 < f_i < f_i + \Delta f/2$, the coefficients

$$\rho_{j,n}(f_i) = \frac{\left|\sum_{i=-Q/2}^{Q/2} X_{j,\left(\frac{n}{2}+n\right)}(f_i) \times X^*_{(j+1),n}(f_i)\right|}{\sqrt{\sum_{i=-Q/2}^{Q/2} \left|X_{j,\left(\frac{n}{2}+n\right)}(f_i)\right|^2 \times \sum_{i=-Q/2}^{Q/2} \left|X^*_{(j+1),n}(f_i)\right|^2}} \tag{6.62}$$

are the normalized cross-correlation coefficients or the coherence estimates between the $N/2$ pairs of sensors that overlap in space. The coefficients in Equation 6.62 are used as weighting factors in Equation 6.61 in order to optimally weigh the good against the bad pairs of sensors during the estimation process of the phase correction factor.

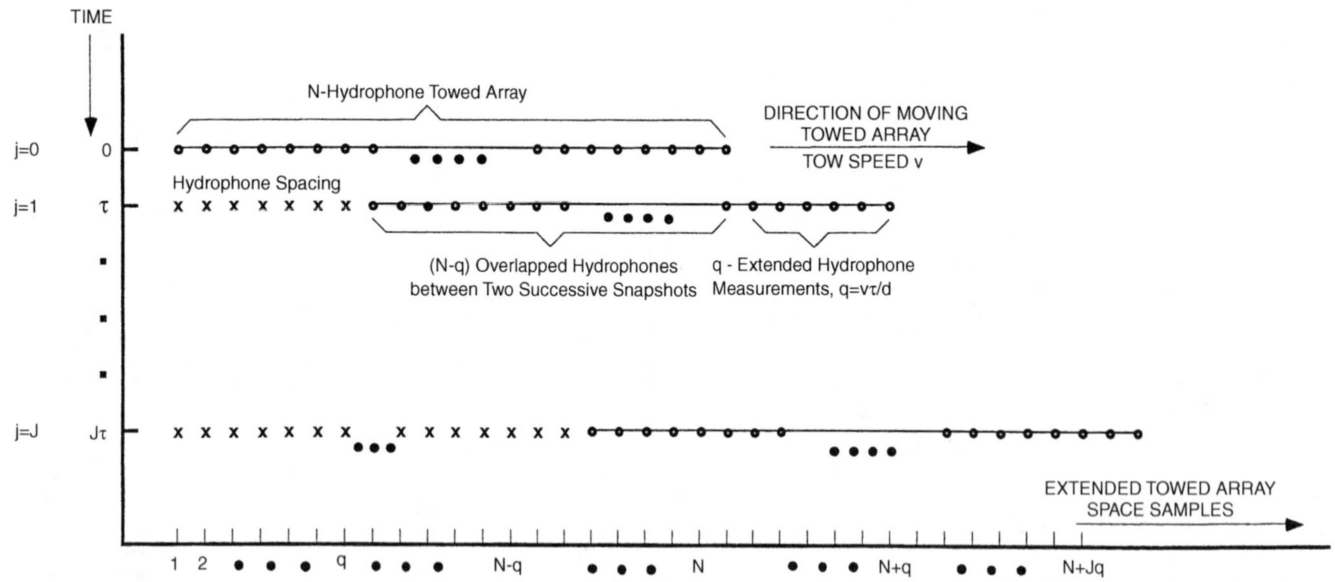

FIGURE 6.12 Concept of the experimental implementation of ETAM algorithm in terms of towed array positions and speed as a function of time and space. (Reprinted by permission of IEEE ©1992.)

The performance characteristics and expectations from the ETAM algorithm have been evaluated experimentally, and the related results have been reported.[1,12,55] The main conclusion drawn from these experimental results is that for narrowband signals or for FM type of pulses from active sonar systems, the overlap correlator in ETAM compensates successfully the speed fluctuations and effectively extends the physical aperture of a line array more than eight times. On the other hand, the threshold value of ETAM is −8 dB re 1-Hz band at the sensor. For values of SNR higher than this threshold, it has been shown that ETAM achieves the theoretical CRLB bounds and it has comparable performance to the maximum likelihood estimation.[12]

6.5.2 Adaptive Beamformers

Despite the geometric differences between the line and circular arrays, the underlying beamforming processes for these arrays, as expressed by Equations 6.11 and 6.12, respectively, are time-delay beamforming estimators, which are basically spatial filters. However, optimum beamforming requires the beamforming filter coefficients to be chosen based on the covariance matrix of the received data by the N sensor array in order to optimize the array response,[15,16] as discussed in Section 6.3.3. The family of algorithms for optimum beamforming that use the characteristics of the noise are called *adaptive beamformers*.[3,17,18–20,22,23] In this section, we will address implementation issues for various partially adaptive variants of the MVDR method and a GSC adaptive beamformer.[1,37]

Furthermore, the implementation of adaptive schemes in real-time systems is not restricted to one method, such as the MVDR technique that is discussed next. In fact, the generic concept of the sub-aperture multi-dimensional array introduced in this chapter allows for the implementation of a wide variety of adaptive schemes in operational systems.[7,25–28] As for the implementation of adaptive processing schemes in active systems, the following issues need to be addressed.

For active applications that include matched filter processing, the outputs of the adaptive algorithms are required to provide coherent beam time series to facilitate the post-processing. This means that these algorithms should exhibit near-instantaneous convergence and provide continuous beam time series that have sufficient temporal coherence to correlate with the reference signal in matched filter processing.[1]

In a previous study,[1] possible improvement in the convergence periods of two algorithms in the sub-aperture configuration was investigated. The GSC[18,44] coupled with the normalized least mean square (NLMS) adaptive filter[45] has been shown to provide near-instantaneous convergence under certain conditions.[1,37] The GSC/NLMS in the sub-aperture configuration was tested under a variety of conditions to determine if it could yield performance advantages and if its convergence properties could be exploited over a wider range of conditions.[1,37] The steered minimum variance (STMV) beamformer is a variant of the MVDR beamformer.[38] By applying narrowband adaptive processing on bands of frequencies, extra degrees of freedom are introduced. The number of degrees of freedom is equal to the number of frequency bins in the processed band. In other words, increasing the number of frequency bins processed decreases the convergence time by a corresponding factor. This is due to the fact that convergence now depends on the observation time bandwidth product, as opposed to observation time in the MVDR algorithm.[38,39]

The STMV beamformer in its original form was a broadband processor. In order to satisfy the requirements for matched filter processing, it was modified to produce coherent beam time series.[1] The ability of the STMV narrowband beamformer to produce coherent beam time series has been investigated in another study.[37] Also, the STMV narrowband processor was implemented in the sub-aperture configuration to produce near-instantaneous convergence and to reduce the computational complexity required. The convergence properties of both the full aperture and sub-aperture implementations have been investigated for line arrays of sensors.[1,37]

6.5.2.1 Minimum Variance Distortionless Response (MVDR)

The goal is to optimize the beamformer response so that the output contains minimal contributions due to noise and signals arriving from directions other than the desired signal direction. For this optimization procedure, it is desired to find a linear filter vector $\overline{W}(f_i, \theta)$, which is a solution to the constrained minimization problem that allows signals from the look direction to pass with a specified gain:[17,18]

$$\text{Minimize: } \sigma_{MV}^2 = \overline{W}^*(f_i, \theta) R(f_i) \overline{W}(f_i, \theta), \text{ subject to } \overline{W}^*(f_i, \theta)\overline{D}(f_i, \theta) = 1 \tag{6.63}$$

where $\overline{D}(f_i, \theta)$ is the conventional steering vector based on Equation 6.24. The solution is given by

$$\overline{W}(f_i, \theta) = \frac{R^{-1}(f_i)\overline{D}(f_i, \theta)}{\overline{D}^*(f_i, \theta)R^{-1}(f_i)\overline{D}(f_i, \theta)} \tag{6.64}$$

Equation 6.64 provides the adaptive steering vectors for beamforming the received signals by the N hydrophone line array. Then in the frequency domain, an adaptive beam at a steering θ_s is defined by

$$B(f_i, \theta_s) = \overline{W}^*(f_i, \theta_s)\overline{X}(f_i) \tag{6.65}$$

and the corresponding conventional beams are provided by Equation 6.25.

6.5.2.2 Generalized Sidelobe Canceller (GSC)

The GSC[44] is an alternative approach to the MVDR method. It reduces the adaptive problem to an unconstrained minimization process. The GSC formulation produces a much less computationally intensive implementation. In general, GSC implementations have complexity $O(N^2)$, as compared to $O(N^3)$ for MVDR implementations, where N is the number of sensors used in the processing. The basis of the reformulation of the problem is the decomposition of the adaptive filter vector $\overline{W}(f_i, \theta)$ into two orthogonal components, $\overline{w}$ and $-\overline{v}$, where $\overline{w}$ and $\overline{v}$ lie in the range and the null space of the constraint of Equation 6.63, such that $\overline{W}(f_i, \theta) = \overline{w}(f_i, \theta) - \overline{v}(f_i, \theta)$. A matrix **C**, which is called signal blocking matrix, may be computed from $\boldsymbol{C}\overline{I} = 0$, where $\overline{I}$ is a vector of ones. This matrix **C** whose columns form a basis for the null space of the constraint of Equation 6.63 will satisfy $\overline{v} = C\overline{u}$, where $\overline{u}$ is defined below by Equation 6.67. The adaptive filter vector may now be defined as $\overline{W} = \overline{w} - C\overline{u}$ and yields the realization shown in Figure 6.13. Then the problem is reduced to the following:

$$\text{Minimize: } \sigma_u^2 = \{[\overline{w} - C\overline{u}]^* R[\overline{w} - C\overline{u}]\} \tag{6.66}$$

which is satisfied by

$$\overline{u}_{opt} = (C^*RC)^{-1}C^*R\overline{w} \tag{6.67}$$

with $\boldsymbol{u}_{opt}$ being the value of the weights at convergence.

The Griffiths-Jim GSC in combination with the NLMS adaptive algorithm has been shown to yield near-instantaneous convergence.[44,45] Figure 6.13 shows the basic structure of the so-called memoryless GSC. The time delayed by $\tau_n(\theta_s)$ sensor time series defined by Equations 6.5 and 6.23 and Figure 6.3 form the pre-steered sensor time series, which are denoted by $x_n(t_i, \tau_n(\theta_s))$. In the frequency domain, these pre-steered sensor data are denoted by $X_n(f_i, \theta_s)$ and form the input data vector for the adaptive scheme in Figure 6.13. On the left-hand-side branch of Figure 6.13, the intermediate vector $\overline{Z}(f_i, \theta_s)$ is the result of the signal blocking matrix **C** being applied to the input $\overline{X}(f_i, \theta_s)$. Next, the vector $\overline{Z}(f_i, \theta_s)$ is an input to the NLMS adaptive filter. The output of the right-hand branch in Figure 6.13 is simply the shaded conventional output. Then the output of this processing scheme is the difference between the adaptive filter output and the "conventional" output:

$$e(f_i, \theta_s) = b(f_i, \theta_s) - \overline{u}^*(f_i, \theta_s)\overline{Z}(f_i, \theta_s). \tag{6.68}$$

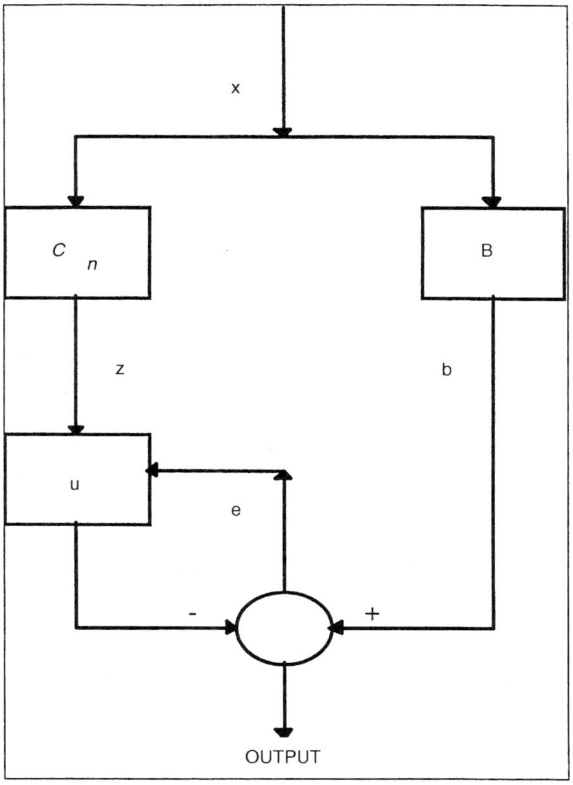

FIGURE 6.13 Basic processing structure for the memoryless GSC. The right-hand-side branch is simply the shaded conventional beamformer. The left-hand-side branch is the result of the signal blocking matrix (constraints) applied to pre-steered sensor time series. The output of the signal blocking matrix is the input to the NLMS adaptive filter. Then, the output of this processing scheme is the difference between the adaptive filter and the conventional output. (Reprinted by permission of IEEE ©1998.)

The adaptive filter, at convergence, reflects the sidelobe structure of any interferers present, and it is removed from the conventional beamformer output. In the case of the NLMS, this adaptation process can be represented by

$$\bar{u}_{k+1}(f_i, \theta_s) = \bar{u}_k(f_i, \theta_s) + \frac{\mu \times e_k^*(f_i, \theta_s)}{\alpha + \bar{X}^*(f_i, \theta_s)\bar{X}(f_i, \theta_s)} \bar{Z}(f_i, \theta_s) \tag{6.69}$$

where k is the iteration number, and α is a small positive number designed to maintain stability. The parameter μ is the convergence controlling parameter or "step size" for the NLMS algorithm.

6.5.2.3 Steered Minimum Variance (STMV) Broadband Adaptive Beamformer

Krolik and Swingler[38] have shown that the convergence time for broadband source location can be reduced by using the space-time statistic called the steered covariance matrix (STCM). This method achieves significantly shorter convergence times than adaptive algorithms that are based on the narrowband cross-spectral density matrix (CSDM)[17,18] without sacrificing spatial resolution. In fact, the number of statistical degrees of freedom available to estimate the STCM is approximately the time-bandwidth product ($T \times BW$) as opposed to the observation time ($T = M/F_s$, F_s being the sampling frequency) in CSDM methods. This provides an improvement of approximately BW, the size of the broadband source bandwidth, in convergence time. The conventional beamformer's output in the frequency domain is shown by Equation 6.25. The corresponding time domain conventional beamformer output $b(t_i, \theta_s)$ is the weighted sum of the steered sensor outputs, as expressed by Equation 6.31. Then the expected broadband beam power, $B(\theta)$, is given by

$$B(\theta_s) = E\{|b(\theta_s, t_i)|\} = \bar{h}^* E\{\bar{x}^*(t_i, \tau_n(\theta))\bar{x}(t_i, \tau_m(\theta))\}\bar{h} \tag{6.70}$$

where the vector $\bar{h}$ includes the weights for spatial shading.[36]

The term

$$\Phi(t_i, \theta_s) = E\{\bar{x}(t_i, \tau_n(\theta_s))\bar{x}^*(t_i, \tau_m(\theta_s))\} \tag{6.71}$$

is defined as the STCM in the time domain and is assumed to be independent of t_i in stationary conditions. The name STCM is derived from the fact that the matrix is computed by taking the covariance of the presteered time domain sensor outputs. Suppose $X_n(f_i)$ is the Fourier transform of the sensor outputs $x_n(t_i)$, and assume that the sensor outputs are approximately band limited. Under these conditions, the vector of steered (or time delayed) sensor outputs $x_n(t_i, \tau_n(\theta_s))$ can be expressed by

$$\bar{x}(t_i, \tau_n(\theta_s)) = \sum_{k=1}^{l+H} T_k(f_k, \theta_s)\bar{X}(f_k)\exp(j2\pi f_k t_i) \tag{6.72}$$

where $T(f_k, \theta)$ is the diagonal steering matrix in Equation 6.73 below, with elements identical to the elements of the conventional steering vector $\bar{D}(f_i, \theta)$:

$$T(f_k, \theta) = \begin{bmatrix} 1 & & \cdots & 0 \\ 0 & d_1(f_k, \theta) & & \\ \vdots & & \ddots & \vdots \\ 0 & & \cdots & d_N(f_k, \theta) \end{bmatrix} \tag{6.73}$$

Then it follows directly from Equations 6.72 and 6.73 that

$$\Phi(\Delta f, \theta_s) = \sum_{k=1}^{l+H} T(f_k, \theta_s)R(f_k)T^*(\theta_s) \tag{6.74}$$

where the index $k = l, l+1, \ldots, l+H$ refers to the frequency bins in a band of interest Δf, and $R(f_k)$ is the CSDM for the frequency bin f_k. This suggests that $\Phi(\Delta f, \theta_s)$ in Equation 6.71 can be estimated from the CSDM, $R(f_k)$ and $T(f_k, \theta)$ expressed by Equation 6.73. In the STMV method, the broadband spatial power spectral estimate $B(\theta_s)$ is given by[38]

$$B(\theta_s) = \left[\bar{I}^* \Phi(\Delta f, \theta_s)^{-1}\bar{I}\right]^{-1}. \tag{6.75}$$

The STMV algorithm differs from the basic MVDR algorithm in that the STMV algorithm yields an STCM that is composed from a band of frequencies and the MVDR algorithm uses a CSDM that is derived from a single frequency bin. Thus, the additional degrees of freedom of STMV compared to those of CSDM provide a more robust adaptive process.

However, estimates of $B(\theta)$ according to Equation 6.75 do not provide coherent beam time series, since they represent the broadband beam power output of an adaptive process. In this investigation,[1] we have modified the estimation process of the STMV matrix in order to get the complex coefficients of $\Phi(\Delta f, \theta_s)$ for all the frequency bins in the band of interest.

The STMV algorithm may be used in its original form to generate an estimate of $\Phi(\Delta f,\theta)$ for all the frequency bands Δf across the band of the received signal. Assuming stationarity across the frequency bins of a band Δf, then the estimate of the STMV may be considered to be approximately the same with the narrowband estimate $\Phi(f_o, \theta)$ for the center frequency f_o of the band Δf. In this case, the narrowband adaptive coefficients may be derived from

$$\bar{w}(f_o, \theta) = \frac{\Phi(f_o, \Delta f, \theta)^{-1}\bar{D}(f_o, \theta)}{\bar{D}^*(f_o, \theta)\Phi(f_o, \Delta f, \theta)^{-1}\bar{D}(f_o, \theta)}, \tag{6.76}$$

The phase variations of $\bar{w}(f_o, \theta)$ across the frequency bins $i = l, l+1, \ldots, l+H$ (where H is the number of bins in the band Δf) are modeled by

$$w_n(f_i, \theta) = \exp[2\pi f_i \Psi(\Delta f, \theta)], \quad i = l, l+1, \ldots, l+H \tag{6.77}$$

where $\Psi_n(\Delta f, \theta)$ is a time-delay term derived from

$$\Psi_n(\Delta f, \theta) = F[w_n(\Delta f, \theta), 2\pi f_o]. \tag{6.78}$$

Then, by using the adaptive steering weights $w_n(\Delta f, \theta)$ that are provided by Equation 6.77, the adaptive beams are formed as shown by Equation 6.65. Figure 6.14 shows the realization of the STMV beamformer and provides a schematic representation of the basic processing steps that include

1. Time series segmentation, overlap, and FFT shown by the group of blocks at the top left-hand side of the schematic diagram in Figure 6.14.
2. Formation of STCM Equations 6.71 and 6.74 shown by the two blocks at the bottom left-hand side of Figure 6.14.
3. Inversion of covariance matrix using Cholesky factorization, estimation of adaptive steering vectors, and formation of adaptive beams in the frequency domain presented by the middle and bottom blocks at the right-hand side of Figure 6.14.
4. Formation of adaptive beams in the time domain through IFFT, discardation of overlap, and concatenation of segments to form continuous beam time series, which is shown by the top right-hand-side block of Figure 6.14.

The various indexes in Figure 6.14 provide details for the implementation of the STMV processing flow in a generic computing architecture. The same figure indicates that estimates of the STCM are based on an exponentially weighted time average of the current and previous STCM, which is discussed in the next section.

6.6 Implementation Considerations

The conventional and adaptive steering vectors for steering angles θ_s, ϕ_s discussed in Section 6.4 are integrated in a frequency domain beamforming scheme, which is expressed by Equations 6.25, 6.28, 6.34, and 6.65. The beam time series are formed by Equation 6.32. Thus, the frequency domain adaptive and conventional outputs are made equivalent to the FFT of the time domain beamforming outputs with proper selection of beamforming weights and careful data partitioning. This equivalence corresponds to implementing FIR filters via circular convolution.[40–42]

Matrix inversion is another major implementation issue for the adaptive schemes discussed in this chapter. Standard numerical methods for solving systems of linear equations can be applied to solve for the adaptive weights. The range of possible algorithms includes

- Cholesky factorization of the covariance matrix $R(f_i)$[17,29] allows the linear system to be solved by backsubstitution in terms of the received data vector. Note that there is no requirement

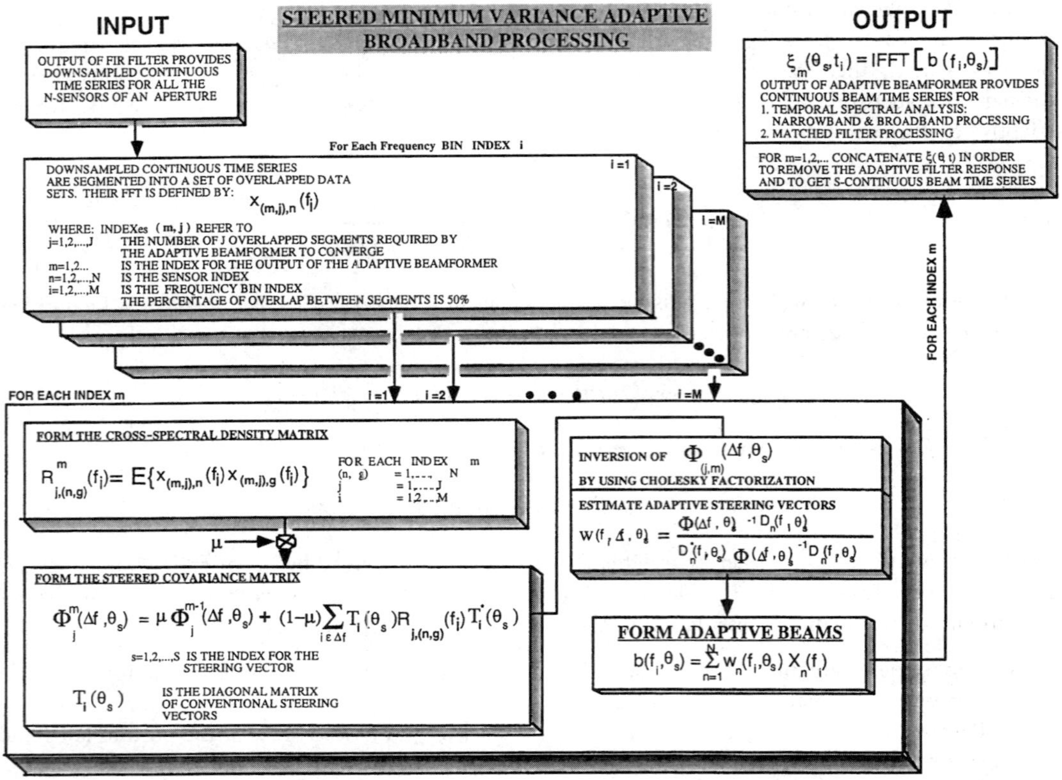

FIGURE 6.14 Realization of the steered covariance adaptive beamformer. The basic processing steps include (1) time series segmentation, overlap, and FFT, shown by the group of blocks at the top left-hand side of the schematic diagram; (2) formation of STCM, shown by the two blocks at the bottom left-hand side; (3) inversion of covariance matrix using Cholesky factorization, estimation of adaptive steering vectors, and formation of adaptive beams in the frequency domain (middle and bottom blocks at the right-hand side); and finally (4) formation of adaptive beams in the time domain through IFFT, discardation of overlap and concatenation of segments to form continuous beam time series (top right-hand-side block). The various indexes provide details for the implementation of the STMV processing flow in a generic computing architecture. (Reprinted by permission of IEEE ©1998.)

to estimate the sample covariance matrix and there is a continuous updating of an existing Cholesky factorization.

- QR decomposition of the received vector $\overline{X}(f_i)$ includes the conversion of a matrix to upper triangular form via rotations. The QR decomposition method has better stability than the Cholesky factorization algorithm, but it requires twice as much computational efforts than the Cholesky approach.
- The SVD (singular value decomposition) method is the most stable factorization technique. It requires, however, three times more computational requirements than the QR decomposition method.

In this implementation study, we have applied the Cholesky factorization and the QR decomposition techniques in order to get solutions for the adaptive weights. Our experience suggests that there are no noticable differences in performance between the above two methods.[1]

The main consideration, however, for implementing adaptive schemes in real-time systems is associated with the requirements derived from Equations 6.64 and 6.65, which require knowledge of second-order statistics for the noise field. Although these statistics are usually not known, they can be estimated from the received data[17,18,23] by averaging a large number of independent samples of the covariance matrixes

$R(f_i)$ or by allowing the iteration process of the adaptive GSC schemes to converge.[1,37] Thus, if K is the effective number of statistically independent samples of $R(f_i)$, then the variance on the adaptive beam output power estimator detection statistic is inversely proportional to $(K - N + 1)$,[17,18,22] where N is the number of sensors. Theoretical suggestions[23] and our empirical observations suggest that K needs to be three to four times the size of N in order to get coherent beam time series at the output of the above adaptive schemes. In other words, for arrays with a large number of sensors, the implementation of adaptive schemes as statistically optimum beamformers would require the averaging of a very large number of independent samples of $R(f_i)$ in order to derive an unbiased estimate of the adaptive weights.[23] In practice, this is the most serious problem associated with the implementation of adaptive beamformers in real-time systems.

Owsley[17,29] has addressed this problem with two important contributions. His first contribution is associated with the estimation procedure of $R(f_i)$. His argument is that, in practice, the covariance matrix cannot be estimated exactly by time averaging because the received signal vector $\bar{X}(f_i)$ is never truly stationary and/or ergodic. As a result, the available averaging time is limited. Accordingly, one approach to the time-varying adaptive estimation of $R(f_i)$ at time t_k is to compute the exponentially time averaged estimator (geometric forgetting algorithm) at time t_k:

$$R^{t_k}(f_i) = \mu R^{t_{k-1}}(f_i) + (1 - \mu)\bar{X}(f_i)\bar{X}^*(f_i) \tag{6.79}$$

where μ is a smoothing factor $(0 < \mu < 1)$ that implements the exponentially weighted time averaging operation. The same principle has also been applied in the GSC scheme.[1,37] Use of this kind of exponential window to update the covariance matrix is a very important factor in the implementation of adaptive algorithms in real-time systems.

Owsley's[29] second contribution deals with the dynamics of the data statistics during the convergence period of the adaptation process. As mentioned above, the implementation of an adaptive beamformer with a large number of adaptive weights in a large array sonar system requires very long convergence periods that will eliminate the dynamical characteristics of the adaptive beamformer to detect the time-varying characteristics of a received signal of interest. A natural way to avoid this kind of temporal stationarity limitation is to reduce the number of adaptive weights requirements. Owsley's[29] sub-aperture configuration for line array adaptive beamforming reduces significantly the number of degrees of freedom of an adaptation process. His concept has been applied to line arrays, as discussed in References 1 and 37. However, extension of the sub-aperture line array concept for multi-dimensional arrays is not a trivial task. In the following sections, the sup-aperture concept is generalized for circular, cylindrical, planar, and spherical arrays.

6.6.1 Evaluation of Convergence Properties of Adaptive Schemes

To test the convergence properties of the various adaptive beamformers of this study, synthetic data were used that included one CW signal. The frequency of the monochromatic signal was selected to be 330 Hz, and the angle of arrival was at 68.9° to directly coincide with the steering direction of a beam. The SNR of the received synthetic signal was very high, 10 dB at the sensor. By definition, the adaptive beamformers allow signals in the look direction to pass undistorted, while minimizing the total output power of the beamformer. Therefore, in the ideal case, the main beam output of the adaptive beamformer should resemble the main beam output of the conventional beamformer, while the side beam outputs will be minimized to the noise level. To evaluate the convergence of the beamformers, two measurements were made. From Equation 6.68, the mean square error (MSE) between the normalized main beam outputs of the adaptive beamformer and the conventional beamformer was measured, and the mean of the normalized output level of the side beam, which is the MSE when compared with zero, was measured. The averaging of the errors was done with a sliding window of four snapshots to provide a time-varying average, and the outputs were normalized so that the maximum output of the conventional beamformer was unity.

6.6.1.1 Convergence Characteristics of GSC and GSC-SA Beamformers

The GSC/NLMS adaptive algorithm, which has been discussed in Section 6.5.2, and its sub-aperture configuration denoted by GSC-SA/NLMS were compared against each other to determine if the use of the sub-aperture configuration produced any improvement in the time required for convergence. Figure 6.15a shows the comparison of the MSE of the main beams of both algorithms for the same step size μ, which is defined in Equation 6.68. The graphs show that the convergence rates of the main beams are approximately the same for both algorithms, reaching a steady-state value of MSE within a few snapshots. The value of MSE that is achieved is dictated by the missadjustment, which depends on μ. The higher MSE produced by the GSC-SA algorithm indicates that the algorithm exhibits a higher misadjustment.

Figure 6.15b shows the output level of an immediate side beam, again for the same step size μ. The side beam was selected as the beam right next to the main beam. The GSC-SA algorithm appears

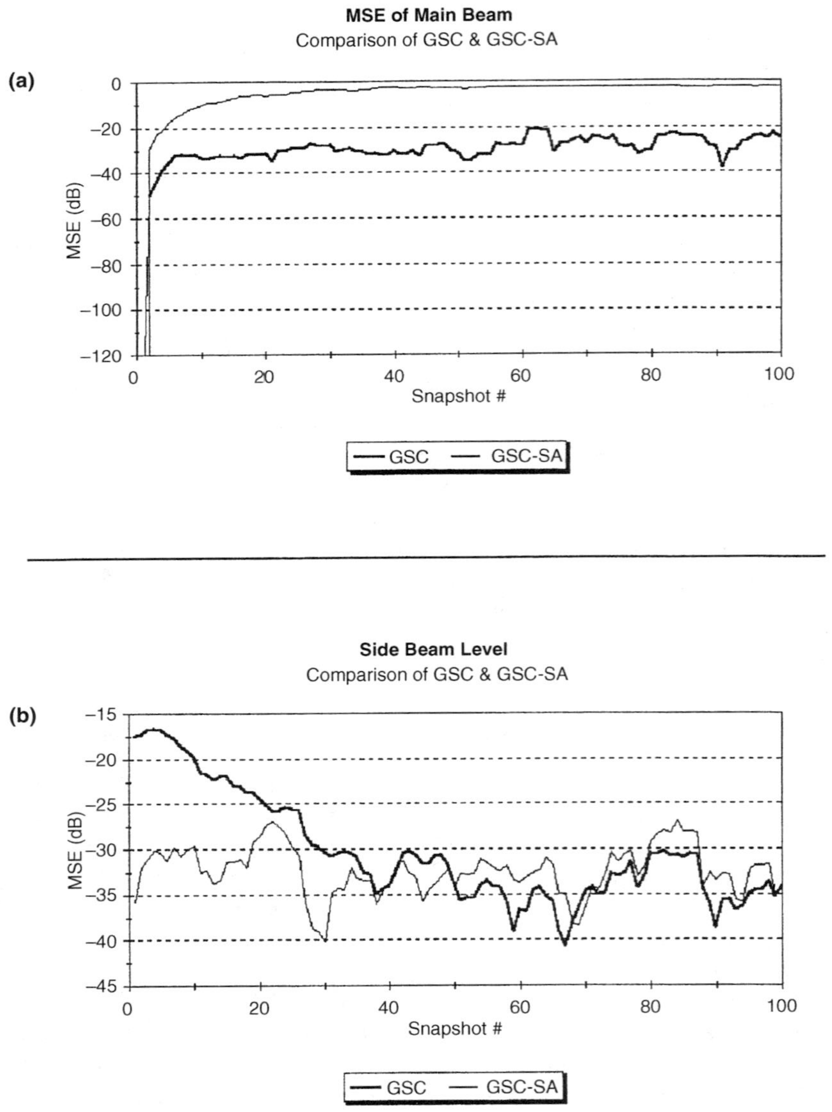

FIGURE 6.15 (a) MSE of the main beams of the GSC/NLMS and the GSC-SA/NLMS algorithms. (b) Sidebeam levels of the GSC/NLMS and GSC-SA/NLMS algorithms. (Reprinted by permission of IEEE ©1998.)

superior at minimizing the output of the side beam. It reaches its convergence level almost immediately, while the GSC algorithm requires approximately 30 snapshots to reach the same level. This indicates that the GSC-SA algorithm should be superior at canceling time-varying interferers. By selecting a higher value for μ, the time required for convergence will be reduced, but the MSE of the main beam will be higher.

6.6.1.2 Convergence Characteristics of STMV and STMV-SA Beamformers

As with the GSC/NLMS and GSC-SA/NLMS beamformers, the STMV and the STMV sub-aperture (STMV-SA) beamformers were compared against each other to determine if there was any improvement in the time required for convergence when using the sub-aperture configuration. Figure 6.16 shows the

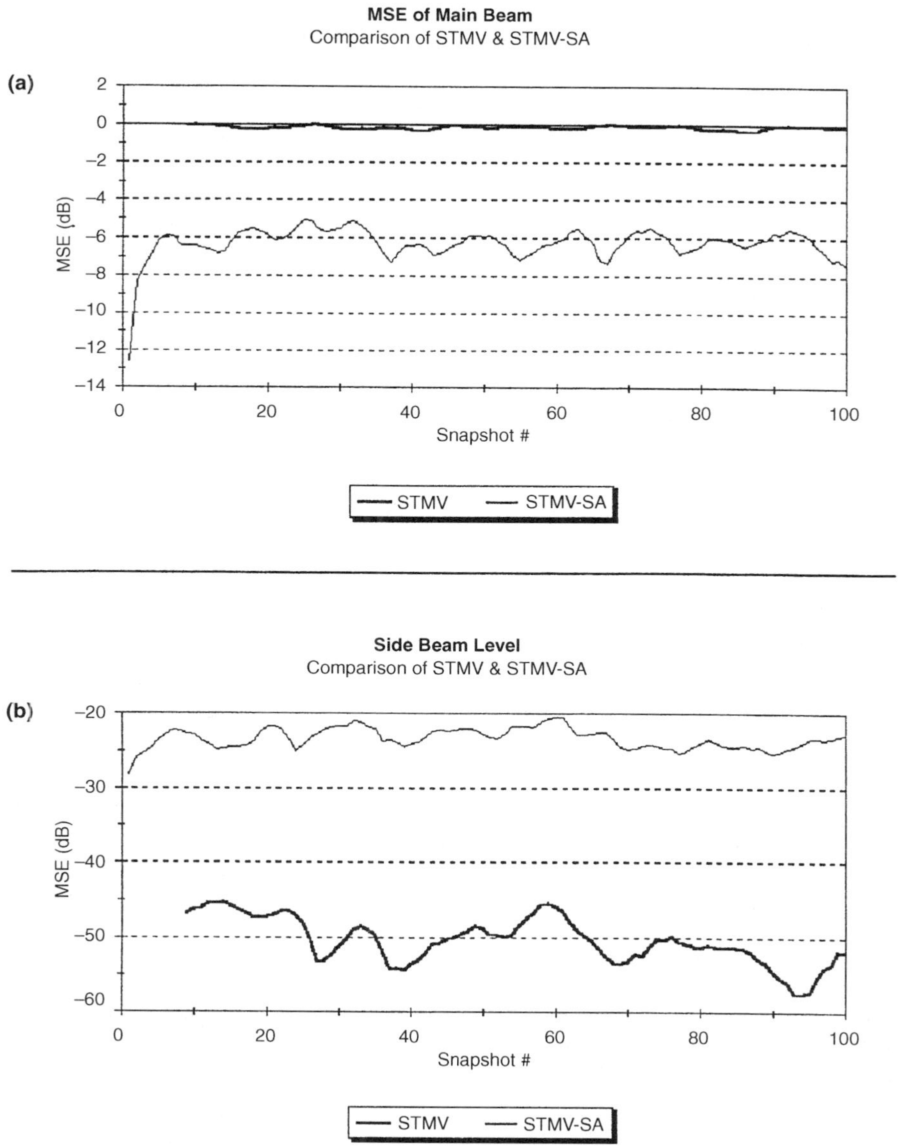

FIGURE 6.16 (a) MSE of the main beams of the STMV and the STMV-SA algorithms. (b) Sidebeam levels of the STMV and STMV-SA algorithms. (Reprinted by permission of IEEE ©1998.)

comparison of the MSE of the main beams of both algorithms. The graph shows that the STMV-SA algorithm reaches a steady state value of MSE within the first few snapshots.

The STMV algorithm is incapable of producing any output for at least eight snapshots as tested. Before this time, the matrices that are used to compute the adaptive steering vectors are not invertible. After this initial period, the algorithm has already reached a steady state value of MSE. Unlike the case of the GSC algorithm, the misadjustment from sub-aperture processing is smaller. Figure 6.16b shows the output level of the side beam for both the STMV and the STMV-SA beamformers. Again, the side beam was selected as the beam right next to the main beam. As before, there is an initial period during which the STMV algorithm is computing an estimate of the STCM and is incapable of producing any output; after that period the algorithm has reached steady state and produces lower side beams than the sub-aperture algorithm.

6.6.1.3 Signal Cancellation Effects of the Adaptive Algorithms

Testing of the adaptive algorithms of this study for signal cancellation effects was carried out with simulations that included two signals arriving from 64° and 69°.[37] All of the parameters of the signals were set to the same values for all the beamformers — conventional, GSC/NLMS, GSC-SA/NLMS, STMV, and STMV-SA. Details about the simulated signal cancellation effects can be found in Reference 37. In the narrowband outputs of the conventional beamformer, the signals appear at the frequency and beam at which they were expected. As anticipated, however, the sidelobes are visible in a number of other beams. The gram outputs of the GSC/STMV algorithm indicated that there is signal cancellation. In each case, the algorithm failed to detect either of the two CWs. This suggests that there is a shortcoming in the GSC/NLMS algorithm when there is strong correlation between two signal arrivals received by the line array. The narrowband outputs of the GSC-SA/NLMS algorithm showed that in this case the signal cancellation effects have been minimized and the two signals were detected only at the expected two beams with complete cancellation of the sidelobe structure. For the STMV beamformer, the grams indicated a strong sidelobe structure in many other beams. However, the STMV-SA beamformer successfully suppresses the sidelobe structure that was present in the case of the STMV beamformer. From all these simulations,[37] it is obvious that the STMV-SA beamformer, as a broadband beamformer, is not as robust for narrowband applications as the GSC-SA/NLMS.

6.6.2 Generic, Multi-Dimensional Sub-Aperture Structure for Adaptive Schemes

The decomposition of the 2-D and 3-D beamformer into sets of line and/or circular array beamformers, which was discussed in Section 6.4.2, provides a first-stage reduction of the numbers of degrees of freedom for an adaptation process. Furthermore, the sub-aperture configuration is considered in this study as a second stage reduction of the number of degrees of freedom for an adaptive beamformer. Then, the implementation effort for adaptive schemes in multi-dimensional arrays is reduced to implementing adaptive processes in line and circular arrays. Thus, a multi-dimensional adaptive beamformer can now be divided into two coherent modular steps which lead to efficient system-oriented implementations.

6.6.2.1 Sub-Aperture Configuration for Line Arrays

For a line array, a sub-aperture configuration includes a large percentage overlap between contiguous sub-apertures. More specifically, a line array is divided into a number of sub-arrays that overlap, as shown in Figure 6.17. These sub-arrays are beamformed using the conventional approach, and *this is the first stage of beamforming*. Then, we form a number of sets of beams with each set consisting of beams that are steered at the same direction, but with each one of them generated by a different sub-array. A set of beams of this kind is equivalent to a line array that consists of directional sensors steered at the same direction, with sensor spacing equal to the space separation between two contiguous sub-arrays and with the number of sensors equal to the number of sub-arrays. *The second stage of beamforming* implements an adaptive scheme on the above kind of set of beams, as illustrated in Figure 6.17.

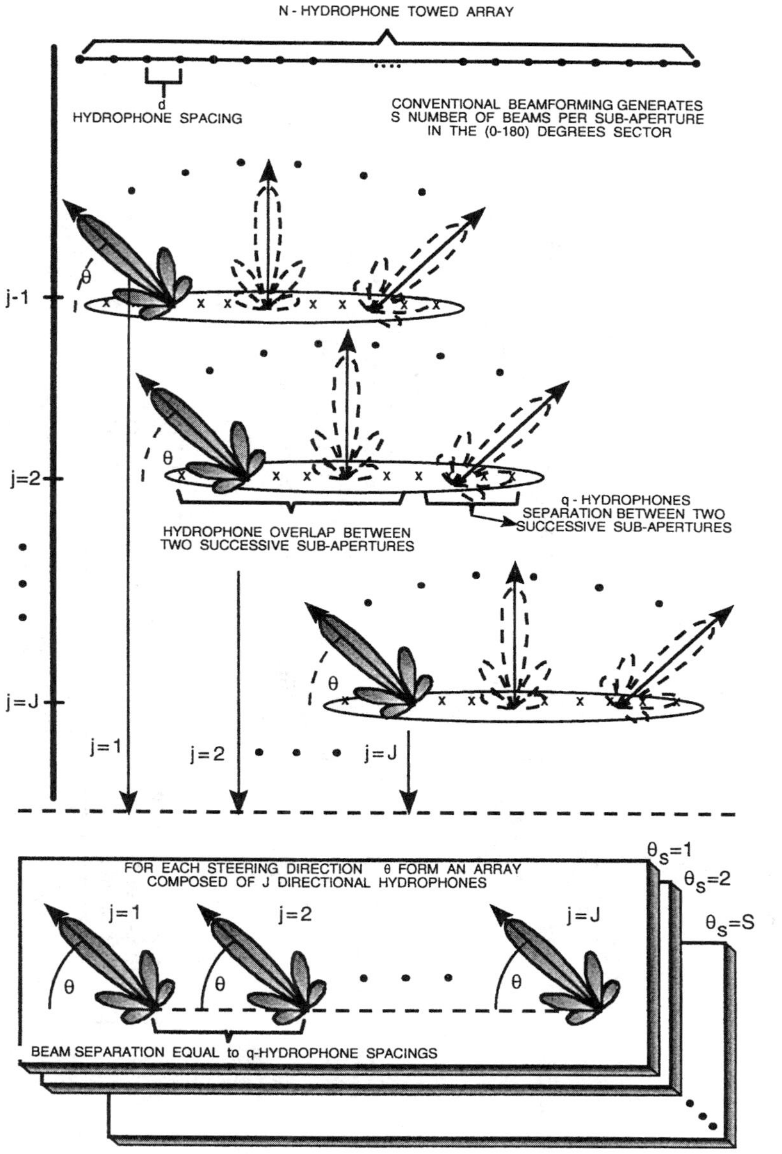

FIGURE 6.17 Concept of adaptive sub-aperture structure for line arrays. Schematic diagram shows the basic steps that include (1) formation of J sub-apertures; (2) for each sub-aperture, formation of S conventional beams; and (3) for a given beam direction θ, formation of line sensor arrays that consist of J number of directional sensors (beams). The number of line arrays with directional sensors (beams) are equal to the number S of steered conventional beams in each sub-aperture. For each line array, the directional sensor time series (beams) are provided at the input of an adaptive beamformer. (Reprinted by permission of IEEE ©1998.)

6.6.2.2 Sub-Aperture Configuration for Circular Array

Consider a circular array with M sensors as shown in Figure 6.18. The first circular sub-aperture consists of the first $M - G + 1$ sensors with $n = 1, 2, \ldots, M - G + 1$, where n is the sensor index and G is the number of sub-apertures. The second circular sub-aperture array consists of $M - G + 1$ sensors with $n = 2, 3, \ldots, M - G + 2$. The sub-aperture formation goes on until the last sub-aperture consists of $M - G + 1$ sensors with $n = G, G + 1, \ldots, M$. In the first stage, each circular sub-aperture is beamformed as discussed in Section 6.4.1.2, and *this first stage of beamforming* generates G sets of beams.

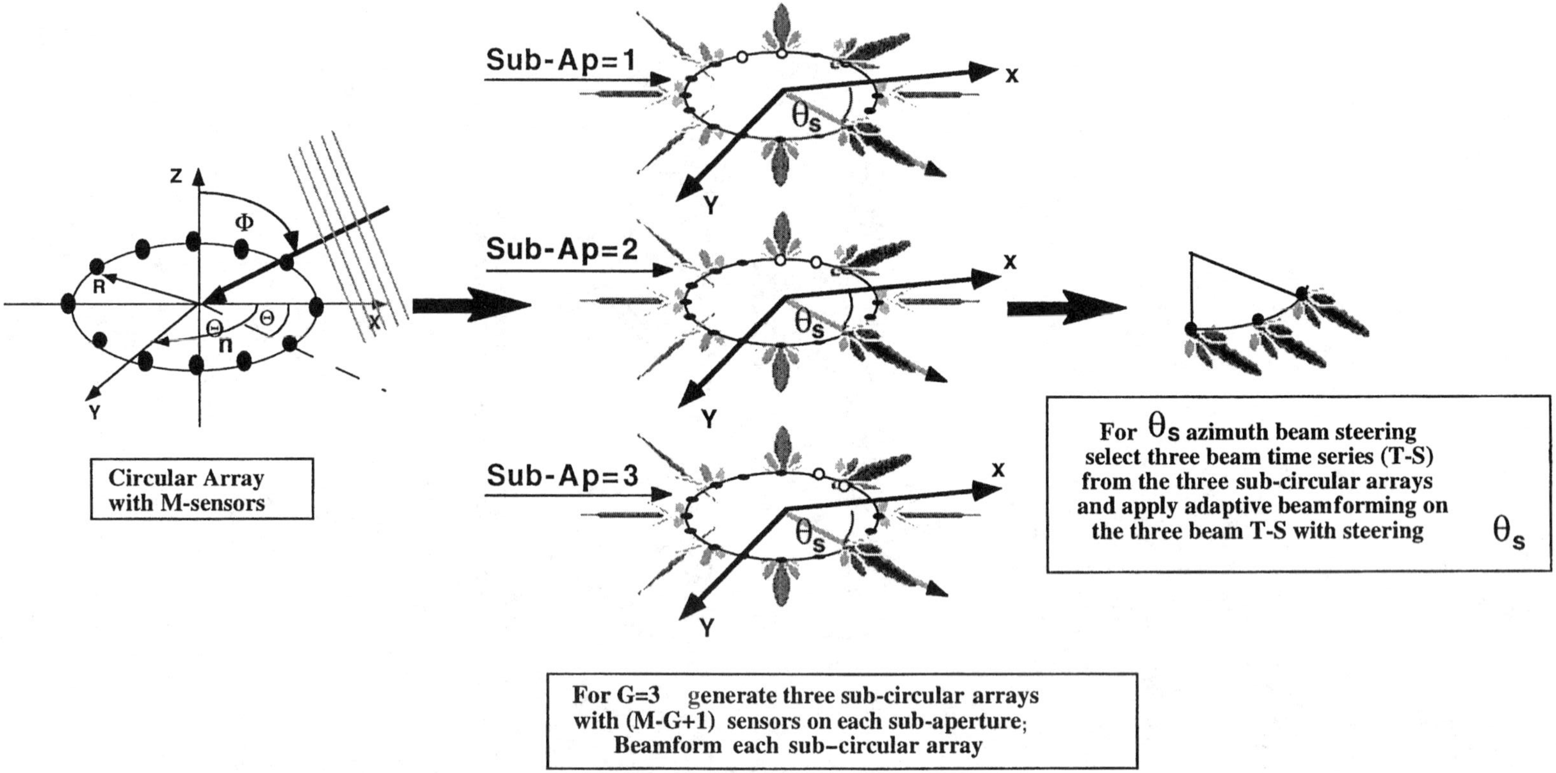

FIGURE 6.18 Concept of adaptive sub-aperture structure for circular arrays, which is similar to that for line arrays shown in Figure 6.17.

As in the previous section, we form a number of sets of beams with each set consisting of beams that are steered at the same direction, but with each one of them generated by a different sub-array. For $G < 5$, a set of beams of this kind can be treated approximately as a line array that consists of directional sensors steered at the same direction, with sensor spacing equal to the space separation between two contiguous sub-arrays and with the number of sensors equal to the number of sub-arrays. *The second stage of beamforming* implements an adaptive scheme on the above kind of set of beams, as illustrated in Figure 6.18, for $G = 3$.

6.6.2.3 Sub-Aperture Configuration for Cylindrical Array

Consider the cylindrical array shown in Figures 6.6 and 6.19 with the number of sensors $\aleph = NM$, where N is the number of circular rings and M is the number of sensors on each ring. Let n be the ring index, m be the sensor index for each ring, and G be the number of sub-apertures. The formation of sub-apertures is as follows:

The *first sub-aperture* consists of the first $(N - G + 1)$ rings, where $n = 1, 2, \ldots, N - G + 1$. In each ring we select the first set of $(M - G + 1)$ sensors, where $m = 1, 2, \ldots, M - G + 1$. However, each ring has M sensors, but only $(M - G + 1)$ sensors are used to form the sub-aperture. These sensors form a cylindrical array cell, as shown in the upper right-hand corner of Figure 6.19.

In other words, the sub-aperture includes the sensors of the full cylindrical array except for $G - 1$ sensors from $G - 1$ rings, which are denoted by small circles in Figure 6.19, which have been excluded in order to form the sub-aperture. Next, the generic decomposition concept of the conventional cylindrical array beamformer, presented in Section 6.4.2.1, is applied to the above sub-aperture cylindrical array cell. For a given pair of azimuth and elevation steering angles $\{\theta_s, \phi_s\}$, the output of the generic, conventional, multi-dimensional sub-aperture beamformer provides the beam time series $b_{g=1}(t_i, \theta_s, \phi_s)$, where the subscript $g = 1$ is the sub-aperture index.

The *second sub-aperture* consists of the next set of $(N - G + 1)$ rings, where $n = 2, \ldots, N - G + 2$. In each ring we select the next set of $(M - G + 1)$ sensors, where $m = 2, \ldots, M - G + 2$. However, each ring has M sensors, but only $(M - G + 1)$ sensors are used to form the sub-aperture. These sensors form the second sub-aperture cylindrical array cell.

Again, the generic decomposition concept of the conventional cylindrical array beamformer, presented in Section 6.4.2.1, is applied to the above sub-aperture cylindrical array cell. For a given pair of azimuth and elevation steering angles $\{\theta_s, \phi_s\}$, the output of the generic conventional multi-dimensional sub-aperture beamformer provides the beam time series $b_{g=2}(t_i, \theta_s, \phi_s)$ with the sub-aperture index $g = 2$.

This kind of sub-aperture formation continues until the *last sub-aperture* which consists of a set of $(N - G + 1)$ rings, where $n = G, G + 1, \ldots, N$. In each ring we select the last set of $(M - G + 1)$ sensors, where $m = G, G + 1, \ldots, M$. Please note also that each ring has M sensors, but only $(M - G + 1)$ sensors are used to form the sub-aperture.

As before, the generic decomposition concept of the conventional cylindrical array beamformer is applied to the last sub-aperture cylindrical array cell. For a given pair of azimuth and elevation steering angles $\{\theta_s, \phi_s\}$, the output of the generic, conventional, multi-dimensional sub-aperture beamformer would provide beam time series $b_{g=G}(t_i, \theta_s, \phi_s)$ with the sub-aperture index $g = G$.

As in the Section 6.6.2.2, we form a number of sets of beams with each set consisting of beams that are steered at the same direction, but with each one of them generated by a different sub-aperture cylindrical array cell. For $G < 5$, a set of beams of this kind can be treated approximately as a line array that consists of directional sensors steered at the same direction, with sensor spacing equal to the space separation between two contiguous sub-aperture cylindrical array cells and with the number of sensors equal to the number of sub-arrays. Then *the second stage of beamforming* implements an adaptive scheme on the above kind of set of beams, as illustrated in Figure 6.19.

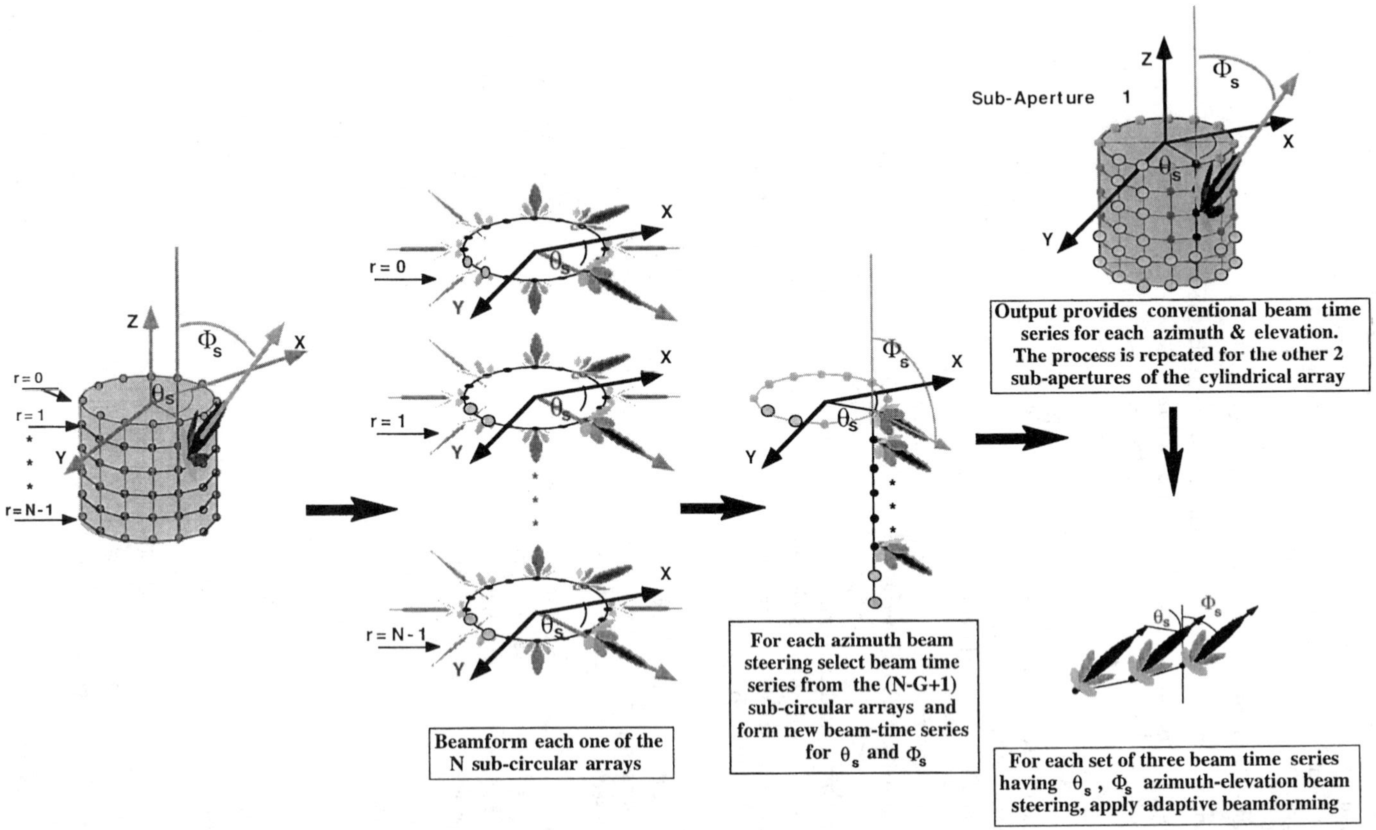

FIGURE 6.19 Coordinate system and geometric representation of the concept of adaptive sub-aperture structure for cylindrical arrays. In this particular example, the number of sub-apertures was $G = 3$. The $\aleph = NM$ sensor cylindrical array beamformer consists of N circular arrays with M being the number of sensors in each circular array. Then, the sub-aperture adaptive structure for cylindrical arrays is reduced to the basic steps of adaptive sub-aperture structures for circular and line arrays as defined in the schematic diagrams of Figure 6.18 and 6.17, respectively. Thus, for a given azimuth θ_s and elevation ϕ_s beam steering and $G = 3$, these steps include (1) formation of a sub-aperture per circular array with $M - G + 1$ sensors, (2) for each sub-aperture formation of S conventional beams, and (3) formation of $N - G + 1$ vertical line sensor arrays that consist of directional sensors (beams). This arrangement defines a circular sub-aperture. The process is repeated to generate two additional sub-aperture circular arrays. The beam output response of the $G = 3$ sub-aperture circular arrays is provided at the input of a line array adaptive beamformer with $G = 3$ the number of directional sensors.

For the particular case shown in Figure 6.19, *the second stage of beamforming* implements an adaptive beamformer on a line array that consists of the $G = 3$ beam time series $b_g(t_i, \theta_s, \phi_s)$, $g = 1, 2, \ldots, G$. Thus, for a given pair of azimuth and elevation steering angles $\{\theta_s, \phi_s\}$ the cylindrical adaptive beamforming process is reduced to an adaptive line array beamformer that includes as input only three beam time series $b_g(t_i, \theta_s, \phi_s)$, $g = 1, 2, 3$ with spacing $\delta = [(R2\pi/M)^2 + \delta_z^2]^{1/2}$, which is the spacing between two contiguous sub-aperture cylindrical cells, where $(R2\pi/M)$ is the sensor spacing in each ring and δ_z is the distance between each ring along z-axis of the cylindrical array. The output of the adaptive beamformer provides one or more adaptive beam time series with steering centered on the pair of azimuth and elevation steering angles $\{\theta_s, \phi_s\}$.

As expected, the adaptation process in this case will have near-instantaneous convergence because of the very small number of degrees of freedom. Furthermore, because of the generic characteristics, the proposed 3-D sub-aperture adaptive beamforming concept may include a wide variety of adaptive techniques such as MVDR, GSC, and STMV that have been discussed in References 1 and 37.

6.6.2.4 Sub-Aperture Configuration for Planar and Spherical Arrays

The sub-aperture adaptive beamforming concepts for planar and spherical arrays are very similar to that of the cylindrical array. In particular, for planar arrays, the formation of sub-apertures is based on the sub-aperture concept of line arrays that has been discussed in Section 6.6.2.1. The different steps of sub-aperture formation for planar arrays as well as the implementation of adaptive schemes on the G beam time series $b_g(t_i, \theta_s, \phi_s)$, $g = 1, 2, \ldots, G$, that are provided by the G sub-apertures of the planar array, are similar with those in Figure 6.19 by considering the composition process for planar arrays shown in Figure 6.7. Similarly, the sub-aperture adaptive concept for spherical arrays is based on the sub-aperture concept of circular arrays that has been discussed in Section 6.6.2.2.

6.6.3 Signal Processing Flow of a 3-D Generic Sub-Aperture Structure

As was stated before, the disussion in this chapter has been devoted to designing a generic sub-aperture beamforming structure that will decompose the computationally intensive multi-dimensional beamforming process into coherent subsets of line and/or circular sub-aperture array beamformers for ultrasound, radar, and integrated active-passive sonar systems. In a sense, the proposed generic processing structure is an extension of a previous effort discussed in Reference 1.

The previous study[1] included the design of a generic beamforming structure that allows the implementation of adaptive, synthetic aperture, and high-resolution temporal and spatial spectral analysis techniques in integrated active-passive line array sonars. Figure 6.20 shows the configuration of the signal processing flow of the previous generic structure that allows the implementation of FIR filters and conventional, adaptive, and synthetic aperture beamformers.[1,40,41,42]

Shown in Figure 6.21 is the proposed configuration of the signal processing flow that includes the implementation of line and circular array beamformers as FIR filters.[40–42] The processing flow is for 3-D cylindical arrays. The reconfiguration of the different processing blocks in Figures 6.20 and 6.21 allows the application of the proposed configuration to a variety of ultrasound, radar, and integrated active-passive sonar systems with planar, cylindrical, or spherical arrays of sensors.

As discussed at the beginning of this section, the output of the beamforming processing block in Figure 6.21 provides continuous beam time series. Then the beam time series are provided at the input of a vernier for passive narrowband/broadband analysis or a matched filter for active applications. This modular structure in the signal processing flow is a very essential processing arrangement in allowing the integration of a great variety of processing schemes such as the ones considered in this chapter. The details of the proposed generic processing flow, as shown in Figure 6.21, are very briefly the following:

- The first block in Figure 6.21 includes the partitioning of the time series from the receiving sensor array, the computation of their initial spectral FFT, the selection of the signal's frequency band of interest via bandpass FIR filters, and downsampling. The output of this block provides continuous time series at a reduced sampling rate.[41,42]

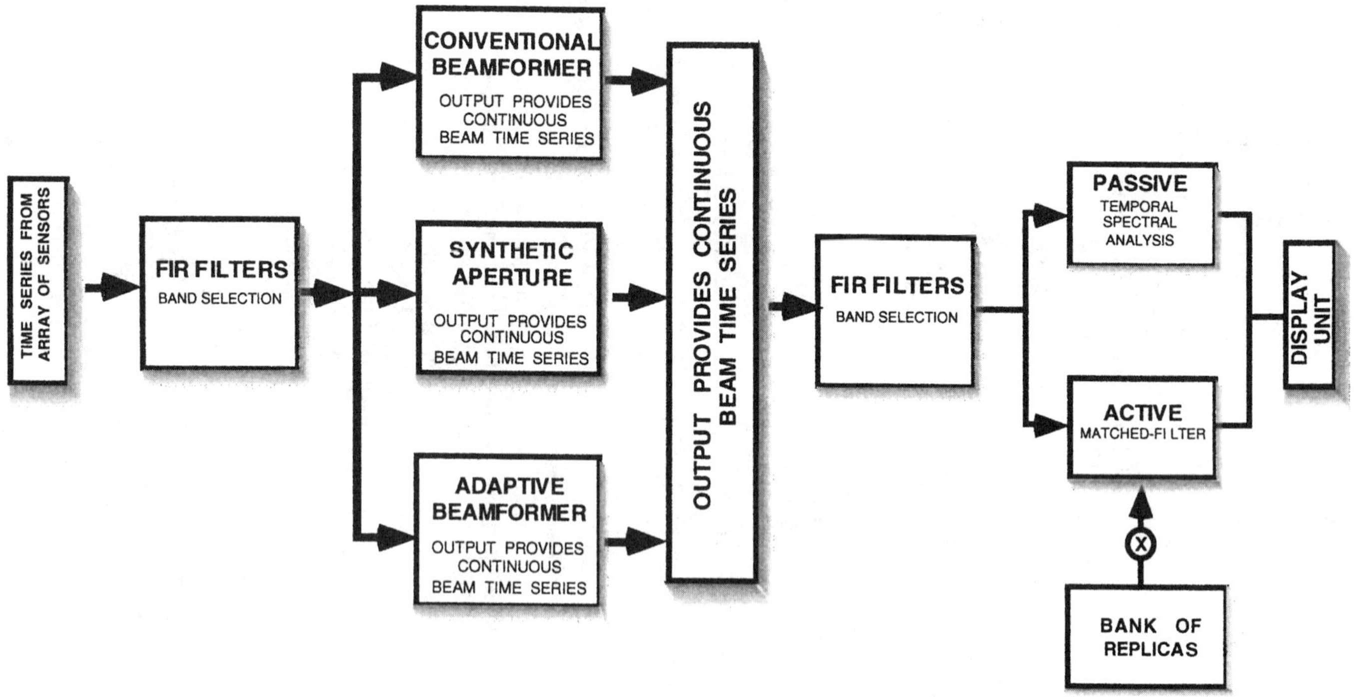

FIGURE 6.20 Schematic diagram of a generic signal processing flow that allows the implementation of conventional, adaptive, and synthetic aperture beamformers in line array sonar and ultrasound systems.

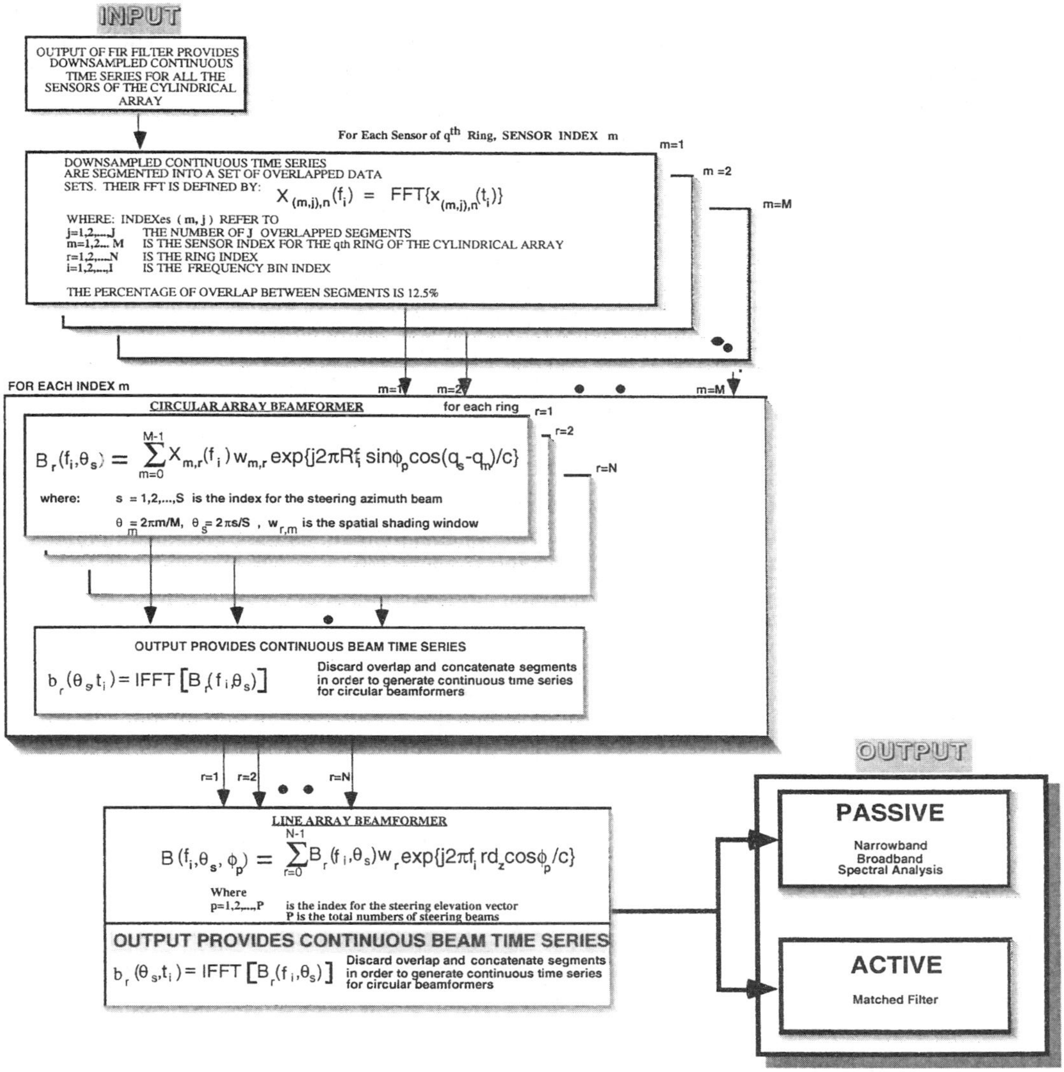

FIGURE 6.21 Signal processing flow of generic structure decomposing the 3-D beamformer for cylindrical arrays of sensors into coherent sub-sets of line and circular array beamformers.

- The second and third blocks titled *Circular Array Beamformer* and *Line Array Beamformer* provide continuous directional beam time series by using the FIR implementation scheme of the spatial filtering via circular convolution.[40] The segmentation and overlap of the time series at the input of each one of these beamformers takes care of the wraparound errors that arise in fast-convolution signal processing operations. The overlap size is equal to the effective FIR filter's length.[41,42]
- The *Active* block is for the processing of echoes for active sonar and radar applications.
- The *Passive* block includes the final processing steps of a temporal spectral analysis.

Finally, data normalization processing schemes are being used to map the output results into the dynamic range of the display devices in a manner which provides a CFAR capability.[34]

In the passive unit, the use of verniers and the temporal spectral analysis (incorporating segment overlap, windowing, and FFT coherent processing) provide the narrowband results for all the beam time series. Normalization and OR-ing are the final processing steps before displaying the output results. Since a beam time sequence can be treated as a signal from a directional sensor having the same AG and directivity pattern as that of the beamformer, the display of the narrowband spectral estimates for all the beams follows the so-called GRAM presentation arrangements, as shown in Figures 6.24 to 6.27. This includes the display of the beam-power outputs as a function of time, steering beam (or bearing), and frequency.[34]

Broadband outputs in the passive unit are derived from the narrowband spectral estimates of each beam by means of incoherent summation of all the frequency bins in a wideband of interest.[34] This kind of energy content of the broadband information is displayed as a function of bearing and time.[1,34,43]

In the active unit, the application of a matched filter (or replica correlator) on the beam time series provides coherent broadband processing. This allows detection of echoes as a function of range and bearing for reference waveforms transmitted by the active transducers of ultrasound, sonar, or radar systems. The displaying arrangements of the correlator's output data are similar to the GRAM displays and include, as parameters, range as a function of time and bearing.[1]

Next, presented in Figure 6.22, is the signal processing flow of the generic adaptive sub-aperture structure for multi-dimensional arrays. The first processing block includes the formation of sub-apertures as discussed in Section 6.6.2. Then the sensor time series from each sub-aperture are beamformed by the generic, multi-dimensional beamforming structure introduced in Section 6.4 and presented in Figure 6.21. Thus, for a given pair of azimuth and elevation steering angles $\{\theta_s, \phi_s\}$, the output of the generic, conventional, multi-dimensional beamformer would provide G beam time series, $b_g(t_i, \theta_s, \phi_s)$, $g = 1, 2, \ldots, G$. The second stage of beamforming includes the implementation of an adaptive beamformer as discussed in Section 6.5.2.

For the synthetic aperture processing scheme, however, there is an important detail regarding the segmentation and overlap of the sensor time series into sets of discontinuous segments. It is assumed here that the received sensor signals are stored as continuous time series. Therefore, the segmentation process of the sensor time series is associated with the tow speed and the size of the synthetic aperture, as discussed in Section 6.5.1.4. So, in order to achieve continuous data flow at the output of the overlap correlator, the N continuous time series are segmented into discontinuous data sets as shown in Figure 6.23. Our implementation scheme in Figure 6.23 considers five discontinuous segments in each data set. This arrangement will provide at the output of the overlap correlator $3N$ continuous sensor time series, which are provided at the input of the conventional beamformer as if they were the sensor time series of an equivalent physical array. Thus the basic processing steps include *time series segmentation, overlap, and grouping of five discontinuous segments, which are provided at the input of the overlap correlator*, as shown by the group of blocks at the top part of Figure 6.23. $T = M/f_s$ is the length in seconds of the discontinuous segmented time series, and M defines the size of FFT. The rest of the blocks provide the indexing details for the formation of the synthetic aperture. These indexes also provide details for the implementation of the segmentation process of the synthetic aperture flow in a generic computing architecture.

The processing arrangements and the indexes in Figure 6.24 provide the details needed for the mapping of this synthetic aperture processing scheme in sonar or ultrasound computing architectures. The basic processing steps include the following:

1. *Time series segmentation, overlap, and grouping of five discontinuous segments, which are provided at the input of the overlap correlator*, are shown by the block at the top part of schematic diagram. Details of this segmentation process are shown also in Figure 6.23.
2. *The main block called* ***ETAM: Overlap Correlator*** *provides processing details for the estimation of the phase correction factor to form the synthetic aperture.*
3. *Formation of the continuous sensor time series of the synthetic aperture is obtained through IFFT, discardation of overlap, and concatenation of segments to form continuous time series*, which is shown by the left-hand-side block.

It is important to note here that the choice of five discontinuous segments was based on experimental observations[10,30] regarding the temporal and spatial coherence properties of the underwater medium. These issues of coherence are very critical for synthetic aperture processing, and they have been addressed in Section 6.4.3.

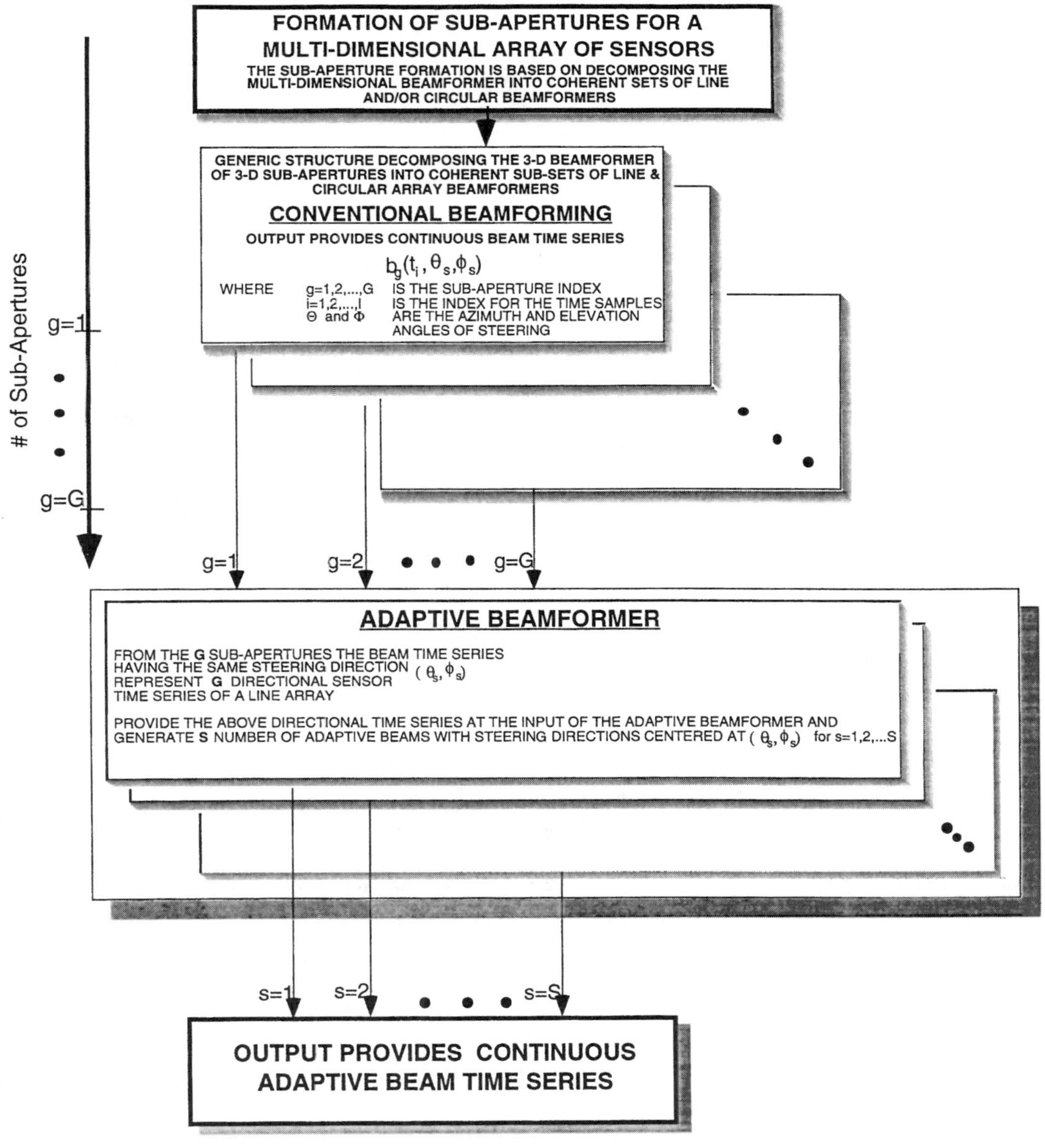

FIGURE 6.22 Signal processing flow of a generic adaptive sub-aperture structure for multi-dimensional arrays of sensors.

6.7 Concept Demonstration: Simulations and Experimental Results

Performance assessment and testing of the generic sub-aperture multi-dimensional adaptive beamforming structure has been carried out with synthetic and real data sets. The data sets include narrowband and hyperbolic frequency modulated (HFM) signals for passive and active applications, respectively. For sonar applications, the frequencies of the passive narrowband signals are taken to be 330 Hz and the active signal consists of HFM pulses with a pulse-width 8 s long, a 100-Hz bandwidth centered at 330 Hz, with a 120-s pulse repetition period, or pulses with a pulse-width 500 μs long, 10-kHz bandwidth

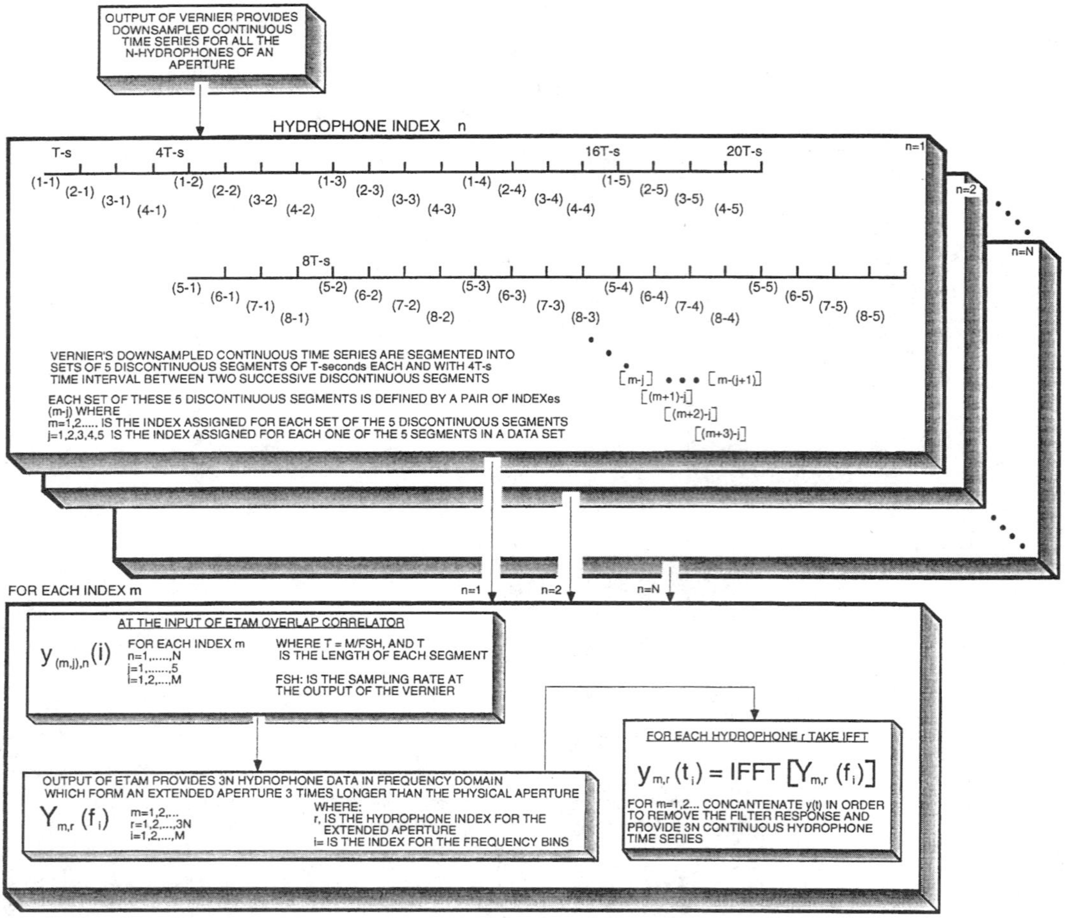

FIGURE 6.23 Schematic diagram of the data flow for the ETAM algorithm and the sensor time series segmentation into a set of five discontinuous segments for the overlap correlator. The basic processing steps include time series segmentation, overlap, and grouping of five discontinuous segments, which are provided at the input of the overlap correlator (shown by group of blocks at the top part of schematic diagram). $T = M/f_s$ is the length in seconds of the discontinuous segmented time series, and M defines the size of FFT. The rest of the blocks provide the indexing details for the formation of the synthetic aperture. These indexes provide details for the implementation of the segmentation process of the synthetic aperture flow in a generic computing architecture. The processing flow is shown in Figure 6.24. (Reprinted by permission of IEEE ©1998.)

centered at 200 kHz, with arbitrary pulse repetition period. For ultrasound applications, the signals consists of FM pulses with a 4-MHz bandwidth centered at 3 MHz. The scope here is to demonstrate that the implementation of adaptive schemes (i.e., GSC and STMV) in real-time systems is feasible. Moreover, it has been shown that the proposed generic configuration of adaptive schemes provides AG improvements when compared with the performance characteristics of the multi-dimensional conventional beamformer. As for active applications, it has been shown that the adaptive schemes of the proposed generic sub-aperture structure achieve near instantaneous convergence, which is essential for active ultrasound, sonar, and radar applications.

The generic adaptive sub-aperture processing structure and the associated signal processing algorithms were implemented in a computer workstation. The memory of the workstation was sufficient to allow processing of long, continuous sensor time series. However, the available memory restricts the number of sensors and the number of steered beams.

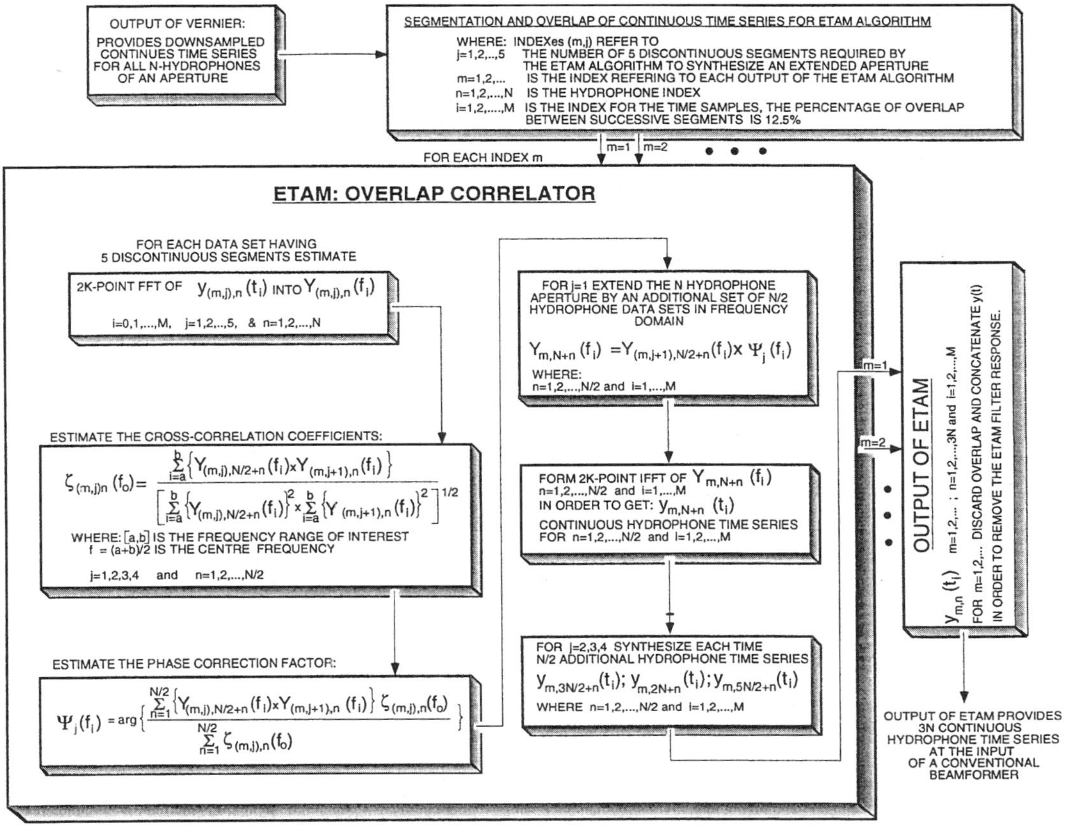

FIGURE 6.24 Schematic diagram for the processing arrangements of the ETAM algorithm. The basic processing steps include (1) time series segmentation, overlap, and grouping of five discontinuous segments, which are provided at the input of the overlap correlator, shown by the block at the top part of schematic diagram (details of the segmentation process are shown also by Figure 6.23); (2) the main block called **ETAM: Overlap Correlator**, which provides processing details for the estimation of the phase correction factor to form the synthetic aperture; and finally, (3) generation of the continuous sensor time series of the synthetic aperture, which are obtained through IFFT, discardation of overlap, and concatenation of segments to form continuous time series (left-hand-side block). The various indexes provide details for the implementation of the synthetic processing flow in a generic computing architecture. (Reprinted by permission of IEEE ©1998.)

Nevertheless, the results are sufficient to demonstrate system-oriented applications of the proposed generic sub-aperture adaptive structure for multi-dimensional arrays of sensors.

6.7.1 Sonars: Cylindrical Array Beamformer

6.7.1.1 Synthetic Data: Passive

A cylindrical array with 160 sensors (16 rings with 10 sensors on each ring) was considered where the distance between rings along z-axis is taken to be equal to the angular spacing between sensors of the rings (i.e., $\delta_z = 2\pi R/M = \delta = 2.09$ m). Continuous sensor time series were provided at the inputs of the generic conventional and adaptive beamformers with the processing flows as shown in Figures 6.21 and 6.22, respectively. The total number of steering beams for both the adaptive and conventional beamformers was 144. For the decomposition process of the generic beamformer, expressed by Equation 6.36, there were 16 beams steered in the angular sector of 0° to 360° for azimuth bearing and 9 beams formed in the angular sector of (0° to 180°) for elevation bearing. Thus, the generic beamformer provided 16 azimuth beams for each of the 9 elevation steering angles, giving a total of 144 beams.

In the upper part of Figure 6.25, the left-hand-side diagram shows the output power of the azimuth beams at the expected elevation bearing of the signal source for the generic, 3-D cylindrical array conventional beamformer; the right-hand side of Figure 6.25 shows the output power of the elevation beams at the expected azimuth angle of the signal source. In both cases, no spatial window has been applied. The results at the left-hand side of the lower part of Figure 6.25 correspond to the azimuth beams for the conventional beamformer with Hamming as a spatial window (dotted line) and the adaptive (solid line) sub-aperture beamformer. In this case, the number of sub-apertures was $G = 3$, with Hamming as a spatial window applied on the sub-aperture conventional circular beamformer.

It is apparent by these results that the spatial shading has significantly suppressed the sidelobe structure of the conventional beamformer and has widened the beam width, as expected.

Moreover, the adaptive beamforming results demonstrate a significant improvement in suppressing the sidelobe structure as compared with the conventional results. The right-hand side of the lower part

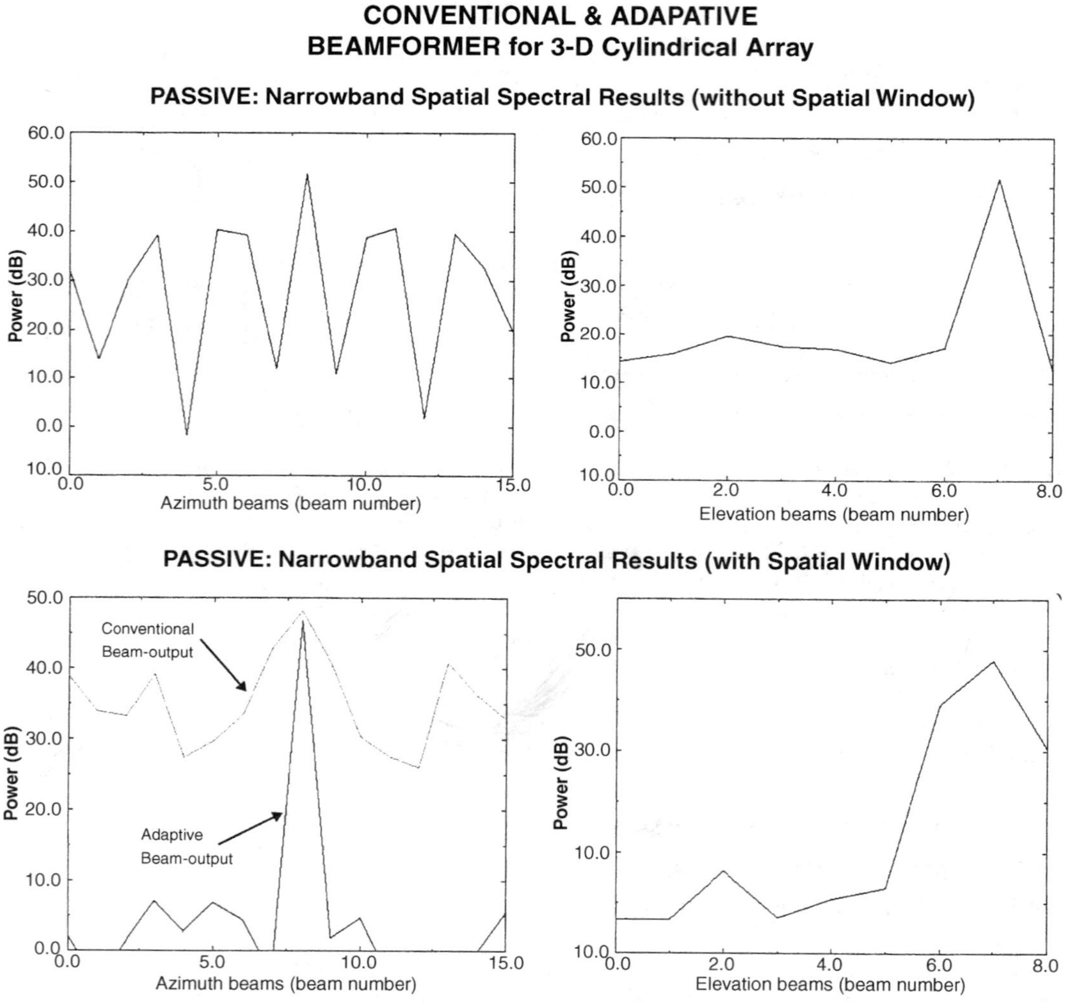

FIGURE 6.25 Passive beamforming results for a cylindrical array. The upper part shows azimuth and elevation bearing response for the proposed generic multi-dimensional beamformer. The lower left-hand side shows beamforming results of the conventional beamformer with a spatial window (dotted line) and adaptive (solid line) beamformers. The lower right-hand side shows elevation bearing response for the conventional cylindrical beamformer with a spatial window.

of Figure 6.25 includes elevation bearing response for the conventional beamformer with spatial shading. At this point, it is important to note that the application of spatial shading on the fully coherent 3-D cylindrical beamformer would have been a much more elaborate process than the one that has been used for the generic multi-dimensional beamformer. This is because the decomposition process for the latter allows two much simpler and separate applications of spatial shading (i.e., one for circular arrays and the other for line arrays), discussed analytically in Section 6.4.2 and in References 7 and 25 to 28.

Figure 6.26 shows the narrowband spectral estimates of the generic, 3-D cylindrical array conventional beamformer with Hamming spatial shading for all the azimuth beams according to the so-called GRAM presentation arrangement, discussed in Section 6.5 and in Reference 34. The GRAMs in Figure 6.26 represent the spectrograms of the output of the azimuth beams steered at the signal's expected elevation bearing. The GRAMs in Figure 6.27 show the corresponding results when the azimuth beams are steered at an elevation angle which is 55° off the expected elevation bearing of the signal. It is obvious from the results of Figure 6.27 that the AG of the conventional beamformer with spatial shading is not very high.

For the same sensor time series, when the adaptive sub-aperture schemes are implemented in the generic multi-dimensional beamformer, the corresponding results are shown in Figure 6.28. When the results of Figure 6.28 are compared with the corresponding conventional results of Figure 6.26, the directional AG improvements of the generic multi-dimensional beamformer become apparent. In this case, the adaptive technique was the sub-aperture GSC. As expected and because of the AG improvements provided by the adaptive scheme, the signal of interest is not present in the GRAMs of Figure 6.29, which provides the azimuth beams steered at an elevation angle which is 55° off the expected elevation bearing of the signal. The results of Figure 6.29 are in sharp constrast with those of Figure 6.27 for the conventional beamformer.

An explanation for the poor angular resolution performance of the conventional beamformer requires interpretation of the results of Figures 6.25 to 6.27. In particular, for the simulated cylindrical array, Figure 6.25 shows that that conventional beamformer with spatial shading has 13 dB sidelobe suppression in azimuth beam steering and approximately 60° beam width in elevation. Furthermore, to improve detection, the power beam outputs shown in the GRAMs of Figures 6.26 and 6.28 have been normalized,[34] since this is a typical processing arrangement for operational sonar displays. However, the detection improvements of the normalization process would enhance the detection of the sidelobe structure shown in Figure 6.25. Thus, the results of Figures 6.26 and 6.27 provide typical angular resolution performance characteristics for sonars deploying cylindrical array beamformers.

In summary, the results of Figures 6.25 to 6.29 and appropriate scaling on the actual array dimensions and the frequency ranges of the signals that have been considered in the simulations may project the performance characteristics for a variety of sonars deploying cylindrical arrays with conventional or adaptive beamformers.

6.7.1.2 Synthetic Data: Active

As discussed before, the configuration of the generic beamforming structure providing continuous beam time series to the input of a matched filter or a temporal spectral analysis unit forms the basis for integrated active or passive sonar applications. However, before the adaptive aperture processing schemes are integrated with a matched filter, it is essential to demonstrate that the beam time series from the outputs of the non-conventional beamformers have sufficient temporal coherence and correlate with the reference signal. For example, if the signal received by a sonar array consists of FM pulses with a pulse repetition period of a few minutes, then questions may be raised about the efficiency of an adaptive beamformer to achieve near-instantaneous convergence in order to provide beam time series with coherent content for the FM pulses. This is because partially adaptive processing schemes require at least a few iterations to converge to a sub-optimum solution.

To address this question, the matched filter and the conventional and adaptive beamformers, shown in Figures 6.21 and 6.22, were tested with simulated data sets including HFM pulses 8-s long with a 100-Hz bandwidth. The pulse repetition period was 120 s. Although this may be considered as a configuration for bistatic active sonar applications, the findings from this experiment can be applied to monostatic active sonar systems as well.

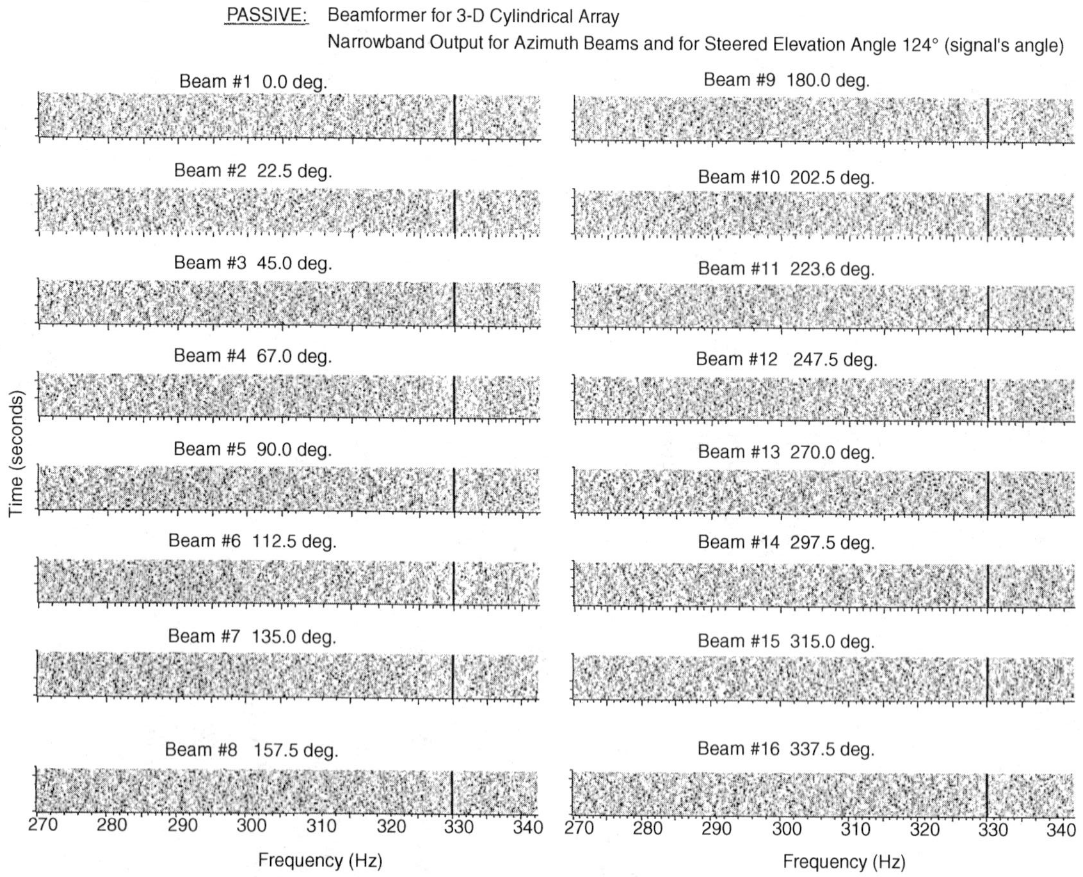

FIGURE 6.26 Narrowband spectral estimates of the generic, 3-D cylindrical array conventional beamformer for all the azimuth beams steered at the signal's expected elevation angle. The 25 windows of this display correspond to the 25 steered beams equally spaced in [1, −1] cosine space. The acoustic field includes two narrowband signals that the very poor angular resolution performance of the conventional beamformer has failed to resolve.

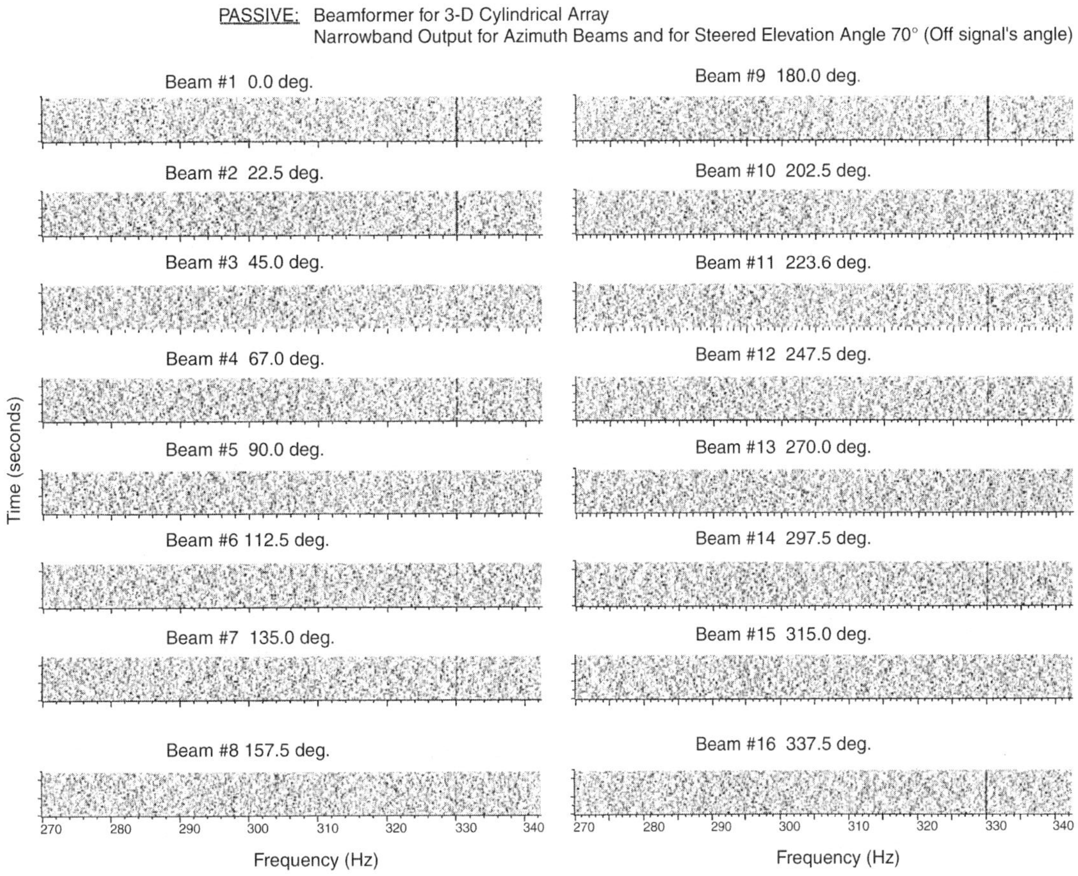

FIGURE 6.27 Narrowband spectral estimates of the generic, 3-D cylindrical array conventional beamformer for all the azimuth beams steered at an elevation bearing which is 55° off the expected signal's elevation angle. The 25 windows of this display correspond to the 25 steered beams equally spaced in [1, –1] cosine space. The acoustic field at this steering does not include signals. However, the very poor sidelobe suppression of the conventional beamformer reveals signals that do not exist at this steering.

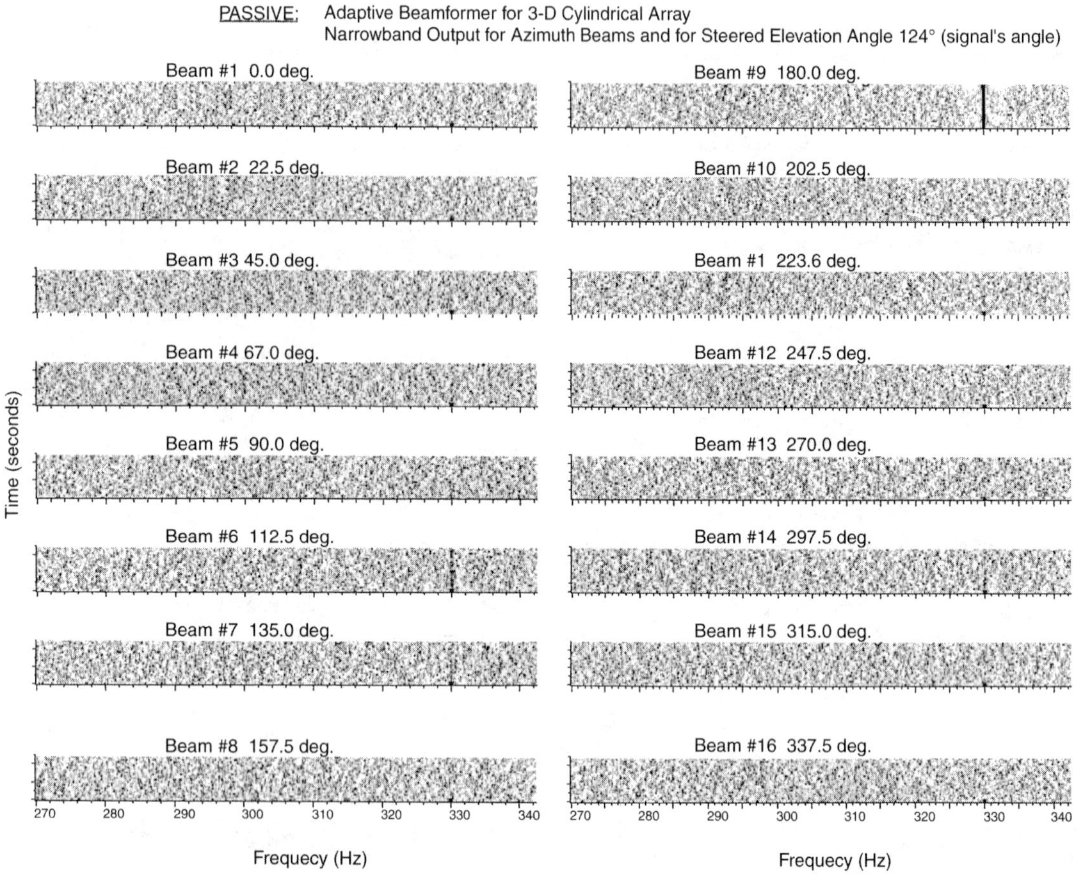

FIGURE 6.28 Narrowband spectral estimates of the generic, 3-D cylindrical array adaptive (GSC) beamformer for all the azimuth beams steered at the signal's expected elevation angle. Input data sets are the same as in Figure 6.24. The 25 windows of this display correspond to the 25 steered beams equally spaced in [1, –1] cosine space. The acoustic field includes two narrowband signals that the very good angular resolution performance of the sub-aperture adaptive beamformer resolves the bearings of the two signals.

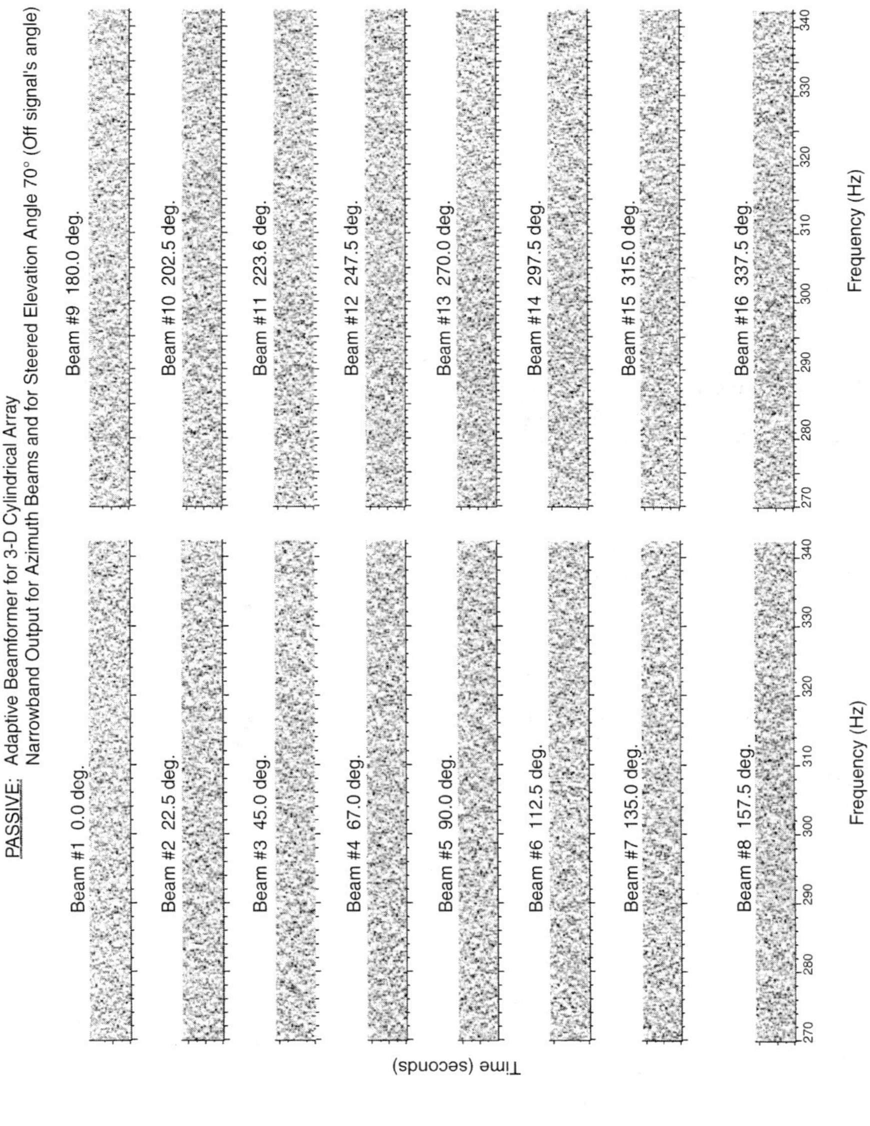

FIGURE 6.29 Narrowband spectral estimates of the generic, 3-D cylindrical array adaptive (GSC) beamformer for all the azimuth beams steered at an elevation bearing which is 55° off the expected signal's elevation angle. Input data sets are the same as in Figure 6.25. The 25 windows of this display correspond to the 25 steered beams equally spaced in [1, −1] cosine space. The acoustic field at this steering does not include signals. Thus, the very good sidelobe suppression of the sub-aperture adaptive beamformer shows that there are no signals present at this steering.

In Figure 6.30, we will present some results from the output of the active unit of the generic signal processing structure. Figure 6.30 shows the output of the replica correlator for the conventional and adaptive beam time series of the sub-aperture GSC and STMV adaptive techniques.[1,37] In this case, the steering angles are the same as those of the data sets shown in Figures 6.26 to 6.29. The horizontal axis in Figure 6.30 represents range or time delay ranging from 0 to 120 s, which is the pulse repetition period. While the three beamforming schemes provide artifact-free outputs, it is apparent from the values of the replica correlator output that the conventional beam time series exhibit better temporal coherence properties than the beam time series of the sub-aperture GSC adaptive beamformer. The significance and a quantitative estimate of this difference can be assessed by comparing the amplitudes of the correlation outputs in Figure 6.30. The replica correlator amplitudes are 12.06, 11.81, and 12.08 for the conventional and the adaptive schemes, GSC-SA (sub-Aperture), and STMV-SA (sub-Aperture), respectively. These results also show that the beam time series of the STMV sub-aperture scheme have temporal coherence properties equivalent to those of the conventional beamformer, which is the optimum case.

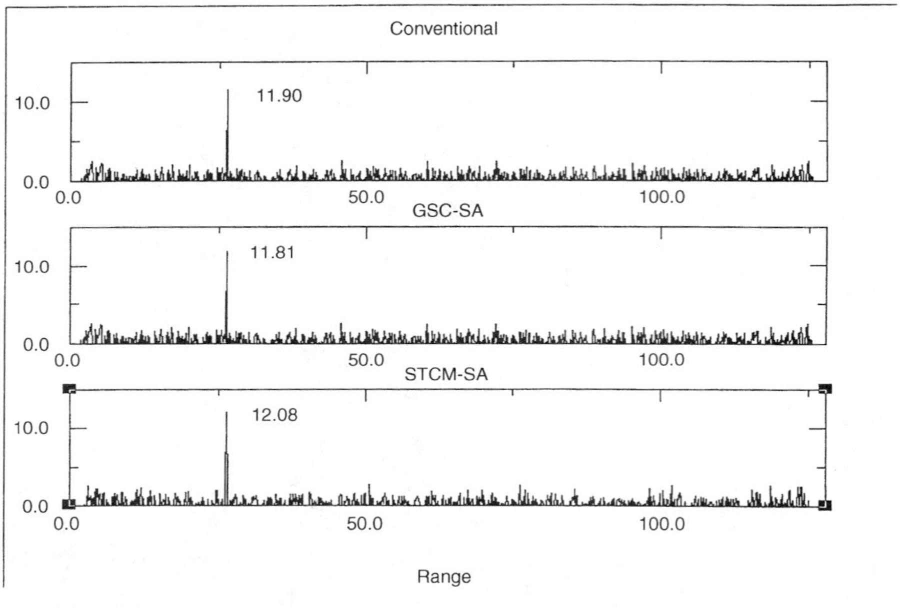

FIGURE 6.30 Output of replica correlator for the conventional and sub-aperture adaptive (GSC, STMV) beam time series of the generic cylindrical array beamformer. The azimuth and elevation beams are steered at the signal's expected bearings, which are the same as those in Figures 6.24 to 6.27.

6.7.1.3 Real Data

The proposed system configuration and the conventional and adaptive signal processing structures of this study were also tested with real data sets from an operational sonar system. Echoes of HFM pulses were received by a cylindrical array having comparable geometric configuration and sensor arrangements as discussed in the simulations. Continuous beam time series were provided at the input of the sub-aperture cylindrical adaptive beamformer with processing flow as defined in Figures 6.21 and 6.22. The total number of steering beams for both the adaptive and conventional beamformers was 144. Figure 6.30 provides the matched filter output results for a single pulse. Figure 6.31a shows the matched filter output of the conventional beamformer, and Figure 6.31b shows the output of the sub-aperture STMV adaptive algorithm.[1,37] The horizontal axis corresponds to the azimuth angular space ranging from 0° to 360°. The vertical axis corresponds to the time delay or range estimate, which is determined by the location of the pick of the matched filter output, and the grayscales show the magnitude of the matched

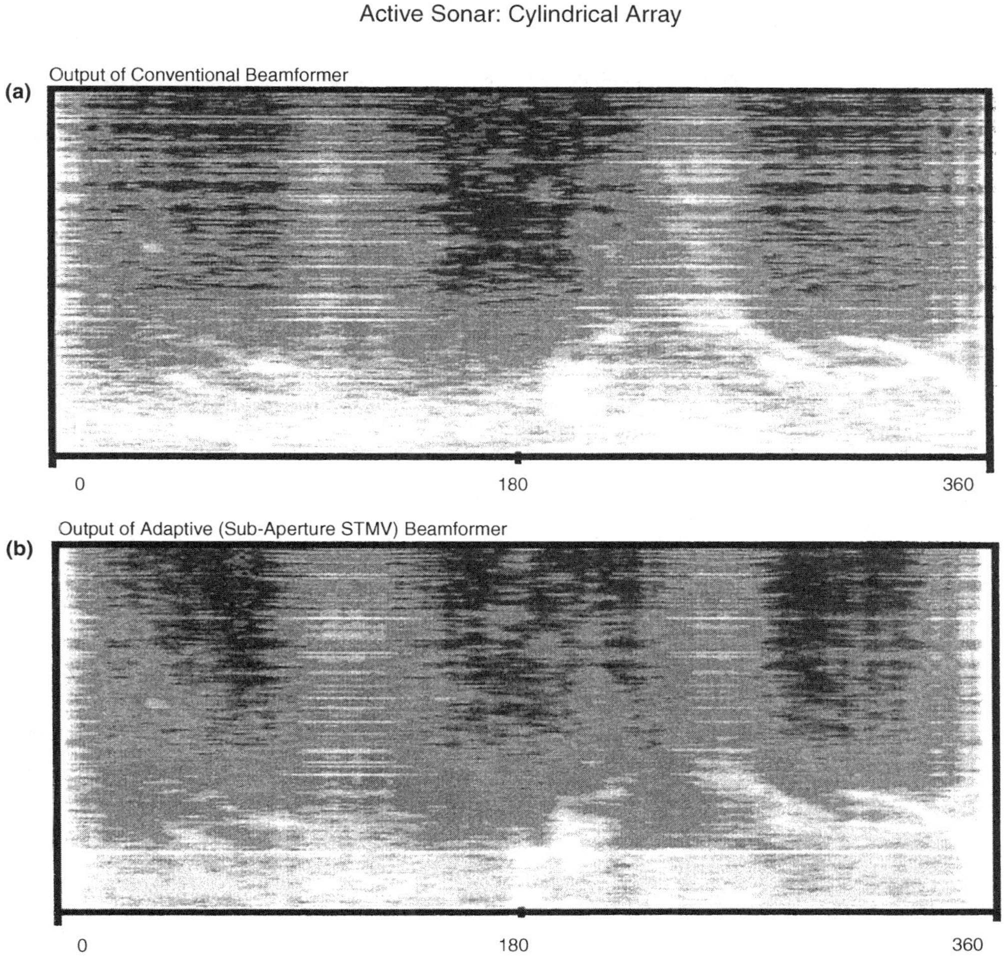

FIGURE 6.31 Shown are the matched filter output of the conventional beamformer (a) and the sub-aperture STMV adaptive algorithm (b) for a cylindrical array. The horizontal axis refers to the angular space covering the bearing range of (0°, 360°). The vertical axis refers to time delay or range estimates of the matched filter, and the grayscales refer to the correlation output. Each vertical color-coded line of Figure 6.31 represents a correlation output of Figure 6.30 for a given bearing angle. The basic difference between the conventional and adaptive matched filter output results is that the improved directionality (or array gain) of the adaptive beam time series localizes the detected HFM pulses and the associated echo returns in a smaller number of beams than the conventional beamformer.

filter output. In a sense, each vertical color-coded line of Figure 6.31 represents the matched filter output (e.g., see Figure 6.30) for a given azimuth steering angle.

Since these are unclassified results provided by an operational sonar, there were no real targets present during the experiments. In a sense, the results of Figure 6.31 present the scattering properties of the medium as they were defined by the received echoes. Although the results of Figure 6.31 are normalized,[34] the amplitudes of the output of the matched filter in Figure 6.31 for the conventional (Figure 6.31a) and adaptive (Figure 6.31b) beam time series were compared before the use of the normalization processing and were found to be approximately the same. Again, these results show that the beam time series of the sub-aperture adaptive scheme have temporal coherence properties equivalent to those of the conventional beamformer; this was also confirmed with simulated data discussed in Section 6.7.1.2.

In summary, the basic difference between the conventional and adaptive matched filter output results is that the improved directionality (or array gain) of the adaptive beam time series localizes the detected HFM pulses and the associated echo returns more accurately than the conventional beamformer.

This kind of AG improvement, provided by the adaptive beamformer, suppresses the reverberation effects during active sonar operations; this is confirmed by the results of Figure 6.31. It is anticipated that the adaptive beamformers will enhance the performance of integrated active-passive and mine-hunting sonars by means of precise detection and localization of echoes that are embedded in reverberation noise fields.

6.7.2 Ultrasound Systems: Line and Planar Array Beamformers

Performance assessment and testing of the generic sub-aperture adaptive beamformers that have been discussed in this chapter have been carried out with the following type of simulated ultrasound data:

- Synthetic data sets that have been generated for an active ultrasound system deploying a line array. The simulated line array sensor time series included the following:
  - Number of sensors, N = 48 sensors.
  - Type of Signal Pulse:

    I. CW: $F_{CW} = 4$ MHz
       Pulse Length: $T_{CW} = 10$ μs

    II. FM: $F_c = 3$ MHz, BW = 4 MHz
       Pulse Length: $T_{FM} = 20$ μs
- Synthetic data sets that have been generated for an active ultrasound system deploying a planar array.

The active CW and FM pulses were the same. The following two sections present the results of these two cases.

6.7.2.1 Synthetic Data Results for Ultrasound Systems Deploying Line Arrays

The results presented in this section are divided into two parts. The first part discusses active CW pulses, which are processed in the same way as in the case of passive narrowband line array sonar applications. The scope here is to evaluate the angular (azimuth) resolution performance of the

- *Sub-aperture adaptive* beamforming
- *Synthetic aperture* beamforming (*ETAM* algorithm)
- Combined beamformer consisting of *synthetic-aperture and adaptive processing*

The impact and merits of these techniques will be contrasted with the angular resolution performance obtained using the conventional beamformer. For details about the synthetic aperture beamformer, the reader is asked to see Reference 1. The synthetic aperture scheme is called the ETAM algorithm and has been tested only with line arrays.[11,12,14,30] Presented in the second part of this section will be similar results from active ultrasound applications with matched filter processing and FM-type pulses.

6.7.2.1.1 Narrowband CW Pulses

Figure 6.32 provides the power of the beam response of the beamformers, as defined in Figures 6.21 and 6.22. The results presented in this section are divided into two parts. The first part discusses active CW pulses, which are processed in the same way as in the case of passive narrowband line array sonar applications. In particular, Figure 6.32 presents the power of the beam response for

- A conventional beamformer on a 48-sensor line array (black color)
- A sub-aperture STMV adaptive beamformer on a 48-sensor line array
- A synthetic aperture (ETAM algorithm) beamformer, which extends the 48-sensor line array into a synthetic 144-sensor array
- A combined synthetic aperture and sub-aperture STMV adaptive beamformer

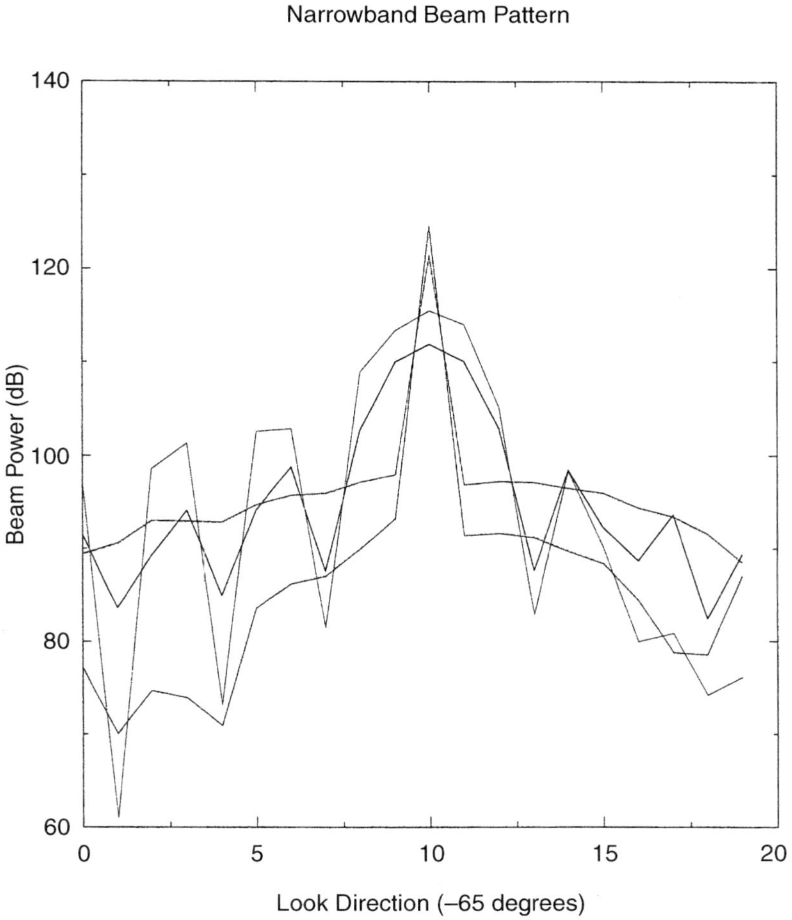

FIGURE 6.32 Output of beamformers [conventional (black), synthetic aperture, sub-aperture STMV adaptive, and combined synthetic-aperture/sub-aperture STMV adaptive] applied on a 48-sensor line array of an ultrasound system with CW pulses of received signals. There are 20 steering beams in a 10° angular sector in the –20° look direction.

At this point, it is important to note that the combined processing operation of a synthetic aperture and sub-aperture STMV adaptive beamformer was a challenging task to test the coherent properties of the synergistic functionality of the generic processing structure shown in the upper part of Figure 6.21. In this case, the output of the synthetic aperture processing scheme provided continuous sensor time series of a synthesized 144-sensor array, which was derived from the simulated 48-sensor real line array. Then, the synthesized 144 continuous sensor time series were provided at the input of the sub-aperture STMV adaptive beamformer.

If the combined signal processing operation between the synthetic aperture and adaptive scheme was not coherent, then the output of the beamformer would have been reduced to a noisy beam pattern.

In summary, the power beam output results of Figure 6.32 demonstrate that the advanced beamformers provide a significantly better beam pattern than that of the corresponding conventional beamformer. However, the sub-aperture beamforming response is not as good as expected. This is because the sub-aperture STMV adaptive scheme, discussed in Reference 1, is basically a broadband beamformer applied on a narrowband type of signals. This is an indication of loss of temporal coherence in the adaptive beam time series. Loss of temporal coherence in the beam time series is caused by non-optimum performance and poor convergence of the broadband adaptive schemes when the input signal is a narrowband CW pulse.

On the other hand, the synthetic aperture beamformer, as narrowband estimator, provides a very good angular resolution response, as expected, because of the narrowband coherent characteristics of the input CW signal. It is equally important to also note that the combined operation of the synthetic aperture and adaptive processing scheme exhibits a superior angular resolution performance and a significant suppression of the sidelobe structure with respect to the other beamformers considered in the investigation.

A common artifact in adaptive beamforming processing applications is the cancellation of a signal of interest when the interfering noise has the same narrowband characteristics as the signal of interest. The results in Figure 6.33 provide a performance assessment of the above algorithms in the presence of two sources radiating the same type of CW signals as defined at the beginning of this section. Furthermore, the beam output results of the beamformers that have been considered in Figure 6.32 are compared with the beam output results of the conventional beamformer applied on a three-times-longer simulated physical line array with 144 sensors.

It is apparent from the results of Figures 6.32 and 6.33 that the generic sub-aperture adaptive beamformers of this investigation improve the angular resolution of an *N* sensor line array by a factor of three. Therefore, these angular resolution improvements are approximately similar with those of the conventional beamformer applied on a three times longer (144 sensors) line array.

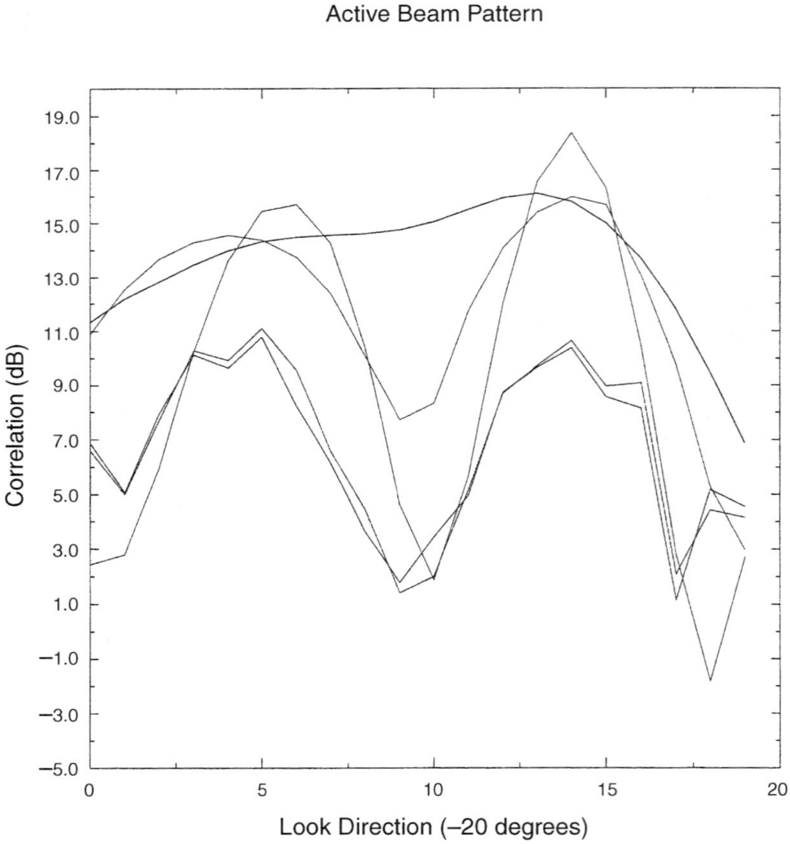

FIGURE 6.33 Output of beamformers [conventional (black), synthetic aperture, sub-aperture STMV adaptive, and combined synthetic-aperture/sub-aperture STMV adaptive] applied on a 48-sensor line array of an ultrasound system with two sources radiating CW pulses. There are 20 steering beams in a 10° angular sector in the –20° look direction. The results are compared with the output of the beam response of a conventional beamformer applied on a 144-sensor line array.

6.7.2.1.2 Broadband FM Pulses

It was discussed in Section 6.6.3 that the configuration of the generic beamforming structure to provide continuous beam time series at the input of a matched filter and a temporal spectral analysis unit forms the basis for ultrasound and integrated passive and active sonar applications. For ultrasound imaging applications, to address this question the matched filter and the sub-aperture adaptive processing scheme, shown in Figures 6.21 and 6.22, were tested with synthetic data sets including FM pulses, as defined at the beginning of Section 6.7.2.

Normalization of the output of a matched filter, such as the results of Figure 6.30, and their display arrangement as GRAMs provide a waterfall display of ranges (depth) as a function of beam steering. Figure 6.34 shows these results for the correlation outputs of the beamformers shown in Figure 6.33. For each beamformer, the horizontal axis includes 20 steering beams in a 10° angular sector centered in the –20° look direction. The vertical axis represents time delay or depth penetration of the signal within a medium of interest. Thus, the detected echoes along the vertical axis of each window of Figure 6.34 represent reflections from simulated objects or organs. For reference, the results are compared with the

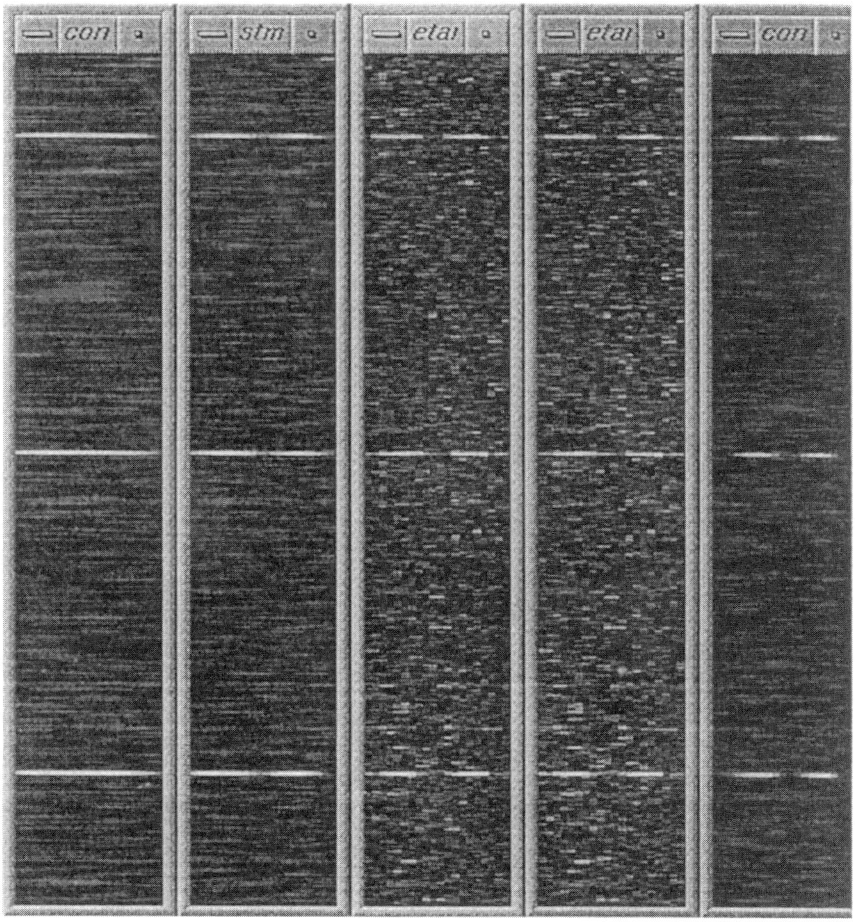

FIGURE 6.34 Outputs of replica correlator for the same beam times series of the beamformers shown in Figure 6.31. For each beamformer, the horizontal axis includes 20 steering beams in a 10° angular sector in the –20° look direction. The vertical axis represents time delay or depth range. The results are compared with the output of the beam response of a conventional beamformer applied on a 144-sensor line array (from the left, the last window of the figure).

outputs of beams of the conventional beamformer for the simulated 144-sensor line array, shown by the window at the left-hand side of Figure 6.34.

In summary, the basic difference between the GRAMs of the adaptive processing with those of the conventional beam time series is that the improved directionality of the sub-aperture adaptive beamformers localizes the detected FM pulses in a smaller number of beams than the conventional beamformer. A quantitative assessment of the improved angular resolution performance of the advanced beamformer is provided by its distinct performance to detect and localize the two simulated echoes that are clearly shown at the left-hand-side window of Figure 6.34, which is the output for the simulated longer array (144 sensors).

Furthermore, it is anticipated that the directional properties of the non-conventional beamformers would suppress the anticipated reverberation levels during active ultrasound or sonar operations. Thus, if there are going to be advantages regarding the implementation of the sub-aperture adaptive beamformers in ultrasound and active mine warefare sonar applications, it is expected that these advantages will include minimization of the impact of the medium's reverberations by means of improved directionality.

6.7.2.2 Synthetic Data Results for Ultrasound Systems Deploying Planar Arrays

Deployment of planar arrays by ultrasound medical imaging systems is gaining increasing popularity because of their advantage to provide 3-D images of organs under medical examination. However, if we consider that a state-of-the-art line array ultrasound system consists of 100 sensors, then a planar array ultrasound system should include at least 10,000 sensors (100×100) in order to achieve the angular resolution performance of a line array system and the additional 3-D image reconstruction capability provided by the elevation beam steering of a planar array. Thus, increased angular resolution in azimuth and elevation beam steering for ultrasound systems means larger sensor arrays with consequent technical and higher cost implications. As discussed in Section 6.5, the alternative is to use synthetic aperture and sub-aperture adaptive beam processing.

In a simulation study, a planar array with 1024 (32×32) sensors was considered that provided continuous sensor time series at the input of the conventional and sub-aperture adaptive beamformers with processing flow similar to that shown in Figures 6.21 and 6.22 for a cylindrical array. As in the case of the cylindrical beamforming results, the power outputs of the beam time series of the conventional and the sub-aperture adaptive techniques implemented in a planar array demonstrated the same performance characteristics with those of Figures 6.25 to 6.34 for the cylindrical array.

6.8 Conclusion

The synthetic and real data results of this chapter indicate that the generic multi-dimensional adaptive concept addresses practical concerns of near-instantaneous convergence, shown in Figures 6.25 to 6.34, for ultrasound and integrated active-passive sonar systems. The performance characteristics of the sub-aperture adaptive beamformer compared with that of the conventional beamformer are reflected as improvements in directional estimates of azimuth and elevation angles and suppression of reverberation effects. This kind of improvement in azimuth and elevation bearing estimates is essential for 3-D ultrasound, sonar, and radar operations.

In summary, a generic beamforming structure has been developed for multi-dimensional sensor arrays that allows the implementation of conventional, synthetic aperture, and adaptive signal processing techniques in integrated active-passive real-time systems. The proposed implementation is based on decomposing the 2-D and 3-D beamforming process in subsets of coherent processes and creating sub-aperture configurations that allow the minimization of the number of degrees of freedom of the adaptive processing schemes. The proposed approach has been applied to line, planar, and cylindrical arrays of sensors where the multi-dimensional beamforming is decomposed into sets of line array and/or circular array beamformers. Moreover, the application of spatial shading on the generic multi-dimensional beamformer is a much simpler process than that of the fully coherent 3-D beamformer. This is because the decomposition process allows two simple and separate applications of spatial shading (i.e., one for circular and the other for line arrays).

The fact that the sub-aperture adaptive beamformers provided AG improvements for CW and HFM signals under a real-time data flow as compared with the conventional beamformer demonstrates the merits of these advanced processing schemes for practical ultrasound, sonar, and radar applications. In addition, the generic implementation scheme of this study suggests that the design approach to provide synergism between the conventional beamformer, the synthetic aperture, and the adaptive processing schemes (e.g., see results in Figures 6.21 and 6.22) is an essential property for system applications.

Although the focus of the implementation effort included only a few adaptive processing schemes, the consideration of other types of spatial filters for real-time ultrasound, sonar, and radar applications should not be excluded. The objective here was to demonstrate that adaptive processing schemes can address some of the challenges that the next-generation ultrasound and active-passive sonar systems will have to deal with in the near future. Once a generic signal processing structure is established, as suggested in the chapter, the implementation of a wide variety of processing schemes can be achieved with minimum efforts for real-time systems deploying multi-dimensional arrays. Finally, the results presented in this chapter indicate that the sub-aperture STMV adaptive scheme addresses the practical concerns of near-instantaneous convergence associated with the implementation of adaptive beamformers in ultrasound and integrated active-passive sonar systems.

References

1. S. Stergiopoulos, Implementation of adaptive and synthetic aperture processing schemes in integrated active-passive sonar systems, *Proc. IEEE*, 86(2), 358–396, February 1998.
2. W.C. Knight, R.G. Pridham, and S.M. Kay, Digital signal processing for sonar, *Proc. IEEE*, 69(11), 1451–1506, 1981.
3. B. Windrow et al., Adaptive antenna systems, *Proc. IEEE*, 55(12), 2143–2159, 1967.
4. A. A. Winder, Sonar System Technology, *IEEE Trans. Sonic Ultrasonics*, SU-22(5), 291–332, 1975.
5. A.B. Baggeroer, Sonar signal processing, in *Applications of Digital Signal Processing*, A.V. Oppenheim, Ed., Prentice-Hall, Englewood Cliffs, NJ, 1978.
6. S. Stergiopoulos and A. T. Ashley, Guest editorial for a special issue on sonar system technology, *IEEE J. Oceanic Eng.*, 18(4), 361–366, 1993.
7. A.C. Dhanantwari, S. Stergiopoulos, and J. Grodski, Implementation of Adaptive Processing in Integrated Active-Passive Sonars Deploying Cylindrical Arrays, Proceedings of Underwater Technology '98, UT'98, Tokyo, Japan, April 1998.
8. W. C. Queen, The directivity of sonar receiving arrays, *J. Acoust. Soc. Am.*, 47, 711–720, 1970.
9. V. Anderson and J. Munson, Directivity of spherical receiving arrays, *J. Acoust. Soc. Am.*, 35 1162–1168, 1963.
10. W. M. Carey and W. B. Moseley, Space-time processing, environmental-acoustic effects, *IEEE J. Oceanic Eng.*, 16, 285–301, 1991; *Progress in Underwater Acoustics*, Plenum Press, New York, 743–758, 1987.
11. S. Stergiopoulos and E. J. Sullivan, Extended towed array processing by overlapped correlator, *J. Acoust. Soc. Am.*, 86(1), 158–171, 1989.
12. S. Stergiopoulos, Optimum bearing resolution for a moving towed array and extension of its physical aperture, *J. Acoust. Soc. Am.*, 87(5), 2128–2140, 1990.
13. E. J. Sullivan, W. M. Carey, and S. Stergiopoulos, Editorial for a special issue on acoustic synthetic aperture processing, *IEEE J. Oceanic Eng.*, 17(1), 1–7, 1992.
14. S. Stergiopoulos and H. Urban, An experimental study in forming a long synthetic aperture at sea, *IEEE J. Oceanic Eng.*, 17(1), 62–72, 1992.
15. G. S. Edelson and E. J. Sullivan, Limitations on the overlap-correlator method imposed by noise and signal characteristics, *IEEE J. Oceanic Eng.*, 17(1), 30–39, 1992.
16. G. S. Edelson and D. W. Tufts, On the ability to estimate narrow-band signal parameters using towed arrays, *IEEE J.Oceanic Eng.*, 17(1), 48–61, 1992.

17. N. L. Owsley, *Sonar Array Processing*, S. Haykin, Ed., Signal Processing Series, A.V. Oppenheim, Series Ed., Prentice-Hall, Englewood Cliffs, NJ, p. 123, 1985.
18. B. Van Veen and K. Buckley, Beamforming: a versatile approach to spatial filtering, *IEEE ASSP Mag.*, 4–24, 1988.
19. H. Cox, R. M. Zeskind, and M. M. Owen, Robust adaptive beamforming, *IEEE Trans. Acoust. Speech Signal Proc.,* ASSP-35(10), 1365–1376, 1987.
20. H. Cox, Resolving power and sensitivity to mismatch of optimum array processors, *J. Acoust. Soc. Am.*, 54(3), 771–785, 1973.
21. A. H. Sayed and T. Kailath, A state-space approach to adaptive RLS filtering, *IEEE SP Mag.*, July, 18–60, 1994.
22. J. Capon, High resolution frequency wavenumber spectral analysis, *Proc. IEEE*, 57, 1408–1418, 1969.
23. T. L. Marzetta, A new interpretation for Capon's maximum likelihood method of frequency-wavenumber spectra estimation, *IEEE Trans. Acoust. Speech Signal Proc.,* ASSP-31(2), 445–449, 1983.
24. S. Haykin, *Adaptive Filter Theory*, Prentice-Hall, Englewood Cliffs, NJ, 1986.
25. A. Tawfik, A. C. Dhanantwari, and S. Stergiopoulos, A Generic Beamforming Structure Allowing the Implementation of Adaptive Processing Schemes into 2-D & 3-D Arrays of Sensors, Proc. MTS/IEEE OCEANS'97, Halifax, Nova Scotia, October 1997.
26. A. Tawfik, A. C. Dhanantwari, and S. Stergiopoulos, A generic beamforming structure for adaptive schemes implemented in 2-D & 3-D arrays of sensors, *J. Acoust. Soc. Am.*, 101(5, Pt. 2), 3025, 1997.
27. S. Stergiopoulos, A. Tawfik, and A. C. Dhanantwari, Adaptive Microphone 2-D and 3-D Arrays for Enhancement of Sound Reception in Coherent and Incoherent Noise Environment, Proc. Inter-Noise'97, OPAKFI H-1027, Budapest, August 1997.
28. A. Tawfik and S. Stergiopoulos, A Generic Processing Structure Decomposing the Beamforming Process of 2-D & 3-D Arrays of Sensors into Sub-Sets of Coherent Processes, Proc. of CCECE'97, Canadian Conference on Electrical & Computer Engineering, St. John's, NF, May 1997.
29. N. L. Owsley, Systolic Array Adaptive Beamforming, NUWC Report 7981, September 1987.
30. S. Stergiopoulos, Limitations on towed-array gain imposed by a non isotropic ocean, *J. Acoust. Soc. Am.*, 90(6), 3161–3172, 1991.
31. A.D. Whalen, *Detection of Signals in Noise*, Academic Press, New York, 1971.
32. D. Middleton, *Introduction to Statistical Communication Theory*, McGraw-Hill, New York, 1960.
33. H. L. Van Trees, *Detection, Estimation and Modulation Theory*, Wiley, New York, 1968.
34. S. Stergiopoulos, Noise normalization technique for beamformed towed array data, *J. Acoust. Soc. Am.*, 97(4), 2334–2345, 1995.
35. S. Stergiopoulos, Influence of Underwater Environment's Coherence Properties on Sonar Signal Processing, Proc. 3rd European Conference on Underwater Acoustics, FORTH-IACM, Heraklion-Crete, V-I, pp. 453–458, 1996.
36. F. J. Harris, On the use of windows for harmonic analysis with discrete Fourier transform, *Proc. IEEE*, 66, 51–83, 1978.
37. A. C. Dhanantwari, Adaptive Beamforming with Near-Instantaneous Convergence for Matched Filter Processing, Master thesis, Departement of Electrical Engineering, Technical University of Nova Scotia, Halifax, NS, Canada, September 1996.
38. J. Krolik and D. N. Swingler, Bearing estimation of multiple broadband sources using steered covariance matrices, *IEEE Trans. Acoust. Speech Signal Process.*, ASSP-37, 1481–1494, 1989.
39. H. Wang and M. Kaveh, Coherent signal-subspace processing for the detection and estimation of angles of arrival of multiple wideband sources, *IEEE Trans. Acoust. Speech Signal Process.*, ASSP-33, 823–831, 1985.
40. A. Antoniou, *Digital Filters: Analysis, Design, and Applications*, 2nd ed., McGraw-Hill, New York, 1993.

41. A. Mohammed, A High-Resolution Spectral Analysis Technique, DREA Memorandum 83/D, Defence Research Establishment Atlantic, Dartmouth, NS, Canada, 1983.
42. A. Mohammed, Novel Methods of Digital Phase Shifting to Achieve Arbitrary Values of Time Delays, DREA Report 85/106, Defence Research Establishment Atlantic, Dartmouth, NS, Canada, 1985.
43. S. Stergiopoulos and A. T. Ashley, An experimental evaluation of split-beam processing as a broadband bearing estimator for line array sonar systems, *J. Acoust. Soc. Am.*, 102(6), 3556–3563, December 1997.
44. L. J. Griffiths and C. W. Jim, An alternative approach to linearly constrained adaptive beamforming, *IEEE Trans. Antennas Propagation*, AP-30, 27–34, 1982.
45. D. T. M. Slock, On the convergence behavior of the LMS and the normalized LMS algorithms, *IEEE Trans. Acoust. Speech Signal Process.*, ASSP-31, 2811–1825, 1993.
46. D. Middleton and R. Esposito, Simultaneous otpimum detection and estimation of signals in noise, *IEEE Trans. Info. Theory*, IT-14, 434–444, 1968.
47. P. Wille and R. Thiele, Transverse horizontal coherence of explosive signals in shallow water, *J. Acoust. Soc. Am.*, 50, 348–353, 1971.
48. P. A. Bello, Characterization of randomly time-variant linear channels, *IEEE Trans. Commun. Syst.*, 10, 360–393, 1963.
49. D. A. Gray, B. D. O. Anderson, and R. R. Bitmead, Towed array shape estimation using kalman filters — theoretical models, *IEEE J. Oceanic Eng.*, 18(4), 543–556, October 1993.
50. B. G. Ferguson, Remedying the effects of array shape distortion on the spatial filtering of acoustic data from a line array of sensors, *IEEE J. Oceanic Eng.*, 18(4), 565–571, October 1993.
51. J. L. Riley and D. A. Gray, Towed array shape estimation using kalman filters — experimental investigation, *IEEE J. Oceanic Eng.*, 18(4), 572–581, October 1993.
52. B. G. Ferguson, Sharpness applied to the adaptive beamforming of acoustic data from a towed array of unknown shape, *J. Acoust. Soc. Am.*, 88(6), 2695–2701, 1990.
53. F. Lu, E. Milios, and S. Stergiopoulos, A new towed array shape estimation method for sonar systems, *IEEE J. Oceanic Eng.*, submitted.
54. N. C. Yen and W. Carey, Application of synthetic-aperture processing to towed-array data, *J. Acoust. Soc. Am.*, 86, 754–765, 1989.
55. S. Stergiopoulos and H. Urban, A new passive synthetic aperture technique for towed arrays, *IEEE J. Oceanic Eng.*, 17(1), 16–25, 1992.
56. V. H. MacDonald and P. M. Schulteiss, Optimum passive bearing estimation in a spatially incoherent noise environment, *J. Acoust. Soc. Am.*, 46(1), 37–43, 1969.
57. G. C. Carter, Coherence and time delay estimation, *Proc. IEEE*, 75(2), 236–255, 1987.
58. C. H. Knapp and G.C. Carter, The generalized correlation method for estimation of time delay, *IEEE Trans. Acoust. Speech Signal Process.*, ASSP-24, 320–327, 1976.
59. D. C. Rife and R. R. Boorstyn, Single-tone parameter estimation from discrete-time observations, *IEEE Trans. Info. Theory*, 20, 591–598, 1974.
60. D. C. Rife and R. R. Boorstyn, Multiple-tone parameter estimation from discrete-time observations, *Bell Syst. Tech. J.*, 20, 1389–1410, 1977.

7

Advanced Applications of Volume Visualization Methods in Medicine

Georgios Sakas
Fraunhofer Institute for Computer Graphics

Grigorios Karangelis
Fraunhofer Institute for Computer Graphics

Andreas Pommert
Institute of Mathematics and Computer Science in Medicine

Abstract

This chapter summarizes the state-of-the-art application of techniques developed over the recent years for visualizing volumetric medical data acquired by modern medical imaging modalities such as computed tomography (CT), magnetic resonance angiography (MRA), magnetic resonance imaging (MRI), nuclear medicine, three-dimensional (3D)-ultrasound, laser confocal microscopy, etc. Although all of the modalities provide "slices of the body," significant differences exist between the image content of each modality. The focus of this chapter is less in explaining algorithms and rendering techniques, but rather to point out their applicability, benefits, and potential in the medical environment.

In the first part, fundamentals of medical image processing and methods for all steps of the volume visualization pipeline from data preprocessing to object display are reviewed, with special emphasis on data structures, segmentation, and surface- and volume-based rendering. Furthermore, volume registration, intelligent visualization, intervention rehearsal, and aspects of image quality are discussed. In the second part, applications are illustrated from the areas of craniofacial surgery, traumatology, neurosurgery, radiotherapy, and medical education. Further, some new applications of volumetric methods are presented: 3D ultrasound, laser confocal data sets, and 3D-reconstruction of cardiological data sets, i.e., vessels as well as ventricle. These new volumetric methods are currently under development, but due to

0-8493-3691-0/01/$0.00+$.50

their enormous application potential they are expected to be clinically accepted within the next years. Finally, in the Appendix, we discuss the most common techniques for medical image pixel brightness transformation (or image grey scale manipulation techniques [GSMT]), enhancement image processing (or image filtering), and image restoration.

7.1 Volume Visualization Principles

7.1.1 Introduction

Medical imaging technology has experienced a dramatic change over the past 25 years. Previously, only X-ray radiographs were available, which showed the depicted organs as superimposed shadows on photographic film. With the advent of modern computers, new *tomographic* imaging modalities like CT, MRI, and positron emission tomography (PET) which deliver cross-sectional images of a patient's anatomy and physiology have been developed. These images show different organs free from overlays with unprecedented precision. Even the 3D structure of organs can be recorded if a sequence of parallel cross-sections is taken. For many clinical tasks like surgical planning, it is necessary to understand and communicate complex and often malformed 3D structures. Experience has shown that the "mental reconstruction" of objects from cross-sectional images is extremely difficult and strongly depends on the observer's training and imagination. For these cases, it is certainly desirable to present the human body as a surgeon or anatomist would see it. The aim of *volume visualization* (also known as *3D imaging*) in medicine is to create precise and realistic views of objects from medical volume data. The resulting images, even though they are of course two-dimensional, are often called *3D images* or *3D reconstructions* to distinguish them from 2D cross sections or conventional radiographs. The first attempts date back to the late 1970s, with the first clinical applications reported on the visualization of bone from CT in craniofacial surgery and orthopedics. Methods and applications have since been extended to other subjects and imaging modalities. The same principles are also applied to sampled and simulated data from other domains, such as fluid dynamics, geology, and meteorology.[53]

7.1.2 Methods

An overview of the volume visualization pipeline is shown in Figure 7.1. After the acquisition of a series of tomographic images of a patient, the data usually undergo some preprocessing for data conversion and possibly image filtering. From this point, one of several paths may be followed.

The dotted line in Figure 7.1 represents an early approach where an object is reconstructed from its contours on the cross-sectional images. All other methods, represented by the solid line, start from a contiguous *data volume*. If required, equal spacing in all three directions can be achieved by interpolation. Like a 2D image, a 3D volume can be filtered to improve image quality. Corresponding to the *pixels* (picture elements) of a 2D image, volume elements are called *voxels* (volume elements).

The next step is to identify the different objects represented in the data volume so that they can be removed or selected for visualization. The simplest way is to binarize the data with an intensity threshold, e.g., to distinguish bone from other tissues in CT. For MRI data especially, however, more sophisticated *segmentation* methods are required.

After segmentation, there is a choice for which *rendering* technique is to be used. The more traditional surface-based methods first create an intermediate surface representation of the object to be shown. It may then be rendered with any standard computer graphics method. More recently, volume-based methods have been developed which create a 3D view directly from the volume data. These methods use the full grey level information to render surfaces, cuts, or transparent and semi-transparent volumes. As a third way, transform-based rendering methods may be used.

Extensions to the volume visualization pipeline not shown in Figure 7.1, but also covered here, include volume registration, intelligent visualization, and intervention rehearsal.

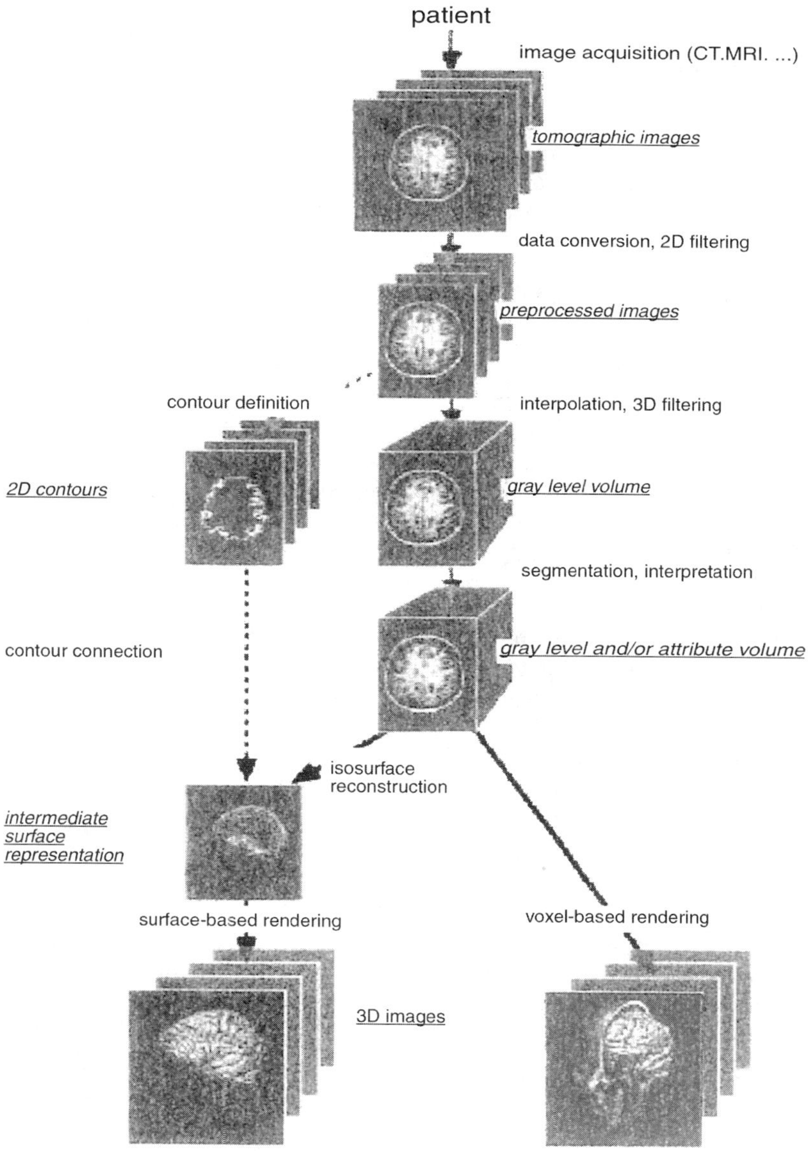

FIGURE 7.1 Overview of the volume visualization pipeline. Individual processing steps may be left out, combined, or reversed in order by a particular method.

7.1.2.1 Preprocessing

The data we consider usually come as a spatial sequence of 2D cross-sectional images. If they are put on top of each other, a contiguous *grey level volume* is obtained. The resulting data structure is an orthogonal 3D array of voxels, each representing an intensity value. This is called the *voxel model*. Many algorithms for volume visualization work on *isotropic* volumes, where the sampling density is equal in all three dimensions. In practice, however, only very few data sets have this property, especially for CT. In these cases, the missing information has to be reconstructed in an *interpolation* step. A quite simple method is linear interpolation of the intensities between adjacent images. Higher order functions such as splines usually give better results for fine details.[68] Shape-based methods are claimed to be superior in certain situations;[7] however, these are dependent on the results of a previous segmentation step.

With respect to later processing steps such as segmentation, it is often desirable to improve the signal-to-noise ratio of the data using image or volume filtering. Well-known *noise filters* are average, median, and Gaussian filters.[90] These methods, however, tend to smooth out small details as well; better results are obtained with *anisotropic diffusion* filters which largely preserve object boundaries.[34]

7.1.2.1.1 Data Structures for Volume Data

There are a number of different data structures for volume data. The most important are the following:

- *Binary voxel model:* Voxel values are either 1 (object) or 0 (no object). This very simple model is not in much use any more. In order to reduce storage requirements, binary volumes may be subdivided recursively into subvolumes of equal value; the resulting data structure is called an *octree*.
- *Grey level voxel model:* Each voxel holds an intensity information. Octree representations have also been developed for grey level volumes.[59]
- *Generalized voxel model:* In addition to an intensity information, each voxel contains *attributes*, describing its membership to various objects and/or data from other sources (e.g., MRI and PET).[47]
- *Intelligent volumes:* As an extension of the generalized voxel model, properties of anatomical objects (such as color, names in various languages, pointers to related information) and their relationships are modeled on a symbolic level.[49,84] This data structure is the basis for advanced applications such as medical atlases discussed later.

7.1.2.2 Segmentation

A grey level volume usually represents a large number of different structures obscuring each other. Thus, to display a particular one, we have to decide which parts of the data we want to use or ignore. Ideally, selection would be done with a command like "show only the brain." This, however, requires that the computer know which parts of the volume constitute the brain and which do not.

A first step toward object recognition is to partition the grey level volume into different regions which are homogeneous with respect to some formal criteria and correspond to real anatomical objects. This process is called *segmentation*. The generalized voxel model is a suitable data structure for representing the results. In a further *interpretation* step, the regions may be identified and labeled with meaningful terms such as "white matter" or "ventricle."

All segmentation methods can be characterized as being either "binary" or "fuzzy," corresponding to the principles of binary and fuzzy logic, respectively.[124] In *binary segmentation*, the question whether a voxel belongs to a certain region is always answered yes or no. This information is a prerequisite, e.g., for creating surface representations from volume data. As a drawback, uncertainty or cases where an object takes up only a fraction of a voxel (*partial volume effect*) cannot be handled properly. For example, a very thin bone would appear with false holes on a 3D image. Strict yes-no decisions are avoided in *fuzzy segmentation*, where a set of probabilities is assigned to every voxel, indicating the evidence for different materials. Fuzzy segmentation is closely related to the so-called volume rendering methods discussed later.

Currently, a large number of segmentation methods for 3D medical images are being developed, which may be roughly divided into three classes: point-, edge-, and region-based methods. The methods described often have been tested successfully on a number of cases; experience has shown, however, that the results should always be used with care.

7.1.2.2.1 Point-Based Segmentation

In point-based segmentation, a voxel is *classified* depending only on its intensity, no matter where it is located. A very simple but nevertheless important example, which is very much used in practice, is *thresholding*: a certain intensity range is specified with lower and upper threshold values. A voxel belongs to the selected class if, and only if, its intensity level is within the specified range.

Thresholding is the method of choice for selecting bone or soft tissue in CT. In volume-based rendering, it is often performed during the rendering process itself so that no explicit segmentation step is required. In order to avoid the problems of binary segmentation, Drebin et al.[24] use a fuzzy maximum likelihood classifier which estimates the percentages of the different materials represented in a voxel, according to Bayes' rule. This method requires that the grey level distributions of different materials be different from each other and known *a priori*. This is approximately the case in musculo-skeletal CT.

Unfortunately, these simple segmentation methods are not suitable if different structures have mostly overlapping or even identical grey level ranges. This situation frequently occurs, e.g., in the case of soft tissues from CT or MRI. The situation is somewhat simplified if multi-parameter data are available, such as T_1- and T_2-weighted images in MRI, emphasizing fat and water, respectively. In this case, individual threshold values can be specified for every parameter. To somewhat generalize this concept, voxels in an n-parameter data set can be considered as n-dimensional vectors in an n-dimensional *feature space*. In *pattern recognition*, this feature space is partitioned into subspaces representing different tissue classes or organs. This is called the *training phase*. In supervised training, the partition is derived from feature vectors which are known to represent particular tissues.[19,35] In unsupervised training, the partition is automatically generated.[35] In the subsequent *test phase*, a voxel is classified according to the position of its feature vector in the partitioned feature space.

With especially adapted image acquisition procedures, pattern recognition methods have successfully been applied to considerable numbers of two- or three-parametric MRI data volumes.[19,35] Quite frequently, however, isolated voxels or small regions are incorrectly classified, such as subcutaneous fat in the same class as white matter. To eliminate these errors, a connected component analysis (see below) is often applied.

A closely related method, based on *neural network* methodology, was developed by Kohonen.[57] Instead of an n-dimensional feature space, a so-called *topological map* of $m \times m$ n-dimensional vectors is used. During the training phase, the map iteratively adapts itself to a set of training vectors which may either represent selected tissues (supervised learning) or the whole data volume (unsupervised learning).[40,115] Finally, the map develops several relatively homogeneous regions which correspond to different tissues or organs in the original data. The practical value of the topological map for 3D MRI data seems to be generally equivalent to that of pattern recognition methods.

7.1.2.2.2 Edge-Based Segmentation

The aim of edge-based segmentation methods is to detect intensity discontinuities in a grey level volume. These edges (in 3D, they are actually surfaces; however, it is common to speak about edges) are assumed to represent the borders between different organs or tissues. Regions are subsequently defined as the enclosed areas. A common strategy for edge detection is to locate the maxima of the first derivative of the 3D intensity function. A method which very accurately locates the edges was developed by Canny.[17] All algorithms using the first derivative, however, share the drawback that the detected contours are usually not closed, i.e., they do not separate different regions properly. An alternative approach is to detect zero-crossings of the second derivative. The Marr-Hildreth operator convolves the input data with the Laplacian of a Gaussian; the resulting contour volume describes the locations of the edges. With a 3D extension of this operator, Bomans et al. segmented and visualized the complete human brain from MRI for the first time.[12] Occasionally, this operator creates erroneous "bridges" between different materials which have to be removed interactively. Also, location accuracy of the surfaces is not always satisfactory.

Snakes[1,52] are 2D image curves which are adjusted from an initial approximation to image features by a movement of the curve caused by simulated forces (Figure 7.2). Image features produce the so-called external force. An internal tension of the curve resists against highly angled curvatures, which makes the Snakes movement robust against noise. After a starting position is given, the Snake adapts itself to an image by relaxation to the equilibrium of the external force and internal tension. To calculate the forces, an external energy has to be defined. The gradient of this energy is proportional to the external force. The segmentation by Snakes is due to its 2D definition performed in a slice-by-slice manner, i.e., the resulting curves for a slice are copied into the neighboring slice and the minimization is started again. The user may control the segmentation process by stopping the automatic tracking if the curves run out of the contours and define a new initial curve.

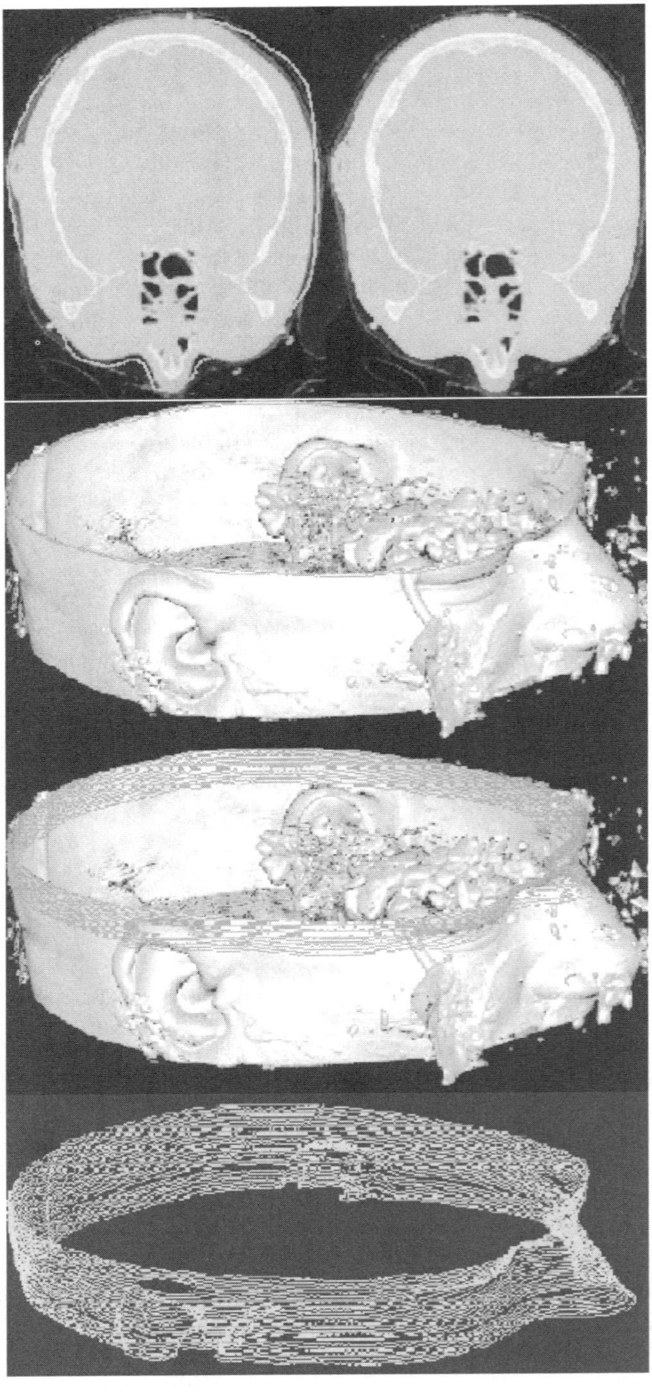

FIGURE 7.2 The principle of segmentation using Snakes.

For this reason, two methods have been applied to enter an initial curve for the Snake. The first is the interactive input of a polygon. Since the Snake contracts due to its internal energy, the contour to be segmented has to be surrounded by this polygon. The second one is a contour tracing method, using an *A* search tree to find the path with minimal costs between two interactively marked points.[76,116] The quality of the result depends on the similarity of two adjacent slices. Normally, this varies within a data set. Therefore, in regions with low similarity, the slices to be segmented by the interactive method must be selected rather tightly.

7.1.2.2.3 ***Region-Based Segmentation***

Region-based segmentation methods consider whole regions instead of individual voxels or edges. Since we are actually interested in regions, this approach appears to be the most natural. Properties of a region are, e.g., its size, shape, location, variance of grey levels, and its spatial relation to other regions.

A typical application of region-based methods is to postprocess the results of a previous point-based segmentation step. For example, a *connected component analysis* may be used to determine whether the voxels, which have been classified as belonging to the same class, are part of the same connected region. If not, some of the regions may be discarded.

A practical interactive segmentation system based on the methods of *mathematical morphology* was developed by Höhne and Hanson.[48] Regions are initially defined with thresholds; the user can subsequently apply simple but fast operations such as *erosion* (to remove small bridges between erroneously connected parts), *dilation* (to close small gaps), connected components analysis, region fill, or Boolean set operations. Segmentation results are immediately visualized on orthogonal cross sections and 3D images, so that they may be corrected or further refined in the next step (Figure 7.3). With this system, segmentation of gross structures is usually a matter of minutes. In Reference 101, this approach is extended to multi-parameter data.

For automatic segmentation, the required knowledge about data and anatomy needs to be represented in a suitable model. A comparatively simply approach is presented by Brummer et al., who use a fixed sequence of morphological operations for the segmentation of brain from MRI.[16] For the same application, Raya and Udupa developed a rule-based system which successively generates a set of threshold values.[86] Rules are applied depending on measured properties of the resulting regions. Bomans generates a set of object hypotheses for every voxel, depending on its grey level.[11] Location, surface-to-volume ratio, etc. of the resulting regions are compared to some predefined values, and the regions are modified accordingly. Menhardt uses a rule-based system which models the anatomy with relations such as "brain is inside skull."[70] Regions are defined as fuzzy subsets of the volume, and the segmentation process is based on fuzzy logic and fuzzy topology.

One of the problems of these and similar methods for automatic segmentation is that the required anatomical knowledge is often represented in more or less ad hoc algorithms, rules, and parameters. A more promising approach is to use an explicit 3D organ model. For the brain, Atata et al. developed an atlas of the "normal" anatomy and its variation in terms of a probabilistic spatial distribution obtained from 22 MRI data sets of living persons.[3] The model was reported as suitable for the automatic segmentation of various brain structures, including white matter lesions. A similar approach is described in Reference 56. Automatic segmentation of cortical structures based on a statistical atlas representing several hundred individuals is presented in Reference 29.

Another interesting idea is to investigate object features in *scale-space*, i.e., at different levels of image resolution. This approach allows irrelevant image detail to be ignored. One such method developed by Pizer et al. considers the symmetry of previously determined shapes, described by medial axes.[129] The resulting ridge function in scale-space is called the *core* of an object. It may be used, e.g., for interactive segmentation, where the user can select, add, or subtract regions or move to larger "parent" or smaller "child" regions in the hierarchy. Other applications such as automatic segmentation or registration are currently being investigated.

In conclusion, automatic segmentation systems are not yet robust enough to be generally applicable to medical volume data. Interactive segmentation, which combines fast operations with the unsurpassed human recognition capabilities, is still the most practical approach.

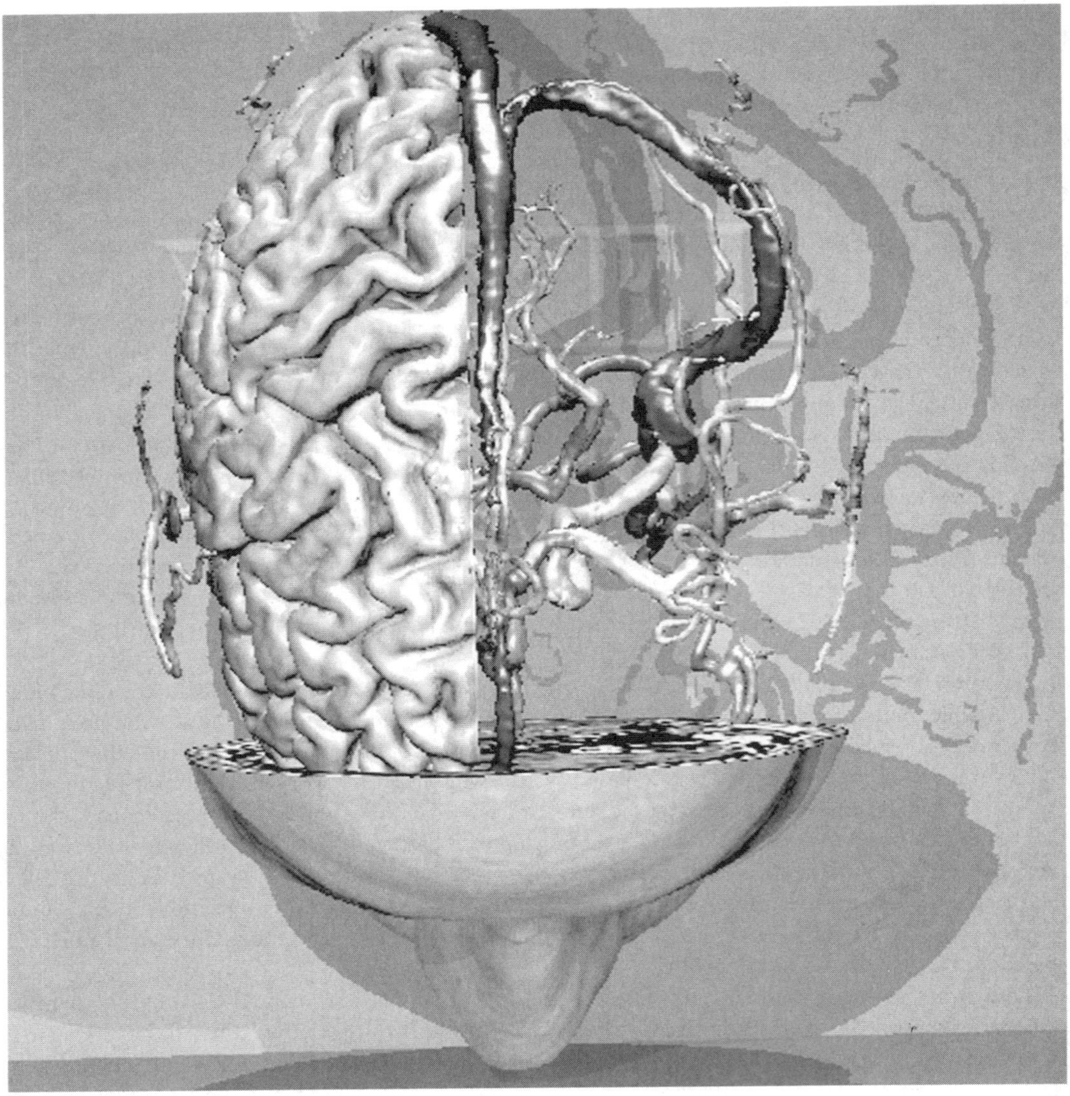

FIGURE 7.3 Results of interactive segmentation of MRI (skin, brain) and MRA (vessels) data.

7.1.2.3 Surface-Based Rendering

The key idea of *surface-based rendering* methods is to extract an intermediate surface description of the relevant objects from the volume data. Only this information is then used for rendering. If triangles are used as surface elements, this process is called *triangulation.* A clear advantage of surface-based methods is the possibly very high data reduction from volume to surface representations. Resulting computing times can be further reduced if standard data structures such as triangle meshes are used with common rendering hard- and software support. On the other hand, the *surface reconstruction* step throws away most of the valuable information on the cross-sectional images. Even simple cuts are meaningless because there is no information about the interior of an object. Furthermore, every change of surface definition criteria, such as thresholds, requires a recalculation of the whole data structure.

An early approach for the reconstruction of the polygonal mesh from a stack of contours is based on the Delauney interpolation developed by Boissinnat.[10] Using this heuristic method, the volume of the

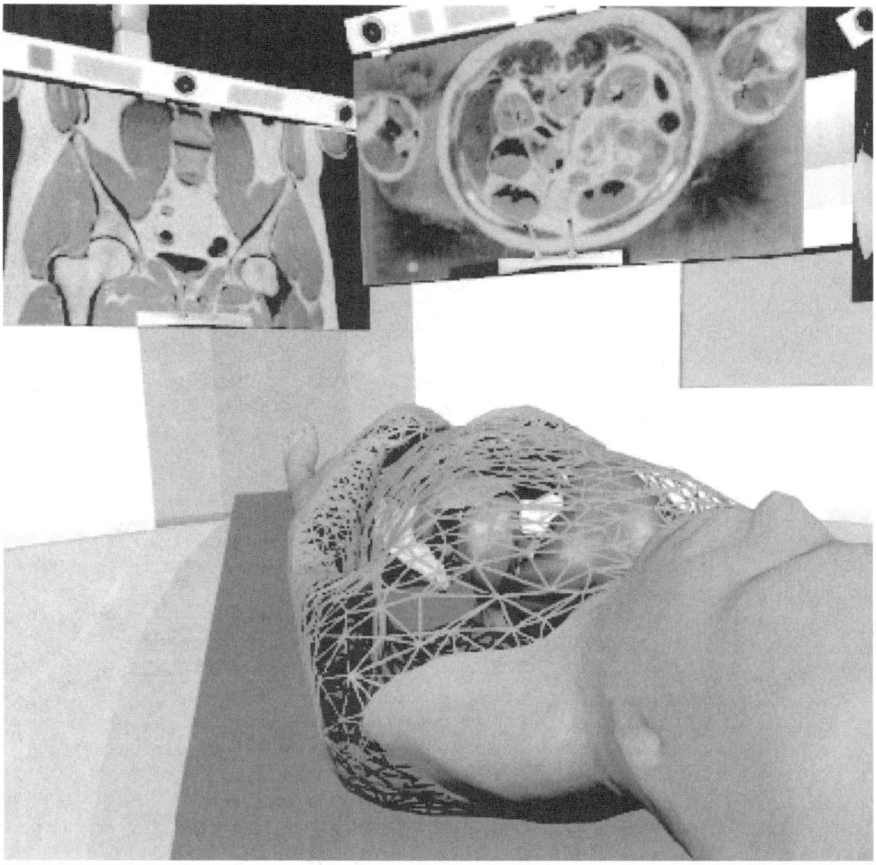

FIGURE 7.4 The virtual patient.

contours is computed by a 3D triangulation which allows an extraction of the surface of the object. Figure 7.4 shows the result of the triangle reduced surface of the Virtual Human Project patient.

A more recent method by Lorensen and Cline, called *Marching Cubes* algorithm, creates an *iso-surface*, representing the locations of a certain intensity value in the data.[65] This algorithm basically considers a cube of $2 \times 2 \times 2$ contiguous voxels. Depending on whether one or more of these voxels are inside the object (i.e., above a threshold value), a surface representation of up to four triangles is placed within the cube. The exact location of the triangles is found by linear interpolation of the intensities at the voxel vertices. The result is a highly detailed surface representation with subvoxel resolution (Figure 7.5). Surface orientations are calculated from grey level gradients. Meanwhile, a whole family of similar algorithms has been developed.[80,118,122]

Applied to clinical data, the Marching Cubes algorithm typically creates hundreds of thousands of triangles. As has been shown, these numbers can be reduced considerably by a subsequent simplification of the triangle meshes, without much loss of information.[103,123] The reduction method can be parameterized and thus allows deriving models of different levels of detail.

7.1.2.3.1 Shading

In general, *shading* is the realistic display of an object based on the position, orientation, and characteristics of its surface and the light sources illuminating it.[32] The reflective properties of a surface are described with an *illumination model* such as the *Phong* model, which uses a combination of ambient light and diffuse (like chalk) and specular (like polished metal) reflections. A key input into these models is the

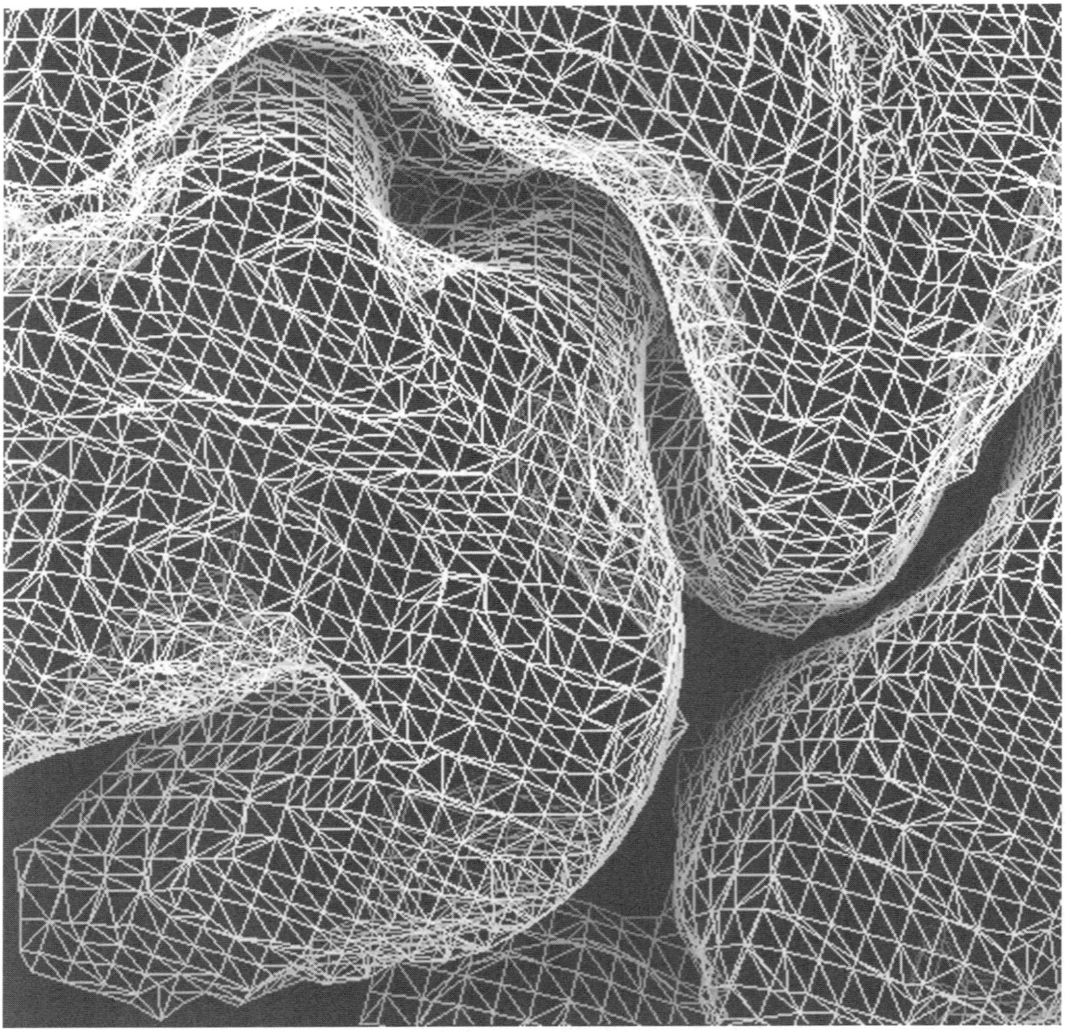

FIGURE 7.5A Triangulated portion of the brain from MRI, created with the Marching Cubes algorithm.

local surface orientation, described by a *normal vector* perpendicular to the surface. The original Marching Cubes algorithm calculates the surface normal vectors from the grey level gradients in the data volume, described later.

7.1.2.4 Volume-Based Rendering

In *volume-based rendering*, images are created directly from the volume data. Compared to surface-based methods, the major advantage is that all grey level information, which has originally been acquired, is kept during the rendering process. As shown by Höhne et al.,[47] this makes it an ideal technique for interactive data exploration. Threshold values and other parameters, which are not clear from the beginning, can be changed interactively. Furthermore, volume-based rendering allows a combined display of different aspects such as opaque and semi-transparent surfaces, cuts, and maximum intensity projections. A current drawback of volume-based techniques is that the large amount of data, which has to be handled, does not allow real-time applications on present-day computers.

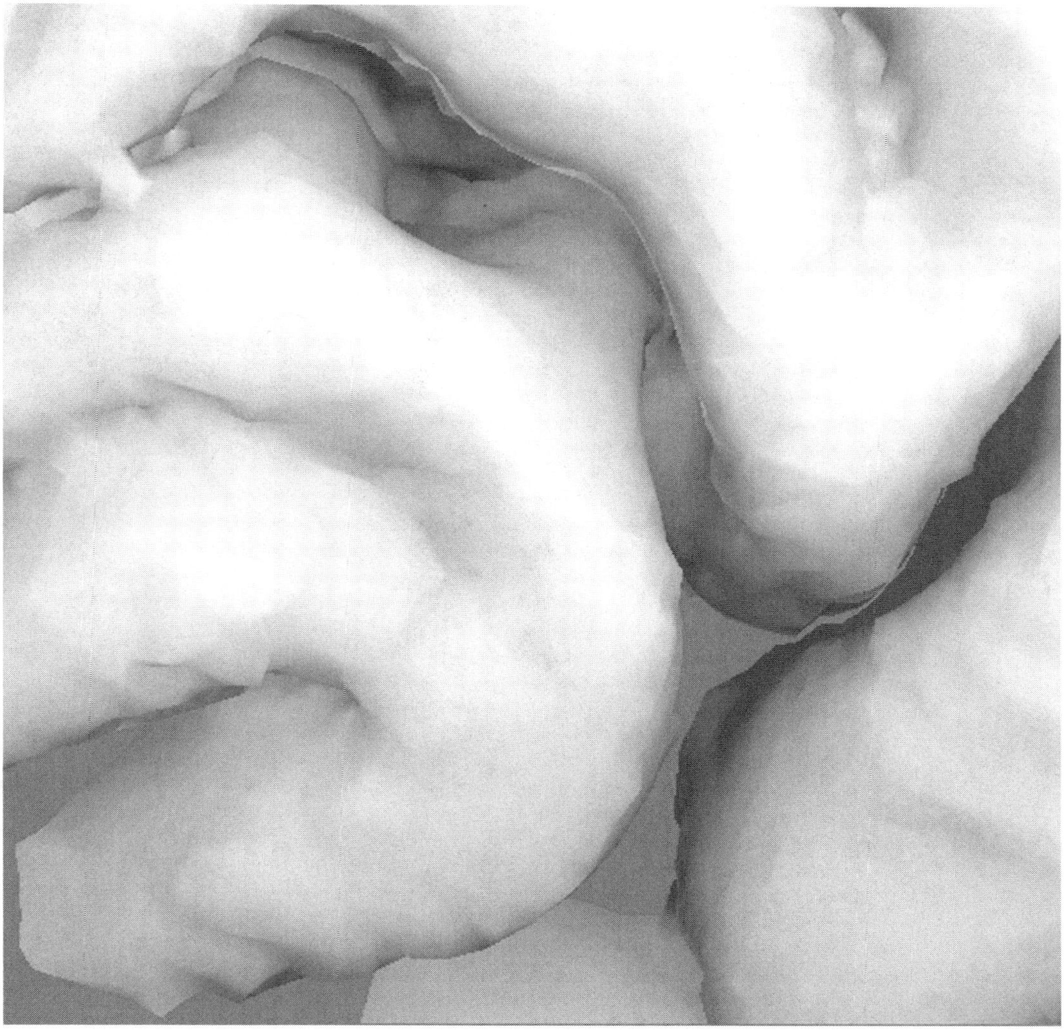

FIGURE 7.5B Shaded portion of the brain from MRI, created with the Marching Cubes algorithm.

7.1.2.4.1 Scanning the Volume

In volume-based rendering, we basically have the choice between two scanning strategies: pixel by pixel (image order) or voxel by voxel (volume order). These strategies correspond to the image and object order rasterization algorithms used in computer graphics.[32] In *image order* scanning, the data volume is sampled on rays along the view direction. This method is commonly known as *ray casting*:[47]

FOR each pixel on image plane DO
    FOR each sampling point on associated viewing ray DO
        compute contribution to pixel

The principle is illustrated in Figure 7.6. Along the ray, visibility of surfaces and objects is easily determined. The ray can stop when it meets an opaque surface. Yagel et al. extended this approach to a full *ray tracing* system which follows the viewing rays as they are reflected on various surfaces.[125] Multiple light reflections between specular objects can thus be handled.

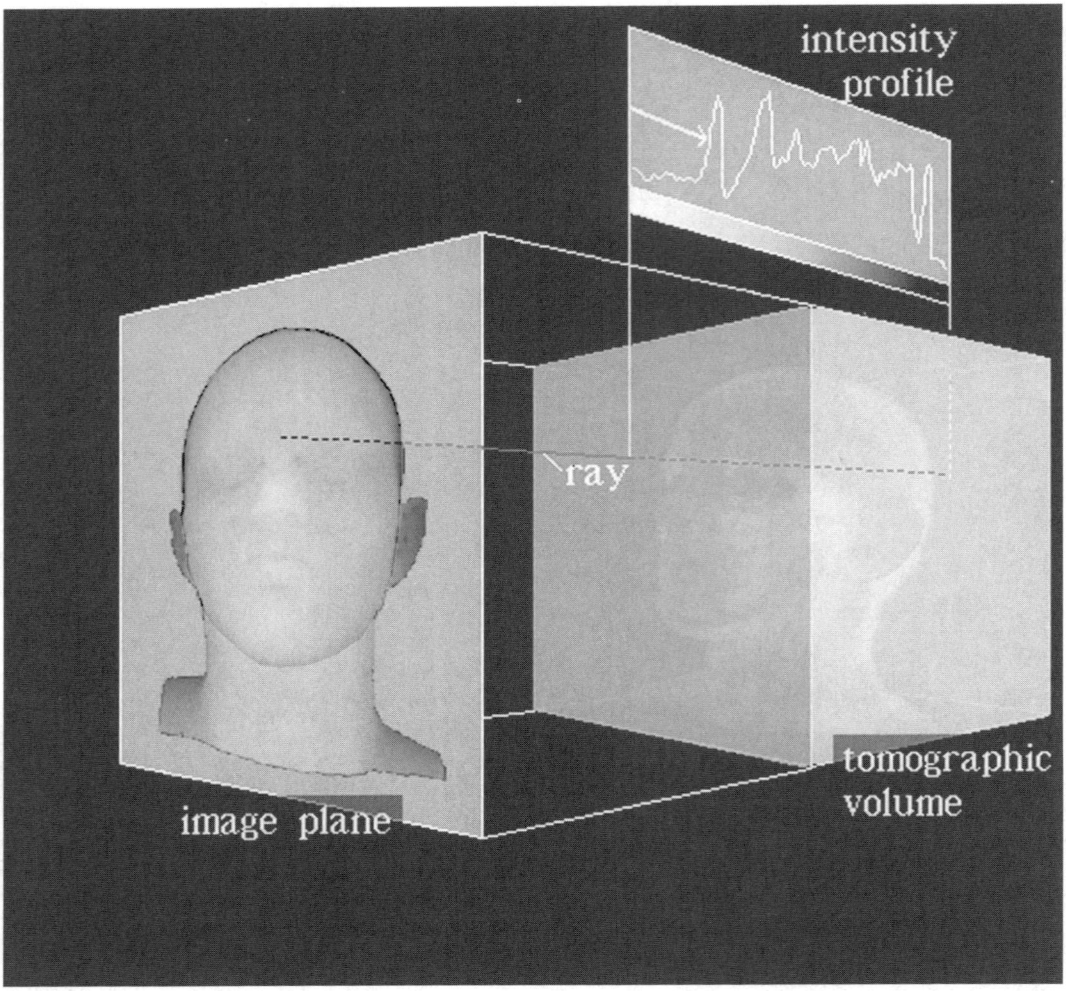

FIGURE 7.6 Principle of ray casting for volume visualization. In this case, the object surface is found using an intensity threshold.

Image order scanning can be used to render both voxel and polygon data at the same time.[63] Image quality can be adjusted by choosing smaller (oversampling) or wider (undersampling) sampling intervals.[62,82] As a drawback, the whole input volume must be available for random access to allow arbitrary view directions. Furthermore, interpolation of the intensities at the sampling points is required. A strategy to reduce computation times is based on the observation that most of the time is spent traversing empty space, far away from the objects to be shown. If the rays are limited to scan the data only within a predefined *bounding volume* around these objects, scanning times are greatly reduced.[4]

In *volume order* scanning, the input volume is sampled along the lines and columns of the 3D array, projecting a chosen aspect onto the image plane in the direction of view:

FOR each sampling point in volume DO

  FOR each pixel projected onto DO

    compute contribution to pixel

The volume can either be traversed in back-to-front (BTF) order from the voxel with maximal to the voxel with minimal distance to the image plane or vice versa in front-to-back (FTB) order. Scanning the input data as they are stored, these techniques are reasonably fast, even on computers with small main memory, and especially suitable for parallel processing. So far, ray casting algorithms still offer a higher flexibility in combining different display techniques. However, volume rendering techniques working in volume order are also available.[121]

7.1.2.4.2 Shaded Surfaces

Using one of the described scanning techniques, the visible surface of an object can be determined with a threshold and/or an object label. Unfortunately, using object labels will introduce a somewhat blocky appearance of the surface, especially when zooming into the scene. An algorithm which solves this problem, based on finding an iso-intensity surface, is presented in Reference 61.

As shown by Höhne and Bernstein,[45] a very realistic and detailed presentation is obtained if the grey level information present in the data is taken into account. Due to the partial volume effect, the grey levels in the 3D neighborhood of a surface voxel represent the relative proportions of different materials inside these voxels. The resulting *grey level gradients* can thus be used to calculate surface inclinations. The simplest variant is to calculate the components of a gradient G for a surface voxel at (i, j, k) from the grey level g of its six neighbors along the main axes as

$$Gx = g(i + 1, j, k) - g(i - 1, j, k)$$

$$Gy = g(i, j + 1, k) - g(i, j - 1, k)$$

$$Gz = g(i, j, k + 1) - g(i, j, k - 1)$$

Scaling G to unit length yields the normal surface. The grey level gradient may also be calculated from all 26 neighbors in a $3 \times 3 \times 3$ neighborhood, weighted according to their distance from the surface voxel.[109] Aliasing patterns are thus almost eliminated. A different approach is to use the first derivative of a higher order interpolation function, such as a cubic spline.[68]

7.1.2.4.3 Cut Planes

Once a surface view is available, a very simple and effective method to visualize interior structures is cutting. When the original intensity values are mapped onto the cut plane, they can be better understood in their anatomical context.[47] A special case is selective cutting, where certain objects are excluded (Figure 7.7).

7.1.2.4.4 Maximum Intensity Projection

For small, bright objects such as vessels from MRA, *maximum intensity projection* (MIP) is a suitable display technique. Along each ray through the data volume, the maximum grey level is determined and projected onto the image plane. The advantage of this method is that neither segmentation nor shading is needed, which may fail for very small vessels. But there are also some drawbacks: as light reflection is totally ignored, MIP does not give a realistic 3D impression. Spatial perception can be improved by rotating the object or by a combined presentation with other surfaces or cut planes.[47]

7.1.2.4.5 Volume Rendering

Volume rendering is the visualization equivalent to fuzzy segmentation. For medical applications, these methods were first described by Drebin et al.[24] and Levoy.[62] A commonly assumed underlying model is that of a colored, semi-transparent gel with suspended low-albedo (low-reflectivity) particles.[9] Illumination rays are partly reflected and change color while traveling through the volume.

Each voxel is assigned a color and an opacity. This opacity is the product of an "object weighting function" and a "gradient weighting function." The object weighting function is usually dependent on the grey level, but it can also be the result of a more sophisticated fuzzy segmentation algorithm. The gradient weighting function emphasizes surfaces for 3D display. All voxels are shaded, using the grey

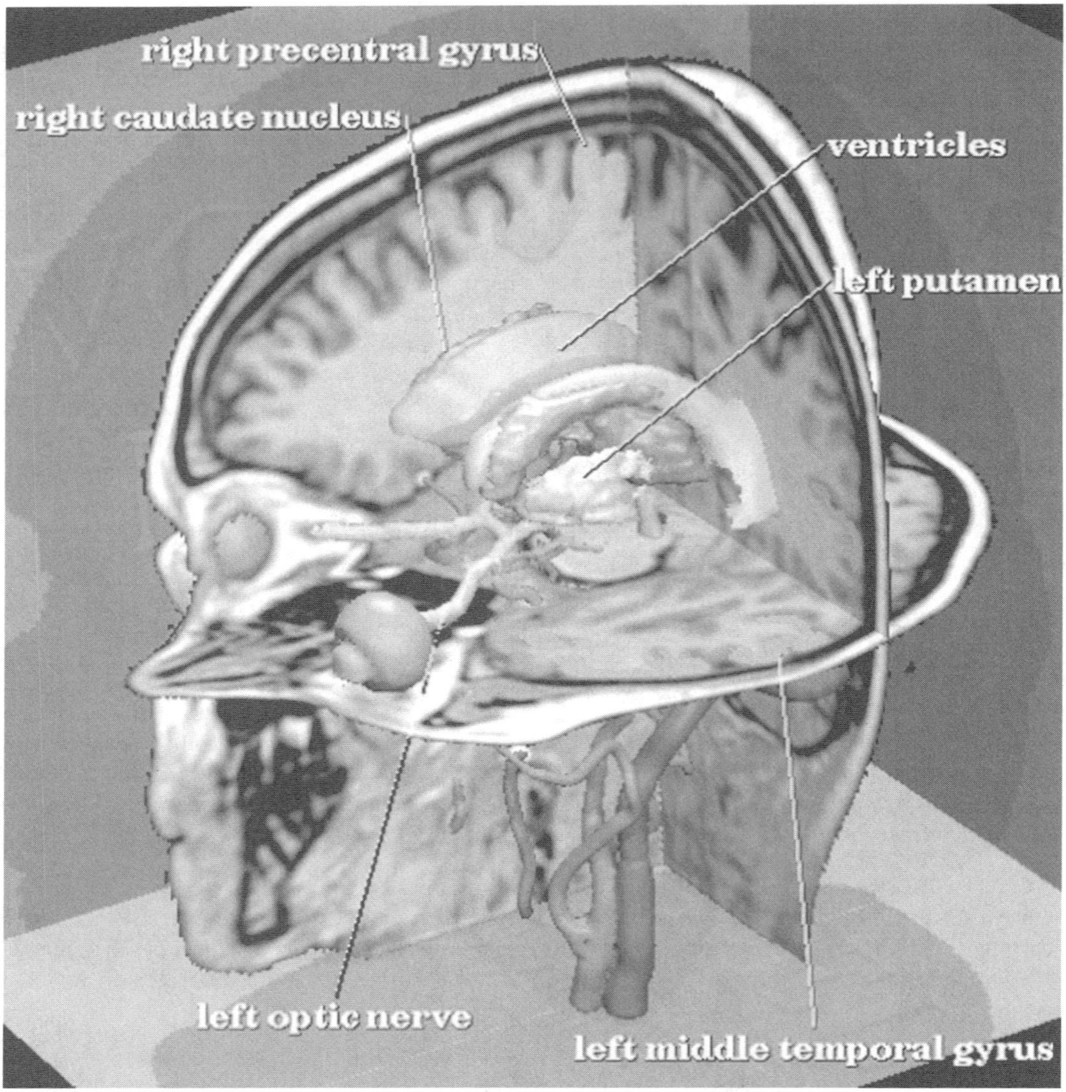

FIGURE 7.7 Brain from MRI. Original intensity values are mapped onto the cut planes.

level gradient method. The shaded values along a viewing ray are weighted and summed up. A somewhat simplified basic equation modeling frontal illumination with a ray casting system is given as follows:

intensity	= intensity of reflected light	
p	= index of sampling point on ray	(0 … max. depth)
l	= fraction of incoming light	(0.0 … 1.0)
α	= local opacity	(0.0 … 1.0)
s	= local shading component	

$$intensity\,(p, l) = \alpha(p) \cdot l \cdot s\,(p) + (1.0 - \alpha(p)) \cdot \text{intensity}\,(p + 1 \cdot (1.0 - \alpha(p)) \cdot l)$$

The total reflected intensity as displayed on a pixel of the 3D image is given as *intensity* (0, 1.0). Since binary decisions are avoided in volume rendering, the resulting images do not show pronounced artifacts such as jagged edges and false holes. On the other hand, a number of superimposed, more or less transparent surfaces are often hard to understand. Spatial perception can be improved by rotating the object. Another problem is the large number of parameters, which have to be specified to define the weighting functions. Furthermore, volume rendering is comparably slow because weighting and shading operations are performed for many voxels on each ray.

7.1.2.5 Transform-Based Rendering

While both surface- and volume-based rendering are operated in a 3D space, 3D images may also be created from other data representations. One such method is *frequency domain rendering*, which creates 3D images in Fourier space, based on the Fourier projection-slice theorem.[112] This method is very fast, but the resulting images are somewhat similar to X-ray images, lacking real depth information.

A more promising approach is *wavelet transforms*. These methods provide a multi-scale representation of 3D objects, with the size of the represented detail locally adjustable. Thus, the amount of data and rendering times may be dramatically reduced. Application to volume visualization is discussed in References 74 and 75.

7.1.2.6 Volume Registration

For many clinical applications, it is desirable to combine information from different imaging modalities. For example, for the interpretation of PET images, which show only physiological aspects, it is important to know the patient's morphology, as shown in MRI. In general, different data sets do not match geometrically. Therefore, it is required to transform one volume with respect to the other. This process is known as registration. The transformation may be defined by using corresponding *landmarks* in both data sets.[113] In a simple case, external markers attached to the patient are available which are visible on different modalities. Otherwise, arbitrary pairs of matching points may be defined. A more robust approach is to interactively match larger features such as surfaces (Figure 7.8), or selected internal features such as the AC-PC line (anterior/posterior commissure) in brain imaging.[100] All these techniques may also be applied in scale-space at different levels of resolution.[73]

In a fundamentally different approach, the results of a registration step are evaluated at every point of the combined volume using *voxel similarity measures*, based on intensity values.[107,120] Starting from a coarse match, registration is achieved by adjusting position and orientation until the mutual information between both data sets is maximized. Since these methods are fully automatic and do not rely on a possibly erroneous definition of landmarks, they are increasingly considered superior. A comparison of various approaches is found in Reference 130.

7.1.2.7 Intelligent Visualization

Knowledge for the interpretation of the 3D images described so far still has to come from the observer. In contrast, the 3D brain atlas VOXEL-MAN/brain shown in Figure 7.9 is based on an *intelligent volume* (see Section 7.1.2.1.1), which has been prepared from an MRI data set.[49,84] It contains spatial and symbolic descriptions in terms of anatomical objects, their relationships, etc. of morphology, function, and blood supply. The brain may be explored on the computer screen in a style close to a real dissection and may be queried at any point. Besides, in education,[20] such atlases are also a powerful aid for the interpretation of clinical images.[100]

If high-resolution cryosections such as those created in the *Visible Human Project* of the National Library of Medicine[106] are used, even more detailed and realistic atlases can be prepared.[101,110] An example image from the VOXEL-MAN Junior/Inner Organs atlas[67] is shown in Figure 7.10. The technical background of the VOXEL-MAN Junior atlases,[20,67] based on precalculated QuickTime VR image matrices, is presented in Reference 104.

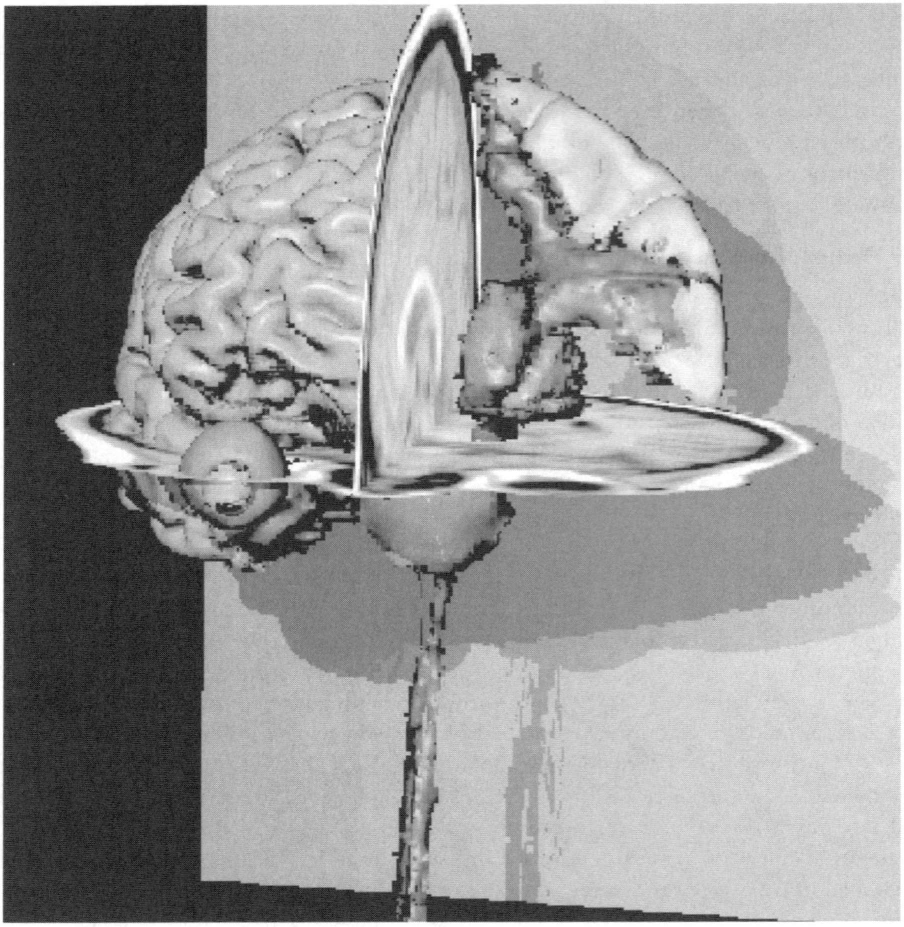

FIGURE 7.8 Volume registration of different imaging modalities MRI, showing morphology, is combined with an FDG-PET, showing the glucose metabolism of a volunteer. Since the entire volume is mapped, the activity can be explored at any location.

7.1.2.8 Intervention Rehearsal

So far, we have focused on merely visualizing the data. A special case is to move the camera inside the patient for virtual endoscopy.[33] Besides, in education, potential applications are in non-invasive procedures, such as interventional radiology (Figure 7.11), gastrointestinal diagnosis, and virtual colonoscopy.

A step further is to manipulate the data at the computer screen for surgery simulation. These techniques are most advanced for craniofacial surgery, where a skull is dissected into small pieces and then rearranged to achieve a desirable shape (Figure 7.12). Several systems have been designed based on the binary voxel model[78,126] or polygon representations.[18] Pflesser et al. developed an algorithm which handles full grey level volumes.[81] Thus, all features of volume-based rendering, including cuts and semi-transparent rendering of objects obscuring or penetrating each other, are available.

Another promising area is the simulation of soft tissue deformation. This may be due to applying force, e.g., using surgical tools,[98] or as a consequence of an osteotomy, modifying the underlying bone structures.[31]

7.1.2.9 Image Quality

For applications in the medical field, it is mandatory to assure that the 3D images show the true anatomical situation or to at least know about their limitations. A common approach for investigating *image fidelity*

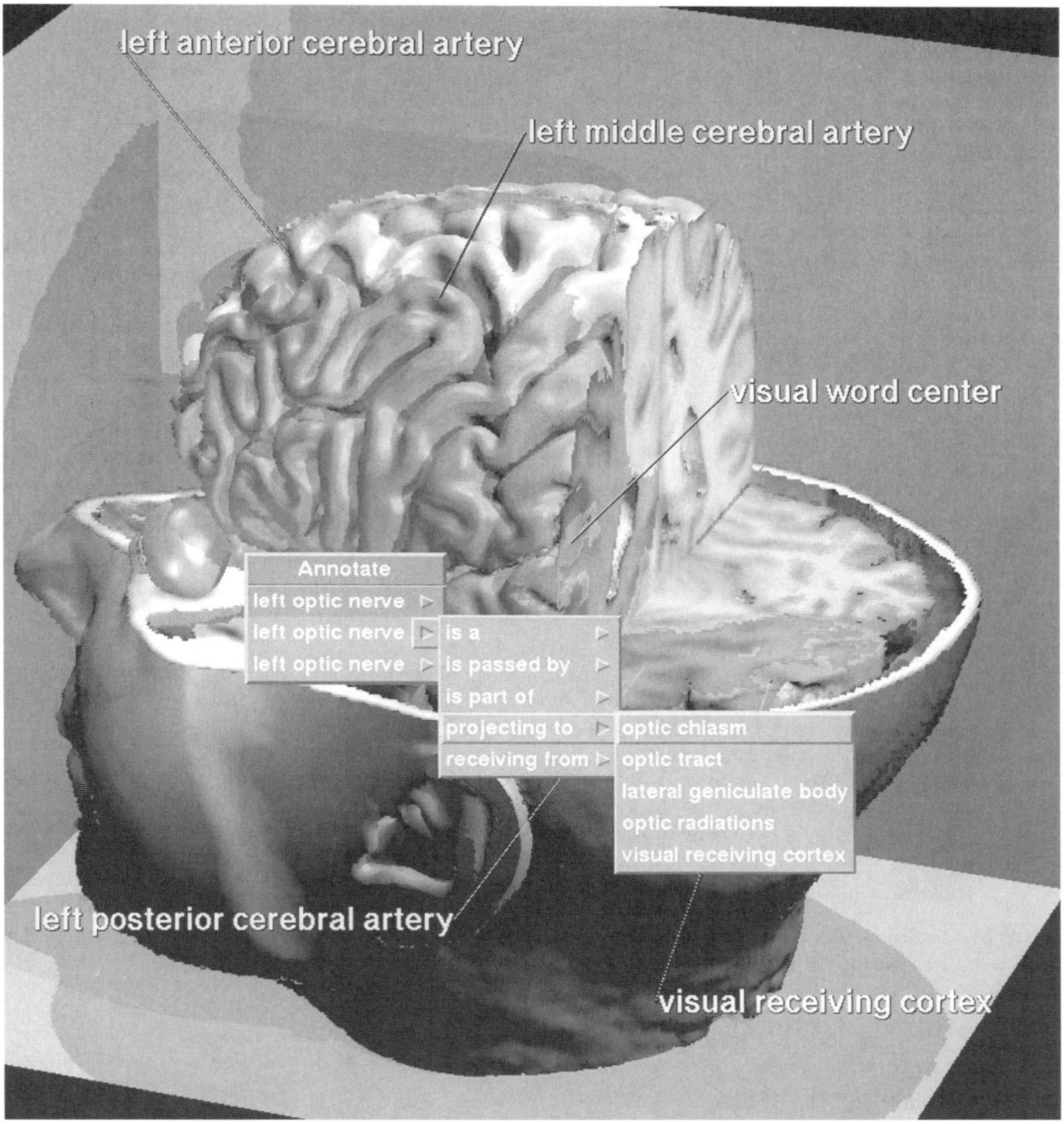

FIGURE 7.9 Exploration of a brain atlas. Arbitrary cutting reveals the interior. The user has accessed information available for the optic nerve concerning functional anatomy, which appears as a cascade of pop-up menus. He has asked the system to colormark some blood supply areas and cortical regions. He can derive from the colors that the visual word center is supplied by the left middle cerebral artery.

is to compare 3D images rendered by means of different algorithms. This method, however, is of limited value since the truth is usually not known. A more suitable approach is to apply volume visualization techniques to simulated data,[66,85,109] and to data acquired from corpses.[25,41,79,83,92] In both cases, the actual situation is available for comparison. In a more recent approach, the frequency behavior of the involved operations is investigated.[68,85]

Another aspect of *image quality* is image utility, which describes whether an image is really useful for a viewer with respect to a certain task. Investigations of 3D image utility in craniofacial surgery may be found in References 2, 105, and 114.

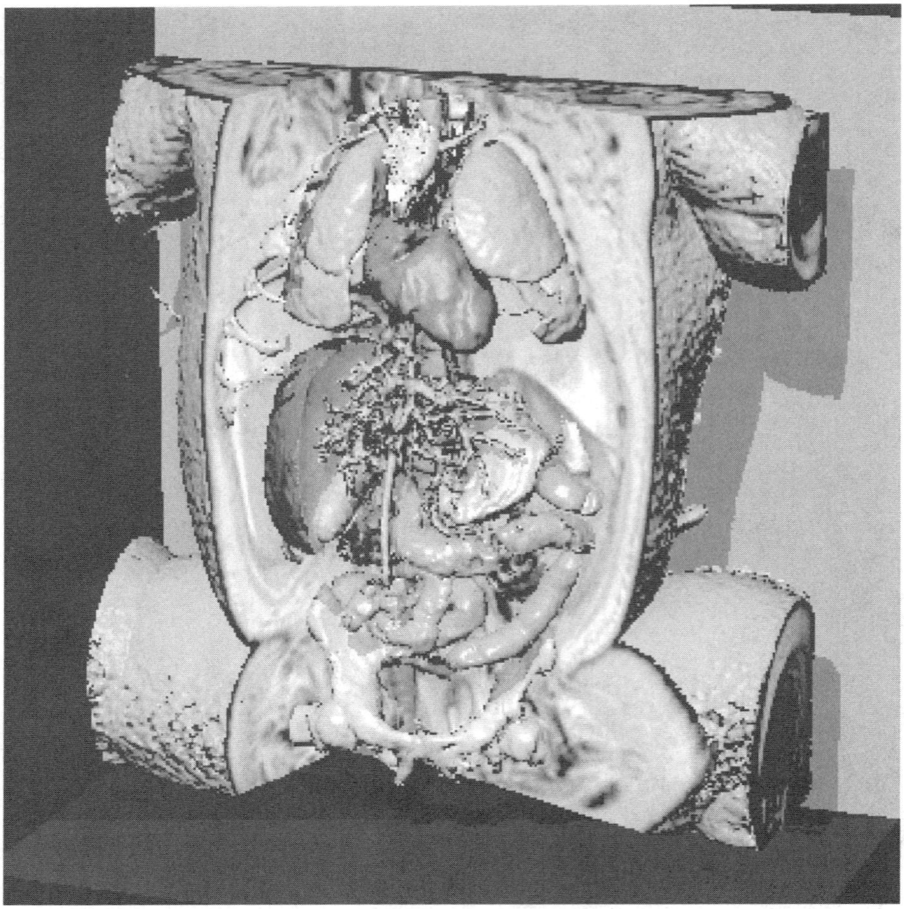

FIGURE 7.10 Dissection of the visible human. Used in a state-of-the-art visualization environment, these data represent a new quality of anatomical imaging.

7.2 Applications to Medical Data

7.2.1 Radiological Data

At first glance, one might expect diagnostic radiology to be the major field of application for volume visualization in medicine. In general, however, this is not the case. One of the reasons is clearly that radiologists are especially skilled in reading cross-sectional images. Another reason is that many diagnostic tasks such as tumor detection and classification can be done based on cross-sectional images. Furthermore, 3D visualization of these objects from MRI requires robust segmentation algorithms which are not yet available. A number of successful applications are presented in Reference 94.

The situation is generally different in all fields where therapeutical decisions have to be made by non-radiologists on the basis of radiological images.[46,127] A major field of application for volume visualization methods is *craniofacial surgery*.[2,22,64,128] Volume visualization not only facilitates understanding of pathological situations, but is also a helpful tool for planning optimal surgical access and cosmetic results of an intervention. A typical case is shown in Figure 7.12. Dedicated procedures for specific disorders have been developed,[105] which are now in routine application.

An application that is becoming more and more attractive with the increasing resolution and specificity of MRI is *neurosurgery planning*. Here, the problem is to choose a proper access path to a

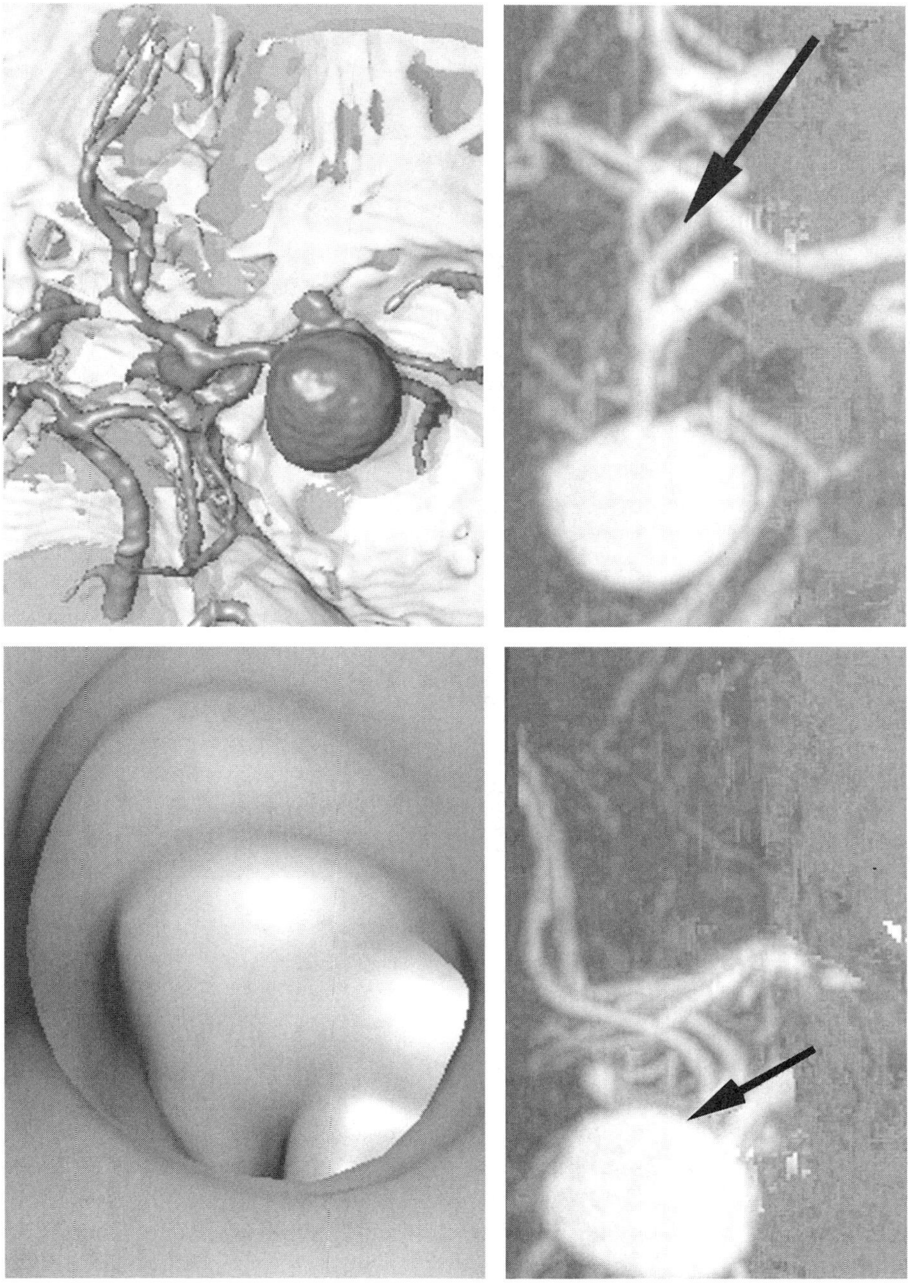

FIGURE 7.11 Virtual catheter examination of a large aneurysm of the right middle cerebral artery, based on CT angiography: (top left) inner structure of the blood vessels as seen from a virtual catheter camera; (top right) 3D overview image showing the blood vessels in relation to the cranial base; (bottom) corresponding MIP images in different orientations. The current position and view direction of the camera are indicated by arrows.

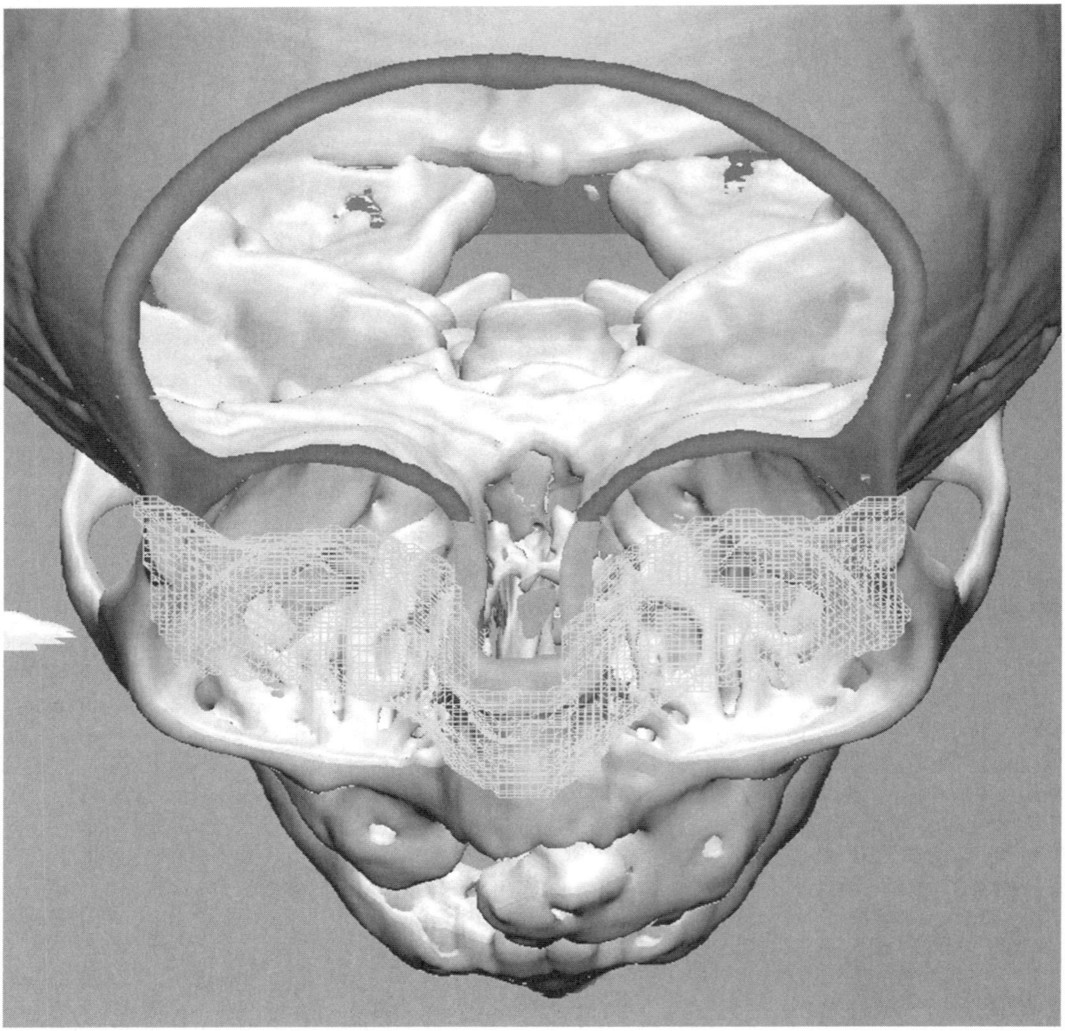

FIGURE 7.12A Craniofacial surgery simulation of a complex congenital malformation of a 7-month-old patient. A CT data set was used to simulate a classic floating-forehead procedure. A frontobasal segment of the skull (shown as a wire mesh) was cut and can be moved in any direction. (Figure 7.12B) Corresponding intra-operative view.

lesion (Figure 7.11). 3D visualization of brain tissue from MRI and blood vessels from MRA before surgical intervention allows the surgeon to find a path with minimal risk in advance.[21,82] In combination with an optical tracking system, the acquired information can be used to guide the surgeon during the intervention.[6] In conjunction with functional information from PET images, localization of a lesion is facilitated (Figure 7.8). The state of the art in *computer-integrated surgery* is presented in Reference 108.

Another important application that reduces the risk of a therapeutical intervention is *radiotherapy planning*. Here, the objective is to focus the radiation as closely as possible to the target volume, while avoiding side effects in healthy tissue and radiosensitive organs at risk. 3D visualization of target volume, organs at risk, and simulated radiation dose allows an iterative optimization of treatment plans.[50,55,99,102]

Applications apart from clinical work include *medical research* (Figure 7.8) and *education* (Figure 7.10). In the current decade of brain research, exploring and mapping brain functions is a major issue

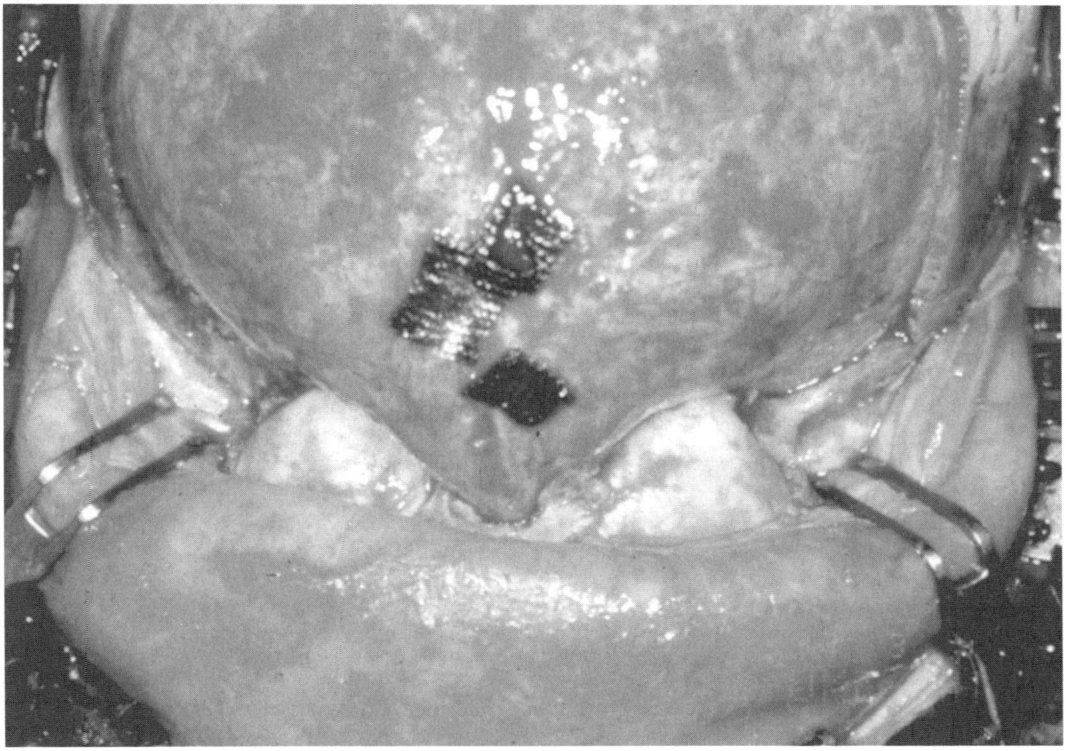

FIGURE 7.12B Craniofacial surgery simulation of a complex congenital malformation of a 7-month-old patient. Corresponding intra-operative view of Figure 7.12A.

(Figure 7.9). Volume visualization methods provide a framework to integrate information obtained from such diverse sources as dissection, functional MRI, or magnetoencephalography.[111]

7.2.4 3D Ultrasound

7.2.4.1 Introduction

3D ultrasound is a very new and interesting application in the area of "tomographic" medical imaging, able to become a fast, non-radiative, non-invasive, and inexpensive volumetric data acquisition technique with unique advantages for the localization of vessels and tumors in soft tissue (spleen, kidneys, liver, breast, etc.). In general, tomographic techniques (CT, MR, PET, etc.) allow for a high anatomical clarity when inspecting the interior of the human body (Figure 7.13).

In addition, they enable a 3D reconstruction and examination of regions of interest, offering obvious benefits (reviewing from any desired angle; isolation of crucial locations; visualization of internal structures; "fly-by;" accurate measurements of distances, angles, volumes, etc.).

The physical principle of ultrasound is as follows:[72] sound waves of high frequency (1 to 15 MHz) emanate from a row of sources that are located on the surface of a transducer which is in direct contact with the skin (Figure 7.14). The sound waves penetrate the human tissue, traveling with a speed of 1450 to 1580 m/s, depending upon the type of tissue. The sound waves are reflected partially if they hit an interface between two different types of tissue (e.g., muscle and bone). The reflected wavefronts are detected by sensors (microphones) located next to the sources on the transducer. The intensity of

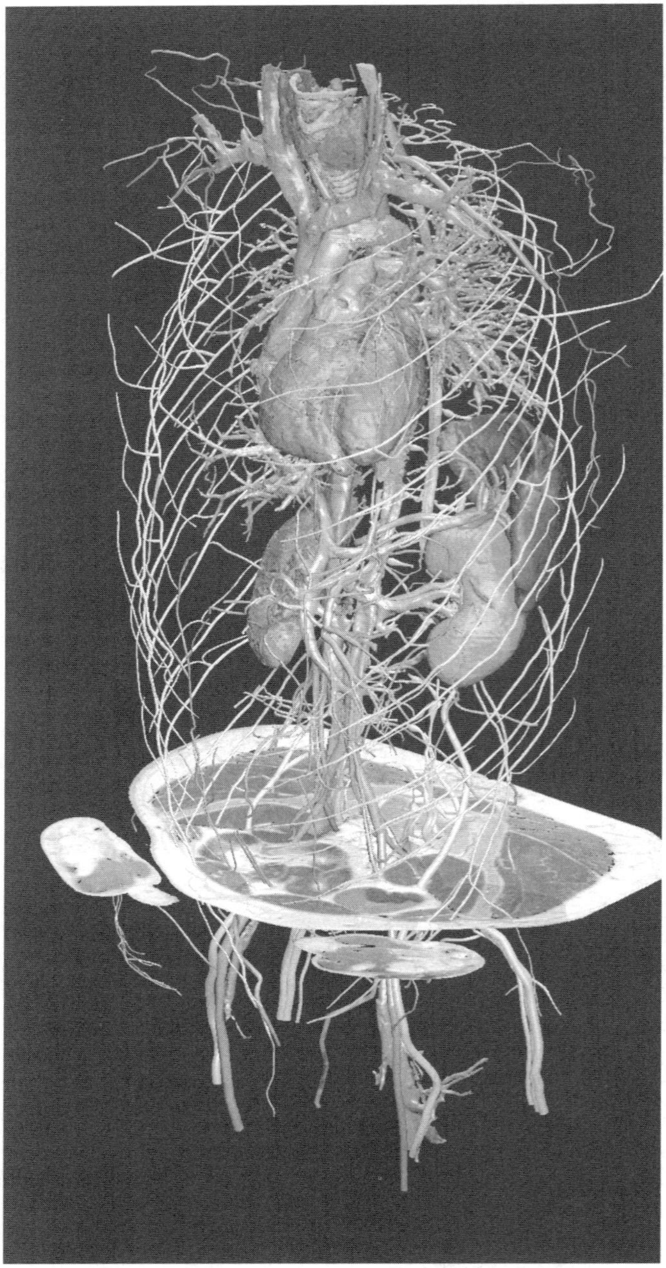

FIGURE 7.13 Gradual transition between surface and MIP visualization of a gamma camera data set of the pelvis. The heart, kidneys, liver, and spleen are visible. Three hemangiomas can be seen in the MIP mode.

reflected energy is proportional to the sound impedance difference of the two corresponding types of tissue and depends on the difference of the sound impendances Z_1 and Z_2:

$$I_r = I_e \cdot \frac{\left(I - \frac{Z_2}{Z_1}\right)}{\left(1 + \frac{Z_2}{Z_1}\right)} \tag{7.1}$$

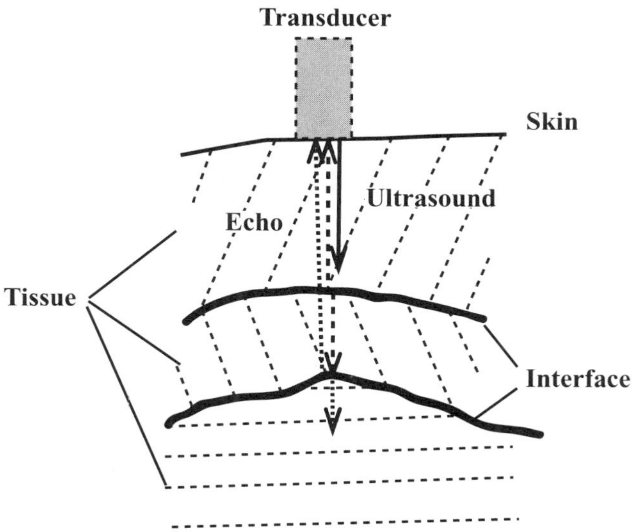

FIGURE 7.14 The principal function of ultrasound.

An image of the interior structure can be reconstructed based upon the total traveling time, the (average) speed, and the energy intensity of the reflected waves. The resulting 3D images essentially represent hidden internal "surfaces." The principle is similar to radar, with the difference being that it uses mechanical instead of electromagnetic waves.

7.2.4.2 Collecting 3D Ultrasound Data

In contrast to the common 2D case where a single image slice is acquired, 3D ultrasonic techniques cover a volume within the body with a series of subsequent image slices. The easiest way to collect 3D ultrasound data is to employ a Kretz Voluson 530 device. This is a commercially available device which allows direct acquisition of a whole volume area instead of a single slice. The principle of the Kretz device is based on a mechanical movement of the transducer during acquisition along a rotational or sweep path (see Figure 7.15)

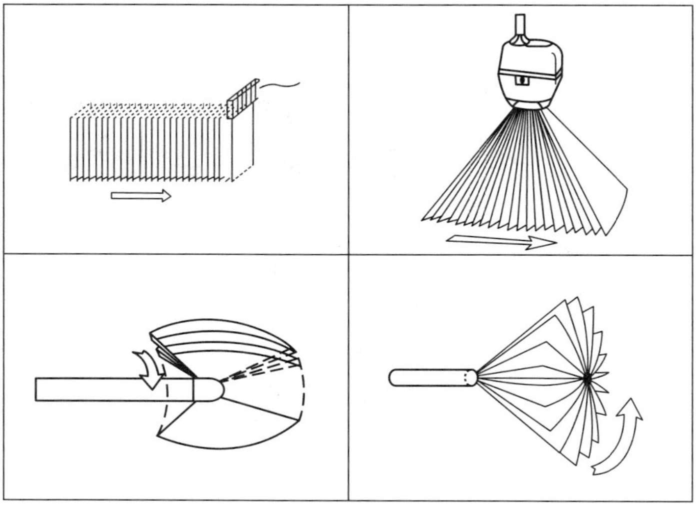

FIGURE 7.15 Different mechanical scanning methods.

The advantage of the Kretz system lies in its high decision and commercial availability. Its disadvantage is that the rather high system price makes it somehow difficult to purchase for physicians. The alternative is a free-hand scanning system which allows the upgrade of virtually any existing conventional (2D) ultrasound system to full 3D-capabilities. Such an update can be done exclusively by external components and hence does not require any manipulation of the existing hardware and software configuration. After the upgrade, the ultrasound equipment can be operated in both the 2D as well as the 3D mode almost simultaneously. Switching from the 2D to the 3D mode requires only a mouse click. As a result, the familiar 2D examination procedure remains unchanged, and the physician can switch on the 3D mode only when this is necessary. The system architecture is illustrated in Figure 7.16. The upgrade requires the employment of two external components:

1. A 6-degrees-of-freedom (6DOF) tracking system for the transducer — Such a tracking system is mounted on the transducer and follows very precisely its position and orientation in 3D space. Thus, each 2D image is associated with corresponding position and orientation coordinates. The physician can now move the transducer free-hand over the region under examination. Commercially, several different types of 6DOF tracking systems exist: mechanical arms, electromagnetic trackers, and camera-based trackers (infrared or visible light).
2. An image digitalization and volume rendering system — This component consists of a frame grabber, a workstation or PC with sufficient memory and processor power, a serial interface, and the usual peripheral devices (monitor, mouse, printer, etc.). The video output of the 2D ultrasound machine is connected to the frame grabber, and the 6DOF tracker is connected to the serial input. Every 2D image presented on the ultrasound screen is digitized in real time and stored together with its corresponding tracker coordinates in the memory. After finishing the scanning procedure, all acquired 2D slices are combined into a 3D volume sample of the examined area. This volume data set is then further processed.

7.2.4.3 Visualization of 3D Ultrasound

One of the major reasons for the limited acceptance of 3D ultrasound to date is the complete lack of an appropriate visualization technique able to display clear surfaces out of the acquired data. The very first approach was to use well-known techniques, such as those used for MRI and CT data, to extract surfaces. Such techniques, reported in more detail in the first part of this chapter, include binarization, iso-surfacing, contour connecting, marching cubes, and volume rendering either as a semi-transparent cloud or as fuzzy gradient shading.[61] Manual contouring is too slow and impractical for real-life applications. Unfortunately, ultrasound images possess several features that cause all these techniques to fail totally. The general appearance of a volume rendered 3D ultrasound data set is that of a solid block covered

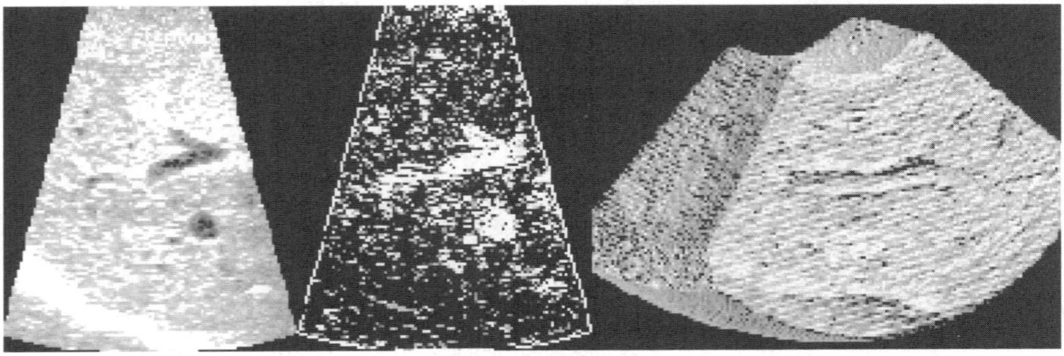

FIGURE 7.16 A grey image of the liver (left), the corresponding opacity values (middle), and a volume rendered data set (right). Note the high opacity values along the interface between data and empty space (middle) causing a solid "curtain" obscuring the volume interior (right).

with "noise snow" (Figure 7.16, right). The most important of these features, as reported in References 95 and 97, are

1. Significant amount of noise and speckle
2. Much lower dynamic range as compared to CT or MR
3. High variations in the intensity of neighboring voxels, even within homogeneous tissue areas
4. Boundaries with varying grey level caused by the variation of surface curvature and orientation to the sound source
5. Partially or completely shadowed surfaces from objects closer and within the direction of the sound source (e.g., a hand shadows the face)
6. The regions representing boundaries are not sharp, but show a width of several pixels
7. Poor alignment between subsequent images (parallel-scan devices only)
8. Pixels representing varying geometric resolutions, depending on the distance from the sound source (fan-scanning devices only)

The next idea in dealing with ultrasound data was to improve the quality of the data during a preprocessing step, i.e., prior to reconstruction, segmentation, and volume rendering. When filtering medical images, a trade-off between image quality and information loss must always be taken into account. Several different filters have been tested: 3D Gaussian for noise reduction, 2D speckle removal for contour smoothing, and 3D median for both noise reduction and closing of small gaps caused by differences in the average luminosity between subsequent images;[95] other filters such as mathematical topology and extended threshold-based segmentation have been tested as well. The best results have been achieved by combining Gaussian and median filters (see Figure 7.17).

However, preprocessing of large data sets (a typical 3D volume has a resolution of 256^3 voxels) requires several minutes of computing, reduces the flexibility to interactively adjust visualization parameters, and aliases the original data. For solving these problems, interactive filtering techniques

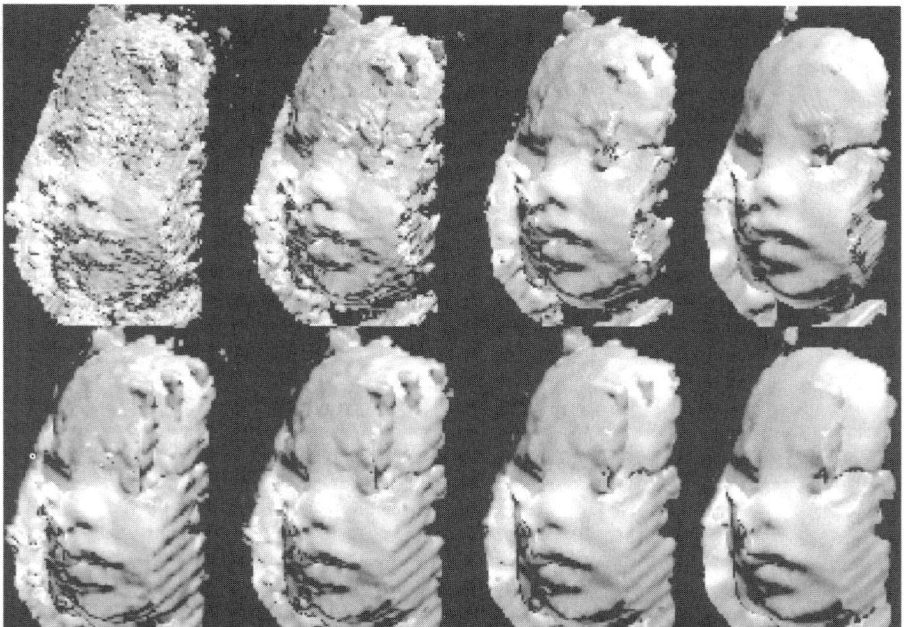

FIGURE 7.17 Volume rendering after off-line 3D median and 3D Gaussian filtering. From left to right: unfiltered and median with a width of 3^3, 5^3, and 7^3. In the lower row, the same data after additional Gaussian filtering with a width of 3^3.

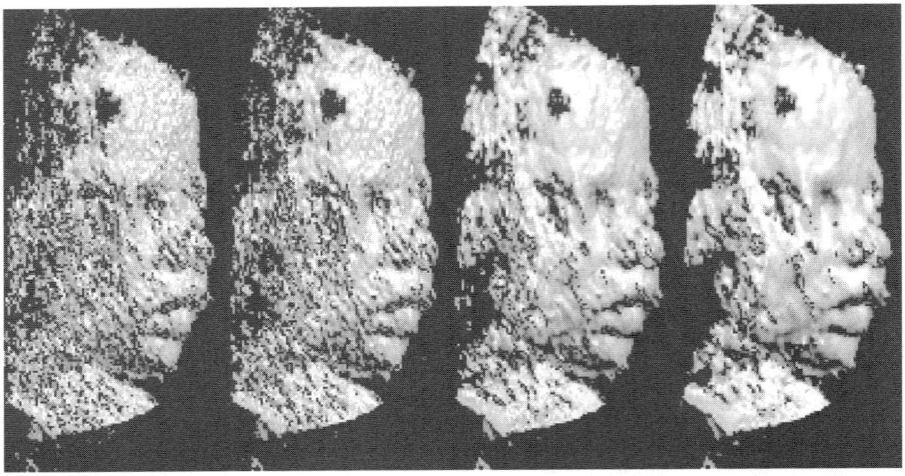

FIGURE 7.18 On-line filtering of the face of a fetus. This filtering is completed in less than 5 s.

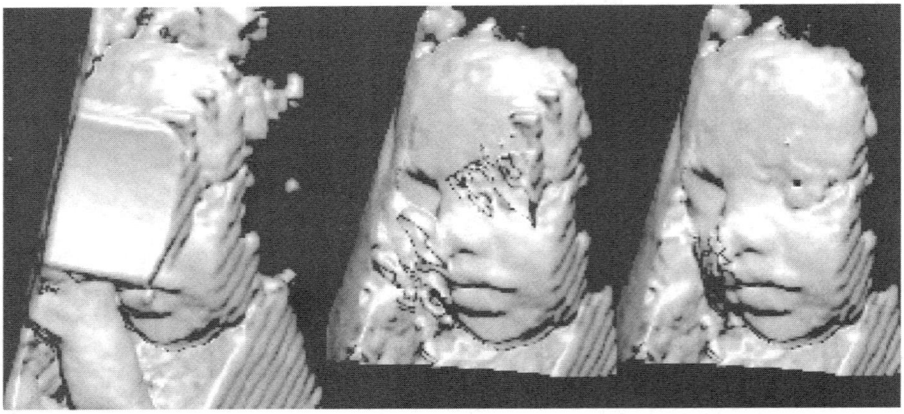

FIGURE 7.19 Fetal face before (left) and after (middle) removing the right hand and the remaining artifacts (right).

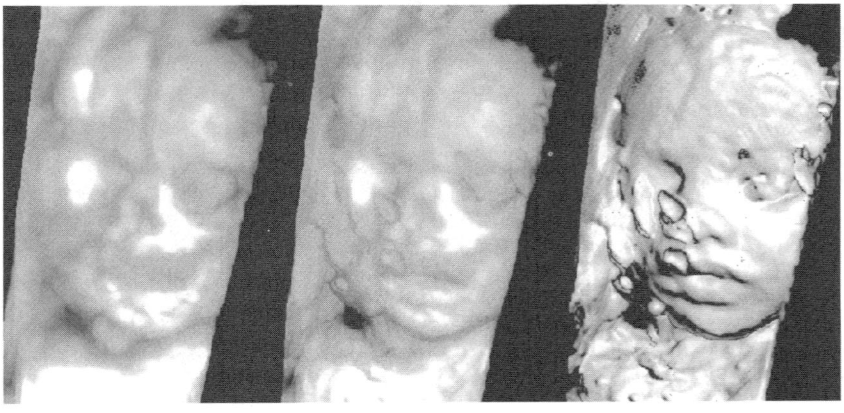

FIGURE 7.20 On-line mixing between surface and MIP models. This operation is performed in real time.

based on multi-resolution analysis and feature extraction have been developed, allowing a user-adjustable, on-line filtering within a few seconds and providing an image quality comparable to the off-line methods[97] (see Figures 7.18, 7.19, and 7.20).

In order to remove artifacts remaining in the image after filtering, semi-automatic segmentation has been applied because of the general lack of a reliable automatic technique. A segmentation can be provided by using the mouse to draw a few crude contours (see Reference 95 for more details). The diagnostic value of surface reconstruction in prenatal diagnosis so far has been seen in the routine detection of small irregularities of the fetal surface, such as cheilo-gnatho-(palato)schisis or small (covered) vertebral defects, as well as in a better spatial impression of the fetus as compared to the 2D imaging. A useful side effect is a psychological one, the pregnant woman gets a plastic impression of the unborn.[5,44] Figure 7.21 compares an image reconstructed from data acquired in the 25th week of pregnancy with a photo of the baby 24 h after birth. The resolution of the data was 256 × 256 × 128 (8 Mbytes); the time for volume rendering one image with a resolution of 300^2 pixels was about 1 s on a Pentium PC.

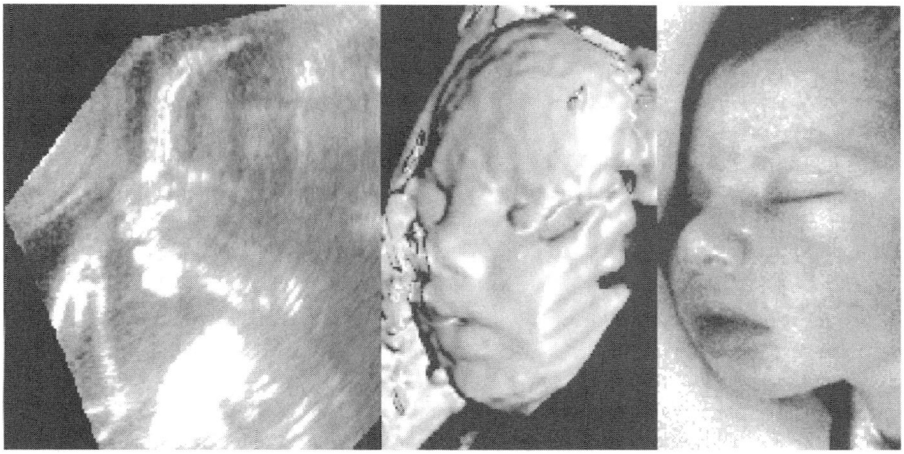

FIGURE 7.21 Comparison of a volume reconstructed from 3D ultrasound data acquired during the 25th pregnancy week (3.5 months before birth) with a photograph of the same baby taken 24 h after birth.

Figure 7.22 shows several other examples of fetal faces acquired at the Mannheim Clinic. It is important to note that these data sets have been acquired under routine clinical conditions, and, therefore, they can be regarded as representative. On average, 80% of the acquired volumes can be reconstructed within ca. 10 min with an image quality comparable to that shown in Figure 7.22. All cases where the fetus was facing the abdominal wall could be reconstructed successfully.

Under clinical aspects, further work should be aimed toward a better distinction and automatic separation of surfaces lying close together and showing relatively small grey scale differences. The reconstruction of surfaces within the fetus, e.g., organs, is highly desirable. Surface properties of organs, but also of pathological structures (ovarian tumors etc.), might give further information for the assessment of the dignity of tumors.

7.2.5 3D Cardiac Reconstruction from 2D Projections

Different imaging modalities are applied in order to acquire medical data. In terms of the human heart, 3D tomographic imaging techniques are not yet suitable for resolving either moving coronary arteries or the changing volume of the heart ventricles.

The golden standard for diagnosis of coronary artery disease or volumetry is X-ray angiography, recently combined with intra-vascular ultrasound (IVUS).[60] The main benefit of this technique is the high spatial and temporal resolution, as well as high image contrast.

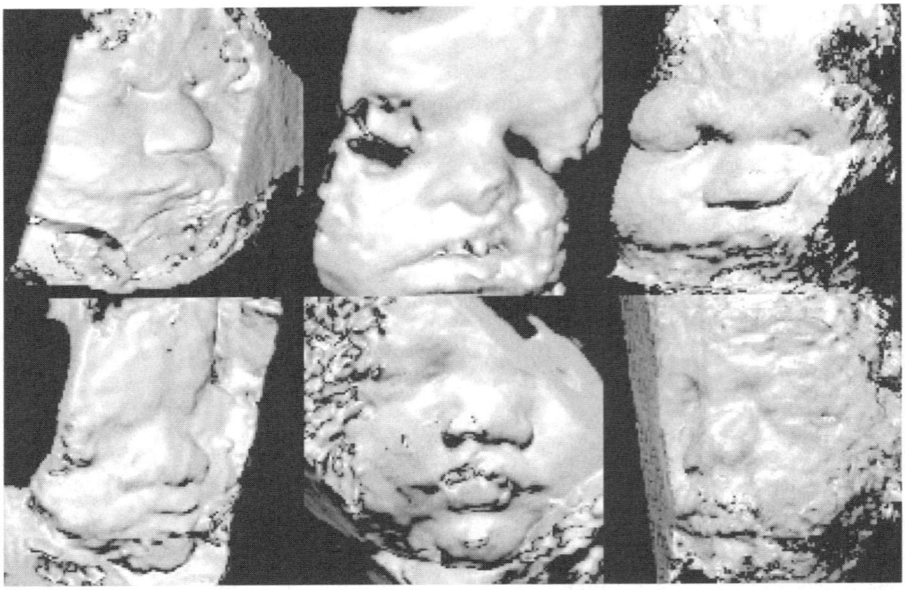

FIGURE 7.22 Six different examples of fetal faces acquired under daily clinical routine conditions.

For treatment planning of angioplasty or bypass surgery or for volumetry, sequences of X-ray images are traditionally acquired and evaluated. Despite the high quality of angiograms, an exact judgment of pathological changes (e.g., stenosis) requires a large amount of experience on the part of the cardiologist.

In order to improve the diagnostic accuracy, 3D reconstruction from 2D coronary angiograms appears desirable.[117] In general, two different approaches can be distinguished. The stereoscopic or multi-scopic determination of ray intersections is a method which makes it necessary to identify correspondent features within different images. If this correspondence is impossible to establish, back-projection techniques[30] are more suitable. The choice of using either the stereoscopic or the back-projection approach mainly depends on the following criteria:

1. **Number of images:** For the stereoscopic approach at least two images are necessary to perform the reconstruction. In order to achieve good results by using back-projection techniques, more than 20 images are necessary.
2. **Relative orientation:** A small, relative orientation results in low accuracy for both stereoscopic and back-projection techniques. Nevertheless, the necessity of a large parallax angle is higher for back-projection techniques.
3. **Morphology:** In order to reconstruct objects which are composed of a number of small, structured parts, stereoscopic techniques are more appropriated. On the other hand, large objects with low structure are easier to reconstruct by back-projection techniques.
4. **Occluding objects:** Occluding objects cause problems when using stereoscopic methods. In contrast, back-projection techniques are able to separate different objects which lie on the same projection ray.

Since the choice of the right technique strongly depends on the current application, both approaches will be described briefly in the following sections.

7.2.5.1 Model-Based Restoration of Teeth

7.2.5.1.1 Introduction

An important goal in image processing of medical images, in addition to the analysis of the images, is the reconstruction and visualization of the scanned objects. The results assist the physician in making a diagnosis and, therefore, contribute to an improved treatment. By means of the reconstruction, it is possible to produce implants made to measure and adjusted to the individual patient's anatomy. Therefore, the object is scanned and afterward reconstructed. Then a high-quality implant is produced using the information of the 3D model. One field of application, where this method gained wide currency, is the restoration of teeth by range images.[13]

7.2.5.1.2 Related Work

The CEREC system, one of the most popular methods for the restoration of teeth, was introduced by Brandestini and colleagues.[13,14] It was developed at the University of Zürich in cooperation with the Brains company (Brandestini Instruments of Switzerland). Today, the CEREC system is developed by Sirona, and in the latest version, it is possible to make crown restorations. Due to the complicated interactive construction process, the surface of the occlusal part of the inlay often has to be modeled manually after the inlay is inserted into the cavity of the prepared tooth.

The Minnesota system, developed from Rekow[87–89] at the University of Minnesota, uses affine transformations of a 3D model tooth to adapt it to the scanned tooth. Instead of covering all kinds of tooth restorations, they consider mainly the production of crowns. The system of Duret et al.,[26,27,28] developed in cooperation with the French company Hennson, works similar to the Minnesota system. They use affine transformations to produce crowns and small bridges.[69] In addition, they use the range image of an intact tooth to determine the occlusal surface of the inlay by 2D free-form deformations based on extracted feature points.

7.2.5.1.3 Occlusal Surface Restoration

Automatic occlusal surface reconstruction for all kinds of tooth restorations is an important ongoing research topic. It is undisputed that an automation of a restoration system is only possible if the typical geometry of teeth is known by the system. One realizable approach is the restoration of the occlusal surface by adapting an appropriate tooth model. Therefore, the starting point for automatic restorations is the theoretical tooth.[37] After positioning and scaling the tooth model, adjustments to the individual patient's anatomy are necessary to get a smooth join of the inlay with the undamaged parts of the real occlusal surface. A description of the method is given in References 37 and 38. The processing pipeline is shown in Figure 7.23.

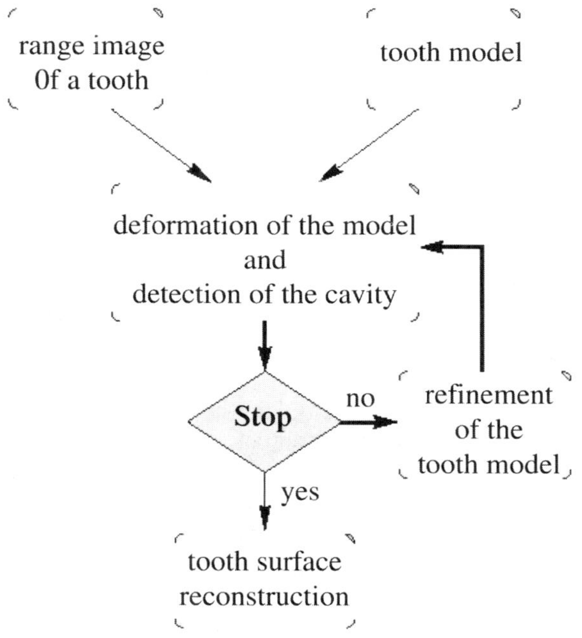

FIGURE 7.23 Occlusal surface restoration using a tooth model (heavy arrows indicate design loop).

7.2.5.1.4 Data Acquisition

For the 3D surface measurements of teeth, we use a new optical 3D laser scanner based on the principle of triangulation. Shaded areas due to steep inclines can be avoided by combining two scans from different directions.[77] The measurement time for 250,000 surface points is about 30 s. The distance between two scanned points is 27 m, and the scanner has a vertical resolution less than 1 m.

7.2.5.1.5 The Geometrically Deformable Model

For the purpose of shape modification, a 3D adaptation of a geometrically deformable model (GDM) is employed. GDMs were introduced by Miller et al.[15,71] for the segmentation and visualization of 2D- and 3D objects. Rückert[91] uses GDMs for the segmentation of 2D medical images. Gürke[39] extends the model by introducing free-function assignment in the control points in order to integrate *a priori* knowledge about the object.

7.2.5.1.6 Optimization Methods

It has been shown that the segmentation process of Miller et al.[15] is very susceptible to noise and artifacts. The reason lies in the usage of the Hillclimbing process. Rückert provided as an improvement the simulated annealing optimization. In theory, this method always finds the global minimum of a function. Due to the enormous complexity of our functions, we decided on the usage of simulated annealing as the optimization method.

7.2.5.1.7 Cavity Detection

In order to avoid a slip of the GDM into the cavity during the adaptation process, there has to be a mechanism to detect the control points of the GDM lying above a cavity. Later, we use this information to calculate the surface of the inlay. We decided on a criterion based on a distance measurement and an adaptive threshold. The threshold depends on the actual mean error calculated in the control points of the GDM.

The detected control points are labeled and removed from the deformation process. In the nonlabelled control points, we store the actual deformation vectors. After we pass all control points in the deformation step, we calculate the deformation vectors for the labeled control points by a weighted Shepard interpolation.

7.2.5.1.8 Results

The images we used in our experiments were captured with a 3D laser scanner at the dental school of Munich and registrated with the Sculptor system 77 at the Fraunhofer Institute for Computer Graphics in Germany.

In Figure 7.24, you can see the range image of a prepared first upper molar (top). For better visibility of details, the rendered image is shown on the bottom.

Figure 7.25 shows the results of the deformations in the different resolutions starting with the initial model and terminating after four refinement steps.

The adaptation process terminates after a sufficient degree of refinement is reached. In our case, we finished after four refinement steps, respectively, five deformation steps. Figure 7.26 shows a 3D view of the reconstructed chewing surface.

Finally, we are able to calculate the inlay by using the information of the range image and the 3D model of the reconstructed occlusal surface. Figure 7.27 shows a volume representation of the inlay.

Figure 7.28 shows the result of a dental restoration of a first lower molar. Following the arrows, you can see, on the left side, the original prepared tooth; then different views of the reconstruction; and finally, the corresponding views of the resulting inlay. The whole process took 21 s on a SPARC 10 workstation, and we obtained an average error of 25 μm.

7.2.5.2 Reconstruction of Coronary Vessels

In this section, a method of reconstructing the 3D appearance of the coronary arteries, based on a sequence of angiograms, acquired by rotating a monoplane system around the heart, will be described. In order to determine the exact phase of the heart cycle for each image, an ECG is recorded

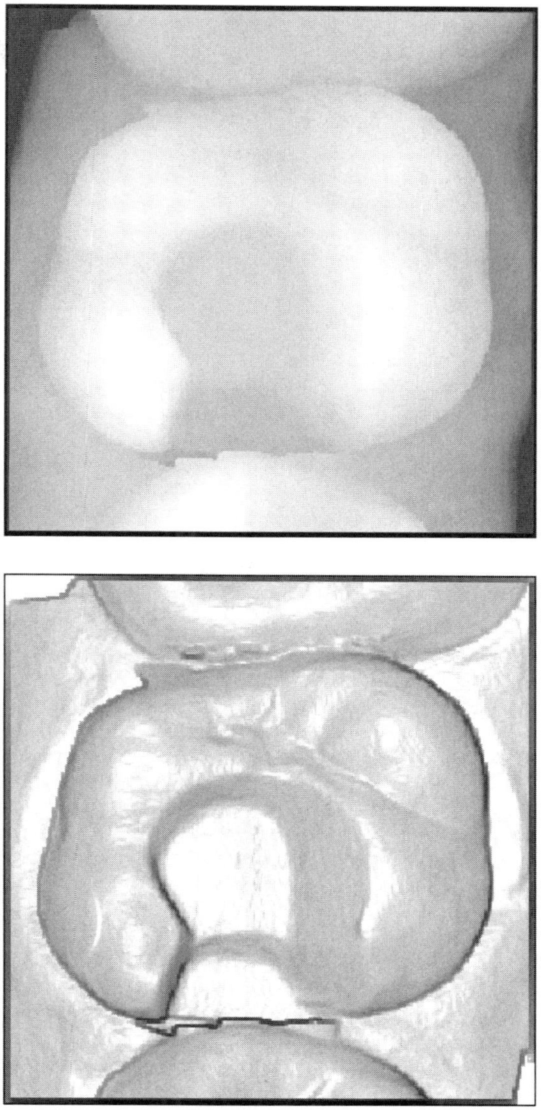

FIGURE 7.24 Range image and corresponding rendered image of a prepared tooth with a cavity.

simultaneously. In order to minimize user interaction and *a priori* knowledge introduced into the reconstruction process,[36] a new method has been developed and implemented. The technique requires a minimum of user interaction limited to the segmentation of vessels in the initial image of each angiographic sequence. The segmentation result is exploited in the entire series of angiograms to track each individual vessel. In contrast to the assumption for 3D reconstruction of objects from multiple projections, coronary arteries are not rigid. Due to the deterministic nature of the mobility of the heart with respect to the phase of the heart motion, distinct images are used, showing the heart at the same phase of the cardiac cycle.

The different processing steps used for reconstructing the 3D geometry of the vessel are shown in Figure 7.29 and are discussed later.[43] In order to separate the vessel tree to be reconstructed, the image has to be segmented. The major drawback of most of the existing segmentation algorithms

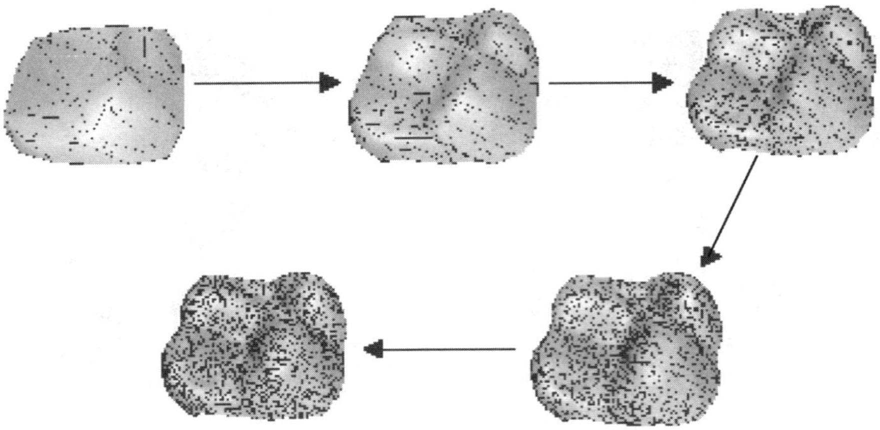

FIGURE 7.25 Result of the adaptation process in the form of triangular meshes.

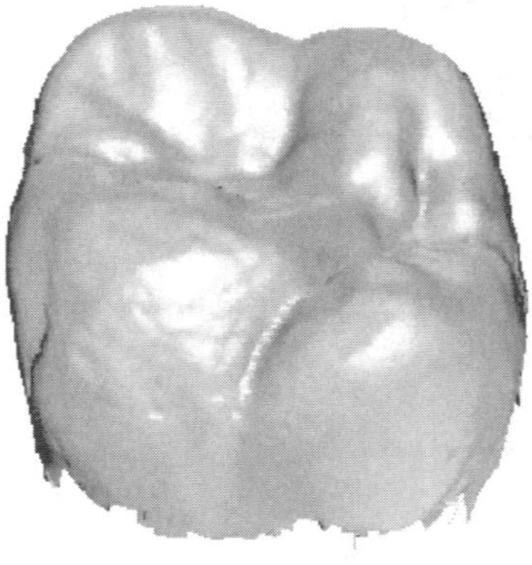

FIGURE 7.26 3D view of the reconstructed occlusal surface.

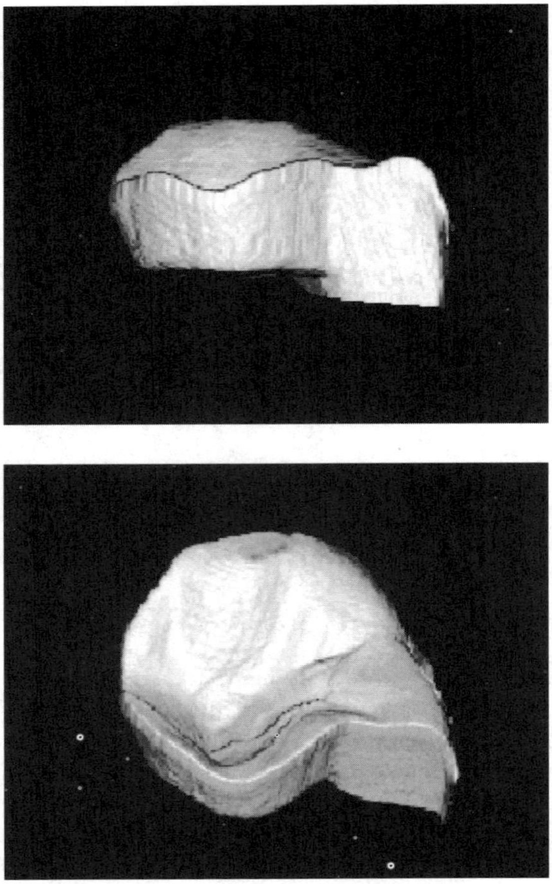

FIGURE 7.27 Volume representation of the calculated inlay.

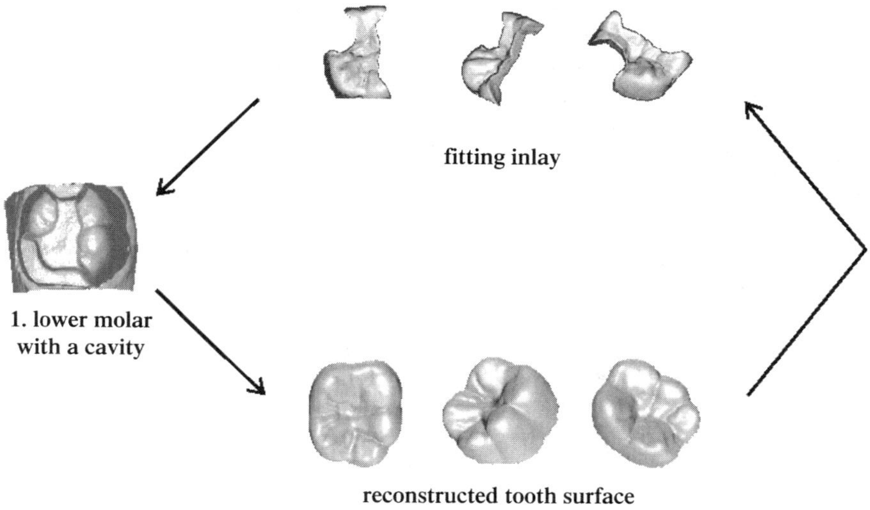

FIGURE 7.28 Dental restoration of a first lower molar.

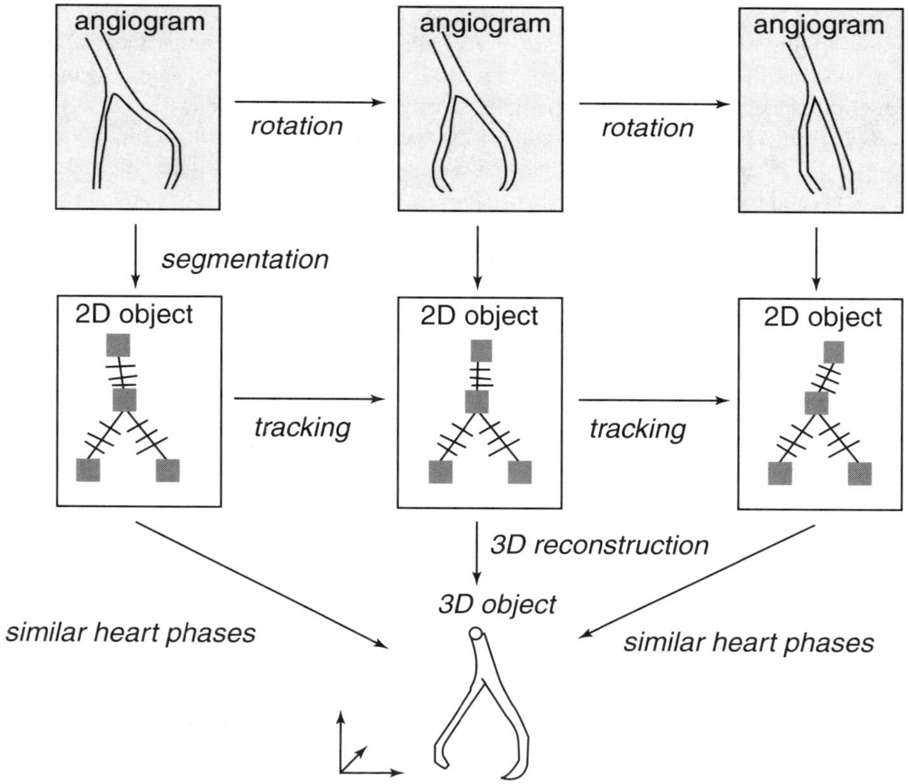

FIGURE 7.29 Processing steps used to reconstruct the 3D geometry of the coronary vessels.

is either a very limited amount of variation in the input data amenable to processing by a fully automatic algorithm or the necessity of extensive user assistance. The approach leads to a compromise in which the user only identifies a very small number of points interactively. The segmentation process is separated into the detection of the vessel centerline and evaluation of the vessel contour. The algorithm works with a cost-minimizing *A* search tree,[76,116] which proves to be robust against noise and may be fully controlled by the user. Snakes track the obtained structure over the angiographic sequence.

Reconstruction is based on the extracted vessel tree structures, the known relative orientation (i.e., the angle) of the projections, and the imaging parameters of the X-ray system. The 3D reconstruction is performed from images of identical heart phases. It begins with the two projections of the same phase, defining the largest angle.

The obtained result is improved afterward by introducing additional views. Applying a 3D optimization techniques the shape of a 3D Snake is adapted according to multiple 2D projections.[42] The obtained 3D structure can be either visualized by performing a volume rendering or, in order to be presented within VR systems, transferred into a polygonal representation.

Besides the 3D geometry of the coronary vessels, the trajectories of distinct points of the vessels are determined during the tracking process. As a result, these trajectories can be used to simulate the movement of the vessel caused during the heartbeat (Figure 7.30, bottom row).

7.2.5.3 Reconstruction of Ventricles

Beside the stereoscopic or multi-scopic feature-based approach, the 3D structure can also be obtained using densitometric information. This technique, also known as the back-projection method, does not need any *a priori* knowledge or image segmentation. Similar to CT, the 3D information is obtained by determining the intensity of a volume element according to the density of the imaged structure. The intensity of each pixel within the angiogram correlates to the amount of X-ray energy, which is received at the image amplifier. This energy depends on the density and the absorption capabilities of the traversed material. As a result, a pixel represents the sum of the transmission coefficients of the different materials which are pierced by the X-ray. For homogeneous material and parallel mono-chromatic X-rays, the image intensity can be described by the rule of *Lambert-Beer*:[8]

$$I = I_0 e^{-\mu \nu d} \tag{7.2}$$

I	= image intensity
I_0	= initial intensity
μ	= absorption coefficient of the structure
ν	= density of the structure
d	= thickness of the structure

If the X-ray travels through a material with varying densities, Equation 7.2 has to be split into parts with constant density. The total amount of transmitted intensity is the sum of these different parts.

$$I = I_0 e - \Sigma_i \mu_i \nu_i d_i \tag{7.3}$$

To improve the image quality, contrast agent is injected during the acquisition process. For this purpose, a catheter is positioned in front of the ventricles (see Figure 7.31).

Applying the back-projection technique, the distribution of the coefficients can be determined. During the acquisition process, the X-ray system is rotated around the center of the heart (see Figure 7.32).

In order to reconstruct the appropriate intensities of the heart, all the images are translated into the center of rotation (see Figure 7.33); thus, according to the amount of images, a cylinder is defined by a

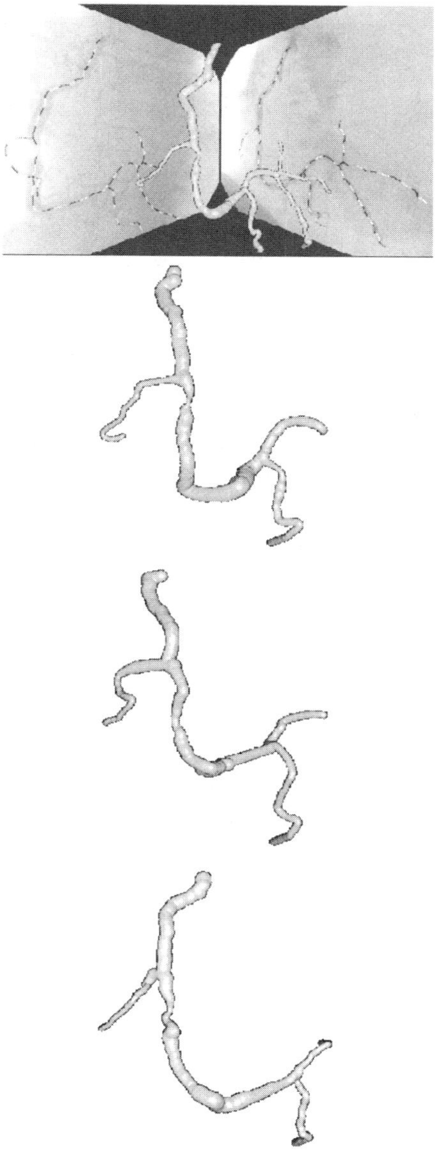

FIGURE 7.30 Reconstructed vessels rendered by InViVo (top: combined presentation of the volume rendered reconstruction result and angiograms; bottom: some frames of the 3D movement simulation).

number of sampling planes. The complete volume of the cylinder can now be determined. Therefore, all the rays starting from the X-ray source and intersecting a distinct voxel are accumulated and weighted according to the intensity of the different planes. Continuing this process for all the voxels of the cylinder, by taking the projection geometry into account by introducing a cone filter,[51] the intensity of each cylinder voxel can be determined. The obtained volume data can be visualized using a volume rendering technique and can be segmented by Snakes (Figure 7.34).

7.2.6 Visualization of Laser Confocal Microscopy Data Sets

Structures in the microscopic scale nerve cells, tissue and muscles, blood vessels, etc. show beautiful, complex, and still mostly unexplored patterns usually with higher complexity than those of organs. In order to understand the spatial relationship and internal structure of such microscopic probes,

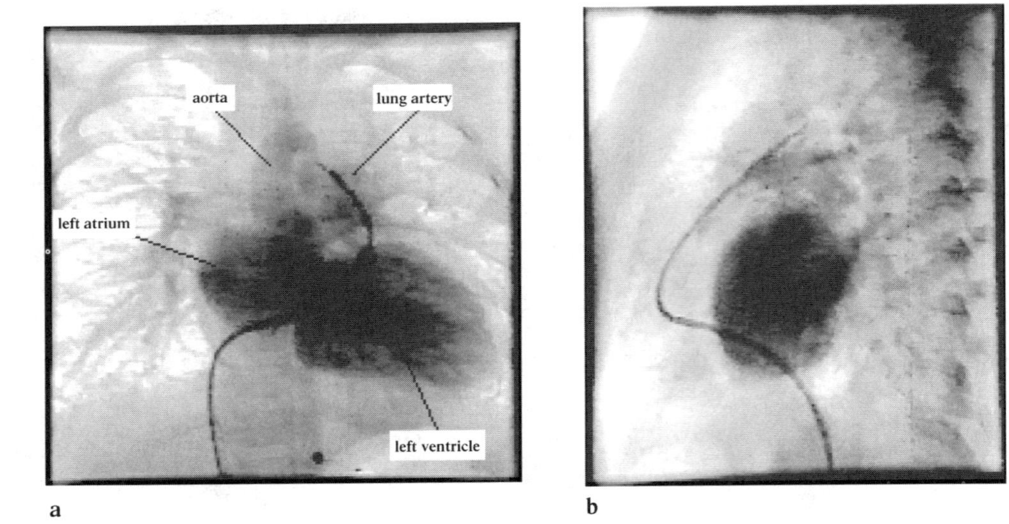

FIGURE 7.31 Angiograms acquired by the biplane X-ray system.

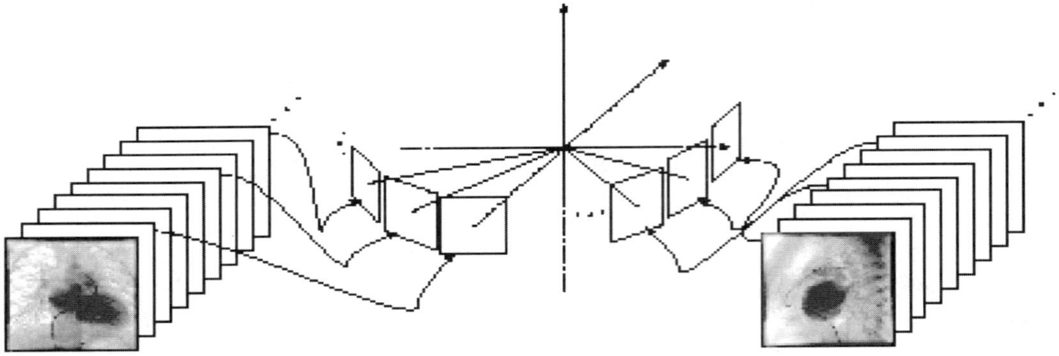

FIGURE 7.32 Acquisition of different angiograms by rotating a biplane X-ray system around the center of the heart.

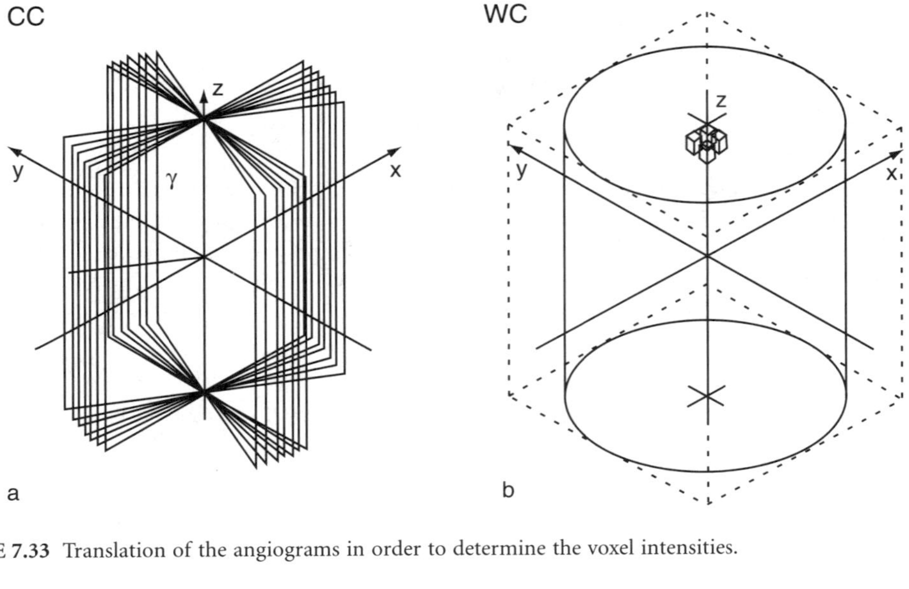

FIGURE 7.33 Translation of the angiograms in order to determine the voxel intensities.

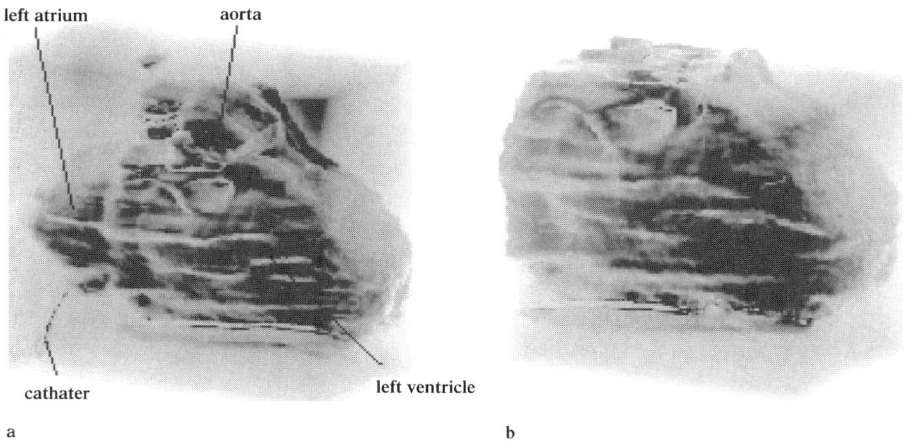

FIGURE 7.34 Volume rendering of the intensities obtained by the back-projection techique.

tomographic series of slices are required in analogy to the tomographies used for organs and other macroscopic structures.

Laser confocal microscopy is a relatively new method, allowing for a true tomographic inspection of microscopic probes. The method operates according to a simple, basic principle.[23]

A visible or ultraviolet laser emission is focused on the first confocal pinhole and then onto the specimen as a diffraction-limited light spot (see Figure 7.35). The primary incident light is then reflected from particular voxel elements or emitted from fluorescent molecules excited within it. Emissions from

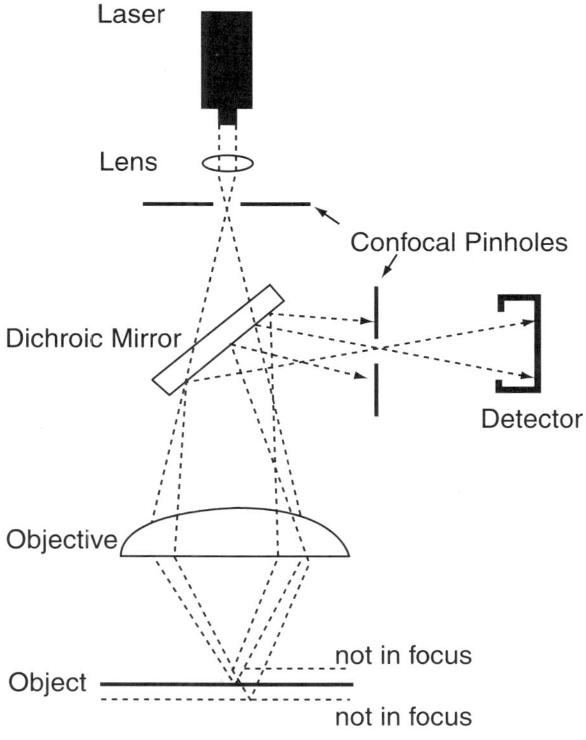

FIGURE 7.35 Principle of laser confocal microscopy.

the object return along the primary laser light pathway and depart from it by lateral reflection from (or passage through, depending on the instrument) a dichroic mirror onto the second confocal pinhole. This aperture is confocal with the in-focus voxel elements in the specimen. The virtual elimination by defocusing of all distal and proximal flanking emissions at this physical point assures that the light passing onto the detector, a sensitive photodetector or camera, is specifically derived from in-focus object voxels with a resolution, e.g., in the Leica instrument, approaching 200 to 400 nm in the x/y and z directions, respectively. In order to image the entire object, the light spot is scanned by a second mirror in the x/y plane in successive z sections by means of a precision stage motor. Rapid scanning preserves fluorescent intensity, but must be reconciled with image quality. The storage, retrieval, and manipulation of light intensity information from the object make static and dynamic 3D imaging possible.

Although not perfect, the new method shows several significant benefits as compared to the traditional procedures. The most important of these benefits are the following: true tomographic method, significant freedom in choosing slice thickness and size, trivial registration of slices, very fast and easy in operation, capable of acquiring *in vivo* cells as well as static or dynamic structures, and non-destructive. Finally, by using different types of laser and fluorophore materials, different spatially overlapping structures can be visualized and superimposed within the same probe.

The data acquired with laser confocal microscopy (LCM) show several characteristics requiring specialized treatment in order to make the method applicable:

1. Typical data sets have a large data size with a resolution of $512^2 \times 64$ pixels. These pixels are colored; thus, a typical RGB data set requires some 50 Mbytes of memory. Obviously, data sets of this size require efficient processing methods.
2. These characteristics of low-contrast, low-intensity gradients and a bad signal-to-noise ratio make a straightforward segmentation between the structures of interest and the background (e.g., by using thresholding, region growing, homogeneity, color differences, etc.) impossible. All these methods listed apply more or less to binary decision criteria whether a pixel/voxel belongs to the structure or not. Such criteria typically fail when used with signals showing the characteristics listed above.
3. Due to unequal resolutions in the plane and the depth directions, a visualization method has to be able to perform with "blocks" or unequal size lengths instead of with cubic voxels. Resampling of the raw data to a regular cubic field will further reduce the signal quality, introduce interpolation artifacts, and generate an even larger data set, probably too large to be handled with conventional computers.
4. Regarding the quality, artifacts have to be avoided as far as possible. Introducing artifacts in an unknown structure will often have fatal effects on their interpretation, since the human observer does not always have the experience for judging the correctness or the fidelity of the presented structures. As an example, an obvious artifact caused by bas parameter settings of the software during the visualization of human anatomy (e.g., of a head) is immediately detected by the observer, since the human anatomy is well known and such artifacts are trivially detected. This is not the case when inspecting an unknown data set.
5. Choosing the "correct" illumination model (e.g., MIP, semi-transparent, surface, etc.) has a significant impact on the clarity and information content of the visualization. Again, due to the lack of experience such a decision is typically much more difficult than in the case of anatomic tomographic data.
6. The speed of visualization becomes the most crucial issue. The visualization parameters have to be adjusted in an interactive, trial-and-error procedure. This can take a very long time if, e.g., after an adjustment the user has to wait for several minutes to see the new result. Furthermore, inspection of new, unknown structures requires rapid changing of directions, illumination conditions, visualization models, etc. Looping and stereo images are of enormous importance for understanding unknown, complicated spatial structures.

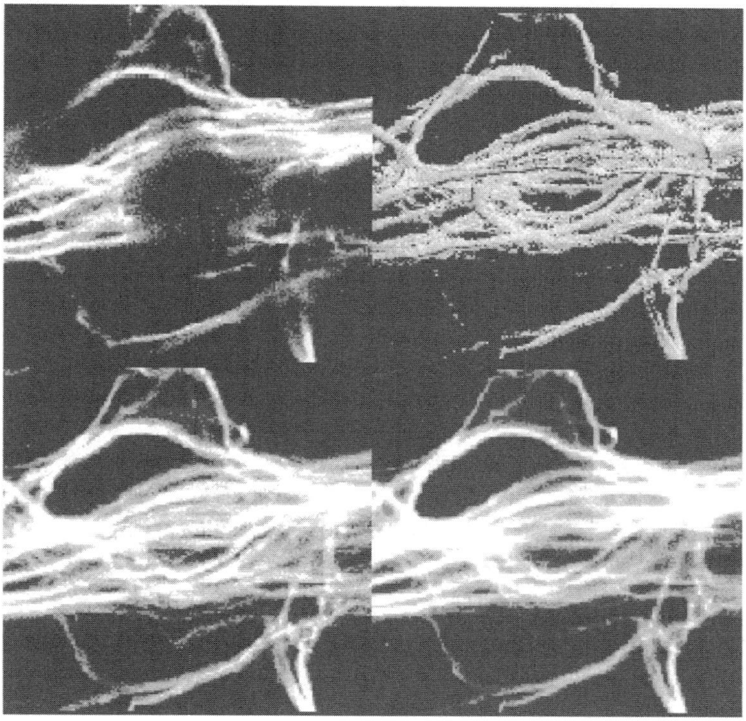

FIGURE 7.36 The wire-like structure of the astrocyte cytoskeleton of the same probe. Resolution of 335 × 306 × 67 voxels with a size of $0.16^2 \times 0.2$ μm. Upper left, a single slice; upper right, surface reconstruction; lower left, MIP; lower right, transmission illumination models.

The main requirement here is to employ a fast volumetric method which allows interactive and intuitive parameter settings during the visualization session. Detailed results of the employed volume visualization are reported in Reference 96. Figures 7.36 and 7.37 present a microscopic preparation of the tubular structure of a cat retina. The data set consist of 335 × 306 × 67 voxels, each with a dimension of $0.16^2 \times 0.2$ μm. The first image presents the extracellular component of the blood vessel. The vessel diameter before the branch point is 19 μm. The second image shows the wire-like structure of the astrocyte cytoskeleton. Both data sets originate from the same probe. In all subsequent images, the difference of the visualization between slicing, MIP, surface, and semi-transparent methods is shown.

Figure 7.38 shows the complicated structure of nerve cells networks. The resolution of the data set is 25 Mbytes ($512^2 \times 100$ voxels). As one can see on Figure 7.38, upper left, single slices are not able to provide full understanding of the complicated topology. The three other images of Figure 7.38 show in much better detail the internal structure of the cell network.

LCM plays a fundamental role for gathering *in vivo* data about not only static, but also dynamic structures, i.e., structures existing typically only within living cells and for a very short period of time (e.g., for a few seconds). Such structures are common in several biological applications. In the case referred to here, we present temporary structures formed by polymerized actin, a structure necessary for cell movements.

Figures 7.39 and 7.40 demonstrate the importance of LCM data visualization for detecting unknown structures. In this case, we studied actin filaments in *Dictyostelium* amoebae with time periods ranging from 10 to 100 s. The data resolution is 512 × 484 × 43 voxels = 10 Mbytes. Note the structure of the surface visible in the "surface volume rendering" image. These structures are hardly visible and, therefore, are difficult to detect when regarding individual slices.

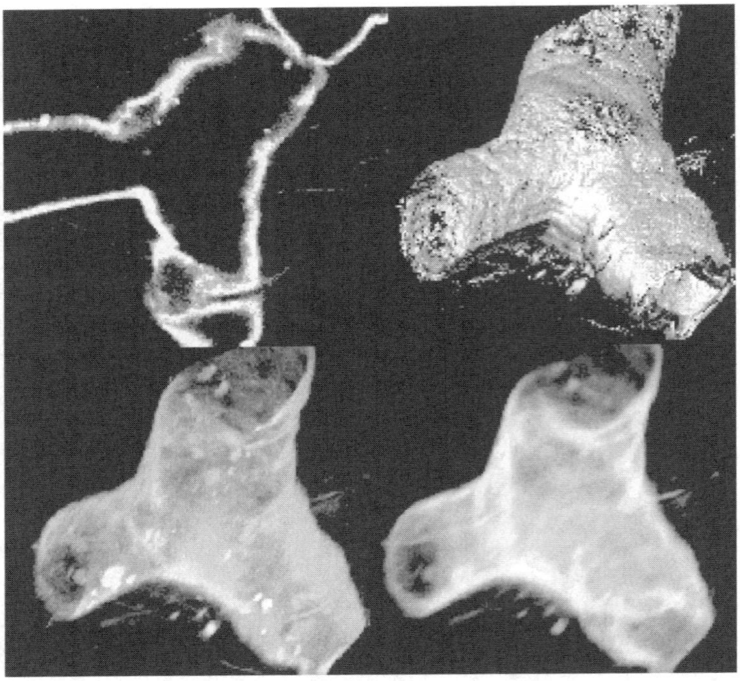

FIGURE 7.37 The extracellular component of a retina blood vessel of a cat. Resolution of 335 × 306 × 67 voxels with a size of $0.16^2 \times 0.2$ μm. Upper left, a single slice; upper right, surface reconstruction; lower left, MIP; lower right, transmission illumination models.

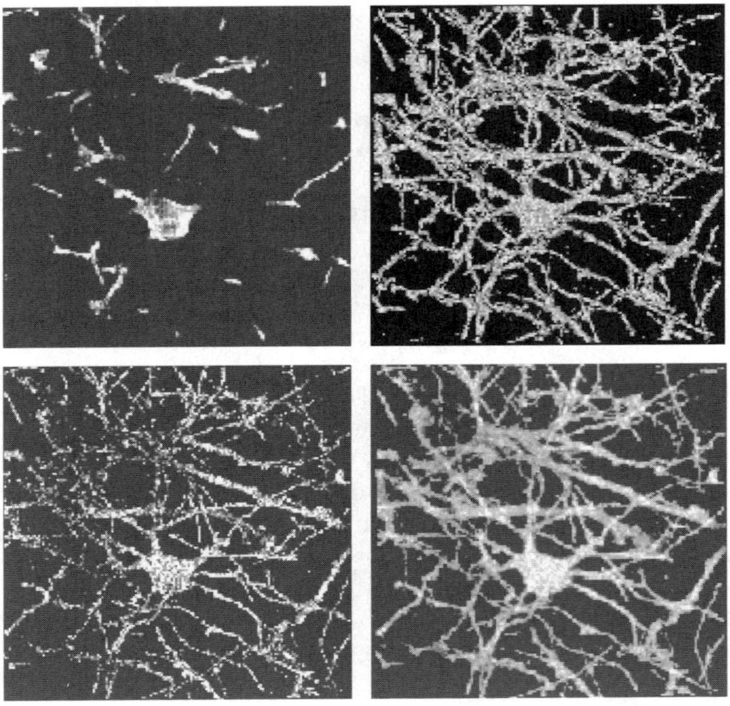

FIGURE 7.38 The complicated topology of nerve cell networks. Resolution 5122 × 100 voxels = 25 Mbytes. Upper left, a single slice; upper right, surface reconstruction; lower left, MIP; lower right, transmission illumination models.

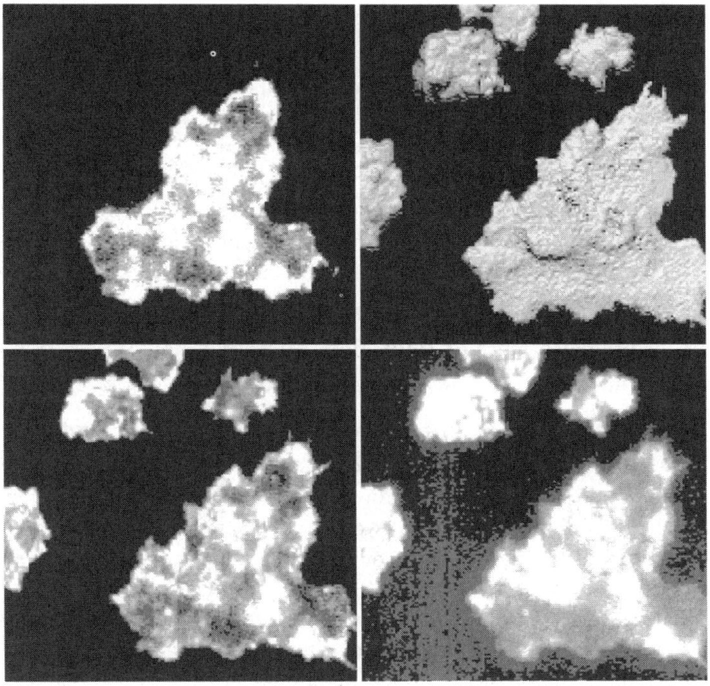

FIGURE 7.39 F-actin structures in *Dictyostelium* amoebae, resolution 512 × 484 × 43 voxels = 10 Mbytes. Upper left, a single slice; upper right, surface reconstruction; lower left, MIP; lower right, transmission illumination models.

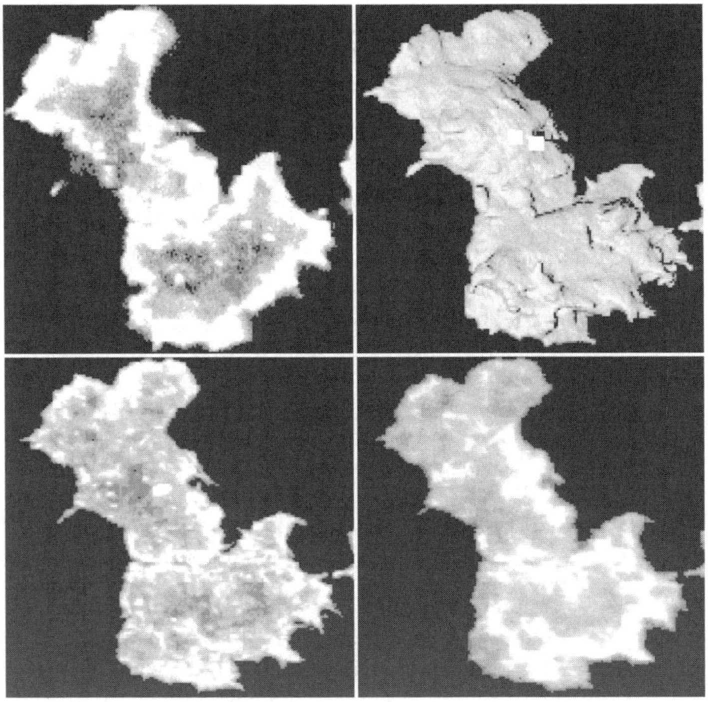

FIGURE 7.40 F-actin structures. Resolution 512 × 484 × 70 voxels = 16.5 Mbytes. Upper left, a single slice; upper right, surface reconstruction; lower left, MIP; lower right, transmission illumination models.

7.2.7 Virtual Simulation of Radiotherapy Treatment Planning

Radiation therapy (RT) is one of the most important techniques in cancer treatment. RT involves several steps which mainly take place in three equipment units: the CT unit, the simulator unit, and the treatment unit.

Before the patient goes to the treatment unit and the actual RT takes place, the treatment plan of the RT must be prepared (RTP). The RTP is performed on the simulator. The simulator machine can perform exactly the same movements and achieve the same position for the patient's RT as the treatment machine, but it uses conventional, diagnostic X-rays instead of high-energy treatment rays.

The conventional simulation of the RT process has several limitations, mainly due to physical movement of the components of the simulator (e.g., gantry and table) and the long period of time the patient must remain in the simulator unit.

7.2.7.1 Current Clinical Routine of Radiation Therapy

Today the general procedure of an RT treatment is the following (see Figure 7.41):

1. Move patient to the simulator. Physicians locate the region of interest (ROI, such as tumor) using the traditional X-ray fluoroscopy (diagnostic imaging).
2. Move patient to the CT scanner. In both rooms, simulator unit and CT scanner, an identical laser system is installed defining the so-called "world coordinates." Physicians place the patient on the CT table in such a way that the CT laser coordinate system matches with the skin markers on the patient. This will assure to recover the patient position from Step 1.
3. Physicians analyze the ROI and define the target volume(s) and the critical organ(s) on each CT slice. Then treatment parameters are selected.
4. Move patient to the simulator again. Physicians simulate the treatment plan to verify its effectiveness using X-rays instead of treatment rays. The treatment field is documented on X-ray films, and skin markers are placed on the patient's body. If the treatment plan is successfully verified, then the patient goes to the treatment unit; otherwise, Step 3 must be resumed.
5. Move patient to the treatment unit. Physicians carry out the actual RT treatment according to the RTP derived during the previous steps.

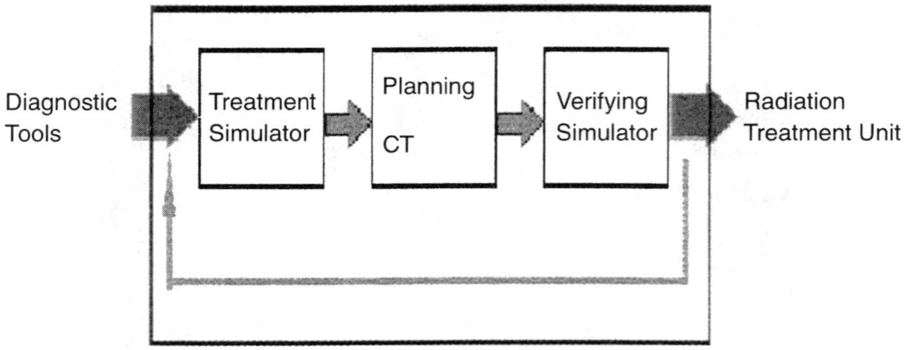

FIGURE 7.41 Clincal RT routine.

7.2.7.2 Proposed "Virtual Simulation"

The general procedure of the virtual radiation treatment planning is the following (see Figure 7.42):

1. Move patient to the CT scanner. Physicians digitize the patient using spiral CT.
2. Transfer patient's CT data to the virtual simulator (VS). Physicians create the therapy plan on the VS using an interactive 3D planning and visualization interface. The VS system supports the aspects: VS interaction, digital reconstructed radiograph (DRR), and visualization, target volumes delineation beam shape design, and orientation determination in 3D space.
3. Move patient to the irradiation machine, where physicians carry out the real treatment on the patient.

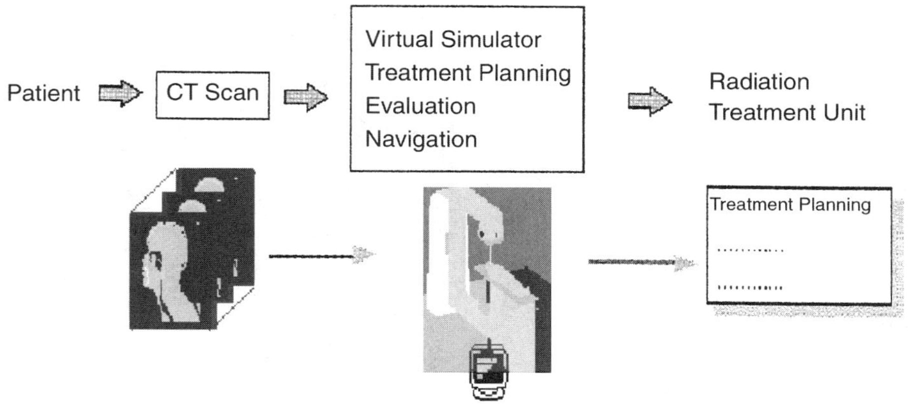

FIGURE 7.42 Virtual RT.

One significant feature of RTP systems (compared with other graphics applications) is that they support two rendering views: "beam's eye view" (BEV) and "observer's eye view" (OEV). In BEV, the patient's image is reconstructed as if the observer eye is placed at the location of the radiation source looking out along the axis of the radiation beam. The BEV window is used to detect the ROI and to define the target volume (tumor). However, BEV alone is insufficient for defining the ROI. Therefore, the OEV is also used as a second indicator to investigate the interaction among treatment beam, target volume (tumor), and its surrounding tissue (see Figure 7.43).

7.2.7.3 Digital Reconstruction Radiograph

In a VS, DRR (or X-ray) images, with which physicians are familiar, are required. In EXOMIO, two kinds of volume illumination methods are supported: DRR images and MIP. An MIP is physically impossible on a real simulator. In contrast to X-rays, the MIP makes the distinction between soft and hard tissues (e.g., bones) easier for the physicians (see Figure 7.44).

7.2.7.4 Registration

The patient's coordinates in different rooms, such as the CT room and the irradiation operation room, must be identical. The patient's position is labeled with several marks on his (or her) skin, at those points where the laser beams are projected onto their skin. These marks define the reference points of radiation iso-center. In EXOMIO, these marks can be seen on the axial slices and on the surface reconstructed model of the patient. In Figure 7.45, marks are displayed in both OEV and slices, allowing the patient to be identically positioned in the CT unit, the treatment unit, and the VS. The RTP parameters (gantry rotation, table position, beam size, etc.) are defined, based on these initial positions.

7.2.8 Conclusions

Medical volume visualization has come a long way from the first experiments to the current, highly detailed renderings. As the rendering algorithms are improved and the fidelity of the resulting images is investigated, 3D images are not just pretty pictures, but a powerful source of information for research, education, and patient care. In certain areas such as craniofacial surgery, volume visualization is increasingly becoming part of the standard preoperative procedures. New applications such as 3D cardiology, 3D ultrasound, and LCM are becoming more and more popular. Further rapid development of volume visualization methods is widely expected.[54]

A number of problems still hinder an even broader use of volume visualization in medicine. First, and most importantly, the segmentation problem is still unsolved. It is no coincidence that volume visualization is most accepted in all areas where clinicians are interested in bone from CT. Especially for MRI,

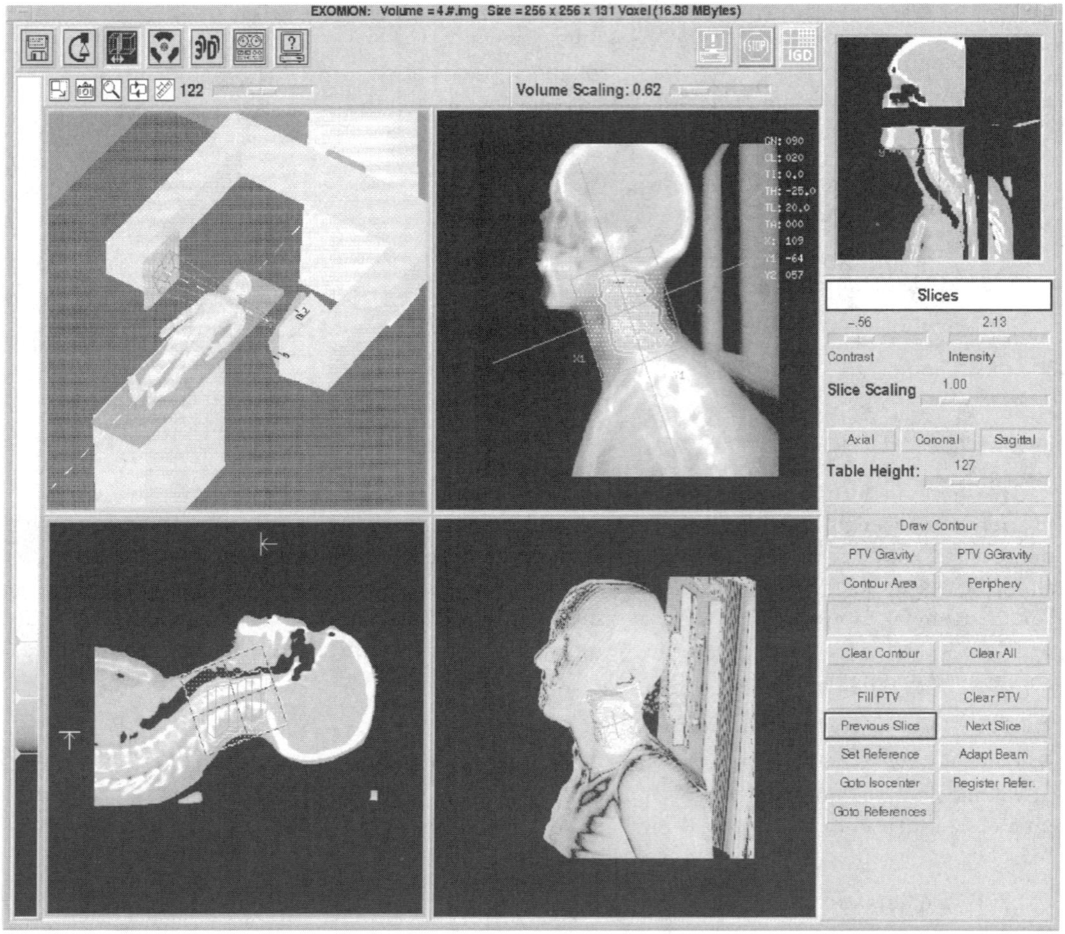

FIGURE 7.43 EXOMIO user interface.

however, automatic segmentation methods are still far from being generally applicable, while interactive procedures are too expensive. As has been shown, there is research in different directions going on; in many cases, methods have already proven valuable for specific applications.

The second major problem is the design of a user interface, which is suitable in a clinical environment. Currently, there is still a large number of rather technical parameters for controlling segmentation, registration, shading, and so on. Acceptance in the medical community will certainly depend heavily on progress in this field.

Third, current workstations are not yet able to deliver 3D images fast enough. For the future, it is certainly desirable to interact with the workstation in real time, instead of just looking at static images or precalculated movies. However, with computing power increasing further, this problem will be overcome in just a few years, even on low-cost platforms.

As has been shown, a number of applications based on volume visualization are becoming operational, such as surgical simulation systems and 3D atlases. Another intriguing idea is to combine volume visualization with virtual reality systems, which enable the clinician to walk around or fly through a virtual patient (Figure 7.46).[58,93] In *augmented reality*, images from the real and virtual world are merged to guide the surgeon during an intervention.[6] Integration of volume visualization with virtual reality and robotics toward *computer-integrated surgery* will certainly be a major topic in the coming decade.[93,108,119]

(a)
(b)

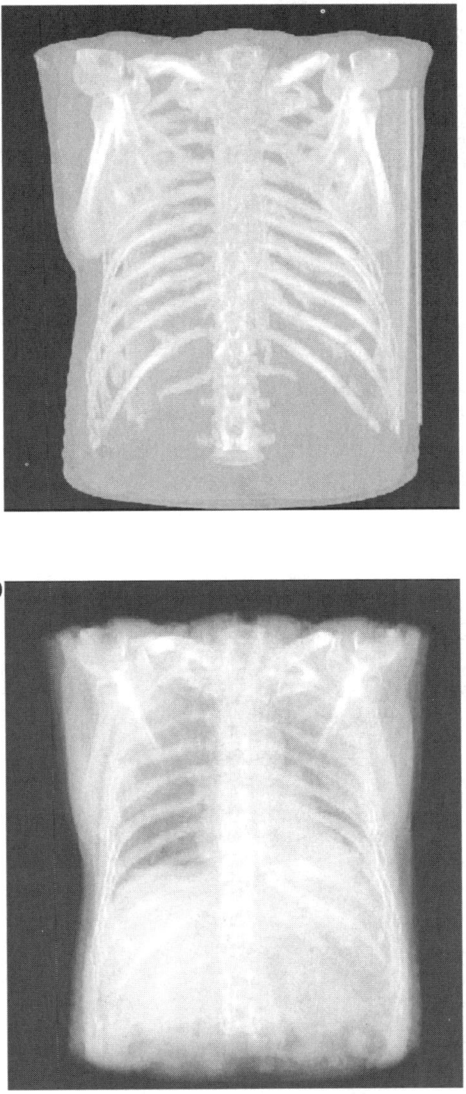

FIGURE 7.44 DRR images: (a) MIP image and (b) X-ray image.

(a)
(b)
(c)

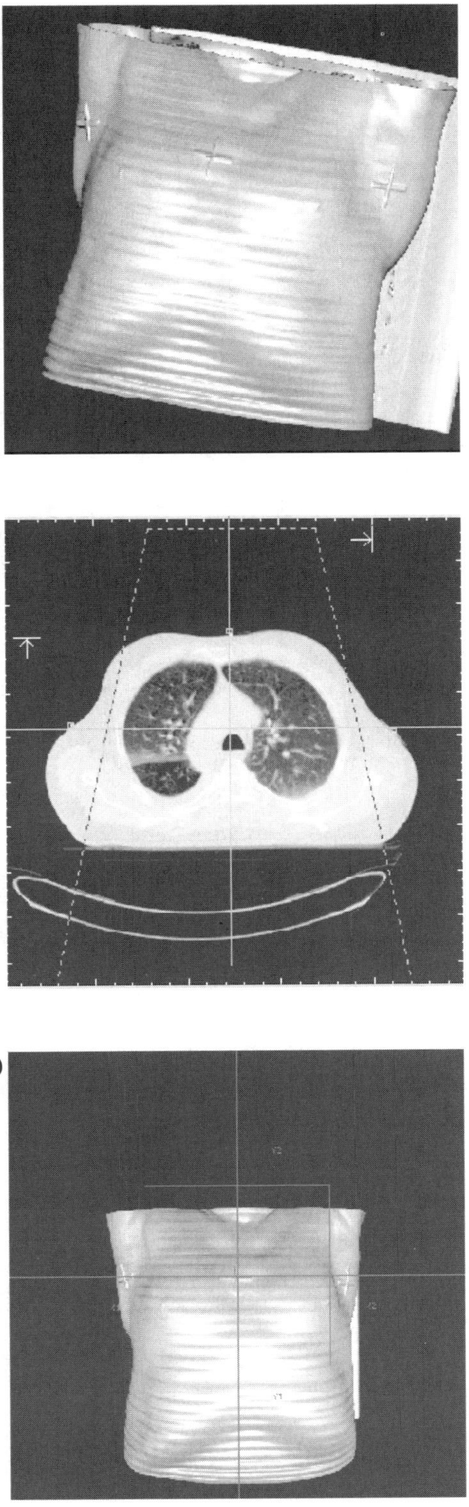

FIGURE 7.45 Marks and patient registration: (a) skin marks (OEV), (b) iso-center in slice, and (c) BEV after register.

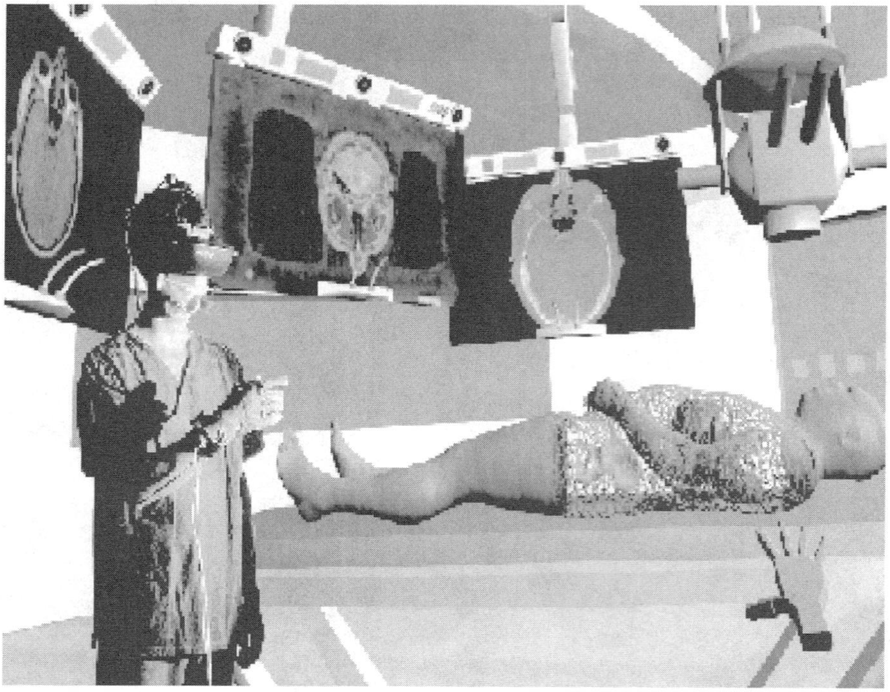

FIGURE 7.46 OP 2000 the operation theatre of the future.

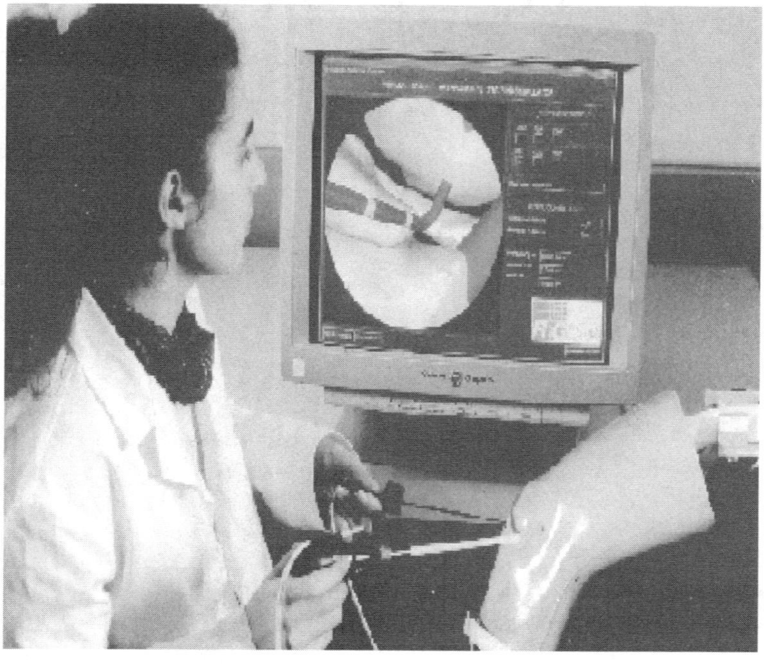

FIGURE 7.47 Virtual arthroscopy using VR and force feedback for training surgical procedures.

Acknowledgments

Georgios Sakas would like to thank Axel Hildebrand, Stefan Großkopf, Jürgen Jäger, Rainer Malkewitz, and Stefan Walter for their work in the different sections of the chapter, and Peter Plath and Mike Vicker from the University of Bremen for the laser confocal microscopy data. Furtherly, data have been provided by Kretztechnik, Visible Human Project, Deutsche Klinik für Diagnostik Wiesbaden; H.T.M. van der Voort, Department of Molecular Biology, University of Amsterdam; Herbert Stüttler; Martin Hoppe, Leica Lasertechnik GmbH, State Hospital of Darmstadt; and Rolf Ziegler (Figure 7.47).

Many of the examples presented in this chapter are based on the work at the IMDM. Andreas Pommert is grateful to Karl Heinz Höhne (director); his colleagues Bernhard Pflesser, Martin Riemer, Thomas Schiemann, Rainer Schubert, and Ulf Tiede; and the students Sebastian Gehrmann, Stefan Noster, Markus Urban (Figure 7.3), and Frank Wilmer (Figure 7.5). Presented applications are in cooperation with Udo Schumacher, Department of Neuroanatomy (Figure 7.10); Christoph Koch, Department of Neuroradiology (Figure 7.11); and Andreas Fuhrmann, Department of Dental Radiology (Figure 7.12). The PET dataset (Figure 7.8) was kindly provided by Uwe Pietrzyk, Max Planck Institute of Neurological Research, Cologne. The Visible Human dataset is courtesy of the National Library of Medicine (Figure 7.10).

References

1. A. Blake and M. Isard. *Active Contours: The Application of Techniques from Graphics, Vision, Control Theory and Statistics to Visual Tracking of Shapes in Motion.* Springer-Verlag, London, 1998.
2. M. E. Alder, S. T. Deahl, and S. R. Matteson. Clinical usefulness of two-dimensional reformatted and three-dimensionally rendered computerized images: Literature review and a survey of surgeons' options. *J. Oral Maxillofac. Surg.*, 53(4), 375–386, 1995.
3. L. K. Atata, A. P. Dhawan, J. P. Broderick, M. F. Gaskil-Shipley, A. V. Levy, and N. D. Volkow. Three-dimensional anatomical model-based segmentation of MR brain images through principal axes registration. *IEEE Trans. Biomed. Eng.*, 42(11), 1069–1078, 1995.
4. R. S. Avila, L. M. Sobierajski, and Arie E. Kaufman. Towards a Comprehensive Volume Visualization System. In Proc. Visualization '92, Boston, MA, pp. 13–20, 1992.
5. K. Baba, K. Stach, and S. Sakamoto. Development of an ultra-sound system for 3d-reconstruction of the foetus. *J. Perinat. Med.*, 17, 19–24, 1989.
6. A. C. F. Colchester, J. Zhao, K. S. Holton-Tainter, C. J. Henri, N. Maitland, P. T. E. Roberts, C. G. Harris, and R. J. Evans. Development and preliminary evaluation of VISLAN, a surgical planning and guidance system using intra-operative video imaging. *Med. Image Anal.*, 1(1), 73–90, 1996.
7. W. Barrett and E. Bess. Interpolation by directed distance morphing. In R. A. Robb, editor. *Visualization in Biomedical Computing 1994*, Proc. SPIE 2359, Rochester, MN, pp. 110–121, 1994.
8. J. Beier. *Automatische Quantifizierung von Koronarstenosen aus angiographischen Röntgenbildern*, Fortschr.-Ber. VDI Reihe 17 Nr.95, VDI-Verlag, Düsseldorf, 1993.
9. J. F. Blinn. Light reflection functions for simulation of clouds and dusty surfaces. *Comput. Graphics*, 16(3), 21–29, 1982.
10. J. D. Boissinat. Surface reconstruction from planar cross-sections. In *Proceedings of IEEE Conference on Computer Vision and Pattern Recognition*, pp. 393–397, June 1985.
11. M. Bomans. Segmentationsverfahren zur 3D-Visualisierung von Kernspintomogrammen des Kopfes: Evaluierung der Standardverfahren und Entwurf und Realisierung eines klinisch einsetzbaren Systems. Dissertation, Fachbereich Informatik, Universität Hamburg, 1994.
12. M. Bomans, K. H. Höhne, U. Tiede, and M. Riemer. 3D-segmentation of MR-images of the head for 3D-display. *IEEE Trans. Med. Imaging*, MI-9(2), 177–183, 1990.
13. M. Brandestini and W. Mörmann. Die CEREC Computer Rekonstruktion. Quint – essenz – Verlag, 1989.
14. M. Brandestini, W. Mörmann, A. Ferru, F. Lutz, and L. Kreijci. Computer machined ceramic inlays. In vitro marginal adaptation. *J. Dent. Res.*, 64, 208, 1985.

15. D. E. Breen, W. E. Lorensen, J. V. Miller, R. M. O'Bara, and M. J. Wozny. Geometrically deformed models: Method for extracting closed geometric models from volume data. *Comput. Graphics*, 25(4), 217–226, July 1991.
16. M. E. Brummer, R. M. Mersereau, R. L. Eisner, and R. R. J. Lewine. Automatic detection of brain contours in MRI datasets. *IEEE Trans. Med. Imaging*, 12, 153–166, 1993.
17. J. Canny. A computational approach to edge detection. *IEEE Trans. Pattern Anal. Mach. Intell.*, PAMI-8(6), 679–698, 1985.
18. E. Keeve, S. Girod, and B. Girod. Craniofacial surgery simulation. In K. H. Höhne, editor. *Visualization in Biomedical Computing, Proc. VBC'96*. Springer-Verlag, Berlin, pp. 541–546, 1996.
19. H. E. Cline, W. E. Lorensen, R. Kikinis, and F. Jolesz. Three-dimensional segmentation of MR images of the head using probability and connectivity. *J. Comput. Assist. Tomogr.*, 14(6), 1037–1045, 1990.
20. K. H. Höhne, editor. *VOXEL-MAN Junior: Interactive 3D Anatomy and Radiology in Virtual Reality Scenes, Part 1: Brain and Skull.* Springer-Verlag Electronic Media, Heidelberg, 1998 (CD-ROM).
21. H. E. Cline, W. E. Lorensen, S. P. Souza, Ferenc A. Jolesz, R. Kikinis, G. Gerig, and T. E. Kennedy. 3D surface rendered MR images of the brain and its vasculature. *J. Comput. Assist. Tomogr.*, 15(2), 344–351, 1991.
22. D. J. David, D. C. Hemmy, and R. D. Cooter. *Craniofacial Deformities: Atlas of Three-Dimensional Reconstruction from Computed Tomography.* Springer-Verlag, New York, 1990.
23. D. M. Shotton. Confocal scanning optical microscopy and its applications for biological specimens. Volume 94, 175–206, 1989.
24. R. A. Drebin, L. Carpenter, and P. Hanrahan. Volume rendering. *Comput. Graphics*, 22(4), 65–74, 1988.
25. R. A. Drebin, D. Magid, D. D. Robertson, and E. K. Fishman. Fidelity of three-dimensional CT imaging for detecting fracture gaps. *J. Comput. Assist. Tomogr.*, 13(3), 487– 489, 1989.
26. F. Duret. Method of Making a Prosthesis, Especially a Dental Prosthesis. Technical Report, United States Patent 4,742,464, 1988.
27. F. Duret, J. I. Blouin, and L. Nahami. Principes de fonctionnement et application techniques de l'empreine optique dans l'exercise des cabinet. *Cah Prothese*, 13, 73, 1985.
28. F. Duret, J. L. Blouin, and B. Duret. CAD/CAM in dentistry. *J. Am. Dent. Assoc.*, 117(11), 715–720, 1988.
29. D. L. Collins, A. P. Zijdenbos, W. F. C. Baare, and A. C. Evans. ANIMAL+INSECT: Improved cortical structure segmentation. In A. Kuba, M. Samal, and A. Todd-Pokropek, editors. *Information Processing in Medical Imaging, Proc. IPMI ´99.* Vol. 1613, Lecture Notes in Computer Science, pp. 210–223, Springer-Verlag, Berlin, 1999.
30. L. A. Feldkamp, L. C. Davis, and J. W. Kress. Practical cone-beam algorithm. *J. Opt. Soc. Am. A*, 1(6), 612–619, 1989.
31. R. M. Koch, M. H. Gross, F. R. Carls, D. F. von Büren, G. Fankhauser, and Y. I. H. Parish. Simulating facial surgery using finite element models. In *Computer Graphics Proceedings*. Annual Conference Series, ACM SIGGRAPH, New Orleans, pp. 421–428, 1996.
32. J. D. Foley, A. van Dam, S. K. Feiner, and J. F. Hughes. *Computer Graphics: Principles and Practice. Second Edition in C.* Addison-Wesley Longman, Amsterdam, 1996.
33. B. Geiger and R. Kikinis. Simulation of endoscopy. In N. Ayache, editor, *Computer Vision, Virtual Reality and Robotics in Medicine, Proc. CVRMed '95.* Vol. 905, Lecture Notes in Computer Science, Springer-Verlag, Berlin, pp. 277–281, 1995.
34. G. Gerig, O. Kübler, R. Kikinis, and F. A. Jolesz. Nonlinear anisotropic filtering of MRI data. *IEEE Trans. Med. Imaging*, 11(2), 221–232, 1992.
35. G. Gerig, J. Martin, R. Kikinis, O. Kübler, M. Shenton, and F. A. Jolesz. Automating segmentation of dual-echo MR head data. In A. C. F. Colchester and D. J. Hawkes, editors. *Information Processing in Medical Imaging, Proc. IPMI '91.* Vol. 511, Lecture Notes in Computer Science, Springer-Verlag, Berlin, pp. 175–187, 1991.

36. S. Großkopf and A. Hildebrand. Three-dimensional reconstruction of coronary arteries from X-ray projections. In P. Lanzer and M. Lipton, editors.*Vascular Diagnostics: Principles and Technology.* Springer-Verlag, Heidelberg, in press.
37. S. Gürke. Generation of Tooth Models for Ceramic Dental Restorations. In The 4th International Conference on Computer Integrated Manufacturing, Singapore, October 1997.
38. S. Gürke. Modellbasierte Rekonstruktion von Zähnen aus intraoralen Tiefenbildern. In *Digitale Bildverarbeitung in der Medizin*, Freiburger Workshop Digitale Bildverarbeitung in der Medizin. Freiburg, Germany, pp. 231–237, March 1997.
39. S. Gürke. Geometrically deformable models for model-based reconstruction of objects from range images. In *Computer Assisted Radiology and Surgery 98.* Tokyo, pp. 824–829, June 1998.
40. S. Haring, M. A. Viergever, and J. N. Kok. A multiscale approach to image segmentation using Kohonen networks. In H. H. Barrett and A. F. Gmitro, editors. *Information Processing in Medical Imaging, Proc. IPMI '93.* Vol. 687, Lecture Notes in Computer Science, Springer-Verlag, Berlin, pp. 212–224, 1993.
41. D. C. Hemmy and P. L. Tessier. CT of dry skulls with craniofacial deformities: Accuracy of three-dimensional reconstruction. *Radiology*, 157(1), 113–116, 1985.
42. A. Hildebrand. Bestimmung Computer-Graphischer Beschreibungsattribute für reale 3D-Objekte mittels Analyse von 2D-Rasterbildern. Ph.D. Thesis, TH Darmstadt, Darmstadt, 1996.
43. A. Hildebrand and S. Großkopf. 3D reconstruction of coronary arteries from X-Ray projections. In *Proceedings of the Computer Assisted Radiology CAR'95 Conference.* Springer-Verlag, Berlin, 1995.
44. W. Hiltman. Die 3d-strukturrekonstruktion aus ultraschall-bildern. October 1994.
45. K. H. Höhne and R. Bernstein. Shading 3D-images from CT using grey level gradients. *IEEE Trans. Med. Imaging*, MI-5(1), 45–47, 1986.
46. K. H. Höhne, M. Bomans, B. Pflesser, A. Pommert, M. Riemer, T. Schiemann, and U. Tiede. Anatomic realism comes to diagnostic imaging. *Diagn. Imaging*, (1), 115–121, 1992.
47. K. H. Höhne, M. Bomans, A. Pommert, M. Riemer, C. Schiers, U. Tiede, and G. Wiebecke. 3D-visualisation of tomographic volume data using the generalized voxel-model. *Visual Comput.*, 6(1), 28–36, 1990.
48. K. H. Höhne and W. A. Hanson. Interactive 3D-segmentation of MRI and CT volumes using morphological operations. *J. Comput. Assist. Tomogr.*, 16(2), 285–294, 1992.
49. K. H. Höhne, B. Pflesser, A. Pommert, M. Riemer, T. Schiemann, R. Schubert, and U. Tiede. A new representation of knowledge concerning human anatomy and function. *Nature Med.*, 1(6), 506–511, 1995.
50. A. Höss, J. Debus, R. Bendl, R. Engenhart-Cabillic, and W. Schlegel. Computerverfahren in der dreidimensionalen Strahlentherapieplanung. *Radiologe*, 35(9), 583–586, 1995.
51. J. Jäger. Volumetric Reconstruction of Heart Ventricles from X-Ray Projections (in German). Dissertation in Technical University Darmstadt, 1996.
52. M. Kass, A. Witkin, and D. Terzopoulos. Snakes: active contour models. *IEEE First Int. Conf. Comput. Vision*, 259–268, 1987.
53. A. Kaufman, editor. *Volume Visualisation.* IEEE Computer Society Press, Los Alamitos, CA, 1991.
54. A. Kaufman, K. H. Höhne, W. Krüger, L. J. Rosenblum, and P. Schröder. Research issues in volume visualisation. *IEEE Comput. Graphics Appl.*, 14(2), 63–67, 1994.
55. M. L. Kessler and D. L. McShan. An application for design and simulation of conformal radiation therapy. In R. A. Robb, editor. *Visualisation in Biomedical Computing 1994.* Proc. SPIE 2359, Rochester, MN, pp. 474–483, 1994.
56. R. Kikinis, M. E. Shenton, D. V. Iosifescu, R. W. McCarley, P. Saiviroonporn, H. H. Hokama, A. Robatino, D. Metcalf, C. G. Wible, C. M. Portas, R. M. Donnino, and F. A. Jolesz. A digital brain atlas for surgical planning, model driven segmentation, and teaching. *IEEE Trans. Visualisation Comput. Graphics*, 2(3), 232– 241, 1996.
57. T. Kohonen. *Self-Organisation and Associative Memory.* 2nd edition. Springer-Verlag, Berlin, 1988.

58. W. Krueger and B. Froehlich. The responsive workbench. *IEEE Comput. Graphics Appl.*, 14(3), 12–15, 1994.
59. D. Laur and P. Hanrahan. Hierarchical splatting: A progressive refinement algorithm for volume rendering. *Comput. Graphics*, 25(4), 285–288, 1991.
60. J. Leugyel, D. P. Greenberg, and R.Ç. Poop Time-dependent three-dimensional intravascular ultrasound. In *Computer Graphics Proceedings*. SIGGRAPH, Los Angeles, pp. 457–464, 1995.
61. U. Tiede, T. Schiemann, and K. H. Höhne. High quality rendering of attributed volume data. In D. Ebert et al., editors. *Proc. IEEE Visualization '98*. IEEE Computer Society Press, Los Alamitos, CA, pp. 255–262, 1998.
62. M. Levoy. Display of surfaces from volume data. *IEEE Comput. Graphics Appl.*, 8(3), 29–37, 1988.
63. M. Levoy. A hybrid ray tracer for rendering polygon and volume data. *IEEE Comput. Graphics Appl.*, 10(2), 33–40, 1990.
64. L.-J. Lo, J. L. Marsh, M. W. Vannier, and V. V. Patel. Craniofacial computer-assisted surgical planning and simulation. *Clin. Plast. Surg.*, 21(4), 501–516, 1994.
65. W. E. Lorensen and H. E. Cline. Marching cubes: A high resolution 3D surface construction algorithm. *Comput. Graphics*, 21(4), 163–169, 1987.
66. M. Magnusson, R. Lenz, and P.-E. Danielsson. Evaluation of methods for shaded surface display of CT volumes. *Comput. Med. Imaging Graphics*, 15(4), 247–256, 1991.
67. K. H. Höhne, editor. *VOXEL-MAN Junior: Interactive 3D Anatomy and Radiology in Virtual Reality Scenes, Part 2: Inner Organs*. Springer-Verlag Electronic Media, Heidelberg, 2000 (CD-ROM).
68. S. R. Marschner and R. J. Lobb. An evaluation of reconstruction filters for volume rendering. In R. D. Bergeron and A. E. Kaufman, editors. *Proc. Visualisation '94*. IEEE Computer Society Press, Los Alamitos, CA, pp. 100–107, 1994.
69. S. Meller, M. Wolf, D. Paulus, M. Pelka, P. Weierich, and H. Niemann. Automatic Tooth Restoration Via Image Warping. In Proceedings of the Computer Assisted Radiology '97 Conference. Berlin, June 1997.
70. W. Menhardt. Iconic fuzzy sets for MR image segmentation. In A. E. Todd-Pokropek and M. A. Viergever, editors. *Medical Images: Formation, Handling and Evaluation*. Vol. 98, NATO ASI Series F, Springer-Verlag, Berlin, pp. 579–591, 1992.
71. J. V. Miller. On gdm's: Geometrically Deformed Models for the Extraction of Closed Shapes from Volume Data. Master's thesis, Rensselaer Polytechnic Institute, Troy, New York, December 1990.
72. R. Millner. Ultraschalltechnik, grundlagen und anwendungen. Physik Verlag, Weinheim, ISBN 3-87664-106-3, 1987.
73. B. S. Morse, S. M. Pizer, and A. Liu. Multiscale medial analysis of medical images. *Image Vision Comput.*, 12(6), 327–338, 1994.
74. S. Muraki. Volume data and wavelet transforms. *IEEE Comput. Graphics Appl.*, 13(4), 50–56, 1993.
75. S. Muraki. Multiscale volume representation by a DOG wavelet. *IEEE Trans. Visualisation Comput. Graphics*, 1(2), 109–116, 1995.
76. P. J. Neugebauer. Interactive segmentation of dentistry range images in CIM systems for the construction of ceramic in-lays using edge tracing. In *Proceedings of the Computer Assisted Radiology CAR'95 Conference*. Springer-Verlag, Berlin, 1995.
77. P. J. Neugebauer. Geometrical cloning of 3d objects via simultaneous registration of multiple range images. In *International Conference on Shape Modeling and Applications 1997*. IEEE Computer Society Press, Los Alamitos, CA, 1997.
78. D. Ney and E. K. Fishman. Editing tools for 3D medical imaging. *IEEE Comput. Graphics Appl.*, 11(6), 63–70, 1991.
79. D. Ney, E. K. Fishman, D. Magid, D. D. Robinson, and A. Kawashima. Three-dimensional volumetric display of CT data: Effect of scan parameters upon image quality. *J. Com-put. Assist. Tomogr.*, 15(5), 875–885, 1991.
80. P. Ning and J. Bloomenthal. An evaluation of implicit surface tilers. *IEEE Comput. Graphics Appl.*, 13(6), 33–41, 1993.

81. B. Pflesser, U. Tiede, and K. H. Höhne. Specification, modelling and visualization of arbitrarily shaped cut surfaces in the volume model. In W. M. Wells et al., editors. *Medical Image Computing and Computer-Assisted Intervention, Proc. MICCAI '98*. Vol. 1496, Lecture Notes in Computer Science, Springer-Verlag, Berlin, pp. 853–860, 1998.
82. A. Pommert, M. Bomans, and K. H. Höhne. Volume visualisation in magnetic resonance angiography. *IEEE Comput. Graphics Appl.*, 12(5), 12–13, 1992.
83. A. Pommert, W.-J. Höltje, N. Holzknecht, U. Tiede, and K. H. Höhne. Accuracy of images and measurements in 3D bone imaging. In H. U. Lemke, M. L. Rhodes, C. C. Jaffe, and R. Felix, editors. *Computer Assisted Radiology, Proc CAR '91*. Springer-Verlag, Berlin, pp. 209–215, 1991.
84. A. Pommert, R. Schubert, M. Riemer, T. Schiemann, U. Tiede, and K. H. Höhne. Symbolic modeling of human anatomy for visualisation and simulation. In R. A. Robb, editor. *Visualisation in Biomedical Computing 1994*. Proc. SPIE 2359, Rochester, MN, pp. 412–423, 1994.
85. A. Pommert and K. H. Höhne. Towards an image quality index in medical volume visualization. In S. K. Mun, editor. *SPIE Medical Imaging 2000: Image Display and Visualization*. San Diego, CA, accepted for publication.
86. S. P. Raya and J. K. Udupa. Low-level segmentation of 3-D magnetic resonance brain images: A rule-based system. *IEEE Trans. Med. Imaging*, MI-9(3), 327–337, 1990.
87. D. E. Rekow. The Minnesota CAD/CAM System Denti-CAD. Technical report, University of Minnesota, 1989.
88. D. E. Rekow. CAD/CAM in dentistry: Critical analysis of systems. In *Computers in Clinical Dentistry*. Quintessence Publishing Co., pp. 172–185, September 1991.
89. D. E Rekow. Method and Apparatus for Modeling a Dental Prosthesis. Technical report, United States Patent 5,273,429, 1993.
90. B. Jahne. *Digital Image Processing: Concepts, Algorithms, and Scientific Applications*. Springer-Verlag, Berlin, 1997.
91. D. Rückert. Bildsegmentierung durch stochastisch optimierte Relaxation eines 'geometric deformable model.' Master's thesis, TU Berlin, 1993.
92. H. Rusinek, M. E. Noz, G. Q. Maguire, A. Kalvin, B. Haddad, D. Dean, and C. Cutting. Quantitative and qualitative comparison of volumetric and surface rendering techniques. *IEEE Trans. Nucl. Sci.*, 38(2), 659–662, 1991.
93. A. Hildebrand, R. Malkewitz, W. Mueller, R. Ziegler, G. Graschew, and S. Grosskopf. *Computer Aided Surgery – Vision and Feasibility of an Advanced Operation Theatre*. Vol. 20, Pergamon Press, Oxford, pp. 825–835, November 1996.
94. P. Bono and G. Sakas, *Special Issue on Medical Visualisation*. Vol. 20, Pergamon Press, Oxford, pp. 759–838, November 1996.
95. L. Schreyer, M. Grimm, and G. Sakas. *Case Study: Visualisation of 3D-Ultrasonic Data*. IEEE Computer Society Press, Los Alamitos, CA, pp. 369–373, October 1994.
96. M. Vicker, P. Plath, and G. Sakas. Case Study: Visualisation of Laser Confocal Microscopy Data. IEEE Computer Society Press, Los Alamitos, CA, pp. 375–380, October 1996.
97. S. Walter and G. Sakas. *Extracting Surfaces from Fuzzy 3D-Ultrasonic Data*. Addison-Wesley, Reading, MA, pp. 465–474, August 1995.
98. T. Schiemann and K. H. Höhne. Definition of volume transformations for volume interaction. In J. Duncan and G. Gindi, editors. *Information Processing in Medical Imaging, Proc. IPMI '97*. Vol. 1230, Lecture Notes in Computer Science 1230, Springer-Verlag, Berlin, pp. 245–258, 1997.
99. T. Frenzel, D. Albers, K. H. Höhne, and R. Schmidt. Problems in medical imaging in radiation therapy. In H. U. Lemke et al., editors. *Computer Assisted Radiology and Surgery, Proc. CAR '97*. Excerpta Medica ICS 1134, Elsevier, Amsterdam, pp. 381–387, 1997.
100. T. Schiemann, K. H. Höhne, C. Koch, A. Pommert, M. Riemer, R. Schubert, and U. Tiede. Interpretation of tomographic images using automatic atlas lookup. In R. A. Robb, editor. *Visualisation in Biomedical Computing 1994*. Proc. SPIE 2359, Rochester, MN, pp. 457–465, 1994.

101. T. Schiemann, U. Tiede, and K. H. Höhne. Segmentation of the visible human for high quality volume based visualisation. *Med. Image Anal.*, 1(4), 263–271, 1997.
102. R. Schmidt, T. Schiemann, W. Schlegel, K. H. Höhne, and K.-H. Hübener. Consideration of time-dose-patterns in 3D treatment planning: An approach towards 4D treatment planning. *Strahlenther. Onkol.*, 170(5), 292–301, 1994.
103. W. J. Schroeder, J. A. Zarge, and W. E. Lorensen. Decimation of triangle meshes. *Comput. Graphics*, 26(2), 65–70, 1992.
104. R. Schubert, B. Pflesser, A. Pommert, K. Priesmeyer, M. Riemer, T. Schiemann, U. Tiede, P. Steiner, and K. H. Höhne. Interactive volume visualization using ``intelligent movies''. In J. D. Westwood, H. M. Hoffman, R. A. Robb, and D. Stredney, editors. *Medicine Meets Virtual Reality. The Convergence of Physical and Informational Technologies, Options for a New Era in Healthcare (Proc. MMVR ´99).* Vol. 62, Studies in Health Technology and Informatics, IOS Press, Amsterdam, pp. 321–327, 1999.
105. R. Schubert, W.-J. Höltje, U. Tiede, and K. H. Höhne. 3D-Darstellungen für die Kiefer- und Gesichtschirurgie. *Radiologe*, 31, 467–473,1991.
106. V. Spitzer, M. J. Ackerman, A. L. Scherzinger, and D. Whitlock. The visible human male: a technical report. *J. Am. Med. Inf. Assn.*, 3(2), 118–130, 1996.
107. C. Studholme, D. L. G. Hill, and D. J. Hawkes. Automated 3-D registration of MR and CT images of the head, *Med. Image Anal.*, 1(2), 163–175, 1996.
108. R. H. Taylor, S. Lavallée, G. C. Burdea, and R. Mösges. *Computer Integrated Surgery: Technology and Clinical Applications*. MIT Press, Cambridge, MA, 1995.
109. U. Tiede, K. H. Höhne, M. Bomans, A. Pommert, M. Riemer, and G. Wiebecke. Investigation of medical 3D-rendering algorithms. *IEEE Comput. Graphics Appl.*, 10(2), 41–53, 1990.
110. U. Tiede, T. Schiemann, and K. H. Höhne. Visualizing the visible human. *IEEE Comput. Graphics Appl.*, 16(1), 7–9, 1996.
111. A. W. Toga and J. C. Mazziotta. *Brain Mapping*. Academic Press, San Diego, 1996.
112. T. Totsuka and M. Levoy. Frequency domain volume rendering. *Comput. Graphics*, 271–278, 1993.
113. P. A. van den Elsen, E.-J. D. Pol, and M. A. Viergever. Medical image matching: A review with classification. *IEEE Eng. Med. Biol. Mag.*, 12(1), 26–39, 1993.
114. M. W. Vannier, C. F. Hildebolt, J. L. Marsh, T. K. Pilgram, W. H. McAlister, G. D. Shackelford, C. J. Offutt, and R. H. Knapp. Craniosynostosis: Diagnostic value of three-dimensional CT reconstruction. *Radiology*, 173, 669–673, 1989.
115. E. Vaske. Segmentation von Kernspintomogrammen mit der topologischen Karte zur 3D-Visualisierung. IMDM Institutsbericht 91/1, Institut für Mathematics und Datenverarbeitung in der Medizin, Universität Hamburg, 1991.
116. G. Vosselman. *Relational Matching*. Springer-Verlag, New York, 1992.
117. A. Wahle, E. Wellnhofer, I. Mugaragu, A. Trebeljahr, H. Oswald, and E. Fleck. Application of accurate 3D reconstruction from biplane angiograms in morphometric analyses and in assessment of diffuse coronary artery disease. In *CAR'95: Computer Assisted Radiology.* Springer Verlag, Berlin, 1995.
118. Å. Wallin. Constructing isosurfaces from CT data. *IEEE Comput. Graphics Appl.*, 11(6), 28–33, 1991.
119. J. D. Westwood, H. M. Hoffman, R. A. Robb, and D. Stredney, editors. *Medicine Meets Virtual Reality. The Convergence of Physical and Informational Technologies, Options for a New Era in Healthcare (Proc. MMVR ´99).* Vol. 62, Studies in Health Technology and Informatics, IOS Press, Amsterdam, 1999.
120. W. M. Wells III, P. Viola, H. Atsumi, S. Nakajima, and R. Kikinis. Multi-modal volume registration by maximization of mutual information. *Med. Image Anal.*, 1(1), 35–51, 1996.
121. L. Westover. Footprint evaluation for volume rendering. *Comput. Graphics*, 24(4), 367–376, 1990.
122. J. Wilhelms and A. van Gelder. Topological considerations in isosurface generation. *Comput. Graphics*, 24(5), 79–86, 1990.

123. I. J. Trotts and B. Hamann and K. I. Joy. Simplification of tetrahedral meshes with error bounds. *IEEE Trans. Visualization Comput. Graphics*, 5(3), 224–237, 1999.
124. P. H. Winston. *Artificial Intelligence*. 3rd edition, Addison-Wesley, Reading, MA, 1992.
125. R. Yagel, D. Cohen, and A. Kaufman. Discrete ray tracing. *IEEE Comput. Graphics Appl.*, 12(5), 19–28, 1992.
126. T. Yasuda, Y. Hashimoto, S. Yokoi, and J.-I. Toriwaki. Computer system for craniofacial surgical planning based on CT images. *IEEE Trans. Med. Imaging*, MI-9(3), 270–280, 1990.
127. F. W. Zonneveld and K. Fukuta. A decade of clinical three-dimensional imaging: A review. Part 2: Clinical applications. *Invest. Radiol.*, 29, 574–589, 1994.
128. F. W. Zonneveld, S. Lobregt, J. C. H. van der Meulen, and J. M. Vaandrager. Three-dimensional imaging in craniofacial surgery. *World J. Surg.*, 13, 328–342, 1989.
129. S. M. Pizer, D. Eberly, D. S. Fritsch, and B. S. Morse. Zoom-invariant vision of figural shape: The mathematics of cores. *Comput. Vision Image Understanding*, 69(1), 55–71, 1998.
130. J. West, J. M. Fitzpatrick, M. Y. Wang, B. M. Dawant, C. R. Maurer, R. M. Kessler, R. J. Maciunas, C. Barillot, D. Lemoine, A. Collignon, F. Maes, P. Suetens, D. Vandermeulen, P. A. van den Elsen, S. Napel, T. Sumanaweera, B. Harkness, P. F. Hemler, D. L. G. Hill, D. J. Hawkes, C. Studholme, J. B. A. Maintz, M. A. Viergever, G. Malandain, X. Pennec, M. E. Noz, G. Q. Maguire, M. Pollack, C. A. Pelizzari, R. A. Robb, D. Hanson, and R. P. Woods. Comparison and evaluation of retrospective intermodality brain image registration techniques. *J. Comput. Assist. Tomogr.*, 21(4), 554–566, 1997.

Appendix

Principles of Image Processing: Pixel Brightness Transformations, Image Filtering, and Image Restoration

A7.1 Introduction

What you see on an image is not the best of what you can get. In most cases, image quality can be further improved. An image "hides" information. To discover all this information and to make it visible, we need to process the image. Image processing is also an important and necessary process for medical imaging. Here, the term "medical imaging" is not limited only to the diagnostic imaging. There are several other applications, such as surgery, oncology, and simulation of medical systems, where image processing techniques are essential.

Image processing or (according to other authors[6,20]) image preprocessing can be separated into pixel brightness transformations (or image GSMT), image enhancement (or image filtering),[11,15] and image restoration.[12,18] In this section, we focus on the most commonly used techniques for medical image GSMT and enhancement. For the rest of the techniques, a number of references are given for further reading.

The main goal of image filtering is to enhance image quality and remove image artifacts such as noise or blurring (Figure A7.1). At the moment, several algorithms can perform these tasks. For smoothing operations filters like Gaussian, median or local averaging masks can be used. Most of the previous operations can be performed either on the 2D image basis or, in the case of volumetric data, in the 3D volume space.[20,21] These filters are applied on the image using multi-dimensional convolution, and they do not demand any further image transformation. Operations such as the Fourier,[7,14,16] wavelet,[6,22,25] and cosine[21] filter the image after it has been transformed.

The most popular and suitable transformation in medical application, including signal processing, used to be and probably still is the Fourier transform. When we transform an image or a signal using the Fourier transform, we mainly create a number of coefficients which represent the image or signal frequency components. One can realize that filtering in the frequency domain is ideal for removing periodic noise.

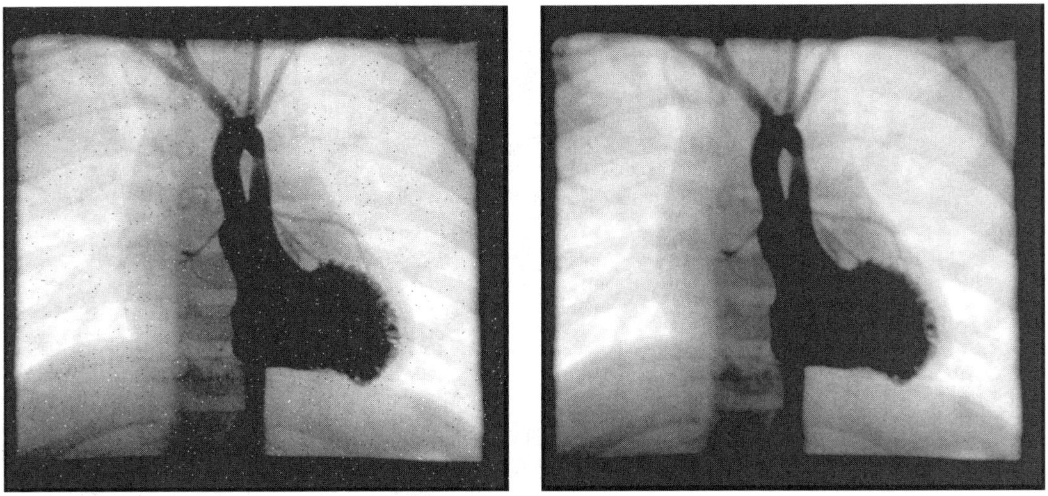

a b

FIGURE A7.1 Random noise (a) and blurring (b) added on a digital angiography image.

Since 1990, the applications of wavelets started to increase rapidly in medicine. Except for image processing,[27,28] the wavelets have been applied also into applications like image compression,[1,3,26] a very important issue in medicine and tele-medicine, since today the digital data sets, especially from acquisition devices like CT, MRI, and PET, are very large.[5] The mathematical and application analysis of wavelets is beyond the scope of this book.

A7.2 2D Convolution Function

A very useful and important operation in image processing is convolution. In case of 2D functions f and h, the convolution is denoted by (f^*h) and is defined by the integral

$$\begin{aligned} G(x, y) &= \int_{-\infty}^{\infty}\int_{-\infty}^{\infty} f(i, j)h(x-i, y-j)didj \\ &= \int_{-\infty}^{\infty}\int_{-\infty}^{\infty} f(x-i, y-j)h(i, j)didj \\ &= (f*h)(x, y) \end{aligned}$$

In digital image processing, where images have a limited domain on the image plane, convolution is used locally. For example, assuming an image $I(x, y)$ and a filtering mask $h(j, k)$ of size $(M \times N)$, their 2D convolution is defined as

$$G(x, y) = \sum_{j=\frac{M}{2}}^{\frac{M}{2}} \sum_{k=\frac{N}{2}}^{\frac{N}{2}} I(x-j, y-k) \cdot h(j, k) \tag{A7.1}$$

There are a number of mathematical properties associated with convolution. Actually, they are the same as the mathematical operation of the multiplication.

1. Convolution is commutative.

$$G = f*h = h*f$$

2. Convolution is associative.

$$G = f*(e*h) = (f*e)*h = f*e*h$$

3. Convolution is distributive.

$$G = f*(e + h) = (f*e) + (f*h)$$

where e, f, h, and G are all images, either continuous or discrete. An extensive description of the convolution theorem can be found in References 7, 9, and 13.

A7.3 The Fourier Transform

Using the Fourier transform, an image can be decomposed into harmonics.[10,14,16] To apply the Fourier transform in image processing, we have to make two assumptions:

1. The image under processing is periodic.
2. The Fourier transform of the periodic function always exists.

The 2D continuous Fourier transform of an image is defined as

$$Fr(u, v) = \int_{-\infty}^{\infty}\int_{-\infty}^{\infty} I(x, y)\exp^{-j2\pi(ux + vy)}dxdy$$

The inverse Fourier transform is

$$I(x, y) = \int_{-\infty}^{\infty}\int_{-\infty}^{\infty} Fr(u, v)\exp^{j2\pi(ux + vy)}dudv$$

To be applicable in medical image processing, the continuous Fourier transform must be converted in a discrete form. Thus, the discrete Fourier transform is equal to

$$Fr(u, v) = \frac{1}{MN}\sum_{x=0}^{M-1}\sum_{y=0}^{N-1} I(x, y)e^{-2\pi j\left(\frac{xu}{M} + \frac{yv}{N}\right)}$$
$$u = 0, 1, \ldots, M-1, v = 0, 1, \ldots, N-1$$

The inverse Fourier transform is given by

$$I(x, y) = \sum_{u=0}^{M-1}\sum_{v=0}^{N-1} Fr(u, v)e^{2\pi j\left(\frac{xu}{M} + \frac{yv}{N}\right)}$$
$$x = 0, 1, \ldots, M-1, y = 0, 1, \ldots, N-1$$

For simplicity, the Fourier transform can be denoted by an operator *Fr*. Thus, the previous equation can be abbreviated to

$$Fr\{I(x, y)\} = Fr(u, v)$$

The result of the Fourier transform is a complex number, which is composed by a real and an imaginary part:

$$Fr(u, v) = Rl(u, v) + jI(u, v)$$

Using the real and the imaginary parts one can compute the values called frequency and the phase spectrum of the image $I(x, y)$. The frequency spectrum is defined as (Figure A7.2)

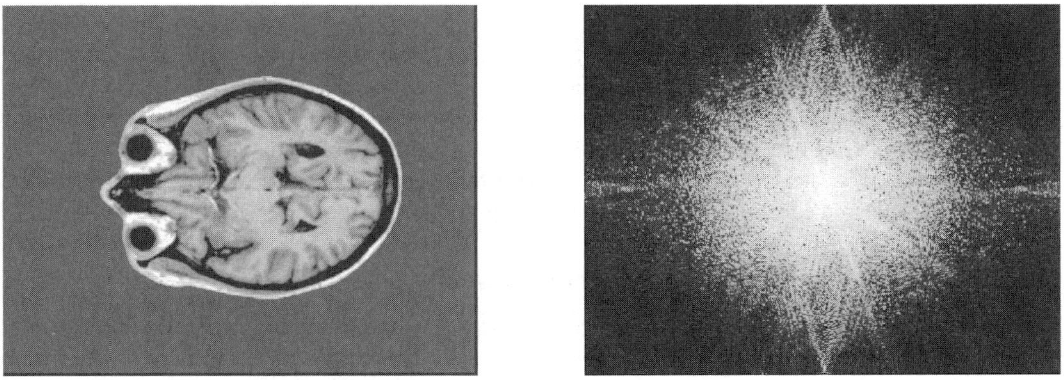

a b

FIGURE A7.2 Head MRI image (a) and its frequency spectrum (b).

$$|Fr(u, v)| = \sqrt{RI^2(u, v)*I^2(u, v)}$$

The phase spectrum is defined as

$$\phi(u, v)0 = \tan^{-1}\left[\frac{I(u, v)}{RI(u, v)}\right]$$

Some interesting properties of the Fourier transform from the image processing point of view are the following:

1. Its relation with the convolution (convolution theorem):

$$Fr\{(I*h)(x, y)\} = Fr(u, v)H(u, v)$$

$$Fr\{I(x, y)h(x, y)\} = (Fr*H)(u, v)$$

2. Shift of the origin, e.g., to the center of the frequency range, in the image and in the frequency domain:

$$Fr\{I(x - u_0, y - v_0)\} = Fr(u, v)e^{-2\pi j\frac{(u_0u + v_0v)}{N}}$$

$$Fr\left\{I(x, y)e^{2\pi j\left(\frac{u_0x + v_0y}{N}\right)}\right\} = Fr(u - u_0, v - v_0)$$

3. Periodicity: The Fourier transform and its inverse are periodic. In practice, if one wants to specify $Fr(x, y)$ in the frequency domain, only one period is enough for that. In mathematical terms, periodicity can be expressed as

$$Fr(u, -v) = Fr(u, N - v),\ I(-x, y) = f(M - x, y)$$

$$Fr(-u, v) = Fr(M - u, v),\ I(x, -y) = f(x, N - y)$$

$$Fr(u_0M + u, v_0N + v) = Fr(u, v),\ I(u_0M + x, v_0N + y) = f(x, y)$$

4. Conjugate symmetry, when $I(x, y)$ contains real values:

$$Fr(u, v) = Fr'(-u, -v) \text{ or}$$

$$|Fr(u, v)| = |Fr'(-u, -v)|$$

with Fr' being a complex Fourier product.

Since the computational effort considering multiplication and addition is proportional to N^2. The fast Fourier transform (FFT) algorithm reduces this effort and makes it proportional to $N \log_2 N$. One way to do this is to compute the terms of $e^{-2\pi j((ux)/N)}$ one time and store it in a table for all subsequent applications. Algorithms for FFT computation can be found in References 9, 13, and 17.

A7.4 Grey Scale Manipulation Techniques

In most medical imaging diagnostic equipment, the obtained images are in grey scale. The grey level look-up table usually does not surpass the 256 values. Therefore, every depth value from the image matrix will be converted to an index to the current look-up table. The observer often must perform a grey scale transformation in order to make more clear specific organs and tissue types. The techniques used to perform this task are called GSMTs.

One can separate these techniques into two categories: (1) histogram techniques and (2) image windowing techniques.[8,13]

A7.4.1 Histogram Techniques

An image histogram provides a general description of the image appearance. It shows the corresponding number of pixels to every grey value of an image. A change of the histogram (histogram modification) shape has as a result a change to the image contrast. A very common histogram technique in medical image processing is the histogram equalization. The aim of this technique is to modify the current histogram so as to have an equal distribution of the pixel's grey level over the whole range of the grey scale values. The benefit of using this technique is the automatic contrast enhancement for grey levels near the maximum of the histogram. The result can be seen in Figure A7.3. The implementation details of the histogram equalization technique can be found in References 7 and 8.

A7.4.2 Image Windowing Techniques

In windowing techniques, only a part of the image depth values is displayed with the available grey values. The term "window" refers to the range of the image depth values which are each time displayed. This window can be moved along the whole range of depth values of the image, each time displaying different

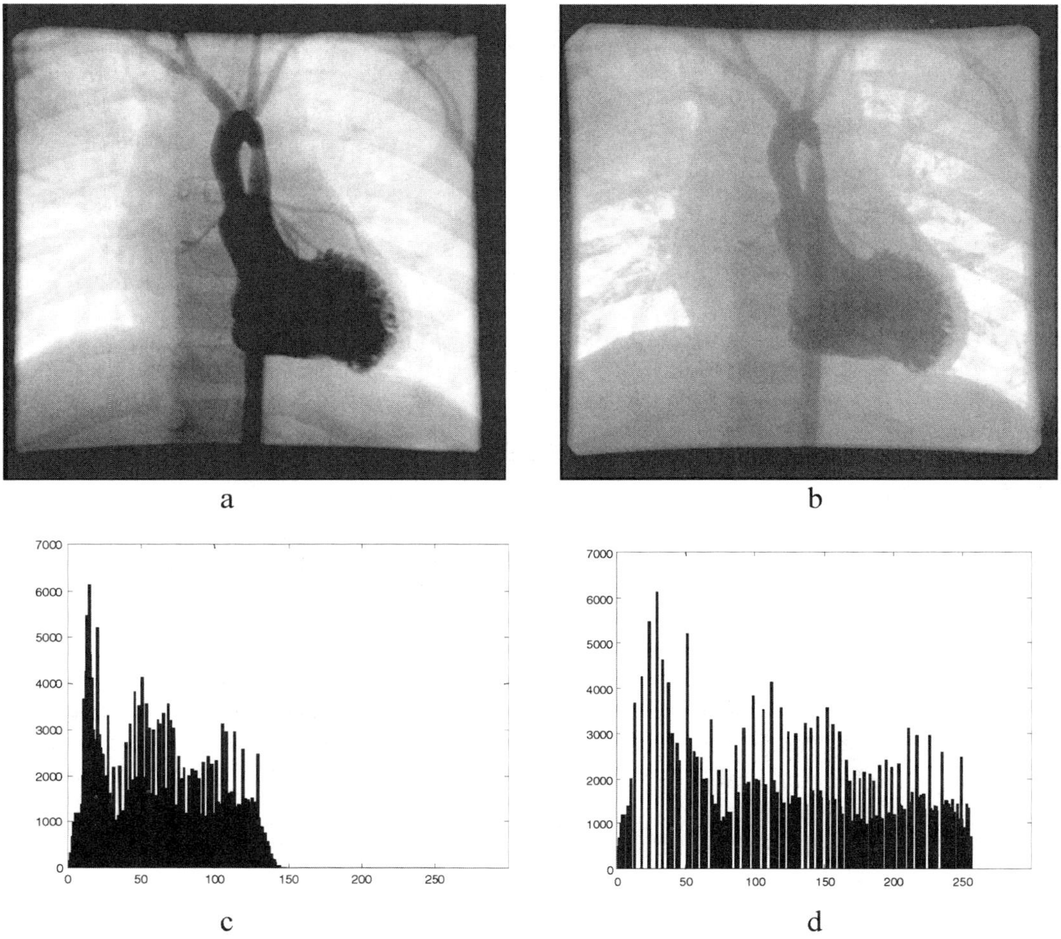

FIGURE A7.3 A digital angiographic image of the heart using injected contrast medium (a) and its histogram (c); the same image after histogram equalization (d).

tissue types in the full range of the grey scale and achieving better image contrast. The new brightness value of the pixel Gv is given by the formula:

$$Gv = \left(\frac{Gv_{max} - Gv_{min}}{We - Ws}\right) \cdot (Wl - Ws) + Gv_{min}$$

where $[Gv_{max}, Gv_{min}]$ is the grey level range, $[Ws, We]$ defines the window width, and Wl defines the window center. This is the simplest case of image windowing. Often, depending on the application, the window might have more complicated forms, such as double window, broken window, or non-linear windows (exponential or sinusoid) (Figure A7.4).

A7.5 Image Sharpening

The sharpening when applied to an image aims to decrease the image blurring and to enhance image edges.[7,8,13] There are two ways to apply these filters on the image:

1. In the spatial domain using the convolution process and the appropriate masks
2. In the frequency domain using high-pass filters

A7.5.1 High-Emphasis Masks

High-emphasis masks are masks with dimensions 3×3, 5×5, 7×7, or even higher, which are applied via convolution (Equation A7.1) to the original image. These masks are the result of the differentiation process.[7,20] Common masks used for high-emphasis filtering are

$$Em_1(j,k) = \begin{bmatrix} 0 & -1 & 0 \\ -1 & 5 & -1 \\ 0 & -1 & 0 \end{bmatrix}, \qquad Em_2(j,k) = \begin{bmatrix} -1 & -1 & -1 \\ -1 & 9 & -1 \\ -1 & -1 & -1 \end{bmatrix}$$

$$Em_3(j,k) = \begin{bmatrix} -1 & -2 & -1 \\ -2 & 13 & -2 \\ -1 & -2 & -1 \end{bmatrix}, \qquad Em_4(j,k) = \begin{bmatrix} 1 & -2 & 1 \\ -2 & 5 & -2 \\ 1 & -2 & 1 \end{bmatrix}$$

The image result after applying these masks is demonstrated in Figure A7.5.

A7.5.2 Unsharp Masking

The unsharp masking is a filter which combines the original image with the result of the image if it is filtered using a Laplacian filter, which we will describe later. First, one should enhance and isolate the edges, e.g., by using the Laplace operators, amplify them, and then add them back to the original image.[29] The results of this filter are similar to the high-emphasis filter.

A7.5.3 High-Pass Filtering

In the frequency spectrum of an image, one cannotice that the image edges and general high variations in grey levels result in high frequencies in the image spectrum. Using a high-pass filter in the frequency domain, we can attenuate the low frequencies without erasing the image edges. Some filter types are given here.

Ideal Filter

The transfer function for a 2D ideal filter is given as

$$H(u,v) = \begin{cases} 0 & \text{if } T(u,v) \le To \\ 1 & \text{if } T(u,v) > To \end{cases}$$

where To is the cut-off distance from the origin of the frequency plane and $T(u, v)$ is equal to

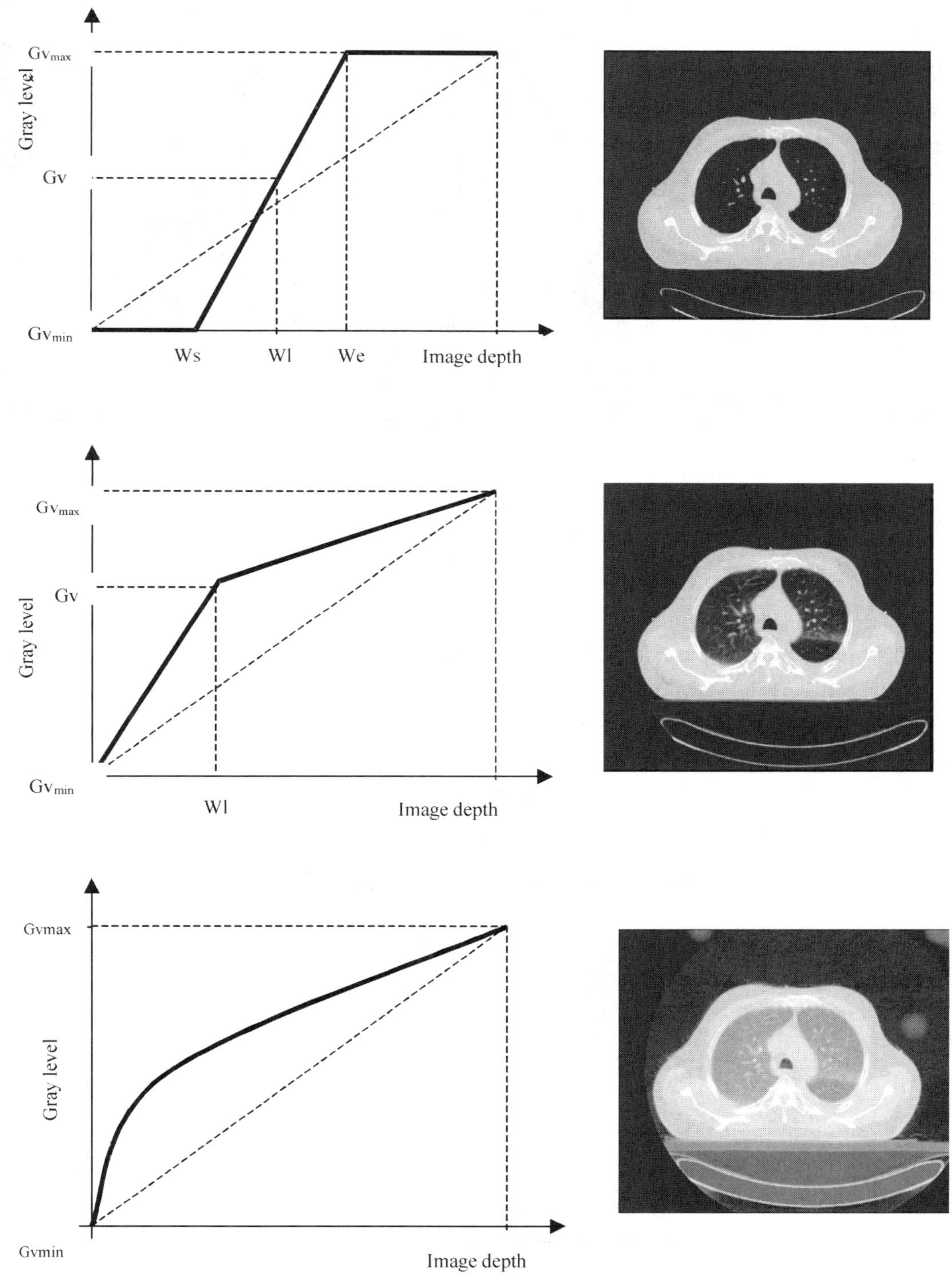

FIGURE A7.4 Examples of image windowing (from top to bottom): linear window, broken window, and non-linear window.

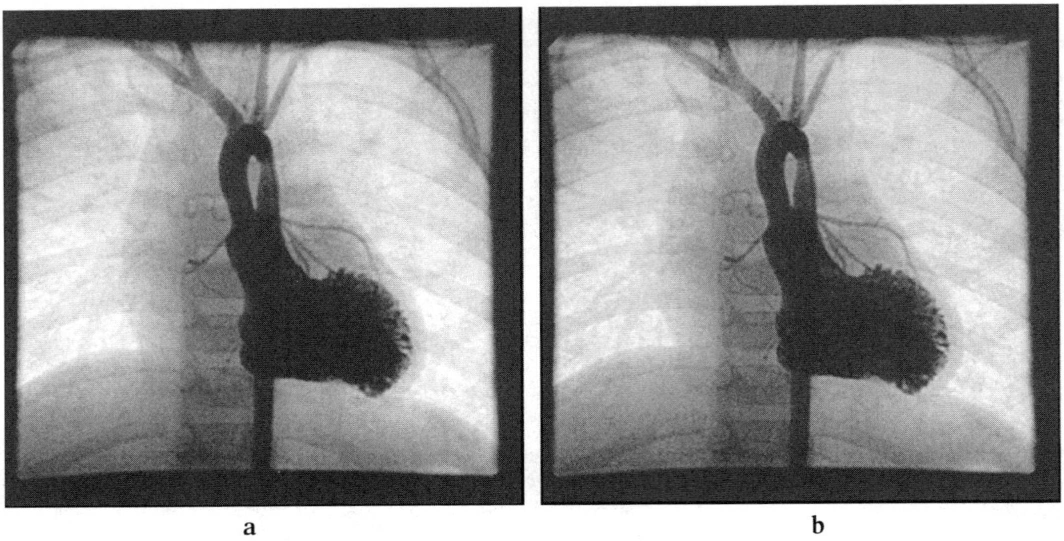

FIGURE A7.5 Removing blurring using high-emphasis masks Em_2 (a) and Em_3 (b). When using the masks Em_1 and Em_4, the image does not improve.

$$T(u, v) = \sqrt{(u^2 + v^2)} \tag{A7.2}$$

Butterworth Filter

Having a cut-off frequency at a distance *To* from the origin, the Butterworth filter of *n* order is defined as

$$H(u, v) = \frac{1}{1 + [To/T(u, v)]^{2n}}$$

Note that when $T(u, v) = To$ it is down to half of the maximum value (Figure A7.6).

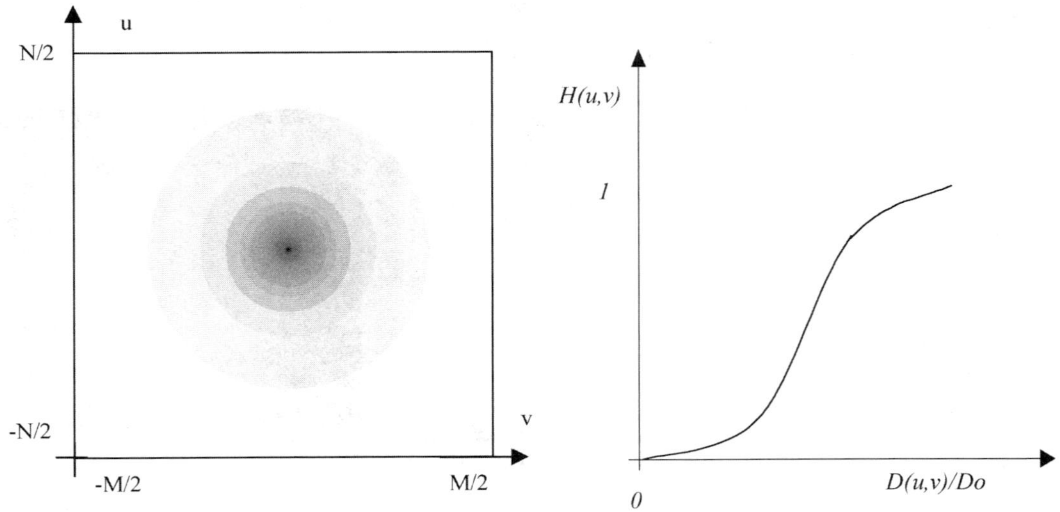

FIGURE A7.6 Butterworth high-pass filter. Shown are the 2D frequency spectrum and its cross section starting from the filter's center.

Exponential Filter

Similar to the Butterworth filter, the high-pass exponential filter can be defined from the relation

$$H(u, v) = e^{-0.347(To/T(u,v))^n}$$

where To is the cut-off frequency and n is the filter order.

The above filters and general the filter in the frequency domain are applied using the equation

$$G(u, v) = H(u, v)Fi(u, v)$$

where $Fi(u, v)$ it the Fourier transform of the image under processing, and $H(u, v)$ is the function which describes the filter. The inverse Fourier transform of the function $G(u, v)$ gives us the sharpened image (Figure A7.7).

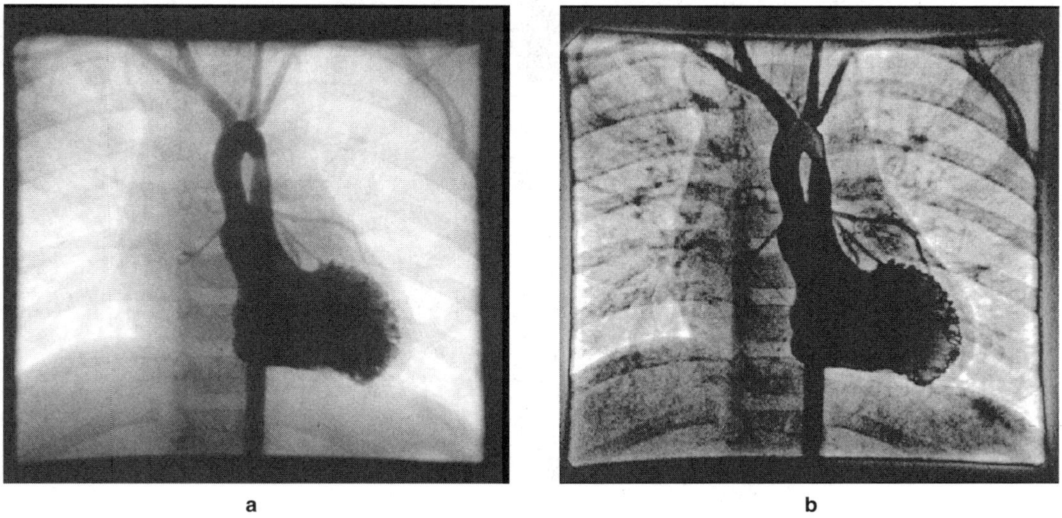

FIGURE A7.7 Filtering example of a digital angiographic image (a) using a high-pass Butterworth filter. In the filtered image (b) the vessels' branches are more distinct.

A7.6 Image Smoothing

Image smoothing techniques are used in image processing to reduce noise. Usually, in medical imaging, the noise is distributed statistically, and it exists in high frequencies. Therefore, one can say that image smoothing filters are low-pass filters. The drawback of applying a smoothing filter is the simultaneous reduction of useful information and mainly detail features, which also exist in high frequencies.

A7.6.1 Local Averaging Masks

Most filters used in the spatial domain use matrices (or masks) which may have dimensions of 3×3, 5×5, 7×7, 9×9, or 11×11. These masks can be applied on the original image using the 2D convolution function. Assuming that $I(x, y)$ is the original image, $E(y, x)$ is the filtered image, and $La(j, k)$ is a mask of size $M = 3, 5, 7, 9, 11$. If we choose M=3, then a 3×3 mask must be applied to each pixel of the image using (Equation A7.1). Different numeric values and different sizes of the masks will have different effects on the image. For example, if we increase the size of the matrix, we will have a more intensive smoothing effect. Usually, 3×3 masks are preferred. By using these masks, not only noise but also useful information will be removed (Figure A7.8). Some types of smoothing masks are

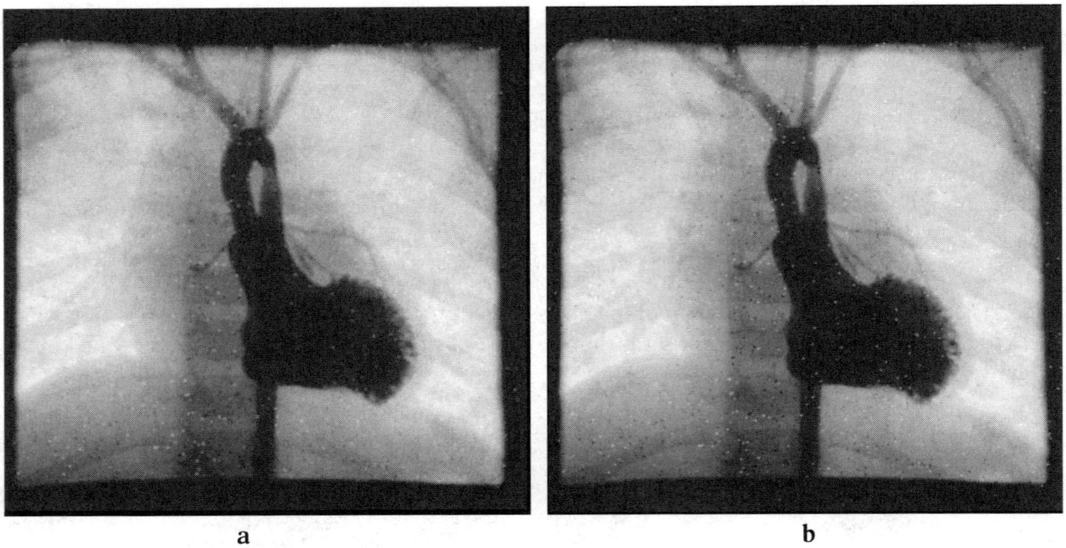

FIGURE A7.8 Filtering noise using the local averaging masks. The results are from (a) La_1 and (b) La_2. The masks La_3 and La_4 have very similar effects on the image.

$$La_1(j,k) = \frac{1}{9}\begin{bmatrix}1 & 1 & 1\\ 1 & 1 & 1\\ 1 & 1 & 1\end{bmatrix}, \qquad La_2(j,k) = \frac{1}{5}\begin{bmatrix}0 & 1 & 0\\ 1 & 1 & 1\\ 0 & 1 & 0\end{bmatrix}$$

$$La_3(j,k) = \frac{1}{10}\begin{bmatrix}1 & 1 & 1\\ 1 & 2 & 1\\ 1 & 1 & 1\end{bmatrix}, \qquad La_4(j,k) = \frac{1}{16}\begin{bmatrix}1 & 2 & 1\\ 2 & 4 & 2\\ 1 & 2 & 1\end{bmatrix}$$

A7.6.2 Median Filter

The median filter is based upon applying an empty mask of size $M \times M$ on the image (as in convolution). While the mask moves around the image, each place P_i of this mask will copy the pixel with the same coordinates (x,y) and also that pixel which is about to be filtered (Figure A7.9).[15,30] Then these collected pixels, which are values of brightness, will be sorted from the lower to the higher value, and their median value will replace the pixel to be filtered, in our example P5 (Figure A7.10). A different approach for efficient median filtering can be found in References 2 and 8.

A7.6.3 Gaussian Filter

The Gaussian filter is a popular filter with several applications, including smoothing filtering. A detailed description of the Gaussian filter can be found in References 8, 15, and 19.

In general, the Gaussian filter is separable:

$$\begin{aligned} h(x,y) &= g_{2D}(x,y) = \left(\frac{1}{\sqrt{2\pi}\sigma}\exp\left(-\frac{x^2}{2\sigma^2}\right)\right)\cdot\left(\frac{1}{\sqrt{2\pi}\sigma}\exp\left(-\frac{y^2}{2\sigma^2}\right)\right) \\ &= g_{1D}(x)\cdot g_{1D}(y) \end{aligned}$$

The Gaussian filter can be implemented in at least three different ways:

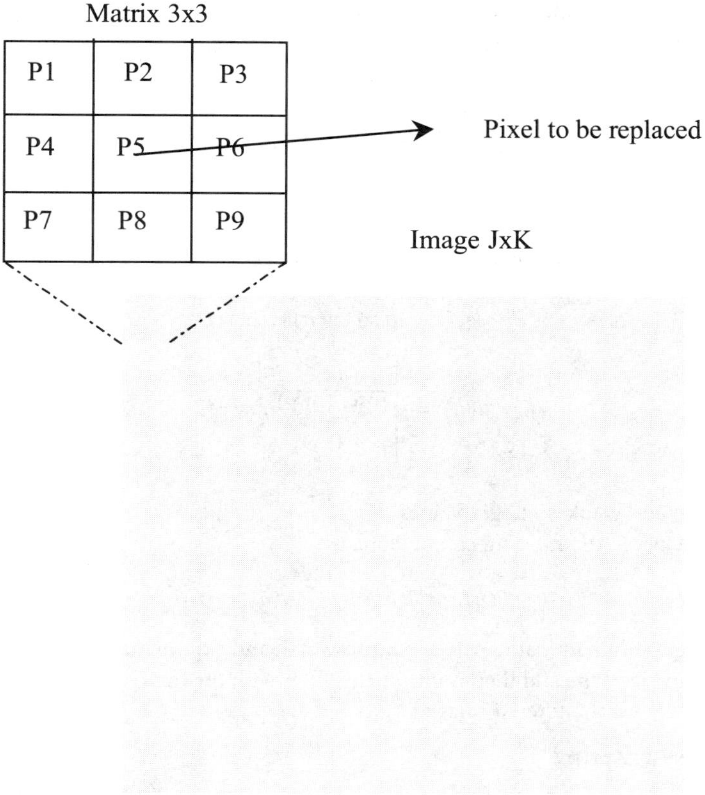

FIGURE A7.9 Applying a 3 × 3 median mask.

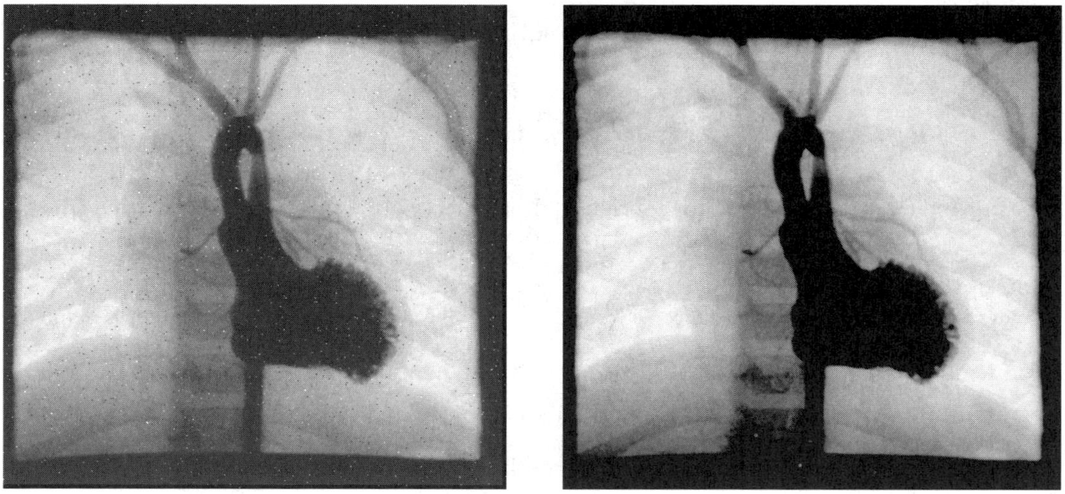

a b

FIGURE A7.10 Filtering the random noise using the median filter. Observe how effective the filter is in this case.

1. Convolution: Using a finite number of samples M of the Gaussian as the convolution kernel — it is common to choose $M = 3\sigma$ or 5σ, where σ is an integer number:

$$g_{1D}(n) = \begin{cases} \frac{1}{\sqrt{2\pi}\sigma} \cdot \exp\left(-\frac{n^2}{2\sigma^2}\right), & |n| \le M \\ 0 & |n| > M \end{cases}$$

2. Repetitive convolution: Using a uniform filter as a convolution kernel:

$$g_{1D} = u(n)*u(n)*u(n),$$

$$u(n) = \begin{cases} \frac{1}{(2M+1)}, & |n| \le M \\ 0 & |n| > M \end{cases}$$

where $M = \sigma$ and σ takes integer values.
In each dimension, the filtering can be done as

$$E(n) = [[I(n)*u(n)]*u(n)]*u(n)$$

3. In the frequency domain: Similar to the Butterworth and exponential filters, one can create a filter using the Gaussian type and then multiply this filter with the image spectrum. The inverse Fourier transform will give the filtered image.

A7.6.4 Low-Pass Filtering

In low-pass filtering, we use same principles as in high-pass filtering. In this case, our aim is to cut the high frequencies, where the noise is usually classified. The benefit of low-pass filtering, compared to spatial domain filters, is that noise with a specific frequency can be isolated and completely cleared from the image. When filtering random noise the drawback is that the edge information will be suppressed. Common filter types are the following:

Ideal Filter
The transfer function for a 2D low-pass ideal filter is given as

$$H(u, v) = \begin{cases} 1 & \text{if } T(u, v) \le To \\ 0 & \text{if } T(u, v) > To \end{cases}$$

where To is the cut-off distance from the origin of the frequency plane and $T(u, v)$ is given from Equation A7.2.

Butterworth Filter
Having a cut-off frequency at distance To from the origin, the Butterworth filter of n order is defined as

$$H(u, v) = \frac{1}{1 + [T(u, v)/To]^{2n}}$$

Note that when $T(u, v) = To$ is down to the half of the maximum value (Figure A7.11).

Exponential Filter
Similar to the Butterworth filter, the low-pass exponential filter can be defined from the relation

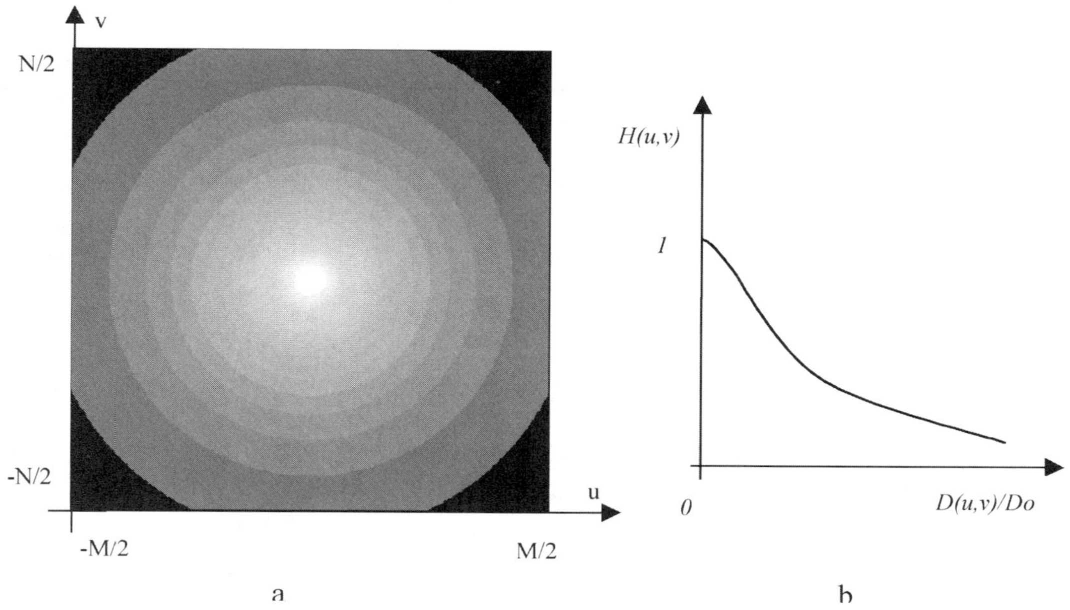

FIGURE A7.11 Butterworth low-pass filter, in a 2D representation of its frequency spectrum (a) and a cross-section of the same filter starting from the filter's center (b).

$$H(u, v) = e^{-0.347(T(u, v)/To)^n}$$

where *To* is the cut-off frequency and *n* is the filter order.

Figure A7.12 is an example of image noise removal using the Butterworth low-pass filter.

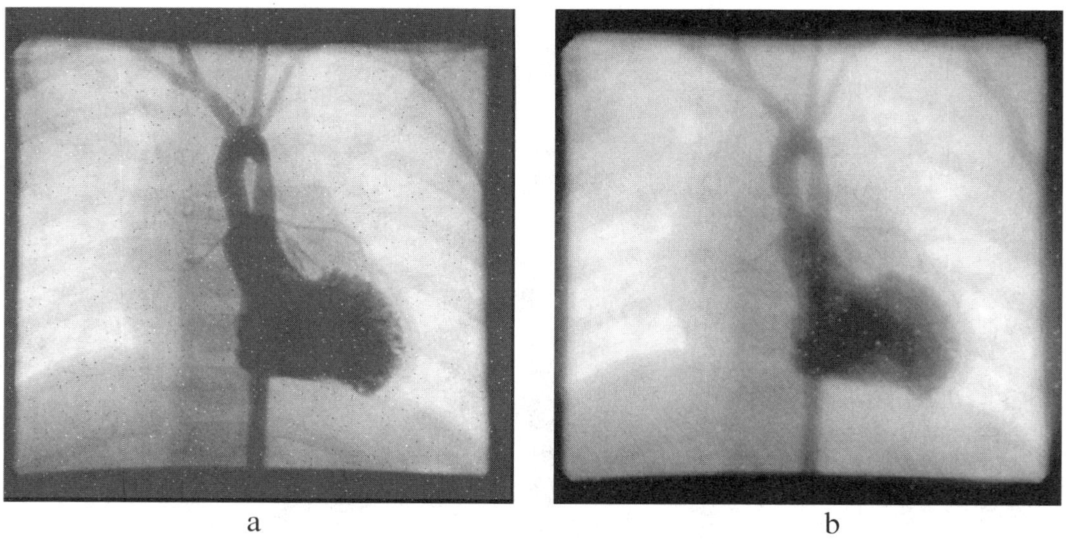

FIGURE A7.12 Removing random noise (a) from a digital angiographic image using a low-pass Butterworth filter. The smoothing effect in the filtered image is obvious.

A7.7 Edge Detection

Edge detection techniques aim to detect the image areas where we have a rapid change of the intensity.[4,8,13] In X-ray diagnostic imaging, these can be areas between bone and soft tissue or soft tissue and air. One can find similar examples in several types of diagnostic images. Here, we describe a number of gradient operators used in medical image edge detection as a mean of 3 × 3 masks. All of the masks can be applied to the original image via 2D convolution.

In general, the gradient magnitude $|grI(x, y)|$ of an image $I(x, y)$ is given as

$$|grI(x, y)| = \sqrt{\left(\frac{\partial I}{\partial x}\right)^2 + \left(\frac{\partial I}{\partial y}\right)^2}$$

A7.7.1 Laplacian Operator

A Laplacian or edge enhancement filter is used to isolate and amplify the edges of the image, but it completely destroys the image information at low frequencies (such as soft tissues). The Laplacian operator is invariant to image rotation and therefore has the same properties to all directions (Figure A7.13) and is calculated as

$$\nabla^2 I(x, y) = \frac{\partial^2 I(x, y)}{\partial x^2} + \frac{\partial^2 I(x, y)}{\partial y^2}$$

Common Laplacian masks are

$$Lp_1(j, k) = \begin{bmatrix} 0 & 1 & 0 \\ 1 & -4 & 1 \\ 0 & 1 & 0 \end{bmatrix}, \qquad Lp_2(j, k) = \begin{bmatrix} 1 & 1 & 1 \\ 1 & -8 & 1 \\ 1 & 1 & 1 \end{bmatrix}$$

$$Lp_3(j, k) = \begin{bmatrix} 1 & 2 & 1 \\ 2 & -12 & 2 \\ 1 & 2 & 1 \end{bmatrix}, \qquad Lp_4(j, k) = \begin{bmatrix} -1 & 2 & -1 \\ 2 & -4 & 2 \\ -1 & 2 & -1 \end{bmatrix}$$

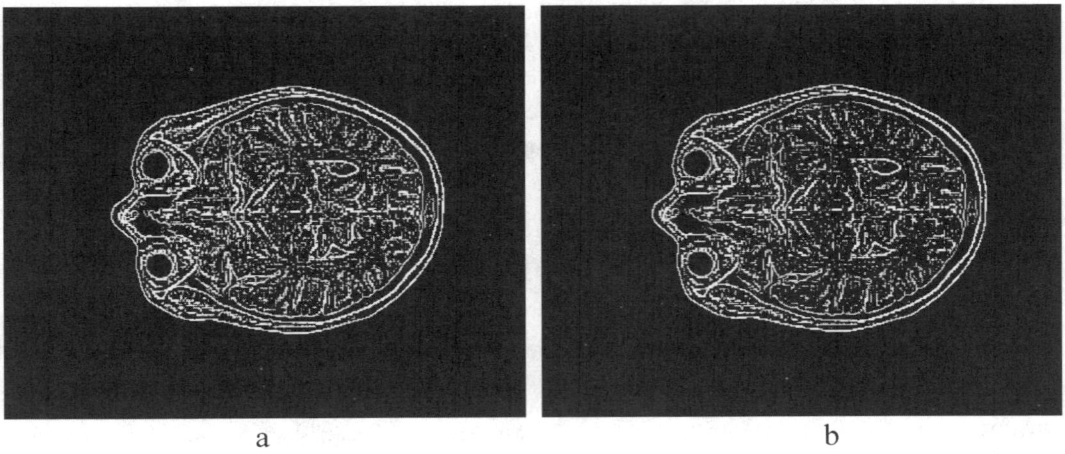

FIGURE A7.13 The Laplacian operator. The results are from (a) Lp_1 and (b) Lp_2.The mask Lp_3 gives a similar result to Lp_1, and the mask Lp_4 deteriorates the internal structures image.

A7.7.2 Prewitt, Sobel, and Robinson Operators

All these operators base their function on the first derivative. For a 3 × 3 mask, the gradient can be estimated for eight different directions. The gradient direction is indicated from the convolution result of greatest magnitude. In contrast to the Laplacian operator, these operators are related to the image orientation, and therefore, the image edges can be enhanced only at one direction for each time. We present here the first four masks from each operator. The rest of the masks are calculated considering the gradient direction we want to check (Figure A7.14).

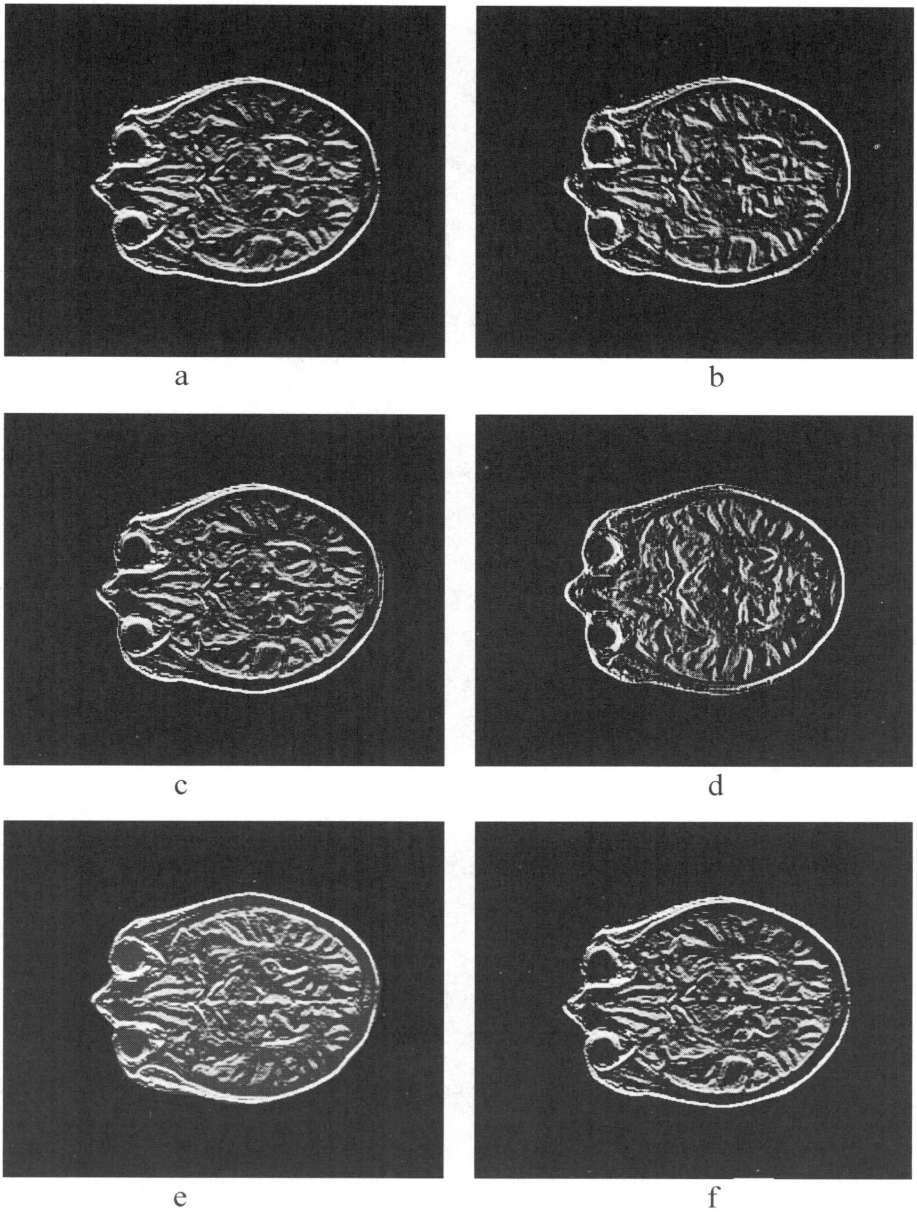

FIGURE A7.14 The Prewitt, Sobel, and Robinson operators applied on the MRI image. In (a) and (b), the results from Pr_1 and Pr_2 are equivalent. In (c) and (d), the results from Rb_1 and Rb_3 are equivalent. In (e) and (f), the results from Sb_0 and Sb_1 are equivalent.

Prewitt

$$Pr_1(j,k) = \begin{bmatrix} 1 & 1 & 1 \\ 0 & 0 & 0 \\ -1 & -1 & -1 \end{bmatrix}, \quad Pr_2(j,k) = \begin{bmatrix} 0 & 1 & 1 \\ -1 & 0 & 1 \\ -1 & -1 & 0 \end{bmatrix}$$

$$Pr_3(j,k) = \begin{bmatrix} -1 & 0 & 1 \\ -1 & 0 & 1 \\ -1 & 0 & 1 \end{bmatrix}, \quad Pr_4(j,k) = \begin{bmatrix} 1 & 0 & -1 \\ 1 & 0 & -1 \\ 1 & 0 & -1 \end{bmatrix}$$

Sobel

$$Sb_1(j,k) = \begin{bmatrix} 1 & 2 & 1 \\ 0 & 0 & 0 \\ -1 & -2 & -1 \end{bmatrix}, \quad Sb_2(j,k) = \begin{bmatrix} 0 & 1 & 2 \\ -1 & 0 & 1 \\ -2 & -1 & 0 \end{bmatrix}$$

$$Sb_3(j,k) = \begin{bmatrix} -1 & 0 & 1 \\ -2 & 0 & 2 \\ -1 & 0 & 1 \end{bmatrix}, \quad Sb_4(j,k) = \begin{bmatrix} 1 & 0 & -1 \\ 2 & 0 & -2 \\ 1 & 0 & -1 \end{bmatrix}$$

Robinson

$$Rb_1(j,k) = \begin{bmatrix} 1 & 1 & 1 \\ 1 & -2 & 1 \\ -1 & -1 & -1 \end{bmatrix}, \quad Rb_2(j,k) = \begin{bmatrix} 1 & 1 & 1 \\ -1 & -2 & 1 \\ -1 & -1 & 0 \end{bmatrix}$$

$$Rb_3(j,k) = \begin{bmatrix} -1 & 1 & 1 \\ -1 & -2 & 1 \\ -1 & 1 & 1 \end{bmatrix}, \quad Rb_4(j,k) = \begin{bmatrix} 1 & 1 & -1 \\ 1 & -2 & -1 \\ 1 & 1 & -1 \end{bmatrix}$$

References

1. M. L. Hilton, B. D. Jawerth, and A. Sengupta. Compressing still and moving images with wavelets. *J. Multimedia Syst.*, 2, 218–227, 1994.
2. T. S. Huang, G. J. Yang, and G. Y. Tang. A fast two-dimensional median filtering algorithm. *IEEE Trans. Acoust. Speech Signal Process.*, ASSP-27, 13–18, 1979.
3. W. R. Zettler, J. Huffman, and D. C. P. Linden. Application of compactly supported wavelets to image compression. In *Image Processing Algorithms and Techniques*. SPIE, Bellingham, WA, pp. 150–160, 1990.
4. D. Marr and E. C. Hildreth. Theory of edge detection. *Proc. R. Soc. London Ser. B.*, 207, 187–217, 1980.
5. A. H. Delany and Y. Bresler. Multiresolution tomographic reconstruction using wavelets. *IEEE Int. Conf. Image Process.*, 1, 830–834, 1994.
6. M. D. Harpen. An Introduction to wavelet theory and application for the radiological physicist. *Med. Physics*, 25, 1985–1993, October 1998.
7. R. C. Gonzalez and P. Wintz. *Digital Image Processing*. 2nd edition, Addison-Wesley, Reading, MA, 1987.

8. M. Sonka, V. Hlavac, and R. Boyle. *Image Processing, Analysis and Machine Vision.* Second Edition, PWS Publishing, 1998.
9. H. J. Nussbaumer, *Fast Fourier Transform and Convolution Algorithms.* 2nd edition, Springer Verlag, Berlin, 1982.
10. A. C. Kak and M. Stanley. *Principles of Computerized Tomographic Imaging.* IEEE, Piscataway, NJ, 1988.
11. J. C. Russ. *The Image Processing Handbook.* 2nd edition, CRC Press, Boca Raton, FL, 1995.
12. H. C. Andrews and B. R. Hunt. *Digital Image Restoration.* Prentice-Hall, Englewood Cliffs, NJ, 1977.
13. R. C. Gonzalez and R. E. Woods. *Digital Image Processing.* Addison-Wesley, Reading, MA, p. 716, 1992.
14. J. W. Goodman. *Introduction to Fourier Optics.* McGraw-Hill Physical and Quantum Electronics Series, McGraw-Hill, New York, p. 287, 1968.
15. K. R. Castleman. *Digital Image Processing.* 2nd edition, Prentice-Hall, Englewood Cliffs, NJ, 1996.
16. H. Stark. *Application of Optical Fourier Transforms.* Academic Press, New York, 1982.
17. T. Pavlidis. *Algorithms for Graphics and Image Processing.* Computer Science Press, New York, 1982.
18. R. H. T. Bates and M. J. McDonnell. *Image Restoration and Reconstruction.* Clarendon Press, Oxford, 1986.
19. I. T. Young and L.J. Van Vliet. Recursive implementation of the Gaussian filter. *Signal Process.*, 44(2), 139–151, 1995.
20. L. Schreyer, M. Grimm, and G. Sakas. *Case Study: Visualization of 3d-Ultrasonic Data.* IEEE, 369–373, October 1994.
21. S. Walter and G. Sakas. *Extracting Surfaces from Fuzzy 3D-Ultrasonic Data.* Addison-Wesley, Reading, MA, pp. 465–474, August 1995.
22. A. Rosenfeld and A. C. Kak. *Digital Picture Processing.* 2nd edition, Academic Press, New York, 1982.
23. K. R. Rao and P. Yip. *Discrete Cosine Transform, Algorithms, Advantages, Applications.* Academic Press, Boston, 1990.
24. C. K. Chui. *An Introduction to Wavelets.* Academic Press, New York, 1992.
25. G. Strang. Wavelet transforms versus Fourier transforms. *Bull. Am. Math. Soc.*, 28, 288–305, 1993.
26. R. Devore, B. Jaeverth, and B. J. Lucier. Image compression through wavelet transform coding. *IEEE Trans. Inf. Theory*, 38, 719–746, 1992.
27. D. Healy and J. Weaver. Two applications of wavelet transforms in MR imaging. *IEEE Trans. Inf. Theory*, 38, 840–860, 1992.
28. R. Devore, B. Jaeverth, and B. J. Lucier. Feature extraction in digital mammography. In A. Aldroubi and M. Unser, Editors. *Waveletsin Medicine and Biology.* CRC Press, Boca Raton, FL, 1996.
29. A. K. Jain. *Fundamentals of Digital Image Processing.* Prentice-Hall, Englewood Cliffs, NJ, 1989.
30. S. G. Tyan. Median filtering, deterministic properties. In T. S. Huang, editor. *Two-Dimensional Digital Signal Processing, Volume II.* Springer-Verlag, Berlin, 1981.

8

Target Tracking

Wolfgang Koch
FGAN Research Institute for Communication, Information Processing, and Ergonomics (FKIE)

Abbreviations

IMM	Interacting multiple models
JPDAF	Joint probabilistic data association filter
KF	Kalman filter
MHT	Multiple hypothesis tracking
MMSE	Minimum mean squared error
NN	Nearest neighbor filter
PDA	Probabilistic data association
PDAF	Probabilistic data association filter

Frequently Used Symbols

$(\ldots)^{\mathrm{T}}$	Transpose
$\lVert\ldots\rVert$	Vector norm
$\det(\ldots)$	Determinant
$\mathbb{E}[\ldots]$	Expectation

0-8493-3691-0/01/$0.00+$.50

∇_x	Gradient with respect to x
$\mathcal{N}(z; x, V)$	Multivariate Gaussian density with mean vector x and covariance matrix V
a_k^2	Received signal strength at time $t_k \to$ SNR
α_ϕ, α_r	Resolution (azimuth, range); $\to P_u$, R_k^u
F_k	System matrix at time $t_k \to \mathrm{Q}_k$
$\lvert G_k \rvert$	Volume of a region containing all relevant reports at time t_k; $\to p_F(n_k)$
h_k	Interpretation hypothesis regarding the origin of a single scan; $\to H_k$, Z_k
H^k, H_n^k	Interpretation history: hypothesis regarding the origin of multiple scans; $\to h_k$, Z^k
H_k, H_k^g	Measurement matrix at time t_k (resolved/group measurement); $\to \mathrm{R}_k$, R_k^g
H_k^u	Unresolved returns: fictitious measurement matrix (relative distance); $\to \mathrm{R}_k^u$
λ_D	Detector threshold; $\to a_k^2$, P_D
λ_G	Threshold for gating; $\to P_c$
m_k, m_k^i	Dynamics model assumed to be in effect at time t_k (i = 1, …, r); $\to M^k$, M_n^k, r
M^k, M_n^k	Model history: hypothesis on a temporal sequence dynamics models assumed; $\to m_k$
μ_{H^k}, $\mu_{H^k}^{M^k}$	Mixture coefficient related to particular histories H^k and M^k
n_k	Number of sensor reports to be processed at time t_k; $\to Z_k$
ν_k	Data innovation vector at time t_k; $\to S_k$
$p_F(n_k)$	Probability of receiving n_k false returns (Poisson); $\to \lvert G_k \rvert$, ρ_F
$p_{m_k m_{k+1}}$, p_{ij}	Model transition probability; $\to m_k$, m_k^i
P_c	Correlation probability; $\to \lambda_G$
P_D, P_{FA}	Detection/false alarm probability; $\to \lambda_D$, snr
$\mathbf{P}_{\dots}^{\dots}$	Covariance matrix related to $\hat{\mathbf{x}}_{\dots}^{\dots}$
P_u	Probability of two objects being unresolved; $\to \mathrm{H}_k^u$, R_k^u
Q_k	Process noise covariance matrix at time t_k; $\to \mathrm{F}_k$
r	Number of models used in the IMM approach
$\mathrm{R}_k^{\dots}$	Measurement noise covariance matrix related to $\mathrm{H}_k^{\dots}$
R_k^u	Measure of the sensor resolution capability; $\to \mathrm{H}_k^u$, P_u
ρ_F	Spatial false return density; $\to p_F(n_k)$
snr, SNR	Mean/instantaneous signal-to-noise ratio
S_k	Innovation covariance matrix at time t_k; $\to \nu_k$
t_k	Instant of time when reports are produced (scan, target revisit, frame time)
T_k	Data innovation interval; $\to t_k$
u_k, v_k	Measurement/process noise vector at time t_k; $\to \mathrm{R}_k$, Q_k
W_k	Kalman Gain matrix; $\to S_k$
x_k	Kinematical state of the object to be tracked at time t_k
$\hat{\mathbf{x}}_k$	Expectation of x_k with respect to the density obtained by filtering, MMSE estimate
$\hat{\mathbf{x}}_{l\mid k}$	Expectation of x_k with respect to the density obtained by pre- ($l > k$) or retrodiction $l > k$
$\hat{\mathbf{x}}_{H_k^k}$, $\hat{\mathbf{x}}_{H^k}^{M^k}$	Expectation of x_k related to particular histories H^k, M^k (filtering)
$\hat{\mathbf{x}}_{H^k}^{M^k}(l\mid k)$	Expectation of x_k related to particular histories H^k, M^k (pre-, retrodiction)
z_k	Individual sensor report to be processed at time $t_k \to Z_k$
Z_k	Set of sensor reports to be processed at time t_k; see n_k, z_k
Z^k	Temporal sequence of data to be processed: (Z_k, n_k, Z_{k-1}, n_{k-1}, …, Z_1, n_1); see n_k, z_k

8.1 Introduction

In many engineering applications, including surveillance, guidance, navigation, robotics, system control, image data processing, or quality management, single stand-alone sensors or sensor networks are used for collecting information on time varying quantities of interest, such as kinematical characteristics and measured attributes of moving objects (e.g., maneuvering air targets, ground vehicles) or, in a more general framework, time varying signal parameters.

More strictly speaking, in these or similar applications, the state of a stochastically driven dynamical system is to be estimated from a series of sensor data sets, also called scans or data frames, which are received at discrete instants of time, being referred to as scan/frame time, target revisit time, or data innovation time. The individual output data produced by the sensor systems considered (sensor reports, observations, returns, hits, plots) typically result from complex estimation procedures which characterize particular waveform parameters of the received signals (*sensor signal processing*). Provided the quantities of interest are related to moving point-source objects or small extended objects (e.g., radar targets), often relatively simple statistical models can be derived from basic physical laws which describe their temporal behavior and thus define the underlying dynamical system. The formulation of adequate dynamics models, however, may be a difficult task in certain applications.

8.1.1 Tracking Systems

For an efficient exploitation of the sensor resources as well as to obtain information not directly provided by the individual sensor reports, appropriate data association and estimation algorithms are required (*sensor data processing*). These techniques result in tracks, i.e., estimates of state trajectories, which statistically represent the quantities or objects considered along with their temporal history. Tracks are initiated, confirmed, maintained, stored, evaluated, fused with other tracks, and displayed by the *tracking system* or data manager. For methodical reasons, the tracking system should be carefully distinguished from the underlying sensor systems, though there may exist close interrelations, such as in the case of multiple-target tracking with an agile-beam radar, raising the problem of sensor management. Evidently, the achievable track accuracy depends on the quality of the sensors involved, the current operating conditions, and the particular scenario considered. Several well-established textbooks provide a general introduction to this practically important subject (see References 1, 2, 3, 5, 7, 8 and, with particular emphasis, the recent monograph in Reference 9).

Figure 8.1 provides a schematic overview of a generic tracking system along with its relation to the underlying sensor system. After passing the detector device, which essentially serves as a means of data rate reduction, the sensor signal processing unit provides estimates of signal parameters characterizing

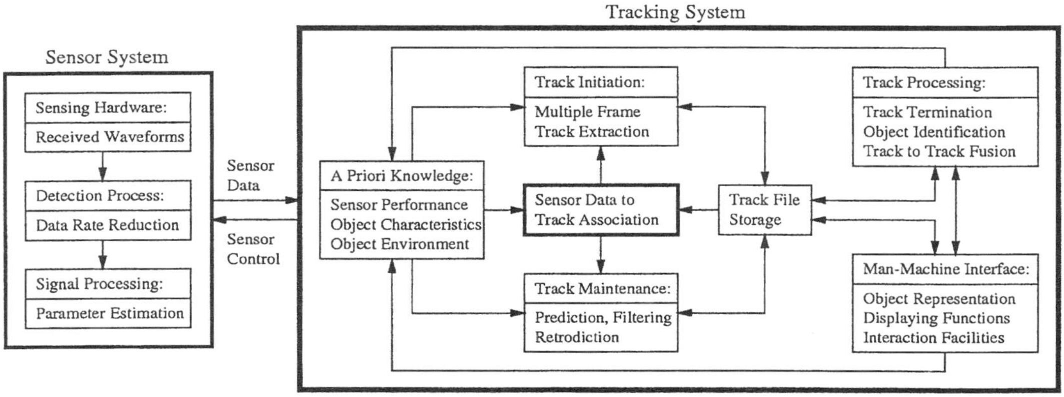

FIGURE 8.1 Generic scheme of a tracking system.

the waveforms received by the sensing hardware (e.g., radar antennas). From these preprocessed estimates, sensor reports are formed, i.e., measured quantities possibly related to the objects of interest, that are input information for the tracking system. In the tracking system itself, all sensor data which can be associated to the already existing tracks are used for track maintenance (prediction, filtering, retrodiction). The remaining non-associated data are processed in order to establish new tentative tracks (track initiation, multiple-frame track extraction). Thus, the plot-to-track association unit plays a key role in any multiple-target tracking system. Evidently, *a priori* knowledge in terms of statistical models of the sensor performance, the object characteristics (including their dynamical behavior), and the object environment is a prerequisite to both track maintenance and track initiation. Track confirmation/termination, object classification/identification, and fusion of tracks representing identical information are performed in the track processing unit. The generic scheme of a tracking system is completed by a man-machine interface with displaying and interaction functions. The available information on the sensor, the objects of interest, and the environment can be specified, updated, or corrected by direct human interaction as well as by the track processor itself, e.g., as a consequence of a successful object classification.

8.1.2 Challenging Conditions

Track maintenance/initiation by processing noise-corrupted returns is by no means trivial if the sensor data are of uncertain origin or if there exists uncertainty regarding the underlying system dynamics. It is this particular topic that is being discussed in this chapter. We focus mainly on four aspects:

1. In general, data association conflicts may arise even for well-separated objects if a false return background is to be taken into account, which can never be completely suppressed by means of signal processing at the individual sensor sites (e.g., random noise, residual clutter, man-made noise). For airborne clutter suppression, see the recent monograph in Reference 10.
2. Even in the absence of unwanted sensor reports, ambiguous correlations between newly received sensor reports and existing tracks are an inherent problem for objects moving closely spaced for some time. This is even more critical in the case of false returns or detections from unwanted objects. In such situations, the identity of the individual object tracks might get lost.
3. Furthermore, closely spaced objects may continuously change from being resolved to unresolved and back again due to the limited resolution capability of every physical sensor, making the data association task even harder. Additional problems arise from sensor returns having a poor quality, due to large measurement errors, low signal-to-noise ratios, or fading phenomena, for instance. Besides that, the scan rates may be low in certain applications, such as long-range air surveillance.
4. Moreover, in a practical tracking task the underlying system dynamics model currently assumed to be in effect might be one particular sample out of a set of several alternatives and, thus, *a priori* unknown. As an example, let us consider a radar application with military air targets. In a given mission, often clearly distinct maneuvering phases can be identified, as even agile targets do not always use their high maneuvering capability. Nevertheless, sudden switches between the underlying dynamics models do occur and are to be taken into account. Tracks must not be lost in such situations.

8.1.3 Bayesian Approach

Many basic ideas and mathematical techniques relevant to the design of tracking systems can be discussed in a unified statistical framework that essentially makes use of Bayes' Rule. The general multiple-object, multiple-sensor tracking task, however, is highly complex and involves rather sophisticated combinatorial and logical considerations that are beyond the scope of this chapter. For a more detailed discussion of the problems involved, see References 11 to 14. Nevertheless, in many applications the task can be partitioned into independent sub-problems of (much) less complexity. Thus, to provide an introduction to statistical tools frequently used in sensor data processing, we follow this approach and analyze practically important examples along with several approximations to their optimal solution being important to practical realization. These examples may serve as elements for developing appropriate tracking systems that meet the requirements of a particular user-defined application.

In a Bayesian view, a tracking algorithm is an iterative updating scheme for conditional probability densities that describe the object states given both the accumulated sensor data and all available *a priori* information (sensor characteristics, object dynamics, operating conditions, underlying scenario). Provided the density iteration, also referred to as the filtering, has been performed correctly, optimal state estimators may be derived related to various risk functions. Under the conditions previously discussed, the densities have a particular formal structure: they are finite mixtures,[15] i.e., weighted sums of individual densities, each of them being related to an individual data interpretation and model hypothesis. Thus, this structure is a direct consequence of the uncertain origin of the sensor data and uncertainty regarding the underlying system dynamics. In the case of well-separated objects without false returns, assuming perfect detection and a single dynamics model, the Bayesian approach reduces to Kalman filtering (see Reference 16, p. 107 ff.).

Bayesian retrodiction is intimately related to filtering in that it provides a backward iteration scheme for calculating the probability densities of the past object states given all information accumulated up to the current scan.[13,17,18] Retrodiction thus proves to be a generalization of smoothing algorithms such as those proposed by Rauch, Tung, and Striebel (see Reference 16, p. 161 ff.).

While in many applications track maintenance and acquisition of sensor data are completely decoupled, the problem of optimal resource allocation arises for more sophisticated sensors like agile-beam radar. Such steered sensors are characterized by various degrees of freedom (for instance, free choice of revisit time, beam position, transmitted energy, detection thresholds). The sensor parameters involved may be varied over a wide range and may be chosen individually for each track. For these sensors, decisions on resource allocations can be adapted to the current lack of information, i.e., the control of basic sensor parameters is taken into the responsibility of the tracking system. In other words, there exists a feedback of tracking information to the sensor system. Thus, track maintenance and data acquisition may be closely interrelated.[19–22]

Tracking algorithms must be initiated by appropriately chosen *a priori* densities (track initiation). This is a relatively simple task provided particular sensor reports are actually valid measurements of the objects to be tracked. For low-observable objects, i.e., targets embedded in a high false return background, however, several frames of observations may be necessary for the detection of all objects of interest moving in the sensors' field of view. By this, a higher level detection process is defined, resulting in algorithms for multiple-frame track extraction (see Reference 23 and the literature cited therein). A more detailed discussion of this practically important aspect, however, is beyond the scope of this chapter.

8.1.4 Sensor Fusion Aspects

Often the input data for the tracking system are provided by a network of homogeneous or heterogeneous sensors that may be co-located on a single platform (aircraft, ship, robot, for instance) or distributed at various sites. The use of a combination of various sensors instead of a single sensor has many advantages: the total coverage of suitably distributed sensors may be much larger and a high-cost sensor might be replaced by a network of low-cost sensors producing the same information. The redundancy provided by sensors with overlapping fields of view results in increased data rates which can be important to tracking low-observable objects. Multiple-sited networks may also provide information that is, on principle, not available by a corresponding single-site sensor (networks of passive sensors, for instance). Moreover, sensor networks are more robust against failure or destruction of individual components. Several well-established textbooks provide general introductions to this practically important subject.[24–26]

If all sensor reports (including information on their source, their position in a common frame of reference, and the related time) are transmitted to a processing center without significant delay, we speak of *centralized* data fusion: at discrete instants of time, a frame of observations is received and processed by the central tracking system. For centralized fusion, being the optimal approach in a theoretical view, it is thus irrelevant if the data are produced by a single-sited sensor or a sensor network. The practical realization of a centralized fusion architecture, however, may be difficult for various reasons, such as the limited capacity of the data links between sensors and fusion center, synchronization problems, or misalignment errors. Therefore, *decentralized* fusion architectures or hybrid solutions have been proposed (see Reference 27, for instance). In this approach, the data of the sensor systems are preprocessed at their individual sites. Hence, the fusion center receives higher level information, i.e., sensor-individual tracks which are to be fused with other tracks resulting in a central track (see Reference 9 and the literature cited therein). The problems arising in the design of operational sensor fusion systems and methods for their efficient solution, however, are beyond the scope of this chapter.

8.2 Discussion of the Problem

As a generic example in this chapter, let us consider a number of radar sensors scanning the vicinity of point-source objects at discrete instants of time t_k. For a single rotating radar, the interval between consecutive target illuminations is constant (also called revisit interval, data innovation interval). For radar networks, this is in general no longer true. Moreover, in the case of agile-beam radar, the tracking system can allocate the sensor at arbitrary instants of time.

8.2.1 Basic Notions

For a more precise discussion, let us consider six sensor reports produced by two closely spaced targets at time t_k (Figure 8.2). This single frame of observations is by no means uniquely interpretable. Among other feasible interpretation hypotheses, the black dots could be assumed to represent valid position measurements of the targets, while all other plots are false (Figure 8.2a). The asterisks indicate the predicted target positions provided by the tracking system. Under certain statistical assumptions discussed later, the target measurements are assumed to be normally distributed about the predictions with a covariance S_k determined by the related state prediction covariance and the measurement error. As any prediction uses assumptions on the underlying system dynamics, both the sensor performance and the dynamics model enter into the statistics of the expected target measurements. The difference ν_k between a measurement and the predicted target position, the *innovation* vector, is assumed to be $N(0, S_k)$ distributed. A natural scalar measure for the deviation

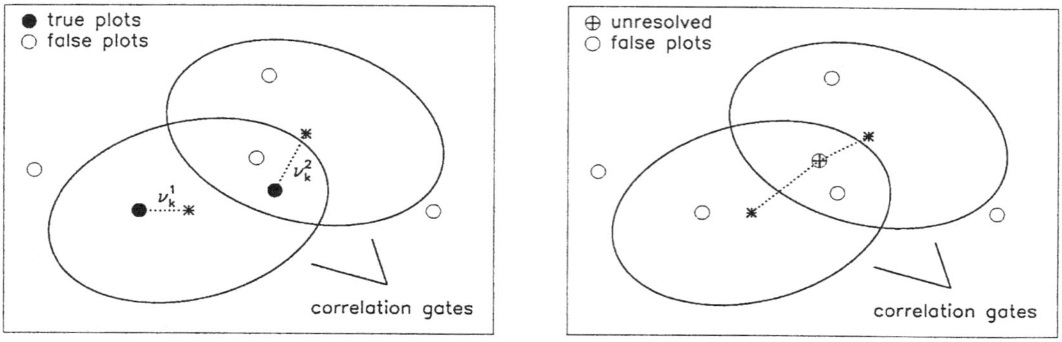

(a) two resolved targets (b) two unresolved targets

FIGURE 8.2 Sensor data of uncertain origin: competing interpretations.

between the predicted and an actually received measurement is thus given by $||\nu_k||^2 = \nu_k^T S_k^{-1} \nu_k$, also called *Mahalanobis norm*. Gating means that only those sensor returns whose innovations are smaller than a certain predefined threshold ($||\nu_k|| < \lambda_G$) are considered for track maintenance. By this criterion, ellipsoid correlation gates are defined containing the valid measurements with a certain *correlation probability* $P_c = P_c(\lambda_G)$.

Competing with the previously discussed data interpretation, however, there exist many more feasible association hypotheses; for instance, the targets could have produced a single unresolved measurement as indicated in Figure 8.2b, with all other plots being false returns. Alternatively, one of both targets might not have been detected or no target detection might have occurred at all, with the gates containing false returns only. The size of the correlation gates and thus the ambiguity of the received sensor data depend on the number of false returns and missing detections to be taken into account, on the measurement errors and data innovation intervals involved, and on the uncertainty regarding the target's maneuvering behavior. As will become clear later, the innovation statistics related to a particular interpretation hypothesis is essential to evaluating its statistical weight.

8.2.2 Ad Hoc Approaches

For dealing with sensor data of uncertain origin, several well-established ad hoc methods exist which are implemented in numerous operational tracking systems. Under benign conditions, *gating* can be sufficient for separating valid target measurements from competing sensor returns. The resulting plot is then processed by *Kalman* filtering (KF) or one of its suboptimal versions (α-β filtering,[1] for instance). In the previous example (Figure 8.2), two sensor reports can be excluded by this measure. Evidently, the gate must be sufficiently large, otherwise the valid plot might be excluded from processing. By *nearest neighbor* (NN) filters,[1] only the measurement having the smallest innovation is processed via KF if competing returns exist in the gates. This approach fails, however, if one of the interpretation hypotheses indicated in Figure 8.2 is true. *(Joint) probabilistic data association* (PDA, JPDA) filters[2] are adaptive mono-hypothesis trackers that show data-driven adaptivity in case of association conflicts.

A more rigorous Bayesian approach capable of handling challenging conditions as sketched in Section 8.1.2 leads to *multiple-hypothesis/multiple-model* filtering discussed later. The ad hoc methods mentioned (KF, NN, PDAF, JPDAF) prove to be limiting cases of this more general approach.

8.3 Statistical Models

A statistical description of what kind of information is provided by the sensor systems is a prerequisite to processing the n_k sensor output data $Z_k = \{z_k^j\}_{j=1}^{n_k}$ (scans, frames) consecutively received at discrete instants of time t_k. Therefore, appropriate statistical models of the sensor performance, the target characteristics (including their expected dynamic behavior), and the underlying operational conditions are required. For the sake of simplicity, our discussion and terminology are confined to point-source objects, small extended objects, possibly unresolved closely spaced objects, or small clusters of such objects; i.e., we consider "small targets" following Oliver Drummond's definition.[28] The underlying statistical models essentially determine the feasible interpretations of the received sensor reports.

8.3.1 Object Dynamics

Although there may exist applications in which adequate dynamics models are difficult to obtain, in many practical cases the dynamic behavior of the objects of interest directly results from simple physical laws. In the subsequent discussion, we consider randomly moving objects in Cartesian coordinates. Their kinematical state x_k at a certain instant of time t_k is defined by the current position of the objects and the related derivatives, such as velocity and acceleration. The order to which temporal derivatives of the target position are taken into account depends on the particular tracking application considered.

In the sequel we refer to linear-Gaussian dynamics models in which the temporal evolution of the state vectors x_k is defined by a discrete time Markov process with transition densities

$$p(x_k|x_{k-1}) = \mathcal{N}(x_k;\, F_k x_{k-1},\, Q_k)^*. \tag{8.1}$$

F_k denotes the system matrix and Q_k denotes the related process or plant noise covariance matrix. Evidently, this model is equivalent to a linear difference equation with additive white Gaussian noise:

$$x_k = F_k x_{k-1} + v_k, \qquad v_k \sim N(0, Q_k). \tag{8.2}$$

In a practical tracking application, however, it might be uncertain which dynamics model out of a set of possible alternatives is currently in effect. Systems with Markovian switching coefficients provide a well-established statistical framework for handling those cases, such as objects characterized by different modes of dynamical behavior. This approach is defined by multiple dynamics models with a given probability of switching between them (IMM, interacting multiple models; see References 4 to 6 and the literature cited therein). Thus, the transition probabilities are part of the modeling assumptions. More strictly speaking, suppose r models given and let m_k be denoting the dynamics model assumed to be in effect at time t_k, $m_k \in \{m_k^i\}_{i=1}^r$, the statistical properties of systems with Markovian switching coefficients are summarized by the following equation:

$$p(x_k, m_k|x_{k-1}, m_{k-1}) = p(x_k|x_{k-1}, m_k)P(m_k|m_{k-1}) \tag{8.3}$$

$$= p_{m_k m_{k-1}}\, \mathcal{N}(x_k;\, F_{m_k} x_{k-1},\, Q_{m_k}). \tag{8.4}$$

Hence, the model switching process is a Markov chain with model transition probabilities $p_{m_k m_{k-1}} = P(m_k|m_{k-1})$, with the individual models being linear Gaussian. For $r = 1$, the linear model (Equation 8.2) results as a limiting case.

If there exists a non-linear relationship between the states at two consecutive scans, $x_k = f_{m_k}(x_{k-1})$ (no plant noise), e.g., in case of a non-Cartesian coordinate system, the function $f_{m_k}(x)$ might be linearized by a first-order Taylor expansion around the estimate $\hat{x}_{k-1}$ that results from sensor data processing up to time t_{k-1} and is provided by the tracking system.[5] Thus, in general, we have to deal with data-dependent system matrices $F_{m_k} \approx \nabla_x f_{m_k}(x)|_{x=\hat{x}_{k-1}}$ which must be approximated at each step of the tracking process.

In many practical applications, the number of individual models possibly being in effect is not r^N (for N objects to be tracked) as might be expected from the previous discussion. For well-separated objects, we essentially have to handle single-target problems. On the other hand, in situations with small clusters of objects moving closely spaced for a while, such as formations, we can assume that all targets obey the same dynamics model at a given time. Otherwise, they would quickly become well separated. A long-living cluster of many closely spaced objects, each in a different maneuvering state, is not realistic.

8.3.1.1 Example: A Simplified Model

Let us consider r simple dynamics models of maneuvering air targets in two spatial dimensions.[1,19] For a given model m_k^i assumed to be in effect at time t_k, let F_k^i, Q_k^i denote the corresponding state transition and plant noise covariance matrices, respectively ($i = 1, \ldots, r$). In the case of an uncorrelated dynamics in both dimensions, F_k^i, Q_k^i are given by block-diagonal matrices

$$F_k^i = \begin{pmatrix} f_k^i & 0 \\ 0 & f_k^i \end{pmatrix},\; Q_k^i = \begin{pmatrix} q_k^i & 0 \\ 0 & q_k^i \end{pmatrix}. \tag{8.5}$$

* $\mathcal{N}(z; x, V) = \det(2\pi V)^{-1/2} \exp(-1/2(z-x)^T V^{-1}(z-x))$, multivariate Gaussian density.

The block matrices for each Cartesian component are characterized by two dynamical parameters and the data innovation interval T_k:

$$f_k^i = \begin{pmatrix} 1 & T_k & \frac{1}{2}T_k^2 \\ 0 & 1 & T_k \\ 0 & 0 & e^{-T_k/\theta_i} \end{pmatrix}, q_k^i = \Sigma_i^2(1 - e^{-2T_k/\theta_i})\begin{pmatrix} 0 & 0 & 0 \\ 0 & 0 & 1 \\ 0 & 0 & 1 \end{pmatrix}. \tag{8.6}$$

In other words, the acceleration process in each Cartesian component is modeled by a stationary Markov process which is defined by $q_k = e^{-T_k/\theta_i} q_{k-1} + \Sigma_i \sqrt{} = (1 - e^{-2T_k/\theta_i})^{1/2} u_k$, $u_k \sim \mathcal{N}(0, 1)$. Due to ergodicity, we obtain for the related expectation and autocorrelation function: $\mathbb{E}[q_k] = 0$, $\mathbb{E}[q_k, q_l] = \Sigma_i^2 e^{-(k-l)T_k/\theta_i}$, $l \le k$. By this the model, parameters Σ_i (*acceleration magnitude*) and θ_i (*maneuver correlation time*) find an intuitively clear interpretation.[19] In an air surveillance application, a worst/best-case modeling may be used with $r = 2$ and $\theta_i = 60$, 30 s, $\Sigma_i = 3$, 30 m/s^2, $i = 1, 2$. For the transition probabilities, $p_{11} = .9$, $p_{22} = .8$, $p_{12} = 1 - p_{11}$, $p_{21} = 1 - p_{22}$ are reasonable choices, i.e., with a probability of 90% the target remains in a slightly maneuvering phase, while it returns to this phase with a probability of 20% when it formerly was strongly maneuvering.

8.3.2 Detection

For resolved objects and a given association hypothesis, each sensor return can be associated to exactly one individual object, whereas for unresolved closely spaced objects a given report may correspond to several objects. Thus, it is reasonable to introduce different detection probabilities for resolved and unresolved objects: P_D, P_D^u. Moreover, the detection process and the production of measurements (fine localization by monopulse processing, for instance [see Reference 29, p. 119 ff.]) are assumed to be statistically independent.

Let G_k denote a region large enough to contain all relevant sensor reports at scan k. False detections or detections produced by unwanted objects are assumed to be equally distributed in G_k and independent from scan to scan. Moreover, let the number of false returns in G_k be Poisson-distributed according to the density

$$p_F(n_k) = \frac{1}{n_k!}(|G_k|\rho_F)^{n_k} e^{-|G_k|\rho_F}, \tag{8.7}$$

where $|G_k|$ denotes the volume of the region G_k and ρ_F denotes the spatial false return density.

8.3.2.1 Example: Swerling-I Targets

In a typical radar application, a simple quadrature detector decides on target detection if the received signal strength exceeds a certain threshold: $a_k^2 > \lambda_D$. For a given fluctuation model of the radar cross section of the targets, the detection probability depends on the mean signal-to-noise ratio (snr) of the sensor and the detector threshold λ_D, $P_D(\text{snr}, \lambda_D) = \int_{\lambda_D}^{\infty} da_k^2 p(a_k^2|\text{snr})$, while the false alarm probability is a function of λ_D alone, $P_{FA} = P_D(1, \lambda_D)$. As for Swerling-I targets a_k^2 is exponentially distributed (see Reference 29, p. 48 ff.) with mean 1 + snr, we directly obtain the well-known relationship: $P_D = P_{FA}^{1/(1+\text{snr})}$ with $P_{FA} = -2\log \lambda_D$. According to the radar range equation (see Reference 29, p. 60), snr depends on the range r_k and the mean radar cross section $\bar{\sigma}$ of the target: $\text{snr} = \text{snr}_0(\sigma/\sigma_0)(r_k/r_0)^{-4}$, with snr_0, $\bar{\sigma}_0$, and r_0 denoting radar parameters. Provided the received signal strength is accessible for the tracking system, it can be used as input information for adaptive threshold control,[9,30] for discriminating of false returns,[23,31] or for phased-array energy management.[22]

8.3.3 Measurements

For a resolved object, let z_k be a bias-free measurement of its kinematical state x_k at time t_k with an additive, normally distributed measurement error:

$$z_k = H_k x_k + w_k,\ w_k \sim N(0, \mathrm{R}_k) \leftrightarrow p(z_k|x_k) = \mathcal{N}(z_k; \mathrm{H}_k x_k, R_k). \tag{8.8}$$

While the *measurement noise* covariance R_k describes the quality of the received sensor measurements, the *measurement matrix* H_k indicates what the measurements can, in principle, say about the state vector x_k. In the case of a non-linear relationship between the target state and the measurement, $z_k = h_k(x_k)$ (no measurement noise), the function $h_k(x)$ may be linearized around the *predicted* target state $\hat{x}_{k|k-1}$: $\mathrm{H}_k \approx \nabla_x h_k(x)|_{x = \hat{x}_{k|k-1}}$.[5] Thus, in general we have to deal with data-dependent measurement matrices. For two *unresolved* objects, let z_k be a bias-free measurement corresponding to the joint state $x_k = (x_k^1, x_k^2)^{\mathrm{T}}$ of both objects, i.e., a measurement of the mean target position (group measurement), which is approximately described by

$$p(z_k|x_k) = \mathcal{N}(z_k; \mathrm{H}_k^g x_k, \mathrm{R}_k^g), \quad \text{with: } \mathrm{H}_k^g x_k = \frac{1}{2}\mathrm{H}_k(x_k^1 + x_k^2). \tag{8.9}$$

8.3.3.1 Example: 2D Radar

Monopulse angle estimation techniques provide approximately bias-free, normally distributed angular measurements. In many practical cases, it is reasonable to assume that the standard deviation of the related measurement error is proportional to the radar beam-width B and inversely proportional to the square root of the *instantaneous* signal-to-noise ratio (SNR) of the target: $\sigma_\phi \propto B/\sqrt{\mathrm{SNR}}$. As SNR is usually unknown, the current signal strength a_k^2 may be treated as an estimate of SNR + 1 (see previous example). Thus, we obtain bias-free estimates of the mean angular error at each scan k: $\hat{\sigma}_\phi^k \propto B/\sqrt{a_k^2 - 1}$. While in principle high precision measurements are available also in range, in many radar systems the range measurement errors are a superposition of errors uniformly distributed in the related range cells. For convenience and without significant degradation of the tracking process, however, in many cases range measurement errors can be assumed to be normally distributed with a standard deviation σ_r which must not be chosen too optimistically. For a more detailed discussion, see Reference 9, Chapter 2 and the literature cited therein. By linearization around the predicted target position $r_{k|k-1}$, $\phi_{k|k-1}$, the time dependent measurement error covariance R_k of the transformed measurements $z_k = r_k(\cos\phi_k, \sin\phi_k)^{\mathrm{T}}$ is thus given by

$$\mathrm{R}_k \approx \begin{pmatrix} \sigma_r^2\cos^2\phi_{k|k-1} + (r_{k|k-1}\sigma_\phi)^2\sin^2\phi_{k|k-1} & (\sigma_r^2 + (r_{k|k-1}\sigma_\phi)^2)\cos\phi_{k|k-1}\sin\phi_{k|k-1} \\ (\sigma_r^2 + (r_{k|k-1}\sigma_\phi)^2)\cos\phi_{k|k-1}\sin\phi_{k|k-1} & \sigma_r^2\sin^2\phi_{k|k-1} + (r_{k|k-1}\sigma_\phi)^2\cos^2\phi_{k|k-1} \end{pmatrix}. \tag{8.10}$$

As a direct consequence, the Cartesian measurement error ellipses typically increase with increasing range. In certain applications, it may be useful to deal with different measurement accuracies, depending on the tracking task under consideration, such as search, acquisition, or high-precision tracking modes in phased-array tracking (see Reference 9).

8.3.4 Resolution

The sensor resolution does not only depend on hardware characteristics, such as the beam- and bandwidth of a radar, but also on the random target fluctuations and the signal processing technique applied (super-resolution techniques,[32] for instance). The probability P_u of two objects being unresolved certainly depends on their relative distance d: $P_u = P_u(d)$. An exact analytical description, however, is not easily obtained. Qualitatively, we expect $P_u = 1$ for $d = 0$, and it will remain large for small values of d. On the other hand, $P_u = 0$ for distances significantly larger than the beam-width. We expect a narrow transient region. In a generic model of the sensor resolution, we may thus describe P_u by a Gaussian-shaped function of the relative distance between the objects, with a "covariance" R_k^u being a quantitative measure of the resolution capability. More strictly speaking,[33,34] let us consider the expression

$$P_u(x_k) = \exp\left\{-\frac{1}{2}(H_k(x_k^1 - x_k^2))^T R_k^{u^{-1}} H_k(x_k^1 - x_k^2)\right\} \tag{8.11}$$

$$= \det(2\pi R_k^u)^{\frac{1}{2}} \mathcal{N}(H_k^u x_k; 0, R_k^u) \text{ with } H_k^u x_k = H_k(x_k^1 - x_k^2). \tag{8.12}$$

$x_k = (x_k^1, x_k^2)^T$ denotes the joint state of the objects. The symmetric, positive definite matrix R_k^u reflects the extension and spatial orientation of ellipsoidal resolution cells and, in general, depends on the object-sensor geometry considered.[34] The Gaussian structure of Equation 8.12 not only significantly simplifies the mathematics involved, but also provides an intuitive interpretation of the resulting processing algorithm: besides the received group measurement, an assumed resolution conflict essentially results in a fictitious measurement of the relative distance between the objects, $H_k(x_k^1 - x_k^2)$, with zero value and a measurement error covariance given by R_k^u.

8.3.4.1 Example: 2D Radar

The range resolution α_r of a radar is essentially determined by the emitted pulse length, while the angular resolution of α_ϕ is limited by the beam-width. With Δr_k and $\Delta\phi_k$ denoting the range and azimuth distance of the targets, we assume $P_u(\Delta r_k, \Delta\phi_k) = e^{-1/2(\Delta r_k/\alpha_r)^2} e^{-1/2(\Delta\phi_k/\alpha_\phi)^2}$. Analogously to the previous example in Cartesian coordinates the resolution matrix R_k^u is approximately

$$R_k^u(r_{k|k-1}^g, \phi_{k|k-1}^g) \approx \begin{pmatrix} \alpha_r^2\cos^2\phi_{k|k-1}^g + (r_{k|k-1}^g\alpha_\phi)^2\sin^2\phi_{k|k-1}^g & (\alpha_r^2 + (r_{k|k-1}^g\alpha_\phi)^2)\cos\phi_{k|k-1}^g\sin\phi_{k|k-1}^g \\ (\alpha_r^2 + (r_{k|k-1}^g\alpha_\phi)^2)\cos\phi_{k|k-1}^g\sin\phi_{k|k-1}^g & \alpha_r^2\sin^2\phi_{k|k-1}^g + (r_{k|k-1}^g\alpha_\phi)^2\cos^2\phi_{k|k-1}^g \end{pmatrix} \tag{8.13}$$

with $r_{k|k-1}^g$ and $\phi_{k|k-1}^g$ denoting the predicted position of the group. For a related approach see Reference 35. As in the previous example, the Cartesian "resolution ellipses" depend on the target range. Suppose we have $\alpha_r = 100$ m and $\alpha_\phi = 1°$, then we expect the resolution in a distance of 50 km to be about 100 m (range) and 900 m (cross range). For military targets in a formation, their mutual distance may well be 200 to 500 m or even less; resolution is thus a real problem in target tracking.[36]

8.3.5 Data Association

The conditional probability density $p(Z_k, n_k|x_k)$, often referred to as likelihood function (see Reference 8, for instance), statistically describes what a single frame of n_k observations $Z_k = \{z_k^j\}_{j=1}^{n_k}$ can say about the joint state x_k of the objects to be tracked. Due to the Total Probability theorem, $p(Z_k, n_k|x_k)$ can be written as a sum over all possible data interpretations h_k, i.e., over all hypotheses regarding the origin of the data set Z_k:

$$p(Z_k, n_k|x_k) = \sum_{h_k} p(Z_k, n_k, h_k|x_k) \tag{8.14}$$

$$= \sum_{h_k} p(Z_k, n_k|h_k, x_k)P(h_k|x_k). \tag{8.15}$$

As shown in the following examples the probability $P(h_k|x_k)$ of h_k being correct as well as the individual likelihood functions $p(Z_k, n_k|h_k, x_k) = p(Z_k|h_k, n_k, x_k)\, P(n_k|h_k)$ directly result from the statistical sensor model previously discussed (Equations 8.7 to 8.9 and 8.12). These considerations make evident that the determination of mutually exclusive and exhaustive data interpretations is a prerequisite to sensor data processing. Though this is, in general, by no means a trivial task, in many practical cases a given multiple-object tracking problem can be decomposed into independent sub-problems of reduced complexity. We thus consider two examples that are practically important but can still be handled more or less rigorously.

8.3.5.1 Example: Well-Separated Objects

For well-separated objects in a cluttered environment, essentially two classes of data interpretations can be identified:[2]

1. h_k^0. The object considered was not detected, all n_k sensor returns in Z_k are false, i.e., assumed to be equally distributed in G_k (one interpretation):

$$p(Z_k, n_k | h_k^0, x_k) = |G_k|^{-n_k} p_F(n_k) \tag{8.16}$$

$$P(h_k^0 | x_k) = (1 - P_D)\,. \tag{8.17}$$

2. h_k^i, $i = 1, \ldots, n_k$. The object was detected, $z_k^i \in Z_k$ is the corresponding measurement, all other sensor returns are false (n_k interpretation hypotheses):

$$p(Z_k, n_k | h_k^i, x_k) = |G_k|^{1-n_k} \mathcal{N}(z_k^i; \mathrm{H}_k x_k, \mathrm{R}_k) p_F(n-1) \tag{8.18}$$

$$P(h_k^i | x_k) = \frac{1}{n_k} P_D\,. \tag{8.19}$$

According to Equation 8.7, the likelihood function $p(Z_k, n_k|x_k)$ is proportional to the sum

$$p(Z_k, n_k | x_k) \propto (1 - P_D)\rho_F + P_D \sum_{i=1}^{n_k} \mathcal{N}(z_k^i; \mathrm{H}_k x_k, \mathrm{R}_k) \tag{8.20}$$

up to a factor $1/n_k \rho_F^{n_k - 1} e^{-|G_k|\rho_F}$ being independent of x_k.

8.3.5.2 Example: Small Object Clusters

For a cluster of two closely spaced objects moving in a cluttered environment, five different classes of data interpretations exist ($x_k = (x_k^1, x_k^2)^\mathsf{T}$):[33]

1. h_k^{ii}, $i = 1, \ldots, n_k$: Both objects were not resolved, but detected as a group; $z_k^i \in Z_k$ represents the group measurement; all remaining returns are false (n_k data interpretations):

$$p(Z_k, n_k | h_k^{ii}, x_k) = |G_k|^{1-n_k} \mathcal{N}(z_k^i; \mathrm{H}_k^g x_k, \mathrm{R}_k^g) p_F(n_k - 1) \tag{8.21}$$

$$P(h_k^{ii} | x_k) = \frac{1}{n_k} P_u(x_k) P_D^u\,. \tag{8.22}$$

2. h_k^{00}: Both objects were neither resolved nor detected as a group; all returns in Z_k are thus assumed to be false (one interpretation hypothesis):

$$p(Z_k, n_k | h_k^{00}, x_k) = |G_k|^{-n_k} p_F(n_k) \tag{8.23}$$

$$P(h_k^{00}|x_k) = P_u(x_k)(1 - P_D^u). \tag{8.24}$$

3. h_k^{ij}, $i, j = 1,\ldots, n_k$, $i \neq j$: Both objects were resolved and detected; z_k^i, $z_k^j \in Z_k$ are the measurements; $n_k - 2$ returns are false ($n_k(n_k - 1)$ interpretations):

$$p(Z_k, n_k|h_k^{ij}, x_k) = |G_k|^{2-n_k}\mathcal{N}(z_k^i;\mathrm{H}_k x_k^1, \mathrm{R}_k)\mathcal{N}(z_k^j;\mathrm{H}_k x_k^2, \mathrm{R}_k)p_F(n_k - 2) \tag{8.25}$$

$$P(h_k^{ij}|x_k) = \frac{1}{n_k(n_k - 1)}(1 - P_u(x_k))P_D^2. \tag{8.26}$$

4. h_k^{i0}, h_k^{0i}, $i = 1, \ldots, n_k$: Both objects were resolved, but only one object was detected; $z_k^i \in Z_k$ is the measurement; $n_k - 1$ returns in Z_k are false ($2n_k$ interpretations):

$$p(Z_k, n_k|h_k^{i0}, x_k) = |G_k|^{1-n_k}\mathcal{N}(z_k^i;\mathrm{H}_k x_k^1, \mathrm{R}_k)p_F(n - 1) \tag{8.27}$$

$$P(h_k^{i0}|x_k) = \frac{1}{n_k}(1 - P_u(x_k))P_D(1 - P_D). \tag{8.28}$$

5. h_k^0: The objects were resolved, but not detected; all n_k plots in Z_k are false (one interpretation):

$$p(Z_k, n_k|h_k^0, x_k) = |G_k|^{-n_k}p_F(n_k) \tag{8.29}$$

$$P(h_k^0|x_k) = (1 - P_u(x_k))(1 - P_D)^2. \tag{8.30}$$

As there exist $(n_k + 1)^2 + 1$ interpretation hypotheses, the ambiguity for even small clusters of closely spaced objects is much higher than in the case of well-separated objects ($n_k + 1$ each). Thus, we expect that only small groups can be handled more or less rigorously. For larger clusters (raids of military aircraft, for instance), a collective treatment[1] seems to be reasonable until the group splits off into smaller sub-clusters or individual objects.

Up to a factor $1/n!\rho_F^{n_k-2}e^{-|G_k|\rho_F}$ independent of x_k (Equation 8.7), the likelihood function of the data,

$$p(Z_k, n_k|x_k) = p(Z_k, n_k, h_k^0|x_k) + \sum_{i,j=0}^{n_k} p(Z_k, n_k, h_k^{ij}|x_k), \tag{8.31}$$

is proportional to the sum

$$\begin{aligned} p(Z_k, n_k|x_k) \propto\ & \rho_F^2(1 - P_D)^2(1 - P_u(x_k)) + \rho_F^2(1 - P_D^u)P_u(x_k) \\ & + P_D^u\rho_F P_u(x_k)\sum_{i=1}^{n_k}\mathcal{N}(z_k^i;\mathrm{H}_k^g x_k, \mathrm{R}_k^g) + \rho_F P_D(1 - P_D)(1 - P_u(x_k))\sum_{i=1}^{n_k}\{\mathcal{N}(z_k^i;\mathrm{H}_k x_k^1, \mathrm{R}_k) \\ & + \mathcal{N}(z_k^i;\mathrm{H}_k x_k^2, \mathrm{R}_k)\} + P_D^2(1 - P_u(x_k))\sum_{\substack{i,j=1\\ i\neq j}}^{n_k}\mathcal{N}(z_k^i;\mathrm{H}_k x_k^1, \mathrm{R}_k)\mathcal{N}(z_k^j;\mathrm{H}_k x_k^2, \mathrm{R}_k). \end{aligned} \tag{8.32}$$

8.4 Bayesian Track Maintenance

In a Bayesian view, tracking algorithms are iterative updating schemes for conditional probability densities $p(x_l|Z^k)$ that represent all available information of the state x_l of an underlying dynamical system at discrete instants of time t_l given the sensor data $Z^k = \{Z_k, n_k, Z_{k-1}, n_{k-1}, \ldots, Z_1, n_1\}$ accumulated up to some time t_k, typically the current scan time. *A priori* information on the system dynamics, the sensor performance, and the environment enters in terms of the probability densities $p(x_k, m_k|x_{k-1}, m_{k-1})$ (Equation 8.4) and $p(Z_k, n_k|x_k)$ (Equations 8.20 and 8.32).

Depending on the time t_l at which an estimate for the state vector x_l is required, the related estimation process is referred to as *prediction* ($t_l > t_k$), *filtering* ($t_l = t_k$), and *retrodiction* ($t_l < t_k$), respectively.[16,137] Being the natural antonym of prediction, retrodiction as a technical term was introduced by O. Drummond: "Retrodiction: The process of computing estimates of states, probability densities, or discrete probabilities for a prior time (or over a period of time) based on data up to and including some subsequent time, typically, the current time" (see Reference 37, p. 255).

Provided the densities $p(x_l|Z^k)$ are calculated correctly, optimal estimators may be derived related to various risk functions, such as minimum mean square estimators, for instance (MMSE, see Reference 5, p. 98).

Figure 8.3 provides a schematic illustration of Bayesian density iteration. The probability densities $p(x_{k-1}|Z^{k-1})$, $p(x_k|Z^k)$, and $p(x_{k+1}|Z^{k+1})$ resulting from filtering at the scan times t_{k-1}, t_k, and t_{k+1}, respectively, are displayed along with the predicted density $p(x_{k+2}|Z^{k+1})$ (Figure 8.3a, forward iteration). At time t_{k-1}, one sensor report has been processed, but no report could be associated to the track at time t_k. Hence, a missing detection according to $P_D < 1$ is assumed. As a consequence of this lack of sensor information, the density $p(x_k|Z^k)$ is broadened, because target maneuvers may have occurred. This, in particular, implies an increased correlation gate for the subsequent scan time t_{k+1}. According to this effect, at time t_{k+1} three correlating sensor reports are to be processed leading to a multi-modal probability density. The multiple modes reflect the ambiguity regarding the origin of the sensor data and also characterize the predicted density $p(x_{k+2}|Z^{k+1})$. By this, the data-driven adaptivity of the Bayesian updating scheme is clearly indicated. In Figure 8.3b the density $p(x_{k+2}|Z^{k+2})$ resulting from processing a single correlating report at t_{t+2} along with the retrodicted densities $p(x_{k+1}|Z^{k+2})$, $p(x_k|Z^{k+2})$, and $p(x_{k-1}|Z^{k+2})$ are shown. Evidently, newly available sensor data significantly improve the estimates of the past states.

8.4.1 Finite Mixture Densities

The tracking problems considered here are inherently ambiguous due to both sensor data of uncertain origin and multiple dynamics models. As in the examples previously discussed, let h_l denote a specific

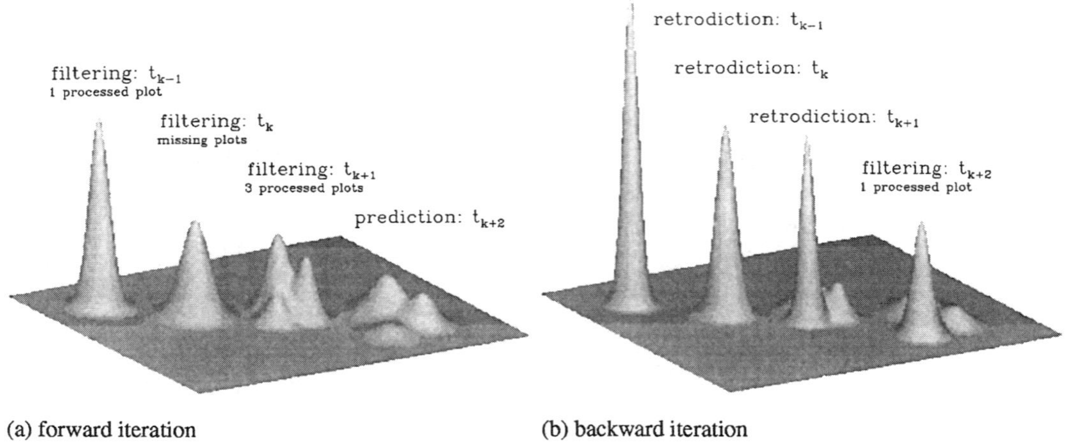

FIGURE 8.3 Scheme of Bayesian density iteration.

interpretation of the sensor data Z_l at scan time t_l taken out of a set of mutually exclusive and exhaustive interpretation hypotheses. Accordingly, the k-tuple $H^k = (h_k, \ldots, h_1)$, consisting of consecutive data interpretations h_l, $1 \le l \le k$, up to the time t_k, is a particular interpretation hypothesis regarding the origin of the accumulated sensor data $Z^k = \{Z_k, n_k, Z_{k-1}, n_{k-1}, \ldots, Z_1, n_1\}$. H^k is thus called an *interpretation history*. For each H^k the related *prehistories* $H^{k-n} = (h_{k-n}, \ldots, h_1)$ provide possible interpretations of sensor data Z^{k-n} accumulated up to scan *k–n*. With $H_n^k = (h_k, \ldots, h_{k-n+1})$, the *recent history*, any H^k can be decomposed in $H^k = (H_n^k, H^{k-n})$.

In close analogy to data interpretation histories, we consider *model histories* $M^k = (m_k, \ldots, m_1)$ along with the related quantities M_n^k and M^{n-k}. As before, m_l, $1 \le l \le k$, denotes the dynamics model assumed to be in effect at a particular time t_l. A model history is thus a hypothesis regarding which dynamics models have been assumed in the past to explain the target behavior.

Due to the Total Probability theorem, the density $p(x_k|Z^k)$ can be written as a sum over all possible interpretation and model histories:

$$p(x_k|Z^k) = \sum_{H^k, M^k} p(x_k, M^k, H^k|Z^k) \tag{8.33}$$

$$= \sum_{H^k, M^k} p(x_k|M^k, H^k, Z^k)P(M^k, H^k|Z^k). \tag{8.34}$$

$p(x_k|Z^k)$ is thus a *finite mixture density*,[15] i.e., a weighted sum of *component densities* $p(x_k|M^k, H^k, Z^k)$ that assume a particular interpretation history H^k and a model history M^k to be true (given the data Z^k). The corresponding *mixing weights* $P(M^k, H^k|Z^k)$ sum up to one. As a consequence of the modeling assumptions previously introduced, the densities $p(x_k|Z^k)$ have a particularly simple structure, as will be shown in the sequel. They are *normal mixtures*, i.e., weighted sums of Gaussian densities:

$$p(x_k|Z^k) = \sum_{H^k, M^k} \mu_{H^k}^{M^k} \mathcal{N}\left(x_k; \hat{\mathbf{x}}_{H^k}^{M^k}, \mathbf{P}_{H^k}^{M^k}\right) \tag{8.35}$$

with $\mu_{H^k}^{M^k} = P(M^k, H^k|Z^k)$ and $\mathcal{N}(x_k; \hat{x}_{H^k}^{M^k}, \mathbf{P}_{H^k}^{M^k}) = p(x_k|M^k, H^k, Z^k)$. The quantities $\hat{x}_{H^k}^{M^k} = \mathbb{E}_{H^k, M^k}[x_k]$ and $\mathbf{P}_{H^k}^{M^k} = \mathbb{E}_{H^k, M^k}[(x_k - \hat{x}_{H^k}^{M^k})(x_k - \hat{x}_{H^k}^{M^k})^{\mathsf{T}}]$ denote the expectation and the related covariance matrix of x_k with respect to the conditional density $p(x_k|M^k, H^k, Z^k)$.

The ambiguity due to the uncertain origin of the data may be treated separately from the ambiguity caused by the underlying IMM modeling:

$$p(x_k|Z^k) = \sum_{H^k} p(x_k|H^k, Z^k)P(H^k|Z^k) \tag{8.36}$$

with

$$p(x_k|H^k, Z^k) = \sum_{M^k} p(x_k|M^k, H^k, Z^k)P(M^k|H^k, Z^k) \tag{8.37}$$

$$P(H^k|Z^k) = \sum_{M^k} P(M^k, H^k|Z^k). \tag{8.38}$$

$p(x_k|Z^k)$ may thus be represented by a mixture with respect to the interpretation hypotheses H^k consisting of component densities $p(x_k|H^k, Z^k)$ that are mixtures themselves. Let $\hat{\mathbf{x}}_{H^k} = \mathbb{E}_{H^k}(x_k)$ denote the expectation of x_k with respect to $p(x_k|H^k, Z^k)$, and $\mathbf{P}_{H^k} = \mathbb{E}_{H^k}[(x_k - \hat{\mathbf{x}}_{H^k})(x_k - \hat{\mathbf{x}}_{H^k})^{\mathsf{T}}]$ denote the related covariance matrix. It is intuitive to call $\{\hat{\mathbf{x}}_{H^l}, \mathbf{P}_{H^l}\}_{l=1}^{k}$ a hypothetical or local track with

$\mu_{M^k} = P(H^k|Z^k)$ denoting its statistical weight. Analogously, $\{\hat{\mathbf{x}}_{H^l}, \mathbf{P}_l\}_{l=1}^{k}$ is a global track where $\hat{x}_l$, $\mathbf{P}_l$ are defined with respect to $p(x_l|Z^l)$.[13]

8.4.2 Prediction

Let us assume $p(x_k|Z^k) = \Sigma_{H^k, M^k} p(x_k, H^k, M^k|Z^k)$ is known at time t_k. Due to *a priori* information on the system dynamics given by $p(x_{k+1}, m_{k+1}|x_k, m_k) = p_{m_{k+1}m_k} \mathcal{N}(x_{k+1}; \mathrm{F}_{m_{k+1}} x_k, \mathrm{Q}_{k+1})$ (Equation 8.4), the future state x_{k+1} at time t_{k+1} can be predicted:

$$p(x_k|Z^k) \xrightarrow[p(x_{k+1},\, m_{k+1}|x_k,\, m_k)]{\text{dynamics model}} p(x_{k+1}|Z^k). \tag{8.39}$$

The predicted density $p(x_{k+1}|Z^k)$ is important to efficiently handling the data association problem (e.g., by individual gating, see later) or decisions on future sensor allocations depending on the current lack of information.[19,22] It is also a prerequisite to the processing of the newly received frame of observations Z_{k+1} at t_{k+1}. Due to the Total Probability theorem and the Markov property of the system dynamics (Equation 8.4), we obtain

$$p(x_{k+1}|Z^k) = \sum_{M^{k+1}} \int dx_k\, p(x_{k+1}, x_k, M^{k+1}|Z^k) \tag{8.40}$$

$$= \sum_{m_{k+1}} \sum_{m_k} \sum_{M^{k-1}} \int dx_k\, p(x_{k+1}, m_{k+1}|x_k, m_k) p(x_k, m_k, M^{k-1}|Z^k). \tag{8.41}$$

As $p(x_k, m_k, M^{k-1}|Z^k) = p(x_k, M^k|Z^k) = \Sigma_{H^k} p(x_k, H^k, M^k|Z^k)$ is assumed to be known, $p(x_{k+1}|Z^k)$ is thus represented by a finite mixture over all possible model histories up to time t_{k+1}.

8.4.2.1 Example: Single Model Prediction

For a Gaussian density $p(x_k|Z^k) = \mathcal{N}(x_k; \hat{x}_k, P_k)$ and a single dynamics model (i.e., $r = 1$), the predicted density $p(x_{k+1}|Z^k)$ is also a Gaussian:

$$\mathcal{N}(x_k; \hat{\mathbf{x}}_k, \mathbf{P}_k) \xrightarrow[\mathrm{F}_{k+1},\, \mathrm{Q}_{k+1}]{} \mathcal{N}(x_{k+1}; \hat{\mathbf{x}}_{k+1|k}, \mathbf{P}_{k+1|k}) \tag{8.42}$$

$$\begin{aligned} \hat{\mathbf{x}}_{k+1|k} &= \mathrm{F}_{k+1}\hat{\mathbf{x}}_k \\ \mathbf{P}_{k+1|k} &= \mathrm{F}_{k+1}\mathbf{P}_k\mathrm{F}_{k+1}^{\mathrm{T}} + \mathrm{Q}_{k+1}. \end{aligned} \tag{8.43}$$

The evaluation of the integral $\int dx_k\, \mathcal{N}(x_{k+1}; \mathrm{F}_{k+1}x_k, \mathrm{Q}_{k+1})\, \mathcal{N}(x_k; \hat{\mathbf{x}}_k, \mathbf{P}_k)$ according to Equation 8.41 results from the following identity that is frequently used and can be proven by a completion of the squares and the matrix inversion lemma (see Reference 5, p. 13 or Reference 8, p. 291; for instance):

$$\mathcal{N}(z; \mathrm{F}x, \mathrm{Q})\, \mathcal{N}(x; \hat{\mathbf{x}}, \mathbf{P}) = \mathcal{N}(z; \mathrm{F}\hat{\mathbf{x}}, S)\, \mathcal{N}(x; \hat{\mathbf{x}} + W(z - \mathrm{F}\hat{\mathbf{x}}), \mathrm{P} - WSW^{\mathrm{T}}) \tag{8.44}$$

with $S = \mathrm{F}\mathbf{P}\mathrm{F}^{\mathrm{T}} + \mathrm{Q}$, $W = \mathbf{P}\mathrm{F}^{\mathrm{T}}S^{-1}$.

8.4.2.2 Example: Innovation Statistics

As a byproduct of Equations 8.43 and 8.44 and under the assumptions of the previous example, the statistics of the expected target measurement z_{k+1} is described by

$$p(z_{k+1}|Z^k) = \int dx_{k+1}\, \mathfrak{N}(z_{k+1}; \mathrm{H}_{k+1}x_k, \mathrm{R}_{k+1})\, \mathcal{N}(x_{k+1}; \hat{\mathbf{x}}_{k+1|k}, \mathbf{P}_{k+1|k}) \tag{8.45}$$

$$= \mathcal{N}(z_{k+1}; \mathrm{H}_{k+1}\hat{\mathbf{x}}_{k+1|k}, S_{k+1}), \text{ with: } S_{k+1} = \mathrm{H}_{k+1}\mathbf{P}_{k+1|k}\mathrm{H}_{k+1}^{\mathrm{T}} + \mathrm{R}_{k+1}. \tag{8.46}$$

In particular, the innovation vector $\nu_{k+1} = z_{k+1} - \mathrm{H}_{k+1}\hat{\mathbf{x}}_{k+1|k}$ is normally distributed with zero mean and the covariance matrix S_{k+1}, which is thus called the *innovation covariance matrix* (see the introductory discussion in Section 8.2.1). According to Equation 8.43, the innovation covariance is determined by both the assumed maneuvering capability of the target and the measurement error statistics of the sensor.

8.4.3 Filtering

The processing of the sensor data $Z_{k+1} = \{z_{k+1}^{i}\}_{j=1}^{n_{k+1}}$ received at time t_{k+1} is basically determined by the probability densities $p(Z_{k+1}, n_{k+1}|x_{k+1})$ (Equations 8.20 and 8.32) and $p(x_{k+1}|Z^k)$ (Equation 8.39). Thus, it depends on the quality of the previous prediction, *a priori* information in terms of the sensor models, and the particular tracking problem considered:

$$p(x_{k+1}|Z^k) \xrightarrow[p(Z_{k+1},\, n_{k+1}|x_{k+1})]{\text{senor model/data}} p(x_{k+1}|Z^{k+1}). \tag{8.47}$$

By this, one step of the tracking loop is completed; $p(x_{k+1}|Z^{k+1})$ is used for the subsequent prediction. The update equations directly result from Bayes' Rule:

$$p(x_{k+1}|Z^{k+1}) = \frac{1}{c_{k+1}} p(Z_{k+1}, n_{k+1}|x_{k+1}) p(x_{k+1}|Z^k) \tag{8.48}$$

with a normalizing constant given by $c_{k+1} = \int dx_{k+1}\, p(Z_{k+1}, n_k|x_{k+1})p(x_{k+1}|Z^k)$.

8.4.3.1 Example: Standard Kalman Filtering

If there is no ambiguity regarding the origin of the data, i.e., $p(Z_k, n_k|x_k) = \mathcal{N}(z_k; \mathrm{H}_k x_k, \mathrm{R}_k)$, and for a single dynamics model, we have $p(x_{k+1}|Z^k) = \mathcal{N}(x_{k+1}; \hat{\mathbf{x}}_{k+1|k}, \mathbf{P}_{k+1|k})$ according to the previous example. Exploiting Equation 8.44, Equation 8.48 directly results in the well-known Kalman filter equations:

$$\mathcal{N}(x_{k+1}; \hat{\mathbf{x}}_{k+1|k}, \mathbf{P}_{k+1|k}) \xrightarrow[\mathrm{H}_{k+1},\, \mathrm{R}_{k+1}]{z_{k+1}} \mathcal{N}(x_{k+1}; \hat{\mathbf{x}}_{k+1}, \mathbf{P}_{k+1}) \tag{8.49}$$

$$\begin{aligned} \hat{\mathbf{x}}_{k+1} &= \hat{\mathbf{x}}_{k+1|k} + W_{k+1}(z_{k+1} - \mathrm{H}_{k+1}\hat{\mathbf{x}}_{k+1|k}) & S_{k+1} &= \mathrm{H}_{k+1}\mathbf{P}_{k+1|k}\mathrm{H}_{k+1}^{\mathrm{T}} + \mathrm{R}_{k+1} \\ \mathbf{P}_{k+1} &= \mathbf{P}_{k+1|k} - W_{k+1}S_{k+1}W_{k+1}^{\mathrm{T}} & W_{k+1} &= \mathbf{P}_{k+1|k}\mathrm{H}_{k+1}^{\mathrm{T}}S_{k+1}^{-1}. \end{aligned} \tag{8.50}$$

In other words, the parameters $\hat{\mathbf{x}}_{k+1}, \mathbf{P}_{k+1}$ defining the density $p(x_{k+1}|Z^{k+1})$ result from a correction of the corresponding prediction result $\hat{\mathbf{x}}_{k+1|k}, \mathbf{P}_{k+1|k}$ by the innovation vector $\nu_{k+1} = z_{k+1} - \mathrm{H}_{k+1}\hat{\mathbf{x}}_{k+1|k}$ and the innovation covariance S_{k+1}, respectively, according to a weighting matrix W_{k+1} (*Kalman Gain* matrix).

8.4.4 Retrodiction

Retrodiction is an iteration scheme for calculating the probability densities $p(x_l|Z^k)$, $l < k$, that describe the past states x_l given all available sensor information Z^k accumulated up to a later scan time $t_k > t_l$, typically the current time:

$$p(x_l|Z^k) \xleftarrow[p(x_l|Z^l)]{\text{dynamics model}} p(x_{l+1}|Z^k),\ l < k. \tag{8.51}$$

The iteration is initiated by the filtering result $p(x_k|Z^k)$ at time t_k. Retrodiction thus describes the impact of newly available sensor data on our knowledge of the past. Adopting the standard terminology,[16] we speak of *fixed-interval* retrodiction. In close analogy to the previous reasoning, an application of the Total Probability theorem yields

$$p(x_l|Z^k) = \sum_{M^k, H^k} \int dx_{l+1} p(x_l|x_{l+1}, M^k, H^k, Z^k) p(x_{l+1}, H^k, M^k|Z^k). \tag{8.52}$$

In Equation 8.52, the density $p(x_{l+1}, M^k, H^k|Z^k)$ is available by the previous step in the retrodiction loop: $p(x_{l+1}|Z^k) = \Sigma_{H^k, M^k} p(x_{l+1}, M^k, H^k|Z^k)$. Due to the Markov property of the system dynamics and Bayes' Rule, the remaining factor yields

$$p(x_l|x_{l+1}, M^k, H^k, Z^k) = \frac{1}{c_{l|k}} p(x_{l+1}|x_l, m_{l+1}) p(x_l|M^l, H^l, Z^l) \tag{8.53}$$

with a normalizing constant given by $c_{l|k} = \int dx_l\, p(x_{l+1}|x_l, m_{l+1})\, p(x_l|M^l, H^l, Z^l)$. With the densities $p(x_{k+1}|x_k, m_{k+1}) = \mathcal{N}(x_{l+1}; \mathrm{F}_{m_{l+1}} x_l, \mathrm{Q}_{m_{l+1}})$ (Equation 8.4) and $p(x_l|M^l, H^l, Z^l) = \mathcal{N}(x_l; \hat{\mathbf{x}}^{M^l}_{H^l}, \mathbf{P}^{M^l}_{H^l})$ (Equation 8.35), the algebraic manipulations and integration can be carried out by exploiting Equation 8.44. This results in the Rauch-Tung-Striebel equations (see the example in Section 8.4.4.1). We thus obtain

$$p(x_l|Z^k) = \sum_{H^k, M^k} \mu^{M^k}_{H^k} \mathcal{N}\left(x_k; \hat{\mathbf{x}}^{M^k}_{H^k}(l|k), \mathbf{P}^{M^k}_{H^k}(l|k)\right). \tag{8.54}$$

Evidently, no sensor data enter into the retrodiction loop. The data processing is completely performed in the successive filtering steps.

A direct consequence of these considerations is the notion of a *retrodicted probability*.[13,37] Due to $H^k = (H^k_n, H^l)$ and the Total Probability theorem, the probability of H^l being correct given the accumulated data up to t_k can be calculated by summing up the weighting factors of all its descendants at time t_k:

$$P(H^l|Z^k) = \sum_{H^k_n} P(H^k_n, H^l|Z^k). \tag{8.55}$$

8.4.4.1 Example: Rauch-Tung-Striebel Smoothing

Under the assumptions of the previous examples (no data ambiguity, a single dynamics model), well-known iteration equations, named after Rauch, Tung, and Striebel (see Reference 16, p. 161 ff.), prove to be a limiting case of Equation 8.51. Equation 8.52 and 8.53 yield, according to Equation 8.44,

$$\mathcal{N}(x_l; \hat{\mathbf{x}}_{l|k}, \mathbf{P}_{l|k}) \xleftarrow[\hat{\mathbf{x}}_l,\, \mathbf{P}_l,\, \hat{\mathbf{x}}_{l+1|l},\, \mathbf{P}_{l+1|l}]{\mathrm{F}_{k+1}} \mathcal{N}(x_{l+1}; \hat{\mathbf{x}}_{l+1|k}, \mathbf{P}_{l+1|k}), \qquad l < k \tag{8.56}$$

$$\begin{aligned} \hat{\mathbf{x}}_{l|k} &= \hat{\mathbf{x}}_l + A_k(\hat{\mathbf{x}}_{l+1|k} - \hat{\mathbf{x}}_{l+1|l}) \\ \mathbf{P}_{l|k} &= \mathbf{P}_l + A_k(\mathbf{P}_{l+1|k} - \mathbf{P}_{l+1|l})A_k^{\mathrm{T}} \end{aligned} \qquad A_k = \mathbf{P}_l \mathrm{F}^{\mathrm{T}}_{l+1} \mathbf{P}^{-1}_{l+1|l}. \tag{8.57}$$

Formally speaking, the Rauch-Tung-Striebel equations are very similar to the Kalman update Equation 8.49, in that the filtering results at time t_l are corrected by a weighted difference between the smoothing and prediction results related to time t_{l+1}. Also, the weighting matrix A_k shows a structure similar to the Kalman Gain matrix.

8.5 Suboptimal Realization

Due to the uncertain origin of the sensor data and uncertainty regarding the underlying system dynamics model, naively applied data processing according to the previous formalism leads to memory explosions: the number of components in the mixture densities $p(x_{k+1}|Z^k)$ and $p(x_{k+1}|Z^{k+1})$ exponentially grow at each prediction and filtering step. Thus, suboptimal approximation techniques are inevitable for any practical realization. Fortunately, the densities resulting from Equation 8.39 and 8.47 are characterized by a finite number of modes that may be large and fluctuating, but do not explosively grow. This is the rationale for adaptive approximation methods that keep the number of mixture components under control without disturbing the density iteration too seriously. In other words, the densities can often be approximated by mixtures with (far) less components. Provided the relevant features of the densities are preserved, the resulting suboptimal algorithms are expected to be close to optimal Bayesian filtering.

The memory explosions due to the IMM modeling of the system dynamics arise in the prediction step, while the uncertain origin of the sensor data affects the calculations in the filtering step. Both cases can be handled separately, leading to IMM-type, PDA- or MHT-type, and hybrid algorithms such as IMM-PDA filtering or IMM-MHT.

8.5.1 Moment Matching

The mean $\hat{\mathbf{x}} = \mathbb{E}[x]$ and covariance $\mathbf{P} = \mathbb{E}[(x - \hat{\mathbf{x}})(x - \hat{\mathbf{x}})^\mathsf{T}]$ of a finite mixture density $p(x) = \sum_i c_i p_i(x)$ can be expressed in terms of its mixing weights c_i and the means $\hat{\mathbf{x}}_i$ and covariances $\mathbf{P}_i$ of the component densities $p_i(x)$:

$$\hat{\mathbf{x}} = \sum_i c_i \int dx\; p_i(x) x = \sum c_i \hat{\mathbf{x}}_i \tag{8.58}$$

$$\mathbf{P} = \sum_i c_i \int dx\; p_i(x)(x - \hat{\mathbf{x}})(x - \hat{\mathbf{x}})^\mathsf{T} = \sum_i c_i \{\mathbf{P}_i + (\hat{\mathbf{x}}_i - \hat{\mathbf{x}})(\hat{\mathbf{x}}_i - \hat{\mathbf{x}})^\mathsf{T}\}. \tag{8.59}$$

Equation 8.59 results from a completion of the squares and $\int \mathbf{dx}\; p_i(x)(x - \hat{\mathbf{x}}_i)(\hat{\mathbf{x}}_i - \hat{\mathbf{x}})^\mathsf{T} = 0$. While $\hat{x}$ is a weighted sum of the individual means $\hat{\mathbf{x}}_i$, for the covariance $\mathbf{P}$ additionally the spread matrix $\sum_i c_i(\hat{\mathbf{x}}_i - \hat{\mathbf{x}})(\hat{\mathbf{x}}_i - \hat{\mathbf{x}})^\mathsf{T}$ appears. It describes the scattering of the individual means $\hat{\mathbf{x}}_i$ around $\hat{\mathbf{x}}$. As the spread matrix is a sum of dyads with positive coefficients, it is positive (semi-) definite and thus implies a "broadening" of the covariance P.

$\hat{\mathbf{x}}$ and $\mathbf{P}$ can be used to approximate a mixture density $p(x)$ via *moment matching*: $p(x) \approx \mathcal{N}(x; \hat{\mathbf{x}}, \mathbf{P})$; i.e., the mixture is replaced by a Gaussian having the same mean and covariance. This second-order approximation is also referred to as *global combining* of the individual mixture components.[13] More generally speaking, a given mixture density can be split into a weighted sum of sub-mixtures such as $(1/(c_i + c_j))(c_i p_i(x) + c_j p_j(x))/c_{ij}$ with the resulting mixing weight $c_{ij} = c_i + c_j$. If sub-mixtures are approximated by a Gaussian via moment matching, we speak of *local combining*.[13] This results in an approximate representation of the original density by a mixture with a reduced number of components. Thus, the "art" of finding good approximations to optimal Bayesian track maintenance primarily consists in identifying appropriate sub-mixtures to be combined without much error.

Figure 8.4 provides a schematic illustration of moment matching. A particular mixture density $p(x) = c_1 p_1(x) + c_2 p_2(x)$ is displayed along with the related mixture components $c_1 p_1(x)$, $c_2 p_2(x)$ (Figure 8.4a). In Figure 8.4b, the mixture $p(x)$ is compared with the Gaussian density $\mathcal{N}(x; \hat{\mathbf{x}}, \mathbf{P})$ with $\hat{\mathbf{x}} = \mathbb{E}_p[x]$, $\mathbf{P} = \mathbb{E}_p[(x - \hat{\mathbf{x}})(x - \hat{x})^\mathsf{T}]$. The bars at the bottom line indicate the relative size of the mixture coefficients c_1, c_2 in this example. Evidently, moment matching can provide a satisfying approximation to a mixture as long as it is unimodal.

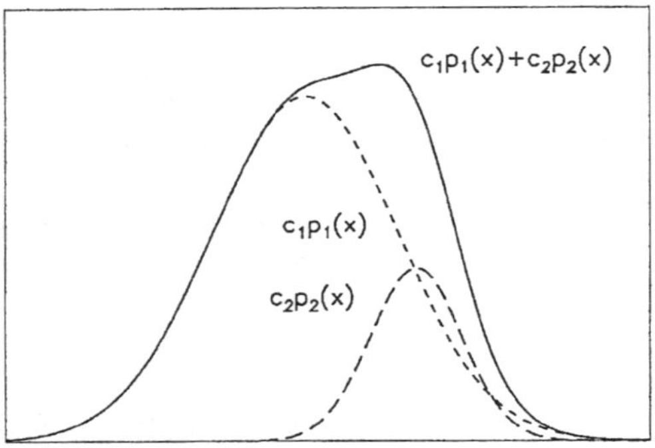

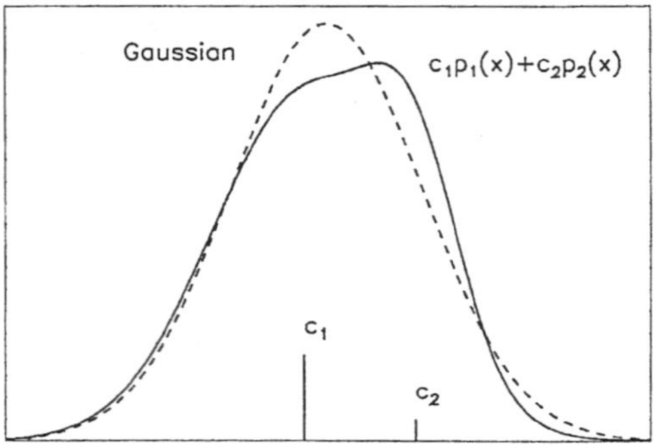

FIGURE 8.4 Scheme of moment matching.

8.5.2 IMM-Type Prediction

For a given interpretation history H^k, the density $p(x_k|H^k, Z^k)$ can be written as a sum over all model histories M^k (Equation 8.37). As the number of mixture components involved is r^k according to the r dynamics models involved, various techniques have been developed for approximately representing $p(x_k|H^k, Z^k)$ by mixtures with a *constant* number of components at each scan k. Among these approximation schemes, the so-called IMM-type and GPB-type algorithms are the most popular (GPB: generalized pseudo-Bayesian). Both approaches are based on the observation that only the recent model histories "n scans back" are actually relevant for predicting the system state (see Reference 5 and the literature cited therein).

With M_n^k denoting a sequence of possible model hypotheses for the last n scans, we are thus looking for approximations of $p(x_k|H^k, Z^k)$ by normal mixtures of the form

$$p(x_k|H^k, Z^k) \approx \sum_{M_n^k} \mu_{H_k}^{M_n^k} \mathcal{N}\left(x_k; \hat{\mathbf{x}}_{H_k}^{M_n^k}, \mathbf{P}_{H_k}^{M_n^k}\right) \tag{8.60}$$

$$\mu_{H_k}^{M_n^k} = P(M_n^k|H^k, Z^k), \quad \mathcal{N}\left(x_k; \hat{\mathbf{x}}_{H_k}^{M_n^k}, \mathbf{P}_{H_k}^{M_n^k}\right) = p(x_k|M_n^k, H^k, Z^k). \tag{8.61}$$

By variation of the parameter n, the quality of the approximation is at the user's disposal and depends on his particular needs and computational resources. In many practically relevant cases, good approximations are provided even by small values of n (1, 2, or 3). Standard probabilistic reasoning in close analogy to Equations 8.40, 8.41, and local combining yields

$$px_{k+1}|H^k, Z^k \approx \sum_{M_n^{k+1}} \sum_{m_{k-n}} \mathrm{p}_{m_{k+1}m_k} \mu_{H_k}^{M_n^k} \int dx_k \mathcal{N}(x_{k+1}; \mathrm{F}_{m_{k+1}} x_k, \mathrm{Q}_{m_{k+1}}) \mathcal{N}\left(x_k; \hat{\mathbf{x}}_{H_k}^{M_n^k}, P_{H_k}^{M_n^k}\right) \tag{8.62}$$

$$= \underbrace{\sum_{M_n^{k+1}} \sum_{m_{k-n}} \mathrm{p}_{m_{k+1}m_k} \mu_{H_k}^{M_n^k} \mathcal{N}\left(x_{k+1}; \mathrm{F}_{m_{k+1}} \hat{x}_{H_k}^{M_n^k}, \mathrm{F}_{m_{k+1}} \mathbf{P}_{H_k}^{M_n^k} \mathrm{F}_{m_{k+1}}^{\mathrm{T}} + \mathrm{Q}_{m_{k+1}}\right)}_{\text{moment matching: approximate by a Gaussian!}} \tag{8.63}$$

$$\approx \sum_{M_n^{k+1}} \mu_{H_k(k+1|k)}^{M_n^{k+1}} \mathcal{N}\left(x_{k+1}; \hat{\mathbf{x}}_{H_k(k+1|k)}^{M_n^{k+1}}, \mathbf{P}_{H_k(k+1|k)}^{M_n^{k+1}}\right) \tag{8.64}$$

where we used $M_n^{k+1} = (m_{k+1}, m_k, M_{n-1}^{k-1}, m_{k-n})$, $M_n^k = (m_k, M_{n-1}^{k-1}, m_{k-1})$. By this approximation scheme, the number of mixture components in $p(x_k|H^k, Z^k)$ is kept constant for each interpretation history H^k during the density iteration.

8.5.2.1 Example: Standard IMM Tracking

If there is no ambiguity regarding the origin of the data, i.e., $p(Z_k, n_k|x_k) = \mathcal{N}(z_k; \mathrm{H}_k x_k, \mathrm{R}_k)$, and for $n = 1$, the densities $p(x_k|Z^k)$ are approximately given by normal mixtures with r components according to the r dynamics models involved. From Equation 8.64 we obtain, due to Equation 8.44, the following prediction formulae:[5]

$$\sum_{i=1}^{r} \mu_k^i \mathcal{N}(x_k; \hat{\mathbf{x}}_k^i, \mathbf{P}_k^i) \xrightarrow[\{\mathrm{F}_{k+1}^i, \mathrm{Q}_{k+1}^i\}_{i=1}^r]{\mathrm{p}_{ij},\ i, j = 1, \ldots, r} \sum_{i=1}^{r} \mu_{k+1|k}^i \mathcal{N}(x_{k+1}; \hat{\mathbf{x}}_{k+1|k}^i, \mathbf{P}_{k+1|k}^i) \tag{8.65}$$

$$\begin{aligned} \mu_{k+1|k}^i &= \Sigma_{j=1}^r \mu_k^{ij} & \mu_k^{ij} &= \mathrm{p}_{ij}\mu_k^j \\ \hat{\mathbf{x}}_{k+1|k}^i &= \frac{1}{\mu_{k+1|k}^i} \Sigma_{j=1}^r \mu_k^{ij} \hat{x}_k^{ij} & \hat{x}_k^{ij} &= \mathrm{F}_{k+1}^i \hat{\mathbf{x}}_k^j \\ \mathbf{P}_{k+1|k}^i &= \frac{1}{\mu_{k+1|k}^i} \Sigma_{j=1}^r \mu_k^{ij} \{P_k^{ij} + (\hat{x}_k^{ij} - \hat{\mathbf{x}}_{k+1|k}^i)(\ldots)^{\mathrm{T}}\} & P_k^{ij} &= \mathrm{F}_{k+1}^i \mathbf{P}_k^j \mathrm{F}_{k+1}^{i\mathrm{T}} + \mathrm{Q}_{k+1}^i. \end{aligned} \tag{8.66}$$

The received measurement z_{k+1} is processed according to Bayes' Rule (Equation 8.48). By a second use of Equation 8.44, one step of the filtering loop is completed:

$$\sum_{i=1}^{r} \mu_{k+1|k}^i \mathcal{N}(x_{k+1}; \hat{\mathbf{x}}_{k+1}^i, \mathbf{P}_{k+1|k}^i) \xrightarrow[\mathrm{H}_{k+1},\ \mathrm{R}_{k+1}]{z_{k+1}} \sum_{i=1}^{r} \mu_{k+1}^i \mathcal{N}(x_{k+1}; \hat{\mathbf{x}}_{k+1}^i, \mathbf{P}_{k+1}^i) \tag{8.67}$$

$$\begin{aligned} \mu_{k+1}^i &= \frac{\mu_{k+1|k}^i}{c_{k+1}} \mathcal{N}(z_{k+1}; \mathrm{H}_{k+1}\hat{\mathbf{x}}_{k+1|k}^i, S_{k+1}^i) & S_{k+1}^i &= \mathrm{H}_{k+1}\mathbf{P}_{k+1|k}^i \mathrm{H}_{k+1}^{\mathrm{T}} + \mathrm{R}_{k+1} \\ \hat{\mathbf{x}}_{k+1}^i &= \hat{\mathbf{x}}_{k+1|k}^i + W_{k+1}^i(z_{k+1} - \mathrm{H}_{k+1}\hat{\mathbf{x}}_{k+1|k}^i) & W_{k+1}^i &= \mathbf{P}_{k+1|k}^i \mathrm{H}_{k+1}^{\mathrm{T}} S_{k+1}^{i-1} \\ \mathbf{P}_{k+1}^i &= \mathbf{P}_{k+1|k}^i - W_{k+1}^i S_{k+1} W_{k+1}^{i\mathrm{T}} \end{aligned} \tag{8.68}$$

$$c_{k+1} = \Sigma_{i=1}^r \mathcal{N}(z_{k+1}; \mathrm{H}_{k+1}\hat{\mathbf{x}}_{k+1|k}^i, S_{k+1}^i)\mu_{k+1|k}^i. \tag{8.69}$$

8.5.3 PDA-Type Filtering

In close analogy to the previous considerations, moment matching can also be used to avoid memory explosion due to the uncertain origin of the sensor data. Global combining of all mixture components in $p(x_k|Z^k) = \Sigma_{H^k} p(x_k|H^k, Z^k)\, p(H^k|Z^k)$ (Equation 8.36) obtained after filtering leads to PDA-type algorithms,[2,5] while local combining of suitably chosen sub-mixtures results in MHT-type algorithms as discussed in the subsequent section.

Let us assume that $p(x_{k+1}|Z^k)$ is available by IMM-type prediction as a normal mixture with respect to the modal histories M_n^{k+1} "n scans back":

$$p(x_{k+1}|Z^k) = \sum_{M_n^{k+1}} \mu_{k+1|k}^{M_n^{k+1}} \mathcal{N}\left(x_{k+1}; \hat{\mathbf{x}}_{k+1|k}^{M_n^{k+1}}, \mathbf{P}_{k+1|k}^{M_n^{k+1}}\right). \tag{8.70}$$

Standard probabilistic reasoning in close analogy to Equation 8.48 and global combining Equations 8.58 and 8.59 yields

$$p(x_{k+1}|Z^{k+1}) \propto \sum_{M_n^{k+1}} p(Z_{k+1}, n_k|x_{k+1})\mu_{k+1|k}^{M_n^{k+1}} \mathcal{N}\left(x_{k+1}; \hat{\mathbf{x}}_{k+1|k}^{M_n^{k+1}}, \mathbf{P}_{k+1|k}^{M_n^{k+1}}\right)$$

$$= \sum_{M_n^{k+1}} \underbrace{\sum_{h_{k+1}} p(Z_{k+1}, n_{k+1}|x_{k+1}, h_{k+1}) P(h_{k+1}|x_{k+1})\mu_{k+1|k}^{M_n^{k+1}} \mathcal{N}\left(x_{k+1}; \hat{\mathbf{x}}_{k+1|k}^{M_n^{k+1}}, \mathbf{P}_{k+1|k}^{M_n^{k+1}}\right)}_{\text{moment matching approximate by a Guassian!}} \tag{8.71}$$

$$\approx \sum_{M_n^{k+1}} \mu_{k+1}^{M_n^{k+1}} \mathcal{N}\left(x_{k+1}; \hat{\mathbf{x}}_{k+1|k}^{M_n^{k+1}}, \mathbf{P}_{k+1|k}^{M_n^{k+1}}\right). \tag{8.72}$$

8.5.3.1 Example: Standard PDA Filtering

For a single object embedded in a false return background, we have up to a normalizing constant independent of x_{k+1}: $p(Z^{k+1}, n_{k+1}|x_{k+1}) \propto (1 - P_D)\rho_F + P_D \Sigma_{j=1}^{n_{k+1}} \mathcal{N}(z_{k+1}^j; H_{k+1}x_{k+1}, \mathrm{R}_{k+1})$ (Equation 8.20). Moreover, let us assume a single dynamics model, i.e., $r = 1$. Following the PDA philosophy, global combining is to be applied to a mixture with $n_{k+1} + 1$ components according to the $n_{k+1} + 1$ possible interpretation hypotheses of the sensor data Z_{k+1}. By exploiting Equation 8.44, we obtain:[2]

$$\mathcal{N}(x_{k+1}; \hat{\mathbf{x}}_{k+1|k}, \mathbf{P}_{k+1|k}) \xrightarrow[P_D,\, \rho_F,\, \mathrm{H}_{k+1},\, \mathrm{R}_{k+1}]{z_{k+1}^j,\, j = 1, \ldots, n_{k+1}} \mathcal{N}(x_{k+1}; \hat{\mathbf{x}}_{k+1}, \mathbf{P}_{k+1}) \tag{8.73}$$

$$\begin{aligned} \hat{\mathbf{x}}_{k+1} &= \hat{\mathbf{x}}_{k+1|k} + W_{k+1}\nu_{k+1} \\ \mathbf{P}_{k+1} &= \mu_{k+1}^0 \mathbf{P}_{k+1|k} + (1 - \mu_{k+1}^0)(\mathbf{P}_{k+1|k} - W_{k+1}S_{k+1}W_{k+1}^{\mathrm{T}}) \\ &\quad + W_{k+1}\{\Sigma_{j=1}^{n_{k+1}} \mu_{k+1}^j (\nu_{k+1}^j - \nu_{k+1})(\nu_{k+1}^j - \nu_{k+1})^{\mathrm{T}}\} W_{k+1}^{\mathrm{T}}. \end{aligned} \tag{8.74}$$

where we used the abbreviations

$$\begin{aligned} S_{k+1} &= \mathrm{H}_{k+1}\mathbf{P}_{k+1|k}\mathrm{H}_{k+1}^{\mathrm{T}} + \mathrm{R}_{k+1}, \quad W_{k+1} = \mathbf{P}_{k+1|k}\mathrm{H}_{k+1}^{\mathrm{T}}S_{k+1}^{-1} \\ \nu_{k+1}^j &= z_{k+1}^j - \mathrm{H}_{k+1}\hat{\mathbf{x}}_{k+1|k}, \quad c_{k+1} = \rho_F(1 - P_D) + P_D\Sigma_{j=1}^{n_{k+1}}\mathcal{N}(\nu_{k+1}^j; 0, S_{k+1}) \\ \mu_{k+1}^0 &= \frac{1}{c_{k+1}}\rho_F(1 - P_D), \quad \mu_{k+1}^j = \frac{1}{c_{k+1}}P_D\mathcal{N}(\nu_{k+1}^j; 0, S_{k+1}), \quad \nu_{k+1} = \Sigma_{j=0}^{n_{k+1}}\mu_{k+1}^j\nu_{k+1}^j \end{aligned} \tag{8.75}$$

Evidently, the PDA update equations for $\hat{\mathbf{x}}_{k+1}$ and $\mathbf{P}_{k+1}$ are closely related to standard KF: $\hat{\mathbf{x}}_{k+1}$ is given by the predicted estimate $\hat{\mathbf{x}}_{k+1|k}$ corrected by the product of the mean innovation ν_{k+1} and the Kalman Gain matrix W_{k+1}. The weighting factors for the individual innovations are data dependent. The Kalman covariance $\mathbf{P}_{k+1|k} - W_{k+1}\, S_{k+1}\, W_{k+1}^{\mathrm{T}}$ appears with a weighting factor that is one for $P_D = 1$ or $\rho_F = 0$. The spread term is responsible for the data-driven adaptivity of PDA filtering. It "broadens" $\mathbf{P}_{k+1}$ depending on the scattering of the individual innovations ν_{k+1}^j around the mean innovation ν_{k+1}.

8.5.3.2 Example: JPDA Filtering for Small Clusters

For two closely space objects embedded in a false return background, the density $p(Z_{k+1}, n_{k+1}|x_{k+1})$ (Equation 8.32) obeys the same formal structure as in the previous example. Due to $P_u(x_{k+1}) = \det(2\pi \mathrm{R}_{k+1}^u)^{1/2}\, \mathcal{N}(\mathrm{H}_{k+1}^u x_{k+1}; 0, \mathrm{R}_{k+1}^u)$ (Equation 8.12) and according to Equation 8.44, it proves to be a normal mixture with positive and negative mixture coefficients summing up to one. For the sake of simplicity, let us consider a single dynamics model, i.e., $r = 1$. Following the PDA philosophy, global combining is to be applied to a mixture with $(n_{k+1} + 1)^2 + 1$ components according to the $(n_{k+1} + 1)^2 + 1$ interpretation hypotheses of the sensor data Z_{k+1}. Let $x_{k+1} = (x^1_{k+1}, x^2_{k+1})^{\mathsf{T}}$ be the joint state of both targets. By processing of the sensor data according to Equation 8.48, we obtain, due to Equation 8.44,[33]

$$\mathcal{N}(x_{k+1};\hat{\mathbf{x}}_{k+1|k}, \mathbf{P}_{k+1|k}) \xrightarrow[\mathrm{H}^g_{k+1},\, \mathrm{R}^g_{k+1},\, \mathrm{H}_{k+1},\, \mathrm{R}_{k+1}]{\{z^j_{k+1}\}^{n_{k+1}}_{j=1},\, P^u_D,\, P_D,\, \rho_F} \begin{cases} \mu^0_{k+1}\mathcal{N}(x_{k+1};\hat{x}^0_{k+1}, P^0_{k+1}) \\ + \Sigma^{n_{k+1}}_{i,j=0}\mu^{ij}_{k+1}\mathcal{N}(x_{k+1};\hat{x}^{ij}_{k+1}, P^{ij}_{k+1}). \end{cases} \tag{8.76}$$

While the individual means $\hat{x}^0_{k+1}$, $\hat{x}^{ij}_{k+1}$ and covariances P^0_{k+1}, P^{ij}_{k+1} result from KF (Equation 8.49), the corresponding weighting factors μ^0_{k+1}, μ^{ij}_{k+1} depend on x_{k+1}:

$$\mu^0_k = (1 - P_u(x_{k+1}))\mu^{*0}_k, \quad \mu^{ii}_k = P_u(x_{k+1})\mu^{*ii}_k, \quad \mu^{ij}_k = (1 - P_u(x_{k+1}))\mu^{*ij}_k, \quad i \neq j \tag{8.77}$$

where we used the abbreviations

$$\begin{aligned} \mu^{*0}_k &= \rho^2_F(1 - P_D)^2, \quad \mu^{*00}_k = \rho^2_F(1 - P^u_D), \quad \mu^{*ii}_k = P^u_D\rho_F\mathcal{N}(z^i_k;\mathrm{H}^g_{k+1}\hat{\mathbf{x}}_{k+1|k}, S^g_{k+1}) \\ \mu^{*i0}_k &= P_D(1 - P_D)\rho_F\mathcal{N}(z^i_k;\mathrm{H}_k\hat{\mathbf{x}}^1_{k+1|k}, S^1_{k+1}), \quad \mu^{*0i}_k \text{ analogously, } i = 1, \ldots, n_{k+1} \\ \mu^{*ij}_k &= P^2_D\mathcal{N}(z^i_k;\mathrm{H}_k\hat{\mathbf{x}}^1_{k+1|k}, S_{k+1})\mathcal{N}(z^i_k;\mathrm{H}_k\hat{\mathbf{x}}^2_{k+1|k}, S_{k+1}),\ i, j = 1, \ldots, n_{k+1},\ i \neq j \end{aligned} \tag{8.78}$$

with $\hat{\mathbf{x}}_{k+1|k} = (\hat{\mathbf{x}}^1_{kj+1|k}, \hat{\mathbf{x}}^2_{kj+1|k})$, and S^g_{k+1}, $S^{1,2}_{k+1}$, S_{k+1} denoting the corresponding innovation covariance matrices. By a second use of Equation 8.44 and global combining (Equations 8.58 and 8.59), we finally obtain

$$\left.\begin{aligned} &\mu^0_{k+1}\mathcal{N}(x_{k+1};\hat{x}^0_{k+1}, P^0_{k+1}) \\ &+ \Sigma^{n_{k+1}}_{i,j=0}\mu^{ij}_{k+1}\mathcal{N}(x_{k+1};\hat{x}^{ij}_{k+1}, P^{ij}_{k+1}) \end{aligned}\right\} \xrightarrow[\mathrm{H}^u_{k+1},\, \mathrm{R}^u_{k+1}]{} \mathcal{N}(x_{k+1};\hat{\mathbf{x}}_{k+1}, \mathbf{P}_{k+1}) \tag{8.79}$$

$$\begin{aligned} \hat{\mathbf{x}}_{k+1} &= \frac{1}{c_{k+1}}\mu^{u0}_{k+1}\hat{x}^{u0}_{k+1} + \frac{1}{c_{k+1}}\Sigma^{n_{k+1}}_{i,j=0}\mu^{uij}_{k+1}\hat{x}^{uij}_{k+1} \\ \mathbf{P}_{k+1} &= \frac{1}{c_{k+1}}\mu^{u0}_{k+1}(P^{u0}_{k+1} + (\hat{x}^{u0}_{k+1} - \hat{\mathbf{x}}_{k+1})(\hat{x}^{u0}_{k+1} - \hat{\mathbf{x}}_{k+1})^{\mathsf{T}}) \\ &\quad + \frac{1}{c_{k+1}}\Sigma^{n_{k+1}}_{i,j=0}\mu^{uij}_{k+1}(P^{uij}_{k+1} + (\hat{x}^{uij}_{k+1} - \hat{\mathbf{x}}_{k+1})(\hat{x}^{uij}_{k+1} - \hat{\mathbf{x}}_{k+1})^{\mathsf{T}}) \end{aligned} \tag{8.80}$$

where we used the abbreviations

$$\begin{aligned} &\hat{x}^{uij}_{k+1} = (1 - W^{uij}_{k+1}\mathrm{H}^u_{k+1})\hat{x}^{ij}_{k+1} \qquad S^{uij}_{k+1} = \mathrm{H}^u_{k+1}P^{ij}_{k+1}\mathrm{H}^{u\mathsf{T}}_{k+1} + \mathrm{R}^u_{k+1} \\ &P^{uij}_{k+1} = P^{ij}_{k+1} - W^{uij}_{k+1}S^{uij}_{k+1}W^{uij\mathsf{T}}_{k+1} \qquad W^{uij}_{k+1} = P^{ij}_{k+1}\mathrm{H}^{u\mathsf{T}}_{k+1}S^{uij-1}_{k+1} \\ &\hat{x}^{u0}_{k+1}, P^{u0}_{k+1}, S^{u0}_{k+1}, W^{u0}_{k+1} \text{ analogously defined} \\ &\mu^{u0}_{k+1} = \mu^0_{k+1}(1 - \mathcal{N}(\mathrm{H}^u_{k+1}\hat{x}^u_{k+1};0, S^{u0}_{k+1})),\ \mu^{uii}_{k+1} = \mu^{ii}_{k+1}\det(2\pi\mathrm{R}^u_{k+1})^{\frac{1}{2}}\mathcal{N}(\mathrm{H}^u_{k+1}\hat{x}^{ii}_{k+1};0, S^{uii}_{k+1}) \\ &\mu^{uij}_{k+1} = \mu^{ij}_{k+1}\left(1 - \det(2\pi\mathrm{R}^u_{k+1})^{\frac{1}{2}}\mathcal{N}(\mathrm{H}^u_{k+1}\hat{x}^{ij}_{k+1};0, S^{uij}_{k+1})\right),\ i \neq j,\ c_{k+1} = \mu^{u0}_{k+1} + \Sigma^{n_{k+1}}_{i,j=0}\mu^{uij}_{k+1}. \end{aligned} \tag{8.81}$$

Within the scope of a second-order statistics and the resolution model (Equation 8.12), Equations 8.76 and 8.79 provide a complete solution to the cluster tracking problem (at least for two objects). In principle, a generalization to small clusters consisting of more than two objects is possible.

8.5.3.3 Example: IMM-PDA Filtering

Finally, a hybrid solution is discussed where IMM prediction is combined with PDA filtering. Let us consider the case $n = 1$ and a single object embedded in a false return background. In close analogy to the previous examples, we obtain[7]

$$\sum_{i=1}^{r} \mu^i_{k+1|k} \mathcal{N}(x_{k+1}; \hat{\mathbf{x}}^i_{k+1|k}, \mathbf{P}^i_{k+1|k}) \xrightarrow[P_D,\, \mathrm{H}_{k+1},\, \mathrm{R}_{k+1}]{\{z^j_{k+1}\}_{j=1}^{n_{k+1}},\, \rho_F} \sum_{i=1}^{r} \mu^i_{k+1} \mathcal{N}(x_{k+1}; \hat{\mathbf{x}}^i_{k+1}, \mathbf{P}^i_{k+1}) \tag{8.82}$$

$$\begin{aligned} \mu^i_{k+1} &= \Sigma_{j=0}^{n_{k+1}} \mu^{ij}_{k+1} \\ \hat{\mathbf{x}}^i_{k+1} &= \hat{\mathbf{x}}^i_{k+1|k} + W^i_{k+1} \nu^i_{k+1} \\ \mathbf{P}^i_{k+1} &= \mu^{i0}_{k+1} \mathbf{P}^i_{k+1|k} + (1 - \mu^{i0}_{k+1})(\mathbf{P}^i_{k+1|k} - W^i_{k+1} S^i_{k+1} W^{i\mathsf{T}}_{k+1}) \\ &\quad + W^i_{k+1} \{\Sigma_{j=1}^{n_{k+1}} \mu^{ij}_{k+1} (\nu^{ij}_{k+1} - \nu^i_{k+1})(\nu^{ij}_{k+1} - \nu^i_{k+1})^{\mathsf{T}}\} W^{i\mathsf{T}}_{k+1} \end{aligned} \tag{8.83}$$

where we used the abbreviations

$$\begin{aligned} S^i_{k+1} &= \mathrm{H}_{k+1} \mathbf{P}^i_{k+1|k} \mathrm{H}^{\mathsf{T}}_{k+1} + \mathrm{R}_{k+1}, \quad W^i_{k+1} = \mathbf{P}^i_{k+1|k} \mathrm{H}^{\mathsf{T}}_{k+1} S^{i^{-1}}_{k+1} \\ \nu^{ij}_{k+1} &= z^j_{k+1} - \mathrm{H}_{k+1} \hat{\mathbf{x}}^i_{k+1|k}, \quad c_{k+1} = \rho_F (1 - P_D) \mu^i_{k+1|k} + P_D \Sigma_{j=1}^{n_{k+1}} \mathcal{N}(\nu^j_{k+1}; 0, S_{k+1}) \mu^i_{k+1|k} \\ \mu^{i0}_{k+1} &= \frac{1}{c_{k+1}} \rho_F (1 - P_D) \mu^i_{k+1|k}, \quad \mu^{ij}_{k+1} = \frac{1}{c_{k+1}} P_D \mathcal{N}(\nu^j_{k+1}; 0, S_{k+1}) \mu^i_{k+1|k}, \quad \nu_{k+1} = \Sigma_{j=0}^{n_{k+1}} \mu^j_{k+1} \nu^j_{k+1}. \end{aligned} \tag{8.84}$$

8.5.4 IMM-MHT-Type Filtering

In case of a more severe false return background or in a multiple-object tracking task with correlation gates overlapping for a longer time, Bayesian track maintenance inevitably leads to densities $p(x_k|Z^k)$ that are characterized by several distinct modes. As this phenomenon is inherent in the uncertain origin of the received data, relevant statistical information would get lost if global combining is applied to such cases. The use of PDA-type filtering is thus confined to a relatively restricted area in parameter space (defined by ρ_F, P_D, for instance).

By *local* combining of suitably chosen sub-mixtures and pruning of irrelevant mixture components, however, memory explosions may be avoided without destroying the multi-mode structure of the densities. Provided this is carefully done with data-driven adaptivity, all statistically relevant information may be preserved while keeping the number of mixture components under control, i.e., the number may be fluctuating and even large in critical situations, but does not explosively grow.[38–41] Evidently, PDA-type filtering is a limiting case of such MHT-type techniques (MHT: multiple hypothesis tracking). As "n scans back" methods defined analogously to IMM-type prediction are non-adaptive, their discussion is omitted here.

8.5.4.1 Individual Gating

In the first step for avoiding unnecessary computational load, sensor data irrelevant for a given track hypothesis are excluded. Let $\hat{\mathbf{x}}_{H^k(k+1|k)}$ be a predicted track hypothesis for a given interpretation history H^k along with its covariance $\mathbf{P}_{H^k(k+1|k)}$. These quantities result from Equation 8.41 by using Equations 8.58 and 8.59. Due to $p(z_{k+1}|\mathrm{H}^k, Z^k) = \int dx_{k+1}\, p(z_{k+1}|x_{k+1})\, p(x_{k+1}|\mathrm{H}^k, Z^k)$, the probability density of the innovation $\nu_{H^k} = z_{k+1} - \mathrm{H}_{k+1} \hat{\mathbf{x}}_{H^k(k+1|k)}$ is approximately given by $p(\nu_{H^k}|\mathrm{H}^k, Z^k) \approx \mathcal{N}(\nu_{H^k}; 0, S_{H^k})$ with $S_{H^k} = \mathrm{H}_{k+1} \mathbf{P}_{H^k(k+1|k)} \mathrm{H}^{\mathsf{T}}_{k+1} + \mathrm{R}_{k+1}$. Individual gating means that only those sensor data are used for continuing a particular track hypothesis H^k whose innovations obey $\nu^{\mathsf{T}}_{H^k} S^{-1}_{H^k} \nu_{H^k} < \lambda_G$. The processing parameter λ_G must be tuned to meet the requirements of a particular application. Evidently, the accuracy of the prediction (depending on the system dynamics model and the previous track hypothesis) and *a priori* information on the sensor performance enter into this decision criterion. Individual gating is a simple measure of preselecting the sensor data. It can be performed for each track hypothesis independently before any further data processing takes place.

8.5.4.2 Pruning Methods

In order to identify insignificant track hypotheses, first for each H^k the weighting factors $\mu_{H^k} = \Sigma_{M_n^k} \mu_{H_k}^{M_n^k}$ (Equation 8.35) are evaluated by processing the sensor data within the gates. This is done before the individual component densities $p(x_{k+1}|H^{k+1}, Z^{k+1})$ (Equation 8.36) and thus the hypothetical tracks $\hat{\mathbf{x}}_{H^k}$, $\mathbf{P}_{H^k}$ are computed. Due to the normalization involved, the size of each weighting factor μ_{H^k} depends on all sensor data in the gates. In contrast to individual gating, pruning is applied after all weighting factors are available. In zero-scan pruning, track hypotheses are deleted that are smaller than a certain predefined threshold: $\mu_{H^k} < p_0$. By this, an additional processing parameter is introduced that must be tuned to meet the requirements of a particular application. The limiting case, where the track hypothesis of highest statistical weight is considered only, is a slightly more general formulation of standard NN filtering.

Delayed or multiple-frame pruning is closely related to this procedure. Here, we consider the retrodicted weighting factors $P(H^{k-n}|Z^k)$ for a past time t_{k-n} given all sensor data up to the current scan k. According to Equation 8.55, they are obtained by summing up the weighting factors μ_{H^k} of all descendants of H^{k-n}. By this procedure, retrospectively, some past hypotheses increase in weight, while others decrease. Analogously to zero-scan pruning, all track hypotheses H^{k-n} along with their descendants up to the current scan are deleted if $P(H^{k-n}|Z^k) < p_0$. Hence, in delayed pruning, the "hard" decision of deleting a hypothesis is based on a broader data basis than in case of zero-scan pruning. In close analogy to NN filtering, a simple limiting case consists of finding the hypothesis H^{*k-n} with the largest retrodicted weight and deleting all other hypotheses along with their descendants. By this method, only the descendants of H^{*k-n} are used to represent the density $p(x_k|Z^k)$.

8.5.4.3 Local Combining

After filtering, a single distinct mode of $p(x_k|Z^k)$ might be a superposition of "similar" mixture components. It is thus reasonable to apply local combining to the sub-mixture producing that mode. Among several realizations, *successive local combining* is particularly simple. Let us start with the mixture component of highest statistical weight, $p_{H_1^k}(x_k) = p(x_k|H_1^k, Z^k)$. In the order of decreasing weighting factors, a component $p_{H_2^k}$ is searched that is "similar" to $p_{H_1^k}$. A very simple scalar criterion for similarity is provided by

$$d(H_1^k, H_2^k) < \kappa_0 \text{ with: } d(H_1^k, H_2^k) = (\hat{\mathbf{x}}_{H_1^k} - \hat{\mathbf{x}}_{H_2^k})^{\mathrm{T}} (\mathbf{P}_{H_1^k} + \mathbf{P}_{H_2^k})^{-1} (\hat{\mathbf{x}}_{H_1^k} - \hat{\mathbf{x}}_{H_2^k}) \tag{8.85}$$

where $\hat{\mathbf{x}}_{H_{1,2}^k}$ and $\mathbf{P}_{H1,2^k}$ denote the mean and covariance with respect to $p_{H1,2^k}(x_k)$, respectively. By this, a third processing parameter κ_0 is introduced (besides λ_G and p_0) that must be tuned to meet the requirements of a particular application. Local combining of $p_{H_1^k}(x_k)$ and $p_{H_2^k}(x_k)$ results in $p_{H_1^{*k}}$ while the corresponding weighting factor increases: $\mu_{H_1^{*k}} = \mu_{H_1^k} + \mu_{H_2^k}$. Then the next similar component is searched in the order of decreasing weighting factors and so on. Having done this, we restart the procedure with the mixture component having the second largest weighting factor. Due to the data-driven adaptivity inherent in this method, MHT-type filtering automatically reduces to PDA-type processing if PDA processing provides good approximation to $p(x_k|Z^k)$.

Objects moving closely spaced for some time may irreversibly loose their identity. When they dissolve again, a unique track-to-target association is impossible. In particular, this means that the component densities $p(x_k, H^k|Z^k)$ and $p(x_k, H^{k\prime}|Z^k)$ are nearly identical if H^k and $H^{k\prime}$ differ only in a permutation of the targets. Thus, it is reasonable to deal with densities that are symmetric under permutations of the individual targets. By this, no statistically relevant information is lost, and the filter performance remains unchanged, while the mean number of hypotheses involved may be significantly reduced.

8.5.5 IMM-Type Retrodiction

In many practical applications, MHT-type filtering results in trees of hypothetical tracks $\{\hat{\mathbf{x}}_{H^l}, \mathbf{P}_{H^l}\}_{l=1}^{k}$ that are simply structured and thus provide satisfying estimates of the state trajectories $\{\hat{\mathbf{x}}_l, \mathbf{P}_l\}_{l=1}^{k}$. Even in a more difficult dense target/dense clutter environment, adaptive pruning and local combining methods remain applicable and preserve all statistically relevant information on the objects of interest. The resulting filtering output may thus consist of a complex and fluctuating but limited hypothesis tree. In consequence, the state estimates $\hat{\mathbf{x}}_k$, $\mathbf{P}_k$ are possibly of a rather poor quality, even if the underlying density is a good approximation to the correct Bayesian density $p(x_k|Z^k)$. As this phenomenon directly results from the current lack of information and data-inherent ambiguity, radical approximation such as in PDA or NN filtering guaranteeing a fixed amount of computing load is not the answer for obtaining more easily interpretable results: track losses would occur. At the expense of some delay, however, a retrodictive analysis of the MHT filtering output may provide significantly improved estimates of the trajectories.

In close analogy to Equation 8.60, we are looking for densities $p(x_l|Z^k) = \Sigma_{H^k} p(x_l, H^k|Z^k)$ in the retrodiction loop (Equation 8.51) which are approximately represented by the same class of functions previously used in the filtering loop:

$$p(x_l|H^k, Z^k) \approx \sum_{M_n^l} \mu_{H_k}^{M_n^l}(l|k)\,\mathcal{N}\left(x_l; \hat{\mathbf{x}}_{H_k}^{M_n^l}(l|k), \mathbf{P}_{H_k}^{M_n^l}(l|k)\right) \tag{8.86}$$

$$\mu_{H_k}^{M_n^l} = P(M_n^l|H^k, Z^k), \quad \mathcal{N}\left(x_l; \hat{\mathbf{x}}_{H_k}^{M_n^l}(l|k), \mathbf{P}_{H_k}^{M_n^l}(l|k)\right) = p(x_l|M_n^l, H^k, Z^k) \tag{8.87}$$

By standard probability reasoning according to Equations 8.52 and 8.53, the density $p(x_l, H^k|Z^k)$ can be written as

$$\begin{aligned} p(x_l, H^k|Z^k) &= \sum_{m_{l+1}} \sum_{M_{n-1}^l} \sum_{m_{l-n}} \int dx_{k+1}\, p(x_l, M_{n-1}^l, m_{l-n}, H^k|x_{l+1}, m_{l+1}, Z^k)\, p(x_{l+1}, m_{l+1}, M_{n-1}^l, H^k|Z^k) \\ &= \underbrace{\sum_{M_n^l} \sum_{m_{l+1}} \int dx_{k+1} \frac{1}{c_{H^k}^l(x_{l+1})} p(x_{l+1}, m_{l+1}|x_l, m_l)\, p(x_{l+1}, M_n^{l+1}|Z^k)\, p(x_l, M_n^l|Z^l)}_{\text{moment matching approximate by a Gaussian!}} \end{aligned} \tag{8.88}$$

$$\approx \sum_{M_n^l} \mu_{H_k}^{M_n^l}(l|k)\,\mathcal{N}\left(x_l; \hat{\mathbf{x}}_{H_k}^{M_n^l}(l|k), \mathbf{P}_{H_k}^{M_n^l}(l|k)\right) \tag{8.89}$$

where we used $M_n^{l+1} = (m_{l+1}, m_l, M_{n-1}^{l-1}, m_{l-n})$. In Equation 8.89, $c_{H^k}^l(x_{l+1})$ denotes a normalizing constant with respect to m_{l-n} and x_l:

$$c_H^l(x_{l+1}) = \sum_{m_{l-n}} \int dx_l\, p(x_{l+1}, m_{l+1}|x_l, m_l)\, p(x_l, M_n^l|Z^l) \tag{8.90}$$

Hence, $p(x_l, H^k|Z^k)$ is completely determined by the retrodicted density at scan $l+1$, the filtering result at scan l, and the underlying dynamics model. By using the approximation $c_H^l(x_{l+1}) \approx c_H^l(\hat{\mathbf{x}}_{H_k}^{M_n^{l+1}}(l+1|k))$[18] and Equation 8.44, the algebraic manipulations and integration can explicitly be carried out, resulting in update formulae for $\mu_{H_k}^{M_n^l}(l|k)$, $\hat{\mathbf{x}}_{H_k}^{M_n^l}(l|k)$, and $\mathbf{P}_{H_k}^{M_n^l}(l|k)$.[18]

8.5.5.1 Example: Standard IMM Retrodiction

If there is no ambiguity regarding the origin of the data and for $n = 1$, the densities $p(x_l|Z^k)$, $l < k$ are approximately given by normal mixtures with r components according to the r dynamics models involved. From Equations 8.88, 8.58, and 8.59, we obtain, due to Equation 8.44, the following prediction formulae:[18]

$$\sum_{i=1}^{r} \mu^i_{l|k} \mathcal{N}(x_l; \hat{\mathbf{x}}^i_{l|k}, \mathbf{P}^i_{l|k}) \xleftarrow[p_{ij},\ \mathbf{F}^i_{l+1},\ \mathbf{Q}^i_{l+1}]{\{\mu^i_l,\ \hat{\mathbf{x}}^i_l,\ \mathbf{P}^i_l\}^r_{i=1}} \sum_{i=1}^{r} \mu^i_{l+1|k} \mathcal{N}(x_{l+1}; \hat{\mathbf{x}}^i_{l+1|k}, \mathbf{P}^i_{l+1|k}),\ l < k \tag{8.91}$$

$$\begin{aligned} \mu^i_{l|k} &= \Sigma^r_{j=1} \mu^{ji}_{l|k} \\ \hat{\mathbf{x}}^i_{l|k} &= \frac{1}{\mu^i_{l|k}} \Sigma^r_{j=1} \mu^{ji}_{l|k} \hat{\mathbf{x}}^{ji}_{l|k} \\ \mathbf{P}^i_{l|k} &= \frac{1}{\mu^i_{l|k}} \Sigma^r_{j=1} v^{ji}_{l|k} \{ P^{ji}_{l|k} + (\hat{x}^{ji}_{l|k} - \hat{\mathbf{x}}^i_{l|k})(\hat{x}^{ji}_{l|k} - \hat{\mathbf{x}}^i_{l|k})^{\mathsf{T}} \} \end{aligned} \tag{8.92}$$

where we used the abbreviations

$$\begin{aligned} \mu^{ji}_{l|k} &= \frac{1}{c_j} p_{ji} \mu^i_l \mu^j_{l+1|k} \mathcal{N}(\hat{\mathbf{x}}^i_{l+1|k}; \hat{x}^{ji}_{l+1|l}, P^{ji}_{l+1|l}) & \hat{x}^{ji}_{l+1|l} &= \mathbf{F}^j_{l+1} \hat{\mathbf{x}}^i_l \\ \hat{x}^{ji}_{l|k} &= \hat{\mathbf{x}}^i_l + A^{ji}_k (\hat{\mathbf{x}}^i_{l+1|k} - \hat{x}^{ji}_{l+1|l}) & P^{ji}_{l+1|l} &= \mathbf{F}^j_{l+1} \mathbf{P}^i_l \mathbf{F}^{j\mathsf{T}}_{l+1} + \mathbf{Q}^j_{l+1} \\ P^{ji}_{l|k} &= \mathbf{P}^i_l + A^{ji}_k (\mathbf{P}^i_{l+1|k} - P^{ji}_{l+1|l}) A^{ji\mathsf{T}}_k & A^{ji}_k &= \mathbf{P}^i_l \mathbf{F}^{j\mathsf{T}}_{l+1} P^{ji^{-1}}_{l+1|l} \\ c_j &= \Sigma^r_{i=1} \mathrm{p}_{ji} \mu^i_l \mu^j_{l+1|k} \mathcal{N}(\hat{\mathbf{x}}^i_{l+1|k}; \hat{x}^{ji}_{l+1|l}, P^{ji}_{l+1|l}) \end{aligned} \tag{8.93}$$

See Reference 42 for a similar solution.

8.6 Selected Applications

The discussion of suboptimal approximations to optimal Bayesian track maintenance is concluded by selected tracking examples with real radar data (ground-based medium- and long-range air surveillance).* For the sake of simplicity, the discussion is confined to single-sited radar; sensor fusion aspects are thus excluded.

8.6.1 JPDA Formation Tracking

Figure 8.5a shows a characteristic detail taken from a set of raw data that was collected at the plot level from a typical 2D L-band, medium-range radar. The antenna is rotating with a scan period of 5 s; the corresponding pulse width is 1 μs, the beam-width 1.5°, and the detection probability about 80%. The data set recorded several formation flights consisting of targets that join, split off, and in the meantime continuously change from being resolved to unresolved and back again. In addition, crossing target situations occur. As in this example, the spatial false return density is low; PDA-type filtering is expected to be applicable.

The sensor model used for tracking is characterized by the following parameters: sensor resolution in range and azimuth α_r = 150 m, α_ϕ = 1.5°, measurement error σ_r = 30 m, σ_ϕ = .2°, measurement error for unresolved returns σ^u_r = 75 m, σ^u_ϕ = .75°, scan period T = 5 s, detection probability $P_D = P^u_D = .8$, and spatial false return density $\rho_F = 10^{-4}/\mathrm{km}^2$. The maneuvering behavior of the targets is modeled by the simple model discussed in Equation 8.6 with $r = 1$, $\theta = 60$ s, and $\Sigma = 10\ \mathrm{m/s}^2$.

---

* The raw data (plot level) were made available by the Defence Research Establishment, Ottawa (DREO) under the auspices of the Germany-Canada Memorandum of Understanding on Radar Technology. The author wishes to thank Martin Blanchette for his kind help and assistance.

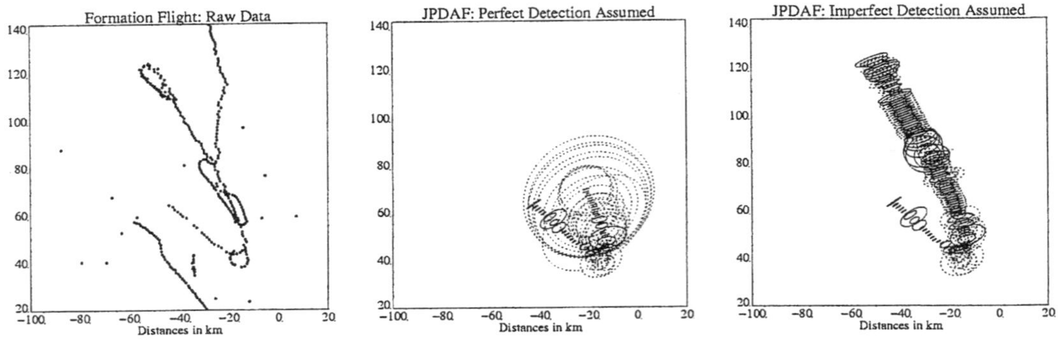

(a) (b) (c)

FIGURE 8.5 JPAF: small cluster tracking.

The example demonstrates that the limited sensor resolution must explicitly be taken into account as soon as the targets begin to move closely spaced. For that purpose, Figures 8.5b and 8.5c show the estimation error ellipses for two targets (solid, dotted) that result from JPDA filtering (Equations 8.76 and 8.79). While in Figure 8.5b wrongly perfect sensor resolution was assumed, i.e., $\alpha_r = \alpha_\phi = 0$, in Figure 8.5c the previous parameters were used. JPDA filtering without considering resolution phenomena evidently fails after a few frames, as indicated by diverging tracking error ellipses. This has a simple explanation: without modeling the finite sensor resolution, an actually produced unresolved plot can only be treated as a single target measurement along with a missed detection. In consequence, the related covariances increase in size. This effect is further intensified by subsequent unresolved returns. If hypotheses related to resolution conflicts are taken into account, however, the tracking remains stable. The error ellipses in Figure 8.5 have been enlarged to make their data-driven adaptivity more visible. The ellipses shrink, for instance, if both targets are actually resolved in a particular scan. The transient enlargement halfway during the formation flight is caused by a crossing target situation. The corresponding track for the third target involved is not displayed.

8.6.2 MHT and Retrodiction

Figure 8.6 shows details from a second data set accumulated over 240 and 290 scans, respectively. Besides many false alarms (probably due to ground clutter), the data recorded two pairs of interceptor aircraft performing an air combat exercise. Such air situations can be relevant in military applications and serve for the purpose of demonstration here.

The detection probability is fairly low (40 to 60%). In addition, rather long sequences of missed detections occur (fading phenomena). The clutter density is about .002/km^2. The data were collected from a rotating S-band, long-range radar. Range and azimuth information was used only; the elevation data were corrupted and thus ignored. The radar is characterized by the following parameters: scan period, 10 s; range accuracy, 350 ft; bearing accuracy, .22°; range resolution, 1600 ft; bearing resolution, 2.4°. For target dynamics, a worst/best-case model as used (i.e., $r = 2$, $\theta_i = 60$, 30 s, $\Sigma_i = 3$, 30 m/s^2, $i = 1, 2$ according to Equation 8.6). The model transition probabilities are chosen as follows: $p_{11} = .9$, $p_{22} = .8$, $p_{12} = 1 - p_{11}$, $p_{21} = 1 - p_{22}$ (see the example in Section 8.3.1). We considered model histories up to the length $n = 3$.

Information on the real target position is crucial for evaluating tracking filters. This is particularly true under conditions where even trained human observers seem unable to assess the filtering output. Here, a secondary radar was used. When primary and secondary radar produced identical information (within a certain correlation gate), the primary plots received an ID number. The target ID served for track assessment exclusively and was not used in the filtering algorithm. The verified primary plots are

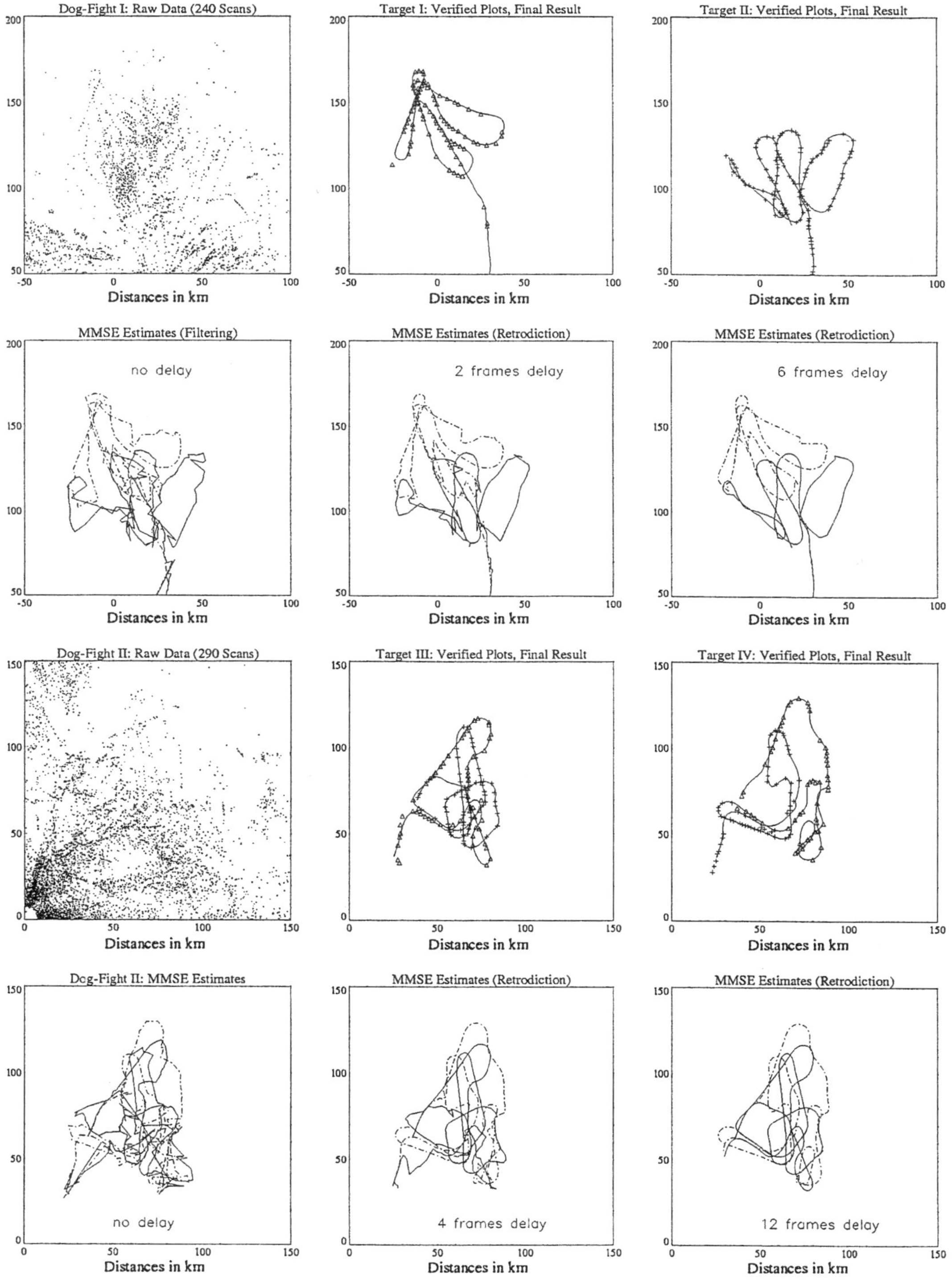

FIGURE 8.6 MHT-type filtering and retrodiction.

indicated by Δ, + in Figure 8.6 along with the final tracking result obtained after processing the raw data (i.e., MHT and retrodiction).

For both scenarios, the minimum mean squared error (MMSE) estimates of the targets' positions are displayed. They were obtained by MHT-type filtering (i.e., no delay). This result apparently seems to be of small value for assessing the air situation. The related variances (often very large) are not indicated. The high inaccuracy observed reflects the complex hypothesis tree resulting from ambiguous data interpretations. Multiple model filtering ($r = 2$) does outperform single-model filtering ($r = 1$, worst case) in some particular situations that are characterized by fewer data association conflicts and at least one non-maneuvering target. Besides those situations, however, the overall impression of the pure filtering result is similar for both cases. By experiment, we found that the consideration of a larger number of dynamics models (i.e., $r > 2$) or longer model histories (i.e., $n > 3$) does not lead to a further improvement in accuracy.

By using MHT retrodiction, even a delay of two frames significantly improves the filtering output. We displayed the MMSE estimates ($r = 1$, worst case) derived from $p(x_l|Z^k)$ (retrodiction loop, Equation 8.51) for $l = 2, 3, 6, 12$. A delay of 6 frames (1 min) provides easily interpretable trajectories, while the maximum gain by retrodiction is obtained after a 12-frame delay. Evidently, the final retrodiction results fit the verified primary plots very well. If IMM retrodiction is used, we essentially obtain the same final trajectory. However, in certain flight phases (not too much false returns, no maneuvers), it is obtained by a shorter delay (about one to three frames less). Under certain circumstances, accurate speed and heading information are available earlier than in a case of a single worst-case model. We also observed some improvement of the achievable results (both filtering and retrodiction) by using model histories longer than 1. If high quality estimates are desired, model histories of some length ($n = 3$) should be considered.

8.6.3. Summary

From our experiments with real radar data, we learned the following lessons:[43,44]

1. IMM-MHT is applicable in situations that are inaccessible to human radar operators.
2. The filter is rather robust and does not critically depend on modeling parameters (within certain limits).
3. Decisive are both its *multiple hypothesis* character allowing tentative alternatives in critical situations and the *qualitatively correct modeling* of all significant effects.
4. Unless properly handled, resolution conflicts can seriously destabilize tracking.
5. Mono-hypothesis approximations to MHT (such as JPDAF) are not applicable in scenarios as considered in Figure 8.6.
6. MHT is highly adaptive, developing its multiple hypothesis character only when needed.
7. Retrodiction provides unique and accurate results from ambiguous MHT output if a small time delay is accepted (some frames).
8. The maximum gain achievable by retrodiction is roughly the same for both worst-case modeling and IMM-MHT.
9. Algorithms employing multiple dynamics models are superior in that the time delays involved are shorter.
10. Finally, it seems notable that a very simplified modeling of the sensor, the target dynamics, and the environment may provide reasonable results if applied to real data.

References

1. Blackman, S., *Multiple-Target Tracking with Radar Applications*, Artech House, Norwood, MA, 1986.
2. Bar-Shalom, Y. and Fortmann, T.E., *Tracking and Data Association*, Academic Press, Orlando, 1988.

3. Bar-Shalom, Y. (Ed.), *Multitarget-Multisensor Tracking: Advanced Applications*, Vols. 1 and 2, Artech House, Dedham, MA, 1990, 1992.
4. Blom, H.A.P. and Bar-Shalom, Y., The Interacting Multiple Model Algorithm for Systems with Markovian Switching Coefficients, *IEEE Trans. Automatic Control*, 33(8), 1988.
5. Bar-Shalom, Y. and Li, X.-R., *Estimation and Tracking: Principles, Techniques, Software*, Artech House, Boston, MA, 1993.
6. Li, X.R., Hybrid Estimation Techniques, in *Control and Dynamic Systems* Vol. 76, C.T. Leondes, Ed., Academic Press, 1996.
7. Bar-Shalom, Y. and Li, X.-R., *Multitarget-Multisensor Tracking: Principles and Applications*, YBS Publishing, Storrs, CT, 1995.
8. Stone, L.D., Barlow, C.A., and Corwin, T.L., *Bayesian Multiple Target Tracking*, Artech House, Norwood, MA, 1999.
9. Blackman, S. and Populi, R., *Design and Analysis of Modern Tracking Systems*, Artech House, Norwood, MA, 1999.
10. Klemm, R., *Space-Time Adaptive Processing*, IEE Publishers, 1998.
11. Sittler, R.W., An Optimal Data Association Problem in Surveillance Theory, *IEEE Trans. Milit. Electronics*, 8(2), 1964.
12. Reid, D.B., An Algorithm for Tracking Multiple Targets, *IEEE Trans. Automatic Control*, 24(6), 1979.
13. Drummond, O.E., Multiple Sensor Tracking with Multiple Frame, Probabilistic Data Association, *SPIE Signal Data Process. Small Targets*, 2561, 322, 1995.
14. Mahler, R.P.S., Multisource, Multitarget Filtering: A Unified Approach, *SPIE Signal Data Process. Small Targets*, 3373, 296, 1998.
15. Titterington, D.M., Smith, A.F.M., and Makov, U.E., *Statistical Analysis of Finite Mixture Distributions*, John Wiley & Sons, New York, 1985.
16. Gelb, A. (Ed.), *Applied Optimal Estimation*, MIT Press, Cambridge, MA, 1974.
17. Koch, W., Retrodiction for Bayesian Multiple Hypothesis/Multiple Target Tracking in Densely Cluttered Environment, *SPIE Signal Data Process. Small Targets*, 2759, 429, 1996.
18. Koch, W., Fixed-Interval Retrodiction Approach to Bayesian IMM-MHT for Maneuvering Multiple Targets, *IEEE Trans. Aerosp. Electron. Syst.*, 36(1), 2000.
19. van Keuk, G. and Blackman, S.S., On Phased Array Radar Tracking and Parameter Control, *IEEE Trans. Aerosp. Electron. Syst.*, 29, 186, 1993.
20. van Keuk, G., Multiple Hypothesis Tracking with Electronically Scanned Radar, *IEEE Trans. Aerosp. Electron. Syst.*, 31, 916, 1995.
21. Kirubarajan, T., Bar-Shalom, Y., Blair, W.D., and Watson, G.A., IMMPDAF Solution to Benchmark for Radar Resource Allocation and Tracking Targets in the Presence of ECM, *IEEE Trans. Aerosp. Electron. Syst.*, 35(4), 1998.
22. Koch, W., On Adaptive Parameter Control for IMM-MHT Phased-Array Tracking, *SPIE Signal Data Process. Small Targets*, 3809, 1999.
23. van Keuk, G., Sequential Track Extraction, *IEEE Trans. Aerosp. Electron. Syst.*, 34, 1135, 1998.
24. Waltz, E. and Llinas, G., *Multisensor Data Fusion*, Artech House, Norwood, MA, 1990.
25. Hall, D.L., *Mathematical Techniques in Multisensor Data Fusion*, Artech House, Norwood, MA, 1992.
26. Goodman, I.R., Mahler, R., and Nguyen, H.T., *Mathematics of Data Fusion*, Kluwer, Dordrecht, 1997.
27. Liggins, M.E. et al., Distributed Fusion Architectures and Algorithms for Target Tracking, Special Issue on Sensor Fusion, *Proc. IEEE*, 85(1), 95, 1997.
28. Drummond, O.E. (Ed.), Introduction, *SPIE Signal Data Process. Small Targets*, 3373, 1998.
29. Bogler, Ph.L., *Radar Principles with Applications to Tracking Systems*, John Wiley & Sons, New York, 1990.
30. Li, X.R. and Bar-Shalom, Y., Detection Threshold Selection for Tracking Performance Optimization, *IEEE Trans. Aerosp. Electron. Syst.*, 30(3), 1994.

31. van Keuk, G., Multihypothesis Tracking Using Incoherent Signal-Strength Information, *IEEE Trans. Aerosp. Electron. Syst.*, 32(3), 1996.
32. Nickel, U., Radar Target Parameter Estimation with Antenna Arrays, in *Radar Array Processing*, S. Haykin, J. Litva, and T.J. Shepherd, Eds., (Springer Series in Information Sciences, Vol. 25, Springer-Verlag, New York, 1993, pp. 47–98.
33. Koch, W. and van Keuk, G., Multiple Hypothesis Track Maintenance with Possibly Unresolved Measurements, *IEEE Trans. Aerosp. Electron. Syst.*, 33(3), 1997.
34. Koch, W., On Bayesian MHT for Formations with Possibly Unresolved Measurements — Quantitative Results, *SPIE Data Process. Small Targets*, 3163, 417, 1997.
35. Chang, K.C. and Bar-Shalom, Y., Joint Probabilistic Data Association for Multitarget Tracking with Possibly Unresolved Measurements and Maneuvers, *IEEE Trans. Automatic Control*, 29(7), 1984.
36. Daum, F.E. and Fitzgerald, R.J., The Importance of Resolution in Multiple Target Tracking, *SPIE Signal Data Process. Small Targets*, 2235, 329, 1994.
37. Drummond, O.E., Target Tracking with Retrodicted Discrete Probabilities, *SPIE Signal Data Process. Small Targets*, 3163, 249, 1997.
38. Salmond, D.J., Mixture Reduction Algorithms for Target Tracking in Clutter, *SPIE Signal Data Process. Small Targets*, 1305, 435, 1990.
39. Pao, L.Y., Multisensor Multitarget Mixture Reduction Algorithms for Tracking, *J. Guidance Control Dynamics*, 17, 1205, 1994.
40. Danchick, R. and Newnam, G.E., A Fast Method for Finding the Exact N-Best Hypotheses for Multitarget Tracking, *IEEE Trans. Aerosp. Electron. Syst.*, 29(2), 1993.
41. Cox, I.J. and Miller, M.L., On Finding Ranked Assignments with Application to Multitarget Tracking and Motion Correspondence, *IEEE Trans. Aerosp. Electron. Syst.*, 31(1), 1995.
42. Helmick, R.E., Blair, W.D., and Hoffman, S.A., Fixed-Interval Smoothing for Markovian Switching Systems, *IEEE Trans. Inform. Theory,* 41(6), 1995.
43. Koch, W., Experimental Results on Bayesian MHT for Maneuvering Closely-Spaced Objects in a Densely Cluttered Environment, *RADAR 97, IEE International Radar Conference*, 729, 1997.
44. Koch, W., Generalized Smoothing for Multiple Model/Multiple Hypothesis Filtering: Experimental Results, ECC 99, European Control Conference, 31.8–3.9.1999, Karlsruhe, Germany.

9

Target Motion Analysis (TMA)

Klaus Becker
FGAN Research Institute for Communication, Information Processing, and Ergonomics (FKIE)

Abbreviations and Symbols

$(\ldots)^{\mathrm{T}}$	Tranpose
$(\dot{\ldots})$	Time derivative
$\lvert\ldots\rvert$	Norm of a vector
$(\ldots)'$	Quantity associated with $\mathbf{r}'_T(t)$
$\nabla_{\mathbf{a}}$	Gradient with respect to $\mathbf{a}$
0_n	n-Dimensional null vector
$\varnothing_n$	$n \times n$ null matrix
α_1, α_2	Weight coefficients in the performance index
$\alpha(t), \tilde{\alpha}(t), \mu(t)$	Scalar functions
β	Vector of exact bearings β_i
$\beta_i = \beta(t_i)$	Exact bearing at time t_i
β^m	Vector of measured bearings β_i^m
$\beta_i^m = \beta^m(t_i)$	Bearing measurement at time t_i
$\beta^m(k+1\|k)$	Measurement prediction
β, ϕ	Angles from the observer to the target
$\Phi(t, t_0)$	Transition matrix
κ	Constant determining V
λ_i	Eigenvalue of $\mathbf{J}$
ν	Doppler-shifted target signal frequency

0-8493-3691-0/01/$0.00+$.50

ν_0	Fixed target signal frequency
ψ^m	Generic measurement vector
σ_β^2	Variance of angle measurement error
ξ_i	Eigenvector of **J**
a	Frequency-TMA ambiguity parameter
$\mathbf{A}, \mathbf{A}_T, \mathbf{A}_{Ob}$	Coefficient matrix of a vector polynomial
$\mathbf{a}$	Generic state parameter
$\hat{\mathbf{a}}$	Estimate of **a**
$\Delta\mathbf{a}$	Estimation error of $\hat{\mathbf{a}}$
$\mathbf{a}_{xO}, \mathbf{a}_{yO}$	Cartesian components of the observer acceleration
ARM	Anti-radiation missile
AWACS	Airborne Warning and Control System
c	Signal velocity
$\mathbf{C}$	Covariance of $\Delta\mathbf{a}$
CR	Cramer Rao
CRLB	Cramer-Rao lower bound
$\mathbf{D}$	Orthogonal transformation
det(...)	Determinant
$E[\ldots]$	Expected value
EKF	Extended Kalman filter
$\mathbf{e}_r$	Unit vector in the direction of LOS
$\mathbf{f}[\mathbf{y}(t_0); t, t_0]$	Solution of the initial value problem in MP coordinates
$\mathbf{F}_k$	Jacobian of **f** at $\mathbf{y}(k\|k)$
$\mathbf{f}_x$	Transformation from MP to Cartesian state
$\mathbf{f}_y$	Transformation from Cartesian state to MP state
$\mathbf{G}$	Filter gain
$\mathbf{H}$	Measurement matrix
$\mathbf{I}_n$	$n \times n$ identity matrix
IR	Infrared
$\mathbf{J}$	Fisher information matrix
$\mathbf{J}_p, \mathbf{J}_v, \mathbf{J}_{pv}$	2×2 partitions of **J**
$J \ldots$	Performance index of the optimal control problem
k	Frequency-TMA ambiguity parameter
K	Number of bearing measurements
$L[\mathbf{x}(t_0); t, t_0]$	Solution of the initial value problem in Cartesian coordinates
LOS	Line-of-sight
MLE	Maximum likelihood estimation
MP	Modified polar
MPEKF	Modified polar EKF
$\mathbf{n}$	Vector of measurement errors n_i

$n_i = n(t_i)$	Measurement error at time t_i
$\mathbf{N}$	Covariance of the measurement vector $\mathbf{n}$
N	Degree of target/observer dynamics
$\mathbf{P}$	Projection operator onto the position space
$\mathbf{P}(k\|k)$	Covariance of $\mathbf{y}(k\|k)$
$\mathbf{P}(k+1\|k)$	Covariance of $\mathbf{y}(k+1\|k)$
$\mathfrak{p}_N$	Class of vector polynomials of a degree less than or equal to N
$p(\beta^m\|\mathbf{x}_{Tr})$	Conditional probability density function
$\mathbf{Q}$	Projection operator onto the velocity space
Q	Quadratic form of the bearing measurement errors
$\mathbf{r}$	Target position relative to the observer
r_x, r_y, r_z	Cartesian components of $\mathbf{r}$
$\mathbf{r}_{Ob}(t)$	Observer trajectory
$\mathbf{r}_T(t)$	Target trajectory
$\mathbf{r}_T^{(i)}$	i^{th} time derivative of $\mathbf{r}_T$
$\mathbf{r}'_T(t)$	Target trajectory leading to the same measurement history as $\mathbf{r}_T(t)$
$\mathbf{t}$	$(N+1)$-dimensional vector consisting of powers of $(t-t_0)$
t ...	Time variable
TMA	Target motion analysis
V	Volume of the concentration ellipsoid
V_n	Volume of the n-dimensional sphere
$\mathbf{w}_{Ob}(t, t_0)$	Non-inertial part of the four-dimensional Cartesian observer state
$\mathbf{x}$	Four-dimensional Cartesian relative state vector
$\mathbf{x}_T$	Four-dimensional Cartesian state of the non-accelerating target
$\mathbf{x}_{Tr} = \mathbf{x}_T(t_r)$	State parameter at t_r
$\mathbf{y}$	MP state vector
$\mathbf{y}(k\|k)$	Estimate of $\mathbf{y}$ at t_k given k measurements
$\mathbf{y}(k+1\|k)$	State prediction of $\mathbf{y}(k\|k)$

9.1 Introduction

This chapter deals with a class of tracking problems that uses passive sensors only. In solving tracking problems, active sensors certainly have an advantage over passive sensors. Nevertheless, passive sensors may be a prerequisite to some tracking solution concepts. This is the case, e.g., whenever active sensors are not a feasible solution from a technical or tactical point of view.

An important problem in passive target tracking is the target motion analysis (TMA) problem. The term TMA is normally used for the process of estimating the state of a radiating target from noisy measurements collected by a single passive observer. Typical applications can be found in passive sonar infrared (IR), or radar tracking systems.

A well-known example is the tracking of a ship by a submarine from passive sonar measurements. Here, the submarine uses a passive system because it does not want to reveal its presence by active transmissions. The measurements are noisy bearings from the radiating acoustic target, which are subsequently processed to obtain an estimate of the target state. In contrast to active sonar, range cannot be measured by the passive system.

Range measurements are also not available under jamming conditions. A fighter that wants to launch a missile against a jammer, however, needs some information on range and, therefore, has to estimate the jammer state. This constitutes an air warfare example of a TMA application.

Another important application is the Airborne Warning and Control System (AWACS), in which, among other things, passive angle measurements to radiating sources are processed for reconnaissance purposes.

TMA techniques are also applied in the field of missile guidance. Some modern anti-radiation missiles (ARM), e.g., exploit the radar transmissions for target state estimation in order to keep a lock-on in case the radar shuts down or operates intermittently for self-protection. Some other modern missiles are equipped with passive radar and/or IR receivers and estimate the target state in order to utilize optimal guidance procedures.

From the definition, passive target localization is a subset of TMA and involves the estimation of position only when the target is stationary. This has been studied in detail in the literature (see, e.g., references 1 and 2 and references cited therein). Conventional TMA, however, typically involves moving targets. This has also been the topic of much research in the literature, and since it will be the topic of this chapter also, the relevant literature will be cited later in a proper context in subsequent sections.

The TMA problem is characterized by the type of measurement extracted from the target signal. Different types induce qualitatively different estimation problems. This point is elaborated in Section 9.2.1, taking angle and frequency measurements as an example.

A peculiarity of passive tracking is the fact that the target may not be observable from the used measurement set. In Section 9.2.2, we separately discuss the observability conditions in the cases of angle and/or frequency measurements. Choosing a general but intuitive method, we can show that fundamental ambiguities exist if no restrictions are imposed on the target motion. It turns out that for the considered types of measurement, target modeling is a prerequisite to ambiguity resolution. Given the target model, the ambiguities can be resolved by suitable observer motions, which depend on the measurement set and the target model as well. For an illustration of this method, the observability conditions are discussed in the case of angle measurements and a three-dimensional Nth-order dynamics target model.

In Section 9.3, we develop steps toward a solution of the TMA problem. Since the steps are the same irrespective of the target model and the type of measurement, the discussion is restricted to the relatively simple, two-dimensional, constant target velocity, bearings-only TMA problem, which is defined in Section 9.3.1. One of the solution steps is a theoretical Cramer-Rao (CR) analysis of the TMA problem. This analysis provides a lower bound on the estimation accuracy, which is valid for any realized estimator, and thus reveals characteristic features of the estimation problem. In Section 9.3.2, the Cramer-Rao lower bound (CRLB) for the specified bearings-only TMA problem is calculated and discussed.

The development of powerful estimation algorithms is another necessary step in solving the TMA problem. In Section 9.3.3, some of the algorithms that have been devised to solve the bearings-only TMA problem are presented, and two of them that have been successfully applied are discussed in more detail, namely, the extended Kalman filter in modified polar coordinates and the maximum likelihood estimator (MLE).

If the observer is free to move, then a further solution step is required. The objective of this step is to find an observer motion that maximizes estimation accuracy. Useful optimality criteria for the resulting optimal control problem can be derived from the CRLB. Some of them are discussed in Section 9.3.4.

9.2 Features of the TMA Problem

9.2.1 Various Types of Measurements

Passive state estimation is based on exploiting the signals coming from the target. In doing so, crucial points are the type and quality of the measurements, which can be extracted from the signal, and their information content about the target. Generally, all measurements are suited for the process of state estimation, which are functions of the target state, e.g., as the angles from the observer to the target, the

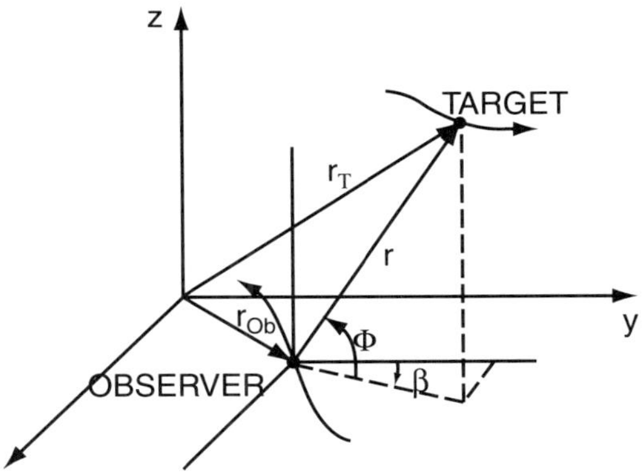

FIGURE 9.1 Target-observer geometry. (Reprinted by permission of IEEE © 1996.)

Doppler-shifted emitter frequencies, time delays, etc. A basic requirement, however, for successful estimation is that the final measurement set contains information on the full emitter state, i.e., that the noise-free measurements can be uniquely assigned to a target state. This point will be elaborated in detail in Section 9.2.2.

The final measurement set thereby used may consist of one single measurement type only, but it may be composed of various types as well. To give an example, let us consider the measurement type *angles* and *Doppler-shifted frequencies* in the three-dimensional scenario illustrated in Figure 9.1. Here, the target is moving along a trajectory $\mathbf{r}_T(t)$. Let us assume that the target emits a signal of constant but unknown frequency ν_0. The observer moving along another trajectory $\mathbf{r}_{Ob}(t)$ (assumed known) receives the signal and tries to estimate the target state from passive measurements of the line-of-sight (LOS) angles β, ϕ and/or of the Doppler-shifted frequency ν. This leads to the three alternative measurement sets:

$$\{\beta(t), \phi(t)\}, \{\nu(t)\}, \{\beta(t), \phi(t), \nu(t)\} \tag{9.1}$$

which are time histories of the LOS angles, the Doppler-shifted frequency, and the combined measurement data, respectively.

In the absence of noise and interference, the angle and frequency measurements satisfy the nonlinear relations

$$\beta(t) = \arctan\frac{r_x(t)}{r_y(t)} \tag{9.2}$$

$$\phi(t) = \arctan\frac{r_z(t)}{\sqrt{r_x^2(t) + r_y^2(t)}} \tag{9.3}$$

$$\nu(t) = \nu_0\left(1 - \frac{\dot{\mathbf{r}}(t)\cdot\mathbf{r}(t)}{cr(t)}\right) = \nu_0\left(1 - \frac{\dot{r}(t)}{c}\right) \tag{9.4}$$

whereas $\mathbf{r}(t) = \mathbf{r}_T(t) - \mathbf{r}_{Ob}(t) = (r_x(t), r_y(t), r_z(t))^T$ is the target position relative to the observer, $r = |\mathbf{r}|$ is its norm, and c is the signal velocity.

The estimation problem and its specific features change with the measurement set. There are target-observer scenarios in which the sensitivities of the measurement Equations 9.2 to 9.4 may be quite different. For example, whereas the orientation of the relative velocity $\dot{\mathbf{r}}$ has a strong effect on frequency, the angles are not affected. That means that maneuvers may lead to a large variation in frequency, while the angle variation is small and vice versa. Simple examples are weaving and spherical relative motions, respectively, as illustrated in Figures 9.2 and 9.3 for two-dimensional motions.

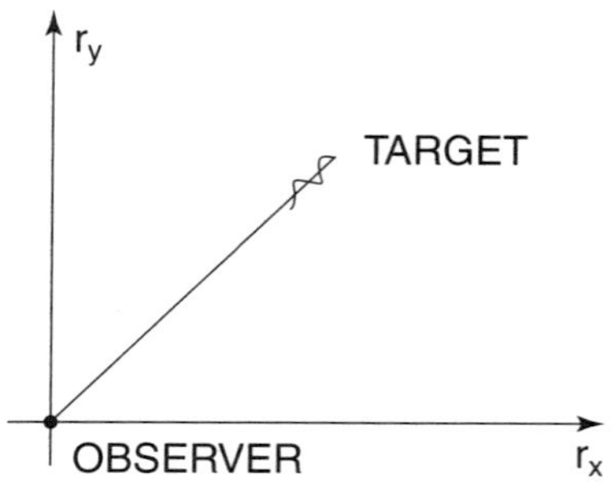

FIGURE 9.2 Weaving motion.

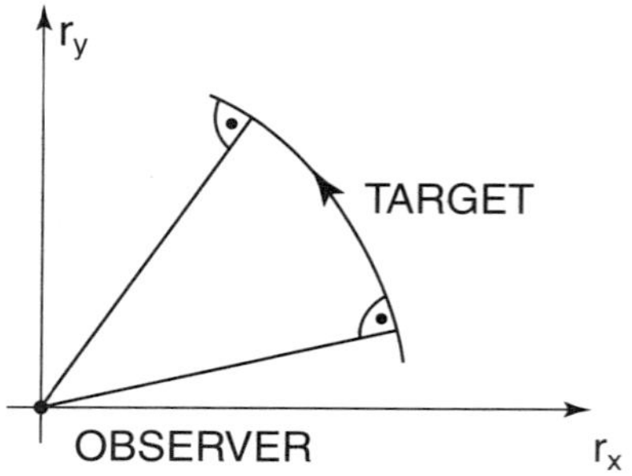

FIGURE 9.3 Spherical motion.

On the other hand, the different formulas may lead to measurement sets with qualitatively different information content. For example, in a straight-line collision course, the angle measurement set provides angle information only, and no information on closing velocity is contained. In contrast, the frequency measurement set provides closing velocity information only (ν_0 assumed known), whereas angle information is not contained. The qualitatively different information content leads to differently oriented estimation error ellipsoids. From this, a significant gain in estimation accuracy may result when the combined set of angle and frequency measurements is processed. This has been verified and discussed in detail in the stationary target case,[2] for example.

Other significant differences resulting from the distinct measurement sets in Equation 9.1 will become apparent in Section 9.2.2.

9.2.2 Observability

As indicated, a basic requirement for passive state estimation is the existence of a unique tracking solution. This leads to the question of observability.[3–11] We shall say that the target state $\mathbf{r}_T(t)$ is observable over the time interval $[t_0, t_f]$ if, and only if, it is uniquely determined by the measurements taken in that interval. Otherwise, it is considered unobservable. The ensuing discussion covers the three specified measurement sets in Equation 9.1. Since observability characteristics can be discerned under ideal conditions, only the noise-free measurements Equations 9.2 to 9.4 need to be considered.

To understand the observability problem emanating from the measurement sets in Equation 9.1 in full detail, it is important to know the transformations that leave the target trajectories compatible with the measurements when no restrictions are imposed on the target motion. Observability analysis for a particular target model then can be done in a systemic way by specializing the general set of compatible trajectories to the model under consideration.[9]

9.2.2.1 Fundamental Ambiguities

Provided that only the LOS angles are measured, it is obviously necessary and sufficient for trajectories $\mathbf{r}'_T(t)$ to lead to the same measurement history as $\mathbf{r}_T(t)$, if at all times the target lies on the LOS defined by the direction of $\mathbf{r}(t)$. Thus,

$$\mathbf{r}'_T(t) = \alpha(t)\mathbf{r}(t) + \mathbf{r}_{Ob}(t) \tag{9.5}$$

where $\alpha(t)$ is an arbitrary scalar function greater than zero, i.e., both $\mathbf{r}_T(t)$ and $\mathbf{r}'_T(t)$ have to be on the same "side" of the observer in order for the LOS angles to be the same. Since $\alpha(t)$ is an arbitrary function, the trajectory $\mathbf{r}'_T(t)$ may be of any shape.

If only frequencies are measured, the trajectories $\mathbf{r}'_T(t)$ and $\mathbf{r}_T(t)$ trivially will lead to the same measurement history if, and only if (cf. Equation 9.4),

$$\nu_0\left(1 - \frac{\dot{r}}{c}\right) = \nu'_0\left(1 - \frac{\dot{r}'}{c}\right) \tag{9.6}$$

where the prime signifies quantities associated with $\mathbf{r}'_T(t)$. Equation 9.6 may be rearranged as $\nu_0\,\dot{r} - \nu'_0\,\dot{r}' = c(\nu_0 - \nu'_0)$. Integrating from t_0 to t and rearranging, we obtain

$$r' = kr + a + c(1-k)(t-t_0) \tag{9.7}$$

with

$$k = \frac{\nu_0}{\nu'_0} \qquad a = r'_0 - kr_0 \tag{9.8}$$

Conversely, it is easy to show that Equation 9.6 follows from Equation 9.7. Thus, a target trajectory $\mathbf{r}'_T(t)$ cannot be distinguished from a trajectory $\mathbf{r}_T(t)$ by frequency measurements in Equation 9.4, if, and only if, the relative distance $r' = |\mathbf{r}'_T - \mathbf{r}_{OB}|$ satisfies Equation 9.7. This is obviously true if, and only if, the trajectories are of the form

$$\mathbf{r}'_T(t) = \mathbf{D}(t)\left[k + \frac{a + c(1-k)(t-t_0)}{r(t)}\right]\mathbf{r}(t) + \mathbf{r}_{Ob}(t) \tag{9.9}$$

where $\mathbf{D}(t)$ is an arbitrary orthogonal transformation.

If LOS angles and frequencies are measured, the compatible trajectories $\mathbf{r}'_T(t)$ necessarily must belong to a common subset of the trajectories defined by Equations 9.5 and 9.9. Since none of the sets in Equations 9.5 and 9.9 is a subset of the other, the intersection will remove some of the arbitrariness in Equations 9.5 and 9.9. Evidently, by the additional angle measurements, the orthogonal transformation in Equation 9.9 becomes the identity transformation, yielding trajectories which are contained in the set defined by Equation 9.5. Therefore, in the case of angle and frequency measurements necessary and sufficient for $\mathbf{r}'_T(t)$ to lead to the same measurement history as $\mathbf{r}_T(t)$ is that $\mathbf{r}'_T(t)$ can be written as

$$\mathbf{r}'_T(t) = \left[k + \frac{a + c(1-k)(t-t_0)}{r(t)}\right]\mathbf{r}(t) + \mathbf{r}_{Ob}(t) \tag{9.10}$$

with $k + (a + c(1-k)(t-t_0))/(r(t) > 0)$.

Equations 9.5, 9.9, and 9.10 show that the true target trajectory is always embedded in a continuum of compatible trajectories if no restrictions are imposed on the class of target motions. As an example, let us consider Equation 9.10. Even if the signal frequency is supposed to be known, i.e., $\nu'_0 = \nu_0$ or $k = 1$, there is still a continuum of compatible trajectories parameterized by $a = r'_0 - r_0$ which lead to the same angle and frequency measurement history, i.e.,

$$\mathbf{r}'_T(t) = \left[1 + \frac{a}{r(t)}\right]\mathbf{r}(t) + \mathbf{r}_{Ob}(t) \tag{9.11}$$

The ambiguity is illustrated in Figure 9.4 for a two-dimensional motion where, besides the true trajectory, some compatible trajectories have been depicted within the observer's coordinate system. According to Equation 9.11, the compatible trajectories result from the true trajectory by a shift of the trajectory points along the instantaneous LOS by an arbitrary but constant amount a.

The relation in Equation 9.11 constitutes a set of compatible trajectories if the measurement data are composed of angle and frequency measurements. In case the measurement set, however, consists of angles or frequencies only, Equations 9.5 and 9.9 introduce additional ambiguities into the curves of Figure 9.4. Whereas Equation 9.5 removes the restriction on shape, Equation 9.9, of course, leaves the shape unchanged, but the curves may be rotated, e.g., by an arbitrary angle.

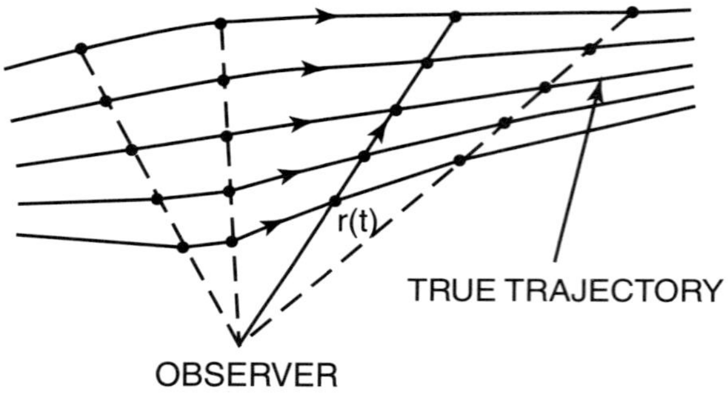

FIGURE 9.4 Compatible trajectories (Equation 9.11); parameter a. (Reprinted by permission of IEEE © 1996.)

The fundamental ambiguities of the target state exhibited in Equations 9.5, 9.9, and 9.10 in the class of unrestricted target motions clearly demonstrate that in TMA the question for observability is of no use unless specific target models are considered. Since the models impose restrictions on the analytical behavior of the target state, the fundamental ambiguities change into specific ones. These can be resolved in general by suitable observer maneuvers.

Note that in this way different target models may lead to completely different observability criteria. Consequently, in case of model mismatch, there may be situations where the target state is observable within the class of modeled motion, but is unobservable within the class of actual motion. The metric embedding of the observed target trajectory in the unobservable ones via Equations 9.5, 9.9, and 9.10 then gives rise to the fear that practical estimation algorithms may suggest convergence even in cases of divergence.

9.2.2.2 Nth-Order Dynamics Target

An illustrative example for a target model is the standard Nth-order dynamics model, i.e., the target motion $\mathbf{r}_T(t)$ can be described over the time interval $[t_0, t_f]$ as a vector polynomial of degree N:

$$\mathbf{r}_T(t) = \sum_{i=0}^{N} \frac{\mathbf{r}_T^{(i)}(t_0)}{i!}(t - t_0)^i \tag{9.12}$$

with $\mathbf{r}_T^{(i)}$ as the ith time derivative. Now, as a result of the model, $\mathbf{r}'_T(t)$ must also be a vector polynomial of degree N. If this condition can only be fulfilled by $\mathbf{r}'_T(t) \equiv \mathbf{r}_T(t)$, then the set of compatible trajectories shrinks to one single element and the state is observable, otherwise, it is not.

To simplify the subsequent discussion, we introduce the class of vector polynomials

$$\mathcal{P}_N = \left\{ \sum_{i=0}^{N} \mathrm{a}_i (t - t_0)^i = \mathbf{At} \right\} \tag{9.13}$$

where $\mathbf{A} = (\mathbf{a}_0, \ldots, \mathbf{a}_N)$ is a arbitrary $3 \times (N + 1)$ matrix of coefficients independent of t and $\boldsymbol{t} = (1, t - t_0, \ldots, (t - t_0)^N)^{\mathrm{T}}$. Obviously, $\mathbf{r}_T(t) \in \mathcal{P}_N$ and $\mathcal{P}_n \subset \mathcal{P}_N$ $(n < N)$, as can be easily verified by a suitable choice of $\mathbf{A}$.

For an illustration of how the method works, let us consider the measurement set $\{\beta(t), \phi(t)\}$. The other measurement sets in Equation 9.1 are discussed in detail in Reference 9. The true target trajectory is described by

$$\mathbf{r}_T(t) = \mathbf{r}(t) + \mathbf{r}_{Ob}(t) \tag{9.14}$$

Subtracting Equation 9.14 from Equation 9.5, the observer motion is eliminated: $\mathbf{r}'_T(t) - \mathbf{r}_T(t) = (\alpha(t) - 1)\mathbf{r}(t)$. Since $\mathbf{r}_T(t) \in \mathcal{P}_N$ and $\mathbf{r}'_T(t) \in \mathcal{P}_N$, the difference also must be in $\mathcal{P}_N$, i.e., $(\mathbf{r}'_T - \mathbf{r}_T) \in \mathcal{P}_N$. So $(\mathbf{r}'_T - \mathbf{r}_T)$ must be of the form $\mathbf{At}$ as in Equation 9.13. From this, it follows that $\mathbf{r}(t)$ can be represented as

$$\mathbf{r}(t) = \tilde{\alpha}(t)\mathbf{At} \tag{9.15}$$

where $\tilde{\alpha} = (\alpha - 1)^{-1}$. This is the necessary and sufficient condition for unobservability in the class of Nth-order dynamics targets.[7–9]

Examples

1. Obviously, the target cannot be observed from angle measurements in case of a constant LOS. This can easily be verified from Equation 9.15 by selecting $\tilde{\alpha}(t) = \mathbf{r}(t)$ and $\mathbf{A} = (\mathrm{e}_r, 0_3, \ldots, 0_3)$, where 0_3 is the three-dimensional null vector and e_r is the constant unit vector in the direction of LOS.
2. For $\tilde{\alpha}(t) \equiv 1$, Equation 9.15 reduces to $\mathbf{r}(t) = \mathbf{At}$, i.e., the target is unobservable, if $\mathbf{r}$ is a polynomial of a degree less than or equal to N. From this, it follows that the target can be observed only if the observer dynamics is of a higher degree than the target dynamics. This is reflected in the well-known fact that a constant velocity target cannot be observed by a stationary or constant velocity observer. The condition of a higher observer dynamics degree, however, is only a necessary but not a sufficient condition. The target may also be unobservable, even then when the observer motion is of a higher degree than the target motion. This is true, e.g., when the higher order terms result in observer displacements in the direction of the instantaneous LOS only, i.e., if the observer trajectory is of the form $\mathbf{r}_{Ob}(t) = \mathbf{A}_{Ob}\mathbf{t} + \mu(t)\mathbf{r}(t)$, where $\mathbf{A}_{Ob}\mathbf{t} \in \mathcal{P}_N$ and $\mu(t)\mathbf{r}(t)$ is the higher order terms observer motion.

Proof

Since $\mathbf{r}_T = \mathbf{A}_T\mathbf{t} \in \mathcal{P}_N$, we have $\mathbf{r} = (\mathbf{A}_T - \mathbf{A}_{Ob})\mathbf{t} - \mu\mathbf{r}$. From this follows $\mathbf{r} = (1 + \mu)^{-1} (\mathbf{A}_T - \mathbf{A}_{Ob})\mathbf{t}$, which is of the form of Equation 9.15.

This proof is illustrated in Figure 9.5 for the example of a constant velocity target.

9.3 Solution of the TMA Problem

9.3.1 Bearings-Only Tracking — A Typical TMA Problem

In the preceding section it has been shown that different types of measurement sets lead to estimation problems different in nature. In this section, we develop the steps toward a solution of the problem.

Since the steps are the same irrespective of the type of measurement, we restrict the discussion to the angles-only problem. Also, the three-dimensional problem is not particularly more enlightening than the two-dimensional one. Therefore, for computational ease, we assume that the target and the observer move in the (x,y)-plane of Figure 9.1. In doing so, the three-dimensional angles-only problem reduces

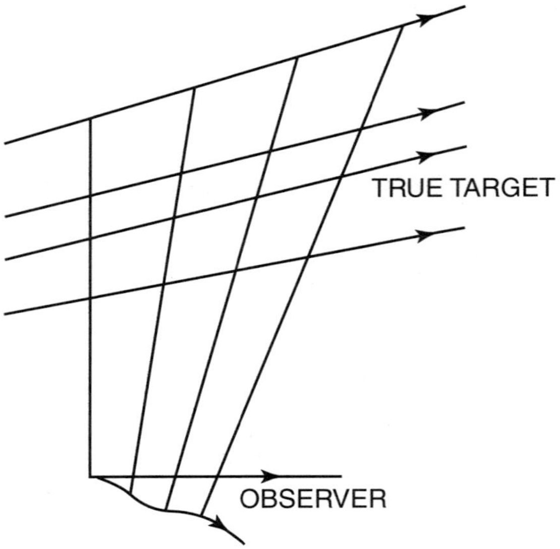

FIGURE 9.5 Constant velocity target ambiguities in case of a constant velocity observer trajectory or its displacement along instantaneous LOS.

to the two-dimensional bearings-only problem. For simplicity reasons, we further assume that the target moves along a straight line with constant velocity. Thus, the goal is to estimate the target position and velocity through noisy bearing measurements and knowledge of the observer motion. Note that since the bearings-only problem is a two-dimensional problem, the position vectors $\mathbf{r}_T(t)$, $\mathbf{r}_{Ob}(t)$, and $\mathbf{r}(t)$ denote two-dimensional vectors in the (x,y)-plane throughout this section.

Let

$$\mathbf{x}_T(t) = (r_{Tx}(t), r_{Ty}(t), \dot{r}_{Tz}, \dot{r}_{Ty})^{\mathrm{T}} \tag{9.16}$$

be the Cartesian four-dimensional, position-velocity vector of the nonaccelerating target. Then a mathematical model of the target can be specified via the linear state equation

$$\dot{\mathbf{x}}_T = \begin{pmatrix} \varnothing_2 & \mathbf{I}_2 \\ \varnothing_2 & \varnothing_2 \end{pmatrix} \mathbf{x}_T \tag{9.17}$$

where $\varnothing_2$ is the 2×2 null matrix, and $\mathbf{I}_2$ is the 2×2 identity matrix. Integrating Equation 9.17, we arrive at the solution

$$\mathbf{x}_T(t) = \Phi(t, t_0)\mathbf{x}_T(t_0) \tag{9.18}$$

Herein is

$$\Phi(t, t_0) = \begin{pmatrix} \mathbf{I}_2 & (t-t_0)\mathbf{I}_2 \\ \varnothing_2 & \mathbf{I}_2 \end{pmatrix} \tag{9.19}$$

the state transition matrix, which relates the state vector (Equation 9.16) at time t to the initial state $\mathbf{x}_T(t_0)$ at time t_0.

The bearing angle defined by the relation in Equation 9.2 is obviously a function of the unknown state vector in Equation 9.16, i.e., $\beta[\mathbf{x}_T(t)]$. Viewed by the observer, the bearing, however, is noise corrupted. Now, let a set of K bearing measurements β_i^m, $i = 1, \ldots, K$, be collected at various times t_i. Then in the presence of additive errors n_i, the measured bearings are given by

$$\beta_i^m = \beta_i + n_i \qquad i = 1, \ldots, K \tag{9.20}$$

where $\beta_i^m = \beta^m(t_i)$ and $\beta_i = \beta[\mathbf{x}_T(t_i)]$. Since $\mathbf{x}_T(t_i) = \Phi(t_i, t_r)\mathbf{x}_T(t_r)$ for an arbitrary reference time t_r (cf. Equation 9.18), the angles β_i can be considered as functions of t_i and of the constant state $\mathbf{x}_{Tr} = \mathbf{x}_T(t_r)$. Hence,

$$\beta[\mathbf{x}_T(t_i)] = \beta_i(\mathbf{x}_{Tr}) \tag{9.21}$$

Identifying β_i^m, β_i, and n_i with the components of vectors, Equation 9.20 can be organized in vector form as

$$\beta^m = \beta(\mathbf{x}_{Tr}) + \mathbf{n} \tag{9.22}$$

The measurement error $\mathbf{n}$ is a K-dimensional multivariate random vector with covariance matrix $\mathbf{N} = E[(\mathbf{n} - E[\mathbf{n}])(\mathbf{n} - E[\mathbf{n}])^{\mathrm{T}}]$, where E[...] denotes the expected value.

In what follows, we assume that the measurement error $\mathbf{n}$ can be adequately described by a multivariate, zero mean, normal probability distribution. Accordingly, the conditional density of β^m, given $\mathbf{x}_{Tr}$, is the multivariate normal density

$$p(\beta^m|\mathbf{x}_{Tr}) = \frac{1}{\sqrt{\det(2\pi\mathbf{N})}}\exp\left\{-\frac{1}{2}[\beta^m - \beta(\mathbf{x}_{Tr})]^{\mathrm{T}}\mathbf{N}^{-1}[\beta^m - \beta(\mathbf{x}_{Tr})]\right\} \tag{9.23}$$

where $\det(2\pi\mathbf{N})$ denotes the determinant of $2\pi\mathbf{N}$. In addition, we assume that the measurements are independent of each other and that the variances are independent of the measurement points, i.e.,

$$\mathbf{N} = \sigma_\beta^2\mathbf{I}_K \tag{9.24}$$

These are reasonable assumptions if the sampling frequency is not too high and if the measurement points are much closer to each other than to the target.

The outlined two-dimensional, single observer, bearings-only problem has been the topic of much research in the past.[12–26] The problem has been solved in detail in a variety of scenarios with different approaches. For the numerical solutions, pertinent plots, and tables, we refer to the cited literature. A discussion of these results is beyond the scope of this more tutorial chapter, which is confined to some theoretical fundamentals only that will be discussed in the subsequent description of the solution steps.

9.3.2 Step 1 — Cramer-Rao Lower Bound

In judging an estimation problem, it is important to know the maximum estimation accuracy that can be attained with the measurements. It is well known that the CRLB provides a powerful lower bound on the estimation accuracy. Moreover, since it is a lower bound for any estimator, its parameter dependence reveals characteristic features of the estimation problem. This and the fact that the optimal performance bound is usually used as an evaluation basis for specific estimation algorithms are the very reasons for the CR analysis to be a viable step in solving the TMA problem.

9.3.2.1 General Case

In its multi-dimensional form, the CR inequality states (see, e.g., References 27 and 28):

Let **a** be an unknown parameter vector of dimension n and let $\hat{\mathbf{a}}\,(\psi^m)$ denote some unbiased estimate of **a** based on the measurements ψ^m. Further, let **C** denote the covariance matrix of the estimation error $\Delta\mathbf{a} = \hat{\mathbf{a}}\,(\psi^m) - \mathbf{a}$ and **J** the Fisher information matrix

$$\mathbf{J} = E[\nabla_{\mathrm{a}} \ln p(\psi^m|\mathbf{a})(\nabla_{\mathrm{a}} \ln p(\psi^m|\mathbf{a}))^{\mathrm{T}}] \tag{9.25}$$

where ∇_{a} is the gradient with respect to **a**. Then the inequality

$$\mathbf{C} \geq \mathbf{J}^{-1} \tag{9.26}$$

holds, meaning $\mathbf{C} - \mathbf{J}^{-1}$ is positive semidefinite.

The relation in Equation 9.26 is the multi-dimensional CR inequality, and $\mathbf{J}^{-1}$ is the CRLB.

Geometrically, the covariance **C** can be visualized in the estimation error space by the concentration ellipsoid[28]

$$\Delta\mathbf{a}^{\mathrm{T}}\mathbf{C}^{-1}\Delta\mathbf{a} = \kappa \tag{9.27}$$

which has the volume

$$V = V_n\sqrt{\kappa^n \det\mathbf{C}} \tag{9.28}$$

Herein V_n is the volume of the n-dimensional unit hypersphere. In these terms, an equivalent formulation of the CR inequality reads:

For any unbiased estimate of **a**, the concentration ellipsoid (Equation 9.27) lies outside or on the bound ellipsoid (Figure 9.6) defined by

$$\Delta\mathbf{a}^{\mathrm{T}}\mathbf{J}\Delta\mathbf{a} = \kappa \tag{9.29}$$

The size and orientation of the ellipsoid (Equation 9.29) can be best described in terms of the eigenvalues and eigenvectors of the positive definite $n \times n$ matrix **J**. To this end, the eigenvalue problem $\mathbf{J}\xi_i = \lambda_i\xi_i$, $(i = 1, \ldots, n)$ has to be solved, where $\lambda_1, \ldots, \lambda_n$ are the eigenvalues of **J** and $\xi_1, \ldots, \xi_n$ are the corresponding eigenvectors. The mutually orthogonal eigenvectors ξ_i coincide with the principal axes of the bound ellipsoid, and the eigenvalues λ_i establish the lengths of the semiaxes via $\sqrt{\kappa/\lambda_i}$.

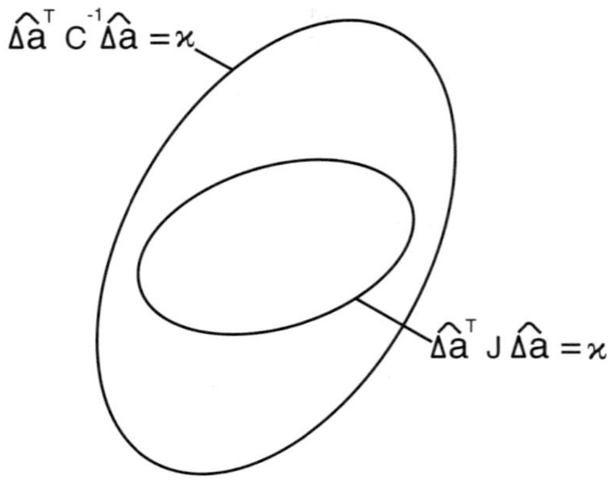

FIGURE 9.6 Geometrical visualization of the CRLB.

9.3.2.2 Bearings-Only Tracking

For bearings-only TMA considered in this section, we have $\mathbf{a} = \mathbf{x}_{Tr}$, $\psi^m = \beta^m$, and the Fisher information matrix is obtained from the conditional density in Equation 9.23. The gradient of the log-likelihood function is

$$\nabla \mathbf{x}_{Tr} \ln p(\beta^m | \mathbf{x}_{Tr}) = \frac{\partial \beta^{\mathrm{T}}}{\partial \mathbf{x}_{Tr}} \mathbf{N}^{-1} [\beta^m - \beta(\mathbf{x}_{Tr})] \tag{9.30}$$

where $\partial\beta/\partial\mathbf{x}_{Tr}$ is the Jacobian matrix of the vector function $\beta(\mathbf{x}_{Tr})$. Inserting Equation 9.30 in Equation 9.25 and taking the expectation, the Fisher information matrix at reference time t_r results in

$$\mathrm{J}(t_r) = \frac{\partial \beta^{\mathrm{T}}}{\partial \mathbf{x}_{Tr}} \mathbf{N}^{-1} \frac{\partial \beta}{\partial \mathbf{x}_{Tr}} \tag{9.31}$$

which enters into the bearings-only tracking bound ellipsoid (cf. Equation 9.29)

$$\Delta \mathbf{x}_{Tr}^{\mathrm{T}} \mathrm{J}(t_r) \Delta \mathbf{x}_{Tr} = \kappa \tag{9.32}$$

The ith row of $\partial\beta/\partial\mathbf{x}_{Tr}$ is calculated from Equation 9.21 by the chain rule

$$\begin{aligned} \frac{\partial \beta_i}{\partial \mathbf{x}_{Tr}} &= \frac{\partial \beta[\mathbf{x}(t_i)]}{\partial \mathbf{x}_T(t_i)} \frac{\partial \mathbf{x}_T(t_i)}{\partial \mathbf{x}_{Tr}} \\ &= \frac{1}{r_i}(\cos\beta_i, -\sin\beta_i, 0, 0)\Phi(t_i, t_r) \end{aligned} \tag{9.33}$$

Denoting $\mathbf{h}_i^{\mathrm{T}} = (\cos \beta_i, -\sin \beta_i, 0, 0)$ and considering that $\mathbf{N}$ is diagonal (cf. Equation 9.24), the matrix in Equation 9.31 takes the particular form

$$\mathrm{J}(t_r) = \frac{1}{\sigma_\beta^2} \sum_{i=1}^{K} \Phi^{\mathrm{T}}(t_i, t_r) \frac{\mathbf{h}_i \mathbf{h}_i^{\mathrm{T}}}{r_i^2} \Phi(t_i, t_r) \tag{9.34}$$

Since $\Phi(t_i, t_m) = \Phi(t_i, t_r)\, \Phi(t_r, t_m)$, it immediately follows that

$$\Phi^{\mathrm{T}}(t_r, t_m) \mathrm{J}(t_r) \Phi(t_r, t_m) = \mathrm{J}(t_m) \tag{9.35}$$

Hence, if we know the information matrix at the reference time t_r, we can calculate from it the information matrix at an arbitrary time by a pre- and post-multiplication with the transition matrix Φ.

According to Equation 9.19, det $\Phi = 1$. Consequently,

$$\det \mathrm{J}(t_r) = \det \mathrm{J}(t_m) \tag{9.36}$$

i.e., det J is invariant under shifts in the reference time. Since the inverse of det J is proportional to the volume of the bound ellipsoid (Equation 9.32) (cf. Equation 9.28), we infer that time translations, of course, may change the orientation and shape of the bound ellipsoid, while the total volume, however, is preserved.

Due to the four-dimensional state parameter vector x_{Tr}, the Fisher information matrix (Equation 9.31) is a 4×4 matrix. The associated bound ellipsoid (Equation 9.32) is a hyperellipsoid in the four-dimensional position-velocity space of mixed dimension and not amenable to direct geometrical interpretation. Only the projections of the hyperellipsoid onto the position and velocity subspaces, respectively, can be visualized and geometrically interpreted.

Generally, the projection of the bound ellipsoid onto a subspace of the state vector corresponds to the subspace estimation error bound when only the subspace components are classified as parameters of real interest and all others as nuisance parameters of no practical interest.[2] Thus, the projection of the hyperellipsoid onto the position subspace is the relevant estimation error bound of a location system, in which the position components of the target are estimated irrespective of its velocity. Obviously, the CRLB on this error is provided by the position space submatrix of $\mathbf{J}^{-1}$ (cf. Equation 9.26), which can be written as

$$\mathbf{P}\mathbf{J}^{-1}\mathbf{P}^{\mathsf{T}} \tag{9.37}$$

where $\mathbf{P}$ denotes the projection operator $\mathbf{P} = (\mathbf{I}_2|\varnothing_2)$ that projects the four-dimensional position-velocity space onto the two-dimensional position subspace. The corresponding estimation error bound (Equation 9.32) is an ellipse given by

$$(\mathbf{P}\Delta\mathbf{x}_{Tr})^{\mathsf{T}}(\mathbf{P}\mathbf{J}^{-1}\mathbf{P}^{\mathsf{T}})^{-1}\mathbf{P}\Delta\mathbf{x}_{Tr} = \kappa \tag{9.38}$$

The information matrix associated to the bound Equation 9.37 is its inverse. Naturally, the information content is affected by the presence of the unknown velocity covariances. Their effect can be calculated in an easy way from the partitioned form of Equation 9.31

$$\mathbf{J} = \left(\begin{array}{c|c} \mathbf{J}_p & \mathbf{J}_{pv} \\ \hline \mathbf{J}_{pv}^{\mathsf{T}} & \mathbf{J}_v \end{array}\right) \tag{9.39}$$

where $\mathbf{J}_p$ and $\mathbf{J}_v$ are the 2 × 2 Fisher information matrices in the case of known velocity and known position components, respectively, and $\mathbf{J}_{pv}$ is the 2 × 2 cross-term block matrix. Now, the difference $\mathbf{J}_p - (\mathbf{P}\mathbf{J}^{-1}\mathbf{P}^{\mathsf{T}})^{-1}$ is the information loss due to the presence of the unknown velocity parameters. Since $(\mathbf{P}\mathbf{J}^{-1}\mathbf{P}^{\mathsf{T}})^{-1}$ is the Schur complement of $\mathbf{J}_p$,[28] the information loss is the positive definite matrix

$$\mathbf{J}_p - (\mathbf{P}\mathbf{J}^{-1}\mathbf{P}^{\mathsf{T}})^{-1} = \mathbf{J}_{pv}\mathbf{J}_v^{-1}\mathbf{J}_{pv}^{\mathsf{T}} \tag{9.40}$$

A similar result holds for the projection of the hyperellipsoid onto the velocity subspace.

9.3.3 Step 2 — Estimation Algorithm

Good estimates of the target state are the ultimate goal in any TMA application. Consequently, powerful estimation algorithms are a very important step in solving the TMA problem.

In this section, the single observer bearings-only tracking problem is considered. Unfortunately, this type of estimation problem is not amenable to a simple solution. First, since observations and states are not linearly related, conventional linear analysis cannot be applied. Second, the measurements provide only directional information on the state and thereby introduce the question of system observability as an important issue into the estimation problem. As discussed in Section 9.2.2, restrictions must be imposed on the observer motion in order to warrant a unique tracking solution. In the considered scenario, e.g., the observer must execute at least one maneuver. But even then, quite realistic target-observer constellations often suffer from poor observability, in which cases TMA proves to be an ill-conditioned estimation problem.

Numerous estimators have been devised for bearings-only TMA. From the implementation viewpoint, the solutions can be loosely grouped into four categories: graphical methods, Kalman filters, explicit methods, and search methods. The graphical solutions are earlier approaches proposed for use without computers. Today, these methods are no longer of any practical value, and they are mentioned here only for completeness reasons. More important are the numerical estimators. Here, a multitude of different algorithms exist. However, in the discussion to follow, we concentrate only on some prominent representatives of each category. For a more complete list of algorithms, see, e.g., References 22 and 26.

- The Kalman filter solutions recursively update the target state estimates. Since the problem is nonlinear, they are basically extended Kalman filters (EKF). Depending on the choice of coordinates, linearizations are necessary for either the state or the measurement equation. Estimators of this kind are
  1. The Cartesian EKF:[12,13] The state equation is linear, whereas the measurement equation is nonlinear. Although the solution can be very good, in many instances, however, it exhibits divergence problems precipitated by a premature convergence of the covariance matrix prior to the first observer maneuver.
  2. The modified polar EKF (MPEKF):[14–16] The measurement equation is linear, but the state equation is nonlinear. The filter is free from premature covariance convergence, since the observable and the unobservable state components are automatically decoupled prior to the first observer maneuver. The performance of the filter is good if initialized properly.
- The explicit methods provide solutions in explicit form as a function of the measurements. The most well-known one is the pseudo-linear estimator (PLE).[17–23] In PLE, the nonlinear measurement equation is replaced with an equation of pseudo-measurements that are derived from the known observer state and the bearing measurements and are linearly related to the target state. Thus, the method of linear least squares can be applied for the explicit solution. Note that because of the linearity PLE may also be implemented in recursive form.[17] Geometrically, the solution minimizes the sum of squared cross-range errors perpendicular to the measured bearing. The PLE method avoids the instability problems of the Cartesian EKF. However, it has not gained widespread acceptance because the estimates are biased whenever noisy measurements are processed.[17,19] The bias can be severe, but modifications of PLE appear to have limited that problem.[22,23]
- The search methods are numerical optimization algorithms which iteratively improve the estimate. They are basically batch methods using the entire measurement set at every iteration. A prominent representative of this group is the MLE. The MLE is the best estimator,[20,22] but it is a computationally expensive solution.

From the performance aspect, MPEKF and MLE are both suitable candidates in a real-time system. They will be discussed in more detail in Sections 9.3.3.1 and 9.3.3.2.

9.3.3.1 The Modified Polar Extended Kalman Filter

It is well known that the system equations often acquire entirely dissimilar properties when expressed in different coordinate systems. In the same way, the performance of an estimator is affected by the choice of coordinates.

The Cartesian formulation of the EKF, though appealing from the computational point of view, was found to be unstable for bearings-only TMA. Therefore, research efforts have focused on alternative coordinates that reduce the problems inherent in Cartesian EKF. Apparently, the premature covariance convergence problem can be avoided by using coordinates whose observable and unobservable components are decoupled in the filter equations prior to the first observer maneuver. Coordinates with these attributes are, e.g., the modified polar (MP) coordinates.

The MP state vector is defined by

$$y(t) = \left(\frac{1}{r(t)}, \beta(t), \frac{\dot{r}(t)}{r(t)}, \dot{\beta}(t)\right)^{\mathrm{T}} \tag{9.41}$$

Herein the last three components are observable without an observer maneuver, while the first component becomes observable only after a maneuver.[14,15] Differentiating Equation 9.41 with respect to time and using the polar coordinate representation for the components of the Cartesian relative position vector $\mathbf{r} = (r\sin\beta, r\cos\beta)^T$, we obtain the state equation in MP coordinates

$$\dot{\mathbf{y}} = \begin{pmatrix} -y_1 y_3 \\ y_4 \\ y_4^2 - y_3^2 - y_1(a_{xO}\sin y_2 + a_{yO}\cos y_2) \\ -2y_3 y_4 - y_1(a_{xO}\cos y_2 - a_{yO}\sin y_2) \end{pmatrix} \tag{9.42}$$

where a_{xO}, a_{yO} are the Cartesian components of the observer acceleration $\ddot{\mathbf{r}}_{Ob}$. The general solution of the nonlinear differential Equation 9.42 can be expeditiously found by solving its Cartesian counterpart

$$\dot{\mathbf{x}} = \begin{pmatrix} \varnothing_2 & \mathbf{I}_2 \\ \varnothing_2 & \varnothing_2 \end{pmatrix} \mathbf{x} - \begin{pmatrix} 0_2 \\ \ddot{\mathbf{r}}_{Ob} \end{pmatrix} \tag{9.43}$$

of the relative state $\mathbf{x} = (\mathbf{r}^T, \dot{\mathbf{r}}^T)^T$ and by making use of the one-to-one transformations

$$\mathbf{x} = \mathbf{f}_x(\mathbf{y}) \qquad \mathbf{y} = \mathbf{f}_y(\mathbf{x}) \tag{9.44}$$

between the Cartesian and MP coordinate system.[15]

The Cartesian state Equation 9.43 can be readily solved. The solution is a linear function of the initial state $\mathbf{x}(t_0)$, and it is given by

$$\begin{aligned} \mathbf{x}(t) &= \Phi(t, t_0)\mathbf{x}(t_0) - \mathbf{w}_{Ob}(t, t_0) \\ &= L[\mathbf{x}(t_0); t, t_0] \end{aligned} \tag{9.45}$$

where

$$\mathbf{w}_{Ob}(t, t_0) = \begin{pmatrix} \int_{t_0}^{t} (t-\lambda) a_{xO}(\lambda)\, d\lambda \\ \int_{t_0}^{t} (t-\lambda) a_{yO}(\lambda)\, d\lambda \\ \int_{t_0}^{t} a_{xO}(\lambda)\, d\lambda \\ \int_{t_0}^{t} a_{yO}(\lambda)\, d\lambda \end{pmatrix} \tag{9.46}$$

and t_0 denotes an arbitrary fixed value of time.

Using the relations in Equations 9.44 and 9.45, the solution of Equation 9.42 can obviously be written in the form of three successive transformations. First, the initial MP state $y(t_0)$ is transformed via Equation 9.44 to its Cartesian counterpart $x(t_0)$, which then is linearly extrapolated via Equation 9.45. Finally, the result $x(t)$ is transformed via Equation 9.44 back to MP coordinates, giving the solution

$$\begin{aligned} \mathbf{y}(t) &= \mathbf{f}[\mathbf{y}(t_0); t, t_0] \\ &= \mathbf{f}_y[\Phi(t, t_0)\mathbf{f}_x[\mathbf{y}(t_0)] - \mathbf{w}_{Ob}(t, t_0)] \end{aligned} \tag{9.47}$$

which is a nonlinear function of $\mathbf{y}(t_0)$. The solution scheme is illustrated in Figure 9.7.

Since $\beta(t)$ is a component of the MP state vector (Equation 9.41), the nonlinear measurement Equation 9.20 becomes linear when expressed in MP coordinates, i.e.,

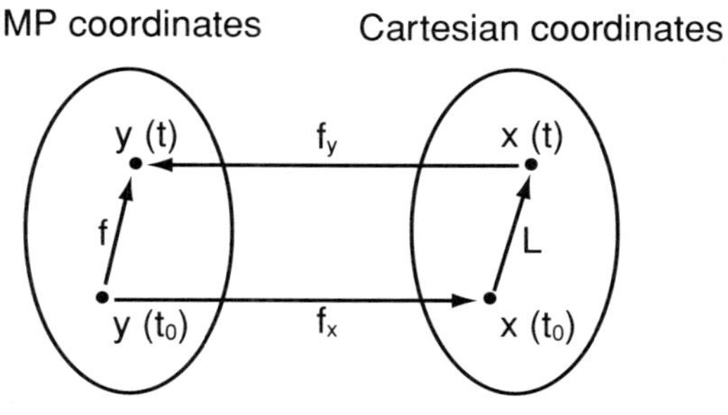

FIGURE 9.7 Solution scheme of the differential Equation 9.42.

$$\beta^m(t) = \mathbf{H}\mathbf{y}(t) + n(t) \tag{9.48}$$

where

$$\mathbf{H} = (0, 1, 0, 0) \tag{9.49}$$

is the measurement matrix.

Equations 9.47 and 9.48 are the MP system equations in continuous form. From these, the discrete time equations readily follow by assigning discrete values to t and t_0. In the simplified index-only time notation, we obtain

$$\mathbf{y}(k+1) = \mathbf{f}[\mathbf{y}(k); t_{k+1}, t_k] \tag{9.50}$$

$$\beta^m(k) = \mathbf{H}\mathbf{y}(k) + n(k) \tag{9.51}$$

Expanding the nonlinear state Equation 9.50 in a Taylor series around the latest estimate $\mathbf{y}(k|k)$ and neglecting higher than first-order terms, we arrive at the linearized state equation

$$\mathbf{y}(k+1) = \mathbf{f}[\mathbf{y}(k|k); t_{k+1}, t_k] + \mathbf{F}_k[\mathbf{y}(k) - \mathbf{y}(k|k)] \tag{9.52}$$

where $\mathbf{F}_k = \partial\mathbf{f}[\mathbf{y}(k|k); t_{k+1}, t_k]/\partial\mathbf{y}(k|k)$ is the Jacobian of the vector function $\mathbf{f}$. Straightforward application of the Kalman filter to the linearized system Equations 9.52 and 9.51 results in the MPEKF. One cycle of the filter is presented in Figure 9.8.

Theoretical and experimental findings have conclusively shown that the performance of the MPEKF is good provided that it is initialized by a proper choice of the initial state estimate $\mathbf{y}(0|0)$ and the initial state covariance matrix $\mathbf{P}(0|0)$.

9.3.3.2 The Maximum Likelihood Estimator

Target tracking becomes increasingly difficult in a scenario of poor observability, for example, as in a long-range scenario. The linearizations at each update in the recursive Kalman filter algorithms may then lead to significant errors in this ill-conditioned estimation problem, whereas the MLE, as a batch algorithm, avoids these linearization error effects.

For the problem specified in Section 9.3.1, the ML estimate is that value of $\mathbf{x}_{Tr}$ which maximizes Equation 9.23. Thus, the ML estimate minimizes the quadratic form

$$\begin{aligned} Q(\mathbf{x}_{Tr}) &= [\beta^m - \beta(\mathbf{x}_{Tr})]^{\mathrm{T}}\mathbf{N}^{-1}[\beta^m - \beta(\mathbf{x}_{Tr})] \\ &= \frac{1}{\sigma_\beta^2}\sum_{i=1}^{K}[\beta_i^m - \beta_i(\mathbf{x}_{Tr})]^2 \end{aligned} \tag{9.53}$$

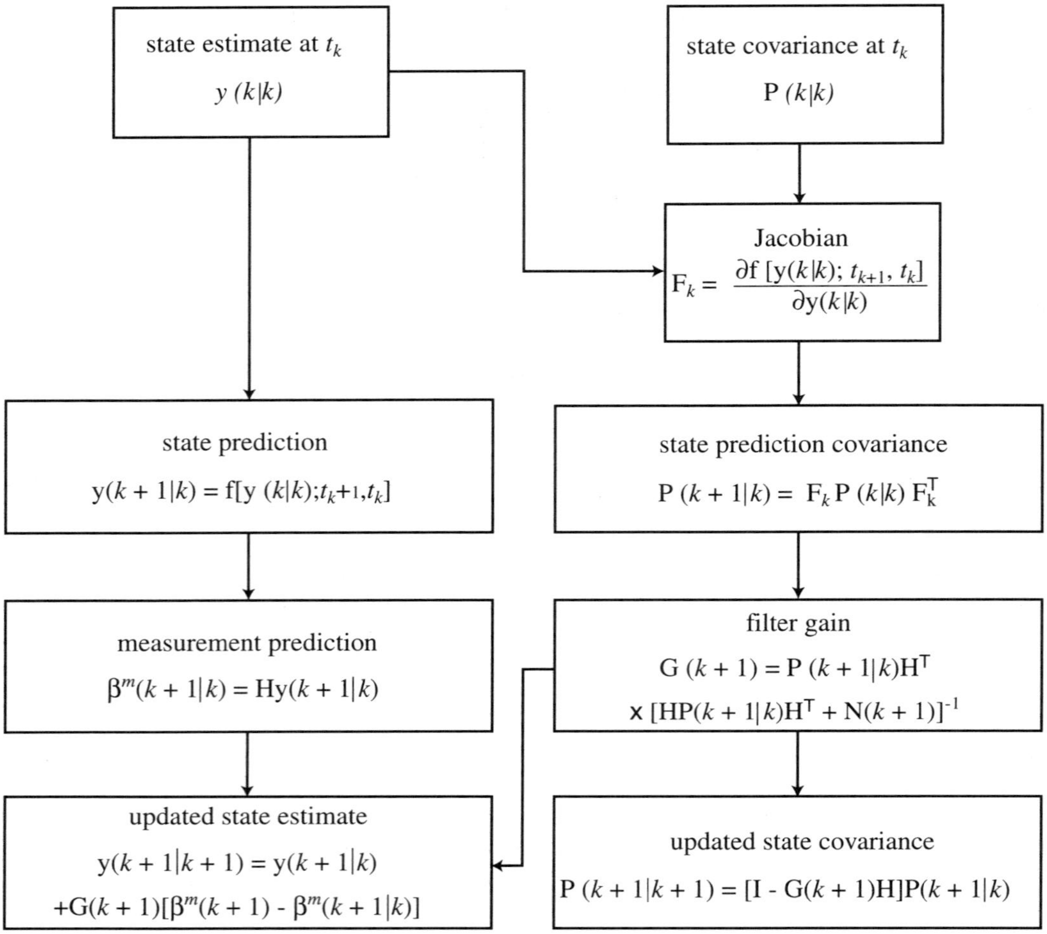

FIGURE 9.8 Flowchart of MPEKF (one cycle).

i.e., in case of a normal distribution the MLE and the least squares estimator with N^{-1} as weight matrix are identical.

Given a proper measurement set, the performance of the MLE is usually very good. In the single observer bearings-only case, the MLE is known to be asymptotically efficient; but analytical closed-form solutions do not exist. To find the minimum of Equation 9.53, a numerical iterative search algorithm is needed. Consequently, application of MLE suffers from the same problems as the numerical algorithms. Suitable optimization algorithms are e.g., the Gauss-Newton and the Levenberg-Marquardt method.[29,30] These methods are easy to implement, but to avoid a possibly large number of time consuming iteration steps, good starting values are usually necessary.

9.3.4 Step 3 — Optimal Observer Motion

In TMA the estimation accuracy highly depends on the target-observer geometry. By changing the geometry, estimation accuracy can increase or decrease, as the case may be. In an application, the target motion is given and cannot be changed; but, if the observer is free to move, the target-observer geometry can be changed by observer maneuvers. This leads to the final step in solving the TMA problem: *Find an optimal observer maneuver which creates a geometry that maximizes estimation accuracy.*

From the discussion in Section 9.3.2, optimality is properly defined by the CRLB. Like the bound, optimality criteria based upon CRLB are independent of the particular estimation algorithm. In the literature, several criteria have been applied. One of them,[31,32] e.g., is maximizing

$$J_{det} = \det \mathbf{J} \tag{9.54}$$

This is an intuitively appealing approach, since it minimizes the volume of the bound ellipsoid (Equation 9.32) (cf. Equation 9.28). The determinant criterion (Equation 9.54) has, on the one hand, the advantages that it is sensitive to the precision of all target state components and that it is independent of the reference time (cf. Equation 9.36). On the other hand, however, it may be disadvantageous, because using volume as a performance criterion may favor solutions with highly eccentric ellipsoids and with large uncertainties in target state components of practical interest.

The eccentricity problem can be alleviated, e.g., by choosing an optimization criterion that minimizes the trace of a weighted sum of the position and velocity lower bounds,[33] i.e.,

$$J_{trace} = \alpha_1 tr\{\mathbf{P}\mathbf{J}^{-1}\mathbf{P}^{\mathrm{T}}\} + \alpha_2 tr\{\mathbf{Q}\mathbf{J}^{-1}\mathbf{Q}^{\mathrm{T}}\} \tag{9.55}$$

Herein $\mathbf{P}$ and $\mathbf{Q}$ are the projection operators onto the position and velocity space, respectively, and α_1, α_2 are weight coefficients to decide whether position or velocity estimation is more important. The trace of the position lower bound $\mathbf{P}\mathbf{J}^{-1}\mathbf{P}^{\mathrm{T}}$ is the sum of its eigenvalues, which according to Section 9.3.2.7 are proportional to the square of the semiaxes of the bound ellipse (Equation 9.38). The same is true for the trace of the velocity lower bound $\mathbf{Q}\mathbf{J}^{-1}\mathbf{Q}^{\mathrm{T}}$ and the bound ellipse in the velocity subspace. Therefore, the optimality criterion (Equation 9.55) penalizes solutions with large semiaxes, and by this, it reduces the possibly highly unbalanced estimation uncertainties resulting from the determinant criterion.

In a localization system the position components of the target state are of main interest. To improve the estimation accuracy of these components, optimality criteria are needed that penalize the position errors above all. To this end, criteria have been proposed that minimize the trace of the position lower bound[34]

$$J_{plb} = tr\{\mathbf{P}\mathbf{J}^{-1}\mathbf{P}^{\mathrm{T}}\} \tag{9.56}$$

or the range variance[22,32]

$$J_r = \frac{\partial r}{\partial \mathbf{x}_{Tr}} \mathbf{J}^{-1} \left(\frac{\partial r}{\partial \mathbf{x}_{Tr}}\right)^{\mathrm{T}} \tag{9.57}$$

respectively.

In finding the optimal observer trajectories for the various optimality criteria, Quasi-Newton optimization procedures[31,33,34] and optimal control theory[32] have been applied. The specific characteristics of the solutions prove to be different for the individual criteria. For the details, we refer to the literature. All solutions, however, involve a trade-off between increasing bearing rate and decreasing range.

9.4 Conclusion

Different types of measurements differ in their functional relation to the target state. As shown, this leads to basically different estimation problems. The differences are reflected in the observability conditions and the estimation accuracy as well. From this, we conclude that a measurement set consisting of different measurement types will result in less restrictive observability conditions and in an improvement of estimation accuracy. Depending on the qualitative differences of the estimates pertinent to the different measurement types, the improvement may be substantial and may justify increased measurement equipment complexity, all the more so as the computational complexity is not severely affected.

In the preceding section, the TMA problem is solved in three consecutive steps. The main step is the development of a powerful algorithm that effectively estimates the target state from the noisy measurements collected by the observer. The performance of the realized algorithm is usually assessed in a simulation. In doing so, typical performance criteria of the estimator are unbiasedness and estimation accuracy. Needed computer power is only a minor point in this context, due to the continually ongoing rapid computer development. Whereas the question of unbiasedness can directly be answered by inspection, the question whether the estimation accuracy is good can only be conclusively answered by a comparison with other estimators or even better by a comparison with an estimation error bound that is independent of any specific estimation algorithm. A bound with this attribute is the CRLB. Its calculation is based on a theoretical measurement model, and because it is a function of the system parameters, an analysis of its parametric dependences reveals characteristic features of the TMA problem under consideration. Since the CRLB on the one hand gives a deep insight into the properties of the estimation problem and on the other hand is used to evaluate particular estimators, the calculation and analysis of the CRLB should always be the first step in solving the TMA problem.

In an application the user should always strive to get the maximum of attainable estimation accuracy. Estimation accuracy can be influenced by the user first via the used algorithm and second, since it is a function of the target-observer geometry, via observer motions as well. The improvement in estimation accuracy via observer motions may be substantial even then, when the estimation accuracy of the used estimator is generally very close to the CRLB. Therefore, if the observer is free to move, a final third step is necessary in the TMA solution process. This step requires the solution of an optimal control problem, in which the observer motion is controlled to achieve the maximum of attainable estimation accuracy. Suitable optimality criteria in the solution of the problem can be derived, e.g., from the CRLB established in the first solution step.

For a better understanding, the individual solution steps have exemplarily been discussed in the relatively simple, constant target velocity, bearings-only TMA problem. Naturally, the complexity of the solution steps increases with the complexity of the estimation problem. For example, the three-dimensional angles-only TMA problem leads to far more complex equations than its two-dimensional bearings-only counterpart. The same is true if the observer has no perfect knowledge of its own state or if the target is allowed to maneuver. But, nevertheless, whatever cases are considered, the solution steps of the pertinent TMA problem are always the same. They differ only in the level of complexity.

References

1. Constantine, J., Airborne passive emitter location (APEL), Proceedings of the Symposium on Electronic Warfare Technology, Brussels, November 1993.
2. Becker, K., An efficient method of passive emitter location, *IEEE Trans. Aerosp. Electron. Syst.*, AES-28, 1091–1004, 1992.
3. Nardone, S.C. and Aidala, V.J., Observability criteria for bearings-only target motion analysis, *IEEE Trans. Aerosp. Electron. Syst.*, AES-17, 162–166, 1981.
4. Shensa, M.J., On the uniqueness of Doppler tracking, *J. Acoust. Soc. Am.*, 70, 1062–1064, 1981.
5. Hammel, S.E. and Aidala, V.J., Observability requirements for three-dimensional tracking via angle measurements, *IEEE Trans. Aerosp. Electron. Syst.*, AES-21, 200–207, 1985.
6. Payne, A.N., Observability conditions for angles-only tracking, in Proceedings of the 22nd Asilomar Conference on Signals, Systems, and Computers, pp. 451–457, October 1988.
7. Fogel, E. and Gavish, M., Nth-order dynamics target observability from angle measurements, *IEEE Trans. Aerosp. Electron. Syst.*, AES-24, 305–308, 1988.
8. Becker, K., Simple linear theory approach to TMA observability, *IEEE Trans. Aerosp. Electron. Syst.*, AES-29, 575–578, 1993.
9. Becker, K., A general approach to TMA observability from angle and frequency measurements, *IEEE Trans. Aerosp. Electron. Syst.*, AES-32, 487–494, 1996.

10. Jauffret, C. and Pillon, D., Observability in passive target motion analysis, *IEEE Trans. Aerosp. Electron. Syst.*, AES-32, 1290–1300, 1996.
11. Song, T.L., Observability of target tracking with bearings-only measurements, *IEEE Trans. Aerosp. Electron. Syst.*, AES-32, 1468–1472, 1996.
12. Kolb, R.C. and Hollister, F.H., Bearings-only target motion estimation, in Proceedings of the 1st Asilomar Conference on Circuits and Systems, pp. 935–946, 1967.
13. Aidala, V.J., Kalman filter behavior in bearings-only tracking applications, *IEEE Trans. Aerosp. Electron. Syst.*, AES-15, 29–39, 1979.
14. Hoelzer, H.D., Johnson, G.W., and Cohen, A.O., Modified Polar Coordinates — The Key to Well Behaved Bearings-Only Ranging, IBM Rep. 78-M19-0001A, IBM Shipboard and Defense Systems, Manassas, VA, 1978.
15. Aidala, V.J. and Hammel, S.E., Utilization of modified polar coordinates for bearings-only tracking, *IEEE Trans. Automat. Control*, AC-28, 283–294, 1983.
16. Van Huyssteen, D. and Farooq, M., Performance analysis of bearings-only tracking algorithm, in SPIE Conference on Acquisition, Tracking, and Pointing XII, Orlando, FL, pp. 139–149, 1998.
17. Lindgren, A.G. and Gong, K.F., Position and velocity estimation via bearing observations, *IEEE Trans. Aerosp. Electron. Syst.*, AES-14, 564–577, 1978.
18. Lindgren, A.G. and Gong, K.F., Properties of a nonlinear estimator for determining position and velocity from angle-of-arrival measurements, in Proceedings of the 14th Asilomar Conference on Circuits, Systems, and Computers, pp. 394–401, November 1980.
19. Aidala, V.J. and Nardone, S.C., Biased estimation properties of the pseudolinear tracking filter, *IEEE Trans. Aerosp. Electron. Syst.*, AES-18, 432–441, 1982.
20. Nardone, S.C., Lindgren, A.G., and Gong, K.F., Fundamental properties and performance of conventional bearings-only target motion analysis, *IEEE Trans. Automat. Control*, AC-29, 775–787, 1984.
21. Holst, J., A Note on a Least Squares Method for Bearings-Only Tracking, Tech. Rep. TFMS-3047, Department of Math. Stat., Lund Institute of Technology, Sweden, 1988.
22. Holtsberg, A., A Statistical Analysis of Bearing-Only Tracking, Ph.D. dissertation, Department of Math. Stat., Lund Institute of Technology, Sweden, 1992.
23. Holtsberg, A. and Holst, J., A nearly unbiased inherently stable bearings-only tracker, *IEEE J. Oceanic Eng.*, 18, 138–141, 1993.
24. Petridis, V., A method for bearings-only velocity and position estimation, *IEEE Trans. Automat. Control*, AC-26, 488–493, 1981.
25. Pham, D.T., Some quick and efficient methods for bearings-only target motion analysis, *IEEE Trans. Signal Process.*, 41, 2737–2751, 1993.
26. Nardone, S.C. and Graham, M.L., A closed-form solution to bearings-only target motion analysis, *IEEE J. Oceanic Eng.*, 22, 168–178, 1997.
27. Van Trees, H.L., *Detection, Estimation, and Modulation Theory, Part 1*, Wiley, New York, 1968.
28. Scharf, L.L., *Statistical Signal Processing*, Addison Wesley, New York, 1991.
29. Gill, P.E., Murray, W., and Wright, M.H., *Practical Optimization*, Academic Press, New York, 1981.
30. Dennis, J.E. and Schnabel, R.B., *Numerical Methods for Unconstrained Optimization and Nonlinear Equations*, Prentice-Hall, Englewood Cliffs, NJ, 1983.
31. Hammel, S.E., Optimal Observer Motion for Bearings-Only Localization and Tracking, Ph.D. thesis, University of Rhode Island, Kingston.
32. Passerieux, J.M. and van Cappel, D., Optimal observer maneuver for bearings-only tracking, *IEEE Trans. Aerosp. Electron. Syst.*, AES-34, 777–788, 1998.
33. Helferty, J.P. and Mudgett, D.R., Optimal observer trajectories for bearings-only tracking by minimizing the trace of the Cramer-Rao lower bound, in Proceedings of the 32nd Conference on Decision and Control, San Antonio, TX, pp. 936–939, December 1993.
34. Helferty, J.P., Mudgett, D.R., and Dzielski, J.E., Trajectory optimization for minimum range error in bearings-only source localization, in OCEAN'93, Engineering in Harmony with Ocean Proceedings, Victoria, BC, Canada, pp. 229–234, October 1993.

II

Sonar and Radar System Applications

10

Sonar Systems*

G. Clifford Carter
Naval Undersea Warfare Center

Sanjay K. Mehta
Naval Undersea Warfare Center

Bernard E. McTaggart
Naval Undersea Warfare Center (retired)

Defining Terms

SONAR: Acronym for "Sound Navigation and Ranging," adapted in the 1940s, involves the use of sound to explore the ocean and underwater objects.

Array gain: A measure of how well an array discriminates signals in the presence of noise as a result of beamforming. It is represented as a ratio of the array SNR divided by the SNR of an individual omnidirectional hydrophone.

Beamformer: A process in which outputs from individual hydrophone sensors of an array are coherently combined by delaying and summing the outputs to provide enhanced SNR.

Coherence: A normalized cross-spectral density function that is a measure of the similarity of received signals and noise between any sensors of an array.

* This work represents a revised version from the CRC Press *Electrical Engineering Handbook*, R. Dorf, Ed., 1993 and from the CRC Press *Electronics Handbook*, J. C. Whitaker, Ed., 1996.

0-8493-3691-0/01/$0.00+$.50

Decibels (dB): Logarithmic scale representing the ratio of two quantities given as 10 $\log_{10}(P_1/P_0)$ for power level ratios and 20 $\log_{10}(V_1/V_0)$ for acoustic pressure or voltage ratios. A standard reference pressure or intensity level in SI units is equal to 1 μPa (1 Pa = 1 N/m^2 = 10 dyn/cm^2).

Doppler shift: Shift in frequency of transmitted signal due to the relative motion between the source and object.

Figure of merit/sonar equation: Performance evaluation measure for the various target and equipment parameters of a sonar system. It is a subset of the broader sonar performance given by the sonar equations, which includes reverberation effects.

Receiver operating characteristics (ROC) curves: Plots of the probability of detection (likelihood of detecting the object when the object is present) vs. the probability of false alarm (likelihood of detecting the object when the object is not present) for a particular processing system.

Reverberation/clutter: Inhomogeneities, such as dust, sea organisms, schools of fish, and sea mounds on the bottom of the sea, which form mass density discontinuities in the ocean medium. When an acoustic wave strikes these inhomogeneities, some of the acoustic energy is reflected and reradiated. The sum total of all such reradiations is called reverberation. Reverberation is present only in active sonar, and in the case where the object echoes are completely masked by reverberation, the sonar system is said to be "reverberation limited."

Sonar hydrophone: Receiving sensors that convert sound energy into electrical or optical energy (analogous to underwater microphones).

Sonar projector: A transmitting source that converts electrical energy into sound energy.

Sound velocity profile (SVP): Description of the speed of sound in water as a function of water depth.

SNR: The signal-to-noise (power) ratios, usually measured in decibels (dB).

Time delay: The time (delay) difference in seconds from when an acoustic wavefront impinges on one hydrophone or receiver until it strikes another.

10.1 Introduction

10.1.1 What Is a Sonar System?

A system that uses acoustic signals propagated through the water to detect, classify, and localize underwater objects is referred to as a sonar system.* Sonars are typically on surface ships (including fishing vessels), submarines, autonomous underwater vehicles (including torpedoes), and aircraft (typically helicopters). A sonar system generally consists of four major components. The first component is a transmitter that (radiates or) transmits a signal through the water. For active sonars, the system transmits energy to be reflected off objects. In contrast, for passive sonar, the object itself is the radiator of acoustic energy. The second component is a receiving array of hydrophones that receives the transmitted (or radiated) signal which has been degraded due to underwater propagation effects, ambient noise, or interference from other signal sources such as surface war ships and fishing vessels. A signal processing subsystem which then processes the received signals to minimize the degradation effects and to maximize the detection and classification capability of the signal is the third component. The fourth component consists of the various displays that aid machine or human operators to detect, classify, and localize sonar signals.

10.1.2 Why Exploit Sound for Underwater Applications?

Acoustic signals propagate better underwater than do other types of energy. For example, both light and radio waves (used for satellite or in-air communications) are attenuated to a far greater degree underwater than are sound waves.** For this reason, sound waves have generally been used to extract information about underwater objects. A typical sonar signal processing scenario is shown in Figure 10.1.

* Also known as sonar.

** There has been some limited success propagating blue-green laser energy in clear water.

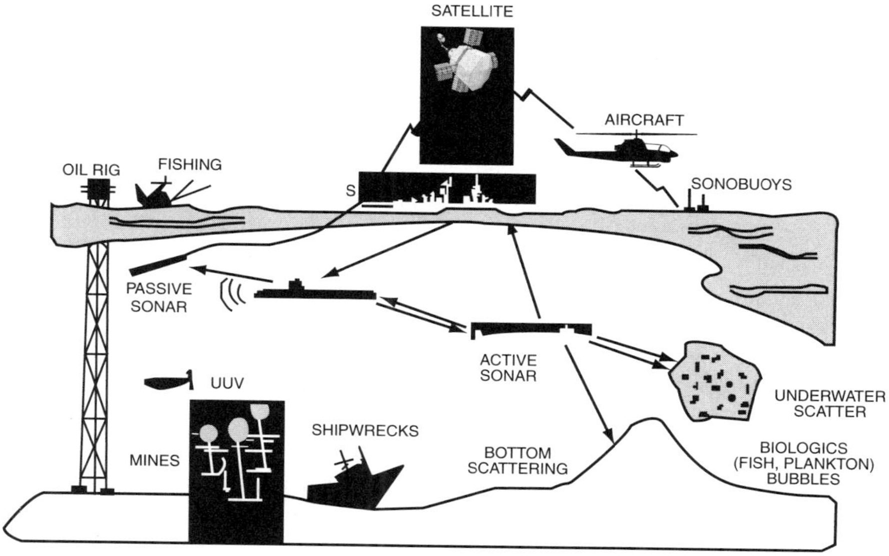

FIGURE 10.1 Active and passive underwater acoustical signal processing.

10.1.3 Background

In underwater acoustics, the metric system has seldom been universally applied and a number of non-metric units are still used: distances of nautical miles (1852 m), yards (0.9144 m), and kiloyards; speeds of knots (nautical miles per hour); depths of fathoms (6 ft or 1.8288 m); and bearing in degrees (0.1745 radian). However, in the past decade, there has been an effort to become totally metric, i.e., to use MKS or standard international units.

Underwater sound signals that are processed electronically for detection, classification, and localization can be characterized from a statistical point of view. When time averages of each signal are the same as the ensemble average of signals, the signals are said to be *ergodic*. When the statistics do not change with time, the signals are said to be *stationary*. The spatial equivalent to stationary is *homogeneous*. For many introductory problems, only stationary signals and homogeneous noise are assumed; more complex problems involve nonstationary, inhomogeneous environments of the type experienced in actual underwater acoustic environments.

Received sound signals have a first-order probability density function (PDF). For example, the PDF may be Gaussian, or in the case of clicking, sharp noise spikes, or crackling ice noise, the PDF may be non-Gaussian. In addition to being characterized by a PDF, signals can be characterized in the frequency domain by their power spectral density functions, which are Fourier transforms of the autocorrelation functions. White signals, which are uncorrelated from sample to sample, have a delta function autocorrelation or a flat (constant) power spectral density. Ocean signals, in general, are much more colorful, and they are neither stationary nor homogeneous.

10.1.4 Sonar

SONAR (sound navigation and ranging), the acronym adapted in the 1940s, is similar to the popular RADAR (radio detection and ranging) and involves the use of sound to explore the ocean and underwater objects.

Passive sonar uses sound radiated from the underwater object itself. The duration of the radiated sound may be short or long in time and narrow or broad in frequency. Transmission through the ocean, from the source to a receiving sensor, is one way.

Active sonar involves transmitting an acoustical signal from a source and receiving reflected echoes from the object of interest. Here, the transmissions from a transmitter to an object and back to a receiving sensor are two way. There are three types of active sonar systems:

- *Monostatic*: In this most common form, the source and receiver can be identical or distinct, but are located on the same platform (e.g., a surface ship or torpedo).
- *Bistatic*: In this form, the transmitter and receiver are on different platforms (e.g., ships).
- *Multistatic*: Here, multiple transmitters and multiple receivers are located on different platforms (e.g., multiple ships).

Passive sonar signals are primarily modeled as random signals. Their first-order PDFs are typically Gaussian; one exception is a stable sinusoidal signal that is non-Gaussian and has a power spectral density function that is a Dirac delta function in the frequency domain. Such sinusoidal signals can be detected by measuring the energy output of narrowband filters. This can be done with fast Fourier transform (FFT) electronics and long integration times. However, in actual ocean environments, an arbitrarily narrow frequency width is never observed, and signals have some finite bandwidth. Indeed, the full spectrum of most underwater signals is quite "colorful." In fact, the signals of interest are not unlike speech signals, except that the signal-to-noise (SNR) ratio is much higher for speech applications than for practical sonar applications.

Received active sonar signals can be viewed as consisting of the results of (1) a deterministic component (known transmit signal) convolved with the medium and reflector transfer functions and (2) a random (noise) component. The Doppler imparted (frequency shift) to the reflected signal makes the total system effect nonlinear, thereby complicating analysis and processing of these signals. In addition, in active systems the noise (or reverberation) is typically correlated with the signal, making detection of signals more difficult.

10.2 Underwater Propagation

10.2.1 Speed/Velocity of Sound

Sound speed, c, in the ocean, in general, lies between 1450 to 1540 m/s and varies as a function of several physical parameters, such as temperature, salinity, and pressure (depth). Variations in sound speed can significantly affect the propagation (range or quality) of sound in the ocean. Table 10.1 gives approximate expressions for sound speed as a function of these physical parameters.

10.2.2 Sound Velocity Profiles

Sound rays that are normal (perpendicular) to the signal acoustic wavefront can be traced from the source to the receiver by a process called ray tracing.* In general, the acoustic ray paths are not straight, but bend in a manner analogous to optical rays focused by a lens. In underwater sound, the ray paths are determined by the *sound velocity profile* (SVP) or the *sound speed profile* (SSP): that is, the speed of sound in water as a function of water depth. The sound speed not only varies with depth, but also varies in different regions of the ocean and with time as well. In deep water, the SVP fluctuates the most in the upper ocean due to variations of temperature and weather. Just below the sea surface is the *surface layer,* where the sound speed is greatly affected by temperature and wind action. Below this layer lies the *seasonal thermocline,* where the temperature and speed decrease with depth and the variations are seasonal. In

* Ray tracing models are used for high-frequency signals and in deep water. Generally, if the depth-to-wavelength ratio is 100 or more, ray tracing models are accurate. Below that, corrections must be made to these models. In shallow water or low frequencies, i.e., when the depth-to-wavelength is about 30 or less, "mode theory" models are used.

TABLE 10.1 Expressions for Sound Speed in Meters Per Second

Expression	Limits	Ref.
$c = 1492.9 + 3\,(T - 10) - 6 \times 10^{-3}(T - 10)^2$	$-2 \bullet T \bullet 24.5°$	1[a]
$\quad - 4 \times 10^{-3}\,(T - 18)^2 + 1.2\,(S - 35)$	$30 \bullet S \bullet 42$	
$\quad - 10^{-2}(T - 18)(S - 35) + D/61$	$0 \bullet D \bullet 1000$	
$c = 1449.2 + 4.6T - 5.5 \times 10^{-2}T^2$	$0 \bullet T \bullet 35°$	2[b]
$\quad + 2.9 \times 10^{-4}T^3 + (1.34 - 10^{-2}T)(S - 35)$	$0 \bullet S \bullet 45$	
$\quad + 1.6 \times 10^{-2}D$	$0 \bullet D \bullet 1000$	
$c = 1448.96 + 4.591T - 5.304 \times 10^{-2}T^2$	$0 \bullet T \bullet 30°$	3[c]
$\quad + 2.374 \times 10^{-4}T^3 + 1.340\,(S - 35)$	$30 \bullet S \bullet 40$	
$\quad + 1.630 \times 10^{-2}D + 1.675 \times 10^{-7}D^2$	$0 \bullet D \bullet 8000$	

Note:
D = depth, in meters;
S = salinity, in parts per thousand; and
T = temperature, in degrees Celsius.

[a] Leroy, C.C., 1969, Development of Simple Equations for Accurate and More Realistic Calculation of the Speed of Sound in Sea Water, *J. Acoust. Soc. Am.*, 46, 216.

[b] Medwin, H., 1975, Speed of Sound in Water for Realistic Parameters, *J. Acoust. Soc. Am.*, 58, 1318.

[c] Mackenzie, K. V., 1981, Nine-term Equation for Sound Speed in the Oceans, *J. Acoust. Soc. Am.*, 70, 807.

Source: Urick, R. J., 1983, *Principles of Underwater Sound*, McGraw-Hill, New York, p. 113. With permission.

the next layer, the *main thermocline*, the temperature and speed decrease with depth, and surface conditions or seasons have little effect. Finally, there is the *deep isothermal layer*, where the temperature is nearly constant at 39°F and the sound velocity increases almost linearly with depth. A typical deep water SVP as a function of depth is shown in Figure 10.2.

If the sound speed is a minimum at a certain depth below the surface, then this depth is called the axis of the underwater sound channel.* The sound velocity increases both above and below this axis. When the sound wave travels through a medium with a sound speed gradient, the direction of travel of the sound wave is bent toward the area of lower sound speed.

Although the definition of shallow water can be signal dependent, in terms of depth-to-wavelength ratio, a water depth of less than 100 m is generally referred to as shallow water. In shallow water, the SVP is irregular and difficult to predict because of large surface temperature and salinity variations, wind effects, and multiple reflections of sound from the ocean bottom.

10.2.3 Three Propagation Modes

In general, there are three dominant propagation modes that depend on the distance or range between the sound source and the receiver.

1. *Direct path:* Sound energy travels in a (nominal) straight-line path between the source and receiver, usually at short ranges.
2. *Bottom bounce path:* Sound energy is reflected from the ocean bottom (at intermediate ranges, see Figure 10.3).
3. *Convergence zone (CZ) path:* Sound energy converges at longer ranges where multiple acoustic ray paths add or recombine coherently to reinforce the presence of signal energy from the radiating/reflecting source.

* Often called the SOFAR, SOund Fixing And Ranging, channel.

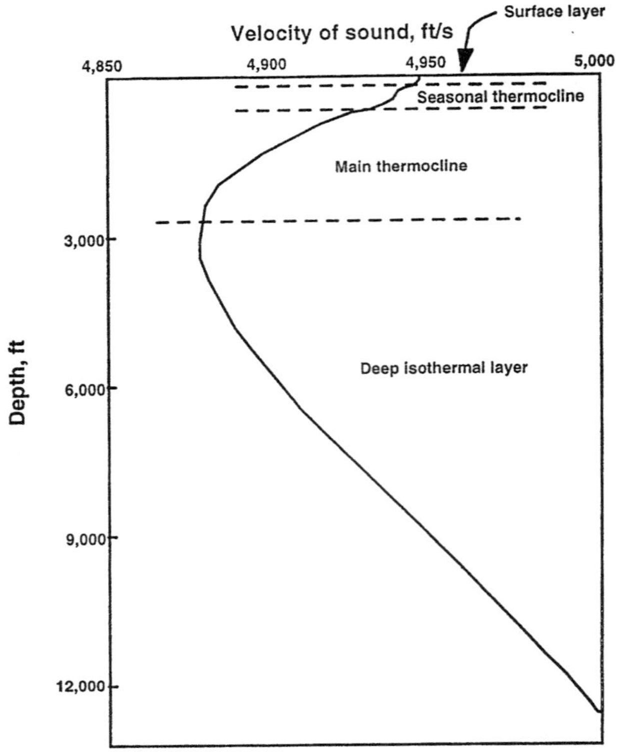

FIGURE 10.2 A typical SVP. (From Urick, R. J., 1983, *Principles of Underwater Sound*, McGraw-Hill, New York, p. 118. With permission.)

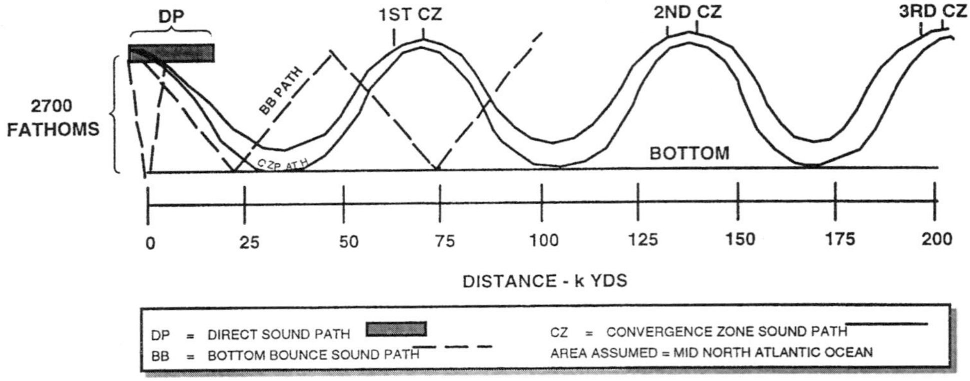

FIGURE 10.3 Typical sound paths between source and receiver. A fathom is a unit of length or depth generally used for underwater measurements. 1 fathom = 6 feet. (From Cox, A.W., 1974, *Sonar and Underwater Sound*, Lexington Books, D.C. Health and Company, Lexington, MA, p. 25. With permission.)

10.2.4 Multipaths

The ocean splits signal energy into multiple acoustic paths. When the receiving system can resolve these multiple paths (or multipaths), then they should be coherently recombined by optimal signal processing to fully exploit the available signal energy for detection.[4] It is also theoretically possible to exploit the geometrical properties of multipaths present in the bottom bounce path by investigation of

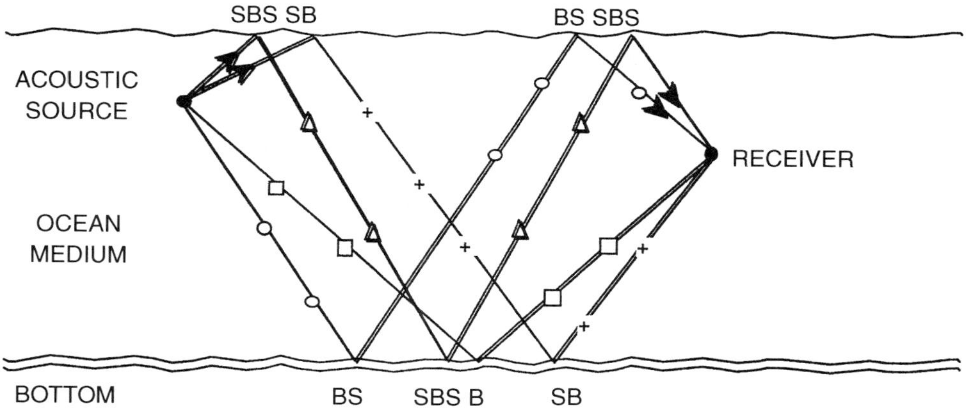

FIGURE 10.4 Multipaths for a first-order bottom bounce propagation model.

a virtual aperture that is created by the different path arrivals to localize the energy source. In the case of a first-order bottom bounce transmission (i.e., only one bottom interaction), there are four paths (from source to receiver):

1. A bottom bounce ray path (B)
2. A surface interaction followed by a bottom interaction (SB)
3. A bottom bounce followed by a surface interaction (BS)
4. A path that first hits the surface, then the bottom, and finally the surface (SBS)

Typical first-order bottom bounce ocean propagation paths are depicted in Figure 10.4.

10.2.5 Sonar System Performance

The performance of sonar systems, at least to first order, is often assessed by the passive and active sonar equations. The major parameters in the sonar equation, measured in decibels (dB), are as follows:

L_S = source level
L_N = noise level
N_{DI} = directivity index
N_{TS} = echo level or target strength
N_{RD} = recognition differential

Here, L_S is the target-radiated signal strength (for passive) or transmitted signal strength (for active), and L_N is the total background noise level. N_{DI}, or *DI*, is the directivity index, which is a measure of the capability of a receiving array to electronically discriminate against unwanted noise. N_{TS} is the received echo level or target strength. Underwater objects with large values of N_{TS} are more easily detectable with active sonar than those with small values of N_{TS}. In general, N_{TS} varies as a function of object size, aspect angle (i.e., the direction at which impinging acoustic signal energy reaches the underwater object), and reflection angle (i.e., the direction at which the impinging acoustic signal energy is reflected off the underwater object). N_{RD} is the recognition differential of the processing system.

The *figure of merit* (FOM), a basic performance measure involving parameters of the sonar system, ocean, and target, is computed for active and passive sonar systems (in decibels) as follows:

For passive sonar,

$$\mathrm{FOM}_P = L_S - (L_N - N_{DI}) - N_{RD}$$

For active sonar,

$$\mathrm{FOM}_A = (L_S + N_{TS}) - (L_N - N_{DI}) - N_{RD}$$

Sonar systems are designed so that the FOM exceeds the signal propagation loss for a given set of parameters of the sonar equations. The amount above the FOM is called the *signal excess*. When two sonar systems are compared, the one with the largest signal excess is said to hold the *acoustic advantage*. However, it should be noted that the set of parameters in the above FOM equations is simplified here. Depending on the design or parameter measurability conditions, parameters can be combined or expanded in terms of such quantities as the frequency dependency of the sonar system in particular ocean conditions, the speed and bearing of the receiving or transmitting platforms, reverberation loss, and so forth. Furthermore, due to multipaths, differences in sonar system equipment and operation, and the constantly changing nature of the ocean medium, the FOM parameters fluctuate with time. Thus, the FOM is not an absolute measure of performance, but rather an average measure of performance over time.

10.2.6 Sonar System Performance Limitations

In a typical reception of a signal wavefront, noise and interference can degrade the performance of the sonar system and limit the system's capability to detect signals in the underwater environment. The effects of these degradations must be considered when any sonar system is designed. The noise or interference could be from a school of fish, shipping (surface or subsurface), active transmission operations (e.g., jammers), or the use of multiple receivers or sonar systems simultaneously. Also, the ambient noise may have unusual vertical or horizontal directivity, and in some environments, such as the Arctic, the noise due to ice motion may produce unusual interference. Unwanted backscatterers, similar to the headlights of a car driving in fog, can cause a signal-induced and signal-correlated noise that degrades processing gain. Some other performance-limiting factors are the loss of signal level and acoustic coherence due to boundary interaction as a function of grazing angle; the radiated pattern (signal level) of the object and its spatial coherence; the presence of surface, bottom, and volume reverberation (in active sonar); signal spreading (in time, frequency, or bearing) owing to the modulating effect of surface motion; biologic noise as a function of time (both time of day and time of year); and statistics of the noise in the medium (e.g., does the noise arrive in the same ray path angles as the signal?).

10.2.7 Imaging and Tomography Systems

Underwater sound and signal processing can be used for bottom imaging and underwater oceanic tomography.[10] Signals are transmitted in succession, and the time delay measurements between signals and measured multipaths are then used to determine the speed of sound in the ocean. This information, along with bathymetry data, is used to map depth and temperature variations of the ocean. In addition to mapping ocean bottoms, such information can aid in quantifying global climate and warming trends.

10.3 Underwater Sound Systems: Components and Processes

In this section, we describe a generic sonar system and provide a brief summary for some of its components. In Section 10.4, we describe some of the signal processing functions. A detailed description of the sonar components and various signal processing functions can be found in Knight et al.,[7] Hueter,[6] and Winder.[13] Figures 10.5 and 10.6 show block diagrams of the major components of a typical active and passive sonar system, respectively. Except for the signal generator, which is present only in active sonar, there are many similarities in the basic components and functions for the active and passive sonar system.

In an active sonar system, an electronic signal generator generates a signal. The signal is then inverse beamformed by delaying it in time by various amounts. A separate projector is used to transmit each of the delayed signals by transforming the electrical signal into an acoustic pressure wave that propagates

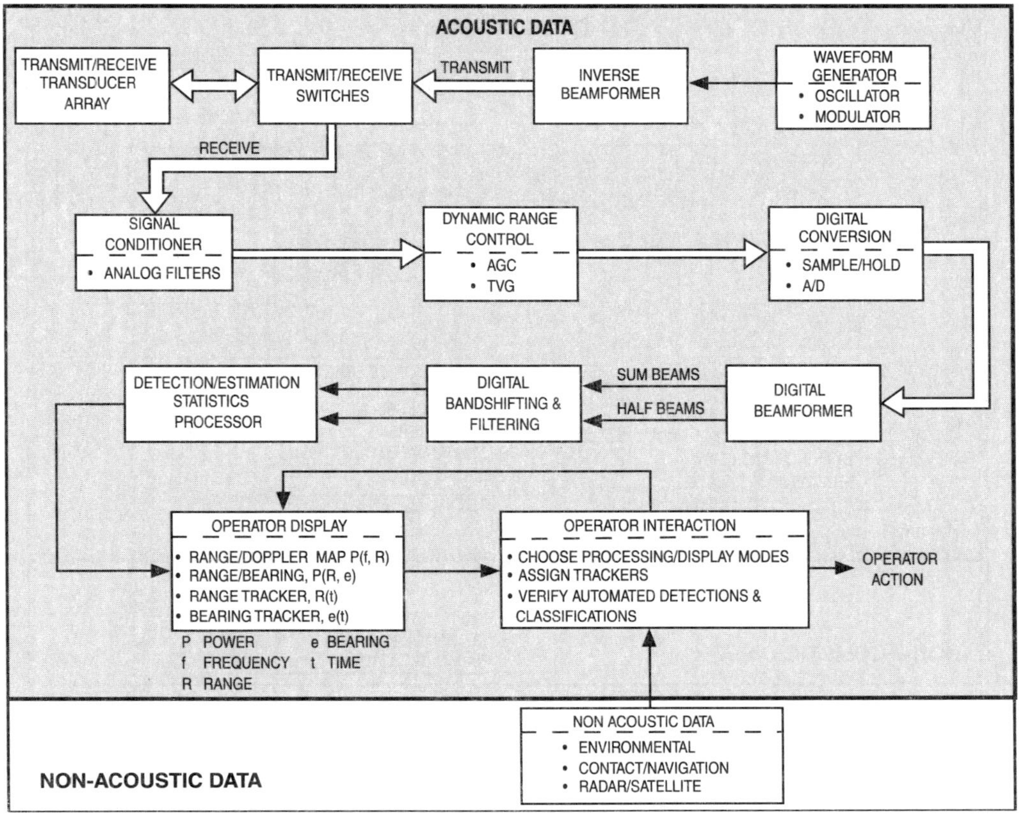

FIGURE 10.5 Generic active sonar system. (Modified Knight, W. C. et al., 1981, Digital Signal Processing for Sonar, *Proc. IEEE,* 69(11), November.)

through water. Thus, an array of projectors is used to transmit the signal and focus it in the desired direction. Depending on the desired range and Doppler resolution, different signal waveforms can be generated and transmitted.

At the receiver (an array of hydrophones), the acoustic or pressure waveform is converted back to an electrical signal. The received signal consists of the source signal (usually the transmitted signal in the active sonar case) embedded in ambient noise and interference from other sources present in water. The signal then goes through a number of signal processing functions. In general, each channel of the analog signal is first filtered in a signal conditioner. It is then amplified or attenuated within a specified dynamic range using an automatic gain control (AGC). For active sonar, we can also use a time-varied gain (TVG) to amplify or attenuate the signal. The signal, which is analog until this point, is then sampled and digitized by analog-to-digital (A/D) converters. The individual digital sensor outputs are next combined by a digital beamformer to form a set of beams. Each beam represents a different search direction of the sonar. The beam output is further processed (bandshifted, filtered, normalized, downsampled, etc.) to obtain detection, classification, and localization (DCL) estimates, which are displayed to the operator on single or multiple displays. Based on the display output (acoustic data) and other nonacoustic data (environmental, contact, navigation, and radar/satellite), the operators make their final decision.

10.3.1 Signal Waveforms

The transmitted signal is an essential part of an active sonar system. The properties of the transmitted signal will strongly affect the quality of the received signal and the information derived from it. The main

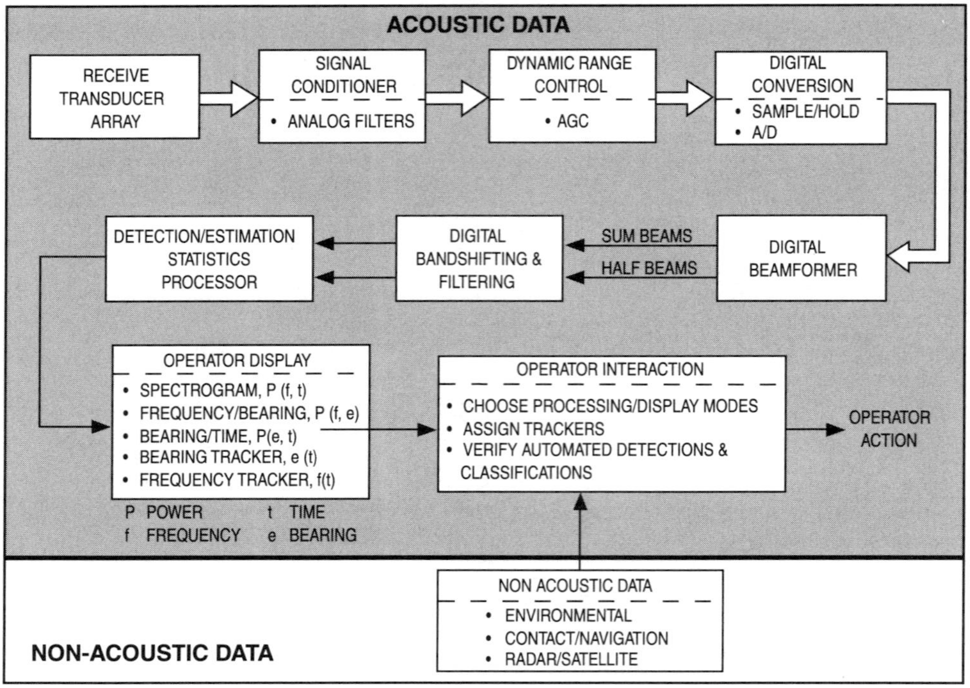

FIGURE 10.6 Generic Passive sonar system. (Modified Knight, W. C. et al., 1981, Digital Signal Processing for Sonar, *Proc. IEEE,* 69(11), November.)

objective in active sonar is to detect a target and estimate its range and velocity. The range and velocity information is obtained from the reflected signal. To show how the range and velocity information is contained in the received signal, we consider a simple case where $s(t)$, $0 \leq t \leq T$ is transmitted. Neglecting medium, noise, and other interference effects, we let the transmitted signal be reflected back from a moving target located at range $R = R_0 + vt$, where v is the velocity of the target. The received signal is then given by

$$r(t) = a(R)s[(1 - b)t - \tau]$$
$$\tau = 2R_0/c$$
$$b = 2v/c$$

where $a(R)$ is the propagation loss attenuation factor* and c is the speed of sound. Measuring the delay τ gives the propagation time to and from the target. We can then calculate the range from the time delay. The received signal is also time compressed or expanded by b. The velocity of the target can be estimated by determining the compression or expansion factor of the received signal. Each signal has different range and Doppler (velocity) resolution properties. Some signals are good for range resolution, but not for Doppler; some for Doppler, but not for range; some for reverberation-limited environments; and some for noise-limited environments.[13] Commonly used signals in active sonar are continuous wave (CW), linear frequency modulation (LFM), hyperbolic frequency modulation (HFM), and pseudo-random noise (PRN) signals. CW signals have been used in sonar for decades, whereas signals like frequency hop codes (FHC) and Newhall waveforms are recently "rediscovered" signals[9] that work well in high-reverberation, shallow water environments.

Some of the most commonly used signals are described below. So far, we have made generalized statements about the effectiveness of signals, which only provide a broad overview. More specifically,

* More generally, $a(R)$ is also a function of frequency.

the signal has properties that depend on a number of factors not discussed here, such as time duration and frequency bandwidth. Rihaczek[9] provides a detailed analysis of the properties and effectiveness of signals. The simplest signal is a rectangular CW pulse, which is a single frequency sinusoid. The CW signal may have high resolution in range (short CW) or Doppler (long CW), but not in both simultaneously. LFM signals are waveforms whose instantaneous frequency varies linearly with time; in HFM signals, the instantaneous frequency sweeps monotonically as a hyperbola. Both these signals are good for detecting low Doppler targets in reverberation-limited conditions. PRN signals, which are generated by superimposing binary data on sinusoid carriers, provide simultaneous resolution in range and Doppler. However, such range resolution may not be as good as LFM alone, and Doppler resolution is not as good as CW alone. An FHC signal is a waveform that consists of subpulses of equal duration. Each subpulse has a distinct frequency, and these frequencies jump or hop in a defined manner. Similar to PRN, FHC also provides simultaneous resolution in range and Doppler. Newhall waveforms (also known as "coherent pulse trains" or "saw-tooth frequency modulation"), which are trains of repeated modulated subpulses (typically HFM or LFM), allow reverberation suppression and low Doppler target detection.

10.3.2 Sonar Transducers

A transducer is a fundamental element of both receiving hydrophones and projectors. It is a device that converts one form of energy into another. In the case of a sonar transducer, the two forms of energy are electricity and pressure, the pressure being that associated with a signal wavefront in water. The transducer is a reciprocal device, such that when electricity is applied to the transducer, a pressure wave is generated in the water, and when a pressure wave impinges on the transducer, electricity is developed. The heart of a sonar transducer is its active material, which makes the transducer respond to electrical or pressure excitations. These active materials produce electrical charges when subjected to mechanical stress and conversely produce a stress proportional to the applied electrical field strength when subjected to an electrical field. Most sonar transducers employ piezoelectric materials, such as lead zirconate titanate ceramic, as the active material. Magnetostictive materials such as nickel can also be used. Figure 10.7 shows a flextensional transducer. In this configuration, the ceramic stack is mounted on the major axis of a metallic elliptical cylinder. Stress is applied by compressing the ellipse along its minor axis, thereby extending the major axis. The ceramic stacks are then inserted into the cylinder, and the stress is released, which places the ceramic stacks in compression. This design allows a small change imparted at the ends of the ceramic stack to be converted into a larger change at the major faces of the ellipse.

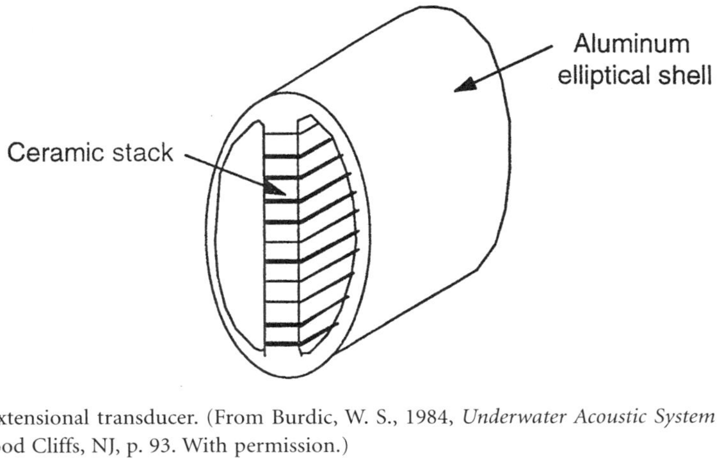

FIGURE 10.7 Flextensional transducer. (From Burdic, W. S., 1984, *Underwater Acoustic System Analysis*, Prentice-Hall, Inc. Englewood Cliffs, NJ, p. 93. With permission.)

10.3.3 Hydrophone Receivers

Hydrophone sensors are microphones capable of operating in water under hydrostatic pressure. These sensors receive radiated and reflected sound energy that arrives through the multiple paths of the ocean medium from a variety of sources and reflectors. As with a microphone, hydrophones convert acoustic pressure to electrical voltages or optical signals. Typical hydrophones are hollow piezoelectric ceramic cylinders with end caps. The cylinders are covered with rubber or polyethylene as water proofing, and an electrical cable exits from one end. These hydrophones are isolation mounted so that they do not pick up stray vibrations from the ship. Most hydrophones are designed to operate below their resonance frequency, thus resulting in a flat or broadband receiving response.

10.3.4 Sonar Projectors (Transmitters)

A sonar projector is a transducer designed principally for transmission, i.e., to convert electrical voltage or energy into sound energy. Although sonar projectors are also good receivers (hydrophones), they are too expensive and complicated (designed for a specific frequency band of operation) for just receiving signals. Most sonar projectors are of a tonpilz (sound mushroom) design with the piezoelectric material sandwiched between a head and tail mass. The head mass is usually very light and stiff (aluminum) and the tail mass is very heavy (steel). The combination results in a projector that is basically a half-wavelength long at its resonance frequency. The tonpilz resonator is housed in a watertight case that is designed so that the tonpilz is free to vibrate when excited. A pressure release material like corprene (cork neoprene) is attached to the tail mass so that the housing does not vibrate, and all the sound is transmitted from the front face. The piezoelectric material is a dielectric and, as such, acts like a capacitor. To ensure an efficient electrical match to driver amplifiers, a tuning coil or even a transformer is usually contained in the projector housing. A tonpilz projector is shown in Figure 10.8.

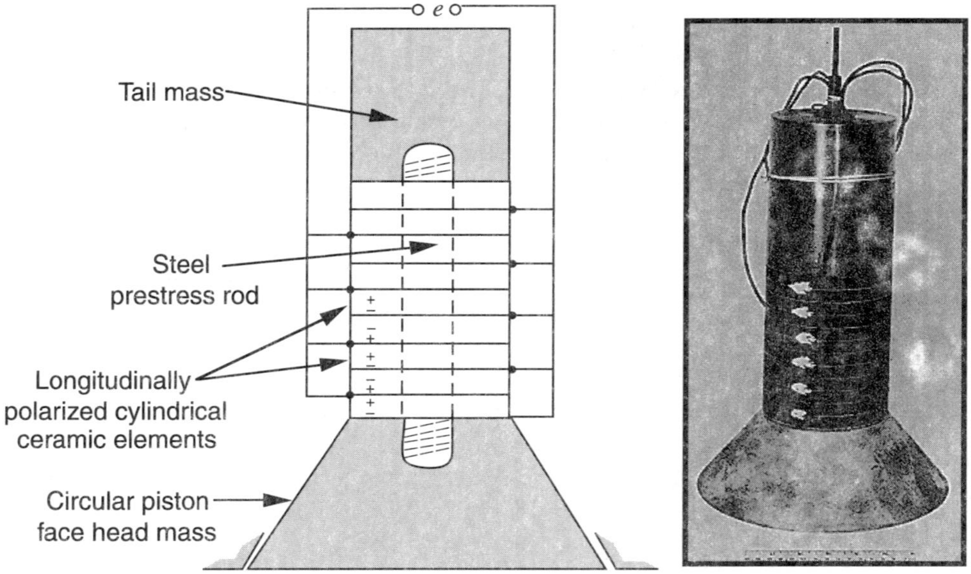

FIGURE 10.8 Tonpilz projector. (Modified from Burdic, W. S., 1984, *Underwater Acoustic System Analysis*, Prentice-Hall, Englewood Cliffs, NJ, p. 92 and Hueter, T. F., 1972, Twenty Years in Underwater Acoustics: Generation and Reception, *J. Acoust. Soc. Am.*, 51 (3, Part 2), March, p. 1032.)

10.3.5 Active Sources

One type of sonar transducer primarily used in the surveillance community is a low-frequency active source. The tonpilz design is commonly used for such projectors at frequencies down to about 2 kHz (a tonpilz at this frequency is almost 3/4 m long). For frequencies below 2 kHz, other types of transducer technology are employed, including mechanical transformers such as flexing shells, moving coil (loud speaker) devices, hydraulic sources, and even impulse sources such as spark gap and air. Explosives are a common source for surveillance, and when used with towed arrays, they make a very sophisticated system.

10.3.6 Receiving Arrays

Most individual hydrophones have an omnidirectional response: that is, sound emanates almost uniformly in all directions. Focusing sound in a particular direction requires an array of hydrophones or projectors. The larger the array, the narrower and more focused is the beam of sound energy, and hence, the more the signal is isolated from the interfering noise. An array of hydrophones allows discrimination against unwanted background noise by focusing its main response axis (MRA) on the desired signal direction. Arrays of projectors and hydrophones are usually designed with elements spaced one-half wavelength apart. This provides optimum beam structure at the design frequency. As the frequency of operation increases and exceeds three-quarter wavelength spacing, the beam structure, although narrowing, deteriorates. As the frequency decreases, the beam increases in width to the point that focusing diminishes. Some common array configurations are linear (omnidirectional), planar (fan shaped), cylindrical (searchlight), and spherical (steered beams). Arrays can be less than 3.5 in. in diameter and can have on the order of 100 hydrophones or acoustic channels. Some newer arrays have even more channels. Typically, these channels are nested, subgrouped, and combined in different configurations to form the low-frequency (LF), mid-frequency (MF), and high-frequency (HF) apertures of the array. Depending on the frequency of interest, one can use any one of these three apertures to process the data. The data are then prewhitened, amplified, and lowpass filtered before being routed to A/D converters. The A/D converters typically operate or sample the data at about three times the lowpass cutoff frequency.

A common array, shown in Figure 10.9a, is a single linear line of hydrophones that makes up a device called a towed array.* The line is towed behind the ship and is effective for searching for low level and LF signals without interference from the ship's self-noise. Figure 10.9b shows a more sophisticated bow array (sphere) assembly for the latest Seawolf submarine.

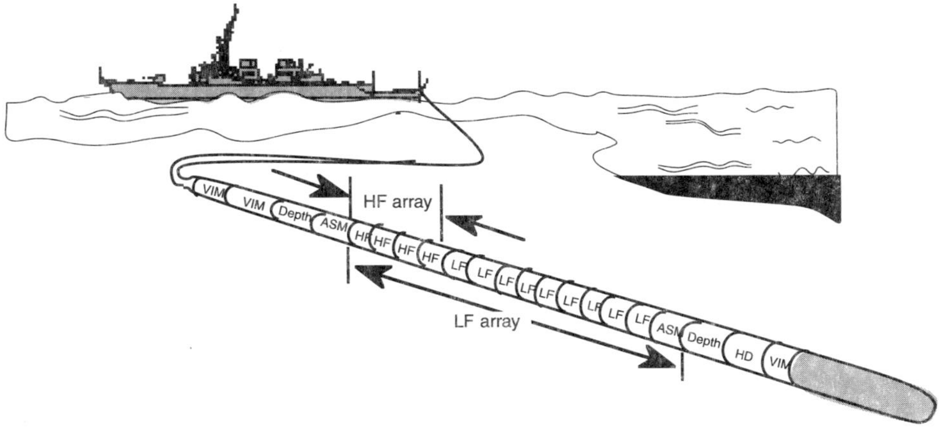

FIGURE 10.9a Schematic of a towed line array. (From Urick, R. J., 1983, *Principles of Underwater Sound*, McGraw-Hill, New York, p. 13. With permission.)

* In the oil exploration business, they are called streamers.

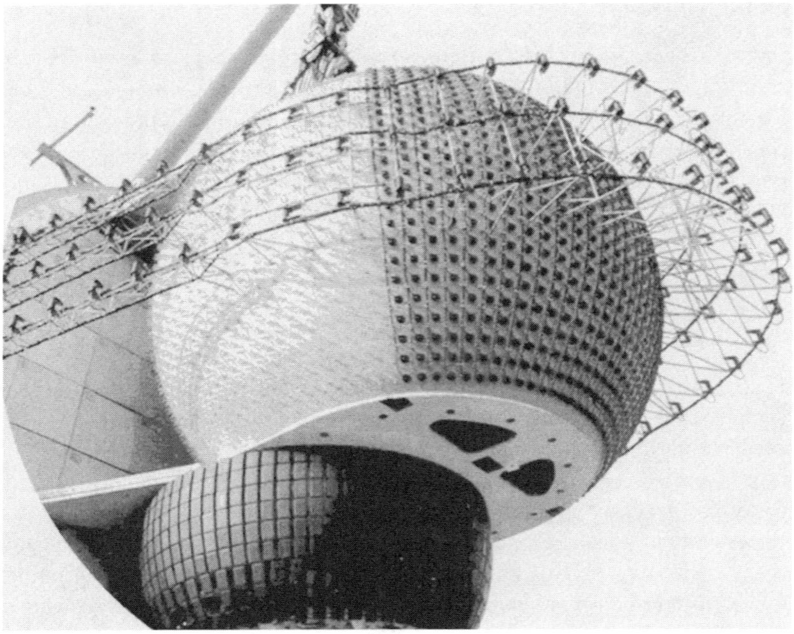

FIGURE 10.9b Bow array assembly for Seawolf submarine (SSN-21). (Taken from the Naval Undersea Warfare Center, Division Newport, BRAC 1991 presentation.)

10.3.7 Sonobuoys

These small expendable sonar devices contain a transmitter to transmit signals and a single hydrophone to receive the signal. Sonobuoys are generally dropped by fixed-wing or rotary-wing aircraft for underwater signal detection.

10.3.8 Dynamic Range Control

Today, most of the signal processing and displays involve the use of electronic digital computers. Analog signals received by the receiving array are converted into a digital format, while ensuring that the dynamic range of the data is within acceptable limits. The receiving array must have sufficient dynamic range so that it can detect the weakest signal, but also not saturate upon receiving the largest signal in the presence of noise and interference. To be able to convert the data into digital form and display it, large fluctuations in the data must be eliminated. Not only do these fluctuations overload the computer digital range capacity, they affect the background quality and contrast of the displays as well. It has been shown that the optimum display background, a background with uniform fluctuations as a function of range, time, and bearing, for detection is one that has constant temporal variance at a given bearing and a constant spatial variance at a given range. Since the fluctuations (noise, interference, propagation conditions, etc.) are time varying, it is necessary to have a threshold level that is independent of fluctuations. The concept is to use techniques that can adapt the high dynamic range of the received signal to the limited dynamic range of the computers and displays. TVG for active sonar and AGC are two popular techniques to control the dynamic range. TVG controls the receiver gain so that it follows a prescribed variation with time, independent of the background conditions. The disadvantage of TVG is that the variations of gain with time do not follow the variations in reverberation. TVG is sufficient if the reverberation is uniform in bearing and monotonically decreasing in range (which is not the case in shallow water). AGC, on the other hand, continuously varies the gain according to the current reverberation or interference conditions. Details of how TVG, AGC, and other gain control techniques such as notch filters, reverberation controlled gain, logarithmic receivers, and hard clippers work are presented by Winder.[13]

10.3.9 Beamforming

Beamforming is a process in which outputs from the hydrophone sensors of an array are coherently combined by delaying and summing the outputs to provide enhanced detection and estimation. In underwater applications, we are trying to detect a directional (single direction) signal in the presence of normalized background noise that is ideally isotropic (nondirectional). By arranging the hydrophone (array) sensors in different physical geometries and electronically steering them in a particular direction, we can increase the SNR in a given direction by rejecting or canceling the noise in other directions. There are many different kinds of arrays that can be beamformed (e.g., equally spaced line, continuous line, circular, cylindrical, spherical, or random sonobuoy arrays). The beam pattern specifies the response of these arrays to the variation in direction. In the simplest case, the increase in SNR due to the beamformer, called the *array gain* (in decibels), is given by

$$AG = 10 \log \frac{SNR_{\text{array(output)}}}{SNR_{\text{single sensor (input)}}}$$

10.3.10 Displays

Advancements in processing power and display technologies over the last two decades have made displays an integral and essential part of any sonar system today. Displays have progressed from a single monochrome terminal to very complicated, interactive, real-time, multiterminal, color display electronics. The amount of data that can be provided to an operator can be overwhelming; time series, power spectrum, narrowband and broadband lofargrams, time bearing, range bearing, time Doppler, and sector scans are just some of the many available displays. Then add to this a source or contact tracking display for single or multiple sources over multiple (50 to 100) beams. The most recent displays provide these data in an interactive mode to make it easier for the operator to make a decision.

For passive systems, the three main parameters of interest are time, frequency, and bearing. Since three-dimensional data are difficult to visualize and analyze, they are usually displayed in the following formats:

Bearing Time: Obtained by integrating over frequency; useful for targets with significant broadband characteristics; also called the BTR for bearing time recorder

Bearing Frequency: Obtained at particular intervals of time or by integrating over time; effective for targets with strong stable spectral lines; also called FRAZ for frequency azimuth

Time Frequency: Usually effective for targets with weak or unstable spectral lines in a particular beam; also called lofargram or sonogram

In active systems, two additional parameters, range and Doppler, can be estimated from the received signals, providing additional displays of range bearing and time Doppler where Doppler provides the speed estimate of the target.

In general, all three formats are required for the operator to make an informed decision. The operator must sort the outputs from all the displays before classifying them into targets. For example, in passive narrowband sonar, classification is usually performed on the outputs from spectral/tonal contents of the targets. The operator uses the different tonal content and its harmonic relations of each target for classification. In addition to the acoustic data information and displays, nonacoustic data such as environmental, contact and navigation information, and radar/satellite photographs are also available to the operator. Figure 10.10 illustrates the displays of a recently developed passive sonar system.

Digital electronics have also had a major impact in sonar in the last two decades and will have an even greater impact in the next decade. As sonar arrays become larger and algorithms become more complex, even more data and displays will be available to the operator. This trend, which requires more data to be processed than before, is going to continue in the future. Due to advancement in technologies, computer processing power has increased, permitting additional signal and data processing. Figure 10.11a shows approximate electronic sonar loads of the past, present, and future sonar systems. Figure 10.11b shows the locations of the controls and displays in relation to the different components of a typical sonar system on a submarine.

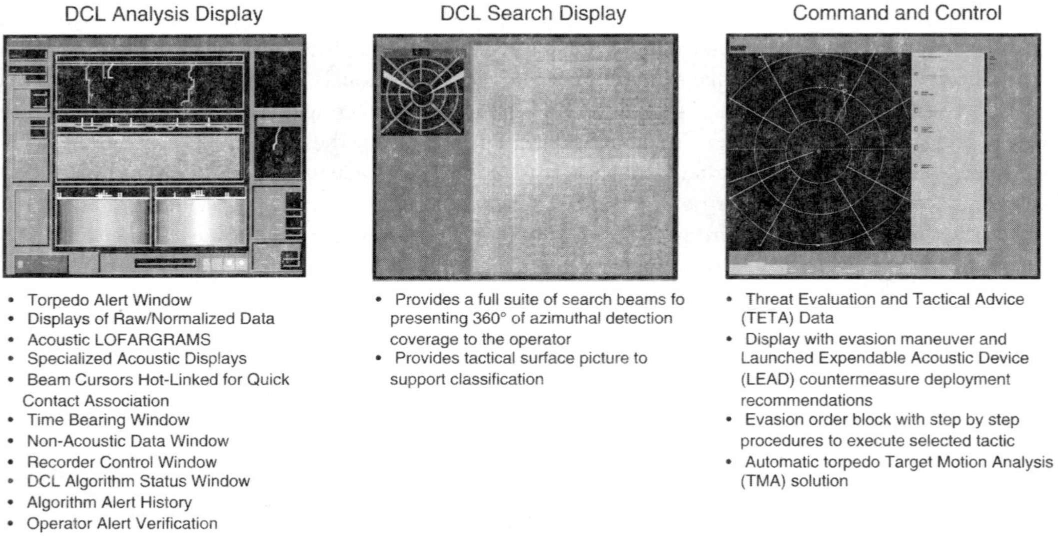
DCL Analysis Display
DCL Search Display
Command and Control
• Torpedo Alert Window
• Displays of Raw/Normalized Data
• Acoustic LOFARGRAMS
• Specialized Acoustic Displays
• Beam Cursors Hot-Linked for Quick Contact Association
• Time Bearing Window
• Non-Acoustic Data Window
• Recorder Control Window
• DCL Algorithm Status Window
• Algorithm Alert History
• Operator Alert Verification
• Provides a full suite of search beams fo presenting 360° of azimuthal detection coverage to the operator
• Provides tactical surface picture to support classification
• Threat Evaluation and Tactical Advice (TETA) Data
• Display with evasion maneuver and Launched Expendable Acoustic Device (LEAD) countermeasure deployment recommendations
• Evasion order block with step by step procedures to execute selected tactic
• Automatic torpedo Target Motion Analysis (TMA) solution

FIGURE 10.10 DCL displays for a passive sonar system. (From Personal communication, E. Marvin, MSTRAP Project, Naval Undersea Warfare Center, Division Newport, Detachment New London, CT.)

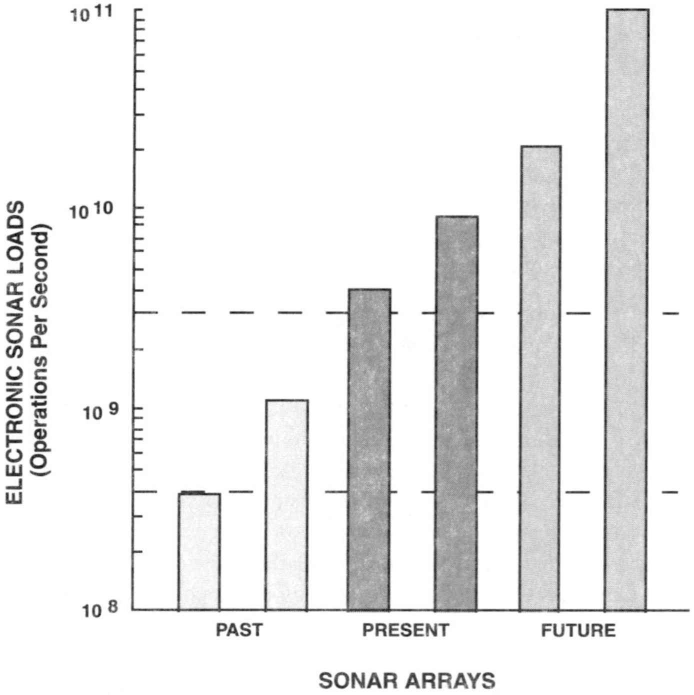
10 11
10 10
10 9
10 8
ELECTRONIC SONAR LOADS (Operations Per Second)
PAST
PRESENT
FUTURE
SONAR ARRAYS

FIGURE 10.11a Electronic sonar load. (From Personal communication, J. Law and N. Owsley, Naval Undersea Warfare Center, Division Newport, Detachment New London, CT.)

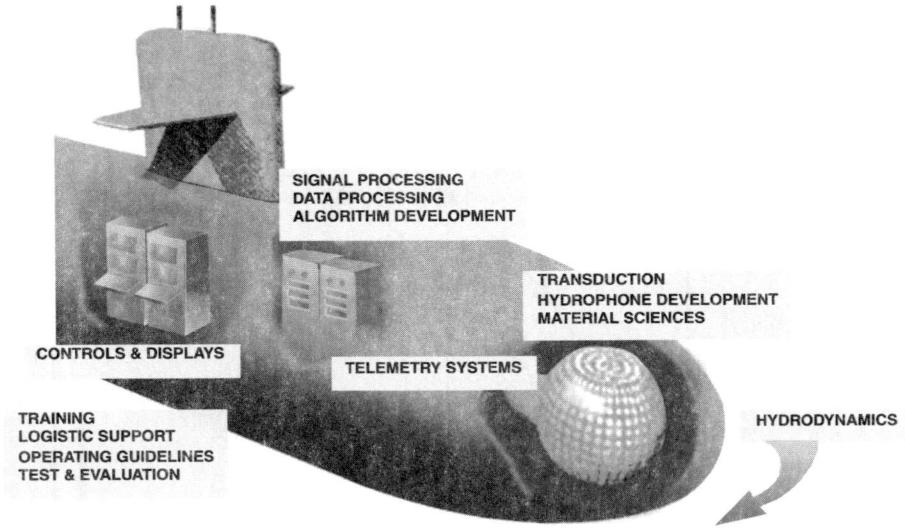

FIGURE 10.11b Control, display, and other components of a submarine sonar system. (From the Naval Undersea Warfare Center, Division Newport, BRAC 1991 presentation.)

10.4 Signal Processing Functions

10.4.1 Detection

Detection of signals in the presence of noise, using classical Bayes or Neyman-Pearson decision criteria, is based on hypothesis testing. In the simplest binary hypothesis case, the detection problem is posed as two hypotheses:

H_0 : Signal is not present (referred to as the null hypothesis).
H_1 : Signal is present.

For a received wavefront, H_0 relates to the noise-only case and H_1 to the signal-plus-noise case. Complex hypotheses (M-hypotheses) can also be formed if detection of a signal among a variety of sources is required.

Probability is a measure, between zero and unity, of how likely an event is to occur. For a received wavefront, the likelihood ratio, L, is the ratio of P_{H_1} (probability that hypothesis H_1 is true) to P_{H_0} (probability that hypothesis H_0 is true). A decision (detection) is made by comparing the likelihood* to a predetermined threshold h. That is, if $L = P_{H_1} / P_{H_0} > h$, a decision is made that the signal is present.

Probability of detection, P_D, measures the likelihood of detecting an event or object when the event does occur. *Probability of false alarm*, P_{fa}, is a measure of the likelihood of saying something happened when the event did NOT occur. Receiver operating characteristics (ROC) curves plot P_D vs. P_{fa} for a particular (sonar signal) processing system. A single plot of P_D vs. P_{fa} for one system must fix the SNR and processing time. The threshold h is varied to sweep out the ROC curve. The curve is often plotted on either log-log scale or "probability" scale. In comparing a variety of processing systems, we would like to select the system (or develop one) that maximizes the P_D for every given P_{fa}. Processing systems must operate on their ROC curves, but most processing systems allow the operator to select where on the ROC curve the system is operated by adjusting a threshold; low thresholds ensure a high probability of detection at the expense of a high false alarm rate. A sketch of two monotonically increasing ROC curves is given in Figure 10.12. By proper adjustment of the decision threshold, we can trade off detection performance for false alarm performance. Since the points (0,0) and (1,1) are on all ROC curves, we can always

* Sometimes the logarithm of the likelihood ratio, called the log-likelihood ratio, is used.

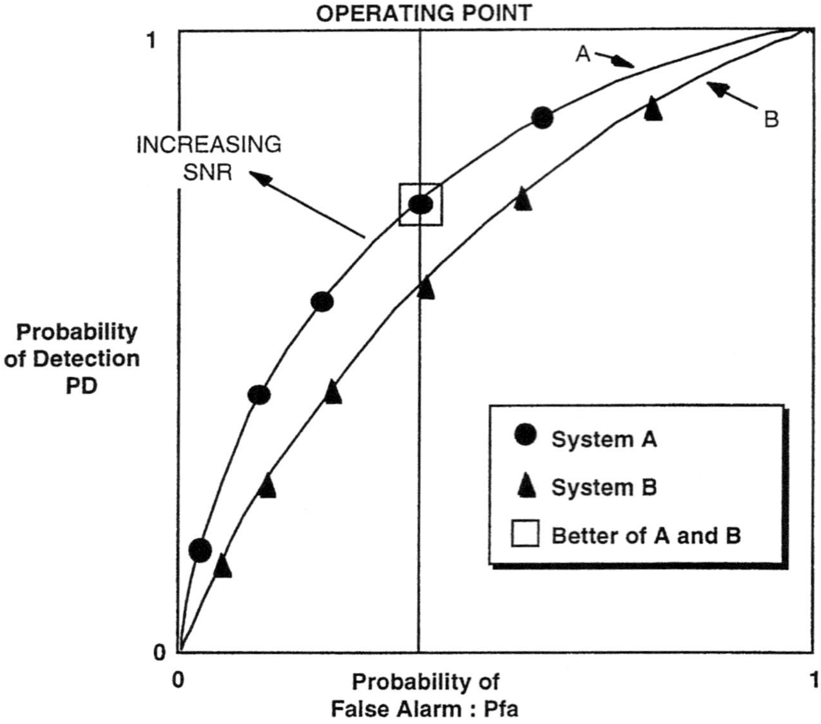

FIGURE 10.12 Typical ROC curves. Note that points (0,0) and (1,1) are on all ROC curves; upper curve represents higher P_D for fixed P_{fa} and hence better performance by having higher SNR or processing time.

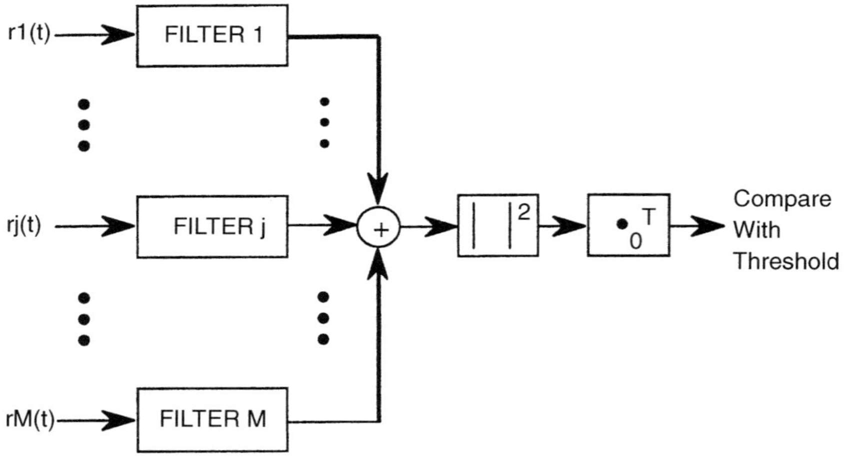

FIGURE 10.13 Log-likelihood detector structure for uncorrelated Gaussian noise in the received signal $r_j(t), j = 1, \ldots, M$.

guarantee 100% P_D with an arbitrarily low threshold (albeit at the expense of 100% P_{fa}) or 0% P_{fa} with an arbitrarily high threshold (albeit at the expense of 0% P_D). The (log) *likelihood detector* is a detector that achieves the maximum P_D for fixed P_{fa}; it is shown in Figure 10.13 for detecting Gaussian signals reflected or radiated from a stationary object. For spiky non-Gaussian noise, clipping prior to filtering improves detection performance.

In active sonar, the filters are matched to the known transmitted signals. If the object (acoustic reflector) has motion, it will induce Doppler on the reflected signal, and the receiver will be complicated by the addition of a bank of Doppler compensators. Returns from a moving object are shifted in frequency by $\Delta f = (2v/c)f$, where v is the relative velocity (range rate) between the source and object, c is the speed of sound in water, and f is the operating frequency of the source transmitter.

In passive sonar, at low SNR, the optimal filters in Figure 10.13 (so-called Eckart filters) are functions of $G_{ss}^{1/2}(f)/G_{nm}(f)$, where f is frequency in Hertz, $G_{ss}(f)$ is the signal power spectrum, and $G_{nn}(f)$ is the noise power spectrum.

10.4.2 Estimation/Localization

The second function of underwater signal processing estimates the parameters that localize the position of the detected object. The source position is estimated in range, bearing, and depth, typically from the underlying parameter of *time delay* associated with the acoustic signal wavefront. The statistical uncertainty of the positional estimates is important. Knowledge of the first-order PDF or its first- and second-order moments, the mean (expected value), and the variance are vital to understanding the expected performance of the processing system. In the passive case, the ability to estimate range is extremely limited by the geometry of the measurements; indeed, the variance of passive range estimates can be extremely large, especially when the true range to the signal source is long when compared with the aperture length of the receiving array. Figure 10.14 depicts direct path passive ranging uncertainty from a collinear array with sensors clustered so as to minimize the bearing and uncertainty region. Beyond the direct path, multipath signals can be processed to estimate source depth passively. Range estimation accuracy is not difficult with the active sonar, but active sonar is not covert, which for some applications can be important.

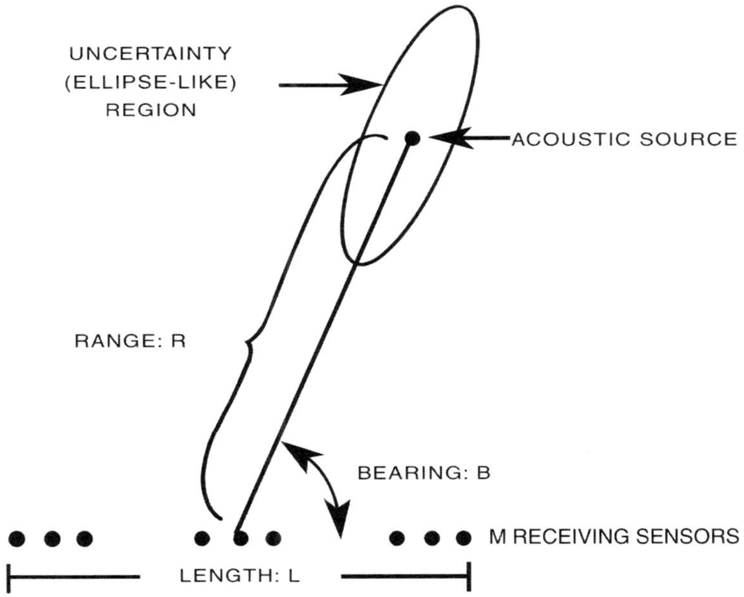

FIGURE 10.14 Array geometry used to estimate source position. (From Carter, G.C., February 1987, Coherence and Time Delay Estimation, *Proc. IEEE,* 75(2), February, p. 251. With permission.)

10.4.3 Classification

The third function of sonar signal processing is classification. This function determines the type of object that has radiated or reflected signal energy. For example, was the sonar signal return from a school of fish or a reflection from the ocean bottom? The action taken by the operator is highly dependent upon this important function. The amount of radiated or reflected signal power relative to the background noise (that is, SNR) necessary to achieve good classification may be many decibels higher than for detection. Also, the type of signal processing required for classification may be different than the type of processing for detection. Processing methods that are developed on the basis of detection might not have the requisite SNR to adequately perform the classification function. Classifiers are, in general, divided into feature (or clue) extractors followed by a classifier decision box. A key to successful classification is feature extraction. Performance of classifiers is plotted (as in ROC detection curves) as the probability of deciding on class A, given A was actually present, or $P(A|A)$, vs. the probability of deciding on class B, given that A was present, or $P(B|A)$, for two different classes of objects, A and B. Of course, for the same class of objects, the operator could also plot $P(B|B)$ vs. $P(A|B)$.

10.4.4 Target Motion Analysis

The fourth function of underwater signal processing is to perform contact or target motion analysis (TMA): that is, to estimate parameters of bearing and speed. Generally, nonlinear filtering methods, including Kalman-Bucy filters, are applied; typically, these methods rely on a state space model for the motion of the contact. For example, the underlying model of motion could assume a straight-line course and a constant speed for the contact of interest. When the signal source of interest behaves like the model, then results consistent with the basic theory can be expected. It is also possible to incorporate motion compensation into the signal processing detection function. For example, in the active sonar case, proper signal selection and processing can reduce the degradation of detector performance caused by uncompensated Doppler. Moreover, joint detection and estimation can provide clues to the TMA and classification processes. For example, if the processor simultaneously estimates depth in the process of performing detection, then a submerged object would not be classified as a surface object. Also, joint detection and estimation using Doppler for detection can directly improve contact motion estimates.

10.4.5 Normalization

Another important signal processing function for the detection of weak signals in the presence of unknown and (temporal and spatial) varying noise is normalization. The statistics of noise or reverberation for oceans typically varies in time, frequency, and/or bearing from measurement to measurement and location to location. To detect a weak signal in a broadband, nonstationary, inhomogeneous background, it is usually desirable to make the noise background statistics as uniform as possible for the variations in time, frequency, and/or bearing. The noise background estimates are first obtained from a window of resolution cells (which usually surrounds the test data cell). These estimates are then used to normalize the test cell, thus reducing the effects of the background noise on detection. Window length and distance from the test cell are two of the parameters that can be adjusted to obtain accurate estimates of the different types of stationary or nonstationary noise.

10.5 Advanced Signal Processing

10.5.1 Adaptive Beamforming

Beamforming was discussed earlier in Section 10.3. The cancellation of noise through beamforming can also be done adaptively, which can improve array gain further. Some of the various adaptive beamforming techniques[7] use Dicanne, sidelobe cancellers, maximum entropy array processing, and maximum-likelihood (ML) array processing.

10.5.2 Coherence Processing

Coherence is a normalized,* cross-spectral density function that is a measure of the similarity of received signals and noise between any sensors of the array. The complex coherence functions between two wide-sense stationary processes x and y are defined by

$$\gamma_{xy}(f) = \frac{G_{xy}(f)}{\sqrt{G_{xx}(f)G_{yy}(f)}}$$

where, as before, f is the frequency in Hertz and G is the power spectrum function. Array gain depends on the coherence of the signal and noise between the sensors of the array. To increase the array gain, it is necessary to have high signal coherence, but low noise coherence. Coherence of the signal between sensors improves with decreasing separation between the sensors, frequency of the received signal, total bandwidth, and integration time. Loss of coherence of the signal could be due to ocean motion, object motion, multipaths, reverberation, or scattering. The coherence function has many uses, including measurement of SNR or array gain, system identification, and determination of time delays.[2,3]

10.5.3 Acoustic Data Fusion

Acoustic data fusion is a technique that combines information from multiple receivers or receiving platforms about a common object or channel. Instead of each receiver making a decision, relevant

TABLE 10.2 Underwater Acoustic Applications

Function	Description
	Military
Detection	Deciding if a target is present or not
Classification	Deciding if a detected target does or does not belong to a specific class
Localization	Measuring at least one of the instantaneous positions and velocity components of a target (either relative or absolute), such as range, bearing, range rate, or bearing rate
Navigation	Determining, controlling, and/or steering a course through a medium (includes avoidance of obstacles and the boundaries of the medium)
Communications	Instead of a wire link, transmitting and receiving acoustic power and information
Control	Using a sound-activated release mechanism
Position marking	Transmitting a sound signal continuously (beacons) or transmitting only when suitably interrogated (transponders)
Depth sounding	Sending short pulses downward and timing the bottom return
Acoustic-speedometers	Using pairs of transducers pointing obliquely downward to obtain speed over the bottom from the Doppler shift of the bottom return
	Commercial Applications
Industrial	Oceanographic
Fish finders/fish herding	Subbottom geological mapping
Fish population estimation	Environmental monitoring
Oil and mineral explorations	Ocean topography
River flow meter	Bathyvelocimeter
Acoustic holography	Emergency telephone
Viscosimeter	Seismic simulation and measurement
Acoustic ship docking system	Biological signal and noise measurement
Ultrasonic grinding/drilling	Sonar calibration
Biomedical ultrasound	

(Modified from Urick, R. J., 1983, *Principles of Underwater Sound*, McGraw-Hill, New York, p. 8, and Cox, A. W., 1974, *Sonar and Underwater Sound*, Lexington Books, D.C. Health and Company, Lexington, MA, p. 2.)

* So that it lies between zero and unity.

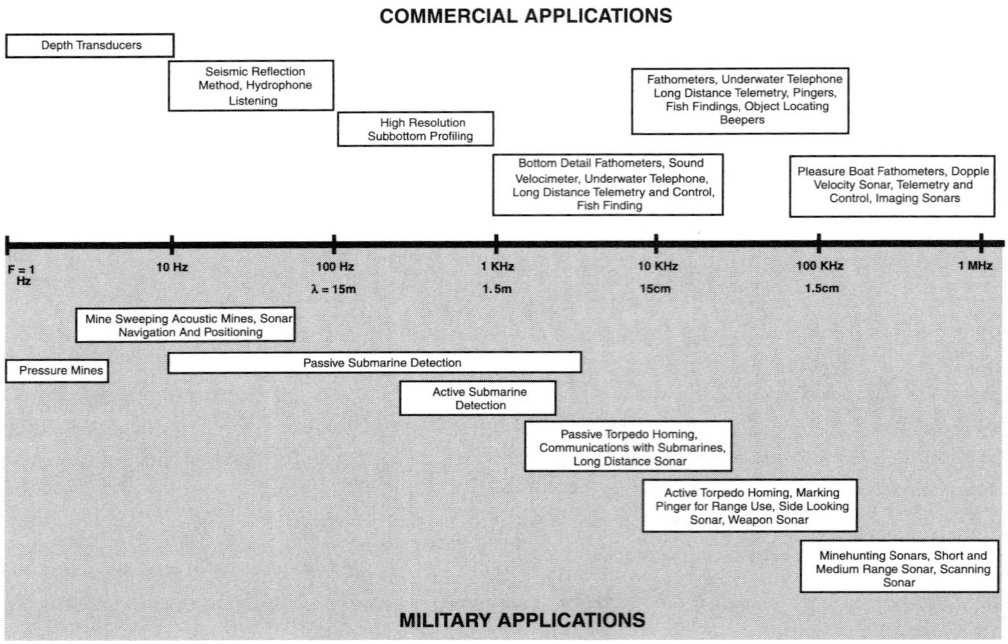

FIGURE 10.15 Sonar frequency allocation. (Modified from Neitzel, E. B., 1973, Civil Uses of Underwater Acoustics, Lectures on Marine Acoustics, AD 156-052.)

information from the different receivers is sent to a common control unit where the acoustic data are combined and processed (hence the name *data fusion*). After fusion, a decision can be relayed or "fed" back to each of the receivers. If data transmission is a concern, due to time constraints, cost, or security, other techniques can be used in which each receiver makes a decision and transmits only the decision. The control unit makes a global decision based on the decisions of all the receivers and relays this global decision back to the receivers. This is called "distributed detection." The receivers can then be asked to re-evaluate their individual decisions based on the new global decision. This process could continue until all the receivers are in agreement or could be terminated whenever an acceptable level of consensus is attained.

An advantage of data fusion is that the receivers can be located at different ranges (e.g., on two different ships), in different mediums (e.g., shallow or deep water, or even at the surface), and at different bearings from the object, thus giving comprehensive information about the object or the underwater acoustic channel.

10.6 Application

Since World War II, in addition to military applications, there has been an expansion in commercial and industrial underwater acoustic applications. Table 10.2 lists the military and nonmilitary functions of sonar along with some of the current applications. Figure 10.15 shows the sonar frequency allocations for military and commercial applications.

Acknowledgment

The authors thank Louise Miller for her assistance in preparing this document.

References

1. Burdic, W. S., 1984, *Underwater Acoustic System Analysis*, Prentice-Hall, Englewood Cliffs, NJ.
2. Carter, G. C., 1987, Coherence and Time Delay Estimation, *Proc. IEEE,* 75(2), February, 236–255.
3. Carter, G. C., Ed., 1993, *Coherence and Time Delay Estimation*, IEEE Press, Piscataway, NJ.
4. Chan, Y. T., Ed., 1989, *Digital Signal Processing for Sonar Underwater Acoustic Signal Processing*, NATO ASI Series, Series E: Applied Sciences, Vol. 161, Kluwer Academic Publishers, Dordrecht.
5. Cox, A. W., 1974, *Sonar and Underwater Sound*, Lexington Books, D.C. Health and Company, Lexington, MA.
6. Hueter, T. F., 1972, Twenty Years in Underwater Acoustics: Generation and Reception, *J. Acoust. Soc. Am.*, 51(3), Part 2, March, 1025–1040.
7. Knight, W. C., Pridham, R. G., and Kay, S. M., 1981, Digital Signal Processing for Sonar, *Proc. IEEE,* 69(11), November, 1451–1506.
8. Oppenheim, A. V., Ed., 1980, *Applications of Digital Signal Processing*, Prentice-Hall, Englewood Cliffs, NJ.
9. Rihaczek, A. W., Ed., 1985, *Principles of High Resolution Radar*, Peninsula Publishing, Los Altos, CA.
10. Spindel, R. C., 1985, Signal Processing in Ocean Tomography, in *Adaptive Methods in Underwater Acoustics*, edited by H.G. Urban, D. Reidel Publishing Company, Dordrecht, pp. 687–710.
11. Urick, R. J., 1983, *Principles of Underwater Sound*, McGraw-Hill, New York.
12. Van Trees, H. L., 1968, *Detection, Estimation, and Modulation Theory*, John Wiley & Sons, New York.
13. Winder, A. A., 1975, II. Sonar System Technology, *IEEE Trans. Sonics and Ultrasonics*, su-22(5), September, 291–332.

Further Information

Journal of Acoustical Society of America (*JASA*); *IEEE Transactions on Signal Processing* (formerly the *IEEE Transactions on Acoustics, Speech and Signal Processing*), and *IEEE Journal of Oceanic Engineering* are professional journals providing current information on underwater acoustical signal processing.

The annual meetings of the International Conference on Acoustics, Speech and Signal Processing, sponsored by the IEEE, and the biannual meetings of the Acoustical Society of America are good sources for current trends and technologies.

"Digital Signal Processing for Sonar," (Knight et al., 1981) and "Sonar System Technology" (Winder, 1975) are informative and detailed tutorials on underwater sound systems. Also, the March 1972 issue of *Journal of Acoustical Society of America* (Hueter) has historical and review papers on underwater acoustics related topics.

11

Theory and Implementation of Advanced Signal Processing for Active and Passive Sonar Systems

Stergios Stergiopoulos
Defence and Civil Institute of Environmental Medicine

University of Western Ontario

Geoffrey Edelson
Sanders, A Lockheed Martin Company

Progress in the implementation of state-of-the-art signal processing schemes in sonar systems has been limited mainly by the moderate advancements made in sonar computing architectures and the lack of operational evaluation of the advanced processing schemes. Until recently, sonar computing architectures allowed only fast-Fourier-transform (FFT), vector-based processing schemes because of their ease of implementation and their cost-effective throughput characteristics. Thus, matrix-based processing techniques, such as adaptive, synthetic aperture, and high-resolution processing, could not be efficiently implemented in sonar systems, even though it is widely believed that they have advantages that can address the requirements associated with the difficult operational problems that next-generation sonars will have to solve. Interestingly, adaptive and synthetic aperture techniques may be viewed by other disciplines as conventional schemes. However, for the sonar technology discipline, they are considered as advanced signal processing schemes because of the very limited progress that has been made in their implementation in sonar systems.

The mainstream conventional signal processing of current sonar systems consists of a selection of temporal and spatial processing algorithms that have been discussed in Chapter 6. However, the drastic

0-8493-3691-0/01/$0.00+$.50

changes in the target acoustic signatures suggest that fundamentally new concepts need to be introduced into the signal processing structure of next-generation sonar systems.

This chapter is intended to address issues of improved performance associated with the implementation of adaptive and synthetic aperture processing schemes in integrated active-passive sonar systems. Using target tracking and localization results as performance criteria, the impact and merits of these advanced processing techniques are contrasted with those obtained using the conventional beamformer.

11.1 Introduction

Several review articles[1–4] on sonar system technology have provided a detailed description of the mainstream sonar signal processing functions along with the associated implementation considerations. The attempt with this chapter is to extend the scope of these articles[1–4] by introducing an implementation effort of non-mainstream processing schemes in real-time sonar systems. The organization of the chapter is as follows.

Section 11.1 provides a historical overview of sonar systems and introduces the concept of the signal processor unit and its general capabilities. This section also outlines the practical importance of the topics to be discussed in subsequent sections, defines the sonar problem, and provides an introduction into the organization of the chapter.

Section 11.2 introduces the development of a realizable generic processing scheme that allows the implementation and testing of *non-linear processing techniques* in a wide spectrum of real-time active and passive sonar systems. Finally, a concept demonstration of the above developments is presented in Section 11.3, which provides real data outputs from an advanced beamforming structure incorporating adaptive and synthetic aperture beamformers.

11.1.1 Overview of a Sonar System

To provide a context for the material contained in this chapter, it would seem appropriate to briefly review the basic requirements of a high-performance sonar system. A *sonar* (SOund, NAvigation, and Ranging) system is defined as a "method or equipment for determining by underwater sound the presence, location, or nature of objects in the sea."[5] This is equivalent to detection, localization, and classification as discussed in Chapter 6.

The main focus of the assigned tasks of a modern sonar system will vary from the detection of signals of interest in the open ocean to very quiet signals in very cluttered underwater environments, which could be shallow coastal sea areas. These varying degrees of complexity of the above tasks, however, can be grouped together quantitatively, and this will be the topic of discussion in the following section.

11.1.2 The Sonar Problem

A convenient and accurate integration of the wide variety of effects of the underwater environment, the target's characteristics, and the sonar system's designing parameters is provided by the sonar equation.[8] Since World War II, the sonar equation has been used extensively to predict the detection performance and to assist in the design of a sonar system.

11.1.2.1 The Passive Sonar Problem

The passive sonar equation combines, in logarithmic units (i.e., units of decibels [dB] relative to the standard reference of energy flux density of rms pressure of 1 μPa integrated over a period of 1 s), the following terms:

$$(S - TL) - (N_e - AG) - DT \geq 0, \qquad (11.1)$$

which define signal excess where

S is the source energy flux density at a range of 1 m from the source.
TL is the propagation loss for the range separating the source and the sonar array receiver. Thus, the term $(S - TL)$ expresses the recorded signal energy flux density at the receiving array.
N_e is the noise energy flux density at the receiving array.
AG is the array gain that provides a quantitative measure of the coherence of the signal of interest with respect to the coherence of the noise across the line array.
DT is the detection threshold associated with the decision process that defines the SNR at the receiver input required for a specified probability of detection and false alarm.

A detailed discussion of the DT term and the associated statistics is given in References 8 and 28 to 30. Very briefly, the parameters that define the detection threshold values for a passive sonar system are the following:

- The time-bandwidth product defines the integration time of signal processing. This product consists of the term T, which is the time series length for coherent processing such as the FFT, and the incoherent averaging of the power spectra over K successive blocks. The reciprocal, $1/T$, of the FFT length defines the bandwidth of a single frequency cell. An optimum signal processing scheme should match the acoustic signal's bandwidth with that of the FFT length T in order to achieve the predicted DT values.
- The probabilities of detection, P_D, and false-alarm, P_{FA}, define the confidence that the correct decision has been made.

Improved processing gain can be achieved by incorporating segment overlap, windowing, and FFT zeroes extension as discussed by Welch[31] and Harris.[32] The definition of DT for the narrow-band passive detection problem is given by[8]

$$DT = 10\log\frac{S}{N_e} = 5\log\left(\frac{d \cdot BW}{t}\right), \tag{11.2}$$

where N_e is the noise power in a 1-Hz band, S is the signal power in bandwidth BW, t is the integration period in displays during which the signal is present, and $d = 2t(S/N_e)$ is the detection index of the receiver operating characteristic (ROC) curves defined for specific values of P_D and P_{FA}.[8,28] Typical values for the above parameters in the term DT that are considered in real-time narrowband sonar systems are $BW = O(10^{-2})$ Hz, $d = 20$, for $P_D = 50\%$, $P_{FA} = 0.1\%$, and $t = O(10^2)$ seconds.

The value of TL that makes Equation 11.1 become an equality leads to the equation

$$\text{FOM} = S - N_e - AG) - DT, \tag{11.3}$$

where the new term "FOM" (figure of merit) equals the transmission loss TL and gives an indication of the range at which a sonar can detect a target.

The noise term N_e in Equation 11.1 includes the total or composite noise received at the array input of a sonar system and is the linear sum of all the components of the noise processes, which are assumed independent. However, detailed discussions of the noise processes related to sonar systems are beyond the scope of this chapter and readers interested in these noise processes can refer to other publications on the topic.[8,33–39]

When taking the sonar equation as the common guide as to whether the processing concepts of a passive sonar system will give improved performance against very quiet targets, the following issues become very important and appropriate:

- During passive sonar operations, the terms S and TL are beyond the sonar operators' control because S and TL are given as parameters of the sonar problem. DT is associated mainly with the design of the array receiver and the signal processing parameters. The signal processing parameters in Equation 11.2 that influence DT are adjusted by the sonar operators so that DT will have the maximum positive impact in improving the FOM of a passive sonar system. The discussion in Section 11.1.2.2 on the active sonar problem provides details for the influence of DT by an active sonar's signal processing parameters.

- The quantity $(N_e - AG)$ in Equations 11.1 and 11.3, however, provides opportunities for sonar performance improvements by *increasing the term AG* (e.g., deploying large size array receivers or using new signal processing schemes) and by *minimizing the term* N_e (e.g., using adaptive processing by taking into consideration the directional characteristics of the noise field and by reducing the impact of the sensor array's self noise levels).

Our emphasis in the sections of this chapter that deal with passive sonar will be focused on the minimization of the quantity $(N_e - AG)$. This will result in new signal processing schemes in order to achieve a desired level of performance improvement for the specific case of a line array sonar system.

11.1.2.2 The Active Sonar Problem

The criterion for sonar system detection requires the signal power collected by the receiver system to exceed the background level by some threshold. The minimum SNR needed to achieve the design false alarm and detection probabilities is called the detection threshold as discussed above. Detection generally occurs when the signal excess is non-negative, i.e., $SE = \text{SNR} - DT \geq 0$. The signal excess for passive sonar is given by Equation 11.1.

A very general active sonar equation for signal excess in decibels is

$$SE = EL - IL - DT, \tag{11.4}$$

in which EL and IL denote the echo level and interference level, respectively. For noise-limited environments with little to no reverberation, the echo and interference level terms in Equation 11.4 become

$$\begin{aligned} EL &= S - TL_1 + TS - TL_2 + \text{AGS} - L_{sp} \\ IL &= NL + AGN, \end{aligned} \tag{11.5}$$

in which TL_1 is the transmission loss from the source to the target, TS is the target strength, TL_2 is the transmission loss from the target to the receiver, L_{sp} denotes the signal processing losses, AGS is the gain of the receiver array on the target echo signal, and AGN is the gain of the receiver on the noise. Array gain (AG), as used in Chapter 6, is defined as the difference between AGS and AGN. All of these terms are expressed in decibels.

In noise-limited active sonar, the SNR, defined as the ratio of signal energy (S) to the noise power spectral density at the processor input (NL) and expressed in decibels, is the fundamental indicator of system performance. Appropriately, the detection threshold is defined as $DT = 10\log(\text{S/NL})$. From the active sonar equation for noise-limited cases, we see that one simple method of increasing the signal excess is to increase the transmitted energy.

If the interference is dominated by distributed reverberation, the echo level term does not change, but the interference level term becomes

$$IL = S - TL'_1 + 10\log(\Omega_s) + S_x - TL'_2 + \text{AGS}' - L'_{sp}, \tag{11.6}$$

in which the transmission loss parameters for the out and back reverberation paths are represented by the primed TL quantities and S_x is the scattering strength of the bottom (dB re m^2), surface (dB re m^2), or volume (dB re m^3). The terms for the gain of the receive array on the reverberation signal and for the signal processing losses are required because the reverberation is different in size from the target and they are not co-located. Ω_s is the scattering area in square meters for the bottom (or surface) or the scattering volume in cubic meters. The scattering area for distributed bottom and surface reverberation at range R is $R\phi((c\tau)/2)$, in which ϕ is the receiver beamwidth in azimuth, c is the speed of sound, and

τ is the effective pulse length after matched filter processing. For a receiver with a vertical beamwidth of θ, the scattering volume for volume reverberation is $(R\phi((c\tau)/2))R\theta$.

The resulting active sonar equation for signal excess in distributed reverberation is

$$\begin{aligned} SE &= (TL'_1 - TL_1) + (TS - 10\log(\Omega_s) - S_x) + (TL'_2 - TL_2) \\ &+ (\mathrm{AGS} - \mathrm{AGS}') - (L_{sp} - L'_{sp}) - DT \end{aligned} \tag{11.7}$$

Of particular interest is the absence of the signal strength from Equation 11.7. Therefore, unlike the noise-limited case, increasing the transmitted energy does not increase the received signal-to-reverberation ratio.

In noise-limited active sonar, the formula for DT depends on the amount known about the received signal.[111] In the case of a completely known signal with the detection index as defined in Section 11.1.2.1, the detection threshold becomes $DT = 10\ \log(d/2\omega t)$, where ω is the signal bandwidth. In the case of a completely unknown signal in a background of Gaussian noise when the SNR is small and the time-bandwidth product is large, the detection threshold becomes $DT = 5\ \log(d/\omega t)$, provided that the detection index is defined as $d = \omega t \cdot (S/NL)^2$.[111] Thus, the noise-limited detection threshold for these cases improves with increasing pulse length and bandwidth.

In reverberation-limited active sonar, if the reverberation power is defined at the input to the receiver as $R = U_R t$ in which U_R is the reverberation power per second of pulse duration, then S/U_R becomes the measure of receiver performance.[112] For the cases of completely known and unknown signals, the detection thresholds are $DT = 10\ \log(d/2\omega_R)$ and $DT = 5\ \log(dt/2\omega_R)$, respectively, with ω_R defined as the effective reverberation bandwidth. Therefore, the reverberation-limited detection threshold improves with increasing ω_R.

Thus, a long-duration, wideband active waveform is capable of providing effective performance in both the noise-limited and reverberation-limited environments defined in this section.

11.2 Theoretical Remarks

Sonar operations can be carried out by a wide variety of naval platforms, as shown in Figure 11.1. This includes surface vessels, submarines, and airborne systems such as airplanes and helicopters. Shown also in Figure 11.1A is a schematic representation of active and passive sonar operations in an underwater sea environment. Active sonar operations involve the transmission of well-defined acoustic signals, which illuminate targets in an underwater sea area. The reflected acoustic energy from a target provides the sonar array receiver with a basis for detection and estimation. The major limitations to robust detection and classification result from the energy that returns to the receiver from scattering bodies also illuminated by the transmitted pulses.

Passive sonar operations base their detection and estimation on acoustic sounds that emanate from submarines and ships. Thus, in passive systems only, the receiving sensor array is under the control of the sonar operators. In this case, major limitations in detection and classification result from imprecise knowledge of the characteristics of the target radiated acoustic sounds.

The depiction of the combined active and passive acoustic systems shown in Figure 11.1 includes towed line arrays, hull-mounted arrays, a towed source, a dipping sonar, and vertical line arrays. Examples of some active systems that operate in different frequency regimes are shown in Figures 11.1B through 11.3C. The low-frequency (LF) sources in Figure 11.1B are used for detection and tracking at long ranges, while the hull-mounted spherical and cylindrical mid-frequency (MF) sonars shown in Figures 11.2A and 11.2B are designed to provide the platform with a tactical capability.

The shorter wavelengths and higher bandwidth attributable to high-frequency (HF) active sonar systems like those shown in Figures 11.3A and 11.3C yield greater range and bearing resolution compared to lower frequency systems. This enables better spatial discrimination, which can be broadly applied, from the geological mapping of the seafloor to the detection and classification of man-made objects. Figure 11.3C shows the geological features of an undersea volcano defined by an HF active sonar. These

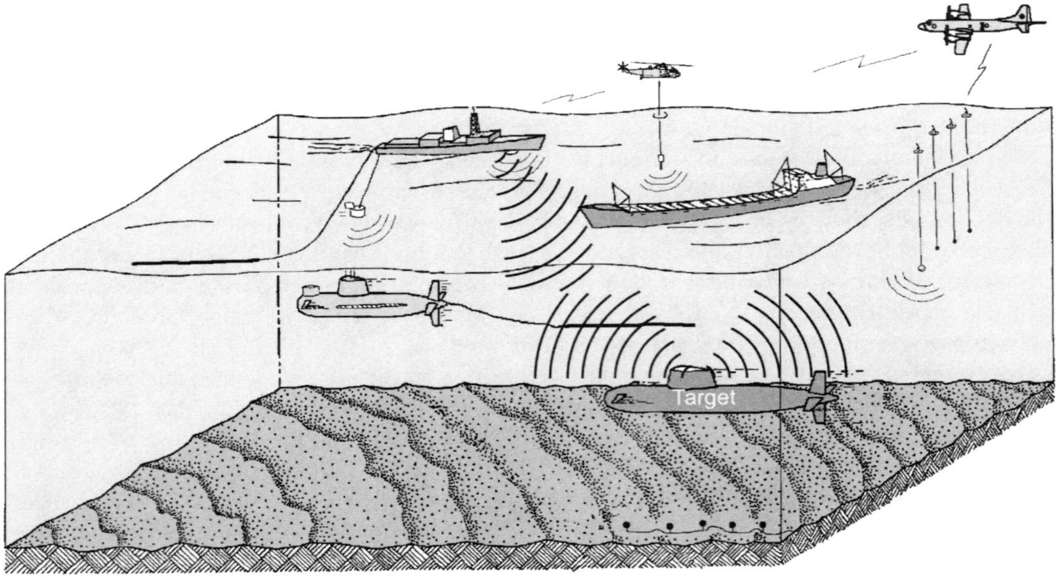

(A)

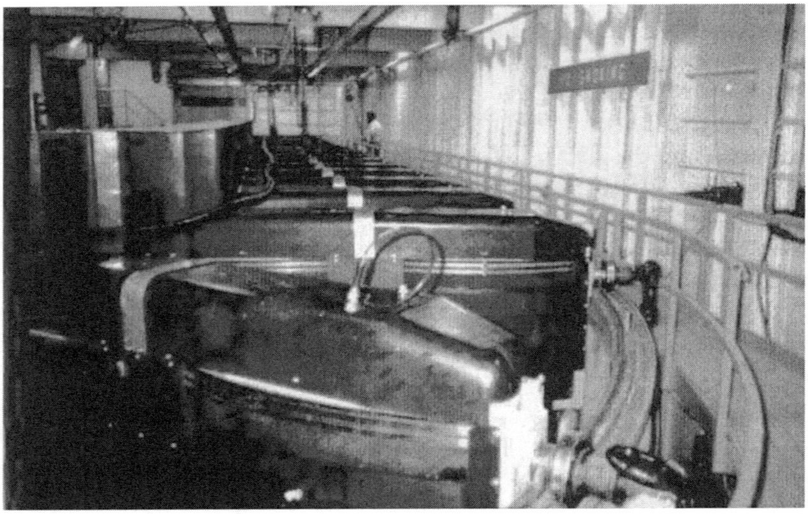

(B)

FIGURE 11.1 (A) Schematic representation for active and passive sonar operations for a wide variety of naval platforms in an underwater sea environment. (Reprinted by permission of IEEE © 1998.) (B) Low-frequency sonar projectors inside a surface ship. (Photo provided courtesy of Sanders, A Lockheed Martin Company.)

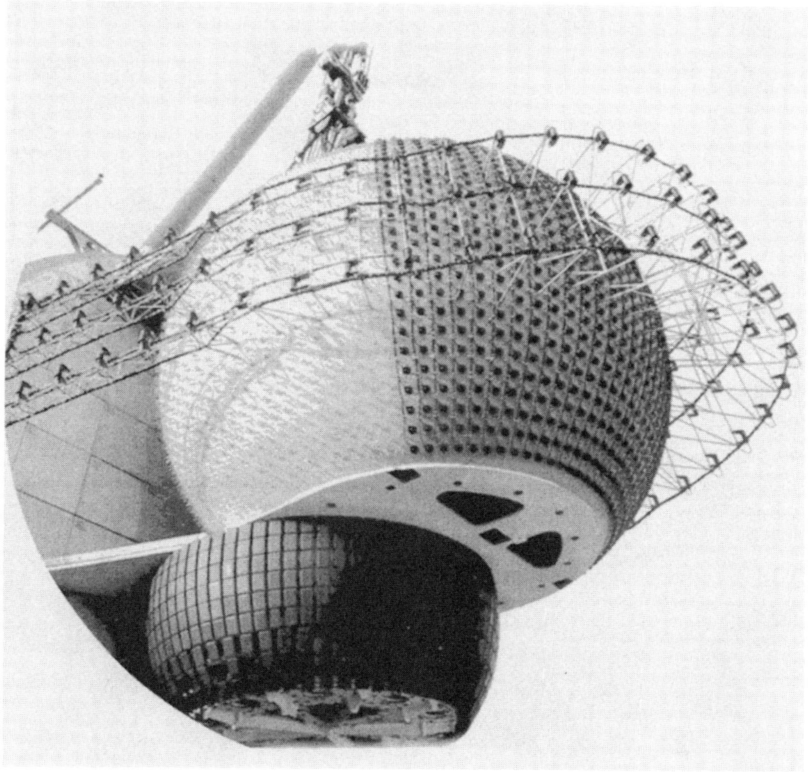

(A)

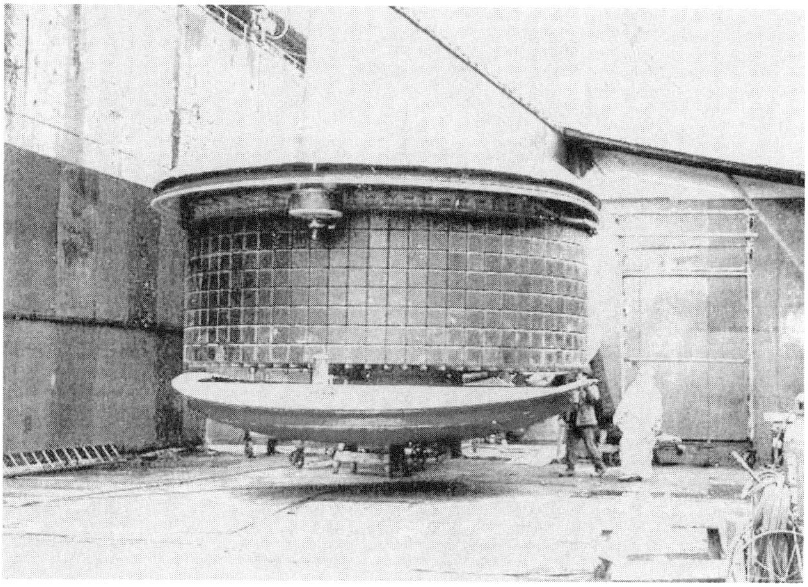

(B)

FIGURE 11.2 (A) A bow-installed, mid-frequency spherical array. (Photo provided courtesy of the Naval Undersea Warfare Center.) (B) A mid-frequency cylindrical array on the bow of a surface ship. (Photo provided courtesy of the Naval Undersea Warfare Center.)

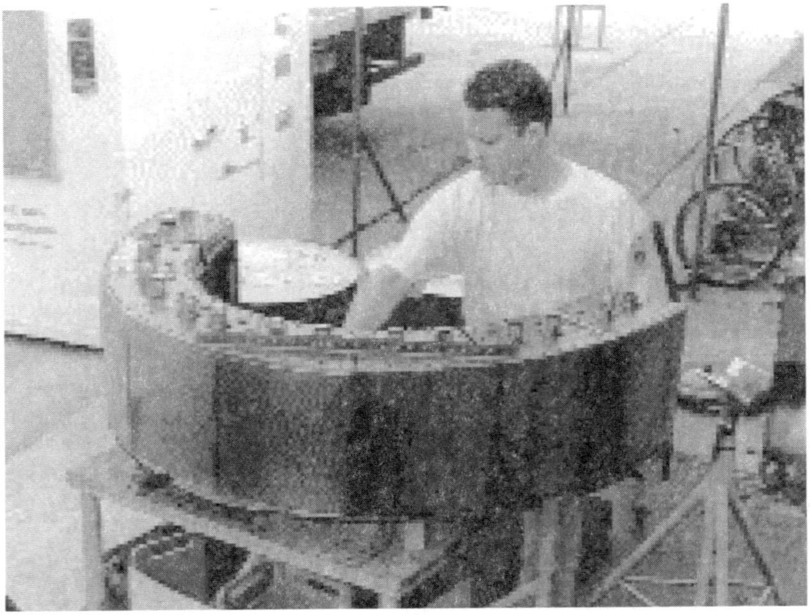

(A)

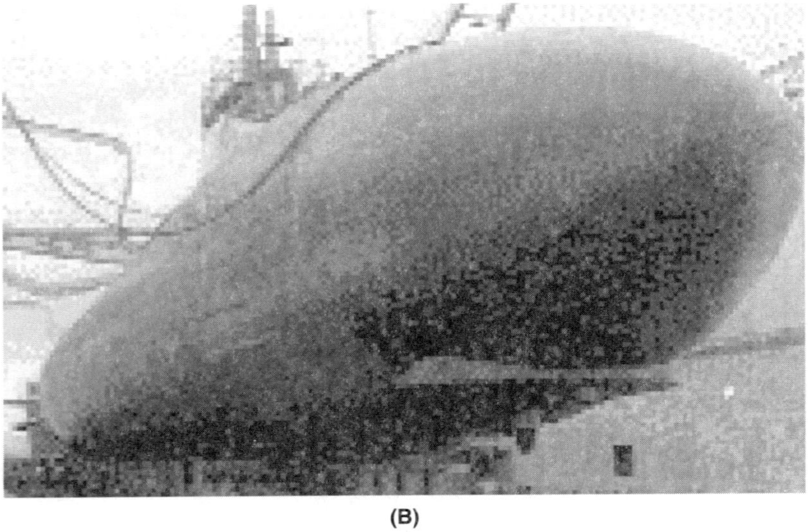

(B)

FIGURE 11.3 (A) Preparation of a high-frequency cylindrical array for installation in a submarine. (Photo provided courtesy of Undersea Warfare Magazine.) (B) High-frequency receiver and projector arrays visible beneath the bow dome of a submarine. (Photo provided courtesy of Undersea Warfare Magazine.) *(continued)*

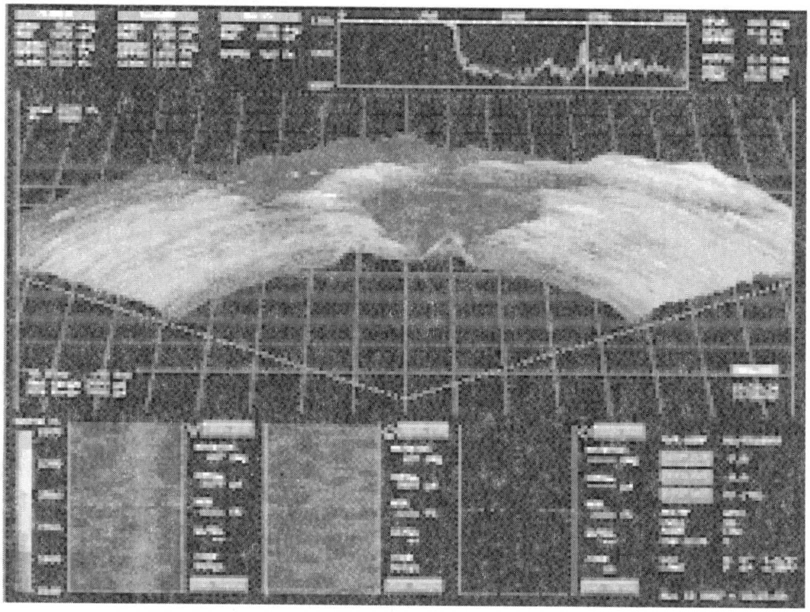

(C)

FIGURE 11.3 (CONTINUED) (C) Output display of a high-frequency sonar system showing the geological features of an undersea volcano. (Photo provided courtesy of Undersea Warfare Magazine.)

spatial gains are especially useful in shallow water for differentiating undersea objects from surface and bottom reverberation. HF arrays have also been used successfully as passive receivers.[113]

The passive sonar concept, in general, can be made clearer by comparing sonar systems with radars, which are always active. Another major difference between the two systems arises from the fact that sonar system performance is more affected than that of radar systems by the underwater medium propagation characteristics. All the above issues have been discussed in several review articles[1–4] that form a good basis for interested readers to become familiar with "main stream" sonar signal processing developments. Therefore, discussions of issues of conventional sonar signal processing, detection, and estimation and the influence of the medium on sonar system performance are briefly highlighted in this section in order to define the basic terminology required for the presentation of the main theme of this chapter.

Let us start with a basic system model that reflects the interrelationships between the target, the underwater sea environment (medium), and the receiving sensor array of a sonar system. A schematic diagram of this basic system is shown in Figure 6.3 of Chapter 6, where sonar signal processing is shown to be two-dimensional (2-D)[1,12,40] in the sense that it involves both temporal and spatial spectral analysis. The temporal processing provides spectral characteristics that are used for target classification, and the spatial processing provides estimates of the directional characteristics (i.e., bearing and possibly range) of a detected signal. Thus, *space-time processing* is the fundamental processing concept in sonar systems, and it has already been discussed in Chapter 6.

11.2.1 Definition of Basic Parameters

This section outlines the context in which the sonar problem can be viewed in terms of models of acoustic signals and noise fields. The signal processing concepts that are discussed in Chapter 6 have been included in sonar and radar investigations with sensor arrays having circular, planar, cylindrical, and spherical geometric configurations. Therefore, the objective of our discussion in this section is to integrate the advanced signal processing developments of Chapter 6 with the sonar problem. For geometrical simplicity and without any loss of generality, we consider here an N hydrophone line array receiver with sensor spacing δ. The output of the nth sensor is a time series denoted by $x_n(t_i)$, where ($i = 1, \ldots, M$) are the time samples for each sensor time series. An * denotes complex conjugate transposition so that $\bar{x}^*$ is the row vector of the received N hydrophone time series $\{x_n(t_i), n = 1, 2, \ldots, N\}$. Then $x_n(t_i) = s_n(t_i) + \varepsilon_n(t_i)$, where $s_n(t_i)$, $\varepsilon_n(t_i)$ are the signal and noise components in the received sensor time series. $\bar{S}$, $\bar{\varepsilon}$ denote the column vectors of the signal and noise components of the vector $\bar{x}$ of the sensor outputs (i.e., $\bar{x} = \bar{S} + \bar{\varepsilon}$). $X_n(f) = \sum_{f=1}^{M} x_n(t_i)\exp(-j2\pi f t_i)$ is the Fourier transform of $x_n(t_i)$ at the signal with frequency f, $c = f\lambda$ is the speed of sound in the underwater medium, and λ is the wavelength of the frequency f. $S = E\{\bar{S}\,\bar{S}^*\}$ is the spatial correlation matrix of the signal vector $\bar{S}$, whose n^{th} element is expressed by

$$s_n(t_i) = s_n[t_i + \tau_n(\theta)], \tag{11.8}$$

E{...} denotes expectation, and

$$\tau_n(\theta) = (n-1)\delta\cos\theta/c \tag{11.9}$$

is the time delay between the first and the nth hydrophone of the line array for an incoming plane wave with direction of propagation θ, as illustrated in Figure 6.3 of Chapter 6.

In this chapter, the problem of detection is defined in the classical sense as a hypothesis test that provides a detection probability and a probability of false alarm, as discussed in Chapter 6. This choise of definition is based on the standard CFAR processor, which is based on the Neyman-Pearson criterion.[28] The CFAR processor provides an estimate of the ambient noise or clutter level so that the threshold can be varied dynamically to stabilize the false alarm rate. Ambient noise estimates for the CFAR processor are provided mainly by noise normalization techniques[42–45] that account for the slowly varying changes in the background noise or clutter. The above estimates of the ambient noise are based upon the average value of the received signal, the desired probability of detection, and the probability of false alarms.

Furthermore, optimum beamforming, which has been discussed in Chapter 6, requires the beamforming filter coefficients to be chosen based on the covariance matrix of the received data by the N sensor array in order to optimize the array response.[46,47] The family of algorithms for optimum beamforming that use the characteristics of the noise are called *adaptive beamformers*,[2,11,12,46–49] and a detailed definition of an adaptation process requires knowledge of the correlated noise's covariance matrix $R(f_i)$. For adaptive beamformers, estimates of $R(f_i)$ are provided by the spatial correlation matrix of received hydrophone time series with the nm^{th} term, $R_{nm}(f,d_{nm})$, defined by

$$R_{nm}(f, \delta_{nm}) = E[X_n(f)X_m^*(f)] \tag{11.10}$$

$R'_\varepsilon(f_i) = \sigma_n^2(f_i)R_\varepsilon(f_i)$ is the spatial correlation matrix of the noise for the ith frequency bin with $\sigma_n^2(f_i)$ being the power spectral density of the noise $\varepsilon_n(t_i)$. The discussion in Chapter 6 shows that if the statistical properties of an underwater environment are equivalent with those of a white noise field, then the *conventional beamformer* (CBF) without shading is the optimum beamformer for bearing estimation, and the variance of its estimates achieve the CRLB bounds. For the narrowband CBF, the plane wave response of an N hydrophone line array steered at direction θ_s is defined by[12]

$$B(f, \theta_s) = \sum_{n=1}^{N} X_n(f)d_n(f, \theta_s), \tag{11.11}$$

where $d_n(f, \theta_s)$ is the nth term of the steering vector $\bar{D}(f, \theta_s)$ for the beam steering direction θ_s, as expressed by

$$d_n(f_i, \theta) = \exp\left[j2\pi\frac{(i-1)f_s}{M}\tau_n(\theta)\right], \tag{11.12}$$

where f_s is the sampling frequency.

The beam power pattern $P(f, \theta_s)$ is given by $P(f, \theta_s) = B(f, \theta_s)B^*(f, \theta_s)$. Then, the power beam pattern $P(f, \theta_s)$ takes the form

$$P(f, \theta_s) = \sum_{n=1}^{N}\sum_{m=1}^{N} X_n(f)X_m^*(f)\exp\left[\frac{j2\pi f\delta_{nm}\cos\theta_s}{c}\right], \tag{11.13}$$

where δ_{nm} is the spacing $\delta(n - m)$ between the nth and m^{th} hydrophones. Let us consider for simplicity the source bearing θ to be at array broadside, $\delta = \lambda/2$, and $L = (N - 1)\delta$ to be the array size. Then Equation 11.13 is modified as[3,40]

$$P(f, \theta_s) = \frac{N^2\sin^2\left[\frac{\pi L\sin\theta_s}{\lambda}\right]}{\left(\frac{\pi L\sin\theta_s}{\lambda}\right)^2}, \tag{11.14}$$

which is the far-field radiation or directivity pattern of the line array as opposed to near-field regions.

Equation 11.11 can be generalized for non-linear 2-D and 3-D arrays, and this is discussed in Chapter 6. The results in Equation 11.14 are for a perfectly coherent incident acoustic signal, and an increase in array size $L = \delta(N - 1)$ results in additional power output and a reduction in beamwidth. The sidelobe structure of the directivity pattern of a receiving array can be suppressed at the expense of a beamwidth increase by applying different weights. The selection of these weights will act as spatial filter coefficients with optimum performance.[4,11,12] There are two different approaches to select these weights: **pattern optimization** and **gain optimization**. For **pattern optimization**, the desired array response pattern $B(f, \theta_s)$ is selected first. A desired pattern is usually one with a narrow main lobe and low sidelobes. The weighting or shading coefficients in this case are real numbers from well-known window functions that modify the array response pattern. Harris' review[32] on the use of windows in discrete Fourier transforms and temporal spectral analysis is directly applicable in this case to spatial spectral analysis for towed line array applications.

Using the approximation $\sin\theta \cong \theta$ for small θ at array broadside, the first null in Equation 11.14 occurs at $\pi L\sin\theta/\lambda = \pi$ or $\Delta\theta \times L/\lambda \cong 1$. The major conclusion drawn here for line array applications is that[3,40]

$$\Delta\theta \approx \lambda/L \text{ and } \Delta f \times T = 1, \tag{11.15}$$

where $T = M/F_s$ is the hydrophone time series length. Both relations in Equation 11.15 express the well-known temporal and spatial resolution limitations in line array applications that form the driving force and motivation for adaptive and synthetic aperture signal processing techniques that have been discussed in Chapter 6.

An additional constraint for sonar applications requires that the frequency resolution Δf of the hydrophone time series for spatial spectral analysis, which is based on FFT beamforming processing, must be

$$\Delta f \times \frac{L}{c} \ll 1 \tag{11.16}$$

to satisfy *frequency quantization* effects associated with the implementation of the beamforming process as **f**inite-**d**uration **i**mpulse **r**esponse (FIR) filters that have been discussed in Chapter 6. Because of the linearity of the conventional beamforming process, an exact equivalence of the frequency domain narrowband beamformer with that of the time domain beamformer for broadband signals can be derived.[64,68,69] The time domain beamformer is simply a time delaying[69] and summing process across the hydrophones of the line array, which is expressed by

$$b(\theta_s, t_i) = \sum_{n=1}^{N} x_n(t_i - \tau_s). \tag{11.17}$$

Since $b(\theta_s, t_i) = \text{IFFT}\{B(f, \theta_s)\}$, by using FFTs and fast-convolution procedures, continuous beam time sequences can be obtained at the output of the frequency domain beamformer.[64] This is a very useful operation when the implementation of adaptive beamforming processors in sonar systems is considered.

When **gain optimization** is considered as the approach to select the beamforming weights, then the beamforming response is optimized so that the output contains minimal contributions due to noise and signals arriving from directions other than the desired signal direction. For this optimization procedure, it is desired to find a linear filter vector $\overline{W}(f_i, \theta)$, which is a solution to the constrained minimization problem that allows signals from the look direction to pass with a specified gain,[11,12] as discussed in Chapter 6. Then in the frequency domain, an adaptive beam at a steering θ_s is defined by

$$B(f_i, \theta_s) = \overline{W}^*(f_i, \theta_s)\overline{X}(f_i), \tag{11.18}$$

and the corresponding conventional beams are provided by Equation 11.11. Estimates of the adaptive beamforming weights $\overline{W}(f_i, \theta)$ are provided by various adaptive processing techniques that have been discussed in detail in Chapter 6.

11.2.2 System Implementation Aspects

The major development effort discussed in Chapter 6 has been devoted to designing a generic beamforming structure that will allow the implementation of adaptive, synthetic aperture, and spatial spectral analysis techniques in integrated active-passive sonar system. The practical implementation of the numerous adaptive and synthetic aperture processing techniques, however, requires the consideration of the characteristics of the signal and noise, the complexity of the ocean environment, as well as the computational difficulty. The discussion in Chapter 6 addresses these concerns and prepares the ground for the development of the above generic beamforming structure.

The major goal here is to provide a **concept demonstration** of both the sonar technology and advanced signal processing concepts that are proving invaluable in the reduction risk and in ensuing significant innovations occur during the formal development process.

Shown in Figure 11.4 is the proposed configuration of the signal processing flow that includes the implementation of FIR filters and conventional, adaptive, and synthetic aperture beamformers. The reconfiguration of the different processing blocks in Figure 11.4 allows the application of the proposed configuration into a variety of active and/or passive sonar systems. The shaded blocks in Figure 11.4 represent advanced signal processing concepts of next-generation sonar systems, and this basically differentiates their functionality from the current operational sonars. In a sense, Figure 11.4 summarizes the signal processing flow of the advanced signal processing schemes shown in Figures 6.14 and 6.20 to 6.24 of Chapter 6.

The **first point** of the generic processing flow configuration in Figure 11.4 is that its implementation is in the frequency domain. The **second point** is that the frequency domain beamforming (or spatial filtering) outputs can be made equivalent to the FFT of the broadband beamformers outputs with proper selection of beamforming weights and careful data partitioning. This equivalence corresponds to implementing FIR filters via circular convolution. It also allows spatial-temporal processing of narrowband and broadband types of signals as well. As a result, the output of each one of the processing blocks in Figure 11.4 provides continuous time series. This modular structure in the signal processing flow is a very essential processing arrangement, allowing the integration of a great variety of processing schemes such as the ones considered in this study. The details of the proposed generic processing flow, as shown in Figure 11.4, are very briefly the following:

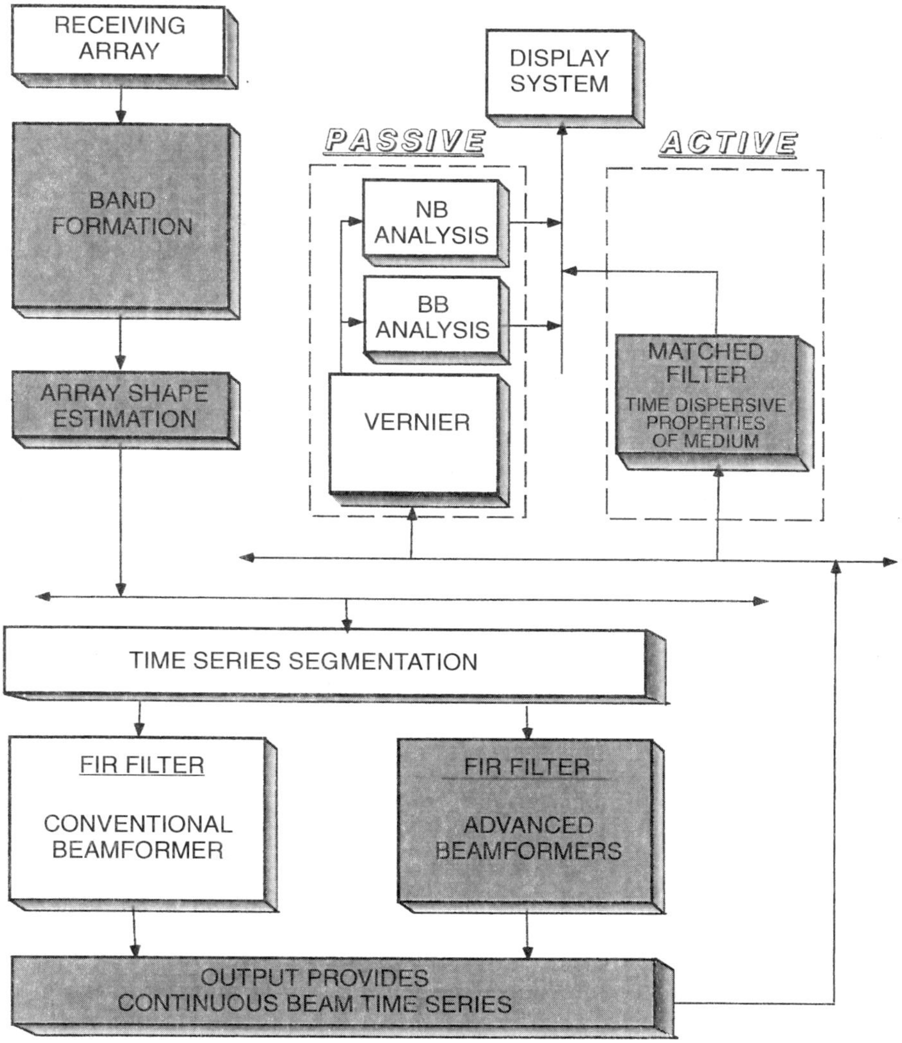

FIGURE 11.4 Schematic diagram of a generic signal processing flow that allows the implementation of non-conventional processing schemes in sonar systems. (Reprinted by permission of IEEE © 1998.)

- The block named as *initial spectral FFT — and Formation* includes the partitioning of the time series from the receiving sensor array, their initial spectral FFT, the selection of the signal's frequency band of interest via bandpass FIR filters, and downsampling.[65–67] The output of this block provides continuous time series at a reduced sampling rate.
- The major blocks including *Conventional Spatial FIR Filtering and Adaptive & Synthetic Aperture FIR Filtering* provide continuous directional beam time series by using the FIR implementation scheme of the spatial filtering via circular convolution.[64–67] The segmentation and overlap of the time series at the input of the beamformers takes care of the wraparound errors that arise in fast-convolution signal processing operations. The overlap size is equal to the effective FIR filter's length.
- The block named *Matched Filter* is for the processing of echoes for active sonar applications. The intention here is to compensate also for the time dispersive properties of the medium by having as an option the inclusion of the medium's propagation characteristics in the replica of the active signal considered in the matched filter in order to improve detection and gain.

- The blocks *Vernier, NB Analysis,* and *BB Analyisis*[67] include the final processing steps of a temporal spectral analysis. The inclusion of the vernier here is to allow the option for improved frequency resolution capabilities depending on the application.
- Finally, the block *Display System* includes the data normalization[42,44] in order to map the output results into the dynamic range of the display devices in a manner which provides a CFAR capability.

The strength of this generic implementation scheme is that it permits, under a parallel configuration, the inclusion of non-linear signal processing methods such adaptive and synthetic aperture, as well as the equivalent conventional approach. This permits a very cost-effective evaluation of any type of improvements during the concept demonstration phase.

All the variations of adaptive processing techniques, while providing good bearing/frequency resolution, are sensitive to the presence of system errors. Thus, the deformation of a towed array, especially during course alterations, can be the source of serious performance degradation for the adaptive beamformers. This performance degradation is worse than it is for the conventional beamformer. So, our concept of the generic beamforming structure requires the integration of towed array shape estimation techniques[73–78] in order to minimize the influence of system errors on the adaptive beamformers. Furthermore, the fact that the advanced beamforming blocks of this generic processing structure provide continuous beam time series allows the integration of passive and active sonar application in one signal processor. Although this kind of integration may exist in conventional systems, the integration of adaptive and synthetic aperture beamformers in one signal processor for active and passive applications has not been reported yet, except for the experimental system discussed in Reference 1. Thus, the beam time series from the output of the conventional and non-conventional beamformers are provided at the input of two different processing blocks, the passive and active processing units, as shown in Figure 11.4.

In the passive unit, the use of verniers and the temporal spectral analysis (incorporating segment overlap, windowing, and FFT coherent processing[31,32]) provide the narrowband results for all the beam time series. Normalization and OR-ing[42,44] are the final processing steps before displaying the output results. Since a beam time sequence can be treated as a signal from a directional hydrophone having the same AG and directivity pattern as that of the above beamforming processing schemes, the display of the narrowband spectral estimates for all the beams follows the so-called LOFAR presentation arrangements, as shown in Figures 11.10 to 11.19 in Section 11.3. This includes the display of the beam-power outputs as a function of time, steering beam (or bearing), and frequency. LOFAR displays are used mainly by sonar operators to detect and classify the narrowband characteristics of a received signal.

Broadband outputs in the passive unit are derived from the narrowband spectral estimates of each beam by means of incoherent summation of all the frequency bins in a wideband of interest. This kind of energy content of the broadband information is displayed as a function of bearing and time, as shown by the real data results of Section 11.3.

In the active unit, the application of a matched filter (or replica correlator) on the beam time series provides coherent broadband processing. This allows detection of echoes as a function of range and bearing for reference waveforms transmitted by the active transducers of a sonar system. The displaying arrangements of the correlator's output data are similar to the LOFAR displays and include, as parameters, range as a function of time and bearing, as discussed in Section 11.2.

At this point, it is important to note that for active sonar applications, waveform design and matched filter processing must not only take into account the type of background interference encountered in the medium, but should also consider the propagation characteristics (multipath and time dispersion) of the medium and the features of the target to be encountered in a particular underwater environment. Multipath and time dispersion in either deep or shallow water cause energy spreading that distorts the transmitted signals of an active sonar, and this results in a loss of matched filter processing gain if the replica has the properties of the original pulse.[1–4,8,54,102,114–115] Results from a study by Hermand and Roderick[103] have shown that the performance of a conventional matched filter can be improved if the reference signal (replica) compensates for the multipath and the time dispersion of the medium. This compensation is a model-based matched filter operation, including the correlation of the received signal with the reference signal (replica) that consists of the transmitted signal convolved with the impulse response of the medium. Experimental results for a one-way propagation problem have shown also that the model-based matched filter approach has improved performance with respect to the conventional matched filter approach by as much as 3.6 dB. The above remarks should be considered as supporting arguments for the inclusion of model-based matched filter processing in the generic signal processing structure shown in Figure 11.4.

11.2.3 Active Sonar Systems

Emphasis in the discussion so far has been centered on the development of a generic signal processing structure for integrated active-passive sonar systems. The active sonar problem, however, is slightly different than the passive sonar problem. The fact that the advanced beamforming blocks of the generic processing structure provide continuous beam time series allows for the integration of passive and active sonar application into one signal processor. Thus, the beam time series from the output of the conventional and non-conventional beamformers are provided at the input of two different processing blocks, the passive and active processing units, as shown in Figure 11.4. In what follows, the active sonar problem analysis is presented with an emphasis on long-range, LF active towed array sonars. The parameters and deployment procedures associated with the short-range active problem are conceptually identical with those of the LF towed array sonars. Their differences include mainly the frequency range of the related sonar signals and the deployment of these sonars, as illustrated schematically in Figure 6.1 of Chapter 6.

11.2.3.1 Low-Frequency Active Sonars

Active sonar operations can be found in two forms. These are referred to as monostatic and bistatic. Monostatic sonar operations require that the source and array receivers be deployed by the same naval vessel, while bistatic or multistatic sonar operations require the deployment of the active source and the receiving arrays by different naval vessels, respectively. In addition, both monostatic and bistatic systems can be air deployed. In bistatic or multi-static sonar operations, coordination between the active source and the receiving arrays is essential. For more details on the principles and operational deployment procedures of multi-static sonars, the reader is referred to References 4, 8, and 54. The signal processing schemes that will be discussed in this section are applicable to both bistatic and monostatic LF active operations. Moreover, it is assumed that the reader is familiar with the basic principles of active sonar systems which can be found in References 4, 28, and 54.

11.2.3.1.1 Signal Ambiguity Function and Pulse Selection

It has been shown in Chapter 6 that for active sonars the optimum detector for a known signal in white Gaussian noise is the correlation receiver.[28] Moreover, the performance of the system can be expressed by means of the ambiguity function, which is the output of the quadrature detector as a function of time delay and frequency. The width of the ambiguity function along the time-delay axis is a measure of the capacity of the system to resolve the range of the target and is approximately equal to

- The duration of the pulse for a continuous wave (CW) signal
- The inverse of the bandwidth of broadband pulses such as linear frequency modulation (LFM), hyperbolic frequency modulation (HFM), and pseudo-random noise (PRN) waveforms

On the other hand, the width of the function along the frequency axis (which expresses the Doppler-shift or velocity tolerance) is approximately equal to

- The inverse of the pulse duration for CW signals
- The inverse of the time-bandwidth product of frequency modulated (FM) types of signals

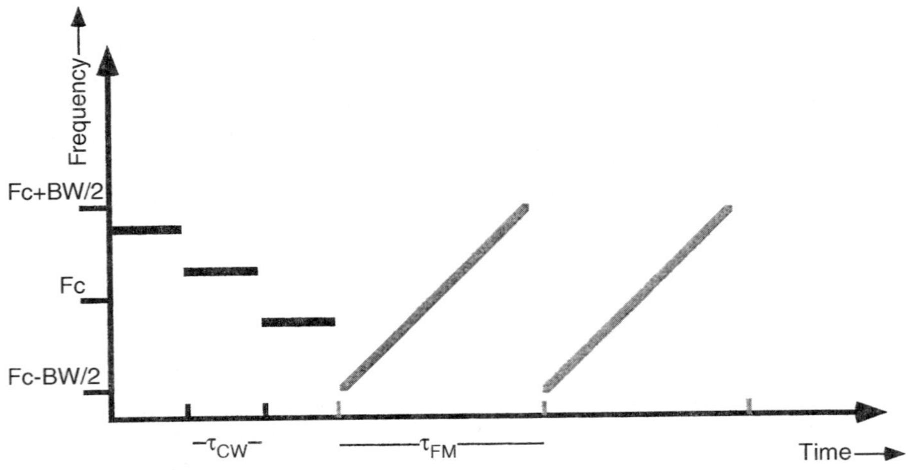

FIGURE 11.5 Sequence of CW and FM types of pulses for an LF active towed array system.

Whalen[28] (p. 348) has shown that in this case there is an uncertainty relation, which is produced by the fact that the time-bandwidth product of a broadband pulse has a theoretical bound. Thus, one cannot achieve arbitrarily good range and Doppler resolution with a single pulse. Therefore, the pulse duration and the signal waveform, whether this is a monochromatic or broadband type of pulse, is an important design parameter. It is suggested that a sequence of CW and FM types of pulses, such as those shown in Figure 11.5, could address issues associated with the resolution capabilities of an active sonar in terms of a detected target's range and velocity. Details regarding the behavior (in terms of the effects of Doppler) of the various types of pulses, such as CW, LFM, HFP, and PRN, can be found in References 8, 28, 54, and 110.

11.2.3.1.2 Effects of Medium

The effects of the underwater environment on active and passive sonar operations have been discussed in numerous papers[1,4,8,41,54,110] and in Chapter 6. Briefly, these effects for active sonars include

- Time, frequency, and angle spreading
- Surface, volume, and bottom scattering
- Ambient and self receiving array noise

Ongoing investigations deal with the development of algorithms for model-based matched filter processing that will compensate for the distortion effects and the loss of matched filter processing gain imposed by the time dispersive properties of the medium on the transmitted signals of active sonars. This kind of model-based processing is identified by the block, *Matched Filter: Time Dispersive Properties of Medium*, which is part of the generic signal processing structure shown in Figure 11.4. It is anticipated that the effects of angle spreading, which are associated with the spatial coherence properties of the medium, will have a minimum impact on LF active towed array operations in blue (deep) waters. However, for littoral water (shallow coastal areas) operations, the medium's spatial coherence properties would impose an upper limit on the aperture size of the deployed towed array, as discussed in Chapter 6.

Furthermore, the medium's time and frequency spreading properties would impose an upper limit on the transmitted pulse's duration τ and bandwidth *Bw*. Previous research efforts in this area suggest that the pulse duration of CW signals in blue waters should be in the range of 2 to 8 s, and in shallow littoral waters in the range of 1 to 2 s long. On the other hand, broadband pulses, such as LFM, HFM, and PRN, when used with active towed array sonars should have upper limits

- For their bandwidth in the range of 300 Hz
- For their pulse duration in the range of 4 to 24 s

Thus, it is apparent by the suggested numbers of pulse duration and the sequence of pulses, shown in Figure 11.5, that the anticipated maximum detection range coverage of LF active towed array sonars should be beyond ranges of the order of $O(10^2)$ km. This assumes, however, that the intermediate range coverage will be carried out by the MF hull-mounted active sonars.

Finally, the effects of scattering play the most important role on the selection of the type of transmitted pulses (whether they will be CW or FM) and the duration of the pulses. In addition, the performance of the matched filter processing will also be affected.

11.2.3.2 Effects of Bandwidth in Active Sonar Operations

If an FM signal is processed by a matched filter, which is an optimum estimator according to the Neyman-Pearson detection criteria, theory predicts[28,110] that a larger bandwidth FM signal will result in improved detection for an extended target in reverberation. For extended targets in white noise, however, the detection performance depends on the SNR of the received echo at the input of the replica correlator.

In general, the performance of a matched filter depends on the temporal coherence of the received signal and the time-bandwidth product of the FM signal in relation to the relative target speed. Therefore, the signal processor of an active sonar may require a variety of matched filter processing schemes that will not have degraded performance when the coherence degrades or the target velocity increases. At this point, a brief overview of some of the theoretical results will be given in order to define the basic parameters characterizing the active signal processing schemes of interest.

It is well known[28] that for a linear FM signal with bandwidth, Bw, the matched filter provides pulse compression and the temporal resolution of the compressed signal is $1/Bw$. Moreover, for extended targets with virtual target length, $T\tau$ (in seconds), the temporal resolution at the output of the matched filter should be matched to the target length, $T\tau$. However, if the length of the reverberation effects is greater than that of the extended target, the reverberation component of bandwidth will be independent in frequency increments, $\Delta Bw > 1/T\tau$.[30,110] Therefore, for an active LF sonar, if the transmitted broadband signal $f(t)$ with bandwidth Bw is chosen such that it can be decomposed into n signals, each with bandwidth $\Delta Bw = Bw/n > 1/T\tau$, then the matched filter outputs for each one of the n signal segments are independent random variables. In this case, called reverberation limited, the SNR at the output of the matched filter is equal for each frequency band ΔBw, and independent of the transmitted signal's bandwidth Bw as long as $Bw/n > 1/T\tau$. This processing arrangement, including segmentation of the transmitted broadband pulse, is called *segmented replica correlator* (SRC).

To summarize the considerations needed to be made for reverberation-limited environments, the area (volume) of scatterers decreases as the signal bandwidth increases, resulting in less reverberation at the receiver. However, large enough bandwidths will provide range resolution narrower than the effective duration of the target echoes, thereby requiring an approach to recombine the energy from time-spread signals. For CW waveforms, the potential increase in reverberation suppression at low Doppler provided by long-duration signals is in direct competition with the potential increase in reverberation returned near the transmit frequency caused by the illumination of a larger area (volume) of scatterers. Piecewise coherent (PC) and geometric comb waveforms have been developed to provide good simultaneous range and Doppler resolution in these reverberation-limited environments. Table 11.1 provides a summary for waveform selection based on the reverberation environment and the motion of the target.

TABLE 11.1 Waveform Considerations in Reverberation

Doppler	Background Reverberation		
	Low	Medium	High
Low	FM	FM	FM
Moderate	FM (CW)	PC (CW, HFM)	CW (PC)
High	CW (HFM)	CW (HFM)	CW

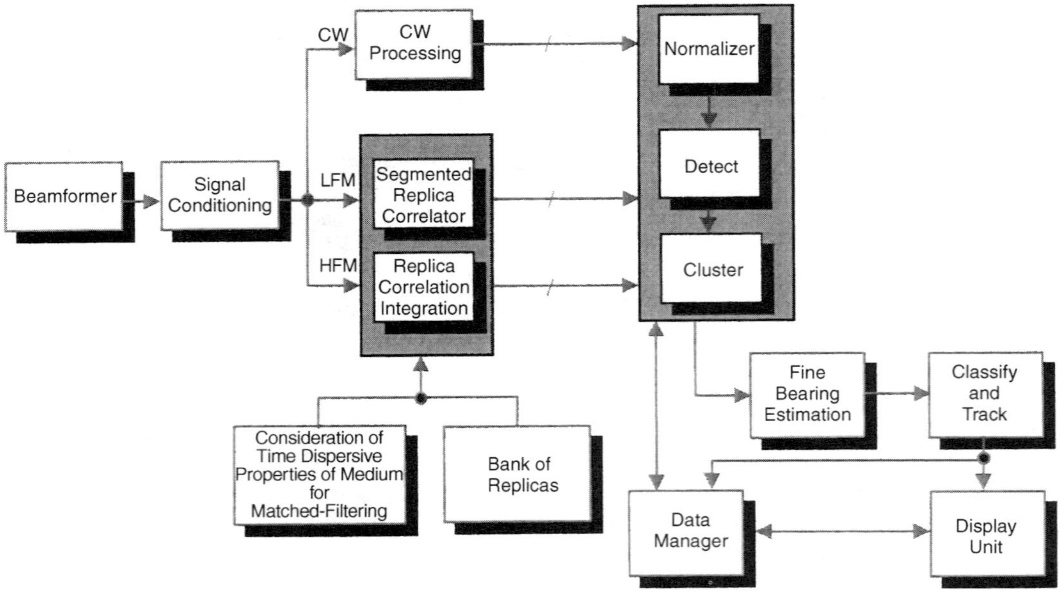

FIGURE 11.6 Active waveform processing block diagram. Inputs to the processing flow of this schematic diagram are the beam time series outputs of the advanced beamformers of Figure 11.4. The various processing blocks indicate the integration of the medium's time dispersive properties in the matched filter and the long or segmented replica correlations for FM type of signals to improve detection performance for noise-limited or reverberation-limited cases discussed in Section 11.2.3.2.

In contrast to the reverberation-limited case, the SNR in the noise-limited case is inversely proportional to the transmitted signal's bandwidth, *Bw*, and this case requires long replica correlation. Therefore, for the characterization of a moving target, simultaneous estimation of time delay and Doppler speed is needed. But for broadband signals, such as HFM, LFM, and PRN, the Doppler effects can no longer be approximated simply as a frequency shift. In addition, the bandwidth limitations, due to the medium and/or the target characteristics, require further processing considerations whether or not a long or segmented replica correlator will be the optimum processing scheme in this case.

It is suggested that a sequence of CW and broadband transmitted pulses, such as those shown in Figure 11.5, and the signal processing scheme, presented in Figure 11.6, could address the above complicated effects that are part of the operational requirements of LF active sonar systems. In particular, the CW and broadband pulses would simultaneously provide sufficient information to estimate the Doppler and time-delay parameters characterizing a detected target. As for the signal processing schemes, the signal processor of an LF active towed array system should allow simultaneous processing of CW pulses as well as bandwidth-limited processing for broadband pulses by means of replica correlation integration and/or segmented replica correlation.

11.2.3.2.1 Likelihood Ratio Test Detectors

This section deals with processing to address the effects of both the bandwidth and the medium on the received waveform. As stated above, the replica correlation (RC) function is used to calculate the likelihood ratio test (LRT) statistic for the detection of high time-bandwidth waveforms.[28] These waveforms can be expected to behave well in the presence of reverberation due to the $1/B$ effective pulse length. Because the received echo undergoes distortion during its two-way propagation and reflection from the target, the theoretical RC gain of 10 logBT relative to a zero-Doppler CW echo is seldom achievable, especially in shallow water where multipath effects are significant. The standard RC matched filter assumes an ideal channel and performs a single coherent match of the replica to the received signal at each point in time. This n^{th} correlation output is calculated as the inner product of the complex conjugate of the transmitted waveform with the received data so that

$$y(n) = \left|\sqrt{2/N}\sum_{i=0}^{N-1} s^*(i)r(i+n)\right|^2. \tag{11.19}$$

One modification to the standard RC approach of creating the test statistic is designed to recover the distortion losses caused by time spreading.[114,115] This statistic is formed by effectively placing an energy detector at the output of the matched filter and is termed "replica correlation integration" (RCI), or long replica correlator (LRC). The RCI test statistic is calculated as

$$y(n) = \sum_{k=0}^{M-1}\left|\sqrt{2/N}\sum_{i=0}^{N-1} s^*(i-k)r(i+n)\right|. \tag{11.20}$$

The implementation of RCI requires a minimal increase in complexity, consisting only of an integration of the RC statistic over a number of samples (M) matched to the spreading of the signal. Sample RCI recovery gains with respect to standard RC matched filtering have been shown to exceed 3 dB.

A second modification to the matched filter LRT statistic, called SRC and introduced in the previous section, is designed to recover the losses caused by fast-fading channel distortion, where the ocean dynamics permit the signal coherence to be maintained only over some period T_c that is shorter than the pulse length.[114,115] This constraint forces separate correlation over each segment of length T_c so that the receiver waveform gets divided into $M_s = T/T_c$ segments, where T is the length of the transmitted pulse. For implementation purposes, M_s should be an integer so that the correlation with the replica is divided into M_s evenly sized segments. The SRC test statistic is calculated as

$$y(n) = \sum_{k=0}^{M_s-1}\left|\sqrt{2M_s/N}\sum_{i=0}^{N/M_s-1} s^*\left(i+\frac{kn}{M_s}\right)r\left(i+n+\frac{kN}{M_s}\right)\right|^2. \tag{11.21}$$

One disadvantage of SRC in comparison to RCI is that SRC does not support multi-hypothesis testing when the amount of distortion is not known *a priori*.[114]

11.2.3.2.2 Normalization and Threshold Detection for Active Sonar Systems

Figure 11.6 presents a processing scheme for active sonars that addresses the concerns about processing waveforms like the one presented in Figure 11.5 and about the difficulties in providing robust detection capabilities. At this point, it is important to note that the block named *Normalizer*[42,44] in Figure 11.6 does not include simple normalization schemes such as those assigned for the LOFAR-grams of a passive sonar, shown in Figure 11.4.

The ultimate goal of any normalizer in combination with a threshold detector is to provide a system-prescribed and constant rate of detections in the absence of a target, while maintaining an acceptable probability of detection when a target is present. The detection statistic processing output (or FFT output for CW waveforms) is normalized and threshold detected prior to any additional processing. The normalizer estimates the power (and frequency) distribution of the mean background (reverberation plus noise) level at the output of the detection statistic processing.

The background estimate for a particular test bin that may contain a target echo is formed by processing a set of data that is assumed to contain no residual target echo components. The decision statistic output of the test bin gets compared to the threshold that is calculated as a function of the background estimate. A threshold detection occurs when the threshold is exceeded. Therefore, effective normalization is paramount to the performance of the active processing flow.

Normalization and detection are often performed using a split window mean estimator.[42,44,45] Two especially important parameters in the design of this estimator are the guard window and the estimation window sizes placed on both sides (in range delay) of the test bin (and also along the frequency axis for CW). The detection statistic values of the bins in the estimation windows are used to calculate the background estimate, whereas the bins in the guard windows provide a gap between the test bin of interest and the estimation bins. This gap is designed to protect the estimation bins from containing target energy if a target is indeed present. The estimate of the background level is calculated as

$$\hat{\sigma}^2 = \frac{1}{K}\sum_{\{k\}} y(k)\,, \tag{11.22}$$

in which $\{y(k)\}$ are the detection statistic outputs in the K estimation window bins. If the background reverberation plus noise is Gaussian, the detection threshold becomes[28]

$$\lambda_T = -\hat{\sigma}^2 \ln P_{\text{fa}}\,. \tag{11.23}$$

The split window mean estimator is a form of CFAR processing because the false alarm probability is fixed, providing there are no target echo components in the estimation bins. If the test bin contains a target echo and some of the estimation bins contain target returns, then the background estimate will likely be biased high, yielding a threshold that exceeds the test bin value so that the target does not get detected. Variations of the split window mean estimator have been developed to deal with this problem. These include (1) the simple removal of the largest estimation bin value prior to the mean estimate calculation and (2) clipping and replacement of large estimation bin values to remove outliers from the calculation of the mean estimate.

Most CFAR algorithms also rely on the stationarity of the underlying distribution of the background data. If the distribution of the data used to calculate the mean background level meets the stationarity assumptions, then the algorithm can indeed provide CFAR performance. Unfortunately, the real ocean environment, especially in shallow water, yields highly non-stationary reverberation environments and target returns with significant multipath components. Because the data are stochastic, the background estimates made by the normalizer have a mean and a variance. In non-stationary reverberation environments, these measures may depart from the design mean and variance for a stationary background. As the non-stationarity of the samples used to compute the background estimate increases, the performance of the CFAR algorithm degrades accordingly, causing

1. Departure from the design false alarm probability
2. A potential reduction in detectability

For example, if the mean estimate is biased low, the probability of false alarm increases. And, if the mean estimate is biased high, the reduction in signal-to-reverberation-plus-noise ratio causes a detection loss.

Performance of the split window mean estimator is heavily dependent upon the guard and estimation window sizes. Optimum performance can be realized when both the guard window size is well matched to the time (and frequency for CW) extent of the target return and the estimation window size contains the maximum number of independent, identically distributed, reverberation-plus-noise bins. The time extent for the guard window can be determined from the expected multipath spread in conjunction with the aspect-dependent target response. The frequency spread of the CW signal is caused by the dispersion properties of the environment and the potential differential Doppler between the multipath components. The estimation window size should be small when the background is highly non-stationary and large when it is stationary.

Under certain circumstances, it may be advantageous to adaptively alter the detection thresholds based on the processing of previous pings. If high-priority detections have already been confirmed by the operator (or by post-processing), the threshold can be lowered near these locations to ensure a higher probability of detection on the current ping. Conversely, the threshold can be raised near locations of low-priority detections to drop the probability of detection. This functionality simplifies the post-processing and relieves the operator from the potential confusion of tracking a large number of contacts.

The normalization requirements for an LF active sonar are complicated and are a topic of ongoing research. More specifically, the bandwidth effects, discussed in Section 11.2.3.2, need to be considered also in the normalization process by using several specific normalizers. This is because an active sonar display requires normalized data that retain bandwidth information, have reduced dynamic range, and have constant false alarm rate capabilities which can be obtained by suitable normalization.

11.2.3.3 Display Arrangements for Active Sonar Systems

The next issue of interest is the display arrangement of the output results of an LF active sonar system. There are two main concerns here. The first is that the display format should provide sufficient information to allow for an unbiased decision that a detection has been achieved when the received echoes include sufficient information for detection. The second concern is that the repetition rate of the transmitted sequence of pulses, such as the one shown in Figure 11.5, should be in the range of 10 to 15 min. These two concerns, which may be viewed also as design restrictions, have formed the basis for the display formats of CW and FM signals, which are discussed in the following sections.

11.2.3.3.1 Display Format for CW Signals

The processing of the beam time series, containing information about the CW transmitted pulses, should include temporal spectral analysis of heavily overlapped segments. The display format of the spectral results associated with the heavily overlapped segments should be the same with that of a LOFAR-gram presentation arrangement for passive sonars.

Moreover, these spectral estimates should include the so-called *ownship Doppler nullification*, which removes the component of Doppler shift due to ownship motion. The left part of Figure 11.7A shows the details of the CW display format for an active sonar as well as the mathematical relation for the ownship Doppler nullification.

Accordingly, the display of active CW output results of an active sonar should include LOFAR-grams that contain all the number of beams provided by the associated beamformer. The content of output results for each beam will be included in one window, as shown at the left-hand side of Figure 11.7B. Frequencies will be shown by the horizontal axis. The temporal spectral estimates of each heavily overlapped segment will be plotted as a series of gray-scale pixels along the frequency axis. Mapping of the power levels of the temporal spectral estimates along a sequence of gray-scale pixels will be derived according to normalization processing schemes for the passive LOFAR-gram sonar displays.

If three CW pulses are transmitted, as shown in Figure 11.5, then the temporal spectral estimates will include a frequency shift that would allow the vertical alignment of the spectral estimates of the three CW pulses in one beam window. Clustering across frequencies and across beams would provide summary displays for rapid assessment of the operational environment.

11.2.3.3.2 Display Format for FM Type of Signals

For FM type of signals, the concept of processing heavily overlapped segments should also be considered. In this case, the segments will be defined as heavily overlapped replicas derived from a long broadband transmitted signal, as discussed in Section 11.2.3.2. However, appropriate time shifting would be required to align the corresponding time-delay estimates from each segmented replica in one beam window. The display format of the output results will be the same as those of the CW signals. Shown at the right-hand side of Figure 11.7A are typical examples of FM types of display outputs. One real data example of an FM output display for a single beam is given in Figure 11.7B. This figure shows replica correlated data from 30 pings separated by a repetition interval of approximately 15 min.

At this point, it is important to note that for a given transmitted FM signal a number of Doppler shifted replicas might be considered to allow for multi-dimensional search and estimation of range and velocity of a moving target of interest.

Thus, it should be expected that during active LF towed array operations the FM display outputs will be complicated and multi-dimensional. However, a significant downswing of the number of displays can be achieved by applying clustering across time delays, beams, and Doppler shift. This kind of clustering

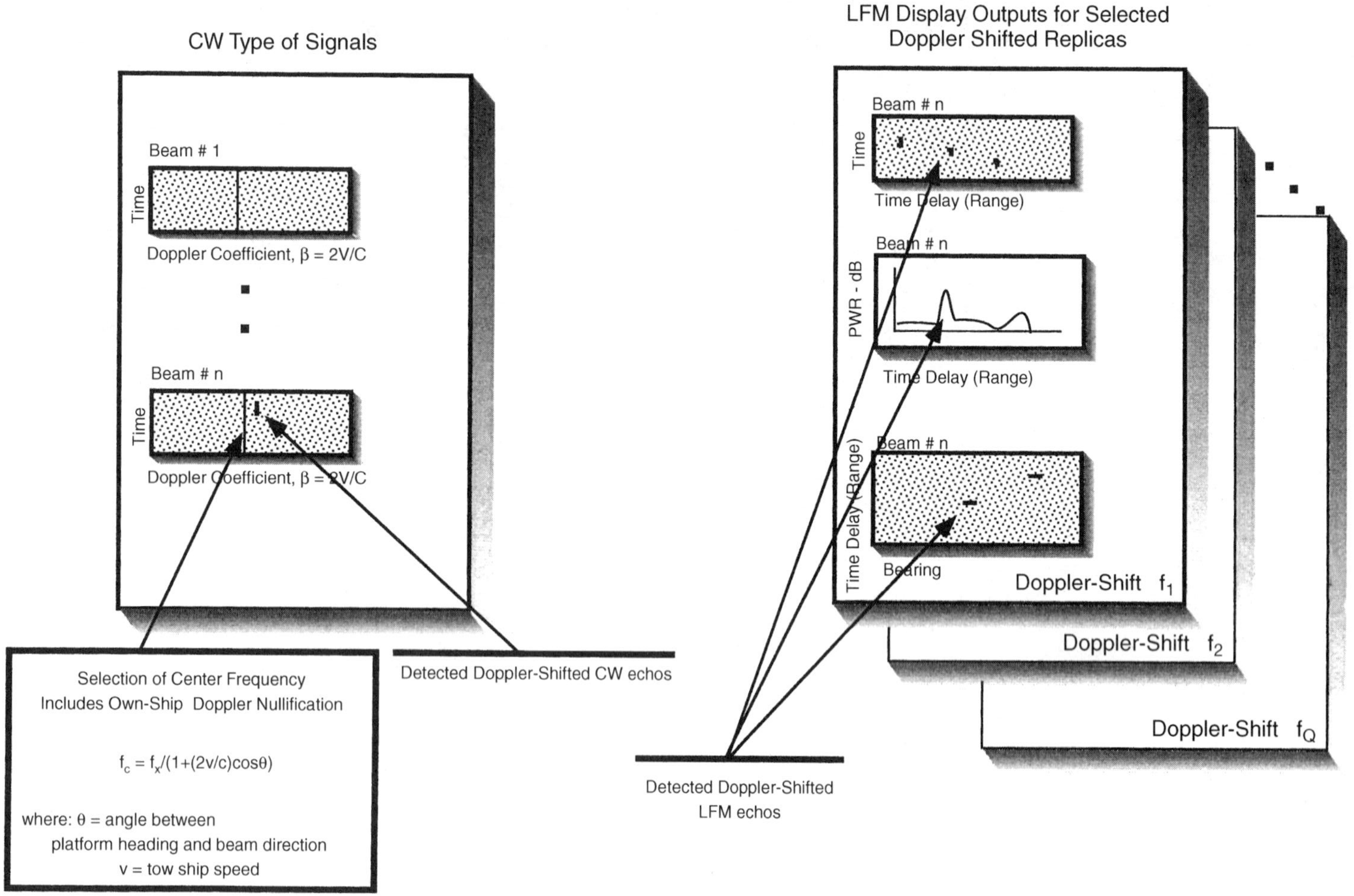

FIGURE 11.7 (A) Display arrangements for CW and FM pulses. The left part shows the details of the CW display format that includes the ownship Doppler nullification. The right part shows the details of the FM type display format for various combinations of Doppler shifted replicas. *(continued)*

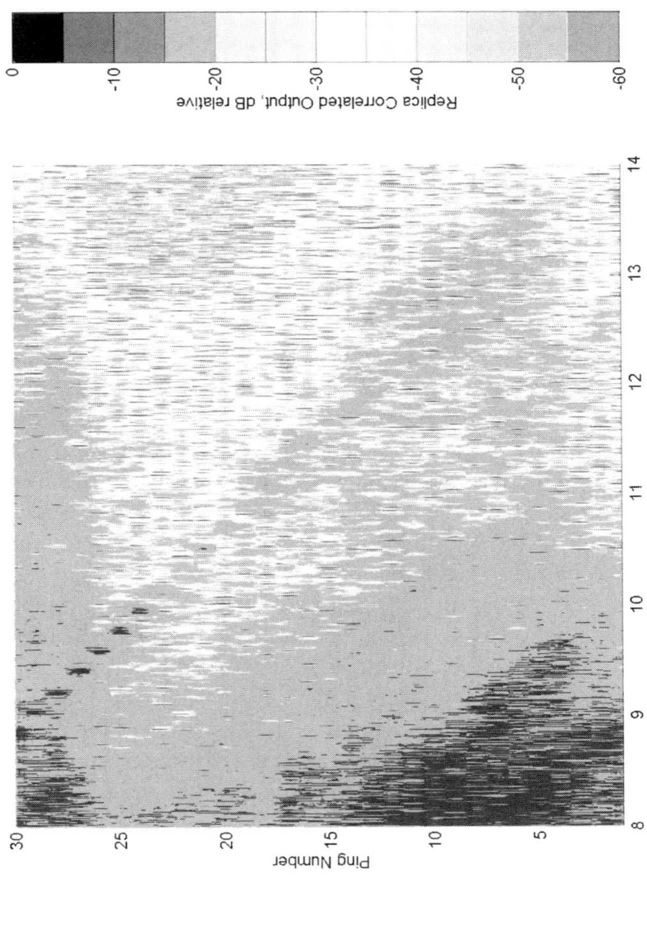

FIGURE 11.7 (CONTINUED) (B) Replica correlated FM data displayed with time on the vertical axis and range along the horizontal axis for one beam. The detected target is shown as a function of range by the received echoes forming a diagonal line on the upper left corner of the display output.

will provide summary displays for rapid assessment of the operational environment, as well as critical information and data reduction for classification and tracking.

In summary, the multi-dimensional active sonar signal processing, as expressed by Figures 11.5 to 11.7, is anticipated to define active sonar operations for the next-generation sonar systems. However, the implementation in real-time active sonars of the concepts that have been discussed in the previous sections will not be a trivial task. As an example, Figures 11.8 and 11.9 present the multi-dimensionality of the processing flow associated with the SRCs and LRCs shown in Figure 11.6. Briefly, the schematic interpretation of the signal processing details in Figures 11.8 and 11.9 reflects the implementation and mapping in sonar computing architectures of the multi-dimensionality requirements of next-generation active sonars. If operational requirements would demand large number of beams and Doppler shifted replicas, then the anticipated multidimensional processing, shown in Figures 11.8 and 11.9, may lead to prohibited computational requirements.

11.2.4 Comments on Computing Architecture Requirements

The implementation of this investigation's non-conventional processing schemes in sonar systems is a non-trivial issue. In addition to the selection of the appropriate algorithms, success is heavily dependent on the availability of suitable computing architectures.

Past attempts to implement matrix-based signal processing methods, such as adaptive beamformers reported in this chapter, were based on the development of systolic array hardware, because systolic arrays allow large amounts of parallel computation to be performed efficiently since communications occur locally. None of these ideas are new. Unfortunately, systolic arrays have been much less successful in practice than in theory. The fixed-size problem for which it makes sense to build a specific array is rare. Systolic arrays big enough for real problems cannot fit on one board, much less one chip, and interconnects have problems. A 2-D systolic array implementation will be even more difficult. So, any new computing architecture development should provide high throughput for vector- as well as matrix-based processing schemes.

A fundamental question, however, that must be addressed at this point is whether it is worthwhile to attempt to develop a system architecture that can compete with a multi-processor using stock microprocessors. Although recent microprocessors use advanced architectures, improvements of their performance include a heavy cost in design complexity, which grows dramatically with the number of instructions that can be executed concurrently. Moreover, the recent microprocessors that claim high performance for peak MFLOP rates have their net throughput usually much lower, and their memory architectures are targeted toward general purpose code.

These issues establish the requirement for dedicated architectures, such as in the area of operational sonar systems. Sonar applications are computationally intensive, as shown in Chapter 6, and they require high throughput on large data sets. It is our understanding that the Canadian DND recently supported work for a new sonar computing architecture called the next-generation signal processor (NGSP).[10] We believe that the NGSP has established the hardware configuration to provide the required processing power for the implementation and real-time testing of the non-conventional beamformers such as those reported in Chapter 6.

A detailed discussion, however, about the NGSP is beyond the scope of this chapter, and a brief overview about this new signal processor can be found in Reference 10. Other advanced computing architectures that can cover the throughput requirements of computationally intensive signal processing applications, such as those discussed in this chapter, have been developed by Mercury Computer Systems, Inc.[104] Based on the experience of the authors of this chapter, the suggestion is that implementation efforts of advanced signal processing concepts should be directed more on the development of generic signal processing structures as in Figure 11.4, rather than the development of very expensive computing architectures. Moreover, the signal processing flow of advanced processing schemes that include both scalar and vector operations should be very well defined in order to address practical implementation issues.

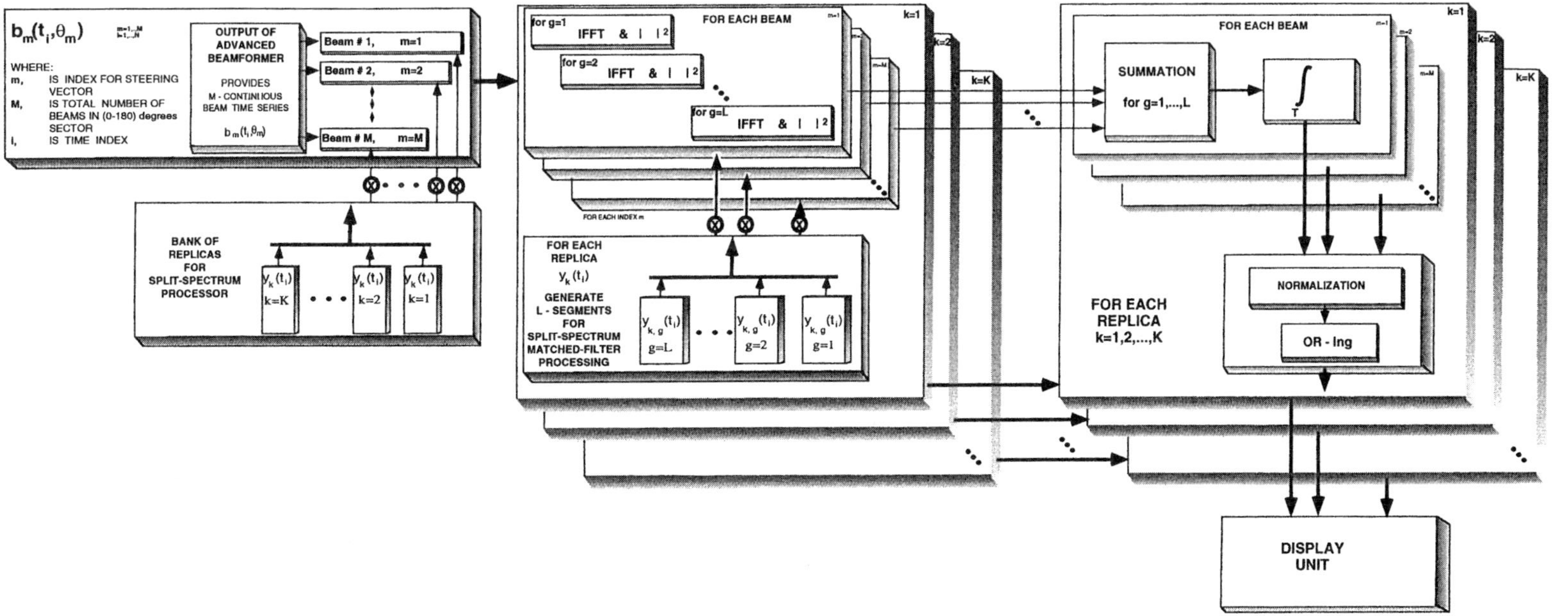

FIGURE 11.8 Signal processing flow of an SRC. The various layers in the schematic diagram represent the combinations that are required between the segments of the replica correlators and the steering beams generated by the advanced beamformers of the active sonar system. The last set of layers (at the right-hand side) represent the corresponding combinations to display the results of the SRC according to the display formats of Figure 11.7.

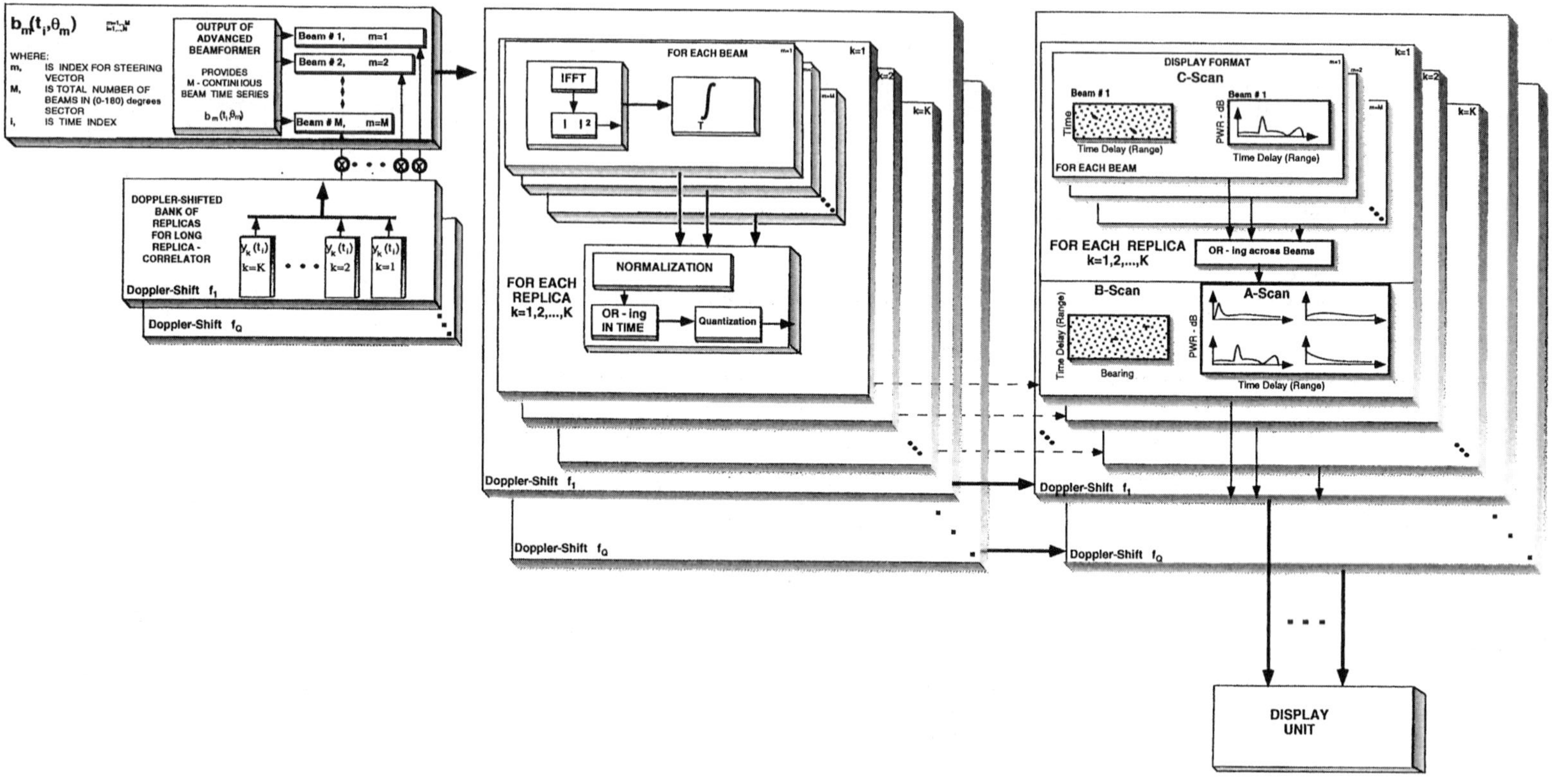

FIGURE 11.9 Processing flow of an LRC. The various layers in the schematic diagram represent the combinations that are required between the Doppler shifted replicas of the LRC and the steering beams generated by the advanced beamformers of the active sonar system. The last set of layers (at the right-hand side) represent the corresponding combinations to display the results of the LRC according to the display formats of Figure 11.7.

In this chapter, we address the issue of computing architecture requirements by defining generic concepts of the signal processing flow for integrated active-passive sonar systems, including adaptive and synthetic aperture signal processing schemes. The schematic diagrams in Figures 6.14 and 6.20 to 6.24 of Chapter 6 show that the implementation of advanced sonar processing concepts in sonar systems can be carried out in existing computer architectures[10,104] as well as in a network of general purpose computer workstations that support both scalar and vector operations.

11.3 Real Results from Experimental Sonar Systems

The real data sets that have been used to test the implementation configuration of the above non-conventional processing schemes come from two kinds of experimental setups. The first one includes sets of experimental data representing an acoustic field consisting of the tow ship's self noise and the reference narrowband CWs, as well as broadband signals such as HFM and pseudo-random transmitted waveforms from a deployed source. The absence of other noise sources as well as noise from distant shipping during these experiments make this set of experimental data very appropriate for concept demonstration. This is because there are only a few known signals in the received hydrophone time series, and this allows an effective testing of the performance of the above generic signal processing structure by examining various possibilities of artifacts that could be generated by the non-conventional beamformers.

In the second experimental setup, the received hydrophone data represent an acoustic field consisting of the reference CW, HFM, and broadband signals from the deployed source that are embodied in a highly correlated acoustic noise field including narrowband and broadband noise from heavy shipping traffic. During the experiments, signal conditioning and continuous recording on a high-performance digital recorder were provided by a real-time data system.

The generic signal processing structure, presented in Figure 11.4, and the associated signal processing algorithms (minimum variance distortionless response [MVDR], generalized sidelobe cancellers [GSC], steered minimum variance [STMV], extended towed array measuremnts [ETAM], matched filter), discussed in Chapter 6, were implemented in a workstation supporting a UNIX operating system and FORTRAN and C compilers, respectively.

Although the CPU power of the workstation was not sufficient for real-time signal processing response, the memory of the workstation supporting the signal processing structure of Figure 11.4 was sufficient to allow above of continuous hydrophone time series up to 3 h long. Thus, the output results of the above generic signal processing structure were equivalent to those that would have been provided by a real-time system, including the implementation of the signal processing schemes discussed in this chapter.

The results presented in this section are divided into two parts. The first part discusses passive narrowband and broadband towed array sonar applications. The scope here is to evaluate the performance of the adaptive and synthetic aperture beamforming techniques and to assess their ability to track and localize narrowband and broadband signals of interest while suppressing strong interferers. The impact and merits of these techniques will be contrasted with the localization and tracking performance obtained using the conventional beamformer.

The second part of this section presents results from active towed array sonar applications. The aim here is to evaluate the performance of the adaptive and synthetic aperture beamformers in a matched filter processing environment.

11.3.1 Passive Towed Array Sonar Applications

11.3.1.1 Narrowband Acoustic Signals

The display of narrowband bearing estimates, according to a LOFAR presentation arrangement, are shown in Figures 11.10, 11.11, and 11.12. Twenty-five beams equally spaced in [1,-1] cosine space were steered for the conventional, the adaptive, and the synthetic aperture beamforming processes. The wavelength λ of the reference CW signal was approximately equal to 1/6 of the aperture size L of the deployed line

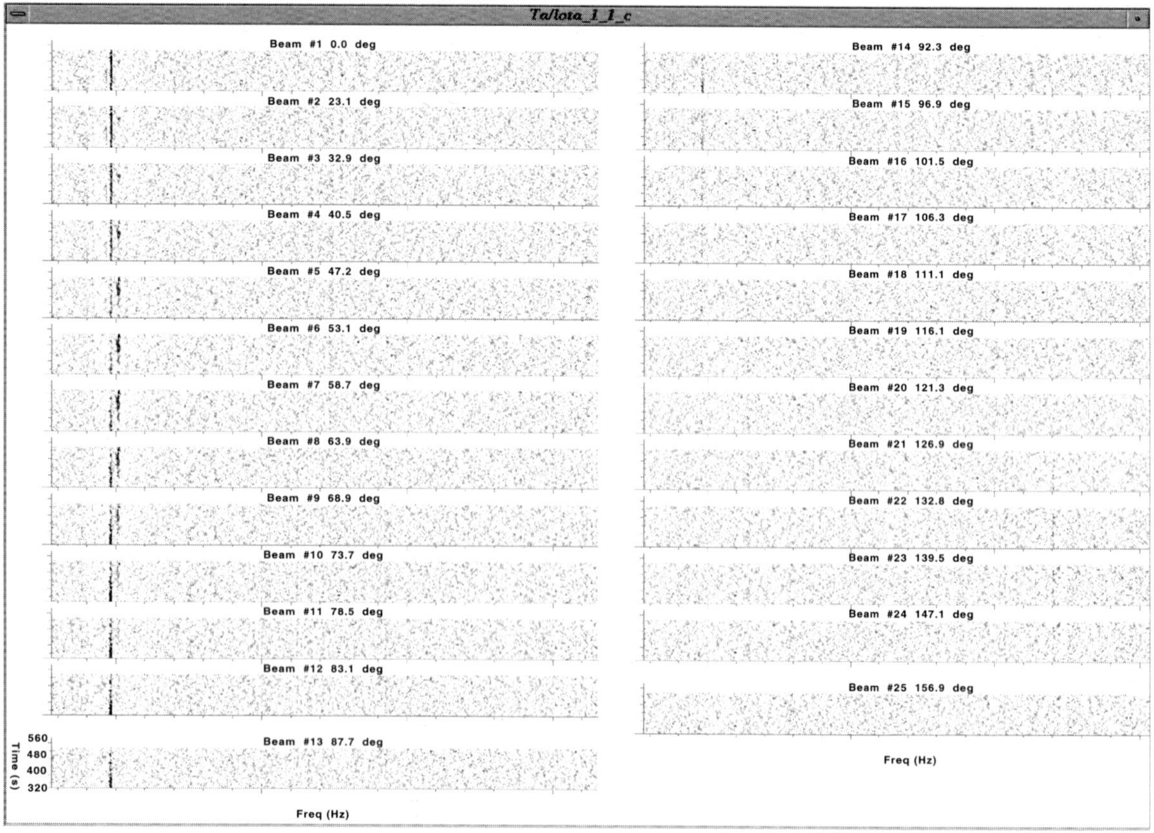

FIGURE 11.10 Conventional beamformer's LOFAR narrowband output. The 25 windows of this display correspond to the 25 steered beams equally spaced in [1, −1] cosine space. The acoustic field included three narrowband signals. Very weak indications of the CW signal of interest are shown in beams #21 to #24. (Reprinted by permission of IEEE ©1998.)

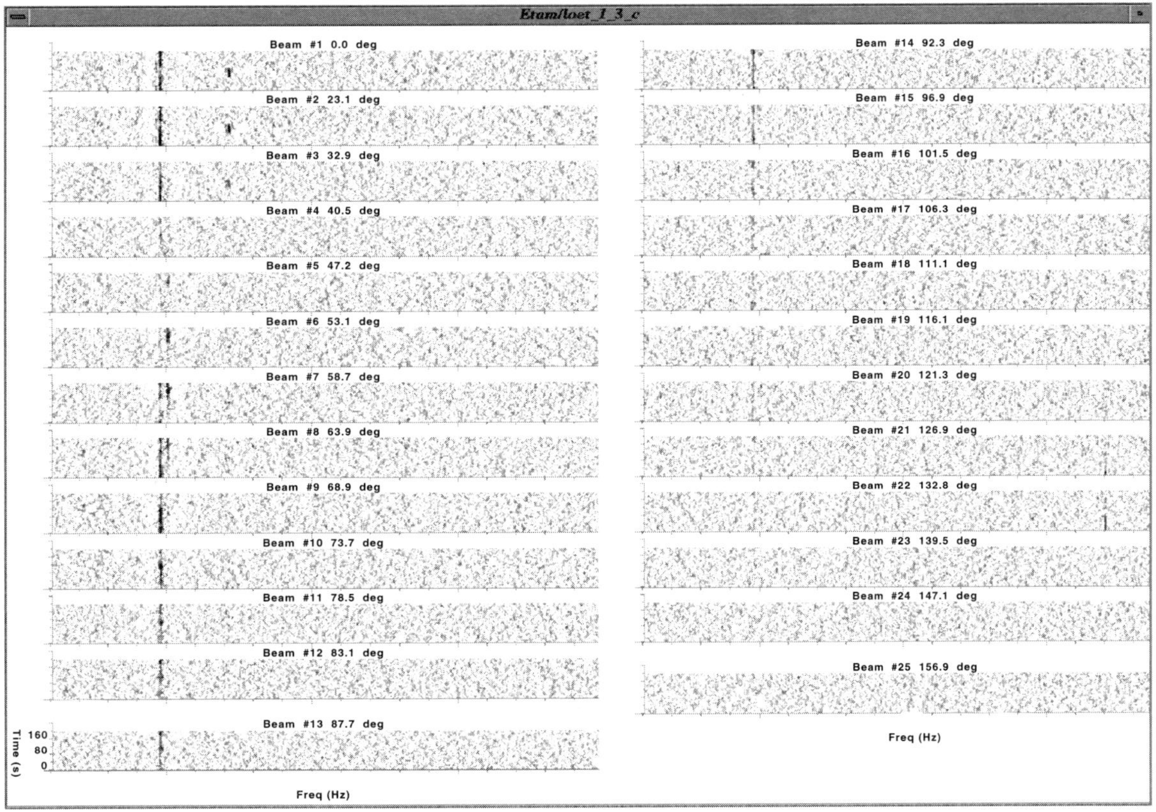

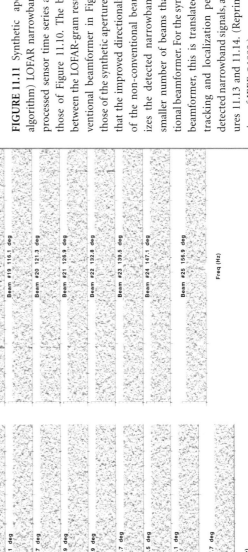

FIGURE 11.11 Synthetic aperture (ETAM algorithm) LOFAR narrowband output. The processed sensor time series are the same as those of Figure 11.10. The basic difference between the LOFAR-gram results of the conventional beamformer in Figure 11.10 and those of the synthetic aperture beamformer is that the improved directionality (array gain) of the non-conventional beamformer localizes the detected narrowband signals in a smaller number of beams than the conventional beamformer. For the synthetic aperture beamformer, this is translated into a better tracking and localization performance for detected narrowband signals, as shown in Figures 11.13 and 11.14. (Reprinted by permission of IEEE ©1998.)

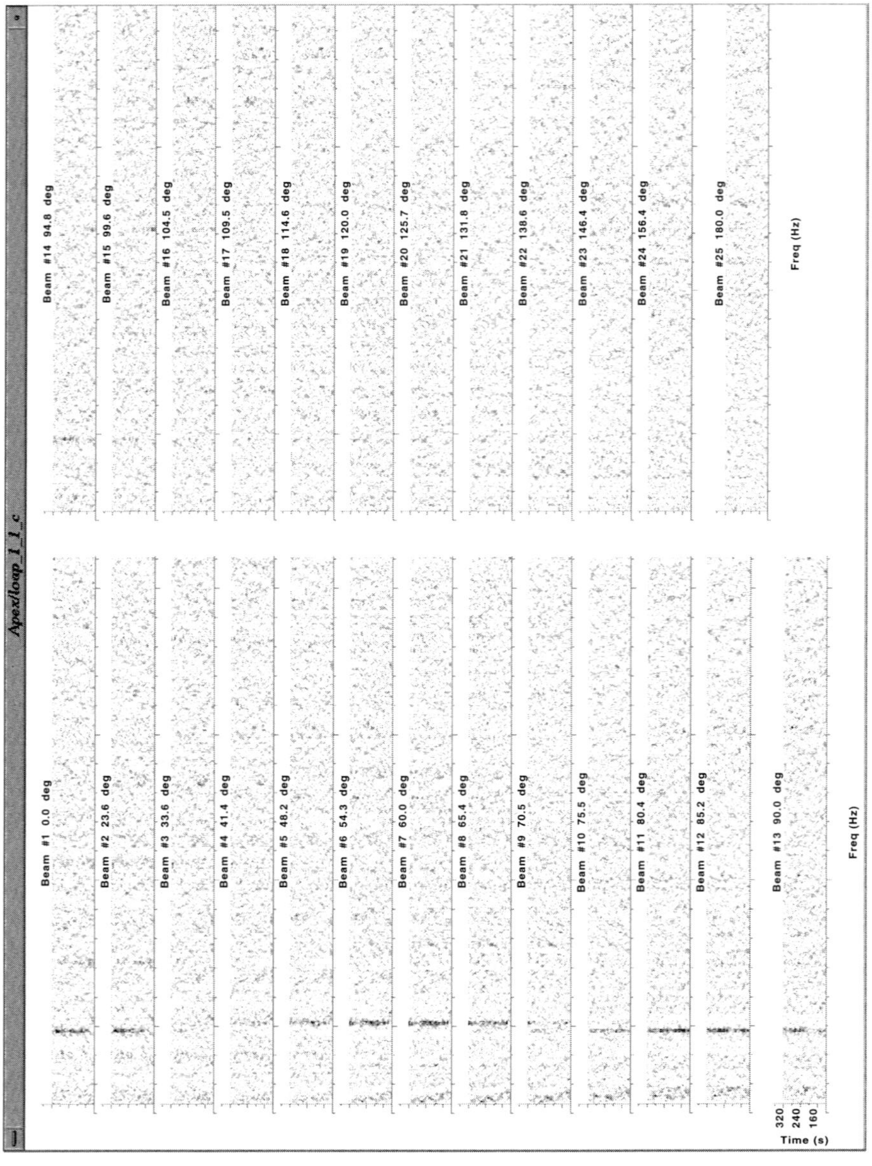

FIGURE 11.12 Sub-aperture MVDR beamformer's LOFAR narrowband output. The processed sensor time series are the same as those of Figures 11.10 and 11.11. Even though the angular resolution performance of the sub-aperture MVDR scheme in this case was better than that of the conventional beamformer, the sharpness of the adaptive beamformer's LOFAR output was not as good as the one of the conventional and synthetic aperture beamformer. This indicated loss of temporal coherence in the adaptive beam time series, which was caused by non-optimum performance and poor convergence of the adaptive algorithm. The end result was poor tracking of detected narrowband signals by the adaptive schemes as shown in Figure 11.15. (Reprinted by permission of IEEE ©1998.)

array. The power level of the CW signal was in the range of 130 dB re 1 μPa, and the distance between the source and receiver was of the order of $O(10^1)$ nm. The water depth in the experimental area was 1000 m, and the deployment depths of the source and the array receiver were approximately 100 m.

Figure 11.10 presents the conventional beamformer's LOFAR output. At this particular moment, we had started to lose detection of the reference CW signal tonal. Very weak indications of the presence of this CW signal are shown in beams #21 to #24 of Figure 11.10. In Figures 11.11 and 11.12, the LOFAR outputs of the synthetic aperture and the partially adaptive sub-aperture MVDR processing schemes are shown for the set of data and are the same as those of Figure 11.10. In particular, Figure 11.11 shows the synthetic aperture (ETAM algorithm) LOFAR narrowband output, which indicates that the basic difference between the LOFAR-gram results of the conventional beamformer in Figure 11.10 and those of the synthetic aperture beamformer is that the improved directionality (AG) of the non-conventional beamformer localizes the detected narrowband signals in a smaller number of beams than the conventional beamformer. For the synthetic aperture beamformer, this is translated into a better tracking and localization performance for detected narrowband signals, as shown in Figures 11.13 and 11.14.

Figure 11.12 presents the sub-aperture MVDR beamformer's LOFAR narrowband output. In this case, the processed sensor time series are the same as those of Figures 11.10 and 11.11. However, the sharpness of the adaptive beamformer's LOFAR output was not as good as the one of the conventional and synthetic aperture beamformer. This indicated loss of temporal coherence in the adaptive beam time series, which was caused by non-optimum performance and poor convergence of the adaptive algorithm. The end result was poor tracking of detected narrowband signals by the adaptive schemes as shown in Figure 11.14.

The narrowband LOFAR results from the sub-aperture GSC and STMV adaptive schemes were almost identical with those of the sub-aperture MVDR scheme, shown in Figure 11.12. For the adaptive beamformers, the number of iterations for the exponential averaging of the sample covariance matrix was approximately five to ten snapshots ($\mu = 0.9$ convergence coefficient of Equation 6.79 in Chapter 6). Thus, for narrowband applications, the shortest convergence period of the sub-aperture adaptive beamformers was of the order of 60 to 80 s, while for broadband applications the convergence period was of the order of 3 to 5 s.

Even though the angular resolution performance of the adaptive schemes (MVDR, GSC, STMV) in element space for the above narrowband signal was better than that of the conventional beamformer, the sharpness of the adaptive beamformers' LOFAR output was not as good as that of the conventional and synthetic aperture beamformer. Again, this indicated loss of temporal coherence in the adaptive beam time series, which was caused by non-optimum performance and poor convergence of the above adaptive schemes when their implementation was in element space.

Loss of coherence is evident in the LOFAR outputs because the generic beamforming structure in Figure 11.4 includes coherent temporal spectral analysis of the continuous beam time series for narrowband analysis. For the adaptive schemes implemented in element space, the number of iterations for the adaptive exponential averaging of the sample covariance matrix was 200 snapshots ($\mu = 0.995$ according to Equation 6.79 in Chapter 6). In particular, the MVDR element space method required a very long convergence period of the order of 3000 s. In cases that this convergence period was reduced, then the MVDR element space LOFAR output was populated with artifacts.[23] However, the performance of the adaptive schemes of this study (MVDR, GSC, STMV) improved significantly when their implementation was carried out under the sub-aperture configuration, as discussed in Chapter 6.

Apart from the presence of the CW signal with $\lambda = L/6$ in the conventional LOFAR display, only two more narrowband signals with wavelengths approximately equal to $\lambda = L/3$ were detected. No other signals were expected to be present in the acoustic field, and this is confirmed by the conventional narrowband output of Figure 11.10, which has white noise characteristics. This kind of simplicity in the received data is very essential for this kind of demonstration process in order to identify the presence of artifacts that could be produced by the various beamformers.

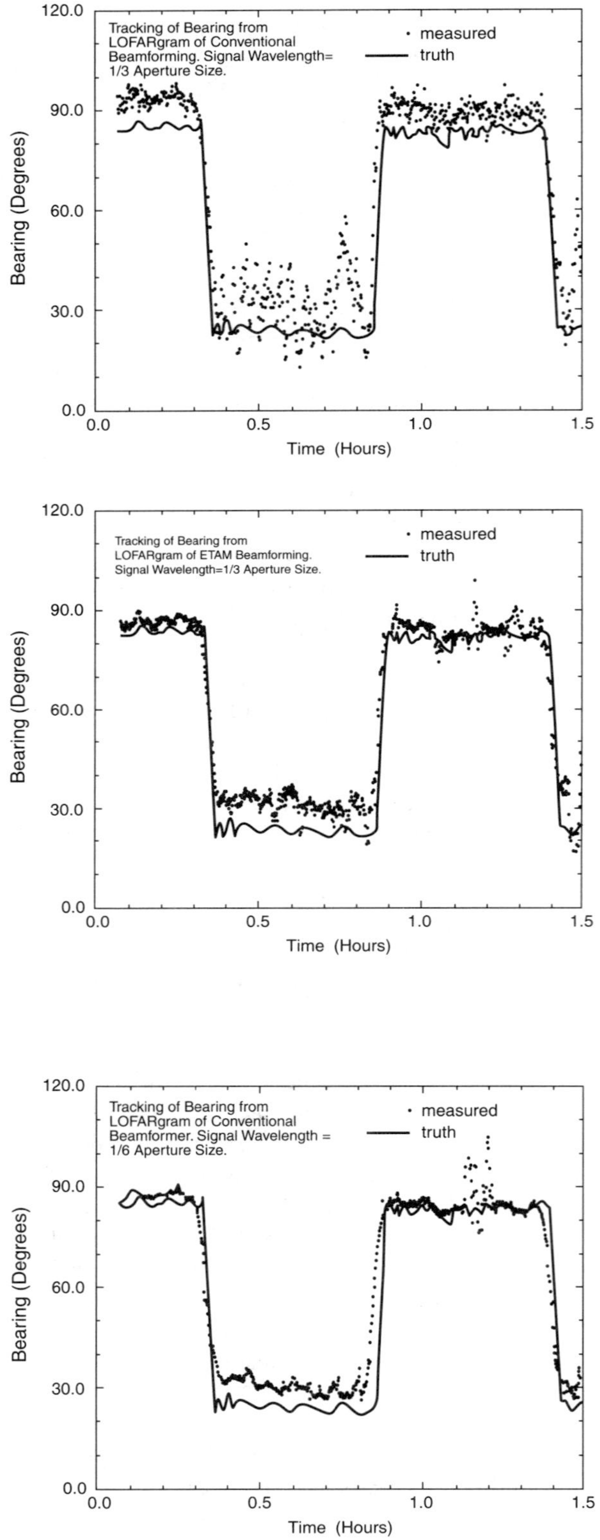

FIGURE 11.13 Signal following of bearing estimates from the conventional beamforming LOFAR narrowband outputs and the synthetic aperture (ETAM algorithm). The solid line shows the true values of source's bearing. The wavelength of the detected CW was equal to one third of the aperture size L of the deployed array. For reference, the tracking of bearing from conventional beamforming LOFAR outputs of another CW with wavelength equal to 1/16 of the towed array's aperture is shown in the lower part. (Reprinted by permission of IEEE ©1998.)

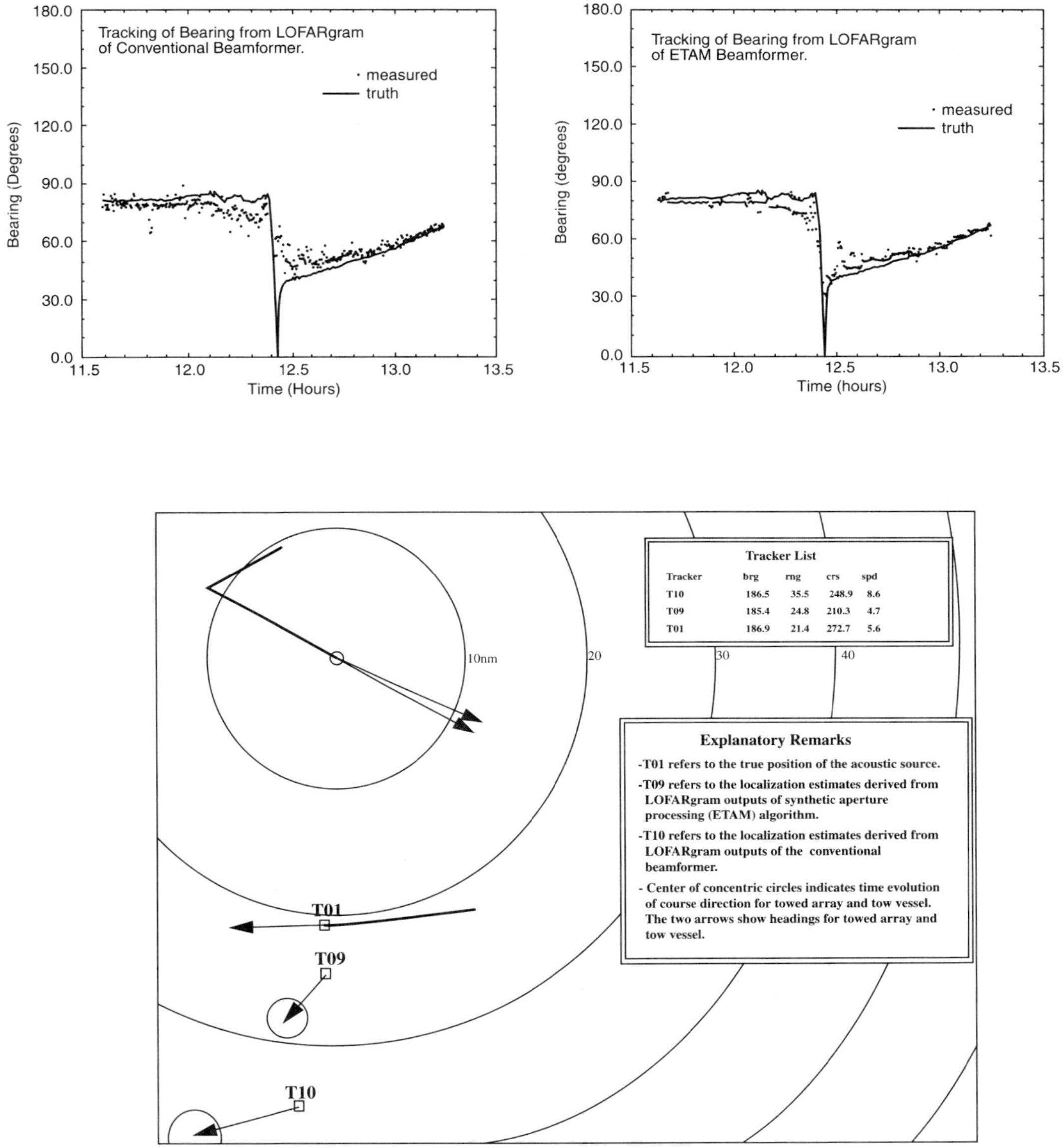

FIGURE 11.14 The upper part shows signal following of bearing estimates from conventional beamforming and synthetic aperture (ETAM algorithm) LOFAR narrowband outputs. The solid line shows the true values of source's bearing. The lower part presents localization estimates that were based on the bearing tracking results shown in the upper part. (Reprinted by permission of IEEE ©1998.)

The narrowband beam power maps of the LOFAR-grams in Figures 11.10 to 11.12 form the basic unit of acoustic information that is provided at the input of the data manager of our system for further information extraction. As discussed in Section 11.3.3, one basic function of the data management algorithms is to estimate the characteristics of signals that have been detected by the beamforming and spectral analysis processing schemes, which are shown in Figure 11.4. The data management processing includes signal following or tracking[105,106] that provides monitoring of the time evolution of the frequency and the associated bearing of detected narrowband signals.

If the output results from the non-conventional beamformers exhibit improved AG characteristics, this kind of improvement should deliver better system tracking performance over that of the conventional beamformer. To investigate the tracking performance improvements of the synthetic aperture and adaptive beamformers, the deployed source was towed along a straight-line course, while the towing of the line array receiver included a few course alterations over a period of approximately 3 h. Figure 11.14 illustrates this scenario, showing the constant course of the towed source and the course alterations of the vessel towing the line array receiver.

The parameter estimation process for tracking the bearing of detected sources consisted of peak picking in a region of bearing and frequency space sketched by fixed gate sizes in the LOFAR-gram outputs of the conventional and non-conventional beamformers. Details about this estimation process can be found in Reference 107. Briefly, the choice of the gate sizes was based on the observed bearing and frequency fluctuations of a detected signal of interest during the experiments. Parabolic interpolation was used to provide refined bearing estimates.[108] For this investigation, the bearings-only tracking process described in Reference 107 was used as a narrowband tracker, providing unsmoothed time evolution of the bearing estimates to the localization process.[105,109] The localization process of this study was based on a recursive extended Kalman filter formulated in Cartesian coordinates. Details about this localization process can be found in References 107 and 109.

Shown by the solid line in Figure 11.13 are the expected bearings of a detected CW signal with respect to the towed array receiver. The dots represent the tracking results of bearing estimates from LOFAR data provided by the synthetic aperture and the conventional beamformers. The middle part of Figure 11.13 illustrates the tracking results of the synthetic aperture beamformer. In this case, the wavelength λ of the narrowband CW signal was approximately equal to one third of the aperture size of the deployed towed array (Directivity Index, DI = 7.6 dB). For this very LF CW signal, the tracking performance of the conventional beamformer was very poor, as this is shown by the upper part of Figure 11.13. To provide a reference, the tracking performance of the conventional beamformer for a CW signal, having a wavelength approximetely equal to 1/16 of the aperture size of the deployed array (DI = 15. dB), is shown in the lower part of Figure 11.13.

Localization estimates for the acoustic source transmitting the CW signal with $\lambda = L/3$ were derived only from the synthetic aperture tracking results, shown in the middle of Figure 11.13. In contrast to these results, the conventional beamformer's localization estimates did not converge because the variance of the associated bearing tracking results was very large, as indicated by the results of the upper part of Figure 11.13. As expected, the conventional beamformer's localization estimates for the higher frequency CW signal $\lambda = L/16$ converge to the expected solution. This is because the system AG in this case was higher (DI = 15. dB), resulting in better bearing tracking performance with a very small variance in the bearing estimates.

The tracking and localization performance of the synthetic aperture and the conventional beamforming techniques were also assessed from other sets of experimental data. In this case, the towing of the line array receiver included only one course alteration over a period of approximately 30 min. Presented in Figure 11.14 is a summary of the tracking and localization results from this experiment. The upper part of Figure 11.14 shows tracking of the bearing estimates provided by the synthetic aperture and the conventional beamforming LOFAR-gram outputs. The lower part of Figure 11.14 presents the localization estimates derived from the corresponding tracking results.

It is apparent from the results of Figures 11.13 and 11.14 that the synthetic aperture beamformer improves the AG of small size array receivers, and this improvement is translated into a better signal tracking and target localization performance than the conventional beamformer.

With respect to the tracking performance of the narrowband adaptive beamformers, our experience is that during course alterations the tracking of bearings from the narrowband adaptive beampower outputs was very poor. As an example, the lower part of Figure 11.15 shows the sub-aperture MVDR adaptive beamformer's bearing tracking results for the same set of data as those of Figure 11.14. It is clear in this case that the changes of the towed array's heading are highly correlated with the deviations of the adaptive beamformer's bearing tracking results from their expected estimates.

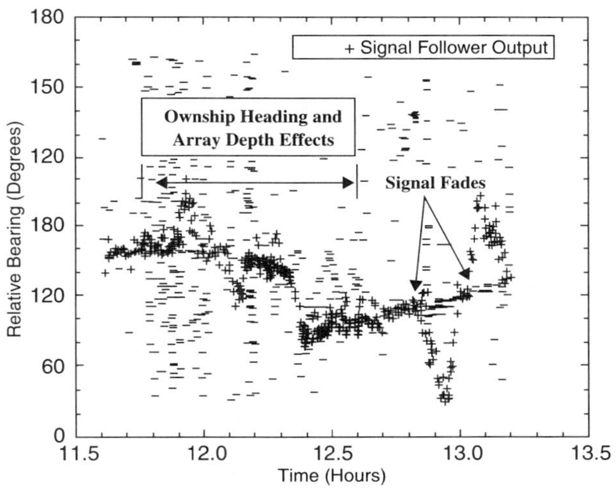

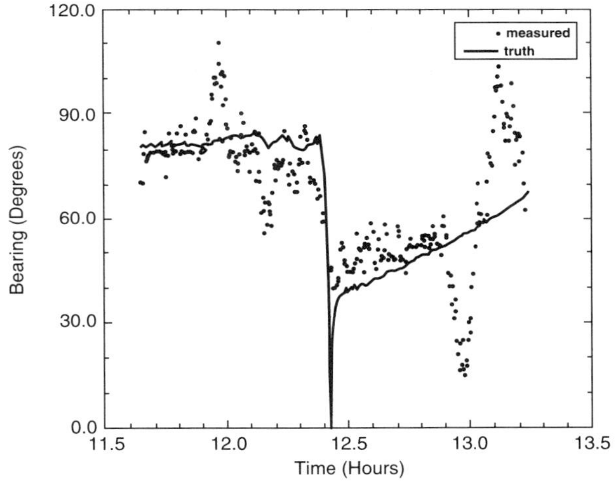

FIGURE 11.15 The upper part provides narrowband bearing estimates as a function of time for the sub-aperture MVDR and for a frequency bin including the signal of interest. These narrowband bearing estimates are for all the 25 steered beams equally spaced in [1, –1] cosine space. The lower part presents tracking of bearing estimates from sub-aperture MVDR LOFAR narrowband outputs. The solid line shows the true values of bearing. (Reprinted by permission of IEEE ©1998.)

Although the angular resolution performance of the adaptive beamformer was better than that of the synthetic aperture processing, the lack of sharpness, fuzziness, and discontinuity in the adaptive LOFAR-gram outputs prevented the signal following algorithms from tracking the signal of interest.[107] Thus, the sub-aperture adaptive algorithm should have provided better bearing estimates than those indicated by the output of the bearing tracker, shown in Figure 11.15. In order to address this point, we plotted the sub-aperture MVDR bearing estimates as a function of time for all the 25 steered beams equally spaced in [1, –1] cosine space for a frequency bin including the signal of interest. Shown in the upper part of Figure 11.15 is a waterfall of these bearing estimates. It is apparent in this case that our bearing tracker failed to follow the narrowband bearing outputs of the adaptive beamformer. Moreover, the results in the upper part of Figure 11.15 suggest signal fading and performance degradation for the narrowband adaptive processing during certain periods of the experiment.

Our explanation for this performance degradation is twofold. First, the drastic changes in the noise field, due to a course alteration, would require a large number of iterations for the adaptive process to converge. Second, since the associated sensor coordinates of the towed array shape deformation had not been considered in the steering vector $\overline{D}(f_i, \theta)$, this omission induced erroneous estimates in the noise covariance matrix during the iteration process of the adaptive processing. If a towed array shape estimation algorithm had been included in this case, the adaptive process would have provided better bearing tracking results than those shown in Figure 11.15.[77,107] For the broadband adaptive results, however, the situation is completely different, and this is addressed in the following section.

11.3.1.2 Broadband Acoustic Signals

Shown in Figure 11.16 are the conventional and sub-aperture adaptive broadband bearing estimates as a function of time for a set of data representing an acoustic field consisting of radiated noise from distant shipping in acoustic conditions typical of a sea state 2–4. The experimental area here is different than that including the processed data presented in Figures 11.10 to 11.15. The processed frequency regime for the broadband bearing estimation was the same for both the conventional and the partially adaptive sub-aperture MVDR, GSC, and STMV processing schemes. Since the beamforming operations in this study are carried out in the frequency domain, the LF resolution in this case was of the order of $O(10^0)$. This resulted in very short convergence periods for the partially adaptive beamformer of the order of a few seconds.

The left-hand side of Figure 11.16 shows the conventional broadband bearing estimates, and the right-hand side shows the partially adaptive broadband estimates for a 2-h-long set of data. Although the received noise level of a distant vessel was very low, the adaptive beamformer has detected this target in time-space position (240°, 6300 s) in Figure 11.16, something that the conventional beamformer has failed to show. In addition, the sub-aperture adaptive outputs have resolved two closely spaced broadband signal arrivals at space-time position (340°, 3000 s), while the conventional broadband output shows an indication only that two targets may be present at this space-time position.

It is evident by these results that the sub-aperture adaptive schemes of this study provide better detection (than the conventional beamformer) of weak signals in the presence of strong signals. For the previous set of data, shown in Figure 11.16, broadband bearing tracking results (for a few broadband signals at bearing 245°, 265°, and 285°) are shown by the solid lines for both the adaptive and the conventional broadband outputs. As expected, the signal followers of the conventional beamformer lost track of the broadband signal with bearing 240° at the time position (240°, 6300 s). On the other hand, the trackers of the sub-aperture adaptive beamformers did not loose track of this target, as shown by the results at the right-hand side of Figure 11.16. At this point, it is important to note that the broadband outputs of the sub-aperture MVDR, GSC, and STMV adaptive schemes were almost identical.

It is apparent from these results that the partially adaptive sub-aperture beamformers have better performance than the conventional beamformer in detecting very weak signals. In addition, the sub-aperture adaptive configuration has demonstrated tracking targets equivalent to the conventional beamformer's dynamic response during the tow vessel's course alterations. For the above set of data, localization estimates based on the broadband bearing tracking results of Figure 11.16 converged to the expected solution for both the conventional and the adaptive processing beam outputs.

Given the fact that the broadband adaptive beamformer exhibits better detection performance than the conventional method, as shown by the results of Figure 11.16 and other data sets which are not reported here, it is concluded that for broadband signals the sub-aperture adaptive beamformers of this study provide significant improvements in AG that result in better tracking and localization performance than that of the conventional signal processing scheme.

At this point, questions may be raised about the differences in bearing tracking performance of the adaptive beamformer for narrowband and broadband applications. It appears that the broadband sub-aperture adaptive beamformers as energy detectors exhibit very robust performance because the incoherent summation of the beam powers for all the frequency bins in a wideband of interest removes the

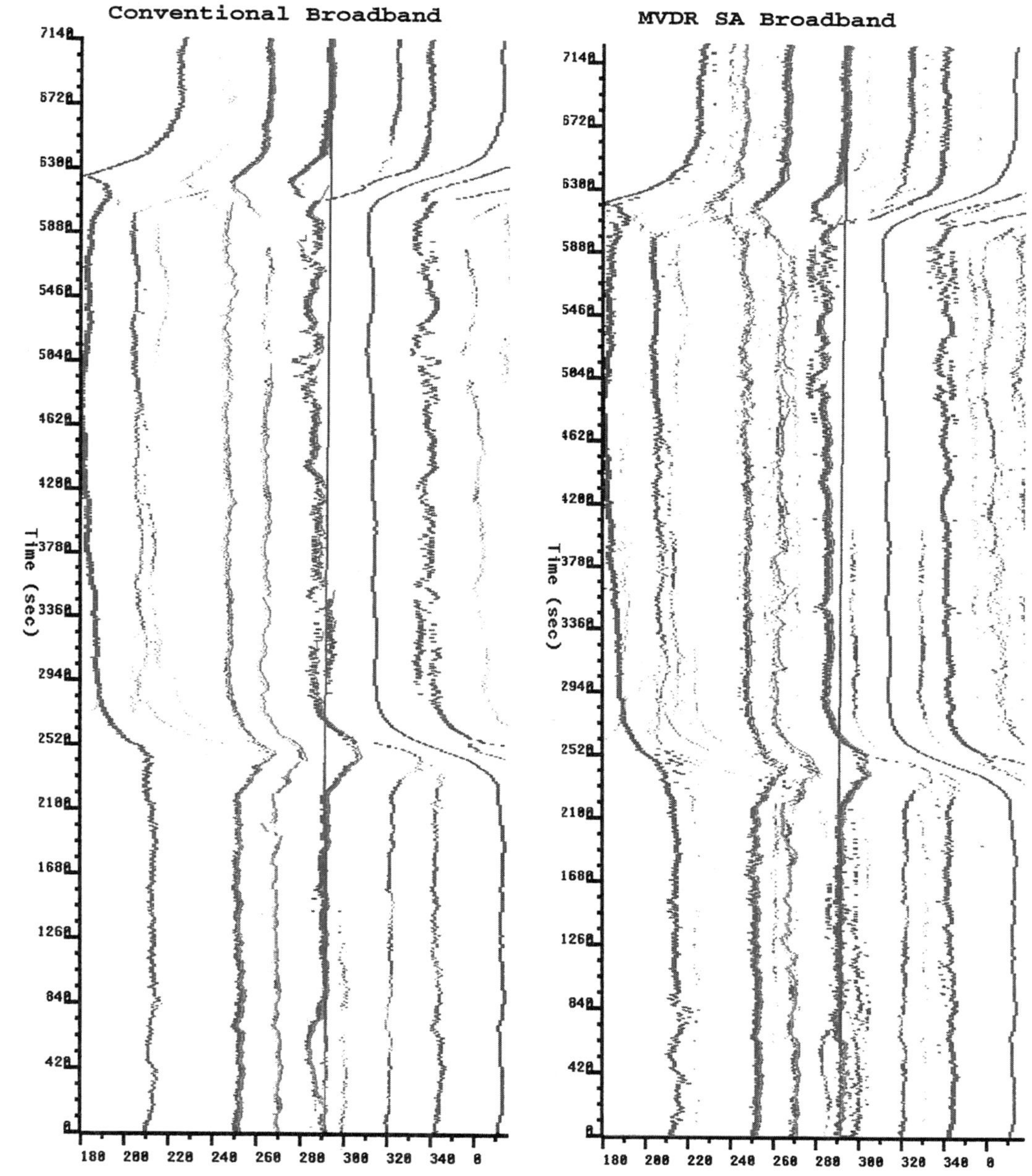

FIGURE 11.16 Broadband bearing estimates for a 2-h-long set of data: left-hand side, output from conventional beamformer; right-hand side, output from sub-aperture MVDR beamformer. Solid lines show signal tracking results for the broadband bearing estimates provided by the conventional and sub-aperture MVDR beamformers. These results show a superior signal detection and tracking performance for the broadband adaptive scheme compared with that of the conventional beamformer. This performance difference was consistent for a wide variety of real data sets. (Reprinted by permission of IEEE ©1998.)

fuzziness of the narrowband adaptive LOFAR-gram outputs, shown in Figure 11.12. However, a signal follower capable of tracking fuzzy narrowband signals[27] in LOFAR-gram outputs should remedy the observed instability in bearing trackings for the adaptive narrowband beam outputs. In addition, towed array shape estimators should also be included because the convergence period of the narrowband adaptive processing is of the same order as the period associated with the course alterations of the towed array operations. None of these remedies are required for broadband adaptive beamformers because of their proven robust performance as energy detectors and the short convergence periods of the adaptation process during course alterations.

11.3.2 Active Towed Array Sonar Applications

It was discussed in Chapter 6 that the configuration of the generic beamforming structure to provide continuous beam time series at the input of a matched filter and a temporal spectral analysis unit forms the basis for integrated passive and active sonar applications. However, before the adaptive and synthetic aperture processing schemes are integrated with a matched filter, it is essential to demonstrate that the beam time series from the output of these non-conventional beamformers have sufficient temporal coherence and correlate with the reference signal. For example, if the received signal by a sonar array consists of FM type of pulses with a repetition rate of a few minutes, then questions may be raised about the efficiency of an adaptive beamformer to achieve near-instantaneous convergence in order to provide beam time series with coherent content for the FM pulses. This is because partially adaptive processing schemes require at least a few iterations to converge to a sub-optimum solution.

To address this question, the matched filter and the non-conventional processing schemes, shown in Figure 11.4, were tested with real data sets, including HFM pulses 8-s long with a 100-Hz bandwidth. The repetition rate was 120 s. Although this may be considered as a configuration for bistatic active sonar applications, the findings from this experiment can be applied to monostatic active sonar systems as well.

In Figures 11.17 and 11.18 we present some experimental results from the output of the active unit of the generic signal processing structure. Figure 11.17 shows the output of the replica correlator for the conventional, sub-aperture MVDR adaptive, and synthetic aperture beam time series. The horizontal axis in Figure 11.17 represents range or time delay ranging from 0 to 120 s, which is the repetition rate of the HFM pulses. While the three beamforming schemes provide artifact-free outputs, it is apparent from the values of the replica correlator output that the conventional beam time series exhibit better temporal coherence properties than the beam time series of the synthetic aperture and the sub-aperture adaptive beamformer. The significance and a quantitative estimate of this difference can be assessed by comparing the amplitudes of the normalized correlation outputs in Figure 11.17. In this case, the amplitudes of the replica correlator outputs are 0.32, 0.28, and 0.29 for the conventional, adaptive, and synthetic aperture beamformers, respectively.

This difference in performance, however, was expected because for the synthetic aperture processing scheme to achieve optimum performance the reference signal is required to be present in the five discontinuous snapshots that are being used by the overlapped correlator to synthesize the synthetic aperture. So, if a sequence of five HFM pulses had been transmitted with a repetition rate equal to the time interval between the above discontinuous snapshots, then the coherence of the synthetic aperture beam time series would have been equivalent to that of the conventional beamformer. Normally, this kind of requirement restricts the detection ranges for incoming echoes. To overcome this limitation, a combination of the pulse length, desired synthetic aperture size, and detection range should be derived that will be based on the aperture size of the deployed array. A simple application scenario, illustrating the concept of this combination, is a side scan sonar system that deals with predefined ranges.

Although for the adaptive beam time series in Figure 11.17 a sub-optimum convergence was achieved within two to three iterations, the arrangement of the transmitted HFM pulses in this experiment was not an optimum configuration because the sub-aperture beamformer had to achieve near-instantaneous convergence with a single snapshot. Our simulations suggest that a sub-optimum solution for the sub-aperture MVDR adaptive beamformer is possible if the active sonar transmission consists of a continuous sequence of active pulses. In this case, the number of pulses in a sequence should be a function of the

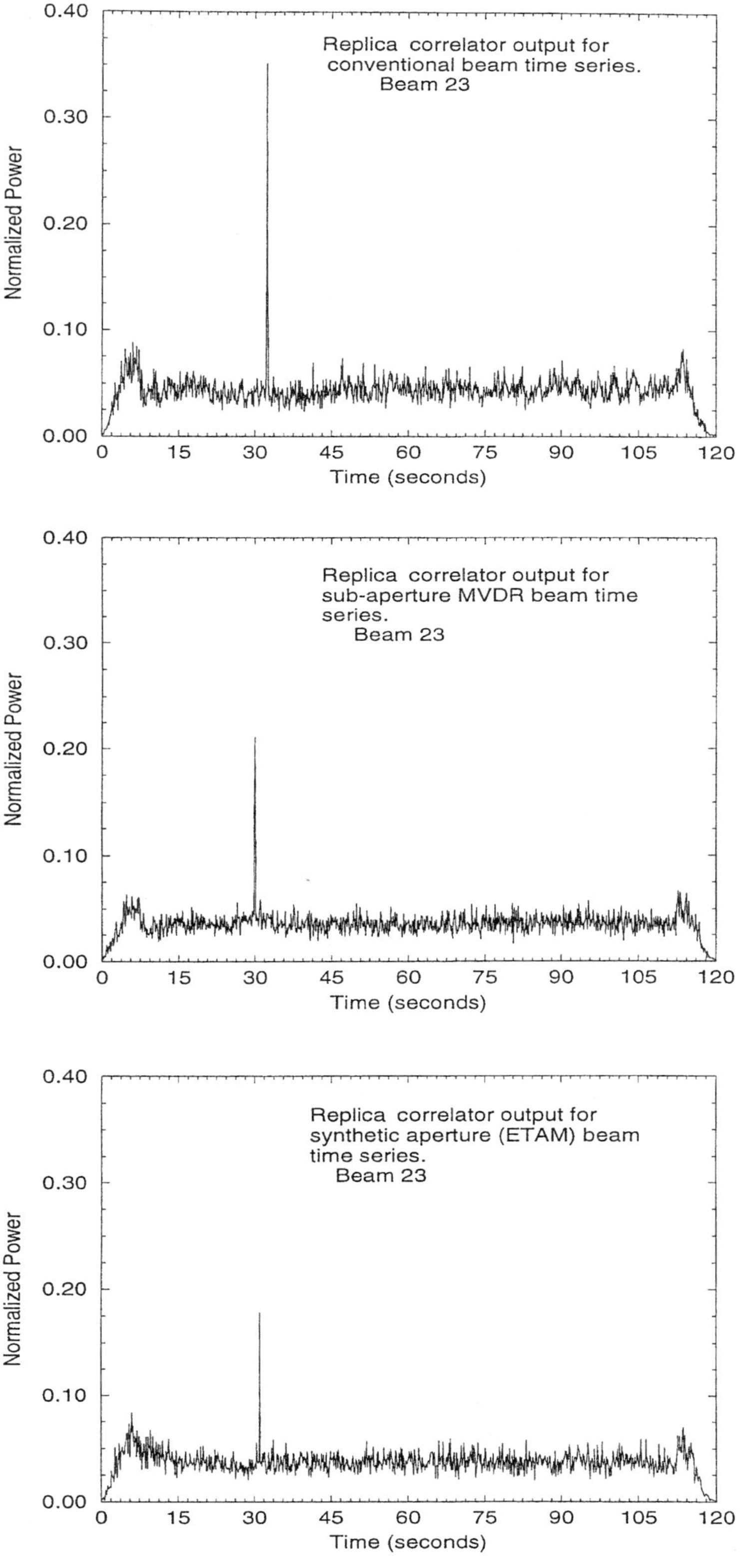

FIGURE 11.17 Output of replica correlator for the beam series of the generic beamforming structure shown in Figures 11.4 and 11.5. The processed hydrophone time series includes received HFM pulses transmitted from the acoustic source, 8-s long with a 100-Hz bandwidth and 120-s repetition rate. The upper part is the replica correlator output for conventional beam time series. The middle part is the replica correlator output for sub-aperture MVDR beam time series. The lower part is the replica correlator output for synthetic aperture (ETAM algorithm) beam time series. (Reprinted by permission of IEEE ©1998.)

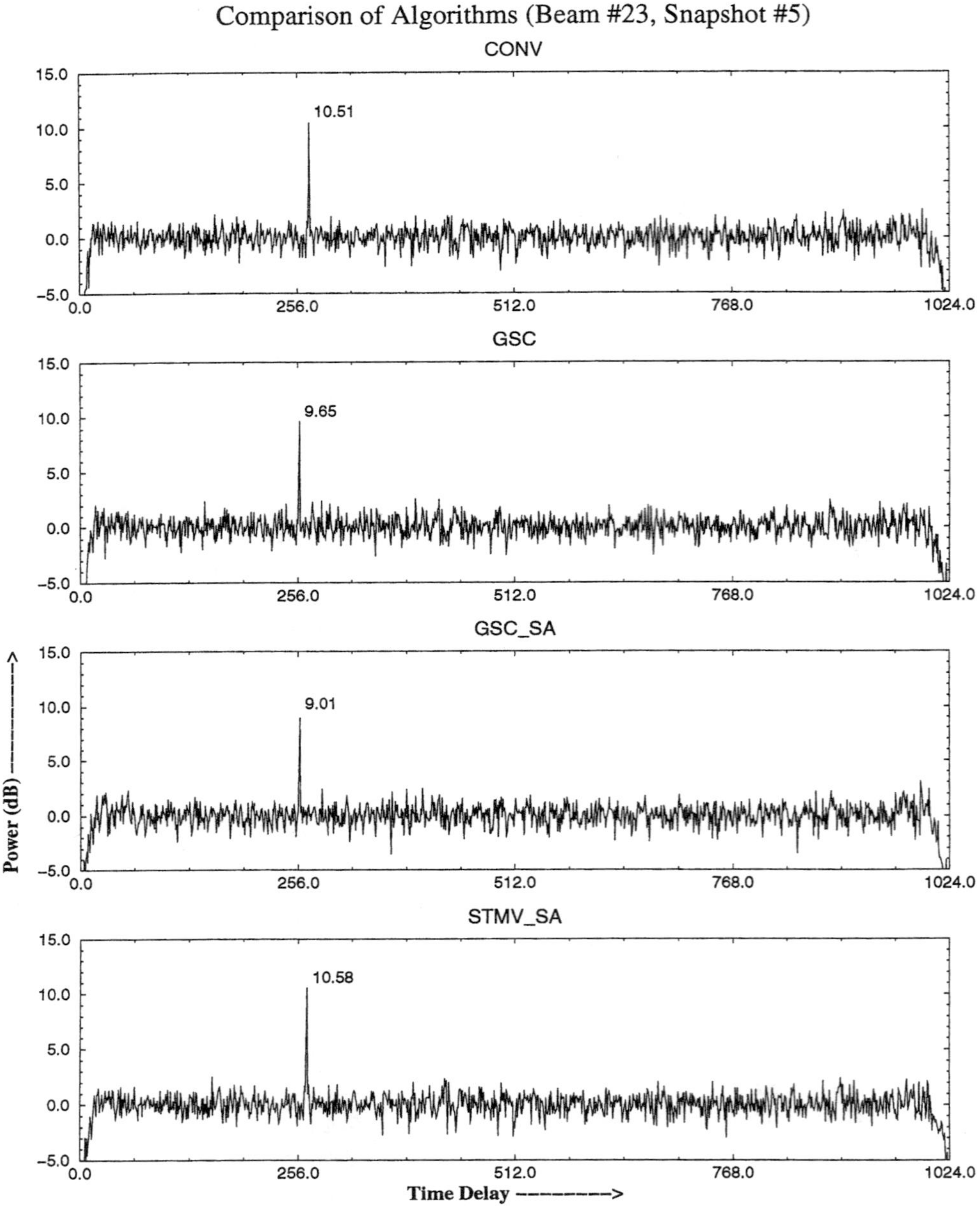

FIGURE 11.18 Output of replica correlator for the beam series of the conventional and the sub-aperture MVDR, GSC, and STMV adaptive schemes of the generic beamforming structure shown in Figure 11.4. The processed hydrophone time series are the same as those of Figure 11.17. (Reprinted by permission of IEEE ©1998.)

number of sub-apertures, and the repetition rate of this group of pulses should be a function of the detection ranges of operational interest.

The near-instantaneous convergence characteristics, however, for the other two adaptive beamformers, namely, the GSC and the STMV schemes, are better compared with those of the sub-aperture MVDR scheme. Shown in Figure 11.18 is the replica correlator output for the same set of data as those in Figure 11.17 and for the beam series of the conventional and the sub-aperture MVDR, GSC, and STMV adaptive schemes.

Even though the beamforming schemes of this study provide artifact-free outputs, it is apparent from the values of the replica correlator outputs, shown in Figures 11.17 and 11.18, that the conventional beam time series exhibit better temporal coherence properties than the beam time series of the adaptive beamformers, except for the sub-aperture STMV scheme. The significance and a quantitative estimate of this difference can be assessed by comparing the amplitudes of the correlation outputs in Figure 11.18. In this case, the amplitudes of the replica correlator outputs are 10.51, 9.65, 9.01, and 10.58 for the conventional scheme and for the adaptive schemes: GSC in element space, GSC-SA (sub-aperture), and STMV-SA (sub-aperture), respectively. These results show that the beam time series of the STMV sub-aperture scheme have achieved temporal coherence properties equivalent to those of the conventional beamformer, which is the optimum case.

Normalization and clustering of matched filter outputs, such as those of Figures 11.17 and 11.18, and their display in a LOFAR-gram arrangement provide a waterfall display of ranges as a function of beam steering and time, which form the basis of the display arrangement for active systems, shown in Figures 11.8 and 11.9. Figure 11.19 shows these results for the correlation outputs of the conventional and the adaptive beam time series for beam #23. It should be noted that Figure 11.19 includes approximately 2 h of processed data. The detected HFM pulses and their associated ranges are clearly shown in beam #23. A reflection from the sidewalls of an underwater canyon in the area is visible as a second echo closely spaced with the main arrival.

In summary, the basic difference between the LOFAR-gram results of the adaptive schemes and those of the conventional beam time series is that the improved directionality of the non-conventional beamformers localizes the detected HFM pulses in a smaller number of beams than the conventional beamformer. Although we do not present here the LOFAR-gram correlation outputs for all 25 beams, a picture displaying the 25 beam outputs would confirm the above statement regarding the directionality improvements of the adaptive schemes with respect to the conventional beamformer. Moreover, it is anticipated that the directional properties of the non-conventional beamformers would suppress the anticipated reverberation levels during active sonar operations. Thus, if there are going to be advantages regarding the implementation of the above non-conventional beamformers in active sonar applications, it is expected that these advantages would include minimization of the impact of reverberations by means of improved directionality. More specifically, the improved directionality of the non-conventional beamformers would restrict the reverberation effects of active sonars in a smaller number of beams than that of the conventional beamformer. This improved directionality would enhance the performance of an active sonar system (including non-conventional beamformers) to detect echoes located near the beams that are populated with reverberation effects. The real results from an active adaptive beamforming (sub-aperture STMV algorithm) output of a cylindrical sonar system, shown in Figure 6.31 of Chapter 6, provide qualitative supporting arguments that demonstrate the enhanced performance of the adaptive beamformers to suppress the reverberation effects in active sonar operations.

11.4 Conclusion

The experimental results of this study were derived from a wide variety of CW, broadband, and HFM types of strong and weak acoustic signals. The fact that adaptive and synthetic aperture beamformers provided improved detection and tracking performance for the above type of signals and under a real-time data flow as the conventional beamformer demonstrates the merits of these non-conventional processing schemes for sonar applications. In addition, the generic implementation scheme, discussed in Chapter 6, suggests that the design approach to provide synergism between the conventional beamformer and the adaptive and synthetic aperture processing schemes could probably provide some answers to the integrated active and passive sonar problem in the near future.

Although the focus of the implementation effort included only adaptive and synthetic aperture processing schemes, the consideration of other types of non-linear processing schemes for real-time sonar applications should not be excluded. The objective here was to demonstrate that non-conventional

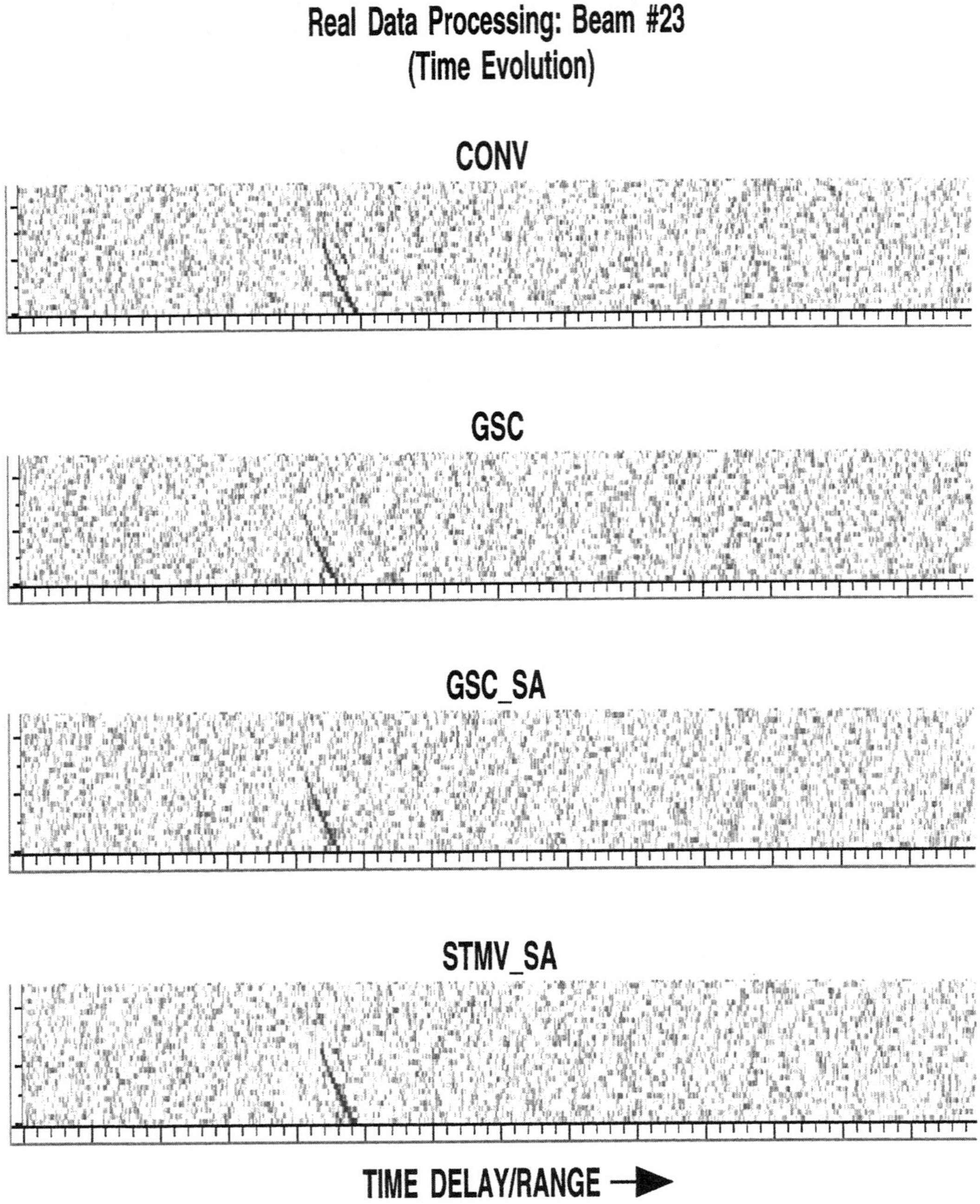

FIGURE 11.19 Waterfall display of replica correlator outputs as a function of time for the same conventional and adaptive beam time series as those of Figure 11.18. It should be noted that this figure includes approximately 2 h of processed data. The detected HFM pulses and their associated ranges are clearly shown in beam #23. A reflection from the side-walls of an underwater canyon in the area is visible as a second echo closely spaced with the main arrival. (Reprinted by permission of IEEE ©1998.)

processing schemes can address some of the challenges that the next-generation active-passive sonar systems will have to deal with in the near future. Once a computing architecture and a generic signal processing structure are established, such as those suggested in Chapter 6, the implementation of a wide variety of non-linear processing schemes in real-time sonar and radar systems can be achieved with minimum efforts. Furthermore, even though the above real results are from an experimental towed array sonar system, the performance of sonar systems including adaptive beamformers and deploying cylindrical or spherical arrays will be equivalent to that of the above experimental towed array sonar. As an example, Figure 6.31 of Chapter 6 reports results from an active sonar system, including adaptive beamformers deploying a cylindrical hydrophone array.

In conclusion, the previous results suggest that the broadband outputs of the sub-aperture adaptive processing schemes and the narrowband synthetic aperture LOFAR-grams exhibit very robust performance (under the prevailing experimental conditions) and that their AG improvements provide better signal tracking and target localization estimates than the conventional processing schemes. It is worth noting also that the reported improvements in performance of the previous non-conventional beamformers compared with that of the conventional beamformer have been consistent for a wide variety of real data sets. However, for the implementation configuration of the adaptive schemes in element space, the narrowband adaptive implementation requires very long convergence periods, which makes the application of the adaptive processing schemes in element impractical. This is because the associated long convergence periods destroy the dynamic response of the beamforming process, which is very essential during course alterations and for cases that include targets with dynamic changes in their bearings.

Finally, the experimental results of this chapter indicate that the sub-aperture GSC and STMV adaptive schemes address the practical concerns of near-instantaneous convergence associated with the implementation of adaptive beamformers in integrated active-passive sonar systems.

References

1. S. Stergiopoulos, Implementation of adaptive and synthetic aperture processing in integrated active-passive sonar systems, *Proc. IEEE,* 86(2), 358–396, 1998.
2. B. Windrow, et al., Adaptive antenna systems, *Proc. IEEE,* 55(12), 2143–2159, 1967.
3. A.A. Winder, Sonar system technology, *IEEE Trans. Sonic Ultrasonics*, SU-22(5), 291–332, 1975.
4. A.B. Baggeroer, Sonar signal processing, in *Applications of Digital Signal Processing*, A.V. Oppenheim, Ed., Prentice-Hall, Englewood Cliffs, NJ, 1978.
5. American Standard Acoustical Terminology S1.1–1960, American Standards Association, New York, May 25, 1960.
6. D. Stansfield, *Underwater Electroacoustic Transducers*, Bath University Press and Institute of Acoustics, 1990.
7. J.M. Powers, Long range hydrophones, in *Applications of Ferroelectric Polymers,* T.T. Wang, J.M. Herbert, and A.M. Glass, Eds., Chapman & Hall, New York, 1988.
8. R.I. Urick, *Principles of Underwater Acoustics*, 3rd ed., McGraw-Hill, New York, 1983.
9. S. Stergiopoulos and A.T. Ashley, Guest Editorial for a special issue on sonar system technology, *IEEE J. Oceanic Eng.*, 18(4), 361–365, 1993.
10. R.C. Trider and G.L. Hemphill, The Next Generation Signal Processor: An Architecture for the Future, DREA/Ooral 1994/Hemphill/1, Defence Research Establishment Atlantic, Dartmouth, N.S., Canada, 1994.
11. N.L. Owsley, *Sonar Array Processing*, S. Haykin, Ed., Signal Processing Series, A.V. Oppenheim Series Editor, Prentice-Hall, Englewood Cliffs, NJ, pp. 123, 1985.
12. B. Van Veen and K. Buckley, Beamforming: a versatile approach to spatial filtering, *IEEE ASSP Mag.*, 4–24, 1988.
13. A.H. Sayed and T. Kailath, A state-space approach to adaptive RLS filtering, *IEEE SP Mag.*, 18–60, July 1994.
14. E.J. Sullivan, W.M. Carey, and S. Stergiopoulos, Editorial special issue on acoustic synthetic aperture processing, *IEEE J. Oceanic Eng.*, 17(1), 1–7, 1992.

15. N.C. Yen and W. Carey, Application of synthetic-aperture processing to towed-array data, *J. Acoust. Soc. Am.*, 86, 754–765, 1989.
16. S. Stergiopoulos and E.J. Sullivan, Extended towed array processing by overlapped correlator, *J. Acoust. Soc. Am.*, 86(1), 158–171, 1989.
17. S. Stergiopoulos, Optimum bearing resolution for a moving towed array and extension of its physical aperture, *J. Acoust. Soc. Am.*, 87(5), 2128–2140, 1990.
18. S. Stergiopoulos and H. Urban, An experimental study in forming a long synthetic aperture at sea, *IEEE J. Oceanic Eng.*, 17(1), 62–72, 1992.
19. G.S. Edelson and E.J. Sullivan, Limitations on the overlap-correlator method imposed by noise and signal characteristics, *IEEE J. Oceanic Eng.*, 17(1), 30–39, 1992.
20. G.S. Edelson and D.W. Tufts, On the ability to estimate narrow-band signal parameters using towed arrays, *IEEE J. Oceanic Eng.*, 17(1), 48–61, 1992.
21. C.L. Nikias and J.M. Mendel, Signal processing with higher-order spectra, *IEEE SP Mag.*, 10–37, July 1993.
22. S. Stergiopoulos, R.C. Trider, and A.T. Ashley, Implementation of a Synthetic Aperture Processing Scheme in a Towed Array Sonar System, 127th ASA Meeting, Cambridge, MA, June 1994.
23. J. Riley, S. Stergiopoulos, R.C. Trider, A.T. Ashley, and B. Ferguson, Implementation of Adaptive Beamforming Processing Scheme in a Towed Array Sonar System, 127th ASA Meeting, Cambridge, MA, June 1994.
24. A.B. Baggeroer, W.A. Kuperman, and P.N. Mikhalevsky, An overview of matched field methods in ocean acoustics, *IEEE J. Oceanic Eng.*, 18(4), 401–424, 1993.
25. R.D. Doolitle, A. Tolstoy, and E.J. Sullivan, Editorial special issue on detection and estimation in matched field processing, *IEEE J. Oceanic Eng.* 18, 153–155, 1993.
26. Editorial special issue on neural networks for oceanic engineering systems, *IEEE J. Oceanic Eng.*, 17, October 1992.
27. A. Kummert, Fuzzy technology implemented in sonar systems, *IEEE J. Oceanic Eng.*, 18(4), 483–490, 1993.
28. A.D. Whalen, *Detection of Signals in Noise*, Academic Press, New York, 1971.
29. D. Middleton, *Introduction to Statistical Communication Theory*, McGraw-Hill, New York, 1960.
30. H.L. Van Trees, *Detection, Estimation and Modulation Theory*, Wiley, New York, 1968.
31. P.D. Welch, The use of fast Fourier transform for the estimation of power spectra: a method based on time averaging over short, modified periodigrams, *IEEE Trans. Audio Electroacoust.*, AU-15, 70–79, 1967.
32. F.J. Harris, On the use of windows for harmonic analysis with discrete Fourier transform, *Proc. IEEE*, 66, 51–83, 1978.
33. W.M. Carey and E.C. Monahan, Guest editorial for a special issue on sea surface-generated ambient noise 20–2000 Hz, *IEEE J. Oceanic Eng.*, 15(4), 265–267, 1990.
34. R.A. Wagstaff, Iterative technique for ambient noise horizontal directionality estimation from towed line array data, *J. Acoust. Soc. Am.*, 63(3), 863–869, 1978.
35. R.A. Wagstaff, A computerized system for assessing towed array sonar functionality and detecting faults, *IEEE J. Oceanic Eng.*, 18(4), 529–542, 1993.
36. S.M. Flatte, R. Dashen, W.H. Munk, K.M. Watson, and F. Zachariasen, *Sound Transmission through a Fluctuating Ocean*, Cambridge University Press, New York, 1985.
37. D. Middleton, Acoustic scattering from composite wind-wave surfaces in bubble-free regimes, *IEEE J. Oceanic Eng.*, 14, 17–75, 1989.
38. W.A. Kuperman and F. Ingenito, Attenuation of the coherent component of sound propagating in shallow water with rough boundaries, *J. Acoust. Soc. Am.*, 61, 1178–1187, 1977.
39. B.J. Uscinski, Acoustic scattering by ocean irregularities: aspects of the inverse problem, *J. Acoust. Soc. Am.*, 86, 706–715, 1989.

40. W. M. Carey and W.B. Moseley, Space-time processing, environmental-acoustic effects, *IEEE J. Oceanic Eng.*, 16, 285–301, 1991; also in *Progress in Underwater Acoustics*, Plenum Press, New York, pp. 743–758, 1987.
41. S. Stergiopoulos, Limitations on towed-array gain imposed by a non isotropic ocean, *J. Acoust. Soc. Am.*, 90(6), 3161–3172, 1991.
42. W.A. Struzinski and E.D. Lowe, A performance comparison of four noise background normalization schemes proposed for signal detection systems, *J. Acoust. Soc. Am.*, 76(6), 1738–1742, 1984.
43. S.W. Davies and M.E. Knappe, Noise Background Normalization for Simultaneous Broadband and Narrowband Detection, Proceedings from IEEE-ICASSP 88, U3.15 pp. 2733–2736, 1988.
44. S. Stergiopoulos, Noise normalization technique for beamformed towed array data, *J. Acoust. Soc. Am.*, 97(4), 2334–2345, 1995.
45. A.H. Nuttall, Performance of Three Averaging Methods, for Various Distributions, Proceedings of SACLANTCEN Conference on Underwater Ambient Noise, SACLANTCEN CP-32, Vol. II, pp. 16–1, SACLANT Undersea Research Centre, La Spezia, Italy, 1982.
46. H. Cox, R.M. Zeskind, and M.M. Owen, Robust adaptive beamforming, *IEEE Trans. Acoust. Speech Signal Process.*, ASSP-35(10), 1365–1376, 1987.
47. H. Cox, Resolving power and sensitivity to mismatch of optimum array processors, *J. Acoust. Soc. Am.*, 54(3), 771–785, 1973.
48. J. Capon, High resolution frequency wavenumber spectral analysis, *Proc. IEEE*, 57, 1408–1418, 1969.
49. T.L. Marzetta, A new interpretation for Capon's maximum likelihood method of frequency-wavenumber spectra estimation, *IEEE Trans. Acoust. Speech Signal Process.*, ASSP-31(2), 445–449, 1983.
50. S. Haykin, *Adaptive Filter Theory*, Prentice-Hall, Englewood Cliffs, NJ, 1986.
51. S.D. Peters, Near-instantaneous convergence for memoryless narrowband GSC/NLMS adaptive beamformers, *IEEE Trans. Acoust. Speech Signal Process.*, submitted.
52. S. Stergiopoulos, Influence of Underwater Environment's Coherence Properties on Sonar Signal Processing, Proceedings of 3rd European Conference on Underwater Acoustics, FORTH-IACM, Heraklion-Crete, V-I, 453–458, 1996.
53. A.C. Dhanantwari and S. Stergiopoulos, Adaptive beamforming with near-instantaneous convergence for matched filter processing, *J. Acoust. Soc. Am.*, submitted.
54. R.O. Nielsen, *Sonar Signal Processing*, Artech House, Norwood, MA, 1991.
55. D. Middleton and R. Esposito, Simultaneous otpimum detection and estimation of signals in noise, *IEEE Trans. Inf. Theory*, IT-14, 434–444, 1968.
56. V.H. MacDonald and P.M. Schulteiss, Optimum passive bearing estimation in a spatially incoherent noise environment, *J. Acoust. Soc. Am.*, 46(1), 37–43, 1969.
57. G.C. Carter, Coherence and time delay estimation, *Proc. IEEE*, 75(2), 236–255, 1987.
58. C.H. Knapp and G.C. Carter, The generalized correlation method for estimation of time delay, *IEEE Trans. Acoust. Speech Signal Process.*, ASSP-24, 320–327, 1976.
59. D.C. Rife and R.R. Boorstyn, Single-tone parameter estimation from discrete-time observations, *IEEE Trans. Inf. Theory*, 20, 591–598, 1974.
60. D.C. Rife and R.R. Boorstyn, Multiple-tone parameter estimation from discrete-time observations, *Bell System Technical J.*, 20, 1389–1410, 1977.
61. S. Stergiopoulos and N. Allcott, Aperture extension for a towed array using an acoustic synthetic aperture or a linear prediction method, *Proc. ICASSP-92*, March 1992.
62. S. Stergiopoulos and H. Urban, A new passive synthetic aperture technique for towed arrays, *IEEE J. Oceanic Eng.*, 17(1), 16–25, 1992.
63. W.M.X. Zimmer, High Resolution Beamforming Techniques, Performance Analysis, SACLANTCEN SR-104, SACLANT Undersea Research Centre, La Spezia, Italy, 1986.
64. A. Mohammed, Novel Methods of Digital Phase Shifting to Achieve Arbitrary Values of Time Delays, DREA Report 85/106, Defence Research Establishment Atlantic, Dartmouth, N.S., Canada, 1985.

65. A. Antoniou, *Digital Filters: Analysis, Design, and Applications*, 2nd ed., McGraw-Hill, New York, 1993.
66. L.R. Rabiner and B. Gold, *Theory and Applications of Digital Signal Processing*, Prentice-Hall, Englewood Cliffs, NJ, 1975.
67. A. Mohammed, A High-Resolution Spectral Analysis Technique, DREA Memorandum 83/D, Defence Research Establishment Atlantic, Dartmouth, N.S., Canada, 1983.
68. B.G. Ferguson, Improved time-delay estimates of underwater acoustic signals using beamforming and prefiltering techniques, *IEEE J. Oceanic Eng.*, 14(3), 238–244, 1989.
69. S. Stergiopoulos and A.T. Ashley, An experimental evaluation of split-beam processing as a broadband bearing estimator for line array sonar systems, *J. Acoust. Soc. Am.*, 102(6), 3556–3563, 1997.
70. G.C. Carter and E.R. Robinson, Ocean effects on time delay estimation requiring adaptation, *IEEE J. Oceanic Eng.*, 18(4), 367–378, 1993.
71. P. Wille and R. Thiele, Transverse horizontal coherence of explosive signals in shallow water, *J. Acoust. Soc. Am.*, 50, 348–353, 1971.
72. P.A. Bello, Characterization of randomly time-variant linear channels, *IEEE Trans. Commun. Syst.*, 10, 360–393, 1963.
73. D.A. Gray, B.D.O. Anderson, and R.R. Bitmead, Towed array shape estimation using Kalman filters — theoretical models, *IEEE J. Oceanic Eng.*, 18(4), October 1993.
74. B.G. Quinn, R.S.F. Barrett, P.J. Kootsookos, and S.J. Searle, The estimation of the shape of an array using a hidden Markov model, *IEEE J. Oceanic Eng.*, 18(4), October 1993.
75. B.G. Ferguson, Remedying the effects of array shape distortion on the spatial filtering of acoustic data from a line array of hydrophones, *IEEE J. Oceanic Eng.*, 18(4), October 1993.
76. J.L. Riley and D.A. Gray, Towed array shape estimation using Kalman Filters — experimental investigation, *IEEE J. Oceanic Eng.*, 18(4), October 1993.
77. B.G. Ferguson, Sharpness applied to the adaptive beamforming of acoustic data from a towed array of unknown shape, *J. Acoust. Soc. Am.*, 88(6), 2695–2701, 1990.
78. F. Lu, E. Milios, and S. Stergiopoulos, A new towed array shape estimation method for sonar systems, *IEEE J. Oceanic Eng.*, submitted.
79. N.L. Owsley, Systolic Array Adaptive Beamforming, NUWC Report 7981, New London, CT, September 1987.
80. D.A. Gray, Formulation of the maximum signal-to-noise ratio array processor in beam space, *J. Acoust. Soc. Am.*, 72(4), 1195–1201, 1982.
81. O.L. Frost, An algorithm for linearly constrained adaptive array processing, *Proc. IEEE*, 60, 926–935, 1972.
82. H. Wang and M. Kaveh, Coherent signal-subspace processing for the detection and estimation of angles of arrival of multiple wideband sources, *IEEE Trans. Acoust. Speech Signal Process.*, ASSP-33, 823–831, 1985.
83. J. Krolik and D.N. Swingler, Bearing estimation of multiple broadband sources using steered covariance matrices, *IEEE Trans. Acoust. Speech Signal Process.*, ASSP-37, 1481–1494, 1989.
84. J. Krolik and D.N. Swingler, Focussed wideband array processing via spatial resampling, *IEEE Trans. Acoust. Speech Signal Process.*, ASSP-38, 1990.
85. J.P. Burg, Maximum Entropy Spectral Analysis, Presented at the 37th Meeting of the Society of Exploration Geophysicists, Oklahoma City, OK, 1967.
86. C. Lancos, *Applied Analysis*, Prentice-Hall, Englewood Cliffs, NJ, 1956.
87. V.E. Pisarenko, On the estimation of spectra by means of nonlinear functions on the covariance matrix, *Geophys. J. Astron. Soc.*, 28, 511–531, 1972.
88. A.H. Nuttall, Spectral Analysis of a Univariate Process with Bad Data Points, via Maximum Entropy and Linear Predictive Techniques, NUWC TR5303, New London, CT, 1976.
89. R. Kumaresan and W.D. Tufts, Estimating the angles of arrival of multiple plane waves, *IEEE Trans. Acoust. Speech Signal Process.*, ASSP-30, 833–840, 1982.

90. R.A. Wagstaff and J.-L. Berrou, Underwater Ambient Noise: Directionality and Other Statistics, SACLANTCEN Report SR-59, SACLANTCEN, SACLANT Undersea Research Centre, La Spezia, Italy, 1982.
91. S.M. Kay and S.L. Marple, Spectrum analysis — a modern perspective, *Proc. IEEE*, 69, 1380–1419, 1981.
92. D.H. Johnson and S.R. Degraaf, Improving the resolution of bearing in passive sonar arrays by eigenvalue analysis, *IEEE Trans. Acoust. Speech Signal Process.*, ASSP-30, 638–647, 1982.
93. D.W. Tufts and R. Kumaresan, Estimation of frequencies of multiple sinusoids: making linear prediction perform like maximum likelihood, *Proc. IEEE*, 70, 975–989, 1982.
94. G. Bienvenu and L. Kopp, Optimality of high resolution array processing using the eigensystem approach, *IEEE Trans. Acoust. Speech Signal Process.*, ASSP-31, 1235–1248, 1983.
95. D.N. Swingler and R.S. Walker, Linear array beamforming using linear prediction for aperture interpolation and extrapolation, *IEEE Trans. Acoust. Speech Signal Process.*, ASSP-37, 16–30, 1989.
96. P. Tomarong and A. El-Jaroudi, Robust high-resolution direction-of-arrival estimation via signal eigenvector domain, *IEEE J. Oceanic Eng.*, 18(4), 491–499, 1993.
97. J. Fawcett, Synthetic aperture processing for a towed array and a moving source, *J. Acoust. Soc. Am.*, 93, 2832–2837, 1993.
98. L.J. Griffiths and C.W. Jim, An alternative approach to linearly constrained adaptive beamforming, *IEEE Trans. Antennas Propagation*, AP-30, 27–34, 1982.
99. D.T.M. Slock, On the convergence behavior of the LMS and the normalized LMS algorithms, *IEEE Trans. Acoust. Speech Signal Process.*, ASSP-31, 2811–1825, 1993.
100. A. C. Dhanantwari, Adaptive Beamforming with Near-Instantaneous Convergence for Matched Filter Processing, Master thesis, Department of Electrical Engineering, Technical University of Nova Scotia, Halifax, N.S., Canada, September 1996.
101. A. Tawfik and S. Stergiopoulos, A Generic Processing Structure Decomposing the beamforming process of 2-D & 3-D Arrays of Sensors into Sub-Sets of Coherent Processes, Proceedings of IEEE-CCECE, St. John's, NF, Canada, May 1997.
102. W.A. Burdic, *Underwater Acoustic System Analysis*, Prentice-Hall, Englewood Cliffs, NJ, 1984.
103. J-P. Hermand and W.I. Roderick, Acoustic model-based matched filter processing for fading time-dispersive ocean channels, *IEEE J. Oceanic Eng.*, 18(4), 447–465, 1993.
104. Mercury Computer Systems, Inc., *Mercury News Jan-97*, Mercury Computer Systems, Inc., Chelmsford, MA 1997.
105. Y. Bar-Shalom and T.E. Fortman, *Tracking and Data Association*, Academic Press, Boston, MA, 1988.
106. S.S. Blackman, *Multiple-Target Tracking with Radar Applications*, Artech House Inc., Norwood, MA, 1986.
107. W. Cambell, S. Stergiopoulos, and J. Riley, Effects of Bearing Estimation Improvements of Non-Conventional Beamformers on Bearing-Only Tracking, Proceedings of Oceans '95 MTS/IEEE, San Diego, CA, 1995.
108. W.A. Roger and R.S. Walker, Accurate Estimation of Source Bearing from Line Arrays, Proceedings of the Thirteen Biennial Symposium on Communications, Kingston, Ont., Canada, 1986.
109. D. Peters, Long Range Towed Array Target Analysis — Principles and Practice, DREA Memorandum 95/217, Defence Research Establishment Atlantic, Dartmouth, N.S., Canada, 1995.
110. J.G. Proakis, *Digital Communications*, McGraw-Hill, New York, 1989.
111. W.W. Peterson and T.G. Birdsall, The theory of signal detectability, *Univ. Mich. Eng. Res. Inst. Rep.*, 13, 1953.
112. J.T. Kroenert, Discussion of detection threshold with reverberation limited conditions, *J. Acoust. Soc. Am.*, 71(2), 507–508, February 1982.
113. L. Moreavek and T.J. Brudner, USS Asheville leads the way in high frequency sonar, *Undersea Warfare*, 1(3), 22–24, 1999.

114. P.M. Baggenstoss, On detecting linear frequency-modulated waveforms in frequency- and time-dispersive channels: alternatives to segmented replica correlation, *IEEE J. Oceanic Eng.*, 19(4), 591–598, October 1994.
115. B. Friedlander and A. Zeira, Detection of broadband signals in frequency and time dispersive channels, *IEEE Trans. Signal Proc.*, 44(7), 1613–1622, July 1996.

12

Phased Array Radars

Nikolaos Uzunoglu
National Technical University of Athens

12.1 Introduction

In radar systems the antenna unit being employed for both transmission and reception has a very important role to fulfill the design requirements and system specifications.

Radar technology developed in 1940 to 1965 was based on the use of antennae providing directional beams incorporating mechanical rotation. Usually, parabolic reflectors were employed to develop directive beam antennae based on the geometrical properties of reflectors. Depending on the requirements, various technologies have been employed, such as parabolic cylinders, paraboloids, offset focus paraboloids, and various types of lenses. One of the difficulties with these antennae has been their three-dimensional structure, which is quite large.[1,2]

An alternative class of antenna has been the use of radiators being constructed as an array of many similar elementary antennae. This provides the ability of reducing the antenna dimensions practically into two dimensions. Such structures are slotted waveguide lines, dipole arrays, microstrip arrays, etc. The resonant nature of the latter type of antennae makes them rather narrowband, and usually, less than 10% frequency bandwidth is achieved.[3]

The fundamental idea of developing electronically controlled beam array antennae was suggested a long time ago, and its use in radar systems was foreseen by several researchers as early as 1940. Despite this fact and the apparent superior properties of non-moving antennae, the realization of phased array radars was delayed for many decades, mainly because of the very high development and maintenance costs. Recently, with the advances in microwave monolithic integrated circuits (MMIC) technology, phased array antennae have started to become feasible at a reasonable cost. The possibility of using alternative technologies, such as hybrid optical microwave techniques, has also increased the possibility of developing low-cost array systems.

In general terms, the use of a phased array provides a significant improvement compared to conventional mechanically related antennae, with the most important benefits being

- Absence of mechanical movements in antenna system
- Very fast search of a given field of view

0-8493-3691-0/01/$0.00+$.50

- Simultaneous tracking of many targets using electronic scanning multiple arrays
- Ability to suppress intentional or unintentional interference
- Electronic countermeasure feature to radar and communication systems using phased array antennae
- Highly flexible control of radiation patterns such as polarization and side lobe levels

Recently, there has been growing interest in using phased array techniques in commercial applications such as mobile and satellite communications. The possibility of achieving space multiplexing is being investigated by various researchers. It is hoped that market-driven demand could facilitate the wide exploitation of phased array technologies, which until recently have remained in the monopoly of military applications.[4,5]

The fundamental theoretical concept of phased array antennae is the exploitation of the superposition of waves radiated by the individual elements of the array's antennae. The ability to control the phase and the amplitude of the waves emitted by each individual element allows the angular movement of the radiated beams. Traditionally, phased arrays have been developed based on the principle of superheterodyne transceiver technology, and because of this, usually phased array antennae have a narrow frequency bandwidth in which they can operate without any significant degradation.

It should be emphasized that phased array techniques are expected to play an essential role in the development of "software radio"; that is, the transceiver units will become a set of high-speed digital signal processing circuits. In this case, the phased array structure can be simplified essentially since all basic functions such as phase shifting and amplitude setting for each element will be implemented with an embedded computational system with parallel processing.

Indeed, we are very close to a time when the whole process of beam steering and radiation pattern synthesis will be carried out through software-programmable, high-speed, digital signal processing circuits.

This new concept, which is applicable for both receiving and transmitting arrays, is an approaching technology which will enhance the use of phased array techniques. It is an entirely new concept and should not to be confused with the traditional phase shifter technology, which usually is under digital control. This entirely new approach combined with the ultra wideband antenna arrays will allow the possibility of using a single aperture antenna for communication, radar, and remote sensing applications on platforms such as ships, airplanes, vehicles, and satellites. Finally, one should mention that this technology is fully compatible with the reduced radar cross-section (RCS) stealth technology required on such platforms.[6]

12.2 Fundamental Theory of Phased Arrays

The fundamental concept in phased array operation is the simultaneous use of many radiating elements and control of the overall array antenna radiation pattern by setting the phase and amplitude in each individual array antenna element properly.

In the present section, the fundamental theory of phased array is presented, assuming each radiating element is operating independently without any mutual interaction with the other elements belonging to the very same antenna array system. This analysis provides the basic characteristics of array antennae, which should be used as a first step in designing new arrays. However, more detailed electromagnetic analysis is needed in designing in detail new arrays and predicting their behavior. This is presented in the next section.

In the following analysis, a continuous wave signal of harmonic type, $\exp(j\omega t)$ time dependence excitation of array, is assumed throughout the study, where ω is the radiation field angular frequency and t is the time variable. According to this, all quantities being used in the analysis are complex numbers and can be considered as Fourier transformation quantities.

The geometry of a generalized array is shown in Figure 12.1 where N active elements distributed in a three-dimensional space are assumed.

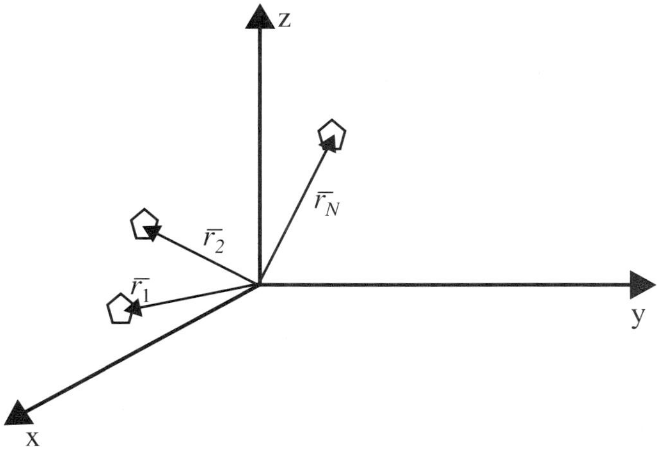

FIGURE 12.1 Three-dimensional array geometry.

Each radiator element position is defined with its position vector $\bar{r}_i(i) = 1, \ldots, \nu$ of a characteristic point on each array element. Assume an arbitrary observation point $\bar{r}$ being at the far field of radiator elements. Based on the knowledge of spherical field radiation from each element, the total electric field $E(\bar{r})$ is computed by using the superposition principle as follows:[7]

$$\bar{E}(\bar{r}) = \Lambda \sum_{i=1}^{N} \bar{f}_i(\hat{r})\alpha_i \frac{e^{-jk|\bar{r}-\bar{r}_i|}}{|\bar{r}-\bar{r}_i|} \tag{12.1}$$

where Λ is a normalization constant depending on radiated power level by the array, k is the free space propagation constant ($k = \omega/c$, where $c = 3 \times 10^8$ m/s is the speed of light in vacuum), $\bar{f}_i(\bar{r})$ is the vector radiation pattern function of the ith element, $\hat{r}$ indicates the unit vector along the observation angle, and the complex numbers $a_i(i = 1, 2, \ldots, N)$ are the impressed excitation amplitudes on each element. It should be emphasized that the amplitudes a_i are not directly controllable by the array signal driving system due to interactions between the elements. This issue is discussed in the next section.

Based on the feet of the "far field" assumption, in Equation 12.1 the phase term can be approximated using the expression:

$$|\bar{r}-\bar{r}_i| = \sqrt{|\bar{r}^2| + |\bar{r}_i^2| - 2\bar{r}\cdot\bar{r}_1} \approx |\bar{r}|\left(1 - \hat{r}\cdot\bar{r}_i\frac{1}{|\bar{r}|}\right) = |\bar{r}| - \hat{r}\bar{r}_i \tag{12.2}$$

while the denominator term can be taken simply as $|\bar{r}-\bar{r}_i| \approx |\bar{r}|$. Then, Equation 12.1 can be rewritten as

$$\bar{E}(\bar{r}) = \Lambda\left(\sum_{i=1}^{N} \bar{f}_i(\hat{r})\alpha_i e^{-jk(\hat{r}\cdot\bar{r}_i)}\right)\frac{e^{-jk|\bar{r}|}}{|\bar{r}|} \tag{12.3}$$

It is evident that the electric field at the position $\bar{r}$ is a spherical wave with amplitude determined by summing wave amplitudes of each array element weighted with a complex number a_i and the phase term $\exp(jk\hat{r}\cdot\bar{r}_i)$. The a_i terms can be written as follows:

$$\alpha_i = |\alpha_i| e^{-j\tau_i\omega} e^{-j\phi_i} \tag{12.4}$$

where $|a_i|$ on each element is the amplitude of the excitation, τ_i is the physical delay prior to element excitation, and ϕ_i is also a phase shift prior to element excitation.

It should be emphasized that in Equation 12.3 one can certainly define an overall phase constant

$$\Phi_i = \tau_i\omega + \phi_i \tag{12.5}$$

and thus replace Equation 12.4 with

$$\alpha_i = |\alpha_i|e^{-j\phi_i} \tag{12.6}$$

However, in phased array systems, it is very important to distinguish the two terms in Equation 12.5. The true time delay $\tau_i\omega$ leads to wideband arrays, while, on the contrary, the control of the $\phi_i\omega$ phases restricts the array bandwidth. Despite the many benefits of "true time delay arrays," until now only a few such arrays have been built because of the involved hardware complexity. Only recently has the use of fiber optics technology provided the possibility of developing such arrays (see Section 12.4). An alternative method has been to construct the array antenna using "subarrays" incorporating "phase control" (second term in Equation 12.5), while each subarray delay is to be driven by a "true time delay" device. This leads to better bandwidth behavior.

12.2.1 Fundamental Array Properties

12.2.1.1 Focusing Properties of Arrays

The basic requirement in phased antenna arrays is to achieve strong directivity of electromagnetic energy in a specific direction. In order to obtain this at a specific operation frequency $\omega = \omega_0$, "constructive interference" is required in a specific direction $\hat{r} = \hat{r}_0$. This could be achieved if in Equation 12.3 the terms under the sum can be added constructively. This requires

$$\alpha_i e^{j\frac{\omega_0}{c}\hat{r}\cdot\bar{r}_i} = |a_i|e^{-j\left(\tau_i\omega_0 + -\frac{\omega_0}{c}\hat{r}_0\cdot\bar{r}_i\right) + j\phi_i\cdot(\omega_0)} = |a_i| \tag{12.7}$$

and therefore, at the ω_0 operation frequency, the phase term should be

$$\frac{\omega_0}{c}\hat{r}_0\cdot\bar{r}_i - \tau_i\omega_0 + \phi_i(\omega_0) = 2\pi n \tag{12.8}$$

where $n = 0, \pm1, \pm2, \pm\ldots$. Then the field amplitude from Equation 12.3 is obtained as

$$\bar{E}(\hat{r}_0) = \Lambda\left(\sum_{i=1}^{N}\bar{f}_i(\hat{r}_0)|a_i|\right)\frac{e^{-jk|\bar{r}|}}{|\bar{r}|} \tag{12.9}$$

Notice that in Equation 12.8 assuming the array elements are "almost identical," $f_i(\hat{r}) \approx f(\hat{r})$ (for $i = 1, 2, \ldots, N$), Equation 12.8 leads to

$$\bar{E}(\hat{r}_0) = \Lambda\bar{f}(\hat{r}_0)\left(\sum_{i=1}^{N}|\alpha_i|\right)\frac{e^{-jk|\bar{r}|}}{|\bar{r}|} \tag{12.10}$$

Examination of the condition in Equation 12.8 leads to the following considerations.

12.2.1.1.1 The Case of "True Time Delay"

$$\phi_i \text{ constant } (i = 1, 2, \ldots, N) \text{ and } (n = 0)$$

In this case, $\tau_i = \hat{r}_0\cdot\bar{r}_i$, and this condition is independent. Equation 12.10 holds for every frequency, provided the individual array radiation pattern $\bar{f}(\hat{r}_0)$ is independent of frequency (ω) or is relatively a slowly varying function of τ. This case corresponds to wideband operation.

12.2.1.1.2 *The Case of "Phase Control"*

In this case τ_i = constant, so at an arbitrary operation frequency ω, Equation 12.3 can be written as

$$\bar{E}(\hat{r}_0) = \Lambda \frac{e^{-j\omega\tau} f(\hat{r}_0)^{e^{-jk|r|}}}{|\bar{r}|} \sum_{i=1}^{N} |\alpha_i| e^{j\frac{\omega - \omega_0}{c} \hat{r}_0 \cdot \bar{r}_i} \tag{12.11}$$

Notice that when $\omega \neq \omega_0$, the summation in Equation 12.11 doesn't represent a "constructive summation" and the array is not any more focused in the direction of $\hat{r} = \hat{r}_0$.

In the case of the phase control array, the electric field at an arbitrary orientation can be computed using Equations 12.3 and 12.8, leading to the following relation:

$$\bar{E}(\hat{r}) = \Lambda \frac{e^{-jk|r|}}{|\bar{r}|} \bar{f}(\hat{r}) \sum_{i=1}^{N} |a_i| \exp\left(j \frac{\omega \hat{r} \cdot \bar{r}_i - \omega_0 \hat{r}_0 \cdot r_i}{c}\right) \tag{12.12}$$

Notice that in Equation 12.12 the electric field is obtained as a product of the individual array element radiation pattern function $\bar{f}(\hat{r})$ and the "array factor" is the summation term.

12.2.1.2 Directivity of Arrays

The definition of the array directivity is the same as in an ordinary antenna, that is,

$$D = \frac{\text{Maximum Power Density}}{\text{Average Power Density}} = \frac{E(\hat{r}_0) \cdot \bar{E}(\hat{r}_0)}{\frac{1}{4\pi} \iint_{4\pi \text{ steradian}} \bar{E}(\hat{r}) \cdot \bar{E}(\hat{r}) d\hat{r}} \tag{12.13}$$

where $d\hat{r}$ is the elementary solid angle, and the surface integral in the denominator is computed on the unit sphere.

12.2.2 Linear Arrays

Consider the case of one-dimensional arrays with an odd number of elements as shown in Figure 12.2, where the number of elements is $2N + 1$ numbered as $i = -N, -N + 1, \ldots, 0, N - 1, N$, and the distance between two array elements is d.

$$\bar{r}_i = \hat{x}(-N + i - 1) \qquad i = 1, 2, \ldots 2N + 1$$

According to Equation 12.12, the array factor when the observation vector is within the x0z plane is computed as follows:

$$A_f = \sum_{i=1}^{2N+1} |a_i| \exp\left[j\left(\frac{\omega}{c}\sin\theta - \frac{\omega_0}{c}\sin\theta_0\right)(i - 1 - N)d\right] \tag{12.14}$$

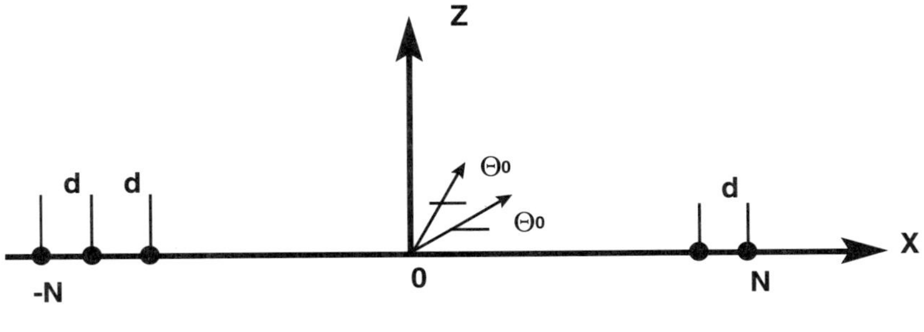

FIGURE 12.2 Linear array geometry.

where $\sin\theta = \hat{r}\cdot\hat{x}$ and $\sin\theta_0 = \hat{r}_0\cdot\hat{x}$ are defined in Figure 12.2.

In the case of uniform excitation, $|a_i| = 1$, and the array factor is computed using the theory of geometric series and leads to the term

$$A_f = \frac{\sin\left(\left(N+\frac{1}{2}\right)\frac{d}{c}(\omega\sin\theta - \omega_0\sin\theta_0)\right)}{\sin\left(\frac{(\omega\sin\theta - \omega_0\sin\theta_0)}{2c}d\right)} \tag{12.15}$$

When the array is operating at $\omega = \omega_0$ (focusing frequency), then Equation 12.15 is

$$A_f = \frac{\sin\left(\left(N+\frac{1}{2}\right)\frac{d}{c}\omega_0(\sin\theta - \sin\theta_0)\right)}{\sin\left(\frac{\omega_0}{2c}d(\sin\theta - \sin\theta_0)\right)} \tag{12.16}$$

At $\theta = \theta_0$, the array factor is equal to

$$A_f(\theta_0) = (2N+1) \tag{12.17}$$

which is independent of the angle θ_0. On substituting Equation 12.15 into Equation 12.12, the electric field of a linear array can be written as

$$\bar{E}(\bar{r}) = \Lambda\frac{e^{-jk|\bar{r}|}}{|\bar{r}|}\bar{f}(\theta)\cdot A_f(\theta) \tag{12.18}$$

Several useful properties of antenna arrays are found using this relation.

12.2.2.1 Grating Lobes

Consider the case of a large number of elements. The variation of $A_f(\theta)$ (when $\omega = \omega_0$) can be drawn as shown in Figure 12.3.

Notice that $A_f(\theta)$ is a periodic function of angle θ.

The angular period is determined by using the condition

$$\frac{\omega_0 d}{2c}(\sin\theta - \sin\theta_0) = \pi n \quad n = \pm 1, \pm 2, \ldots \quad \text{or} \quad \sin\theta_n = \frac{2n}{k_0 d} + \sin\theta_0 \tag{12.19}$$

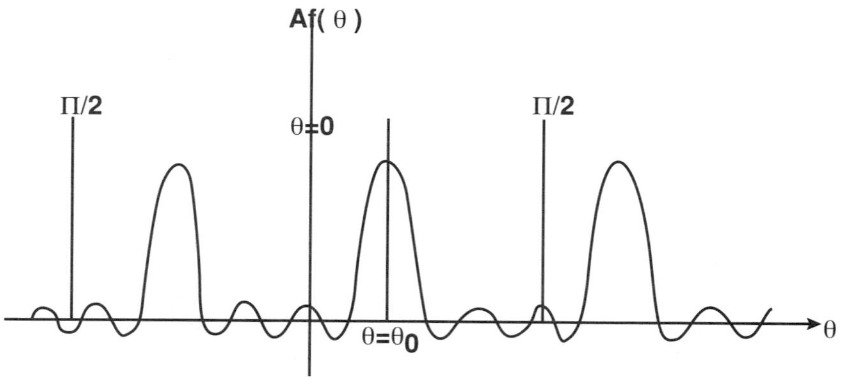

FIGURE 12.3 Array factor dependence to angle θ.

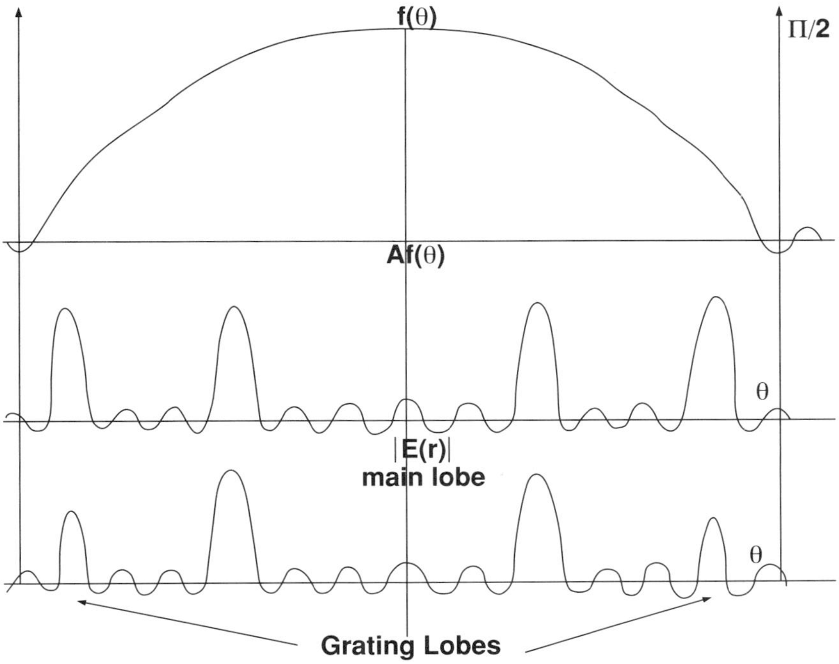

FIGURE 12.4 Radiation pattern of a linear array.

where $k_0 = \omega_0/c$. Depending on the value of $k_0 d$, we might have any one of the many peak values of the $A_f(\theta)$ functions. When $k_0 d << 1$, then the only valid solution is $\theta = \theta_0$, while if $k_0 d >> 1$, then many peaks could appear within the angular region $-\pi/2 < \theta < \pi/2$ (see Figure 12.3).

Consider what happens when the individual element radiation pattern function $\bar{f}(\theta)$ (Equation 12.18) is multiplied with $A_f(\theta)$. The picture shown in Figure 12.4 is obtained for the overall radiation pattern.

12.2.2.2 Side Lobe Level

In the case of a uniform excitation array ($|a_i| = 1$), the array factor given in Equation 12.16 shows that that array's first side lobe level is easily computed to be –13 dB. This is because of the uniform array excitation. Using a non-uniform excitation of the array elements can decrease the side lobe level. Several cases such as the binomial or the Chebyshev type distribution could be used. An example is the case of binomial distribution which is defined by the following equation:

$$|a_i| = \binom{2N+1}{i-1} \xi^{2N+1-(i-1)} \tag{12.20}$$

where ξ is a constant to be specified for $i = 1, 2, 3, \ldots, (2N + 1)$. On substituting Equation 12.20 into Equation 12.14, the array factor is found to be

$$\begin{aligned} A_f(\theta) &= \sum_{i=1}^{2N+1} \binom{2N+1}{i-1} \xi^{2N+1-(i-1)} e^{j\left(\frac{\omega}{c}\sin\theta_0 - \frac{\omega_0}{c}\sin\theta_0\right)(i-1-N)d} \\ &= \left\{\xi + \exp\left[j\left(\frac{\omega}{c}\sin\theta - \frac{\omega_0}{c}\sin\theta_0\right)\right]\right\}^{2N+1} \exp\left[-j\left(\frac{\omega}{c}\sin\theta - \frac{\omega_0}{c}\sin\theta_0\right)\mathrm{N}d\right] \end{aligned} \tag{12.21}$$

12.2.2.3 Array Bandwidth

Upon returning to Equation 12.15, in the case of uniform excitation array, and assuming $\omega = \omega_0 + \delta\omega$ with $\delta\omega/\omega_0 << 1$, the angular shift $\delta\theta$ of the main beam axis can be computed easily by taking into account the condition

$$(\omega_0 + \delta\omega)\sin(\theta_0 + \delta\theta) = \omega_0 \sin\theta_0 \tag{12.22}$$

and after some simple algebra,

$$\delta\theta = -\left(\frac{\delta\omega}{\omega_0}\right)\frac{1}{\cos\theta_0}$$

which shows that at $\theta_0 = 0$ the angle shift is minimum. This shifting property is being used extensively in practical arrays in the development of electronic scanning arrays by frequency shifting.

12.2.3 Two- and Three-Dimensional Arrays

The fundamental theory presented earlier allows the computation of array properties in two and three dimensions. The most common case is the planar two-dimensional array shown in Figure 12.5, where the array elements are placed on the x-y plane.

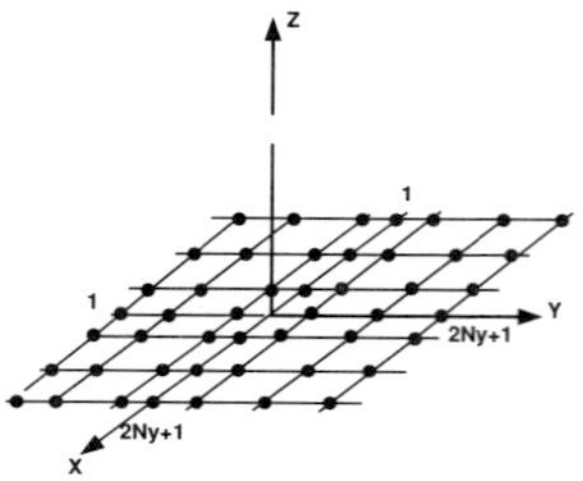

FIGURE 12.5 Two-dimensional array.

In this case also, the numbering of array elements is similar to the case of the linear array. The total number of array elements is $(2N_x + 1)(2N_y + 1)$. Assuming a uniform excitation of array elements and going through an analysis similar to the case of Equations 12.14 to 12.16, the electric field is obtained as

$$\begin{aligned}\bar{E}(\bar{r}) &= \Lambda\frac{e^{-jk|\bar{r}|}}{|\bar{r}|}\bar{f}(\hat{r})\\ &\frac{\sin\left(\left(N_x + \frac{1}{2}\right)\frac{d}{c}(\omega\sin\theta\cos\phi - \omega_0\sin\theta_0\cos\phi_0)\right)}{\sin\left(\frac{\omega\sin\theta\cos\phi - \omega_0\sin\theta_0\cos\phi_0}{2c}\right)}\\ &\frac{\sin\left(\left(N_y + \frac{1}{2}\right)\frac{d}{c}(\omega\sin\theta\sin\phi - \omega_0\sin\theta_0\sin\phi_{00})\right)}{\sin\left(\left(\frac{d}{2c}(\omega\sin\theta\sin\phi - \omega_0\sin\theta_0\sin\phi_0)\right)\right)}\end{aligned} \tag{12.23}$$

where $\hat{r} = \hat{x}\cos\phi\sin\theta + \hat{y}\sin\phi\sin\theta + \hat{z}\cos\theta$ is the unit vector along the observation direction and $\hat{r}_0$ is the corresponding main beam direction. Similar conclusions, as in the case of linear arrays, can be deduced using Equation 12.17. The case of non-planar, known as conformal, arrays can also be computed.

12.3 Analysis and Design of Phased Arrays

In order to analyze and design real-world arrays, rigorous computational electromagnetism techniques are required, instead of using trial-and-error efforts in constructing phased arrays, which is an extremely costly approach.

During the last decades, much effort has been spent in developing computational techniques to model complex structures. The most known techniques are[8]

1. Method of moments (MOM)
2. Finite difference time domain (FDTD) and transmission line method (TLM)
3. Finite element (FE)
4. Method of auxiliary sources (MAS)

In the following, the basics of the MOM are presented. It is important to stress the fact that most analysis techniques are based on the infinite array approximation to compute the field distribution on the array elements. Then the relation between near and far field elements is used to compute the radiation pattern of the array antenna by taking into account a finite number of array elements. This approximation has a very solid basis in many cases, but because of the negligence of not considering boundary array elements being taken, significant effects on secondary array properties could be observed, such as side lobe level and polarization properties of radiated waves. Furthermore, rigorous techniques are required to take the interaction between the array elements into account. It is important to realize that in the modeling of an array.

12.3.1 Statement of the Boundary Value Problem

Consider the surface S in free space which can take any arbitrary shape and on which the active array elements are placed. Each element is driven with a transmission line. Usually, the surface S is a perfect electric conductor, and this assumption will also be adopted here. Array antenna elements, which are individually fed by transmission lines, are placed on this surface.

The array elements are considered to be identical. This is the most commonly encountered case in practice. Furthermore, the phased array antenna could have in its vicinity a passive structure such as, in some cases, a reflector or lens. The latter case is quite common in quasi-optic type phased array designs.

Integral equation techniques treated by various types of MOM are considered to be the most accurate among the computational algorithms.

Green's function is employed widely in developing the formulation of the corresponding boundary problems. Green's function is the response function of an electromagnetic structure that is excited by an elementary source such as an infinitesimal electric or magnetic type dipole source.[7] The simplest case of Green's function is in free space, which has well-known properties. If the studied structure incorporates a substructure which has a canonical shape such as a sphere, an infinite circular cylinder (either of conductor or dielectric), etc., then Green's function can be computed by taking into account the "canonical shape object" boundary conditions and by applying the method of separation of variables based on the method established by Sommerfeld. Although the new Green's function in terms of its expression is much more complicated as compared to the corresponding free space, the obtained numerical solution is significantly more efficient in comparison with the case when the free space in Green's function is employed. Consider the geometry of Figure 12.6 where a generalized phased array system is shown.

The interaction between the transmission lines and the array elements is assumed to be only through their direct connection, and no leakage phenomena are assumed in the volume hosting the transmission lines. The electromagnetic fields arising from the array antenna can be described in terms of the following generalized Kirchoff–Chu integrals.

$$\bar{E}(\bar{r}) = \sum_{l=1}^{N}\left(\iint_{S_1}(\bar{G}^{ee}(\bar{r}|\bar{r}')\bar{J}_1^{e}(\bar{r}') + \bar{G}^{em}(\bar{r}|\bar{r}')\bar{J}_1^{m}(\bar{r}'))\right)ds \tag{12.24}$$

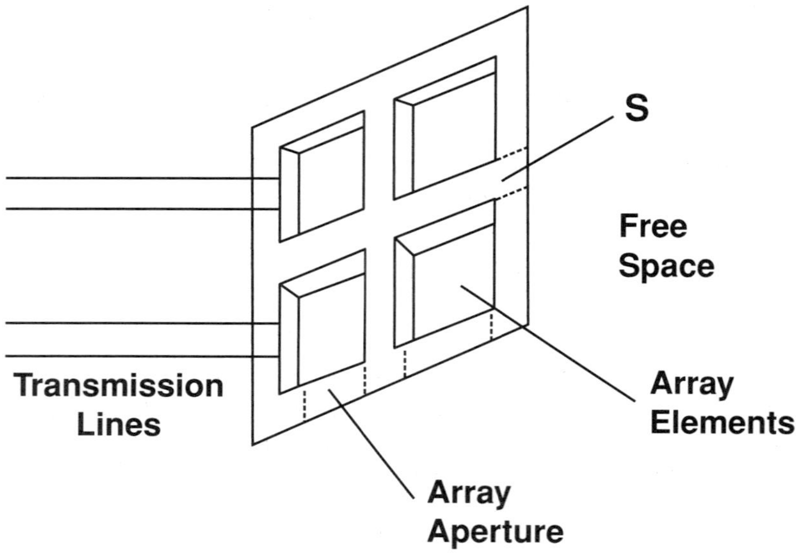

FIGURE 12.6 Phased array geometry.

$$\bar{H}(\bar{r}) = \sum_{l=1}^{N}\left(\iint_{S_l}(\bar{G}^{me}(\bar{r}|\bar{r}')\bar{J}_l^e(\bar{r}') + \bar{G}^{mm}(\bar{r}|\bar{r}')\bar{J}_l^m(\bar{r}'))\right)ds \tag{12.25}$$

where $\bar{J}_l^e(\bar{r}')$ and $\bar{J}_l^m(\bar{r}')$ are the electric and magnetic current distributions on the lth array element surface (S_l), while $\bar{G}^{ee}(\bar{r}|\bar{r}')$, $\bar{G}^{em}(\bar{r}|\bar{r}')$ are the dyadic (3 × 3 tensor) Green's functions corresponding to the electric and magnetic fields generated by elementary electric and magnetic dipoles, respectively, in the presence of the surface S (see Figure 12. 6), which is usually a conductor surface.

Only in limited cases (planar, spherical, or cylindrical surfaces) can the $\bar{G}^{ee}$, $\bar{G}^{em}$ dyadics be computed and amenable to the analytic solution. If this is not possible, then the corresponding free space in Green's dyadic could be used at the expense of the complexity of the solution, since the induced electric currents on the supporting surface S should also be taken into account.

Notice that in Equations 12.24 and 12.25 the source terms $\bar{J}_l^e(\bar{r}')$ and $\bar{J}_l^m(\bar{r}')$ are related to electric and magnetic fields on the surface S_l with the following relations:

$$\bar{J}_l^e(\bar{r}') = \hat{n}_s x \bar{H}(\bar{r}')$$

$$\bar{J}_l^m(\bar{r}') = -\hat{n}_s x \bar{E}(\bar{r}')$$

where $\hat{n}_s$ is the unit normal vector on the surface S_l. Upon moving the observation point $\bar{r}$ (which appears in Equations 12.18 and 12.19) to an arbitrary array element surface S_l, and taking the vector product of both equations, a set of homogeneous integral equations are obtained in terms of the unknown surface electric and magnetic fields on the $S_1, S_2, \ldots, S_n$ array elements. However, this system of equations is not sufficient to be solved in itself since no excitation is taken into account and the effects of transmission lines are not considered in Equations 12.18 and 12.19. To this end, on each transmission line an integral equation is written as follows:

$$\bar{E}_l(\bar{r}) = \bar{E}_l^{inc}(\bar{r}) + \iint_{S_l}(\bar{g}^{ee}(\bar{r}|\bar{r}')\cdot\bar{J}_l^e(\bar{r}') + \bar{g}^{em}(\bar{r}|\bar{r}')\cdot\bar{J}_l^m(\bar{r}'))ds \tag{12.26}$$

$$\bar{H}_1(\bar{r}) = \bar{H}_1^{inc}(\bar{r}) + \iint_{S_1}(\bar{g}^{me}(\bar{r}|\bar{r}')\cdot\bar{J}_1^{e}(\bar{r}') + \bar{g}^{mm}(\bar{r}|\bar{r}')\cdot\bar{J}_1^{m}(\bar{r}'))ds \qquad (12.27)$$

where $\bar{E}_1(\bar{r})$, $\bar{H}_1(\bar{r})$ are the electric and magnetic fields insidethe l^{th} transmission line and $\bar{E}_1^{inc}(\bar{r})$, $\bar{H}_1^{inc}(\bar{r})$ are the corresponding incident guided waves feeding the array elements.

On placing the observation point $\bar{r}$ in Equations 12.26 and 12,27 and requesting the continuity of the tangential field array element surfaces, the final set of integral equations are obtained with the following form:

$$\sum_{l'=1}^{N}\iint_{S_{1'}}\begin{pmatrix}\bar{r}^{ee}(\bar{r}_1,\bar{r}_{1'}) & \bar{r}^{em}(\bar{r}_1,\bar{r}_{1'})\\ \bar{r}^{me}(\bar{r}_1,\bar{r}_{1'}) & \bar{r}^{mm}(\bar{r}_1,\bar{r}_{1'})\end{pmatrix}\begin{pmatrix}\bar{J}_{1'}^{e}(r'_1)\\ \bar{J}_{1'}^{m}(r'_1)\end{pmatrix} = \begin{pmatrix}\bar{V}^{e}(\bar{r}_1)\\ \bar{V}^{m}(\bar{r}_1)\end{pmatrix} \qquad (12.28)$$

where $\bar{r}^{ee}$, $\bar{r}^{em}$, $\bar{r}^{me}$, and $\bar{r}^{mm}$ are known 3 × 3 dyadics, and $\bar{V}^{1}$ and $\bar{V}^{m}$ are known terms related to incident waves $\bar{E}_1^{inc}$ and $\bar{H}_1^{inc}$ on the first waveguide. Notice that Equation 12.28 should be satisfied on each $l = 1, 2, \ldots, N$ array element surface.

12.3.2 Solution of the *N* × *N* System of Equations

Following the setting up of the fundamental integral Equation 12.28 one has to find a solution to this. It is clear that as the number of array elements N increases, the corresponding solution complexity rises. A significant reduction of complexity is obtained if the array is assumed to be of an infinite number of elements with spatial periodicity. This topic is discussed in the following section.

In order to explain the proposed solution based on the Galerkin technique in the following, a general approach is presented.

In the first place, the unknown surface current distributions $\bar{J}_\ell^{e}$, $\bar{J}_\ell^{m}$ are described in terms of known vector functions such as

$$\bar{J}_{1'}^{e}(\bar{r}'_1) = \sum_{q=1}^{M^e} c_q^{1'}\bar{\psi}_q^{1'}(\bar{r}'_e) \qquad (12.29)$$

$$\bar{J}_{1'}^{m}(\bar{r}'_1) = \sum_{q=1}^{M^m} d_q^{1'}\bar{\varphi}_q^{1'}(\bar{r}'_e) \qquad (12.30)$$

where $c_q^{1'}$, $d_q^{1'}$ are unknown coefficients on the lth element. The proper selection of the $\bar{\psi}_q^{1'}$ and $\bar{\varphi}_q^{1'}$ functions should be based on physical reasoning such as

- In case of waveguide aperture, the array elements' use of a guiding model function
- In case of printed microstrip elements (dipole or patch geometries) pulse or global domain functions including the edge conditions

According to the MOM technique, Equations 12.29 and 12.30 are substituted into Equation 12.28, then the inner product of this equation is computed by the vector $\bar{\psi}_p^{1}(\bar{r}_e)\bar{\phi}_p^{1}(\bar{r}_e)$ and, finally, integrated on the surface S_1, giving the system of equations.

The numerical inversion of this set of equations provides an approximate solution to the phased array boundary value problem. By increasing the number of describing functions (M^e, M^m), convergence is exhibited in computing the aperture fields on the array elements provided the functions $\bar{\psi}_q^{\ell'}$ and $\bar{\varphi}_q^{\ell'}$.

Assuming that the array elements aperture fields are known, the radiation pattern at the far field region is computed easily by substituting Equations 12.29 and 12.30 into Equations 12.24 and 12.25 and then taking the limiting case of $|\bar{r}| \to +\infty$ for the observation point.

The previous method, being rigorous, foresees all phenomena involved in the array operation. Of course, as in all electromagnetic simulations, the array size that can be simulated is restricted by the RAM available for storage. This technique has been already applied by several authors in analyzing arrays.

Important phenomena uncovered through the previous analysis are

- Array blindness, which occurs because of the cumulative effects of mutual coupling between array elements and can be explained as a surface wave excited in the array surface
- Boundary array element effects which increase the side lobe level of radiation patterns

12.3.2.1 Infinite Array Theory

In cases of large size arrays with spatial periodicity, the behavior of elements not at the array boundary is very close to the case in which the array elements were infinite.

Assuming a periodic excitation of array elements, the current distribution on the array element surfaces can then be explained by Fourier series being periodic functions. On introducing these expansions into Equation 12.28, the corresponding equation is reduced into a "single array" integral equation which is solved by applying a similar scheme as in Equations 12.23 to 12.25.

12.4 Array Architectures

The traditional approach of developing phased arrays in radars in the early stages has been the use of a single powerful power source, while a single receiver was used at the output of a combining network receiving inputs from the phased array elements. The basic architecture is shown in Figure 12.7.

The inherent drawback in this architecture is the excessive power loss occurring by both transmitting and receiving at combining (power divider) and transmission lines. This type of architecture is suitable for conventional search radars and limited angular measurement accuracy of targets because of the diffraction limited scheme employed in measuring either azimuth or elevation angles. In order to improve the angle measurement of the radar system, monopulse radar techniques have been introduced into radar antennae. In Figure 12.8, the architecture of a passive phased array for a single axis, monopulse angle measurement is shown.

On each symmetric pair of elements, there is a hybrid circuit providing the individual sum (Σ) and the difference (Δ) channels which are combined into two outputs providing the overall sum and the difference outputs. In order to measure both azimuth and elevation angles, a two-level architecture is employed. In this type of beamforming system in hybrid circuits, excessive loss is suffered. A significant improvement of overall system sensitivity of the order of 10 dB can be achieved by using transmit/receive units at the front end of each array element. This is accomplished by replacing the single transmitter and

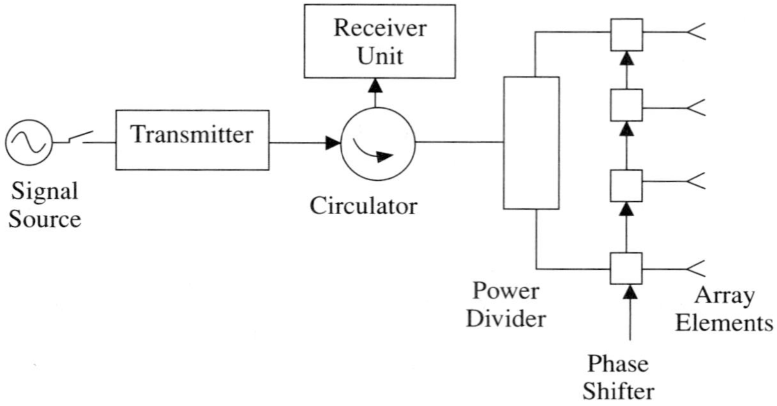

FIGURE 12.7 First generation array architecture.

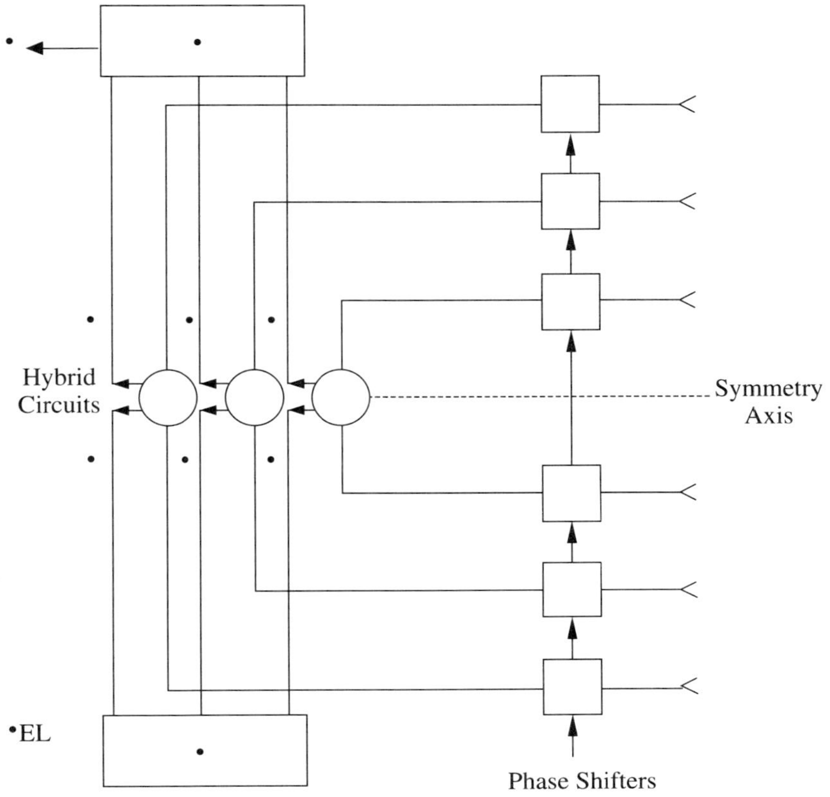

FIGURE 12.8 Monopulse angle measurement phased array.

receiver with a large number of low-power transmitters and low-noise receiver units in a distributed architecture. In addition to the improvement of system sensitivity, the overall system reliability is also improved. In the former case, the failure of the single transmitter loads into a total failure, while in the latter case, a graceful performance is obtained depending on the number of failed arrays.

12.5 Conclusion

A brief review of phased arrays used in radar systems is presented. Fundamental theories of analyzing and designing phased arrays are examined. Present technological limitations are reviewed, and the possibility of developing fully digital arrays is suggested based on multi-gigabit signal processing elements.

References

1. J. E. Reed, The AN/FPS-85 Radar System, *Proc. IEEE*, 57, 324–335, 1969.
2. M. I. Skolnik, Fifty years of radar, *Proc. IEEE*, 73, 182–197, 1985.
3. C. Balanis, *Antenna Theory: Analysis and Design*, 2nd ed., John Wiley & Sons, New York, 1982.
4. J. J. Schuss, J. Upton, B. Myers, T. Sikine, A. Rohwer, P. Makridakos, R. Fransois, l. Wardlle, and R. Smith, The IRIDIUM main mission antenna concept, *IEEE Trans. Antennae Propagation*, 47, 416–424, 1999.
5. L. M. Buckler, The Use of Phased Array Radars at Civilian Airports, *Proc. IEEE 1996 Int. Symp. on Phased Array Systems and Technology*, pp. 334–339, 15–18 October, 1996.

6. C. Hemmi, R. T. Dover, F. German, and A. Vespa, Multifunction wide-band array design, *IEEE Trans. Antennae Propagation*, 47, 425–431, 1999.
7. D. S. Jones, *Theory of Electromagnetism*, Pergamon Press, Oxford.
8. N. K. Uzunoglu, K. S. Nikita, and D. Kaklamani, Eds., *Applied Computational Electromagnetics: State of the Art and Future Trends*, NATO ASI SERIES (961049), Springer-Verlag, New York, 2000.
9. H. Zmunda amd E.N. Toughlien, Photonic aspects of modern radar, in *The Aertech House Opto-electronics Library*, Aertech House, 1994, p. 550.
10. B. L. Anderson, S. A. Collins, C. A., Klein, and S. B. Brown, Highly Parallel Optical True Time Delay Device for Phased Array Antennas, Proc. 6th Annu. ARPA Symp. on Photonic Systems for Antenna Applications, Monterey, CA, 1996.
11. J. B. L. Rao, D. P. Patel, and U. Kricherstey, Voltage controlled ferroelectric lens phased array, *IEEE Trans. Antennae Propagation*, 47, 458–468, 1999.

III

Medical Imaging System Applications

13

Medical Ultrasonic Imaging Systems

John M. Reid
Drexel University

Thomas Jefferson University

University of Washington

Abbreviations

a.c.	Alternating current
A/D	Analog to digital converter
AGC	Automatic gain control; also called time controlled gain (TCG)
cm	Centimeters
CPU	Central processing unit
D	Dimension
dB	Decibels
d.c.	Direct current
D/A	Digital to analog converter
F	Numerical aperture
f	Frequency
FFT	Fast Fourier transform
I	In-phase
MHz	Megahertz
p.r.f.	Pulse repetition rate
Q	Quadrature
ROM	Read only memory
r.f.	Radio frequency

0-8493-3691-0/01/$0.00+$.50

x	A Distance, range
θ	Angular beamwidth
λ	Wavelength, = c/f

13.1 Introduction

Ultrasound has been called the most widely used imaging modality in medicine. No demonstrable harm has resulted from its diagnostic use, and there are strict regulatory limits on the output. These systems are widely used for primary diagnosis as well as for screening, monitoring, and follow-up procedures. The equipment is movable — some even can be carried by hand — and it can have high frame rates. The many features of these systems have been summarized by Greenleaf.[1] The large installed base and experienced personnel almost guarantee that any improvement in image information will benefit patients. Since the frame rates are high enough to follow changes in real time, and the scan head transducers are easily movable, the examinations are very interactive. Many different scan planes can be imaged by moving the small scan heads shown in Figure 13.1. The motions of muscles, organs, and even of blood flow can be seen.

The static images in publications, such as those shown here, give a very poor impression of the picture on the display. On the display, the heart beats, muscles and tendons move, and blood flows in pulses. An example of a static image is shown in Figure 13.2. In the actual examination, the pulsating blood flow in the descending aorta of a fetus *in utero* can be seen, including a branch to the kidney. A fetus will respond by moving visibly to stimuli so that a neurological examination can be carried out as well. However, experienced personnel are needed to acquire and to save the particular images regarded as diagnostic. This requirement is a current weakness of ultrasound imaging.

These systems offer many opportunities for improvements through signal processing. The speed of operation and the extraction of diagnostic information are two promising areas of work. The phase, as well as the amplitude, of the received signals is available, and most modern systems are fully digital, offering software control of both transmitting and receiving functions (within the limits of the hardware). The received digital signals can be made available for external processing.

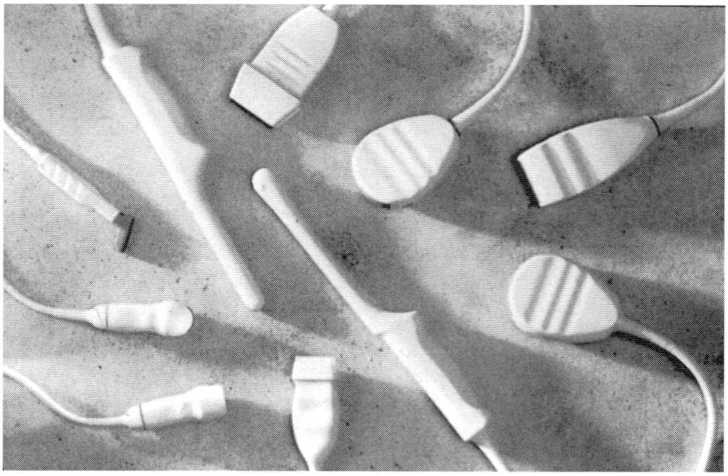

FIGURE 13.1 Representative ultrasonic scan heads (array transducers). Clockwise from top: linear array, curvilinear abdominal array, high frequency linear, another curved abdominal, curved intravaginal, two phased arrays, highly curved array, side-looking linear array for intraoperative use, and microcurved intravaginal array. (Courtesy of ATL Ultrasound.)

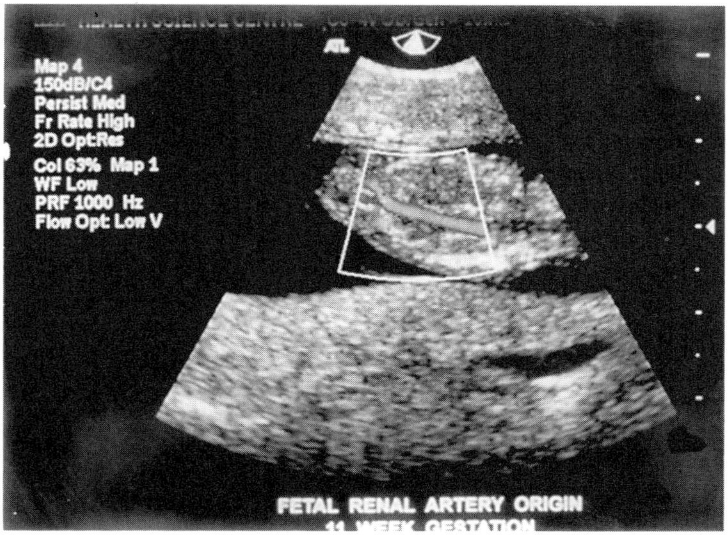

FIGURE 13.2 Ultrasonic image of a fetus *in utero*. The central portion is a color flow image of the fetal aorta. In the actual examination, fetal motion and the pulsating blood flow are seen. (Courtesy of ATL Ultrasound.)

Medical ultrasound systems, in common with radar and sonar systems, are active in that interrogating signals are transmitted by a transducer: in our case, an electro-acoustic transducer. Small changes in the elastic properties and density of tissue elements reflect the energy back to the receiving transducer. The resolution of the system is set by the dimensions of the emitted and received sound beams in both space and time. The overall system response is the product of the separate transmit and receive beams, which need not be identical. The propagation factors must also be included in the response. These factors include tissue attenuation and the diffractive properties of transducer geometry. Medical ultrasound systems have some unique features and limitations that set them apart from radar and sonar systems. At the same time, the differences open new avenues and opportunities for problem solving.

The major differences are

1. High and frequency-dependent attenuation of tissue leads to a restriction on the highest usable frequency, hence the depth of imaging and resolution.
2. The tissue of interest can be in the near field of practical transducer apertures, so tight focusing is possible.
3. Anatomy sets serious restrictions on access. Bone and lung are more difficult to penetrate than other tissues.
4. Many targets are not discrete objects, so we must map *areas* of echo density and texture.
5. Unresolvable scattering structures give false discrete targets and mask the echoes from small structures through the "speckle" artifact.[2]
6. Only sound speed sets a limit on frame rate, which can be high enough to stop all motions encountered, even in the heart.

Signal processing in this environment can be viewed as having three goals:

1. Faster and better image formation; avenues include selecting transmitting and receiving elements (beamforming), controlling the transmitted signals, and then processing returns from individual sensor elements in hardware or software
2. Image enhancement as a second step to increase image quality and to lessen the need for highly trained operators
3. Extraction of diagnostic information about tissue, sometimes called tissue characterization, and incorporation of this information into the image[3,4]

Current systems perform the first step to make images and can also do scan conversion to permit display on raster scan monitors. In the second step, enhancement processing, such as interpolation of additional image lines, speckle reduction, and edge enhancement, is done before images are displayed. Details of these processes are highly proprietary.

The third step includes several processing steps that go beyond basic image formation. One example is to use the Doppler effect and its relatives to assess blood flow and tissue motion, which are important functional measures.[5] Other efforts to measure elasticity and to derive statistical properties of the scattered energy are showing promise. These added operations require calculations that slow the imaging process with current methods. This added information can be integrated into the images with selection algorithms and are often shown as an area of color substitution within the basic gray scale, or B-mode, image.

13.2 System Fundamentals

13.2.1 Resolution

The lateral resolution is set by the dimensions of the sound beam from the transducer. Consider the beam from an acoustic aperture, shown at the left in Figure 13.3. This beam can be either that transmitted or the area of sensitivity of the same aperture as a receiver. The wavefronts spread out to the right and form the far field. The lateral resolution is thus poorer at large distances, but improves with a larger aperture. Better resolution closer to the transducer is possible for a smaller aperture. If a small receiving aperture is expanded with time after transmit, this "expanding aperture" system can have a better overall lateral resolution than a system with fixed aperture.

1.3

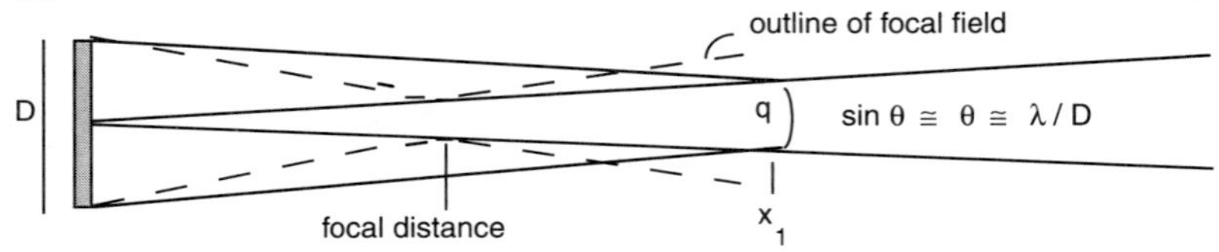

FIGURE 13.3 Schematic cross-section of the approximate extent of the transmitted field from (or the receiving sensitivity of) a transducer aperture of dimension *D*. The far field spreads out with the divergence angle, θ, at long distances from the aperture. The minimum field size occurs at x_1, the transition between the near and far fields. The dotted line shows an assumed focal field. Focusing can only narrow the field to the same divergence angle, θ.

The beamwidth at a range x, along the diverging beam, is given approximately by $\lambda x/D$, which is λ times the instantaneous "*F*" number of the aperture. In a typical expanding aperture imaging system with a constant receiving *F* number of 2 or 3, the lateral resolution can be constant and equal to 2 or 3λ. This requires that the aperture, Figure 13.2, be both expanded and focused. At an assumed 4 MHz frequency and a sound speed of 1540 m/s, the lateral resolution is less than 1 mm. The range resolution can be set to be the same as the lateral resolution. This gives a bit more than 100 resolution elements in either direction. Since both resolution and penetration vary inversely with frequency, their ratio is nearly constant, so that the relative resolution is the same in any ultrasound image. The frequency of operation must be chosen to suit the application. Very high absolute resolution is possible at high frequencies, but the penetration is less. For example, the acoustic microscope can approach optical resolution by operating at gigaHertz frequencies.[6]

In the previous example, the resolution was improved by focusing at all ranges. However, focusing is only effective in narrowing the beam within the near field of the transducer aperture, since focusing narrows the field to have the same angular width as exists in the far field from the same aperture. In an array transducer system, focusing on the receiver can be adjusted electronically to track the echo region as the pulse propagates. By this "dynamic focusing," the number of resolution elements can be increased to several hundred if the anatomy allows a sufficiently large aperture. In some patients, the variable sound speed in tissues, mainly inclusions of fat, can upset the precise timing needed to form very narrow beams. This results in beam distortions that can misplace structures in the image or can fill normally echo-free cysts with low-level echoes. This makes the important cystic vs. solid tumor decision more difficult. The correction of these "phase aberrations" is an active area of research.

13.2.2 Scanning and Transducers

The transducer is a critical component of the imaging system.[7] Its size, resonant frequency, and bandwidth determine the resolution and penetration. Many types have been developed to provide desired images in a wide variety of medical applications. These include both single element types needed for high frequencies, and the multi-element arrays shown in Figure 13.1.

The basic scanning array, sometimes called a one-dimensional (1-D) array, is shown in Figure 13.4. The array is formed by slicing the radiating structure into independent elements with widths of the order of the center frequency wavelength in tissue or water. Electrical connections are made to each element. The transmitting and receiving beams of ultrasound are formed and scanned by the associated electronic circuits.[8]

Because of the Fourier transform relationship between the radiating aperture size and the beamwidth, the resolution is better in the scanning plane, Figure 13.4, than in the transverse, or cross-plane, direction. Efforts are being made to reduce this effect, which can produce confusing echoes from structures that are out of the image plane. The poorer transverse resolution also results in the reception of more clutter signal energy, which makes the speckle artifact worse.

A current development is the use of elevation focusing of arrays to narrow the thickness of the imaged slice of tissue. A simple, weak focusing lens of a rubber with a sound speed less than that of tissue has been used for this purpose for some time, since it also provides a curved front surface that is comfortable for the patient. With weak focusing, the focal region in the elevation plane is cigar shaped, since then the far field contracts toward the axis at long ranges.

Better focusing in elevation can result from dividing each array element into several sections, as shown in Figure 13.5. Separate connections to these parts of the element allow expanding the aperture using switches (called a 1.25 D array) or allow focusing in elevation using delays (called a 1.5 D array). A more

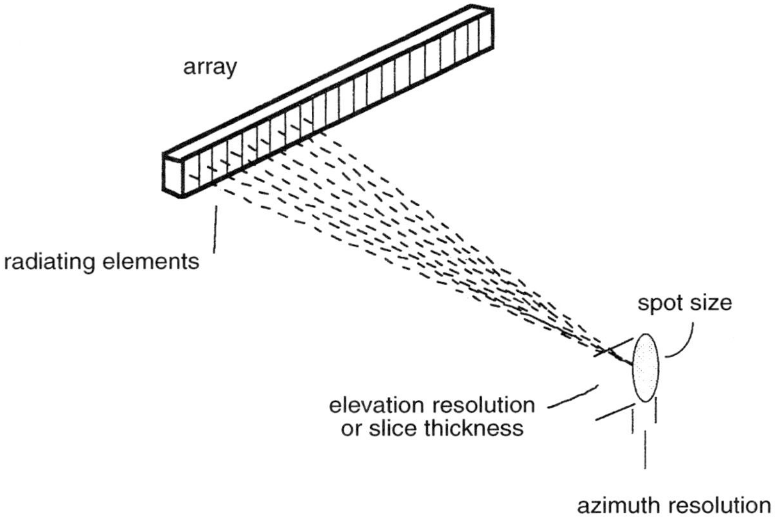

FIGURE 13.4 Drawing of a multi-element array transducer showing the path of sound beams from individual elements forming the scanning spot. Electronic control of time delays for excitation and reception form the scanned and focused spot as shown.

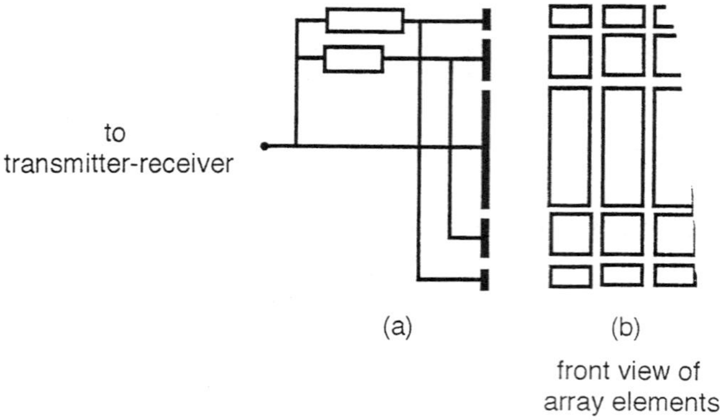

FIGURE 13.5 Elevation focusing with subdivided array elements. Control of focus is accomplished by using switches or delay elements within the blocks. (From Reid, J. M. and Lewin, P. A., *Ultrasonic Transducers, Imaging*, Vol. 22, Encyclopedia of Electrical and Electronics Engineering, Webster, J., Ed., Wiley, New York, 1999. With permission.)

finely divided array element could be programmed to scan over a limited range in elevation. This is called a 1.75 D array.[8] The ultimate solution to the coarse focusing in elevation is the fully two-dimensional (2-D) array, with equal numbers of elements in azimuth and in elevation. In this case, the beam can be directed freely over a three-dimensional (3-D) region of tissue to collect a 3-D data set.

The goal of 3-D data collection is being approached in three ways. These are by post-processing of individual scan images; sweeping a 2-D scan plane in the transverse direction by manual or mechanical means and storing the data in real time; or using a true 2-D array. One true 3-D system of this last type exists, and more may be expected.[9] They are the most useful for imaging complicated 3-D structures such as the heart and the fetus, in which the plane "slice" that is imaged by any one scan cuts through complicated structures and is difficult to interpret. In these cases the goal is to display a slice at an orientation that shows the desired structures. A suitably diagnostic scan plane often can be found after searching the 3-D data set. The examination takes only a few minutes for data collection and removes the need for most repeat scans.

The current array transducers use several scan formats, as shown in Figure 13.6, and the system may be required to accommodate more than one.

The rectilinear scan (Figure 13.6A) is used in small parts and some abdominal imaging. The beam only needs to be translated, not phased, to form this format. The dotted lines show regions where the transmit beam can be focused. Since these structures move only slowly, repeat scans can be made with different fixed positions for the transmit focus, and then the image is assembled from the focal regions of the scans.

The scan from a curvilinear array (Figure 13.6B) increases the field of view over that of the rectilinear scan and also requires only simple translation of the beam. Expanding aperture and dynamic focus operations are still possible, but the number of elements that can contribute to any one scan line is limited by the curvature.

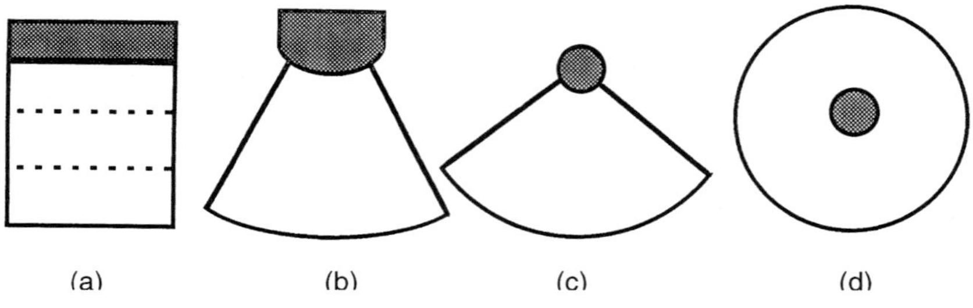

FIGURE 13.6 Image formats: (a) rectilinear scan, (b) scan from a curvilinear array, (c) sector scan, (d) rotational scan. See text. (From Reid, J. M. and Lewin, P. A., *Ultrasonic Transducers, Imaging*, Vol. 22, Encyclopedia of Electrical and Electronics Engineering, Webster, J., Ed., Wiley, New York, 1999. With permission.)

The sector scan (Figure 13.6C) is used with a phased array for imaging the heart or any structure where the imaging "window" is limited by anatomy. Here, mechanical rotation or time delays are used to scan the angle. Some very strongly curved linear arrays can produce scans in this format.

The fully rotational scan (Figure 13.6D) is used for intravascular imaging, either by mechanical rotation of a single element or by a circular phased array.

13.3 Tissue Properties' Influence on System Design

13.3.1 Attenuation

The tissue environment is one of relatively high and frequency-dependent attenuation, which has a profound effect on system design. These properties have been tabulated in some detail.[10] Average soft tissue has a one-way loss of approximately 1 dB/cm for each megahertz of frequency, which ranges from about 1 to 40 MHz, depending on the application. Lower values of attenuation that are sometimes found in the literature were adopted to make conservative estimates of safety from heating and cavitation, but they do not govern the imaging properties. The presence of attenuation means that the frequency cannot be selected arbitrarily, since both the penetration and the resolution depend on the choice.

At 4 MHz, the loss over a 10-cm range could be over 80 dB, plus enough reserve sensitivity to handle 30- to 60-dB reflection loss, or maybe 140 dB. A receiver gain function that increases with time after transmitting is needed to avoid overload in the receiver. This is not just a problem with analog systems.

In a digital system, the problem is to stay in range of the analog to digital (A/D) converter. A dynamic range of over 140 dB needs over 23 bits. Modern systems need digitization rates to 20 MHz or more, and this hardware combination is now impractical. Therefore, a controlled gain function (automatic gain control, AGC) is still needed in the receiver to compensate for attenuation. Currently, the rate is set by the operator. The amplifiers are followed by anti-aliasing filters and 8 to 12 bit flash or pipeline A/D converters. The converters must be located in the channel connected to each active receiving element. Since one bit is lost with alternating current (a.c.) signals and another is lost to noise, the total dynamic range from the remaining bits could still be insufficient to handle the expected 60-dB dynamic range from echo variations alone. Since the outputs of many individual A/D converters are added together in the usual delay and sum beamformer, a longer word length results, which can help to meet the requirement.

13.3.2 Speed

The frame rate is limited only by the speed of sound, c. Fortunately, soft tissues have nearly the same sound speed. Fat alone deviates from the average by about 10%. Frequency dispersion is similarly low. At an average speed of 1540 m/s the ranging time, $2x/c$, is 13 μsec/cm. The ranging time for radar is 12.2 μsec per nautical mile. So, considering the different distance scales, the timing electronics and the processor speeds are similar. Our pulse repetition rate (p.r.f.) is a few kilohertz for abdominal imaging. Imaging time is proportional to the number of actual lines in the image. Sixty frames per second is about the theoretical maximum for a 128-line system with a 10-cm range. A lower rate is needed for retrace allowance, but single scan planes can be imaged rapidly enough to stop tissue motion.

For 3-D imaging, another strategy must be used to keep the frame rate high enough for the heart. The needed multiple channels have been formed by using receiving channels that synthesize multiple receive beams for each pulse transmission. One 3-D system has a rate of 22 full 3-D data sets per second and uses 16 simultaneous receiving channels.[9]

13.3.3 Structure

Tissue structure can be viewed as being heterogeneous on any size scale. This leads to the production of echoes throughout the volumes we wish to image. The dimensions of the resolution cells can be viewed as forming a sample volume from which both desired and undesired, or clutter, signals are received. Since the clutter elements seem to be randomly distributed, their echoes add with random phases and lead to the formation of speckle, just as in radar and sonar. The amplitude peaks of the video speckle signal can mask small structures and break up organ outlines.

The dimensions of the speckles are the size of the resolution elements in the image plane and, therefore, are useful in judging the resolution of a system. Expanding aperture systems strive for a constant speckle size, but the speckles do get larger at the maximum range after the full aperture is reached. Another use for speckle is to calculate the speed of an echo from a moving structure independently of its direction. The differences in speckle location from image to image are mapped as a velocity field. Correlation of one image with another can be done on either the radio frequency (r.f.) or the video signal.

Both frequency and spatial compounding have been used to reduce the speckle artifact with modest success. The full 3-D systems have an advantage in being able to produce a very small sample volume, since the out-of-plane resolution can be made the same as that in the image plane. This results in less speckle energy. Many statistical methods of reducing speckle have been investigated without much success. They usually require signals from regions that are larger than the smallest possible resolution elements. The processing tends to blur the edges and structures.

13.4 Imaging Systems

13.4.1 Single Channel

The basic principles are the same for single and multiple channel systems, and will be described first. Originally, mechanical scanning was used in single channel analog systems. A typical functional block diagram is shown in Figure 13.7. These systems are still used at the highest frequencies where array fabrication is not possible because the saw kerfs used to subdivide array elements remove too much material. These single transducer systems are used in the acoustic microscope, scanning inside small blood vessels, and for scanning the skin and the eye.

The transmitter emits short, wideband pulses of only a few cycles. With modern, high-bandwidth, transducer materials, the resolution goal of a few wavelengths is easily met; the repetition rate is a few kilohertz, chosen to be low enough to avoid "second time around" echoes. A duplexer is used to protect the receiver so that it can recover full gain rapidly after the overload from the transmitter. Passive circuits are used here, since active switches have too much spurious output from switching transients.

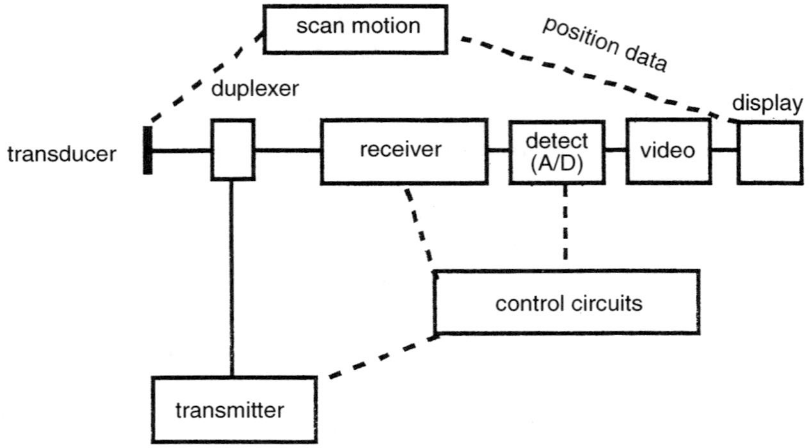

FIGURE 13.7 Single channel system block diagram. This mechanical scanning system is used with high frequencies where array fabrication is impractical.

The r.f. portion of the receiver has two main functions:

1. To amplify the weak echoes with a gain that increases with time (AGC) to compensate for attenuation
2. To filter to select the signal bandwidth, to prevent aliasing, and to reject noise

Post-detection digital techniques are now common in high-frequency systems and are also used in low-end equipment. The A/D converters are located just following the detector shown in Figure 13.7 and operate at the lower frequencies of the video signals. The introduction of digital techniques allows digital image storage, which also provides a scan converter operation for the TV raster display. The associated software control allows some signal processing to be done digitally. Examples are interpolation of image lines, sweep integration, contrast adjustment, and edge enhancement. The digital output from these systems is only the detected and filtered video display signal.

In the common scanner designs meant for observing blood flow, the Doppler shift is extracted at a single location in the image and is interpreted as the flow velocity component in the direction of the sound beam. The spectral display shows the direction of flow and can indicate the presence of turbulence as well. The receiver incorporates demodulation to baseband in in-phase (I) and quadrature (Q) video channels; see Figure 13.8. Sample and hold circuits select signal amplitudes at one range at the p.r.f., and filter to remove strong signals that result from motion of the vessel walls and to correct the frequency response. This complex video signal is fed to a fast Fourier transform (FFT) analyzer to find the Doppler shift at that range. The resulting Doppler spectrum is usually presented as a function of time along with an image showing the location where the Doppler shifted signal originates; see Figure 13.9.[5]

The phase is sampled at the p.r.f., which limits the maximum unambiguous frequency shift in each direction to half the p.r.f. Since the systems are directional this is a total frequency span equal to the p.r.f. Since the zero frequency can be offset by a small change in the reference frequency, the upper limit of Doppler shift measurable without aliasing equals the p.r.f.

Since there may also be strong signals from the vessel walls, a high-pass "wall filter" is needed before the A/D converter to remove them. This filter cutoff frequency may be selectable so that the operator can reduce its effect when observing low flows. In some designs, a low cutoff frequency may remain when the control is set to "zero."

13.4.2 Multi-Channel Systems, Arrays

Figure 13.10 shows the basic electronic system used with modern aperture arrays. The active elements of the array used for transmitting and receiving are selected by the multiplexer. Switching is done during the retrace time to avoid transients. The number of receiving and transmitting channels may be the same

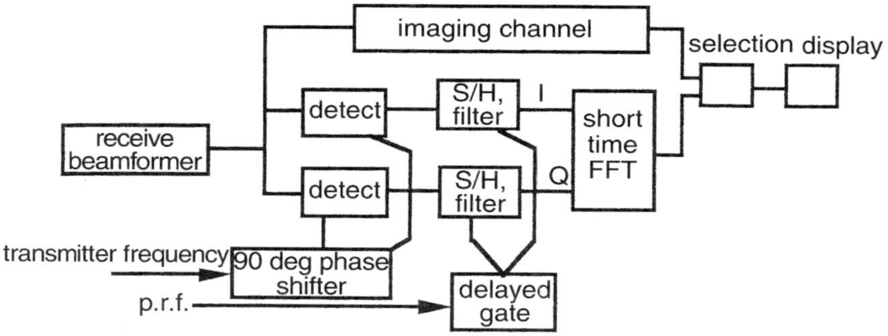

FIGURE 13.8 Block diagram of a spectral Doppler processor using quadrature detection to derive in-phase (I) and quadrature (Q) components for fast Fourier transform processing. See Figure 13.9 for an example of the display.

(a)
(b)
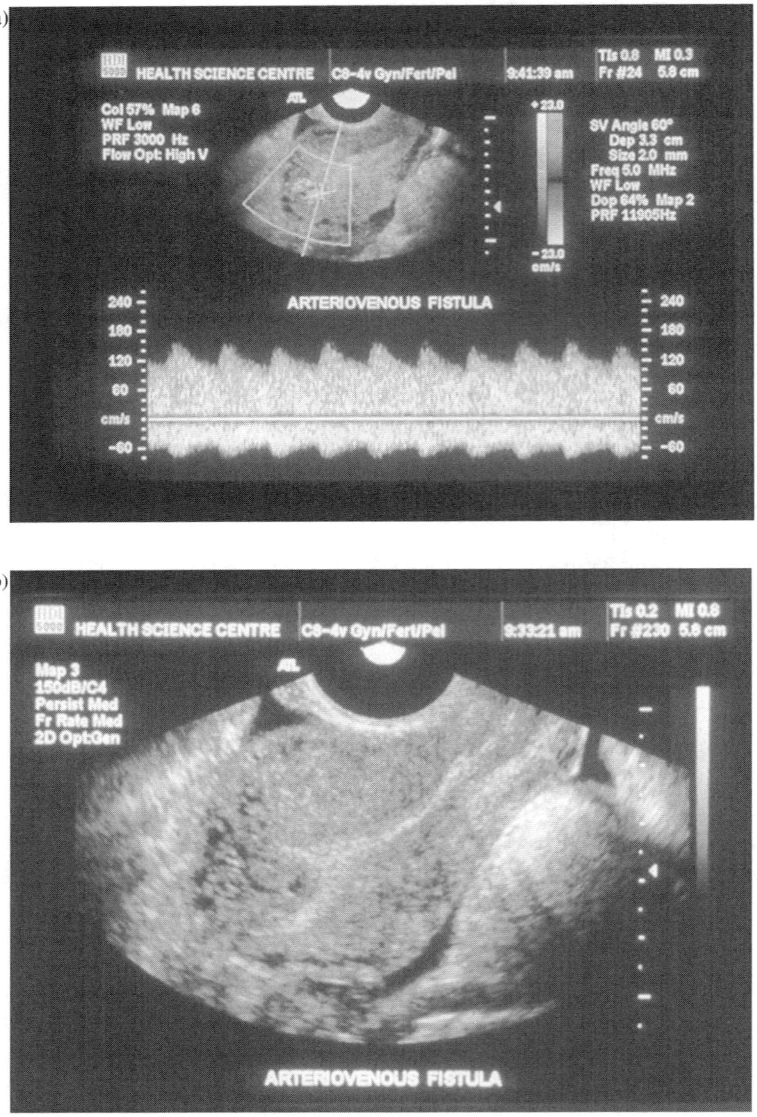

FIGURE 13.9 Scans of a uterine growth: (a) Doppler scans of a suspicious region at the head of the uterus, and (b) original B-mode scan. The flow sector in (a) shows very high-speed flow, as referenced by the scale at the right. The Doppler spectrum from a sample volume placed within a high-flow region of the image is shown at the bottom as a function of time. The frequencies, converted to flow velocities by reference to the direction of the line segment in the image, show a uniform distribution of flow speed in both directions from near zero to the maximum. This indicates simultaneous flow in both directions, which is characteristic of tangled vascular beds or turbulence. (Courtesy of ATL Ultrasound.)

or less than the number of transducer elements. The transducer is connected to the transmitter and the receiver through their respective beamformers; see Figures 13.11 and 13.12. The beamformers introduce any time delays needed for focusing and adjust the amplitudes (apodization) to reduce sidelobes in the radiation pattern. Grating lobes, which result from a too-wide array element spacing, are eliminated by using a fine pitch in the array or are reduced by differing transmitting and receiving beam patterns in which the grating lobes do not overlap.

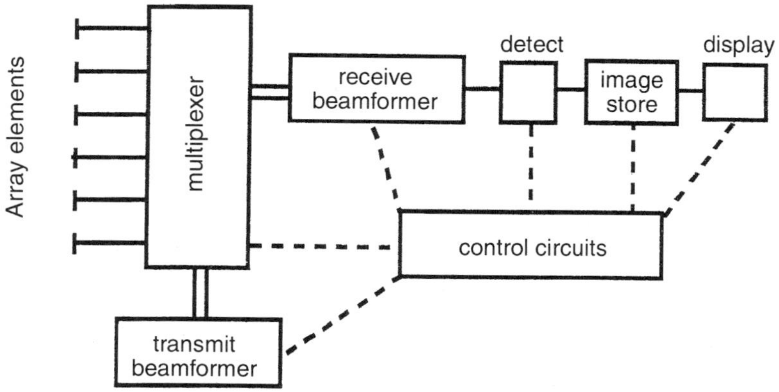

FIGURE 13.10 Array system block diagram. A multiplexer is needed to connect the transmitting and receiving channels to the array elements.

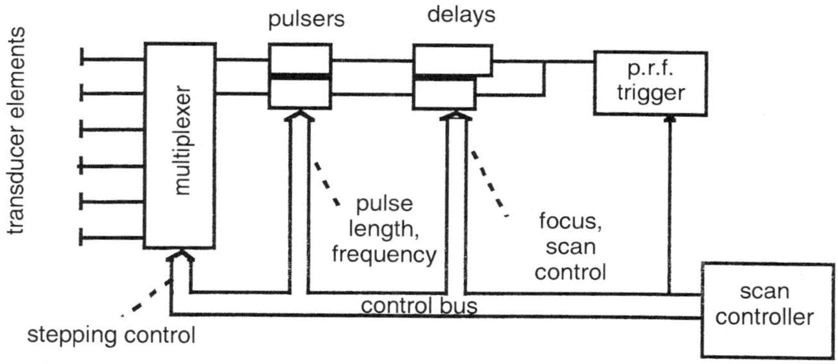

FIGURE 13.11 Detail of the transmit beamformer of Figure 13.10. Note that simple trigger delays are used for focus and scanning.

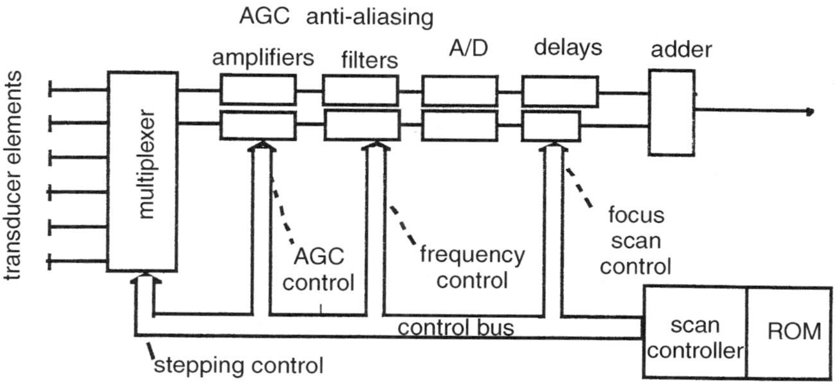

FIGURE 13.12 Detail of the receive beamformer of Figure 13.10. Here, wideband or digital delays are used.

In the basic linear array, the multiplexer determines which elements will transmit and receive, and it changes which elements are active to scan. If the array has a flat surface, the scan format in Figure 13.6a results. In a curvilinear array the curvature of the front face provides angular sweeping of the beams, and the scan formats of Figure 13.6b or 13.6c result.

The beamformers for so-called phased arrays do beam steering and focusing by introducing suitable *time delays*.[8] Phasing is a narrowband method that is only useful for fine, or vernier, delays in this system.

The array receiver beamformer is shown in more detail in Figure 13.12. It can adjust the focusing delays with time after transmission (dynamic focusing). This technique tracks the received focal spot as it moves through the tissue at the speed of sound to remove any depth of field problem with the focus and attempts to maintain a constant F number for the system.

Each channel of the beamformer, Figure 13.12, contains all the needed receiver elements except for the detector and video processors. Many AGC amplifiers are needed to keep the A/D converters within range. The amplifiers also contain duplexers and anti-aliasing filters. The A/D converters must handle twice the r.f. The processing in the array systems must provide some filtering to remove low frequencies introduced by the differing direct current (d.c.) offsets in the A/D converters. This may be done either in hardware or software. Again, quadrature detection, as shown in Figure 13.8 but operating on all of the received signals, can use A/D converters operating at just over twice the signal bandwidth rather than at the highest r.f. The problem remains that it is quite expensive to provide more than 64 or 128 channels. So-called "synthetic aperture" systems avoid this problem.

The synthetic aperture approach was introduced in commercial systems to produce systems with the performance of many channels without actually having to build them. Here, this term means time sharing of a number of transmitter and receiver channels to synthesize the electronic system. The full physical aperture is already provided by the transducer. More advanced synthetic aperture processing is being developed and is a promising field, but better hardware is needed.[11]

The basic block diagram illustrating a fully digital system suitable for synthetic aperture operation is shown in Figure 13.13. Two important new functions have been added. First, the data collected by firing the smaller number of transducer elements and collecting the returns from the active receive channels are stored. The signals can be at the original r.f. or in I and Q channels if downconverted to baseband by Q detection. Thus, a large, high-speed memory is needed. Then all of the original functions needed for image formation are done in a processor using software.

The transmitter beamformer structure is not changed, but the receiver is a bit more simple; see Figure 13.14. The output of the A/D converters now goes directly to the data memory. Filtering for the d.c. offsets is still needed. The functions of the processor, Figure 13.14, can include selecting the data from a selected number of elements for expanding the aperture, performing apodization, and adjusting the timing of the data lines for scanning and focusing. Since the r.f. data are available, Doppler processing in software can be performed with this same hardware.

New types of processing are now possible in this system. Theoretically, a new focused transmitter firing can be done for each image pixel, and the echo data collected from each focused transducer element, so that focusing of the full array can be done on transmit as well as on receive. This truly confocal system has been called the "gold standard" beamformer. For an "N" element array only $N^2/2$ firings are needed

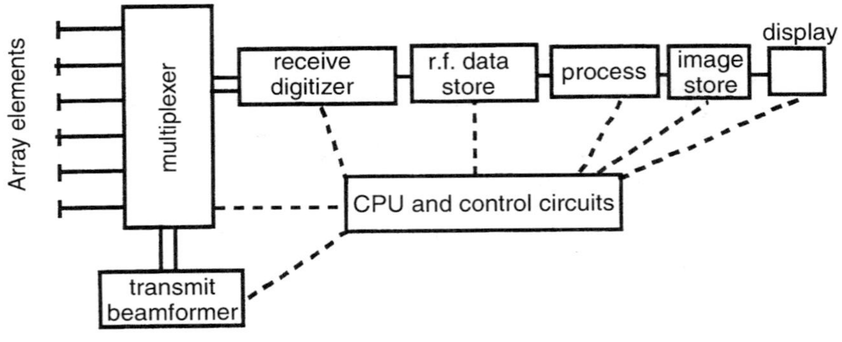

FIGURE 13.13 Block diagram of a fully digital system suitable for synthetic aperture operation.

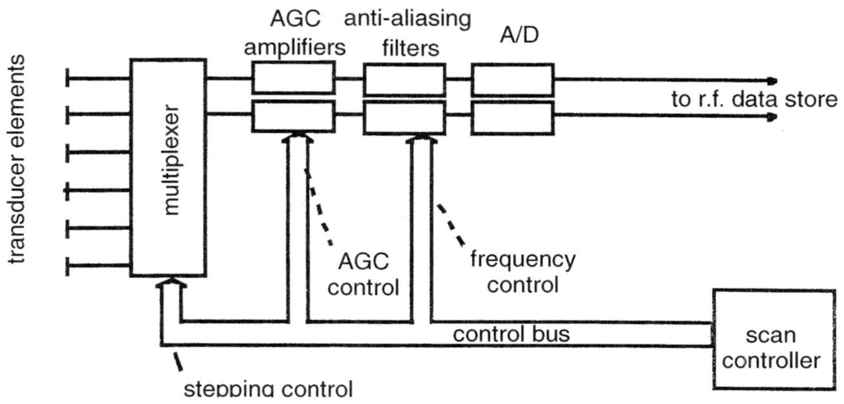

FIGURE 13.14 Detail of receive digitizer for the system in Figure 13.13. All beamforming and processing steps are done with software.

since reciprocity holds.[8] In applying any form of signal processing on data from an ultrasound system, it should be recognized that significant processing may be inherent in the many possible types of operation. Collaboration with the manufacturer generally is needed to be able to determine what has been done and how it may affect data acquisition.

13.4.3 Doppler and Advanced Systems

When the I and Q signals from Q detection are present, Doppler shift extraction can be done for all pixels in the image, since phase information is preserved. With suitable algorithms the rate of change of phase can be calculated at any range. From this information, the mean Doppler frequency and other parameters, such as its variance, can be calculated and displayed as velocity and turbulence, respectively. The mean Doppler frequency is usually found from the phase of the complex autocorrelation function as

$$\bar{\omega} = \frac{1}{\Delta T}\tan^{-1}\frac{\sum_{i=1}^{n} I(i)Q(i-1) - Q(i)I(i-1)}{\sum_{i=1}^{n} I(i)I(i-1) + Q(i)Q(i-1)} \tag{13.1}$$

where the index, i, refers to n successive image lines spaced ΔT apart.[12]

Since the scan line data also includes the much stronger tissue echoes, it is necessary to precede the Doppler processor by a stationary echo canceller, similar to those used in radar. This serves the same function as the wall filter in single channel systems.

Digital control of the display is also needed to allow selection of the region where the Doppler output, or color flow, signals are seen in shades of color to distinguish them from the gray tissue image. This region is usually smaller than the full image to allow time for additional processing, which requires several transmit bursts. The selection of which data set to display is done on the basis of amplitude, since the echoes from red cells are quite weak.

The scan image in Figure 13.9a is an example of this display. The B-mode image of the uterus without the Doppler component is shown in Figure 13.9b. The added information clearly shows the presence of the high-speed, turbulent flow that is found at arterio-venous fistulas.

An alternate method of Doppler display is called power Doppler.[5] Only the magnitude or presence of the Doppler signal is displayed, not its frequency, and the signal is sampled only at its peak in the cardiac cycle. The image requires a few heartbeats to develop fully and is useful for imaging regions of low-speed

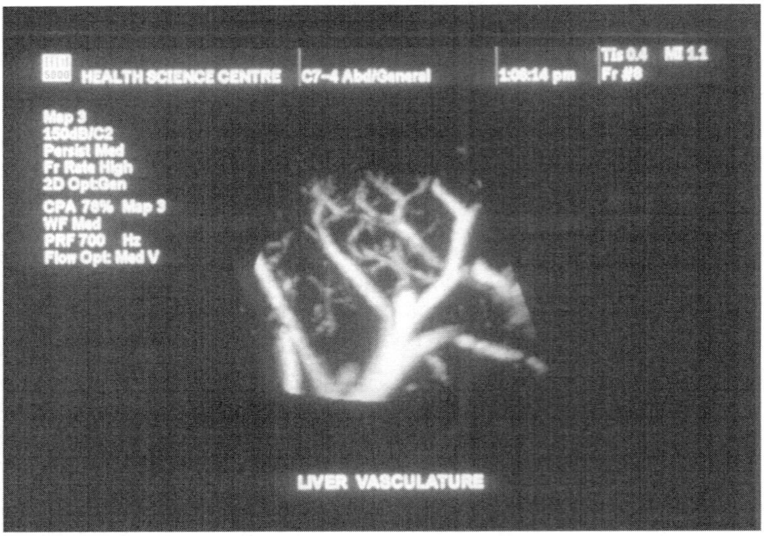

FIGURE 13.15 Power Doppler image of flow in the liver in small vessels. (Courtesy of ATL Ultrasound.)

flow in small vessels and for revealing non-echogenic plaques on vessel walls. Figure 13.15 shows how power Doppler can delineate the vessels in the liver.

Algorithms for displaying low flows take advantage of the high-scanning speed of electronic scanning. Following radar practice, the early mechanical scanners collected the Doppler data from a small number of firings in each look direction before moving to the next, slowing the scan rate. This method collects enough data for a number of images equal to the number of transmitter firings, which the electronically scanned array can produce. With the faster imaging rate, the pixel sampling rate decreases to p.r.f./n, where n is the number of look directions, and the sample window is wider.

A third type of Doppler processing is proving to be useful. In this "tissue Doppler," the wall filter or stationary echo canceller is removed, and the display then shows the echoes from moving tissues in color. The motions of the myocardium can be seen better in this display. The display is also useful in assessing plaques and blockages in arteries.

Imaging using the nonlinear properties of tissue and contrast agents has recently been introduced. Three types of operation are available: harmonic, subharmonic, and pulse inversion imaging. The first uses higher harmonics of the transmitted frequency. The transmitter operates at a given frequency, and the receiver is tuned to the second harmonic. Nonlinear propagation of finite amplitude ultrasound produces the harmonics which increase in strength with depth. The resulting images are superior to those obtained by imaging at the transmitted frequency for reasons that are not fully understood. Since a lower transmit frequency can be used, image distortions caused by sound speed variations are lessened, while the better resolution for the harmonic components improves the image. The depth of penetration is a bit less with current embodiments, but the image quality and diagnostic utility have impressed the clinicians.

The second type of nonlinear imaging uses the reflections at half of the transmitted frequency: the subharmonic. Since tissues produce a very low level of subharmonic energy, the contrast echoes are more easily seen in a tissue background.

A particularly effective form of nonlinear imaging can also be done if the hardware permits. In this "pulse inversion" imaging, the polarity of the transmitted pulse is changed between two transmit firings.[13] The received pulse trains from sequential firings are added in the receiver, so that the fundamental frequencies in the echoes are cancelled and only the nonlinear components are used for imaging. This method can include the subharmonics.

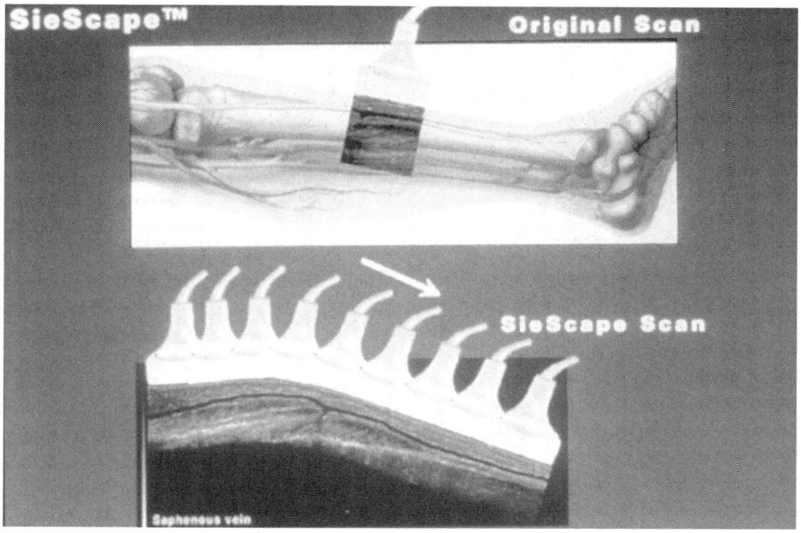

FIGURE 13.16 Schematic showing the SieScape™ process of assembling a large field of view image from individual images taken by sliding the transducer along the leg. The image of the saphenous vein at the bottom shows an open flow channel. This scan was not confined to a plane, but followed the path of the vein. (Courtesy of Siemens Medical Systems, Inc., Ultrasound group.)

These nonlinear methods offer great promise for use with contrast agents. Ultrasound contrast agents contain small gas bubbles, which are nonlinear scatterers and produce strong reflections as well. The echoes from the fluid in the very smallest vessels are often undetectable without the use of contrast. Harmonic methods are capable of enhancing the contrast echo since the bubbles are more nonlinear than the tissues.

Coded pulse transmissions and suitable receiver processing are also being done in at least one commercial system. This allows the sensitivity to be increased without increasing the peak power and hence the possibility of cavitation.[14] With this system, deeper penetration is possible for a given frequency. The echoes from red cells can be seen on the image. A selection algorithm is needed to control the regions of the image where echoes from blood are shown, just as in Doppler or color flow imaging.

One disadvantage of ultrasonic imaging is the small region that is imaged. It is difficult to include "landmarks" in the image which allow persons other that the operator to interpret the image. It is also difficult to perform registration with images from other imaging modalities. It may even be difficult to compare ultrasonic images from adjacent regions for diagnostic purposes. It has been possible to use image data itself to allow adjacent scans to be connected to form a larger image.[15] The principle is shown in Figure 13.16. An algorithm based on "fuzzy logic" is used to connect the separate images without the need for any mechanical constraint of the scan head. Another example of this SieScape™* processing is shown in Figure 13.17, which shows how a necrotic abdominal wall mass can be shown not to involve the liver.

13.5 Conclusion

For our use in doing signal processing, we now have systems with stored digitized data that can be made accessible to us. Software control of processing is built into the machines and can be used for many purposes. Many new schemes are now possible, including creating multiple receiver beams for each conventional firing to increase the frame rate and special processing to correct for phase aberrations and

* SieScape is a trademark of Siemens Medical Systems, Inc., Ultrasound group, Issaquah, WA.

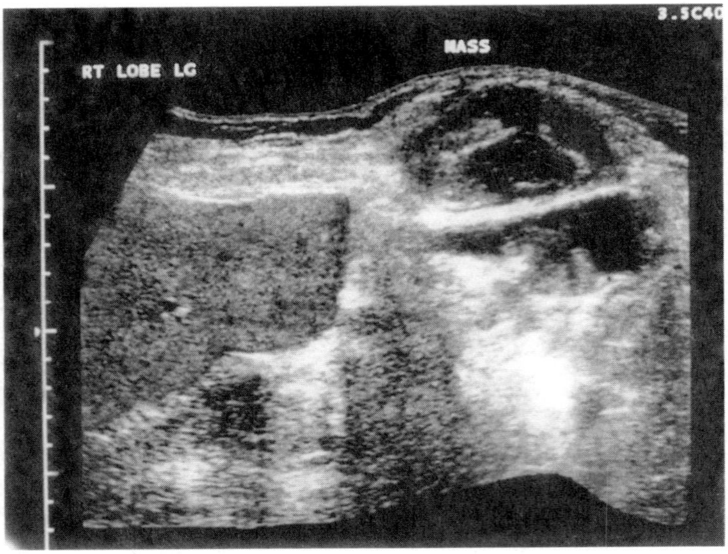

FIGURE 13.17 This clinical SieScape™ image shows the relationship of a necrotic metastatic abdominal wall mass to the liver all on one image. (Courtesy of Siemens Medical Systems, Inc., Ultrasound group.)

speckle degradation of images. In the most modern systems, some control of the transmit waveform on each element is possible, although care must be used to stay within safety guidelines. The received signals from each element are sent to a digital memory, since many image forming and enhancing operations are now done in software. The manufacturers can make these data available for processing, but close cooperation is necessary.

We now have the opportunity to apply even more advanced methods of signal processing to improve the utility of these systems. This includes increasing operating speed, lessening dependence on trained operators, improving image quality, and increasing diagnostic accuracy.

References

1. Greenleaf, J. F., Acoustical medical imaging instrumentation, in *Encyclopedia of Acoustics,* Vol. 4, Crocker, M. J., Ed., John Wiley & Sons, New York, 1997, chap. 144.
2. Wagner, R. F., Smith, S. W., Sandrik, J. M., and Lopez, H., Statistics of speckle in ultrasound B scans, *IEEE Trans. Sonics Ultrason.*, 30, 156, 1983.
3. Greenleaf, J. F., *Tissue Characterization with Ultrasound,* Vols. I and II, CRC Press, Boca Raton, FL, 1986.
4. Shung, K. K. and Thieme, G. A., *Ultrasonic Scattering in Biological Tissues,* CRC Press, Boca Raton, FL, 1993.
5. Evans, D. H., *Doppler Ultrasound: Physics, Instrumentation and Clinical Applications,* John Wiley & Sons, New York, 1989.
6. Lemons, R. F. and Quate, C. F., Acoustic microscopy, in *Physical Acoustics; Principles and Methods,* Vol. 14, Mason, W. P. and Thurston, R. N., Eds., Academic Press, New York, 1979.
7. Reid, J. M. and Lewin, P. A., *Ultrasonic Transducers, Imaging,* Vol. 22, *Encyclopedia of Electrical and Electronics Engineering,* Webster, J., Ed., Wiley, New York, 1999, 664.
8. Thomenius, K., Evolution of ultrasound beamformers, *Proc. 1996 IEEE Ultrasonics Symp.*, IEEE Press, San Antonio, TX, 1996, 1615.
9. Light, E. D., Davidsen, R. E., Fiering, J. O., Hruschka, T. A., and Smith, S. W., Progress in 2-D arrays for real time volumetric imaging, *Ultrason. Imaging,* 20, 235, 1998.

10. Goss, S. A., Frizzell, L. A., and Dunn, F., Ultrasonic absorption and attenuation in mammalian tissues, *Ultrasound Med. Biol.*, 5, 181, 1979.
11. Hazard, C. R. and Lockwood, G. R., Theoretical assessment of a synthetic aperture beamformer for real-time 3-D imaging, *IEEE Trans. Ultrason. Ferroelectr. Frequency Control,* 46, 972, 1999.
12. Kasai, C., Namekawa, K., Koyano, A., and Omoto, R., Real-time, two-dimensional blood flow imaging using an autocorrelation technique, *IEEE Trans. Sonics Ultrason.*, SU-32, 458, 1985.
13. Simpson, D. H., Chin, C. T., and Burns, P. N., Pulse inversion Doppler: a new method for detecting nonlinear echoes from microbubble contrast agents, *IEEE Trans. Ultrason. Ferroelectr. Frequency Control,* 46, 372, 1999.
14. Haider, B., Lewin, P. A., and Thomenius, K., Pulse elongation and deconvolution filtering for medical ultrasound imaging, *IEEE Trans. Ultrason. Ferroelectr. Frequency Control,* 45, 98, 1998.
15. Weng, L., Tirumalai, A. P., Lowery, C. M., Nock, L. F., Gustafson, D. E., Von Behren, P. L., and Kim, J. H., U.S. extended-field-of-view imaging technology, *Radiology,* 203(3), 877, 1997.

14

Basic Principles and Applications of 3-D Ultrasound Imaging

Aaron Fenster
University of Western Ontario

Donal B. Downey
University of Western Ontario

14.1 Introduction

The discovery of X-rays by Röntgen over 100 years ago introduced a new way to visualize the interior of the human body. The radiographic image produced by the use of X-rays displayed a shadow of the

0-8493-3691-0/01/$0.00+$.50

3-dimensional (3-D) structures within the body in the form of a 2-dimensional image (2-D). Since all 3-D information is lost to the physician by this approach, many attempts have been made to develop imaging techniques in which 3-D information within the body is preserved in a recorded image. The introduction of computed tomography (CT) in the early 1970s revolutionized diagnostic radiology, as 3-D information could be preserved and presented to the physician as a series of 2-D image slices of the body. In addition, computers became central in the processing and display of images used in radiology. The wide availability of true digital 3-D anatomical information stimulated the field of 3-D image processing and visualization for a variety of applications in diagnostic radiology.[1–4]

The development of ultrasound imaging is more recent than X-ray imaging and followed the pioneering work of Wild and Reid in the 1950s. The medical use of ultrasound progressed slowly, from A-mode systems, producing oscilloscope traces of acoustic reflections, to B-mode grey-scale images of the anatomy. These developments led to systems producing real-time images of the anatomy and blood flow. Over the past four decades, the image quality of medical ultrasound images has advanced from low-resolution, bistable images to images with sub-millimeter resolution and sufficient grey scale to differentiate subtle lesions. Medical ultrasound imaging is now an important and often indispensable imaging modality in disease diagnosis, image-guided biopsy, therapy, and obstetrics. Because it is not invasive and, due to the improved image quality and blood flow information, ultrasonography is progressively achieving a greater role in radiology, cardiology, and image-guided surgery and therapy. The major advantages of ultrasonography are

- Ultrasound transducers are small and easily manipulated, allowing the generation of real-time tomographic images at orientations and positions controlled by the user.
- The ultrasound image has sufficient resolution (0.2 to 2.0 mm) to display many structures within the body.
- Ultrasound instrumentation is inexpensive, compact, and mobile.
- Ultrasound imaging can provide real-time images of blood velocity and flow, allowing the physician to map vascular structures ranging in size from arteries to angiogenic tumor vessels.

In spite of these advantages, ultrasonography still suffers from several limitations, which are being addressed by researchers and imaging companies. In this chapter, we discuss a recent advance, the development of 3-D ultrasound imaging, which is achieving rapid growth in overcoming one limitation of ultrasonography.

14.2 Limitations of Ultrasonography Addressed by 3-D Imaging

In conventional 2-D ultrasonography, an experienced diagnostician manipulates the ultrasound transducer, mentally transforms the 2-D images into a lesion or anatomical volume, and makes the diagnosis or performs an interventional procedure. Although this is extremely useful, it leads to a limitation resulting primarily from the use of a spatially flexible 2-D imaging technique to view 3-D anatomy/pathology. This is particularly important in image-guided therapeutic procedures, as the process of guiding needles into the proper location and following changes during the procedure is limited by the 2-D restrictions of conventional ultrasonography. Specifically, 3-D ultrasonography addresses the following limitations:

- Since conventional ultrasound images are 2-D, the diagnostician must mentally transform multiple images to form a 3-D impression of the anatomy/pathology during the diagnostic examination or during the interventional procedure. This process is time-consuming and inefficient, but, more important, variable and subjective, potentially leading to incorrect decisions in diagnosis, planning, and delivery of therapy.
- Diagnostic (obstetrics) and therapeutic decisions (staging and planning) often require accurate measurements of organ or tumor volume. Current volume measurement techniques use only

simple measures of the width in two views and assume an idealized shape to calculate volume. This volume measurement technique potentially leads to inaccuracy and operator variability.

- It is difficult to place the 2-D image plane at a particular location within an organ and even more difficult to find the same location again later. Thus, 2-D ultrasound is not well suited for planning or monitoring therapeutic procedures or for performing quantitative prospective or follow-up studies.
- The patient's anatomy or position sometimes restricts the orientation of the 2-D transducer needed to obtain the optimal image plane. Visualization of the anatomy is hindered, preventing accurate diagnosis of the patient's condition and monitoring of interventional procedures.

The goal of 3-D ultrasound imaging is to overcome these limitations by providing the diagnostician or therapist with views of the anatomy in 3-D, thereby reducing the variability of conventional ultrasound techniques.

Since conventional ultrasonography is inherently tomographic like CT or magnetic resonance (MR), the information it provides can be reconstructed into 3-D images to be used for 3-D visualization. However, unlike CT and MR imaging, in which the images are usually acquired at a slow rate as a stack of parallel slices, ultrasound provides 2-D images at a high rate (15 to 60 images per second). In addition, the relative orientation of each image in the sequence is arbitrary and under the user control. The high rate of image acquisition, arbitrary orientation of the images, and problems imposed by the ultrasound image (e.g., speckle, shadowing, and distortions) provide unique problems to overcome and opportunities to be exploited in extending conventional ultrasonography from its 2-D presentation of images to 3-D and dynamic 3-D (i.e., 4-D).

Over the past two decades, many developments of the various types of 3-D imaging techniques have taken advantage of ultrasound imaging positioning flexibility and data acquisition speed.[5–11] These approaches have focused on reconstructing 3-D images by integrating transducer position information with the acquired 2-D ultrasound images. Because of the enormous demands on computational speed needed to produce near real-time images and on development of low-cost systems, most attempts did not succeed. Only in the last few years has computer technology and visualization techniques progressed sufficiently to make 3-D ultrasonography viable. Review articles and two books have described progress in the development of 3-D ultrasound imaging for use in radiology and echocardiology.[6,8,11–16] These articles and books provide extensive lists of references and show that there have been numerous attempts at producing 3-D ultrasound systems by many investigators.

In this chapter, we review the various approaches that investigators have pursued in the development of 3-D ultrasound imaging systems and discuss the uses of these techniques. We also show example images demonstrating their use and report on the performance of these systems.

14.3 Scanning Techniques for 3-D Ultrasonography

14.3.1 Introduction

Except for the development of 2-D arrays for direct 3-D ultrasound imaging, all the approaches make use of conventional 1-D ultrasound transducers producing 2-D ultrasound images. The major differences in the various 3-D ultrasonography approaches are in the specific methods used to locate the position of the 2-D ultrasound image within the tissue volume under investigation. The ultrasound transducer's position must be known accurately and precisely so that high-quality 3-D images are produced without distortions. Clearly, the choice of the scanning technique used to produce the 3-D image is crucial in producing accurate representation of the anatomy.

The most current 3-D ultrasound imaging systems are based on commercially available transducers whose position is accurately controlled by a mechanical device or monitored by a position and orientation sensing device. Position data may be obtained from stepping motors in the transducer's scan head, a translation or rotation mechanical device, or a position and orientation sensor that may be electromagnetic, acoustic, or mechanical. These approaches result in an acquired sequence of 2-D images that may be arranged in the pattern of a wedge, in a series of parallel slices, in a cone obtained by rotation around a central axis, or in arbitrary orientations. These approaches are discussed in detail in the subsequent sections.

To produce 3-D images without distortions, three factors must be considered and optimized:

- The scanning technique must be rapid or gated to avoid artifacts due to involuntary, respiratory, and/or cardiac motion.
- The locations and orientations of the acquired 2-D images must be known accurately and precisely, and the devices must be calibrated correctly to avoid geometric distortions and measurement errors.
- The scanning technique must be easy to use so that it does not interfere with the patient exam or be inconvenient to use.

Four different scanning approaches have been pursued: mechanical scanners, sensed free-hand techniques, free hand without location sensing, and 2-D arrays.

14.3.2 Mechanical 3-D Scanning Devices

In this approach, a sequence of 2-D images is acquired rapidly while the conventional transducer is scanned over the anatomy using a variety of mechanical motorized techniques. Since mechanical means are used to move the conventional transducer over the anatomy in a precise predefined manner, the relative position and orientation of each 2-D image in the sequence are known accurately. The angular or spatial interval between the digitized 2-D images is precomputed and is usually made adjustable to minimize the scanning time while optimally sampling the volume. Thus, with proper calibration, it is possible to avoid geometric distortions and inaccuracies.

Before efficient, high-speed video image digitizers were readily available, the 2-D images were stored on videotape for later processing. However, in current systems, either the images are stored in their original digital format in the ultrasound system's computer memory, or the video output from the ultrasound machine is digitized immediately and stored in an external computer memory. After a sequence of 2-D images has been acquired, either the ultrasound machine's computer or an external computer reconstructs the 3-D image using the predefined geometric information, which relates the digitized 2-D images to each other.

Different mechanical scanning approaches have been developed, with each requiring that the conventional transducer be mounted in a special assembly and made to rotate or translate by a motor. When the motor is activated under computer control, the transducer rotates or translates rapidly over the region being examined. The mechanical assemblies that have been developed vary in size from small 3-D probes that house the mechanical mechanism within the transducer housing to ones in which the conventional transducer is mounted on an external fixture.

The integrated mechanical 3-D probes, which house the mechanism within the housing, are usually easy for the operator to use, but are larger and heavier than conventional probes. Since these integrated probes are specially designed for 3-D imaging, they require the purchase of a special 3-D ultrasound machine that can interface to them.

The mechanical scanning approaches that make use of external fixtures are generally bulkier than the integrated probes, but they can be adapted to any conventional ultrasound machine's transducers, obviating the need to purchase a special purpose 3-D ultrasound machine. Thus, improvements in image quality and developments of new imaging techniques (e.g., harmonic imaging) by the ultrasound machine manufacturers can also be achieved in 3-D. In addition, since the external fixture approach can house a transducer from any ultrasound machine, the 3-D capability can be made available in conjunction with most ultrasound machines.

Researchers and commercial companies have developed different types of mechanical scanning assemblies for 3-D ultrasonography. These can be divided into three basic types of motion, as shown schematically in Figures 14.1, 14.4, 14.5, and 14.7:

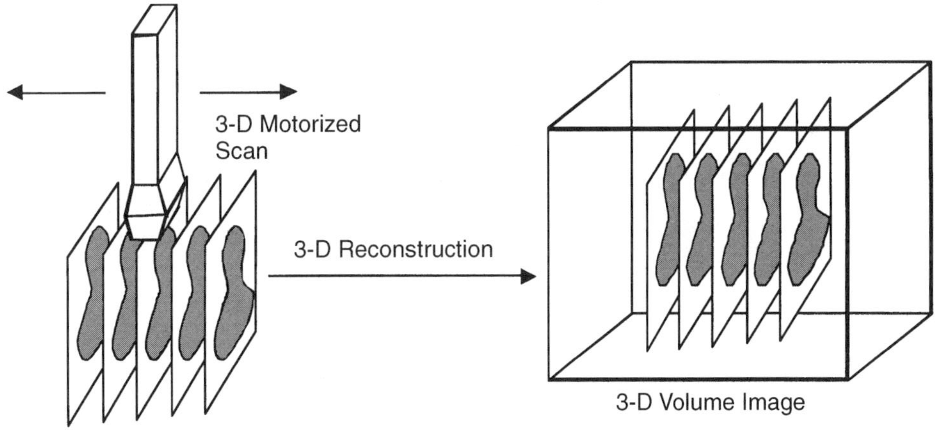

FIGURE 14.1 Schematic diagram showing the linear scanning approach, in which a series of parallel 2-D images is collected and used to reconstruct the 3-D image.

- Linear
- Tilting
- Rotation

14.3.2.1 Mechanical 3-D Scanning: Linear

In this approach, the mechanical mechanism causes the transducer to translate linearly over the patient's skin, so that the 2-D images in the sequence are parallel to each other. This is accomplished by mounting the conventional transducer in an assembly that houses a motor and drive mechanism as shown in Figure 14.1. When the motor is activated, the drive mechanism moves the transducer in a direction that is parallel to the skin surface. While the transducer is moved over the anatomy, 2-D images are acquired at regular spatial intervals that sample the anatomy appropriately. The ability to vary the sampling interval allows imaging of the anatomy in 3-D with an appropriate sampling interval dictated by the particular elevational resolution of the transducer.

Since the acquired 2-D images are parallel to each other (see Figure 14.1) with a known spatial sampling interval, the parameters required for the reconstruction can be pre-computed, and the 3-D image can be obtained immediately after performing a linear scan.[17]

Since the resolution in the acquired 2-D images is different in the axial and lateral directions, and the elevational resolution varies with depth, the resolution in the resulting reconstructed 3-D image will not be isotropic. However, the simplicity of the scanning motion makes interpretation of resolution variation more easily understood. In the planes of the 3-D image corresponding to the original 2-D images, the resolution is unchanged. Because of the poor elevational resolution of conventional 1-D transducers (or even 1.5-D transducers), the resolution in the scanning direction is worse. For optimal resolution in 3-D, a transducer with good elevational resolution should be used.

The mechanical linear scanning approach has been used successfully in many vascular imaging applications using the B-mode of various organs:[18–22] color Doppler for imaging the carotid arteries,[19,23–27] Doppler power imaging,[19,20,24] tumor vascularity,[20,28–30] and test phantoms.[24,31] An example of the linear scanning approach with an external fixture is shown in Figures 14.2 and 14.3.

14.3.2.2 Mechanical 3-D Scanning: Tilting

In this scanning geometry, the transducer is tilted by the mechanical assembly about an axis that is parallel to the face of the transducer, as shown in Figure 14.4. The tilting axis can be either at the face of the transducer, producing a set of 2-D images that intersect at the face, or above the transducer's axis, causing the set of 2-D planes to intersect above the skin. This mechanical scanning approach has been achieved with an external fixture as well as with an integrated 3-D probe approach. In either case, this scanning approach allows the transducer face or the 3-D probe housing to be placed at a single location on the patient's skin. The use of the external fixture causes the transducer to pivot on the point of contact on the skin (Figure 14.4). The integrated 3-D probe approach allows the housing to be placed against the skin, while the transducer is tilted and slides against the housing to produce an angular sweep.

The tilting approach for the acquisition of the 2-D images at constant angular intervals results in a set of 2-D image planes arranged in fan-like geometry. With the appropriate choice of angular interval and the number of 2-D images to be acquired, this approach can sweep out a large region of interest. This approach has been used successfully in abdominal imaging applications.[23,32–36]

The mechanical 3-D tilting approach lends itself to compact designs for both integrated 3-D probes and external fixtures. Kretztechnik (Zipf, Austria) and Aloka (Korea) have developed special 3-D ultrasound systems with integrated 3-D probes for use in abdominal and obstetrical imaging.

This scanning approach can also be used with endocavity transducers, such as transesophageal (TE) and transrectal (TR) transducers (Figures 14.5 and 14.6). In this approach, a side-firing linear transducer array is used with either the external fixture or an integrated 3-D probe. When the motor is activated, the transducer rotates about its long axis while 2-D images are acquired. After a rotation of about 80° to 110°, the acquired 2-D images are arranged in fan-like geometry, similar to that obtained with the tilting scan used for abdominal imaging. This approach is successful in prostate imaging (see

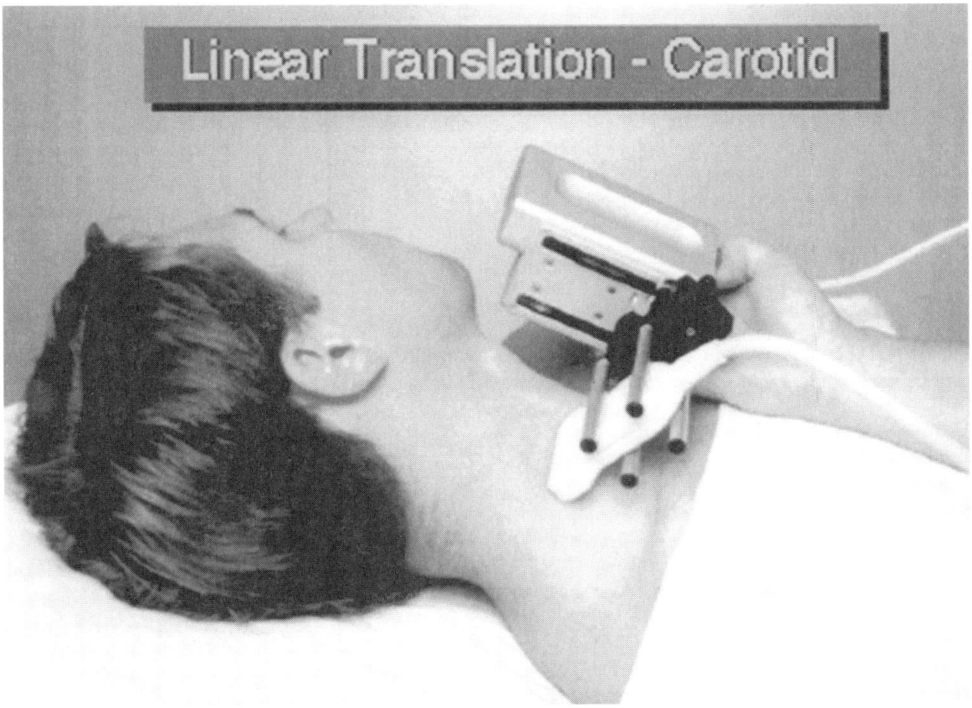

FIGURE 14.2 Photograph of a linear scanning mechanical assembly being used to image the carotid arteries.

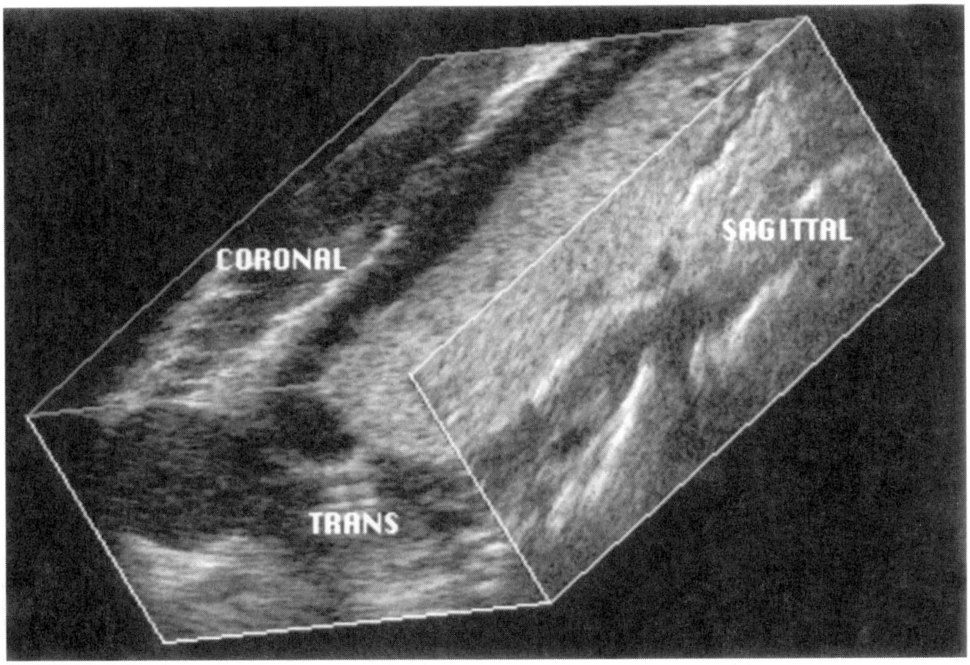

FIGURE 14.3 3-D image of a patient's thyroid. The image has been "sliced" to reveal the coronal, sagittal, and transaxial views of the thyroid using the multi-planar reformatting visualization approach.

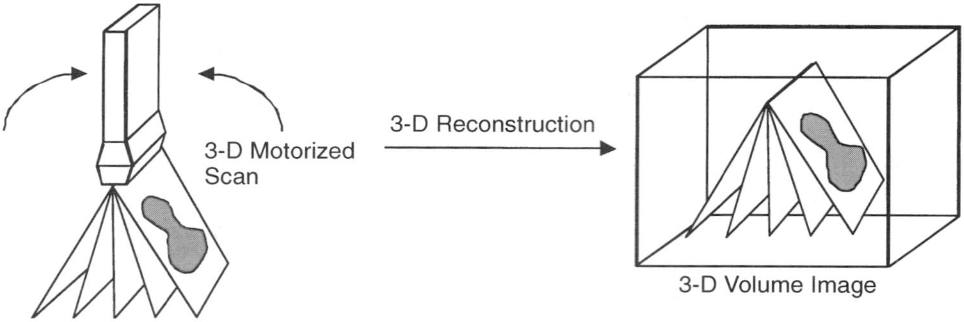

FIGURE 14.4 Schematic diagram showing the tilting scanning approach, in which a series of 2-D images is collected as the transducer is tilted and then reconstructed into a 3-D image.

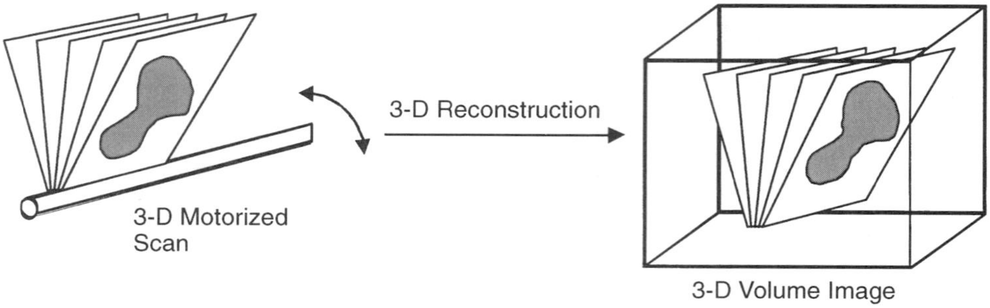

FIGURE 14.5 Schematic diagram of the tilt scanning approach used with a side-firing TR transducer. This approach is being used to produce 3-D images of the prostate.

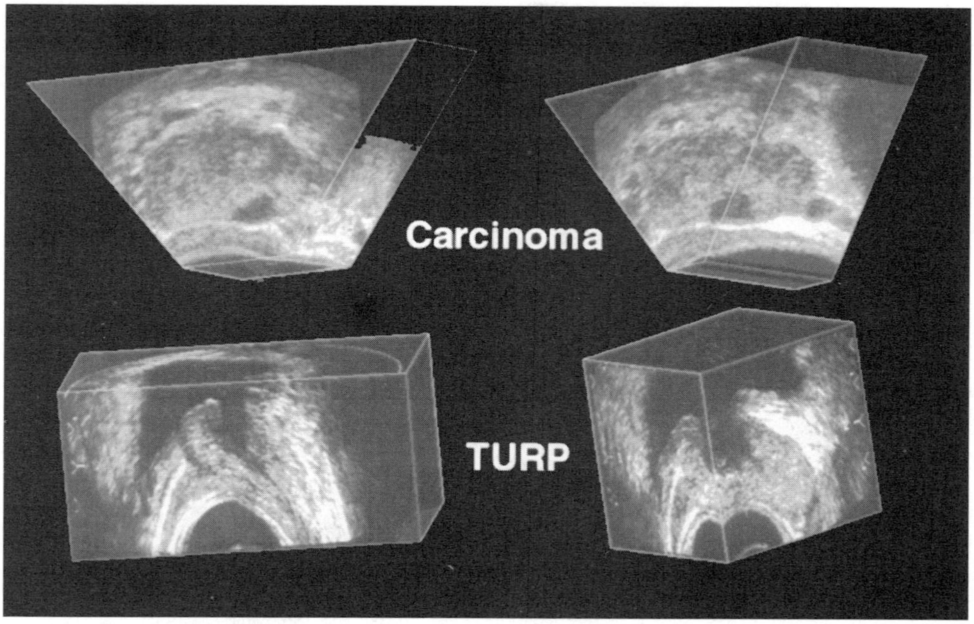

FIGURE 14.6 3-D images of a prostate with carcinoma. The top two images were obtained with the tilt scanning approach using a side-firing TR transducer. The image has been sliced to reveal a tumor (dark lesion) and its relationship to the seminal vesicle. The bottom two images were obtained with the rotational approach using a end-firing TR transducer. The 3-D prostate image has been sliced to display the TURP (transurethral resection procedure) deficit.

Figure 14.6)[19,20,23,37,38] and 3-D ultrasound-guided cryosurgery and brachytherapy[17,40,41] and has been commercialized by Life Imaging Systems, Inc. (London, Canada)

The main advantage of the tilt scanning approach is that the scanning mechanism can be made compact to allow easy hand-held manipulation. In addition, using a suitable choice of scanning angle and angular interval, the scanning time can be made short. With reduced resolution, Aloka has exhibited the acquisition of three volumes per second. However, with a typical angular interval of 1° or less and a scanning angle of about 90°, the scanning time can range from 3 to 9 s, depending on the ultrasound image frame rate (which depends on the number of focal zones and the depth of imaged tissue). Since the set of planes is acquired in a predefined geometry, again, many of the geometric parameters can be precalculated, allowing short 3-D reconstruction times.

Because of the variations in the axial, lateral, and elevational resolution of the conventional ultrasound transducer, the resolution in the 3-D image will not be isotropic. Since the 2-D images are acquired in a fan-like geometry, the distance between acquired planes will increase with depth. Near the transducer the sampling distance will be small, while far from the transducer the distance will be large. However, the angular sampling interval can be optimized by matching the change in sampling distance with depth to the elevational resolution degradation of 2-D ultrasound transducers with depth. Nonetheless, the resulting resolution in the 3-D image will be worse in the scan (tilting) direction, due to the combined effects of elevational resolution and sampling, with the resolution degrading with distance from the transducer.

14.3.2.3 Mechanical 3-D Scanning: Rotational

In this scanning geometry, the mechanical mechanism rotates the transducer by more than 180° around an axis that is perpendicular to the ultrasound array and bisects it as shown in Figure 14.7. This approach allows the axis of rotation to remain fixed as the transducer rotates, causing the acquired images to sweep out a conical volume in a propeller-like fashion. As with the tilting mechanical scanning approach, the angular rotational interval between acquired 2-D images is adjustable and remains fixed during the scan, resulting in a set of acquired 2-D images arranged as shown in Figure 14.7. Also similar to the tilting scanning approach, the rotation scanning can be implemented using both the external fixture and the integrated 3-D probe approach.

Since the rotational scanning approach causes the images to intersect along the central rotational axis, the spatial sampling will be highest near this axis and lowest away from it. The combined effects of the changing sampling with distance from the axis of rotation and the degradation of elevational resolution with depth will cause the resolution in the 3-D image to vary in a complicated manner. The resolution will be best near the transducer and will degrade away from the transducer due to the degradation of both lateral and elevational resolution with depth in the original 2-D images. In addition, improper angular sampling may impose additional resolution degradation away from the central scanning axis.

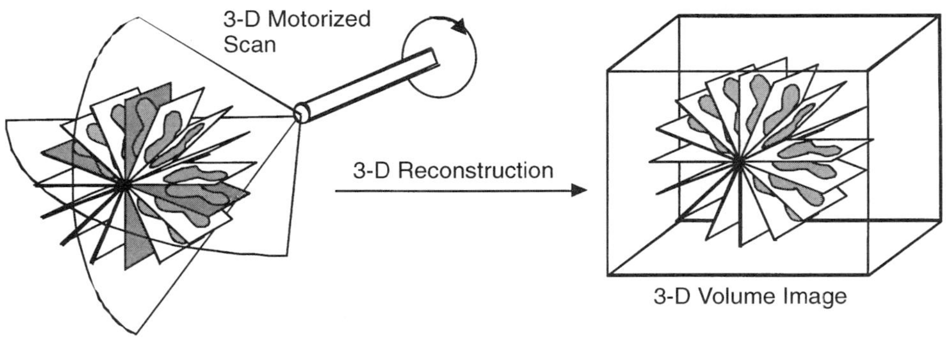

FIGURE 14.7 Schematic diagram of the rotational scanning approach used with an end-firing transrectal transducer. This approach has been used in endovaginal and TR imaging.

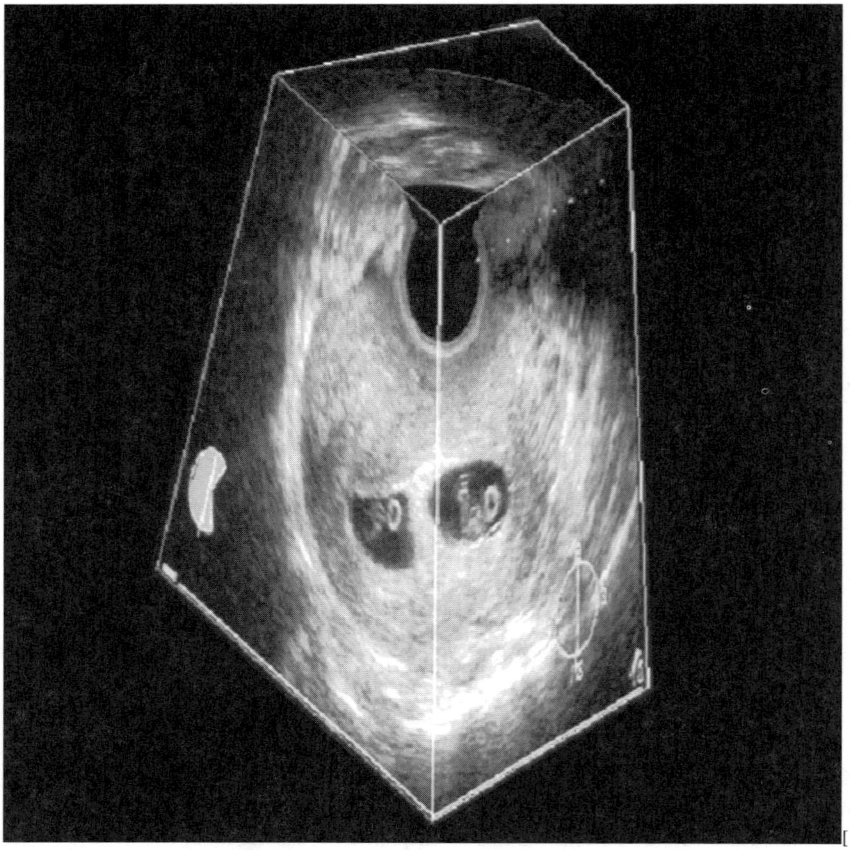

FIGURE 14.8 3-D image of a pregnant uterus with twins. The image has been sliced to reveal the two gestational sacs. This image was obtained using the rotational scanning approach with an end-firing endovaginal transducer.

This approach has been used successfully with end-firing transducers in TR imaging of the prostate and endovaginal imaging of the uterus (Figure 14.8). Downey reported on the use of rotational scanning for imaging the prostate with 200 images (320 × 320 pixels each) acquired in 13 s over a 200° angular sweep.[20]

Since the acquired images intersect along the rotational axis in the center of the 3-D image, any motion of the axis of rotation or of the patient during the acquisition will cause artifacts in a reconstructed 3-D image. For example, if motion were to occur, the images acquired at 0° and 180° would not be mirror images of each other. In addition to the sensitivity to motion, the relative geometry of the acquired images must be accurately known. Thus, if the transducer array is placed into its housing in a way that causes an image tilt or offset from the rotation axis, then 3-D image distortions will occur. These geometrical non-idealities must be known, and the reconstruction must compensate for these to avoid artifacts in the center of the 3-D image.

14.3.3 Sensed Free-Hand 3-D Scanning

The mechanical scanning approach to 3-D ultrasonography described earlier offers speed and accuracy; however, the bulkiness and weight of the scanning mechanisms are inconvenient to use at times. In addition, the restricted scanning geometry hinders scanning of large structures in the body. Many researchers have attempted to overcome this problem by developing various free-hand scanning techniques that do not require a mechanical fixture and yet provide information on the position and orientation of the transducer.

In this sensed free-hand 3-D scanning approach, a position and orientation sensor is attached to the transducer. The transducer is then held by the operator in the usual manner and moved over the anatomy, while a computer is used to record the conventional 2-D images generated by the ultrasound machine together with their position and orientation. Thus, the operator is free to manipulate the transducer and select the optimal orientation which best displays the anatomy and accommodates complex surfaces of the patient.

Since the operator is free to manipulate the transducer, the scanning geometry is not predefined as with the mechanical scanning approaches. Therefore, the relative positions and orientation of the digitized 2-D images must be accurately known if the information is to be used to reconstruct the 3-D image without distortions and inaccuracies. In addition, the operator must move the transducer in a manner that allows adequate spatial sampling without significant gaps and overlaps. This can be achieved by training operators to scan the anatomy in a constant velocity that is appropriate for the acquisition frame rate or with computer prompts generated by the analysis of the sensed position and orientation data. Over the past two decades, several sensed free-hand 3-D scanning approaches have been developed which use four basic position sensing techniques:

- Articulated arms
- Acoustic tracking
- Magnetic field tracking
- Image-based information

14.3.3.1 Sensed Free-Hand 3-D Scanning: Articulated Arms

In this free-hand 3-D scanning approach, the ultrasound transducer is mounted on a multiple jointed mechanical arm system. The relative rotations of the arms, provided by the potentiometers located at the joints of the movable arms, are used to calculate the position and orientation of the transducer. This arrangement allows the operator to manipulate the transducer and scan the desired patient anatomy, while the computer records the 2-D images and the relative rotations of all the arms. The 3-D image can then be reconstructed using the calculated positions and orientations of the transducer associated with the recorded 2-D images. To avoid distortions and inaccuracies in the reconstructed 3-D image, the potentiometers must provide accurate and precise relative rotation information, and the arms must not flex. By keeping the arms as short as possible and reducing the number of movable joints, sufficient accuracy can be achieved. Thus, increased precision and accuracy in the reconstructed 3-D image can be achieved at the expense of reduction in scanning flexibility and reduction in the size of the scanned volume.

14.3.3.2 Sensed Free-Hand 3-D Scanning: Acoustic Tracking

Acoustic ranging was one of the first free-hand scanning methods used to produce 3-D ultrasound images. In this approach, three sound emitting devices, such as spark gaps, were mounted on the transducer, and an array of fixed microphones was mounted above the patient. While the transducer was moved freely over the patient's anatomy, 2-D ultrasound images were recorded, while the sound emitting devices were activated and the microphones continuously received sound pulses. The position and orientation of the transducer, and hence the recorded images, were then calculated from knowledge of the speed of sound in air and the time of flight of the sound pulses from the emitters on the transducers to the fixed microphones. The fixed microphones were placed over the patient to provide unobstructed lines of sight to the emitters and sufficiently close to allow detection of the sound pulses with good signal to noise. Corrections to the speed of sound with the room's temperature and humidity were required to avoid distortions in the 3-D image.[5,42–48]

14.3.3.3 Sensed Free-Hand 3-D Scanning: Magnetic field Sensors

The most popular and successful technique to measure the ultrasound transducer's position and orientation free-hand 3-D scanning approach makes use of a six degree-of-freedom magnetic field sensor.

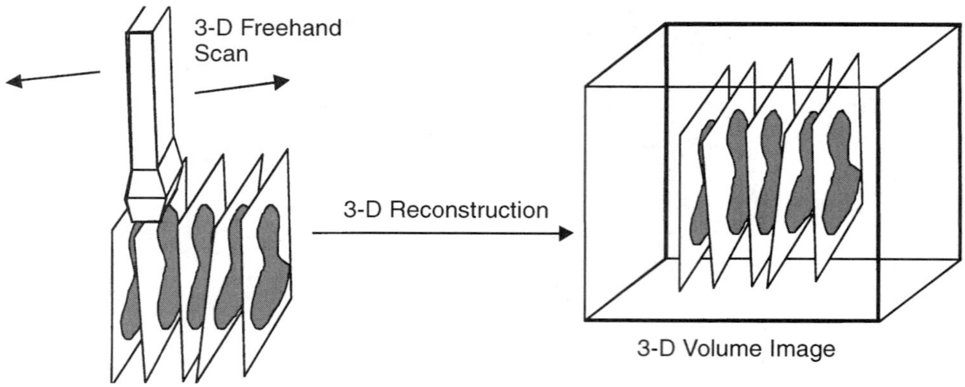

FIGURE 14.9 Schematic diagram showing the free-hand scanning approach. The 2-D images are arranged without a regular geometry. For accurate 3-D reconstruction, the position and orientation of each plane must be known.

This technique (Figure 14.9) uses a transmitter, which produces a spatially varying magnetic field, and a small receiver containing three orthogonal coils to sense the magnetic field strength. By measuring the strength of three components of the local magnetic field, the ultrasound transducer's position and orientation can be calculated.

Magnetic field sensors are small and unobtrusive, allowing for less constraining tracking of position and orientation than the previously described approaches. However, a number of effects can combine to produce distortions in the reconstructed 3-D images.

- The accuracy of the measured position and orientation can be compromised by electromagnetic interference from sources such as CRT monitors, alternating current (AC) power cabling, and some electrical signals from ultrasound transducers.
- Ferrous and highly conductive metals will distort the magnetic field, causing errors in the recorded geometric information.
- Errors in the determination of the location of the moving transducer will occur if the magnetic field sampling rate is not sufficiently high and the transducer is moved too quickly. This image "lag" artifact can be overcome with a sampling rate of about 100 Hz or higher.

By ensuring that the environment is free of electrical interference and metals, and with the appropriate sampling rate, high-quality 3-D images can be obtained, as shown in Figure 14.10. Two companies are currently producing magnetic positioning devices of sufficient quality for 3-D ultrasound imaging: the Fastrack by Polhemus and Flock-of-Birds by Ascension Technologies. Investigators have used these devices successfully in many diagnostic applications, such as echocardiography, obstetrics, and vascular imaging.[15,23,49–60]

14.3.3.3 Sensed Free-Hand 3-D Scanning: Tracking Based on Image Information

The free-hand scanning techniques described previously require a sensor to measure the relative positions of the acquired images. Measurement of the relative position and orientation of the acquired 2-D images can also be obtained by using the well-known phenomenon of ultrasound speckle decorrelation. When a source of coherent energy interacts with scatterers, the reflected spatial energy pattern will vary and appear as a speckle pattern. Image speckle is characteristic of ultrasound images and is used in making images of the velocity of moving blood.[61] For this application, two sequential signals generated by scattering of ultrasound from stationary red blood cells will be correlated, i.e., the speckle pattern will be identical. However, if the red blood cells are moving, the signal will be decorrelated. The degree of decorrelation is proportional to the distance the cells have moved.

The same principle can be used in free-hand 3-D ultrasound scanning. If two images are acquired from the same location, then the speckle pattern will be the same, i.e., the correlation will be maximum.

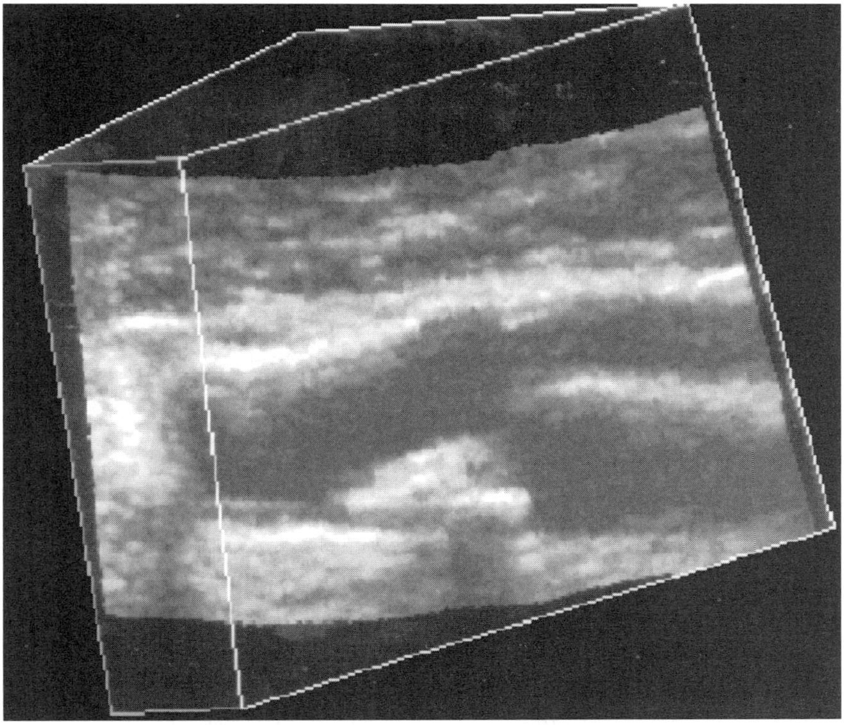

FIGURE 14.10 3-D B-mode image of the carotid arteries obtained with the free-hand scanning approach using a magnetic field sensing device. The 3-D image has been sliced to reveal an atherosclerotic plaque at the entrance of the internal carotid artery. The plaque is heterogeneous with calcified regions casting a shadow in the common carotid artery.

However, if one of the images is moved with respect to the first, then the degree of decorrelation will be proportional to the distance between the two images.[62] The exact relationship relating the degree of decorrelation to distance will depend on many transducer parameters, including the number of focal zones, the degree of focusing, and the tissue depth. Thus, obtaining an accurate determination of the separation between two images requires careful determination of the relationship between distance and the degree of speckle decorrelation for various transducer parameters.

The free-hand 3-D scanning approach does not guarantee that the transducer motion will generate images that are parallel to each other. Thus, a complicated strategy must be adopted to be able to determine if the transducer has been tilted or rotated. Typically, the acquired 2-D images are subdivided into smaller regions, allowing similar regions in adjacent images to be cross-correlated. In this manner, a pattern of decorrelation values is generated for the array of sub-regions and is used to generate a pattern of translation vectors. These are then analyzed to give not only the relative position, but also the orientation between adjacent images, which are used in the 3-D reconstruction algorithm.

14.3.4 Free-Hand Scanning without Position Sensing

Although the sensed free-hand scanning approach allows the transducer to be manipulated in the usual manner, it nevertheless requires that a position and orientation sensing device be added to the transducer or the position and orientation must be calculated from information in the sequence of the acquired 2-D images. An alternative approach involves manipulating the transducer over the patient in a regular manner without any sensing device and reconstructing the 3-D image assuming a predefined scanning geometry. Since no position or orientation information is recorded during the transducer's motion, the transducer must be moved with constant linear or angular velocity so that the acquired images are

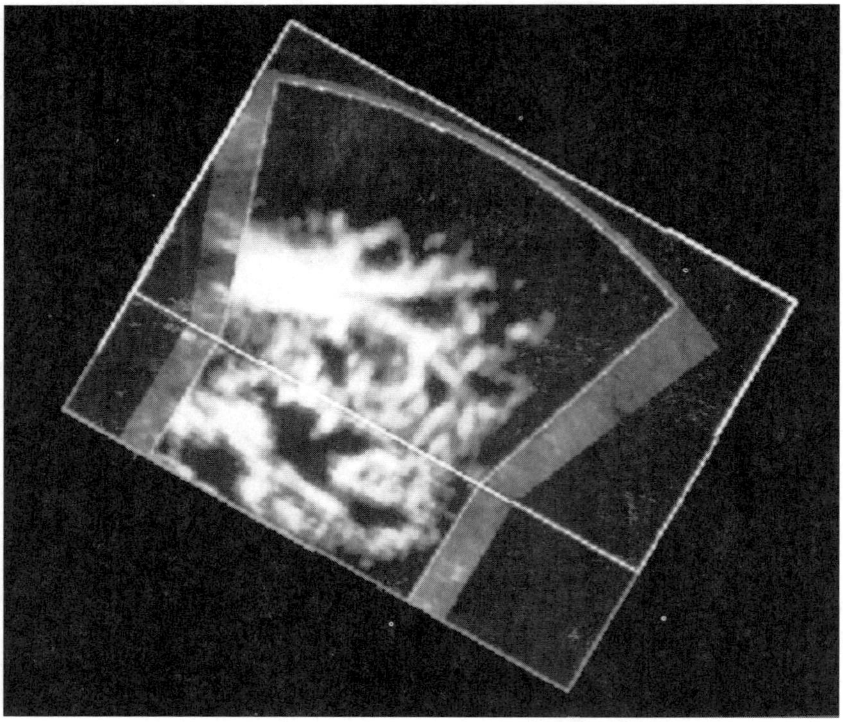

FIGURE 14.11 3-D power Doppler image of a kidney. The image was obtained with the free-hand scanning approach without any positioning sensing. The image has been sliced to demonstrate that excellent 3-D reconstructions can be obtained of vascular structures. However, the geometry may not be correct, and measurements should not be made with images obtained with this approach.

obtained with as regular spacing as possible. To avoid 3-D image distortion, the total linear scanning distance or angular sweep must also be assumed or estimated.

This approach to 3-D scanning may be achieved by moving the transducer over the skin in a linear motion, while the 2-D images are recorded at regular temporal intervals. If the transducer's motion is uniform and steady, then, with the knowledge of the transducer's velocity and distance scanned, a very good 3-D image may be reconstructed, as shown in Figure 14.11.[19] Since this approach does not guarantee that the image geometry or image distances are correct, the 3-D images obtained using this approach should not be used for measurements.

14.3.5 Dynamic 3-D Ultrasonography Using 2-D Arrays

All the techniques described previously make use of conventional 1-D arrays. To obtain a 3-D image requires that a planar beam of ultrasound be mechanically or manually swept over the anatomy. A better approach would keep the transducer stationary, but use electronic steering to sweep out the ultrasound beam over the desired volume. This has been accomplished with the use of 2-D arrays, which produce a broad ultrasound beam covering the entire volume under examination. Investigators have described different 2-D array designs, but the one developed at Duke University in Raleigh, NC for real-time 4-D echocardiography has achieved the greatest success and has been used for dynamic imaging of the heart.[63–70]

The transducer used in dynamic 3-D imaging is composed of a 2-D phased array of elements which are used to transmit a broad beam of ultrasound diverging away from the array and sweeping out pyramidal volumes. The returned echoes are detected by the 2-D array and then processed to display in real-time multiple planes from the volume. These planes can be manipulated interactively to allow the user to explore the whole volume under investigation.

2-D arrays are being used to produce real-time echocardiographic 3-D images by displaying about 20 3-D views per second of the beating heart. Although this approach has the potential to be the dominant method to produce 3-D ultrasound images, it still suffers from a number of limitations that must be overcome before it becomes widespread in radiology and cardiology. These problems are related to the cost and low yield resulting from the manufacture of numerous small elements and the connection and assembly of many electronic leads. This can be alleviated with the use of sparse arrays, which have already produced useful images. Since the 2-D arrays are small, measuring about 2 × 2 cm in size, the volume imaged is relatively small and suited for cardiac imaging, but not yet for larger organs.

14.4 Reconstruction of the 3-D Ultrasound Images

The reconstruction process of a 3-D representation of the scanned anatomy is achieved by placing the acquired 2-D images into the 3-D image at their correct relative position. The 3-D reconstruction process can be carried out during the acquisition of the 2-D images (with multiple processors) or after the 2-D ultrasound images have been acquired. The relative position and orientation of the 2-D images are used either from knowledge of the predefined scanning geometry (mechanical scanners) or from the sensors used in the free-hand scanning approach. The 3-D image can be reconstructed using two distinct methods:

- Feature-based 3-D reconstruction
- Voxel-based 3-D reconstruction

14.4.1 Feature-Based 3-D Reconstruction

In this method, the desired features in the patient's anatomy must be identified first, and only these are reconstructed. Thus, the sequence of 2-D images must be analyzed first and then desired features in each acquired 2-D image can be identified and classified. For example, in echocardiographic or obstetrical images, the boundaries of fluid filled regions are outlined manually or by using an automated computerized method. The boundaries of the desired structures are assigned different labels and colors, and other structures are eliminated. The boundaries of the desired structures are then represented by meshes and displayed with surface viewing software. This approach has been used extensively in echocardiographic imaging to identify the boundaries of the ventricles either manually or automatically. From the 3-D surface model descriptions of the heart chambers, the complex motion of the left ventricle can be viewed with a computer workstation. Changes in the ventricular volume during the cardiac cycle may be measured as well. A similar approach has been used also to reconstruct the vascular lumen in 3-D using intravascular ultrasound (IVUS) imaging.

Since the 3-D image has been reduced to description of surfaces of a few structures, inexpensive computer workstations and software can be readily used for optimized viewing. In addition, this approach artificially increases the contrast between different structures by the manual or automatic identification and classification step, leading to improved appreciation of the 3-D anatomy. It also enables efficient manipulation of the 3-D image processing, including "fly-through," using inexpensive computer display hardware.

Although this approach has been used successfully in CT and MR imaging, it has disadvantages when used in 3-D ultrasonography.

- Since the identification and classification process results in only simple descriptions of anatomical boundaries, important image information, such as subtle tissue features and texture, is lost at the initial phase of the reconstruction and cannot be recovered at a later time.

- The boundary identification step is tedious and time-consuming if done manually or is subject to errors if done automatically by a computer. Both approaches are also subject to variability in the final reconstructed boundaries.
- The classification process artificially accentuates the contrast between different structures, distorting or misrepresenting subtle image features.

To avoid image artifacts and generation of false information, the classification process must be accurate and reproducible. The classification stage can be accurate and reproducible if the image contrast is high, as found at the boundary of fluid filled regions. However, the classification is very difficult in situations where the image contrast is small, as found between tumors and tissue or even between different soft tissues. Because of the tedious and potentially erroneous aspects of the boundary classification and identification process, this method for reconstruction of 3-D ultrasound images is not often used. It is useful though in situations where volume measurements of fluid filled regions are needed.

14.4.2 Voxel-Based 3-D Ultrasound Reconstruction

The more common approach to reconstruction of 3-D ultrasound images is based on the use of the sequence of acquired 2-D images to build a voxel-based volume (i.e., 3-D grid of picture elements). The reconstruction process involves placing each pixel in the acquired 2-D images in its correct location in the 3-D grided volume. Clearly, the location and orientation of each image in the sequence of acquired 2-D images is necessary, as well as the dimensions of the pixels in the acquired 2-D images. If the 2-D acquired images do not intersect a particular region (voxel or group of voxels) in the 3-D image, then their values (color or grey scale) must be calculated by interpolation. This can be accomplished using the values from neighboring voxels in the same acquired image and/or neighboring 2-D images.

Since the arrangement of the acquired 2-D images in both rotational and tilting mechanical 3-D scanning approaches can be described in cylindrical coordinates, the reconstruction process involves the transformation of the acquired data from cylindrical coordinates to Cartesian. For example, if the ultrasound transducer axis, and hence the rotation axis, is designated as the z-axis, the reconstruction of every x-y plane is the same and consists of mapping the source image pixel values $P(r, \theta, z)$ in cylindrical coordinates to the reconstructed image pixel values $P'(x, y, z)$ in Cartesian coordinates. Thus, for each value of z, the polar coordinates (r, θ) of each Cartesian grid point are calculated using Equation 14.1.

$$\begin{aligned} r &= (x^2 + y^2) \\ \theta &= \arctan(x/y) \end{aligned} \tag{14.1}$$

The image values of $P'(x, y)$ are obtained by bilinear interpolation from the values $P(r, \theta)$ of its nearest neighbors in the appropriate source image. Therefore, the pixel value of each reconstructed 3-D image point is interpolated from four source image pixel values from the same x-y plane. By using a precomputed look-up table of interpolation weights, which is applied repeatedly for each successive z value, the 3-D image can be rapidly reconstructed from the set of 2-D images.[38]

Since the voxel-based reconstruction approach preserves all the original information, the original 2-D images can always be generated as well as new 2-D views not found in the original set of acquired 2-D images. However, if the scanning process does not sample the volume properly, leaving gaps between acquired images with a distance greater than half the elevational resolution, the interpolation process will fill the gap with information that does not represent the true anatomy. In this situation, erroneous information is introduced, and the resolution in the reconstructed 3-D image will be degraded. To sample the volume properly and to avoid gaps, large data files will be generated, requiring efficient 3-D viewing software. Typical data files can range from 16 to 96 MB, depending on the application.[17,71]

Because the voxel-based reconstruction approach maintains all the image information, the 3-D image can be reviewed repeatedly using a variety of rendering techniques. For example, the operator can scan through the data (i.e., view the complete 3-D image) and then choose the rendering technique that best displays the desired features. The operator may then use segmentation and classification algorithms to measure volumes and segment boundaries or to perform various volume-based rendering operations. If the process has not achieved the desired results, then the operator can return to the original 3-D image and attempt a different procedure.

14.5 Sources of Distortion in 3-D Ultrasound Imaging

The 3-D ultrasound imaging systems making use of multiple 2-D images require accurate knowledge of the relative positions and orientations of the individual 2-D images. Any errors in the mechanical mechanism or tracking sensors will lead to 3-D image distortion and measurement errors. In Sections 14.5.1 to 14.5.3, we discuss some of these sources of distortions.

14.5.1 Mechanical 3-D Scanning: Linear

The acquired 2-D images are constrained to be parallel in the mechanical linear scanning approach. The transducer is oriented along the x-axis and aimed in the y-direction. For Doppler 3-D imaging, the transducer is tilted by an angle θ about the x-axis and swiveled by an angle ϕ about the y-axis. The transducer is then scanned in the z-direction in steps of size d to acquire the set of 2-D images necessary for the 3-D reconstruction. To avoid 3-D image distortions, three geometrical parameters must be known accurately: the distance d and the tilt angle θ between the acquired 2-D images and the scanning direction, and the swivel angle ϕ. Any error in these parameters will result in image distortions and, hence, error in volume measurements.

In their paper on geometrical distortions in linearly scanned 3-D ultrasound images, Cardinal et al. showed that the error V in an arbitrary volume V due to errors d in d, θ in θ, and ϕ in ϕ:

$$V/V = \mathbf{E} \cdot \mathbf{U} \tag{14.2}$$

where **E** is the error vector due to errors in the three parameters and **U** is the unit vector normal to the acquired 2-D images:

$$\mathbf{E} = (-\Delta\phi/\cos\phi\cos\theta,\ \Delta\theta/\cos\theta,\ \Delta d/d\cos\phi\cos\theta)^{T} \tag{14.3}$$

$$\mathbf{U} = (\sin\phi\cos\theta,\ -\sin\theta,\ \cos\phi\ \cos\theta)^{T} \tag{14.4}$$

The calculated errors in volume measurements for various errors in the three parameters are tabulated in Table 14.1. This table shows that the percentage error in volume measurements will be less than 5% if the systematic error in the distance between acquired images, d, is less than 0.01 mm; the error in the tilt angle is less than 4°, and the swivel angle is less than 8°.

TABLE 14.1 Relative Errors in Volume Measurements of Cubical Objects in a 3-D Image

$\Delta\phi$ (°)	$\Delta\theta$ (°)	Δd (mm)	$\Delta V/V$ (%)
8	0	0	+0.98
4	0	0	+0.24
0	4	0	−3.65
0	2	0	−1.92
0	0	0.02	+8.70
0	0	0.01	+4.17

The 3-D image was obtained with the linear scanning approach as a function of the errors in the swivel angle ($\Delta\phi$), the tilt angle, and the interslice distance ($\Delta\theta$) for nominal parameters $\theta = 0°$, $\theta = 30°$, and $d = 0.25$ mm.

14.5.2 Mechanical 3-D Scanning: Tilting

An analysis of the linear, area, and volume measurement errors due to errors in calibration of tilting 3-D scanning mechanisms (Figure 14.4) has been reported by Tong et al.[72] In the mechanical tilting 3-D scanning approach, the transducer is tilted through a total scan angle θ about its face by mechanically rotating the probe about its axis, while N 2-D images are digitized at equally spaced angular intervals, ($\theta = N\Delta\theta$). If the region digitized in each 2-D image is of size $X \times Y$ with area $A = XY$, the reconstructed 3-D image will be in an annular sector of a cylinder with an average radius of R, subtending an angle θ about the rotation axis. The inner radius of the sector will be R_0, as shown in Figure 14.12. Thus, reconstruction of the 3-D image without distortions requires that the two geometrical parameters R_0 and θ must be known accurately.

To estimate the distortion in the reconstructed 3-D image, we consider two possible errors in the geometrical parameters. If the tilting mechanism's actual geometrical parameters are R_0 and θ, but the calibration procedure measured these parameters erroneously as $\Delta R_0 + R$ and $\theta + \Delta\theta$, then the average radius of the annular sector would be $\Delta R + R$, with an area $A + \Delta A$. Using these parameters, Tong et al.[72] showed that the relative error $\Delta A/A$ is approximately

$$\Delta A/A = (1 + \Delta R/R)\,(1 + \Delta\theta/\theta) - 1 \tag{14.5}$$

Using the error in the reconstructed area ΔA, the error in any volume measurement can be estimated. Summing cross-sectional organ areas in multiple parallel image slices and then multiplying the sum measurement by the interslice separation give the measurement of the volume of the organ. Since the interslice separation is known accurately, the relative volume error $\Delta V/V$ in an arbitrary volume V will be the same as the relative area error $\Delta A/A$ in an arbitrary area A, i.e., approximately

$$\Delta V/V = \Delta A/A = (\Delta R/R) + (\Delta\theta/\theta) + (\Delta R/R)(\Delta\theta/\theta) \tag{14.6}$$

where R is the mean radius of the measured volume from the axis of rotation. Using typical values of the two geometrical parameters, we can estimate the required accuracy of their measurement.

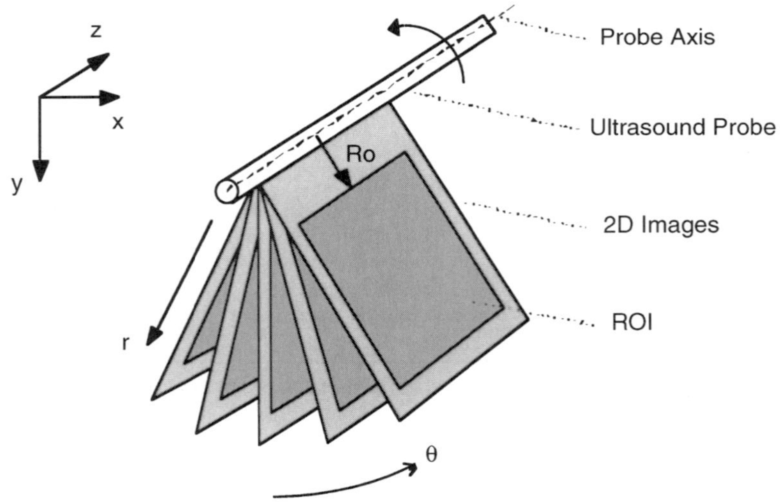

FIGURE 14.12 Schematic diagram illustrating the geometry used for 3-D scanning using the tilt scanning approach with a side-firing TR transducer. A region of interest (ROI) is digitized in each of a series of conventional 2-D images that are acquired while the side-firing transducer is rotated about its axis through an angle θ. The ROI is shown as the shaded area with an inner radius of R_0.

- If the location of the axis of rotation is known exactly, then the percentage error in a volume measurement will equal that in θ. To ensure a volume error of less than 5% for a scanning angle of $\theta = 100°$ with a total number of acquired 2-D images $N = 100$, the accumulated error in angle θ must be less than 5. Since the angular errors between individual 2-D images add, the angular interval between images must be known accurately to 0.05.
- Similarly, if the scan angle, θ, is known exactly, the percentage error in volume will equal that in R. Then to ensure a volume error of less than 5% for $R = 10$ mm requires that ΔR be less than 0.5 mm.

Tong et al.[72] provided a detailed analysis of the shape, distance, area, and volume distortions that arise due to these errors in the mechanical tilting approach, as well as a calibration method for virtually eliminating them, and a method for calculating the mean radius R of a volume.

14.5.3 Sensed Free-Hand 3-D Scanning

In the free-hand 3-D scanning techniques, the movement of the ultrasound transducer is not constrained. Therefore, to avoid distortions and geometrical measurement errors, the position and orientation of each acquired 2-D image must be known accurately and precisely. Random errors in the position and orientation sensor will lead to local distortions appearing as shifts between adjacent images. To avoid these types of distortions, the position and orientation measurements must be obtained with good signal to noise, thereby minimizing random errors. Global distortions of the reconstructed 3-D image will lead to errors in measurement of distance, area, and volume. To avoid these global errors, accurate calibration of the position and orientation sensing devices is the most important requirement. Random errors are important in regard to image quality, but they tend to "average out" over the distances and volumes to be measured. However, systematic errors will not average out, leading to distortions and measurement errors.

A number of researchers have reported on techniques for calibrating free-hand 3-D scanning systems based on magnetic field tracking. Detmer et al.[50] described a technique which made use of a small bead and cross-wire calibration and demonstrated that the root-mean-square (RMS) error of the calibration at a tissue depth of 60 mm ranged from 2.1 to 3.5 mm. This error depended on the distance from the magnetic field transmitter to the small receiver that was mounted on the ultrasound transducer. Leotta et al.[54] reported on two calibration methods for the magnetic positioning sensing device and found an rms calibration error of about 2.5 mm. In a later paper, Leotta et al.[54] examined an upgraded version of the magnetic positioning device, and found an rms calibration error ranging from 0.5 to 2.5 mm, depending on tissue depth.

These reports demonstrated that free-hand systems using the magnetic field sensing devices have errors greater then those exhibited by mechanical scanners. Nonetheless, their size and ease of use make the free-hand 3-D scanning approach based on magnetic field tracking an attractive alternative to the bulkier mechanical scanning approach.

14.6 Viewing of 3-D Ultrasound Images

3-D ultrasound images can be viewed interactively using a number of 3-D visualization techniques. The quality of the acquired 2-D ultrasound images, the image acquisition characteristics of a 3-D ultrasound system, and the 3-D reconstruction method are crucial in determining the quality of the final 3-D image. However, the rendering technique chosen to display the 3-D image to the user also plays an important and, at times, dominant role in determining the information transmitted to the operator. There are many techniques and commercially available software modules for displaying 3-D images. These techniques are still being actively investigated by many investigators and commercial companies and are being reported on at international conferences. In the past few years, investigators and commercial companies have developed various display techniques for 3-D ultrasound imaging that can be divided into three broad classes: surface-based, multi-planar, and volume-based rendering. The optimal choice of the rendering technique is generally determined by the clinical application and is often under control of the user.

14.6.1 Surface-Based Rendering (SR)

The most common technique used to display 3-D images is based on visualization of surfaces of structures or organs. Since the surfaces and boundaries between organs and structures must be identified, a segmentation and classification step precedes rendering. In this first step, the operator or the algorithm analyzes the 3-D image and determines the structure to be rendered. Boundaries of anatomical structures can be identified by the operator using manual contouring,[73–75] or by algorithms using simple thresholding techniques or more complex statistical and geometric properties of parts of the 3-D image. Segmentation of boundaries in 2-D and 3-D medical images is still a very active area of research, and the methods for 3-D ultrasound imaging are still being investigated. Once the tissues or structures are classified and their boundaries are identified, they can be represented by a wire frame or mesh, and the surface texture mapped with an appropriate color and texture can be to represent the anatomical structure.

Surface rendering using wire frames was first used in 3-D ultrasound imaging because it is the simplest and does not require advanced computer workstations. This approach has been used for displaying the fetus,[42,76–78] various abdominal structures,[79–81] and endocardial and epicardial surfaces of the heart.[46,74,82–86]

To improve the display of the surfaces and understanding of the anatomy, the wire frames or other, more complex representations of the surfaces are texture mapped, shaded, and illuminated. Depth cues can be added in the display so that topography and 3-D geometry are more easily comprehended. Since understanding of a 3-D object from a single perspective is difficult, automatic rotation or user-controlled motion is used to allow the operator to view the anatomy from different perspectives.

14.6.2 Multi-Planar Rendering of 3-D Ultrasound Images (MPR)

This technique requires that either a 3-D voxel-based image be reconstructed first or planes be extracted at any arbitrarily orientation from the originally acquired images. Two approaches are typically used to view the 3-D images. In the first approach, computer user-interface tools are provided to the operator to allow selection of single or multiple planes, including oblique from the 3-D image. Often, three perpendicular planes are displayed on the screen simultaneously, with screen cues as to their relative orientation and intersection. This method presents familiar 2-D images to the operator and allows the operator to orient the planes optimally for the examination.[10,87–90] With appropriate interpolation of the original acquired 2-D images, these planes may appear similar to the images that would be obtained by conventional 2-D ultrasound imaging.

In the second approach, the 3-D image is presented as a polyhedron representing the boundaries of the reconstructed volume. Each face of the polyhedron is rendered with the appropriate 2-D ultrasound image for that plane.[91] User-interface tools are provided to allow the operator to rotate the polyhedron and obtain the desired orientation of the 3-D image. User-interface tools are also provided to allow the user to move the faces of the polyhedron in or out parallel to the original face or to move the face obliquely to the original. At each new location of the face, the appropriate ultrasound image is texture mapped in real time. In this way, the operator always has 3-D image-based cues relating the plane being manipulated to the rest of the anatomy.[11,23,26,38] Figure 14.13 shows examples of the use of this approach in displaying anatomy in 3-D.

14.6.3 Volume-Rendering (VR)

Both the surface and MPR techniques reduce the display of 3-D information to a display of 2-D information in the form of complex or planar surfaces. Since our visual senses are best suited for viewing and interpretation of surfaces, the user easily understands images produced by these two approaches. However, surface and planar rendering techniques present only a small part of the complete 3-D image at one time, and software tools are needed to allow the user to explore all the data.

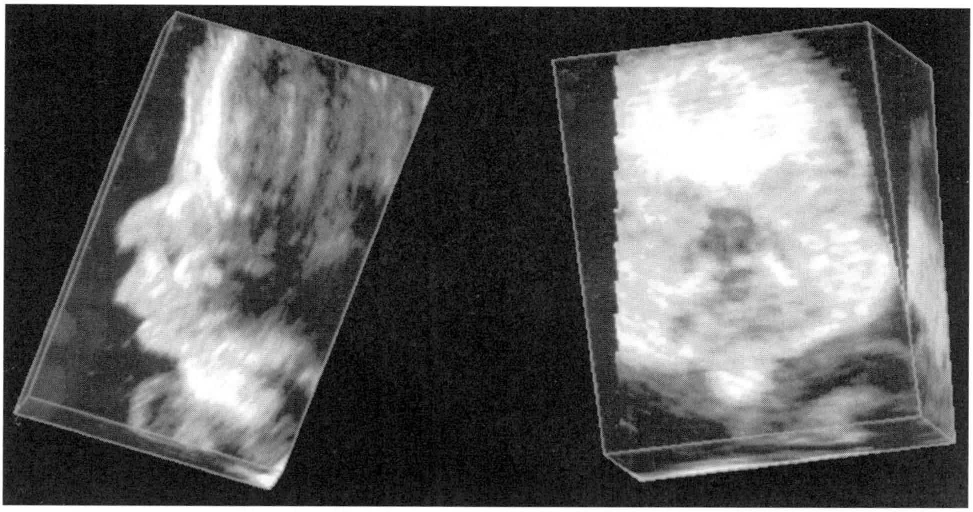

FIGURE 14.13 3-D image of a fetal face displayed using the multi-planar viewing approach. The 3-D image is presented as a polyhedron with the faces painted with the appropriate 2-D image. The polyhedron can be sliced and the faces tilted to reveal the appropriate 2-D image in relation to the other faces. Here, the 3-D image has been "sliced" and oriented to display a profile and en face view of the fetal face.

VR is an alternative technique, presenting the entire 3-D image to the viewer after it has been projected onto a 2-D plane. The most common approach used in 3-D ultrasound imaging is based on the ray-casting techniques,[92–94] which project a 2-D array of rays through the 3-D image (see Figure 14.14). Each ray intersects the 3-D image along a series of voxels, which are weighted and summed to achieve the desired rendering result. If desired structures in the 3-D image have been segmented and classified, the voxels can be weighted and colored appropriately to achieve translucent or opaque

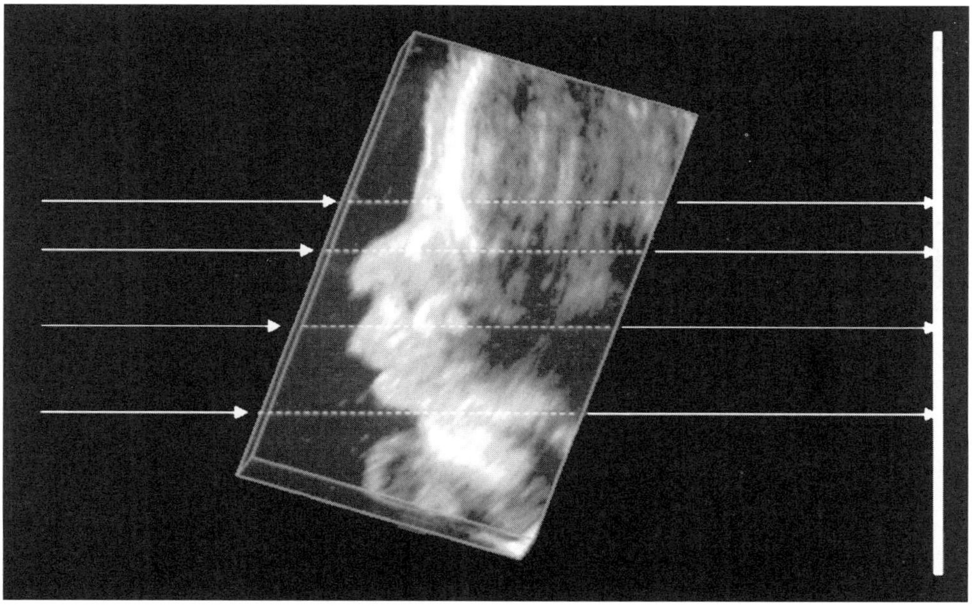

FIGURE 14.14 Image shows the ray-casting approach, in which an array of rays is cast through the 3-D image, projecting the 3-D image onto a plane producing a 2-D image. The rays can be made to interact with the 3-D image data in different ways to produce different types of renderings.

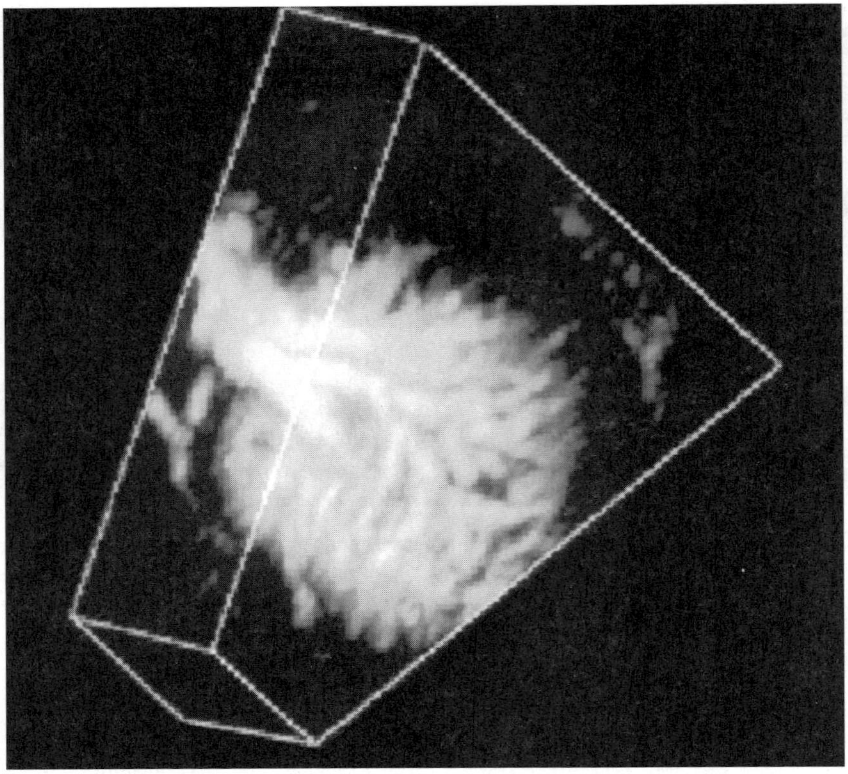

FIGURE 14.15 The 3-D image of the kidney has been rendered using an MIP algorithm with the ray-casting approach.

representation of the structures. Another common approach displays only the voxels with the maximum (or minimum) intensity along each ray to form a "maximum intensity projection" (MIP) image as shown in Figure 14.15.

A common approach used in 3-D ultrasound imaging is the translucency/opacity rendering approach, in which the voxels along each ray are weighted according to Equation 14.7.

$$V(i) = V(i-1)(1-a(i)) + c(i)a(i) \tag{14.7}$$

where $V(i)$ is the value of the ray exiting from the ith voxel, and $V(i-1)$ is the value of the ray exiting from the $(i-1)^{th}$ voxel and entering the ith voxel. The parameters "c" and "a" are chosen to control the specific desired rendering approach, where a controls the opacity and c is a modifying parameter chosen to control the luminance of the voxel that can be based on the local gradient. For example, the ith voxel is transparent, and the ray is transmitted through that voxel unchanged when $a(i) = 0$; if $a(i) = 1$, then the voxel is opaque or luminescent depending on the value of c. Typically, the values of a are summed progressively until the sum reaches 1 and the value V is displayed.

The volume rendering techniques preserve all the 3-D information, but projects it (after nonlinear processing) onto a 2-D plane for viewing. Like radiographic X-ray projection, depth information is lost by the projection of the 3-D information onto a 2-D image. Depth cues can be added in the rendering process (e.g., shading) or the image can be rocked back and forth in real time to achieve motion parallax. Nontheless, this rendering approach results in images that are difficult to interpret and is therefore best suited for simple anatomical structures in which clutter has been removed or is not present. Investigators have demonstrated success, particularly in displaying fetal[10,15,56,58,95,96] and vascular anatomy.[19,97] An example of the volume rendering approach is shown in Figure 14.16.

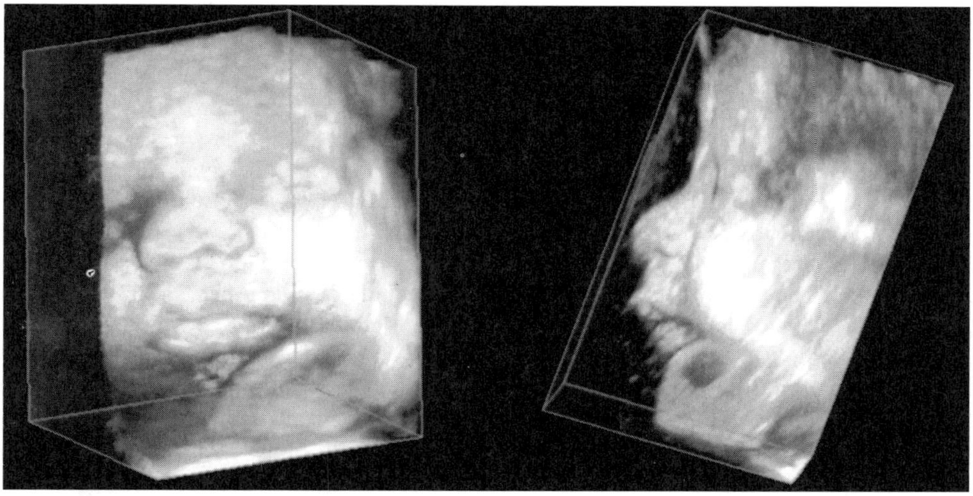

FIGURE 14.16 The 3-D image of the fetal face has been rendered using a translucency rendering algorithm with the ray-casting approach. In this image, the amniotic fluid has been made transparent, and tissues have been made transparent or opaque depending on the voxel intensity as described in Equation 14.7.

14.7 3-D Ultrasound System Performance

14.7.1 Introduction

3-D ultrasound imaging is ideally suited for providing quantitative measurements such as volume, area, and distance in complex anatomical structures. While qualitative visual assessment is valuable, quantitative data provide a more accurate basis for decision making and comparisons against previous studies or reference data.

In most applications, measuring the organ or lesion volume is the goal rather than its diameter or perimeter. However, conventional ultrasound imaging provides only a 2-D image, from which the 3-D volume must be estimated. Nevertheless, techniques have been developed that provide useful estimates of organ volume from 2-D ultrasonography. Many studies have been published on the accuracy and reproducibility of these techniques for various applications, such as the estimate of gestational age, the effects of growth retardation, or prostate volume.

Although techniques based on 2-D ultrasonography have been useful, the estimation of the volume suffers from many deficiencies that 3-D ultrasound addresses. The main deficiency relates to the fact that direct measurement of volumes is not possible with conventional 2-D methods since it is not possible to relate directly measurements in adjacent tomographic slices. Since 3-D ultrasound imaging provides a complete view of the structure, rather than just a few selected planes, it should provide more accurate estimates of volume, especially for complex structures. In addition, 3-D ultrasound also provides volume measurements with reduced variability, since the whole structure is available rather than two or three user-selected planes.

14.7.2 Distance Measurement

Since conventional ultrasound images are tomographic in nature, measurement of length within the 2-D image has always been possible. The accuracy and precision of the measurement depend on the frequency of the transducer, the field of view, and the limiting resolution of the ultrasound machine. Distance measurement requires that the operator select two points. With 3-D ultrasonography, these points need not lie on the same original acquired image. They can instead be located anywhere in the volume, since all the acquired images have been geometrically located relative to each other by the 3-D reconstruction process. Thus, for the distance measurement to be accurate, the reconstruction process must not distort the geometry.

The accuracy of distance measurements in 3-D ultrasonography has been evaluated by imaging a test phantom made of a grid of wires with known separation.[98] The phantom was composed of four layers of 0.25 mm diameter surgical wires, with eight parallel wires per layer. Each wire layer was separated from its neighbor by 10.00 mm, and each wire was also separated from its neighbors in the layer by 10.00 mm. The wire phantom was immersed in a bath composed of a 7% glycerol solution (1540 m/s speed of sound) and then imaged with the 3-D ultrasound system.

These accuracy measurements were made with a 3-D transrectal ultrasound (TRUS) system developed in our laboratory. The system made use of a mechanical tilting scanning mechanism which rotated a TRUS transducer with a side-firing linear array (Figure 14.5) around its long axis. The 3-D images were reconstructed using 100 2-D images which were collected over 60°. The separations between wire layers were then measured in the 3-D image for two orientations of the phantom test (i.e., direction of wires along the X- and Z-axis). The results of the measurement of separations between adjacent wires are tabulated in Table 14.2. The results of this study showed that with proper calibration of the scanning geometry, 3-D ultrasound images system can provide distance measurements with an error of about 1.0%.

TABLE 14.2 Measurements of Distances Between Adjacent Wires

Direction of Wire Axis	Direction of Measurement	Measured Distance Mean ± SD (mm)
Z	x	10.11 ± 0.11
Z	y	10.10 ± 0.06
X	y	10.12 ± 0.17
X	z	10.07 ± 0.29

Note: Measurements were obtained with a wire phantom made of four layers of wires 10.00 mm apart and four wires in each layer also 10.00 mm apart.

14.7.3 Cross-Sectional Area Measurement Using 3-D Ultrasound

Because the 3-D image can be sliced in any orientation using the MPR approach, any desired cross-section of the organ may be obtained. For measurement of the cross-sectional area, the desired area can then be outlined either manually, using planimetric techniques,[98,99] or automatically, using segmentation algorithms. The area of the region is measured by counting the pixels in the outlined region and multiplying by the pixel area.

14.7.4 Volume Measurements: Theoretical Derivation of Variability

Since the measurement of volume using a 3-D image can be obtained by the planimetric method, the variability of the measurement can be estimated by summing the variability of measurements of individual cross-sectional areas of the object. Thus, the volume of a solid of length "*a*" is measured by the summation of cross-sectional areas in each slice k, where $0 < k \leq n$, multiplied by the interslice distance Δx. For simplicity, we assume that the cross-sectional region in each slice is circular with a radius r_k. The volume of the solid in each slice can be given by $V_k = \pi r_k^2 .\Delta x$. Thus, the volume of the entire solid is computed as[39]

$$V = \pi\Delta x \sum_{k=1}^{n} r_k^2 \Rightarrow \pi \int_0^a r_x^2 dx \tag{14.8}$$

The variance in measurement of the volume of each slice is

$$\sigma_{V_k}^2 = (2\pi r_k \Delta x)^2 \sigma_r^2 = (4\pi\Delta x)\pi\Delta x . r_k^2 . \sigma_r^2 \tag{14.9}$$

where σ_r^2 is the variance in measurement of the radius in each cross-sectional slice. The variance in measurement of the volume of the entire solid is the sum of variances in measurement of the volume in each slice.

$$\sigma_V^2 = 4\pi\Delta x.\left(\pi\Delta x\sum_{k=1}^{n} r_k^2\right).\sigma_r^2 = 4\pi\frac{a}{n}.V.\sigma_r^2 \tag{14.10}$$

Dividing σ_V by the volume of the solid, we obtain the fractional variability in measurement of the volume of the general solid, as given by

$$\frac{\sigma_V}{V} = \sqrt{\frac{4\pi a\sigma_r^2}{nV}} \tag{14.11}$$

Thus, the fractional variability in volume measurement using the planimetric method can be reduced with increasing the number of 2-D images used to measure the cross-sectional area. In addition, the fractional variability depends on the standard deviation of measuring the cross-sectional area. This value depends on the resolution of the 2-D images, since identification of the margins of the object will be made difficult if the edges are not clearly visible.

To estimate the variability of an object with a volume of 64 cm^3 and a length of 4 cm, we consider that it has been cut into eight slices, 0.5 mm apart. We consider that the object has been outlined in each slice with a standard deviation in estimating the edge of 1 mm. Using these values and Equation 14.11, the fractional variability of the volume estimate will be 3.1%.

14.7.5 Volume Measurements: Experimental Measurements with Balloons

The volume of an organ in a 3-D image may be measured by a manual technique called manual planimetry. In this technique, the multi-planar viewing is used to slice the 3-D image into a series of uniformly spaced parallel 2-D images. In each 2-D image, the cross-sectional area of the organ is manually outlined on the computer screen using a mouse, trackball, or other pointing device. The areas on these slices are then summed and multiplied by the interslice distance to obtain an estimate of the organ volume.

To evaluate the accuracy of volume measurements using 3-D ultrasound, our 3-D TRUS system was used to image five balloons filled with different known volumes of 7% glycerol solution. The volumes of the balloons were than measured using the planimetry method from the 3-D image, and the results were compared to the true volumes in Table 14.3.[98]

TABLE 14.3 Volume Measurement Accuracy Obtained by Imaging Balloons Filled with Different Volumes of Water/Glycerol Solution

True Volume (cm^3)	Measured Volume Mean ± SD (cm^3)	Error (%)
23.14	22.81 ± 0.33	1.4
35.79	35.49 ± 0.72	0.8
41.69	41.27 ± 0.65	0.1
49.66	49.84 ± 0.62	0.4
65.84	66.31 ± 1.29	0.7

The volume measurements shown in Table 14.3 have an rms error of 0.9% and an rms precision of 1.7%. Also, a least-squares regression of measured vs. true volume resulted in a best-fit line with a slope of 1.0004 ± 0.0039 and a correlation coefficient of 0.99997. These results demonstrate that volume measurements using 3-D ultrasonography can be very accurate and precise when imaging structures that are easy to manually segment.

14.7.6 Volume Measurements: Prostates *In Vitro*

Volume measurement of water-filled balloons using 3-D ultrasound images produces idealized and presumably the most accurate measurements. To determine the accuracy of volume measurement in a more realistic situation, we compared the measurement of prostate volume using 3-D TRUS to their true volumes. The volumes of six fixed prostates, harvested from fresh cadavers, were measured by water displacement in a graduated cylinder.[99]

A 3-D mechanical tilting transducer assembly (Figure 14.5) was used to produce 3-D images of the prostates. The prostate volumes were then measured by manual planimetry using a similar technique to the balloon volume measurements. Each prostate was sliced into 20 to 30 transaxial slices 2 to 5 mm apart, and the boundary of the prostate in each slice was outlined. The volume was obtained by summing the area thickness products of each slice. A linear regression of measured vs. true volume yielded a slope of 1.006 ± 0.007. The accuracy (rms deviation from the line of identity) of the measurements was 2.6%, and the precision (rms deviation from the best-fit line) was 2.5%.

These *in vitro* studies using balloons and cadaveric specimens clearly show that, by using a properly calibrated 3-D ultrasound imaging system under ideal laboratory conditions, the accuracy of 3-D volume estimates can be very good, with errors of less than 5%.

14.7.7 Volume Measurements: Choice of Interslice Distance

Volume estimation requires that the 3-D image of the organ be sliced into individual parallel slices and the boundary of the anatomical structure be outlined in every slice. The accuracy of the volume measurement will depend on the number of slices in which the area is measured and the complexity of the volume. As the interslice distance is reduced, the accuracy will increase, but the length of time needed for manual planimetry will also increase. While a large interslice distance will be less time-consuming, the volume estimate may be in error if the structure is complex.

Irregular or complex structures require the use of smaller interslice distances. However, an interslice distance less than the spacing of the acquired 2-D images used to reconstruct the 3-D image or less than the elevational resolution of the transducer will not result in increased accuracy. Thus, it is difficult to generalize the choice of interslice distance due to varying sizes and shapes of different organs and the use of different transducers.

To examine the effect of the choice of the interslice distance in a specific application (3-D TRUS imaging of the prostate), the volume of one cadaveric prostate was repeatedly measured in a 3-D TRUS image by the planimetry method in the transaxial, sagittal, and coronal planes. Interplane distance ranged from 1 to 15 mm.[99] The results showed that the measured prostate volume was constant (within the precision of the volume estimation) up to an interslice distance of 8 mm. An interslice distance greater than 8 mm resulted in an underestimation of the prostate volume.

14.7.8 Intra- and Inter-Observer Variability in Measuring Prostate Volume

Currently, prostate volume estimation with 2-D ultrasound is done by measuring the height (H), width (W), and length (L) of the prostate from two selected orthogonal views and by estimating the prostate volume (V) as that of the corresponding ellipsoid, i.e., $V = (\pi/6)HWL$. Since the prostate is not ellipsoidal, the choice of the appropriate orthogonal views and which three chords to use in measuring H, W, and L in the selected views is not clear-cut. It is largely dependent on observer preference, leading to high interobserver variability in volume estimation. Even for a single observer with a single set of images, the choice is still arbitrary, leading to high intraobserver variability. However, with 3-D ultrasound, the entire prostate is available for volume measurement. Thus, any arbitrary-shaped volume can be measured, and the only observer variability is in deciding the boundary of the prostate.

Eight observers participated in a study to compare the variability of prostate volume measurements made with 2-D and 3-D techniques. Four observers were experienced radiologists and four were technicians or graduate students. The prostates of 15 patients were scanned *in vivo*, reconstructed, and then measured. The volume of each prostate was measured four times by each observer.

- Twice via the HWL method, using transverse and sagittal 2-D image cross-sections obtained from the 3-D image to measure H, W, and L
- Twice via the 3-D TRUS method, using manual planimetry with an interslice spacing of 4 mm

The analysis of variance (ANOVA) was used to assess the intra- and interobserver variability of prostate volume measurements made via the HWL and 3-D ultrasound methods.[100] In particular, the intra- and interobserver standard errors of measurement SEM_{intra} and SEM_{inter} were used to characterize the variability of volume measurements. These were also expressed in terms of the minimum volume changes V_{intra} and V_{inter} that can be detected with a given confidence level in successive measurements.

The ANOVA results shown in Table 14.4 demonstrate that the 3-D prostate volume method provides results that are less variable. Using the 3-D method, the intraobserver SEM is reduced by about a factor of 3, and the interobserver SEM is reduced by about a factor of 2.[100]

TABLE 14.4 Results of the Intra- and Interobserver Study Comparing Prostate Volume Measurements Using 3-D Ultrasound and the Conventional HWL Method

	SEM/Volume (%)		ΔV/Volume (%)	
	3-D	HWL	3-D	HWL
Intra	5.1	15.5	14	43
Inter	11.4	21.9	32	61

Note: Values of the minimum relative prostate volume change that can be detected between successive prostate volume measurements at the 95% level on confidence. These values are given by $\Delta V = 2.77$ SEM.

The values for the SEMs are often interpreted in terms of the minimum volume change (ΔV) that can be confidently detected between successive measurements at the 95% level of confidence, which is given by $\Delta V = 2.77$ SEM. The values for minimum detectable prostate volume change are also tabulated in Table 14.4. The intraobserver variability values show that if one observer makes repeated measurements of prostate volume, then the volume must change by 43% before it can be confidently detected by the HWL method, but by only 14% before it can be confidently detected by the 3-D ultrasound method. The interobserver variability values show that if two different observers measure the same prostate, then the prostate must change by 61% before it can be confidently detected by the HWL method, but by only 32% by the 3-D method.

14.8 Use of 3-D Ultrasound in Brachytherapy

14.8.1 Identification of Seeds

3-D ultrasound imaging has an important role to play in ultrasound-guided brachytherapy, particularly in the preimplantation phase of the procedure when the dose plan is determined. Its role could be greatly expanded to the postimplantation verification phase if brachytherapy seeds could be accurately detected and their locations determined. Consequently, we conducted a study to determine the variability and accuracy of localizing brachytherapy seeds in 3-D ultrasound images using a tissue-mimicking phantom.

The ultrasound phantom was made of 3% by weight agar gel to which was added to 7% solution of glycerol containing 50 μm cellulose particles. To simulate the rectum and accommodate the TRUS transducer, a 2.5 cm diameter acrylic rod was suspended in the molten mixture, which was allowed to solidify. When the rod was removed, the resulting channel simulated the rectum.

Twenty gold brachytherapy seeds, approximately 1 mm in diameter and 3 mm long, were inserted into the phantom test in a fan pattern at varying depths. After all the seeds were inserted, 3-D ultrasound images of the phantom test were acquired using the 3-D TRUS system with a tilting scanning approach.

Table 14.5. Values of the Standard Error of Measurements for Localizing Brachytherapy Seeds in 3-D Ultrasound Images of an Agar Phantom

Plane	SEM_{inter} (mm)	SEM_{intra} (mm)	Δ_{inter} (mm)	Δ_{intra} (mm)
Coronal	0.49	0.43	1.35	1.18
Sagittal	0.33	0.38	0.91	0.83
Transaxial	0.39	0.34	1.08	0.95

Note: The values plotted are the mean of the SEMs for the three coordinates (*x*, *y*, *z*) determined by "cutting" into the 3-D image from the 3 directions (coronal, sagittal, transaxial). Also tabulated are the minimum detectable changes in the coordinate.

Seven observers measured the Cartesian coordinates of all the seeds using the 3-D viewing software. Each observer "cut" into the 3-D image to reveal sagittal sections of the prostate and measured the (x,y,z) coordinates of the seeds in that revealed view. This procedure was repeated twice for each observer. An ANOVA was performed to determine the standard error of measurement and the minimum detectable change in the coordinates. The results are shown in Table 14.5.

These results show that under ideal conditions, such as those found when imaging agar phantom tests, the location of the seeds can be detected at the 95% confidence level to better than 1 mm when localizing the seeds in the sagittal plane.

Two factors affect the measurement of the seed localization. The first is the size of the seed, resulting in uncertainty in identifying the center. The second is the loss of resolution due to the poor elevation resolution of the ultrasound transducer. This affects the appearance of the seeds in the rotational direction causing the seeds to smear with increasing depth.

14.9 Trends and Future Developments

In the past few years, computer technology and visualization techniques became sufficiently advanced to allow the migration of 3-D ultrasound imaging from the research laboratory to the clinical examination room. However, a number of advances in 3-D ultrasound must still be made for it to become a widespread routine clinical tool.

14.9.1 Computer-Aided Exams

For 3-D imaging to be accepted as a routine tool, diagnosticians must accept the viewing of images on computer workstations rather than the ultrasound machine. This change in clinical practice requires education and training, leading to a cultural change in medical diagnosis. To fully exploit the 3-D image and the potential expanded utility of this type of viewing, the diagnostician must accept the paradigm shift of performing the exam with the aid of a computer.

14.9.2 User-Interface Improvements

To permit optimal use of the 3-D ultrasound system and to avoid interference with efficient patient management, diagnosticians and sonographers must be able to control the viewing of the 3-D image in an intuitive manner without a difficult learning curve. Since viewing of 3-D images is new to most diagnosticians, the user-interface design is crucial for clinical acceptance. Thus, the 3-D system's user interface must not intimidate the user with an extensive selection of icons and multi-level menus.

14.9.3 Speed of Operation

To enhance patient management and to minimize waiting time for the 3-D image, it must be acquired rapidly, reconstructed immediately, and controlled interactively. Ultimately, the acquisition of the 3-D image should be sufficiently fast to provide at least about five volumes per second. The resulting 3-D images must be presented to the viewer immediately after they have been acquired. Interactive viewing of the 3-D image can already be achieved by using computationally intensive VR techniques. All advances in increase in speed should be achieved without incurring additional costs in hardware.

14.9.4 Applications in Image-Guided Therapy and Surgery

It is now recognized that 3-D ultrasound imaging has an important role to play in ultrasound-guided therapy such as prostate cryosurgery and brachytherapy. Its role could be greatly expanded if a number of developments were achieved in coupling the 3-D image acquisition and display to therapy planning and monitoring. For example, advances in 3-D organ segmentation are required to allow efficient delineation of the organ from the 3-D ultrasound image for use in the therapy planning process. In most therapy applications, a needle or optical fiber is introduced into the tissue to deliver the therapy (e.g., cryosurgery, brachytherapy, PDT). Techniques must be developed that allow for real-time guidance of the needle/fiber into the tissue and techniques to validate their positions in the organ. Techniques must be developed to allow real-time monitoring of the therapy in situations in which tissue changes can be observed with ultrasound. These changes must be monitored in 3-D and with sufficient geometric accuracy to provide warning to the physician of possible errors in the procedure. In image-guided prostate brachytherapy, the radioactive seeds should be accurately detected and their locations determined from the 3-D ultrasound image. These advances will all contribute to the development of intraoperative 3-D ultrasound-guided therapy procedures that would reduce the requirements for multiple patient visits and multiple imaging modalities.

References

1. E.K. Fishman, D. Magid, D.R. Ney, E.L. Chaney, S.M. Pizer, J.G. Rosenman, et al., Three-dimensional imaging, *Radiology*, 181, 321–337, 1991.
2. R.A. Robb and C. Barillot, Interactive display and analysis of 3-D medical images, *IEEE Trans. Med. Imaging*, 8, 217–226, 1989.
3. M.W. Vannier, J.L. Marsh, and J.O. Warren, Three-dimensional CT reconstruction for craniofacial surgical planning and evaluation, *Radiology*, 150, 179–184, 1984.
4. T.R. Nelson, D.B. Downey, D.H. Pretorius, and A. Fenster, *Three-Dimensional Ultrasound*, Lippincott/Williams & Wilkins, Philadelphia, 1999.
5. J.F. Brinkley, S.K. Muramatsu, W.D. McCallum, and R.L. Popp, *In vitro* evaluation of an ultrasonic three-dimensional imaging and volume system, *Utrason. Imaging*, 4, 126–139, 1982.
6. A. Fenster and D.B. Downey, 3-D ultrasound imaging: a review, *IEEE Eng. Med. Biol.*, 15, 41–51, 1996.
7. A. Ghosh, N.C. Nanda, and G. Maurer, Three-dimensional reconstruction of echocardiographic images using the rotation method, *Ultrasound Med. Biol.*, 8, 655–661, 1982.
8. J.F. Greenleaf, M. Belohlavek, T.C. Gerber, D.A. Foley, and J.B. Seward, Multidimensional visualization in echocardiography: an introduction, *Mayo Clin. Proc.*, 68, 213–219, 1993.
9. D.L. King, A.S. Gopal, P.M. Sapin, K.M. Schroder, and A.N. Demaria, Three-dimensional echocardiography, *Am. J. Card Imaging*, 3, 209–220, 1993.
10. T.R. Nelson and D.H. Pretorius, Three-dimensional ultrasound of fetal surface features, *Ultrasound ObstetGynecol.*, 2, 166–174, 1992.
11. R.N. Rankin, A. Fenster, D.B. Downey, P.L. Munk, M.F. Levin, and A.D. Vellet, Three-dimensional sonographic reconstruction: techniques and diagnostic applications, *Am. J. Radiol.*, 161, 695–702, 1993.
12. M. Belohlavek, D.A. Foley, T.C. Gerber, T.M. Kinter, J.F. Greenleaf, and J.B. Seward, Three- and four-dimensional cardiovascular ultrasound imaging: a new era for echocardiography, *Mayo Clin. Proc.*, 68, 221–240, 1993.

13. A. Fenster, 3-D sonography — technical aspects, in *Echoenhancers and Transcranial Color Duplex Sonography*, U. Bogdahn, G. Becker, and F. Schlachetzki, Eds., Blackwell Science, Oxford, 1998, pp. 121–140.
14. K.A. Spaulding, M.E. Kissner, E.K. Kim, D.H. Pretorius, S.C. Rose, K. Garroosi, and T.R. Nelson, Three-dimensional gray scale ultrasonographic imaging of the celiac axis: preliminary report, *J. Ultrasound Med.*, 17, 239–248, 1998.
15. T.R. Nelson and T.T. Elvins, Visualization of 3D ultrasound data, *IEEE Comp. Graphics Applic.*, November, 50–57, 1993.
16. E.O. Ofili and N.C. Nanda, Three-dimensional and four-dimensional echocardiography, *Ultrasound Med. Biol.*, 20, 669–675, 1994.
17. D.B. Downey, J.L. Chin, and A. Fenster, Three-dimensional U.S.-guided cryosurgery, *Radiology*, 197(P), 539, 1995.
18. D.H. Blankenhorn, H.P. Chin, S. Strikwerda, J. Bamberger, and J.D. Hestenes, Common carotid artery contours reconstructed in three dimensions from parallel ultrasonic images, *Radiology*, 148, 533–537, 1983.
19. D.B. Downey and A. Fenster, Vascular imaging with a three-dimensional power Doppler system, *Am. J. Radiol.*, 165, 665–668, 1995.
20. D.B. Downey and A. Fenster, Three-dimensional power Doppler detection of prostatic cancer, *Am. J. Radiol.*, 165, 741, 1995.
21. J.F. Greenleaf, Three-dimensional imaging in ultrasound, *J. Med. Syst.*, 6, 579–589, 1982.
22. R.H. Silverman, M.J. Rondeau, F.L. Lizzi, and D.J. Coleman, Three-dimensional high-frequency ultrasonic parameter imaging of anterior segment pathology, *Ophthalmology*, 102, 837–843, 1995.
23. A. Fenster, S. Tong, S. Sherebrin, D.B. Downey, and R.N. Rankin, Three-dimensional ultrasound imaging, *SPIE Phys. Med. Imaging*, 2432, 176–184, 1995.
24. Z. Guo and A. Fenster, Three-dimensional power Doppler imaging: a phantom study to quantify vessel stenosis, *Ultrasound Med. Biol.*, 22, 1059–1069, 1996.
25. P.A. Picot, D.W. Rickey, R. Mitchell, R.N. Rankin, and A. Fenster, Three-dimensional colour Doppler imaging of the carotid artery, *SPIE Proc. Image Capture Formatting Display*, 1444, 206–213, 1991.
26. P.A. Picot, D.W. Rickey, R. Mitchell, R.N. Rankin, and A. Fenster, Three-dimensional colour Doppler imaging, *Ultrasound Med. Biol.*, 19, 95–104, 1993.
27. D.H. Pretorius, T.R. Nelson, and J.S. Jaffe, 3-Dimensional sonographic analysis based on color flow Doppler and gray scale image data: a preliminary report, *J. Ultrasound Med.*, 11, 225–232, 1992.
28. J.C. Bamber, R.J. Eckersley, et al., Data processing for 3-D ultrasound visualization of tumour anatomy and blood flow, *SPIE*, 1808, 651–663, 1992.
29. P.L. Carson, X. Li, J. Pallister, A. Moskalik, J.M. Rubin, and J.B. Fowlkes, Approximate quantification of detected fractional blood volume and perfusion from 3-D color flow and Doppler power signal imaging, in *1993 Ultrasonics Symposium Proceedings*, IEEE, Piscataway, NJ, 1993, pp. 1023–1026.
30. D.L. King, D.L.J. King, and M.Y. Shao, Evaluation of in vitro measurement accuracy of a three-dimensional ultrasound scanner, *J. Ultrasound Med.*, 10, 77–82, 1991.
31. Z. Guo, M. Moreau, D.W. Rickey, P.A. Picot, and A. Fenster, Quantitative investigation of in vitro flow using three-dimensional colour Doppler ultrasound, *Ultrasound Med. Biol.*, 21, 807–816, 1995.
32. A. Delabays, N.G. Pandian, Q.L. Cao, L. Sugeng, et al., Transthoracic real-time three-dimensional echocardiography using a fan-like scanning approach for data acquisition: methods, strengths, problems, and initial clinical experience, *J. CV Ultrasound Allied Tech.*, 12, 49–59, 1995.
33. D.B. Downey, D.A. Nicolle, and A. Fenster, Three-dimensional orbital ultrasonography, *Can. J. Ophthalmol.*, 30, 395–398, 1995.
34. D.B. Downey, D.A. Nicolle, and A. Fenster, Three-dimensional ultrasound of the eye, *Admin. Radiol. J.*, 14, 46–50, 1995.

35. O.H. Gilja, N. Thune, K. Matre, T. Hausken, S. Odegaard, and A. Berstad, In vitro evaluation of three-dimensional ultrasonography in volume estimation of abdominal organs, *Ultrasound Med. Biol.*, 20, 157–165, 1994.
36. C. Sohn, W. Stolz, M. Kaufmann, and G. Bastert, Die dreidimensionale ultraschalldarstellung benigner und maligner brusttumoren — erste klinische erfahrungen, *Geburtshilfe Frauenheilkd.*, 52, 520–525, 1992.
37. T.L. Elliot, D.B. Downey, S. Tong, C.A. Mclean, and A. Fenster, Accuracy of prostate volume measurements in vitro using three-dimensional ultrasound, *Acad. Radiol.*, 3, 401–406, 1996.
38. S. Tong, D.B. Downey, H.N. Cardinal, and A. Fenster, A three-dimensional ultrasound prostate imaging system, *Ultrasound Med. Biol.*, 22, 735–746, 1996.
39. S.K. Nadkarni, D.R. Boughner, M. Drangova, and A. Fenster, Three-dimensional echocardiography: assessment of inter- and intra-operator variability and accuracy in the measurement of left ventricular cavity volume and myocardial mass, *Phys. Med. Biol.*, 45, 1255–1273, 2000.
40. J.L. Chin, D.B. Downey, M. Mulligan, and A. Fenster, Three-dimensional transrectal ultrasound guided cryoablation for localized prostate cancer in nonsurgical candidates: a feasability study and report of early results, *J. Urol.*, 159, 910–914, 1998.
41. G.M. Onik, D.B. Downey, and A. Fenster, Three-dimensional sonographically monitored cryosurgery in a prostate phantom, *J. Ultrasound Med.*, 15, 267–270, 1996.
42. J.F. Brinkley, W.D. McCallum, S.K. Muramatsu, and D.Y. Liu, Fetal weight estimation from lengths and volumes found by three-dimensional ultrasonic measurements, *J. Ultrasound Med.*, 3, 163–168, 1984.
43. D.L. King, D.L. King, Jr., and M.Y.C. Shao, Three-dimensional spatial registration and interactive display of position and orientation of real-time ultrasound images, *J. Ultrasound Med.*, 9, 525–532, 1990.
44. D.L. King, M.R. Harrison, D.L. King, Jr., et al., Ultrasound beam orientation during standard two-dimensional imaging: assessment by three-dimensional echocardiography, *J. Am. Soc. Echocardiogr.*, 5, 569–576, 1992.
45. R.A. Levine, M.D. Handschumacher, A.J. Sanfilippo, A.A. Hagege, P. Harrigan, J.E. Marshall, and A.E. Weyman, Three-dimensional echocardiographic reconstruction of the mitral valve, with implications for the diagnosis of mitral valve prolapse, *Circulation*, 80, 589–598, 1989.
46. W.E. Moritz, A.S. Pearlman, D.H. McCabe, D.H. Medema, M.E. Ainsworth, and M.S. Boles, An ultrasonic technique for imaging the ventricle in three dimensions and calculating its volume, *IEEE Trans. Biomed. Eng.*, BME-30, 482–491, 1983.
47. J.M. Rivera, S.C. Siu, M.D. Handschumacher, J.P. Lethor, and J.L. Guerrero, Three-dimensional reconstruction of ventricular septal defects: validation studies and in vivo feasibility, *J. Am. Coll. Cardiol.*, 23, 201–208, 1994.
48. J.L. Weiss, L.W. Eaton, et al., Accuracy of volume determination by two-dimensional echocardiography: defining requirements under controlled conditions in the ejecting canine left ventricle, *Circulation*, 67, 889–895, 1983.
49. F. Bonilla-Musoles, F. Raga, N.G. Osborne, and J. Blanes, Use of three-dimensional ultrasonography for the study of normal and pathologic morphology of the human embryo and fetus: preliminary report, *J. Ultrasound Med.*, 14, 757–765, 1995.
50. P.R. Detmer, G. Bashein, T. Hodges, K.W. Beach, et al., 3D ultrasonic image feature localization based on magnetic scanhead tracking: in vitro calibration and validation, *Ultrasound Med. Biol.*, 20, 923–936, 1994.
51. U. Ganapgthy and A. Kaufman, 3D acquisition and visualization of ultrasound data. Visualization in biomedical computing, *SPIE*, 1808, 535–545, 1992.
52. T.C. Hodges, P.R Detmer, D.H Burns, K.W. Beach, and D.E. Strandness, Jr., Ultrasonic three-dimensional reconstruction: in vitro and in vivo volume and area measurement, *Ultrasound Med. Biol.*, 20, 719–729, 1994.

53. S.W. Hughes, T.J.D. Arcy, D.J. Maxwell, et al., Volume estimation from multiplanar 2D ultrasound images using a remote electromagnetic position and orientation, *Ultrasound Med. Biol.*, 22, 561–572, 1996.
54. D.F. Leotta, P.R Detmer, and R.W. Martin: Performance of a miniature magnetic position sensor for three-dimensional ultrasound imaging, *Ultrasound Med. Biol.*, 23, 597–609, 1997.
55. O.H. Gilja, P.R. Detmer, J.M. Jong, D.F. Leotta, X.N. Li, K.W. Beach, R. Martin, and D.E. Strandness, Jr., Intragastric distribution and gastric emptying assessed by three-dimensional ultrasonography, *Gastroenterology*, 113, 38–49, 1997.
56. T.R. Nelson and D.H. Pretorius, Visualization of the fetal thoracic skeleton with three-dimensional sonography: a preliminary report, *Am. J. Radiol.*, 164, 1485–1488, 1995.
57. R. Ohbuchi, D. Chen, and H. Fuchs, Incremental volume reconstruction and rendering for 3D ultrasound imaging, *SPIE Visualization Biomed. Comput.*, 1808, 312–323, 1992.
58. D.H. Pretorius and T.R. Nelson, Prenatal visualization of cranial sutures and fontanelles with three-dimensional ultrasonography, *J. Ultrasound Med.*, 13, 871–876, 1994.
59. F.H. Raab, E.B. Blood, et al., Magnetic position and orientation tracking system, *IEEE Trans. Aerosp. Electron. Syst.*, AES-15, 709–717, 1979.
60. M. Riccabona, T.R. Nelson, D.H. Pretorius, and T.E. Davidson, Distance and volume measurement using three-dimensional ultrasonography, *J. Ultrasound Med.*, 14, 881–886, 1995.
61. B.H. Friemel, L.N. Bohs, K.R. Nightingale, and G.E. Trahey, Speckle decorrelation due to two-dimensional flow gradients, *IEEE Trans. Ultrason. Ferroelectr. Frequency Control*, 45, 317–327, 1998.
62. T.A. Tuthill, J.F. Krucker, J.B. Fowlkes, and P.L. Carson, Automated three-dimensional US frame positioning computed from elevational speckle decorrelation, *Radiology*, 209, 575–582, 1998.
63. D.P. Shattuck, M.D. Weinshenker, S.W. Smith, and O.T. von Ramm, Explososcan: a parallel processing technique for high speed ultrasound imaging with linear phased arrays, *J. Acoust. Soc. Am.*, 75, 1273–1282, 1984.
64. S.W. Smith, H.G. Pavy, Jr., and O.T. von Ramm, High-speed ultrasound volumetric imaging system. Part I. Transducer design and beam steering, *IEEE Trans. Ultrason. Ferroelec. Frequency Control*, 38, 100–108, 1991.
65. S.W. Smith, G.E. Trahey, and O.T. von Ramm, Two dimensional arrays for medical ultrasound, *Ultrason. Imaging*, 14, 213–233, 1992.
66. J.E. Snyder, J. Kisslo, and O.T. von Ramm, Real-time orthogonal mode scanning of the heart. I. System design, *J. Am. Coll. Cardiol.*, 7, 1279–1285, 1986.
67. D.H. Turnbull and F.S. Foster, Beam steering with pulsed two-dimensional transducer arrays, *IEEE Trans. Ultrason. Ferroelec. Frequency Control*, 38, 320–333, 1991.
68. O.T. von Ramm and S.W. Smith, Real time volumetric ultrasound imaging system, *SPIE*, 1231, 15–22, 1990.
69. O.T. von Ramm, S.W. Smith, and H.G. Pavy, Jr., High-speed ultrasound volumetric imaging system. II. Parallel processing and image display, *IEEE Trans. Ultrason. Ferroelec. Frequency Control*, 38, 109–115, 1991.
70. O.T. von Ramm, H.G. Pavy, Jr., S.W. Smith, and J. Kisslo, Real-time, three-dimensional echocardiography: the first human images, *Circulation*, 84(Suppl. 2), II-685, 1991.
71. J.L. Chin, D.B. Downey, G. Onik, and A. Fenster, Three-dimensional prostate ultrasound and its application to cryosurgery, *Tech. Urol.*, 2, 187–193, 1996.
72. S. Tong, H.N. Cardinal, D.B. Downey, and A. Fenster, Analysis of linear, area, and volume distortion in 3D ultrasound imaging, *Ultrasound Med. Biol.*, 24, 355–373, 1998.
73. M. Neveu, D. Faudot, and B. Derdouri, Recovery of 3D deformable models from echocardiographic images, *SPIE*, 2299, 367–376, 1994.
74. G. Coppini, R. Poli, and G. Valli, Recovery of the 3-D shape of the left ventricle from echocardiographic images, *IEEE Trans. Med. Imaging*, 14, 301–317, 1995.
75. S. Lobregt and M.A. Viergever: a discrete dynamic contour model, *IEEE Trans. Med. Imaging*, 14, 12–24, 1995.

76. C. Sohn, J. Grotepab, and W. Swobodnik, Moglichkeiten der 3 dimensionalen ultraschalldarstellung, *Ultraschall,* 10, 307–313, 1989.
77. C. Sohn and G. Rudofsky, Die dreidimensionale ultraschalldiagnostik — ein neues verfahren fur die klinische routine?, *Ultraschall Klin Prax*, 4, 219–224, 1989.
78. J.F. Brinkley, W.D. McCallum, S.K. Muramatsu, and D.Y. Liu, Fetal weight estimation from ultrasonic three-dimensional head and trunk reconstructions: evaluation in vitro, *Am. J. Obstet. Gynecol.*, 144, 715–721, 1982.
79. C. Sohn and J. Grotepass, Representation tridimensionnelle par ultrason, *Radiol. J. CEPUR*, 9, 249–253, 1989.
80. C. Sohn, J. Grotepab, W. Schneider, A. Funk, G. Sohn, P. Jensch, H. Fendel, W. Ameling, and H. Jung, Erste untersuchungen zur dreidimensionalen darstellung mittels ultraschall, *Z. Geburtshilfe Perinatd.*, 192, 241–248, 1988.
81. C. Sohn, A new diagnostic technique: three-dimensional ultrasound imaging, *Ultrason. Int. 89 Conf. Proc.,* 1148–1153, 1989.
82. D.G. Fine, D. Sapoznikov, M. Mosseri, and M.S. Gotsman, Three-dimensional echocardiographic reconstruction: qualitative and quantitative evaluation of ventricular function, *Comput. Methods Programs Biomed.*, 26, 33–44, 1988.
83. J.V. Nixon, S.I. Saffer, K. Lipscomb, and C.G. Blomqvist, Three-dimensional echoventriculography, *Am. Heart J.*, 106, 435–443, 1983.
84. R.W. Martin, G. Bashein, P.R. Detmer, and W.E. Moritz, Ventricular volume measurement from a multiplanar transesophageal ultrasonic imaging system: an *in vitro* study, *IEEE Trans. Biomed. Eng.*, 37, 442–449, 1990.
85. D.T. Linker, W.E. Moritz, and A.S. Pearlman, A new three-dimensional echocardiographic method of right ventricular volume measurement: in vitro validation, *J. Am. Coll. Cardiol.*, 8, 101–106, 1986.
86. H. Sawada, J. Fujii, K. Kato, M. Onoe, and Y. Kuno, Three dimensional reconstruction of the left ventricle from multiple cross sectional echocardiograms value for measuring left ventricular volume, *Br. Heart J.*, 50, 438–442, 1983.
87. N. Zosmer, D. Jurkovic, E. Jauniaux, K. Gruboeck, C. Lees, and S. Campbell, Selection and identification of standard cardiac views from three-dimensional volume scans of the fetal thorax, *J. Ultrasound Med.*, 15, 25–32, 1996.
88. E.O. Gerscovich, A. Greenspan, M.S. Cronan, L.A. Karol, and J.P. McGahan, Three-dimensional sonographic evaluation of developmental dysplasia of the hip: preliminary findings, *Radiology,* 190, 407–410, 1994.
89. H.C. Kuo, F.M. Chang, et al., The primary application of three-dimensional ultrasonography in obstetrics, *Am. J. Obstet. Gynecol.*, 166, 880–886, 1992.
90. D. Kirbach and T.A. Whittingham, 3D ultrasound — the kretztechnik voluson approach, *Euro. J. Ultrasound,* 1, 85–89, 1994.
91. R.A. Robb, *Three-Dimensional Biomedical Imaging: Principles and Practice*, VCH Publishers, New York, 1995.
92. H.K. Tuy and L.T. Tuy, Direct 2-D display of 3-D objects, *IEEE Comp. Graphics Appl.*, 4, 29–33, 1984.
93. M. Levoy, Volume rendering, a hybrid ray tracer for rendering polygon and volume data, *IEEE Comp. Graphics Appl.*, 10, 33–40, 1990.
94. M. Levoy, Efficient ray tracing of volume data, *ACM Trans. Graphics*, 9, 245–261, 1990.
95. T.R. Nelson, D.H. Pretorius, M. Sklansky, and S. Hagen-Ansert, Three-dimensional echocardiographic evaluation of fetal heart anatomy and function: acquisition, analysis, and display, *J. Ultrasound Med.*, 15, 1–9, 1996.
96. D.H. Pretorius and T.R. Nelson, Fetal face visualization using three-dimensional ultrasonography, *J. Ultrasound Med.*, 14, 349–356, 1995.
97. D. Fine, S. Perring, J. Herbetko, C.N. Hacking, et al., Three-dimensional (3D) ultrasound imaging of the gallbladder and dilated biliary tree: reconstruction from real-time B-scans, *Br. J. Radiol.*, 64, 1056–1057, 1991.

98. S. Tong, D.B. Downey, H.N. Cardinal, and A. Fenster, A three-dimensional ultrasound prostate imaging system, *Ultrasound Med. Biol.*, 22, 735–46, 1996.
99. T.L. Elliot, D.B. Downey, S. Tong, C.A. Mclean, and A. Fenster, Accuracy of prostate volume measurements in vitro using three-dimensional ultrasound, *Acad. Radiol.*, 3, 401–406, 1996.
100. S. Tong, H.N. Cardinal, R.F. McLoughlin, D.B. Downey, and A. Fenster, Intra- and inter-observer variability and reliability of prostate volume measurement via 2D and 3D ultrasound imaging, *Ultrasound Med. Biol.*, 24, 673–681, 1998.

15

Industrial Computed Tomographic Imaging

Harry E. Martz, Jr.
University of California

Daniel J. Schneberk
University of California

15.1 Introduction

X-rays are the most commonly employed and technologically important form of ionizing radiation for industrial or nondestructive evaluation (NDE) computed tomography (CT) imaging applications. It is the penetrating nature of X-rays and the ease of interpretation that are most utilized in industrial applications. These same characteristics led to their discovery by Wilhelm Röntgen in 1895. On the 100-year anniversary of Röntgen's work, *Physics Today* published a special issue[1] summarizing the important developments in X-ray science and applications in the past century.

The radiation imaging method of NDE enjoys an advantage over many other NDE methods in that it is inherently pictorial and interpretation is to some extent intuitive. Analyzing and interpreting the images require skill and experience, but the casual user of radiation imaging services can easily recognize the item being imaged and can often recognize flaws and defects without expert interpretation. For example, in an X-ray radiograph's image of a hand (Figure 15.1), you do not have to be an expert radiographer to distinguish hard tissue (bone) from soft tissue (flesh). This simple example reveals the potential of the radiographic technique to inspect or characterize objects (e.g., humans or turbine blades) for internal details.

X-rays can be used to yield data in one (1D), two (2D), or all three (3D) spatial dimensions. CT combines many 1D or 2D projection radiography measurements of X-ray attenuation with computer reconstruction methods to produce a 3D image of an object under study (Figures 15.2 and 15.3). Within two decades of Röntgen's discovery, the mathematical possibility of computer reconstruction was developed by Radon.[2] However, practical applications awaited the advent of digital computers and electronic

0-8493-3691-0/01/$0.00+$.50

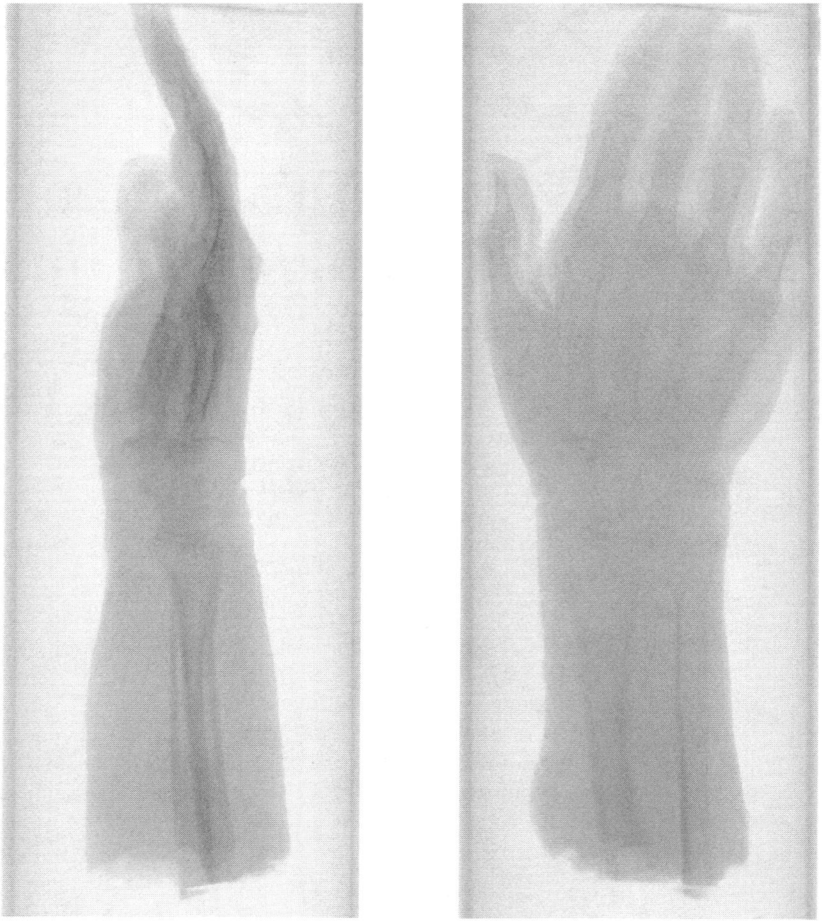

FIGURE 15.1 Radiographs of a women's hand at two angles, 0° and 90°. In these radiographs, both the hard (bone) and soft (flesh) tissues are observable.

detection methods. In the early 1970s, CT was developed as a tool for medical diagnostic imaging,[3] and its inventors Cormack and Hounsfield were awarded the Nobel Prize in medicine in 1979. Soon thereafter, medical CT scanners entered the market.* The medical CT systems were based on linear-array detectors.[4,5] By the mid 1980s, CT was being widely developed and applied for industrial and scientific purposes.[6] Industrial systems began employing area-array detectors, e.g., charge-coupled device (CCD) cameras. This led to large gains in data acquisition speed. By the 1990s, new reconstruction methods,[7–9] widespread availability of faster computers, and new detector technology continued to expand CT as a practical and powerful NDE method.[10,11]

Radiation sources other than X-rays are employed in certain circumstances. Gamma (γ)-rays are electromagnetic radiation of nuclear origin as opposed to X-rays, which are of electronic origin. The physical laws governing transport and detection of X-rays and γ-rays are the same. γ-rays tend to be of higher energy than X-rays, and it is quite common to encounter usage of the terms X-ray and γ-ray based on energy rather on the correct definition. Neutrons,[12,13] protons,[14] and electrons offer useful imaging characteristics, occasionally complementary to X-rays.

* In medical CT an attenuation unit, called Hounsfield, is used. It is defined relative to the attenuation of water. The Hounsfield is of little use in industrial CT.

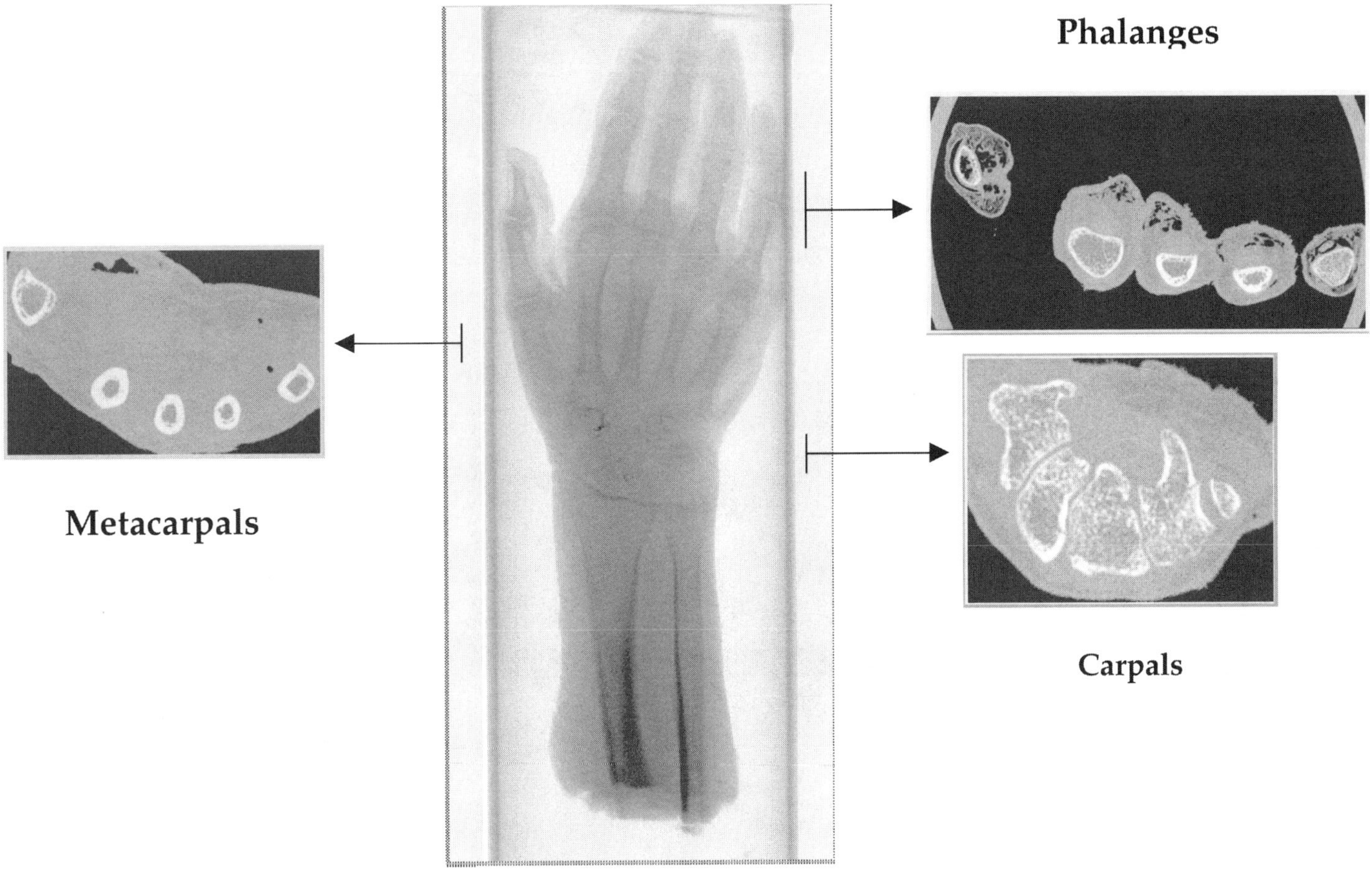

FIGURE 15.2 Radiograph (center) with some exemplar CT cross-sectional images (left and right) of a woman's hand. The true 3D nature of CT is revealed by combing the contiguous cross-sectional CT images.

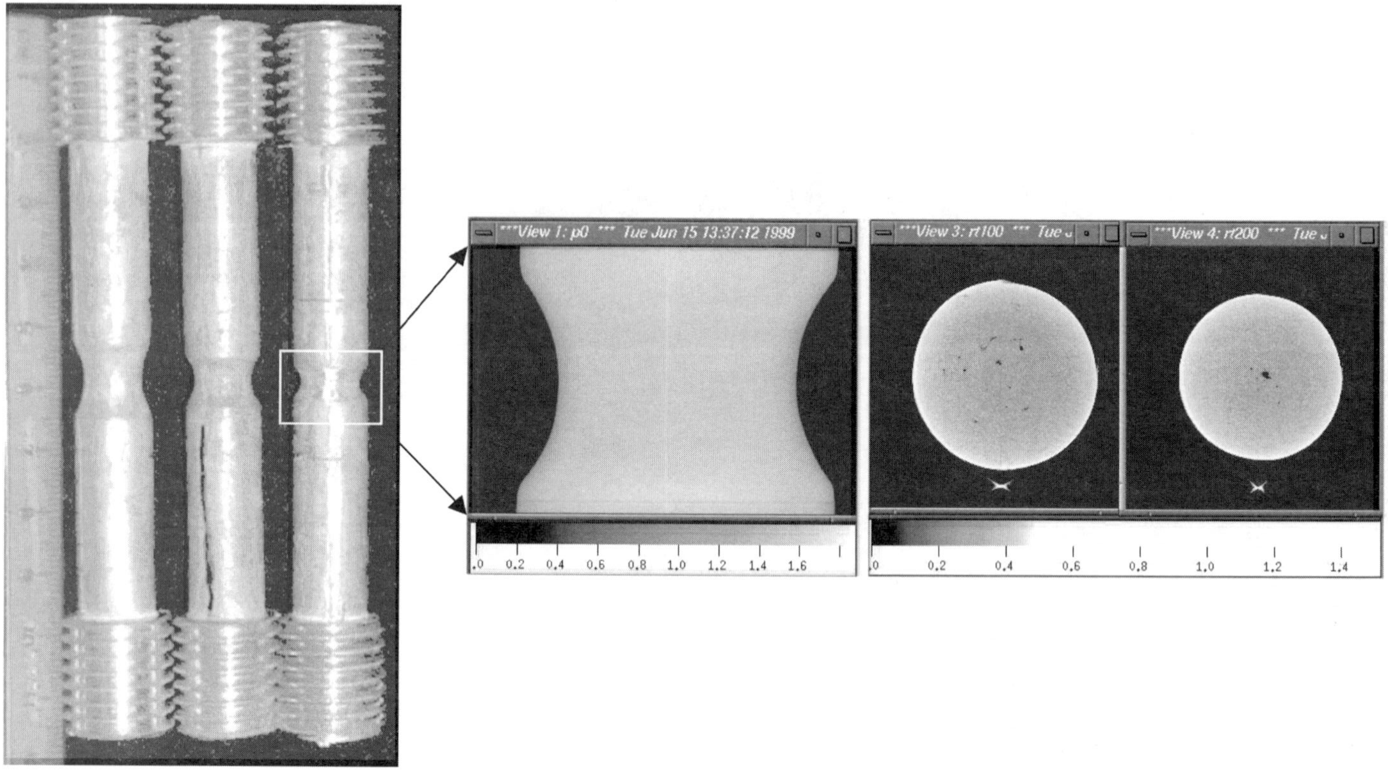

FIGURE 15.3 Photograph of three different notched, magnesium tensile bars (left). A digital radiograph of one notched tensile bar with a wire fiducial (middle). Two tomograms at different locations within the tensile bar (right). The tomograms easily reveal a pore or void (dark hole) within the center of the tensile bar. The bright spot at the bottom is the fiducial wire.

The most common application for radiation NDE methods is detection of internal structures. Radiation methods are suitable for sensing changes in composition and/or density. They are often applied to confirm the presence or configuration of internal features and components. They are especially applicable to finding voids, inclusions, and open cracks and are often the method of choice for verification of internal assembly details. For example, castings are susceptible to internal porosity and cracks. Figure 15.4 shows an example of porosity detection in cast magnesium tensile specimens. Radiography (2D) is limited in utility for detecting cracks. For a crack to affect the transmission of radiation, there must be an opening resulting in a local absence of material. A closed crack is not detectable using radiation. In addition, even when the crack has a finite opening, it will generally only be detectable in a radiograph at certain orientations. CT (3D) overcomes this orientation

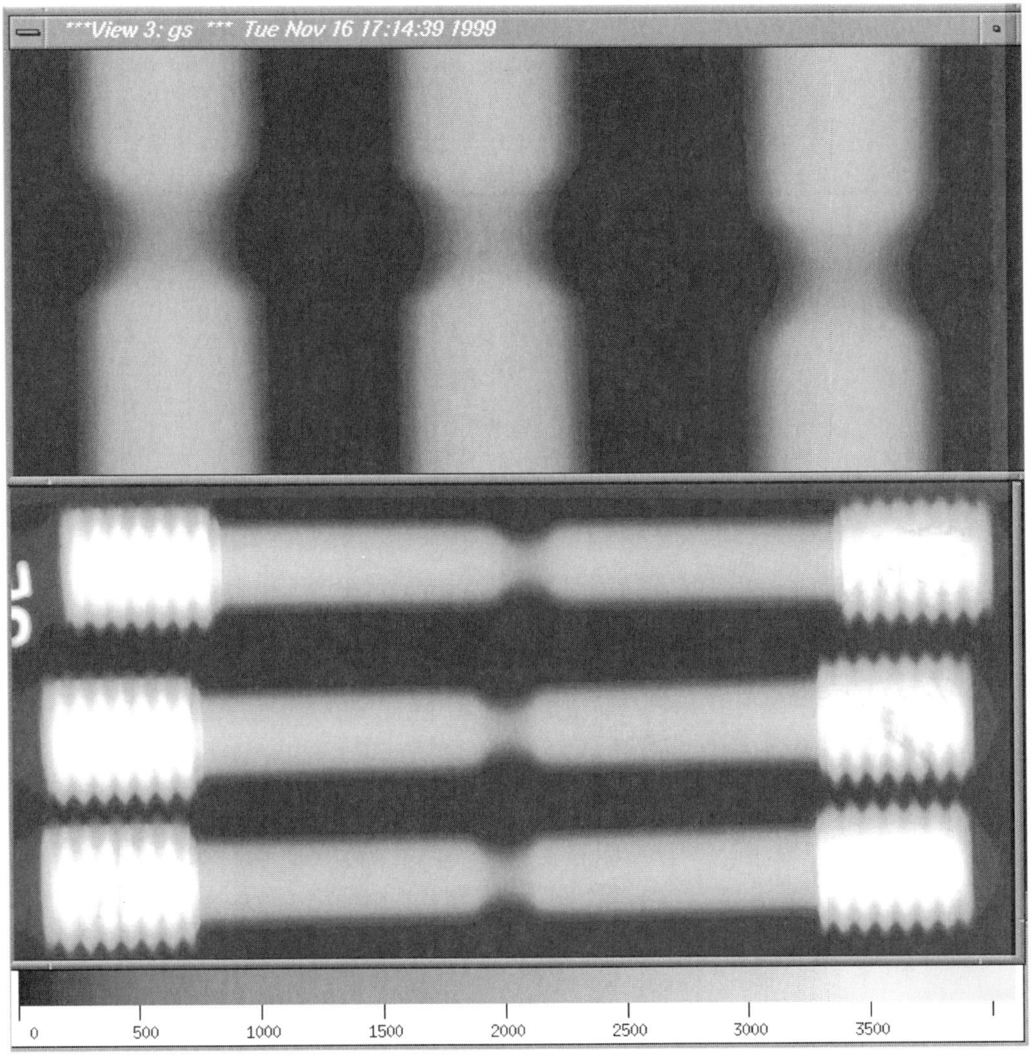

FIGURE 15.4 Digitized film radiographs of three magnesium notched tensile bars. The upper images are blown up to show more detail of the tensile bars given in the lower radiograph. In the upper central tensile bar near the middle, the dark spot in the notch is a pore (dark area). In the lower radiograph, the middle tensile bar has a crack (dark area) in the grip on the right.

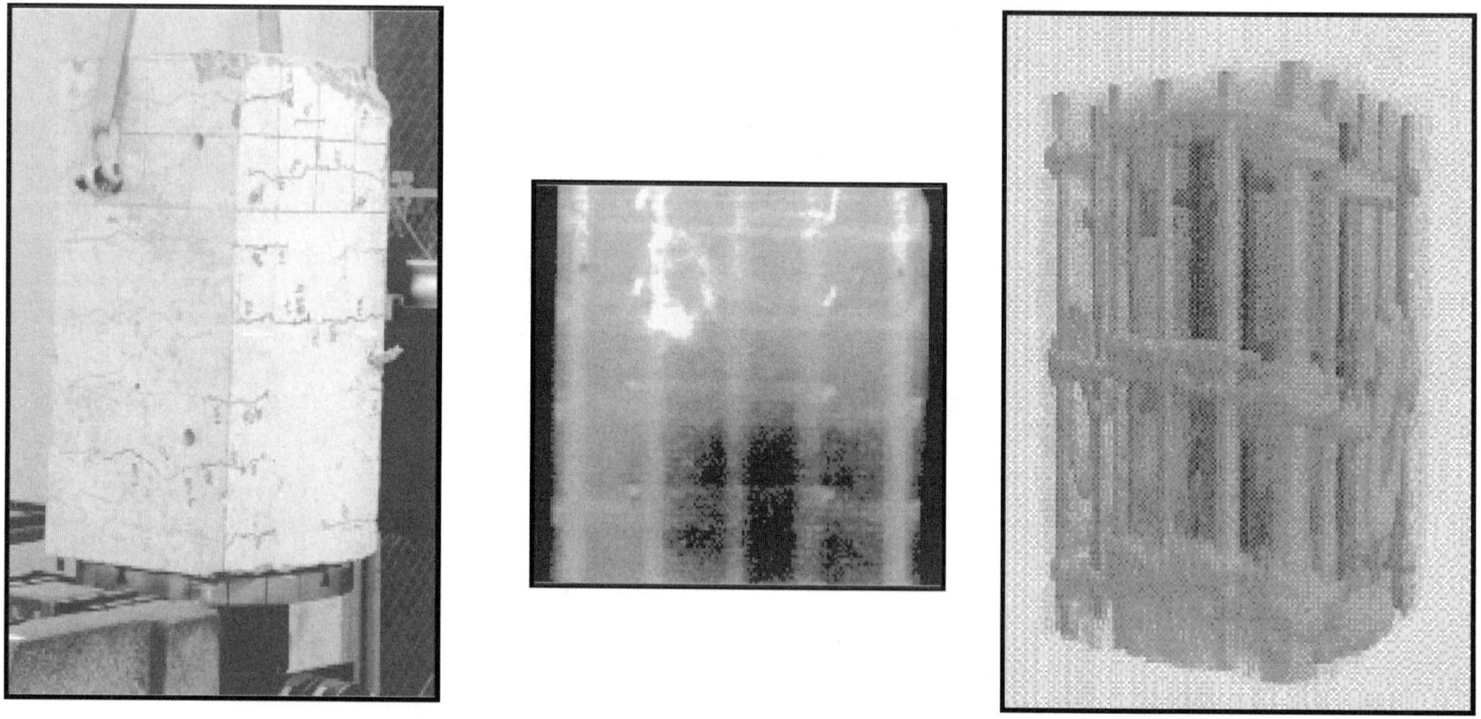

FIGURE 15.5 Photograph of a large California highway reinforced concrete column (provided by CALTRANS) (left). Radiograph of the column shows internal details, which are difficult to interpret due to the superposition of the concrete and rebar (middle); 3D rendered image of tomograms of the column (right). The concrete is transparent, and the rebar is opaque. When these data are rotated in a computer, the internal details are revealed in all three dimensions.

problem and is a suitable method for crack detection/characterization when a finite opening exists (e.g., see Figure 15.3). In this study, CT is being used to understand the growth of porosity cavities as the specimen undergoes plastic deformation.[15]

CT excels where information is needed in three spatial dimensions, such as reinforced concrete columns (Figure 15.5) and turbine blades (Figure 15.6). Indeed, it has replaced destructive sectioning methods in many instances. Because complex equipment is required and the time required for data acquisition can be substantial, CT is generally applied to a limited number of samples or to process development rather than 100% inspection on a production line. It also tends to be justified only for high-value components or instances involving issues such as safety. We know of X-ray CT being applied to objects ranging in size from 5×10^{-6} m[16] to 2 m.[17]

The disadvantages of radiation methods are that they can be cumbersome to apply because they use radiation and often high voltage. This creates concerns for personnel safety, regulatory oversight, use of heavy shielding materials, and the requirement for capable operators.

Medical CT operates under a number of constraints, including the amount of radiation or dose per scan, and sometimes it is difficult to keep the patient or organs from moving. However, medical CT imaging has the advantage of a limited range of materials (typically soft and hard tissues) for inspection. On the other hand, CT of industrial objects does not include constraints on dose, up to the point of "activating" materials in components and assemblies. This is only a problem for very high-energy X-rays. The dose requirement for medical CT leads to limitations on imaging performance and spatial resolution. Obtaining better image quality on hard tissue or bone scans requires higher source energy. Alternatively, obtaining greater spatial resolution requires more collimation, smaller detectors, or, for some scan geometries, more data acquisition. All of these circumstances result in more X-ray dose delivered to the patient or greater discomfort to the patient (has to sit still longer).

The range of objects for industrial CT is much larger than medical CT. The latter is the human, while industrially, we have inspected objects as small as 1 mm and as large as 100 cm in outer diameter. The industrial materials can range from low X-ray attenuating plastics to highly attenuating plutonium. Consequently, industrial CT includes a greater variety of techniques and energy ranges. For example, we have configured a suite of scanners to inspect a wide range of industrial problems at LLNL (see Figure 15.7). Notice the small area covered by medical CT scanners compared with the CT scanners required at LLNL. Distinct effects are part of the different energy ranges, and these effects sort into different types of artifacts in radiographic and tomographic images. X-ray scatter generates its own set of features in the CT image which can mask features in the object. Examples include the inner edge of turbine blades and the scattering observed in a simple hollow cylinder at interfaces as shown in Figure 15.8. Related to the higher energy range is the presence of more scattered signal from components and shielding in the scanner itself. To describe and characterize the different effects and the impact on data acquisition and processing, we need to discuss a few fundamental concepts.

15.2 CT Theory and Fundamentals

A CT reconstructed image is the product of the acquired data and the entire image forming process. Figure 15.9 contains one outline of the different phases for CT imaging. We will use these phases as a guide for describing the different modalities and aspects of imaging performance for industrial CT scanners.

15.2.1 Electromagnetic Radiation Interactions with Matter

The interaction of radiation with matter can be simplified by separating radiation into two types: electromagnetic and particles. Here, we only describe the interaction of electromagnetic radiation with matter. For greater details, we refer the reader to several textbooks such as Evans,[18] Knoll,[19] or Barrett and Swindell.[20]

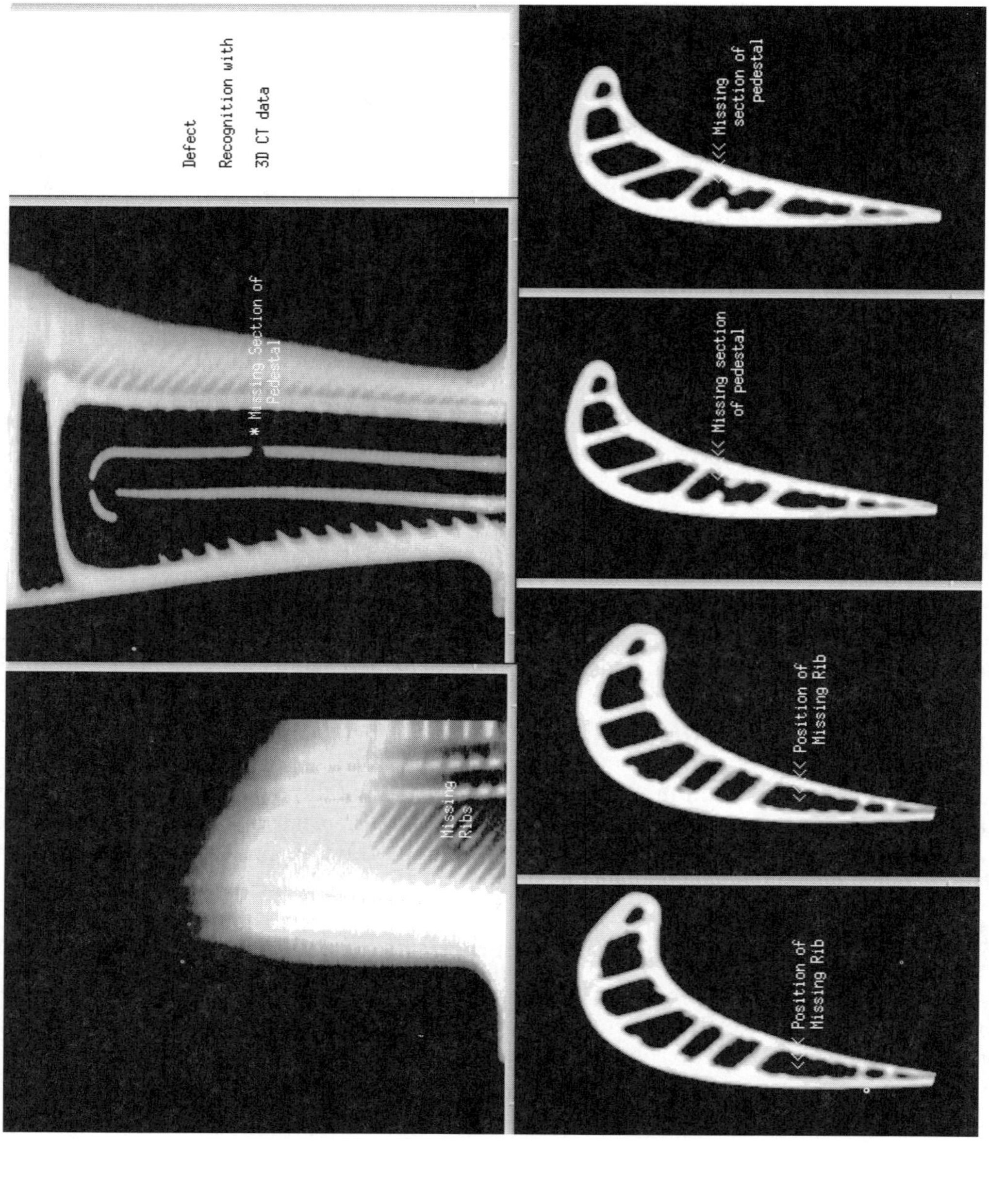

FIGURE 15.6 Representative CT slices of a nickel alloy turbine blade reveals missing pedastal sections and ribs as labeled.

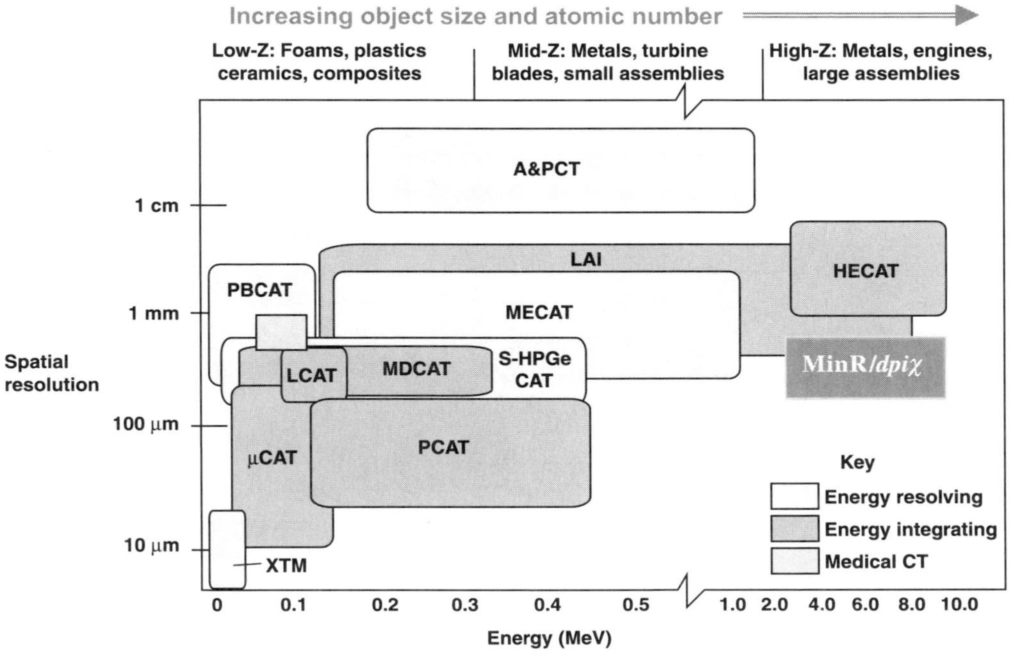

FIGURE 15.7 Spatial resolution performance vs. X-ray energy for LLNL CT scanners. Possible applications are given along the top of the figure.

The attenuation of electromagnetic radiation, e.g., X-rays passing through matter, involves three mechanisms, each accompanied by secondary processes. The three mechanisms are photoelectric effect, Compton scattering, and pair production. In the photoelectric effect, an X-ray dissipates its entire energy by knocking out an electron from an atom. In Compton scattering, the X-ray imparts some energy to an electron, but survives with a lower energy and different direction. In pair production, an X-ray is absorbed to create an electron-positron pair. In order for pair production to occur, the incident X-ray must have an energy equal to or greater than the rest mass of an electron-positron pair. This energy is 1.02 MeV, since the rest mass of each is 0.511 MeV.

The probability of each of these interaction mechanisms varies with atomic number and X-ray energy. The photoelectric effect is more dominant at low X-ray energy and high atomic number. Compton scattering is dominant at energies around 1 MeV, especially for materials with a low atomic number. Pair production becomes dominant around 5 MeV for material with a high atomic number (and at somewhat higher energy for lighter elements). A more complete discussion of these processes can be found in classic books by Evans[18] and Heitler.[21]

The different attenuation mechanisms combine in any radiographic or projection image to produce a signal from a variety of secondary processes. For example, a photoelectric electron can interact with the surrounding material and create secondary X-rays. Likewise, scattered X-rays can undergo photoelectric attenuation, and the resultant electron again may create an X-ray. At lower energies the secondary processes are not prominent. However, at medium- to high-energy regimes, the secondary and ancillary processes can dominate imaging quality.

Most radiographic imaging theory treats radiation as if it traveled in straight lines and disregards the effects of scatter and secondary X-ray production (the result of photoelectric interactions, Compton scattering, and pair production) in recovery of the image. The attenuation arising from the separate processes is usually lumped together and treated as total attenuation. However, for any particular scanner in the field, the photons from these other interactions are a regular part of the detected signal. To describe the impact of these different sources of measured signal, we introduce an accounting model for the

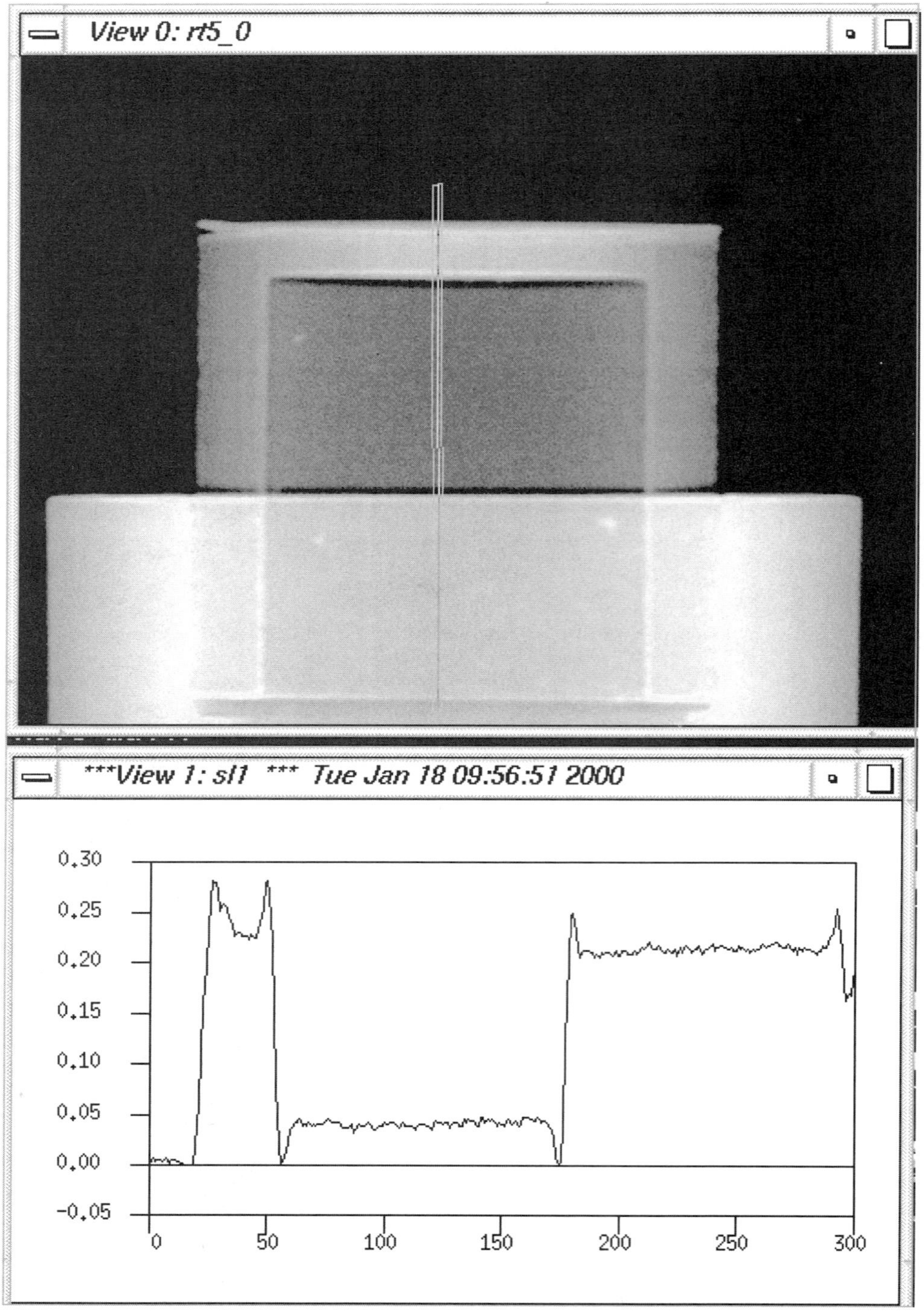

FIGURE 15.8 Digital radiogaphic (ln I_0/I) image of a hollow, 1-mm diameter, berilium cylinder held within a lucite cylinder (top). The data were acquired at 20 kV, 0.47 mA, and 180 sec per projection. Line out from the top digital radiographic image shows the edge artifacts caused by X-ray scattering even at this low energy (bottom).

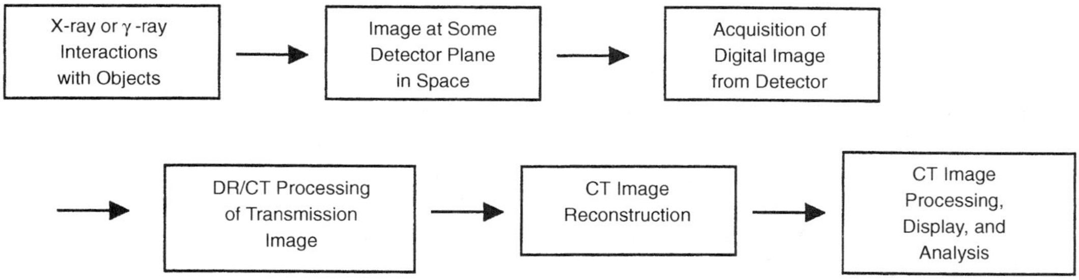

FIGURE 15.9 Representative flow diagram of the phases required in CT imaging.

photons in space which will lay the foundation for discussing different aspects of imaging performance in fielded scanners.

The X-ray beam is the starting point for all radiographic scanners. The properties of the beam, its shape and energy content, the physical mechanisms of X-ray attenuation, and the geometry of the object with respect to the beam are crucial determinants of imaging quality. The beam generates an image in space, and the detector samples the X-ray photons of this beam. Prior to the introduction of any detector or detector/collimation scheme, a number of noise sources are present in the transmission measurements. For any detector position (some 3D position on the other side of the object from the X-ray beam), the total number of X-ray photons transmitted through the object, N_T, at a particular detector position divide into different types as follows:

$$N_T[S(E), L] = N_P[S(E), L] + N_S[S(E), L]; \tag{15.1}$$

$$N_S[S(E), L] = N_{Sbk}[S(E), L] + N_{Sobj}[S(E), L]. \tag{15.2}$$

The schema in Equations 15.1 and 15.2 decomposes the total photon flux from an X-ray source with energy spectrum $S(E)$, along a ray path L, into primary photons, N_P and scattered photons N_S. Further, there are two types of scattered photons: background scatter, N_{Sbk}, and object scatter, N_{Sobj}. Most X-ray measurements have a means for taking into account the radiation field independent of the object, N_{T0}, given by

$$N_{T0}[S(E), L] = N_{P0}[S(E), L] + N_{Sbk}[S(E), L]. \tag{15.3}$$

Typically, this measurement contains some background scattered photons from the supporting fixtures in the X-ray scanner or in the detection hardware, as well as the primary photons launched by the X-ray source.

The performance of imaging techniques using penetrating radiation follows from the properties and the proportions of the different classes of X-ray photons in the image. The "primary" X-ray photons account for the performance of the transmission image. These are the "straight-line" projections through the materials, which best conform to the "idealized" ray path implicit to the CT reconstruction algorithms to be presented subsequently. Spatial resolution of the primary photons is bounded by the spot size blur[22] and, in the case of heavily collimated systems, the effective width of the source and detector apertures. The contrastive properties of the primary photons have the distribution of an X-ray gauge with standard deviation equal to the inverse of the square root of N.[23]

The photons scattered in the object are more difficult to interpret. The line or path of the scattered photons is energy dependent, usually forward peaked, and always broader than the path of the primary photons. Object scatter can vary throughout the image due to the lengths and types of material in the specimen. High attenuating sections of an object that are adjacent to low attenuating elements will generate substantial signals for the image of the low attenuating sections (see Figure 15.8). In radiography, the contrast in the low attenuating sections is simply compromised, and this effect is independent of the detector. For CT reconstructed images, scatter results in "streak" artifacts along the directions of the longest chord lengths or highest attenuation.[24] The spatial variant character of object scatter makes removal of this source of artifacts difficult.

"Background scatter" is the fluence that arises from the detector and its environment, e.g., cabinet and room hardware, independent of the object. If there was no detector or cabinet hardware, there would be no photons of this type, but then there would be no image. Consequently, we include this category as always part of any acquired image, even though the photons are present due to the detector itself. The detector hardware includes all the aspects of collimation, scatter from adjacent parts of the cabinet/detector enclosure, fluence from the fixturing supporting the object, as well as sources of scatter within the detector (blooming in the scintillator etc). Background scatter fills up the detector dynamic range with counts that do not carry information about the object. Consequently, contrast is always compromised by this source of signal.

The photons arising in the course of X-ray attenuation mechanisms can be organized into classes, and the different classes do not have the same properties for imaging the object function. However, as we will discuss, the theory of digital radiography (DR)/CT is based on the idealized ray-path model of attenuation. Imaging artifacts are the result of this mismatch between theory and actual measurement, which varies with each particular scanner. Keeping track of the sources of artifact content is important in industrial CT, where the applications utilize the full scope of the source energies available. Also, knowing how different types of systems result in different types of artifacts is useful for finding the right system for a particular application.

Acquiring DR/CT data involves some mechanism for digitizing the image in space, which involves its own set of issues. While the performance of DR/CT systems can be accounted for by analyzing the properties of the different types of photons, images and inspections are built on the measured intensities, $I[S(E), L]$ and $I_o[S(E), L]$ — the digitized versions of N_T and N_{T0}. Digitization of the photon flux has two aspects: (1) a change into a more measurable energy deposition (e.g., X-rays converted by a scintillator into visible light or X-rays converted by high-purity germanium crystal into current) and (2) some readout of the energy deposition into a digitized quantity (bit depth). The issues for converting the X-ray energy pivot about the treatment of the incoming X-ray energy spectrum. Scintillators differ in their sensitivity to different energy ranges and their ability to record much of the signal for high-energy X-rays. On occasions, the energy-windowing character of X-ray detectors is an advantage. For CT, the choice of scintillator directly affects the amount of "beam hardening" in the reconstructed image. This effect can be more significant for industrial CT, where the energies are substantially higher than in medical CT. The act of digitizing the signal results in some noise added to the signal, and this can be more significant for systems involving signal amplification as part of the digitization, for example, image intensifier (II) detectors. Independent of other considerations lower noise and higher bit depth will provide the most faithful acquisition of X-ray image in space.

15.2.2 Theory of Tomography and Attenuation

Tomography systems measure the effects of an object on an incident beam or "ray" that is transmitted through the object along a straight path. Fundamentally all that happens is that the radiation is absorbed or scattered by an object under study. Both of these modes attenuate the radiation as seen by the detector. For example, in X-ray or γ-ray CT, the data measured are the photon intensities of the incident beam (i.e., without the object), $I_o[S(E), L]$, and the transmitted beam, $I[S(E), L]$, that was attenuated by the object along each ray path, L, for a photon energy spectrum $S(E)$, as shown in Figure 15.10. We now present an overview of the models, machinery, and choices for industrial CT image processing and reconstruction.

15.2.2.1 Transmission CT 3D Imaging

Tomography is the imaging of a cross-sectional portion of an object with all other planes eliminated. CT is the process used to create such a cross-sectional image using a computer for image reconstruction. CT is just the measurements of radiographs over several angles around the object to be inspected. Algorithms

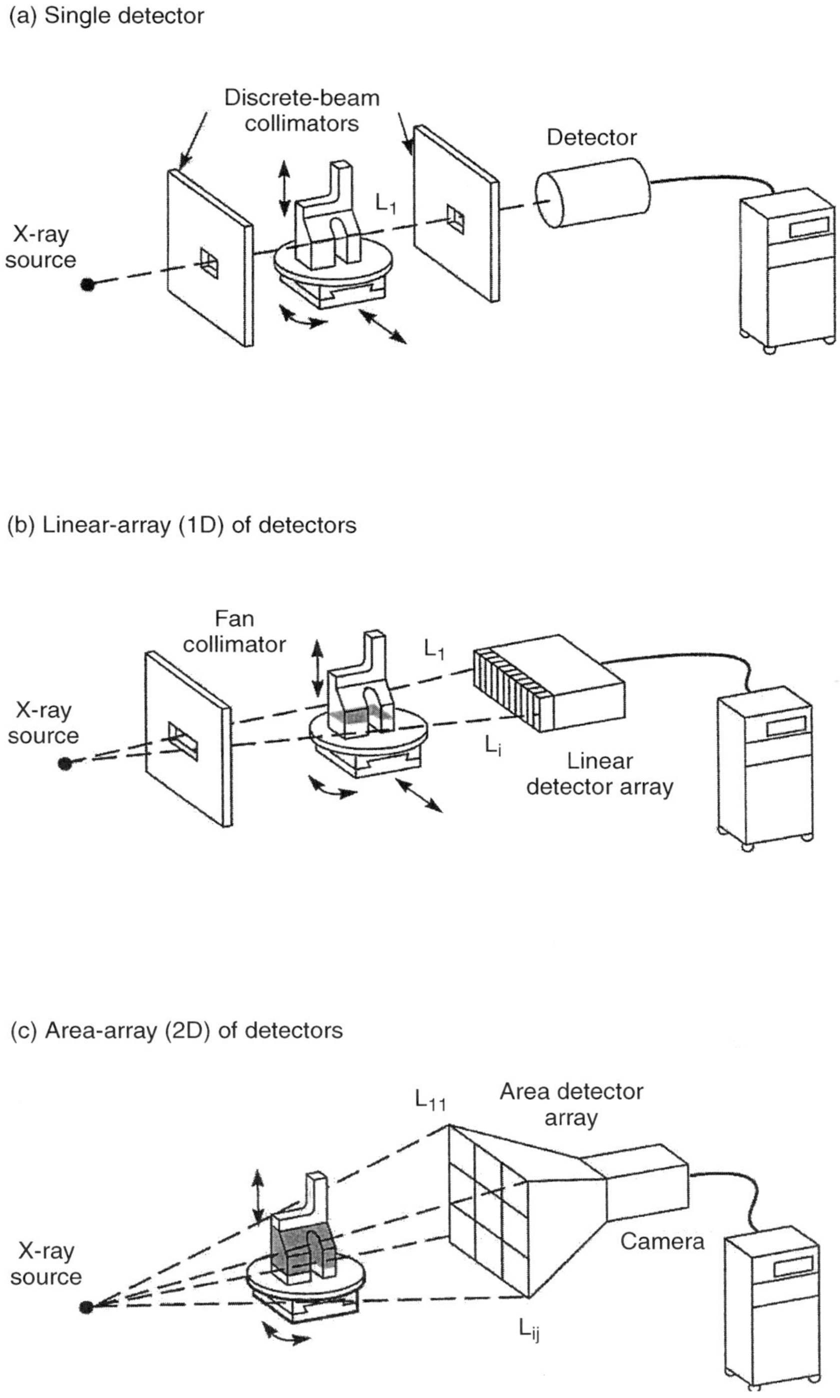

FIGURE 15.10 Data acquisition geometries used in industrial CT imaging. (a) First-generation, discrete-beam translate/rotate scanner configuration using a single detector; (b) second-generation, well-collimated, fan-beam configuration using a linear detector array; (c) third-generation, cone-beam configuration using a area-array detector. Second- and third-generation scanners alike can acquire data using either fan or cone beams. They differ in that third-generation scanners do not require horizontal translations, acquiring data by rotating the object or source/detector only.

are used to reconstruct a 2D or 3D representation (or image) of the object from these different views or angles around the part. The quantity that is reconstructed in CT is the value of some function, $f[S(E), \boldsymbol{x}]$, for some volume element or voxel at location $\boldsymbol{x} = (x, y, z)$ within the object. The reconstruction algorithms require line integrals, also called ray sums, for many ray paths L, which are defined as

$$g[S(E), L] = \int_L f[S(E), x]du, \tag{15.4}$$

where du is the incremental distance along L. For X- and γ-ray CT, these ray paths are determined from the intensity measurements using the Beer's law relationship:

$$g[S(E), L] = \ln\left[\frac{I_0[S(E), L]}{I[S(E), L]}\right]. \tag{15.5}$$

These ray sums over many paths are needed to reconstruct $f[S(E), \boldsymbol{x}]$.

Conventionally, industrial and medical CT use an X-ray machine source with a wide energy spectrum and a current-integrating detector that integrates the energy deposited by photons over all energies. The resultant attenuation image is given by

$$f[S(E), x] = \int_{S(E)} \mu[\rho(x), Z(x), E]dE \text{ (polyenergetic)}, \tag{15.6}$$

where μ is the linear attenuation coefficient, which is a function of volume density, ρ, the atomic number, Z, and energy, E. Note that the resultant attenuation is integrated over the entire energy spectrum. Some special CT scanners differ from the conventional scanners in that they discriminate between photons of different energies or their use of monenergetic sources. In this case, the resultant image is given by

$$f(E, \boldsymbol{x}) = \mu[\rho(\boldsymbol{x}), Z(\boldsymbol{x}), E] \text{ (monoenergetic)}. \tag{15.7}$$

The results are thus a discrete quantitative measurement of the linear attenuation coefficient at one energy E; i.e., there is no integration over the energy spectrum $S(E)$. For a single material object, plugging Equation 15.7 into Equation 15.4 yields

$$g[E, L] = \int_L \mu du = \mu\ell, \tag{15.8}$$

where ℓ is the thickness of the object along the ray path L. It is useful to note that the CT image reconstruction process accounts for the integral over the path length and results in images of μ for each voxel.

Here, it is instructive to show the differences between monoenergetic and polyenergetic sources and their effects on tomographic imaging. A good example of this is given below:

Example: Monoenergetic vs. Polyenergetic Imaging

As mentioned previously, monoenergetic imaging techniques produce the best data with the least number of artifacts. Monoenergetic and polyenergetic CT imaging of a mock explosive was used to study how each technique could be used to distinguish between the different materials that make up the mock explosive and, thus, in the future for real explosives characterization. Component materials of mock plastic binder explosive (PBX) do not have the same X-ray attenuation as PBX. However, the microstructure morphology of mock PBX approximates PBX.

Two CT systems were used to quantify the structure of a 2-mm diameter sample of mock (inert nonhazardous material) PBX.[25] These systems mainly differed in their X-ray source. The XTM used a 15-keV monochromatic synchrotron radiation source and a scintillator lens coupled to a CCD camera detector.[26,27] The mock PBX sample was scanned with 5 μm spatial resolution. This spatial resolution is determined from the full-width half-maximum (FWHM) value of the edge-spread function. The KCAT X-ray tomographic system uses a polychromatic X-ray machine source and a scintillator lens coupled to a CCD camera detector. KCAT was used to scan the mock PBX sample using a polyenergetic of 60-keV peak with 10-μm FWHM spatial resolution.

Figure 15.11 shows CT tomograms or cross-sectional slices of the 2-mm outer diameter sample from each of the two tomographic systems. KCAT data are shown in Figure 15.11A, and XTM data are shown in Figure 15.11B. The slices are representative slices taken at different locations on the sample. Because of higher spatial resolution and higher contrast sensitivity, the XTM slice in Figure 15.11B provides sharper edges and better contrast of the component structures than the KCAT slice. Histograms of linear attenuation values for each of the systems are shown in Figures 15.11C and 15.11D. Attenuation values in the XTM histogram data show four distinct attenuation regions (phases) in the sample. The KCAT data do not show separate attenuation regions (phases) in the histogram.

Mock PBX consists of four materials (phases): talc, Kel-F, cyanuric acid, and voids. Expected attenuation values for these component materials at discrete energies ranging from 10 to 100 keV are

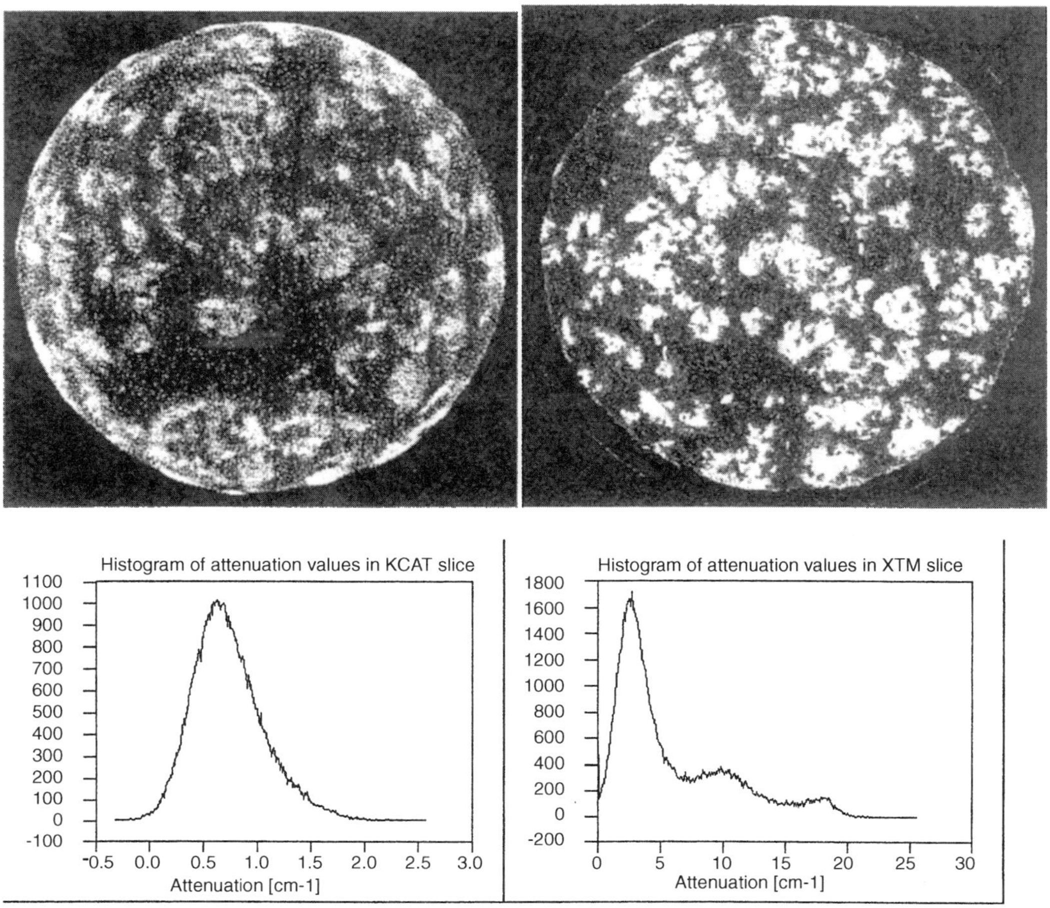

FIGURE 15.11 KCAT and XTM representative slices show the difference in the systems: (top left) KCAT slice and (top right) XTM slice. Histograms of attenuation values in (bottom left) KCAT slice and (bottom right) XTM slice show the difference in contrast sensitivity from the two systems. The XTM system distinguishes four phases.

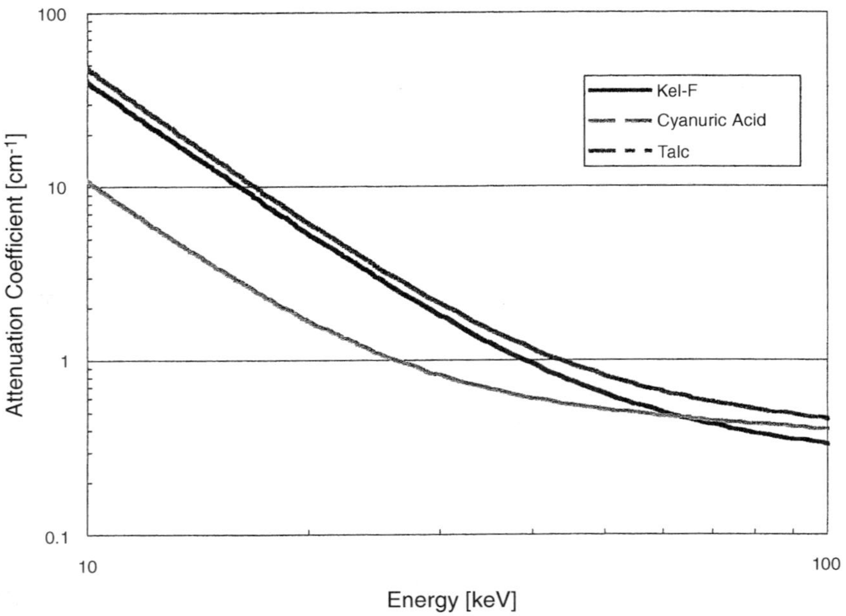

FIGURE 15.12 Calculated linear attenuation coefficient vs. energy for three of the four components of PBX are plotted for comparison to the experimental results. Attenuation is calculated from chemical compositions and density where Kel-F = $H_2C_8F_{11}C_{13}$ @ 2.02, cyanuric acid = $H_{10}C_3N_3O_2$ @ 2.50, and talc = $H_2O_{12}Si_4Mg_3$ @ 2.78.

calculated and presented in Figure 15.12. For the monochromatic XTM energy of 15 keV, expected attenuation values are 14 cm^{-1} for talc, 12 cm^{-1} for Kel-F, 3.4 cm^{-1} for cyanuric acid, and 0 cm^{-1} for void space. These values are based on 100% density. It should be noted that *in situ* density of the component materials can deviate substantially from 100% density. Consequently, *in situ* attenuation values can be different from the expected attenuation values. The four peaks in the XTM histogram correlate well with the attenuation values for each of the three materials and voids as expected.

Although the KCAT slice in Figure 15.11B shows some morphological features of the microstructure, distinct phases are not readily identifiable in the histogram in Figure 15.11C. Unlike the monochromatic XTM source, the polychromatic KCAT source energy ranges from a few kiloelectronvolts up to 60 keV. Figure 15.12 shows the range of attenuation values for each of the component materials. Contrast sensitivity between the phases is diminished by the integration of attenuation values over the polychromatic source, and some are diminished by the poorer spatial resolution of the KCAT system. As a result, the phases in the sample are difficult to identify based on their measured attenuation values alone.

Micro-CT was assessed for use in the 3D characterization of the mock PBX microstructure. Comparison of two micro-CT systems indicates that a monochromatic source with 5-μm FWHM spatial resolution provides sufficient spatial resolution and contrast sensitivity to characterize and identify the mock PBX microstructure. Our initial attempt to use a polychromatic source at 10-μm spatial resolution allows imaging of the microstructure, but does not easily permit identification of phases in mock PBX.

15.2.3 Image Reconstruction Algorithms

In this section, we briefly describe the algorithms required to obtain tomographic images from radiographic projections as a function of angle. This knowledge is not necessary to do CT imaging, however, it is very useful to have a basic understanding of CT algorithms to get the most out of CT imaging.

Several books describe the theory and algorithms used for CT image reconstruction in great detail. Some good references are Herman,[5] Kak and Slaney,[24] Barrett and Swindel,[20] and Macovski.[4] In this chapter, we briefly provide an overview of the fundamentals for conversion of DR projections to CT images. The most common form of CT imaging is the reconstruction of 2D cross-sectional images from 1D or linear-array radiographic projections.

15.2.3.1 Transmission CT Algorithms

The ray path, L, is simplified if we consider a single 2D x-y plane fixed along the longitudinal axis, z, of an object as shown in Figure 15.13. We can treat a single discrete X- or γ-ray beam in that plane as a line or ray path defined by s, the distance between the ray path and the $(x - y)$ origin, and θ, the angle of the s axis from the x axis. The transmitted (or active) beam intensity $I(E, s, \theta)$ for this ray path at a fixed single energy, E, is

$$I(E, s, \theta) = I_0(E, s, \theta)\exp\left[-\iint \mu(E, x, y)\delta(x\cos\theta + y\sin\theta - s)dxdy\right], \tag{15.9}$$

where $\mu(E, x, y)$ is the spatial distribution of the linear attenuation coefficients at energy E, I_0 is the intensity of the incident beam, and δ is the Dirac delta function. The equation for the ray path is $x\cos\theta + y\sin\theta - s = 0$. As mentioned earlier the argument of the exponential is known as a ray sum, $g(E, s, \theta)$, as shown in Equation 15.4, and in this case, it is equal to

$$g(E, s, \theta) = \ln\left[\frac{I_0(E, s, \theta)}{I(E, s, \theta)}\right] = \iint \mu(E, x, y)\delta(x\cos\theta + y\sin\theta - s)dxdy, \tag{15.10}$$

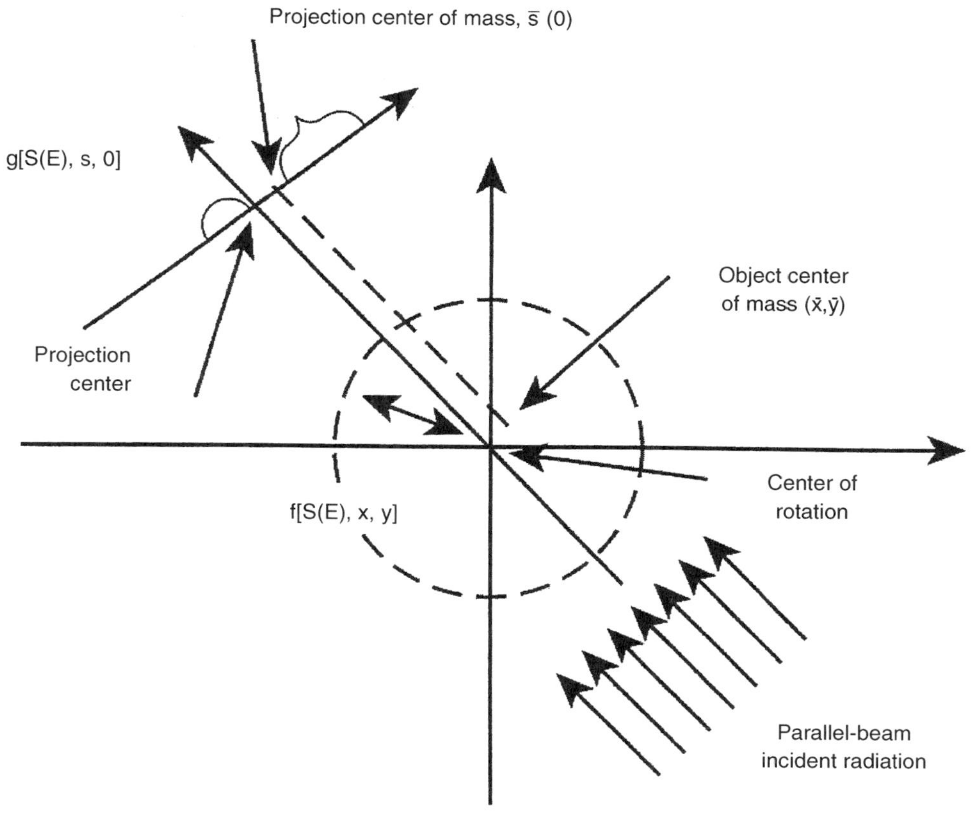

FIGURE 15.13 Diagram of the parallel-projection CT geometry.

It is useful to note that the CT ray sum is analogous to a simple γ-ray transmission gauge experiment.

The set of ray sums at all values of s at a given elevation (slice) for a fixed E and θ is called a "projection." A complete set of parallel-beam projections at all θ (from 0 to 180° or 360°) for a fixed E is called a "sinogram." From measurements of I and I_0, a complete transmission CT sinogram can be determined; and various methods have been devised to reconstruct μ, the linear attenuation coefficient, with filtered backprojection (FBP) being the most common.[28] Therefore, μ is the parameter determined by image reconstruction of the X- or γ-ray transmission CT measurements at a selected energy value and voxel. All voxels for a given plane are called the slice plane or tomogram. For any object, the attenuation due to the object's contents, whether heterogeneous or homogeneous, is accurately measured in the third dimension by measuring sinograms at different z planes (or elevations) of the object and subsequent image reconstruction.

The monoenergetic case is atypical. For the polyenergetic case, which is more common medically and industrially, the ray sums must be integrated over the entire energy spectrum of the source used. This case is given by

$$g[S(E), s, \theta] = \ln\left[\frac{I_0[S(E), s, \theta]}{I[S(E), s, \theta]}\right] = \int_{s(E)}\iint W(E)S(E)\mu(E, x, y)\delta(x\cos\theta + y\sin\theta - s)\,dx\,dy\,dE, \quad (15.11)$$

where W is the quantum efficiency of the detector. This problem can be expanded to the third dimension as well by acquiring several 2D radiographic projections as a function of angle and image reconstruction. For the parallel-beam case, each row in the radiograph is reconstructed into a cross-sectional view separately. By stacking the individual cross-sectional slices, you have a 3D tomogram of the entire object.

Cone-beam data acquisition and image reconstruction methods are the most direct means for obtaining 3D CT of objects and assemblies. In this modality, the area-array detector projection images are input directly into the CT image reconstruction code. For a 3D volume, scanning speed is increased from the more efficient use of the source output. Cone-beam scanning can benefit from a reduced scatter component which occurs with increasing cone angle.[29] However, large cone angles require more data acquisition, more precise alignments, and somewhat different reconstruction algorithms.[7,8,30] Furthermore, large cone angles put more emphasis on small X-ray spot sizes, which are not equally available across the energy ranges needed for some industrial objects. Recent advances in area-array detectors and computer speed have recommended cone-beam CT in more industrial applications. We expect this trend to continue. Cone-beam CT image algorithms are summarized by Smith.[7]

15.2.3.2 Single Photon Emission CT (SPECT) Algorithms

Note that transmission CT does *not* measure the presence or identity of any radioisotope, source strength, or activity within an object. The radioactivity of an object is measured by emission or passive computed tomography (PCT). This technique is common in the medical community, but not so common in the industrial community. Two emission methods exist. One is called single photon emission CT (sometimes called SPECT) imaging and the other is positron emission tomography (PET).

The SPECT technique is very important for a particular industrial problem, the waste drum nondestructive assay problem, presented later. PCT is used to measure the identity and location of γ-ray emitting radionuclides within an object. The ray sum for PCT or SPECT imaging, $g_\gamma(E, s, \theta)$, is defined by[28]

$$g_\gamma(E, s, \theta) = I_e(E, s, \theta) = \iint \mathbf{p}(E, x, y)a(E, x, y, x, \theta)\delta(x\cos\theta + y\sin\theta - s)\,dx\,dy, \quad (15.12)$$

where I_e is the passively emitted counts measured at each ray sum position and

$$a(E, x, y, s, \theta) = \exp\left[-\int_x^{detector}\int_y \mu(E, x', y')\delta(x'\cos\theta + y'\sin\theta - s)\,dx'\,dy'\right] \quad (15.13)$$

is the half-line attenuation integral from the (x, y) position to the detector position defined by (s, θ); and $\mathbf{p}(E, x, y)$ are the photons of energy E emitted per unit volume per unit time for each voxel within an object.

A single photon-emitted ray sum is the integrated radioisotope activity, modified by one or a multiple of exponential attenuations, along the path from the source position within the drum to the detector. The influence of the term $a(E, x, y, s, \theta)$ depends on the magnitude and distribution of the attenuations within the object. Unlike the medical SPECT problem, for example, the waste drum problem, the attenuations are typically large and nonhomogeneous for most energies emitted. To obtain the most accurate results from the PCT measurements, the energy-dependent attenuations must be determined from transmission or active CT measurements. The more commonly used assumption of a constant attenuation coefficient is inadequate for accurate measurements of inhomogeneous waste matrices.

15.2.3.3 Coupling Active and Passive CT

Coupling the active CT (ACT) and PCT modes allows accurate and quantitative attenuation corrections to be determined specific to the location of any radioactivity detected. That is, once the attenuation caused by the object (e.g., a waste drum) and geometry of the CT scanner is accounted for, an accurate measurement of the emitted photons, **p**, from a radioisotope is determined. The radioisotopic activity from a particular radioisotope γ-ray, j, is determined as follows:

$$C_j(E) = \sum_i p_i(E), \tag{15.14}$$

where $C_j(E)$ is the total photons per unit time obtained from the sum of all the voxels at energy E for the reconstruction of the PCT data corrected by the ACT attenuation map. Once the total photons are obtained, the activity A_j is obtained from

$$A^j(mCi) = \frac{C_j(E)}{t\varepsilon(E)\beta_j k}, \tag{15.15}$$

where t is the ray sum integration time, $\varepsilon(E)$ is the detector efficiency* at the particular energy E of the emitted γ-ray measured, β_j is the branching ratio for this particular γ-ray, and k is the constant 3.7×10^7 disintegrations per second per milliCurie. Finally, the measured activity is converted to a specific gram value mass, m, using

$$m_j(g) = \frac{A^j}{A^j_{sp}}, \tag{15.16}$$

where m_j is the mass in grams of radioisotope j that has a specific activity given by A^j_{sp}.

15.2.4 Fundamentals of the CT Measurement Equipment

Before we discuss applications of industrial CT, it is useful to point out some fundamental limitations of the measurement equipment used to capture the projection image in space. For the uncollimated image in space, the spatial properties have a lower blur limit related to the size of the X-ray spot and a higher blur limit convolved with the scatter field generated by the object and detector hardware. The contrastive properties are best for the primary photons and have the distribution of an X-ray gauge with standard deviation equal to the inverse of the square root of N.[23] More primary photons will increase the contrastive performance of any system. All types of scattered photons contribute photons that do not carry information on straight lines through the object and at best inflate the number of counts in a

* This does not include the solid angle since it is accounted for in the image reconstruction and assay algorithm.

detector location without improving contrast. At worst, scattered photons carry information on adjacent parts of the object or assembly and mask the changes in material along the straight-line location.

The physical configuration of the detector fundamentally affects the proportions of the above quantities detected for X-ray imaging. Indeed, the detector performs a sampling of the X-ray image for both the "intensity" of the spectrum and the spatial landscape of the transmission through the object. Additionally, the internal configuration of the X-ray detector determines important practical performance properties. For any fielded detector, the speed of the acquisition and the efficiency with which the source fluence is detected (or quantum efficiency [QE]) are two useful detector properties. For 3D applications, system speed is also a function of source utilization, or source efficiency. Figure 15.10 contains sketches of the three general types of source detector configurations for which there are many specific modalities.

It is the perspective of this chapter that there is no one detector suited for all purposes. The best detector for the application is the choice that best meets the specific goals, which involve all kinds of practical considerations. However, in order to clearly describe the strengths and weaknesses of specific modalities, we categorize imagers on five properties: (1) N_P/N_S; (2) detector blur; (3) number of internally scattered photons; (4) source efficiency; and (5) QE. Together, these properties define important trade-offs that result from the different detector choices. These trade-offs are summarized in Table 15.1.

TABLE 15.1 Trade-Offs that Result from the Different Detector Choices

Detector	N_p/N_s	Detector Blur	N_s	Quantum Efficiency	Source Efficiency
Area array					
Film	Low	Small	Low	Low	Medium
Flat panel	Low	Small/Medium	Medium	High	High
Camera/scintillator w/small cone	Low	Medium/High	High	Low	High
Camera/image intensifier w/large cone	Medium	High	High	High	High
Linear array					
Slit collimated w/septa	High	Small	Low	High	Low
Slit w/o septa	Medium	Medium/Small	Low	Medium	Low
Single detector					
Single detector w/spectroscopy	Highest	Smallest	Smallest	Medium	Lowest

The greater range of choices for industrial scanners provides opportunities and trade-offs for a paraticular application. For data quality, the best choice is the spectroscopy DR/CT-based systems.[31] This is the experimental analog of imaging with the primary photons. Sources of scatter are removed by collimation or from the energy resolution in the detector. However, the impact on system speed is dramatic (factors of a thousand). Slit collimated linear detector array (LDA) systems offer the next best alternative for acquiring DR/CT data with the greatest proportion of primary photons.[4] The difficulty is when the focus of the application is on obtaining higher spatial resolution and/or full 3D inspection data. In this case, the use of area-array detectors can improve system speed by a factor of 20. However, all types of area-array detectors include varying amounts of scatter blur, and this subtracts directly from the dynamic range of the system.[32]

Contrastive performance for any scanner is limited by the system dynamic range, which varies by the scanner and application. The dynamic range of the system is not simply the number of bits in the detector or the number of bits subtracting the readout noise. Rather, dynamic range in a particular area of the image is bit depth minus readout noise minus the background scatter signal in the system. For certain medium-energy systems, the proportion of background scatter signal can be as high as 20% of the detected fluence, reducing the effective dynamic range greatly. These kinds of considerations are important for scanner selection and for interpreting scans of objects with high scatter fractions.

Depending on the system, different procedures have been developed to correct for some of the effects in the measured signals resulting from the particular scanner configuration. The goal of these procedures is to process the transmission data into the ray-path model, which is the basis for the image reconstruction. Industrial CT scanners can be organized into just how many processing steps are applied to transmission

TABLE 15.2 Preprocessing Steps for Various CT Scanner Configurations

	Scanner Configuration		
Step	Single	Linear	Area
Restoring ray-path geometry			X[a]
Correcting glare and blur			X[a]
Correcting for dark current		X	X
Correcting for detector nonlinearity			X[a]
Balancing detectors		X	X
Normalizing I_0	X	X	X
Calculating ray sums	X	X	X

[a] This is required to correct for curved area detector arrays.

data prior to reconstruction (see Table 15.2). Only minimal processing is needed for single detector spectroscopy systems, since they are the closest physical realization of the ideal ray path. LDA require a bit more processing, but much less than area detector arrays. It is typical for LDA scanners to involve some detector balancing or detector linearity corrections. Area arrays involve at least the application of some detector balancing correction and, in some cases, involve a correction for spatial distortions. Each one of these corrections can create scanning artifacts (ring artifacts or spatial distortions) in itself and can mask features in the object, depending on the location of the feature.

Fundamental issues for applications of CT to industrial objects involve this recognition of artifact content in different CT scanners. All of the different types of scanners can provide quality inspection data for a particular application. At the same time, different applications are better suited for different types of scanners. As indicated in a previous section, some artifacts (e.g., object scatter) are part of the transmission image in space independent of the detection scheme. A fully optimized scanner still has some artifact content, and sometimes these artifacts compromise inspection performance. At best, they are benign features within the image. Also, the detector configuration can contribute its own set of artifacts that mask the imaging of object features. We cannot underestimate the importance of scanner characterization through the use of CT phantoms (well-known 3D objects). This is the main means for distinguishing artifacts from features within the object.

To summarize our perspective on industrial DR/CT scanners, we have found the graphic in Figure 15.14 to be helpful. The different industrial DR/CT scanners can be plotted on a four axis scale as in Figure 15.14. At the four axes are spatial resolution, contrast resolution, energy resolution, and system

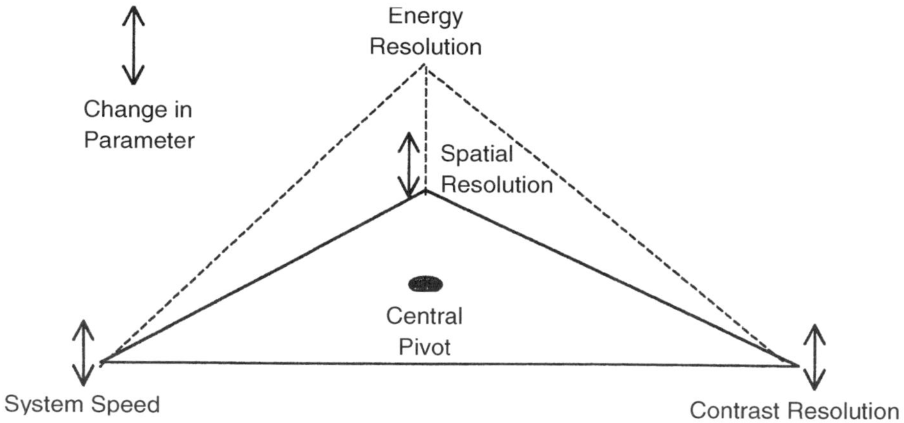

FIGURE 15.14 Three parameters — spatial resolution, contrast resolution, and system speed—placed at edges of a triangle with a central pivot graphically reveal that increasing any two parameters decreases the third. Adding a fourth parameter, energy resolution, extends the parameter space in ways that are not as straightforward to describe. We do know that an increase in energy resolution can result in an increase in contrast resolution, but with a likely decrease in spatial resolution and system speed.

speed. It is our experience that improvement in any two of the parameters results in a worsening of at least one of the other two. More collimation will improve image contrast, and with smaller apertures better spatial resolution can be obtained. However, the cost in system speed can be prohibitive for even the hottest X-ray source. Obtaining more scanning speed usually sacrifices image contrast and resolution. Reducing the energy spectrum for the scanning makes for slower scans for most conventional X-ray sources. Making trade-offs will be required. The point is to know your application well enough to make the best choice.

We now turn to inspection applications of using some of the variety of scanning modalities current in industrial DR/CT. In each case, the choice for the particular scanner was informed by the needs of the application.

15.3 Selected Applications

In this section, we have selected several applications that highlight the many different DR/CT scanners and CT reconstructed images representative of industrial CT. We start with linear-array detector applications since they are the most common; second, we discuss area-array application studies; and end with single detector energy discriminating applications.

15.3.1 A Linear-Array, Nonenergy Discriminating, Detector-Based DR and CT

Most of the medical- and industrial-based DR and CT systems are linear-array-based scanners. These scanners usually use CBP for image reconstruction. We have developed a linear-array-based CT scanner at LLNL called LCAT.[33] LCAT consists of a 160-kV, 1.9-mA X-ray machine source with a 0.2-mm spot size and a Thompson LDA. The detector array has 1024 detector elements with a 0.45-mm pitch and 0.5-mm high. The effective detector size is controlled by geometric magnification up to 3X can be used. LCAT has been used for a wide range of applications. We have selected two application studies: one is the characterization of an asphalt sample, and the second uses CT results to generate finite element analysis meshes to study prosthetic implant designs.

15.3.1.1 Material Analysis of Asphalt

The Federal Highway Administration (FHWA) was interested in learning how they could use CT to better characterize asphalt. To meet this end, the FHWA provided us with an asphalt sample ~102 mm in diameter. The sample is shown in Figure 15.15. The objective was to quantify components such as the amount of asphalt (i.e., tar), aggregate, and porosity within the sample. LCAT operating at the maximum source voltage and current, i.e., 160 kV and 1.9 mA, was used to acquire 1D projection images of the asphalt sample. The projection data were reconstructed by a CBP algorithm into a tomogram of 225 × 225 × 225-μm^3 voxels. A representative CT image is shown in Figure 15.16.

The X-ray attenuation image was segmented by two different methods to quantify the amount of the materials within the asphalt. One of the methods of segmentation is discussed elsewhere.[34] The results of the segmentation are shown in Figures 15.17 to 15.20. A composite of the different segments is provided in Figure 15.21. The volume fraction of the different components obtained by analysis of the segmented data is summarized in Table 15.3. These data are useful for understanding the constituents of the asphalt and to help understand the fabrication process and predict performance of asphalt road beds. These results were supplied to the FHWA for further analysis.

15.3.1.2 Improved Prosthetic Implant Design

Human joints are commonly replaced in cases of damage from traumatic injury, rheumatoid diseases, or osteoarthritis. Frequently, prosthetic joint implants fail and must be surgically replaced by a procedure that is far more costly and carries a higher mortality rate than the original surgery. Poor understanding of the loading applied to the implant leads to inadequate designs and ultimately to failure of the prosthetic.[75]

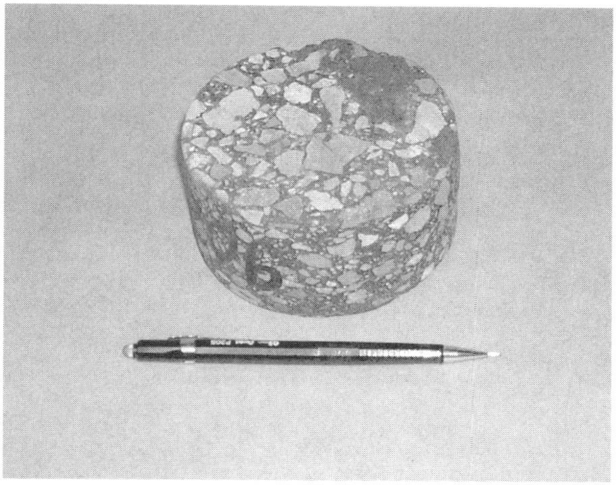

FIGURE 15.15 Photograph of the asphalt sample provide by the FHWA.

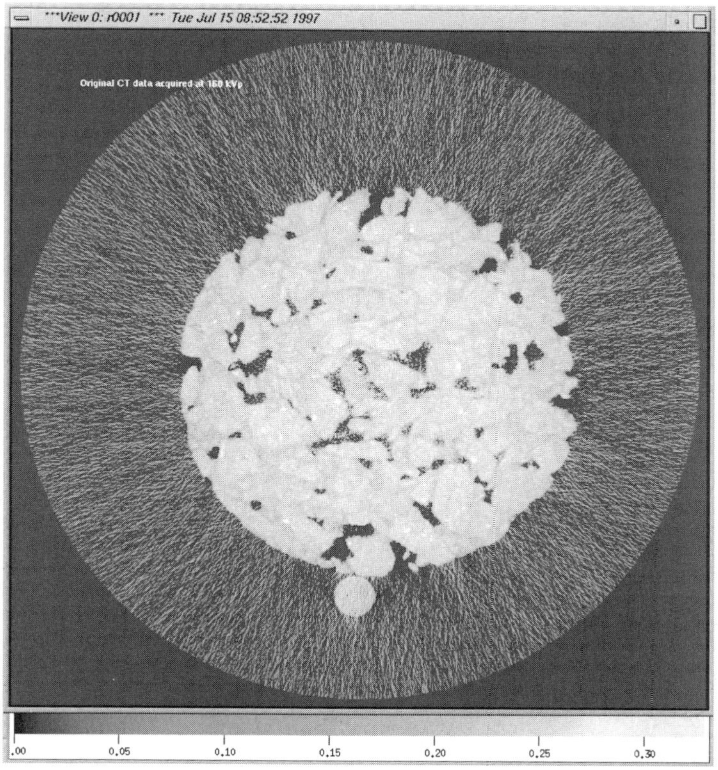

FIGURE 15.16 Representative CT image of the asphalt sample. The little disk at the bottom of the sample is a cylinder filled with water.

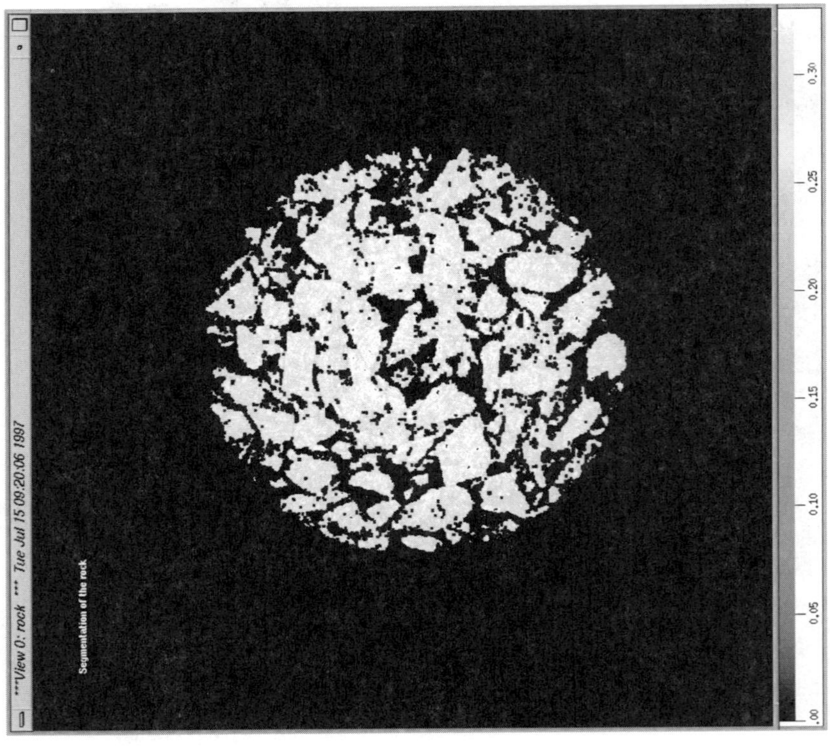

FIGURE 15.17 Segmentation results of the rock within the asphalt sample for the slice plane shown in Figure 15.16.

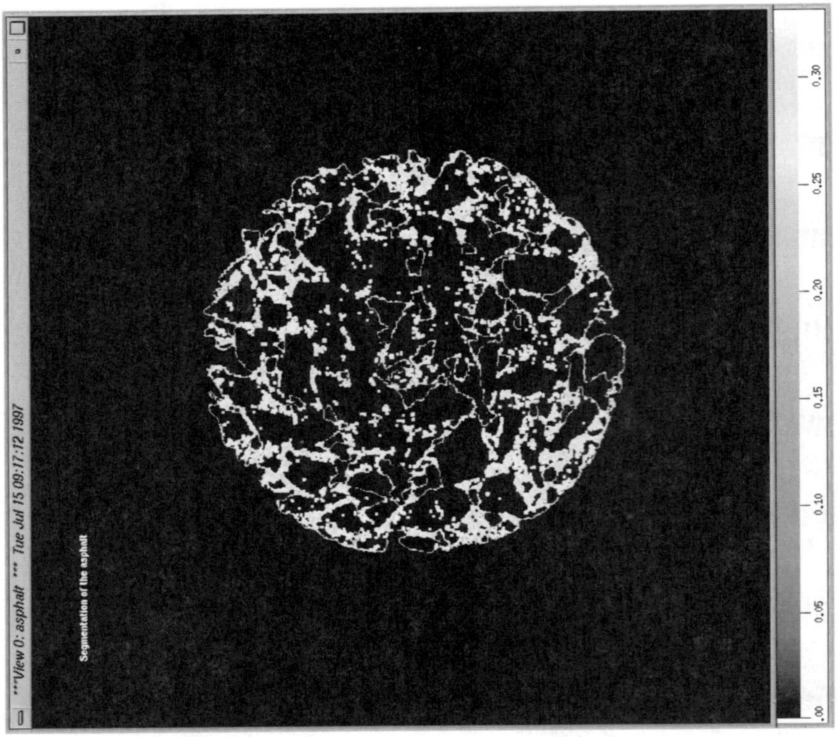

FIGURE 15.18 Segmentation results of the asphalt or tar within the asphalt sample for the slice plane shown in Figure 15.16.

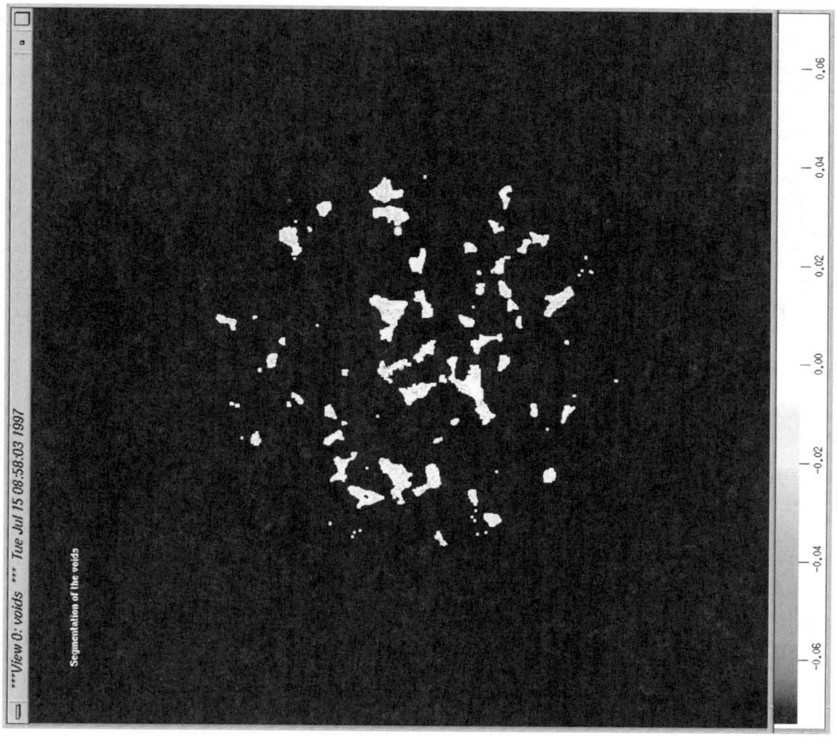

FIGURE 15.19 Segmentation results of the air, including cracks and voids, within the asphalt sample for the slice plane shown in Figure 15.16.

FIGURE 15.20 Segmentation results of apparent metal or high atomic number inclusions within the asphalt sample for the slice plane shown in Figure 15.16.

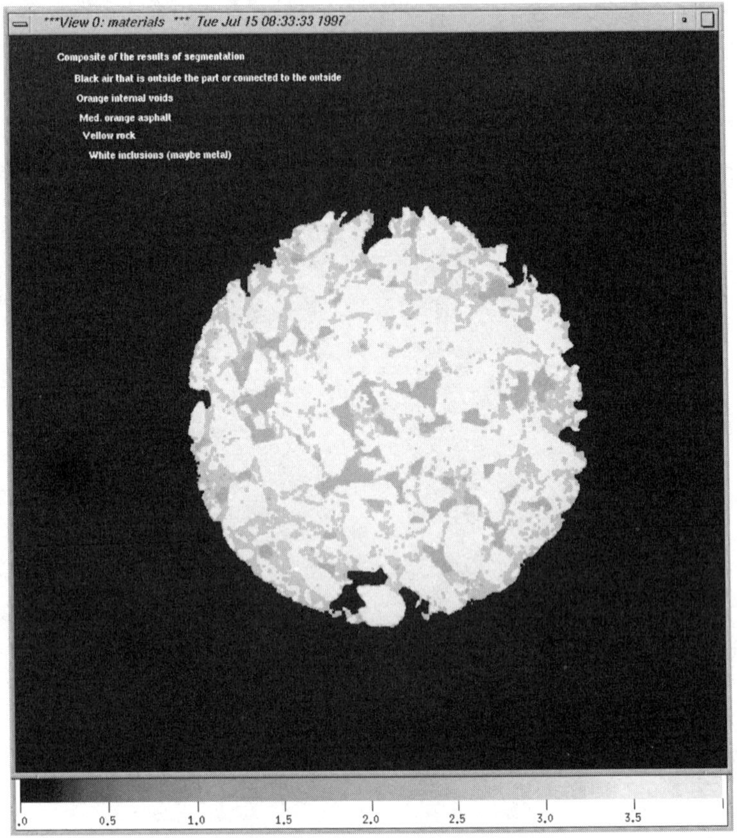

FIGURE 15.21A composite of the different segments within the asphalt sample as labeled.

TABLE 15.3 Volume Fraction in Percent for the Different Materials within the Asphalt Sample

Material	Percent of total volume
Aggregate (rock)	49
Asphalt (tar)	29
Voids & cracks	16
Inclusions	6

One approach to prosthetic joint design offers an opportunity to evaluate and improve joints before they are manufactured or surgically implanted. The modeling process begins with CT data (Figure 15.2). These data are acquired using LCAT and a magnification of almost 3× resulting in ~160 × 160 × 160-μm^3 voxels. The CT data are used to segment the internal hard tissues (bone) as shown in Figure 15.22.[38] A 3D surface of each joint structure of interest is created from the segmented data set (Figure 15.23). An accurate surface description is critical to the validity of the model. The marching cubes algorithm is used to create polygonal surfaces that describe the 3D geometry of the structures (typically bone) identified in the scans. Each surface is converted into a 3D volumetric, hexahedral, finite-element mesh that captures its geometry. The meshed bone structure for the woman's hand is shown in Figure 15.23. Boundary conditions determine initial joint angles and ligament tensions as well as joint loads. Finite-element meshes are combined with boundary conditions and material models. The analysis consists of a series of computer simulations of human joint and prosthetic joint behavior.

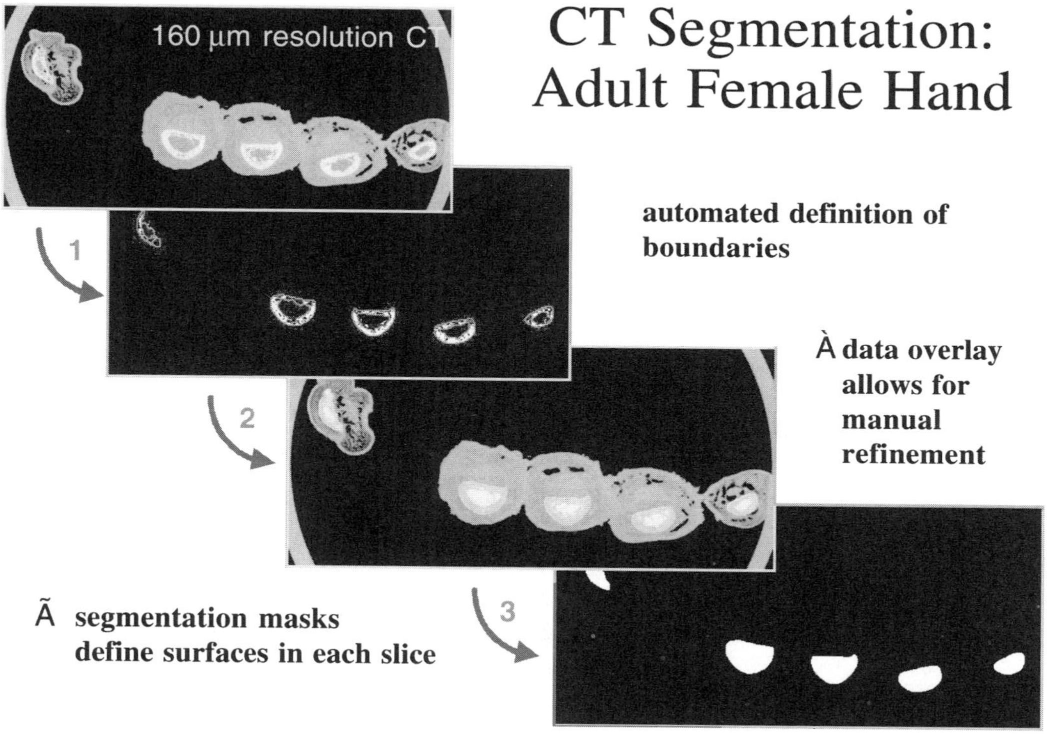

FIGURE 15.22 From left to right: one representative CT slice; automatically defined boundaries for this CT slice; semi-automated segmentation results overlaid onto its CT slice; final surface segmentation results for this slice. After all of the 1084 slices are segmented, they are investigated for interconnectivity and result in a 3D segmentation of the bone surfaces within the woman's hand.

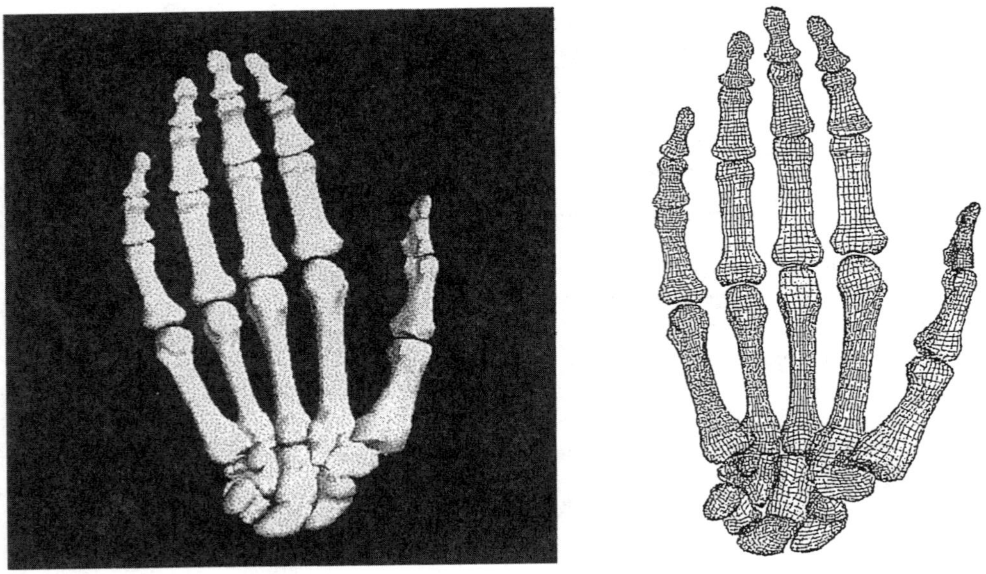

FIGURE 15.23 A 3D surface-rendered image of the bone structure from the CT data of the woman's hand (left); 3D hexahedral finite-element mesh of the segmented bone structure (right).

The simulations provide qualitative data in the form of scientific visualization and quantitative results, such as kinematics and stress-level calculations as shown in Figure 15.24. In human joints, these calculations help us understand what types of stresses occur in daily use of hand joints. This provides a baseline for comparison of the stresses in the soft tissues after an implant has been inserted into the body. Similarly, the bone–implant interface stresses and stresses near the articular surfaces can be evaluated and used to predict possible failure modes.

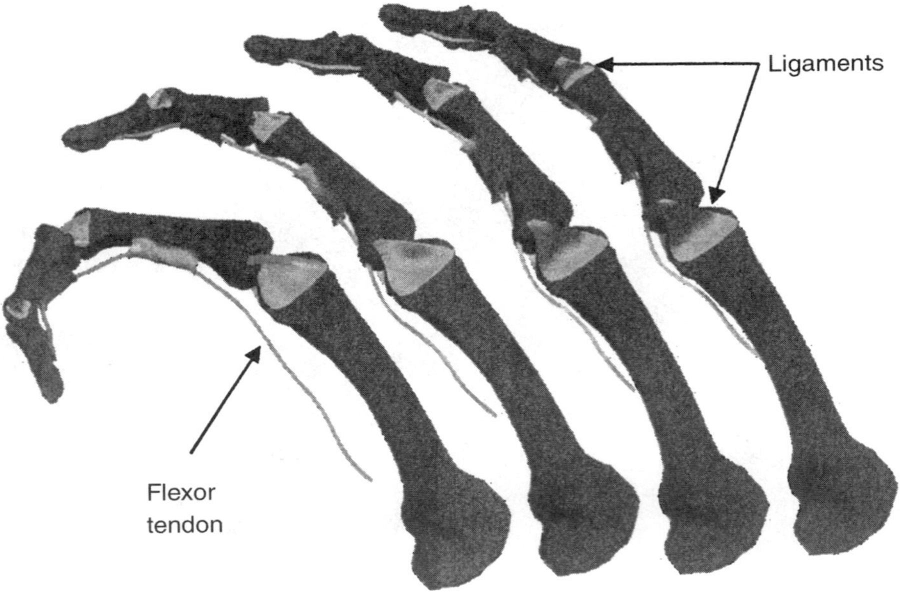

FIGURE 15.24 3D model of the bone structure generated from CT data of the index finger. Different stages of finger flexion were studied by finite-element analysis as shown. The different gray tones within the tendons and ligaments indicate stresses during flexion.

Results from the finite-element analysis are used to predict failure and to provide suggestions for improving the design. Multiple iterations of this process allow the implant designer to use analysis results to incrementally refine the model and improve the overall design. Once an implant design is agreed upon, a prototype is made using computer-aided manufacturing techniques. The resulting implant can then be laboratory tested and put into clinical trials.[49]

15.3.2 Area-Array Nonenergy Discriminating, Detector-Based DR and CT

We have developed several area-array CT scanners at LLNL. These scanners are mainly based on a scintillator lens coupled to CCD cameras as the X-ray detector.[32,45,58,70] Recently, we have been investigating the use of new amorphous-silicon[44] and amorphous-selenium[55] flat panel array detector technology for DR and CT imaging. The area-array scanners developed at LLNL have been used for a wide range of applications.[35,39] We have selected three application studies of different area-array CT scanners.

15.3.2.1 3D CT Analysis of Porosity in Ball Grid Arrays

High-speed I/O switching in electronic circuit boards involves arrays of small solder joints in a complex wiring scheme — commonly called Ball Grid Arrays (BGA). The soldering operations often involve porosity in the solder joint. While the mere existence of a small amount of porosity is not particularly problematic, the presence of porosity in certain parts of the joint volume can compromise functionality. The variety of knobs and settings in the machine soldering process make certain "tuning" choices important for product yield and quality.[52]

Radiography of the BGA (see Figure 15.25) is a typical part of normal operations in this type of circuit board manufacture. However, the radiography is not able to pinpoint the depth of the porosity in the ball joint. Laminographic systems have been fielded with varying amounts of success for inspecting the porosity at different depths in the circuit board. We acquired full 3D CT reconstructed volumes of a number of BGA for evaluation of the usefulness of CT for this type of inspection.

Since the emphasis was upon the 3D aspects of the array, we employed a camera/scintillator glass area detector scanner called PCAT.[70] Also, in order to get a large sample of the balls, we used a 5.08-cm field of view to include 121 balls in the same scan. The board was scanned "standing up" (see Figure 15.25), resulting in a high aspect ratio for the rotate-only scan. We used 300 kV to penetrate the long aspect of the board, which also included enough fidelity to image the porosity in the balls for the short aspect of the board. We used a rotation-only scan of 360 views over 360° and reconstructed

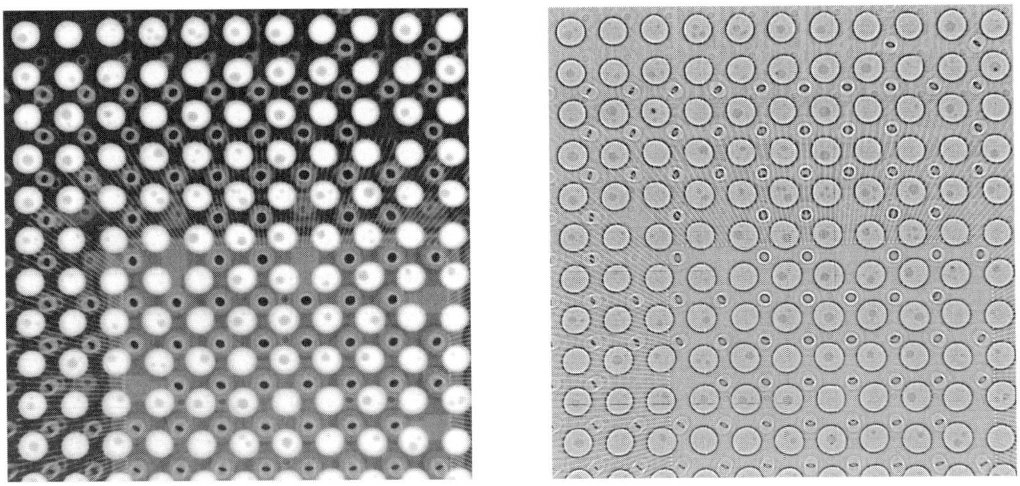

FIGURE 15.25 Representative DR of a BGA in a circuit board (left); results of an unsharp mask transform of the DR image (right).

the data with a standard convolution backprojection method. The goal of this scan is to unravel the position of the porosity in the thickness of each of the balls and quantify the changes in porosity in all three dimensions.

Different slices through the 3D scanned BGA volume are shown in Figures 15.26 and 15.27. The tendency of the porosity to be located closest to the fiberglass portion of the circuit board is shown in Figure 15.26. This is a single, horizontal cross-sectional slice for a single BGA row of balls. Since this scan is 3D, we can obtain slices through the volume in any direction. Different vertical slices along the plane of the board (Figure 15.27) show the tendency of the porosity to be located close to the circuit board.

The 3D volumetric data enable a number of calculations, which are difficult to perform with radiographic or limited laminographic data. A segmentation technique was applied to each slice of the volume in the plane of the board, isolating the total solder ball joint volume within each slice plane. Within each ball volume, a threshold was applied to identify the porous sections of the balls. The total porous volume was divided by the total ball volume to obtain a fraction porosity per slice in the 3D cube of data. Figure 15.28 contains the resultant plot from this calculation. Notice the high fraction of porosity at the slices closest to the board base and the lack of porosity in the middle of the ball joints. These kinds of data can provide crucial information for understanding the parameters required to minimize porosity in the ball joint welding process.

15.3.2.2 3D CT Analysis of Metal Powder Filling Methods

Powdered metallurgy has a number of important advantages over many types of casting processes, chiefly in terms of part cost. Casting processes require a substantial investment in supporting systems (metal furnaces, metal transport, subsequent trim and extensive machining operations, etc.). Powdered metallurgy is much simpler, only requiring systems for powder transport and filling prior to pressing and possible sintering. Parts emerge from the powdered metal press in near net shape, requiring little subsequent processing of any kind. While there are more and more applications for powdered-metal parts, the range of application for powdered metallurgy would increase dramatically if parts could be made with greater toughness and more uniform density.

The principal challenge for powdered metallurgy is configuring the filling process to eliminate voids and density gradients. It has been shown that in down-hole scans of soil and rocks, packing flaws form in the regular layering of soil on layer of soil. These flaws are extremely resistant to changes by high-pressure packing forces.[66] In the same way, packing flaws in powdered-metal parts form at the point of filling and leave voids in the part even after large pressing forces are applied. To get a 3D understanding of the powdered-metal fill process, we configured a fixture for filling a 7.6-cm diameter cylindrical volume with powered metal.

We performed two experiments with an iron-based powder, which has substantial density and atomic number. The cylinder volume was filled using a "simple fill" method from a container of powder passing over the cylinder opening. We performed the fill on the rotational stage of the PCAT scanner so that we did not have to move the configuration. Three metal and three plastic beads were mixed into the powder as indicators of scan quality. We used PCAT to acquire 360 2D projections over 360° by a third-generation, rotate-only scan of the cylinder with powdered metal. Due to the substantial X-ray attenuation of the iron powder in the cylinder, an energy of 450 kV was employed to obtain adequate signal through the object. After the initial scan, the powdered-metal volume was changed by insertion of two different diameter plastic rods into the cylinder of powder. Another scan identical to the first scan was acquired. The experiments were designed to determine changes in porosity and powdered-metal particle packing or density.

Both sets of CT data were reconstructed using the stack wise convolution backprojection method. Each CT 3D volume was reconstructed into ~100 × 100 × 100 μm³ voxels. Subsequently, the two 3D CT scans were subtracted from each other to highlight the changes in packing density. The subtracted image revealed the dimensions of the plastic rods. The fidelity in the subtracted image is a measure of our ability to identify changes in the packing of the powder volume with this dual-scan CT method.

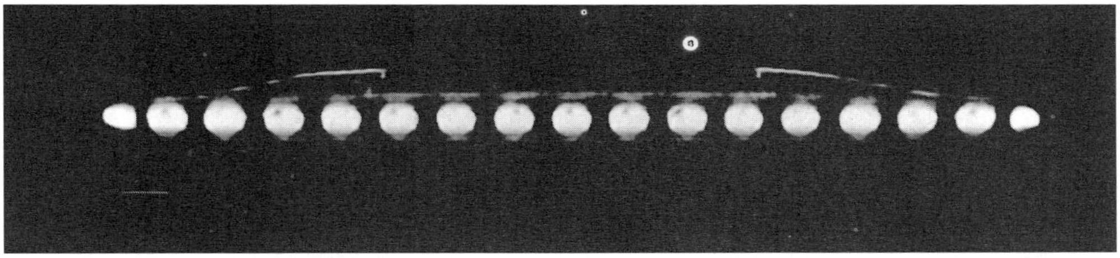

FIGURE 15.26 Horizontal cross-sectional CT slice of the BGA reveals the location of porosity within each solder ball.

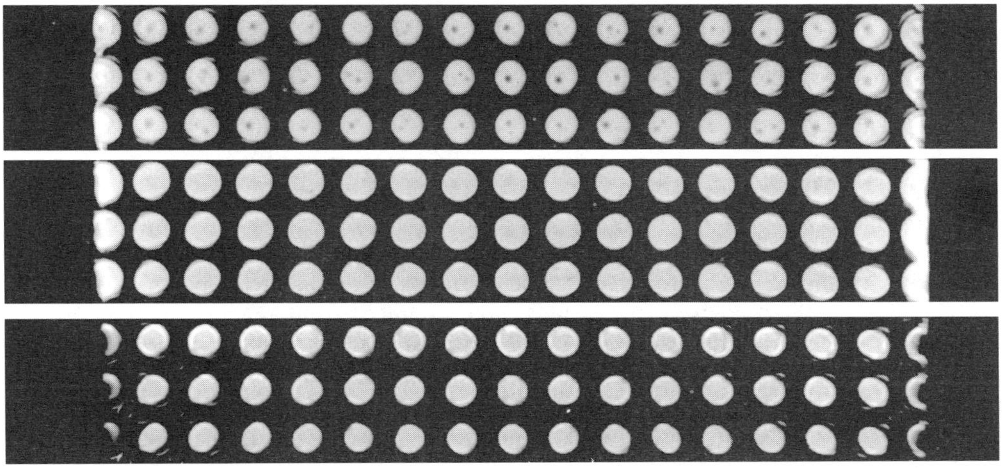

FIGURE 15.27 Three vertical CT slices of the BGA. The slice plane locations from top to bottom, respectively, are near to far from the fiberglass portion of the circuit board. Note that near the fiberglass the solder balls have more porosity.

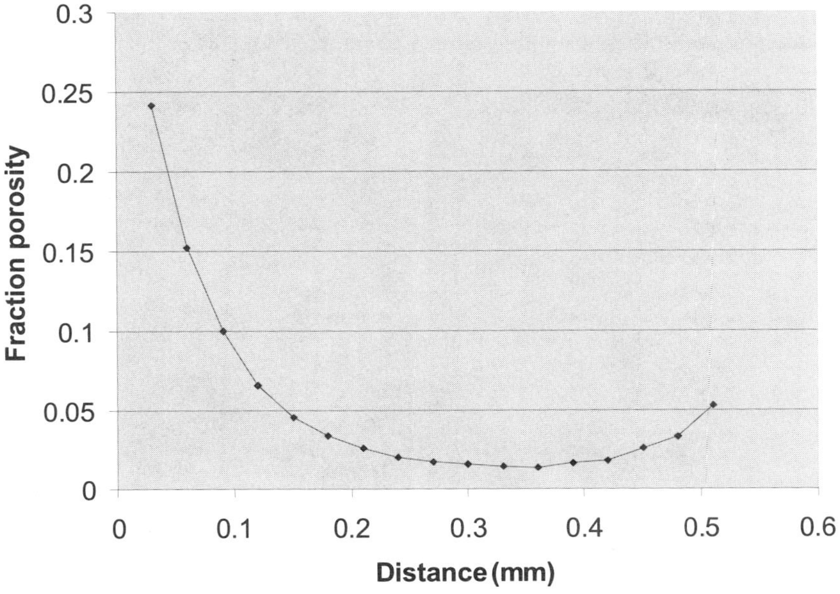

FIGURE 15.28 Plot of fraction of porosity for all the balls to the total volume of balls vs. distance from circuit board fiberglass base.

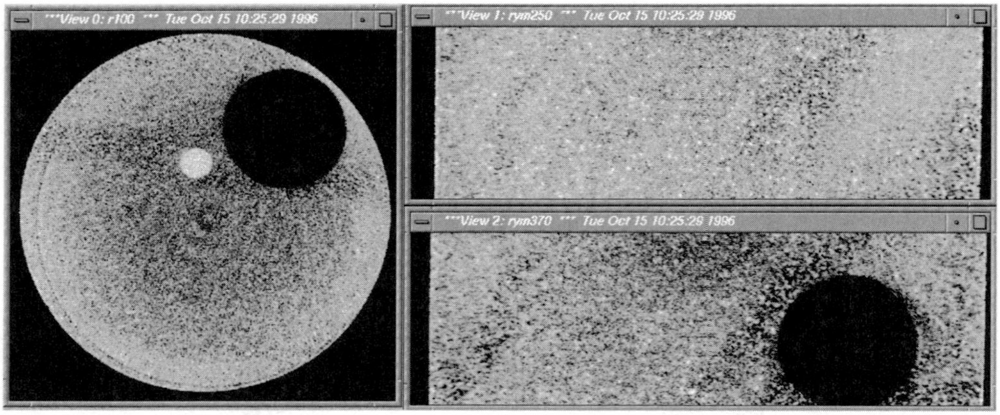

FIGURE 15.29 Horizontal (left) and vertical (right) cross-sectional slices through the powdered-metal cylinder volume reveals the powdered-metal (gray), steel (white), and plastic (black) beads.

Representative CT slices through the cylindrical volume along different axes are shown in Figure 15.29. These images show the differences in X-ray attenuation. The darker areas indicate material with lower density and/or atomic number — for example, small void spaces within the powder — while the lighter areas indicate material with higher density and/or atomic number within the image — for example, the steel beads. Also included in the ensemble of slices are cross sections of the steel (white) and a plastic (black) bead. Notice the attenuation of the steel ball is greater than the powder because of density, but the powder is more attenuating than the plastic due to density and atomic number.

A single CT cross-sectional slice through the powder before and after inserting the plastic rods is shown in Figure 15.30. Also included in this particular slice is a void space that occurred as a result of the filling process. Notice the small changes in the character of the void space even after the insertion of the rod. This shows how small changes in the powder packing have little result on features small distances away from the change in packing density.

To obtain a clear picture of the extent of the changes stemming from the insertion of the rods, the two 3D CT scans were subtracted from each other. The subtracted image of the two single slices shown

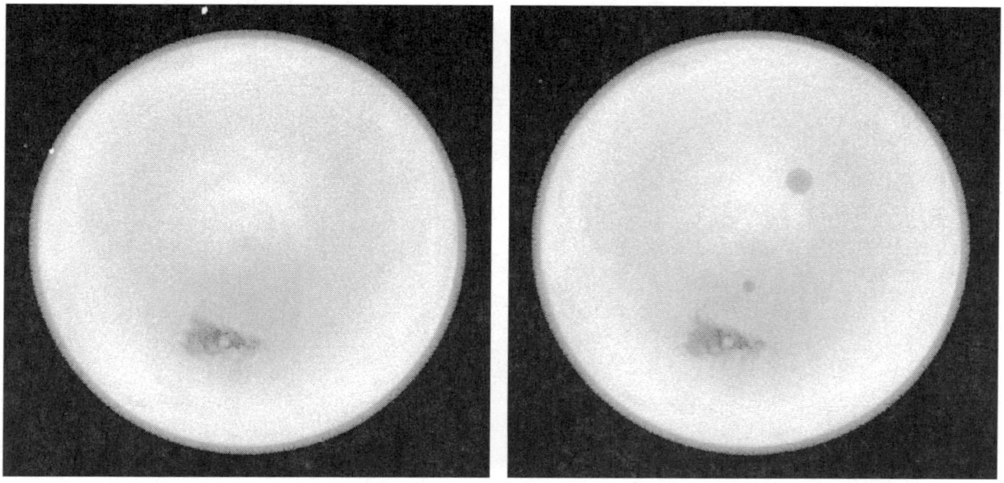

FIGURE 15.30 Horizontal slices of the powdered-metal cylinder volume before (left) and after (right) insertion of two different diameter plastic rods.

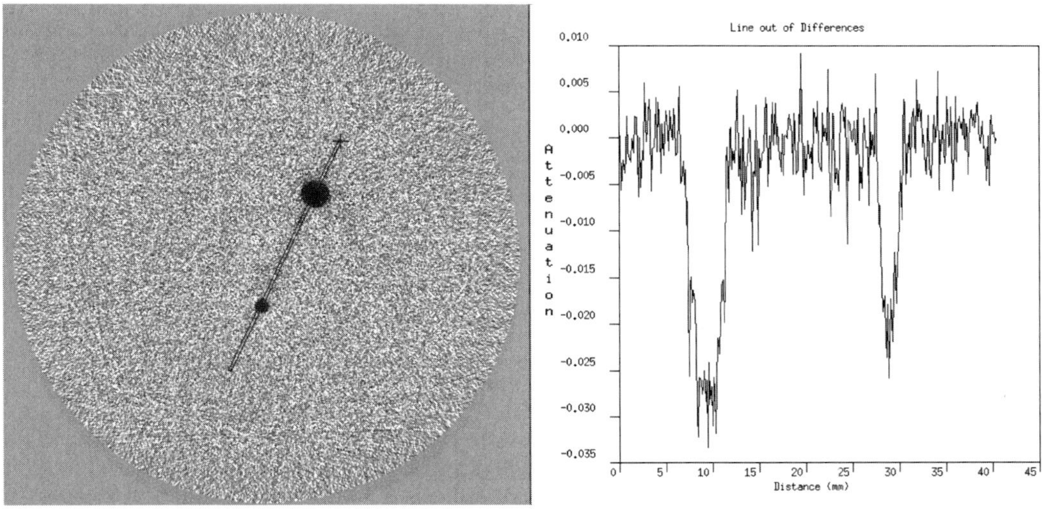

FIGURE 15.31 Difference image for the two single cross-sectional slices shown in Figure 15.30 (left); profile for the line shown within the image to the left (right).

in Figure 15.30 is shown in Figure 15.31. The packing void in the lower middle-left section of the powder volume (Figure 15.30) is not shown within the difference image given in Figure 15.31. This indicates that the insertion of the rods did not disturb the particles around the packing void to any appreciable degree.

Different vertical slice images through the subtracted volume are given in Figure 15.32. The images show local changes in packing density within the vicinity of and down the length of the rods. Line outs from the vertical slices are also shown in Figure 15.32. In the case of the large rod, the particle density decreased directly around the rod. We expect that the particles displaced by the rod were pushed to the top and to the side in the cylinder volume. The motion used to insert the large rod was a series of pushes in the powder. In the case of the small rod the local density increased. This rod was shoved into the powder in a single thrust.

These scans show the kind of 3D detail that can be obtained from CT scanning of the powder fill process. The subtraction method can reveal 3D changes in packing density, provided scans can be

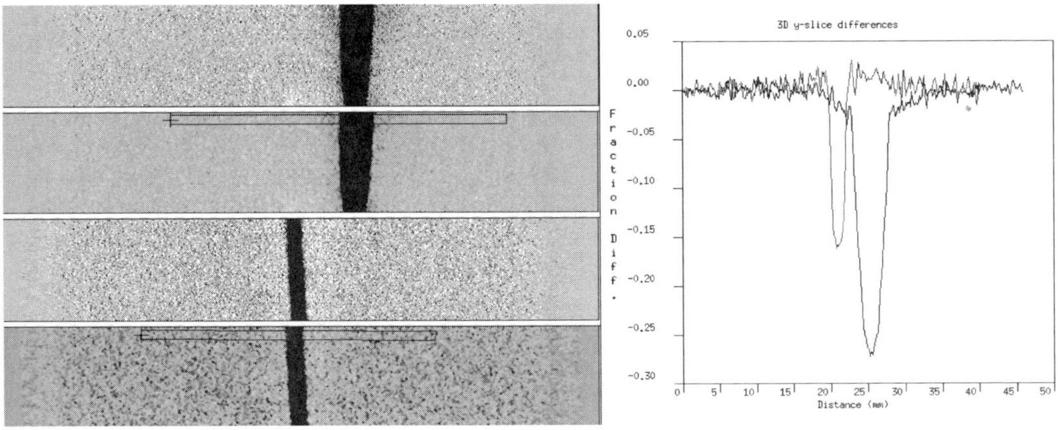

FIGURE 15.32 Vertical CT slices (left) through the rod inserts and line outs (right) showing changes in local density near the top of each rod.

registered between powder fill operations. The changes in density can be viewed in all three dimensions and compared to expectations or possibly to 3D models of the powder fill process.

15.3.2.4 Inspection of Bridge Pins

Bridge pins were used in the hanger assemblies for some multi-span steel bridges built prior to the 1980s, and are sometimes considered fracture critical elements of a bridge. For example, a bridge pin failure was the cause of the 1983 collapse of the Mianus River Bridge that resulted in the deaths of 3 people. Bridge pins (Figure 15.33) are typically made of steel, ~20 cm long with a 7.6-cm wide barrel and threaded ends of ~6 cm outer diameter. The pins typically fail at the shear plane, the region where the shear forces from the hanger plate are transmitted to the web plate and where the majority of flaws so far are detected.

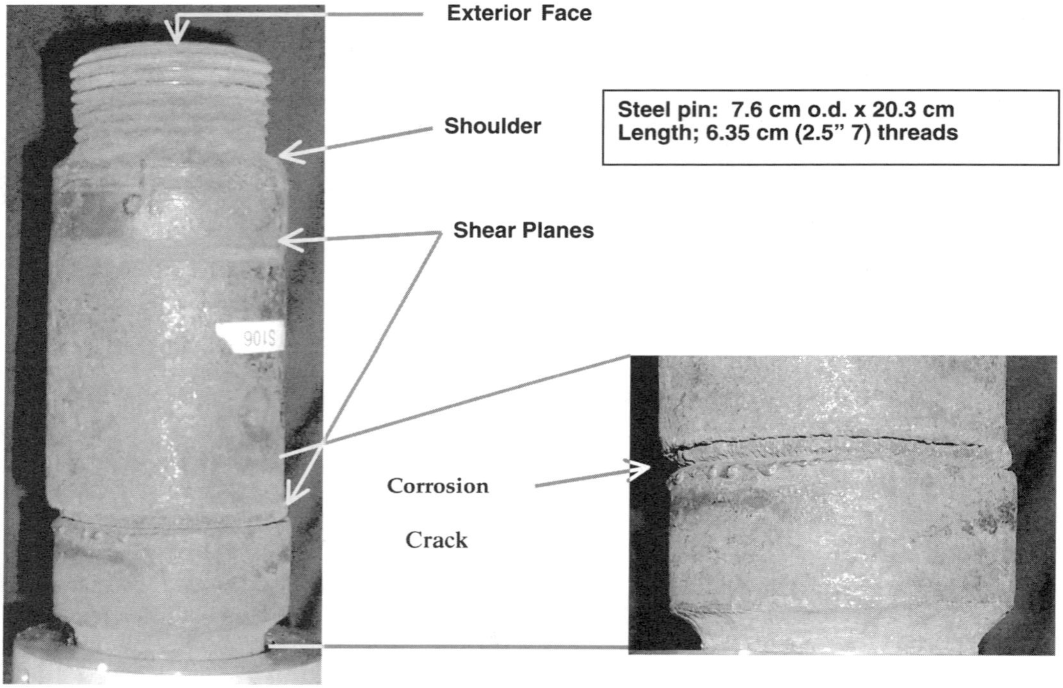

FIGURE 15.33 Photographs of a surface-cracked bridge pin. Details of the pin are labeled in the figure.

The FHWA recently acquired several pins that were removed from a bridge in Indiana following an ultrasonic field inspection. Some of these pins were cracked, and this provided the opportunity to better understand the failure mechanism(s) of bridge pins using laboratory NDE techniques. Visual inspection and laboratory inspection techniques such as ultrasonic C-scans and film radiography were performed by the FHWA. Qualitatively, these laboratory methods yielded similar results as found in the field inspections.

At LLNL, we applied DR and CT characterization methods to further study the bridge pins.[76] A 9-MV LINAC X-ray source was combined with a new amorphous-silicon, flat panel detector[77] to acquire digital radiographs of three pins (see Figure 15.34) and CT projections of one of the severely cracked pins. Other research by Dolan et al.[44] has verified the higher spatial resolution that can be obtained by the use of the amorphous-silicon technology. Scatter is reduced in the amorphous-silicon detector due to the thin scintillator selected. This is analogous to high-energy X-ray film methods.

The CT data consisted of 180 projections over 180° and were reconstructed into several hundred tomograms (cross-sectional images) with a volume element (voxel) size of ~127 × 127 × 127 μm³. Two tomograms are shown in Figure 15.35, and three tomograms are shown in Figure 15.36. It is useful to

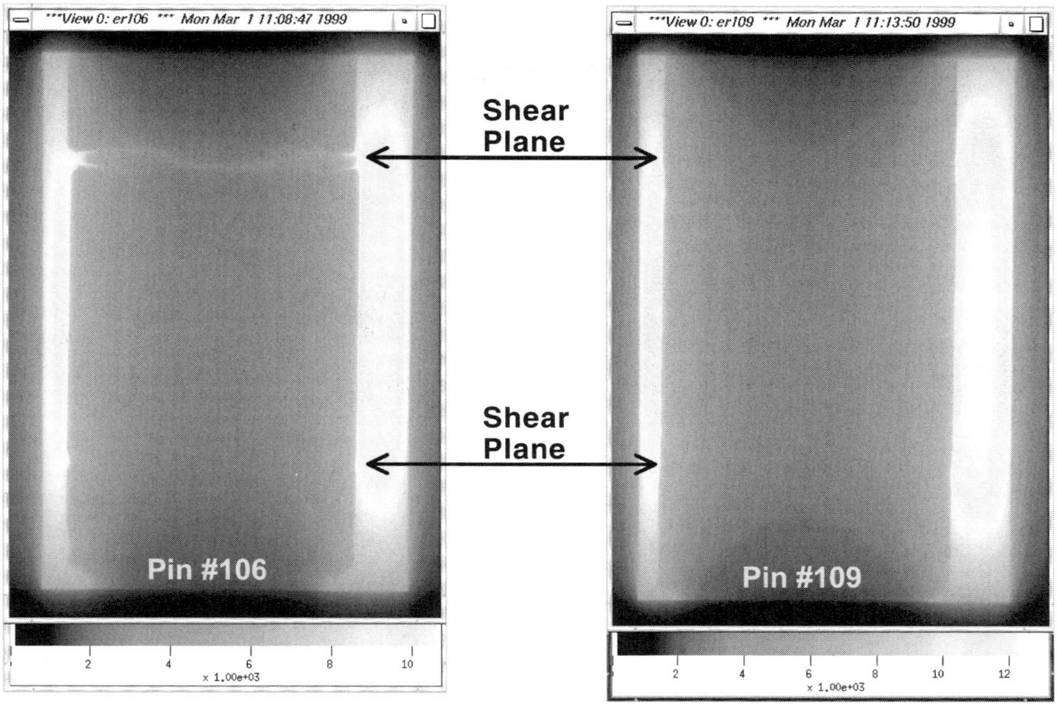

FIGURE 15.34 Representative digital radiographs of two bridge pins. The one on the left is clearly cracked. The one on the right does not appear to have any cracks.

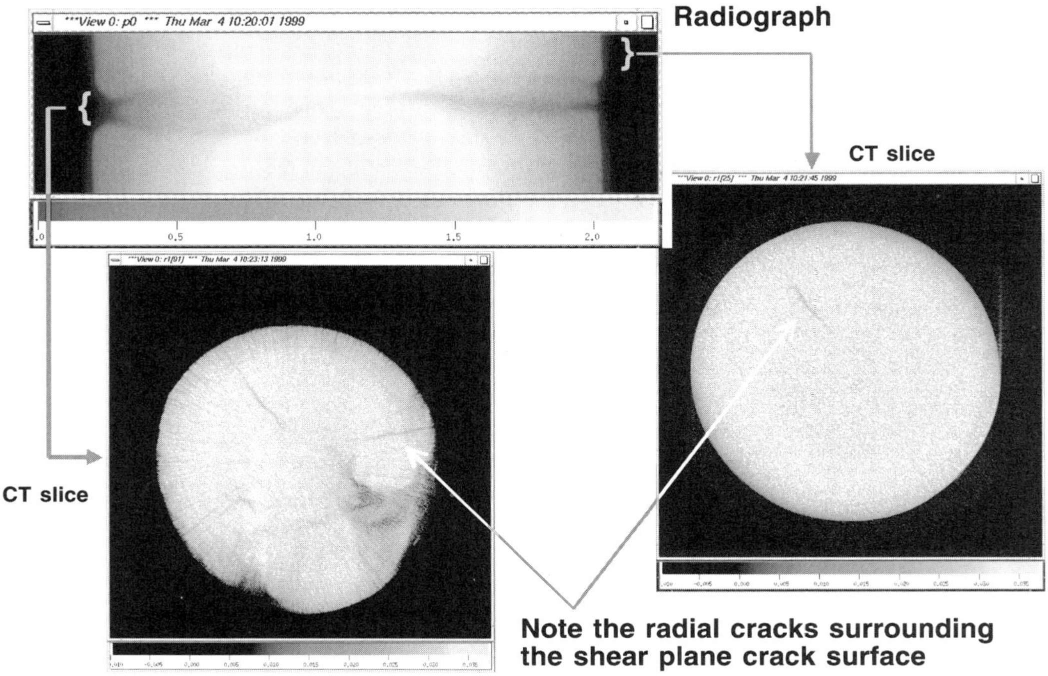

FIGURE 15.35 Representative tomograms for two planes within the cracked bridge pin #106.

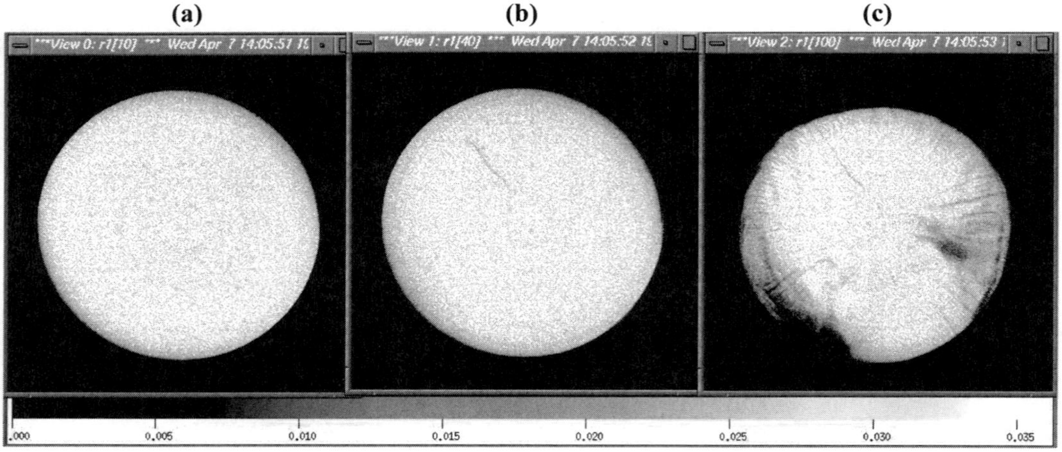

FIGURE 15.36 Tomograms along the longitudinal axis of a cracked bridge pin. The tomograms are (a) far from, (b) near, and (c) within the cracked shear plane region. The gray scale relates gray tones to relative attenuation in mm^{-1} units.

point out that far from the shear plane, Figure 15.36a, there are no apparent cracks or voids in the tomograms. Near the shear plane, Figure 15.36b, a crack appears within the pin at ~11 p.m. Within the shear plane, Figure 15.36c, we observe several cracks. The large surface-breaking cracks from 6 to 8 p.m. in Figure 15.36c are observed visually and ultrasonically. Several smaller radial cracks are revealed in the tomograms that are not observed either visually or in the ultrasonics data. These small radial cracks, Figure 15.36c, are only in the shear plane region and most likely lead to the larger cracks, which most likely cause pin failure.

CT reveals many more internal defects within the bridge pins than those observed in visual, radiographic, or ultrasonic inspection. For example, we observe large and small radial cracking and are able to measure crack sizes in all three spatial dimensions. To first order the field and laboratory ultrasonic inspection results were qualitatively corroborated by the large cracks observed in the X-ray tomograms. This provides some level of confidence in the ultrasonic field inspection method. Further work is required to quantitatively corroborate and correlate the X-ray tomographic data with the field and laboratory ultrasonic inspection results, as well as to obtain a better understanding of bridge pin failure mechanisms and to improve filed inspections reliability in predicting bridge pin failure.

15.3.3 Single, Energy Discriminating Detector-Based CT

Single, energy discriminating, detector-based CT scanners are unique to industrial CT imaging applications. These are the most quantitative and artifact-free CT scanners. However, they are very slow when compared to linear- and area-array detector scanners. They are often used for research to characterize objects when you need the most quantitative and artifact-free CT data available and can afford the time to acquire the data. At LLNL, we have developed several single, energy discriminating, detector-based CT scanners[31,67] to characterize materials[54,59,60,65] and to nondestructively radioassay waste drums.[68] Here, we describe three applications that range from a DC motor through explosives to waste drums.

15.3.3.1 Material Analysis of a DC Motor

Monoenergetic CT imaging of a motor was used to study how this technique could be used to distinguish between the different materials that make up a DC motor. This may be useful, for example, if you needed to cut the motor into smaller pieces and to do so in an efficient manner to yield pieces that could be recycled and some that can be disposed of properly. What we found is that a motor contains several materials from plastics to lead.

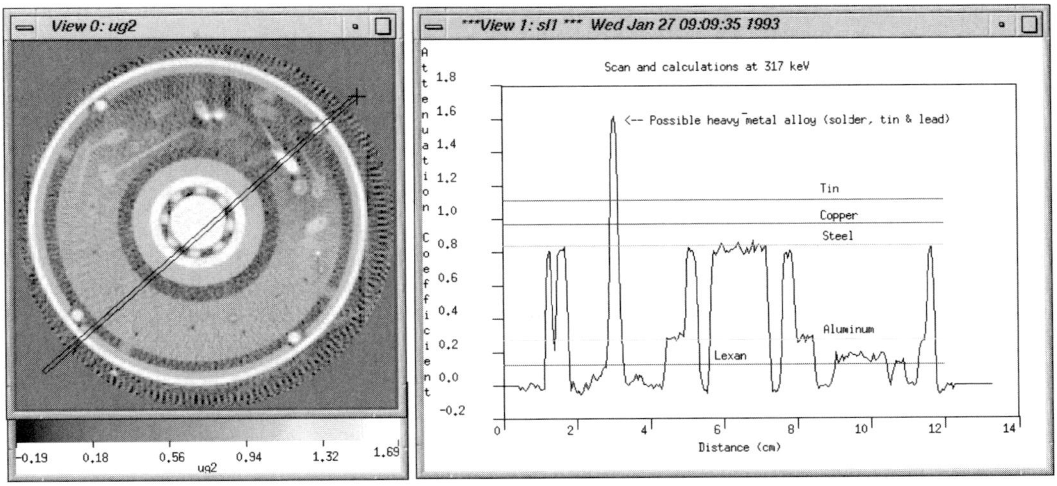

FIGURE 15.37 Materials contained in a DC motor (from plastics to lead). The line out is overlaid with horizontal lines that represent the tabulated linear attenuation coefficient for several materials.

Many of these materials are shown in Figure 15.37. These data were acquired with an ^{192}Ir radioisotope and an energy discriminating (high-purity germanium) detector. The results for the 317-keV X-ray radiation peak reveal the material that make up the motor. The line out in Figure 15.37 is overlaid with horizontal lines that represent the "theoretical-tabulated" linear attenuation coefficient[42] at 317 keV for several materials as labeled. As expected, the housing, bearings, and race are made of steel. There are also aluminum (just around the race) and plastic parts (circuit board). Another analysis method is to plot a histogram of the pixels within the slice plane in Figure 15.37 as shown in Figure 15.38. This reveals the amount of materials by a simple integration of the total number of voxels under each peak. For monoenergetic CT, there is a simple correlation of the "theoretical/tabulated" and CT measured linear attenuation coefficients, which is not the case for polyenergetic sources. However, due to the time required to obtain this pristine data, they are not often used.

15.3.3.2 Material Analysis of High Explosives

CT techniques have been investigated to characterize high-explosive (HE) materials and to improve the current manufacturing process flow. To meet this objective, we obtained a series of CT data of several explosives. Some of these CT experiments are discussed here and in more detail elsewhere.[57,58,65] These

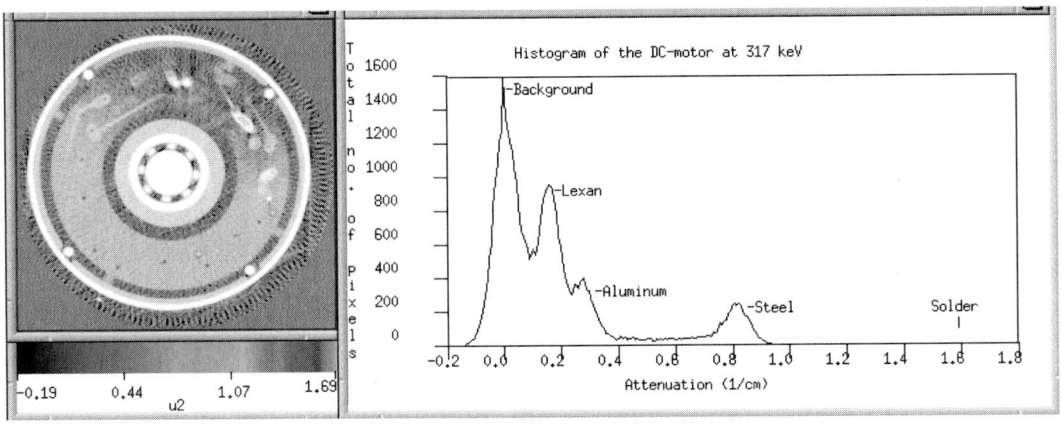

FIGURE 15.38 Histogram of the DC motor within the slice plane shown to the left in Figure 15.37.

CT inspections were focused on determining HE material uniformity: density, chemical composition, and void/crack content. If X-ray CT can adequately replace the current techniques, overall safety and process flow time and costs could be significantly reduced.

PBX9502 material properties result from its two constituents and the pressing process. Using Kouris' et al. chemical formula mixture rule,[51] effective-Z is calculated to be 7.1 for TATB and 12 for Kel-F. The PBX9502 pellet bulk proportions are 95% TATB and 5% Kel-F, resulting in an effective-Z of 7.6. Prior to the formulation, the density of TATB is ~1.938 g/cm^3 and ~2.02 g/cm^3 for Kel-F. For the Phase I study, the pressed PBX9502 density value is ~1.9 g/cm^3.

Different sets of explosive pellet pressings were fabricated. The explosives discussed here were cylindrical in shape and differed only in length. One set consisted of four pellets ~1.27 cm diameter and ~1.27 cm long, referred to as the small pellets. A set of mock (a material that simulates explosive materials, but is not explosive) pellets of this dimension was also pressed. The second set consisted of three pellets ~1.27 cm diameter and ~4 cm long, referred to as the medium pellets. For the small pellets, all four were placed within the experimental apparatus at once for scanning. This allowed CT experiments to be conducted on all four small pellets simultaneously, while only one medium pellet was studied at a time.

A single detector, pencil-beam computerized axial tomography (PBCAT) scanner[31] was used to nondestructively evaluate both the small and medium PBX9502 pellets in two dimensions. The PBCAT scanner uses the first-generation ("translate/rotate") scan geometry. In this study, an ~63-mCi ^{109}Cd (energy levels: 22, 25, and 88 keV) radioisotopic source, a high-purity intrinsic-germanium detector, γ/X-ray spectroscopy electronics, and a multi-channel analyzer were used to acquire the sinogram data. All PBCAT reconstructed images are presented as values of absolute linear attenuation coefficients (cm^{-1}).

One single medium pellet was scanned at two slice planes using PBCAT.[57] Only the slice plane at the pellet center, perpendicular to its longitudinal axis, will be discussed. This sinogram data set consisted of 90 projections at 2° intervals over 180° and 40 ray sums with a 1-mm^2 source and detector aperture and 1/2-mm translational overlap, resulting in a spatial resolution of ~1 mm. The total data acquisition time was 4 days. Photon counting statistics, on the average, were ~1 and ~3% for the 22 and 25 keV ray sum data, respectively. The medium pellet sinogram data were reconstructed into 40 x 40 pixel images (Figure 15.39).

A qualitative inspection of and comparison among the CT image data reveals no major anomalies (i.e., cracks, voids, and/or inclusions) within each PBX9502 pellet or from pellet to pellet at the slice plane(s) chosen. On the other hand, a CT study of mock HE pellets using PBCAT revealed a major anomaly (an inclusion) in one mock pellet (Figure 15.40). The corresponding line out shows the amplitude (1.4 cm^{-1}) and size (1 mm^2) of the inclusion present.

Overall, the linear attenuation coefficient images are uniform (Figure 15.39). The average small pellet PBCAT data resulted in mean and σ linear attenuation coefficient values of 1.1 ± 0.1 and 0.9 ± 0.2 cm^{-1} at 22 and 25 keV, respectively. The medium pellet's PBCAT data resulted in mean and σ values of 1.12 ± 0.03 and 0.87 ± 0.7 cm^{-1} at 22 and 25 keV, respectively. These data agree within experimental error with the individual values and with the predicted values 1.11 ± 0.03 and 0.86±0.03.[57]

Three conclusions are drawn from the CT nondestructive inspection of PBX9502: (1) both small and medium pellets contain a high degree of internal uniformity, as measured by the linear attenuation coefficient for the PBCAT data; (2) the uniformity is maintained from pellet to pellet; and (3) in spite of this overall internal uniformity there appear to be small deviations due to changes in Z and/or ρ (discussed in detail below).

For the PBCAT data, a method has been determined to separate the measurement and reconstruction errors from the Z and/or ρ deviations.[57] Using this method, a detailed analysis of the CT data (Figure 15.39) shows that a small proportion of the linear attenuation coefficient pixel-to-pixel deviations is not due to a combination of measurement and reconstruction errors; thus, we attribute the deviations to be due to subtle changes in Z and/or ρ.

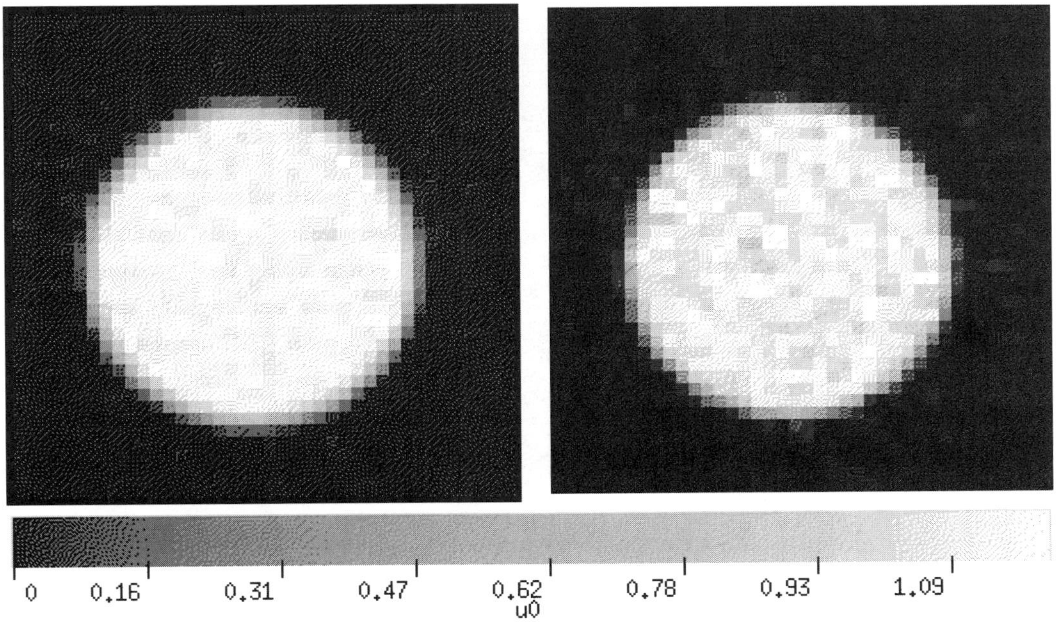

FIGURE 15.39 Reconstructed images from the PBCAT 22 (left) and 25 (right) keV sinogram data for the medium pellet. The gray scale is in cm^{-1}.

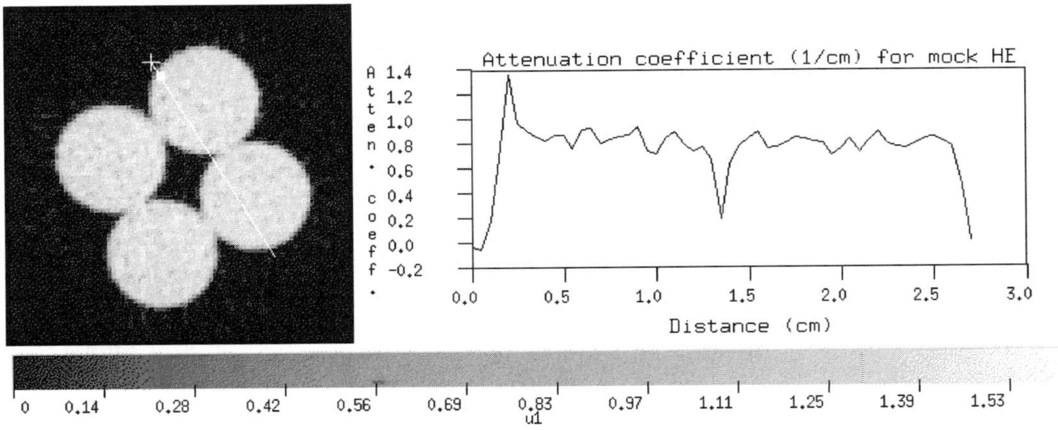

FIGURE 15.40 A reconstructed image from the PBCAT 22-keV sinogram data of mock HE pellets (left). The location of an extracted line out is shown by the white line in the image. The line out (averaged over three adjacent pixels) is plotted to the right.

Uniformity of the of the attenuation coefficient notwithstanding, there could be variation in the distribution of Z and ρ. In PBX9502, changes in Z are a result of a variation in the distribution of TATB ($Z = 7.1$) and Kel-F ($Z = 12$) within the pressed pellet. Density variations can be a result of nonuniformities in the pressing process or from higher amounts of Kel-F, which has a slightly higher density than TATB. Quantitative multiple energy data generated by PBCAT for the medium pellet were used to calculate *effective-Z, weight fraction,* and *density* images (see Figure 15.41).[59]

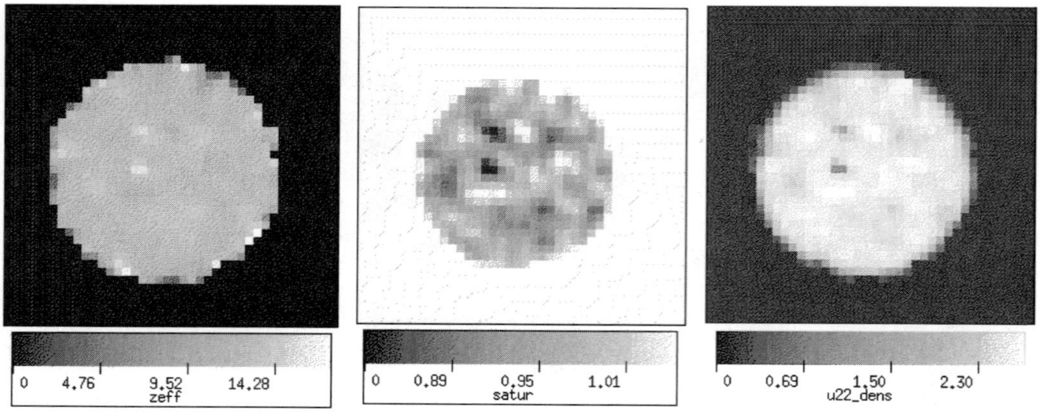

FIGURE 15.41 Resultant effective-Z (left), TATB weight fraction (middle), and density (right) images for the medium pellet. The gray scale is given in effective-Z, TATB weight fraction, and density (g/cm^3) units, respectively.

At the slice plane chosen, the experimentally determined mean effective-*Z* is 7.8 ± 0.5, which is within experimental error of the predicted value, 7.6. The variations in the effective-*Z* image are due solely to changes in the PBX9502 constituents (TATB and Kel-F) and not upon density. This is just one component of the linear attenuation coefficient pixel-to-pixel variations (Figure 15.39). The weight fraction coefficient calculations involve both tabulated attenuation coefficients and nominal density values for TATB and Kel-F. The experimentally determined TATB weight fraction mean is 0.94 ± 0.03 (Figure 15.41). The expected TATB weight fraction is 0.95. The experimental and expected values are well within experimental error. The pixel-to-pixel variation in the weight fraction image tracks with the variation in the effective-*Z* image; i.e., higher weight fraction values for Kel-F correspond with higher effective-Z pixels.

The density image for the medium pellet also is presented in Figure 15.41. The experimentally determined slice plane mean density value is 1.9 ± 0.2 g/cm^3 and is within the wet/dry bulk density measured value of 1.904 g/cm^3. This agreement provides some credibility to the density values on a pixel-to-pixel basis within the density image. There is considerably more variation in the pixel-to-pixel density values than in the linear attenuation coefficient images (Figure 15.39).

Comparison of all three images in Figure 15.41 reveals two small regions in the top left quarter (near the center) of each image that stand out. For both regions, the effective-*Z* values are high and the TATB weight fraction values are low, implying a large concentration of Kel-F, but the density values are unexpectedly low. The low density values can be explained by either TATB and/or Kel-F with a large void fraction. The effective-Z and weight fraction results suggest that the small regions are Kel-F with a larger than normal void fraction.

Overall, the linear attenuation coefficient data reveal that the PBX9502 is very uniform. However, there are a few variations in these data that are larger than the calculated error bars (determined from a combination of measurement and reconstruction errors). The experimentally determined effective-Z, weight fraction, and density results suggest the following:

- The mean effective-Z value is slightly higher than, but within error of, the predicted value.
- The Kel-F weight fraction value is slightly higher than, but within error of, the formulated value.
- The linear attenuation coefficient pixel-to-pixel variations are a function of a nonuniform distribution of TATB, Kel-F, and density.

In summary, a comparison between CT and the current manufacturing inspection techniques reveals that CT may be able to replace most, if not all, of the current techniques used. If the small and medium pellet results are scalable to full-size explosives, then CT is a likely candidate for an automated high explosives inspection system. It should be pointed out that CT can be used to determine gradients not only in density, but in chemical constituents as well. Therefore, it may be necessary to reinvestigate the current bulk density specification with respect to wet/dry measurement techniques.

15.3.3.3 γ-Ray Nondestructive Radioassay for Waste Management

Before drums of radioactive or mixed (radioactive and hazardous) waste can be properly stored or disposed of, the contents must be known. Hazardous and "nonconforming" materials (such as free liquids and pressurized containers) must be identified, and radioactive sources and strengths must be determined. Opening drums for examination is expensive, mainly because of the safety precautions that must be taken. Current nondestructive methods of characterizing waste in sealed drums are often inaccurate and cannot identify nonconforming materials.*

Traditional NDA measurement errors are related to nonuniform measurement responses associated with unknown radioactive source and waste-matrix-material distributions. These errors can be reduced by the application of imaging techniques that better measure the spatial locations of sources and matrix attenuations.[46,47,53] LLNL has developed an emerging γ-ray NDA technology, called active and passive computed tomography (A&PCT), that identifies and accurately quantifies all detectable radioisotopes in closed containers of wastes, regardless of their classification: low level, transuranic, or mixed waste.[61]

The A&PCT technology uses two separate measurements. The first is an active interrogation of the drum using an external radioactive source(s), and the second is a passive measurement of the radioactive emissions from the drum. The results of these two measurements are combined to produce an attenuation corrected γ-ray assay of the drum. The γ-ray A&PCT method[61] involves (1) data acquisition,[68] (2) γ-ray spectral analysis,[43] and (3) Image reconstruction and assay.[50]

Currently, there are two operational single detector A&PCT systems Equation 15.12. One is located at LLNL and is used for research (see Figure 15.42). The A&PCT technology has been transferred to a private sector company, Bio-Imaging Research, Inc. (BIR). BIR has built a mobile Waste Inspection Tomography (WIT) trailer that contains the other single detector A&PCT technology.[36,37] The single detector A&PCT system performance has been determined by several open and blind tests of well-known radioactive sources within a variety of waste matrices in addition to several actual waste drums.[61] At LLNL, WIT characterized drums that contained smaller containers with solidified chemical wastes; at RFETS,** WIT measured drums with low-density combustible matrices; at INEEL,*** WIT characterized graphite-, glass-, and metal-matrix drums, lead-lined drums with combustibles, and very dense sludge drums. The plutonium mass within these drums ranged from 1 to 70 g.

Recently, in order to increase throughput, a multiple detector A&PCT system was jointly developed by LLNL and BIR. This system uses a multiple collimated aperture (*width*: 5.7 cm; *height*: 5.7 cm; *length*: 12.7 cm) with septa for six, high-purity germanium detectors (coaxial, ~65% relative efficiency) and six ^{152}Eu (~7 mCi) external radioactive sources. It is housed within a mobile land/sea container and is called the WIT container. The drum manipulator can handle up to 365 kg. The system can be operated in two different modes: (1) collimated γ-ray scanner (CGS) or (2) A&PCT. The CGS mode is used to determine the A&PCT slice positions and data acquisition time. The acquisition time is a function of radioactivity and matrix absorption. A&PCT is used to accurately determine the isotopics of radioactive source(s) and their activity.

There is a well-defined series of open and blind tests required of NDA measurement systems used to characterize transuranic waste prior to shipment to the Waste Isolation Pilot Plant.[64,74] A DOE sponsored performance demonstration program has built a set of well-known radioactive standards that can be inserted into a set of drums with well-known mock waste matrices. We have used these drums and standards to determine the preliminary precision (% relative standard deviation), accuracy (% recovery), and bias for the six-detector A&PCT system.

* References for contemporary research, development, application, and implementation of NDE and NDA systems are given in Proceedings of the NDA/NDE Waste Characterization Conferences.[63]

** Rocky Flats Environmental Technology Site, Rocky Flats, CO.

*** Idaho National Engineering and Environmental Laboratory, Idaho Falls, ID.

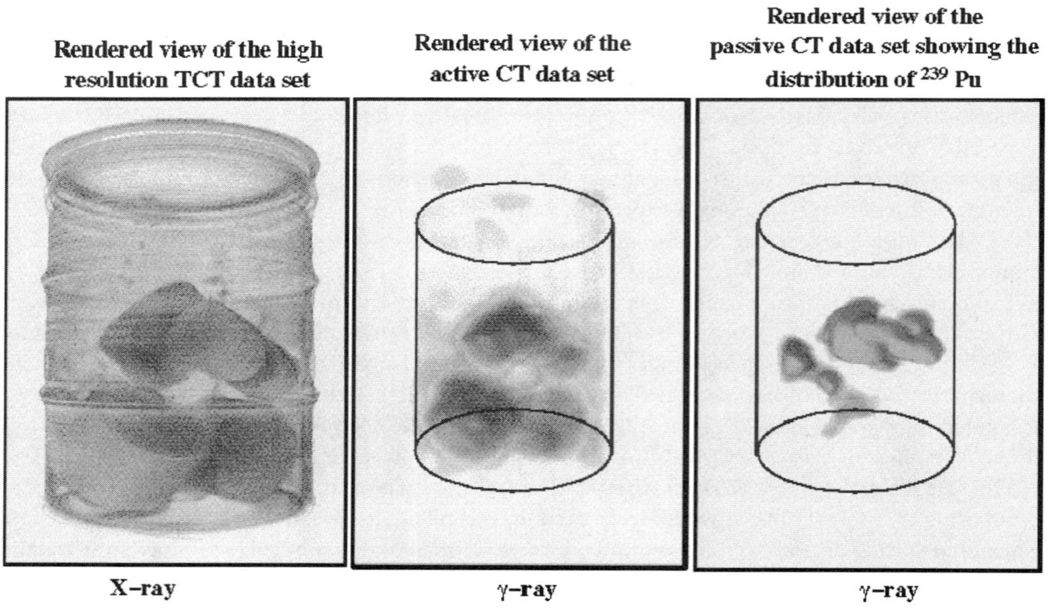

FIGURE 15.42 Representative 3D rendered CT images of an LLNL transuranic waste drum. High spatial (2-mm voxels) with no energy resolution X-ray CT image at 4 MeV reveals the relative attenuation caused by the waste matrix (left). Low spatial (50-mm voxels) with high-energy resolution active γ-ray CT image at 411 keV of the same drum reveals the quantitative attenuation caused by the waste matrix (middle). Low spatial (50-mm voxels) with high-energy resolution PCT image at 414 keV reveals the location and distribution of radioactive ^{239}Pu in the drum. A&PCT was used to determine that this drum contained 3 g of weapons-grade plutonium (right).

Several 15 replicate studies were completed for three of the four required activity ranges (nominal compliance values are 0.1, 1.0, 10, and 160 g weapons-grade plutonium). The latter three ranges were measured by acquiring CGS and A&PCT data for three separate placements of radioactive standards within an empty matrix drum. The standards had a total mass of 0.93, 9.3, and 33.48 g of ^{239}Pu positioned within the drum and required 4, 0.75, and 0.5 h total (CGS and A&PCT) assay time, respectively. The performance results are summarized in Table 15.4.

TABLE 15.4 Nondestructive Radio-Assay Performance Requirements and Measurement Results

	Nominal radioactivity range	0.04 to 0.4 a-Ci	0.4 to 4.0 a-Ci	> 4.0 a-Ci
	Nominal ^{239}Pu amount	0.94 g	9.4 g	150.4 g
	^{239}Pu amount measured	0.93 g	9.3 g	33.48 g
Performance Requirements	Precision	15%	10%	5%
	Low accuracy	50%	75%	75%
	High accuracy	150%	125%	125%
	Low bias	35%	67%	67%
	High bias	300%	150%	150%
Measured Values	^{239}Pu measured average (15 reps)	0.84 g	9.12 g	32.33 g
	Standard deviation	0.09 g	0.47 g	1.40 g
	% Relative standard deviation	9.45% (PASS)	5.02% (PASS)	4.17% (PASS)
	% Recovery	90.19% (PASS)	98.03% (PASS)	96.57% (PASS)
	Bias low	40.63% (PASS)	67.3% (PASS)	67.07 (PASS)
	Bias high	294.37% (PASS)	149.7% (PASS)	149.93 (PASS)

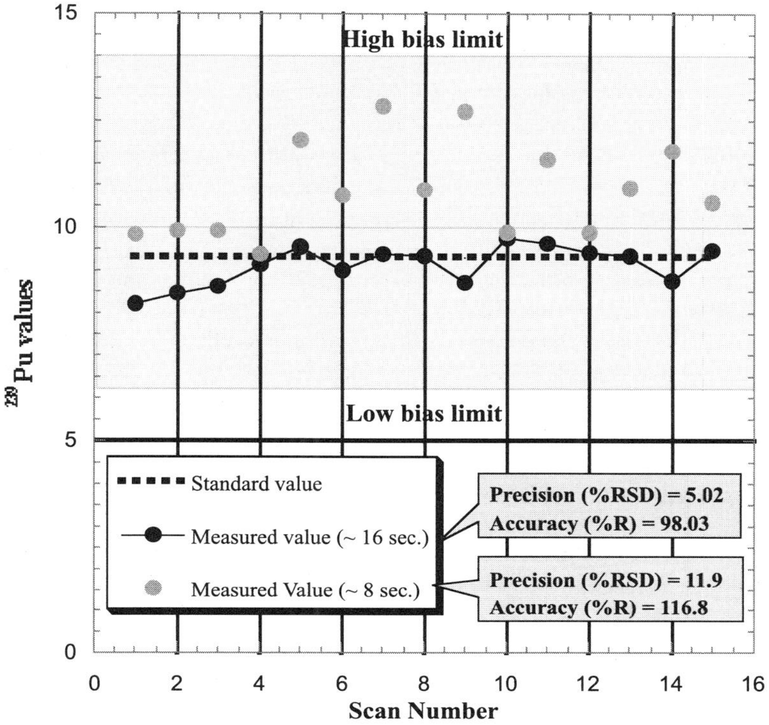

FIGURE 15.43 Six-detector A&PCT performance results for the empty matrix drum with 9.3 g ^{239}Pu. These data reveal that a reduction of 2× in data acquisition time — 16 and 8 s is the ray sum acquisition time — nearly meets performance. New research shows that we meet performance.

Research is being conducted to increase throughput, for example, trading off accuracy and precision for decreased data acquisition time, as shown in Figure 15.43. These data reveal that we can increase throughput by almost 2× for most drums and still be able to meet the performance requirements. New research has shown that, on average, we can meet the required performance for the three ranges in 2, 0.5, and 0.5 h, respectively. These and other data reveal that we have a lower limit data acquisition time of ~0.5 h/drum. This is mainly limited by data transfer rates.

15.4 Summary

Industrial CT encompasses a broad range of scanning modalities and applications. The greater range of energy, varied types of objects, and stringent spatial resolution requirements have produced a wide variety of scanners and techniques. Industrial CT has been used to produce 3D data for many different types of objects from a wide array of different types of scanners. We have described many of the scanner types and applications for those types. As the access to CT scanning is increased, new applications will emerge.

15.5 Future Work

In the last 5 years, we have seen barriers to the application of industrial CT applications eroding steadily. New, flat panel amorphous-silicon and amorphous-selenium detectors developed for medical imaging are just being applied for industrial imaging. These new detectors have extended the field of view for simple area detector rotation-only scanning. At the same time, multiple row linear-array detectors have been introduced for helical scanning.[73] Recently introduced motion control hardware has higher performance at lower cost. While still a formidable task, more and more desktop computers can manage the load of 3D CT data sets. Partly due to the emergence of 3D visible light and laser systems, more and more commercial software can analyze and manipulate 3D data sets. Different CAD/CAM/FEA packages now recognize 3D CT data as a possible input for meshing and modeling phenomena.

Simultaneous with the greater access to CT scanning and CT data sets is the increased need for more detailed investigations of complex objects and assemblies. For inspections requiring fine detail, destructive testing is hard pressed to inspect the part without possibly cutting away the feature of interest. This is especially true for high-value parts. Parts and assemblies fail in three dimensions, and sometimes only 3D inspection techniques will provide the information required.

We anticipate future work in industrial CT to continue in the following five areas: (1) extensions of CT to more industrial objects, assemblies, and material and process characterization; (2) more emphasis on CT as a means for obtaining 3D data as a basis for computer-aided design (CAD) and manufacture (CAM) and finite-element analysis (FEA) tasks; (3) algorithm research in the area of model-based CT reconstructions;[56] (4) new detector technologies for traditional and helical scanning; and (5) brighter and smaller focal spot X-ray sources.

Acknowledgments

The authors want to thank Diane Chinn, Ken Dolan, Jerry Haskins, John Kinney, Clint Logan, Derrill Rikard, Pat Roberson, Kenn Morales, Earl Updike, and Amy Waters for their help in acquiring, reconstructing, and analyzing the data presented in this chapter. We also thank Stergios Stergiopoulos for asking us to contribute a chapter to this handbook. This work is performed under the auspices of the U.S. Department of Energy by the LLNL under contract W-7405-ENG-48.

References

1. Special issue celebrating the cententary of Röntgen's discovery of X-rays, *Phys. Today,* 48, 11, Nov. 1995.
2. J. Radon, Über die Bestimmung von Funktionen durch ihre Integralwerte längs gewisser Mannigfaltigkeiten (On the determination of functions from their integrals along certain manifolds), *Ber. Sächsische Akad. Wiss. Leipzig Math Phys. Kl.,* 69, 262–267, 1917.
3. G. N. Hounsfield, Computerized transverse axial scanning (tomography). I. Description of system, *Br. J. Radiol.*, 46, 1016–1022, 1973.
4. A. Macovski, *Medical Imaging Systems*, Prentice Hall, Englewood Cliffs, NJ, 1983.
5. G. T. Herman, *Image Reconstruction from Projections: The Fundamentals of Computerized Tomography,* Academic Press, New York, 1980.
6. *Proceedings of ASNT Topical Conference on Industrial Computerized Tomography,* Seattle, WA, July 25–27, 1989, American Society for Nondestructive Testing, Columbus, OH.
7. B. D. Smith, Cone-beam tomography: recent advances and a tutorial review, *Opt. Eng.,* 29(5), 524–534, 1990.
8. P. Grangeat, Analysis d'un Systeme d'Imagerie 3D par Reeconstruction á Partir de Radiographies X en Géométrie Conique, Ph.D. thesis, l'Ecole Nationale Superieure des Telecommunications, Grenoble, France, 1987.
9. L. A. Feldkamp, L. C. Davis, and J. W. Kress, Practical cone-beam algorithm, *J. Opt. Soc. Am.*, 1, 612–619, 1984.
10. J. Goebbels, U. Zscherpel, and W. Bock, Eds., Proceedings International Symposium on Computerized Tomography for Industrial Applications and Image Processing in Radiology, Berlin, March 15–17, 1999.
11. Paper Summaries of the 1991 Industrial Computed Tomography II Topical Conference, San Diego, CA, May 20–24, 1991, American Society for Nondestructive Testing, Columbus, OH.

12. H. Berger, Ed., *Practical Applications of Neutron Radiography and Gauging*, STP 586, American Society for Testing and Materials, Philadelphia PA, 1976.
13. H. Berger, *Neutron Radiography–Methods, Capabilities and Applications*, Elsevier, Amsterdam, 1965.
14. A. E. Pontau, A. J. Antolak, D. H. Morse, A. A. Ver Berkmoes, J. M. Brase, D. W. Heikkinen, H. E. Martz, and I. D. Proctor, Ion microbeam microtomography, *Nucl. Instr. Methods*, B40/41, 646, 1989.
15. A. M. Waters, H. Martz, K. Dolan, M. Horstemeyer, D. Rikard, and R. Green, Characterization of Damage Evolution in an AM60 Magnesium Alloy by Computed Tomography, Proceedings of Nondestructive Characterization of Materials™ IX, R. E. Green, Jr., Ed., Sydney, Australia, June 28-July 2, 1999, *AIP Conference Proceedings 497*, Melville, NY, pp. 616–621.
16. W. S. Haddad, I. McNulty, J. E. Trebes, E. H. Anderson, R. A. Levesque, and L. Yang, Ultrahigh-resolution X-ray tomography, *Science*, 266, 1213, 1994.
17. P. D. Tonner and J. H. Stanley, Supervoltage computed tomography for large aerospace structures, *Mat. Eval.*, 12, 1434, 1992.
18. R. D. Evans, *The Atomic Nucleus*, McGraw Hill, New York, 1955.
19. G. F. Knoll, *Radiation Detection and Measurement,* John Wiley & Sons, New York, 1989.
20. H. H. Barrett and W. Swindell, *Radiological Imaging*, Academic Press, New York, 1981.
21. W. Heitler, *The Quantum Theory of Radiation*, Dover Publications, New York, 1984.
22. A. A. Harms and A. Zeilinger, A new formation of total unsharpness in radiography, *Phys. Med. Biol.*, 22(1), 70–80, 1977; L. E. Bryant and P. McIntire, *Nondestructive Testing Handbook, Second Edition, Volume 3, Radiography and Radiation Testing*, American Society for Nondestructive Testing, Columbus, OH, 1985.
23. L. Grodzins, Optimum energies for X-ray transmission tomography of small samples, *Nucl. Instr. Methods*, 206, 541–545, 1983.
24. A. C. Kak, and M. Slaney, *Principles of Computerized Tomographic Imaging*, IEEE Press, New York, 1987.
25. D. Chinn, J. Haskins, C. Logan, D. Haupt, S. Groves, J. Kinney, and A. Waters, Micro-X-Ray Computed Tomography for PBX Characterization, Lawrence Livermore National Laboratory, Livermore, CA, Engineering Research, Development and Technology, UCRL-53868–98, February 1999.
26. J. H. Kinney, D. L. Haupt, M. C. Nichols, T. M. Breunig, G. W. Marshall, and S. J. Marshall, The X-ray tomographic microscope: 3-dimensional perspectives of evolving microstructure, *Nucl. Instr. Methods Phys. Res. A*, 347, 480–486, 1994.
27. J. H. Kinney and M. C. Nichols, X-ray tomographic microscopy using synchrotron radiation, *Ann. Rev. Mater. Sci.*, 22, 121–152, 1992.
28. T. F. Budinger, G. T. Gullberg, and R. H. Huesman, Emission computed tomography, in *Image Reconstruction from Projections Implementation and Applications*, G.T. Herman, Ed., Springer-Verlag, New York, p. 147, 1979.
29. M. G. Light, D. J. Schneberk, and F. Bray, Turbine Blade Internal Structure and Defects NDE, Technical Operating Report, Southwest Research Institute, San Antonio, TX, May 1993.
30. K. C. Tam, Computation of radon data from cone beam data in cone beam imaging, *J. Nondes. Eval.*, 17(1), 1–15, 1998.
31. H. E. Martz, G. P. Roberson, D. J. Schneberk, and S. G. Azevedo, Nuclear-spectroscopy-based, first-generation, computerized tomography scanners, *IEEE Trans. Nucl. Sci.*, 38, 623, 1991.
32. H. E. Martz, D. J. Schneberk, G. P. Roberson, and S. G. Azevedo, Computed Tomography, Lawrence Livermore National Laboratory, Livermore, CA, UCRL-ID-112613, September 1992.
33. K. W. Dolan, J. J. Haskins, D. E. Perskins, and R. D. Rikard, X-ray Imaging: Digital Radiography, Lawrence Livermore National Laboratory, Livermore, CA, Engineering Research, Development and Technology, Thrust Area Report, UCRL 53868–93, 1993.
34. S. K. Sengupta, IMAN-3D: A Software Tool-Kit for 3-D Image Analysis, Lawrence Livermore National Laboratory, Livermore, CA, Engineering Research, Development and Technology, UCRL 53868–98, 1998.

35. S. G. Azevedo, H. E. Martz, D. J. Schneberk, and G. P. Roberson, Quantitative Measurement Tools for Digital Radiography and Computed Tomography Imagery, Lawrence Livermore National Laboratory, Livermore, CA, UCRL-53868–94, 1994.
36. R. T. Bernardi and H. E. Martz, Jr., Nuclear Waste Drum Characterization with 2 MeV X-Ray and Gamma-Ray Tomography, presented at Proceedings of the SPIE's 1995 International Symposium on Optical Science, Engineering, and Instrumentation, San Diego, CA, July 13–14, 1995, Vol. 2519.
37. R. T. Bernardi, Field Test Results for Radioactive Waste Drum Characterization with Waste Inspection Tomography (WIT), presented at the 5th Nondestructive Assay and Nondestructive Examination Waste Characterization Conference, Salt Lake City, UT, January 14–16, 1997, INEL CONF-970126, pp. 107–115.
38. P.-L. Bossart, H. E. Martz, H. R. Brand, and K. Hollerbach, Application of 3D X-ray CT data sets to finite element analysis, in *Review of Progress in Quantitative Nondestructive Evaluation*, D. O. Thompson and D. E. Chimenti, Eds., Plenum Press, New York, 15, 1996, pp. 489–496.
39. H. R. Brand, D. J. Schneberk, H. E. Martz, P.-L. Bossart, and S. G. Azevedo, Progress in 3-D QuantitativeDR/CT, Lawrence Livermore National Laboratory, Livermore, CA, UCRL-53868–95.
40. J. T. Bushberg, J. A. Seibert, E. M. Leidholdt, Jr., and J. M. Boone, *The Essential Physics of Medical Imaging*, Williams & Wilkins, Baltimore, MD, 1994.
41. Cormack and Hounsfield, Nobel Prize for X-ray science, in *Nobel Prize in Physiology or Medicine 1979*, R. B. Fenner, S. Picologlou, and G. K. Shenoy, Eds., Advanced Photon Light Source at Argonne National Laboratory, March 20–26, 1999, pp. 71–77.
42. D. E. Cullen et al., Tables and Graphs of Photon-Interaction Cross Sections from 10 eV to 100 GeV Derived from the LLNL Evaluated-Photon-Data Library (EPDL), Lawrence Livermore National Laboratory, Livermore, CA, UCRL-50400, Vol. 6, Rev. 4, 1989.
43. D. DeLynn and D. Decman, Transuranic Isotopic Analysis Using Gamma Rays, presented at the 6th Nondestructive Assay Waste Characterization Conference, Salt Lake City, UT, November 17–19, 1998, pp. 243–258.
44. K. W. Dolan, C. M. Logan, J. J. Haskins, R. D. Rikard, and D. Schneberk, Evaluation of an Amorphous-Silicon Array for Industrial X-Ray Imaging, Lawrence Livermore National Laboratory, Livermore, CA, Engineering Research, Development and Technology, UCRL 53868–99, February 2000.
45. K. W. Dolan, H. E. Martz, J. J. Haskins, and D. E. Perkins, Digital Radiography and Computed Tomography for Nondestructive Evaluation of Weapons, Lawrence Livermore National Laboratory, Livermore, CA, Engineering Research, Development and Technology, Thrust Area Report, UCRL-53868–94, 1994.
46. T. Q. Dung, Calculation of the systems Kc error and correction factor in gamma waste assay system, *Ann. Nucl. Energy*, 24(1), 33–47, 1997.
47. R. J. Estep, T. H. Prettyman, and G. A. Sheppard, Tomographic gamma scanning to assay heterogeneons radioactive waste, *Nucl. Sci. Eng.*, 118, 145–152, 1994.
48. D. M. Goodman, E. M. Johansson, and T. W. Lawrence, On Applying the Conjugate Gradient Algorithm to Image Processing Problems, in *Multivariate Analysis: Future Directions*, C. R. Rao, Ed., Elsevier Science Publishers, New York, 1993, chap. 11.
49. K. Hollerbach and A. Hollister, Computerized prosthetic modeling, *Biomechanics*, September, 31–38, 1996.
50. J. A. Jackson, D. Goodman, G. P. Roberson, and H. E. Martz, An Active and Passive Computed Tomography Algorithm with a Constrained Conjugate Gradient Solution, presented at the 6th Nondestructive Assay Waste Characterization Conference, Salt Lake City, UT, November 17–19, 1998, pp. 325–358.
51. K. Kouris, N. M. Spyrou, and D. F. Jackson, Materials analysis using photon attenuation coefficients, in *Research Techniques in Nondestructive Testing, Vol. VI*, R. S. Sharpe, Ed., Academic Press, New York, 1982.
52. J. Lau, *Ball Grid Array Technology*, McGraw-Hill, New York, 1995.

53. F. Lévai, Z. S. Nagy, and T. Q. Dung, Low Resolution Combined Emission-Transmission Imaging Techniques for Matrix Characterization and Assay of Waste, presented at the 17th Esarda Symposium, Aachen, May 1995, pp. 319–323.
54. C. M. Logan, G. P. Roberson, D. L. Weirup, J. C. Davis, I. D. Proctor, D. W. Heikkinen, M. L. Roberts, H. E. Martz, D. J. Schneberk, S. G. Azevedo, A. E. Pontau, A. J. Antolak, and D. H. Morse, Computed Tomography of Replica Carbon, ASNT's Industrial Computed Tomography Conference II, Topical Conference Paper Summaries, San Diego, CA, May 20–24, 1991, p. 61.
55. C. Logan, J. Haskins, K. Morales, E. Updike, D. Schneberk, K. Springer, K. Swartz, J. Fugina, T. Lavietes, G. Schmid, and P. Soltani, Evaluation of an Amorphous Selenium Array for Industrial X-ray Imaging, Lawrence Livermore National Laboratory, Livermore, CA, Engineering NDE Center Annual Report, UCRL-ID-132315, 1998.
56. H. E. Martz, Jr., D. M. Goodman, J. A. Jackson, C. M. Logan, M. B. Aufderheide, III, A. Schach von Wittenau, J. H. Hall, and D. M. Sloan, Quantitative Tomography Simulations and Reconstruction Algorithms, Lawrence Livermore National Laboratory, Livermore, CA, Engineering Research, Development and Technology, UCRL 53868–99, February 2000.
57. H. E. Martz, G. P. Roberson, M. F. Skeate, D. J. Schneberk, S. G. Azevedo, and S. K. Lynch, High Explosives (PBX9502) Characterization Using Computerized Tomography, Lawrence Livermore National Laboratory, Livermore, CA, UCRL-ID-103318, 1990.
58. H. E. Martz, S. G. Azevedo, D. J. Schneberk, M. F. Skeate, G. P. Roberson, and D. E. Perkins, Computerized Tomography, Lawrence Livermore National Laboratory, Livermore, CA, UCRL-53868–90, October 1991.
59. H. E. Martz, D. J. Schneberk, S. G. Azevedo, and S. K. Lynch, Computerized tomography of high explosives, in *Nondestructive Characterization of Materials IV*, C. O. Ruud et al., Eds., Plenum Press, New York, pp. 187–195, 1991.
60. H. E. Martz, D. J. Schneberk, G. P. Roberson, and P. J. M. Monteiro, Computed tomography assessment of reinforced concrete, *Nondestr. Test. Eval.*, 8–9 1035–1047, 1992.
61. H. E. Martz, G. P. Roberson, D. C. Camp, D. J. Decman, J. A. Jackson, and G. K. Becker, Active and Passive Computed Tomography Mixed Waste Focus Area Final Report, Internal report Lawrence Livermore National Laboratory, Livermore, CA, UCRL-ID-131695, November 1998.
62. H. E. Martz, The role of nondestructive evaluation in life cycle management, in *Frontiers of Engineering: Reports on Leading Edge Engineering from 1997 NAE Symposium on Frontiers of Engineering*, National Academy Press, Washington, D.C., pp. 56–71, 1998.
63. Papers presented at the 6th Nondestructive Assay and Nondestructive Examination Waste Characterization Conference, Salt Lake City, UT, November 17–19, 1998; 5th Nondestructive Assay and Nondestructive Examination Waste Characterization Conference, Salt Lake City, UT, January 14–16, 1997; 4th Nondestructive Assay and Nondestructive Examination Waste Characterization Conference, Salt Lake City, UT, October 24–26, 1995.
64. Performance Demonstration Program Plan for Nondestructive Assay for the TRU Waste Characterization Program, U.S. Department of Energy, Carlsbad Area Office, National TRU Program Office, CAO-94–1045, Revision 1, May 1997.
65. D. E. Perkins, H. E. Martz, L. O. Hester, G. Sobczak, and C. L. Pratt, Computed Tomography Experiments of Pantex High Explosives, Lawrence Livermore National Laboratory, Livermore, CA, UCRL-CR-110256, April 1992.
66. C. M. Prince and R. Ehrlich, Analysis of spatial order sandstones. I. Basic principles, *Math. Geo.*, 22(3), 333–359, 1990.
67. G. P. Roberson, H. E. Martz, D. J. Schneberk, and C. L. Logan, Nuclear-Spectroscopy Computerized Tomography Scanners, 1991 ASNT Spring Conference, Oakland, CA, March 18–21, 1991, p. 107.
68. G. P. Roberson, H. E. Martz, D. J. Decman, J. A. Jackson, D. Clark, R. T. Bernardi, and D. C. Camp, Active and Passive Computed Tomography for Nondestructive Assay, 6th Nondestructive Assay and Nondestructive Examination Waste Characterization Conference, Salt Lake City, UT, November 17–19, 1998, pp. 359–385.

69. J. Rowlands and S. Kasap, Amorphous semiconductors usher in digital X-ray imaging, *Phys. Today,* 50–11, 24, 1997.
70. V. Savona, H. E. Martz, H. R. Brand, S. E. Groves, and S. J. DeTeresa, Characterization of static- and fatigue-loaded carbon composites by X-ray CT, in *Review of Progress in Quantitative Nondestructive Evaluation*, D. O. Thompson and D. E. Chimenti, Eds., Plenum Press, New York, pp. 1223–1230, 15, 1996.
71. V. D. Scott and G. Love, *Quanitative Electron-Probe Microanalysis,* Ellis Horwood, Chichester, 1983.
72. P. K. Soltani, D. Wysnewski, and K. Swartz, Amorphous Selenium Direct Radiography for Industrial Imaging, Presented at Computerized Tomography for Industrial Application and Image Processing in Radiology, Berlin, Germany, March 15–17, 1999, DGAfP Proceedings BB 67-CD.
73. K. Taguchi and H. Aradate, Algorithm for image reconstruction in multi-slice helical CT, *Med. Phys.*, 25(4), 550–561, 1998.
74. Transuranic Waste Characterization Quality Assurance Program Plan, U.S. Department of Energy, Carlsbad Area Office, National TRU Program Office, CAO-94–1010, Interim Change, November 15, 1996.
75. M. C. H. van der Meulen, *Frontiers of Engineering: Reports on Leading Edge Engineering from 1997 NAE Symposium on Frontiers of Engineering*, Third Annual Symposium on Frontiers of Engineering, National Academy Press, Washington, D.C., pp. 12–15, 1998.
76. A. M. Waters, H. E. Martz, C. M. Logan, E. Updike, and R. E. Green, Jr., High Energy X-ray Radiography and Computed Tomography of Bridge Pins, Proceedings of the Second Japan-US Symposium on Advances in NDT, Kahuku, Hawaii, June 21–25, 1999, pp. 433–438.
77. R. L. Weisfield, M. A. Hartney, R. A. Street, and R. B. Apte, New amorphous-silicon image sensor for X-ray diagnostic medical imaging applications, *SPIE Med. Imaging, Phys. Med. Imaging*, 3336, 444–452, 1998.

16

Organ Motion Effects in Medical CT Imaging Applications

Ian Cunningham
University of Western Ontario

Stergios Stergiopoulos
Defence and Civil Institute of Environmental Medicine

University of Western Ontario

Amar Dhanantwari
Defence and Civil Institute of Environmental Medicine

16.1 Introduction

X-ray computed tomography (CT) was developed in the 1960s and early 1970s as a method of producing transverse tomographic (cross-sectional) images of the human body.* It has since become widely accepted as an essential diagnostic tool in medical centers around the world.

A summary of CT imaging and reconstruction concepts has been described by Martz and Schneberk in Chapter 15. Recall that a CT image is essentially a tomographic "map" of the calculated X-ray linear attenuation coefficient, $\mu(x, y)$, expressed as a function of position (x, y) in the patient. The attenuation coefficient is a function of X-ray energy, and CT images are produced using a spectrum of X-ray energies between approximately 30 and 140 keV, although this varies slightly with manufacturer and sometimes with the type of examination being performed. In order to provide a consistent scale of image brightness for medical uses between vendors and scan parameters, CT images are calibrated and expressed in terms of "CT number" (CT#) in "Hounsfield" units (HU) defined as

* Read Webb[58] for an interesting historical account of the development of CT.

0-8493-3691-0/01/$0.00+$.50

$$CT\# = \frac{\mu_t - \mu_w}{\mu_w} \times 1000\text{HU} \tag{16.1}$$

where μ_t and μ_w are the attenuation coefficients of tissue in a specified image pixel and in water, respectively. In this way, air corresponds to a CT# of –1000 HU, water to 0 HU, and bone to approximately 1000 HU. Most soft tissues of medical interest are in the range of approximately –20 (fat) to 60 HU (muscle) for all CT systems. A change of 1 HU corresponds to a 0.1% change in the attenuation coefficient. The standard deviation image noise in most CT scanners is approximately 3 to 8 HU, depending on the scan protocol. A CT image is thus a map of the linear attenuation coefficient expressed in HU.

16.1.1 CT Systems

Images are calculated from a large number of measurements made of X-ray transmission through the body within a specified plane. Figure 16.1 is an illustration of a typical medical CT installation. The important system components include (1) a gantry containing an X-ray tube and an array of detector elements that rotate about the patient, (2) a patient table to position the patient within the gantry, and (3) a control console with associated computer hardware to control the data acquisition process and perform image reconstruction.

As CT scanners evolved, different configurations for measuring X-ray transmission through the patient were adopted.[1] The first scanners used what is known as first-generation geometry consisting of an X-ray tube and a single X-ray detector. The tube and detector were translated across the patient, taking parallel projections. The tube and detector were then rotated by a small angle (typically ≤ 1°), and another translation was performed. In such a manner, projections through the subject were measured for angles spanning 180°.

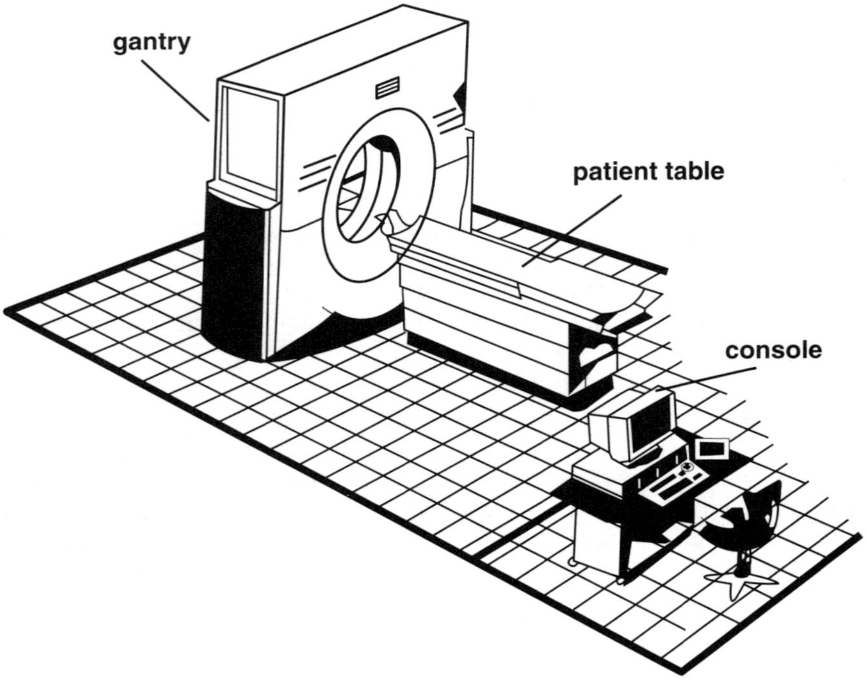

FIGURE 16.1 Schematic illustration showing a typical medical CT installation consisting of the gantry, patient table, and control console. (Courtesy of Picker International Inc.)

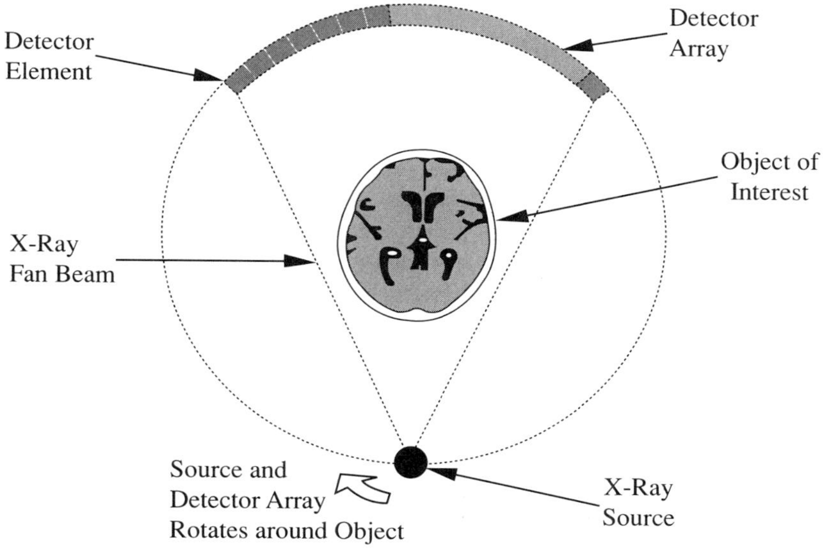

FIGURE 16.2 Typical arrangement of a third-generation X-ray CT.

Today, several configurations are in clinical use. Third-generation CT scanners use a diverging "fan beam" of X-rays produced at the focal spot of an X-ray tube that encompasses the patient, as illustrated in Figure 16.2. X-ray transmission is measured using an array of detector elements, which eliminates the need for a translation motion, and the X-ray source and detector array simply rotate about the patient. Modern, third-generation CT scanners complete one rotation in less than 1 s.

Fourth-generation CT scanners make use of a stationary circular ring of detectors that surround the patient within the gantry and an X-ray source that moves on a circular track inside the ring. At each X-ray source position, projection data are measured by the opposing arc of detectors. Scan times are similar to those of third-generation geometry, and both systems are widely used at present.

Slip-ring scanners transfer signals and power to the rotating part of the gantry through brush couplings that slide along stationary conductive rings. This allows for multiple continuous rotations of the gantry without the need to rewind cables and leads to the development of "spiral" or "helical" scanning in which the patient moves through the gantry with a continuous motion while the X-ray tube rotates about the patient.

16.1.2 The Sinogram

Transmission of X-rays along path L through a patient is described by the integral expression

$$I = I_o e^{-\int_L \mu(l)dl} \tag{16.2}$$

where I is the measured intensity, I_o is the X-ray intensity measured in the absence of the patient, and $\mu(l)$ is the X-ray linear attenuation coefficient at position l. The detector array is used to obtain a measure of X-ray transmission $T = I/I_o$. The data acquisition process for a single tomographic image is typically 0.5 to 1.0 s, during which the source and detectors generally rotate a full circle about the patient. Approximately 500,000 to 1,000,000 transmission measurements, called projections, are used to reconstruct a single image.

The set of projection data acquired during a CT scan can be presented as a grey-scale image of the relative attenuation coefficient $\ln(1/T)$ as a function of θ_i, the angular position of the X-ray source during

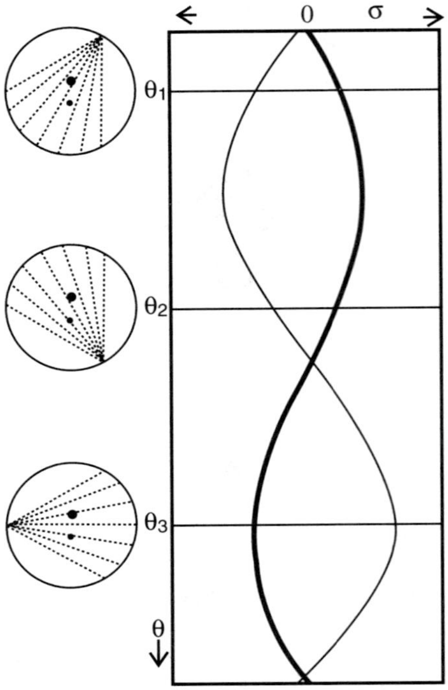

FIGURE 16.3 Sinogram data representation: each line consists of projections measured at one source angular position θ and many angles σ.

the ith projection, and σ_n, the angle of the nth detector element within the fan beam. This representation is called a sinogram, as illustrated in Figure 16.3. Each horizontal line displays one fan beam of projection data acquired at a particular source angle. The projections of each point in the image plane trace out a quasi-sinusoidal curve when using fan-beam geometry. In parallel-beam geometry, the curves are true sinusoids, hence the name sinogram, as shown in Figure 16.3.

16.1.3 Image Reconstruction

Several methods of reconstructing CT images have been proposed over the years and described by many authors,[1–4] including iterative algebraic techniques, the direct Fourier transform technique, and convolution-backprojection. The direct Fourier method is perhaps the simplest method conceptually. It is based on the central-section theorem, which states that the one-dimensional (1-D) Fourier transform of a projection at angle θ is equal to a line through the two-dimensional (2-D) Fourier transform of the image at the same angle. This relationship is illustrated in Figure 16.4. When a sufficient number of projections are acquired, the complete 2-D Fourier transform is interpolated from the samples, and the image is obtained as the inverse Fourier transform. Although the technique is conceptually simple, it is computationally complex.

The reconstruction technique of greatest practical importance is known as convolution backprojection (or filtered backprojection). Backprojection refers to the distribution of projections back across the image plane in the direction from which they were measured.

The data acquisition process for first-generation, fan-beam CT systems is depicted in Figure 16.5. The projection measurements $\{p_n(r_n, \theta_i); n = 1, \ldots, N\}$are defined as line integrals of the attenuation coefficient through the object $f(x, y)$. For a given detector n and projection angle θ_i, the projections are given by

$$p_n(r_n, \theta_i) = \iint f(x, y)\delta(x\cos\theta_i + y\sin\theta_i - r_n)dxdy \tag{16.3}$$

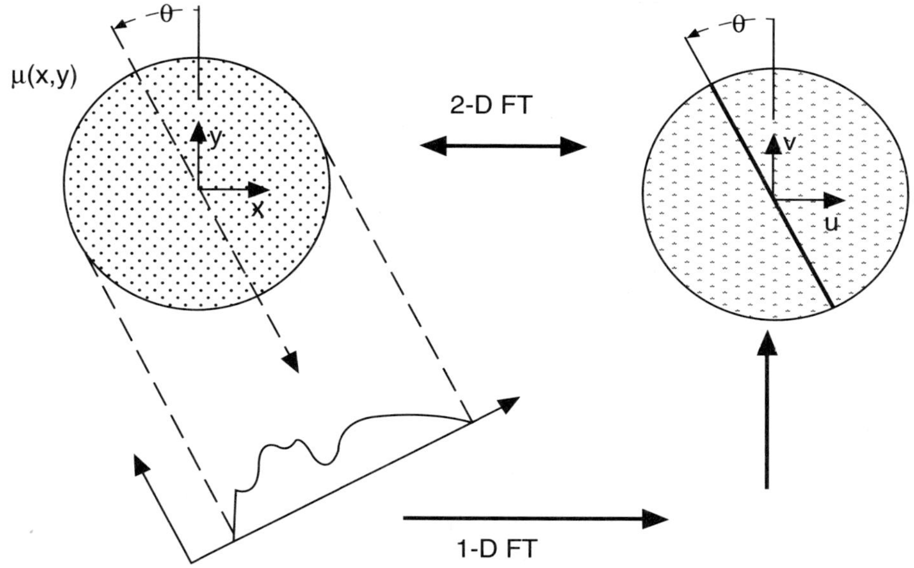

FIGURE 16.4 Central-slice theorem: Fourier transform of a set of projections of μ taken at angle θ equals a line in the 2-D Fourier transform of the image oriented at angle θ.

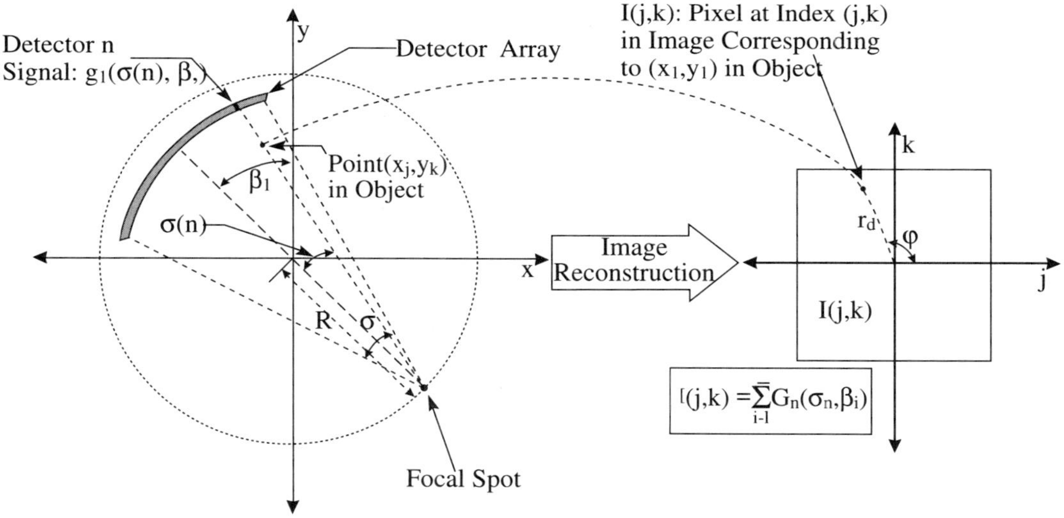

FIGURE 16.5 Schematic diagram of projection function for CT X-ray imaging systems. (Reprinted by permission of IEEE © 2000.)

The angular increment between two projections of the X-ray scanner is $\Delta\theta = 2\pi/M$, where M is the number of projections taken during the period T required for one full rotation of the source around the object. The transformations $r_n = R\sin\sigma_n$ and $\theta_i = \sigma_n + \beta_i$ are required to account for the geometry of the fan beam as shown in Figure 16.5. The projection function is then defined by

$$g_n(\sigma_n, \beta_i) = p_n\{[r_n = R\sin\sigma_n], [\theta_l = \sigma_n + \beta_i]\} \tag{16.4}$$

where g_n is the signal from the nth element of the detector array.

In the image reconstruction process, the pixel $I(j, k)$ in the actual image, shown at the right of Figure 16.5, corresponds to the Cartesian point (x_j, y_k) in the CT scan plane. The pixel value $I(j, k)$ is given by

$$I(j,k) = \sum_{i=1}^{M} G_n(\sigma_n, \beta_i), \tag{16.5}$$

where $G_n(\sigma_n, \beta_i)$ is the filtered version of the projection $g_n(\sigma_n, \beta_i)$ that has been adjusted to account for geometric effects. The angle σ_n defines the detector that samples the projection through a point (x_j, y_k) for a given projection angle β_i and is provided by Equation 16.6, where (r_d, φ) is the polar representation of (x_j, y_k):

$$\sigma_n = Tan^{-1}\left[\frac{r_d \sin(\varphi - \beta_i)}{R + r_d \cos(\varphi - \beta_i)}\right] \tag{16.6}$$

16.2 Motion Artifacts in CT

For an image to be reconstructed successfully from a data set using convolution backprojection, the data set must be complete and consistent. For a data set to be complete, projection data must be acquired over a sufficient range and with uniform angular spacing. No angular views should be missing. In parallel-beam geometry, projection data must be acquired over a range of 180°. In fan-beam geometry, data must be acquired over a range of 180° plus the fan angle.

Obtaining a consistent projection data set also requires that the patient not move during the entire data acquisition process, which may last 15 to 50 s for a multi-slice study. If the examination includes the chest or upper abdomen, the patient must hold their breath, and it may be necessary for multi-slice studies to be acquired in several single-breath-hold sections.

Motion during a scan can be in three dimensions and generally results in artifacts that appear as streaks or distorted semi-transparent structures in the general vicinity of the motion. It may be caused by a variety of reasons. Young or infirm patients are often restless and may not be cooperative or able to hold their breath. On occasion, general sedation may be required, and in extreme cases, muscle paralysis has been used.[5] Injection of a vascular contrast agent may result in involuntary patient motion[6–8] or blood flow artifacts.

16.2.1 Clinical Implications

These artifacts may obscure diagnostically important details or, in some circumstances, give a false indication of an unrelated condition. Respiratory artifacts may cause a ghosting of pulmonary nodules[9] or interfere with the assessment of interstitial lung disease.[10] Cardiac motion may produce an artifact that can be misdiagnosed as an aortic dissection.[11–15] Although such cases can be controversial, an additional angiographic examination may be required to confirm or exclude this diagnosis. Transesophageal ultrasound (echocardiography) can also be used to help determine whether a dissection is present. Cardiac and respiratory motion artifacts may hinder the visualization and clinical scoring of coronary calcifications, an important component of disease assessment and risk management.[16,17] Figure 16.6 illustrates an example of a cardiac motion artifact in a trauma victim that could be misinterpreted as a dissection (separation of the arterial wall) of the ascending aorta. In particular, the curvilinear shape of the artifact, in this example running approximately parallel to the aortic wall, and the fact that the artifact is restricted to the interior of the vessel, are suggestive of dissection and complicate the diagnosis. The ability to properly diagnose an aortic rupture or dissection is critical in the examination of many clinical settings.

16.3 Reducing Motion Artifacts

A seemingly straightforward approach to reducing motion artifacts is to minimize the data acquistion time. However, the ability to rotate a conventional X-ray tube rapidly about the patient is limited by

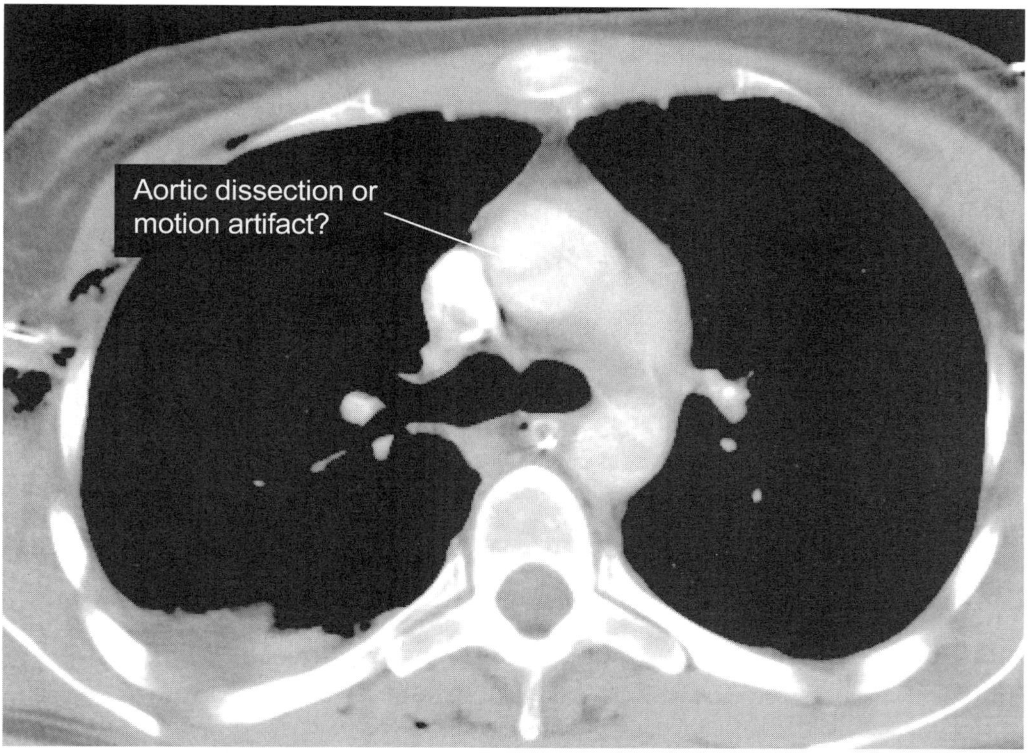

FIGURE 16.6 A CT image with a cardiac motion artifact in the ascending aorta that mimics an aortic dissection.

several factors including large forces that would be exerted on bearings supporting the rotating anode and the maximum output exposure rate. It is unlikely that acquistion times significantly less than 0.5 s will be achieved without major design changes to X-ray tubes. For this reason, many alternative methods have been developed for the purpose of reducing motion artifacts, and each has been successful under particularly circumstances. In this chapter, a brief summary is presented of (1) established methods, particular underscanning; (2) fast-acquisition methods using high-speed scanners; (3) respiratory gating; and (4) electrocardiographic (ECG) gating. Somewhat more detail is given in a description of new image processing methods for ECG gating.

16.3.1 Established Methods

It is often true that motion is relatively continuous, in which case the maximum discrepancy caused by motion occurs between the beginning and the end of a scan, which is the longest time span between two projection views.[18] These views are considered to be less reliable, and Pelc and Glover[19] developed a method of minimizing motion artifacts by applying a weighting factor to minimize the contributions of the most inconsistent projections. They showed that the weighting factor was a function of view angle and fan angle. Motion artifact reduction using this "underscanning" method has gained wide acceptance for routine clinical use.

In general, the extent of the artifact is dependent on the direction of motion. When the motion is in a direction parallel to the X-ray beam, the motion does not cause a change in the measured X-ray transmission and has no effect on the projection values. However, when the motion is in a perpendicular direction, it is likely that the motion will affect projection measurements. By choosing the first and last views to be obtained when motion is parallel to the X-ray beam, motion artifacts can be reduced. Respiratory motion tends to be in the vertical direction, and hence artifacts can sometimes be reduced by starting a scan with the X-ray tube either directly above or below the patient.

16.3.2 Fast Acquisition

Early work by Alfidi et al.[41] suggested that effective scan times of approximately 50 ms are required to reduce artifacts to an acceptable level. There are two approaches to meeting this 50-ms criterion. The first involves using high-speed CT scanners with very short scanning times with respect to the period of the cardiac cycle. The second requires ECG gating to synchronize the data acquisition process with the beating heart.

The earliest high-speed scanner was the dynamic spatial reconstructor (DSR) at the Mayo Clinic.[20,21] It employed 14 X-ray tubes, fired in rapid succession, to reduce scan times to 10 ms. While successful at demonstrating the need for high-speed volume CT systems for cardiac imaging and other applications, it was primarily a research tool and was never commercialized.

An alternative approach to using conventional X-ray tubes was the development of the electron-beam CT (EBCT).[17,22–24] It uses an electron beam deflected within the gantry to produce X-rays from a focal spot that rotates about the patient. Projection data are acquired in approximately 50 ms, which is fast enough to avoid both respiration and cardiac motion artifacts. It has been used successfully for measuring ventricular mass and border definition (essential for quantification of ventricular anatomy and function),[17] for detection of thrombi,[25] and for the management of stroke patients.[26] It provides an accurate method for scoring calcification of the coronary arteries.[16,17] However, EBCT is relatively complex and has not yet demonstrated images that can compete with the quality of conventional CT systems for non-cardiac applications.

16.3.3 Respiratory Gating

Respiratory gating has been used with an algorithm to predict when a motionless period is about to occur, which is then used to trigger an acquisition.[27–30] Adaptive prediction schemes were developed to accommodate variable respiration patterns.[31] These methods were successful at reducing motion artifacts, but they increase the data acquistion times for multi-slice studies, increasing the probability of misregistration between slices.

Active breathing control methods have also been developed that can be used to control a patient's respiratory motion in an attempt to ensure reproducibility of respiratory motion,[32] but they are generally inappropriate for use in diagnostic procedures.

16.3.4 ECG Gating

The earliest attempts at ECG gating used non-slip-ring scanners.[33] Projection data was acquired only during diastole when the heart is moving the least. In order to keep scan times to within 20 ms, a small number of angles were used. Morehouse et al.[34] introduced retrospective gating in 1980 with a technique that involved measuring projection data continuously during four rotations of the X-ray source and selecting the projection data acquired during a specified window of the cardiac cycle. The technique resulted in missing views, causing artifacts of a different nature. These artifacts were reduced by the techniques of reflection, augmentation, and interpolation.[35,36] Joseph and Whitley[37] proposed that a small number of views may be adequate, provided that the heart can be isolated and centered in the field of view. Johnson et al.[38] performed 8 scans per level on 32 contiguous levels of the heart, and reformatted the images into an early three-dimensional (3-D) data set. Although the images were improved over previous ECG-gating attempts, imaging time approached 1 h. In 1983, Moore et al.[27] used prospective ECG gating to determine the optimal start time for each rotation, later including coincident-ray considerations.[39] They were able to bring the time resolution down to 100 ms.

The myocardium of the heart behaves like a spring that is restrained from contracting by an electric potential, controlled mainly by concentrations of sodium and potassium ions. When discharged, or depolarized, these concentrations change, and the myocardium contracts vigorously. Repolarization causes the fibers to lengthen again, allowing the heart to fill with blood. It is the motion of the ions during depolarization and repolarization that provides the electrical signal detected by the ECG (Figure 16.7). In an ECG

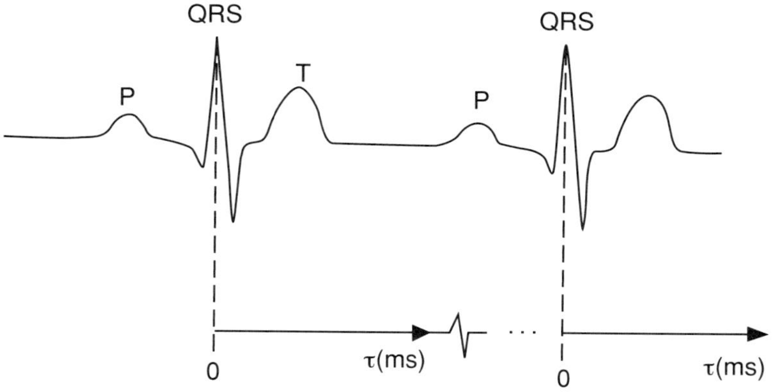

FIGURE 16.7 Components of an ECG waveform.

waveform, the *P* wave corresponds to atrial depolarization, the QRS complex corresponds to ventricular depolarization, and the *T* wave corresponds to ventricular repolarization. The period of inactivity between the *P* wave and the QRS complex is caused by the delay in conduction through the AV node. In the term "cardiac phase," τ is used loosely here to represent the time following the *R* wave (ms) within one cycle.

16.3.5 Single-Breath-Hold ECG Gating

The development of slip-ring CT scanners made possible single-breath-hold ECG gating. Nolan[40] used a continuous rotation of the X-ray tube and acquisition of an ECG waveform for 12 to 16 s during a single breath hold. They used a "data space" diagram as shown in Figure 16.8 to represent the relationship between the cardiac phase τ and the angular position of the X-ray focal spot for each rotation. After multiple rotations, data space is occupied by a series of diagonal lines (Figure 16.9). Once the data space is filled for all source angles covering 180° plus fan angle and any specified cardiac phase, a complete set of projection data exists, and an image can be reconstructed. In practice, it is generally necessary to reconstruct an image without a complete data set.

An improvement on this technique is to select data from each source position spanning 360° that is closest in phase to the desired cardiac phase. Coincident-ray replacement[39] was used to compress the 360° sinogram into one spanning only 180° plus the fan angle.

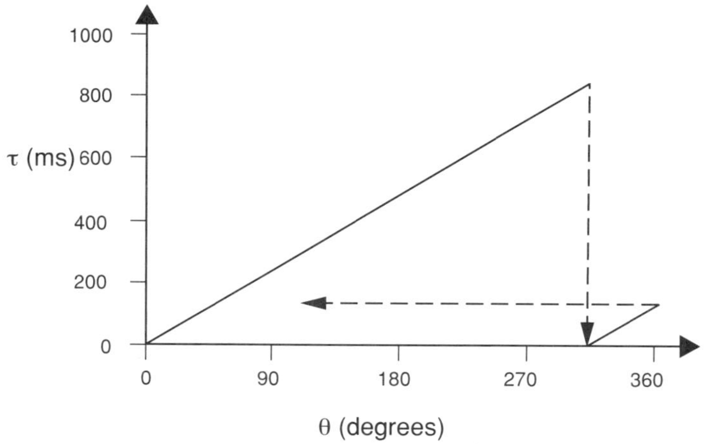

FIGURE 16.8 Filling of data space showing the relationship between the cardiac phase and the source position for one rotation.

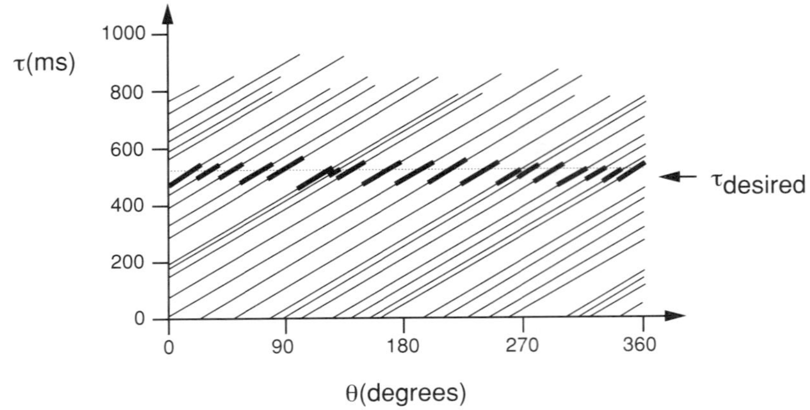

FIGURE 16.9 An image can be reconstructed when projection data have been acquired for source positions covering 180° plus fan angle for any specified cardiac phase. This generally results in a residual temporal inconsistency.

An image of a cardiac phantom simulating ventricle expansion is shown in Figure 16.10. The motion artifact, characterized by distortion of the circular phantom and background streaks, is significantly reduced in the ECG-gated image. Images such as this can be reconstructed for each of approximately 16 different cardiac phases. Using an ECG study of normal volunteers, Nolan suggested that a residual temporal inconsistency of 50 ms could be achieved with 12 to 16 rotations of the X-ray tube for approximately 95% of a normal population if the operator could choose prospectively between two rotation speeds.[40]

The ECG-gating methods described here require some degree of reproducibility of cardiac motion. While this is potentially a limitation in some circumstances, most of the variability in the cardiac cycle is due to variations in the P-QRS interval (Figure 16.7). By synchronizing data acquisition and reconstruction relative to τ, the time following the *R* peak, very little variability is observed from one cardiac cycle to the next.

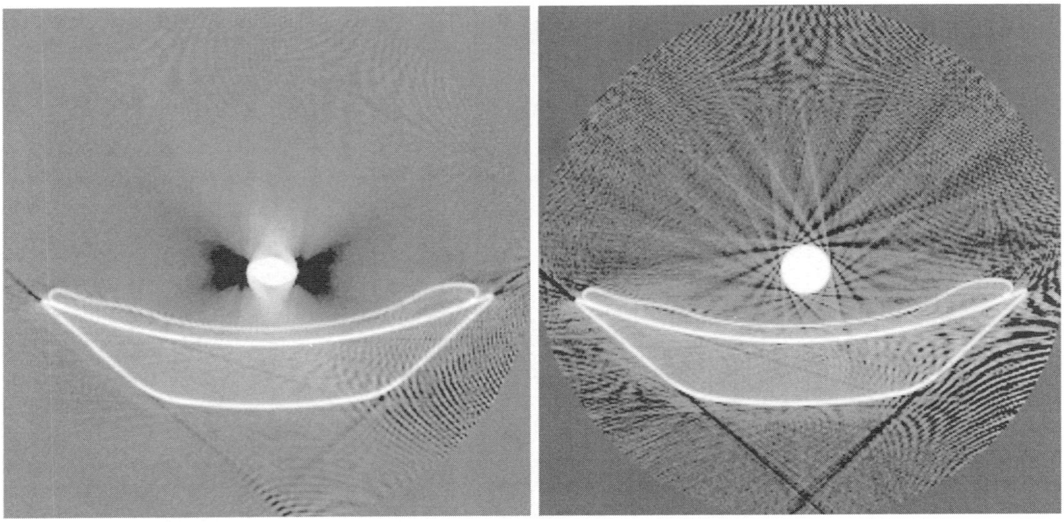

FIGURE 16.10 Comparison of a conventional CT scan of a cardiac phantom simulating ventricle expansion with an ECG-gated scan acquired in 16 s.

16.4 Reducing Motion Artifacts by Signal Processing — A Synthetic Aperture Approach

Early work by Alfidi et al.[41] suggested that effective scan times of approximately 50 ms are required to reduce artifacts to an acceptable level. This is consistent with the scan times of EBCT and with the residual temporal inconsistency of the ECG-gated technique of Nolan.[40] However, Ritchie et al.[42] have suggested more recently that 50 ms is not fast enough for many applications of clinical importance, and additional methods are required for motion correction. It is unlikely that scan times can be reduced significantly, but there may be additional benefits from sophisticated new image processing techniques.

Several mathematical techniques have been proposed as solutions to this problem. In some specific cases, 3-D reconstructions have been used to assist in distinguishing motion artifacts from physical dissections in the descending aorta.[43] Most methods require a simple model of organ motion, such as a translational, rotational, or linear expansion.[44] In most situations of practical importance, such simplifications are not very useful. A more general technique attempts to iteratively suppress motion effects from projection data[45] by making assumptions regarding the spectral characteristics of the motion. However, this depends on knowing some properties of the motion *a priori* and requires a number of iterations to converge. This is generally undesirable for CT imaging, as it results in additional radiation doses to the patient from the X-ray exposure.

Motion artifacts have been reduced in magnetic resonance imaging (MRI) for chest scans by first defining the motion with a parametric model and then adapting the reconstruction algorithm to correct for the modeled motion.[46] Ritchie[47] attempted to address cardiac and respiratory motion in CT using a pixel-specific backprojection algorithm that was conceptually influenced by this MRI approach. Motion of a frame of reference was specified by making an estimate of where each pixel location was when each projection was acquired. These maps then formed the basis of a backprojection algorithm that reconstructed each pixel in a frame of reference that moved according to the information provided by the maps. The method requires manual efforts to describe the motion of each pixel and is therefore not practical for routine clinical use at present.

The problem of motion artifacts has also been addressed in other types of real-time imaging systems such as radar satellites and sonars.[48,49] In this case, it was found that application of synthetic aperture processing increases the resolution of a phased array imaging system as well as corrects for the effects of motion. Reported results showed that the problem of correcting motion artifacts in sonar synthetic aperture applications is centered on the estimation of a phase correction factor. This factor is used to compensate for the phase differences between sequential sensor array measurements in order to coherently synthesize the spatial information into a synthetic aperture.

Dhanantwari et al.[50,51] described a synthetic aperture approach to correct for motion artifacts in CT, which is described here in more detail. Their approach consists of three components:

1. Detection of changes in the CT projection data caused by organ motion using a spatial overlap correlator approach, resulting in a "motion" sinogram that reflects changes in the projection data
2. Use of an adaptive interference canceller approach to isolate the effects of organ motion using the motion sinogram and the conventional sinogram corrupted by motion to make an estimate of a "stationary" sinogram
3. Use of a "coherent sonogram synthesis" technique that identifies through a replica correlation process the segments of the continuous sinograms that have identical phases of the motion effects

These are described in more detail in the following sections.

16.4.1 Spatial Overlap Correlator to Identify Motion Effects

The spatial overlap correlator[48] makes use of two X-ray sources that rotate about the patient separated by a very small time delay δ, where $\delta = T/M$, with T as the total acquisition time for one slice (typically 1 s) and M as the number of angular projections. Source #1 trails source #2, so that if t_0 is the starting time of one source rotation, a view acquired by source #2 at time $t = t_0 + n\delta$ will be sampled again by

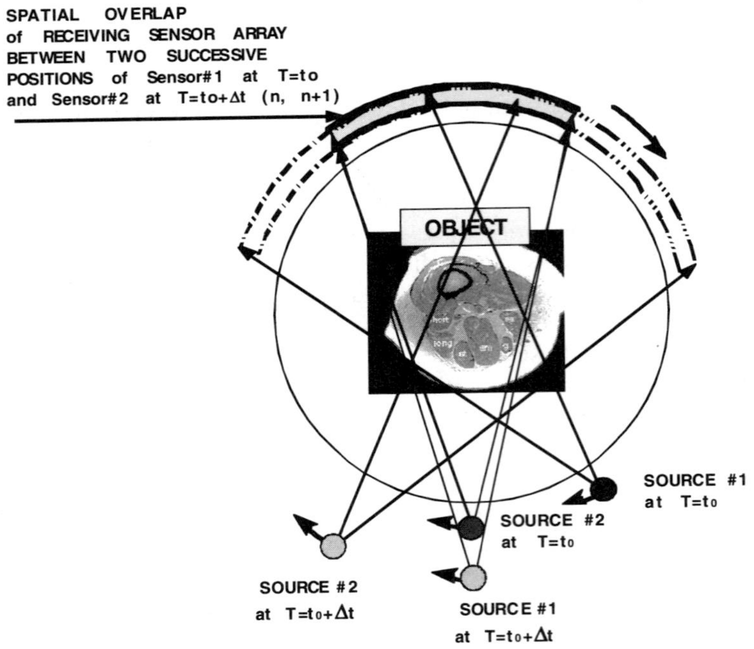

FIGURE 16.11 Two-source concept of spatial overlap correlator for CT X-ray imaging systems. (Reprinted by permission of IEEE © 2000.)

source #1 at time $t = t_0 + (n + 1)\delta$, as illustrated in Figure 16.11. Comparison of any two spatially overlapping measurements will provide a measurement that is associated with any organ motion that may have occurred during the elapsed time δ. For X-ray CT systems, differences between the two sets of samples are caused by organ motion or system noise.

Figure 16.12 shows a graphical representation of the above 2-D space-time sampling process. The vertical axis shows the times associated with the angular positions of the source-array receiver shown on the horizontal axis. Line segments along the diagonal represent the measurements of an X-ray CT scanner. Darker line segments show positions of the first source-detector pair and lighter lines show the second pair. Image reconstruction algorithms work best when the object is stationary, corresponding to horizontal lines of Figure 16.12.

In the following, let the projection measurement be given by $g_n(\sigma(n), \beta(t), t)$, given as a function of the fan angle $\sigma(n)$ on the detector arc, projection angle $\beta(t)$, and time t. The projection measurement for the first and second detector-source pair is given by

$$\{g_{n_{s1}}(\sigma(n_{s1}), \beta(t), t), (n_{s1} = q, q+1, \ldots, N)\} \tag{16.7}$$

$$\{g_{n_{s2}}(\sigma(n_{s2}), \beta(t), t+\delta), (n_{s2} = 1, 2, \ldots, N-q)\} \tag{16.8}$$

where there are $N - q$ overlapping detectors. The source locations are identical for both acquisitions, and hence the projection angles for each are given by $\beta(t) = \beta_{s2}(t) = \beta_{s1}(t+\delta)$. In Figure 16.12, these spatially overlapping measurements are depicted by the pair of lines overlapping in angular space, but in two successive time moments. The difference between the two data acquisitions for a given spatial location defined by Equations 16.7 and 16.8 is

$$\begin{aligned}\Delta g_n(\sigma(n), \beta(t), t) &= g_{n_{s2}}(\sigma(n_{s2}), \beta(t), t+\delta) \\ &- g_{n_{s1}}(\sigma(n_{s1}), \beta(t), t) \text{ for } (n = 1, 2, \ldots, N-q)\end{aligned} \tag{16.9}$$

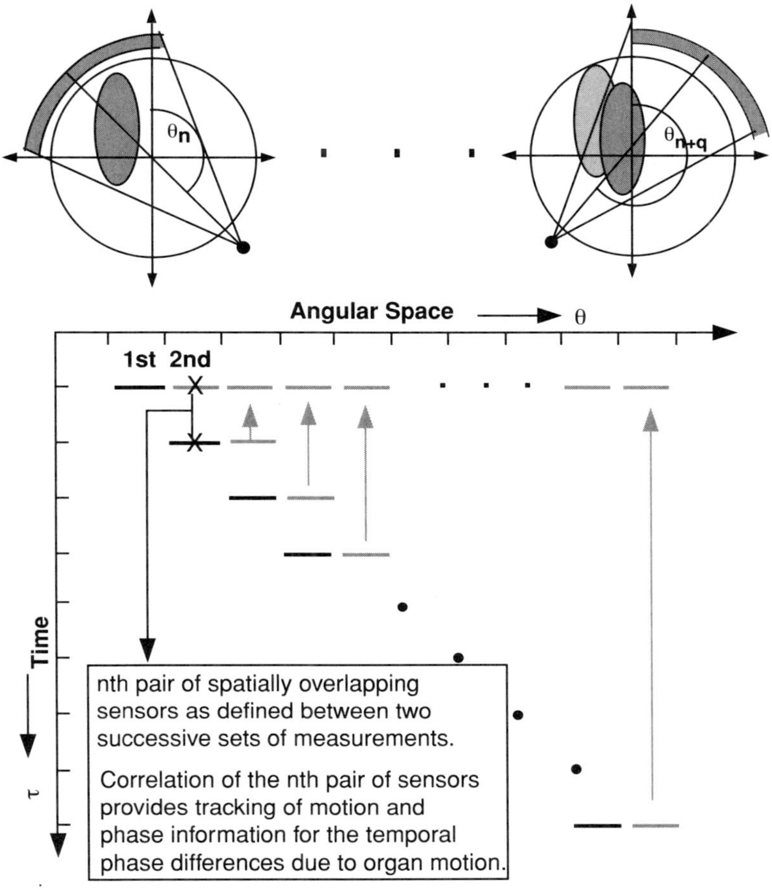

FIGURE 16.12 Graphical representation of the space-time sampling process of the spatial overlap correlator for CT X-ray imaging systems. (Reprinted by permission of IEEE © 2000.)

where it is necessary that $\sigma(n) = \sigma(n_{s1}) = \sigma(n_{s2})$ to ensure spatial overlap.

With reference to Equations 16.3 and 16.4, the time dependent projection measurement for a fan-beam X-ray CT scanner is given by

$$g_n(\sigma(n), \beta(t), t) = \iint f(x, y, t)\delta\{x\cos[\sigma(n) + \beta(t)] + y\sin[\sigma(n) + \beta(t)] - R\sin\sigma(n)\}dxdy \quad (16.10)$$

Therefore, the time dependent projection measurement may be rewritten as

$$\Delta g_n(\sigma(n), \beta(t), t) = \iint [f(x, y, t) - f(x, y, t + \delta)] \{\delta\{x\cos[\sigma(n) + \beta(t)]\} + y\sin[\sigma(n) + \beta(t)] - R\sin\sigma(n)\}dxdy \quad (16.11)$$

where $f(x, y, t) - f(x, y, t + \delta)$ indicates differences within the image plane caused by motion. It is clear that a stationary object will result in a zero output from the spatial overlap correlator.

If the projection measurement $g_n(\sigma(n), \beta(t), t)$ consists of both stationary and moving components, $g_{ns}(\sigma(n), \beta(t), t)$ and $g_{nm}(\sigma(n), \beta(t), t)$, respectively, then

$$g_n(\sigma(n), \beta(t), t) = g_{ns}(\sigma(n), \beta(t), t) + g_{nm}(\sigma(n), \beta(t), t) \quad (16.12)$$

and

$$\Delta g_n(\sigma(n), \beta(t), t) = g_{nm_{s2}}(\sigma(n_{s2}), \beta(t), t+\delta) - g_{nm_{s1}}(\sigma(n_{s1}), \beta(t), t) \text{ for } (n = 1, 2, \ldots, N-q) \quad (16.13)$$

If the motion is oscillatory, then the motion may be represented as

$$g_{nm_{s2}}(\sigma(n), \beta(t), t) = \sin(2\pi f_0 t) \quad (16.14)$$

$$g_{nm_{s1}}(\sigma(n), \beta(t), t) = \sin(2\pi f_0 (t+\delta)) \quad (16.15)$$

$$\sin(2\pi f_0 t) - \sin(2\pi f_0(t+\delta)) = -2\left[\sin(\pi f_0 \delta)\cos\left(2\pi f_0\left(t+\frac{\delta}{2}\right)\right)\right] \quad (16.16)$$

$$\Delta g_n(\sigma(n), \beta(t), t) = -2\left[\sin(\pi f_0 \delta)\cos\left(2\pi f_0\left(t+\frac{\delta}{2}\right)\right)\right] \quad (16.17)$$

Non-periodic motion may be considered as being piecewise periodic, where the motion is broken into small periodic segments. In such a case, the scale factor $\sin(\pi f_0 \delta)$ will vary as f_0 varies, while the piecewise periodic signal

$$\cos\left(2\pi f_0\left(t+\frac{\delta}{2}\right)\right)$$

will continue to track the motion. The resulting time series from the spatial overlap correlator will yield a signal that tracks the organ motion. This signal will be scaled depending on the frequency of the organ motion.

Although motion is assumed to be periodic for mathematical evaluations, the spatial overlap correlator is capable of tracking any form of organ motion, including transients if they are present, limited by sampling considerations as the sources rotate about the patient.

16.4.1.1 Simulations

A simulation study illustrates operation of the spatial overlap correlator using a Shepp-Logan phantom with the dark lower ellipse on the left-hand side of the phantom undergoing deformation. The sinogram from the standard data acquisition process is shown on the left of Figure 16.13. The right-hand side of Figure 16.13 shows motion tracking by the spatial overlap correlator. It is expected that the spatial overlap correlator will track the boundaries of the deforming ellipse, since that is where the motion occurs.

Reconstruction using these sinograms is shown in Figure 16.14. The image on the left of Figure 16.14 corresponds to the sinogram on the left of Figure 16.13 and shows the reconstructed phantom with the motion artifacts caused by the dark lobe on the left-hand side moving. The image on the right of Figure 6.14 corresponds to the sinogram on the right of Figure 16.13. The image shows only the moving object and no indication of any of the stationary objects. Since the motion of the object is tracked in an incremental fashion, the image shows the changes that occur along the boundary of the ellipse.

16.4.1.2 Hardware Implementation

Some currently available X-ray CT scanners (Siemens, GE, and Elscint) use a dual focal spot technique that can be modified for implementation of the spatial overlap correlator. Conventional projections are identified as the spatial locations $\beta(t)$, $\beta(t + \delta)$, $\beta(t + 2\delta)$, and so on. The second focal spot position is adjusted so that it will coincide with location $\beta(t)$ at time $t + \delta/2$. In this fashion, the two-source concept of the spatial overlap correlator may be achieved.

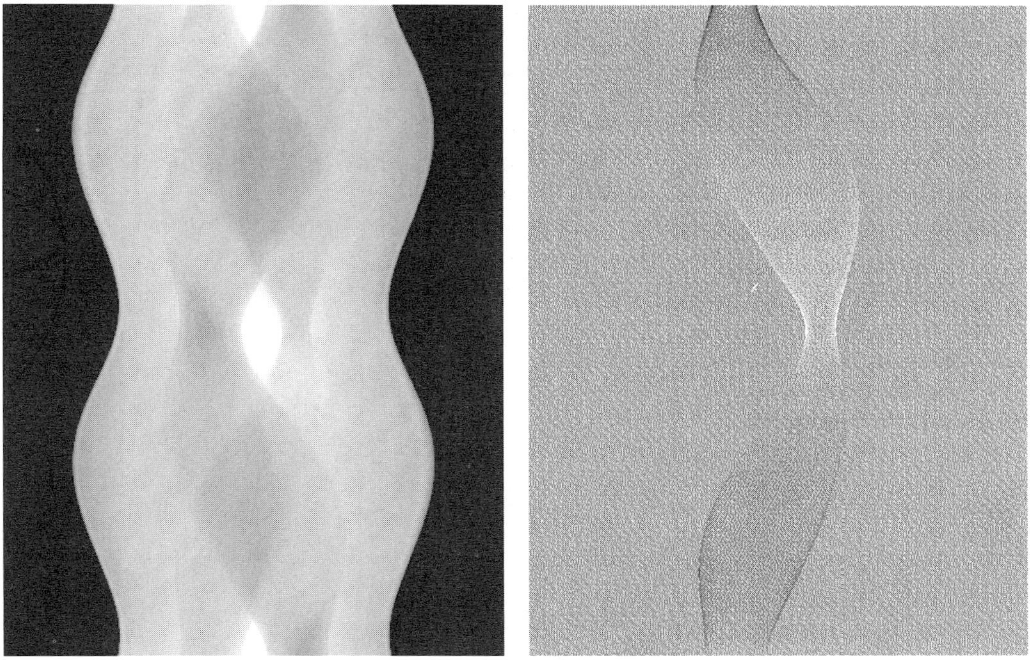

FIGURE 16.13 Simulated sinograms of the Shepp-Logan phantom (left conventional, right hardware spatial overlap correlator). A reconstructed image is in Figure 16.14.

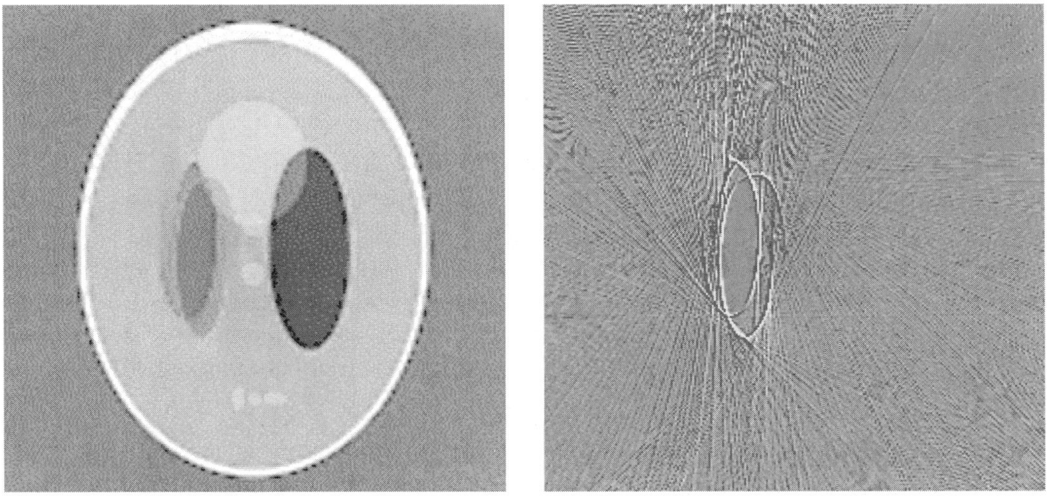

FIGURE 16.14 Reconstructed images from standard X-ray CT data acquisition and the hardware spatial overlap correlator.

However, with third-generation scanners, there is also a shift of the detector array over the time interval between projections from source #1 and #2, $\delta/2$, that does not correspond to an integer number of detector elements. This prevents proper alignment without an expensive hardware modification.

16.4.1.3 Software Implementation

A second approach to implementing the spatial overlap correlator makes use of the rotation of a single X-ray source.[50] Rather than using two projections from different sources that differ in time by the small value δ, two projections from the same source over two rotations that differ in time by $T = M\delta$ are used. With this approach, the subtraction process is sensitive to motions on a time scale of approximately 1 s and requires a minimum of two full gantry rotations. As such, this approach is not sensitive to transient motions. In addition, the assumption that a signal be viewed as being piecewise periodic is often acceptable for motions sampled on an interval of approximately 1 ms, but less likely to be acceptable when the same motion is sampled on an interval of approximately 1 s. However, cardiac motions are remarkably periodic, and good results are still obtained.

Figure 16.15 shows an image obtained from the software spatial overlap correlator for the example shown in Figure 16.14. While sampling at intervals of approximately 1 s does not do as well as sampling at intervals of approximately 1 ms, the method still identifies the important motion components.

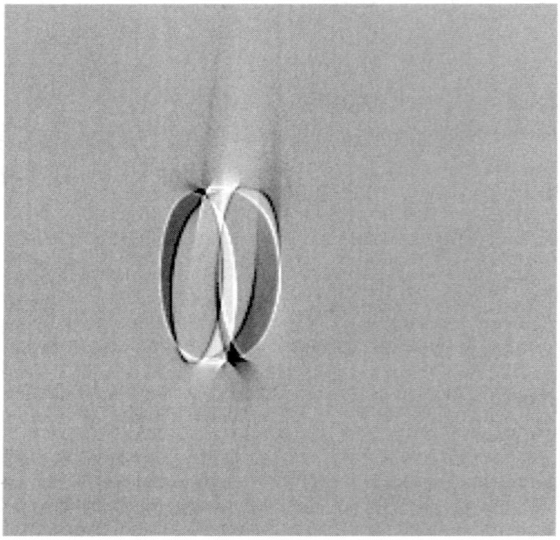

FIGURE 16.15 Reconstructed imagtes of motion using the software spatial overlap correlator.

16.4.2 Adaptive Processing to Remove Motion Artifacts

The adaptive interference canceller (AIC) approach is used to remove the motion artifacts identified by the spatial overlap correlator. It has been used extensively for isolation of signals that were originally measured in the presence of noise.[49,52] The data sequence from each detector is treated as a time series. The sequence from the conventional CT acquisition is treated as the signal including unwanted interference due to motion effects, and the sequence from the spatial overlap correlator is treated as the interference in the AIC processing scheme. Both the AIC and the coherent sinogram synthesis (CSS) methods, which are discussed in Section 16.4.3, require that data be acquired over a number of revolutions of the X-ray source. In the case of the AIC algorithm, the number of rotations is a function of the convergence rate of the adaptive algorithm and requires at least two rotations. For the CSS method, this number is simply a function of desired image quality. There is a direct relationship between the length of the data sequence acquired and the exactness of the synthesized sinogram to the desired sinogram of a stationary object and the speed of the object's motion. Both AIC and CSS algorithms may be implemented with the hardware spatial overlap correlator, but only the CSS algorithm can be used with the software spatial overlap correlator.

At this point, the physical significance of the incremental tracking of organ motion by the hardware spatial overlap correlator is analyzed. Also of importance is the relationship between this incremental motion tracked and the standard X-ray CT measurements. First, the ideal case where the motion artifacts are readily available for removal is described, then the relationship between such a system and the hardware spatial overlap correlator (HSOC) data acquisition scheme is developed.

Let the projections for a single sinogram of M projections acquired without motion be represented as

$$g_{n_{ct}}(\sigma(n_{ct}), \beta(t_j), t_j), (n_{ct} = 1, \ldots, M), (j = 1, \ldots, M) \tag{16.18}$$

This is a sinogram for a stationary object. Let the projections for a single sinogram for the same object, but with motion effects present, be given by

$$g_{n_{mov}}(\sigma(n_{mov}), \beta(t_j), t_j), (n_{mov} = 1, \ldots, N), (j = 1, \ldots, M) \tag{16.19}$$

The difference between these two sinograms provides information about the accumulated motion over the data acquisition period, given by

$$\begin{aligned} \Delta f_n(\sigma(n), \beta(t_j), t_j) &= g_{n_{mov}}(\sigma(n_{mov}), \beta(t_j), t_j) - g_{n_{ct}}(\sigma(n_{ct}), \beta(t_j), t_j); \\ &(n = n_{ct} = n_{mov} = 1, \ldots, N), (j = 1, \ldots, M) \end{aligned} \tag{16.20}$$

Alternatively stated, the optimum sinogram is the difference between the sinogram with organ motion effects included and that of only the organ motion effects, given by

$$\begin{aligned} g_{n_{ct}}(\sigma(n_{ct}), \beta(t_j), t_j) &= g_{n_{mov}}(\sigma(n_{mov}), \beta(t_j), t_j) - \Delta f_n(\sigma(n), \beta(t_j), t_j); \\ &(n = n_{ct} = n_{mov} = 1, \ldots, N), (j = 1, \ldots, M) \end{aligned} \tag{16.21}$$

Thus, an artifact-free image can be reconstructed if an estimate of $\Delta f_n(\sigma(n), \beta(t_j), t_j)$ can be obtained and subtracted from the sinogram measured in the presence of motion.

From Equation 16.9, the measurements obtained from the HSOC are written in discrete form as

$$\begin{aligned} \Delta g_n(\sigma(n), \beta(t_j), t_j) &= g_{n_{s2}}(\sigma(n_{s2}), \beta(t_j), t_j + \delta) \\ &- g_{n_{s1}}(\sigma(n_{s1}), \beta(t_j), t_j) \text{ for } (n = 1, 2, \ldots N), (j = 1, \ldots, M) \end{aligned} \tag{16.22}$$

It is evident that the measurement described by Equation 16.9 is that of any motion that occurred over the time interval from t_j to $t_j + \delta$ from view angle $\beta(t_j)$. This means that over the complete data acquisition period of $M\delta$ seconds, the spatial overlap correlator will sample the motion M times at the angles $\beta(t_j)$ $(j = 1, \ldots, M)$. Therefore, Equation 16.22 can be rewritten as

$$\Delta g_n(\sigma(n), \beta(t_j), t_j) = \left(\frac{g_{n_{s2}}(\sigma(n_{s2}), \beta(t_j), t_j + \delta) - g_{n_{s1}}(\sigma(n_{s1}), \beta(t_j), t_j)}{\delta}\right)\delta \tag{16.23}$$

Recall that $\Delta g_n(\sigma(n), \beta(t_j), t_j)$ is simply a measure of the motion present, since the constant terms due to the stationary components disappear:

$$\begin{aligned} \Delta g_n(\sigma(n), \beta(t_j), t_j) &= g_{nm_{s2}}(\sigma(n_{s2}), \beta(t_j), t_j + \delta) \\ &- g_{nm_{s1}}(\sigma(n_{s1}), \beta(t_j), t_j) \text{ for } (n = 1, 2, \ldots N), (j = 1, \ldots, M) \end{aligned} \tag{16.24}$$

It follows directly that integration of the time series derived from the HSOC gives the compound motion at the mth projection:

$$\begin{aligned} \int_{t_0}^{t_{m-1}} \Delta g_n(\sigma(n), \beta(t_j), m\delta)dt &= \sum_{j=1}^{M} \left\{ g_{nm_{s2}}^{t_{j+1}}(\sigma(n), \beta(t_j), t_j + \delta) - g_{nm_{s1}}^{t_j}(\sigma(n), \beta(t_j), t_j) \right\} \\ &\text{for } (n = 1, \ldots, N), (m = 1, \ldots, M) \end{aligned} \tag{16.25}$$

In Equation 16.25, the projection index j specifies the time t_j, and the detector index n specifies the detector fan angle $\sigma(n)$. For simplicity, $g_n^{t_j}(\sigma(n), \beta(t_j), t_j)$ is expressed as g_n^j.

The HSOC makes measurements continuously over the time period $M\delta$, but at each moment the measurement corresponds to different projection numbers, or view angles β. Rewriting Equation 16.25 leads to Equation 16.26, where the factors ρ_m, $(m = 1, \ldots, M)$ compensate for the views being from different angles:

$$\int_{t_0}^{t_{m-1}} \Delta g_n(\sigma(n), \beta(t_j), t_j)dt = (g_{nm_{s2}}^{j=1} - g_{nm_{s1}}^{j=0})\rho_1 + (g_{nm_{s2}}^{j=2} - g_{nm_{s1}}^{j=1})\rho_2 + \ldots + (g_{nm_{s2}}^{j=m} - g_{nm_{s1}}^{j=m-1})\rho_m \quad (16.26)$$

The factors $(g_{nm_{s2}}^{j=m} - g_{nm_{s1}}^{j=m-1})$ represent measurements from the HSOC as described in Equation 16.24. Alternatively, this may be represented in an incremental form as in Equation 16.27, with the references to source positions $s1$ and $s2$ omitted but assumed. This relationship suggests that there is no interdependency between the individual ρ_m, $(m = 1, \ldots, M)$.

$$\int_{t_0}^{t_{m-1}} \Delta g_n(\sigma(n), \beta(t_j), t_j)dt = \int_{t_0}^{t_{m-2}} \Delta g_n(\sigma(n), \beta(t_j), t_j)dt + (g_{nm}^{m-1} - g_{nm}^{m-2})\rho_{m-1} \quad (16.27)$$

Defining the function $\Delta h_n^m(\sigma(n), \beta(t_j), t_j)$ to be an estimate of the function $\Delta f_n(\sigma(n), \beta(t_j), t_j)$, of Equation 16.20 and using the definition in Equation 16.28, the relationship between HSOC measurements and the desired organ motion effect measurements may be expressed as

$$\Delta h_n^m(\sigma(n), \beta(t_j), t_j) = \int_{t_0}^{t_{m-1}} \Delta g_n(\sigma(n), \beta(t_j), t_j)dt \quad (16.28)$$

$$\Delta h_n^m(\sigma(n), \beta(t_j), t_j) = \Delta h_n^{m-1}(\sigma(n), \beta(t_j), t_j) + (g_n^{m-1} - g_n^{m-2})\rho_{m-1} \quad (16.29)$$

The initial condition is $\rho_1 = 1$. In cases where the motion is independent of view angle, as would be the case for a deforming object at the center of the field of view, $\rho_m = 1$ $(m = 1, \ldots, M)$. In such a case, Equation 16.26 (or Equation 16.27) reduces to Equation 16.30:

$$\int_{t_0}^{t_{m-1}} \Delta g_n(\sigma(n), \beta(t_j), t_j)dt = g_n^{j=m} - g_n^{j=0} \quad (16.30)$$

Because of Equation 16.31, Equation 16.30 may be represented as two separate sinograms.

$$g_n(\sigma(n), \beta(t), t) = \iint f(x, y, t)\delta\{x\cos[\sigma(n) + \beta(t)] + y\sin[\sigma(n) + \beta(t)] - R\sin\sigma(n)\}dxdy \quad (16.31)$$

The first sinogram is the standard X-ray CT sinogram with motion artifacts present, defined by Equation 16.32. The second corresponds to no motion and defined in Equation 16.33.

$$g_n^{j=M}(\sigma(n), \beta(t), t) = \iint f(x, y, t)\delta\{(x, y), (\sigma(n), \beta(t), R(t))\}dxdy \quad (16.32)$$

$$g_n^{j=0}(\sigma(n), \beta(t_0), t_0) = \iint f(x, y, t_0)\delta\{(x, y), (\sigma(n), \beta(t_0), R(t_0))\}dxdy \quad (16.33)$$

The first sinogram is derived over the complete data acquisition period, whereas the second sinogram is derived from the time that the first projection is taken; in effect, all motion has been frozen. Comparing Equations 16.30, 16.32, and 16.33 with Equation 16.20, it is evident that the sinograms are identical in the two representations. Specifically, these relationships are defined as

$$g_n^{j=0}(\sigma(n), \beta(t_0), t_0) = g_{n_{ct}}(\sigma(n), \beta(t), t) \tag{16.34}$$

$$g_n^{j=M}(\sigma(n), \beta(t_M), t_M) = g_{n_{mov}}(\sigma(n), \beta(t), t) \tag{16.35}$$

As a result, for the case where the factors ρ_m=1, (m = 1, …, M) the estimate of $\Delta f_n(\sigma(n), \beta(t), t_j)$ is given by

$$\Delta f_n(\sigma(n), \beta(t_j), t_j) = \int_{t_0}^{t_j} \Delta g_n(\sigma(n), \beta(t), t)dt \tag{16.36}$$

For X-ray CT applications, this integral expression is not a simple problem because the scalar factors ρ_m, (m = 1, …, M), introduce dc offsets. The impact of these factors on the integration process of Equation 16.27 may be removed by means of a normalization process[53] or non-linear adaptive processing.[52,54,55] Moreover, from Equation 16.36, the derivative of the ideal set of measurements, $\Delta f_n(\sigma(t_j), \beta(t_j), t_j)$, which define the difference between sinograms corresponding to projections with motion effects present and those acquired without motion effects, should predict the measurements of the spatial overlap correlator:

$$\Delta g_n(\sigma(t_j), \beta(t_j), t_j) \approx \frac{d(\Delta f_n(\sigma(n), \beta(t_j), t_j))}{dt}\Delta t \tag{16.37}$$

This relationship suggests that the HSOC provides the derivative of the motion effects, and it follows directly that measurements of the spatial overlap correlator form the basis of a new processing scheme to remove motion artifacts associated with the CT data acquisition process. In particular, motion effects are defined by temporal integration of the HSOC measurements, as given by

$$\Delta f(\sigma(n), \beta(t_j), t_j) \approx \sum_{j=1}^{J} \Delta g_n(\sigma(n), \beta(t_j), t_j)\rho_j, \text{ for } (n = 1, 2, \ldots, N) \tag{16.38}$$

In practical terms, this suggests the possibility of generating two types of sinograms. The first is the standard X-ray CT measurements $g_{n_{mov}}(\sigma(n), \beta(t_j), t_j)$, with organ motion effects present. The second is the HSOC measurements $\Delta g_n(\sigma(n), \beta(t_j), t_j)$, which provide estimates of $\Delta f_n(\sigma(n), \beta(t_j), t_j)$ as described in Equation 16.38.

In the case of non-linear effects, which require a normalization process,[53] or an estimation scheme for the scalar factors ρ_m, alternative optimum estimates of $\Delta f_n(\sigma(n), \beta(t_j), t_j)$ can be provided by an AIC process with inputs as the sinograms $g_{n_{mov}}(\sigma(n), \beta(t_j), t_j)$ and $\Delta g_n(\sigma(n), \beta(t_j), t_j)$, as discussed in the next section.

16.4.2.1 Adaptive Interference Cancellation

The concept of adaptive interference cancellation is particularly useful for isolating signals in the presence of additive interferences.[52,54,55] The AIC scheme is shown in Figure 16.16. The detector signal with interference due to motion is represented as $y(j\delta) = s(j\delta) + n(j\delta)$, where $s(j\delta)$ and $n(j\delta)$ are the signal and interference components, respectively. In an AIC system with performance feedback, it is essential

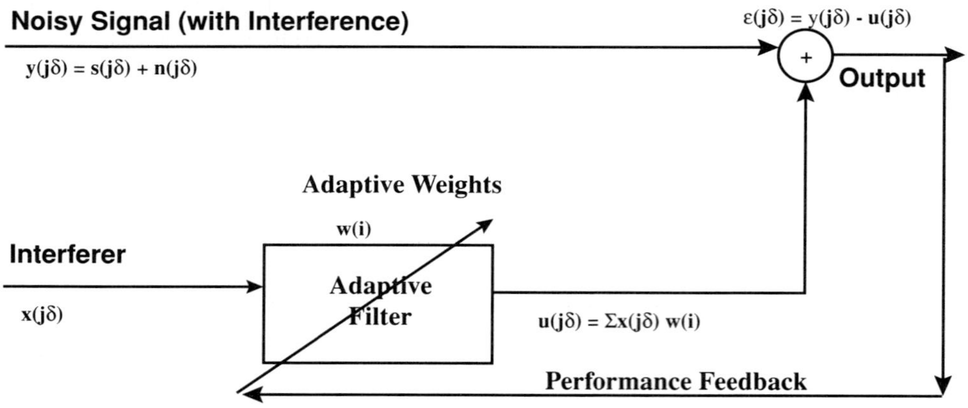

FIGURE 16.16 Concept of an AIC. (Reprinted by permission of IEEE © 2000.)

that the interference component $x(j\delta)$ is either available or measured simultaneously with the received noisy signal $y(j\delta)$.[52]

We assume that an adaptation process with performance feedback provides the weight vector $w(i)$ and through linear combination generates estimates of the interference $n(j\delta)$. In general, the adaptation process includes a minimization of the mean square value of the error signal defined by the performance feedback. Optimization by this criterion is common in many adaptive and non-adaptive applications. In the case of the X-ray CT system, measurements provided by the spatial overlap correlator $\Delta g_n(\sigma(n), \beta(t_j), t_j)$ form the basis of the interference estimates for the adaptation process, while the standard X-ray CT projection measurements $\Delta g_{mov}(\sigma(n), \beta(t_j), t_j)$ represent the noisy signal.

We also assume that interference measurements at the input of the adaptation process in Figure 16.16 are provided by the input $\bar{X}$ with terms $[x(\delta), x(2\delta), \ldots, x(L\delta)]^T$. Furthermore, the output of the adaptation process, $u(j\delta)$, is a linear combination of the input measurements $\bar{X}$ and the weight adaptive coefficients $[w(1), w(2), \ldots, w(L)]^T$ of the vector $\bar{W}$. In general, the weights $\bar{W}$ are adjusted so that the system descends toward the minimum of the surface of the performance feedback.[52,56] The output of the AIC is given by Equation 16.39, where $\bar{Y}$ is the input vector of the interference measurements, $y(j\delta)$ for $j = 1, 2, \ldots, M$ and M is the maximum number of samples to be processed. Since M is generally larger than L, segments of the inputs of length L are selected in a sliding window fashion and processed.

$$\bar{\varepsilon} = \bar{Y} - \bar{X}^T\bar{W} \tag{16.39}$$

The AIC concept is based on the minimization of Equation 16.39 in the least mean square sense, giving

$$E_{\min}[(\bar{\varepsilon})^2] = E_{\min}[(\bar{S} + \bar{N} - [\bar{X}^T\bar{W}])^2] \tag{16.40}$$

When $E[(\bar{\varepsilon})^2]$ is minimized, the signal power $E[\bar{S}]$ is unaffected, and the term $E[\bar{N} - \bar{X}^T\bar{W}]$ is minimized. Thus, the output $u(j\delta)$ of an adaptive filter with L adaptive weights $(w_1, w_2, \ldots, w_L)$ and for an interference input vector $x(j\delta)$, of arbitrary length, at time $j\delta$ are given by Equation 16.41, where the output of the adaptive filter depends on some history of the interference. The number of past samples required is determined by the length of the adaptive filter.

$$u(j\delta) = \sum_{i=1}^{L} w_i^{j\delta} \times x((j+i-L)\delta) \tag{16.41}$$

The weights of the adaptive filter are adjusted at each time interval, and the update method depends on the adaptive algorithm. If μ is the adaptive step size, the update equation for the least mean square (LMS) adaptive filter is given as[52,55,57]

$$w_i^{(j+1)} = w_i^{j\delta} + (\mu \times x((j+1-L)\delta)) \times u(j\delta), (i = 1, 2, \ldots, L) \tag{16.42}$$

Similarly, for the normalized least mean square (NLMS) algorithm, the update equation is given by Equation 16.43. In this update equation, λ is the adaptive step size parameter, α is a parameter included for stability, and $|n|$ is the Euclidean norm of the vector input interference vector $[x((j+1-L)\delta), x((j+2-L)\delta), \ldots, x((j)\delta)]$.

$$w_i^{(j+1)\delta} = w_i^{j\delta} + \left(\frac{\lambda}{\alpha + |n|} \times x((j+i-L)\delta) \times u(j\delta)\right), (i = 1, 2, \ldots, L) \tag{16.43}$$

Figure 16.17 shows the AIC processing structure that has been modified to meet the requirements of the X-ray CT motion artifact removal problem. The CT sinogram can then be expressed as

$$\bar{g}_{n_{CT}} = \bar{g}_{n_{mov}} - \underline{P_n}\overline{\Delta g_n} \tag{16.44}$$

where vectors $\bar{g}_{n_{CT}} = [g_{n_{CT}}(t_1), g_{n_{CT}}(t_2), \ldots, g_{n_{CT}}(t_M)]^T$ and $\bar{g}_{n_{mov}} = [g_{n_{mov}}(t_1), g_{n_{mov}}(t_2), \ldots, g_{n_{mov}}(t_M)]^T$ are defined for each one of the detector elements $n = 1, 2, \ldots, N$, of the CT detector array. The vecto $\overline{\Delta g_n} = [\Delta g_n(t_1), \Delta g_n(t_2), \ldots, \Delta g_n(t_M)]^T$ represents the HSOC measurements, which represent the interference measurements. The matrix $\underline{P_n}$ includes the adaptive weights defined by

$$\underline{P_n} = \begin{bmatrix} \rho_1 & 0 & 0 & 0 \\ \rho_1 & \rho_2 & 0 & 0 \\ \cdot & \cdot & \cdot & \cdot \\ \rho_1 & \rho_2 & \cdot & \rho_M \end{bmatrix} \tag{16.45}$$

The adaptive filter weights $w(j)$ are replaced with ρ_m, $(m = 1, \ldots, M)$. Recall also that $\rho_1 = 1$. The optimization process by the AIC processor now includes optimizing ρ_m, to reduce the effects of the

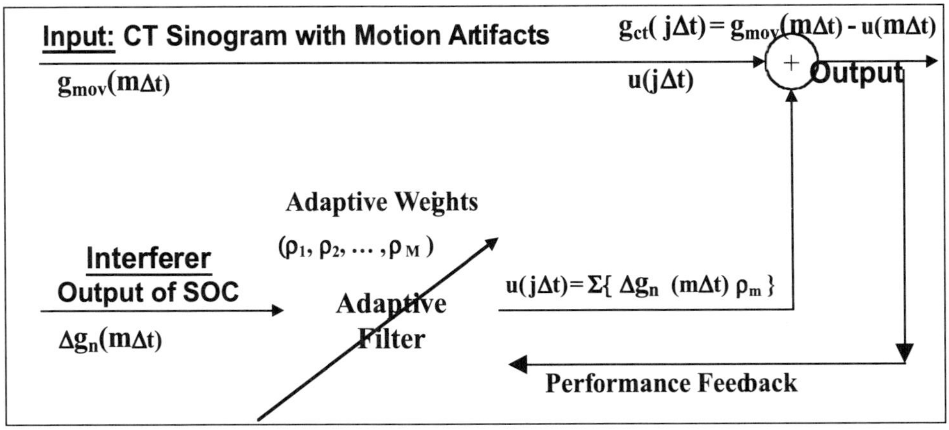

FIGURE 16.17 Concept of an AIC and spatial overlap correlator for removing motion artifacts in CT medical imaging applications. (Reprinted by permission of IEEE © 2000.)

motion artifacts. Thus, the adaptive algorithms form the basis of an iterative estimation and accumulation process for the terms $(g_n^{m-1} - g_n^{m-2})\rho_{m-1}$ that defines the non-linear temporal integration of the HSOC output, as defined in Equation 16.27. The output of the adaptive filter $u(j\delta)$ is a predictive estimate of the term $\Delta f_n(\sigma(n), \beta(t_j), t_j)$ of Equation 16.20. The output of this AIC process provides predictive estimates for sinograms that have been corrected for motion artifacts according to the information provided by the measurements of the spatial overlap correlator.

16.4.3 Coherent Sinogram Synthesis from Software Spatial Overlap Correlator

An alternative approach is to assemble a sinogram using the CSS method, which uses the time series produced by either version of the spatial overlap correlator. When the object, or heart in the case of cardiac X-ray CT imaging, is at a specified phase in its motion cycle, the CSS technique defines this as the phase of interest. It then isolates every subsequent time moment during the data acquisition period when the object is again at that phase of its motion cycle. A number of projections are selected at each of these time moments and assembled into a sinogram. Figure 16.18 depicts the complete process using the curves obtained from using the software spatial overlap correlator (SSOC) scheme for the purpose of illustration. The process is identical with the time series from the HSOC.

16.4.3.1 Phase Selection by Correlation

A sliding window correlation process is used to isolate the moments during the data acquisition that include the same information and phase as the phase of interest. In general, the cross-correlation coefficient, CC_{sr}, between two time series, s_i and r_i of length L, is given by Equation 16.46.[52]

$$CC_{sr} = \frac{\sum_{i=1}^{L} s_i r_i}{\sqrt{\sum_{i=1}^{L} s_i^2 \sum_{i=1}^{L} r_i^2}} \tag{16.46}$$

where s_i is the signal from the spatial overlap correlator and r_i is the replica kernel, which is a short segment of the motion cycle extracted from s_i. The sliding window correlation technique uses a subset of the signal near the phase of the interest as the replica kernel and then correlates this replica with segments of the continuous signal to compute a time varying correlation function. The segments of the signal used in the cross-correlation function are selected in a sliding window fashion. The time varying cross-correlation function CC_i is given by Equation 16.47, where L is the length of the segment used in the cross-correlation function and N is the length of the complete time series from the spatial overlap correlator.

$$CC_i = \frac{\sum_{j=-L/2}^{L/2} s_{i+j} r_{j+\frac{L}{2}}}{\sqrt{\sum_{j=-L/2}^{L/2} s_{i+j}^2 \sum_{j=-L/2}^{L/2} r_{j+\frac{L}{2}}^2}}, \quad i = \frac{L}{2}, \frac{L}{2}+1, \ldots, N-\frac{L}{2} \tag{16.47}$$

The time moments at which the maxima of the correlation function are sufficiently close to 1 are considered as the time moments at which the phase of interest reoccurs. The time moments directly define a projection number, since the projections are acquired sequentially at a known sampling rate. With all of the projections that define a phase of interest known, the sinogram may now be assembled.

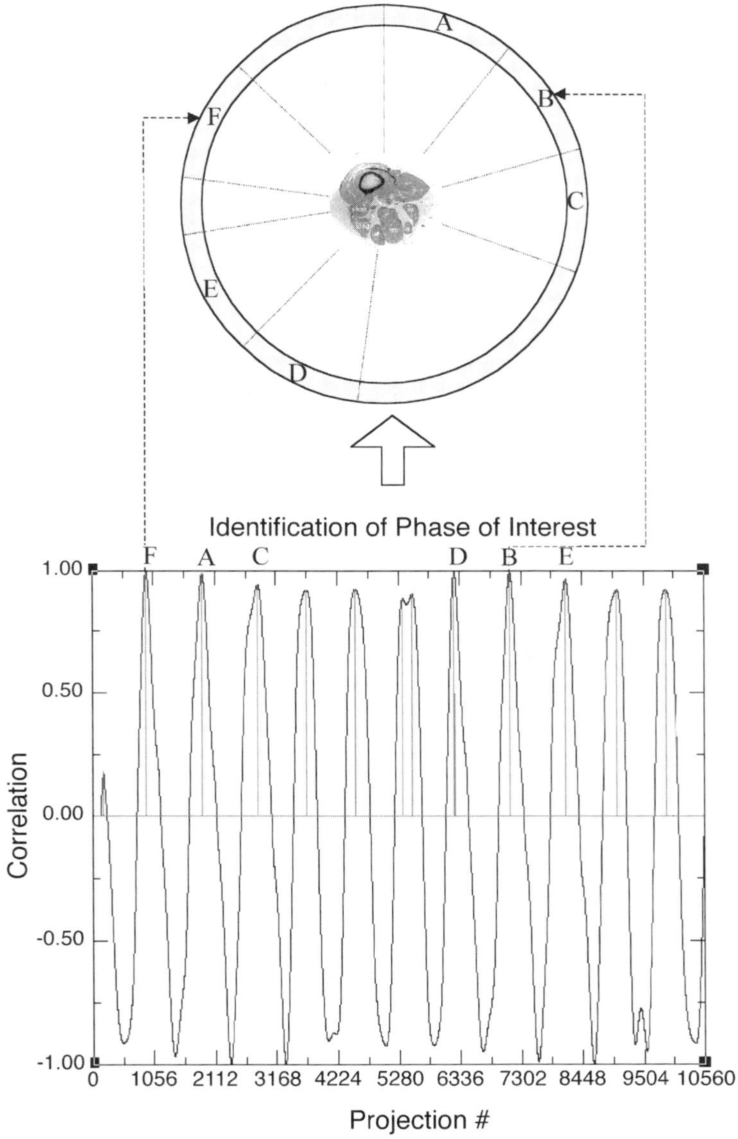

FIGURE 16.18 Operation of the CSS method. (Reprinted by permission of IEEE © 2000.)

16.4.3.2 Assembling the Coherent Sinogram

In the final stage, segments of the continuous sinogram are selected and used to assemble a sinogram for a single image. The selection criterion is to use projections that occur when the organ is at the desired point in its motion cycle. In other words, the time moments at which the level or correlation approaches one in the curve of Figure 16.18 are the time moments that are used as a selection criterion to identify the projections for coherent synthesis of the sinogram. The curve in the lower segment of Figure 16.18 shows the time moments selected because of the sufficiently high level of correlation at these times. These time moments correspond to angular locations of projections, as shown in the upper panel of Figure 16.18. The angular locations they map to are determined by the physical locations of the X-ray source and detector array at the time the organ is at the desired point in its phase.

Since the X-ray source rotates about the patient, spatial locations distributed on a circle are sampled repeatedly. A projection number P maps to a physical projection number p, where P is referenced to the entire data acquisition period and p is referenced to the physical location on the circle along which data are acquired. The relationship between P and p is given by

$$P = nN + p; 0 \le P \le \infty; 0 \le p \le N; n = 1, 2, \ldots, \infty \tag{16.48}$$

In effect, the segments chosen are synchronized to the organ motion cycle, and valid projections are only extracted when the organ is exactly at the desired point in its motion cycle. The difference is that data are acquired continuously, and not all projections are used in generating a single image. However, there is the need for additional data for the interpolation process, and there may be a need for images from a number of phases. Therefore, although not all of the data are used for a single image, the complete processing scheme requires all of the data acquired to produce a complete image set.

Under ideal conditions, one view would be taken each time the organ reaches the desired point in its motion cycle, and after N cycles of motion, a complete sinogram would be available. Since there is no physical synchronization between the data acquisition process and the organ motion, there may be a repetition of some views, while other views may be missing.

16.4.3.3 Interpolation to Complete the Sinogram

Interpolation is used to account for any missing angular segments data acquired by the X-ray CT system. The first option is to take data from one complete revolution and use those data as a basis for the final image. The idea is to try to improve the image that would be produced by this standard X-ray CT sinogram. Using this sinogram as a starting point, projection windows are selected in the same manner as described in the previous section. Whenever a suitable projection window is found, it is used to overwrite the original projections of the sinogram.

The second option synthesizes a new sinogram. This method does not limit the number of projections in a projection window. Rather, windows are allowed to be as large as necessary to fill in the entire sinogram. The center of the initial window is defined as the desired point, and missing projections are filled in using appropriate projections from the data acquired by choosing projections as close as possible to the desired point. This method is preferred when the original image quality is poor, since it attempts to create a completely new image.

16.4.4 Signal Processing Structure for an AIC

A block diagram representation of the signal processing structure that implements the AIC and CSS schemes are shown in Figures 16.19 and 16.20, respectively. Both structures consists of three major blocks: the data acquisition system, the signal processor, and the display functionality.

The data acquisition system requires specialized CT hardware to support the HSOC. This means that the data acquisition system will effectively provide two data streams, with both streams consisting of samples from the same spatial locations, but at different times on the order of δ.

16.4.5 Phantom Experiment

The phantom shown in Figure 16.21, consisting of a hollow Plexiglas™ cylinder and an inner solid Teflon™ cone, was constructed to demonstrate the motion-artifact reduction potential of the synthetic aperture approach. The cone moves back and forth through the image plane in the gantry, simulating an expanding and contracting ventricle. Seven metal wires were placed on the outside of the cone to simulate arterial calcifications. A conventional CT image of this phantom when stationary is shown in Figure 16.22. This image has no motion artifacts and represents the target image for any motion-artifact reduction method. A conventional CT image of the phantom operating with a period of 0.6 s, simulating a heart rate of 100 beats per minute, is shown Figure 16.23. Severe motion artifacts are evident.

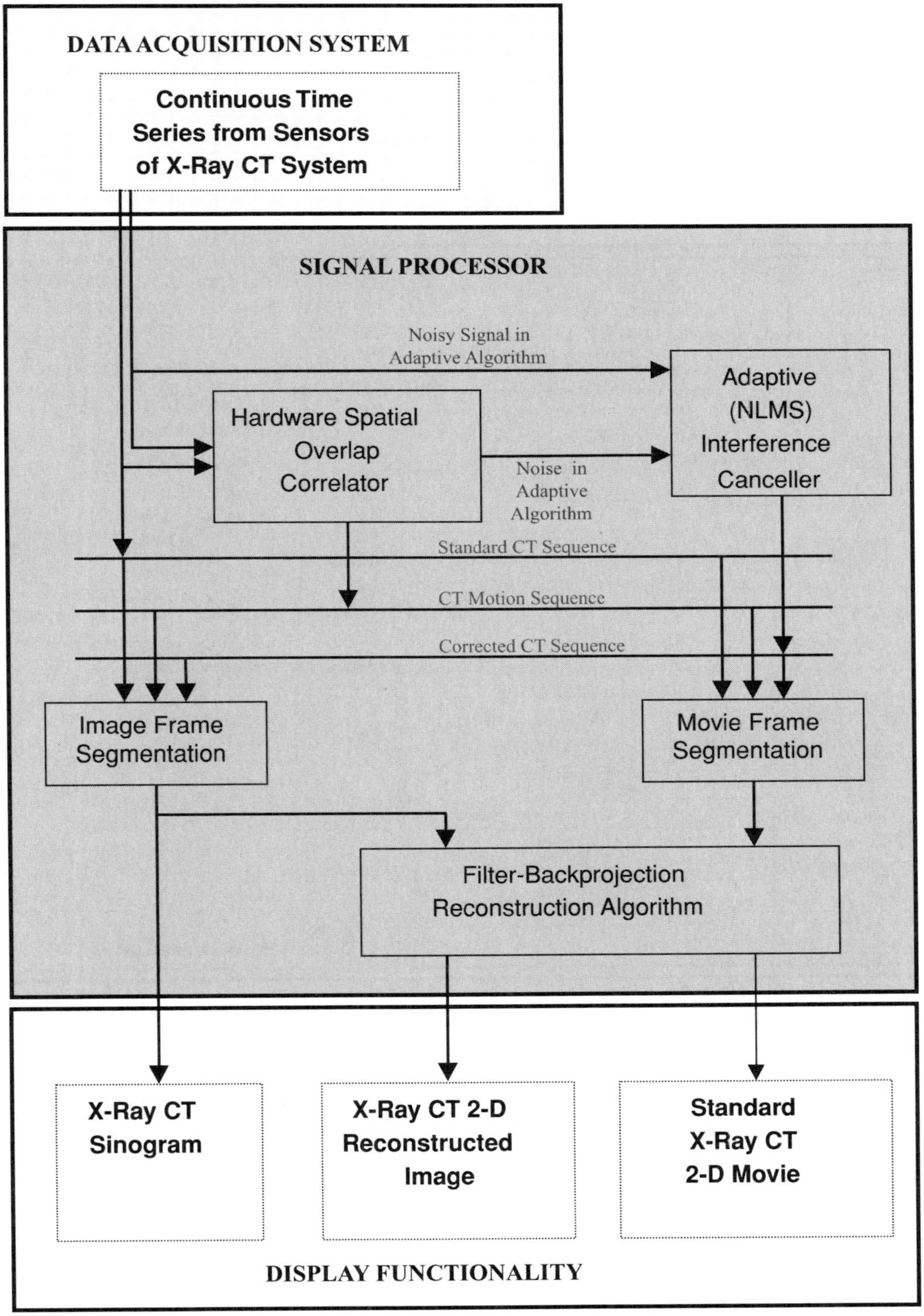

FIGURE 16.19 Signal processing structure for AIC implementation.

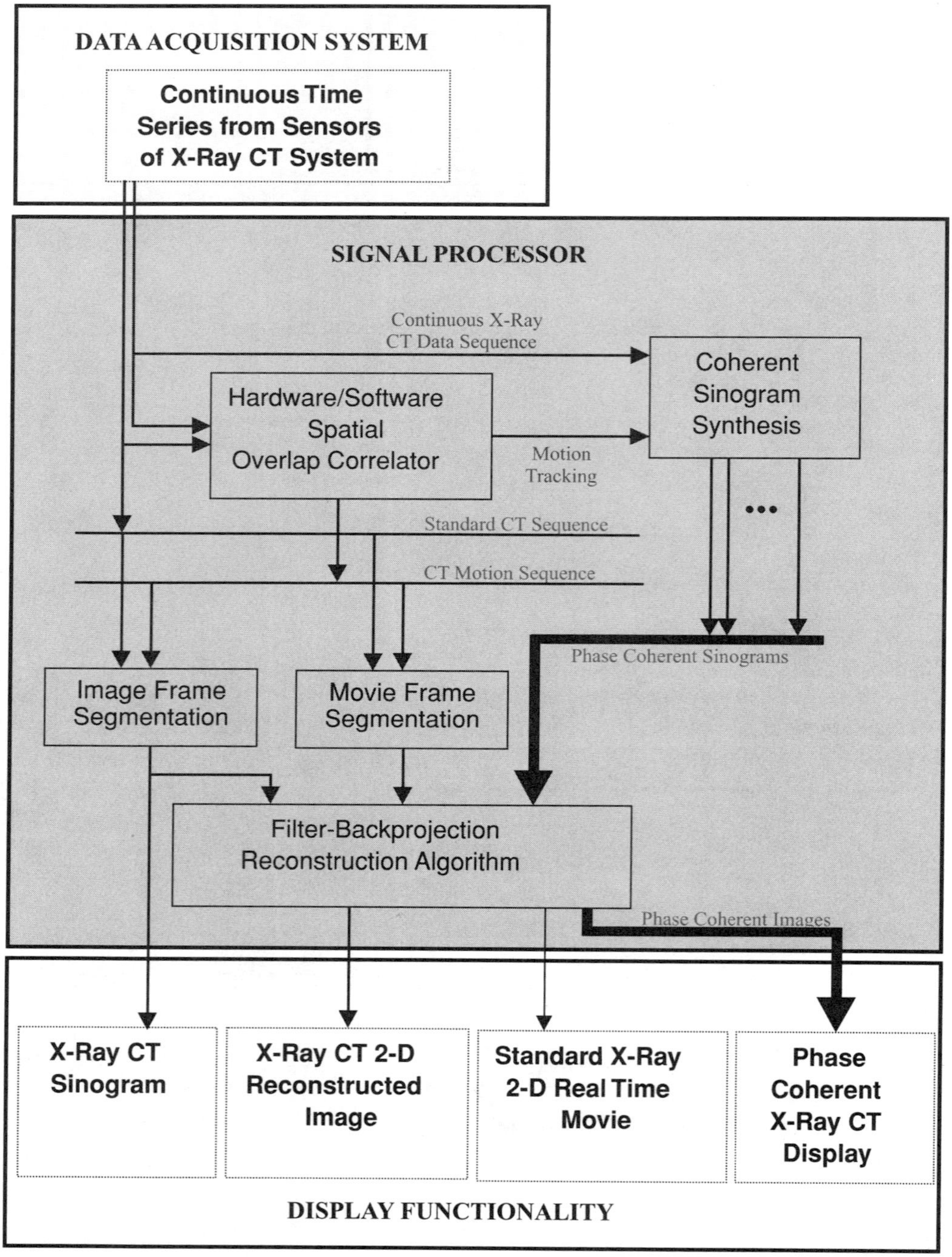

FIGURE 16.20 Signal processing structure for CSS implementation.

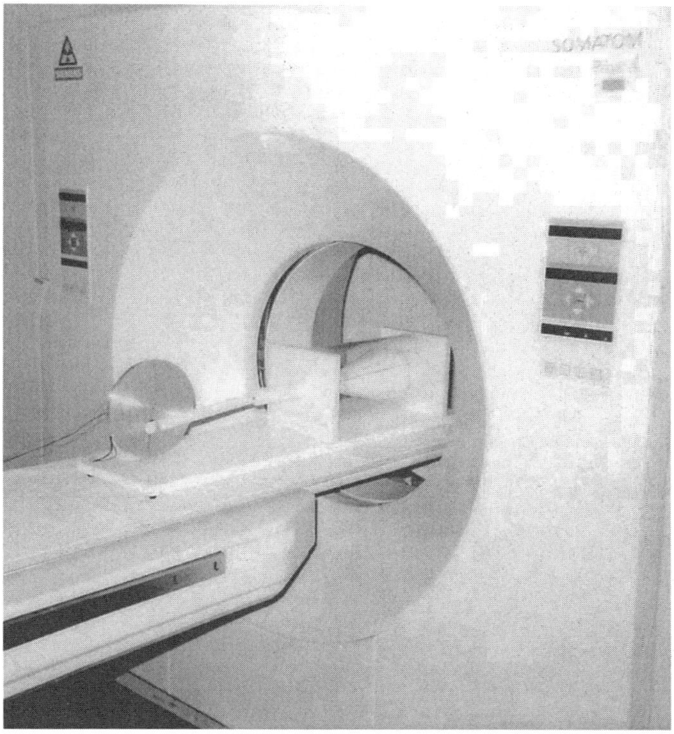

FIGURE 16.21 Experimental phantom in a CT scanner simulating an expanding ventricle.

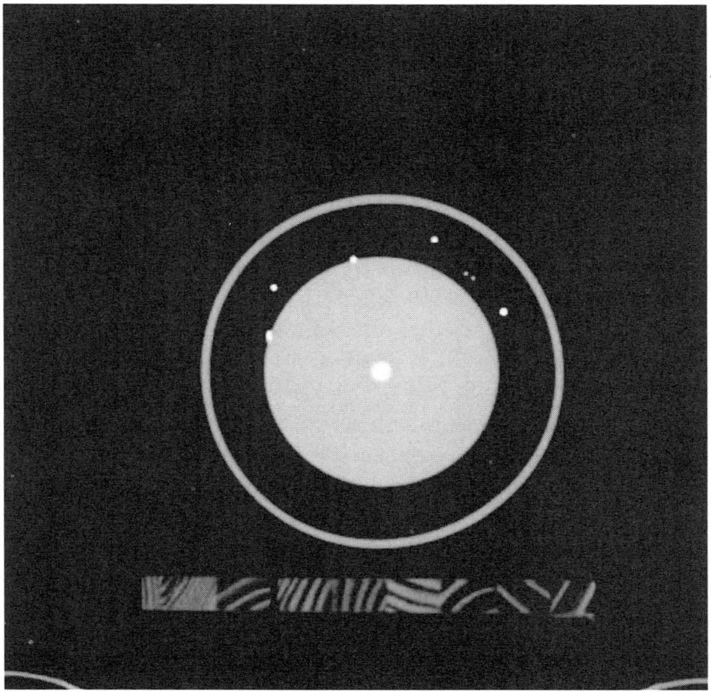

FIGURE 16.22 Conventional CT image of a stationary phantom. (Reprinted by permission of IEEE © 2000.)

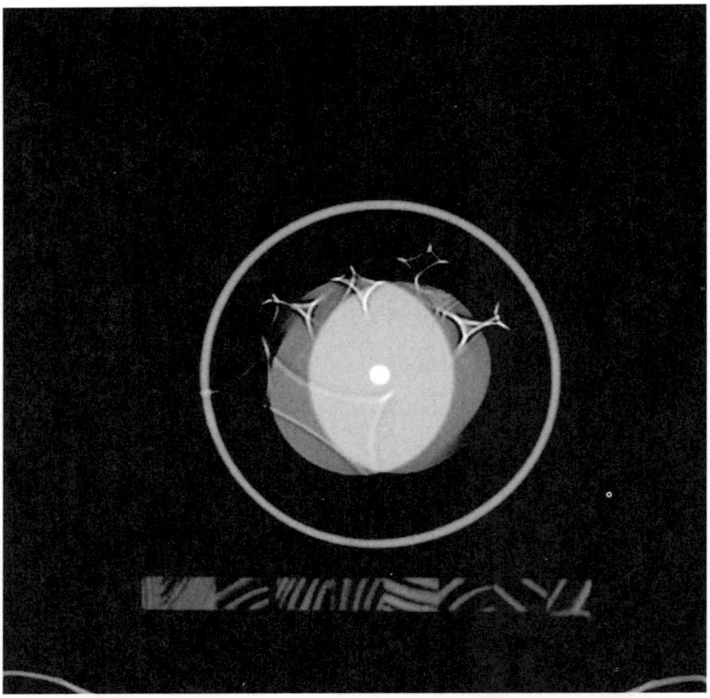

FIGURE 16.23 Conventional CT image of a moving phantom showing severe motion artifacts. (Reprinted by permission of IEEE © 2000.)

A sequence of six CSS/SSOC images were generated at equally spaced points covering one cycle of the phantom's motion. Figure 16.24 shows images obtained using projection data acquired over 9 s. Motion artifacts are reduced in all images relative to the conventional image, showing expansion and contraction of the simulated ventricle. Artifacts exist due to motion of the simulated calcifications, but they are relatively minor in the top-right image corresponding to end diastole where motion is the least. Figure 16.25 shows a sequence of images obtained over a period of 3 s. Residual motion artifacts are more pronounced, but all images are superior to the conventional image.

16.4.6 Human Patient Results

The CSS/SSOC method was evaluated in a cardiac study of a middle-aged female. The patient's heart rate was approximately 72 beats per minute. CT projection data were acquired during multiple rotations of the X-ray tube. No restriction was placed on the patients' breathing. The conventional CT image is shown in Figure 16.26. There is no indication of any calcification in the arteries of the heart in this image, and the effect of respiratory motion is evident. The chest walls and sternum are not clearly defined, appearing as dual images.

Figure 16.27 corresponds to the synthesized sinogram output of the SSOC and CSS processes. In this case, motion artifacts due to breathing effects have been removed, as indicated by the clarity of the image near the area of the sternum. This was achieved with the CSS process by selecting segments of the sinogram corresponding to the same phases of the heart and breathing motion cycles. Since the period of the breathing motion is long (2 to 3 s) compared to that of the heart's periodic motion (0.5 to 1 s), another method to remove the breathing motion effects from the SSOC time series is by applying a band pass filter on the SSOC time series.

Another improvement of diagnostic importance, in this case, is the better and brighter definition compared to the conventional CT image (Figure 16.26) of the bright region corresponding to a coronary calcification in the top-right area of the heart. Overall quality of the image in Figure 16.27 is superior to the conventional CT image in Figure 16.26.

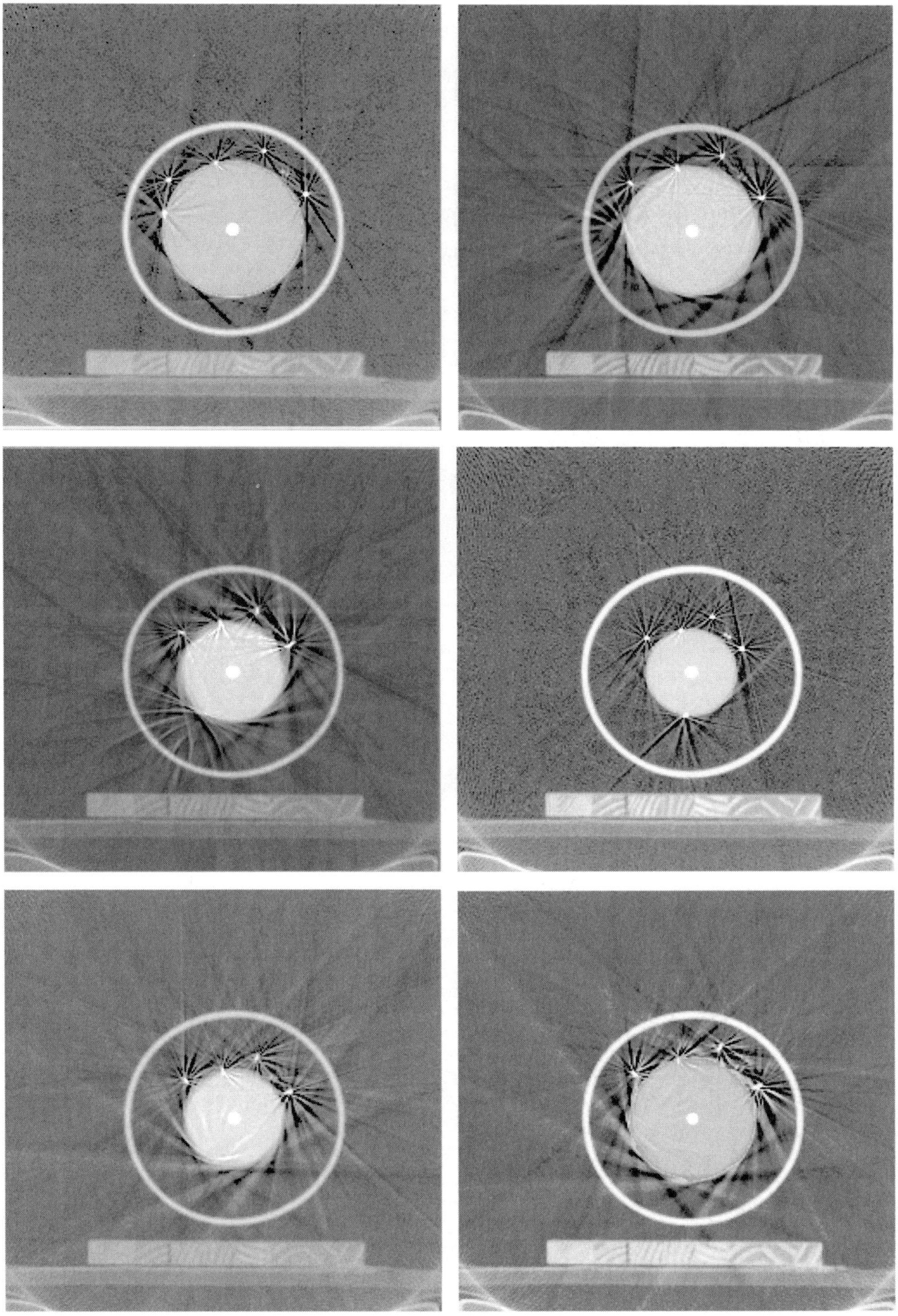

FIGURE 16.24 Resulting images of the CSS/SSOC method with a phantom motion cycle period of 0.6 and 9 s of total data acquisition. (Reprinted by permission of IEEE © 2000.)

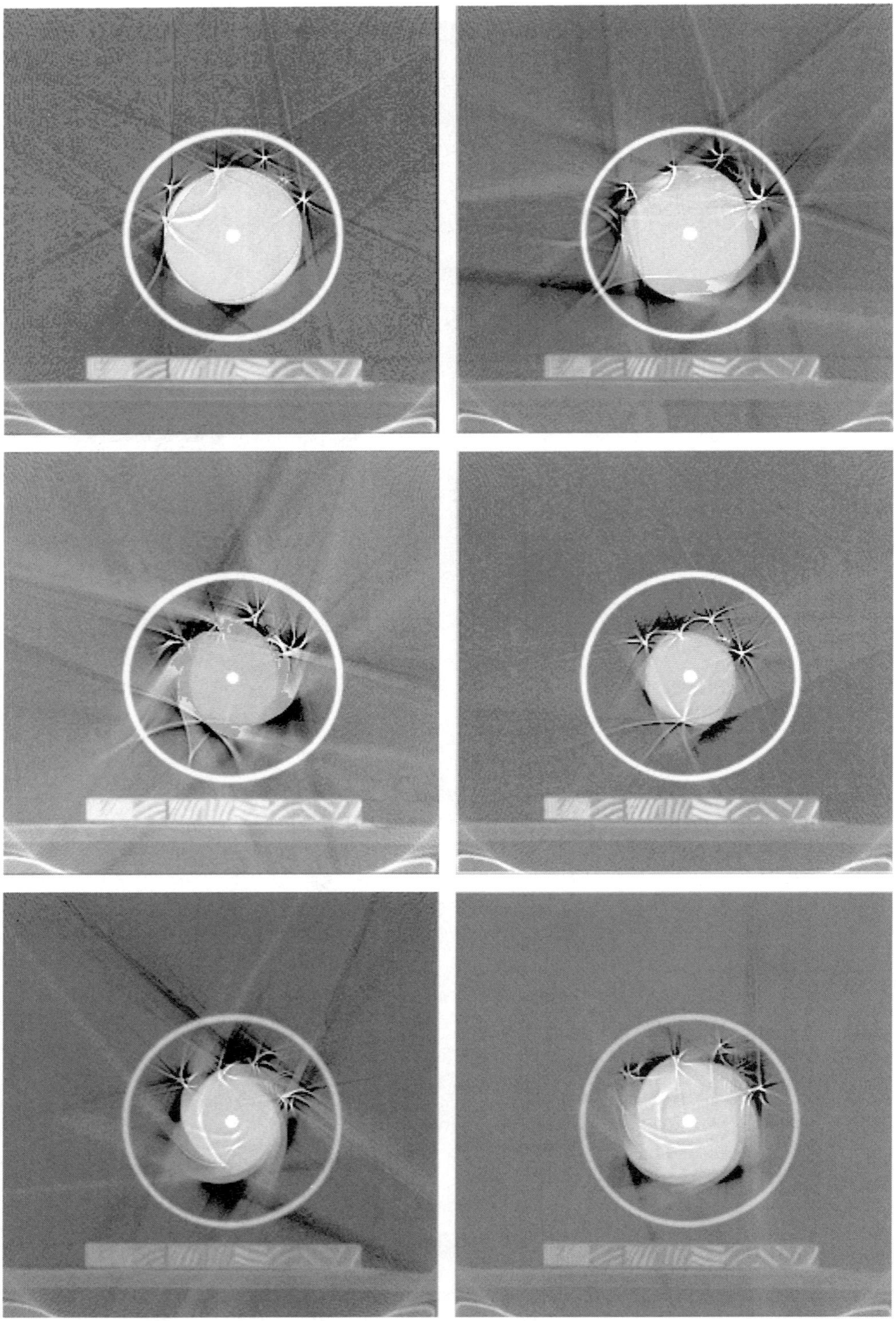

FIGURE 16.25 Resulting images from the CSS/SSOC method with a phantom motion cycle period of 0.6 and 3 s of total data acquisition.

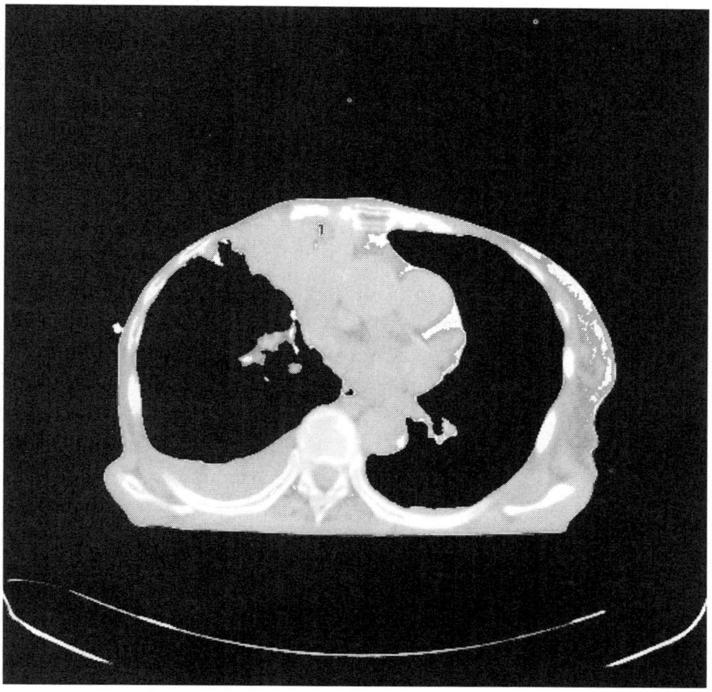

FIGURE 16.26 Conventional CT image including cardiac and respiratory motions. (Reprinted by permission of IEEE © 2000.)

16.5 Conclusions

The problem of motion artifacts is well known in X-ray CT systems. Some types of motion are generally considered to be controllable for most patients, such as respiratory motion and patient restlessness, although there are many exceptions. Other forms of motion are not controllable, such as cardiac motion and blood flow. Patient motion during the data acquistion process results in an inconsistent set of projection data and a degradation of image quality due to motion artifacts. These artifacts appear in the form of streaking, distortions, and blurring. Anatomical regions that are moving appear distorted, while nearby stationary structures may be corrupted by overlying artifacts. These artifacts may result in inaccurate or misleading diagnoses.

Numerous techniques have been developed as partial solutions to this problem. They include (1) simple methods that restrict controllable forms of motion, (2) techniques that modify the data acquisition and reconstruction processes to minimize motion effects, and (3) more generalized techniques that perform data acquisition in a conventional fashion and reduce the effects of motion using retrospective signal processing.

The synthetic aperture signal processing approach has been treated in detail. It provides a measure of organ motion from an analysis of the projection data using a spatial overlap correlator. This information is then used to estimate the effects of this motion and to provide tomographic images with suppressed motion artifacts.

Two implementations of the spatial overlap correlator are described. The first is a hardware implementation. While it provides the best estimate of organ motion, it generally requires a hardware modification to conventional CT systems. The second is a software implementation that provides an approximate measure of organ motion. It works very well with most physiologic motions and can be implemented on existing slip-ring CT systems.

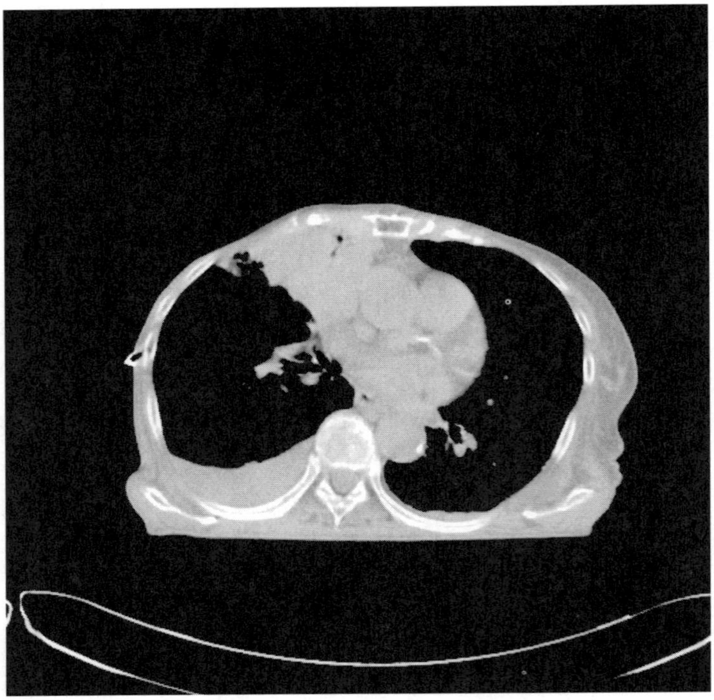

FIGURE 16.27 Corrected image from the SSOC/CSS motion correction scheme. (Reprinted by permission of IEEE © 2000.)

Two motion-correction schemes are described. The first is an AIC that uses organ motion information from the HSOC as an "interference" signal. Motion artifacts are suppressed by removing this interference from the conventional CT sinogram. The second scheme is called a CSS method. It uses organ motion as determined by either the HSOC or SSOC to generate "phase-coherent" sinograms. These are estimated motion-free sinograms from which motion-free images are reconstructed for any specified moment in the motion cycle. Multiple images can be reconstructed to generate "cine-loop" movies of the patient, showing both cardiac and respiratory motions.

The effectiveness of the CSS method along with the spatial overlap correlator was demonstrated with a simulation study, experiments using a moving phantom, and a clinical patient study. The clinical images showed significantly reduced motion artifacts in the presence of both cardiac and respiratory motions.

References

1. Brooks, R.A. and Di-Chiro, G., Principles of computer assisted tomography (CAT) in radiographic and radioisotopic imaging, *Physics in Medicine & Biology,* 21, 689–732, 1976.
2. Edelheit, L.S., Herman, G.T., and Lakshminarayanan, A.V., Reconstruction of objects from diverging x rays, *Medical Physics,* 4, 226–31, 1977.
3. Herman, G.T., Advanced principles of reconstruction algorithms, in Newton, T.H. and Potts, D.G. (Eds.), *Radiology of the Skull and Brain,* C.V. Mosby, St. Louis, 1981, Ch. 110-II.
4. Macovski, A., Basic concepts of reconstruction algorithms, in Newton, T.H. and Potts, D.G. (Eds.), *Radiology of the Skull and Brain,* C.V. Mosby, St. Louis, 1981, Ch. 110-I.
5. Hutchins, W.W., Vogelzang, R.L., Fuld, I.L., and Foley, M.J., Utilization of temporary muscle paralysis to eliminate CT motion artifact in the critically ill patient, *Journal of Computer Assisted Tomography,* 8(1), 181–183, 1984.

6. Stockberger, S.M., Jr., Hicklin, J.A., Liang, Y., Wass, J.L., and Ambrosius, W.T., Spiral CT with ionic and nonionic contrast material: evaluation of patient motion and scan quality, *Radiology*, 206(3), 631–636, 1998.
7. Stockberger, S.M., Jr., Liang, Y., Hicklin, J.A., Wass, J.L., Ambrosius, W.T., and Kopecky, K.K., Objective measurement of motion in patients undergoing spiral CT examinations, *Radiology*, 206(3), 625–629, 1998.
8. Foley, W.D., Contrast-enhanced hepatic CT and involuntary motion: an objective assessment [editorial], *Radiology*, 206(3), 589–591, 1998.
9. Luker, G.D., Bae, K.T., Siegel, M.J., Don, S., Brink, J.A., Wang, G., and Herman, T.E., Ghosting of pulmonary nodules with respiratory motion: comparison of helical and conventional CT using an in vitro pediatric model, *American Journal of Roentgenology*, 167(5), 1189–1193, 1996.
10. Mayo, J.R., Muller, N.L., and Henkelman, R.M., The double-fissure sign: a motion artifact on thin-section CT scans, *Radiology*, 165(2), 580–581, 1987.
11. Qanadli, S.D., El Hajjam, M., Mesurolle, B., Lavisse, L., Jourdan, O., Randoux, B., Chagnon, S., and Lacombe, P., Motion artifacts of the aorta simulating aortic dissection on spiral CT, *Journal of Computer Assisted Tomography*, 23(1), 1–6, 1999.
12. Loubeyre, P., Grozel, F., Carrillon, Y., Gaillard, C., Guyard, F., Pellet, O., and Minh, V.A., Prevalence of motion artifact simulating aortic dissection on spiral CT using a 180 degree linear interpolation algorithm for reconstruction of the images, *European Radiology*, 7(3), 320–322, 1997.
13. Duvernoy, O., Coulden, R., and Ytterberg, C., Aortic motion: a potential pitfall in CT imaging of dissection in the ascending aorta, *Journal of Computer Assisted Tomography*, 19(4), 569–572, 1995.
14. Mukherji, S.K., Varma, P., and Stark, P., Motion artifact simulating aortic dissection on CT [letter; comment], *American Journal of Roentgenology*, 159(3), 674, 1992.
15. Burns, M.A., Molina, P.L., Gutierrez, F.R., and Sagel, S.S., Motion artifact simulating aortic dissection on CT [see comments], *American Journal of Roentgenology*, 157(3), 465–467, 1991.
16. Kaufman, R.B., Sheedy, P.F., Breen, J.F., Kelzenberg, J.R., Kruger, B.L., Schwartz, R.S., and Moll, P.P., Detection of heart calcification with electron beam CT: interobserver and intraobserver reliability for scoring quantification, *Radiology*, 190, 347–352, 1994.
17. Lipton, M.J. and Holt, W.W., Value of ultrafast CT scanning in cardiology, *British Medical Bulletin*, 45, 991–1010, 1989.
18. Hsieh, J., Image artifacts, causes, and correction, in Goldman, L.W. and Fowlkes, J.B. (Eds.), *Medical CT and Ultrasound: Current Technology and Applications*, Advanced Medical Publishing for the American Association of Physicists in Medicine, Madison, WI, 1995, pp. 487–518.
19. Pelc, N.J. and Glover, G.H., Method for reducing image artifacts due to projection measurement inconsistencies, U.S. patent #4,580,219, 1986.
20. Ritman, E.L., Physical and technical considerations in the design of the DSR, and high temporal resolution volume scanner, *American Journal of Roentgenology*, 134, 369–374, 1980.
21. Ritman, E.L., Fast computed tomography for quantitative cardiac analysis — state of the art and future perspectives, *Mayo Clin Proceedings*, 65, 1336–1349, 1990.
22. Boyd, D.P., A proposed dynamic cardiac 3D densitometer for early detection and evaluation of heart disease, *IEEE Transactions in Nuclear Science*, 2724–2727, 1979.
23. Boyd, D.P. and Lipton, M.J., Cardiac computed tomography, *Proceedings of the IEEE*, 198–307, 1983.
24. Lipton, M.J., Brundage, B.H., Higgins, C.B., and Boyd, D.P., Clinical applications of dynamic computed tomography, *Progress Cardiovascular Disease*, 28(5), 349–366, 1986.
25. Nakanishi, T., Hamada, S., Takamiya, M., Naito, H., Imakita, S., Yamada, N., and Kimura, K., A pitfall in ultrafast CT scanning for the detection of left atrial thrombi, *Journal of Computer Assisted Tomography*, 17, 42–45, 1993.
26. Helgason, C.M., Chomka, E., Louie, E., Rich, S., Zajac, E., Roig, E., Wilbur, A., and Brundage, B.H., The potential role for ultrafast cardiac computed tomography in patients with stroke, *Stroke*, 20, 465–472, 1989.

27. Moore, S.C., Judy, P.F., Garnic, J.D., Kambic, G.X., Bonk, F., Cochran, G., Margosian, P., McCroskey, W., and Foote, F., Prospectively gated cardiac computed tomography, *Medical Physics,* 10, 846–855, 1983.
28. Kalender, W., Fichie, H., Bautz, W., and Skalej, M., Semiautomatic evaluation procedures for quantitative CT of the lung, *Journal of Computer Assisted Tomography,* 15, 248–255, 1991.
29. Crawford, C.R., Goodwin, J.D., and Pelc, N.J., Reduction of motion artifacts in computed tomography, *Proceedings of the IEEE Engineering in Medicine and Biological Society,* 11, 485–486, 1989.
30. Ritchie, C.J., Hsieh, J., Gard, M.F., Godwin, J.D., Kim, Y., and Crawford, C.R., Predictive respiratory gating: a new method to reduce motion artifacts on CT scans, *Radiology,* 190(3), 847–852, 1994.
31. Hsieh, J., Generalized adaptive median filters and their application in computed tomography, *Applications of Digital Image Processing XVII. Proceedings of the SPIE,* 662–669, 1994.
32. Wong, J.W., Sharpe, M.B., Jaffray, D.A., Kini, V.R., Robertson, J.M., Stromberg, J.S., and Martinez, A.A., The use of active breathing control (ABC) to reduce margin for breathing motion, *International Journal of Radiation Oncology, Biology, Physics,* 44(4), 911–919, 1999.
33. Sagel, S.S., Weiss, E.S., Gillard, R.G., Hounsfield, G.N., Jost, R.G.T., Stanley, R.J., and Ter-Pogossian, M.M., Gated computed tomography of the human heart, *Investigative Radiology,* 12 563–566, 1977.
34. Morehouse, C.C., Brody, W.R., Guthaner, D.F., Breiman, R.S., and Harell, G.S., Gated cardiac computed tomography with a motion phantom, *Radiology,* 134(1), 213–217, 1980.
35. Nassi, M., Brody, W.R., Cipriano, P.R., and Macovski, A., A method for stop-action imagingof the heart using gated computed tomography, *IEEE Transactions in Biomedical Engineering,* 28, 116–122, 1981.
36. Cipriano, P.R., Nassi, M., and Brody, W.R., Clinically applicable gated cardiac computed tomography, *American Journal of Roentgenology,* 140, 604–606, 1983.
37. Joseph, P.M. and Whitley, J., Experimental simulation evaluation of ECG-gated heart scans with a small number of views, *Medical Physics,* 10, 444–449, 1983.
38. Johnson, G.A., Godwin, J.D., and Fram, E.K., Gated multiplanar cardiac computed tomography, *Radiology,* 145, 195–197, 1982.
39. Moore, S.C. and Judy, P.F., Cardiac computed tomography using redundant-ray prospective gating, *Medical Physics,* 14, 193–196, 1987.
40. Nolan, J.M., Feasibility of ECG Gated Cardiac Computed Tomography, MSc thesis, University of Western Ontario, 1998.
41. Alfidi, R.J., MacIntyre, W.J., and Haaga, J.R., The effects of biological motion on CT resolution, *American Journal of Roentgenology,* 127, 11–15, 1976.
42. Ritchie, C.J., Godwin, J.D., Crawford, C.R., Stanford, W., Anno, H., and Kim, Y., Minimum scan speeds for suppression of motion artifacts in CT, *Radiology,* 185(1), 37–42, 1992.
43. Posniak, H.V., Olson, M.C., and Demos, T.C., Aortic motion artifact simulating dissection on CT scans: elimination with reconstructive segmented images, *American Journal of Roentgenology,* 161(3), 557–558, 1993.
44. Srinivas, C. and Costa, M.H.M., Motion-compensated CT image reconstruction, *Proceedings of the IEEE Ultrasonics Symposium,* 1, 849–853, 1994.
45. Chiu, Y.H. and Yau, S.F., Tomographic reconstruction of time varying object from linear time-sequential sampled projections, *Proceedings of the IEEE* Conference on Acoustic, Speech and Signal Processing, 1, V307–V312, 1994.
46. Hedley, M., Yan, H., and Rosenfeld, D., Motion artifacts correction in MRI using generalized projections, *IEEE Transactions in Medical Imaging,* 10(1), 40–46, 1991.
47. Ritchie, C.J., Correction of computed tomography motion artifracts using pixel-specific back-projection, *IEEE Transactions in Medical Imaging,* 15(3), 333–342, 1996.
48. Stergiopoulos, S., Optimum bearing resolution for a moving towed array and extension of its physical aperture, *Journal of the Accoustical Society of America,* 87(5), 2128–2140, 1990.
49. Stergiopoulos, S., Implementation of adaptive and synthetic aperture processing in real-time sonar systems, *Proceedings of the IEEE,* 86(2), 358–396, 1998.

50. Dhanantwari, A.C. and Stergiopoulos, S., Spatial overlap correlator to track and adaptive processing to correct for organ motion artifacts in X-ray computed tomography medical imaging systems, *IEEE Transactions in Medical Imaging*, submitted.
51. Dhanantwari, A.C., Synthetic Aperture and Adaptive Processing to Track and Correct for Motion Artifacts in X-Ray CT Imaging Systems, Ph.D. thesis, University of Western Ontario, 2000.
52. Widrow, B. and Steams, S.D., *Adaptive Signal Processing*, Prentice-Hall, Engelwood Cliffs, NJ, 1985.
53. Stergiopoulos, S., Noise normalization technique for broadband towed array data, *Journal of the Accoustical Society of America*, 97(4), 2334–2345, 1995.
54. Widrow, B., Glover, J.R., McCool, J.M., Kaunitz, J., Williams, C.S., Hearn, R.H., Zeidler, J.R., Dong, E., Jr., and Goodlin, R.C., Adaptive noise cancelling: principles and applications, *Proceedings of the IEEE*, 63(12), 1692–1716, 1975.
55. Haykin, S., *Adaptive Filter Theory*, Prentice-Hall, Engelwood Cliffs, NJ, 1986.
56. Chong, E.K.P. and Zak, S.H., *An Introduction to Optimization*, John Wiley & Sons, New York, 1996.
57. Slock, D.T.M., On the convergence behavior of the LMS and the normalized LMS algorithms, *IEEE Transactions on Signal Processing*, 41(9), 2811–2825, 1993.
58. Webb, S., *From the Watching of Shadows: The Origins of Radiological Tomography*, Adam Hilger, New York, 2000.

17

Magnetic Resonance Tomography — Imaging with a Nonlinear System

Arnulf Oppelt
Siemens Medical Engineering Group

17.1 Introduction

Since its introduction to clinical routine in the early 1980s magnetic resonance imaging (MRI) or tomography (MRT) has developed to a preferred imaging modality in many diagnostic situations due to its unparalleled soft tissue contrast, combined with high spatial resolution, and its capability to generate images of slices in arbitrary orientation or even of entire volumes. Furthermore, the possibility to display blood vessels, to map brain functions, and to analyze metabolism is widely valued.

Magnetic resonance (MR) is the phenomenon according to which particles with an angular and a magnetic moment precess in a magnetic field, thereby absorbing or emitting electromagnetic energy. This effect is called electron spin resonance (ESR) or electron paramagnetic resonance (EPR) for unpaired electrons in atoms, molecules, and crystals and nuclear magnetic resonance (NMR) for nuclei. ESR was discovered in 1944 by the Russian scientist Zavoisky,[1] but until now has not yet gained any real significance for medical applications. NMR was observed independently in 1945 by Bloch et al.[2] at Stanford University in California and by Purcell et al.[3] in Cambridge, MA. The Nobel Prize for physics was awarded in 1952 to these two groups.

0-8493-3691-0/01/$0.00+$.50

In 1973, Lauterbur[4] described how magnetic field gradients could be employed to obtain images similar to those recently generated with X-ray computed tomography. The limits placed on spatial resolution by the wavelength in the imaging process with waves are circumvented in MRI by superposing two fields. With the aid of a radio frequency (rf)-field in the megahertz (MHz) range and a locally variable static magnetic field, the sharp resonance absorption of hydrogen nuclei in biological tissue is used to obtain the spatial distribution of the nuclear magnetization. Contrary to other imaging modalities in medicine such as X-rays or ultrasound, imaging with NMR employs a nonlinear system. The signal used to construct an image does not depend linearly on the rf-energy applied to generate it and can be influenced in a very wide range by the timing of the imaging procedure. In the following, we will summarize the basic principles of MR and the concepts of imaging.

17.2 Basic NMR Phenomena

All atomic nuclei with an odd number of protons or neutrons, i.e., roughly two thirds of all stable atomic nuclei, possess an intrinsic angular momentum or spin. This is always coupled with a magnetic dipole moment, which is proportional to the angular momentum. As a consequence, these particles align in an external magnetic field. As in matter, many atomic nuclei exist, e.g., 1 mm^3 water contains 6.7 10^{19} hydrogen nuclei, a small but measurable angular momentum per unit volume and an associated macroscopic magnetization with results proportional to the external magnetic flux density and inversely proportional to temperature.

At thermal equilibrium, the nuclear magnetization of a sample with nuclear spins is aligned parallel to an applied magnetic field. However, if this parallel alignment is disturbed, e.g., by suddenly changing the direction of the field, a torque acts on the magnetic moment of the sample. According to the law of conservation of angular momentum, this torque causes a temporal change of the angular momentum of the sample, resulting in a precession of the magnetization with the (circular) Larmor frequency

$$\omega = \gamma B_z. \tag{17.1}$$

This precession (NMR) can be detected by measuring the alternating voltage induced in a coil wound around the sample. For hydrogen nuclei or protons, which represent the most frequently occurring nuclei in nature, a sharp resonance frequency of 42.577 MHz is observed at a magnetic flux density of 1 *T*.

In an NMR experiment, the precession of the nuclear magnetization is often stimulated by disturbing the alignment of the nuclear magnetization parallel to the static magnetic field by an rf-field having a frequency similar to the Larmor frequency. It is provided by a coil wound around the sample with its field axis orthogonal to the static magnetic field. This coil can also be used for signal detection. The linearly polarized rf-field in this coil can be thought of as the superposition of two circularly polarized fields rotating in opposite directions. Thus, there is always the same direction of rotation as the Larmor precession, and in this reference frame (the rotating frame), there is a constant magnetic flux density B_1. The resulting torque causes the nuclear magnetization to precess around the axis of the B_1 field in the rotating frame in the same way as around the static magnetic field B_Z in the laboratory frame. The combined precession movement around the static and the rf-field in the laboratory system causes the tip of the nuclear magnetization vector to execute a spiral path on the surface of a sphere (Figure 17.1). The contra-rotating rf-field, having twice the Larmor frequency in the rotating frame, acts as a perturbation and is effectively averaged out.

Under the influence of the rotating rf-magnetic flux density B_1, an angle α between the static magnetic field and the nuclear magnetization emerges, which is proportional to the duration t of the rf-field:

$$\alpha = \gamma B_1 t. \tag{17.2}$$

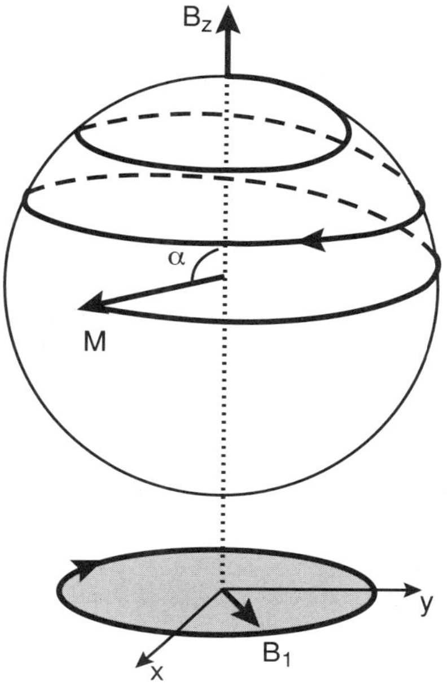

FIGURE 17.1 Motion of the nuclear magnetization vector M_0 under the influence of a static magnetic field B_z and a circularly polarized rf-field B_1 with Larmor frequency γB_0. The initial position of M_0 is parallel to B_z; after time t, M_0 is oriented with an angle $\alpha = \gamma B_1 t$ along the direction of B_z. (The direction of precession depends on whether the angular and the magnetic moment of the nuclei are parallel or antiparallel. For protons, a clockwise rotation follows.)

For reasons of simplicity, we shall consider in the following the transverse components in the rotating frame, i.e., the coordinate system rotating with ω_L in the laboratory system. With a B_1 field that confines an angle φ with the x-axis, the nuclear magnetization attains the components (Figure 17.2).

$$\begin{aligned} M_x &= -M_o \sin\alpha \sin\varphi \\ M_Y &= M_o \sin\alpha \cos\varphi \\ M_Z &= M_o \cos\alpha. \end{aligned} \tag{17.3a}$$

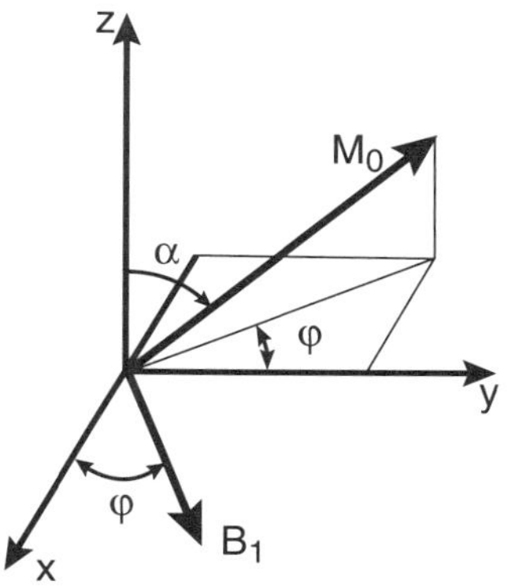

FIGURE 17.2 Tilting of the magnetization M_0 by B_1 field in the rotating frame.

The x and y components of the transverse magnetization can be combined into the complex quantity

$$\begin{aligned} M_\perp &= M_X + iM_Y \\ i &= \sqrt{-1}, \end{aligned} \tag{17.4}$$

giving

$$M_\perp = iM_0 \sin\alpha e^{i\varphi}. \tag{17.3b}$$

Switching off the rf-pulse after the nuclear magnetization is aligned orthogonally to the static magnetic field ($\alpha = 90°$, hence a 90° pulse) induces the maximum signal in the sample coil.

17.3 Relaxation

Relaxation describes the effect that the precession of the nuclear magnetization decays with time; the original state of equilibrium is reestablished, with the magnetization aligned parallel to the static magnetic field. This phenomenon is described by two separate relaxation time constants, T_1 and T_2, in which the equilibrium states, M_0 and 0, respectively, of the nuclear magnetization M_z parallel and $M_\perp$ perpendicular to the static magnetic field are obtained again; T_1 is always $\geq T_2$. Longitudinal relaxation is associated with the emission of energy to the surroundings, i.e., the lattice in which the nuclei are embedded, and is therefore also referred to as spin-lattice relaxation. Transverse relaxation is caused by collisions of the nuclear spins and, thus, is often referred to as spin-spin relaxation. In the latter case, since the longitudinal component of the magnetization remains unchanged, the energy of the nuclear ensemble does not change; only the relationship of the phases between the individual spins is lost. T_1 results from an energy effect, and T_2 results from an entropy effect.

The behavior of the nuclear magnetization in an external magnetic field B_Z undergoing relaxation was described by Bloch et al.[5] by adding empirical terms to the classical law of motion conservation:

$$\begin{aligned} \frac{dM_z}{dt} &= \gamma(\vec{M} \times \vec{B})_Z + \frac{(M_0 - M_Z)}{T_1} \\ \frac{dM_\perp}{dt} &= \gamma(\vec{M} \times \vec{B})_\perp - \frac{M_\perp}{T_2}. \end{aligned} \tag{17.5}$$

It is the wide range of relaxation times in biological tissue that makes NMR so interesting in medical diagnostics. T_1 is of the order of magnitude of several 100 ms, while T_2 is in the range 30 to 100 ms. T_1 of biological tissue decreases, when temperature increases. This effect is investigated with MRI to probe temperature changes in the human body.

17.4 NMR Signal

From Bloch's equations (Equation 17.5), one obtains for the precessing nuclear magnetization after a 90° pulse around the x-axis in the rotating frame

$$M_\perp(t) = iM_0 e^{\frac{-t}{T_2}} \tag{17.6}$$

for the transverse component and

$$M_Z(t) = M_0\left(1 - e^{\frac{-t}{T_1}}\right) \tag{17.7}$$

for the longitudinal component.

The precessing two components of tranverse magnetization can be measured independently in the laboratory system with two induction coils oriented perpendicular to each other; this is named a free induction decay (FID). An oscillating signal with Larmor frequency ω_L is observed that follows from Equation 17.6 by multiplication with $e^{-i\omega_L t}$.

The frequency dependence of the transverse magnetization is given by its Fourier transformation (Figure 17.3). The imaginary part of the Fourier transformation describes the so-called absorption line

$$M_y(\omega) = M_0 \frac{T_2}{1 + \omega^2 T_2^2}, \tag{17.8}$$

and the real part the dispersion line

$$M_x(\omega) = M_0 \frac{-\omega T_2^2}{1 + \omega^2 T_2^2}. \tag{17.9}$$

Instead of taking the Fourier transformation of the FID, it is also possible to measure absorption and dispersion directly by recording the change of resistance and inductance of the signal coil surrounding the sample as a function of frequency (or as a function of the flux density of the static magnetic field). Such continuous wave (cw) methods, however, are much slower than pulse methods and are therefore hardly used anymore.

The full width at half maximum of the absorption and the distance between the extreme points of the dispersion line are given by the transverse relaxation time

$$\Delta\omega_{1/2} = \frac{2}{T_2}. \tag{17.10}$$

Protons in distilled water exhibit a transverse relaxation time $T_2 \approx 1$ s. The measurement of T_2 through the FID, however, is only possible in very homogenous magnetic fields. In practice, the static magnetic

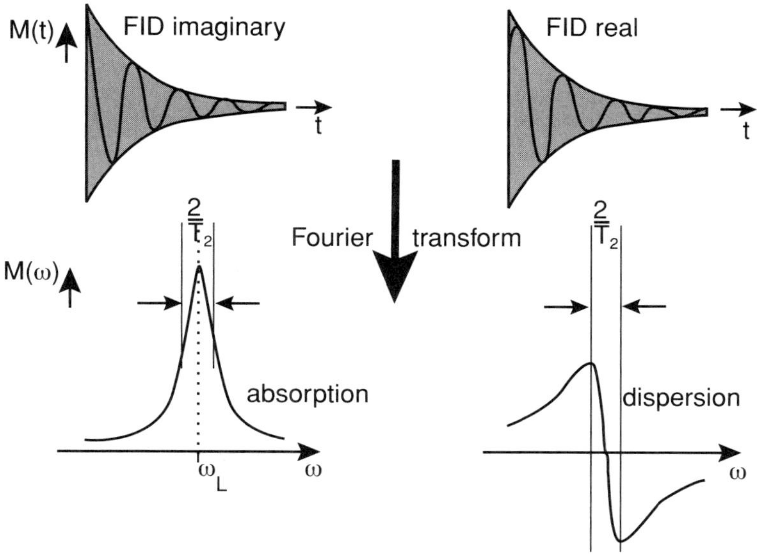

FIGURE 17.3 FID after a 90° pulse with its Fourier transform, representing the NMR absorption and dispersion line. In the laboratory system, resonance is oberserved at $\omega = \omega_L$, compared to $w = 0$ in the rotating frame.

field varies in space, resulting in different precession frequencies of the nuclear magnetization. Because of destructive interference, a shortened FID is observed, resulting in an inhomogeneously broadened resonance line. The line shape depends on the spatial distribution of the static magnetic field deviations ΔB_z over the entire sample. Reference is often made to an effective transverse relaxation time

$$\frac{1}{T_2^*} = \frac{1}{T_2} + \frac{\gamma \Delta B}{2} \tag{17.11}$$

which, however, can only coarsely describe the effect of magnetic field inhomogeneities since the signal clearly no longer decays exponentially.

The signal loss in an inhomogeneous static magnetic field can be recovered by means of a refocusing or a 180° rf-pulse.[6] The diverging transverse magnetization after a 90° pulse due to field inhomogeneities converges again, since the 180° pulse reverses the order of the spins (Figure 17.4). Thus, slowly precessing spins which have lagged behind now move ahead and realign themselves with the faster precessing spins after the interval between the two rf-pulses. A spin echo is observed, the amplitude of which is determined by transverse relaxation.

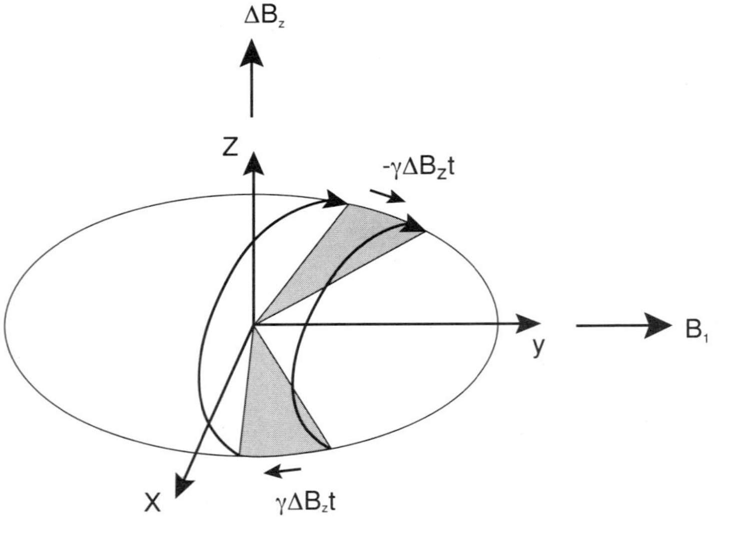

FIGURE 17.4 Dephasing by the angle $\gamma \Delta B_z t$ to transverse nuclear magnetization in the rotating frame due to field inhomogeneities is reversed with a 180° rf-pulse. A spin echo is created.

In this context, we will mention another type of echo, the so called stimulated echo.[6] When applying two 90° pulses instead of a 90°/180° pair, an echo also occurs, but only with half the amplitude resulting from using a 180° pulse. The missing magnetization is stored along the z-axis, thereby undergoing longitudinal relaxation. It can be tilted again into the transverse plane with a third 90° pulse and can manifest itself as a stimulated echo with a distance from the third 90° pulse corresponding to that of the first two 90° pulses. The amplitude of the stimulated echo is determined by the longitudinal relaxation.

Relaxation is not the only mechanism that affects the amplitudes of the echoes. Especially the molecules of liquids move stochastically (diffuse) during the time between excitation and observation of the echo from one position in the inhomogeneous static magnetic field to another; in accordance with the field difference, the nuclei precess there with a different Larmor frequency and thus no longer contribute fully to the echo amplitude.[7] The influence of spin diffusion can be enhanced by applying a strong, magnetic field gradient pulse symmetrically before and after the refocusing rf-pulse. In MRI, such types of experiments are of interest because diffusion is anisotropic due to tissue microstructure. This offers the possibility to get information, e.g., about the fiber structure of tissue. Diffusion depends on temperature and therefore provides an alternative to the precise measurement of T_1 for noninvasive of the monitoring temperature *in vivo*.

The gyromagnetic ratio determining the Larmor frequency of nuclei in the static magnetic field is a fixed constant for each nuclear species. In NMR experiments with nuclei embedded in different molecules, however, slightly different resonance frequencies are observed. This effect is caused by the molecular electrons responsible for chemical bonding. These electrons screen the static magnetic field, with the result that the atomic nucleus "sees" different magnetic fields (chemical shift) depending on the nature of the chemical bond. In a molecular complex, often several resonance lines attributable to individual groups of molecules are observed. Quantitatively, the chemical shift is usually given in parts per million (ppm) relative to a reference line.

Besides the chemical shift, a fine splitting of the MR lines is also frequently observed. This is caused by the magnetic interaction (spin-spin coupling) between the nuclei, which again acts indirectly via the valence electrons. Therefore, in chemistry, molecular structure is often investigated amenable to study with NMR spectroscopy.

17.5 Signal-to-Noise Ratio

To obtain biological or medical information from a living being with MR, it is necessary to attribute the recorded nuclear magnetization to the site of its origin. Before discussing methods of spatial resolution, however, we will turn first to the fundamental restriction for such measurements. The signal induced by the precessing nuclear magnetization in the pick-up coil around the sample must compete with the noise generated by the thermal motion of the electrons in the coil and the Brownian molecular motion in the object under investigation, i.e., the human body.

Noise being generated thermally in the coil provided to pick up the NMR signal according to Nyquist is given by

$$U_{\text{Noise}} = \sqrt{\frac{2}{\pi} k_B (R_{\text{Coil}} T_{\text{Coil}} + R_{\text{Sample}} T_{\text{Sample}}) \Delta\omega}\,, \tag{17.12}$$

$\Delta\omega$ = detection bandwidth for the signal

R_{Coil} = resistance of the signal coil without sample

R_{Sample} = contribution of the sample to the resistance of the signal coil

T_{Coil} = temperature of signal the coil

T_{Sample} = temperature of the sample

While with small samples of a few cubic millimeters, as commonly investigated in the laboratory, the noise contribution $R_{\text{Coil}} T_{\text{Coil}}$ dominates, this is not true for samples as large as the human body. With a conductive sample, the resistance of the signal coil can be derived from the power distributed in the sample by a mean rf-field B_1 that would be generated by a current i in that coil:[8]

$$\begin{aligned} P_{\text{Sample}} &= R_{\text{Sample}} i^2 \\ &= \frac{1}{4} \sigma B_1^2 \omega^2 \int r_\perp^2 \, dv, \end{aligned} \tag{17.13}$$

$r_\perp$ = radius coordinate orthogonal to the field axis of the signal coil

σ = electrical conductivity of the sample

ω = signal frequency, i.e., Larmor frequency

When the rf-field B_1 is replaced by the "field per unit current B_i," which describes the dependence of the magnetic field on the geometry of the coil, Equation 17.13 gives the resistance due to the coupling to the sample.

The voltage that is induced in the signal coil by the precessing magnetization in a volume element (voxel) Δv of the sample after a 90° pulse is given by

$$U_{\text{Signal}} = B_i \omega_L M_0 \Delta v, \tag{17.14}$$

where $M_0 \propto (B_Z/T)$ is the transverse nuclear magnetization.

The signal-to-noise (S/N) ratio follows after a succession of steps omitted here:

$$\frac{U_{\text{Signal}}}{U_{\text{Noise}}} = \sqrt{\frac{\pi}{24k_B^3}}\gamma N_v \frac{\omega_L}{T_{\text{Sample}}\sqrt{\frac{1}{2\mu_o}\frac{T_{\text{Coil}}V_{\text{Sample}}}{\eta Q \omega_L} + \frac{\sigma}{4}T_{\text{Sample}}\int r_\perp^2 dv}} \frac{\Delta v}{\Delta \omega}, \tag{7.15}$$

where N_V is the density of H^1 nuclei and μ_o is the permeability in vacuum.

Thus the "filling factor" η, which is a measure of the ratio of the sample volume to the signal coil volume, and the "coil quality factor" Q, which gives the ratio of the energy stored in the coil to the energy loss per oscillation cycle, have been introduced.

When smaller volume elements in a large sample are to be resolved, a worse S/N ratio is obtained. Since the nuclear magnetization increases with the Larmor frequency, the S/N ratio improves with increasing strength of the static magnetic field. For small values of the "moment of inertia" $\int r_\perp^2 dv$ (i.e., small samples) and for small filling and coil quality factors, the S/N ratio is proportional to $\omega^{3/2}$; for a high coil quality factor or a low coil temperature, it is directly proportional to the Larmor frequency, i.e., the flux density of the static magnetic field.

Since the temperature cannot be lowered with living samples, once the NMR apparatus is set up (i.e., static magnetic field and antennas are chosen), the S/N ratio can only be influenced by voxel size Δv and bandwidth $\Delta\omega$. Repeating the NMR experiment n times and adding the single signals together results in an S/N ratio improvement by a factor of $\sqrt{n}$. However, since a reduction in spatial resolution is as undesirable as a longer time of measurement, these two choices are normally avoided.

17.6 Image Generation and Reconstruction

To derive an MR signal from a localized small volume within a larger object, at least one of the two fields (i.e., the static magnetic field and the rf-field) required for the NMR measurement has to vary over space. It has been proposed that a sharp maximum or minimum be generated in space for these fields so that the MR signal observed would originate mainly from that region.[9] However, along with the technical difficulties of generating sufficiently sharp field extrema, to yield information from other regions would require the movement of the area sensitive to MR through the object under investigation, leading to a long time of measurement, when each voxel has to be measured several times in order to obtain a sufficient S/N ratio.

The utilization of the signal from the entire object rather than from only a single voxel is achieved with the use of magnetic field gradients G, for which the Larmor frequency varies linearly along one direction in space, giving in the laboratory system

$$\omega_L = \gamma(B_Z + Gr) \tag{17.16a}$$

and in the rotating frame

$$\omega = \gamma G r. \tag{17.16b}$$

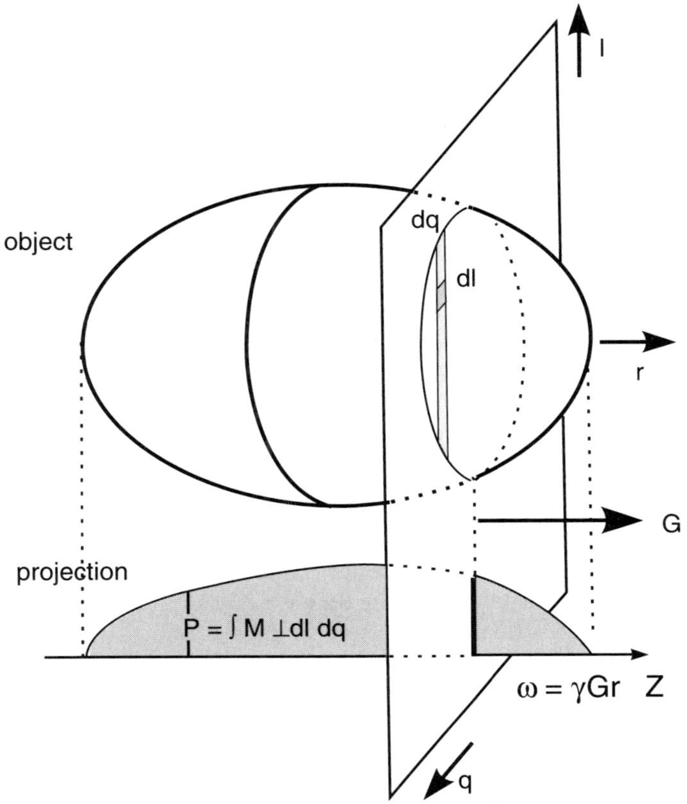

FIGURE 17.5 The NMR signal amplitude as a function of frequency for an object in a linear magnetic field gradient, representing the plane integral of the transverse magnetization (number of spins) in planes orthogonal to the field gradient (projection).

The amplitude of the NMR signal as a function of frequency then corresponds to the sum of all spins in the planes orthogonal to the direction of the magnetic field gradient,[14] i.e., the projection of the nuclear magnetization (Figure 17.5)

$$P(r, \varphi, \vartheta) = \int M_{\perp}(x, y, z)\,dl\,dq\,. \tag{17.17}$$

Normally, the NMR signal is measured as a function of time rather than a function of frequency, and the projection can then be obtained from the Fourier transformation of the time-dependent NMR signal

$$P(\omega, \varphi, \vartheta) = \frac{1}{2\pi}\int M_{\perp}(t)e^{i\omega t}dt\,. \tag{17.18}$$

Lauterbur's original suggestion was to collect a set of projections onto different gradient directions φ, ϑ and reconstruct an image of the nuclear magnetization with the same methods as in X-ray computed tomography.[10,11] It emerges, however, that a modification of his proposal by Kumar et al.[12] offers greater flexibility and simpler image reconstruction. Their method is now used routinely in MRI.

For simplicity, we will restrict ourselves for the moment to a two-dimensional (2d) object. The sample is excited with an rf-pulse so that transverse nuclear magnetization, $M_{\perp}(x, y)$, is generated in the rotating frame. The phase of the magnetization is then made to vary along the y-direction with a gradient G_y switched on for a time T_y:

$$M_{\perp}(x, y, G_y T_y) = M_{\perp}(x, y)e^{-i\gamma G_y y T_y}. \tag{17.19}$$

Then a gradient in the x-direction, G_X, is applied, and the NMR signal, which is the integral of the magnetization precessing differently over the object, is recorded as a function of the time *t*:

$$M_{\perp}(G_y, t) = \iint dxdy\, M_{\perp}(x, y)e^{-i\gamma G_y y + B_0 T_y} e^{-IG_x xt}. \tag{17.20}$$

When longitudinal magnetization has been reestablished due to longitudinal relaxation, again transverse magnetization is generated with an rf-pulse, and the encoding procedure is repeated. Thus, with a variety of gradients, G_y, a 2d set of signals as a function of G_y and the recording time *t* is obtained. One can consider this signal set as an interferogram. A 2d Fourier transformation yields the distribution of the local transverse nuclear magnetization:

$$M_{\perp}(x, y) = \frac{1}{(2\pi)^2}\int_{-T_x}^{T_x}\int_{-G_y^{max}}^{G_y^{max}} M_{\perp}(G_y, t)e^{i\gamma(yT_y G_y + xG_x t)}dG_y dt. \tag{17.21}$$

Since the spatial distribution of the transverse nuclear magnetization is given by the Fourier transformation of a 2d data set, it is usual to view these data as being acquired in Fourier or *k*-space (Figure 17.6b) with spatial frequency coordinates

$$k_x = \gamma G_x t \text{ and } k_y = \gamma T_y G_y \tag{17.22a}$$

or in the more general case of time-dependent gradients

$$k_x(t) = \gamma\int_0^t G_x(t)dt \text{ and } k_y(t) = \gamma\int_0^t G_y(t)dt \tag{17.22b}$$

The MR image signal is sampled along parallel lines in Fourier space that are addressed by a combination of rf- and gradient pulses (Figure 17.6a). When the scanning trajectory crosses the axis $k_y = 0$, a maximum signal named gradient echo is obtained. The Fourier transformation of the amplitudes (real and imaginary part) along a line through the center of the Fourier space results in the projection (Equation 17.17) of the investigated object onto the direction of this line. This is known as the central slice or projection slice theorem.

In digital imaging, an object is sampled with image elements (pixels) of size Δx, Δy, corresponding to the spatial sampling frequencies

$$k_x^s = \frac{2\pi}{\Delta x} \text{ and } k_y^s = \frac{2\pi}{\Delta y}. \tag{17.23}$$

According to the sampling theorem, the object can then be completely reconstructed, when it does not contain spatial frequencies higher than the (spatial) Nyquist frequency, which is half the sampling frequency

$$k_x^{max} \le \frac{k_x^s}{2} \text{ and } k_y^{max} \le \frac{k_y^s}{2}. \tag{17.24a}$$

If the object contains information at spatial frequencies larger than the Nyquist frequency, truncation or aliasing leads to typical artifacts; e.g., sharp intensity borders in an object are displayed with parallel lines (Gibbs ringing). So the pixel size has to be chosen according to the spatial resolution (see below).

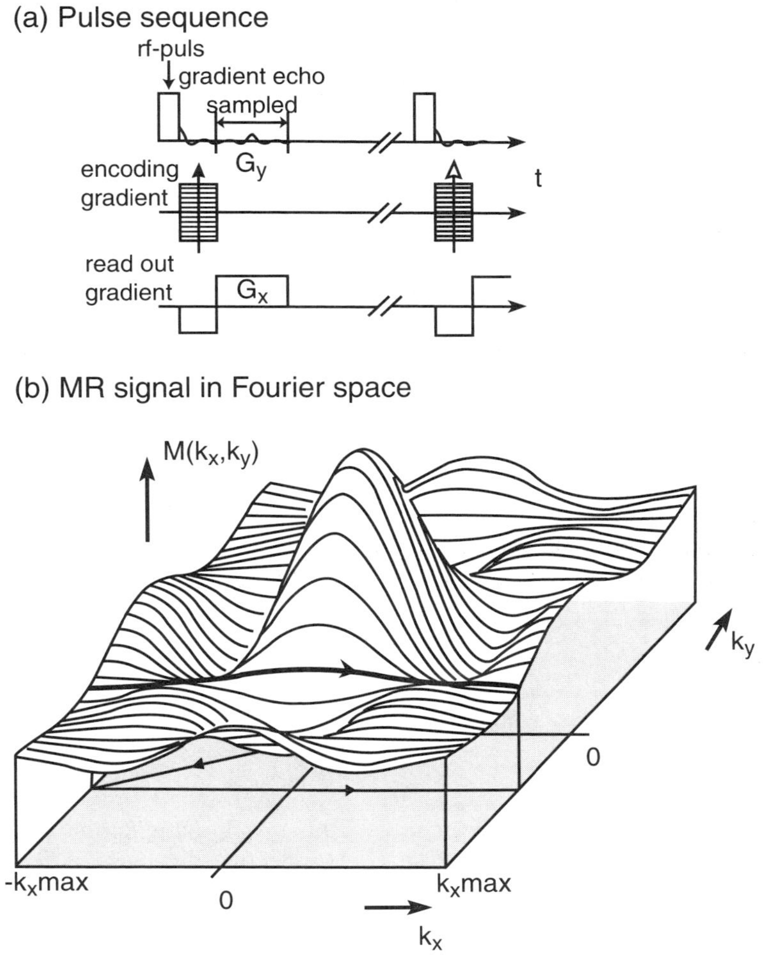

FIGURE 17.6 Scanning 2d object in k-space: the negative lobe of G_y and the encoding gradient G_y address the Fourier line to be recorded (b). The NMR signal is sampled during the positive lobe of gradient G_x (a); the preceding negative lobe ensures sampling to start always at $-k_y^{max}$.

The spatial frequency interval necessary to image an object with size $\pm x^{max}$ and $\pm y^{max}$ is given accordingly by

$$\Delta k_x = \frac{\pi}{x^{max}} \text{ and } \Delta k_y = \frac{\pi}{y^{max}}. \tag{17.24b}$$

Hence, in order to image an object with diameter $2y^{max}$ in y-direction (and $2x^{max}$ in x-direction) with the required spatial resolution Δy, $N_y = (2y^{max}/\Delta y)$ phase encoding steps are necessary, during which the gradient $-G_y^{max} \leq G_y \leq G_y^{max}$ is stepped through, whereby each time $N_X = (2x^{max}/\Delta x)$ samples have to be taken of the NMR signal. Usually, N_Y and N_X are chosen to be a power of two in order to employ the fast Fourier transform (FFT) algorithm for image reconstruction.[13] As during the image procedure, each voxel in the object under investigation is measured N_xN_y times compared to a (hypothetical) sequential scan, an improvement in the S/N of $\sqrt{N_xN_y}$ results related to an identical measurement time.

Spatial resolution depends on the magnetic field gradient strength. To derive a condition for the gradient G_X in the readout direction, we assume two spins separated by the distance d_X. In order to distinguish the two points, the frequency difference due to the gradient must be greater than that due to the natural line width and static magnetic field inhomogeneities:

$$\gamma G_x d_x \geq \frac{2}{T_2} + \gamma \Delta B_z . \tag{17. 25a}$$

This condition is equivalent to

$$T_x < T_2^* , \tag{17.25b}$$

which ensures that the loss of signal intensity due to the signal decay in a voxel caused by transverse relaxation and field inhomogeneities remains acceptable. Pixel size then should be chosen to be about

$$\Delta x \approx \frac{1}{2} d_x \text{ and } \Delta y \approx \frac{1}{2} d_y \tag{17.26}$$

in order to avoid truncation.

Since signal loss due to field inhomogeneities during the encoding interval can be recovered with a 180° pulse, the duration of the encoding gradient has only to be smaller than T_2 rather than T_2^*:

$$T_Y < T_2 , \tag{17.27a}$$

hence

$$\gamma G_Y d_Y \geq \frac{2}{T_2} . \tag{17.27b}$$

An additional restriction for spatial resolution in MRI is given by self-diffusion. Phase variation over a pixel due to diffusion of the water molecules has to be smaller than that caused by the gradient leading to the constraint

$$d_{x,y} > \sqrt[3]{\frac{D}{\gamma G_{x,y}}} , \tag{17.28}$$

where D is the diffusion coefficient (e.g., in tissue $D = 10^{-5} - 10^{-6}$ cm²/s). Though this restriction can be neglected in normal imaging experiments, it is of importance for MR microscopy.

When reconstructing the image from the MR signal set with the 2d FFT (Equation 17.21), it is usual to display the magnitude

$$|M_{\perp}(x, y)| = \sqrt{M_x^2(x, y) + M_y^2(x, y)} \tag{17.29}$$

in order to get rid of phase factors mixing the real and imaginary parts of the magnetization that might arise from sampling delays and the phase of the exciting rf-pulses. Because transverse relaxation poses a multiplicative term

$$e^{-\frac{t}{T_2(x, y)}}$$

on the acquired signal, the local image signal according to Equation 17.29 has to considered as being convoluted with the magnitude of the absorption and dispersion NMR line described by Equations 17.8 and 17.9. Blurring due to this effect is avoided when the condition in Equation 17.25 is observed.

The imaging principle described can be easily extended to three dimensions by adding and stepping through a gradient G_z during the encoding phase. However, this requires a longer time of measurement. Since information for the complete three-dimensional (3d) object is not always required, a 2d object is often generated from the 3d object by selective excitation.

17.7 Selective Excitation

To obtain an image from a slice through a 3d object, a gradient perpendicular to that slice is applied during excitation with the rf-pulse. In this way, the spins are tilted only in a plane, where the precession frequency is identical with the pulse frequency (selective excitation), whereby the bandwidth of the rf-pulse determines the thickness of the excited slice. Assuming a spin system with nuclear magnetization M_0 being exposed to a magnetic field gradient G_z and to a "long" amplitude-modulated rf-pulse $B_1(t)$ of duration $2T_z$, transverse magnetization is generated according to Equation 17.3b that precesses during the time $2T_z$ in the gradient G_z. The distribution of the transverse magnetization along the z-direction can be approximated in the rotating frame as

$$M_{\perp}(z, 2T_z) = iM_0 \sin|\alpha(\omega)| e^{i\varphi(\omega)} e^{-i\omega 2T_z} \tag{17.30}$$

where $\omega = \gamma G_z z$.

The flip angle $|\alpha(\omega)|$ is determined by the spectral amplitude of the rf-pulse $B_1(t)$ given by its Fourier transformation

$$\alpha(\omega) = \gamma \int B_1(t) e^{i\omega t} dt, \tag{17.31a}$$

and the azimuth $\varphi(\omega)$ of the axis around which the magnetization is tilted (measured against the x-axis in the rotating frame) follows from

$$\varphi(\omega) = \arctan\left(\frac{\mathrm{Im}(\alpha(\omega))}{\mathrm{Re}(\alpha(\omega))}\right). \tag{17.31b}$$

In order to obtain a rectangular distribution of the transverse magnetization along the slice thickness d at position z_0, the shape of the rf-pulse must be selected so as to give a rectangular frequency distribution with spectral width $\Delta\omega = \gamma G_z d$ and a center frequency $\omega_0 = \gamma G_z z_o$ (recalling that we are looking at the spins from the frame rotating with Larmor frequency ω_L, so that in the laboratory system a pulse with frequency $\omega_0 + \omega_L$ must be applied to the coil surrounding the sample). Since an rf-pulse with the shape of a sinc (i.e., $(\sin \pi x)/(\pi x)$) function has such a rectangular spectrum, $B_1(t)$ is chosen to be a sinc function modulated with the center frequency ω_0:

$$B_1(t) = B_1(T_z) \operatorname{sinc}\left(\frac{\Delta\omega}{2\pi}(t - T_z)\right) e^{-i\omega_0(t - T_z)} \tag{17.32}$$

where

$\Delta\omega = \gamma G_z d$

$\omega_0 = \gamma G_z z_0$

$0 \le t \le 2T_Z$

$2T_z > \dfrac{2\pi}{\gamma G_z d}$ = pulse duration

The sinc pulse is restricted here to a duration of $2T_Z$ in order to obtain a selective rf-pulse of finite length and shifted by the interval T_Z in order to expose the spins to the signal part left of the pulse maximum as well; it should be adjusted to extend over several sinc oscillations in order to approximate a rectangular slice profile. With the phase of the rf-pulse chosen to be aligned along the x-axis in the rotating frame, for the transverse magnetization, it follows that

$$\begin{aligned} M_{\perp}(z, 2T_z) &= iM_0 \sin\left(\gamma \left|\int B_1(t) e^{i\omega t} dt\right|\right) e^{-iT_z\omega} \\ &= iM_0 \sin\alpha_0 \operatorname{rect}\left(\frac{z - z_0}{d}\right) e^{-i\gamma G_z z T_z} \end{aligned} \tag{17.33}$$

$$\operatorname{rect}(x) = 1 \text{ for } -1/2 < x < 1/2\,,\ \operatorname{rect}(x) = 0 \text{ for } |x| > 1/2$$

$$\alpha_o \approx 2\pi \frac{B_1(T_z)}{G_z d}\,.$$

An oscillating function of the transverse nuclear magnetization along the slice thickness then results (Figure 17.7). The selective rf-pulse has flipped each spin in the slice addressed into the transverse plane, but in its own rotating frame. Since the resonance frequency changes over the slice thickness due to the applied gradient, the transverse nuclear magnetization is twisted. Almost no signal can be observed in a following FID, since the effective value of the spiral-shaped nuclear magnetization cancels out. Reversing

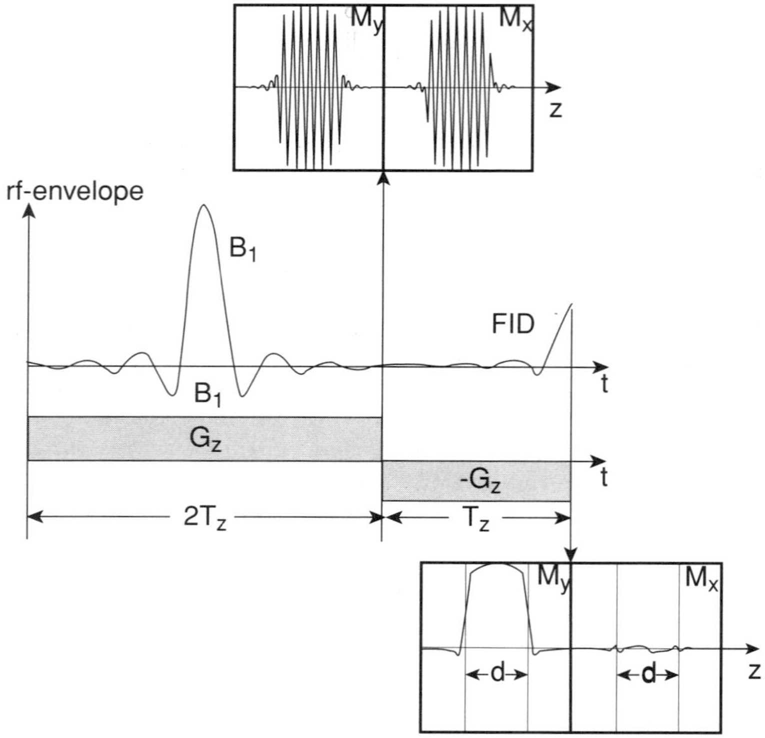

FIGURE 17.7 x and y component of the transverse nuclear-magnetization in the rotating frame after excitation with an rf-pulse of duration $2T_z$ in a magnetic field gradient Gz that is applied along the y-axis. The twisted transverse magnetization resulting immediately after the selective pulse realigns in a refocusing interval of duration T_z with a reversed gradient. The remaining oscillations at the edges of the refocused magnetization originate from the truncation of the sinc pulse. To minimize them, the sinc pulse has been multiplied with a Hanning function.

the polarity of the field gradient during a period T_Z following the rf-pulse[14] refocuses all those spins, resulting in an FID corresponding to the full transverse magnetization in the excited slice (Figure 17.7).

It should be mentioned that Equation 17.33 is only an approximation, since Equation 17.3 and hence Equation 17.30 is only valid for those nuclei for which the frequency of the rf-pulse equals the Larmor frequency. Taking exactly into account the Bloch equations requires numerical methods. It can be seen that a residual nuclear magnetization $M_y(z)$ remains even after refocusing, which can be minimized, however, by tuning the refocusing interval and by modified rf-pulse shapes.

For the necessary strength of the slice selection gradient, a similar argumentation holds as for the encoding gradient, i.e., the duration can be chosen according to T_2 rather than T_2^*, when a 180° pulse is used instead of gradient reversal for refocusing; however, the slice profile is influenced by the static magnetic field inhomogeneities (i.e., becomes curved) when the field variations over the slice thickness due to the presence of the gradient are comparable with or less than those due to the field inhomogeneities.

17.8 Pulse Sequences

For imaging with MR, the object to be investigated must be exposed to a sequence of rf- and gradient pulses. Many modifications of these pulse sequences have been designed to optimize the experiment with respect to special problems such as tissue contrast, display of flow, diffusion, susceptibility, or data acquisition time.

For an image, transverse magnetization has to be generated and phase encoded. If the repetition time of the rf-pulses necessary is made short with respect to the longitudinal relaxation time T_1 in order to speed up imaging, after some rf-pulses a dynamic equilibrium or steady state is established at which the nuclear magnetization is the same after each rf-pulse, i.e., in a distance t from the rf-pulse one observes

$$\vec{M}(nT_R + t) = \vec{M}((n+1)T_R + t) = \vec{M}(t) = \begin{pmatrix} M_x(t) \\ M_y(t) \\ M_z(t) \end{pmatrix}, \tag{17.34}$$

where T_R is the repetition time and n is the number of rf-pulses.

The magnetization immediately after the rf-pulse $\vec{M}(0)$ follows from that directly before $\vec{M}(T_R)$ from the rotation by tip angle α caused by the rf-field as

$$\vec{M}(0) = Q(\alpha)\vec{M}(T_R), \tag{17.35}$$

with the rotation matrix

$$Q(\alpha) = \begin{pmatrix} 1 & 0 & 0 \\ 0 & \cos\alpha & \sin\alpha \\ 0 & -\sin\alpha & \cos\alpha \end{pmatrix}. \tag{17.36}$$

On the other hand, because of precession and relaxation between the rf-pulses, $\vec{M}(T_R)$ and $\vec{M}(0)$ are related by

$$\vec{M}(T_R) = R(T_R)\vec{M}(0) + M_0 \begin{pmatrix} 0 \\ 0 \\ 1 - e^{-\frac{T_R}{T_1}} \end{pmatrix}, \tag{17.37}$$

whereby the matrix $R(T_R)$ describes precession and relaxation in the rotating frame

$$R(T_R) = \begin{pmatrix} \cos\omega t\, e^{-\frac{T_R}{T_2}} & \sin\omega t\, e^{-\frac{T_R}{T_2}} & 0 \\ \sin\omega t\, e^{-\frac{T_R}{T_2}} & \cos\omega t\, e^{-\frac{T_R}{T_2}} & 0 \\ 0 & 0 & e^{-\frac{T_R}{T_1}} \end{pmatrix}, \tag{17.38}$$

where ω is the precession frequency in the rotating frame.

In the following, we will analyze the steady-state magnetization for two cases relevant in MR imaging. First, we shall assume that directly before the rf-pulse no transverse magnetization has remained:

$$\begin{aligned} M_x(T_R) &= M_y(T_R) = 0 \\ M_x(0) &= M_z(T_R)\cos\alpha \\ M_z(T_R) &= M_0 - (M_0 - M_z(0))e^{-\frac{T_R}{T_1}}. \end{aligned} \tag{17.39}$$

The transverse magnetization immediately after the rf-pulse then follows to

$$M_y(0) = M_z(T_R)\sin\alpha = M_0\sin\alpha\frac{\left(1 - e^{-\frac{T_R}{T_1}}\right)}{1 - \cos\alpha e^{-\frac{T_R}{T_1}}} \tag{17.40}$$

and reaches a maximum for the so-called Ernst angle

$$\cos\alpha_{\text{opt}} = e^{-\frac{T_R}{T_1}}. \tag{17.41}$$

For a given repetition time and flip angle, the NMR signal intensity is determined by the local longitudinal relaxation time T_1. Since at short repetition times with small flip angles a large signal can still be obtained, this pulse sequence is often referred to as FLASH (fast low angle shot).[15] With FLASH imaging, it is assumed that the phase memory of the transverse nuclear magnetization has been lost at the end of the repetition interval; since this is not true when the repetition interval is shorter than the transverse relaxation time, spoiling gradient pulses are applied at the end of each interval in order to prevent the emergence of coherent image artifacts or a stochastically varying jitter is added to the repetition time.[16]

Next, we shall assume that between $M(0)$ and $M(T_R)$ a relation

$$M_y(T_R) = -M_y(0)e^{-\frac{T_R}{T_2}} \tag{17.42}$$

exists, which can either be assured by adjusting the frequency of the rf-pulses in the rotating frame to

$$\omega = \pm\frac{\pi}{T_R} \tag{17.43}$$

or alternating their phase between adjacent pulses by 180° (and keeping $\omega = 0$).

Then for the other magnetization components,

$$\begin{aligned} M_y(0) &= M_y(T_R)\cos\alpha + M_z(T_R)\sin\alpha \\ M_z(0) &= -M_y(T_R)\sin\alpha + M_z(T_R)\cos\alpha \\ M_z(T_R) &= M_0 - (M_0 - M_z(0))e^{-\frac{T_R}{T_1}} \end{aligned} \tag{17.44}$$

and the transverse magnetization immediately after the rf-pulse yields

$$M_y(0) = \frac{M_0\left(1 - e^{-\frac{T_R}{T_1}}\right)\sin\alpha}{\left(1 - e^{-T_R\left(\frac{1}{T_1}+\frac{1}{T_2}\right)}\right) + \left(e^{-\frac{T_R}{T_2}} - e^{-\frac{T_R}{T_1}}\right)\cos\alpha}. \tag{17.45}$$

For $(T_R/T_{1,2}) << 1$,

$$M_y(0) = \frac{M_0\sin\alpha}{\left(1 + \frac{T_1}{T_2}\right) + \left(1 - \frac{T_1}{T_2}\right)\cos\alpha} \tag{17.46}$$

is obtained, which reduces for $\alpha = 90°$ to

$$M_y(0) = \frac{M_0}{1 + \frac{T_1}{T_2}}. \tag{17.47}$$

Such a sequence is referred to as true FISP (fast imaging with steady precession)[17] or TRUFI (true fast imaging) and has to be constructed with completely balanced gradients (Figure 17.8). Signal intensity is determined by the T_1:T_2 ratio. For fluids such as water, where $T_1 \approx T_2$, a signal equivalent to half of the maximum nuclear magnetization can be obtained using 90° pulses with rapid pulse repetition. However, the prerequisite for such a strong signal is that no dephasing of transverse magnetization due to field inhomogeneities occurs in an image voxel, i.e.,

$$G_{x,y,z}\Delta x, y, z >> \Delta B, \tag{17.48}$$

where $\Delta x, y, z$ are the lengths of the edges.

If this condition cannot be maintained, image artifacts will arise. However, in order to obtain a steady state, it is not necessary that the net precession angle or phase of the transverse magnetization by 180° between the rf-pulses as assumed in Equation 17.43, a constant phase between the pulse is sufficient. Such a pulse sequence, dubbed FISP, reverses only the encoding gradient G_Y at the end of the repetition interval and is much more insensitive to field inhomogeneities. Signal behavior is somewhere between FLASH and TRUFI.

Without preparation gradients (i.e., the gradient pulses applied before the data are sampled) in x- and z-directions one would observe in a steady-state sequence a focused magnetization before and after the rf-pulse.[16] Graphically, one can consider the signal after the rf-pulse as an FID and before as one half of an echo with an amplitude reduced by the factor $e^{-(TR/T2)}$ compared to the FID. In FISP, one is using the steady-state FID to obtain a projection, but one can also utilize the steady-state half echo signal to get an FISP-like image with additional T_2 weighting (though this is not a strong effect at the short repetition times applied in steady-state sequences). In this case, the time course of the imaging sequence has to be reversed, therefore being dubbed PSIF (Figure 17.9). It is even possible to combine FISP and PSIF in a single sequence giving two images differing in T_2 contrast (DESS, double echo in the steady state). In

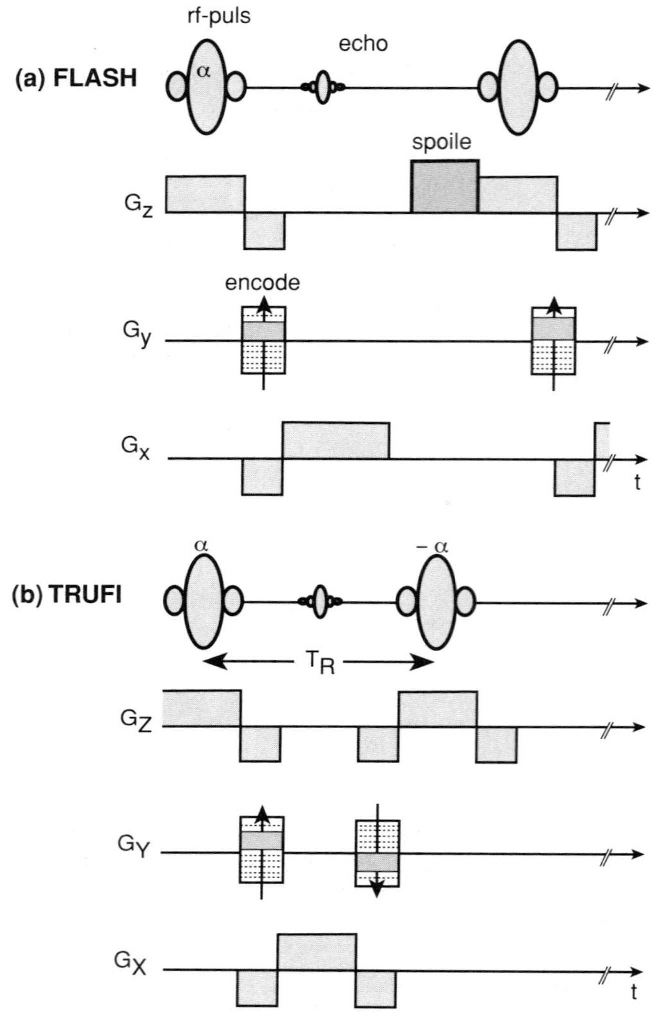

FIGURE 17.8 Examples of steady-state sequences: at FLASH, the transversal magnetization at the end of the repetition interval has to be destroyed, e.g., with a spoiler gradient (a); at true FISP (TRUFI), it is rewound to the state immediately at the end of the rf-pulse (b).

very homogeneous fields or with very strong gradients using a short repetition time, the TRUFI sequence can be set up, where the FISP and the PSIF signals are superimposed.

With steady-state sequences — referred to as gradient echo in contrast to spin echo sequences (see below), because the signal (or echo) is formed without refocusing rf-pulses — fast image acquisition within less than 1 s is possible. To enhance signal contrast between different tissues, the nuclear equilibrium magnetization can be inverted with a 180° rf-pulse before the imaging sequence is started.[18] Thus, during the fast imaging experiment the longitudinal magnetization undergoes relaxation back to its equilibrium state, producing image contrast with respect to tissues having different longitudinal relaxation times T_1.

Because gradient echo sequences are so fast, they are very well suited for 3d data acquisition. Either the rf-pulses are applied nonselectively, i.e., without a slice selection gradient, or a very thick slice is excited. Spatial resolution is then achieved by successively encoding the nuclear magnetization with the gradients G_z and G_y in the y- and z-directions and reading out the signal in the projection gradient G_x.

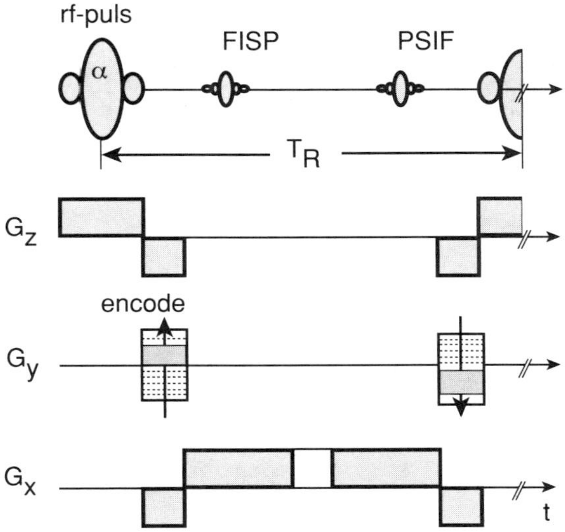

FIGURE 17.9 DESS sequence employing a combination of FISP and PSIF.

Since each volume element is repeatedly measured according to the number of phase encoding steps, the S/N ratio improves considerably. Image reconstruction is performed with a fast 3d Fourier transformation, resulting in a block of images that can be displayed as slices through the three main coordinate axes. Image postprocessing also allows the display of images in arbitrary projections (multi-planar reformatting = MPR).

With so-called spin echo sequences utilizing an additional 180° pulse, there is greater flexibility with respect to the manipulation of image contrast than in gradient echo sequences, though generally at the expense of acquisition time. Spin echo sequences are also much more stable against static field inhomogeneities of the static magnetic field, since dephasing of transverse magnetization during the encoding interval is reversed; compared to gradient echo sequences, signal loss is less. In this context, we want to mention that in the direction of the phase encoding gradient field, inhomogeneities cause no image distortions.

In the standard spin echo imaging sequence, slice selective 90° and 180° rf-pulses are used with encoding gradients between them (Figure 17.10); the echo signal is read out in the projection gradient. Two parameters are available for signal manipulation, the sequence repetition time T_R and the echo time T_E. The use of a long echo time allows transverse relaxation of the spin system before signal acquisition, whereas rapidly repeating the pulse sequence prevents longitudinal magnetization from reestablishing. This effect is called saturation. The signal intensity in a picture element is given by

$$M_{\perp}(x, y) = M_0(x, y)\left(1 - e^{\frac{-T_R}{T_1(x, y)}}\right) e^{\frac{-T_E}{T_2(x, y)}}. \tag{17.49}$$

The repetition time and the echo time can be adjusted so that the image contrast due to different types of tissue is determined by either M_o, T_1, or T_2. Short values of T_E and T_R give T_1-weighted images, while a long T_E and a short T_R give spin density or M_o-weighted images, and long values of both T_E and T_R give T_2-weighted images. Thus, in MR, the contrast-to-noise ratio is determined and can be changed in a wide range by the pulse sequence. This is a unique feature for this image modality and cannot not be obtained by retrospective filtering or postprocessing.

Contrast between adjacent anatomic structures can be further enhanced by means of contrast agents. Since the addition of other magnetic moments increases magnetic interactions during the collisions

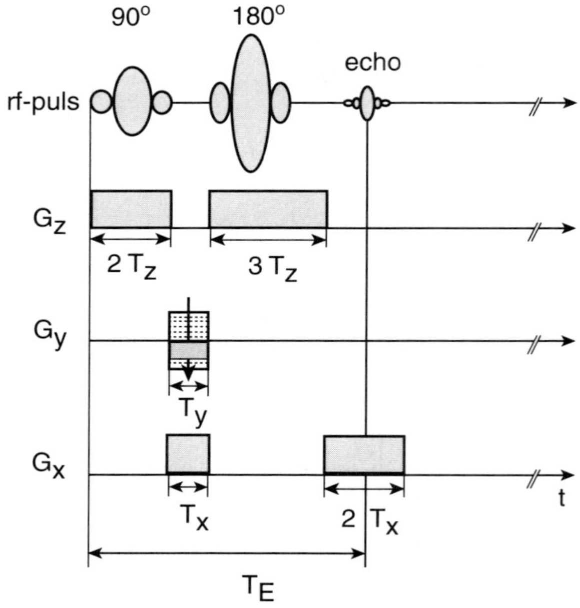

FIGURE 17.10 Standard spin echo imaging sequence. The compensation of the twisted transversal magnetization caused by the selective 90° pulse is achieved here with the prolonged, slice selecting gradient pulse after the selective 180° pulse.

between fluid molecules, a paramagnetic agent dispersed in the tissue accelerates the relaxation of excited spins, longitudinally as well as transversely. A common contrast agent is Gd-DTPA (gadolinium diethylenetriaminepentaacetic acid).[19] Administered to the blood stream, the contrast agent will accumulate at various levels in tissue due to the different microvascular structures.

The standard spin echo imaging sequence can be modified in several ways. At long repetition times, images of several different slices can be acquired in a shorter acquisition time than for a single slice, when the different slices are addressed during the waiting interval. To obtain information on transverse relaxation, the NMR signal can be recovered several times by repeating the 180° pulses. Thus, several images are reconstructed with varying T_2 weighting for a single slice; the transverse relaxation time can be calculated in each pixel and even displayed as an image.

When each echo of a multi-echo sequence is encoded in the y-direction with a gradient pulse, several lines in Fourier space are recorded during one pulse sequence, significantly reducing the time of measurement (TSE, turbo spin echo); it is even sufficient to scan only half of the Fourier space (HASTE, half Fourier acquired single shot turbo spin echo). When the 180° pulses are omitted and the polarity of the projection gradient is alternatingly reversed[20] to generate gradient echoes, a complete image can be acquired in less than 100 ms (EPI, echo planar imaging).

Since rapid switching of strong magnetic field gradients is not a simple technical task, in EPI the encoding gradients can be kept switched on during the total time of data acquisition and a sine wave oscillating readout gradient can be employed (Figure 17.11a). The resulting trajectory in Fourier space is shown in Figure 17.11b.

17.9 Influence of Motion

If the object to be imaged or parts of it are moving during data acquisition, artifacts occur, since a moving volume element acquires another phase in the applied gradient fields than if it were resting. Image reconstruction then attributes the position of that voxel to other origins that give rise to typical image distortions such as blurring or mirror images. However, depending on the type of movement, data

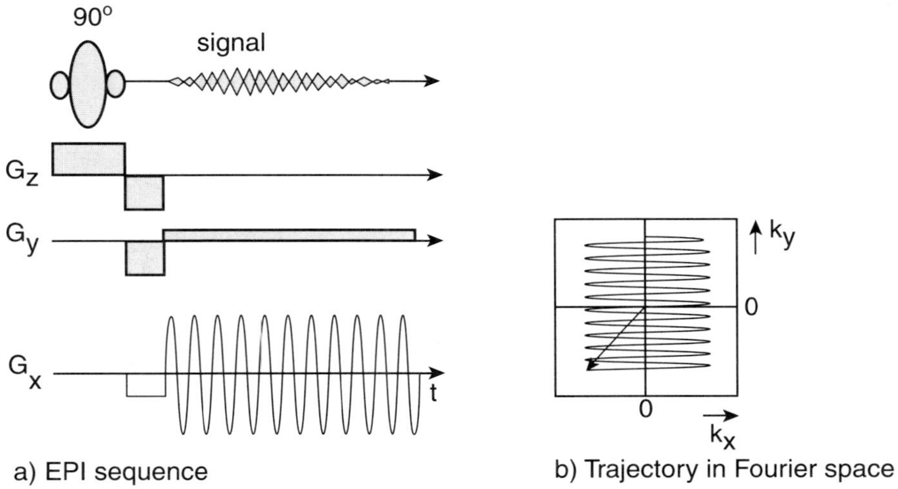

FIGURE 17.11 EPI employing a sine wave readout gradient (a) and the according trajectory in Fourier space (b).

acquisition strategies can be developed to avoid those artifacts or even to get information on parameters of the motion as, e.g., in the case of flow on the velocity distribution.

In principle, a moving object can be described as a four-dimensional (4d) distribution of transverse magnetization with three coordinates in space and one in time. Image artifacts can then be explained as an uncomplete data sampling process in the 4d space with time and space or their Fourier conjugates frequency and spatial frequency as coordinates. For illustrative reasons, we will restrict our discussion in the following to a 2d distribution $M(x, y, t)$ of transverse magnetization moving in time as it might be generated by selective excitation. In Fourier or k-space, such an object is described by

$$M(k_x, k_y, t) = \iint M(x, y, t) e^{-i(k_x x + k_y y)} dxdy, \tag{17.50}$$

putting up a 3d space consisting of the familiar two coordinates k_x, k_y of spatial frequency and the time coordinate t. When a moving object is successively sampled line by line in Fourier space, the k, t–space is crossed along a tilted plane (Figure 17.12)

$$\begin{aligned} q(k_x, k_y, t) &= ak_x(t) + bk_y(t) + t \\ &= \frac{k_x}{\gamma G_x}\cos\alpha + \frac{k_y}{\gamma G_y}\sin\beta + t = 0 \end{aligned} \tag{17.51}$$

with angles α and β with respect to the k_x- and k_y-axes (scaled with the factors $\gamma G_{x,y}$ to give them the same dimension as the t- axis) that pass the k_x, k_y, $t = 0$ center of origin.

Reconstructing the image from the samples on this plane must lead to artifacts, since only sampling along the $q(k_x, k_y, t = 0)$ plane would result in a reconstruction of the object $M(x, y)$ at $t = 0$ and only a complete scan of k, t-space would reconstruct the complete time course $M(x, y, t)$ of the object.

The nature of these artifacts is revealed when one considers the moving object not in x, y, t-space, but in x, y, ω-space:

$$M(x, y, \omega) = \frac{1}{2\pi}\int M(x, y, t) e^{-i\omega t} dt, \tag{17.52}$$

in which the spatial dependence of the harmonics of the moving object is displayed. Since x, y, ω are Fourier conjugates to k_x, k_y, t, one can apply the projection slice theorem, which states that the data $M(t)$

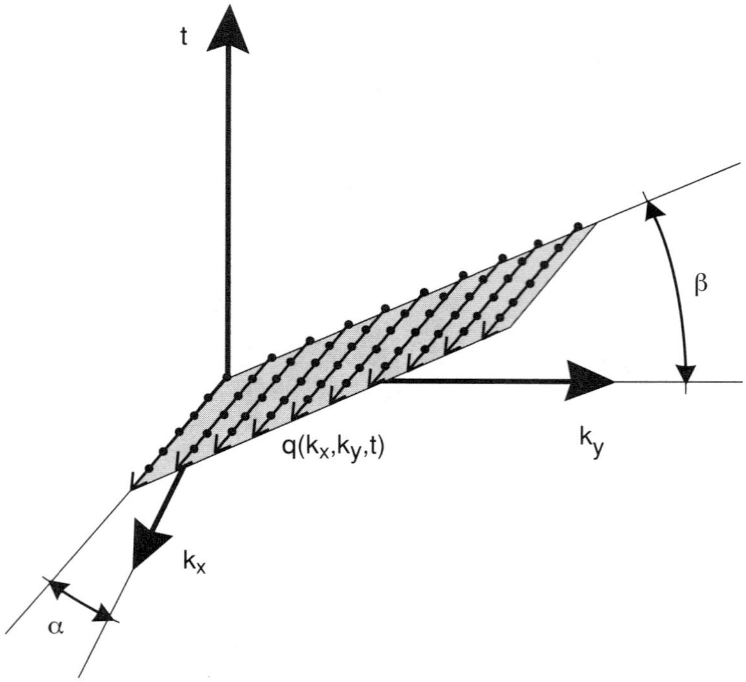

FIGURE 17.12 A moving object is scanned along a tilted plane in Fourier space.

sampled along the plane $q(k_x, k_y, t)$ through the center of Fourier (i.e., k_x, k_y, t) space characterized by the angles α and β correspond to the projection of the data in the original (i.e., x, y, ω) space on a plane

$$p(x, y, \omega) = \gamma G_x x \cos\alpha + \gamma G_y y \cos\beta + \omega = 0 \tag{17.53}$$

with the same direction α and β. For example, in the case of an object moving periodically, the occurrence of replication or ghosting is explained by the projection of the spectral island occurring in the x, y, ω plane onto $p(x, y, \omega)$[21] (Figure 17.13).

Of course, no motion artifacts will occur if the tilting angles of planes $q(k_x, k_y, t)$ and $p(x, y, t)$ are α, $\beta = 0$, i.e., no movement would occur during scanning. This implies the application of very rapid pulse sequences using gradient echoes as steady-state sequences or EPI. Unfortunately, rapid imaging often results in a low S/N ratio and/or low contrast. Imaging moving parts and organs of the human body with high contrast, e.g., with a spin echo sequence, requires long acquisition times due to the time interval between the phase encoding steps. Ghosting can be avoided in this case when data acquisition is triggered or gated by the movement so that the single phase encoding steps always occur at the same position of the object. Triggering is a prospective method, which puts restrictions on the pulse repetition time, while gating works retrospectively, and with proper reordering of the acquired phase encoding steps, it allows in principle a complete scan of the k_x, k_y, t volume in case of a periodically moving object. For imaging of the heart, trigger and gating pulses can be provided by an electrocardiogram (ECG) run simultaneously during the MRI investigation, while respiratory gating asks for a pressure transducer. An alternative is the use of navigator echoes, which are created during the repetition intervals. A scanning line is laid through the organ along the main movement direction by selective 90° and 180° pulses each with a different gradient direction, and the spin echo is read out in the perpendicular gradient. Fourier transformation of the echo monitors the position of the organ on the projection line, from which a gating signal can be derived.

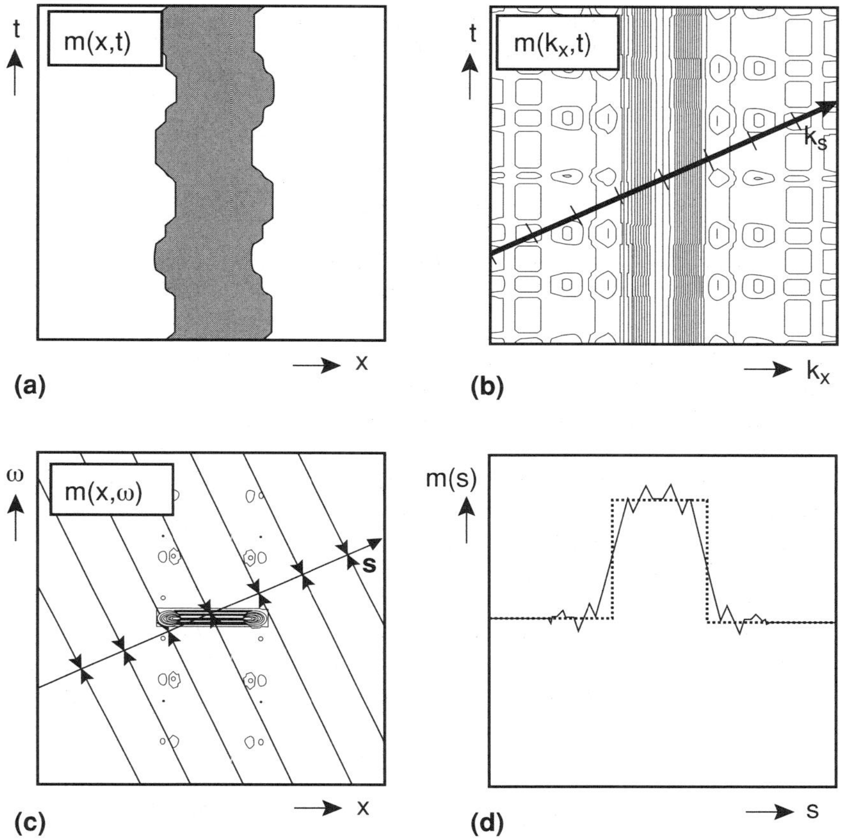

FIGURE 17.13 Illustrating the influence of movement at the example of an oscillating one-dimensional (1d) object (a). In the spatial frequency domain, the moving edges of the object lead to blurring and harmonics (c). With MRI, the object is sampled along an inclined line in the time-spatial frequency domain (b). The imaged transversal magnetization (d) is yielded after Fourier-transforming these samples and represents the projection in the space frequency domain on the direction of this line (d).

17.10 Correction of Motion during Image Series

The sensitivity of gradient echo sequences as steady-state sequences or EPI to magnetic field inhomogeneities can be utilized to image effects in the human body which respond sensitively to changes in magnetic susceptibility. Thus, e.g., the perfusion of the cortex varies with the performing of certain actuating or perceptive tasks. If a certain area of the brain becomes activated, e.g., the visual cortex, when the person under study sees light flashes, the local oxygen requirement increases. The circulatory system reacts by increasing the local blood supply even more than necessary. Consequently, the activation process results in an increased oxygen content of the venous blood flow. Oxygen is transported by hemoglobin in the red blood cells. Oxygenated hemoglobin (HbO_2) is diamagnetic, while deoxygenated hemoglobin (Hb) is paramagnetic due to the presence of four unpaired electrons. These different magnetic susceptibilities lead to different signal intensities in a properly weighted sequence. So Hb acts as a natural contrast agent, its effect varying with the oxygen supply and the utilization of the brain cells. Blood oxygen level dependent (BOLD) contrast can thus be studied with a susceptibility-sensitive sequence (functional MRI[21]).

In functional MRI (fMRI), an analysis of voxel time courses is performed in order to detect changes in the regional blood oxygenation of the brain. Enhanced blood supply happening upon stimulation of certain areas of the eloquent cortex is detected from signal changes between images acquired with and without stimulus applied. By calculating the correlation coefficient

$$CC = \frac{\sum_t (X_t - \bar{X}) \sum_t (Y_t - \bar{Y})}{\sqrt{\sum_t (X_t - \bar{X})^2 \sum_t (Y_t - \bar{Y})^2}} \tag{17.54}$$

where X_t is the time course of a given voxel during no stimulation and Y_t is the time course of the same voxel during stimulation, brain stimulation is assumed, when CC exceeds a threshold defined from the probability one is willing to accept for an accidental correlation (t-test).

Though in fMRI such fast image sequences, e.g., EPI, are used that one can assume no motion during data acquisition, the head might move between acquiring one image and another. This movement can occur even when the head is fixed and might result from an involuntary reaction on the applied stimuli.[23] Due to partial volume effects, motion shifts even less than a voxel size can lead to differences in signal intensity misleading the BOLD effect. Efforts to record head motion with MR or optical monitored fiducials fixed to the patient's head or with orbital navigator echoes were only partially successful, because there is not a tight enough correlation between the motion of the brain and the scalp. An algorithm that detects if motion has occurred between two images and corrects for it with postprocessing, therefore, seems to be a better solution.

The region of interest in a reference image is used to define a set of voxels to form a vector $\vec{X} = X_1, \ldots, X_n)$ and a vector $\vec{Y}$ from the corresponding region in the actual image that might have moved with respect to the reference image and therefore, is to be corrected. A linear transformation

$$\vec{Y} = \vec{X} + A\vec{p} \tag{17.55}$$

is assumed between both images, where the parameters p_j describe the motion parameters (three rotational, three translational) and the transformation matrix contains the derivatives $(\partial X_i / \partial p_j)$ of voxel intensity X_i with respect to parameter p_j. The coefficients of matrix **A** are known, as they can be determined for the object under investigation by exposing the reference image to virtual motions.

The parameter vector p describing motion between the reference and the actual image can be estimated in first order from the Moore-Penrose or pseudo-inverse

$$A^+ = (A^T A)^{-1} A^T \tag{17.56}$$

of the transformation matrix to be

$$\vec{p} = A^+(\vec{Y} - \vec{X}). \tag{17.57}$$

Then the values of p_i obtained are used for a coordinate transformation (describing shift and rotation) of the actual image to yield a corrected image $\vec{Y}_1$. If a transformed voxel falls between the points of the sampling grid of the reference image, its intensity can be distributed into the neighboring grid points, e.g., with a linear or with a sinc interpolation. This procedure is often referred to as regridding.

With $\vec{Y}_1$ it can again be checked if a shift or rotation still exists with respect to the reference image. Then the described algorithm can be repeated until no further changes in $\vec{p}$ are observed. Procedures of this kind be can assumed as sensitive to motion down to some 10 μm.[24]

17.11 Imaging of Flow

Flow-related phenomena are used in magnetic resonance angiography (MRA). Two effects have to be considered, namely, time of flight and phase changes. In standard spin echo sequences, it is often observed that the NMR signal of flowing nuclei is enhanced at low and reduced at high flow velocities, compared with stationary tissue.[25] The increase in intensity (sometimes referred to as a paradoxical phenomenon) is explained by introducing nuclei to the imaged slice which are not magnetized from previous excitations, while the signal void at high velocities occurs because part of the nuclei leave the slice between excitation and the echo measurement.

Time of flight effects can be utilized to create images similar in their appearance to those produced in X-ray angiography. With nonselective excitation pulses or pulses exciting a thick slice, a rapidly repeating 3d gradient echo sequence is set up, which gives a weak signal. Flow introduces fully relaxed spins to the imaged volume, giving rise to a stronger signal. Signal intensity increases with flow velocity. Contrary to a spin echo sequence, no signal void is observed at gradient echo sequences because the refocusing gradient effects all spins, those staying in the excited volume and those moving out. In order to visualize the vascular structure, the method of maximum intensity projection (MIP) is often used.[26] In image postprocessing, the acquired 3d image volume can be regarded as illuminated with parallel rays. Along each ray, the image element with the highest signal intensity is searched and displayed in the projection plane.

Phase-sensitive MRA makes use of the fact that moving spins acquire different transverse phases, according to their velocity in magnetic field gradients, than stationary spins. When the volume element is sufficiently small, so that it contains only spins with similar flow velocities, the flow can be quantitatively measured by analyzing the signal phase (phase contrast angiography[27]). In comparison with time of flight, phase contrast angiography is well suited to slow flow velocities.

When, on the other hand, a volume element large enough to contain spins with a variety of velocities is chosen, this will give only a weak signal because the individual contributions of the different spins cancel.

The transverse phase angle of an ensemble of stationary and moving spins follows (in the rotating frame) from the integration of Equation 17.16b:

$$\varphi(t) = \gamma \int_0^t G(t) r(t) dt. \tag{17.58}$$

Assuming uniform flow in a constant gradient,

$$r(t) = r_o + v_t, \tag{17.59}$$

hence

$$\varphi(t) = \varphi_o + \gamma G r_o t + \frac{1}{2}\gamma G v t^2. \tag{17.60}$$

As the phases of the stationary spins are recovering following the application of a bipolar gradient, the same happens at a uniform flow velocity with two bipolar gradients back to back (Figure 17.14). The signal loss due to dephasing spins in the readout gradient, e.g., in a gradient echo sequence, can therefore be avoided by applying a bipolar gradient pulse before the data readout phase. Such an imaging sequence is often referred to as a motion-compensated or flow-rephased pulse sequence. Subtraction between a flow-dephased and a flow-rephased image cancels out the signal from the stationary tissue, leaving only the vascular structure.[28] For this reason, the method can be viewed as magnitude contrast angiography.

Though flow-sensitive MRI enables MRA without contrast agents, investigation time can be rather long and spatial resolution limited. Recently, contrast enhanced (ce) MRA has been introduced where relaxation time shortening contrast agents are administered intravenously. The high contrast-to-noise ratio yielded, e.g., allows one to follow the contrast bolus from the arterial to the venous phase through the whole body, finding arterial and venous occlusions.

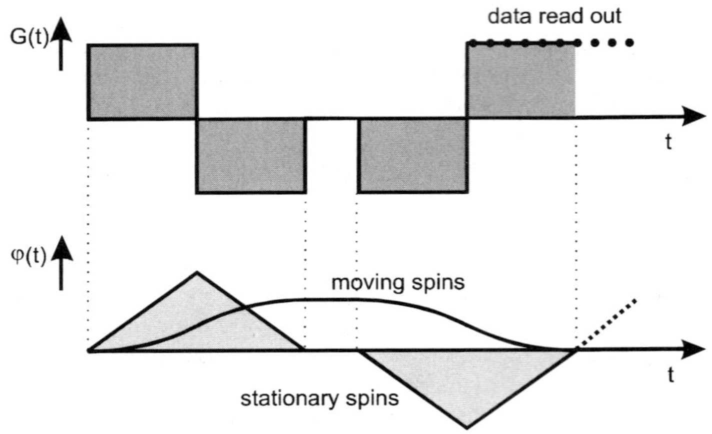

FIGURE 17.14 Gradient pulses for phase and motion compensation and time dependence of the phase stationary and moving spins. Applying a bipolar gradient pulse before the data readout phase minimizes the influence of uniform flow.

17.12 MR Spectroscopy

The signal displayed in MR images derives mostly from the hydrogen nuclei in water and fat. The different chemical shift of fat with respect to water can cause typical image artifacts, e.g., because slightly different slices are excited with a selective rf-pulse. Also, fat and water appear shifted with respect to each other in the direction of the readout gradient. Although these effects can be masked using stronger slice selection and readout gradients, this requires a greater rf-power and a larger bandwidth, in turn leading to an increase in image noise. Therefore, various pulse sequences have been developed to obtain pure fat or pure water images. Such sequences are employed when the strong signal of one compound obscures the weak signal of the other.

One possibility is to suppress the fat or the water signal pulse before beginning the imaging sequence, using an initial 90° pulse having exactly the frequency of the undesired compound and dephasing (spoil) the transverse magnetization of this compound with a strong gradient pulse. The subsequent imaging sequence then acts only on the desired compound.[29] When applied to suppress the signal of fat, this method is often referred to as fat saturation because fat does no longer gives a signal.

In some cases, pure fat and water images are generated using two pulse sequences with different readout delays. The difference is chosen so that the magnetization of fat and water is parallel in one case and antiparallel in the other case.[30] Adding and subtracting the signals from the two sequences yield either a pure fat or a pure water image. The technique of identifying metabolites by employing echo times that generate defined in-phase or opposite-phase alignment of the transverse nuclear magnetization is called spectral editing. It can be used, e.g., to separate lactate from lipid resonances in proton spectroscopy.

Of special interest, however, is the detection of metabolites either by the NMR of hydrogen ^{1}H or of other nuclei, such as phosphorus ^{31}P. Due to the very low concentration of metabolites, their NMR signal is much weaker than that of water and fat. To obtain a sufficient S/N ratio, it is therefore necessary to work with lower spatial resolution and longer acquisition times than with normal imaging.

Measuring chemical shifts requires a static magnetic field with a high field flux density (>1T) and very good homogeneity. Either the spectrum of a single volume element can be measured (single voxel spectroscopy) or the spatial distribution of spectra in the object can be acquired (chemical shift imaging).

Since the intensity of the water signal can be several orders of magnitude greater than that of the signals from the metabolites, it must be suppressed, e.g., with a narrow bandwidth 90° pulse and a spoiler gradient. The volume selection sequence, e.g., a selective 90° pulse with a gradient in the x-direction and two selective 90° or 180° pulses with a gradient in the y- and z-directions, then acts only on the metabolites, the spectral lines of which result from a Fourier transformation of the FID.[31]

Although NMR spectroscopy in living beings at first seems to be very attractive since it should permit immediate insight into cell metabolism, its clinical importance has remained limited up to now. Compared with nuclear medicine, where radioactive tracers with very low concentration can be detected, NMR is a very insensitive method. This restricts its application concerning components other than fat and water to the detection of volume elements with a size of several cubic centimeters and to metabolites of limited biological or medical usefulness.

17.13 System Design Considerations and Conclusions

For MRI, processor-based Fourier NMR spectrometers used routinely in analytical chemistry have been adapted to the size of a human patient, and components and software have been structured to meet imaging requirements. Figure 17.15 shows a block diagram of an MR imager. Of special importance is the system, or host computer which controls all components of the system often applying digital processors themselves, and acts as the interface to the user. The mighty software has to control the system, run the imaging sequences, perform image reconstruction, interact with the user, perform archiving, and perform increasingly morepost processing tasks.

The magnet is by far the most important (and expensive) component. The optimum field strength for MRI is still a matter of controversy. Flux densities above 0.5 *T* in a volume suitable for patient investigations can only be obtained with superconducting magnets, while permanent and resistive magnets with iron yoke flux return paths are applied at field strengths <0.5 *T*. Though clinical MR systems are restricted to flux densities not larger than 1.5 *T*, experimental instruments apply magnets up to 8 *T*. High magnetic fields are often employed in order to obtain a better S/N ratio, but as the rf-power required for spin excitation increases with B_z^2 (Equation 17.13), limitations are placed on image sequences in order not to generate excessive heat in the patient. The different contrast behavior of tissue and possibilities to better distinguish metabolites by spectroscopy are reasons to explore very high magnetic fields. Here, the influence of decreasing penetration of the rf-field into the sample (skin effect) and possible dielectric resonances occurring at tissue borders are new experimental challenges.

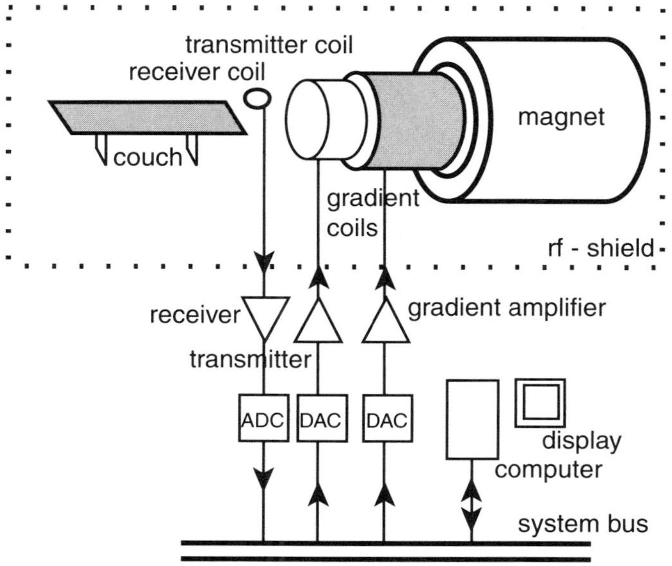

FIGURE 17.15 Block diagram of an MRI apparatus (ADC = analog-to-digital converter, DAC = digital-to-analog converter).

Other than by the static magnetic field, the S/N ratio can be influenced with the rf-antennas. Using arrays of small coils rather than a single large coil leads to a significant improvement[32] because a small coil picks up much less noise from the body.

The gradient strength is another important factor affecting the performance of an MRI system. In order to enable short image acquisition times with steady state sequences or by acquiring several lines in Fourier space, as in TSE or EPI, strong gradients are also necessary for sequences sensitive to flow, perfusion, or diffusion. The switching time for the gradients must always be as short as possible, since during switching intervals undesired signal decay occurs.

The gradient strength and switching time are ultimately limited by the electrophysiology of the patient, who will eventually experience stimulation of the peripheral nerves; even though painful, they are harmless since they already occur at much lower values than would have any effect on the cardiac system.[33] Gradient strengths up to 40 (mT/m) with a rise time of $5(\mu s/(mT/m))$ are now in clinical use.

17.14 Conclusion

Since the first availability of commercial instruments at the beginning of the 1980s, clinical MR has expanded rapidly in terms of both medical applications and the number of units installed. First considered to be an expensive method to create images of inferior quality, it has since established itself as a clinical tool for diagnosis in previously inconceivable applications, and the potential of the method is still not exhausted. MRI has led to the first large-scale industrial application of superconductivity and has brought about a far greater public awareness of a physical effect previously known only to a handful of scientists.

Up to now, the growth and spectrum of applications of MR have exceeded all predictions. The most recent development is that of rendering brain functions visible. Cardiac MR can display coronaries and analyze perfusion of the myocardium and hemodynamics of the heart. Thus, MRI is entering the domain of nuclear medicine.

An interesting new application of MRI is its use as an imaging modality during minimal invasive procedures such as rf-ablation, interstitial laser therapy, or high intensity focused ultrasound. With temperature-sensitive sequences, the development of temperature and tissue damage can be checked during heating and destroying of diseased tissue. The sensitivity of MRI to flow helps the physician to stay away from vessels during an intervention. MRI is also used for image-guided surgery, e.g., resection of tumors in the brain. Special open systems have been designed for such purposes, and dedicated nonmagnetic surgery tools have already been developed.

References

1. Zavoisky, E., *J. Phys. USSR*, 9, 211, 1945.
2. Bloch, F.W., W.W. Hansen, and M. Packard, Nuclear induction, *Phys. Rev. (L)*, 69, 127, 1946.
3. Purcell, E.M., H.C. Torrey, and R.V. Pound, Resonance absorption by nuclear magnetic moments in a solid, *Phys. Rev. (L)*, 69, 37, 1946.
4. Lauterbur, P.C., Image formation by induced local interactions: Examples employing nuclear magnetic resonance, *Nature*, 242, 469, 1955.
5. Bloch, F., W.W. Hansen, and M. Packard, The nuclear induction experiment, *Phys. Rev.*, 70, 474, 1946.
6. Hahn, E., Spin echos, *Phys. Rev.*, 80, 580, 1950.
7. Carr, H.Y. and E.M. Purcell, Effects of diffusion on free precession in nuclear magnetic resonance experiments, *Phys. Rev.*, 94, 630, 1954.
8. Hoult, D.I. and P.C. Lauterbur, The sensitivity of the Zeugmatographic experiment involving human samples, *J. Magn. Reson.*, 343, 425, 1979.
9. Abe, Z., K. Tanaka, K. Hotta, and M. Imai, Noninvasive measurements of biological information with application of nuclear magnetic resonance, in *Biological and Clinical Effects of Low Magnetic and Electric Fields*, Charles C Thomas, Springfield, IL, 1974, pp. 295–317.

10. Cormack, A.M., Representation of a function by its line integrals, with some radiological applications. I, *J. Appl. Phys.*, 34(1), 2722, 1963; 35(2), 2908, 1964.
11. Hounsfield, G.N., J. Ambrose, J. Perry, et al., Computerized transverse axial scanning (tomography), *Br. J. Radiol.*, 46(1, 2), 1016, 1973.
12. Kumar, A., D. Welti, and R. Ernst, NMR Fourier Zeugmatography, *J. Magn. Reson.*, 18, 69, 1975.
13. Cooley, J.W. and J.W. Tukey, An algorithm for machine calculation of complex Fourier series, *Math. Computation,* 19, 297, 1965.
14. Hoult, D.I., Zeugmatography: A criticism of the concept of a selective pulse on the presence of a field gradient, *J. Magn. Reson.*, 26, 165, 1977.
15. Haase, A., J. Frahm, D. Matthaei, W. Hänicke, and K. Merboldt, FLASH imaging: Rapid nmr imaging using low flip angle pulses, *J. Magn. Reson.*, 67, 217, 1986.
16. Freeman, R. and H.D.W. Hill, Phase and intensity anomalies in Fourier transform NMR, *J. Magn. Reson.*, 4, 366, 1971.
17. Oppelt, A., R. Graumann, H. Barfuß, H. Fischer, W. Hartl, and W. Schajor, FISP — eine neue schnelle Pulssequenz für die Kernspintomographie, *Electromedica*, 54, 15, 1986.
18. Haase, A., D. Matthaei, R. Bartkowski, E. Dühmke, and D. Leibfritz, Inversion recovery snapshot FLASH MRI: Fast dynamic T1 contrast, *J. Comput. Assist. Tomogr.*, 13, 1036, 1989.
19. Weinmann, H.J., R.C. Brasch, W.R. Press, and G.E. Wesbey, Characteristics of gadolinium-DTPA complex: A potential NMR contrast agent, *Am. J. Radiol.*, 143, 619, 1984.
20. Mansfield, P., Multi planar image formation using NMR spin echos, *J. Phys. C*, 10, L55, 1977.
21. Lauzon, M.L. and B.K. Rutt, Generalized k-space analysis and correction of motion effects in MR imaging, *Magn. Reson. Med.*, 30, 438, 1993.
22. Ogawa, S., D.W. Tank, R. Menon, J.M. Ellermann, S.G. Kim, H. Merkle, and K. Ugurbil, Intrinsic signal changes accompanying sensory stimulation: Functional brain mapping with magnetic resonance imaging, *Proc. Natl. Acad. Sci. U.S.A.*, 89, 5951, 1992.
23. Hajnal, J.V., R. Myers, A. Oatridge, J.E. Schwieso, I.R. Young, and G.M. Bydders, Artifacts due to stimulus correlated motion in functional imaging of the brain, *Magn. Reson. Med.*, 31, 283–291, 1994.
24. Friston, K.J., S.R. Williams, R. Howard, R.S.J. Frackowiak, and R. Turner, Movement-related effects in fMRI time-series, *Magn. Reson. Med.*, 35, 346–355, 1996.
25. Crooks, L., P. Sheldon, L. Kaufmann, W. Rowan, and T. Millert, Quantification of obstruction in vesssels by nuclear magnetic resonance (NMR), *IEEE Trans. Nucl. Sci.*, NS-29, 1181, 1982.
26. Koenig, H.A. and G.A. Laub, The processing and display of three-dimensional data in magnetic resonance imaging, *Electromedica*, 56, 42, 1988.
27. Dumoulin, C.L., S.P. Souza, M.F. Walker, and W. Wagle, Three-dimensional phase contrast angiography, *Magn. Reson. Med.*, 9, 139, 1989.
28. Laub, G.A. and W.A. Kaiser, MR angiography with gradient motion refocussing, *J. Comput. Assist. Tomogr.*, 12, 377, 1988.
29. Haase, A., J. Frahm, W. Hänicke, and D. Matthaei, 1H NMR chemical shift selective (CHESS) imaging, *Phys. Med. Biol.*, 30, 341, 1985.
30. Dixon, W.T., Simple proton spectroscopic imaging, *Radiology,* 153, 189, 1986.
31. Frahm, J., H. Bruhn, M.L. Gyngell, K.D. Merboldt, W. Hänicke, and R. Sauter, Localized high-resolution protron NMR spectroscopy using stimulated echos: initial applications to human brain in vivo, *Magn. Reson. Med.*, 9, 79, 1989.
32. Roemer, P.B., W.A. Edelstein, C.E. Hayes, S.P. Souza, and O.M. Mueller, The NMR phased array, *Mag. Reson. Med.*, 16, 192, 1990.
33. Budinger, T.F., H. Fischer, D. Hentschel, H.E. Reinfelder, and F. Schmitt, Physiological effects of fast oscillating magnetic field gradients, *J. Comput. Assist. Tomogr.*, 15, 909, 1991.

18

Functional Imaging of Tissues by Kinetic Modeling of Contrast Agents in MRI

Frank S. Prato
University of Western Ontario

Charles A. McKenzie
Beth Israel Deaconess Medical Center and Harvard Medical School

Rebecca E. Thornhill
University of Western Ontario

Gerald R. Moran
University of Western Ontario

18.1 Introduction

18.1.1 Rationale

It has long been recognized that the more fundamental and mechanistic the information encoded in a medical image, the lower the spatial resolution. Nuclear medicine images, which show tissue function and/or biochemistry, have typical object resolution that exceeds 1 cm. Conversely, the anatomical location of a bone fracture can be resolved to within fractions of a millimeter using X-ray imaging. Magnetic resonance imaging (MRI) of the heart muscle typically achieves a resolution of 8 mm^3 voxels with an acceptable signal-to-noise ratio (SNR). Via magnetic resonance spectroscopy (MRS), it is currently possible to interrogate the energetics of heart muscle, but, to achieve an acceptable SNR, one must pay a serious penalty in spatial resolution. ^{31}P MRS of the heart often requires voxels >12 cm^3 in order to compensate for the relatively low concentrations and gyromagnetic ratio of phosphorus metabolites.[1] There is a tremendous clinical and research need for improved spatial resolution and SNR in functional/biochemical medical imaging, especially for the heart and brain. In the heart tissue, biochemical status and function can change across the left ventricular wall over a distance of 1 mm, while present nuclear medicine (functional) imaging methods are limited to 1 to 2 cm spatial resolution.

0-8493-3691-0/01/$0.00+$.50

In the brain, biochemical and functional changes in health and disease occur within sub-millimeters, yet functional/biochemical imaging modalities such as single photon emission computed tomography (SPECT) or positron emission tomography (PET) can measure brain blood flow to a spatial resolution of only 1 cm^3.

Our hypothesis is that MRI is flexible enough and rich enough with respect to the number of parameters which determine final image SNR so that it can be used to produce functional/biochemical images of spatial resolution superior to that of current modalities. However, this will require (1) fast volume imaging of the changing MRI signal and (2) the development of mathematical models that relate this dynamic signal to underlying function and/or biochemistry. Note that we have decided to focus on MRI rather than MRS, which measures biochemistry directly because the intrinsic difference in tissue sample size to maintain SNR is far too great for MRS to compete effectively with other modalities such as nuclear medicine.

The first truly significant breakthrough has been the use of MRI to measure changes in brain blood flow with a spatial resolution of a few cubic millimeters and temporal resolution of tens of milliseconds. This has been a significant advance over nuclear medicine techniques such as SPECT and PET, where resolution is no better than 1 cm^3. This was achieved by recognizing that the nuclear magnetic resonance (NMR) signal in a region decreased as blood oxygenation levels increased in tissue with increased blood flow.[2] Although this technique is limited as a physiological measurement, i.e., a qualitative measure of change in brain blood flow,[3] its development has had significant impact on neuroscience research and clinical neurology. Unfortunately, this technique has been primarily limited to the brain, and other techniques will be needed for other organs such as the heart. The success of brain blood flow imaging using endogenous contrast agents suggests, however, that exogenous MRI contrast agents introduced into the blood may be equally successful for the determination of hemodynamic parameters elsewhere in the body.

However, the measurement of functional/biochemical tissue parameters via the injection of exogenous contrast agents must overcome numerous hurdles before it can reach widespread research and clinical use.

18.1.2 Overall Approach

The approach is to inject an exogenous MRI contrast agent and use MRI to determine the concentration of that contrast agent as a function of time. Unlike nuclear medicine where metabolically active agents can be injected, the relatively high concentrations needed in MRI require that biologically inert agents be used.[4] Both intra- and extravascular compounds can be used, owing to the variety of molecular sizes available.[5–7] As a result, once concentration (contrast agent) time curves are measured, one can calculate a number of contrast agent-dependent parameters such as distribution volume (V_b) and permeability-surface area products (e.g., flow across capillary membranes). Other, independent, strictly physiological parameters, such as tissue blood flow and tissue blood volume, can also be determined.

18.2 Contrast Agent Kinetic Modeling

If $A(t)$ represents the concentration of a contrast agent as a function of time in an artery going to a tissue and $T(t)$ represents the concentration in that tissue at any time t, then these can be related as

$$T(t) = A(t) * I(t) \tag{18.1}$$

where $*$ represents a convolution and $I(t)$ represents the tissue curve if $A(t)$ were a delta (Δ) function; $I(t)$ is often called the tissue impulse residue function.[8] The function $I(t)$ is dependent on tissue and contrast agent parameters. Most of the contrast agent tracer kinetic theory is associated with solving Equation 18.1 for $I(t)$. Once solved, parameters such as tissue blood flow, blood volume, and permeability-surface area product can be calculated.[9] The goal of functional MRI is to measure $T(t)$ and $A(t)$ with sufficient temporal and spatial resolution and SNR such that the derived physiological parameters are determined with sufficient spatial resolution, precision, and accuracy. For example, in cardiac imaging, spatial resolution should cover the left myocardium with a minimum in-plane resolution of 1 mm for slices 5 to 8 mm thick. A parameter such as blood flow should be determined to an accuracy and precision of 0.2 $ml \cdot min^{-1} \cdot g^{-1}$ if one wishes to

1. Estimate absolute flow when hibernating myocardium is supected[10]
2. Estimate perfusion reserve in patients with significant coronary artery disease[11]

There are many approaches to the problem of tracer kinetic modeling, however, most will fall under two categories: compartment models or distributed parameter models. Compartment models attempt to mimic tissue by dividing the problem into several compartments between which the tracer can be distributed. With this type of approach, the tracer concentration is a function of time only. Distributed parameter models compartmentalize the tissue as well; however, typically within one compartment the concentration is allowed to be a function of position as well as time. In the Tissue Homogeneity Model, for instance, the tissue is modeled by two concentric cylinders with the plasma being represented by the innermost cylinder.[9] This innermost volume will have a concentration gradient as the tracer moves through the tissue with time, whereas the other compartment will have tracer concentrations that are approximated as a function of time only.

18.3 Measurement of Contrast Agent Concentration

18.3.1 MRI Contrast and Tissue Contrast Agent Concentration

Unlike nuclear medicine or X-ray computed tomography (CT), the change in the MRI signal caused by a contrast agent is not related directly to the contrast agent's tissue concentration. In MRI, the signal is a complicated function of a number of sample parameters, such as the relaxation parameters T_1, T_2, and T_2^* and the equipment parameters TR, TE, and α. In general, MRI contrast agents work by changing the relaxation parameters (T_1, T_2, or T_2^*) in the sample in a *predictable* fashion. Extravascular contrast agents such as gadolinium diethylenetriaminepentaacetic acid (Gd-DTPA) change T_1 in a concentration-dependent way, while intravascular agents such as Gd-DTPA tagged to albumen change T_2^*.[12] A typical approach has been to assume that

$$[\text{Gd-DTPA}](t) = k\Delta R_1(t) \tag{18.2}$$

where $\Delta R_1(t) = R_{1\ \text{contrast}}(t) - R_{1\ \text{pre-contrast}}$, with $R_1 = {}^1/_{T1}$; $R_{1\ \text{contrast}}(t)$ is the $T1$ relaxation at time t after the initial arrival of Gd-DTPA; and $R_{1\ \text{pre-contrast}}$ is the $T1$ relaxation of the tissue before the arrival of contrast.[13]

A limiting assumption of Equation 18.2 is that the tissue/organ in question can be represented by a single relaxivity at all concentrations of the contrast agent. This assumption may break down, particularly at high concentrations of injected contrast agents.[14]

It has been shown separately[15] that $\Delta R_1(t)$, i.e., $R_{1\ \text{contrast}}(t)$, should be determined once every 1 to 2 s for accurate calculation of physiological tissue parameters via Equation 18.1. However, there is at least one exception to this. It has been shown that, in some cases, V_b can be measured after tissue contrast agent concentrations have reached steady state using constant infusion protocols. In such cases, rapid measurements are not needed. It should be noted that this constant infusion approach is limited and cannot be used to measure important physiological parameters such as blood flow and capillary permeability.[16]

18.3.2 Determinants of $R_{1\ \text{contrast}}(t)$

The task is to measure $R_{1\ \text{contrast}}(t)$ once per second with sufficient resolution (e.g., 1 m × 1 mm × 8 mm thick) over the entire tissue/organ in question. In order to understand how we can measure some of these intrinsic magnetic resonance (MR) parameters, we must first understand the origins of the signal received in an MRI experiment and how that signal can be affected by these parameters.

18.3.2.1 Signal Equation

The signal received from a voxel of a sample in an MRI experiment is proportional to the net transverse magnetization of that voxel. In a two-dimensional (slice selective) experiment, the net signal is the complex valued summation of the signal from every voxel in the slice of interest (assuming the receiver has uniform sensitivity over the slice). If we separate the transverse magnetization into its amplitude (M_{xy}) and phase (ϕ), the signal is[1]

$$s(t) = \int_x \int_y M(x, y, t) e^{-i\phi(x, y, t)} dx dy \tag{18.3}$$

The time evolution of the phase depends on the Larmor relation, where the rotational frequency $\omega = \gamma B$, so

$$\phi(x, y, t) = \gamma \int_0^t B(x, y, \tau) d\tau \tag{18.4}$$

In a typical imaging sequence, B will consist of the static main field B_0 and gradient fields in the x and y directions so that $B(x, y, t) = B_0 + G_x(t)x + G_y(t)y$. Over the imaging volume, the gradient fields cause the phase to vary as a linear function of position.

$$\begin{aligned} \phi(x, y, t) &= \gamma\left(\int_0^t B_0 d\tau + x\int_0^t G_x(\tau) d\tau + y\int_0^t G_y(\tau) d\tau\right) \\ &= \omega_0 t + 2\pi k_x(t) x + 2\pi k_y(t) y \end{aligned} \tag{18.5}$$

where

$$k_i(t) = \frac{\gamma}{2\pi}\int_0^t G_i(\tau) d\tau \qquad i = x, y \tag{18.6}$$

Thus, the gradient fields control the phase of the magnetization as a function of position. The term $\omega_0 t$ describes the Larmor frequency rotation of the magnetization. This term can be eliminated, since the signal is normally measured using quadrature phase sensitive detection, which removes the $\omega_0 t$ term by demodulation of the signal at the frequency ω_0. If we substitute the demodulated Equation 18.5 into Equation 18.3 we get

$$s(t) = \int_x \int_y M(x, y, t) e^{-i2\pi k_x(t) x} e^{-i2\pi k_y(t) y} dx dy \tag{18.7}$$

Assuming $M(x, y, t)$ is time invariant, the two-dimensional (2D) Fourier transform (FT) of $M(x, y)$ is

$$m(k_x, k_y) = \int_x \int_y M(x, y) e^{-i2\pi k_x x} e^{-i2\pi k_y y} dx dy \tag{18.8}$$

where k_x and k_y are in units of spatial frequency (typically 1/mm), which means that at any time t, $s(t)$ is the 2D FT of $M(x, y)$ at some spatial frequency [i.e., $s(t) = m(k_x(t), k_y(t))$]. This implies that we can solve for $M(x, y)$ by taking the inverse 2D FT of $s(t)$.

The signal $s(t)$ corresponds directly to a trajectory through spatial frequency space (k-space) so that $s(t) = s(k_x(t), k_y(t))$. This equation implies that if M does vary in time, it also varies as a function of spatial frequency ($M(x, y, t) = M(x, y, k_x(t), k_y(t))$). In that case, Equation 18.7 is no longer the FT of $M(x, y)$, so it would no longer be possible to solve for $M(x, y)$ by taking the inverse FT of $s(t)$.

In an imaging experiment the continuous function $s(t)$ is sampled at discrete values of k_x and k_y, in which case Equation 18.7 becomes

$$S(k_x, k_y) = \sum_y \left(\sum_x M(x, y, k_x, k_y) e^{-i\frac{2\pi}{Nk_x}k_x x} \right) e^{-i\frac{2\pi}{Nk_y}k_y y}$$
$$x, k_x = -N_{k_x}/2 + 1 \rightarrow N_{k_x}/2 \qquad y, k_y = -N_{k_y}/2 + 1 \rightarrow N_{k_y}/2 \tag{18.9}$$

where N_{kx} and N_{ky} are the total number of samples of k_x and k_y acquired for that imaging sequence. Again, if $M(x, y, k_x, k_y)$ is independent of time (that is M does not vary during the sampling of k-space), then Equation 18.9 is identical to the 2D discrete FT of $M(x, y)$.

The actual value of M depends on $\lambda(x, y)$, which is a set of variables such as M_0 (fully relaxed longitudinal magnetization); T_1 and T_2 (which depend only on x and y); and $t(k_x, k_y)$, a function that describes the time dependence of the acquisition in k-space. M can be re-expressed as $M(\lambda(x, y), t(k_x, k_y))$, in which case Equation 18.9 becomes:

$$S(k_x, k_y) = \sum_y \left(\sum_x M(\lambda(x, y), t(k_x, k_y)) e^{-i\frac{2\pi}{N_{k_x}}k_x x} \right) e^{-i\frac{2\pi}{N_{k_y}}k_y y} \tag{18.10}$$

In most imaging sequences, t is assumed to be a constant independent of k_x and k_y. When this is true, $M(x, y)$ does not change during data acquisition, so it is a function of $\lambda(x, y)$ only and it is equal to the 2D discrete FT (DFT) of $S(k_x, k_y)$. There is not enough data in a single collection of $S(k_x, k_y)$, however, to solve for $\lambda(x, y)$ (e.g., T_1 and T_2). To do that, several $S(k_x, k_y)$ must be collected, each having a different dependence on t.

18.3.2.2 T_1 Measurement

18.3.2.2.1 Inversion Recovery

The behavior of the magnetization $\mathbf{M}$ ($=M_x\mathbf{i} + M_y\mathbf{j} + M_z\mathbf{k}$) is described by a phenomenological equation known as the Bloch equation:[1,17]

$$\frac{d\mathbf{M}}{dt} = \mathbf{M} \times \gamma\mathbf{B} - \frac{M_x\mathbf{i} + M_y\mathbf{j}}{T_2} - \frac{(M_z - M_0)\mathbf{k}}{T_1} \tag{18.11}$$

where **i**, **j**, and **k** are unit vectors in the x, y, and z directions; **B** is the strength of the applied magnetic fields; γ is the gyromagnetic ratio; $\mathbf{M}_0$ is the equilibrium magnetization of the sample; T_1 is the time constant of the spin-lattice relaxation; and T_2 is the time constant of the spin-spin relaxation.

In the absence of RF and gradient magnetic fields, $\mathbf{B} = \mathbf{B}_0\mathbf{k}$. In this case, if we look only at the z component of **M**, the Bloch equation reduces to

$$\frac{dM_z}{dt} = \frac{M_z - M_0}{T_1} \tag{18.12}$$

If we solve Equation 18.12 we get

$$M_z(t) = M_0\left(1 - \beta e^{-\frac{t}{T_1}}\right) \tag{18.13}$$

where the value of β depends on $M_z(0)$.

Given that $M_z = M_0$ in an undisturbed sample, and since T_1 describes the time rate of change of M_z, it is necessary to perturb the magnetization of the sample so that we can observe how M_z changes with time as it returns to its equilibrium value of M_0. This is often done by using a 180° RF pulse to invert M_z (i.e., $M_z(0) = -M_0$). In this case, $\beta = 2$ in Equation 18.13. This technique, whereby T_1 is determined by following the recovery of M_z after it is inverted, is called inversion recovery (IR).

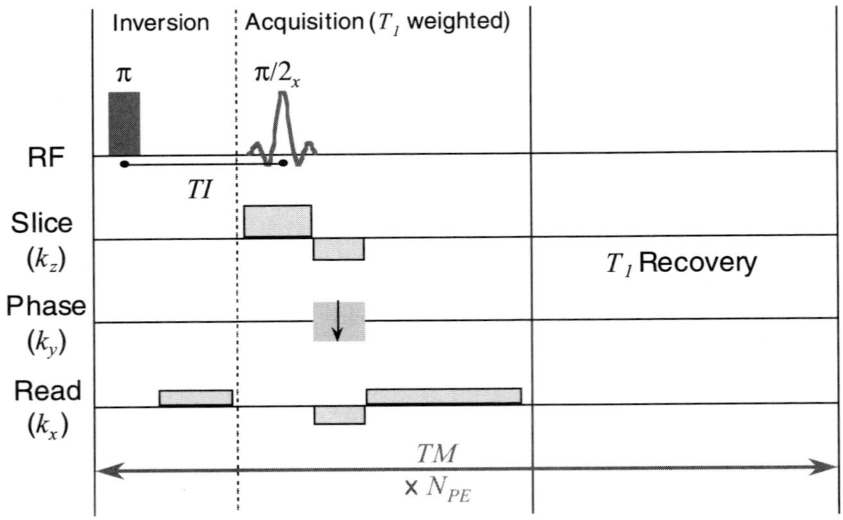

FIGURE 18.1 Diagram of an IR pulse sequence. This sequence must be repeated for each *TI* that is acquired.

As mentioned at the beginning of Section 18.3.2.1, the signal in an MRI experiment depends on the transverse magnetization M_{xy}, not the longitudinal magnetization M_z. Thus, in order to measure T_1, it is necessary to rotate some portion of M_z into the xy plane and measure the resulting transverse magnetization. If a 90° RF pulse is applied to the sample some time after the inversion pulse (called the inversion time [*TI*]), there will have been some recovery of M_z toward M_0. If M_{xy} is measured immediately after the 90° pulse, $M_{xy}(TI) = M_z(TI)$, we will have obtained one measurement of $M_z(t)$. Figure 18.1 illustrates an IR sequence.

There are two unknowns in Equation 18.13, so two (or preferably more) samples of M_z must be obtained at different *TI*s ($N_{TI} \geq 2$). Since the RF pulse used to rotate M_z into the xy plane disturbs the longitudinal magnetization, it is not possible to obtain more than one sample of M_z after an inversion pulse and still use Equation 18.13 to determine T_1. A second measurement of M_z can only be made after M_z has recovered back to M_0. Full recovery is generally assumed to have occurred after waiting for five T_1s[18] after the 90° sampling pulse. This means that the measurement repetition time (*TM*) can be no faster than $5T_1 + TI$. Since T_1 in biological tissues is on the order of 1 s at 1.5 *T*,[19] *TM* must generally be >5 s.

In order to produce a map of the T_1 distribution of an imaged object, a series of images must be acquired, where Equation 18.13 determines the pixel signal intensity. From the perspective of Equation 18.10, this means that each image must be acquired so that $\lambda(x, y)$ only depends on M_0 and T_1, and $t(k_x, k_y)$ is a constant (*TI*), but is varied between images. In this case,

$$M(\lambda(x, y), t(k_x, k_y)) = M_0(x, y)\left(1 - 2e^{\frac{-TI}{T_1(x, y)}}\right) \tag{18.14}$$

Thus after a 2D DFT of Equation 18.10, we see that the signal intensity of each pixel will be described by Equation 18.13. Fitting Equation 18.13 to the signal intensities of each pixel in the series of images of the object produces a T_1 map.

In a typical imaging experiment, at least 64 phase encodes will be collected ($N_{PE} = 64$), which means that the total measurement time for a single point on the T_1 recovery curve will be ~5.3 min (N_{PE} * *TM*). Considering that multiple images must be collected, each with a different *TI*, the acquisition time (*TA*) needed to measure T_1 with the IR method will be $N_{TI} * N_{PE} * TM$, which can easily exceed

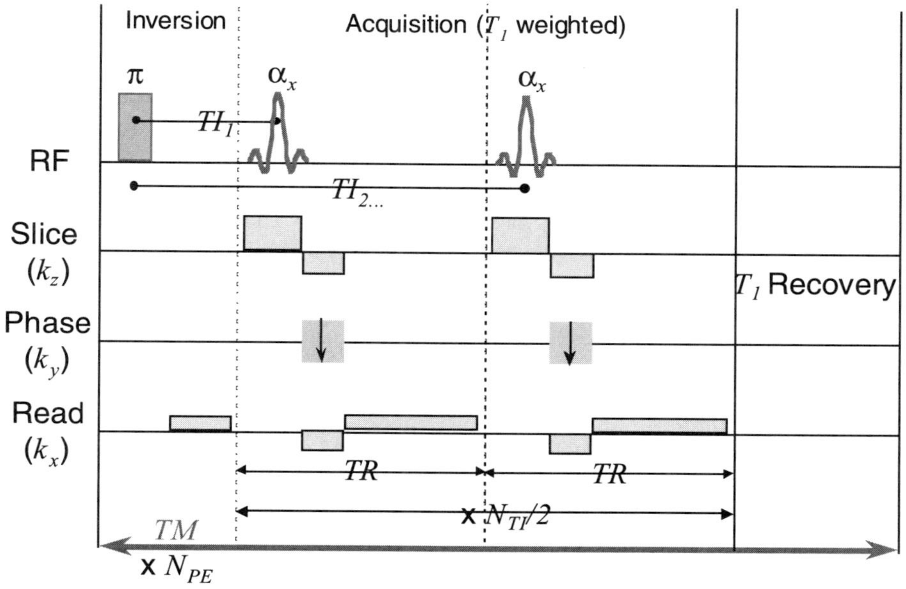

FIGURE 18.2 Diagram of a look-locker pulse sequence. A single repetition of this sequence produces all N_{TI} samples of the T_1 recovery.

40 min, clearly making it inappropriate for routine clinical use, let alone the dynamic imaging needed for kinetic modeling.

18.3.2.2.2 Look-Locker

If a method could be found that would allow for multiple samples of M_z in a single measurement period (*TM*), the total time needed to measure T_1 would be significantly reduced compared to IR. A number of such techniques based on snapshot fast low angle shot (FLASH),[20–23] inversion prepared echo-planar imaging (IR-EPI),[24] and stimulated echo imaging[25–27] have been developed. However, Crawley and Henkleman[28] showed that, in terms of noise per unit time, sequences based on the method of Look and Locker (hereafter LL)[29] were superior to any of the other fast T_1 mapping sequences. Indeed, for a given total imaging time T_1, maps derived from LL or IR sequences were shown to have the same ratio of error in the measured T_1 to the true T_1.

LL[29] showed that a train of small flip angle RF pulses applied to a sample with a constant inter-pulse interval (*TR*) will yield a series of signals which will be proportional to the longitudinal magnetization M_z. By accounting for the effects of the RF pulses on the evolution of M_z, it is possible to determine T_1 from the data acquired during a single pulse train.

A number of T_1 mapping methods are based on the LL method.[30–33] An example of these sequences is shown Figure 18.2. In these sequences, all N_{TI} samples of M_z are acquired in a single *TM*. One phase encode line for each of the N_{TI} images is acquired per *TM*, so the time needed to produce a single T_1 map would be $N_{PE} * TM$, reducing *TA* by a factor of N_{TI}(~5 min for LL vs. >40 min for IR). The reduced *TA* for LL T_1 measurements will reduce the SNR of the data and therefore increase the noise in a single LL T_1 map relative to an IR map. However, it has been shown that, given equal measurement times, there is little penalty for using LL instead of IR, since both sequences have the same ratio of error in the measured T_1 to the true T_1.[28]

The *TR* of the LL pulse trains can, in theory,* be very short (<5 ms), allowing for a large number of samples and wide temporal coverage of the recovery of M_z. This allows for the accurate determination

* In References 29 to 32, TR was never less than 40 ms. However, the state of the art has progressed since the LL imaging sequences were published. Currently, *TR*s less than 5 ms are feasible.[19]

of T_1 over a wide range of T_1s, and offers the possibility of quantifying multi-component T_1 relaxation. The reasonably short *TA*s for the LL sequences makes mapping of T_1 practical in a clinical setting; however, these sequences are still two orders of magnitude too long for dynamic imaging.

One recent innovation in LL T_1 measurement techniques, the LL-EPI sequence,[34] deserves special mention. Like the LL sequences mentioned above, the LL-EPI sequence acquires several samples of M_z during a single periodic train of RF pulses. However, instead of collecting a single phase encode line after each RF pulse, an EPI sequence is used to collect an entire image. Now, instead of having to repeat the RF pulse train N_{PE} times, all the images needed to determine T_1 could be collected in a single pulse train. This makes LL-EPI a true one-shot T_1 mapping technique with a *TA* of approximately 3 s, making it capable of dynamic imaging or of imaging rapidly moving objects.[35] However, the use of EPI to collect each sample of the T_1 relaxation curve limits the *TR* of the RF pulse train to greater than 50 ms. This restriction limits the maximum number of samples M_z that can be obtained and will limit the minimum T_1 that can be quantified accurately. Further, a 3 s *TA* time is not fast enough for cardiac applications, as the heart is stationary for mere fractions of a second.

18.3.2.3 Fast Acquisition Relaxation Mapping (FARM)

All of the T_1 mapping sequences discussed so far rely on determining T_1 from the signal intensity of a number of images produced from the samples of the relaxation curve. These images must be reconstructed using the 2D DFT, and so (from Equation 18.10) $t(k_x, k_y)$ must be a constant. If $t(k_x, k_y)$ changes as a function of frequency or phase encode position, $M(x, y)$ will vary in time and not equal the 2D DFT of $S(k_x, k_y)$. In this case, a 2D DFT of $S(k_x, k_y)$ will produce an image, but one with artifacts.

However, as long as $\lambda(x, y)$ in Equation 18.10 does not depend on k_x and k_y, then given enough data it is possible to solve for the variables in λ by exploiting how $t(k_x, k_y)$ interacts with $\lambda(x, y)$. This type of quantitative image reconstruction is more complicated, mathematically, than the standard 2D DFT, yet it affords greater acquisition flexibility, for example, increasing the temporal efficiency of quantitative imaging.

18.3.2.3.1 T_1 *FARM*

The principles of the T_1 FARM technique were discussed previously by Tong and Prato.[36] Briefly, this method involves two gradient echo *k*-space data acquisitions separated by a single inversion pulse. The first acquisition occurs at steady state during a train of α pulses; the second acquisition follows a 180° pulse and acquires data during the return to steady state. An iterative reconstruction algorithm combines the two *k*-space data sets to form a $1/T_1$ map. The following discusses the details of the data acquisition and reconstruction techniques used in T_1 FARM.

According to LL,[29] when using a train of RF pulses applied with a constant inter-pulse period of *TR*, M_z will evolve according to

$$M_z(nTR) = M_0\frac{1-e^{-TR/T_1}}{1-e^{-TR/T_1}\cos(\alpha)}(1-(e^{-TR/T_1}\cos(\alpha))^n) + M_1(e^{-TR/T_1}\cos(\alpha))^n \tag{18.15}$$

where M_0 is the undisturbed magnetization induced by the B_0 field, M_1 is the magnetization just after the first RF pulse in the train, n is the number of RF pulses applied since the beginning of the experiment, and α is the flip angle of the RF pulses. After many RF pulses ($n = \infty$), $M_z(nTR)$ will reach an equilibrium value M_{eq}:[36]

$$M_{eq} \equiv M_0\frac{1-e^{-TR/T_1}}{1-e^{-TR/T_1}\cos(\alpha)} \tag{18.16}$$

so Equation 18.15 simplifies to

$$M_z(\infty) = M_{eq} \tag{18.17}$$

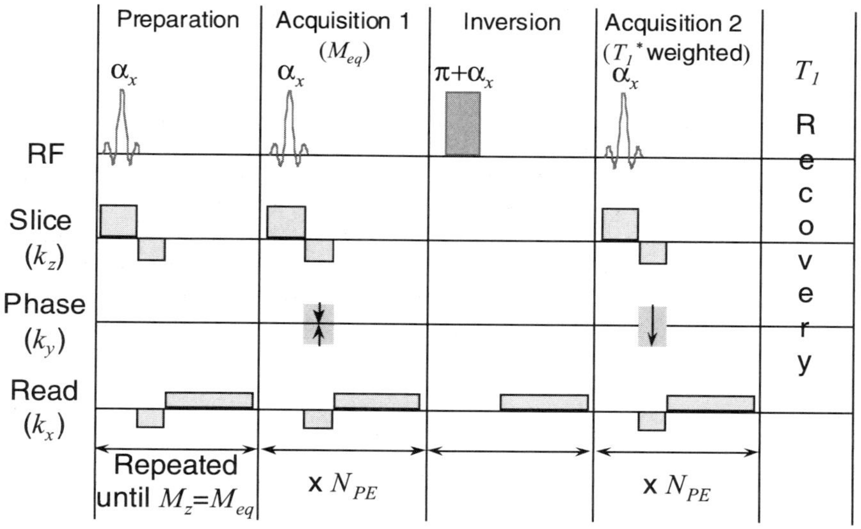

FIGURE 18.3 Schematic diagram of the T_1 FARM pulse sequence. The preparation was repeated until M_z had reached its steady state value of M_{eq}. A delay time with no gradient or RF activity was inserted between the end of the T_1 weighted measurement and the start of the next segment to allow M_z to recover toward M_{eq}.

Before data acquisition in T_1 FARM, the longitudinal magnetization is prepared by a series of rapidly applied (2 ms < TR < 10 ms) small flip angle (3° < α < 19°) RF pulses (Figure 18.3). The preparation pulses are repeated until n in Equation 18.15 is effectively infinity and the preparation has driven M_z to M_{eq}.

The preparation is immediately followed by acquisition of the first of two k-space data sets ($S_1(k_x, k_y)$) with a short gradient echo (GE) sequence (Figure 18.3). The gradients and RF pulses applied during data acquisition are identical to those used for preparation of M_z, except that now gradients are included in each TR period to spatially (Fourier) encode the data in the k_y (phase encode) direction. Each k_y line of data (phase encode line) is acquired during the GE caused by the gradients in the k_x (frequency encode) direction. The frequency encode gradient spatially encodes the data in the k_x direction.

One phase encode line (k_y) is collected after each α pulse. Due to the preparation, $M_z = M_{eq}$ for the entire first acquisition, so M_{xy} will be constant with respect to k_y in this acquisition with $M_{xy} = M_z \sin(\alpha)$. The echo time ($TE$) and time to acquire a frequency encode line (TF) can be made short enough (TE = 1.45 ms and TF = 0.8 ms)[37] that the effect of T_2^* signal decay can be neglected. Thus, we can also assume that M_{xy} is constant as a function of frequency encode position (k_x). This means that (for Equation 18.10) $\lambda(x, y) = M_{eq}(x, y)e^{i\phi(x,y)}\sin(\alpha)$ (ϕ, the complex phase of the magnetization, is needed to account for any phase offset in the experimental data) and $t(k_x, k_y)$ is a constant.

From Equation 18.10, $M(\lambda(x,y), t(k_x, k_y)) = M_{eq}(x,y)\, e^{i\phi(x,y)} \sin(\alpha)$, so S_1 (the M_{eq} data set) is described by

$$S_1(k_x, k_y) = \sum_y \left(\sum_x M_{eq}(x, y) e^{i\phi(x,y)} e^{-i\frac{2\pi}{N_f}k_x x} \right) e^{-i\frac{2\pi}{N_{PE}}k_y y} \tag{18.18}$$

where N_f is the number of frequency encodes and N_{PE} is the number of phase encodes. (The factor $\sin(\alpha)$ which comes from the relation $M_{xy} = M_z \sin(\alpha)$ has been absorbed into M_{eq} here.) Since there is no dependence of $M_{eq}(x, y)$ or $\phi(x, y)$ on k_x or k_y, we can determine those quantities by performing a 2D DFT on S_1.

Immediately after the collection of S_1, an RF pulse with flip angle 180 + α is applied, and a second k-space data set (S_2) is collected in an identical manner to S_1. The 180 + α pulse is equivalent to a 180°

inversion pulse followed by an α sample pulse. The α pulse is necessary to maintain the equilibrium set up by the preparation and M_{eq} acquisition. This near-inversion pulse that precedes the data acquisition means that $M_1 = -M_{eq}$ so Equation 18.15 becomes

$$M_z(nTR) = M_{eq}(1-(e^{-TR/T_1}\cos(\alpha))^n) - M_{eq}(e^{-TR/T_1}\cos(\alpha))^n \quad (18.19)$$

Note that n is now the number of RF pulses applied since the inversion pulse. After the inversion pulse, the longitudinal magnetization starts to recover toward equilibrium. Because S_2 is acquired with a train of α pulses, the recovery of the longitudinal magnetization is disturbed each time an RF pulse tips a portion of M_z into the transverse plane. This means that M_z does not recover at the rate T_1, but at an apparent relaxation rate of T_1^* (Figure 18.4). Note that M_z is recovering from $-M_{eq}$ to M_{eq} and not from $-M_0$ to M_0. The relationship between T_1^* and T_1, α, and TR is[32]

$$e^{-\frac{TR}{T_1^*}} = e^{-\frac{TR}{T_1}}\cos(\alpha) \quad (18.20)$$

If we substitute Equation 18.20 into Equation 18.19 we get

$$M_z(nTR) = M_{eq}\left(1-2e^{-nTR/T_1^*}\right) \quad (18.21)$$

This is similar to Equation 18.14, with T_1 replaced by T_1^* and M_0 replaced by M_{eq}. In Equation 18.14, the phase encode coordinate k_y is increasing with n, so $k_y = n - N_{PE}/2$. Note that M_z now contains time (nTR) explicitly, so for Equation 18.10 $t(k_x, k_y)$ is no longer a constant, but is equal to nTR, and $M(x, y)$ is a function of $M_{eq}(x, y)$, $\phi(x, y)$ and $T_1^*(x, y)$. S_2 (the T_1^* data set) is described by

$$S_2(k_x, k_y) = \sum_y\left(\sum_x M_{eq}(x, y)e^{i\phi(x, y)}\left(1-2e^{-nTR/T_1^*}\right)e^{-i\frac{2\pi}{N_f}k_x x}\right)e^{-i\frac{2\pi}{N_{PE}}k_y y} \quad (18.22)$$

There are three unknowns in S_2: $M_{eq}(x, y)$, $\phi(x, y)$, and $T_1^*(x, y)$. We determined M_{eq} and $\Delta\phi$ from S_1, so they can be substituted into Equation 18.22, leaving the single unknown T_1^*. Because $t(k_x, k_y)$ is not a constant, but depends on the phase encode position k_y, it is not possible to solve for T_1^* simply by taking a 2D DFT of S_2. However, $t(k_x, k_y)$ is constant in the frequency encode (k_x) direction so a one-dimensional (1D) DFT can be performed on S_2 (Equation 18.22) with respect to k_x to give

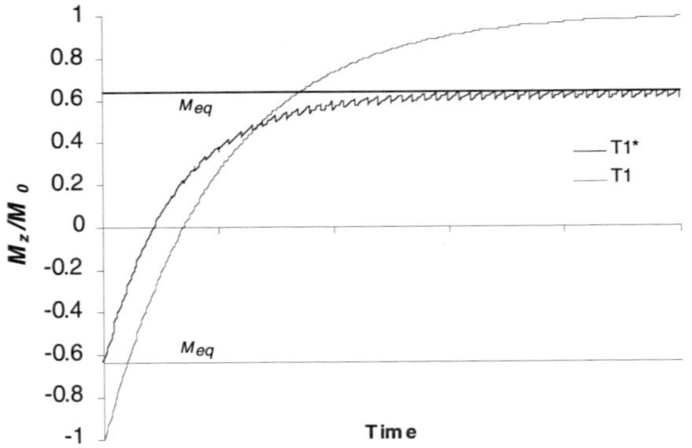

FIGURE 18.4 Plots of the recovery longitudinal magnetization M_z with (T_1^*) and without (T_1) the application of periodic α pulses.

$$s_2(x, k_y) = \sum_y M_{eq}(x, y)e^{i\phi(x, y)}\left(1 - 2e^{-nTR/T_1^*}\right)e^{-i\frac{2\pi}{N_{PE}}k_y y} \tag{18.23}$$

Each column x in Equation 18.23 is independent, so solutions can be found one column at a time. After substituting $\phi(m, n)$ and $M_{eq}(m, n)$ into Equation 18.23, the following function can be constructed:[38]

$$F(x) \equiv \sum_{k_y} |\Delta(x, k_y)|^2 \tag{18.24}$$

where Δ is the difference between the left (measured) and right (calculated) sides of Equation 18.23.

A single arbitrary initial value is assigned to each T_1^* in a column x and substituted into Equation 18.24. A nonlinear least squares minimization routine[38,39] varies $T_1^*(x)$ until $F(x)$ is minimized. Once Equation 18.24 has been minimized for all x a complete T_1^* map is constructed. At this point, assuming TR and α are known, Equation 18.20 can be used to convert the T_1^* map back to a T_1 map. Note, however, that α need not be known for contrast concentration determination. If we take the natural logarithms of Equation 18.20, we notice that R_1^* differs from R_1 by a constant:

$$\frac{1}{T_1^*} = \frac{1}{T_1} - \frac{\ln(\cos(\alpha))}{TR} \tag{18.25}$$

Thus, $\Delta R_1^* = \Delta R_1$ and Equation 18.2 can be rewritten as

$$[\text{Gd-DTPA}](t) = k\Delta R_1^*(t) \tag{18.26}$$

and the Gd-DTPA concentration-time curve can be determined directly from the R_1^* maps.

18.4 Application of T_1 Farm to Bolus Tracking

For an unconventional MRI approach like FARM to become accepted, it must be able to.

1. Provide R_1^* maps of sufficient spatial and temporal resolution.
2. Map R_1^* accurately and precisely over the range of values anticipated for the arterial input and tissues curves.
3. Exceed the functionality that would otherwise be achieved by conventional, i.e., FT, MRI methods.

18.4.1 Dynamic Range of T_1 FARM

Experiments were performed comparing the dynamic range of T_1 FARM and saturation recovery turboFLASH (SrTFL). As can be seen in Figure 18.5, the dynamic range of T_1 FARM exceeds that of SrTFL by a factor of 3, implying that if a bolus tracking application is limited to a lower than optimal dose of contrast agent, using T_1 FARM will allow a threefold increase in administered dose. Of course, a threefold increase, to be useful, must result in superior SNR in the arterial input and tissue curves. Although a recent publication suggests that this is true,[37] further research is needed. Significant extension of the dynamic range of T_1 FARM is dependent on two constraints: one physiological and one due to NMR properties.

Greater concentrations of extracellular agents, such as Gd-DTPA, will result in even shorter T_1s. This, in turn, will require faster sampling (i.e., faster acquisition times). Previous work[40] suggests that *TR*s as short as 1 ms would still provide sufficient SNR in T_1 FARM acquisition. This will require, however, significant increases in gradient slew rates, perhaps those approaching nerve stimulation limits.[41] Alternatively, rather than reducing TR it should be possible to combine FARM with parallel MRI techniques such as SMASH,[42] which reduce the number of gradient phase encode steps required to produce an

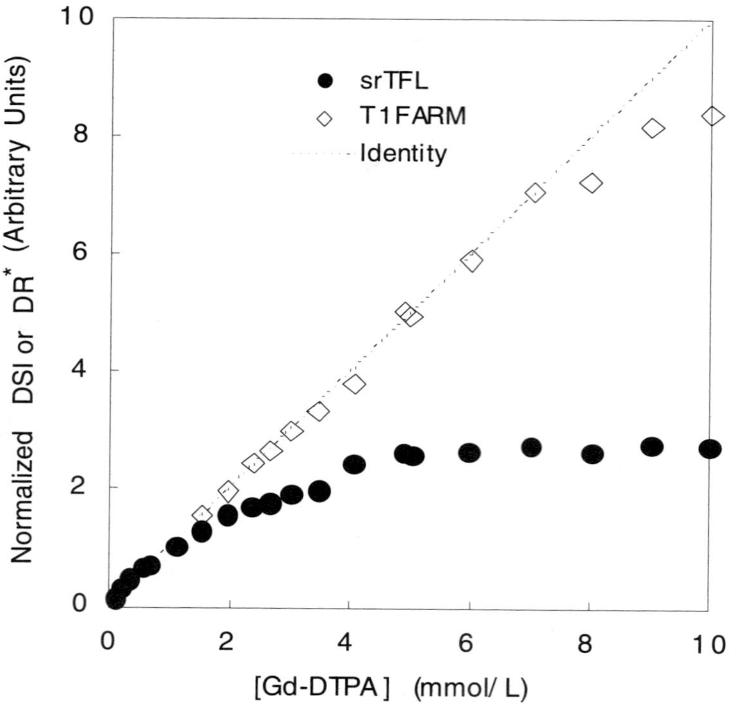

FIGURE 18.5 Comparison of normalized signal changes in SrTFL or T_1 FARM vs. concentration of Gd-DTPA. ΔSI and $\Delta R_1{}^*$ were normalized to their values at 1.15 mmol/l.

image. Indeed, it has been demonstrated that FARM becomes more precise at high contrast concentrations as the number of phase encode steps is reduced.[40]

The SNR in MRI images varies as the square root of the total imaging time. Whether faster imaging is achieved through increased demands on the gradient hardware or through application of parallel imaging techniques, the SNR of the acquired data will necessarily decrease. This decrease in SNR will set a fundamental limit on the maximum possible temporal resolution of any sequence used for bolus tracking. We have done some theoretical calculations, similar to the work of Crawley and Henkelman,[28] which show that given a fixed amount of measurement time, T_1 FARM will produce T_1 maps with twice the SNR of any other currently available T_1 mapping technique. Therefore, due to its superior efficiency, T_1 FARM has better intrinsic SNR than other sequences and thus is more amenable to rapid imaging as it can better tolerate the reduction in SNR inherent in the faster imaging.

In addition to increasing the constraints on pulse sequence design, greater Gd-DTPA concentrations will probably result in a breakdown of fast exchange assumptions (i.e., the validity of Equation 18.2). Consider, for instance, that the water within a given tissue exists in many different chemical environments or compartments. In these different environments (intravascular extracellular space, extravascular extracellular space, etc.), water molecules will give rise to different relaxation times (T_1 and T_2). The MRI signal of the entire tissue, however, most often decays with a single exponential relaxation time. This is due to the rapid motion of the water molecules between these physical compartments. Due to this motion, the water molecules "sample" the relaxation times of other compartments, resulting in an average relaxation time, one that is weighted by the volumes of the compartments. This process is called chemical exchange, and this motional limit is the fast exchange regime.

When a contrast agent is introduced, the relaxation rates within the compartments are increased, and this situation becomes more complicated. Woessner gives an excellent review of this phenomenon.[43] The relaxation rate may be increased or, equivalently, the relaxation time T_1 may be decreased to the degree that the exchange rate is not considered rapid on a T_1 time scale. In other words, the time that the water molecule resides in a compartment is now long enough so that the water protons can be fully relaxed, and they do not sample the chemical environment of the after compartments. This motional regime is the slow exchange limit.

In the slow and the intermediate exchange regimes, the relaxation behavior of the entire sample can no longer be adequately represented by a single exponential relaxation time, but rather is described by a weighted sum of the contributions originating from within each of the compartments.[14] In this context, the breakdown of Equation 18.2 can be made clear. If the fast exchange approximation is no longer a valid assumption, then the notion of a single relaxation time for the tissue is no longer valid. Clearly, what is needed in this case is a treatment of the individual compartments and a determination of how the relaxation times change within each compartment by using Equation 18.2.

Work is currently under way in which a tissue (myocardium) is divided up into many compartments in a form similar to the tissue homogeneity model.[9] Using a tracer kinetic modeling approach, the concentration of a bolus injection of Gd-DTPA will be tracked through the tissue.[44] The resulting contrast enhanced relaxation times are calculated for each compartment, and a theoretical MRI signal, specific to two pulse sequences used for bolus tracking, SrTFL and T_1 FARM, will then be produced which can correctly account for the chemical exchange.

Further calculations and experiments are needed to determine the error introduced by the assumptions implicit in Equation 18.2 before we know the limitations imposed by this effect. These issues are important because increased dynamic range would allow greater doses of contrast agent to be given, resulting in improved SNR in tissue curves. Better tissue curves would be very welcome and, in fact, they are crucial to the precision of blood flow measurements.[45]

18.4.2 Cardiac T_1 FARM Concentration-Time Curves

As indicated in Section 18.1, a major goal is to measure functional parameters such as blood flow in the left ventricular heart muscle, i.e., left myocardium. Figure 18.6 shows a left myocardial concentration-time curve that was generated with T_1 FARM. The temporal resolution was one map every 1.5 s with the following pulse sequence parameters: TR = 2.3 ms, TE = 1.45 ms, α = 10°, 64 phase encode steps, slice

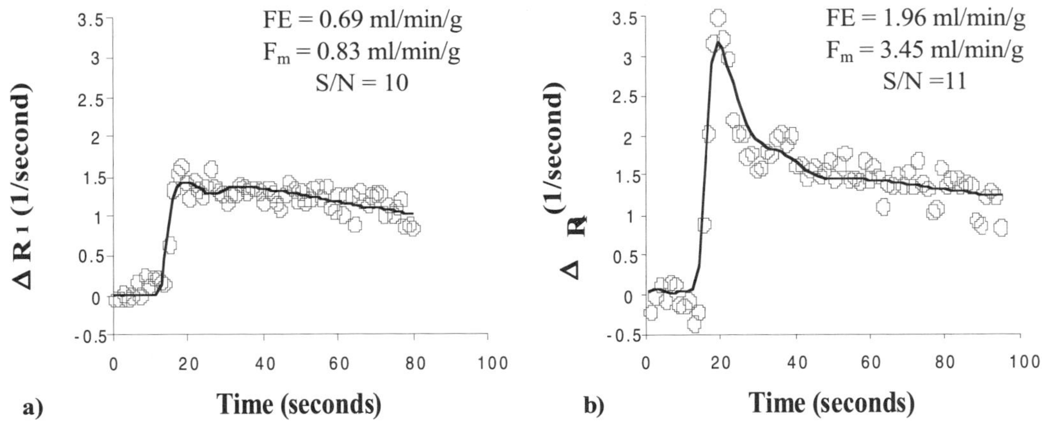

FIGURE 18.6 Sample ΔR_1-time curves. The T_1 FARM data were acquired (a) at rest and (b) during pharmacological stress for a region of interest in the left ventricular heart muscle. The signal to noise (S/N) was (a) 10 and (b) 11. The solid line was estimated by fitting the data to the Modified Kety model (Equation 18.27), a two-parameter tracer kinetic model. The myocardial blood flow extraction fraction (FE) product estimated using the model was (a) 0.69 and (b) 1.96 ml/min/g. The actual myocardial blood flow (F_m) determined using radioactive microspheres was (a) 0.83 and (b) 3.45 ml/min/g.

thickness 10 mm, 150 volume selective (50 mm slice thickness) preparation pulses, total acquisition time of 300 ms. These data were fit with the Modified Kety model, a two-parameter tracer kinetic model:

$$[\text{Gd-DTPA}]_t(t) = FE\int_0^t [\text{Gd-DTPA}]_b(\tau)e^{-\frac{FE}{\lambda}(t-\tau)}d\tau \qquad (18.27)$$

where $[\text{Gd-DTPA}]_t(t)$ is the concentration of Gd-DTPA in the tissue, F is regional blood flow (ml/min/g), E is the extraction efficiency (unitless; E is the fraction of tracer that diffuses from blood into tissue during a single pass through the tissue), $[\text{Gd-DTPA}]_b$ is the concentration of Gd-DTPA in the blood (input function), and λ is the "partition coefficient" of Gd-DTPA (ml/g; equivalent to V_b/[1-Hematocrit]). Note that blood flow values, obtained by these fits, were comparable to actual values (determined invasively). The average SNR determined from a number of myocardial tissue curves was 7:1.[45]

18.4.3 Bolus Tracking with Other Imaging Methods

Until the development of T_1 FARM, T_1 measurements required MR acquisition times that were prohibitively long: concentration-time curves, for the heart, demand rapid image acquisition times (i.e., high temporal resolution). In fact, before the advent of LL-EPI (see Section 18.3.2.2.2), *no* organ could be assessed adequately with MR bolus tracking. Responding to the demands of bolus tracking, fast-GE sequences (e.g., turboFLASH) were introduced. Today, the most common sequence used in bolus tracking is the inversion recovery or saturation recovery turboFLASH sequence (IrTFL or SrTFL). In the case of SrTFL, it is assumed that the signal intensity is proportional to R_1. This approach has suffered from two severe limitations: (1) the linear relationship between signal intensity and R_1 breaks down for moderate concentrations of contrast agents; and (2) as M_{xy} is changing during the image acquisition, the 2D DFT reconstructed images have considerable artifacts; therefore, spatial resolution may be limited.[36] Note that other, recently developed sequences such as fast imaging with steady processing (FISP) will suffer from similar limitations.[46]

18.4.4 Future Enhancements and Modifications of T_1 FARM

The performance of T_1 FARM in cardiac bolus tracking bodes well for the technique. Functional imaging of the myocardium is the greatest challenge facing functional MRI. In large part, this difficulty is owed to the temporal limitations imposed by the cardiac cycle. The heart cycle averages 800 ms in duration in humans. If acquisition time exceeds 100 ms at end diastole or 50 ms at other times, blurred images will result. In addition, since the T_1 FARM acquisition is actually two acquisitions (M_{eq} and R_1 weighted data sets), its temporal resolution will be less than that of sequences such as SrTFL by approximately a factor of 2. Further, the dynamic range of T_1 FARM, although greater than that of conventional imaging methods, could still be improved to allow even greater injected doses to improve the tissue SNR curves. Finally, the FARM sequences we have developed are single slice, whereas multi-slice acquisitions are needed to cover the heart properly. Simulations have shown that the deviation of the slice profile from a rectangular function reduces the SNR by as much as a factor of 2.

In summary, the T_1 FARM method could be improved by

1. Reducing acquisition times, thereby improving temporal resolution.
2. Improving slice selection to improve SNR and allow for multi-slice acquisition.

The reduction in acquisition times may very well allow greater doses to be injected because the short T_1^* recoveries can be better sampled.

We have been concentrating on the changes in MRI signal intensity associated with R_1 produced by extracellular contrast agents. The FARM method is a general approach which can be applied to other NMR parameters such as T_2 and T_2^* as indicated by the generality of Equation 18.10 and a recent publication on T_2 FARM.[47] In the future, it may be possible to use the change in T_2^* to follow intravascular contrast agents such as Gd-DTPA-albumin or Gadomer 17. Additional problems will have to be addressed, however; specifically, the relationship of contrast agent concentration and the change in T_2^* or T_2, as no simple equations (analogous to Equation 18.2) have been validated for intravascular contrast agents.[48]

18.5 Summary

There is a tremendous clinical need to produce functional maps (e.g., blood flow maps of the heart muscle) with spatial resolution in excess of that afforded by nuclear medicine methods. With the recent advent of multi-slice X-ray CT, it is anticipated that cardiac functional imaging will be attempted using this method. Even if successful, however, the application will be limited by concerns over the relatively high X-ray doses which are needed to produce concentration-time curves with sufficient temporal resolution (approximately 1 s).[15] Attempts to use MRI have been limited to conventional imaging approaches, in which these maps are calculated from conventional images formed using FT reconstruction methods. When reconstruction must be performed using FT, severe constraints are put on the nature of data acquisition. For example, in conventional MR, the magnetization in the x,y plane must be kept constant during the acquisition for the entire k_y data set. The successful use of T_1 FARM to generate concentration-time curves of the myocardium shows, however, that if reconstruction is not limited to FT methods, the evolution of NMR relaxation can be allowed during the acquisition of a single data set. The modest success of a FARM method by a single MR group has opened a window to new ways to image function with MR. With the improvement in computer power, which allows realistic non-Fourier reconstruction times, it is now time to aggressively pursue this alternative form of MRI.

Acknowledgments

Over the T_1 FARM development years, the Canadian Medical Research Council has supported this work, in part. The authors would like to thank Mr. John Parr for assistance in the preparation of this manuscript, Ms. Deanna Bellamy for providing the blood myocardial concentration-time curves, Mrs. Jane Sykes for assistance in animal experiments, and Dr. Dick J. Drost for technical advice.

References

1. Nishimura, D.G., *Principles of Magnetic Resonance Imaging,* Department Electrical Engineering, Stanford University, Stanford, CA, 1996.
2. Ogawa, S., Lee, T.M., Kay, A.R., and Tank, D.W., Brain magnetic resonance imaging with contrast dependent on blood oxygenation, *Proc. Natl. Acad. Sci. U.S.A.*, 87(24), 9868, 1990.
3. Buxton, R.B. and Frank, L.R., A model for the coupling between cerebral blood flow and oxygen metabolism during neural stimulation, *J. Cereb. Blood Flow Metab.,* 17, 64, 1997.
4. Weinmann, H.J., Laniado, M., and Mutzel, W., Pharmacokinetics of Gd-DTPA/dimeglumine after intravenous injection into healthy volunteers, *Physiol. Chem. Phys. Med. NMR,* 16, 167, 1984.
5. Judd, R.M., Reeder, S.B., and May-Newman, K., Effects of water exchange on the measurement of myocardial perfusion using paramagnetic contrast agents, *Magn. Reson. Med.,* 41, 334, 1999.
6. Demsar, F., Shames, D.M., Roberts, T.P.L., Stiskal, M., Roberts, H.C., and Brasch, R. C., Kinetics of MRI contrasts agents with a size ranging between Gd-DTPA and albumin-Gd-DTPA: use of cascade Gd-DTPA-24 polymer, *Electro-Magnetobiol.,* 17, 283, 1998.
7. Adam, G., Muhler, A., Spuntrup, E., Neuerburg, J.M., Kilbinger, M., Bauer, H., Fucezi, L., Kupper, W., and Gunther, R.W., Differentiation of spontaneous canine breast tumors using dynamic magnetic resonance imaging with 24-gadolonium-DTPA-cascade-polymer, a new blood-pool agent: preliminary experience, *Invest. Radiol.,* 31, 267, 1996.
8. Bassingthwaighte, J.B. and Holloway, G.A., Estimation of blood flow with radioactive tracers, *Semin. Nucl. Med.,* 6(2), 141, 1976.

9. St. Lawrence, K.S. and Lee, T.Y., An adiabatic approximation to the tissue homogeneity model for water exchange in the brain. I. Theoretical derivation, *J. Cereb. Blood Flow Metab.*, 18(12), 1365, 1998.
10. Canty, J.M. and Fallavollita, J.A., Resting myocardial flow in hibernating myocardium: validating animal models of human pathophysiology, *Am. J. Physiol.*, 277(1,2), H417, 1999.
11. Wilson, R.F., Marcus, M.L., and White, C.W., Prediction of the physiologic significance of coronary arterial lesions by quantitative lesion geometry in patients with limited coronary artery disease, *Circulation,* 75(4), 723, 1987.
12. Schmiedl, U., Ogan, M., Paajanen, H., Marotti, M., Crooks, L.E., Brito, A.C., and Brasch, R.C., Albumin labeled with Gd-DTPA as an intravascular, blood pool-enhancing agent for MR imaging: biodistribution and imaging studies, *Radiology,* 162, 205, 1987.
13. Donohue, K.M., Burstein, D., Manning, W.J., and Gray, M.L., Studies of Gd-DTPA relaxivity and proton exchange rates in tissue, *Magn. Reson. Med.*, 32, 66, 1994.
14. Donohue, K.M., Wisskoff, R.M., and Burstein, D., Water diffusion and exchange as they influence contrast enhancement, *J. Magn. Reson. Imaging,* 7, 102, 1997.
15. Henderson, E., Rutt, B.K., and Lee, T.Y., Temporal sampling requirement for the tracer kinetic modeling of breast tissue, *Magn. Reson. Med.*, 16, 1057, 1998.
16. Tong, C.Y., Prato, F.S., Wisenberg, G., Lee, T.Y., Carroll, E., Sandler, D., and Wills, J., Techniques for the measurement of the local myocardial extraction efficiency for inert diffusible contrast agents such as gadopentate dimeglumine, *Magn. Reson. Med.*, 30(3), 332, 1993.
17. Bloch, F., Nuclear induction, *Phys. Rev.*, 70, 460, 1946.
18. Fukushima, E. and Roeder, S.B.W., *Experimental Pulse NMR: A Nuts and Bolts Approach,* Addison-Wesley, Reading, MA, 1981.
19. Bottomley, P.A., Foster, T.H., Argersinger, R.E., and Pfeifer, L.M., A review of normal tissue hydrogen NMR relaxation times and relaxation mechanisms from 1–100 MHz: dependence on tissue type, NMR frequency, temperature, species, excision, and age, *Med. Phys.*, 11, 425, 1984.
20. Haase, A., Snapshot FLASH MRI. Applications to T_1, T_2, and chemical shift imaging, *Magn. Reson. Med.*, 13, 77, 1990.
21. Nekolla, S., Gneiting, T., Syha, J., Deichmann, R., and Haase, A., T_1 Maps by *k*-space reduced snapshot-FLASH MRI, *J. Comput. Assist. Tomogr.*, 16(2), 327, 1992.
22. Deichmann, R. and Haase, A., Quantification of T_1 values by snapshot-FLASH NMR imaging, *J. Magn. Reson.*, 96, 608, 1992.
23. Blüml, S., Schad, L.R., Stepanow, B., and Lorenz, W.J., Spin-lattice relaxation time measurement by means of a turboFLASH technique, *Magn. Reson. Med.*, 30, 289, 1993.
24. Ordidge, R.J., Gibbs, P., Chapman, B., Stehling, M.K., and Mansfield, P., High-speed multislice T_1 mapping using inversion-recovery echo-planar imaging, *Magn. Reson. Med.*, 16, 238, 1990.
25. Haase, A. and Frahm, J., NMR imaging of spin-lattice relaxation using stimulated echoes, *J. Magn. Reson.*, 65, 481, 1985.
26. Mareci, T.H., Sattin, W., Scott, K.N., and Bax, A., Tip-angle reduced T_1 imaging, *J. Magn. Reson.*, 67, 55, 1986.
27. Franconi, F., Seguin, F., Sonier, C.B., Le Pape, A., and Akoka, S., T_1 mapping from spin echo and stimulated echoes, *Med. Phys.*, 22, 1763, 1995.
28. Crawley, A.P. and Henkelman, R.M., A comparison of one-shot and recovery methods in T_1 imaging, *Magn. Reson. Med.*, 7, 23, 1988.
29. Look, D.C. and Locker, D.R., Time saving in measurement of NMR and EPR relaxation time, *Rev. Sci. Instrum.*, 41, 250, 1970.
30. Hinson, W.H. and Sobol, W.T., A new method of computing spin-lattice relaxation maps in magnetic resonance imaging using fast scanning protocols, *Med. Phys.*, 15(4), 551, 1988.
31. Brix, G., Schad, L.R., Deimling, M., and Lorenz, W.J., Fast and precise T_1 imaging using a TOMROP sequence, *Magn. Reson. Imaging,* 8, 351, 1990.

32. Kay, I. and Henkelman, R.M., Practical implementation and optimization of one-shot T_1 imaging, *Magn. Reson. Med.,* 22, 414, 1991.
33. Gowland, P. and Leach, M.O., Fast and accurate measurements of T_1 using a multi-readout single inversion recovery sequence, *Magn. Reson. Med.,* 26, 79, 1992.
34. Gowland, P. and Mansfield, P., Accurate measurement of T_1 *in vivo* in less than 3 seconds using echo-planar imaging, *Magn. Reson. Med.,* 30, 351, 1993.
35. Freeman, A.J., Gowland, P.A., and Mansfield, P., Optimization of the ultrafast look-locker echo-planar imaging T_1 mapping sequence, *Magn. Reson. Imaging,* 16, 765, 1998.
36. Tong, C.Y. and Prato, F.S., A novel fast T_1 mapping method, *J. Magn. Reson. Imaging,* 4, 701, 1994.
37. McKenzie, C.A., Pereira, R.S., Prato, F.S., Chen, Z., and Drost, D.J., Improved contrast agent bolus tracking using T_1 FARM, *Magn. Reson. Med.,* 41, 429, 1999.
38. Chen, Z., Prato, F.S., and McKenzie, C.A., T_1 fast acquisition relaxation mapping (T_1 FARM): an optimized reconstruction, *IEEE Trans. Med. Imaging,* 17, 155, 1998.
39. Marquardt, D., An algorithm for least squares estimation of nonlinear parameters, *SIAM J. Appl. Math.,* 11, 431, 1963.
40. McKenzie, C.A., Prato, F.S., Thornhill, R.T., and Drost, D.J., T_1 fast acquisition relaxation mapping (T_1 FARM): optimised data acquisition, *Magn. Reson. Imaging,* 18, 129, 2000.
41. Reilly, J.P., Peripheral nerve stimulation by induced electric currents: exposure to time-varying magnetic fields, *Med. Biol. Eng. Comput.,* 27(2), 101, 1989.
42. Sodickson, D.K. and Manning, W.J., Simultaneous acquisition of spatial harmonics (SMASH): fast imaging with radiofrequency coil arrays, *Magn. Reson. Med.,* 38, 591, 1997.
43. Woessner, D.E., Brownian motion and its effects in NMR chemical exchange and relaxation in liquids, *Concepts Magn. Reson.,* 8(6), 397, 1996.
44. Moran, G.R. and Prato, F.S., An estimate of the error introduced by ^{1}H exchange in bolus tracking, in *Programs and Abstracts from the 8th Scientific Meeting and Exhibition of the ISMRM,* Denver, CO, 2000.
45. Bellamy, D., Pereira, R.S., McKenzie, C., Prato, F.S., Sykes, J., and Wisenberg, G., Estimation of myocardial blood flow using tracer kinetic modeling and a fast T_1-mapping method, in *Programs and Abstracts from the 8th Scientific Meeting and Exhibition of the ISMRM,* Denver, CO, 2000.
46. Jerosch-Herold, M., Huang, H., and Wilke, N., Magnetization prepared TrueFISP myocardial perfusion imaging, in *Programs and Abstracts from the 7th Scientific Meeting and Exhibition of the ISMRM,* Philadelphia, PA, 1999.
47. McKenzie, C.A., Chen, Z., Drost, D.J., and Prato, F.S., Fast acquisition of quantitative T2 maps, *Magn. Reson. Med.,* 41(1), 208, 1999.
48. Tian, G., Shen, J.F., Dai, G., Sun, J., Xiang, B., Luo, Z., Somorjai, R., and Deslauriers, R., An interleaved T_1-T_2^* imaging sequence for assessing myocardial injury, *J. Card. Magn. Reson.,* 1(2), 145, 1999.

19

Medical Image Registration and Fusion Techniques: A Review

George K. Matsopoulos
National Technical University of Athens

Konstantinos K. Delibasis
National Technical University of Athens

Nikolaos A. Mouravliansky
National Technical University of Athens

19.1 Introduction

Registration between two- (2-D) or three-dimensional (3-D) images is a common problem encountered when more than one image of the same anatomical structure is obtained, but taken at different times, either when using different imagery or when performing dynamic studies. In all cases, the information present in the different images must be combined to produce fused or parametric images. For efficient fusion of information, the different images have to be registered. Several registration and fusion techniques have been proposed in the literature, with varying degrees of user intervention, and the most representative will be discussed in this chapter.

The chapter is divided in two main sections: medical image registration and medical image fusion. The medical image registration section is concerned with the introduction of a generic medical image registration scheme based on global transformations. This scheme consists of the identification of common features in the images to be registered, the definition of geometrical transformation, the selection of an appropriate measure of match between the reference and transformed image, and the application of various optimization techniques for the determination of the transformation parameters with respect to the measure of match. A detailed description of an elastic deformation model combined with the proposed generic registration scheme will be described in detail to refine the accuracy of the registration results. In the application section, qualitative and quantitative results will be presented for registering retinal images (2-D case) and computed tomography (CT)-magnetic resonance (MR) human heads (3-D case). In both cases, the superiority of the automatic registration methods against the manual one will be proven.

The medical image fusion section is concerned with the introduction of various techniques commonly used to combine information from different modalities after the application of the medical image registration process. These techniques will be divided into two main categories: fusion of information at

0-8493-3691-0/01/$0.00+$.50

the image level and at the object level. Finally, results of the information fusion will be qualitatively assessed from various modalities.

19.2 Medical Image Registration

19.2.1 Literature Review

The process of image registration can be formulated as a problem of optimizing a function that quantifies the match between the original (reference) and the transformed image. Several image features have been used for the matching process, depending on the modalities used, the specific application, and the implementation of the transformation. Comprehensive surveys of medical image registration can be found in References 1 and 2 in terms of image modalities and employed techniques. According to the matching features, the medical image registration process can be divided into three main categories: point-based, surface-based, and volume-based methods.

Point-based registration involves the determination of the coordinates of corresponding points in different images and the estimation of geometrical transformation using these corresponding points. Such points can be defined using external markers placed on the patient's skin before the acquisition,[3–4] stereotactic frames,[5–6] and landmarks.[8] External markers have been used to register single photon emission computed tomography (SPECT)-MR images with the affine transformation method[3] and CT-MR data with 3-D global transformations.[7] Landmarks, as another point-based approach, were placed in the images by experts by means of software in order to equivalently define and register anatomical areas.[8–10] A landmark-based method to register CT-MR images, using Singular Value Decomposition, is described in Reference 8. A more advanced deformation based on the Thin Plate Splines interpolation model is reported in Reference 11. The low resolution along the transverse axis, the small number of corresponding markers, as well as possible inaccuracies in their placement during the acquisition from each modality, render these methods as manual, resulting in inaccuracies and inconsistencies.

Surface-based registration involves the determination of the surfaces of the images to be matched and the minimization of a distance measure between these corresponding surfaces. The surfaces are generally represented as a set of large number of points obtained by segmenting contours in contiguous image slices. There may be a requirement for the points to be triangulated. The difference between point- and surface-based registration algorithms is that point correspondence is defined by the user for the former, whereas it is automatic for the latter. Pelizzari et al.[12] introduced the idea of using surfaces to register brain images by obtaining a rigid body transformation, which when applied to "hat" coordinates (points that belong to the skin surface of the scan with the lower resolution) minimizes a residual that is the mean square distance between hat points and "head" surface (a stack of skin contours from the higher resolution scan) using an optimization technique described by Press et al.[13] The Euclidean distance between a point of an image and the closest surface point is used in Reference 14 as the closest point projection rule, whereas the integer approximations of the Euclidean distance as well as its highly computation cost required were improved by using the well-known chamfer distance transform.[15] This method was then applied to medical image registration.[16–17] Besl and McKay[18] presented a general purpose registration technique called the "Iterative Closest Point method," which was extended and implemented toward medical applications in References 19 and 20. A method designed to register preoperative CT images to cerebral surface points acquired intraoperatively from ultrasound images was presented in Reference 21, along with qualitative results and accuracy comparisons using marker-based methods.

Volume-based registration involves the optimization of a quantity measuring the similarity of all geometrically corresponding voxel pairs, considering some predefined features. Multiple volume-based algorithms have been proposed in Reference 22 and 23, optimizing a measure of the absolute difference between image intensities of corresponding voxels within overlapping parts in a region of interest. These methods were based on the assumption that the two images are linearly correlated, which is not the general case. Cross-correlation of feature images derived from the original image data has been applied to CT-MR modeling using geometrical features, such as edges[24] and ridges,[25,26] or using specially designed intensity transformations.[27] Misregistration was measured by the dispersion of the 2-D histogram of the image intensities of corresponding voxel pairs, which was assumed to be minimal in the registered position. The criterion of Studholme et al.[28] required segmentation of the images or delineation of specific histogram regions, while the criterion of Woods et al.[29] was based on an additional assumption concerning relationships between the gray values in the different modalities to reduce the complexity. In References 30 and in an extended work,[31] the mutual information (MI) registration criterion was introduced, measuring the statistical dependence between two random variables of the amount of information that one variable contains about the other. The MI of the image intensity values of corresponding voxel pairs was maximal if the images were geometrically aligned. A comparative study between surface- and volume-based registration algorithms was recently published,[32] indicating that the volume-based techniques seemed to give more accurate and reliable results when the CT-MR images were registered and slightly more accurate results for the positron emission tomography (PET)-MR images. The reason lies with the fact that surface-based registration methods require well-defined corresponding surfaces prior to registration.

Most of the aforementioned registration algorithms use rigid transformations to account for global misalignment. When merging brain image data or matching anatomical representations such as atlases, elastic matching techniques seem most appropriate to record local shape differences.[33,34] Bookstein[35] used a user-defined set of landmarks and the Thin Plate Splines interpolation model to interpolate the displacement function over the whole image. Approaches based on deformation models include similarity-based methods, polynomial transformations, and contour matching. An elastic deformation method,[36] based on the cross-correlation coefficients, was applied iteratively to CT images until a match was made with a predefined atlas model. Low degree polynomial transformations applied to register MR images caused a polynomial warping,[37] whereas 2-D CT-MR chest images were registered using the dynamic elastic method which matched corresponding contours and led to a global translation vector for coarse alignment and local residual shifts along the contours.[38] Furthermore, a contour registration technique was presented, consisting of a boundary homothetic mapping together with an elastic shape deformation model based on active models.[39] The advantageous implementations of the elastic deformation models avoid the application of the geometrical distortion correction operation, due to the imaging modality required prior to registration. Several imaging techniques, such as the MRI system, cause local geometrical distortions in their field of view. In Reference 32, where surface- and volume-based registration methods were evaluated, MR images were corrected from static field inhomogeneities using various techniques.[40,41]

Current trends on multimodal medical image registration include the improvement of the registration accuracy as well as the implementation of various registration schemes involving approaches from different registration process categories (point-, surface-, and volume-based methods). Fitzpatrick et al.[42] derived approximation expressions for the "Target Registration Error" (TRE) and for the expected squared alignment error of an individual fiducial marker. They found that the expected registration accuracy is worse when the fiducial points are closely aligned, thus providing the surgeons with the appropriate guidance in placing the fiducial markers before surgery and increasing the accuracy of point-based guidance systems. Also, it was revealed in Reference 42 that fiducial alignment alone should not be trusted as an indicator of registration success. The efficiency of a weighted geometrical feature algorithm was demonstrated by registering CT and MR volume head images using fiducial points, surfaces, and various weighted combinations of points and surfaces, according to References 20 and 43. In Reference 44, a combined registration scheme based on the use of image features with the affine transformation and an elastic deformation model was applied on various 2-D and 3-D head data, improving the registration performance by obtaining global and local brain deformations.

19.2.2 Generic Medical Image Registration Scheme

A generic medical image registration scheme may consist of the selection of the image features that will be used during the matching process, the definition of a measure of match (MOM) that quantifies the spatial matching between the reference and the transformed image, and the application of an

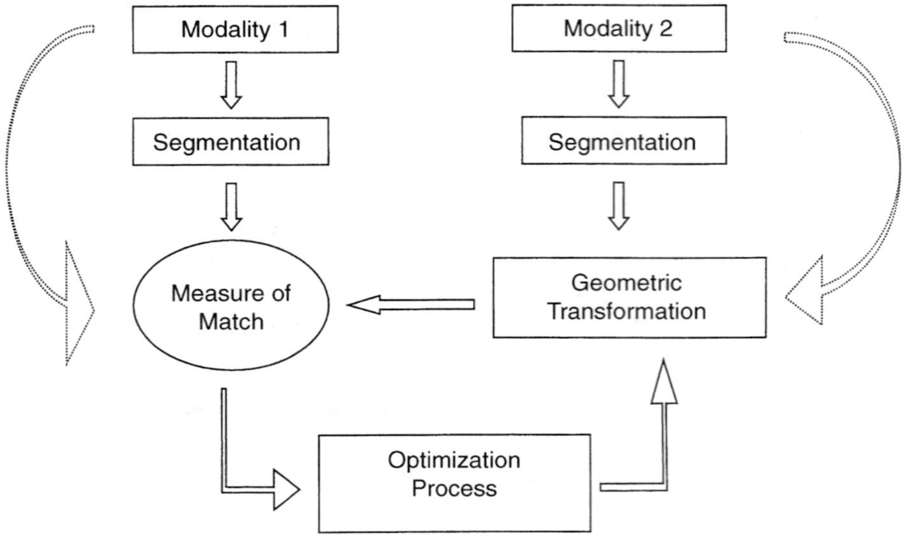

FIGURE 19.1 Generic medical image registration scheme.

optimization technique that determines the independent parameters of the transformation model employed, according to the MOM. The proposed generic medical image registration scheme is shown in the Figure 19.1.

19.2.2.1 Transformation Models

Several transformation models may be employed for the global medical image registration.[1] The most representative include the affine, the bilinear, the rigid, and the projective transformations.

1. *The affine transformation* can be decomposed into a linear transformation and a simple translation. It can be shown that this transformation maps straight lines into straight lines, whereas it preserves parallelism between lines. In the 3-D case, it can be mathematically expressed as follows:

$$\begin{pmatrix} x' \\ y' \\ z' \end{pmatrix} = \begin{pmatrix} a_1 & a_2 & a_3 \\ b_1 & b_2 & b_3 \\ c_1 & c_2 & c_3 \end{pmatrix} \begin{pmatrix} x \\ y \\ z \end{pmatrix} + \begin{pmatrix} dx \\ dy \\ dz \end{pmatrix} \tag{19.1}$$

   The affine transformation is completely defined by nine independent parameters (a_i, b_i, c_i) for $i = 1, 2, 3$ and *dx*, *dy*, and *dz*, whereas, in the 2-D case, the affine transformation is defined by six independent parameters.
2. *The bilinear transformation* is the simplest polynomial transformation, which maps straight lines from the original image to curves. It can be expressed for the 3-D case as

$$\begin{aligned} x' &= \alpha_0 + \alpha_1 x + \alpha_2 y + \alpha_3 z + \alpha_4 xy + \alpha_5 yz + \alpha_6 zx + \alpha_7 xyz \\ y' &= b_0 + b_1 x + b_2 y + b_3 z + b_4 xy + b_5 yz + b_6 zx + b_7 xyz \\ z' &= c_0 + c_1 x + c_2 y + c_3 z + c_4 xy + c_5 yz + c_6 zx + c_7 xyz \end{aligned} \tag{19.2}$$

   The bilinear transformation is completely defined by 24 independent parameters (a_i, b_i, c_i) for $i = 0, 1, 2, \ldots, 7$. For the 2-D case, eight independent parameters completely define the bilinear transformation.

3. *The rigid transformation* preserves the distance between any two points in the transformed image. The mathematical expression of the transformation for the 3-D case is given by

$$\begin{pmatrix} x' \\ y' \\ z' \end{pmatrix} = \begin{pmatrix} 1 & 0 & 0 \\ 0 & \cos(t_x) & -\sin(t_x) \\ 0 & \sin(t_x) & \cos(t_x) \end{pmatrix} \begin{pmatrix} \cos(t_y) & 0 & \sin(t_y) \\ 0 & 1 & 0 \\ -\sin(t_y) & 0 & \cos(t_y) \end{pmatrix} \begin{pmatrix} \cos(t_z) & -\sin(t_z) & 0 \\ \sin(t_z) & \cos(t_z) & 0 \\ 0 & 0 & 1 \end{pmatrix} \begin{pmatrix} x \\ y \\ z \end{pmatrix} + \begin{pmatrix} dx \\ dy \\ dz \end{pmatrix} \quad (19.3)$$

   where t_x, t_y, and t_z represent the rotation angles and dx, dy, and dz represent the translation displacements along the x, y, and z axes, respectively.
4. *The projective transformation* maps any straight line in the original image onto a straight line in the transformed image. Parallelism is not preserved. The mathematical expression of the transformation for the 2-D case is given by

$$\begin{pmatrix} u \\ v \\ w \end{pmatrix} = \begin{pmatrix} a_{11} & a_{12} & a_{13} \\ a_{21} & a_{22} & a_{23} \\ a_{31} & a_{32} & a_{33} \end{pmatrix} \begin{pmatrix} x \\ y \\ 1 \end{pmatrix}, \begin{pmatrix} x' \\ y' \end{pmatrix} = \begin{pmatrix} \frac{u}{w} \\ \frac{v}{w} \end{pmatrix} \quad (19.4)$$

where w represents the extra homogeneous coordinate and u and v are dummy variables. The projective transformation is mainly employed in the 2-D case, strongly resembles the bilinear, and is completely defined by nine independent parameters ($a_{11}, \ldots, a_{33}$).

19.2.2.2 Matching Features — Measure of Match

Several image features have been used for the matching process. The use of external *skin markers* is a well-documented and relatively simple approach. They are placed physically on the patient prior to the acquisition of the image. The use of four skin markers for registering CT-MRI using 3-D global rigid transformation was reported in Reference 7. A similar application on CT images for stereotactic surgery was described in Reference 45. A combination of skin markers with affine transformation appears in Reference 46, considering SPECT and MR images, whereas SPECT and PET images were registered with CT images using four skin markers and a least square fit method in Reference 47.

Landmarks placed in the images by an expert to identify anatomically equivalent areas in the pair of images is another approach. Usually, the small number of landmarks as well as possible inaccuracies in their placement renders these methods semi-automatic and able to estimate simple transforms. Landmarks and Singular Value Decomposition methods were applied in CT-MR images,[8] whereas registering 2-D myocardial perfusion images required high order polynomial transformations.[48] Landmarks and a global affine transformation were employed for CT-SPECT registration in Reference 49, whereas a more advanced transformation (Thin Plane Splines) was used in Reference 11 for CT-MR images.

Matching the *principal axes* can be suitable for registering specific anatomical structures, but the resulting transform is usually approximate using rigid transform,[50] or affine transform,[36] requiring further local transformations. Considering the object of interest as a collection of N voxels, the principal axes of the object are defined as the eigenvectors of the covariance matrix of the position vectors (x^1, x^2, x^3) of the object voxels, which is defined as follows:

$$C_{ij} = \frac{1}{N}\sum_{k=1}^{N}(x_k^i - \bar{x}^i)(x_k^j - \bar{x}^j) \quad (19.5)$$

where $i, j = 1, 2, 3$. The eigenvectors of the covariance matrix can be calculated using the method of Singular Value Decomposition.[13]

In cases of 3-D images, a match between *anatomical structures* is calculated (surface to surface or points to surface). For instance, in the case of registering CT with MR images, a match between the head and hat is estimated, with the hat extracted from the outer contours of the MR modality and the head being the surface generated from the outer contours of the CT image. The "head-hat" match, combined with rigid or affine transformation in References 12 and 50, was used in registering CT and magnetic resonance imaging (MRI) with PET.

Several match measures have been employed, depending on the application as well as the implementation of the transform. In cases of manual registration, the match is quantified by the *distance* between selected pairs of points or markers. In the case of matching the principal axes, the required translation is determined by the centroids of the two structures, whereas scale and rotation are determined by the two systems of principal axes. For the head-hat match, the *distance transform* is often used.[1,14]

Correlation methods are suitable for both 2-D and 3-D image intramodality registration. Generally, the correlation coefficient between two images, $I_A(i, j)$ and $I_B(i, j)$, is defined for the 2-D case according to the following equation:

$$I_A(i,j) \otimes I_B(i,j) = \frac{\sum_{i,j}[(I_A(i,j) - \bar{I}_A)(I_B(i,j) - \bar{I}_B)]}{\sqrt{\sum_{i,j}(I_A(i,j) - \bar{I}_A)^2 \sum_{i,j}(I_B(i,j) - \bar{I}_B)^2}} \tag{19.6}$$

where $\bar{I}_A$ and $\bar{I}_B$ are the corresponding average values for the images.

Correlation techniques map regions of pixels on a best-fit basis and, therefore, require minimal user interaction. These methods have been combined with very simple transformations (translations only in Reference 51 or translation and rotation in Reference 52). The use of global affine transform with cross correlation implemented in the spatial frequency domain was also reported in Reference 53, whereas curved transformations for digital subtraction angiography was described in Reference 54.

The MI criterion is also employed for registering multimodal medical images. MI is a concept measuring statistical dependence between two random variables or the amount of information that one variable contains about the other. The MI registration criterion, proposed in Reference 31, states that the MI of the image intensity values of corresponding voxel pairs is maximized if the images are geometrically aligned according to the following equation:

$$MI = P(x,y)\log(P(x,y)) - P(x)\log(P(x)) - P(y)\log(P(y)) \tag{19.7}$$

where $P(x)$ is the probability of an image pixel having value equal to x and $P(x, y)$ is the double histogram of the two registered images, defined as the probability of a pixel (i, j) having value of y in image B, given that the same pixel has a value of x in image A, according to the following equation:

$$P(x,y) = P(I_B(i,j) = y | I_A(i,j) = x)$$

19.2.2.3 Determination of the Transformation Parameters Using Global Optimization Techniques

The determination of the transformation parameters strongly depends on the objective function, as well as on the medical images to be registered. In the case of matching pairs of internal landmarks, the parameters may be calculated using a closed form solution based on the Singular Value Decomposition method, assuming that the number of landmarks is sufficient. Furthermore, the *search-based methods* provide an alternative, based on the optimization of an MOM between the reference and the transformed images, with respect to the transformation parameters. If the MOM is well behaved (is continuous and has only one extreme), simple gradient-based methods (steepest descent, conjugate gradient method[13]) or the Downhill Simplex method[55] may suffice.

However, if the MOM has multiple extremes, presents discontinuities, or cannot be expressed analytically, as in the case of common clinical images, brute force-based exhaustive search is a method that guarantees successful determination of the parameters. An exhaustive search in the case of an affine transformation model was proposed in order to register remote sensing images, assuming that some of the unknown parameters could be estimated one at a time, in a serial manner, thus converting the multidimensional search into a sequence of optimization problems of lower dimensionality.[56] Under this assumption, the authors were forced to narrow the range of the parameters in order to accelerate the execution of the program. However, in cases of non-trivial transformations with many independent parameters, an exhaustive search is not possible.

The most attractive solution for search methods using non-trivial transformations is based on global optimization techniques. This work considers the use of two global optimization techniques, simulated annealing (SA)[57] and genetic algorithms (GAs).[58] The transformation models employed in the case of 2-D medical images introduce 6, 8, and 9 independent parameters (corresponding to the affine, bilinear, and projective transformation models, respectively), whereas in the 3-D case the transformation models employed require up to 24 independent parameters, corresponding to the bilinear transformation, defined over a wide range of values to achieve robustness. This fact, combined with the presence of multiple local extremes of the objective function, necessitates the use of global optimization techniques.

19.2.2.3.1 Simulated Annealing (SA)

SA presents an optimization technique that can process cost functions with arbitrary degrees of nonlinearities and arbitrary boundary conditions. SA has been successfully applied to image processing,[59] molecular biology,[60] biological signal processing,[61] neural network training,[62] etc.

A function of a system's state, $f(\vec{x})$, described by the vector $\vec{x} = (x_i{:}i = 1, \ldots, D)$, is minimized (respectively maximized) using a process called annealing. The system searches for optimal values of the function $f(\vec{x})$, while adding to the function a noise component whose magnitude is a descending function of time. For every minimization problem, a quantity T, known as "temperature," is defined as a descending function of time k, and the function $T(k)$ is called the annealing schedule. A standard definition of the annealing schedule is the following:[37,38]

$$T(k) = \frac{T_0}{\ln k} \tag{19.8}$$

where k is a time index and T_0 is a starting value large enough to cover the search space. An initial guess, $\vec{x}_0$, is made for the set of the unknown parameters, and the function $f(\vec{x})$ is evaluated. The $\vec{x}_0$ is then modified by $d\vec{x}$ and $f(\vec{x})$ is reevaluated. The temperature T affects the system, since $d\vec{x}$ is affected by the T.

Faster annealing schemes, such as the following, have been shown to behave optimally in several cases.[63]

- Very fast SA (VFSA), where

$$T(k) = \frac{T_0}{k} \tag{19.9}$$

- Exponential SA (ESA), where

$$T(k) = T_0 \exp(k(c-1)) \tag{19.10}$$

In the case of a multidimensional cost function, a different temperature T could be assigned to each of the independent parameters of the transformation model used. This type of SA is called adaptive simulated annealing (ASA).[64]

19.2.2.3.2 Genetic Algorithms (GAs)

GAs are global optimization methods inspired by Darwinian evolution.[58] The method starts by creating a population of random solutions of the optimization problem. A solution to the problem usually consists of the values of the independent parameters of the function to be optimized; this function is called the *objective function*. These values often are converted to binary and concatenated to a single string, called *chromosome* or individual, although *real encoding* is also used. The method treats each individual as an organism to which it assigns a measure of *fitness* or ability to survive. Each individual's fitness is estimated by the value of the objective function calculated over the values of the parameters that are stored in the individual. Using the axiom of *survival of the fittest*, pairs of fit individuals are selected to recombine their encoded parameters to produce offspring. The most basic genetic operators that act on the individuals are *crossover* and *mutation*, although a number of others have been proposed.[58] In this way, a new generation of solutions is produced that replaces the previous one. The selection pressure will direct the evolution of the population toward fitter states, which is equivalent to optimizing the objective function. This process is formulated in pseudocode as follows:

```
initialize the first generation of n individuals randomly
  while (termination_condition is false)
  {
      calculate the objective function of the n individuals
      select N/2 pairs of individuals
      apply crossover and mutation operator to produce offspring
      replace the current generation by the n offspring
  }
```

The convergence theorem for GAs can be expressed as the following. Researchers introduced the concept of *schema* to mathematically formalize GAs. In the case of binary encoding of the unknown parameters, a schema is a binary string with some of the 0s and 1s replaced by wild characters (*). A schema is thought of as a hyperplane of the problem's parameter space. A schema is contained into a chromosome if it can be produced by replacing 0s and 1s of the chromosome by *s. When decoding and evaluating a chromosome a large number of schemata are sampled at the same time — a phenomenon called *implicit parallelism*. GAs achieve function optimization since the selection operator statistically ensures that *building blocks* (low order schemata with small number of non-*s), which appear in fit chromosomes, propagate through generations. If the fitness f of a given schema H is greater than the average fitness $\bar{f}$ of the current population, then the number of chromosomes containing this schema in the next generation $m(H, t+1)$ is expected to be greater than the number of chromosomes containing the schema in the current generation $m(H, t)$:

$$m(H, t+1) \geq m(H, t)\frac{f(H)}{\bar{f}}\left(1 - p_c\frac{\delta(H)}{L-1} - o(H)p_{mut} - \ldots\right) \tag{19.11}$$

where $\delta(H)$ and $L(H)$ and $o(H)$ are schema-related quantities and p_c, p_{mut} are the probabilities for crossover and mutation. Toward the end of the search, when the population is enriched with building blocks, the crossover operator combines them to produce solutions of the optimization problem.

19.2.3 Medical Image Registration Refinement Using Elastic Deformation

A refinement of the aforementioned generic medical image registration scheme will be extensively presented by applying sequentially an elastic deformation method based on the Thin Plate Splines (TPS) deformation model. Elastic matching based on TPS deformation is considered an important method toward accurate automatic registration of medical images. It assumes that one of the images is made of an elastic material, while the other image serves as a reference. A number of pairs of corresponding points are defined between the two images. The method of TPS deformation assumes a mathematical function that transforms the one image with respect to the reference. This function has known values only at the corresponding points (the displacement vectors between the pairs of the corresponding points) and, therefore, requires interpolation for each voxel of the transformed image. These pairs of corresponding points are often referred to as "interpolant" points. The interpolation produces a displacement field defined for every voxel of the transformed image, achieving a match between the interpolant points. The overall elastic deformation can be controlled by a bending factor, thus resulting in a trade-off between exact interpolant point matching and the degree of image warping. Application of this approach in 2-D medical image registration can be found in References 11 and 65. These applications are considered as manual registrations, since the interpolant points are defined by user intervention. Techniques that perform automatic selection of interpolant points belong to the realm of automatic registration.

The proposed deformation model uses the points of the external surfaces of the image data as interpolant points. The correspondence between the interpolant points of the two surfaces is performed automatically using a Kohonen neural network. TPS interpolation is subsequently employed for elastic warping of the image volume using the correspondence between the interpolant points.

19.2.3.1 Self Organizing Maps (SOMs): An Overview

The Self-Organizing Map (SOM) is a neural network algorithm that uses a competitive learning technique to train itself in an unsupervised manner. Kohonen first established the relevant theory and explored possible applications.[66] The Kohonen model comprises a layer of neurons m, usually one or two dimensional, each neuron of which is connected to each input vector (data point) $\vec{x} \in \Re^n$ with a weight vector $\vec{w} \in \Re^n$. Each time a data point is input into the network, only the neuron j, whose weight vector resembles most the input vector, is selected to fire, according to the following rule:

$$j = \arg \min_{i=1}^{m} \left(\left\| \vec{x} - \vec{w}_i \right\|^2 \right) \tag{19.12}$$

The firing neuron j and its neighboring neurons i have their weight vectors $\vec{w}$ modified according to the following rule:

$$\vec{w}_i(t+1) = \vec{w}_i(t) + h_{ij}(t)(\vec{x}(t) - \vec{w}_i(t)) \tag{19.13}$$

where $h_{ij}(t) = h(\|\vec{r}_i - \vec{r}_j\|, t)$ is a kernel defined on the neural network space as a function of the distance $(\|\vec{r}_i - \vec{r}_j\|)$ between the firing neuron j and its neighboring neurons i, as well as the time, defined as the number of iterations t. This kernel has the approximate shape of the "Mexican hat" function, which in its discrete form has maximum value at interneuron distance in the case of $i = j$, whereas its value drops in a Gaussian manner as the distance increases. The width of this function decreases monotonically with time t. In this way, convergence to the global optimum is attempted during the early phases of the self-training process, whereas gradually the convergence becomes more local as the size of the kernel decreases. Each time a new signal is fed into the network, the neurons compete and the one with the weight vector closest to the signal, according to Equation 19.12, is selected to fire. The firing neuron adjusts its weight vector so that it better matches the incoming signal. Its neighboring neurons modify their weight vectors so they also resemble the input signal, but less strongly, depending on their distance from the winner. This learning mechanism is completely defined by the $h_{ij}(t)$ kernel, as described earlier. The previous process achieves spatial coherence within the network of neurons as far as the input signal is concerned. This property is equivalent to clustering, since after self-training the neurons form clusters which reflect the input signal (data points).

19.2.3.2 Surface Matching Based on the Kohonen Model

The 3-D elastic deformation method is comprised of the following steps:

1. Surface Triangulation Based on Kohonen Model

The size of a rectangular grid of neurons is set to a desirable value, and their weight vectors are initialized to the coordinates of the points of a surface, which is homeomorphic to the surface to be triangulated. Usually, for closed objects, convex or concave with no holes, the initializing shape is a sphere centered on the object's center of mass. First the network triangulates the surface, which is to be deformed, as described in References 67 to 69. After the process of triangulation, the weighting vectors of the initialized neurons have such values so that the corresponding wire frame closely follows the surface to be deformed. A similar approach is applied for the reference surface.

2. Automatic Definition of Corresponding Points

The search for corresponding points is based on the concept of replicating the topology of the triangulated surface S_1 on the input layer of a Kohonen neural network model. One neuron is assigned to each node of the triangulated surface. The connections between the neurons are identical with the connections of the wire frame (lattice) of the triangulated surface. No connection between two neurons is allowed if the two corresponding nodes are not directly connected on the triangulated surface. The initial weight vector of the neurons, according to Equation 19.13, is the Cartesian coordinates of the corresponding wire frame nodes in the 3-D space, $\vec{w} \in \Re^3$, as described in the previous step. Note that the traditional configuration of the neuron layer is a 2-D orthogonal grid consisting of $N_1 x N_2$ neurons, where each neuron is connected to a constant number of neurons.

The training of the network is performed by presenting the network with the coordinates of randomly selected points sampled from the reference surface, S_2. The kernel that controls the learning rate of the network is defined as follows. The spatial part consists of a Gaussian function of the Euclidean distance $d_{rq} = ||r - q||$ between the winner neuron r and the neighboring one q. This kernel is formulated as follows:

$$h_{rq}(\|r-q\|, t) = h_{rq}(d_{rq}, t) = \varepsilon_i \left(\frac{\varepsilon_f}{\varepsilon_i}\right)^{t/t_{\max}} e^{\frac{-d_{rq}}{\sigma^2(t)}} \tag{19.14}$$

where ε_i and ε_f are positive initial and final parameters defined experimentally and $t_{\max}$ corresponds to the maximum number of iterations. The standard deviation $\sigma(t)$ of the Gaussian is defined as $\sigma(t) = \sigma_i(\sigma_r/\sigma_i)^{t/t_{\max}}$, where σ_i and σ_f are the initial and final values of $\sigma(t)$.

The lateral interactions between the winning and the neighboring neurons are confined to a 3 × 3 region of neurons throughout the network training. The definition of the distance function d_{rq} between the two neurons r and q is defined as Manhatan distance (the sum of the absolute differences of the neuron coordinates) within the 3 × 3 neuron neighborhood. To ensure that the deformed surface, produced by the weight vectors of the Kohonen neural network, is closed at the boundary neurons of the orthogonal grid, it is essential that these neurons interact with each other. Therefore, the definition of a 3 × 3 neighborhood around a boundary neuron is modified so that it includes neurons, which are symmetrical to the winning one with respect to the middle grid axis.

The convergence of the Kohonen neural network results in a deformed surface (S'_1). Each node of surface S'_1 corresponds to a neuron of the Kohonen network, whose initial weighting vector (x_0, y_0, z_0) equals the initial Cartesian coordinates of this node. In the deformed surface, this node is displaced to new coordinates, equal to the final weighting vector (x_1, y_1, z_1). The new position always coincides with a node of the reference surface, S_2. The points (x_0, y_0, z_0) and (x_1, y_1, z_1) constitute a pair of interpolant points. Thus, the Kohonen neural network model discovers the correspondence between a number of pairs of interpolant points lying on the two surfaces equal to the total number of neurons of the network (typically a few thousands). Figure 19.2 demonstrates schematically the process of establishing the corresponding points, based on the Kohonen neural network, for the CT-MR surfaces. Figure 19.2a shows a random point correspondence, defined by the initial weight vectors, whereas in Figure 19.2b the correct point correspondence is obtained after the Kohonen neural network training.

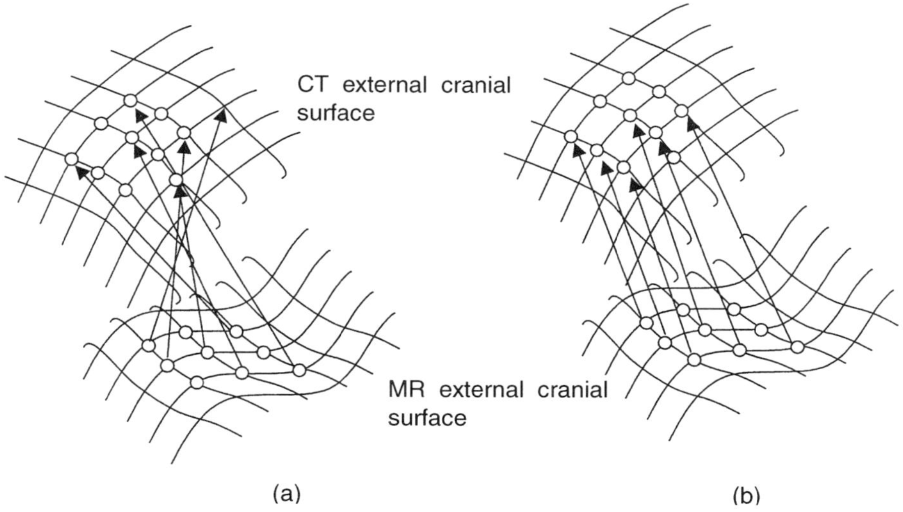

FIGURE 19.2 Correspondence of interpolant points of two surfaces (e.g., CT-MR skin surfaces) using the Kohonen neural network: (a) before training and (b) after Kohonen network convergence.

3. The 3-D Thin Plate Splines Warping

The elastic deformation of a 3-D deformed image with respect to the corresponding reference image is based on the calculation of a displacement field (DF) function, defined over the volume image, which will be deformed. A DF is a 3-D vector function which assigns a spatial displacement, $\delta\vec{r}$, to each of the MR image voxels (x, y, z):

$$DF(\vec{r}) = DF(x, y, z) = \delta\vec{r}(x, y, z) \tag{19.15}$$

The value of the *DF* function is known for each position $\vec{r}_{ij}$ of the points of the wire frame (interpolant points) of the skin surface of the deformed image prior to deformation. The coordinates of the interpolant points coincide with the initial weighting vector of the neurons of the Kohonen network. The value of the *DF* function is defined according to the following equation:

$$DF(\vec{r}_{ij}) = w_{ij} - r_{ij}, \text{ for } i = 1, \ldots, N_1 \text{ and } j = 1, \ldots, N_2 \tag{19.16}$$

where $\vec{w}_{ij}$ is the final value of the weighting vector of the (i, j) neuron calculated by the Kohonen network. The problem of 3-D image registration is equivalent to calculating the *DF* function for every voxel of the image prior to deformation. The calculation of the *DF*, as a function of spatial position, given its value at specific positions, indicates a problem of multivariate interpolation. Assuming the TPS interpolation method, the value of the displacement field at any position $\vec{r}(x, y, z)$ of the deformed image volume is given by

$$DF(\vec{r}) = \sum_{i=1}^{N_1}\sum_{j=1}^{N_2} \vec{\alpha}_m g(\|\vec{r} - \vec{r}_{ij}\|) + p(\vec{r}) \tag{19.17}$$

where $g{:}\Re^+ \rightarrow \Re$ is a univariate function, m is a dummy variable defined as $m = m(i, j) = (j - 1)N_1 + i$, and $p(\vec{r}) = \vec{c}_0 + \vec{c}_1 x + \vec{c}_2 y + \vec{c}_3 z$ is a first-degree polynomial which is added because pure radial sums cannot realize the affine transformation. $\vec{c}_n = (\vec{c}^x_n, \vec{c}^y_n, \vec{c}^z_n)$, with $n = 0, 1, 2, 3$ and $\vec{a}_m = (a^x_m, a^y_m, a^z_m)$ are the function parameters to be estimated. The definition of Equation 19.17 has been proven to be an effective tool for the problem of multivariate interpolation.[65] The coefficients $\vec{a}_m$ and $\vec{c}_n$ are calculated by solving the three systems of linear equations (one for each Cartesian axis) of size $(M + 4) \times (M + 4)$, where $M = N_1 \times N_2$:

$$GA = DF \tag{19.18}$$

where $A = (\vec{a}_1, \vec{a}_2, \ldots, \vec{a}_M, \vec{c}_0, \vec{c}_1, \vec{c}_2, \vec{c}_3)^T$ and $\vec{a}_m = (a^x_m, a^y_m, a^z_m)$, $m = 1, \ldots, M$

$$G = \begin{bmatrix} g_{11} & g_{12} & \cdots & g_{1M} & 1 & x_1 & y_1 & z_1 \\ g_{21} & g_{22} & \cdots & g_{2M} & 1 & x_2 & y_2 & z_2 \\ \vdots & \vdots & \cdots & \vdots & \vdots & \vdots & \vdots & \vdots \\ g_{M1} & g_{M2} & \cdots & g_{MM} & 1 & x_M & y_M & z_M \\ 1 & 1 & \cdots & 1 & 0 & 0 & 0 & 0 \\ x_1 & x_2 & \cdots & x_M & 0 & 0 & 0 & 0 \\ y_1 & y_2 & \cdots & y_M & 0 & 0 & 0 & 0 \\ z_1 & z_2 & \cdots & z_M & 0 & 0 & 0 & 0 \end{bmatrix} \tag{19.18a}$$

where $g_{mn} = g(\|\vec{r}_{ij} - \vec{r}_{kl}\|)$, with $i, k = 1, \ldots, N_1$; $j, l = 1, \ldots, N_2$; and $m, n = 1, \ldots, M$. The index m of the element of the G matrix and the indexes i, j of the interpolant points are connected as in the following:

$$i = mod\left(\frac{m-1}{N_1}\right) + 1 \text{ and } j = div\left(\frac{m-1}{N_1}\right) + 1 \tag{19.18b}$$

where *mod* and *div* are the functions returning the modulus and the integer part of the division of two real numbers. Similar relationships hold for the index n with k and l indexes. The triplet (x_m, y_m, z_m), $m = 1, \ldots, M$ are the Euclidean coordinates of the interpolant point $\vec{r}_{ij}$, where the indexes i and j are calculated according to Equation 19.18b.

Finally, $DF = (DF_1, DF_2, \ldots, DF_M, 0, 0, 0, 0)^T$ and $DF_m = (\vec{w}_{ij} - \vec{r}_{ij})$, where the indexes i, j, and m are connected according to Equation 19.18b.

Several approaches have been reported in the literature for the g function, for which the three linear systems of Equation 19.18 have a unique solution.[65] The shifted log function is selected as the g function according to the following equation:

$$g(t) = \log(t^2 + c^2)^{\frac{1}{2}} \text{ for } c^2 \geq 1 \text{ (\textit{shifted} log)} \tag{19.19}$$

The DF function can be defined according to Equation 19.17 for each dimension as follows:

$$DF(\vec{r}) = (DF_x(\vec{r}), DF_y(\vec{r}), DF_z(\vec{r})) \tag{19.20}$$

where

$$DF_x(\vec{r}) = \sum_{i=1}^{N_1} \sum_{j=1}^{N_2} a^x_m g(\|\vec{r} - \vec{r}_{ij}\|) \tag{19.21}$$

$$DF_y(\vec{r}) = \sum_{i=1}^{N_1} \sum_{j=1}^{N_2} a^y_m g(\|\vec{r} - \vec{r}_{ij}\|) \tag{19.22}$$

$$DF_z(\vec{r}) = \sum_{i=1}^{N_1} \sum_{j=1}^{N_2} a^z_m g(\|\vec{r} - \vec{r}_{ij}\|) \tag{19.23}$$

where $m = m(i, j) = (j - 1)N_1 + i$.

In References 11 and 57, for any transform $T:\Re^3 \rightarrow \Re^3$, the quantity

$$J(T) = \iiint_{\Re^3} ((T_{xx})^2 + (T_{yy})^2 + (T_{zz})^2 + (T_{xy})^2 + (T_{yz})^2 + (T_{zx})^2)dxdydz \tag{19.24}$$

is a measure of the total spatial bending, where the double subscripts denote double partial derivatives. Bookstein[11] and Arad et al.[65] introduced a bending factor λ, which balances the requirement for exact matching of the interpolant points, whereas the total spatial bending, $J(DF_x) + J(DF_y) + J(DF_z)$, is minimal. The introduction of the bending factor λ requires the G matrix (Equation 19.18) to be redefined so that the M first main diagonal elements of matrix G are set to λ: $g_{mm} = \lambda$, $m = 1, \ldots, M$.

When the bending factor is zero, $\lambda = 0$, the calculated DF results in an exact match of the interpolant points at the expense of a possibly high total spatial bending measurement, $J(DF)$, according to Equation 19.24, of the image space (pure TPS interpolation). When λ approaches positive infinity, the sum of square distances between actual and calculated positions of the interpolant points is minimized, while the total spatial bending measurement, $J(DF)$, is kept below a threshold value which depends on λ, or, equivalently, the geometric transformation approaches the affine one. The value of the bending factor λ is estimated by visual inspection of the deformed image. Figure 19.3 confirms these arguments for a synthetic grid image containing six (6) interpolant points (crossed bullets in the figure). Four interpolant

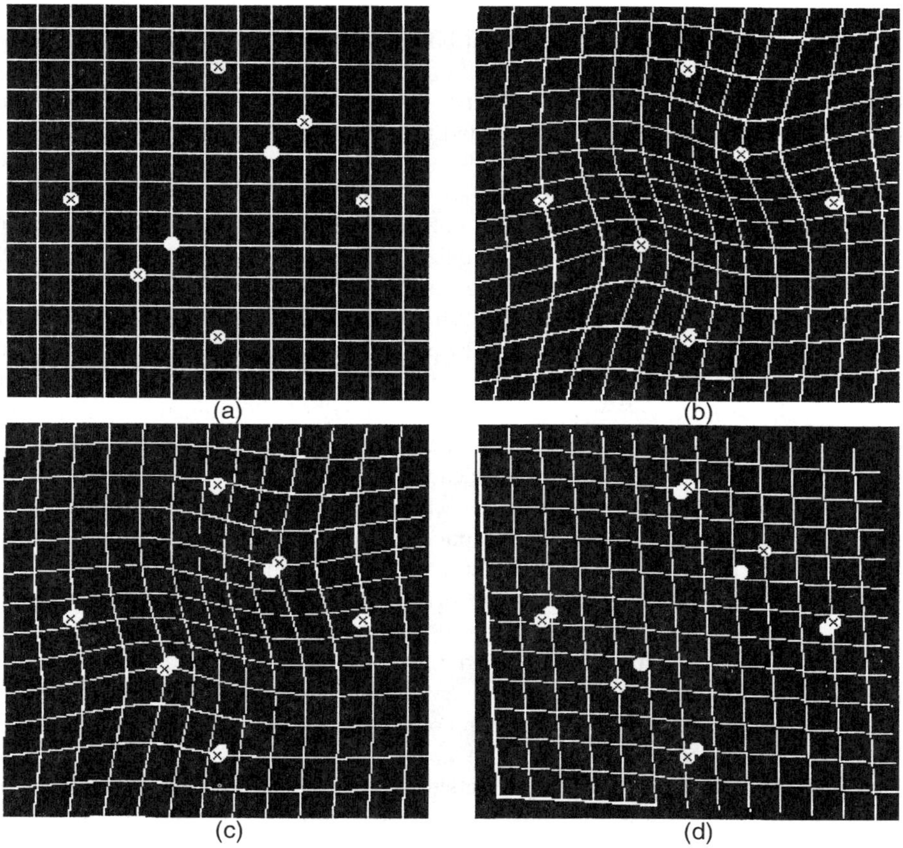

FIGURE 19.3 Effect of the bending factor λ on image warping. (a) The synthetic grid image showing the initial (marked with "x") and final positions of the interpolant points; (b) for $\lambda = 0.0001$ (pure TPS Interpolation); (c) $\lambda = 0.001$; (d) $\lambda = 0.1$ (affine transformation approximation).

points remain fixed, whereas the other two interpolant points move to the opposite direction, as shown in Figure 19.3a. The warped images are shown in Figures 19.3b to 19.3d for increasing values of the bending factor λ corresponding from the pure TPS interpolation method (Figure 19.3a) to the affine transformation approximation (Figure 19.3d).

19.2.4 Application of Medical Image Registration

19.2.4.1 2-D Case: Automatic Retinal Image Registration

Ophthalmologists commonly compare red-free (RF) retinal images, which are reference images taken without intravenous injection of a dye while illuminating the retina with a green light, with the corresponding fluoroscein angiography (FA) or indocyanine green chorioangiography (ICG) image of the patient acquired at different times. This task presents great difficulty due to the misalignment of the retinal images caused by the geometry during the acquisition at different times and to possible progression of various diseases. The relative study of retinal images enhances the information on the reference RF images by superimposing useful information contained in FA or ICG retinal images, and it is considered an important step toward a carefully directed laser treatment, a process that is commonly used in the clinical practice.

A fundus camera is the instrument for acquiring FA and ICG retinal images. In the case of retinal images, the objects of interest are the retinal vessels: arteries and veins. In Reference 70, it was shown that the registration algorithm operates more efficiently if the vessels are segmented, using at least a crude segmentation. Therefore, image preprocessing is required prior to registration. The preprocessing involves two main processes: (1) vessel enhancement and border suppression and (2) vessel detection. These processes are performed on all retinal images with the addition of an inversion procedure applied only to the RF images in order to emulate the appearance of FA and ICG images. A MOM corresponding to the correlation function was employed to quantify the spatial matching between the reference and the transformed image. A comparative study of the three transformation models (affine, bilinear, and projective) in conjunction with the two global optimization techniques (SA and GAs), as well as the Downhill Simplex method (DSM) as a widely applied local optimization technique, was performed on 26 retinal image pairs.[70] In Figure 19.4a, the two global optimization methods, SA and GAs, and the DSM, using the bilinear transformation, are compared by plotting the values of the MOM against the number of function evaluations for a retinal pair. In Figure 19.4b, the three transformation models, affine, bilinear, and projective, using the GAs, are also compared by plotting the MOM against the number of function evaluations. The MOM is averaged for ten different program executions. It becomes evident from Figure 19.4a that although the DSM converges very fast (e.g., typically in 300 function evaluations), it was usually trapped in local maxima, failing to locate the global one. It can also be observed that the GAs performed better than the SA with a significantly higher convergence rate. The convergence to the optimal value of the objective function, using the GAs, was achieved for a fewer number of function evaluations than the SA, which required the total number of function evaluations to approximate the optimal value. Furthermore, the superiority of the affine and bilinear transformations in Figure 19.4b is evident for the retinal image pair, both in terms of numerical performance and convergence rate.

Retinal registration results can be seen in Figure 19.5. Regions of interest (ROI) from the FL and ICG images, as rectangular areas, are superimposed on the corresponding RF images and the results are shown in Figures 19.5a and 19.5b, respectively. The accuracy of the registration is evident by the continuity of the retinal vessels along the borders of the superimposed ROI. The automatic registration scheme (GAs with bilinear transformation) was also applied on a pathologic pair of RF and FL images. Figures 19.5c and 19.5d show the registered FL image of the patient with three boundaries traced by the expert: one corresponding to the fovea (best fitting ellipse) and the other two to the areas of the abnormality (free drawings). These boundaries are easily identified only in the FL image and are of extreme importance in directing the laser treatment procedure. Also, the proposed methodology was also compared with the manual method using the affine transformation, confirming the advantage of the automatic method in terms of accuracy and consistency against the manual one.

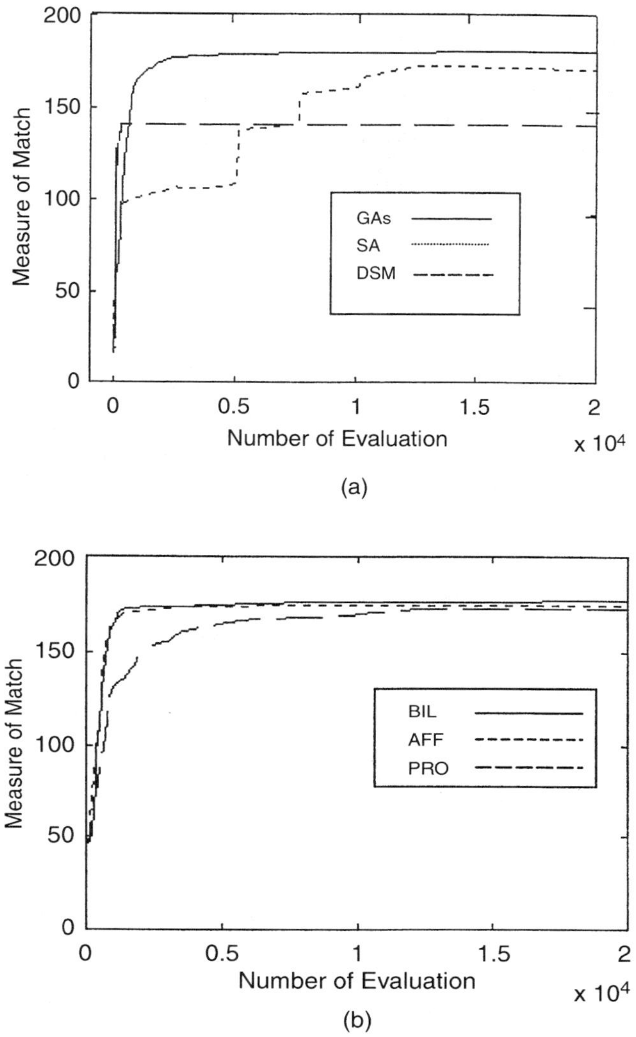

FIGURE 19.4 Performance of the automatic registration technique for a retinal image pair: (a) GAs (continuous line) vs. SA (dotted line) vs. DSM (dashed line) in conjunction with the bilinear transformation model; and (b) performance of the three transformation models (affine [AFF, dotted line] vs. bilinear [BIL, continuous line] vs. projective [PRO, dashed line]) with GAs.

19.2.4.2 3-D Case: Automatic CT-MRI Registration

X-ray CT and MR brain images were acquired from seven patients (seven pairs of data) at the Strahlenklinik of the Stadtische Kliniken Offenbach in Germany. The CT images were acquired using a Siemens Somatom Plus scanner. Each CT slice contained 256 × 256 pixels, using 12 bits per pixel, with no interslice spacing or slice overlap for all CT data sets. The T1-weighted MR images were acquired using a Siemens SP 1.5-Tesla scanner. Each MR slice contained 256 × 256 pixels, using 8 bits per pixel, with constant interslice spacing for each MR data set.

In the case of the pair of CT and MR brain images, a match had been achieved between the skin CT and MR surfaces. The preprocessing step involved the removal of the background noise and the extraction and triangulation of the skin surfaces of the CT and MR images using a modified Marching Cubes algorithm.[71] A coarse registration was then performed by translating the points of the MR surface so that their center of mass coincided with CT surface's center of mass and by scaling the

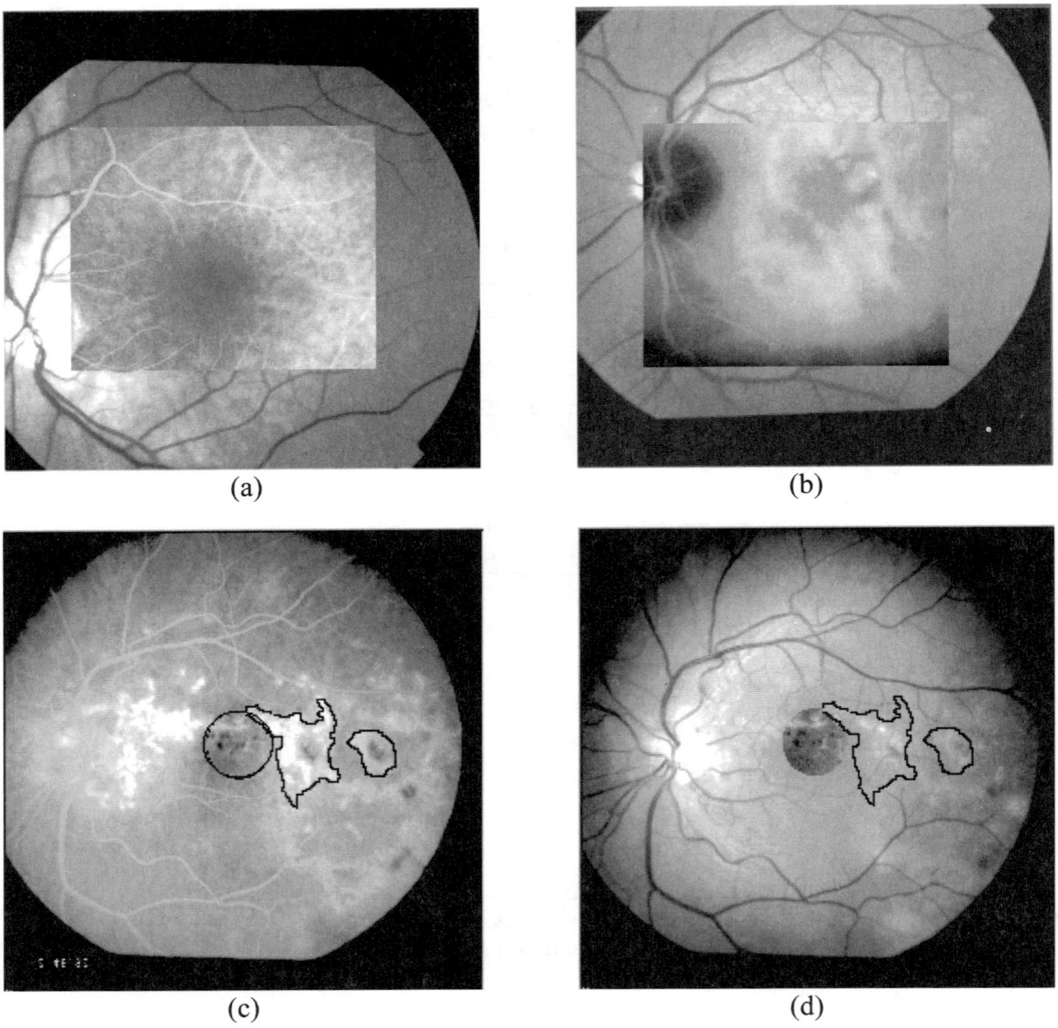

FIGURE 19.5 Accuracy of the automatic registration: (a) superposition of an area of the registered FL image on the corresponding RF image; (b) superposition of an area of the registered ICG image on the corresponding RF image; (c) a registered pathological FL image with three boundaries traced by an expert; and (d) superposition of the boundaries on the corresponding RF image.

MR surface points coordinates to compensate for the difference of voxel size between the two modalities. The final step of the preprocessing consisted of the production of the distance map from the CT surface, *DM*(*CT*). The distance map is a discrete space in which each voxel holds a value equal to its Euclidean distance from the closest node of the CT skin surface.[14] The distance map accelerates the process of matching two surfaces consisting of N nodes each, since it reduces the problem's complexity from $O(N^2)$ to $O(N)$. After preprocessing of the CT and MR data, the registration proceeded as an optimization of the function of MOM over the parameters of the selected geometric transformation model. The MOM was defined as the average Euclidean distance between the CT and MR skin surfaces.[72]

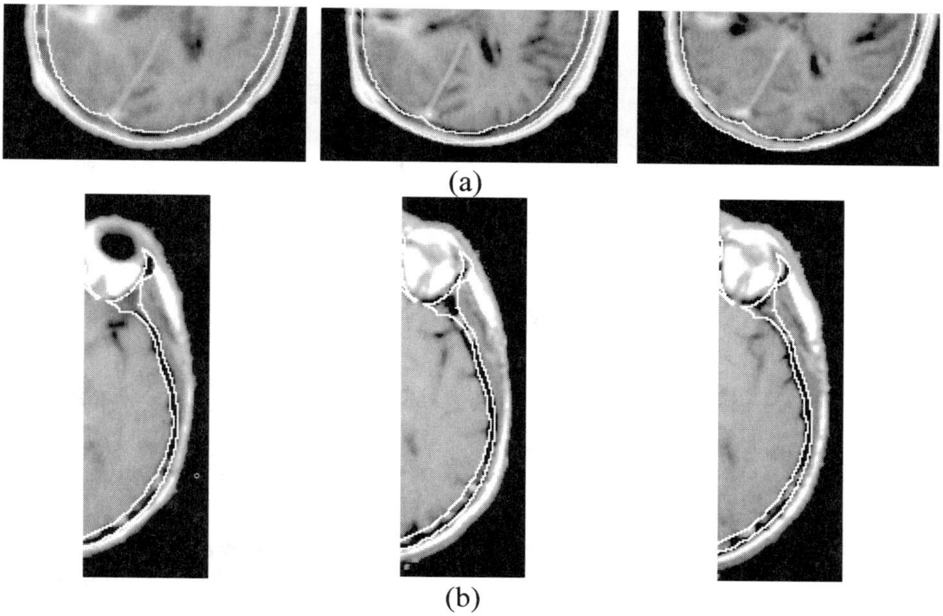

FIGURE 19.6 Magnified areas of the CT bone contours (outer and inner) superimposed on the corresponding MR slice for two image pairs (a and b). The three images on each row, from left to right, correspond to the automatic elastic refinement method, the GAs in conjunction with the affine transformation surface-based method, and the affine-based manual method.

Three registration methods had been assessed: (1) as the surface-based method, GAs in conjunction with the affine transformation; (2) as the automatic registration method, GAs and affine combined with the proposed elastic deformation method (TPS model), also called the elastic registration refinement method; and (3) as a manual method, the affine transformation in conjunction with a number of markers.[72] The accuracy of the medical registration methods is visually demonstrated in Figure 19.6, where CT skull contours are superimposed on magnified areas of the corresponding MR transverse sections for two CT-MR image pairs (Figures 19.6a and 19.6b). The application of the GAs-affine method and the application of the affine-based manual method result in slight inaccuracies in the placement of the CT skull contours, which often invade the brain, and the zygomatic bone at the vicinity of the eye, and also fail to locate accurately the superior sagittal sinus. These inaccuracies were corrected by the combination of the GAs-affine method and the elastic registration refinement method described in the above section.

The performance of two automatic registration methods against the manual one, in terms of the MOM, is quantitatively assessed for all CT-MR image pairs, as shown in Figure 19.7. The averaged skin surface distance (MOM) for the seven pairs of CT and MR brain data was obtained, in millimeters, and the values of MOM for both automatic registration methods were averaged over ten independent executions for all image pairs to compensate for the stochastic (randomized) nature of the optimization method, whereas for the manual method, the values of MOM were averaged over three trials for each pair. It can be observed from Figure 19.7 that the values of the surface distance of the elastic registration refinement method were systematically lower than those of the other automatic registration method when using the affine transformation with GAs only. These results clearly show a definite registration refinement for all pairs of CT-MR data, due to the application of the elastic deformation method. Furthermore, it can be observed that the manual registration method based on the affine transformation performed significantly worse than the two automatic methods. It is also evident that the performance of the manual method was strongly influenced by the number of markers placed.

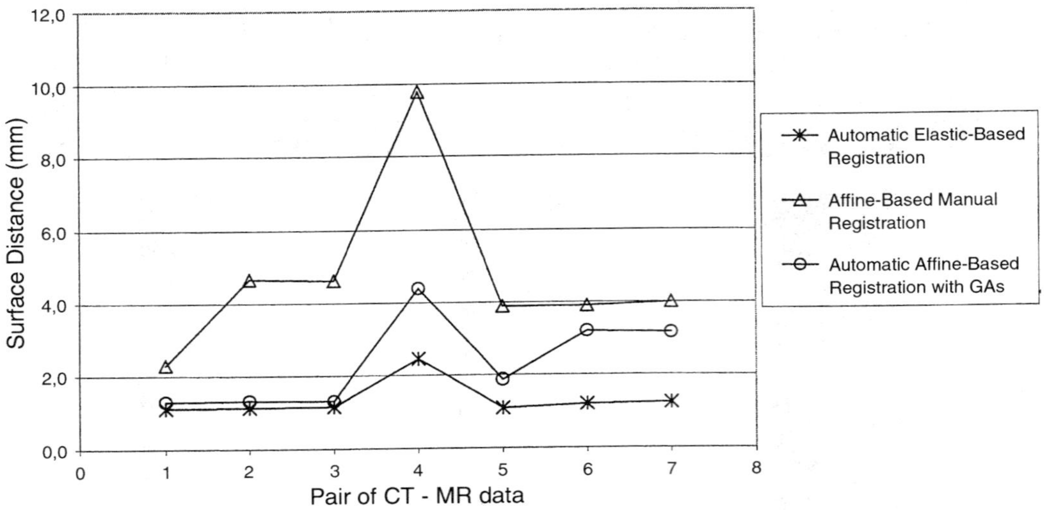

FIGURE 19.7 Performance of (a) the automatic elastic-based refinement method, (b) the automatic GAs with affine surface-based method, and (c) the affine-based manual method, in terms of surface distance (MOM), in millimeters, for all CT and MR image pairs.

19.3 Medical Image Fusion

Imaging the same parts of human anatomy with the different modalities, or with the same modality at different times, provides the expert with a great amount of information which must be combined in order to become diagnostically useful. Medical image fusion is a process that combines information from different images and displays it to the expert so that its diagnostic value is maximized.

Although much attention has been drawn to the process of medical image registration, medical image fusion, as a prerequisite of the process for image registration, is not extensively explored in the literature, mainly because it is considered a straightforward step. A generic approach of the medical image fusion can be seen in Figure 19.8, where all the necessary steps are highlighted.

In this section, the concept of the medical image fusion process, as well as the most representative techniques reported, will be revised and broadened to include several methods to combine diagnostic information.

19.3.1 Fusion at the Image Level

19.3.1.1 Fusion Using Logical Operators

This technique of fusing information from two images can be twofold. The reference image, which is not processed, accommodates a segmented region of interest from the second registered image. The simplest way to combine information from the two images is by using a logical operator, such as the XOR operator, according to the following equation:

$$I(x, y) = I_A(x, y)(1 - M(x, y)) + I_B(x, y)M(x, y) \tag{19.25}$$

where $M(x, y)$ is a Boolean mask that marks with 1s every pixel, which is copied from image B to the fused image $I(x, y)$.

Moreover, in certain cases, it is desirable to simply delineate the outline of the object of interest from the registered image and to position it in the coordinate system of the reference image. An example of this technique is the extraction of the boundary of an object of interest, such as a tumor from a MR image and the overlay on the coordinate system of the registered CT image of the same patient. Figure 19.9

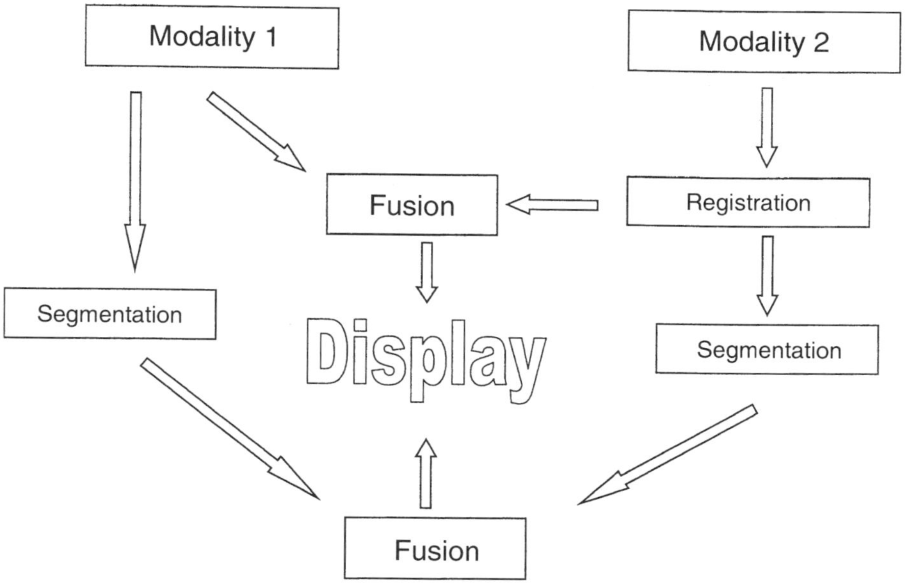

FIGURE 19.8 Generic medical image fusion scheme.

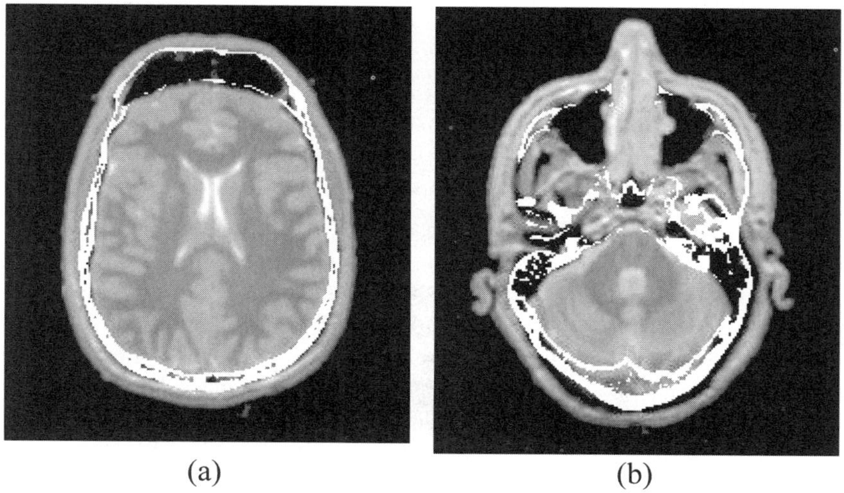

FIGURE 19.9 Fused images using the XOR operator: CT information (bone structures) (a) and (b) superimposed on two registered MR transverse slices.

demonstrates the fusion technique using the logical operators on the same patient data, where CT information (bone structures) is superimposed on two registered MR transverse slices.

19.3.1.2 Fusion Using a Pseudocolor Map

According to this fusion technique, the registered image is rendered using a pseudocolor scale and is transparently overlaid on the reference image.[73] There are a number of pseudocolor maps available, defined algorithmically or by using psychophysiologic criteria. A pseudocolor map is defined as a correspondence of an (R, G, B) triplet to each distinct pixel value. Usually, the pixel value ranges from 0 to 255, and each of the elements of the triplet varies in the same range, thus producing a true color effect. The color map is called "pseudo" because only one value is acquired for each pixel during the acquisition of the image data.

Two of the pseudocolor maps that are defined by psychophysiologic criteria are the *geographic color map* and the *hot body color map.* The (R, G, B) triplet values are defined as a function of the original pixel value according to the following equation:

$$(R, G, B) = (R(pixel\_value), G(pixel\_value), B(pixel\_value)) \tag{19.26}$$

The *RGB Cube* color map is a positional color map, which maps the original values to a set of colors that are determined by traversing the following edges of a cube:

$$B(0, 0, 1) \rightarrow (1, 0, 1) \rightarrow R(1, 0, 0) \rightarrow (1, 1, 0) \rightarrow G(0, 1, 0) \rightarrow (0, 1, 1) \rightarrow B(0, 0, 1) \tag{19.27}$$

This configuration sets up six color ranges, each of which has $N/6$ steps, where N is the number of colors (distinct pixel values) of the initial image. If N is not a factor of 6, the difference is made up at the sixth range.

The *CIE Diagram* color map follows the same philosophy, thus traversing a triangle on the *RGB Cube* in the following manner:

$$R(1, 0, 0) \rightarrow G(0, 1, 0) \rightarrow B(0, 0, 1) \tag{19.28}$$

Another useful color map for biomedical images is the *Gray Code.* This color map is constructed by a three-bit gray code, where each successive value differs by a single bit. A number of additional color maps, such as *HSV Rings* and *Rainbow,* are constructed by manipulating the color in the HSV coordinate system (Hue/Saturation/Value coordinate system). The conversion from HSV to RGB is straightforward and well documented in the literature. Figure 19.10 demonstrates the fusion using the RGB Cube pseudocolor map of a SPECT image on the corresponding CT image of the same patient.

19.3.1.3 Clustering Algorithms for Unsupervised Fusion of Registered Images

Fusion of information on the image level can be achieved by processing both registered images in order to produce a fused image with an appropriate pixel classification. The key quantity in this technique is the double histogram $P(x, y)$ of the two registered images, which is defined as the probability of a pixel (i, j) having a value of y in image B, given that the same pixel has a value of x in image A:

$$P(x, y) = P(I_B(i, j) = y | I_A(i, j) = x) \tag{19.29}$$

This quantity is very closely related to the entropy of the two images and can be very useful for effective tissue classification/segmentation, since it utilizes information from both (registered) images rather than one. The concept of the double histogram can be generalized in the $\Re^n$ space fusing n registered images. Dynamic studies from single photon emission CT (SPECT) are examples of such cases. In this general case, the n-dimensional histogram is defined as follows:

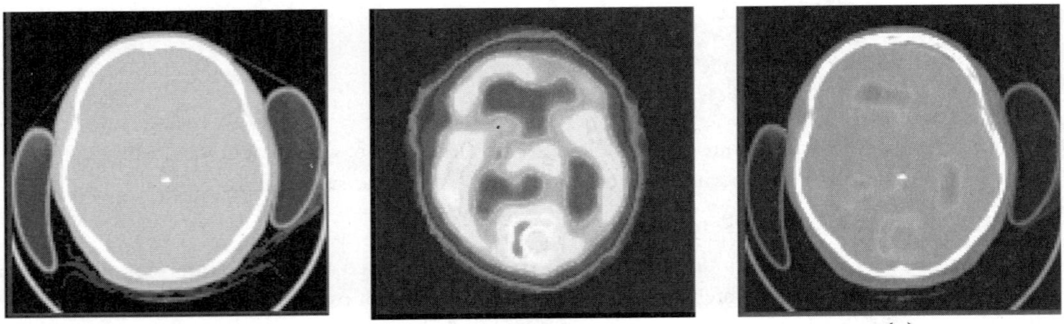

(a) (b) (c)

FIGURE 19.10 Fused image using the RGB Cube pseudocolor map: (a) reference CT image, (b) registered SPECT image in pseudocolor, and (c) fusion result.

$$H_n(v_1, v_2, \ldots, v_n) = P(I_n(i, j) = y|I_1(i, j) = v_1; I_2(i, j) = v_2; \ldots; I_{n-1}(i, j) = v_{n-1}) \quad (19.30)$$

Fusing the registered images to produce an enhanced or segmented/classified image becomes equivalent to partitioning the n-dimensional histogram to a desired number of classes using a clustering algorithm.

The goal of clustering is to reduce the amount of data by categorizing similar data items together. Such grouping is pervasive in the way humans process information. Clustering is hoped to provide an automatic tool for constricting categories in data feature space. Clustering algorithms can be divided into two basic types: *hierarchical* and *partitional*.[74] Hierarchical algorithms are initialized by a random definition of clusters and evolve by splitting large inhomogeneous clusters or merging small similar clusters. Partitional algorithms attempt to directly decompose the data into a set of disjoint clusters by minimizing a measure of dissimilarity between data points in the same clusters, while maximizing dissimilarity between data points in different clusters.

19.3.1.3.1 The K-Means Algorithm

The K-means algorithm is a partitional clustering algorithm, which is used to distribute points in feature space among a predefined number of classes.[75] An implementation of the algorithm applied in pseudocode can be summarized as follows:

Initialize the centroids of the classes, $i = 1, \ldots, k$ at random $\vec{m}_i\ (0)$

$t = 0$

repeat

for the centroids $\vec{m}_i\ (t)$ of all classes

locate the data points whose Euclidean distance from m_i is minimal

set $\vec{m}_i\ (t + 1)$ equal to the new center of mass of x_i, $\vec{m}_i\ (t + 1) = \dfrac{\vec{x}_i}{n_i}$

$t = t + 1$

until($|\vec{m}_i\ (t - 1 - \vec{m}_i\ (t)| \leq$ error, for all classes i)

This algorithm is applied to the n-dimensional histogram of the images to be fused, as defined in Equation 19.30. The centroids of the classes are selected randomly in the n-dimensional histogram, and the algorithm evolves by moving the position of the centroids so that the following quantity is minimized:

$$E = \sum_j (\vec{x}_j - \vec{m}_{c(\vec{x}_j)})^2 \quad (19.31)$$

where j is the number of data points index and $\vec{m}_{c(\vec{x}_j)}$ is the class centroid closest to data point $\vec{x}_j$. The above algorithm could be employed to utilize information from pairs of images, such as CT, SPECT, MRI-T1, MRI-T2, and functional MRI, to achieve tissue classification and fusion. For tissue classification, nine different classes are usually sufficient, corresponding to background, cerebrospinal fluid, gray and white matter, bone material, abnormalities, etc. If fusion is required, then 64 or more classes could be defined, providing enhancement of the fine details of the original images. In Figure 19.11, a fused image using the K-means algorithm for different classes of tissue is produced by fusing CT and MR images of the same patient.

19.3.1.3.2 The Fuzzy K-Means Algorithm

The Fuzzy K-means algorithm (FKM) is a variation of the K-means algorithm, with the introduction of fuzziness in the form of a membership function. The membership function defines the probability with

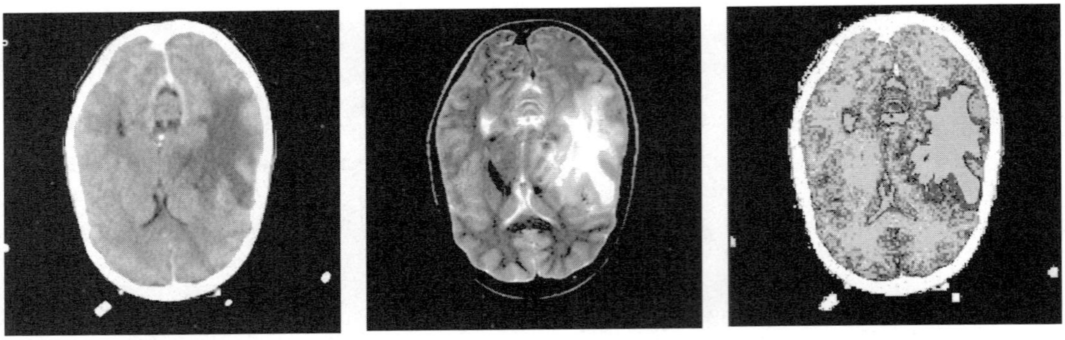

(a) CT image (b) MR image (c) Fused Image

FIGURE 19.11 Fused image using the *K*-means algorithm for different classes of tissue: (a) reference CT image, (b) registered MRI image of the same patient, and (c) fusion result.

which each image pixel belongs to a specific class. It is also applied on the n-dimensional histogram of the images to be fused, as in the previous method. The FKM algorithm and its variations have been employed in several types of pattern recognition problems.[75,76]

If a number of n unlabeled data points is assumed, which have to be distributed over k clusters, a membership function u can be defined, such that

$$u:(j, \vec{x}) \rightarrow [0, 1] \tag{19.32}$$

where $1 \leq j \leq k$ is an integer corresponding to a cluster. The membership function assigns to a data point x a positive, less than or equal to one, probability u_{jx}, which indicates that point x belongs to class j. To meet computational requirements, the membership function is implemented as a 2-D matrix, where first index indicates the cluster and the second index indicates the value of the data point. If the membership function is to express mathematical probability, the following constraint applies:

$$\sum_{i=1}^{n} u_{ij} = 1, \forall cluster\ j \tag{19.33}$$

The FKM algorithm evolves by minimizing the following quantity:

$$E = \sum_{j=1}^{k}\sum_{i=1}^{n} u_{ij}^{p}(\vec{x}_i - \vec{m}_j)^2 \tag{19.34}$$

where the exponential p is a real number greater than 1 and controls the fuzziness of the clustering process. The FKM algorithm can be described in pseudocode as follows:

t = 0

initialize the matrix u randomly

repeat

calculate the cluster centroid using the formula: $m_j(t) = \dfrac{\sum_{i=1}^{n} u_{ij}^{p}\vec{x}_i}{\sum_{i=1}^{n} u_{ij}^{p}}, j = 1, \ldots, k$

calculate the new values for u: $u_{ij}(t) = \left(\sum\left(\frac{|\vec{x}_i - \vec{m}_j|}{|\vec{x}_i - \vec{m}_j|}\right)^{\frac{2}{p-1}}\right)^{-1}$

$t = t + 1$

until $(|\vec{m}_i\,(t-1) - \vec{m}_i\,(t)| \le error$ *or* $|u_{ij_i}\,(t-1) - u_{ij}(t)| \le error)$

19.3.1.3.3 Self Organizing Maps (SOMs)

The SOM is a neural network algorithm which uses a competitive learning technique to train itself in an unsupervised manner. A comprehensive overview can be found in Section 19.2.3.1. According to Reference 66, the Kohonen model comprises a layer of neurons m, usually one dimensional (1-D) or 2-D. The size of the network is defined by the purpose of the fusion operation. If it is desirable to simply fuse information to enhance fine detail visibility, a large number of neurons are required, typically 8 × 8 to 16 × 16. If it is known *a priori* that a small number of tissue classes exist within the images, then the number of neurons of the network is set equal to the number of classes. The resulting image is a segmented image, which classifies each pixel to one of these classes. A number up to 9 is commonly the most meaningful choice in this case. Each neuron is connected to the input vector (data point) $\vec{x} \in \Re^n$ with a weight vector $\vec{w} \in \Re^n$. In the case of image fusion, the input signal has a dimensionality equal to the number of images to be fused, whereas the signal itself is comprised of the values of the images at a random pixel (i, j):

$$\vec{x} = \{I_1(i,j), I_2(i,j), \ldots, I_n(i,j)\} \tag{19.35}$$

It becomes obvious that the proposed method can fuse an arbitrary number of images (assuming that registration has been performed for all of the images). However, the commonest used case is for $n = 2$.

The above process achieves spatial coherence within the network of neurons as far as the input signal is concerned. This property is equivalent to clustering, since after self-training the neurons form clusters which reflect the input signal (data points). Figure 19.12 shows two fused images by applying the SOM algorithm on the same patient data imaged by a Gatholinium (Gd)-enhanced MRI (Figure 19.12a) and MRI-T2 (Figure 19.12b). The upper right fused image (Figure 19.12c) is obtained using 256 classes, whereas the lower right image (Figure 19.12c) is classified using 9 classes.

19.3.1.4 Fusion to Create Parametric Images

It is often necessary to fuse information from a series of images of a dynamic study to classify tissues according to a specific parameter. The result of the fusion process is the creation of a *parametric* image, which visualizes pixel by pixel the value of the diagnostically useful parameter.[77] The required classification is performed by thresholding the parametric image at an appropriate level.

A typical example of this process is the gated blood pool imaging performed for the clinical assessment of the ventricular function. A sequence of images — typically 16 per cardiac cycle — is acquired of the cardiac chamber of interest, which usually is the left ventricle. The assessment of ventricular function is performed by producing parametric images, visualizing several critical parameters:

- The *amplitude image* measures the change of volume as function of time at a pixel by pixel basis, calculating the ejection fraction (EF) for every pixel in the sequence of images.
- The *phase image* visualizes the synchronization of cardiac contraction at a pixel by pixel basis.

The first parameter is valuable in assessing dyskinetic cardiac tissue due to infarcts, whereas the second parameter is a strong indication of fatal conditions such as fibrillation. Both images are calculated by means of Fourier analysis, using only the first few harmonics. In this way, information of a large number of images is fused into a single parametric image, thus enabling the expert to quantify clinical evaluation beyond any qualitative inaccuracies.

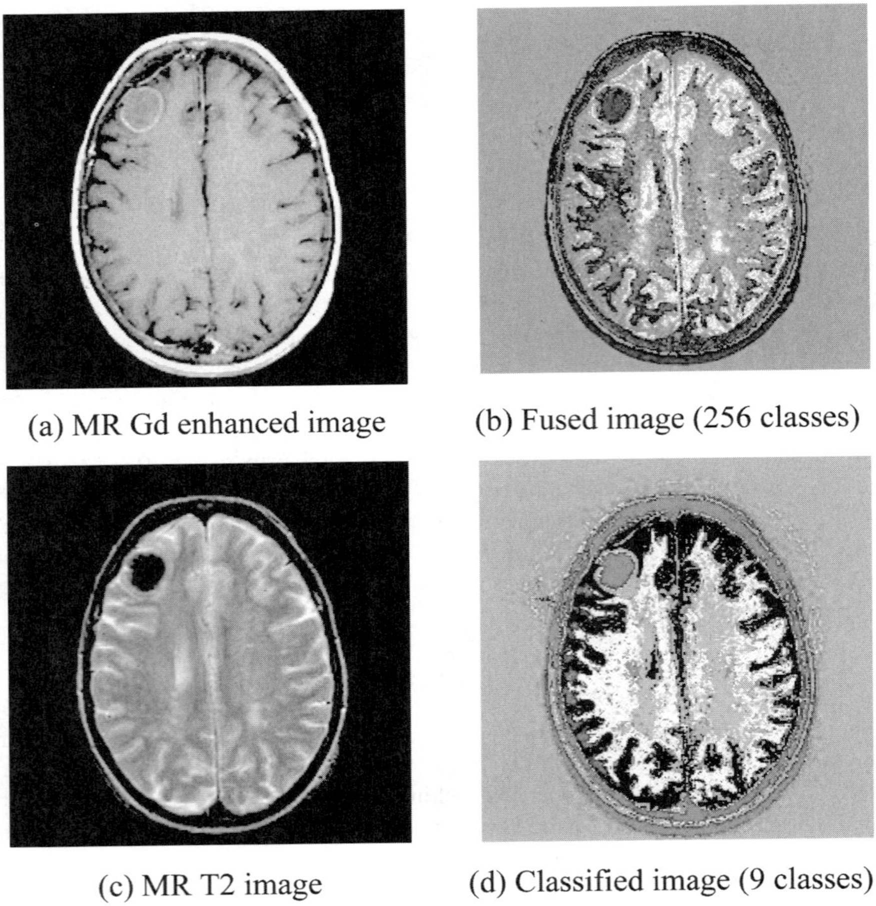

FIGURE 19.12 Fused images by applying the SOM algorithm on the same patient data imaged by (a) Gd-enhanced MRI and (c) MRI-T2. Fused image (b) is obtained using 256 classes, whereas image (d) is classified using 9 classes.

19.3.2 Fusion at the Object Level

Fusion at the object level involves the generation of either a spatio-temporal model or a 3-D textured object of the required object. Segmentation and triangulation algorithms must be performed prior to the fusion process.

Fusion at the object level is demonstrated by fusing temporal information on the spatial domain. Four-dimensional (4-D) MR cardiac images, such as *gradient echo*, are obtained, providing both anatomical and functional information about the cardiac muscle.[78] The left ventricle from each 3-D image, covering the whole cardiac phase, is first segmented, and the produced endocardiac surface is then triangulated. The radial displacement between the end-diastolic and end-systolic phase at each point of the triangulated surface is mapped on the surface and coded as pseudocolor, thus producing a textured object (Figure 19.13) which visualizes information of myocardial wall motion during the entire cardiac cycle.[79]

19.4 Conclusions

In this chapter, a comprehensive review of the most widely used techniques for automatic medical image registration and medical image fusion has been presented.

In terms of medical image registration, three main categories have been reported: the point-based registration, usually rendered as manual, and the surface- and volume-based registrations, rendered as

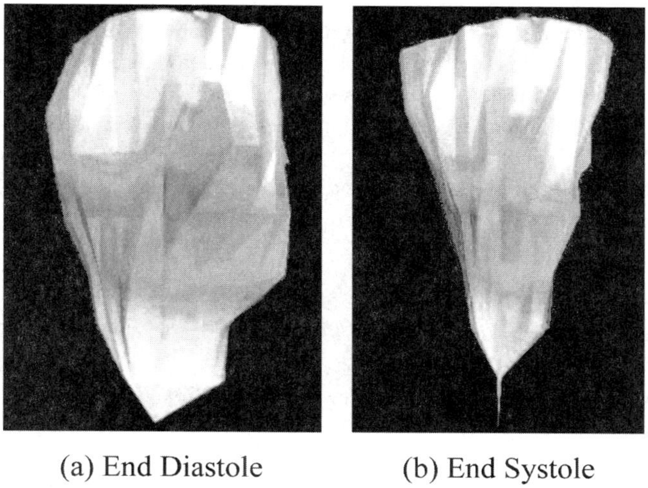

FIGURE 19.13 Fusion at the object level: the end-diastolic (a) and end-systolic (b) phases captured from the animated VRML files of a normal left ventricle.

automatic. A generic automatic registration scheme was then presented, which consisted of the selection of the image features that were used during the matching process, the definition of a MOM that quantified the spatial matching between the reference and the transformed image, and the application of an optimization technique that determined the independent parameters of the transformation model employed according to the MOM. The scheme formulates the problem of the medical image registration as an optimization problem employing SA and GA as two global optimization techniques, in conjunction with affine, bilinear, rigid, and projective transformation models. An elastic deformation model based on the use of the Kohonen neural network and the application of an elastic 3-D image warping method, using the TPS method, has been presented to capture local shape differences. This method may be applied sequentially to any combination of transformations and optimization techniques highlighted in the generic medical image registration scheme.

The aforementioned medical registration methods have been successfully applied on 2-D and 3-D medical data. For the 2-D retinal registration, three optimization methods (DSM, SA, and GA) have been evaluated in terms of accuracy and efficiency for retinal image registration. Results have been presented in the form of evolution of the optimization with the number of function evaluations averaged over a number of independent executions to compensate for the stochastic nature of the techniques. Results have also been presented in the form of the best achieved registration, both numerically and visually, for a sufficient number of data applied on three different retinal images: RF, FA, and IGC images. The experimental results showed the superiority of GA as a global optimization process in conjunction with the bilinear transformation model.

For the 3-D CT-MRI head registration, a global transformation to account for global misalignment and an elastic deformation method to capture local shape differences have been proposed. Since the registration is a surface-based method, it strongly depends on the initial segmentation step to obtain the CT-MR surfaces. The external skin surface has been selected as the common anatomical structure in both modalities for this specific application. The global transformation has been based on the novel implementation of the affine transformation method in conjunction with the GAs, as a global optimization process. The proposed elastic deformation method has been applied sequentially to the global transformation. Two novelties have been addressed through the application: (1) the introduction of the Kohonen neural network model to define an automatic one-to-one point correspondence between the transformed and the reference surfaces (the so-called interpolant points) and (2) the extended implementation of the TPS Interpolation method in three dimensions. The qualitative and the quantitative results for all CT-MR image pairs show the advantageous performance of the proposed automatic methods, in terms of surface distance (MOM), against the manual method.

Medical image fusion techniques have also been presented to increase diagnostic value from the combination of information from medical images of different modalities. The fusion process is required prior the application of a medical image registration scheme. Two main categories of the medical image fusion have been addressed: fusion at the image level and fusion at the object level. The most representative techniques of these categories have been revised and broadened throughout the chapter to include several methods, although more extensive research toward medical image fusion is required.

References

1. P. A. Van den Elsen, E. J. Pol, and M. A. Viergever, Medical image matching: a review with classification, *IEEE Eng. Med. Biol.*, 12(1), 26–39, 1993.
2. C. R. Maurer and J. M. Fitzpatrick, A review of medical image registration, in *Interactive Image Guided Neurosurgery*, R. J. Maciunas, Ed., Am. Assoc. Neurological Surgeons, Park Ridge, IL, pp. 17–49, 1993.
3. D. J. Hawks, D. L. G. Hill, and E. C. M. L. Bracey, Multi-modal data fusion to combine anatomical and physiological information in the head and the heart, in *Cardiovascular Nuclear Medicine and MRI*, J. H. C. Reiber and E. E. Van der Wall, Eds., Kluwer Academic Publishers, Dordrecht, the Netherlands, pp. 113–130, 1992.
4. C. R. Maurer, J. M. Fitzpatrick, M. Y. Wang, R. L. Galloway, R. J. Maciunas, and G. G. Allen, Registration of head volume images using implentable fiducial markers, *IEEE Trans. Med. Imaging*, 16, 447–462, 1997.
5. W. E. L. Grimpson, G. J. Ettinger, S. J. White, T. Lozano-Perez, W. M. Wells, and R. Kikinis, An automatic registration method for frameless stereotaxy, image guided surgery, and enhanced reality visulization, *IEEE Trans. Med. Imaging*, 15, 129–140, 1996.
6. V. Morgioj, A. Brusa, G. Loi, E. Pignoli, A. Gramanglia, M. Scarcetti, E. Bomburdieri, and R. Marchesini, Accuracy evaluation of fusion of CT, MR, and SPECT images using commercially available software packages (SRS Proto with IFS), *Int. J. Radiat. Oncol. Biol. Phys.*, 43(1), 227–234, 1995.
7. P. Clarysse, D. Gibon, J. Rousseau, S. Blond, C. Vasseur, et al., A computer-assisted system for 3-D frameless localization in stereotaxic MRI, *IEEE Trans. Med. Imaging*, 10, 523–529, 1991.
8. D. L. G. Hill, D. J. Hawkes, J. E. Crossman, M. J. Gleeson, T. C. S. Cox, et al., Registration of MR and CT images for skull base surgery using point-like anatomical features, *Br. J. Radiol.*, 64, 1030–1035, 1991.
9. C. Evans, T. M. Peters, D. L. Collins, C. J. Henri, S. Murrett, G. S. Pike, and W. Dai, 3-D correlative imaging and segmentation of cerebral anatomy, function and vasculature, *Automedica*, 14, 65–69, 1992.
10. R. Amdur, D. Gladstone, K. Leopold, and R. D. Hasis, Prostate seed implant quality assessment using MR and CT image fusion, *Int. J. Oncol. Biol. Phys.*, 43, 67–72, 1999.
11. F. L. Bookstein, Principal warps: thin-plate splines and the decomposition of deformation, *IEEE Trans. Pattern Anal. Mach. Intell.*, 11, 567–585, 1989.
12. C. A. Pelizzari, G. T. Y. Chen, D. R. Spelbring, R. R. Weichselbaum, and C. T. Chen, Accurate three-dimensional registration of CT, PET and/or MR images of the brain, *J. Comput. Assist. Tomogr.*, 13, 20–26, 1989.
13. W. Press, B. Flannery, S. Teukolsky, and W. Vetterling, *Numerical Recipes in C*, 2nd edition, Cambridge University Press, London, 1992.
14. D. Kozinska, O. J. Tretiak, and J. Nissanov, Multidimensional alignment using Euclidean distance transform, *Graphical Models Image Proces.*, 59, 373–387, 1997.
15. G. Borgerfors, Multidimensional chamfer matching: a tree edge matching algorithm, IEEE *Trans. Pattern Anal. Mach. Intell.*, 10, 849–865, 1988.

16. M. Van Herk and H. M. Kooy, Automatic three-dimensional correlation of CT-CT, CT-MRI, and CT-SPECT using chamfer matching, *Med. Phys.*, 21, 1163–1178, 1994.
17. M. Jiang, R. A. Robb, and K. J. Molton, A new approach to 3-D registration of multimodality medical image by surface matching, *Visualization in Biomedical Computing*, Proc. SPIE-1808, Int. Soc. Opt. Eng., Bellingham, WA, pp. 196–213, 1992.
18. P. J. Besl and N. D. McKay, A method for registration of 3-D shapes, *IEEE Trans. Pattern Anal. Mach. Intell.*, 14, 239–256, 1992.
19. A. Collignon, D. Vanderaeulen, P. Suetens, and G. Marshal, Registration of 3-D multimodality medical images using surfaces and point landmarks, *Pattern Recogn. Lett.*, 15, 461–467, 1994.
20. A. Maurer, G. B. Aboutanos, B. M. Dawant, R. J. Maciunas, and J. M. Fitzpatrick, Registration of 3-D images using weighted geometrical features, *IEEE Trans. Med. Imaging*, 15, 836–849, 1996.
21. J. L. Herring, B. M. Dawant, C. R. Maurer, D. M. Muratore, G. L. Galloway, and J. M. Fitzpatrick, Surface-based registration of CT images to physical space for image guided surgery of the spliene: a sensitivity Study, *IEEE Trans. Med. Imaging*, 17, 743- 752, 1998.
22. J. Y. Chianos and B. J. Sallivan, Coincident bit counting — a new criterion for image registration, *IEEE Trans. Med. Imaging*, 12, 30–38, 1992.
23. T. Radcliffe, R. Rajapekshe, and S. Shaler, Pseudocorrelation: a fast, robust, absolute, gray level image alignment algorithms, *Med. Phys.*, 41, 761–769, 1994.
24. J. B. A. Maintz, P. A. van den Elsen, and M. A. Viergever, Comparison of feature based matching of CT and MR brain images, in *Computer Vision, Virtual Reality, and Robotics in Medicine*, N. Ayache, Ed., Springer-Verlag, Berlin, pp. 219–228, 1995.
25. J. B. A. Maintz, P. A. van den Elsen, and M. A. Viergever, Evaluation of ridge seeking operators for multimodality medical image matching, *IEEE Trans. Pattern Anal. Mach. Intell.*, 18, 353–365, 1996.
26. P. A. Van den Elsen, J. B. A. Maintz, E. J. D. Pol, and M. N. Viergever, Automatic registration of CT and MR brain images using correlation of geometrical features, *IEEE Trans. Med. Imaging*, 14, 384–396, 1995.
27. P. A. Van den Elsen, E. J. D. Pol, T. S. Samanaweera, P. F. Hemler, S. Napel, and S. R. Adler, Grey value correlation techniques used for automatic matching of CT and MRI brain and spine images, *Visualization Biomed. Comput. 1994*, 2359, 227–237, 1994.
28. C. Studholme, D. L. G. Hill, and D. J. Hawkes, Automated registration of truncated MR and CT datasets of the head, *Proc. Br. Mach. Vision Assoc.*, 27–36, 1995.
29. R. P. Woods, J. C. Mazziotta, and S. R. Cherry, MRI — PET registration with automated algorithm, *J. Comput. Assist. Tomogr.*, 97, 536–546, 1993.
30. A. Collignon, F. Maes, D. Delaere, D. Vandermeulen, P. Suetens, and G. Marshal, Automated multimodality image registration based on information theory, in *Information Processing in Medical Imaging 1995*, Y. Bizais, C. Barillot, and R. Di Paola, Eds., Kluwer Academic Publishers, Dordrecht, the Netherlands, pp. 263–274, 1995.
31. F. Maes, A. Collignon, D. Vandermeulen, G. Marchal, and P. Suetens, Multimodality image registration by maximization of mutual information, *IEEE Trans. Med. Imaging*, 16, 167–198, 1997.
32. J. West, J. M. Fitzpatrick, M. Y. Wang, B. M. Dawant, C. R. Maurer, R. M. Kassler, and R. J. Maciunas, Retrospective intermodality registration techniques for images of the head: surface-based versus volume-based, *IEEE Trans. Med. Imaging*, 18, 147–150, 1999.
33. D. Lemoine, C. Barillot, C. Cibaud, and E. Pasqualinin, A 3-D CT stereoscopic deformation method to merge multimodality images and atlas datas, in *Proc. of Computer Assisted Radiology (CAR)*, Springer-Verlag, Berlin, pp. 663–668, 1991.
34. J. C. Gee, M. Reivich, and R. Bajscy, Elastically deforming a 3-D atlas to match anatomical brain images, *J. Comput. Assist. Tomogr.*, 17, 225–236, 1993.
35. L. Bookstein, Thin Plate Splines and the atlas problem for biomedical image, in *Lectures Notes in Computer Sciences, Information Processing in Medical Imaging*, Vol. 511, A. C. F. Colchester and D. J. Hawkes, Eds., Springer-Verlag, Berlin, pp. 326–342, 1991.

36. R. Bajcsy and S. Kovacic, Multiresolution elastic matching, *Comput. Vision Graph Image Process.*, 46, 1–29, 1989.
37. M. Singh, R. R. Brechner, and V. W. Henderson, Neuromagnetic localization using magnetic resonance images, *IEEE Trans. Med. Imaging*, 11, 125–134, 1984.
38. M. Moshfeghi, Elastic matching of multimodality medical images, *Graphical Models Image Process.*, 53, 271–282, May 1991.
39. C. Davatzikos, J. L. Prince, and R. N. Bryan, Image registration based on boundary mapping, *IEEE Trans. Med. Imaging*, 15, 112–115, 1996.
40. H. Chang and J. M. Fitzpatrick, A technique for accurate magnetic resonance imaging in the presence of field inhomogeneities, *IEEE Trans. Med. Imaging*, 11, 319–329, 1992.
41. J. Michiels, H. Bosmans, P. Pelgrims, D. Vandermeulen, J. Gybels, G. Marshal, and P. Suetens, On the problem of geometric distortion of MR images for stereostatic neurosurgery, *Magn. Reson. Imaging*, 12, 749–764, 1994.
42. J. M. Fitzpatrick, J. B. West, and C. R. Maurer, Predicting error in rigid-body point-based registration, *IEEE Trans. Med. Imaging*, 17, 694–702, 1998.
43. C. R. Maurer, R. J. Maciunas, and J. M. Fitzpatrick, Registration of head CT images to physical space using a weighted combination of points and surfaces, *IEEE Trans. Med. Imaging*, 17, 753–761, 1998.
44. C. Barillot, B. Gibaud, J. C. Gee, and D. Lemoine, Segmentation and fusion of multimodality and multisubjects data for the preparation of neurosurgical procedures, in *Medical Imaging: Analysis on Multimodality 2-D/3-D Image*, L. Beolchi and M. H. Kuhn, Eds., IOS Press, pp. 70–82, 1995.
45. L. Schad, R. Boesecke, W. Schlegel, G. Hartmann, V. Sturm, et al., Three-dimensional image correlation of CT, MR and PET studies in radiotherapy treatment planning of brain tumors, *J. Comput. Assist. Tomogr.*, 11(6), 948–954, 1987.
46. D. Hawks, D. L. G. Hill, and E. C. M. L. Bracey, Multi-modal data fusion to combine anatomical and physiological information in the head and the heart, in *Cardiovascular Nuclear Medicine and MRI*, J. H. C. Reiver and E. E. Van der Wall, Eds., Kluwer Academic Publishers, Dordrecht, the Netherlands, pp. 113–130, 1992.
47. K. S. Arun, T. S. Huang, and S. Blostein, Least-squares fitting of two 3-D sets, *IEEE Trans. Pattern Anal. Mach. Intell.*, 9(5), 698–700, 1987.
48. M. Singh, W. Frei, T. Shibata, G. Huth, and N. Telfer, A digital technique for accurate change detection in nuclear medical images — with application to myocardial perfusion studies using thallium-201, *IEEE Trans. Nucl. Sci.*, 26(1), 565–575, 1979.
49. E. Kramer, M. Noz, J. Sanger, A. Megibaw, and G. Maguire, CT-SPECT fusion to correlate radiolabeled monoclonal antibody uptake with abdominal CT findings, *Radiology*, 172(3), 861–865, 1989.
50. K. Toennies, J. Udupa, G. Herman, I. Wornom, and S. Buchman, Registration of 3D objects and surfaces, *IEEE Comput. Graphics Appl.*, 10(3), 52–62, 1990.
51. E Peli, R. Augliere, and G. Timberlake, Feature-based registration of retinal images, *IEEE Trans. Med. Imaging*, 6, 272–278, 1987.
52. L. Junck, J. G. Moen, G. D. Hutchins, M. B. Brown, and D. E. Kuhl, Correlation methods for the centering, rotation and alignment of functional brain images, *J. Nucl. Med.*, 31(7), 1220–1226, 1990.
53. A. Appicella, J. H. Nagel, and R. Duara, Fast multimodality image matching, in *Ann. Int. Conf. IEEE Eng. Med. Biol. Soc.*, Vol. 10, IEEE Comp Soc., Los Alamos, pp. 414–415, 1988.
54. A. Venot and V. Leclerc, Automated correction of patient motion and gray values prior to subtraction in digitized angiography, *IEEE Trans. Med. Imaging*, 3(4), 179–186, 1984.
55. S. L. Jacoby, J. S. Kowalik, and J. T. Pizzo, *Iterative Methods for Nonlinear Optimization Problems*, Prentice-Hall, Englewood Cliffs, NJ, 1972.
56. C. Fuh, and P. Maragos, Motion displacement estimation using an affine model for image matching, *Opt. Eng.*, 30, 881–887, 1991.

57. E. Aarts and Van Laardhoven, *Simulated Annealing: Theory and Practice*, John Wiley & Sons, New York, 1987.
58. D. Goldberg, *Genetic Algorithms in Optimization, Search and Machine Learning*, Addison-Wesley, Reading, MA, 1989.
59. H. Haneishi, T. Masuda, N. Ohyama, T. Honda, and J. Tsujiuchi, Analysis of the cost function used in simulated annealing for CT image reconstruction, *Appl. Opt.*, 29(2), 259–265, 1990.
60. D. E. Palmer, C. Pattaroni, K. Nunami, R. K. Chadha, M. Goodman, T. Wakamiya, K. Fukase, S. Horimoto, M. Kitazawa, H. Fujita, A. Kubo, and T. Shiba, Effects of dehydroalanine on peptide conformations, *J. Am. Chem. Soc.*, 114(14), 5634–5642, 1992.
61. L. Ingber, Statistical mechanics of neocortical interactions: a scaling paradigm applied to electroencephalography, *Phys. Rev.*, A 44(6), 4017–4060, 1991.
62. M. S. Kim and C. Guest, Simulated annealing algorithm for binary phase only filters in pattern classification, *Applied Opt.*, 29(8), 1203–1208, 1990.
63. L. Ingber, Very fast simulated re-annealing, *J. Math. Comput. Modelling*, 12(8), 967–973, 1989.
64. L. Ingber, Simulated annealing practice versus theory, *J. Math. Comput. Modelling*, 18(11), 29–57, 1993.
65. N. Arad, N. Dyn, D. Reisfeld, and Y. Yeshurun, Image warping by radial basis functions: application to facial expressions, *J. Comput. Vision Graphics Image Process.*, 56, 161–172, 1994.
66. T. Kohonen, Self organized formation of topologically correct feature maps, *Biol. Cybernetics*, 43, 59–69, 1982.
67. J. Camp, B. Cameron, D. Blezek, and R. Robb, Virtual reality in medicine and biology, *Future Generation Comput. Syst.*, 14, 91–108, 1998.
68. T. Martinetz and K. Schulten, A neural gas network learns topologies, in *Artificial Neural Networks*, T. Kohonen et al., Eds., North-Holland, Amsterdam, pp. 397–402, 1991.
69. R. A. Robb and D. P. Hanson, The ANALYZE software system for visualization analysis in surgery simulation, in *Computer Integrated Surgery*, S. Lavalie et al., Eds., MIT Press, Cambridge, MA, 1993.
70. G. K. Matsopoulos, N. A. Mouravliansky, K. K. Delibasis, and K S. Nikita, Automatic registration of retinal images with global optimization techniques, *IEEE Trans. Inf. Technol. Bioeng.*, 3, 47–60, 1999.
71. K. K. Delibasis, G. K. Matsopoulos, N. A. Mouravliansky, and K. S. Nikita, Efficient implementation of the marching cubes algorithm for rendering medical data, *Lecture Notes in Computer Science, High Performance Computing and Networking*, in Proc. 7th Int. Conf., HPCN 1593, P. Sloot, M. Bubak, A. Hoekstra and B. Hertzberger, Eds., Springer-Verlag, Berlin, pp. 989–999, 1999.
72. G. K. Matsopoulos, K K. Delibasis, N. Mouravliansky, and K. S. Nikita, Unsupervised learning for automatic CT — MR image registration with local deformations, European Medical & Biological Engineering Conference, EMBEC'99, Vienna, November 1999.
73. J. Gomes, L. Darsa, B. Costa, and L. Velho, *Warping and Morphing of Graphical Objects*, Morgan Kaufman Publishers, Inc., San Francisco, CA, 1998.
74. A. K. Jain and R. C. Dubes, *Algorithms for Clustering Data*, Prentice-Hall, Englewood Cliffs, NJ, 1988.
75. R. Rezaee, C. Nyqvist, P. van der Zwet, E. Jansen, and J. Reiber, Segmentation of MR images by a fuzzy C-means algorithm, *Comput. Cardiol.*, 21–24, 1995.
76. J. Bezdek, Partition structures: a tutorial, in *The Analysis of Fuzzy Information*, B. Raton, Ed., CRC Press, Boca Raton, 1987.
77. P. Sharp, H. Gemmel, and F. Smith, *Practical Nuclear Medicine*, IRL Press, Oxford, 1989.
78. E. Wall, Magnetic resonance in cardiology: which clinical questions can be answered now and in the near future?, in *What's New in Cardiovascular Imaging*, J. Reiber and E. Wall, Eds., Kluwer Academic Publishers, Dordrecht, the Netherlands, pp. 197–206, 1998.

79. K. K. Delibasis, N. Mouravliansky, G. K. Matsopoulos, K. S. Nikita, and A. Marsh, MR functional cardiac imaging: segmentation, measurement and WWW based visualization of 4D data, *Future Generation Comput. Syst.*, 15, 185–193, 1999.

20

The Role of Imaging in Radiotherapy Treatment Planning

Dimos Baltas
Strahlenklinik, Städtische Kliniken Offenbach

National Technical University of Athens

Natasa Milickovic
Strahlenklinik, Städtische Kliniken Offenbach

Christos Kolotas
Strahlenklinik, Städtische Kliniken Offenbach

Nikolaos Zamboglou
Strahlenklinik, Städtische Kliniken Offenbach

National Technical University of Athens

20.1 Introduction

The aim of radiation therapy is to deliver a prescribed dose of radiation to the target volume, while sparing surrounding healthy tissues and organs. In other words, the aim is to achieve the maximum therapeutic effect with the minimum risk of complications. Different technical methods and devices can be used,[1] depending on their availability and on the level of expertise of the radiation therapy staff.

It is common practice in modern radiation oncology that the decision for the treatment of a specific patient is a result of a multidisciplinary cooperation (pathologists, surgeons, and medical and radiation oncologists) based on clinical, histopathological, and imaging data. After the decision for radiation therapy is made, the next step is the planning of the therapy itself: the treatment planning step.

It is obvious that an accurate description of the target volume in three dimensions is a fundamental task in radiation therapy treatment planning, definitively influencing the outcome of the patient. We can postulate that the accuracy of the radiotherapy treatment planning depends directly on the accuracy of the delineation of the tumor and on the definition of the target volume.

Although radiotherapy treatment planning is mainly computed tomography (CT) based, other imaging modalities can provide complementary anatomical information for the tumor location. Magnetic resonance imaging (MRI) is actually more sensitive and can provide more specific information for brain tumors, head and neck localizations, and soft tissues. Other imaging modalities such as positron emission spectrometry (PET), single photon emission computed tomography (SPECT), and ultrasound can have an important contribution for some tumor localization and the definition of tumor extension including detection of the spread of lymph nodes.[2,3] Establishment of widely applicable techniques for image correlation for all these modalities could have a significant contribution to improvements in the treatment

0-8493-3691-0/01/$0.00+$.50

planning and, subsequently, to the treatment results. Furthermore, for some specific treatment planning procedures, it is of interest to correlate imaging data from different time points (time series) during the history of the cancer case. Such cases are the correlation of preoperative or prechemotherapy to post-treatment imaging data (data acquired during the radiotherapy treatment planning procedure).

In summary, imaging plays a central role in the treatment planning procedure in modern radiation therapy. We will focus our chapter on some examples of three-dimensional (3D) treatment planning procedures.

20.2 The Role of Imaging in the External Beam Treatment Planning

We can identify the following main tasks in the clinical treatment planning procedure:

- Identification and accurate definition of the 3D planning target volume (PTV)
- Design of a conformal treatment technique (treatment plan), where the 3D dose distribution is adjusted to the shape of the PTV
- Accurate 3D dose calculations taking into consideration the real tissue characteristics (inhomogeneities)
- Evaluation of the 3D dose distribution using effective tools
- Accurate treatment delivery that can be realized when using patient immobilization tools and verification systems able to visualize, under real treatment conditions, the position of the radiation fields to the patient anatomy (portal imaging devices)

The final achievable treatment accuracy, which also defines the success of treatment, depends on the accuracy levels that can be realized in the above mentioned treatment planning steps.

The accuracy of definition of the 3D PTV, the first member in the chain of treatment planning procedure, will predefine the quality and level of accuracy of the following steps. That is why defining target volumes in a consistent and accurate way is the most important and critical task for the radiation oncologist in this procedure. In addition, this task is also the most time-consuming.

According to ICRU 50,[1] the following explains the procedure of defining PTV. First the gross tumor volume (GTV) has to be identified. This is the palpable or visible part of the tumor. Consequently, the clinical target volume (CTV) is defined as the volume that has to be treated adequately in order to achieve the aim of therapy, which is cure or palliation. CTV is built from GTV by adding adequate margins to account for subclinical or microscopic extensions of the malignancy. PTV is further built on CTV by adding geometrical margins around the CTV. These margins have to consider (1) movements of tissues inside CTV, (2) variations in size and shape of the tissues contained by the CTV, and (3) day-to-day variations in patient set-up and alignment.

The margins applied to CTV for deriving PTV depend on the localization and the immobilization facilities available for these locations and are generally clinic dependent. Usually applied values are isotropic margin values of 5 to 10 mm.

In practice, the GTV is identified using CT or magnetic resonance (MR) imaging modalities. In contrast to PTV, the margin used to derive CTV from GTV depend on the ability to identify borders between the tumor (GTV) and its neighboring tissues using imaging modalities like CT, MR, SPECT, or PET and on the clinician's knowledge of the disease process. In some cases, multimodality imaging can assist the clinician in deriving CTV very effectively.

An accurate definition of PTV requests the availability of clinician-friendly, fast, flexible, multimodality imaging supporting computerized systems. In the following discussion, tools developed to simplify the PTV definition, as well as to assist clinicians in evaluating and effectively visualizing treatment techniques, will be presented.

20.2.1 Digitally Reconstructed Radiographs (DRRS)

In conventional radiation therapy, X-ray images (films or fluoroscopy) made under the same geometrical conditions as in the therapy unit (usually a accelerator) are made using a dedicated X-ray machine called a simulator. This machine is able to *simulate* any geometrical parameter such as movement, rotations, and radiation field that can be selected in the real treatment machine and to visualize the therapeutic beam applying diagnostic X-ray energies (50 to 120 keV). This machine enables high image quality and anatomical resolution, which are difficult to achieve with the real, high energy therapeutic beam. These simulation images are then used as references for verification of the treatment delivery. Furthermore, such X-ray images are also applied for an initial localization of the PTV and the anatomical region of the patient that has to be treated. In past decades, tumor localization was mainly done on the simulator machine. This was clearly the influence of the radiology-oriented education of the physicians and also a result of the missing CT or MR imaging equipment at the clinics; even if such systems existed, they were under the control of the radiologists (diagnostic). At that time, CT was an instrument for diagnosis and not for treatment planning.

Modern radiation therapy is imaging based. CT scanning is the basic modality for the dosimetric computerized treatment planning for reasons that will be explained later in this chapter. CT scanners are more and more available in radiation oncology departments and are now accepted as essential instruments in the hands of radiation oncologists for planning and optimizing radiation therapy. Computer calculated radiographs based on CT volumetric information of the patient, the so-called DRRs, play a continuously increasing role in the treatment planning procedure. DRRs replace the conventional X-ray films and can be calculated for any beam configuration and also for that which cannot be simulated in a real simulator. In the following, some physical background for the methods of calculating DRR from CT imaging data will be described and briefly explained.

20.2.1.1 Measurement of the Hounsfield Unit (HU)-Electron Density Curve

20.2.1.1.1 Interaction of X-Ray Irradiation with Matter

When a beam of X-rays passes through matter, its intensity is reduced by an amount that is determined by the physical properties, thickness, density, and atomic number of the material through which the beam passes. The variations of these properties between the different kinds of human tissues lead to different attenuation values and differences in the amount of absorbed energy by the beam. In the case of conventional X-ray radiography, these differences create the anatomical details visible in the radiographic image.

Photons are absorbed when they collide with electrons, which means that the absorption of photons depends on the concentration of electrons within the material, the so-called electron density of the material. The number of electrons per cubic centimeter, ρ_e, is a very important dosimetric parameter of a material and is given by

$$\rho_e[e^-/cm^3] = \rho N(Z/A)$$

where N is Avogadro's number ($N = 6.0221367\ 10^{23}\ mol^{-1}$), Z is the atomic number, A is the atomic weight, and ρ is the mass density of the material.

The relation (Z/A) is approximately 0.5. (Z/A) varies less than 20% for the human tissues and is essentially constant, except for hydrogen (H) that does not have any neutron and has a (Z/A) ratio equal to 1. So, the only factor that could significantly change the electron density is the mass density of the material ρ. For mixtures, ρ is calculated as the weighted sum of the densities of the various elements of the mixture, where the weights are equal to the corresponding relative concentrations.

The effective atomic number for a compound or a mixture is given by

$$\bar{Z} = \sqrt[2.94]{a_1 Z_1^{2.94} + a_2 Z_2^{2.94} + a_3 Z_3^{2.94} + \ldots},$$

where $a_1, a_2, a_3, \ldots, a_n$ are the fractional contents of electrons belonging to elements $Z_1, Z_2, Z_3, \ldots, Z_n$, respectively. The exponent 2.94 is derived experimentally from the relationship between X-ray interactions and the atomic number. From this definition, for example, $Z_{air} = 7.64$, $Z_{water} = 7.42$, $Z_{muscle} = 7.46$, and $Z_{bone}(\text{femur}) = 14$[4].

Water is taken as the reference material for dosimetric purposes and calculations. The presence of different materials is commonly expressed as heterogeneity. The influence of heterogeneity on dosimetric parameters can be described using correction factors, which depend on the electron density of the specific material expressed relative to that of water, ρ_e^w. The accuracy of such heterogeneity corrections strongly depends on the accuracy of the estimation of ρ_e^w. There is a need for an accurate estimation of the ρ_e^w values for the human tissues from their measured HU numbers using a CT scanner. This will allow for a voxel-by-voxel correction and an accurate voxel-by-voxel calculation of the energy absorption.

In the following, we describe such a procedure of establishment of the relationship between HUs and ρ_e^w: $\rho_e^w = f(\text{HU})$. Such dependencies can be stored as tables in modern 3D treatment planning systems and are then used for an accurate calculation of the 3D dose distribution.

20.2.1.1.2 *Experimental Estimation of the HU↔ρ_e^w Relationship*

Computed axial transmission tomography imaging, the CT scanning, is based on taking a large number of one-dimensional (1D) attenuation profiles of a body slice from many different directions and reconstructing the anatomical detail within the slice. Digitalization is applied during the reconstruction process, resulting in a matrix where each element, pixel, is defined with the CT numbers expressed in HU. This specific value for each pixel is related to the density of the tissue in the corresponding matrix element. We use this property of CT scanning to establish a relationship between the CT numbers and their corresponding linear attenuation coefficients and relative electron densities. These data can further be used either for dosimetric purposes or in the process of DRR calculation.

As the chemical compounds of different tissues are well known from the literature,[5–10] the next step is to make a calibration of the CT scanning machine to obtain the curve describing the dependence of relative electron density on the Hounsfield number. This calibration was made using the 33-cm-diameter phantom, RMI Electron* Density Phantom model 465[11–14] (Figure 20.1). This phantom has 20 2.8-cm-diameter holes, into which 6 rods made of commercially available plastics and 11 rods simulating different tissues with known elemental composition and electron densities are inserted.[11–14] The remaining 3 rods are made of solid water, the same material as the bulk of the phantom. In Table 20.1, data for the 16 different materials of the RMI Electron Density Phantom model 465 are given. Mass density ρ, relative electron density ρ_e^w, and HU values for seven typical human tissues or compounds are given in Table 20.2.

The relationship between relative electron densities and Hounsfield numbers is described by a density conversion function, HU to ρ_e^w. By scanning the described phantom, the HU values of each rod with known ρ_e^w can be obtained, and these values are then used to establish a scanner-specific density conversion function.

1. For known relative electron density of material and unknown CT number,

$$\begin{aligned}
\text{if}(\tilde{n}_{ei}^w \le 1.001) \Rightarrow \text{HU}_i &= 1018.608 \cdot \tilde{n}_{ei}^w - 1012.1429 \\
\text{if}(1.001 < \tilde{n}_{ei}^w \le 1.083) \Rightarrow \text{HU}_i &= 541.2442 \cdot \tilde{n}_{ei}^w - 527.635 \\
\text{if}(\tilde{n}_{ei}^w > 1.083) \Rightarrow \text{HU}_i &= 1953.923 \cdot \tilde{n}_{ei}^w - 2051.391
\end{aligned}$$

2. For known CT number of material and unknown relative density coefficient,

$$\begin{aligned}
\text{if}(\text{HU}_i \le 7) \Rightarrow \tilde{n}_{ei}^w &= 0.00098 \cdot \text{HU}_i - 0.994 \\
\text{if}(7 < \text{HU}_i \le 58) \Rightarrow \tilde{n}_{ei}^w &= 0.00185 \cdot \text{HU}_i - 0.975 \\
\text{if}(\text{HU}_i > 58) \Rightarrow \tilde{n}_{ei}^w &= 0.00051 \cdot \text{HU}_i - 1.05
\end{aligned}$$

* Radiation Measurements Inc., Middleton, WI.

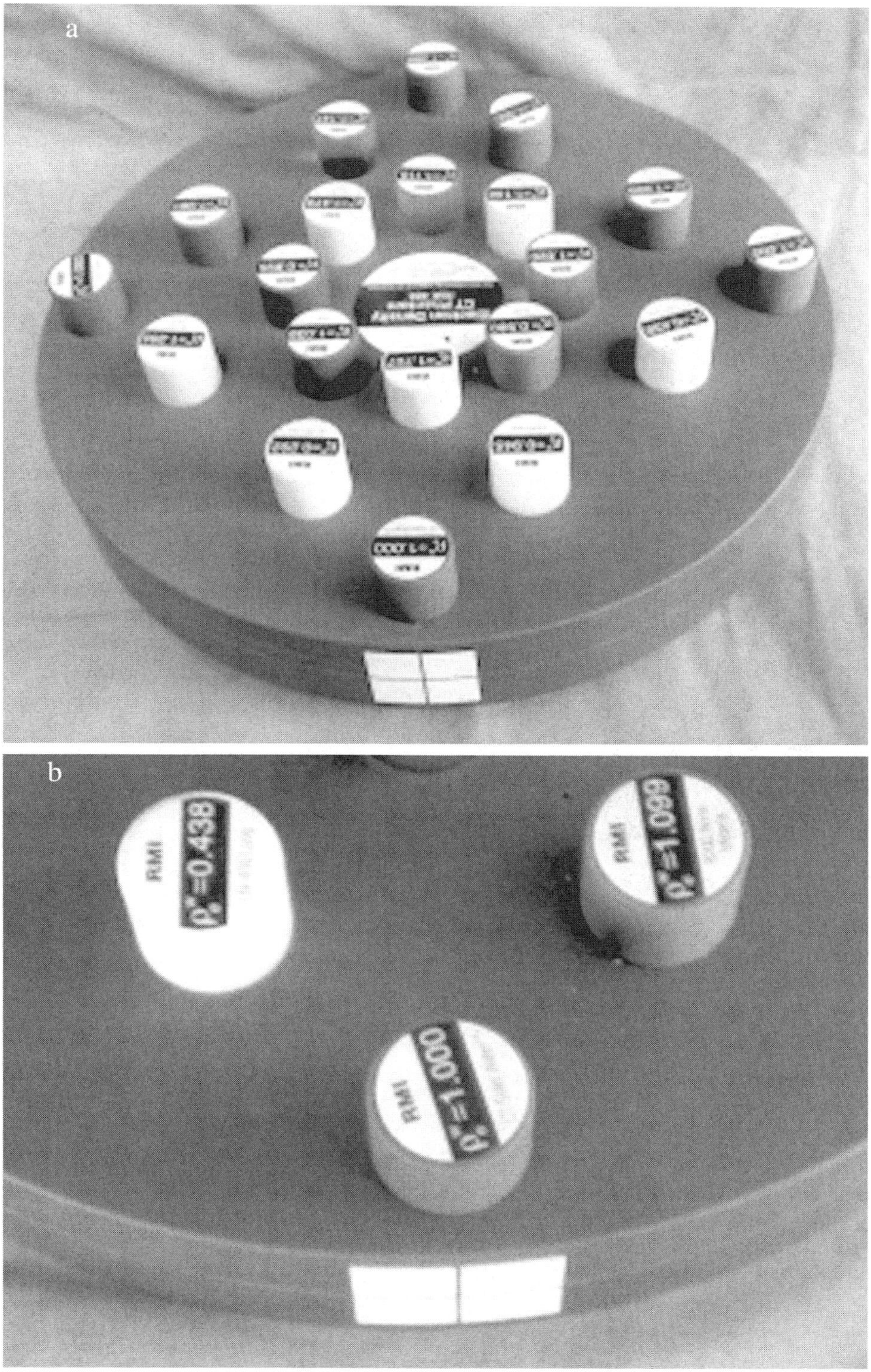

FIGURE 20.1 Photos (a, b) and CT scan (c) of an RMI Electron Density Phantom model 465 that was used for establishing a scanner-specific density function. *(continued)*

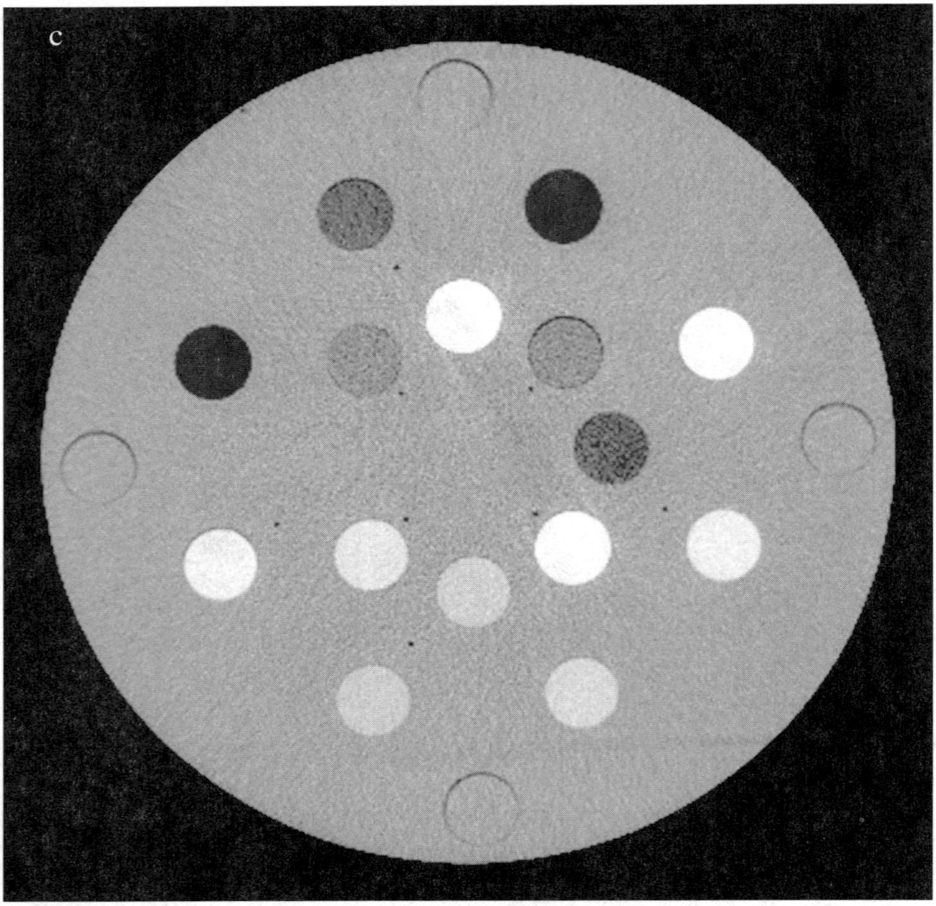

FIGURE 20.1 (CONTINUED) Photos (a, b) and CT scan (c) of an RMI Electron Density Phantom model 465 that was used for establishing a scanner-specific density function.

TABLE 20.1 Mass Density ρ, Electron Density ρ_e^w, and CT Numbers for Different Materials of the RMI Electron Density Phantom Model 465

Material Type	ρ (g/cm³)	ρ_e^w (e⁻/cm³)	CT Number (HU)
LN300 (lung)	0.300	0.292	–706.3 ± 27.3
LN450 (lung)	0.450	0.438	–560.9 + –16.6
AP6 (adipose)	0.920	0.895	–86.8 ± 18.1
AV/BR/SRI (breast)	0.990	0.980	–20.3 ± 15.1
CT Solid Water™	1.015	1.000	0 ± 0.5
CB3 (resin)	1.020	1.020	10.9 ± 13.7
BRN/SR2 (brain)	1.045	1.039	30.9 ± 15.2
LVI (liver)	1.080	1.050	94.5 ± 17.2
Inner bone	1.120	1.081	60.3 ± 17.9
B200 bone mineral	1.145	1.099	238 ± 19.6
CB4 (resin)	1.150	1.116	115.4 ± 15.2
CB2 (10% $CaCO_3$)	1.170	1.142	181.9 ± 13.6
Acrylic	1.180	1.147	138.4 ± 13.1
CB2 (30% $CaCO_3$)	1.340	1.285	469.5 ± 17.2
CB2 (50% $CaCO_3$)	1.560	1.473	840.3 ± 19.7
SB3 (cortical bone)	1.840	1.707	1273.1 ± 19.8

TABLE 20.2 Mass Density ρ, Relative Electron Density ρ_e^w, and Measured CT Numbers (HU) for Different Human Tissues and Compounds

Material Type	ρ (g/cm^3)	ρ_e^w (e$^-$/cm^3)	CT Number (HU)
Water	1.000	1	6.46
Muscle	1.060	0.973–0.994	−21.04/0.35
Bone	1.090–1.920	0.980–1.781*	−11.87/1428
Lung	0.260–1.050	0.258–1.042	−749.34/36.5
Fat	0.920–0.940	~0.925	−70
Air	0.0012	0.0001	−1011.12/−1012
Liver	1.050–1.070	0.994–1.000	0.35/6.46

Figure 20.2 shows the resulting calibration curve for our Somatom Plus 4 CT scanner.*

The relative electron density conversion function varies between CT scanners, and the scanner-specific density function must be derived for each scanner. The scanner-specific parameters, such as the kVp, beam quality, beam hardening, filter, and reconstruction algorithm, can affect the CT number of each volume element (voxel) because the CT number is related to the attenuation coefficient of the tissue in the voxel. Periodical checks of the scanner-specific function $\rho_e^w = f(\mathrm{HU})$ are recommended.

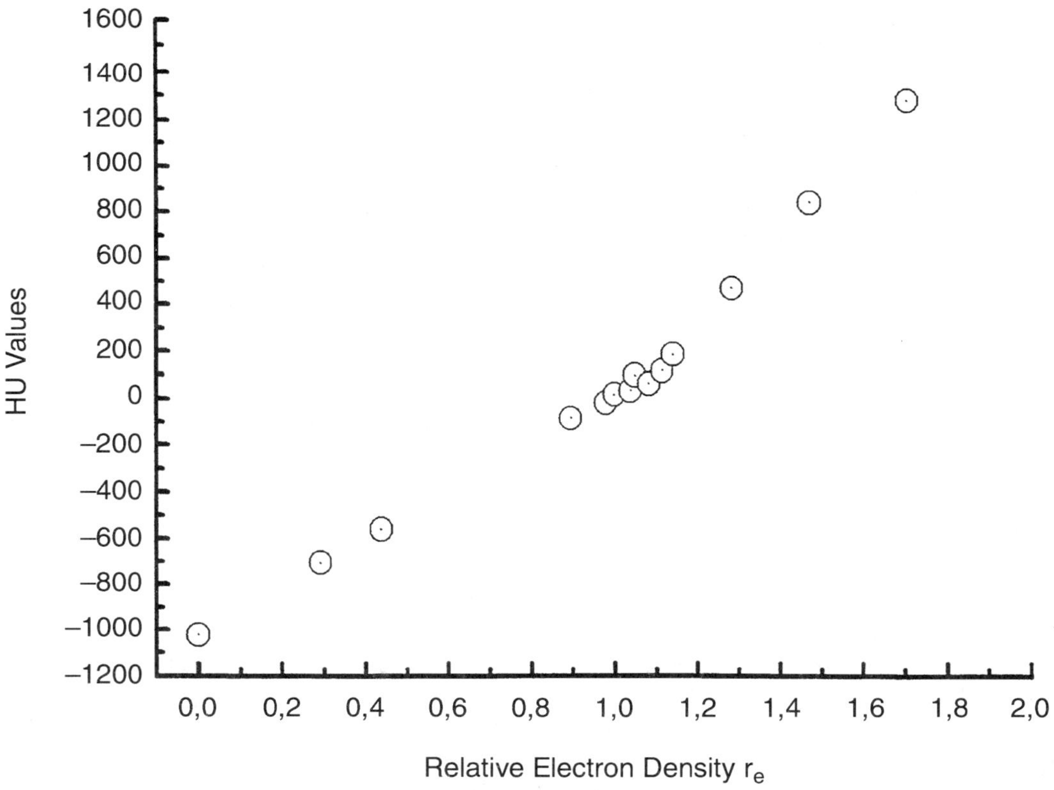

FIGURE 20.2 A calibration curve relating the CT number (HU) to the relative electron density (ρ_e). *(conti-*

* Siemens, Medizinische Technik, Erlangen, Germany.

20.2.1.1.3 Experimental Estimation of the HU↔μ_e^w Relationship

The whole range of CT numbers, [−1500 ÷ 4500] HU, is subdivided in 23 subranges defined by materials of known relative electron density and chemical composition. Using the XCOM[14] program, we then calculate the linear attenuation coefficients of those 23 materials for the selected accelerator voltage *V*. The linear attenuation coefficient μ of material *x* having the CT number $HU_x \in [HU_a, HU_b]$, where $[HU_a, HU_b]$ is one of 23 subranges, for voltage *V* is calculated by linear interpolation between the μ_a and μ_b:

$$i_x = \frac{HU_b - HU_x}{HU_b - HU_a} \cdot i_a + \frac{HU_x - HU_a}{HU_b - HU_a} \cdot i_b \,.$$

In this way, inhomogeneity correction factors can be included in dose calculations, improving the accuracy of dosimetric treatment planning.

The linear attenuation coefficient μ_x for the applied CT voltage is given by:[15,16]

$$\mu_x = [(HU_x/1000) + 1] \cdot \mu_w,$$

where μ_w is the linear attenuation coefficient of water. This established relationship is used in the calculation process of DRR. A brief discussion of this procedure is given in the following section.

20.2.1.2 Estimation of Parameters for Calculation of DRRs

Volume visualization of scalar data, or volume rendering, is the process of generating a two-dimensional (2D) image from a scalar function defined over some region of a 3D space. In radiology, this is used to produce DRRs from the 3D data set formed of patient CT images.

We will present an algorithm for pseudo-simulation[17,18] of the X-ray passing from the accelerator source through the human tissue on the DRR image for selected accelerator energy. The importance of scattered radiation in dosimetry must not be underestimated, but in DRR calculation, we get much better image quality by calculation of only primary energy attenuation. To do this, we establish the relationship $HU_x \leftrightarrow \mu_x$, $x = 1, \ldots, 6000$, for all HU∈ [−1500 ÷ 4500], for all accelerator voltages of interest. These data are stored as a set of look-up tables.

The basic DRR calculation algorithm is presented in Figure 20.3, and resulting images for six representative energies are shown in Figure 20.4. The loss of details in DRR images can be noticed as the accelerator voltage increases. This is caused by the decrease of difference between the linear attenuation coefficients of different human materials with the increase of accelerator voltage[4,17,18] (Figure 20.5). This effect occurs for the following reasons. The photoelectric cross section varies with atomic number as Z^4 and with photon energy approximately as E^{-3}. In this way, the photoeffect plays an important role up to the energy of 150 keV. Between 150 keV and 24 MeV, the Compton effect, which does not depend on the atomic number Z, is dominant and causes images acquired in this energy region to have very pure contrast between the different materials.[4]

20.2.1.3 Clinical Examples of DRRs Application

In Figures 20.6 and 20.7, several options that are available for working with DRRs for a case of prostate cancer are demonstrated. In addition to the features available in the X-ray films, (1) control of the patient position based on the bony structures, (2) visualization of some normal tissues using contrast medium, and (3) field geometry visualization using the DRR of technology the oncological volumes GTV/CTV and PTV can also be projected and visualized on the images. This increases the efficiency of the localization and verification procedures. The images resulting from superimposing the divergent beam geometry on the patient anatomy are called DRR-based beam eye views (BEVs).

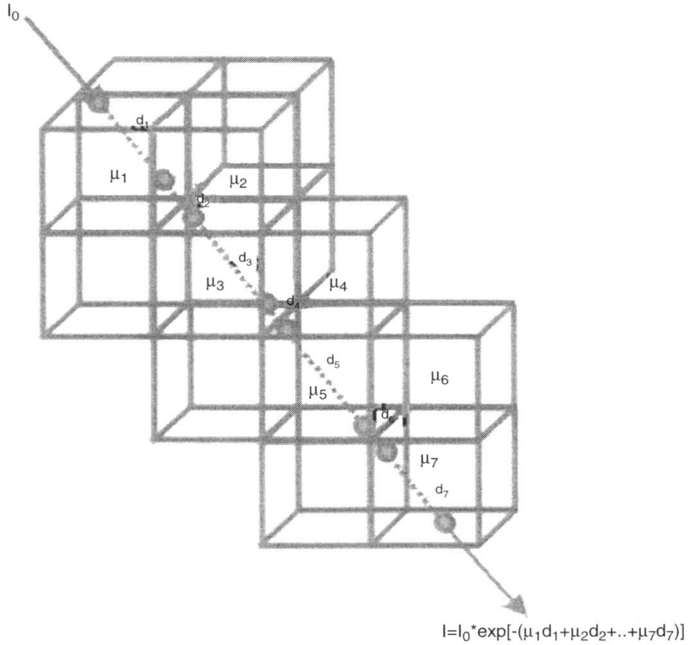

FIGURE 20.3 Ray tracing process presentation. I_0 is the initial ray energy, and I is an attenuated ray energy after the ray tracing through the volumetric data set is done. d_i is distance between the ray entrance and exit point in the voxel i, and μ_i is the linear attenuation coefficient of voxel i.

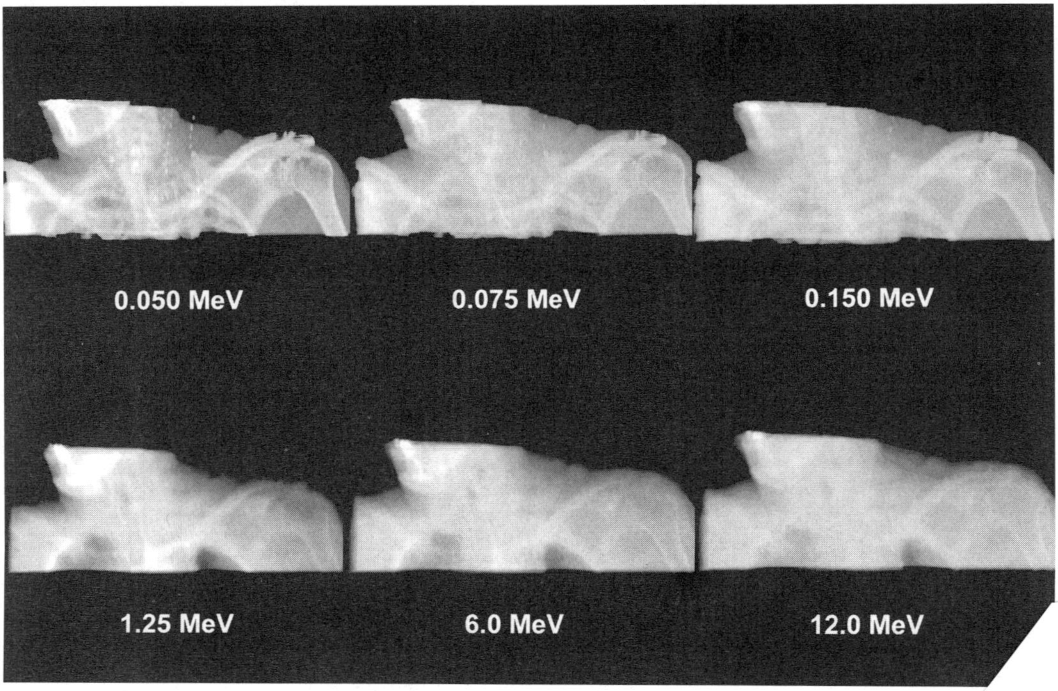

FIGURE 20.4 DRRs calculated for the different accelerator voltages.

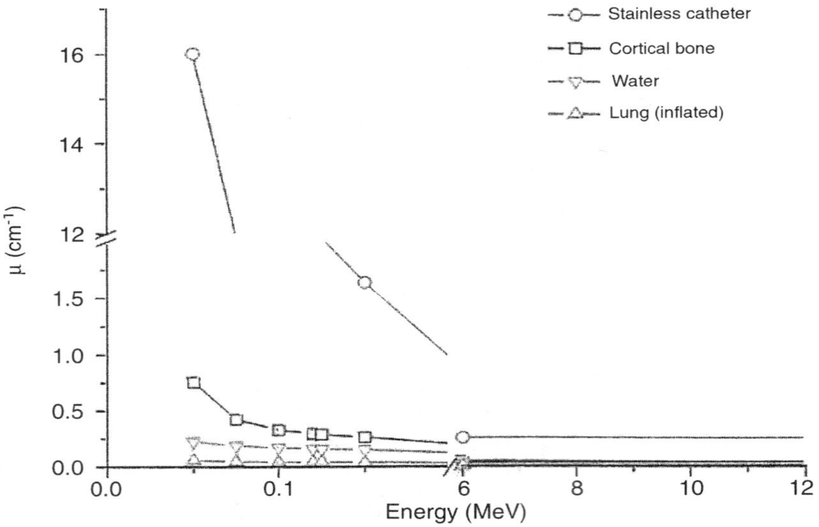

FIGURE 20.5 Region of lower energies obviously produces a better contrast for different materials (stainless catheter, cortical bone, water, and inflated lung, for example).

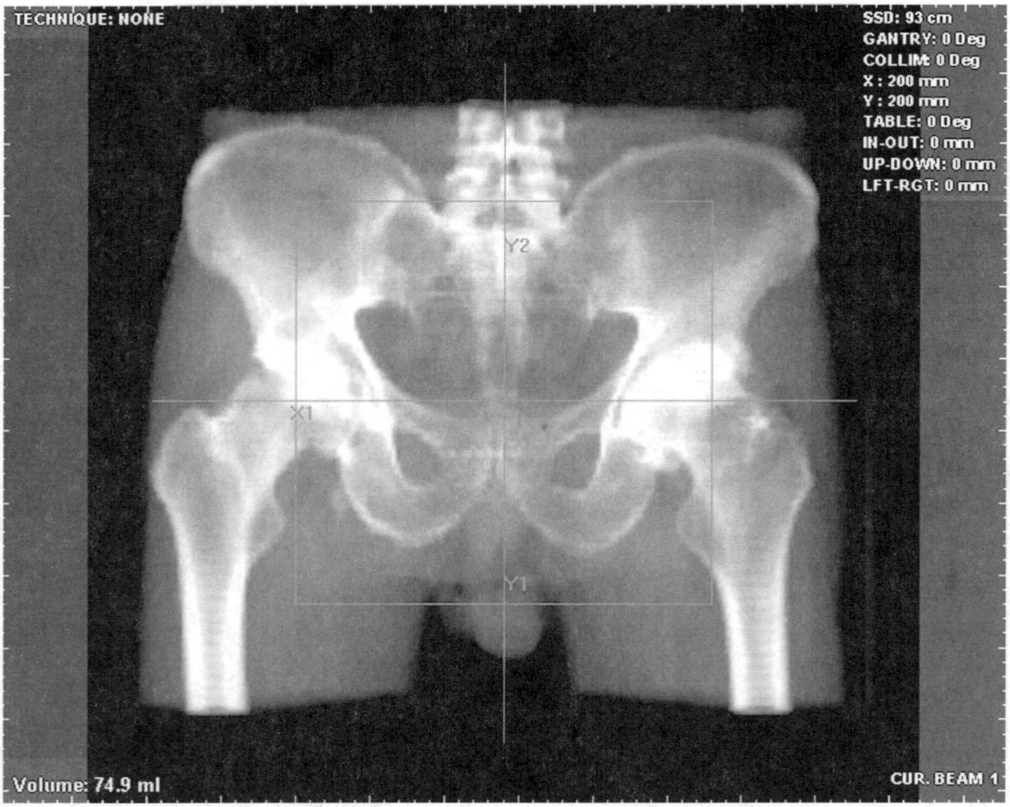

FIGURE 20.6 High quality DRRs calculated from a CT study using 3-mm slice thickness and 3-mm interslice distance for a prostate cancer case. These images were calculated using the EXOMIO CT-based simulation system (MedlinTec GmbH, Bochum, Germany). Original anatomy information available from CT and an arbitrary field configuration are visualized. Also, the GTV as delineated by the radiation oncologist on the CT images is also projected.

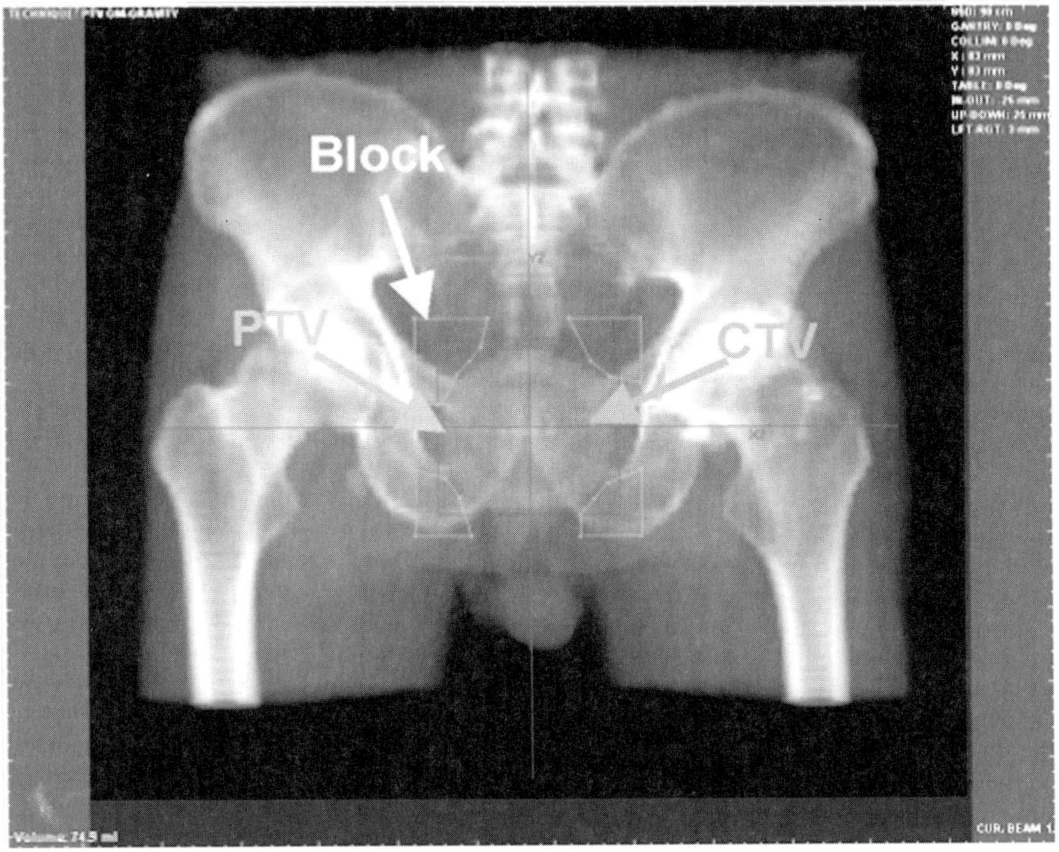

FIGURE 20.7 DRRs for the same case as in Figure 20.6. The GTV/CTV and PTV are extracted from the CTV by applying an isotropic margin of 10 mm and are superimposed to the computer-calculated radiological image. Planning or simulating a beam for the defined PTV using customized blocks (beam formers) is also presented.

20.2.2 Imaging Tools for PTV Definition

Traditionally, the PTV is delineated by the physician stepping from slice to slice (transaxial images). Modern software systems offer the possibility to first define the GTV/CTV contours on the CT images and then to extract the PTV by adding a margin around the GTV/CTV as discussed previously. Currently, there are also commercial systems available offering features for the delineation of the GTV/CTV on any plane: transaxial or reconstructed coronal and sagittal images. This simplifies and speeds up significantly the contouring procedure. In the case of regularly shaped tumor volumes, GTV/CTV can be adequately delineated by entering its contours on a total of 5 to 8 images: 3 to 5 transaxial images and a coronal and a sagittal image. In Figures 20.8 and 20.9, we demonstrate such features for the PTV definition using CT imaging for the case of a prostate cancer patient.

20.2.3 CT-Based Simulation

Conventional simulation was introduced in the 1960s. This technology offered a significant impro' ment in the accuracy and, consequently, the quality of radiotherapy treatment planning. Toda' and non-expensive CT scanners are available. This makes it possible for more and more radio' institutions to integrate such scanners into their daily treatment planning routine. Th' availability of CT enables the realization of treatment planning procedures based compl

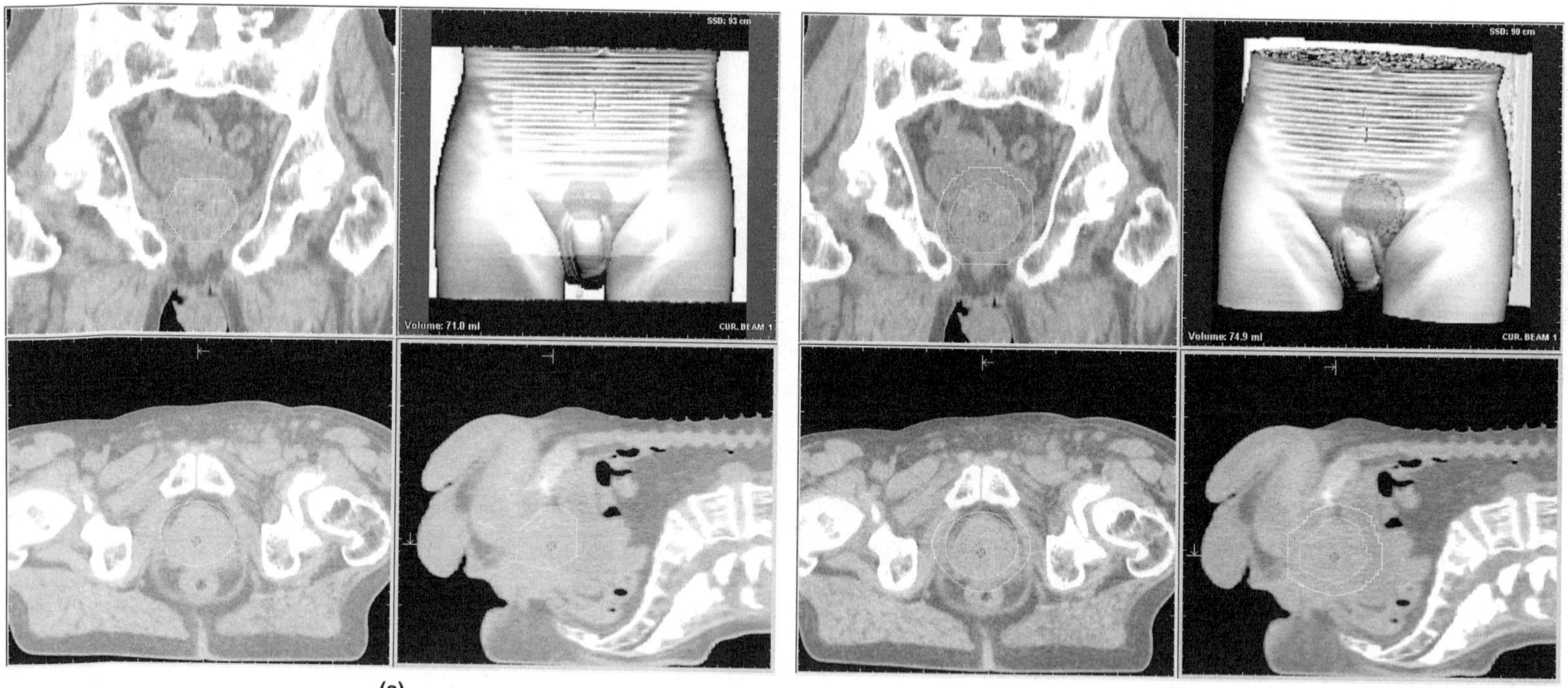

(a) (b)

FIGURE 20.8 Delineation of the GTV/CTV for the prostate cancer case using CT imaging and the EXOMIO software package. The CT study was carried out using a spiral CT: SOMATOM Plus $4^{\#}$ scanner and CT transaxial reconstructed images with 3-mm slice thickness and 3-mm slice distance so that a complete coverage of the related anatomical volume is guaranteed. GTV/CTV contouring is possible on transaxial images as well as on coronal and sagittal reconstructed images. On each image, the contours from the other two planes are projected (perpendicular lines). (a) EXOMIO can then reconstruct the 3D GTV/CTV volume from at least three such contour lines entered. The contours of the final reconstructed GTV/CTV are shown on all axial, coronal, and sagittal images. The GTV/CTV is also superimposed on the 3D surface rendering, visualizing the position/location of this volume in the patient body. (b) Derivation of the PTV by applying an isotropic 3D margin of 10 mm to the GTV/CTV volume working with the EXOMIO system. The GTV/CTV contours delineated by the physician and the resulting PTV contours are shown on the coronal (upper left), CT slice (lower left), and sagittal (lower right) images as well as on the 3D surface rendered image (upper right).

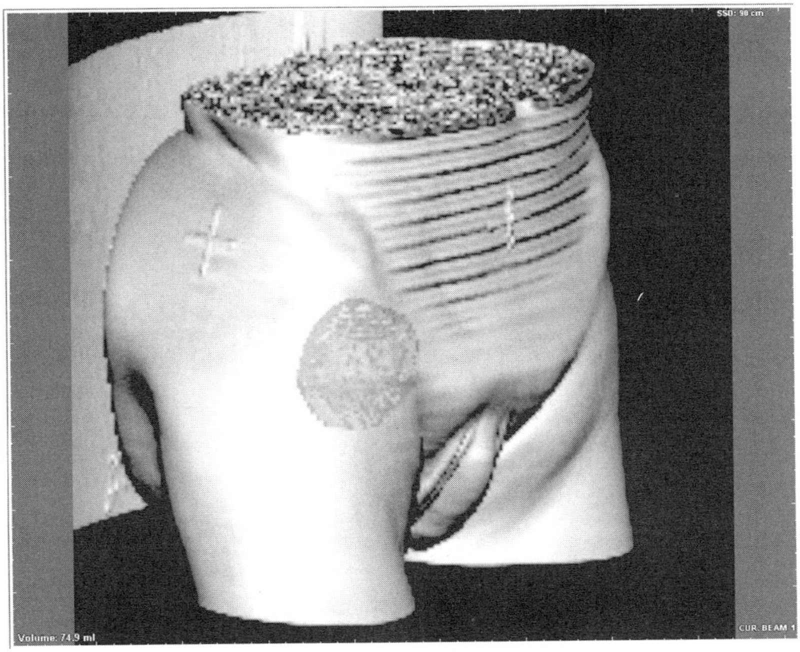

FIGURE 20.9 3D surface rendering of the CT data with superimposed projection of the GTV/CTV and PTV volumes for the prostate cancer case.

imaging.[19–24] This CT-based localization and simulation procedure replaces in stages the conventional hardware-based simulation. Developed CT-based simulation software systems offer features such as

- Effective GTV/CTV and PTV delineation
- 3D visualization of patient anatomy, including oncological volumes such as GTV/CTV, PTV, and critical structures/organs (3D surface rendering, 3D contour visualization)
- 3D visualization of beam geometry (single or multibeam arrangements)
- Realistic and interactive 3D visualization of a simulator or treatment machine (virtual simulator, virtual machine)
- Real-time interactive user interface for all machine-related parameters
- Display of high resolution, real-time BEVs
- Display of high resolution, real-time DRRs

Figures 20.10 to 20.13 demonstrate these features for the clinical case of a prostate cancer patient, using the EXOMIO* CT-based simulation system.[25] Using modern imaging tools, a very realistic and effective physician working environment is realized.

20.3 Introduction to Imaging-Based Brachytherapy Treatment Planning

20.3.1 General Remarks

Modern 3D brachytherapy treatment planning is imaging based.[26–33] A frequently used imaging mod is CT. Efforts are made to also include MR and ultrasound imaging in the brachytherapy tre

* MedInTec GmbH, Bochum, Germany.

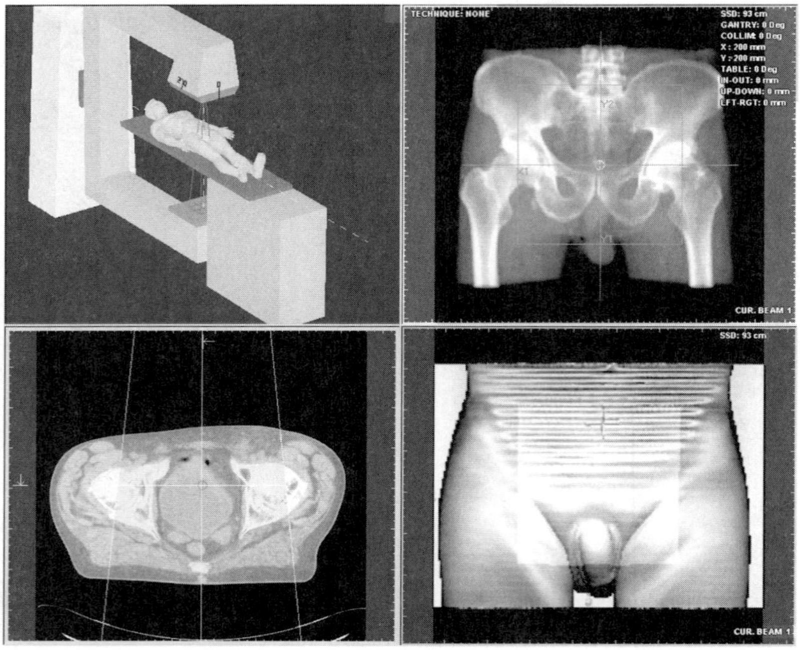

FIGURE 20.10 One of the possible screen configurations of EXOMIO: room view of the virtual simulator including the table and the patient in position (upper left); BEV with superimposed DRR for an arbitrary beam configuration (upper right); CT image window, where the area of the slice is covered by the beam (lower left); and 3D surface rendering window (lower right). Here, the projection of the field on the patient skin is visualized. This corresponds to the light field of the real simulator as visible on the patient skin.

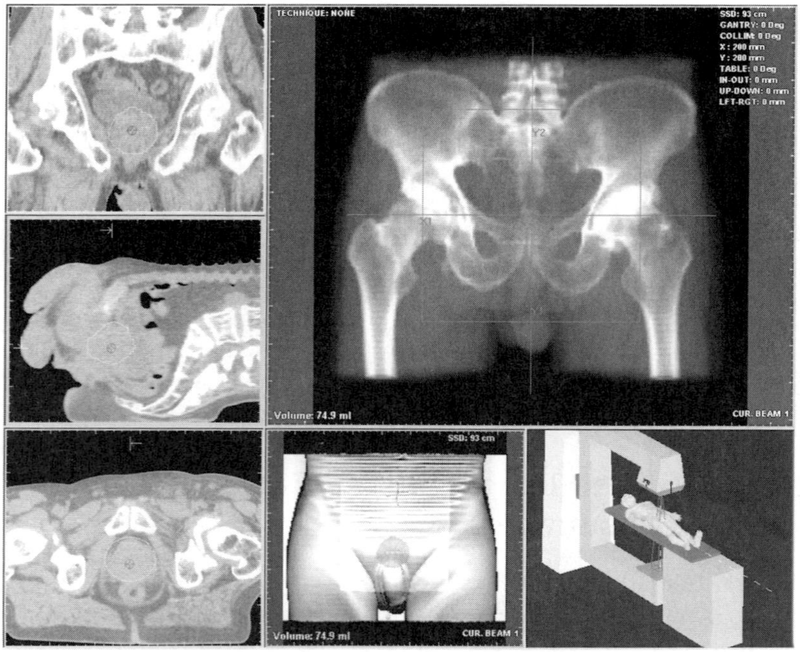

FIGURE 20.11 Another screen configuration of EXOMIO. In addition to the view windows shown in Figure 20.10, a coronal and sagittal reconstructed image window are included. The delineated GTV/CTV is shown on different ˜T images, BEV, and 3D surface rendering window.

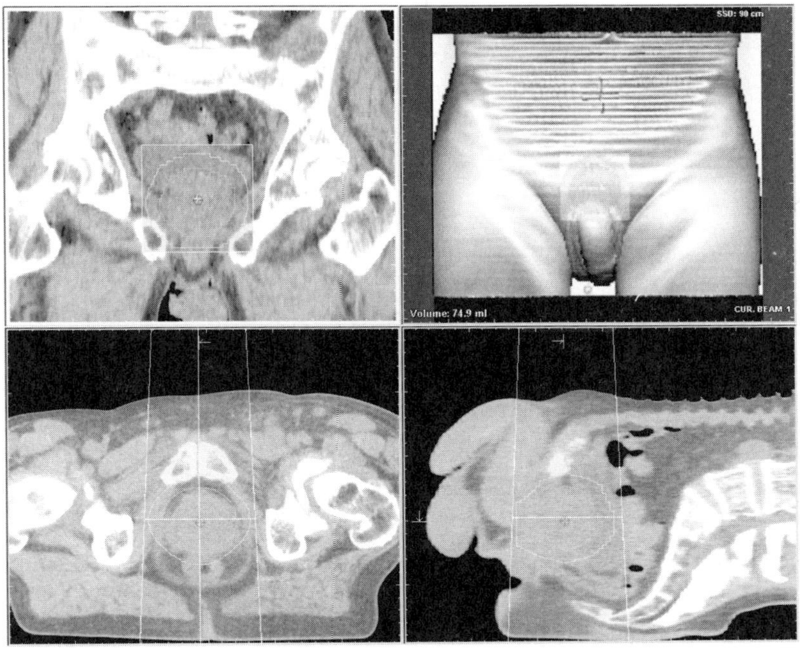

FIGURE 20.12 Automatic field adjustment on the PTV: EXOMIO screen arrangement with transaxial, coronal, and sagittal image view windows and a 3D surface rendering window. All GTV/CTV, PTV (margins around GTV/CTV), and field areas are visualized in the 3D surface rendering window. How closely the field edges fit the PTV can be defined by the planner interactively.

planning procedure. Even if MR could have some advantages with reference to tissue resolution and classification (brain and gynecological localizations), its use is limited mainly because of the long acquisition times, much longer than for CT. Current developments in 3D ultrasound imaging open new horizons for the routine integration of this imaging modality in the brachytherapy procedure. In this section we will be focused on the CT imaging modality. The integration of imaging in the brachytherapy treatment planning enables clinicians to define the PTV and the relevant critical structures, as in external beam radiotherapy. This information can then be used to adjust the dose distribution so that the clinical aims are fulfilled.

The available implementations of CT-based brachytherapy planning systems consider the manual CT-based reconstruction of implanted catheters and the dose optimization regarding the PTV. The reconstruction accuracy depends mainly on the CT imaging parameters such as the slice thickness and interslice distance. In addition, accuracy also depends on graphical resolution and the observational ability of the user. The overall reconstruction accuracy of a CT-based treatment planning system can be as low as 1.0 mm.[19] In clinical brachytherapy, the most time-consuming and error-sensitive part of the treatment planning procedure is catheter reconstruction. This is because the number of catheters can be very large, as in the case of interstitial brachytherapy, where greater than 30 catheters are possible. Imaging-based, 3D treatment planning methods can significantly reduce the time required for the treatment planning process when compared to the use of projectional reconstruction methods (PRM) using radiographs. Even so, a significant part of the treatment planning time is still spent in the reconstruction of the catheters. For a large number of catheters and for complicated catheter geometries, the manual catheter reconstruction method can account for more than half of the treatment planning time.

From the analysis of 35 clinical implants, Tables 20.3 and 20.4, the manual catheter recor procedure took an average of 41% of the total treatment planning time (range of 21 to times do not include those for image processing and contouring. The reconstructio

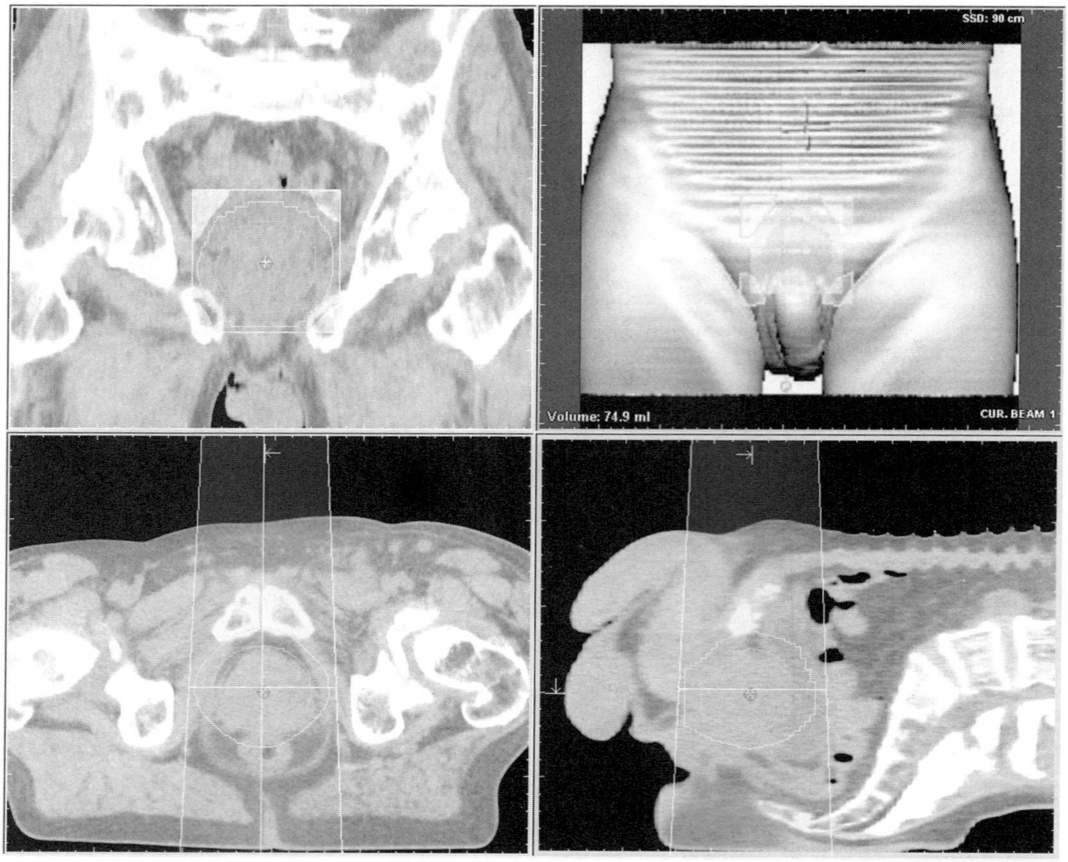

FIGURE 20.13 Block design in EXOMIO. The shape of the blocks is visualized on any image, including BEV with DRR and 3D surface rendering views.

TABLE 20.3 Imaging Parameters for 35 Clinical Implants Analyzed

Images	Matrix (pixels × pixels)	Slice Thickness (mm)	Interslice Distance (mm)	Catheters	Plastic	Metallic
37 ± 10[a] [15, 57][b]	512 × 512	3.0	3.0–5.0	6 ± 3 [3, 13]	10	25

[a] ±1 S.D.
[b] Range.

TABLE 20.4 Time Analysis for the Different Components of the Treatment Planning Procedure for 35 Clinical Implants

Imaging and Contouring (min)	Reconstruction (min)	Source Dwell Position Selection (min)	Optimization (min)	Evaluation (min)	Total (min)
16.3 ± 7.3[a]	13.1 ± 10.5	5.1 ± 3.5	5.7 ± 4.8	5.8 ± 2.5	46.0 ± 19.1
[7.5, 38.0][b]	[4.5, 50.1]	[1.5, 15.0]	[2.0, 30.0]	[1.5, 12.0]	[21.7, 102.9]
(36 ± 9)%	(26 ± 11)%	(11 ± 5)%	(20 ± 10)%	(14 ± 7)%	29.6 ± 14.5
—	(41 ± 14)%[c]	(17 ± 7)%	(26 ± 11)%	(22 ± 11)%	[12.7, 76.1]

[a] ±1 S.D.
[b] Range.
[c] Values excluding the duration of imaging and contouring procedure.

average of 13.1 min (range of 4.5 to 50.1 min). The reconstruction time per catheter was an average of 142.4 s (range of 42.9 to 312 s). The reconstruction time per catheter per CT image was an average of 4.1 s (range of 1.2 to 10.0 s). The selection of the source dwell positions took an average of 17% of the treatment planning procedure (range of 6 to 33%). The mean time spent on this procedure is 5.1 min (range of 1.5 to 15 min), which corresponds to a mean time of 56 s per implanted catheter (range of 22.5 to 100.0 s).

These results demonstrate that there is an obvious need for methods for automatic reconstruction of catheters and automatic selection of the correct source dwell positions that can decrease significantly the duration of the dose treatment planning procedure and increase its safety and reliability. In the following section, we will refer to three developed applications for 3D brachytherapy treatment planning.

20.3.2 CT-Based Catheter Autoreconstruction

Algorithms have been developed[17,34,35] for the automatic reconstruction and recognition of metallic and plastic implanted catheters (Figures 20.14 and 20.15). The process is based on postimplantation acquired CT images with the catheters *in situ* in their final positions. This includes the relevant patient anatomy, target volume(s), organs at risk, and the catheters. Catheter searching is made on a sequence of CT slices and is based on the HU of the catheter material, catheter outer diameter, interslice distance, slice thickness, and geometry of the catheter shape on the CT slices. If there is no patient motion during CT data acquisition, there is virtually no error in the autoreconstruction process.

These algorithms overcome a number of difficulties which arise when a large number of catheters are present. These include situations with intersecting catheters and with loop techniques. The time required for the catheter reconstruction process using our autoreconstruction method is significantly reduced. The accuracy of autoreconstruction is at least as high as the classical, manual slice-by-slice method.

The accuracy and time analysis have been done for 30 different clinical implants and 1 phantom implant with 3 looped plastic catheters. The accuracy analysis is subdivided into two arms: (1) the geometrical difference defined as the geometrical shift between the manually and automatically reconstructed catheter describing points on each transaxial image and (2) the source dwell position difference defined as the geometrical shift between the corresponding dwell positions generated by the manual and automatic catheter reconstruction procedure.

An analysis was made for the dwell positions produced every 2.5 mm, starting from a given catheter tip. The mean differences over all catheters and catheter describing points (arm a) or source dwell positions (arm b) in each catheter are calculated. The catheter describing point-based error analysis shows mean geometrical errors varying from 0.36 ± 0.25 to 1.12 ± 0.35 mm with a mean value over all 31 implants of 0.67 ± 0.36 mm, whereas source dwell position-based analysis gave mean geometrical errors varying from 0.38 ± 0.22 to 1.41 ± 0.44 mm with a mean value of 0.87 ± 0.36 mm.

The reconstruction time analysis presented next shows that our algorithm is extremely time-efficient. In 27 of 30 clinical cases (90%), no manual intervention by the user was needed during the autoreconstruction-based process. For these 27 cases, the catheter reconstruction with our algorithm was 25.7 times (mean value) faster than the manual reconstruction. The mean time needed for our autoreconstruction method was 21.4 s, compared to 547.2 s for the corresponding manual procedure.

In the case of the phantom implant with three looped plastic catheters, no manual intervention by the user was needed during the autoreconstruction-based process. The reconstruction time with our algorithm was 26.1 times faster than the manual reconstruction: 22 s compared to 574.2 s.

For the three cases where manual intervention was required, the catheter reconstruction based autoreconstruction algorithm was 8.2 times faster (mean value) than the corresponding man struction. The mean reconstruction time with our method, including the intermediate man tion, was 81.7 s, compared to 739.8 s for the manual procedure.

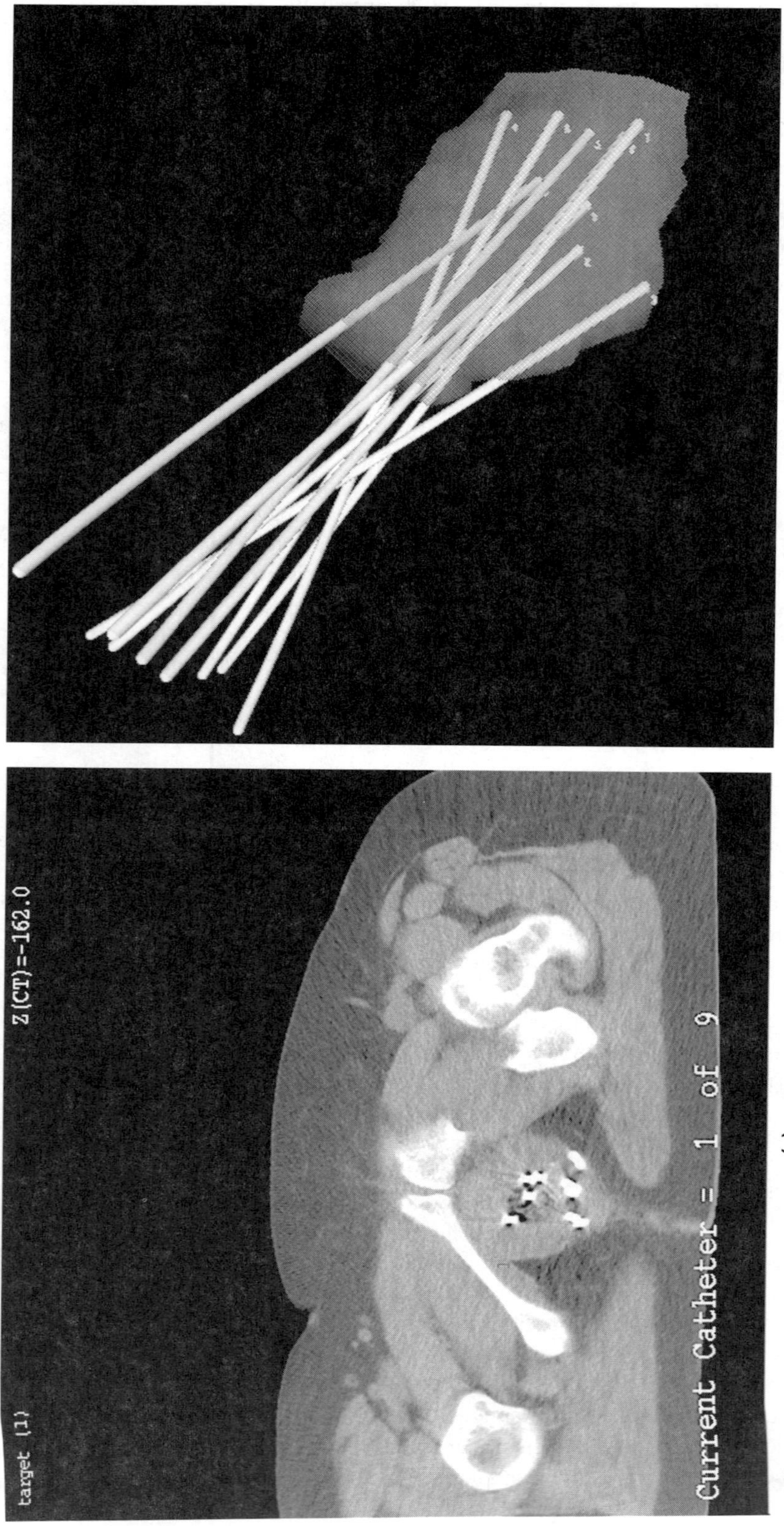

FIGURE 20.14 Cervix implant with nine metallic trocar point needles: (a) representative CT image showing the catheter areas; (b) 3D view of the autoreconstructed catheters and the PTV volume.

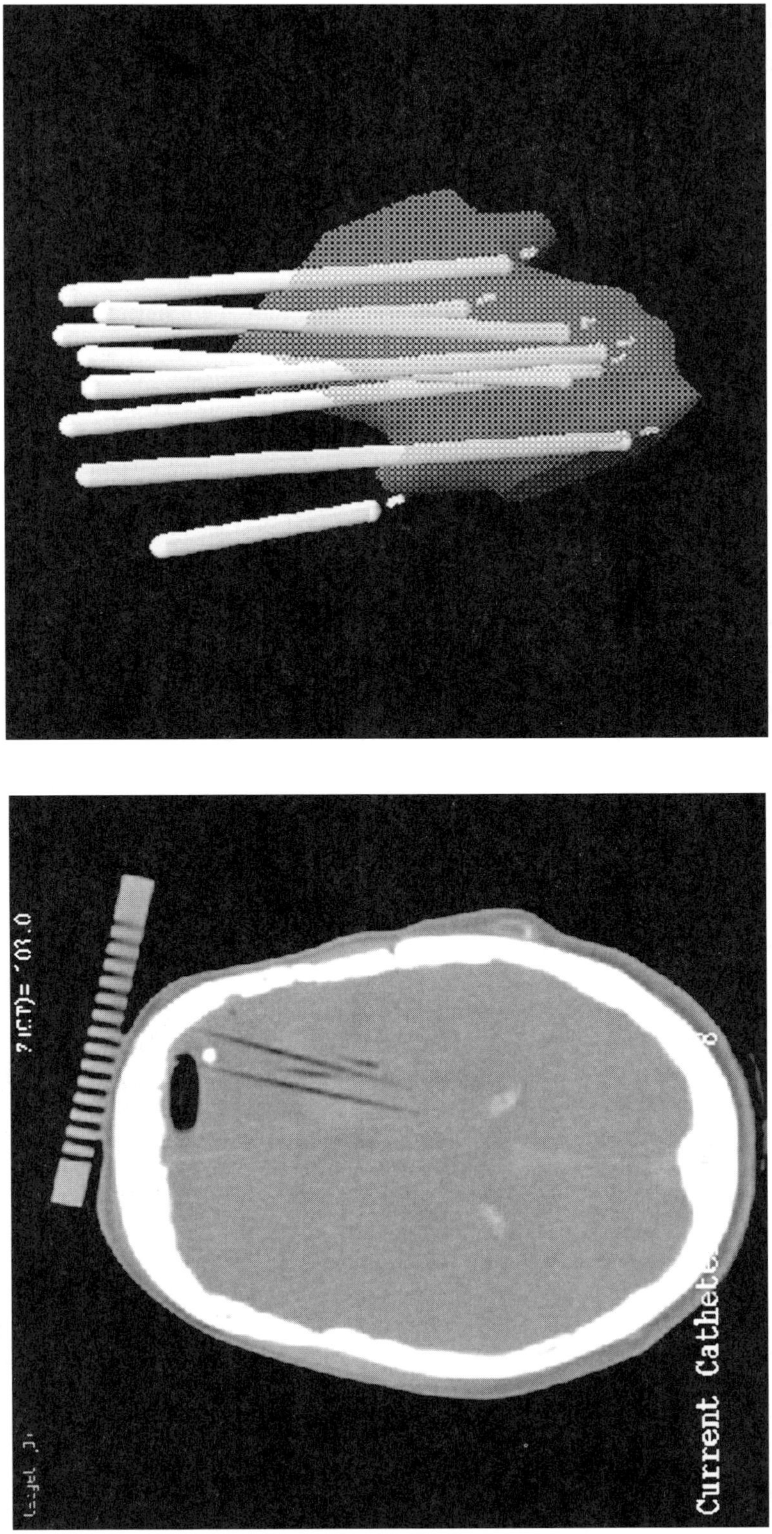

FIGURE 20.15 Brain implant with eight plastic flexible needles: (a) representative CT image showing the catheter areas and the template; (b) 3D view of the autoreconstructed catheters and the PTV volume.

20.3.3 DRR-Based Catheter Reconstruction

Efficient tools have been developed for catheter reconstruction based on CT-reconstructed radiographs (CTRR) calculated from 3D volume constructed from CT slices.[17,18] Until now and for PRM methods, catheter reconstruction was done from true X-ray radiographs.[36–39] This could cause errors in catheter reconstruction because of patient motion between two radiograph acquisitions and because of geometrical inaccuracies of the X-ray system. Additionally, catheters or parts of them can be covered by bones.

Two different methods have been developed for the catheter reconstruction: (a) matching of corresponding points supported by navigation tools, Figure 20.16a, and (b) automatic matching of corresponding points and catheters on the CTRRs (general polynomial method without the need for defining the correspondence).

As two CTRRs are used for catheter reconstruction, we do not have the problem of reconstruction errors resulting from patient motion, as is the case when making two classical radiographs. Navigation tools guide the user through the whole process, and the real-time 3D visualization offers the opportunity to have effective control of the reconstruction results, looking at it in the 3D view window or at any CT slice (Figure 20.16).

Some advantages of this approach to catheter reconstruction are given here. (1) There is no additional patient irradiation. This enables calculation of the necessary number of DRRs to find the two best views. (2) Use of *anatomical* and *catheter filters* enables reconstruction of the catheters that would otherwise be whole or partly obscured by bones and soft tissue (Figure 20.17). (3) There is no need for the radiological film acquisition and digitalization that significantly speeds up the process of catheter reconstruction. (4) Possible patient motion during the CT slice acquisition is insignificant compared to the motion present between the two radiological films acquisition. (5) There is no need for matching between two DRRs, as in the case of radiological films, as DRRs are produced from the same volumetric data set. (6) There are simple solutions for the ill-defined cases that can occur and cannot be solved if working with classical radiological films.

The developed CTRR-based reconstruction methods can further help in the case of classical gynecological brachytherapy applications to identify bony structures and reference points as purposed by ICRU 38.[40]

20.3.4 CT–MR Image Registration

It is of great importance for radiation oncologists to be able to accurately delineate the GTV, CTV, and PTV for the high effective and high dose brachytherapy treatments. Therefore, the combination of different imaging modalities can improve the accuracy of the tumor localization and PTV definition and, consequently, improve the accuracy of the brachytherapy treatment planning, including the catheter implantation procedure. This is especially true for tumor localization in brain, prostate, and gynecological tumors (endometrium and cervix carcinomas).[41–45] Usually, CT and MR imaging are combined for prostate and gynecological cancer, where in the case of brain tumors efforts are made to combine the benefits of CT and MR imaging regarding anatomic and spatial resolution with the metabolic information available from SPECT or PET imaging. The obtained information could assist to separate necrosis from vital tumor. Since CT and MR are the most common imaging modalities available in modern radiation oncology centers, we will discuss here the use of complementary CT and MR imaging modalities for brachytherapy treatment planning for brain tumors.

In the case of brachytherapy treatment planning we have a unique situation in that imaging has to be used to (1) identify GTV, CTV, and PTV; (2) assist during planning of catheter position and realization of catheter implantation; and (3) identify and reconstruct implanted catheters. CT is actually the most reliable imaging modality regarding geometrical accuracy. Furthermore, it is insensitive to influences from catheter material. This is the reason why catheter reconstruction is proffered to be done based on CTs. In the case of brain implants, where plastic catheters are implanted into the GTV using stereotactic- or template-based methods,[46] the treatment planning is based on postimplantation CT images, where for the PTV definition either preimplantation or postimplantation MRI is additionally considered. The combination of the two imaging modalities is called image fusion and aims to determine the geometrical relationship between the two imaging studies. In this way, the information available in one study can been transferred to the other study.

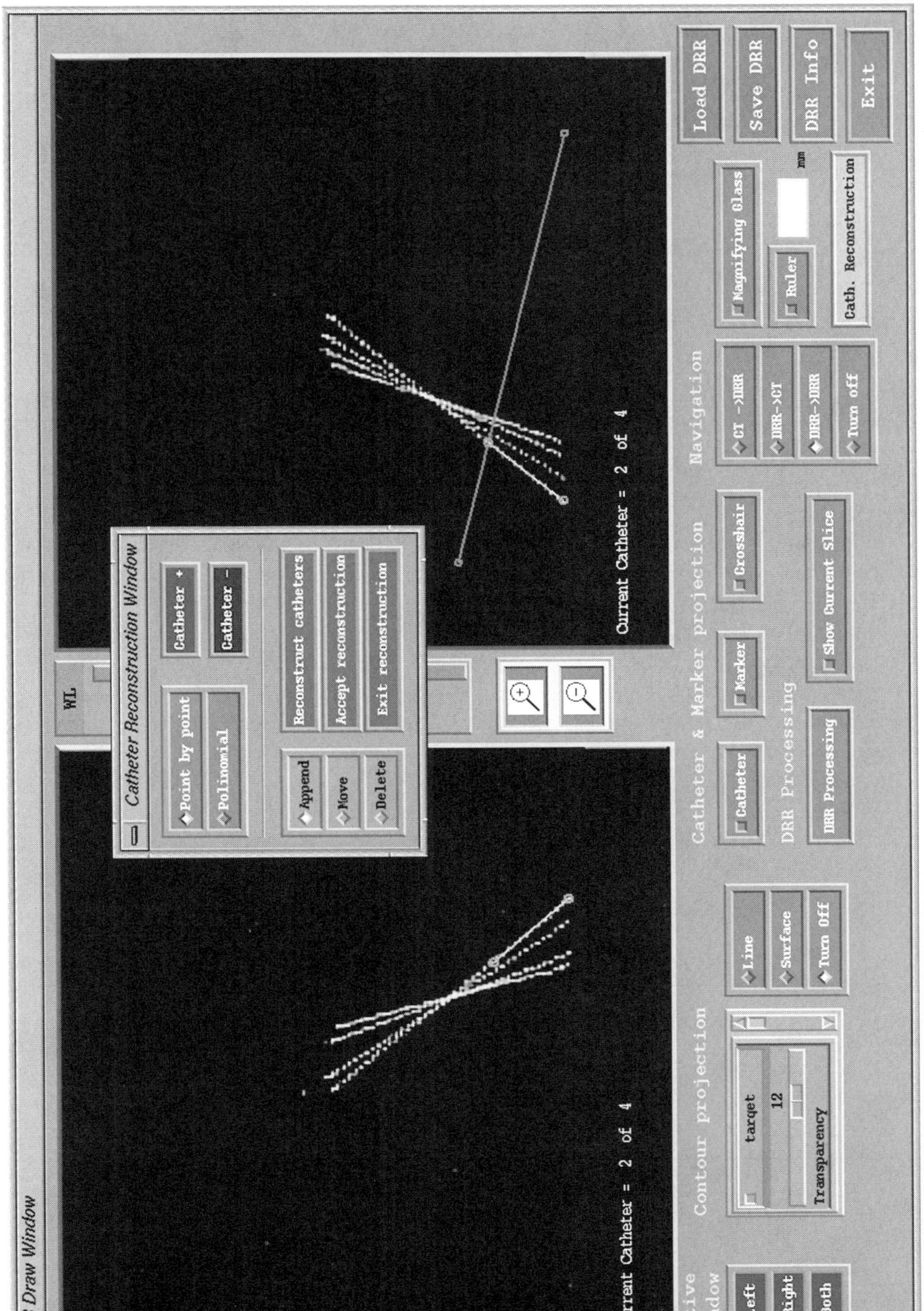

FIGURE 20.16 (a) Two DRRs calculated from the same volumetric data set and with the same isocenter coordinate are presented. Points are user-defined catheter points on both DRRs. The navigation tool from one to the other DRR is presented. (b) After catheter reconstruction, results can be seen in 3D and 2D (on CT slice) views.

(continued)

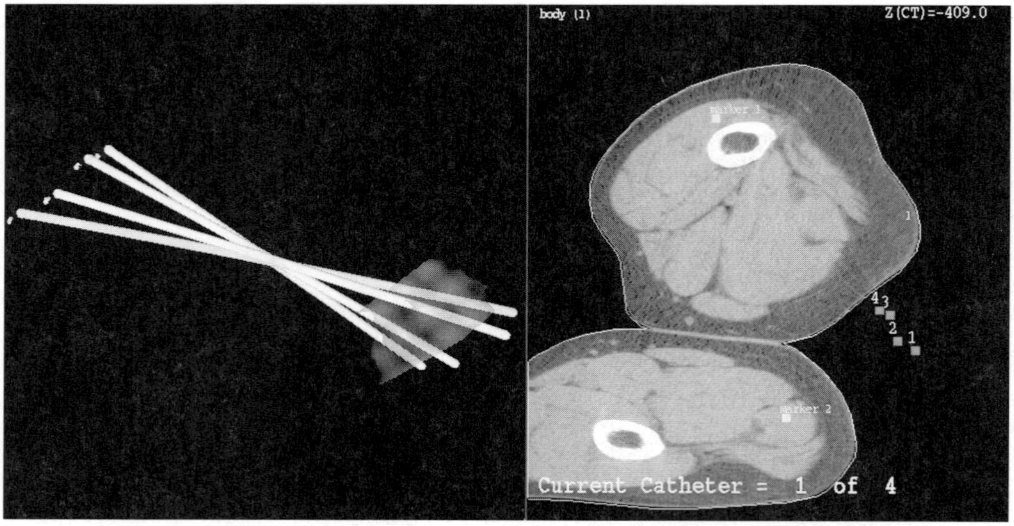

FIGURE 20.16 (CONTINUED) (b) After catheter reconstruction, results can be seen in 3D and 2D (on CT slice) views.

Since for head localizations the anatomy can be assumed to be more or less rigid (rigid body approximation), the geometrical relationship consists only of translation, rotation, and scaling to account for the differences in the pixel/voxel size between the two image studies. The elements of the transformation matrix can be simply estimated by matching anatomical landmarks or markers added to the patient anatomy.[47–50] These markers can be embedded in a stereotactic frame if this is used for the catheter implantation, or can be small plastic spheres that are fixed on the head skin in the case of template-based implantation techniques. In both of these techniques, the markers are filled with an adequate liquid (solution) that makes them visible in CT as well as in MR studies. Both frame-based and free positioned markers have to be used in both imaging studies to build the data set for calculating the transformation matrix.

We present here a case of a paranasal sinus tumor where the CT-MR image fusion has been realized using markers. Figure 20.19 demonstrates a representative CT and MR image where one of the used markers is clearly visible in both of the two images. There is a clear advantage for the MRI, since no artifacts caused by teeth are observed, as is the case for the CT imaging.

Figure 20.20 demonstrates the superiority of MRI to CT imaging with reference to GTV localization and PTV definition for the same case as in Figure 20.18. In CT, even if contrast medium were used, the GTV cannot be clearly identified. In contrast, the GTV is clearly identified in MRI using gadolinium-based contrast medium. The markers are clearly identified in both imaging modalities (Figure 20.19).

The catheter reconstruction was based on CT.[26] The resulting dose distribution can then be visualized in MR images, and their conformity in relation to PTV can be evaluated (Figure 20.20).

20.4 Conclusion

Imaging plays an essential role in radiation therapy treatment planning. 3D imaging-based computer optimized treatment planning allows the delivery of conformal dose distribution. Multimodality imaging enables more accurate identification of tumors resulting in conformity of the dose distribution relative to the "real" target volume. A considerable amount of research and development has to be done in order to increase the reliability, user friendliness, and speed of such multimodality procedures. Virtualization of conventional treatment planning processes also taking advantage of state-of-the-art 3D visualization tools and computing facilities will further increase the efficiency of radiation therapy planning. Further development in the field of verification of treatment delivery will additionally increase the clinical acceptance of such systems.

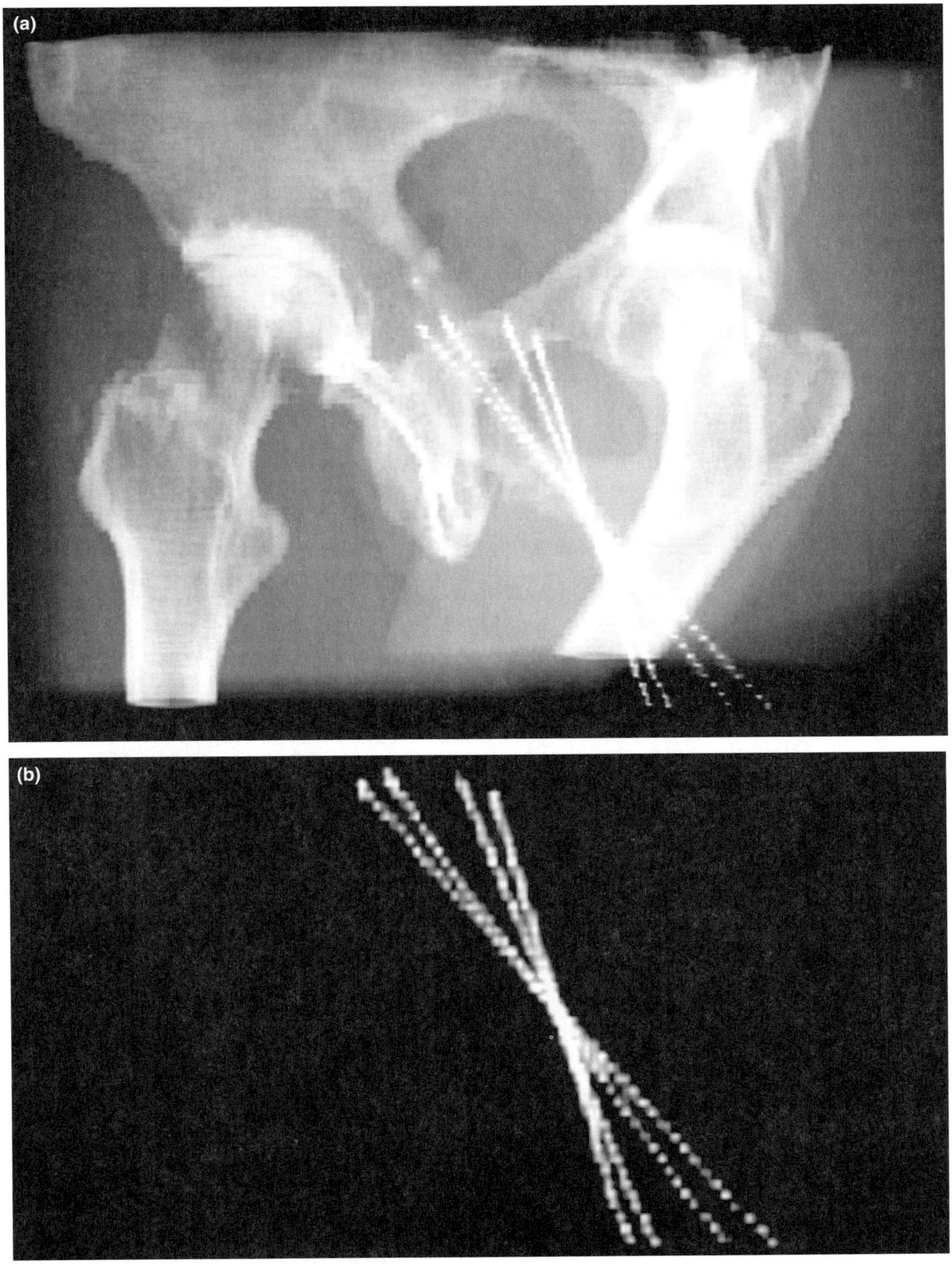

FIGURE 20.17 Application of catheter filters in the DRR calculation process. The gain obtained in the use of catheter filters is obvious. (a, c) DRR images calculated from the postimplanted volumetric data set. (b, d) DRR images calculated using the catheter metallic and plastic filters, respectively. Notice that the plastic catheters in (c) are completely obscured.

(*continued*)

(c)

(d)

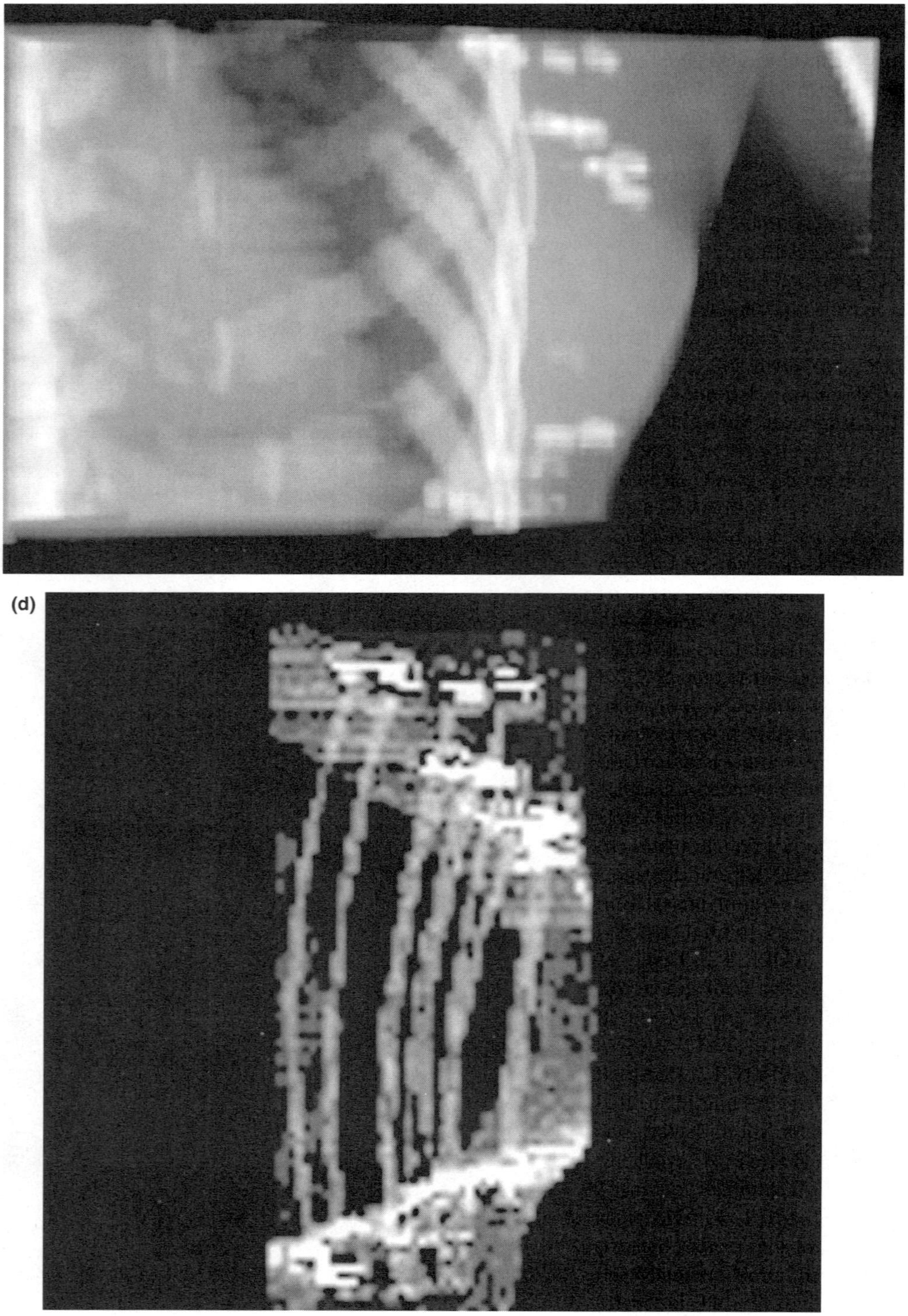

FIGURE 20.17 (CONTINUED) Application of catheter filters in the DRR calculation process. The gain obtained in the use of catheter filters is obvious. (a, c) DRR images calculated from the postimplanted volumetric data set. (b, d) DRR images calculated using the catheter metallic and plastic filters, respectively. Notice that the plastic catheters in (c) are completely obscured.

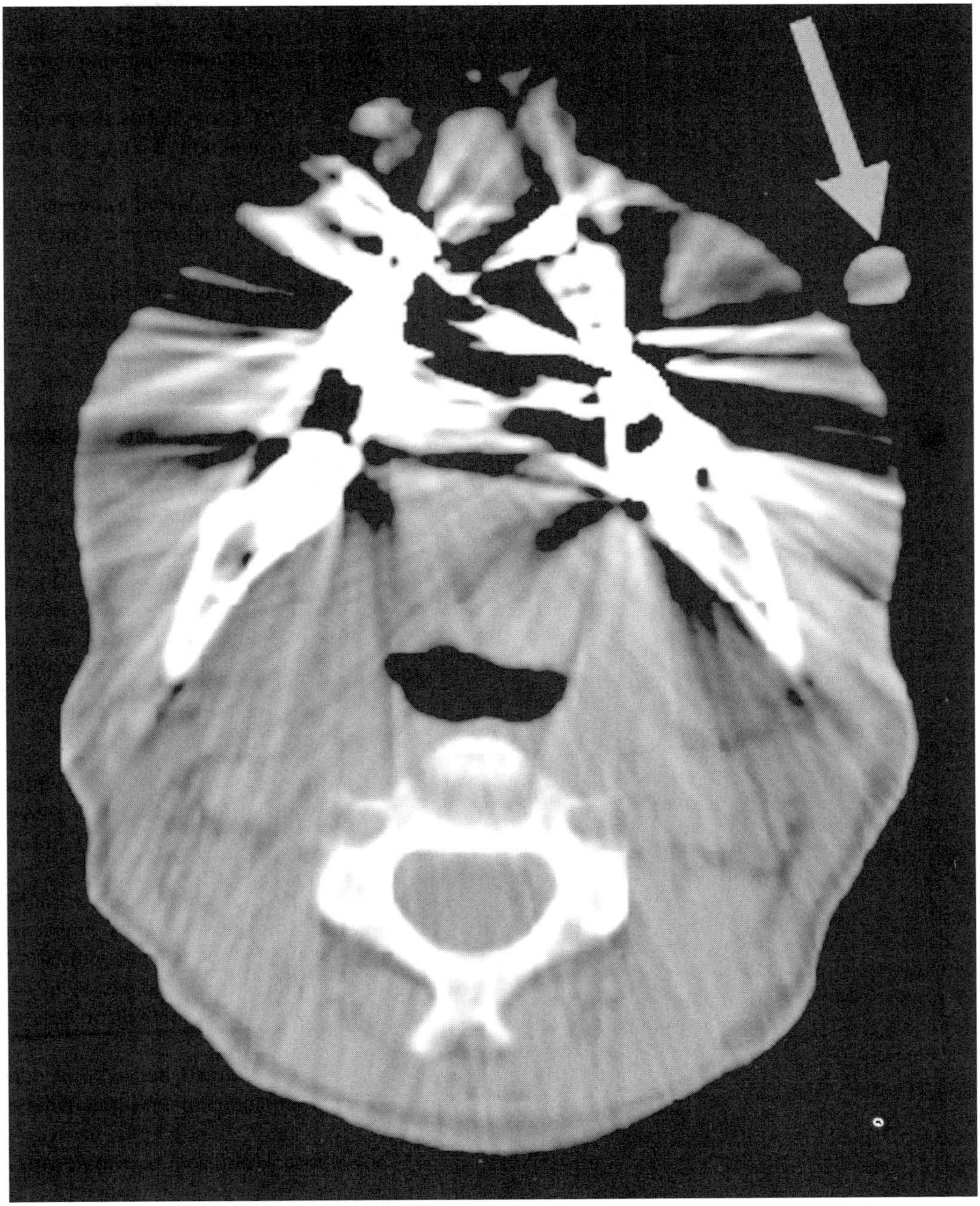

FIGURE 20.18a Representative CT (a) and MR (b) images for the clinical example of a paranasal sinus tumor. One of the six used markers is clearly visible in both images. There are no artifacts observed in the MR image. These artifacts caused by teeth scrawl all over the CT image. There is a clear advantage of the MRI for identifying GTV.

(continued)

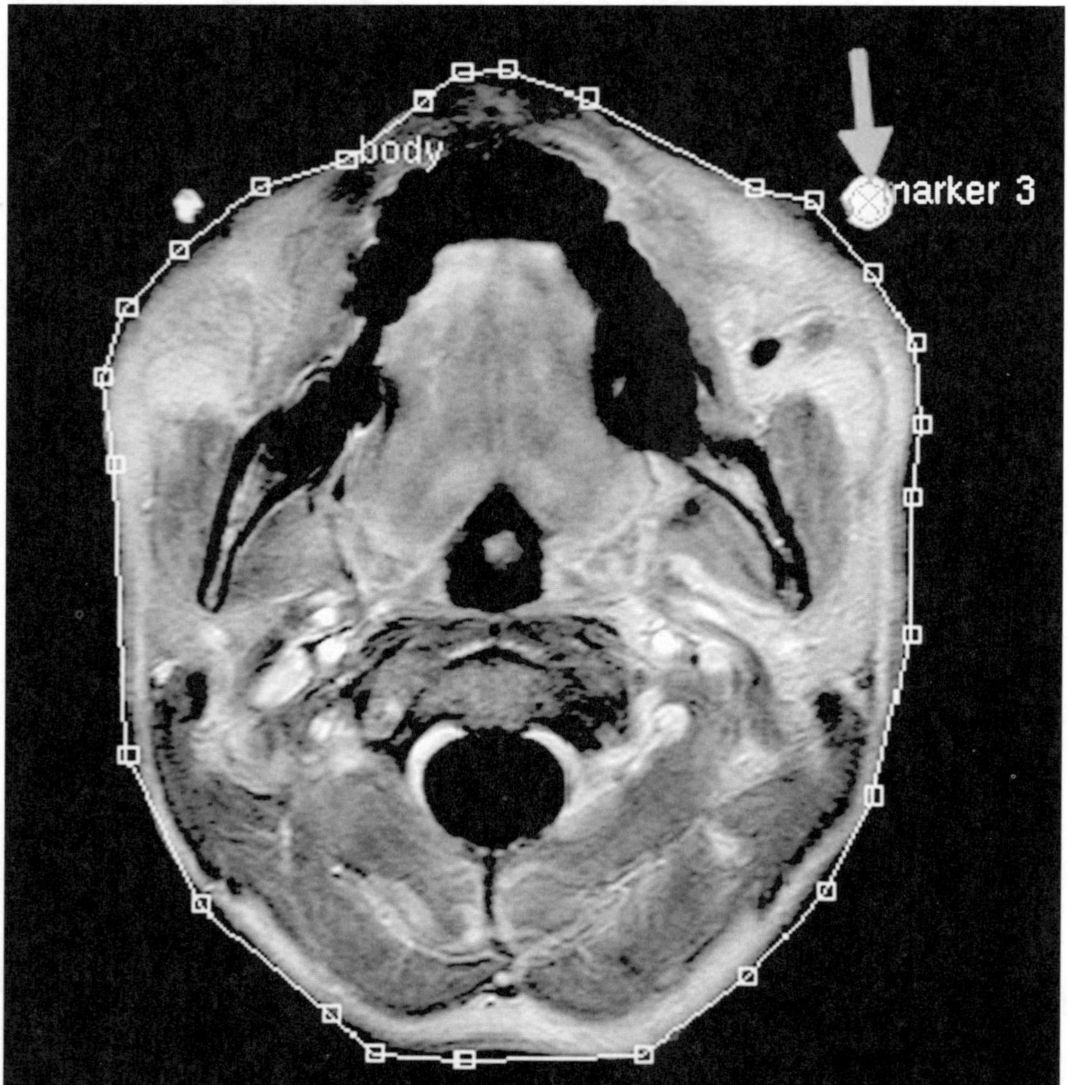

FIGURE 20.18b (CONTINUED) Representative CT (a) and MR (b) images for the clinical example of a paranasal sinus tumor. One of the six used markers is clearly visible in both images. There are no artifacts observed in the MR image. These artifacts caused by teeth scrawl all over the CT image. There is a clear advantage of the MRI for identifying GTV.

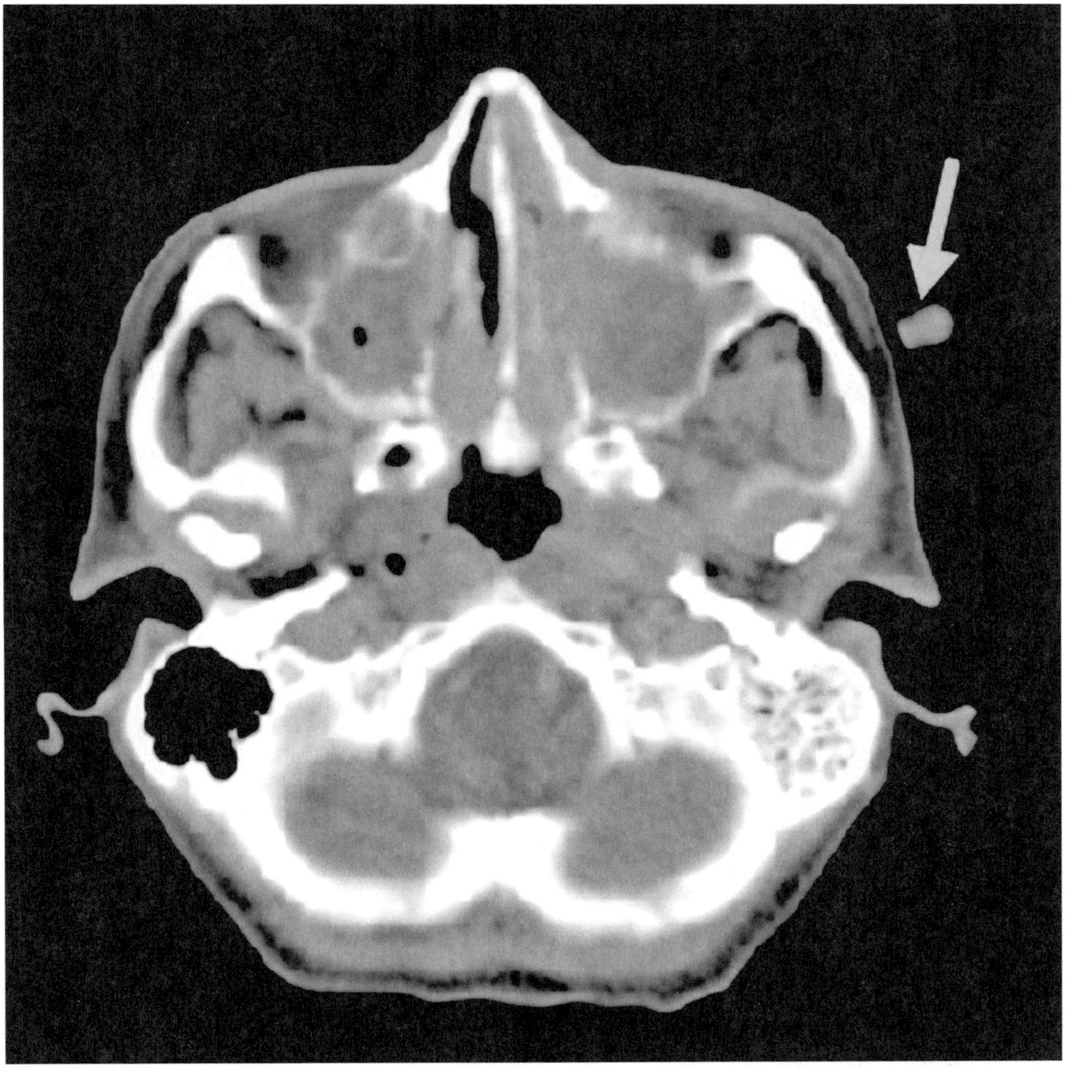

FIGURE 20.19a The same patient as in Figure 20.18. Here, CT images (a, c) and MR images (b, d) in the tumor region are demonstrated. For both imaging studies, adequate contrast medium has been used. MRI with gadolinium-based contrast medium is obviously more proper for the localization of GTV and PTV definition (see light gray and white marked region in (b) and (d) images). In the CT images, it is very difficult to distinguish between normal tissue and tumor. Some of the markers are clearly identified in both imaging modalities (a and b). Parts of the inserted plastic catheters (tubes with an outer diameter of 1.9 to 2.0 mm) are visible as dark lines/curves on both CT (c) and MR (d) images. *(continued)*

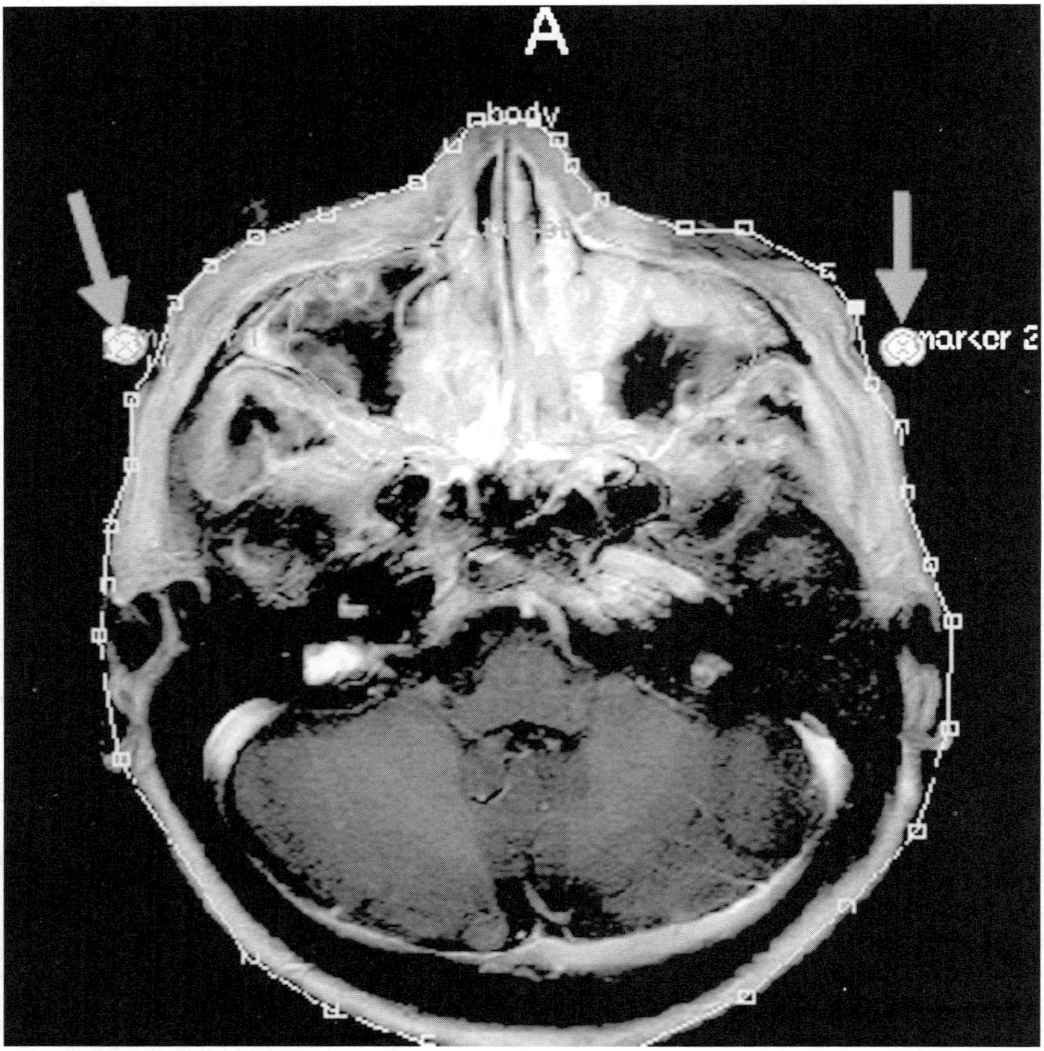

FIGURE 20.19b (CONTINUED) The same patient as in Figure 20.18. Here, CT images (a, c) and MR images (b, d) in the tumor region are demonstrated. For both imaging studies, adequate contrast medium has been used. MRI with gadolinium-based contrast medium is obvious more proper for the localization of GTV and PTV definition (see light gray and white marked region in (b) and (d) images). In the CT images, it is very difficult to distinguish between normal tissue and tumor. Some of the markers are clearly identified in both imaging modalities (a and b). Parts of the inserted plastic catheters (tubes with an outer diameter of 1.9 to 2.0 mm) are visible as dark lines/curves on both CT (c) and MR (d) images. *(continued)*

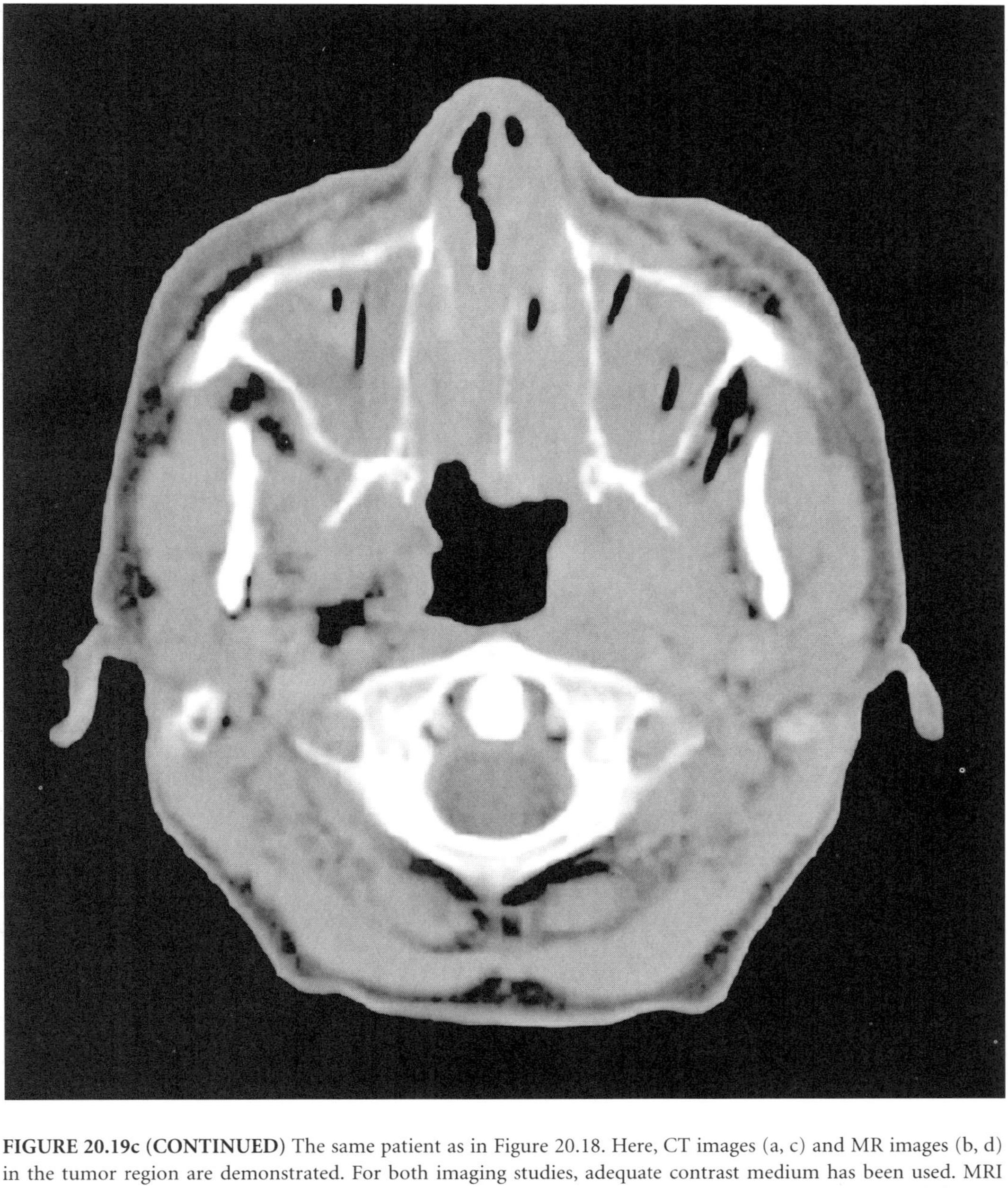

FIGURE 20.19c (CONTINUED) The same patient as in Figure 20.18. Here, CT images (a, c) and MR images (b, d) in the tumor region are demonstrated. For both imaging studies, adequate contrast medium has been used. MRI with gadolinium-based contrast medium is obvious more proper for the localization of GTV and PTV definition (see light gray and white marked region in (b) and (d) images). In the CT images, it is very difficult to distinguish between normal tissue and tumor. Some of the markers are clearly identified in both imaging modalities (a and b). Parts of the inserted plastic catheters (tubes with an outer diameter of 1.9 to 2.0 mm) are visible as dark lines/curves on both CT (c) and MR (d) images. *(continued)*

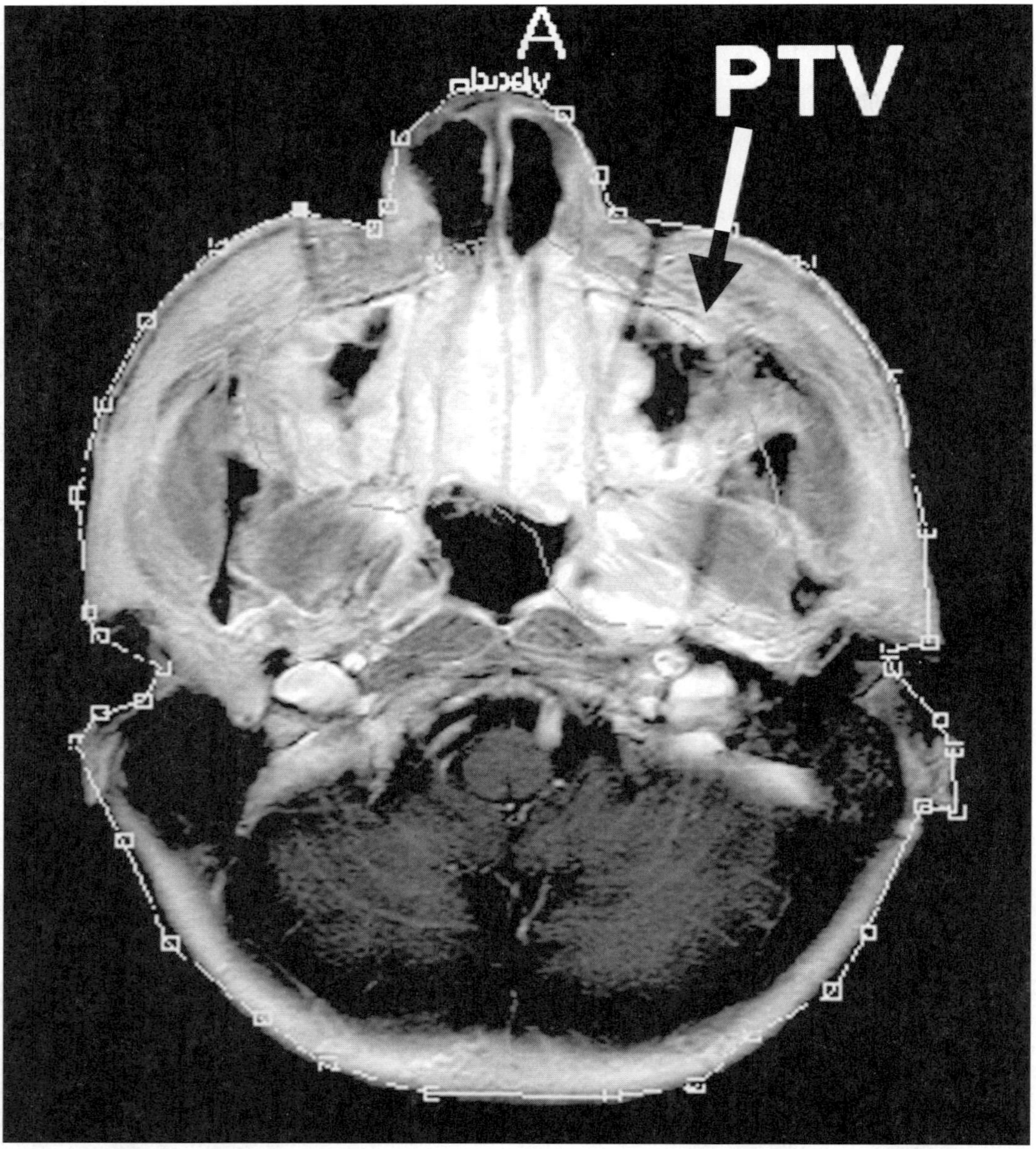

FIGURE 20.19d (CONTINUED) The same patient as in Figure 20.18. Here, CT images (a, c) and MR images (b, d) in the tumor region are demonstrated. For both imaging studies, adequate contrast medium has been used. MRI with gadolinium-based contrast medium is obvious more proper for the localization of GTV and PTV definition (see light gray and white marked region in (b) and (d) images). In the CT images, it is very difficult to distinguish between normal tissue and tumor. Some of the markers are clearly identified in both imaging modalities (a and b). Parts of the inserted plastic catheters (tubes with an outer diameter of 1.9 to 2.0 mm) are visible as dark lines/curves on both CT (c) and MR (d) images.

(continued)

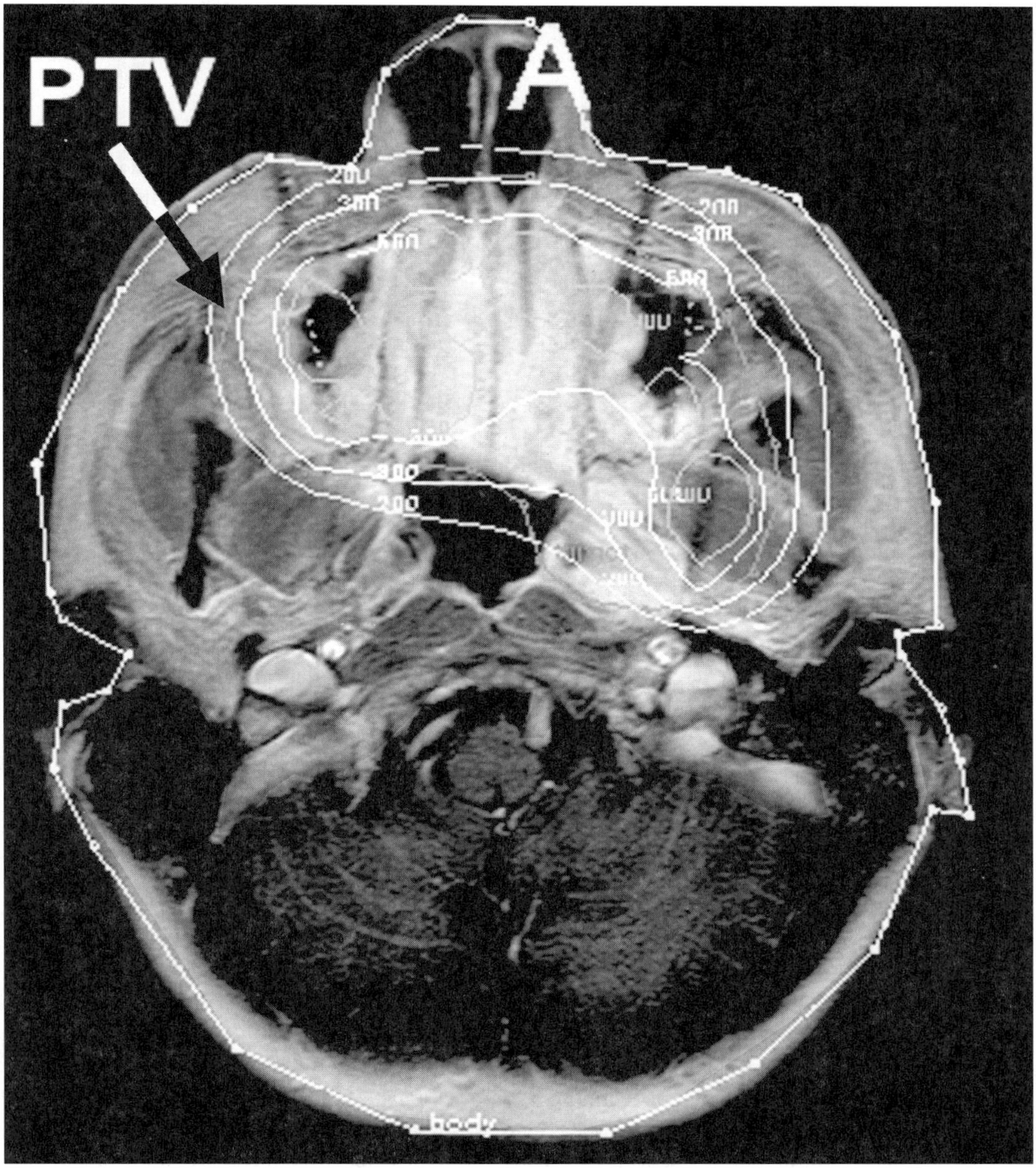

FIGURE 20.20 Patient anatomy reproduced by MR with the delineated PTV and the calculated isodose lines. This MR image is the same as in Figure 20.19d. The catheter reconstruction was based on CT. After dose optimization, the resulting dose distribution was projected on the MR images using the transformation matrix derived, as explained in the text, from the six markers positioned and fixed on the head of the patient.

Future treatment planning systems will make maximal use of the advantages of 3D multimodality imaging, virtualization, and telematics tools. The treatment planning of the 21st century will be based on accurate target definition and completely automatic optimization procedures, making use of the whole spectrum of hardware facilities (multileafs etc.). Imaging-based real-time treatment delivery verification systems will enable clinicians to have effective control of the realized treatment accuracy.

References

1. International Commission on Radiation Units and Measurements, Prescribing, Recording, and Reporting Photon Beam Therapy, ICRU Report 50, Bethesda, MD, 1993.
2. Austin-Seymour, M., Chen, G.T.Y., Rosenman, J., Michalski, J., Lindsley, K., and Goitein, M., Tumor and target delineation: current research and future challenges, *Int. J. Radiat. Oncol. Biol. Phys.*, 33, 1041–1052, 1995.
3. Sailer, S.L., Rosenman, J.G., Soltys, M., Cullip, T.J., and Chen, J., Improving treatment planning accuracy through multimodality imaging, *Int. J. Radiat. Oncol. Biol. Phys.*, 35, 117–124, 1996.
4. Johns, H.E. and Cunningham, J.R., *The Physics of Radiology*, 4th ed., Charles C Thomas, Springfield, IL, 1983.
5. Attix, F.H., Energy-absorption coefficients for γ-rays in compounds and mixtures, *Phys. Med. Biol.*, 29, 869–871, 1984.
6. Constantinou, C., Tissue Substitutes for Particulate Radiations and Their Use in Radiation Dosimetry and Radiotherapy, Ph.D. thesis, London University, 1974.
7. Hubbell, J.H. and Seltzer, S.M., Tables of X-Ray Mass Attenuation Coefficients and Mass Energy-Absorption Coefficients 1 keV to 20 MeV for Elements Z=1 to 92 and 48 Additional Substances of Dosimetric Interest, NISTIR 5632, U.S. Department of Commerce, May 1995.
8. ICRP, Report of the Task Group on Reference Man, ICRP Publication 23, 1975.
9. International Commission on Radiation Units and Measurements, Tissue Substitutes in Radiation Dosimetry and Measurement, ICRU Report 44, Bethesda, MD, 1989.
10. International Commission on Radiation Units and Measurements, Photon, Electron, Proton and Neutron Interaction Data for Body Tissues, ICRU Report 46, 1992.
11. Constantinou, C. and Harrington, J., An electron density calibration phantom for CT based treatment planning computers, *Med. Phys.*, 19-2, 325–327, 1992.
12. Constantinou, C., Harrington, J., Cadieux, R.A., Choi, S., and Vasa, M.A., Dosimetry at lung-muscle and lung-bone interfaces using 6 MV and 10 MV X-rays and 12–18 MeV electrons (Abstract), *Int. J. Radiat. Oncol. Biol. Phys.*, 15(Suppl. 1), 244, 1988.
13. Mackie, T.R., El-Khatib, E., Battista, J., Scringer, J., van Dyk, J., and Cunningham, J.R., Lung dose corrections for 6 and 15 MV X-rays, *Med. Phys.*, 12(3), 327, 1985.
14. Berger, M.J. and Hubbell, J.J.H., XCOM: Photon Cross Sections on a Personal Computer, NBSIR 87-3597, 1987.
15. Battista, J.J., Ph.D. thesis, University of Toronto, Canada, 1977.
16. Battista, J.J. and Bronskill, M.J., Compton scatter imaging of transverse sections: an overall appraisal and evaluation for radiotherapy planning, *Phys. Med. Biol.*, 26-1, 81–99, 1981.
17. Milickovic, N., Three Dimensional CT Based Reconstruction Techniques in Modern Brachytherapy Treatment Planning, Ph.D. thesis, National Technical University of Athens, December 1999.
18. Milickovic, N., Baltas, D., Giannouli, S., Uzunoglu, N., and Zamboglou, N., Use of CT-reconstructed radiographs for catheter reconstruction in brachytherapy treatment planning, *Med. Biol. Eng. Comput.*, November 1999.
19. Michalski, J.M., Purdy, J.A., Harms, W., and Matthews, J.W., The CT-simulation 3D treatment planning process, 3D conformal radiotherapy, *Front. Radiat. Ther. Oncol.*, 29, 43–56, 1996.
20. Rosenman, J., Where will 3D conformal radiation therapy be at the end of the decade? 3D conformal radiotherapy, *Front. Radiat. Ther. Oncol.*, 29, 264–271, 1996.

21. Chen, G.T.Y., Pelizzari, C.A., and Vijayakumar, S., Imaging: the basis for effective therapy, 3D conformal radiotherapy, *Front. Radiat. Ther. Oncol.*, 29, 31–42, 1996.
22. Purdy, J.A., 3D radiation treatment planning: a new era, 3D conformal radiotherapy, *Front. Radiat. Ther. Oncol.*, 29, 1–16, 1996.
23. Vanuytsel, L. and Weltens, C., Imaging techniques for radiotherapy planning, *Oncol. Prac.*, 2, 18–21, 1999.
24. Butker, E.K., Helton, D.J., Keller, J.W., Hughes, L.L., Crenshaw, T., and Davis, L.W., A totally integrated simulation technique for three field breast treatment using a CT simulator, *Med. Phys.*, 23(10), 1809–1814, 1996.
25. Cai, W., Sakas, G., and Karangelis, G., Volume Interaction Techniques in the Virtual Simulation of Radiotherapy Treatment Planning, International Conference on Computer Graphics and Vision (Graphicon), Moscow, 1999.
26. Tsalpatouros, A., Baltas, D., Kolotas, C., van der Laarse, R., Koutsouris, D., Uzunoglu, N.K., and Zamboglou, N., CT-based software for 3-D localization and reconstruction in stepping source brachytherapy, *IEEE Trans. Med. Imaging,* 1, 229–242, 1997.
27. Zamboglou, N., Kolotas, C., Baltas, D., Martin, T., Rogge, B., Strassman, G., Tsalpatouros, A., and Vogt, H.G., Clinical evaluation of CT based software in treatment planning for interstitial HDR brachytherapy, in *Brachytherapy for the 21st Century*, Spencer, B.L. and Mould, R.F., Eds., Nucletron B. V., 1998, pp. 312–326.
28. Kolotas, C., Birn, G., Baltas, D., Fogt, H.G., Martin, T., and Zamboglou, N., CT guided template technique interstitial brachytherapy, in new developments, in *Interstitial Remote Controlled Brachytherapy*, Zamboglou, N., Ed., Zuckschwerdt, New York, 1997, pp. 143–152.
29. Zamboglou, N., Interstitial brachytherapy possibilities, in *New Developments in Interstitial Remote Controlled Brachytherapy*, Zamboglou, N., Ed., Zuckschwerdt, New York, 1997, pp. 174–180.
30. Baltas, D., Kolotas, C., Geramani, K., Mould, R.F., Ioannidis, G., Kekchidi, M., and Zamboglou, N., A conformal index (COIN) to evaluate implant quality and dose specification in brachytherapy, *Int. J. Radiat. Oncol. Biol. Phys.*, 40, 515–524, 1998.
31. Martel, M.K. and Narayana, V., Brachytherapy for the next century: use of image-based treatment planning, *Radiat. Res.,* 150(Suppl.), 178–188, 1998.
32. Vicini, F.A., Jaffray, D.A., Horwitz, E.M., Edmundson, G.K., DeBiose, D.A., Kini, V.R., and Martinez, A.A., Implementation of 3D-virtual brachytherapy in the management of breast cancer: a description of a new method of interstitial brachytherapy, *Int. J. Radiat. Oncol. Biol. Phys.*, 40, 629–635, 1998.
33. Kini, V.R., Edmundson, G.K., Vicini, F.A., Jaffray, D.A., Gustafson, G., and Martinez, A.A., Use of three-dimensional radiation therapy planning tools and intraoperative ultrasound to evaluate high dose rate prostate brachytherapy implants, *Int. J. Radiat. Oncol. Biol. Phys.*, 40, 629–635, 1998.
34. Milickovic, N., Giannouli, S., Baltas, D., Lahanas, M., Uzunoglu, N., Kolotas, C., and Zamboglou, N., Catheter reconstruction in CT based brachytherapy treatment planning, *Med. Phys.*, 27(5), 1047–1057, 2000.
35. Milickovic, N., Giannouli, S., Baltas, D., Lahanas, M., Zamboglou, N., and Uzunoglu, N., A new algorithm for autoreconstruction of catheters in CT based brachytherapy treatment planning, *IEEE Trans. Biomed. Eng.*, to be published.
36. Kassaee, A. and Altschuler, M.D., Semiautomated matching and seed position location for implanted ribbons, *Med. Phys.*, 21, 643–650, May 1994.
37. Li, S., Pelizzari, C.A., and Chen, G.T.Y., Unfolding patient motion with biplane radiographs, *Med. Phys.*, 21, 1427–1433, May 1994.
38. Li, S., Chen, G.T.Y., Pelizzari, C.A., Reft, C., Roeske, J.C., and Lu, Y., A new source localization algorithm with no requirement of one-to-one source correspondence between biplane radiographs, *Med. Phys.*, 23, 921–927, June 1996.

39. Metz, C.E. and Fencil, L.E., Determination of three-dimensional structure in biplane radiography without prior knowledge of the relationship between the two views: theory, *Med. Phys.*, 16, 45–51, January/February 1989.
40. International Commission on Radiation Units and Measurements, Dose and Volume Specification for Reporting Intracavitary Therapy in Gynecology, ICRU Report No. 38, Bethesda, MD, 1985.
41. Narayana, V., Roberson, P.L., Pu, A.T., Sandler, H., Winfield, R.H., and McLaughlin, P.W., Impact on differences in ultrasound and computed tomography volumes on treatment planning of permanent prostate implants, *Int. J. Radiat. Oncol. Biol. Phys.*, 37(5), 1181–1185, 1997.
42. Roach, M., Faillace-Akazawa, P., Malfatti, C., Holland, J., and Hricak, H., Prostate volumes defined by magnetic resonance imaging and computerized tomographic scans for three dimensional conformal radiotherapy, *Int. J. Radiat. Oncol. Biol. Phys.*, 35(5), 1011–1018, 1996.
43. Pelizzari, C.A., Chen, G.T.Y., Spelbring, D.R., et al., Accurate three dimensional registration of CT, PET, and/or MR images of the brain, *J. Comput. Oncol. Assist. Tomogr.*, 13, 20–26, 1989.
44. Kessler, M.L., Pitluck, S., Petti, P., et al., Integration of multimodality imaging data for radiotherapy treatment planning, *Int. J. Radiat. Oncol. Biol. Phys.*, 21, 1653–1667, 1991.
45. Zhang, J., Levesque, M.F., Wilson, C.L., et al., Multimodality imaging of brain structures for stereotactic surgery, *Radiology*, 175, 435–441, 1990.
46. Kolotas, C., Birn, G., Baltas, D., Rogge, B., Ulrich, P., and Zamboglou, N., CT guided interstitial high dose rate brachytherapy for recurrent malignant gliomas, *Br. J. Radiol.*, 72, 805–808, 1999.
47. Amdur, R.J., Gladstone, D., Leopold, L.A., and Harris, R.D., Prostate seed implant quality assessment using MR and CT image fusion, *Int. J. Radiat. Oncol. Biol. Phys.*, 43(1), 67–72, 1999.
48. Mongioj, V., Brusa, A., Loi, G., Pignoli, E., Gramaglia, A., Scorsetti, M., Bombardieri, E., and Marchesini, R., Accuracy evaluation of fusion of CT, MR, and SPECT images using commercially available software packages (SRS Plato and IFS), *Int. J. Radiat. Oncol. Biol. Phys.*, 43(1), 227–234, 1999.
49. Kremser, C., Plangger, C., Bösecke, R., Pallua, A., Aichner, F., and Felber, S., Image registration of MR and CT images using a frameless fiducial marker system, *Magn. Reson. Imaging*, 15(5), 579–585, 1997.
50. Soltanian-Zadeh, H. and Widham, J.P., A multiresolution approach for contour extraction from brain images, *Med. Phys.*, 24(12), 1844–1853, December 1997.

Index

D

E

F

G

H

I

J

K

L

M

S

T

U

V

W

X

Y